Translation Mechanisms and Control

A subject collection from *Cold Spring Harbor Perspectives in Biology*

OTHER SUBJECT COLLECTIONS FROM *COLD SPRING HARBOR PERSPECTIVES IN BIOLOGY*

Cytokines

Circadian Rhythms

Immune Memory and Vaccines: Great Debates

Cell–Cell Junctions, Second Edition

Prion Biology

The Biology of the TGF-β Family

Synthetic Biology: Tools for Engineering Biological Systems

Cell Polarity

Cilia

Microbial Evolution

Learning and Memory

DNA Recombination

Neurogenesis

Size Control in Biology: From Organelles to Organisms

Mitosis

Glia

Innate Immunity and Inflammation

The Genetics and Biology of Sexual Conflict

The Origin and Evolution of Eukaryotes

Endocytosis

SUBJECT COLLECTIONS FROM *COLD SPRING HARBOR PERSPECTIVES IN MEDICINE*

RAS and Cancer in the 21st Century

Enteric Hepatitis Viruses

Bone: A Regulator of Physiology

Multiple Sclerosis

Cancer Evolution

The Biology of Exercise

Prion Diseases

Tissue Engineering and Regenerative Medicine

Chromatin Deregulation in Cancer

Malaria: Biology in the Era of Eradication

Antibiotics and Antibiotic Resistance

The p53 Protein: From Cell Regulation to Cancer

Aging: The Longevity Dividend

Epilepsy: The Biology of a Spectrum Disorder

Molecular Approaches to Reproductive and Newborn Medicine

The Hepatitis B and Delta Viruses

Intellectual Property in Molecular Medicine

Retinal Disorders: Genetic Approaches to Diagnosis and Treatment

Translation Mechanisms and Control

A subject collection from *Cold Spring Harbor Perspectives in Biology*

EDITED BY

Michael B. Mathews
Rutgers New Jersey Medical School

Nahum Sonenberg
McGill University

John W.B. Hershey
University of California, Davis

COLD SPRING HARBOR LABORATORY PRESS
Cold Spring Harbor, New York • www.cshlpress.org

Translation Mechanisms and Control

A subject collection from *Cold Spring Harbor Perspectives in Biology*
Articles online at www.cshperspectives.org

Executive Editor	Richard Sever
Managing Editor	Maria Smit
Senior Project Manager	Barbara Acosta
Permissions Administrator	Carol Brown
Production Editor	Diane Schubach
Production Manager/Cover Designer	Denise Weiss
Publisher	John Inglis

Front cover artwork: The cover assemblage reflects 50+ years of protein synthesis research. The cartoon of a polysome in action is superimposed on a page from Marshall Nirenberg's tabulation of data elucidating the genetic code. Following the historic discovery that polyuridylate (poly U) stimulates phenylalanine incorporation into protein (Nirenberg and Matthaei 1961; *Proc Natl Acad Sci* **47:** 1588), a specific tRNA-binding assay was developed (Leder and Nirenberg 1964; *Proc Natl Acad Sci* **52:** 420). The assay measured the ability of synthetic trinucleotides (listed on the left) to stimulate the binding to ribosomes of tRNAs that had been labeled with an individual amino acid (listed across the top). The results gave codon assignments for all but the three stop codons and confirmed that the code is degenerate (i.e., more than one codon can specify a particular amino acid). Polysomes and messenger RNA were discovered at around the same time. Since then, the components, functioning, and regulation of the translation system have been investigated in depth as discussed in this volume, including detailed mechanistic and kinetic aspects (Sokabe and Fraser 2019; see p. 143). We thank Jonathan Weissman, Masaaki Sokabe, Christopher Fraser, Tsafi Pe'ery and Cheyenne Moorman for their contributions to the cover. (The background, excerpted from the Marshall W. Nirenberg Papers, is sourced from the public domain courtesy of the U.S. National Library of Medicine.)

Library of Congress Cataloging-in-Publication Data

Names: Mathews, Michael, editor.
Title: Translation mechanisms and control / edited by Michael B. Mathews, Rutgers New Jersey Medical School, Nahum Sonenberg, McGill University, and John W.B. Hershey, University of California, Davis.
Description: Cold Spring Harbor, New York : Cold Spring Harbor Laboratory Press, [2019] | A subject collection from Cold Spring Harbor Perspectives in Biology. | Includes bibliographical references and index.
Identifiers: LCCN 2018012350 | ISBN 9781621821861 (hardcover : alk. paper)
Subjects: LCSH: Genetic translation.
Classification: LCC QH450.5 .T195 2019 | DDC 572.8/845--dc23
LC record available at https://lccn.loc.gov/2019012350

All World Wide Web addresses are accurate to the best of our knowledge at the time of printing.

For a complete catalog of all Cold Spring Harbor Laboratory Press publications, visit our website at www.cshlpress.org.

Contents

Contents

Preface

PROTEIN SYNTHESIS AND TRANSLATIONAL CONTROL are central to life as we know it. Proteins are synthesized through the translation of genetic information by an elaborate machinery that is present in nearly all cells in copious amounts. The machinery itself is expensive to produce and to run. It needs to function accurately and it has to respond to regulation over a broad dynamic range and in countless developmental and environmental circumstances. Translation is not the only step in the gene expression pathway that can be regulated, but it is one of the most important. Because of its sensitivity and flexibility as well as the immediacy of its actions on biological events, translational control has widespread influence at the molecular, cellular, and organismal levels. It modulates physiological and pathophysiological processes, and mutations in components of the translation system engender profound consequences leading to a spectrum of genetic diseases. Many antibiotics, including some of the earliest and most often prescribed, target ribosomes and other components of the system, which offer numerous opportunities for clinical intervention. For all these reasons and more, the field is a rich and fertile one for basic and applied investigation.

The book in your hands is the fifth in the series. When the first one, simply entitled *Translational Control*, was published in 1996, we hoped it would be useful and well-read. We knew that the field was fast-moving and, therefore, expected that the book might serve its purpose for a few years before being overtaken by new discoveries. But none of us contemplated the pace and scale of the advances seen over the following 20 years. Many of these were discussed in the next three volumes—*Translational Control of Gene Expression* (2000), *Translational Control in Biology and Medicine* (2007), and *Protein Synthesis and Translational Control* (2012)—and the present volume brings the series up to date. Authorities in the field provide reviews that survey the remarkable developments made in the understanding of protein synthesis and its regulation over the last five years, up to late 2017/early 2018.

Progress has been made in the dimensions of time and space, and mechanistic details continue to unfold in prokaryotes as well as eukaryotes. New insights have been gained as single-cell, single-molecule, and kinetic approaches have been brought to bear. Although most of the components of the translation system have been known for several decades and many of its regulatory processes have been outlined, new factors and regulators continue to be discovered and characterized. Further, the functions of some known but incompletely defined components have become better appreciated. The structures and the modes of action of ribosomes and of protein–protein and ribonucleoprotein complexes are being visualized and examined at ever-finer levels of resolution through the application of techniques such as cryoelectron microscopy and ribosome profiling. At the same time, the field has continued to expand in breadth and complexity. Fresh light is illuminating the relationships between translation and messenger RNA degradation and their partition into subcellular compartments and particles. Answers to long-standing questions, such as the roles of modified bases in RNA, have begun to emerge, yet variations of RNA regulatory elements and structures continue to proliferate and new questions arise that challenge time-honored assumptions. As the details of established mechanisms are revealed at increasingly intricate levels, unorthodox and entirely novel mechanisms of translation initiation have been uncovered and are being analyzed. These advances have increased understanding of processes operating in areas such as cancer, development, neuroscience, and virus infection, and have brought therapeutic interventions within reach. In sum, this is an exciting time in the field; the last few years have yielded much new knowledge but have also shown how much more remains to be learned.

This book would not have been possible without the efforts of many people. First and foremost, we thank the authors for generating the content, chapters that are scholarly, thoughtful, and thought-provoking. We are greatly indebted to Barbara Acosta for her optimism, patience, and tireless attention to our submissions; to Diane Schubach and the production team at CSHLP for ensuring the quality of the final product; and to John Inglis and Richard Sever for their continued support of this series. Finally, we are, as ever, grateful to Cheyenne Moorman and Tsafi Pe'ery for their inestimable help and encouragement.

MICHAEL B. MATHEWS
NAHUM SONENBERG
JOHN W.B. HERSHEY

Principles of Translational Control

John W.B. Hershey,[1] Nahum Sonenberg,[2] and Michael B. Mathews[3]

[1]Department of Biochemistry and Molecular Medicine, University of California, Davis, California 95616

[2]Goodman Cancer Research Centre and Department of Biochemistry, McGill University, Montreal, Quebec H3A 1A3, Canada

[3]Department of Medicine, Rutgers New Jersey Medical School, Newark, New Jersey 07103

Correspondence: jwhershey@ucdavis.edu

Protein synthesis involves a complex machinery comprising numerous proteins and RNAs joined by noncovalent interactions. Its function is to link long chains of amino acids into proteins with precise sequences as encoded by the genome. Regulation of protein synthesis, called translational control, occurs both at a global level and at specific messenger RNAs (mRNAs). To understand how translation is regulated, knowledge of the molecular structures and kinetic interactions of its components is needed. This review focuses on the targets of translational control and the mechanisms employed.

Proteins are vital constituents of all cells and are frequent products for secretion. Regulation of their synthesis contributes importantly to cell homeostasis, survival, differentiation, and many other functions. The process of protein synthesis is expensive, consuming much of a cell's energy. Consequently, it must be regulated to conserve energy. Equally important, cells need to make proteins in the right amounts, at the right time, and in the right place. To understand how cells deal with these challenges, an understanding of the mechanisms of protein synthesis and translational control is essential. Translational control is an important contributor to regulating gene expression and many disease states occur because of its dysfunction. This review examines the general principles of translational control.

THE PROTEIN SYNTHESIS PATHWAY

Protein synthesis involves the formation of peptide bonds that link the carboxyl group of one amino acid to the amino group of another to form a chain. Peptide bond formation is a relatively simple chemical process involving a nucleophilic attack by an amino group of one amino acid on the carboxyl group of another. Activation of the carboxyl group by the formation of an ester or anhydride facilitates the reaction, and requires energy. Such peptide bond formation reactions can then occur spontaneously. In theory, a simple enzyme should be capable of catalyzing peptide bond formation. However, the synthesis of proteins in cells involves a very complex machinery comprising ribosomes, numerous RNAs and proteins, and, of course, amino acids and energy. Such complexity is needed because the attachment of amino acids must occur in such a way that the resulting sequence of amino acids in the protein is generated with high fidelity according to the genetic code. This sequence information comes from the messenger RNA (mRNA), which helps select the next amino acid and aligns it with the nascent peptide chain.

A growing peptide can, in theory, add the next amino acid either to its amino terminus or its carboxyl terminus. Chemists developed a nonphysiological method called solid-phase protein synthesis that involves attachment of the carboxyl group of the first amino acid to a solid support, followed by the sequential addition of activated amino acids, leading to elongation at the amino terminus of the nascent peptide chain (Merrifield 1964). In contrast, cells synthesize proteins by first generating an activated amino acid attached to a specific transfer RNA (tRNA) through an ester bond. A specific activated amino acid (aa-tRNA) is selected through interaction between a codon in the mRNA and the anticodon in its tRNA. Bound to an adjacent upstream codon in the mRNA is the nascent peptide, attached to its tRNA through its carboxy terminal ester bond. This activated nascent peptide (in the peptidyl-tRNA) is transferred to the amino group of the adjacent aa-tRNA on the mRNA (for details of this mechanism, see Dever et al. 2018; Rodnina 2018). Thus, native protein synthesis involves elongation at the carboxyl terminus of the nascent peptide. In contrast to an amino-terminal extension mechanism, carboxyl-terminal extension involves the binding of the peptidyl-tRNA and incoming aa-tRNA to adjacent codons in the mRNA. Therefore, each reaction in the multiple cycles of elongation occurs similarly at the same place on the ribosome, enabling protein synthesis to occur with high fidelity. A brief description of the process of protein synthesis follows.

The translation pathway in cells can be broken into four stages: initiation, elongation, termination, and ribosome recycling. In eukaryotes, the major eukaryotic cellular translation components are two ribosomal subunits, 40S and 60S, composed of four RNAs and over 80 proteins; an mRNA; aminoacyl-tRNAs (aa-tRNAs) formed by attaching each of the 20 amino acids to its specific tRNAs through an energy-consuming reaction catalyzed by their specific aminoacyl-tRNA synthetases; protein factors that interact with the other components; and GTP and ATP, providing energy through their hydrolysis.

Protein synthesis on most mRNAs begins by forming a complex containing the 40S small ribosomal subunit and the initiator methionyl-tRNA$_i$ (Met-tRNA$_i$), which then binds to the 5′ end of the mRNA. The resulting complex, called the preinitiation complex (PIC), also contains many protein factors (initiation factors) as well as GTP and ATP. The PIC then scans down the mRNA through an energy-driven process until the anticodon in the Met-tRNA$_i$ base-pairs with the initiator codon (usually AUG) in the mRNA. This interaction causes termination of PIC scanning, the hydrolysis of bound GTP, and ejection of many of the initiation factors, enabling junction with the 60S ribosomal subunit to form an 80S initiation complex competent to enter the elongation phase of protein synthesis. An important feature of the initiation pathway is that no covalent bond is made or broken, other than those involved in the charging of methionine to its tRNA and the hydrolysis of GTP and ATP. Thus, formation of an 80S initiation complex mainly involves noncovalent interactions between the numerous proteins and RNAs. A detailed description of the eukaryotic initiation pathway is found in Merrick and Pavitt (2018).

Following initiation complex formation, with the Met-tRNA$_i$ bound to the P (peptidyl) site of the 80S ribosome, a new aa-tRNA is selected and binds into the ribosomal A (aminoacyl) site through base-pairing between its anticodon and the codon in the mRNA just downstream of the initiator AUG. The ribosome catalyzes the transfer of the methionyl group from its tRNA to the amino group of the aa-tRNA in the A site, resulting in the formation of the first peptide bond. The peptidyl-tRNA in the A site, together with the mRNA, move along the ribosome into the P site (called translocation). Then the mRNA codon in the A site

Cite this article as *Cold Spring Harb Perspect Biol* doi: 10.1101/cshperspect.a032607

binds a new aa-tRNA, again by base-pairing. The repetitive aa-tRNA binding, peptide bond formation, and translocation steps, called the elongation phase of protein synthesis, are promoted by proteins called elongation factors that transiently associate with the components involved. Besides breaking the aa-tRNA ester bond to form the peptide bond and the hydrolysis of GTP, elongation consists of numerous noncovalent interactions that enable the alignments on the ribosome to produce a proper amino acid sequence in the growing peptide chain. This pathway is described in detail in Dever et al. (2018).

When the ribosome encounters an A-site mRNA codon of UAA, UGA, or UAG, no aa-tRNA recognizes it. Instead, a protein release factor binds into the A site and catalyzes the hydrolysis of the nascent peptidyl-tRNA in the P site, resulting in termination of protein synthesis. Additional proteins then interact with the ribosome to promote the ejection of the tRNA and mRNA and dissociation of the ribosome into its subunits (a process called ribosome recycling). Termination and recycling are described in detail in Hellen (2018).

Most of the principal features of the protein synthesis pathway are common to all domains of life, especially the structure of the ribosome responsible for aa-tRNA binding to mRNA codons and peptide bond formation. The components and mechanism of initiation may differ significantly, however. For example, the scanning mechanism for start-site selection in eukaryotes is replaced in prokaryotes by a mechanism involving base-pairing between ribosomal RNA and mRNA (reviewed by Rodnina 2018). Alternative initiation mechanisms are also exploited in some eukaryotic systems (see Kwan and Thompson 2018) and their viruses (Stern-Ginossar et al. 2018), often for regulation.

FEATURES OF TRANSLATIONAL CONTROL

The cellular level of a protein is determined by a balance between its rates of synthesis and degradation. The synthesis of a specific protein requires transcription of its mRNA, a process that is highly regulated (transcriptional control).

mRNAs may also undergo several critical modifications at their termini (capping and polyadenylation), as well as internally (splicing, methylation, etc.). In addition to transcription, mRNA stability also affects mRNA levels. But mRNA levels alone do not determine the amounts of proteins in cells, as some mRNAs are not actively translated, or are translated poorly, being bound to proteins, called messenger ribonucleoproteins (mRNPs), or sequestered in stress granules (SGs) or processing bodies (PBs) (Ivanov et al. 2018). In addition, translation rates are not always uniform among different mRNA species or along individual mRNAs (Biswas et al. 2018; Ingolia et al. 2018).

The synthesis rate of a specific protein is determined by the number of ribosomes translating its mRNA together with the overall rate of peptide bond formation. The number of translating ribosomes is determined by the number of active mRNAs, the coding length of the mRNA, the rate of ribosome attachment to the mRNA (initiation), and rate of elongation. The average rate of elongation in eukaryotic cells is normally quite fast (often about five amino acids per second), whereas the initiation step is usually rate-limiting (around one every 5 seconds for strongly translated mRNAs) (Palmiter 1975). However, if elongation rates are greatly reduced, this will affect initiation rates as well, as the ribosome must move out of the initiation region of the mRNA to enable the next ribosome to initiate. A ribosome bound at the initiator codon must move downstream at least five codons (~1 second when fast) before a new initiating ribosome can bind there.

How are the rates of global protein synthesis measured? A classical method is to measure the rate of radioactive amino acid incorporation into proteins. This can be applied to measuring both global rates and specific protein rates, the latter generally involving immunoprecipitation. Stable isotope labeling with amino acids in cell culture (SILAC) enables quantification of newly synthesized proteins by mass spectrometry (Schwanhausser et al. 2011). A third method is to determine the number of active ribosomes and their elongation/termination rate. The elongation/termination rate is equal to the average size of synthesized proteins divided by the ribosome transit time, measured by radioactive amino acid labeling followed by centrifugation at different times to separate released proteins from ribosome-bound nascent proteins (Fan and Penman 1970). To determine the rate of initiation for a protein, measuring its polysome size divided by the ribosome transit time can be employed (Palmiter 1975). Polysome profiling, a method involving separation of polysome sizes by sucrose gradient centrifugation, is used to detect changes in initiation rates, as the rate of initiation greatly affects polysome size. However, elongation rates also affect polysome size, so measurement of the elongation rate also is needed. Ribosome profiling (Ingolia et al. 2018), which entails a global analysis of ribosome-protected mRNA segments, also provides information about global and specific protein synthesis rates, as well as fine details of ribosomal progression along an mRNA, and is now frequently employed in studying translational control.

Most mechanisms of translational control in cells affect the rate of initiation. By regulating initiation, rates of protein synthesis can be quickly altered and can affect specific mRNAs differently. The rates of peptide bond formation (elongation) and termination/recycling also can affect the rate of ribosome attachment. If all nascent peptides are completed as full-length protein products, then the initiation rate defines the overall protein synthesis rate during steady state protein synthesis. Because both short and long mRNA open reading frames (ORFs) may initiate at similar rates, resulting, respectively, in light versus heavy polysomes, both large and small proteins can, in principle, be made at similar rates.

What determines the initiation rate of an mRNA, and how is this regulated? Specific mRNAs can initiate with different efficiencies because of their sequences or secondary structures that affect interactions with the translational machinery. *Trans*-acting factors (proteins, small RNAs, riboswitches/ligands) that bind specific mRNAs can enhance or inhibit the recruitment of ribosomes (see Breaker 2018; Duchaine and Fabian 2018). The cellular levels of available ribosomes, initiation factors, and

Met-tRNA$_i$ influence initiation rates. The initiation factors also can be modified posttranslationally, for example, by phosphorylation (see Proud 2018; Wek 2018), and the extent of such modifications can affect factor activities and initiation rates.

Protein synthesis is one of the most energy-consuming pathways in cell metabolism (Rolfe and Brown 1997); two high-energy phosphate bonds are used to activate the amino acid by producing aa-tRNA and two to generate the peptide bond (elongation), making four per amino acid incorporated. Although six high-energy phosphate bonds are broken to synthesize one mRNA codon, the codon is translated multiple times, making protein synthesis much more expensive than mRNA transcription. To regulate the translation rate by slowing elongation and/or termination is expensive, as the translating ribosome is less productive. Such slower rates would require more ribosomes to generate the same amounts of protein, and ribosomes are large and very expensive to make. However, inhibition of elongation can rapidly decrease protein synthesis and thereby conserve energy when the cell's energy level drops suddenly (Wek 2018). Modest reductions in elongation rates (e.g., by phosphorylation of elongation factors; Dever et al. 2018; Proud 2018) or rare codon usage, are sometimes used to adjust translation rates to coordinate cotranslation protein secretion and folding. Thus, the rate of global or specific protein synthesis can be determined by the sum of all of these possible regulatory mechanisms. If a number of different mechanisms affect the same mRNA at the same time, the individual effect of each mechanism could be small, difficult to detect, yet nevertheless important.

REGULATORY MECHANISMS

The targets for regulation at initiation are very numerous, as it is not a covalent bond reaction that is being regulated, but rather the many noncovalent interactions required to form initiation complexes in the correct conformation. An important target for the regulation of global translation is eukaryotic initiation factor (eIF)2,

Cite this article as *Cold Spring Harb Perspect Biol* doi: 10.1101/cshperspect.a032607

whose phosphorylation inhibits its recycling by eIF2B and thus its ability to bind Met-tRNA$_i$, an interaction involved in the translation of nearly all mRNAs (Merrick and Pavitt 2018; Wek 2018). Another target is eIF4E, the cap-binding protein, which is involved in selecting an mRNA and also promotes the scanning process through stimulation of the RNA helicase activity of eIF4A. Its down-regulation by binding eIF4E-binding proteins (4E-BPs) (Merrick and Pavitt 2018; Proud 2018; Robichaud et al. 2018) can affect mRNAs differently, as some mRNAs require high eIF4E activity, whereas others do not. Many other targets of initiation are described in the scientific literature. Elongation rates also can be affected by codon usage, tRNA levels, and elongation factor modifications (Dever et al. 2018; Proud 2018).

Posttranslational Modifications of Translational Machinery Proteins

The most common method for regulating protein synthesis is through phosphorylation of the proteins involved. Phosphorylation levels are controlled by protein kinases and phosphatases whose activities are responsive to conditions in the cell. As mentioned above, the most important targets are eIF2 and the 4E-BPs (Merrick and Pavitt 2018). In addition, other initiation and elongation factors are phosphorylated and contribute to translational control (Proud 2018; Wek 2018). To rigorously establish a correlation between phosphorylation and translational control, it is important to identify the phosphorylation site and measure the extent of phosphorylation. To establish that a phosphorylation actually causes a change in translational efficiency, mutant forms of the target protein that mimic or prevent phosphorylation are frequently tested, as are inhibition or knockout of specific protein kinases.

mRNA Structure and Complexes

mRNAs have unique sequences, both in their coding and noncoding regions. The lengths and sequences of the 5' and 3' untranslated regions (UTRs) differ and can affect interactions with the translational machinery. Alternate promoters and alternative splicing can generate different mRNAs that encode the same protein, but their translational efficiencies may not be identical. Cellular mRNAs can be modified (Peer et al. 2018) and can bind various proteins to form mRNPs (Mitchell and Parker 2014) that may differ in their abilities to initiate translation. mRNAs also interact with small RNAs (microRNAs) that affect initiation (Duchaine and Fabian 2018). A current challenge is to determine the precise structure of an mRNA, especially as found in vivo when bound with cellular proteins.

mRNA Levels

mRNA levels obviously play a role in determining the amounts of protein produced by the translational machinery. As mentioned earlier, mRNA levels are determined by their rates of transcription and processing, but also by their rates of transport into the cytoplasm (Biswas et al. 2018) and by their rates of degradation that may be coupled to or influenced by their translation (Heck and Wilusz 2018), by their sequences (Karousis and Mühlemann 2018), as well as by microRNAs (Duchaine and Fabian 2018). mRNAs also can be sequestered in PBs and SGs (Ivanov et al. 2018) where they are not being translated. Finally, mRNAs may bind other proteins to form mRNPs, which undergo either enhanced or reduced translation. Because stimulating eukaryotic gene expression through transcriptional control requires a considerable amount of time to elevate mRNA levels, and in some cases to transport them to specific cellular regions (e.g., in neurons), translational control can be much more rapid.

FUTURE PROSPECTS

Although much is known about the mechanism of protein synthesis and its control, there are aspects of our understanding that need to be improved. The structure of ribosomes and protein factors provide much needed information (Jobe et al. 2018), but the structures of initiation complexes remain elusive. What is the actual structure of a native mRNA, especially its 5'UTR, and how is it altered by the binding of proteins or small RNAs? Another area concerns the kinetics of the various noncovalent interactions involved in initiation (Prabhakar et al. 2018; Sokabe and Fraser 2018). What is the order of binding of the various components? What are their rates of binding and dissociation, and how are these rates affected by their levels, modifications, and order of binding? How is a given mRNA selected from the pool of mRNAs in the cytoplasm, as this is one of the most important reactions for translational control? A thorough knowledge of the mechanisms and rates of protein synthesis is needed to better understand how translational control contributes to the regulation of gene expression.

REFERENCES

*Reference is also in this collection.

* Biswas J, Liu Y, Singer RH, Wu B. 2018. Fluorescence imaging methods to investigate translation in single cells. *Cold Spring Harb Perspect Biol* doi: 10.1101/cshperspect. a032722.

* Breaker RR. 2018. Riboswitches and translation control. *Cold Spring Harb Perspect Biol* doi: 10.1101/cshperspect. a032797.

* Dever TE, Dinman JD, Green R. 2018. Translation elongation and recoding in eukaryotes. *Cold Spring Harb Perspect Biol* doi: 10.1101/cshperspect.a032649.

* Duchaine TF, Fabian MR. 2018. Mechanistic insights into microRNA-mediated gene silencing. *Cold Spring Harb Perspect Biol* doi: 10.1101/cshperspect.a032771.

Fan H, Penman S. 1970. Regulation of protein synthesis in mammalian cells. II: Inhibition of protein synthesis at the level of elongation. *Proc Natl Acad Sci* **87:** 328–332.

* Heck AM, Wilusz J. 2018. The interplay between the RNA decay and translation machinery in eukaryotes. *Cold Spring Harb Perspect Biol* doi: 10.1101/cshperspect. a032839.

* Hellen CUT. 2018. Translation termination and ribosome recycling in eukaryotes. *Cold Spring Harb Perspect Biol* doi: 10.1101/cshperspect.a032656.

* Ingolia NT, Hussmann JA, Weissman JS. 2018. Ribosome profiling: Global views of translation. *Cold Spring Harb Perspect Biol* doi: 10.1101/cshperspect.a032698.

* Ivanov P, Kedersha N, Anderson P. 2018. Stress granules and processing bodies in translational control. *Cold Spring Harb Perspect Biol* doi: 10.1101/cshperspect.a032813.

* Jobe A, Liu Z, Gutierrez-Vargas C, Frank J. 2018. New insights into ribosome structure and function. *Cold Spring Harb Perspect Biol* doi: 10.1101/cshperspect. a032615.

* Karousis ED, Mühlemann O. 2018. Nonsense-mediated mRNA decay begins where translation ends. *Cold Spring Harb Perspect Biol* doi: 10.1101/cshperspect.a032862.

* Kwan T, Thompson SR. 2018. Noncanonical translation initiation in eukaryotes. *Cold Spring Harb Perspect Biol* doi: 10.1101/cshperspect.a032672.

* Merrick WC, Pavitt GD. 2018. Protein synthesis initiation in eukaryotic cells. *Cold Spring Harb Perspect Biol* doi: 10.1101/cshperspect.a033092.

Merrifield RB. 1964. Solid-phase peptide synthesis. 3: An improved synthesis of bradykinin. *Biochemistry* **3:** 1385–1390.

Mitchell SF, Parker R. 2014. Principles and properties of eukaryotic mRNPs. *Mol Cell* **54:** 547–558.

Palmiter R. 1975. Quantitation of parameters that determine the rate of ovalbumin synthesis. *Cell* **4:** 189–197.

* Peer E, Moshitch-Moshkovitz S, Rechavi G, Dominissini D. 2018. The epitranscriptome in translation regulation. *Cold Spring Harb Perspect Biol* doi: 10.1101/cshperspect. a032623.

* Prabhakar A, Puglisi EV, Puglisi JD. 2018. Single-molecule florescence applied to translation. *Cold Spring Harb Perspect Biol* doi: 10.1101/cshperspect.a032714.

* Proud CG. 2018. Phosphorylation and signal transduction pathways in translational control. *Cold Spring Harb Perspect Biol* doi: 10.1101/cshperspect.a033050.

* Robichaud N, Sonenberg N, Ruggero D, Schneider RJ. 2018. Translational control in cancer. *Cold Spring Harb Perspect Biol* doi: 10.1101/cshperspect.a032896.

* Rodnina MV. 2018. Translation in prokaryotes. *Cold Spring Harb Perspect Biol* doi: 10.1101/cshperspect.a032664.

Rolfe DF, Brown GC. 1997. Cellular energy utilization and molecular origin of standard metabolic rate in mammals. *Physiol Rev* **77:** 731–758.

Schwanhausser B, Busse D, Li N, Dittmar G, Schuchhardt J, Wolf J, Chen W, Selbach M. 2011. Global quantification of mammalian gene expression control. *Nature* **473:** 337–342.

* Sokabe M, Fraser CS. 2018. Toward a kinetic understanding of eukaryotic translation. *Cold Spring Harb Perspect Biol* doi: 10.1101/cshperspect.a032706.

* Stern-Ginossar N, Thompson SR, Mathews MB, Mohr I. 2018. Translational control in virus-infected cells. *Cold Spring Harb Perspect Biol* doi: 10.1101/cshperspect. a033001.

* Wek RC. 2018. Role of eIF2α kinases in translational control and adaptation to cellular stress. *Cold Spring Harb Perspect Biol* doi: 10.1101/cshperspect.a032870.

Protein Synthesis and Translational Control: A Historical Perspective

Soroush Tahmasebi,[1,4] Nahum Sonenberg,[1] John W.B. Hershey,[2] and Michael B. Mathews[3]

[1]Department of Biochemistry and Goodman Cancer Research Center, McGill University, Montreal, QC H3A 1A3, Canada

[2]Department of Biochemistry and Molecular Medicine, University of California, School of Medicine, Davis, California 95616

[3]Department of Medicine, Rutgers New Jersey Medical School, Newark, New Jersey 07103

Correspondence: mathews@njms.rutgers.edu

Protein synthesis and its regulation are central to all known forms of life and impinge on biological arenas as varied as agriculture, biotechnology, and medicine. Otherwise known as translation and translational control, these processes have been investigated with increasing intensity since the middle of the 20th century, and in increasing depth with advances in molecular and cell biology. We review the origins of the field, focusing on the underlying concepts and early studies of the cellular machinery and mechanisms involved. We highlight key discoveries and events on a timeline, consider areas where current research has engendered new ideas, and conclude with some speculation on future directions for the field.

Proteins account for the largest fraction of the macromolecules in a cell, are important components of the extracellular milieu, and fulfill multiple roles—enzymatic, structural, transport, regulatory, and other—in all organisms. Their synthesis, through the translation of genetic information encoded in messenger RNA (mRNA), requires extensive biological machinery and demands delicate and sophisticated regulation (Hershey et al. 2018). Protein synthesis is modulated quantitatively, and in time and space, through a network of stimuli, responses, and interactions collectively referred to as translational control. A large proportion of the resources of cells and organisms is devoted to translation and translational control, as discussed previously in terms of genetics, bioenergetics, and cell biology (Mathews et al. 2007).

Here we summarize the beginnings of the field and outline the pathway that led to our current understanding of the processes of protein synthesis and translational control, placing landmark discoveries on a timeline (Fig. 1). The field is, and always has been, a broad one. It originated in studies of topics ranging from virus infection to embryology and development, and has grown to encompass learning, memory, and genetic disease (Tahmasebi et al. 2018), as well as therapeutic intervention, among other biomedical areas. It continues to diversify and develop with the advent of approaches of increasing depth and precision. At the same time, it continues to generate fresh concepts and present challenges to well-accepted paradigms.

---

[4]Present address: Department of Pharmacology, University of Illinois at Chicago, Chicago, Illinois 60612.

Cite this article as *Cold Spring Harb Perspect Biol* doi: 10.1101/cshperspect.a035584

A brief history of
protein synthesis and translation
control (1951–2017)

- GTP required for PS
- 1st human disease (SCD) linked to a change in AA sequence of a protein
- tRNA
- AA-tRNA
- Adaptor hypothesis and central dogma
- AA-tRNA synthetase
- Cycloheximide inhibits PS in euk. cells
- Nobel Prize - insulin sequence

- 1st protein sequence
- 1st cell-free translation system
- 1st human cell line

- Antibiotics inhibit PS in bacteria

- Ribosomes
- Iron stimulates PS in rabbit reticulocytes

| 1951 | 1952 | 1953 | 1954 | 1955 | 1956–8 | 1959 | 1960 |

- Nobel Prize - partition chromatography
- Insulin stimulates PS in isolated rat diaphragm
- Plasma and AAs stimulate PS in cell-free system

- Nobel Prize - protein structure
- ATP is required for PS
- RNA tie club

- Puromycin inhibits PS
- Ribosomes composed of two unequal subunits
- Ribosomes are the site of PS

- Sucrose gradient velocity sedimentation

- mRNA is information carrier
- Messenger hypothesis
- Messenger-dependent prok. cell-free translation system
- Streptomycin proposed to target ribosomes
- Fertilization stimulates PS in sea urchin eggs

- fMet-tRNA$_f$ mediates initiation of PS
- Synthesis of GMP-PCP
- Prok. elongation factors, EF-Tu, EF-Ts, EF-G
- Completion of the genetic code
- Wobble hypothesis
- Prok. initiation factors, IF1, IF2, IF3

- Translation of 1st mammalian mRNA in vitro
- 1st ribosome-binding site on mRNA

| 1961 | 1962–3 | 1964–5 | 1966 | 1967–8 | 1969 | 1970 |

- Inhibition of PS causes memory loss
- Polysomes
- Tape mechanism
- Euk. elongation factors, eEF-1, eEF-2
- Ribosomal RNA is made in nucleolus
- tRNA mediates mRNA decoding
- GFP was purified
- Nobel Prize - structure of hemoglobin

- Stop codons
- Poliovirus infection inhibits host cell PS
- Ribosomes read mRNA in 5′ to 3′ direction
- 1st RNA sequence
- Heme/Fe^{+2} availability controls globin synthesis
- Ribosome three-site model
- fMet-tRNA$_f$
- Chloramphenicol interacts with ribosomes

- Nobel Prize - genetic code
- Primary sequence and secondary structure of mRNA regulate translation initiation
- Reconstitution of 30S ribosomal subunit
- Release factor
- GTP role in initiation of peptide synthesis
- Mitochondria contain ribosomes

- SDS-PAGE
- Met-tRNA$_i$
- Initial purification of euk. initiation factors
- Messenger-dependent euk. cell-free translation system

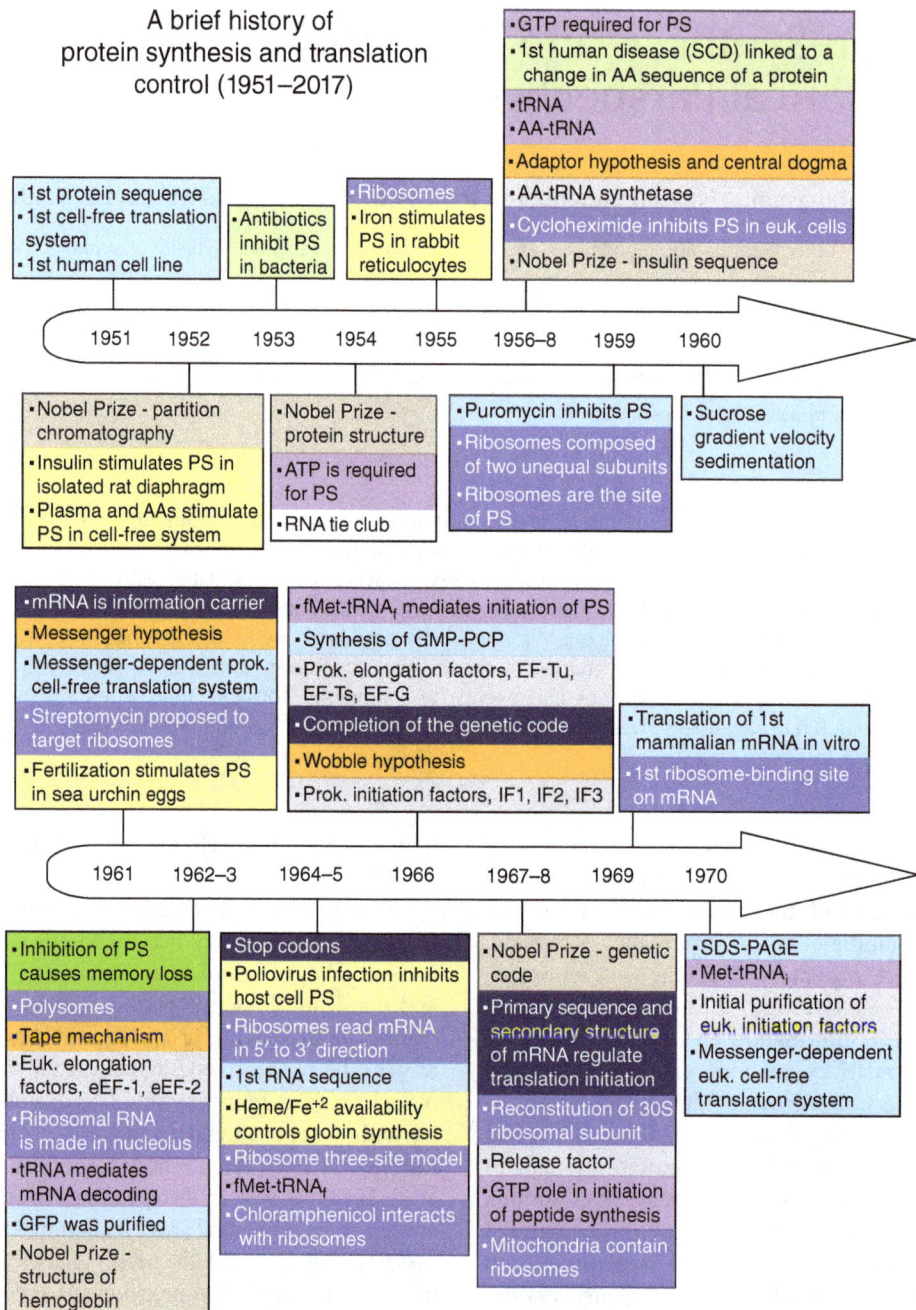

Figure 1. Timeline of discoveries in the fields of protein synthesis (PS) and translational control (1951–2017). Principal advances are shown according to the year of publication, and are color coded by topic (color key and abbreviation definitions are at the end of the figure). Some other relevant events are also noted.

Cite this article as *Cold Spring Harb Perspect Biol* doi: 10.1101/cshperspect.a035584

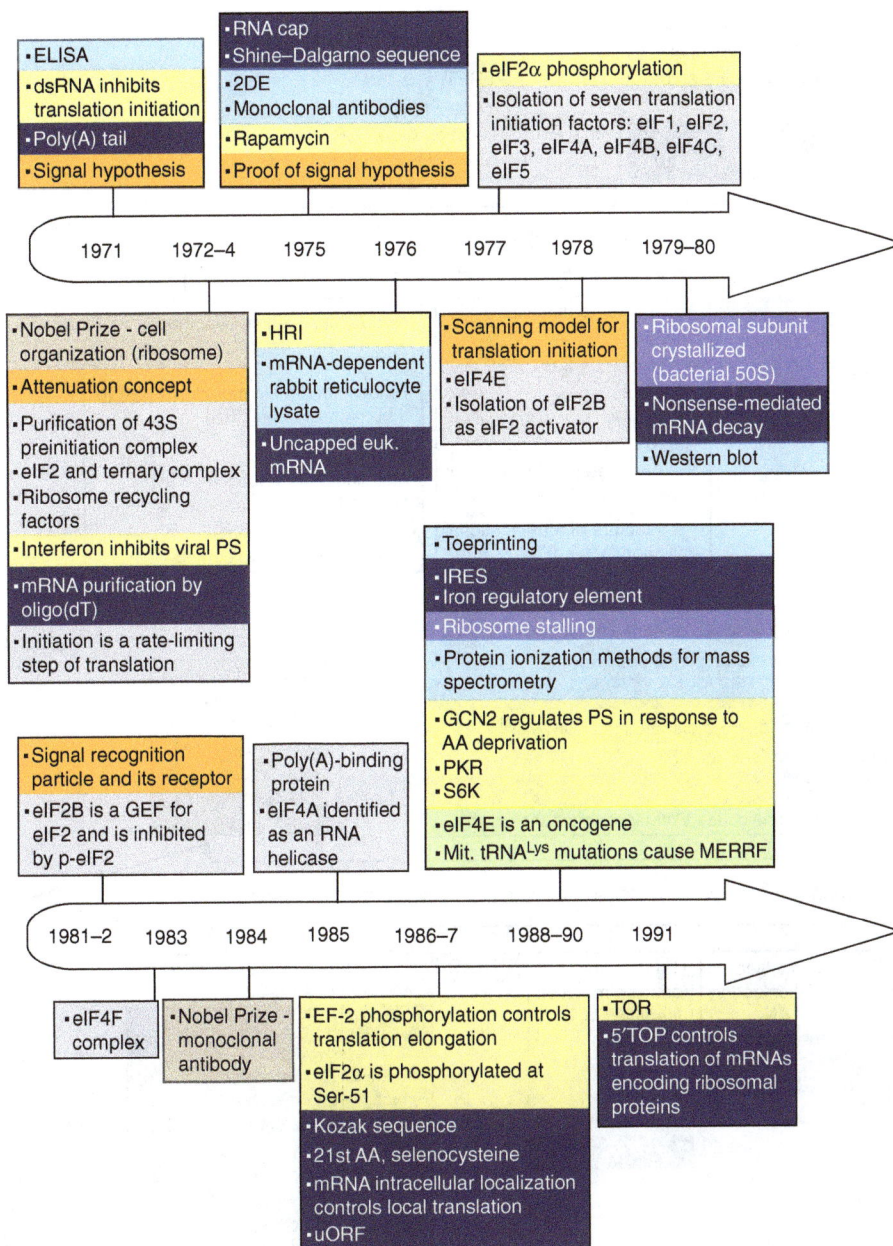

Figure 1. *Continued.*

TRANSLATION TIMELINE

Before the early 1950s, most protein synthesis research addressed physiological questions and the findings were largely descriptive in nature (e.g., Daly and Mirsky 1952). Theories of pro-

tein synthesis via enzyme assembly and peptide intermediates were entertained along with template theories (Campbell and Work 1953), and no component of the translation system was known (ribosomes included [Palade 1955]). In this era, studies of protein synthesis were per-

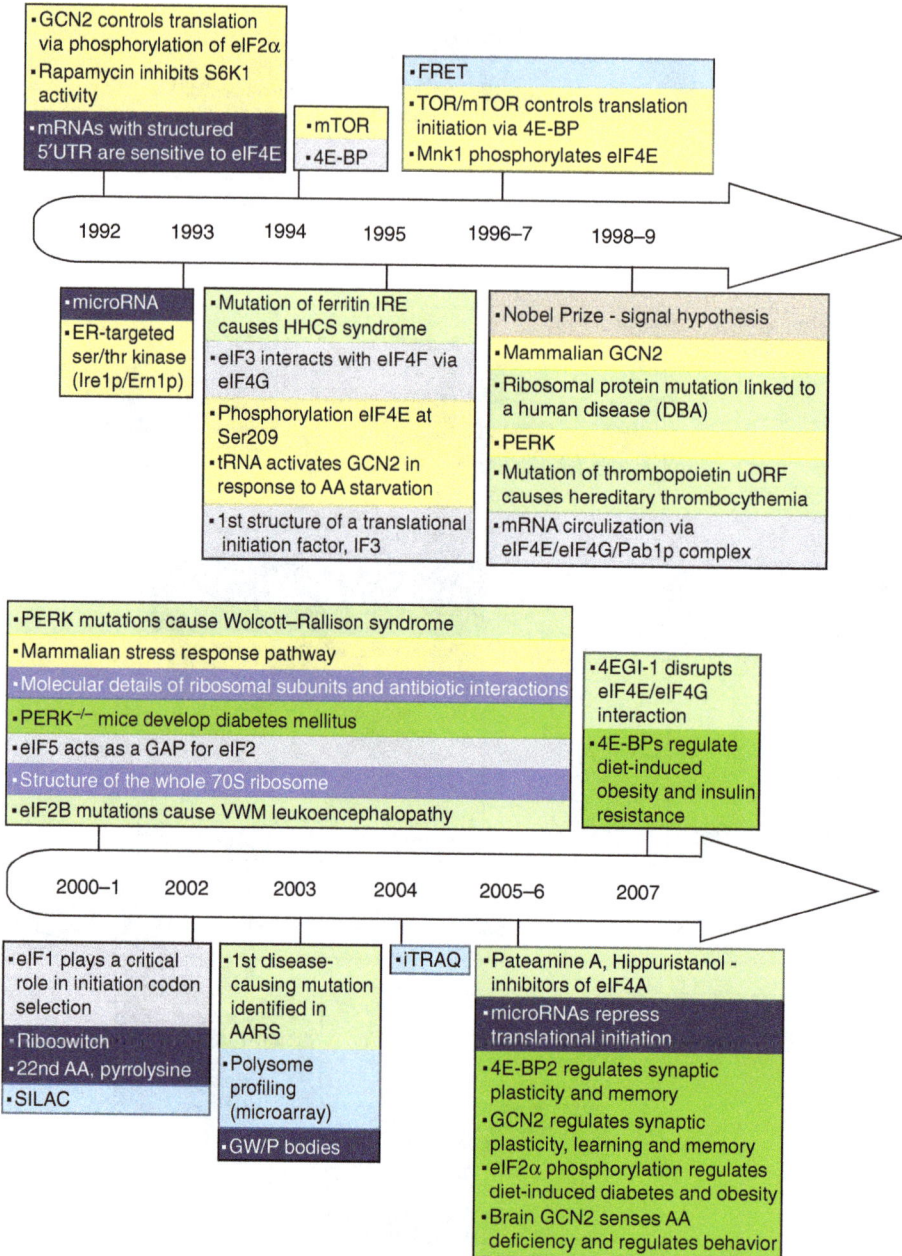

Figure 1. *Continued.*

formed on tissues or tissue slices or in whole animals.

During the decade of the 1950s, the field underwent a transformation with the development of cell-free systems, fundamental discoveries (e.g., of transfer RNA [tRNA] and ribosomes), and crucial technical advances (such as radiolabeled amino acids, sucrose gradient centrifugation, and inhibitors). Subsequent decades were dominated by themes, concepts, and discoveries that furthered the field in different ways. Foundational discoveries (of polysomes and

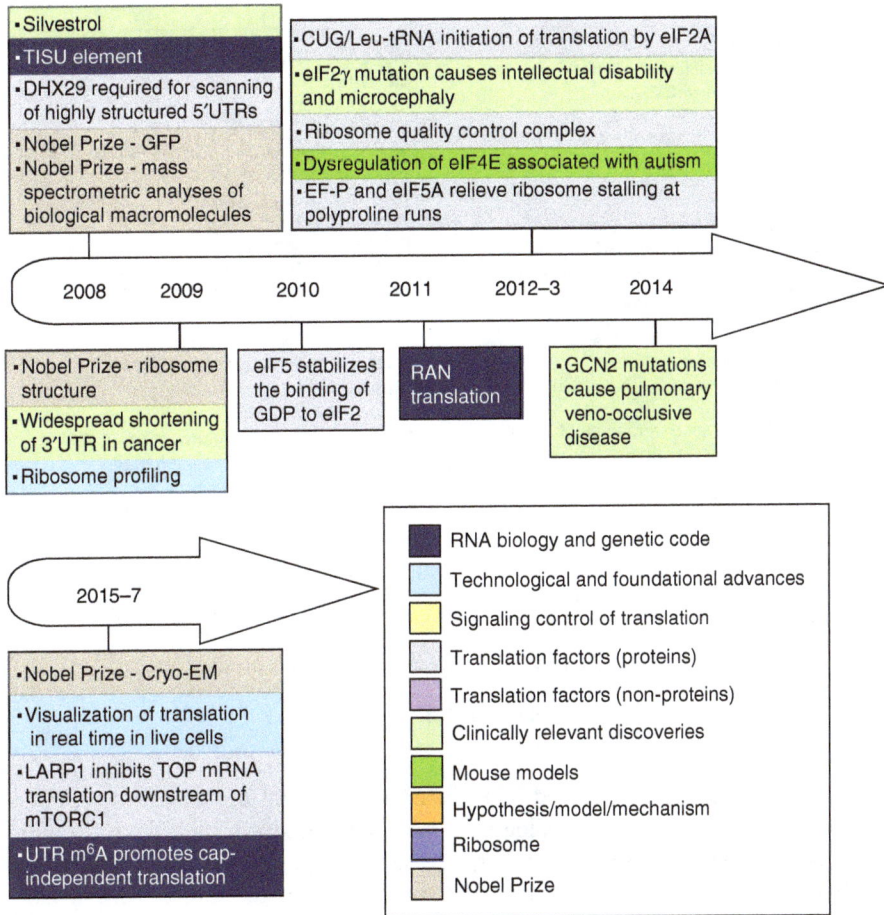

Figure 1. *Continued.*

Principal discoveries are recorded chronologically and thematically (distinguished by color coding) in Figure 1. The timeline illustrates the fact that in this field, as in others, research did not always progress in a systematic and orderly fashion. Some discoveries developed slowly or even lay fallow for many years; the roles of eukaryotic initiation factors (eIFs) afforded several examples of this. Other discoveries, such as the poly(A) tail of eukaryotic mRNA, immediately spawned far-reaching advances in multiple areas of mRNA characterization, isolation, metabolism, and translation. Many advances benefitted from studies in greater depth that were enabled by new techniques (e.g., gel electrophoresis, blotting, toeprinting, ribosome profiling)

mRNA, for example) continued to be made in the 1960s, together with the seminal elucidation of the genetic code. The 1970s saw rapid growth of the field, including the characterization of most of the components of the eukaryotic translation system and the beginning of mechanistic studies. Mechanistic and regulatory themes dominated the 1980s, and the appreciation of regulatory pathways expanded rapidly in the 1990s. All of these themes contributed to the current activity in the first decades of the 21st century, when much research sought—and still seeks—to explain physiological and pathophysiological responses of the translation system and to develop therapeutics for treatment of genetic and acquired diseases and infections.

Abbreviations

AA: Amino acid
AARS: Aminoacyl tRNA synthetase
AA-tRNA: Aminoacyl-tRNA
Cryo-EM: Cryoelectron microscopy
DBA: Diamond–Blackfan anemia
dsRNA: Double-strand RNA
ER: Endoplasmic reticulum
Euk.: Eukaryotic
eIFs: Eukaryotic initiation factors
ELISA: Enzyme-linked immunosorbent assay
FRET: Fluorescence resonance energy transfer
GAP: GTPase-activating protein
GCN2: General control nonderepressible 2
GEF: Guanine nucleotide exchange factor
GFP: Green fluorescent protein
GMP-PCP: Guanosine-5′-[(β,γ)-methyleno] triphosphate
GW/P bodies: Glycine- and tryptophan-rich cytoplasmic/processing bodies
HHCS: Hereditary hyperferritinemia cataract syndrome
HRI: Heme-regulated inhibitor
IRE: Iron regulatory element
IRES: Internal ribosome entry site
iTRAQ: Isobaric tags for relative and absolute quantitation
LARP1: La-related protein 1

m⁶A: N^6-methyladenosine
MERRF: Myoclonic epilepsy with ragged red fibers
Mit.: Mitochondrial
PERK: Protein kinase R (PKR)-like endoplasmic reticulum kinase
PKR: Protein kinase R
Prok.: Prokaryotic
PS: Protein synthesis
RAN: Repeat-associated non-ATG
S6K: Ribosomal protein S6 kinase
SCD: Sickle cell disease
SDS-PAGE: Sodium dodecyl sulfate polyacrylamide gel electrophoresis
SILAC: Stable isotope labeling with amino acids in cell culture
TISU: Translation initiator of short 5′UTR
TOP: Terminal oligopyrimidine tract
TOR: Target of rapamycin
uORF: Upstream open reading frame
UTR: Untranslated region
VWM: Vanishing white matter
2DE: Two-dimensional gel electrophoresis
4E-BP: Eukaryotic translation initiation factor 4E-binding protein

Figure 1. *Continued.*

and technological developments in related fields (cloning, reverse genetics, X-ray crystallography). These drove research forward, providing fresh insights and mechanistic understanding of increasing clarity and detail. Striking examples came from structural investigations of ribosomes and components of the translation system with which they interact (Jobe et al. 2018). Concurrently, investigations sometimes in related fields of study, led to a steady stream of unforeseen observations, including unorthodox initiation mechanisms (Meyer et al. 2015; Zhou et al. 2015; Kwan and Thompson 2018; Zu et al. 2018) and novel concepts such as ribosomal heterogeneity (Sauert et al. 2015; Genuth and Barna 2018) and the ribosomal concentration model (Mills and Green 2017; Khajuria et al. 2018), which expanded the scope of the field.

DEVELOPMENT OF THE PROTEIN SYNTHESIS FIELD

Biochemical investigations of protein synthesis began when concepts that are now nearly axi-

omatic were still uncertain. The view of proteins as unique linear arrays of just 20 amino acid residues was about to be established with the publication of the first protein sequence (the insulin B chain [Sanger and Tuppy 1951]), and mechanisms of protein synthesis involving the reversal of proteolysis or phosphorylated intermediates were entertained (Zamecnik 1969). Radioactive isotopes had begun to revolutionize many areas of biomedical science in the late 1940s, and radiolabeled amino acids came into use as tracers around 1950. Researchers synthesized them from simple labeled compounds such as formaldehyde or sodium cyanide as a first step in their experiments (e.g., Borsook et al. 1952), until they became commercially available in the latter part of the decade. Enabled by this profound technical advance, investigations proliferated rapidly and biochemistry ran ahead of genetics until the advent of cloning and the systematic exploitation of the yeast system that began to make their mark in the 1980s.

Siekevitz and Zamecnik (1951) produced a cell-free preparation from rat liver that incorpo-

Cite this article as *Cold Spring Harb Perspect Biol* doi: 10.1101/cshperspect.a035584

rated amino acids into protein and showed that energy was required in the form of ATP and GTP (Zamecnik and Keller 1954; Keller and Zamecnik 1956). The translation system was refined by stages and resolved into subfractions including a microsomal fraction that contained ribosomes attached to intracellular membrane fragments (Zamecnik 1960). Pulse-chase experiments showed that ribosomes are the site of protein synthesis, not an easy task in bacterial cells where protein synthesis is very rapid: the assembly of a protein chain on a ribosome was estimated to take only 5–10 seconds (McQuillen et al. 1959). It is salutary to recall that this was accomplished in advance of an understanding of the central role of RNA in the flow of genetic information to protein, before the visualization of polysomes, and well before the first RNA sequence was completed (Holley et al. 1965). Amid many concepts (Crick 1959), one idea posited that each ribosome is dedicated to the synthesis of a single protein, the "one gene–one ribosome–one protein" hypothesis. Early in the 1960s, however, polysomes were observed and their function appreciated in light of the messenger hypothesis and the "tape mechanism" of translation discussed below (Marks et al. 1962; Warner et al. 1962, 1963; Arlinghaus and Schweet 1963; Gierer 1963; Goodman and Rich 1963; Nakamoto et al. 1963; Noll and Wettstein 1963; Wettstein et al. 1963). Technical advances in electron microscopy (EM) and high-speed centrifugation made vital contributions during this phase of the field's development.

The role of aminoacyl-tRNA was established in the late 1950s. An intermediate, activated state of amino acids was first detected (Hultin and Beskow 1956), then characterized (Hoagland et al. 1958, 1959) and recognized as the physical manifestation of the adaptor RNA predicted on theoretical grounds (Crick 1958). Once its function had been realized, the name transfer RNA replaced the term "soluble" RNA (sRNA). Chemical modification of the amino acid moiety of cysteine-charged tRNA^{Cys} (to alanine) confirmed that the RNA component is responsible for decoding the template (Chapeville et al. 1962). Thus, fidelity of information transfer from nucleic acid to protein rests in part

on the aminoacyl-tRNA synthetases. One of these, the valine-specific enzyme from *Escherichia coli*, was arguably the first macromolecular component of the protein synthetic apparatus to be characterized (Berg and Ofengand 1958), and additional synthetases soon followed.

Numerous enzymes catalyzing and facilitating the several steps in protein synthesis were steadily purified over the years, with an intense burst of activity in the 1960s and 1970s. In advance of full authentication of their purity and function, these proteins were provisionally called "factors," a term that has stuck. Although many of the factors have been known for almost half a century, the activities of some of them remained obscure or debatable until recently (e.g., EF-P and its homolog eIF5A) (Kang and Hershey 1994; Aoki et al. 1997; Doerfel et al. 2013; Ude et al. 2013), whereas others are still emerging (e.g., eIF2A and eIF2D) (Komar et al. 2005; Ventoso et al. 2006; Dmitriev et al. 2010; Starck et al. 2012; Kearse and Wilusz 2017), and new ones with specialized functions, such as the internal ribosome entry site (IRES) *trans*-acting factors (ITAFs), are being discovered (King et al. 2010; Lee et al. 2017).

The messenger RNA concept revolutionized thinking about gene expression in all cells. Genetics and bacteriophage biology, as well as biochemistry, played key parts in the genesis and confirmation of the messenger hypothesis (see, for example, Cobb 2015). Jacob and Monod (1961) hypothesized the existence of an unstable intermediate between the DNA of the gene and the ribosome, which could be related to the RNA produced in phage T2-infected cells (Volkin and Astrachan 1956; Nomura et al. 1960). On this view, the ribosome and other components of the protein synthesis machinery constitute a relatively stable decoding and synthetic apparatus that is programmed by unstable mRNA. This was soon confirmed in bacteria (Brenner et al. 1961; Gros et al. 1961) and bacterial cell-free systems. The discovery that poly(U) can direct the synthesis of polyphenylalanine in vitro (Nirenberg and Matthaei 1961) was a transformative event, spearheading the elucidation of the genetic code by the mid-1960s. The wobble hypothesis, which rationalizes features of the code's redundancy and its decoding by tRNAs, was published by Crick (1966). In higher cells, the existence of a class of rapidly labeled RNA, heterogeneous in size and with distinct chromatographic properties, was recognized. Its essential feature as an informational intermediary were confirmed and messenger-dependent eukaryotic cell-free translation systems appeared at the end of the decade (Laycock and Hunt 1969; Lockard and Lingrel 1969; Mathews and Korner 1970).

Building on these foundations, mechanisms explaining important aspects of translation in both prokaryotes and eukaryotes emerged in the 1970s. The question of initiation site selection was largely accounted for in prokaryotes by the Shine–Dalgarno sequence base-pairing with 16S ribosomal RNA (Shine and Dalgarno 1975). A solution to the problem in eukaryotes came later, in the 1980s, with recognition of cap-dependent scanning and the Kozak consensus sequence (Kozak 1978, 1986, 1987; Kozak and Shatkin 1978), followed by identification of IRES-dependent mechanisms (Jang et al. 1988; Pelletier and Sonenberg 1988). The optimal codon hypothesis explained how a bias in the usage of synonymous codons can influence the production levels of individual genes (Ikemura 1981) and the signal hypothesis (discussed below) accounted for protein transport into the endoplasmic reticulum (ER). Investigations of mRNA translation and metabolism were greatly facilitated by the discovery in the early 1970s of the terminal hallmarks found on most eukaryotic mRNAs, 5′ caps (Adams and Cory 1975; Both et al. 1975; Furuichi and Miura 1975; Furuichi et al. 1975a,b; Perry and Kelley 1975; Wei and Moss 1975) and 3′ poly(A) tails (Darnell et al. 1971a,b; Edmonds et al. 1971; Lee et al. 1971).

By the end of the 1980s, as a result of the identification and purification of most of the components of the translation system and the reconstitution of their activities in vitro, the pathway of protein synthesis had been defined and was well understood in outline (Hershey et al. 2018). Subsequent detailed analyses led to an in-depth understanding of many of the mechanisms of initiation, elongation, and termination in bacteria (Rodnina 2018) and in eukaryotic cells, with indispensable contributions

from yeast genetics and biochemistry (Dever et al. 2018; Hellen 2018; Merrick and Pavitt 2018). This has allowed questions of regulation to be addressed at ever-increasing levels of sophistication.

ORIGINS OF TRANSLATIONAL CONTROL

The idea of regulation at the transcriptional level flowed naturally from the messenger concept. Jacob and Monod (1961) wrote that "the synthesis of individual proteins may be provoked or suppressed within a cell, under the influence of specific external agents, and … the relative rates at which different proteins are synthesized may be profoundly altered, depending on external conditions." They recognized that such regulation "is absolutely essential to the survival of the cell," although the notion that it could be exerted at the translational level was not a principal focus in the bacterial field. Still, the seeds of the concept that gene expression can be regulated by the efficiency of protein synthesis emerged early, and some from work in bacterial systems.

Among the first observations of regulation at this level of gene expression were those made in rabbit reticulocytes (Borsook et al. 1952; Kruh and Borsook 1956). The term "translational control" itself was used in 1963 with respect to the differential expression of proteins from the RNA genome of MS2 phage in an *E. coli* cell-free translation system (Ohtaka and Spiegelman 1963). The concept spread rapidly into other areas of research, to the extent that less than 10 years later the presence of translationally silent mRNA that is activated on fertilization of sea urchin eggs was referred to as a "classical conclusion" (Humphreys 1971). After the early studies on phages, much of the focus was on eukaryotic systems. A virtue of translation as a site of regulation is that it affords a rapid response to external stimuli without invoking nuclear pathways for mRNA synthesis, processing, and transport. Correspondingly, the first cases recognized were mostly ones in which it was evident or simple to establish that transcription and other nuclear events were not responsible. To illustrate how the evidence for translational control arose, we briefly describe four paradigms, together with an early example of translational control at the level of elongation.

Sea Urchin Eggs

The eggs of sea urchins and other invertebrates synthesize protein at a very low rate but are triggered to incorporate amino acids within a few minutes of fertilization with little or no concomitant RNA synthesis (Hultin 1961; Nemer 1962; Gross et al. 1964). The first wave of increased translation, lasting several hours, is not blocked by inhibiting transcription (Gross et al. 1964) because the eggs contain preexisting mRNAs in a masked form that are not translated until fertilization. In principle, the limitation could be caused by a deficiency in the translational machinery, but there is little evidence to support this possibility (Humphreys 1969). For example, egg ribosomes can translate added poly(U) even though they display little intrinsic protein synthetic activity (Nemer 1962; Wilt and Hultin 1962). Deproteinized egg RNA can be translated in a cell-free system (Maggio et al. 1964; Monroy et al. 1965) and cytoplasmic messenger ribonucleoprotein (mRNP) particles were observed (Spirin and Nemer 1965). Because the assembly of masked mRNP complexes must take place during oogenesis, the sea urchin system exemplifies a reversible process of mRNA repression and activation. Current understanding of the diverse translational control processes operative during embryonic development and stem-cell differentiation in the adult are described by Teixeira and Lehmann (2018).

Mammalian Reticulocytes

It was taken for granted that protein synthesis (mainly hemoglobin) in mammalian reticulocytes, which are enucleate immature red blood cells, would be regulated at the translational level. In the intact rabbit reticulocyte, the synthesis of heme parallels that of globin (Kruh and Borsook 1956) and globin synthesis is controlled by the availability of heme or ferrous ions (Bruns and London 1965). Regulation by heme occurs in the reticulocyte lysate (Lamfrom and Knopf 1964), the forerunner of the messenger-dependent translation system of Pelham and Jackson (1976) and some coupled transcription-translation systems. When globin synthesis is inhibited in cells or extracts, the polysomes dissociate to monosomes (Hardesty et al. 1963; Waxman and Rabinowitz 1966), arguing that regulation impacts translation initiation. The effects of heme deprivation are mediated by the protein kinase HRI (heme-regulated inhibitor, EIF2AK1) and are mimicked by unrelated stimuli, including addition of glutathione disulfide (Kosower et al. 1971) or double-stranded RNA mediated by PKR (protein kinase R, EIF2AK2) (Ehrenfeld and Hunt 1971; Kosower et al. 1971). Regulation extends to all mRNAs in the reticulocyte lysate (Mathews et al. 1973), implying that a general mechanism of translational control is being invoked. This mechanism centers on the phosphorylation of the α subunit of eIF2, which results in reduced levels of ternary complex (eIF2•GTP•Met-tRNA$_i$) and impaired loading of the 40S ribosomal subunit with Met-tRNA$_i$ (Farrell et al. 1977). Considerable attention has been given to the family of eIF2 kinases, which confer sensitivity to a wide range of stimuli. In addition to HRI and PKR, PERK (PKR-like ER kinase, EIF2AK3) and GCN2 (general control nonderepressible 2, EIF2AK4) are activated by ER stress and uncharged tRNA, respectively (Merrick and Pavitt 2018; Wek 2018). PKZ, a PKR-like eIF2α kinase in fish, is activated by Z-DNA and can also inhibit translation (Bergan et al. 2008; Liu et al. 2013; Taghavi and Samuel 2013).

Physiological Stimuli

Cells and tissues of higher organisms regulate the expression of individual genes or classes of genes at the translational level in response to a wide variety of stimuli and conditions. Examples include responses to hormones (Eboué-Bonis et al. 1963; Garren et al. 1964; Martin and Young 1965; Tomkins et al. 1965) and ions (Drysdale and Munro 1965); changes in cell state, such as mitosis (Steward et al. 1968; Hodge et al. 1969; Fan and Penman 1970) and differentiation (Heywood 1970); and stress resulting from heat shock (McCormick and Penman

1969), treatment with noxious substances, and the incorporation of amino acid analogs (Thomas and Mathews 1984). Although these findings strengthened the view that such control is widespread and important, proof that it was exerted at the translational level was sometimes challenging in nucleated cells, let alone in tissues and whole organisms. One approach to this issue took advantage of selective inhibitors of transcription or translation, such as actinomycin D and cycloheximide, but the results were liable to be complicated by indirect or side effects of the drugs in complex systems. The rapidity of a response could also provide suggestive evidence for an effect at the translational level. Compelling data often came from investigations of the underlying biochemical processes, for example, by demonstrating changes in polysome profiles or initiation factor phosphorylation states. Several methods can provide rigorous evidence (Hershey et al. 2018) and ribosome profiling is a powerful and increasingly popular modern approach (Ingolia et al. 2009, 2018).

Cell growth is dependent on protein synthesis and translational control mediated by the mammalian or mechanistic target of rapamycin (mTOR), a protein kinase that lies at the nexus of numerous regulatory pathways. In the early 1990s, genetic screening for rapamycin-resistant genes in budding yeast uncovered TOR as a major regulator of cell growth (Heitman et al. 1991). A few years later, it was established that protein synthesis is a major downstream target (Barbet et al. 1996), and that mTOR controls translation initiation through phosphorylation of eIF4E-binding proteins (4E-BPs) (Beretta et al. 1996). eIF4E is the mRNA cap-binding protein (Sonenberg et al. 1979) required for cap-dependent initiation, and its activity is prevented by dephosphorylated 4E-BP. mTOR phosphorylates 4E-BP, releasing eIF4E, and allowing cap-dependent translation. The control of several other translation factors and regulators is also linked to mTOR activity (Proud 2018). mRNAs that harbor a $5'$-terminal oligopyrimidine (TOP) motif (Levy et al. 1991) were the first ones found to be translationally suppressed by rapamycin (Jefferies et al. 1994; Terada et al. 1994). The TOP mRNA class includes those encoding ribosomal proteins and elongation factors, consistent with the importance of mTOR in ribosome biogenesis, cell growth, and cancer (Proud 2018; Robichaud et al. 2018). The exact mechanism of TOP mRNA translation regulation remained elusive for many years, but in recent years the LARP1 protein has been shown to mediate this effect (Fonseca et al. 2015; Lahr et al. 2017).

Virus-Infected Cells

The small RNA phages, MS2 and its relatives, provided some of the first evidence for translational control, as well as the first clear case of a mechanism specific for the synthesis of an individual protein. The phage genome encodes four polypeptides (the maturation, coat, and lysis proteins and RNA replicase) that are initiated individually and are produced at dissimilar rates. Several regulatory interactions among them are now known. One was revealed by the observation that a nonsense mutation early in the cistron coding for phage coat protein down-regulates replicase synthesis (Lodish and Zinder 1966); passage of ribosomes through a critical region of the coat protein cistron melts the long-range RNA structure and allows replicase translation. A second nonsense mutation leads to overproduction of the replicase because the coat protein acts as a repressor of replicase translation, and the binding of phage coat protein to the hairpin structure containing the replicase AUG is a well-characterized RNA–protein interaction (Witherell et al. 1991). Subsequent studies have disclosed numerous translational control mechanisms in phages and in viruses infecting eukaryotes (Breaker 2018; Stern-Ginossar et al. 2018).

Cellular mRNAs are also subject to translational control during infection with many viruses (Stern-Ginossar et al. 2018). Inhibition of cellular mRNA translation, an aspect of host cell shutoff, may begin before the onset of viral protein synthesis and without any apparent interference with cellular mRNA production or stability. In poliovirus infection, the shutoff of host-cell translation can be complete within 2 hours after infection and is followed by a wave of viral protein synthesis (Summers et al. 1965).

Cite this article as *Cold Spring Harb Perspect Biol* doi: 10.1101/cshperspect.a035584

In the first phase, polysomes break down without any effect on translation elongation or termination (Penman and Summers 1965; Summers and Maizel 1967). In the second phase, virus-specific polysomes form (Penman et al. 1963). The cellular mRNA remains intact and translatable in vitro (Leibowitz and Penman 1971), evidence that initiation has become selective for viral mRNA. Translational inhibition extends to mRNAs produced by other viruses in a double infection (Ehrenfeld and Lund 1977), indicative of a general effect that later work ascribed to modification of the cap-binding complex, eIF4F. Cleavage of the eIF4G subunit of this complex prevents cap-dependent initiation on cellular mRNAs but does not interfere with initiation on the viral mRNA, which occurs by internal ribosome entry (Kwan and Thompson 2018). Viruses have evolved many such specialized mechanisms (Stern-Ginossar et al. 2018), some of which have led to the identification of parallel mechanisms in uninfected cells themselves.

Secretory Pathway

Protein synthesis is regulated predominantly at the level of initiation, consistent with the principle that it is more efficient to govern a pathway at its outset than to interrupt it midstream with the ensuing accumulation of intermediates and logjam of recyclable components. Nevertheless, well-characterized cases do occur later in the translational pathway, at the elongation and termination level (Dever et al. 2018). Proteins destined for secretion or retention in cell membranes are made on polysomes attached to the ER. In the early 1970s, it began to seem likely that ribosomes become associated with cell membranes only after protein synthesis has been initiated (Lisowska-Bernstein et al. 1970; Rosbash 1972) and what came to be called a signal peptide was found on secreted proteins (Milstein et al. 1972; Devilliers-Thiery et al. 1975). These findings lent substance to the signal hypothesis that proposed that an amino-terminal sequence might be responsible for secretion (Blobel and Sabatini 1971). The development of cell-free systems enabled biochemical dissection of the se-

cretory pathway (Blobel and Dobberstein 1975) leading to the discovery of the signal recognition particle (SRP), a ribonucleoprotein, and its receptor on the ER (Walter and Blobel 1981; Walter et al. 1981; Gilmore et al. 1982a,b; Meyer et al. 1982). The SRP also interacts with the ribosome such that the binding of the SRP to a nascent signal peptide causes translational arrest that is relieved when the ribosome docks with its ER receptor. This mechanism ensures cotranslational protein export and prevents the accumulation of secretory proteins in an improper subcellular compartment (the cytosol). From another perspective, this mechanism also represents an example of mRNA localization achieved by controlling its translation, distinct from several other methods used by cells to compartmentalize translation (Buxbaum et al. 2015; Biswas et al. 2018).

WHAT OF THE FUTURE?

In the past seven decades, combined genetic, biochemical, cell biological, pharmacological and structural approaches have uncovered the major components involved in protein synthesis, their interactions, and many of the sophisticated regulatory processes that adjust protein synthesis to developmental and environmental demands. Technological improvements such as cryoelectron microscopy (cryo-EM), real-time single-molecule microscopy, DNA and RNA sequencing, mass spectrometry, and rapid kinetic analysis now provide the opportunity to interrogate translation on spatial and temporal scales in ways that were not possible before. This increased resolution promises to bring the field from studies of cells in bulk to organelle-specific and even single-mRNA levels in vivo. Together with the increasingly detailed understanding of the role of translation in physiology and disease pathogenesis (Tahmasebi et al. 2018), there is optimism that therapeutic relief from acquired and genetic diseases may be on the horizon.

Yet the immense complexity of the translation system continues to pose new challenges and many uncharted areas remain. Little is known, for example, about protein synthesis and translational control in archaea, or in chloroplasts and mitochondria, even though mitochondrial ribosomes were reported half a century ago (Kuntzel and Noll 1967). Novel mechanisms can be confidently predicted to be uncovered by study of these organisms and organelles, as well, perhaps, from the giant viruses of *Acanthamoeba* that encode components of the translation apparatus (Bekliz et al. 2018; Stern-Ginossar et al. 2018). Even in well-studied areas, growing appreciation of the flexibility and complexity of the translational apparatus points to the likelihood that more mechanistic variety will be found than is currently appreciated. Discoveries such as repeat-associated non-ATG (RAN) translation (Zu et al. 2011, 2018) and ribosomal heterogeneity (Kondrashov et al. 2011; Robichaud et al. 2018) exemplify how much is still to be learned. Although it would be rash to be specific, advances in RNA biology—including base modifications and epitranscriptomics (Peer et al. 2018), and regulation by circular and noncoding RNAs (Chekulaeva and Rajewsky 2018) and microRNAs (Duchaine and Fabian 2018), as well as antisense and interfering RNAs—present new avenues for basic research and open fresh possibilities for bench-to-bedside translation.

REFERENCES

Reference is also in this collection.

Adams JM, Cory S. 1975. Modified nucleosides and bizarre 5′-termini in mouse myeloma mRNA. *Nature* **255:** 28–33.

Aoki H, Dekany K, Adams SL, Ganoza MC. 1997. The gene encoding the elongation factor P protein is essential for viability and is required for protein synthesis. *J Biol Chem* **272:** 32254–32259.

Arlinghaus RF, Schweet R. 1963. A ribosome-bound intermediate in polypeptide synthesis. *Biochem Biophys Res Commun* **11:** 92–96.

Barbet NC, Schneider U, Helliwell SB, Stansfield I, Tuite MF, Hall MN. 1996. TOR controls translation initiation and early G1 progression in yeast. *Mol Biol Cell* **7:** 25–42.

Bekliz M, Azza S, Seligmann H, Decloquement P, Raoult D, La Scola B. 2018. Experimental analysis of mimivirus translation initiation factor 4a reveals its importance in viral protein translation during infection of *Acanthamoeba polyphaga. J Virol* **92:** e00337.

Beretta L, Gingras AC, Svitkin YV, Hall MN, Sonenberg N. 1996. Rapamycin blocks the phosphorylation of 4E-BP1 and inhibits cap-dependent initiation of translation. *EMBO J* **15:** 658–664.

Berg P, Ofengand EJ. 1958. An enzymatic mechanism for linking amino acids to RNA. *Proc Natl Acad Sci* **44:** 78–86.

Bergan V, Jagus R, Lauksund S, Kileng O, Robertsen B. 2008. The Atlantic salmon Z-DNA binding protein kinase phosphorylates translation initiation factor 2α and constitutes a unique orthologue to the mammalian dsRNA-activated protein kinase R. *FEBS J* **275:** 184–197.

* Biswas J, Liu Y, Singer RH, Wu B. 2018. Fluorescence imaging methods to investigate translation in single cells. *Cold Spring Harb Perspect Biol* doi: 10.1101/cshperspect.a032722.

Blobel G, Dobberstein B. 1975. Transfer of proteins across membranes. II: Reconstitution of functional rough microsomes from heterologous components. *J Cell Biol* **67:** 852–862.

Blobel G, Sabatini DD. 1971. Ribosome-membrane interaction in eukaryotic cells. In *Biomembranes* (ed. Manson LA), pp. 193–195. Plenum, New York.

Borsook H, Deasy CL, Haagensmit AJ, Keighley G, Lowy PH. 1952. Incorporation in vitro of labeled amino acids into proteins of rabbit reticulocytes. *J Biol Chem* **196:** 669–694.

Both GW, Furuichi Y, Muthukrishnan S, Shatkin AJ. 1975. Ribosome binding to reovirus mRNA in protein synthesis requires 5′ terminal 7-methylguanosine. *Cell* **6:** 185–195.

* Breaker RR. 2018. Riboswitches and translation control. *Cold Spring Harb Perspect Biol* doi: 10.1101/cshperspect.a032797.

Brenner S, Jacob F, Meselson M. 1961. An unstable intermediate carrying information from genes to ribosomes for protein synthesis. *Nature* **190:** 576–581.

Bruns GP, London IM. 1965. The effect of hemin on the synthesis of globin. *Biochem Biophys Res Commun* **18:** 236–242.

Buxbaum AR, Haimovich G, Singer RH. 2015. In the right place at the right time: Visualizing and understanding mRNA localization. *Nat Rev Mol Cell Biol* **16:** 95–109.

Campbell PN, Work TS. 1953. Biosynthesis of proteins. *Nature* **171:** 997–1001.

Chapeville F, Lipmann F, von Ehrenstein G, Weisblum B, Ray WJ, Benzer S. 1962. On the role of soluble ribonucleic acid in coding for amino acids. *Proc Natl Acad Sci* **48:** 1086–1092.

* Chekulaeva M, Rajewsky N. 2018. Roles of long noncoding RNAs and circular RNAs in translation. *Cold Spring Harb Perspect Biol* doi: 10.1101/cshperspect.a032680.

Cobb M. 2015. Who discovered messenger RNA? *Curr Biol* **25:** R526–R532.

Crick FHC. 1958. On protein synthesis. *Symp Soc Exp Biol* **12:** 138–163.

Crick FH. 1959. Biochemical activities of nucleic acids. The present position of the coding problem. *Brookhaven Symp Biol* **12:** 35–39.

Crick FH. 1966. Codon–anticodon pairing: The wobble hypothesis. *J Mol Biol* **19:** 548–555.

Daly MM, Mirsky AE. 1952. Formation of protein in the pancreas. *J Gen Physiol* **36:** 243–254.

Cite this article as *Cold Spring Harb Perspect Biol* doi: 10.1101/cshperspect.a035584

Darnell JE, Philipson L, Wall R, Adesnik M. 1971a. Polyadenylic acid sequences: Role in conversion of nuclear RNA into messenger RNA. *Science* **174:** 507–510.

Darnell JE, Wall R, Tushinski RJ. 1971b. An adenylic acid-rich sequence in messenger RNA of HeLa cells and its possible relationship to reiterated sites in DNA. *Proc Natl Acad Sci* **68:** 1321–1325.

* Dever TE, Dinman JD, Green R. 2018. Translation elongation and recoding in eukaryotes. *Cold Spring Harb Perspect Biol* doi: 10.1101/cshperspect.a032649.

Devilliers-Thiery A, Kindt T, Scheele G, Blobel G. 1975. Homology in amino-terminal sequence of precursors to pancreatic secretory proteins. *Proc Natl Acad Sci* **72:** 5016–5020.

Dmitriev SE, Terenin IM, Andreev DE, Ivanov PA, Dunaevsky JE, Merrick WC, Shatsky IN. 2010. GTP-independent tRNA delivery to the ribosomal P-site by a novel eukaryotic translation factor. *J Biol Chem* **285:** 26779–26787.

Doerfel LK, Wohlgemuth I, Kothe C, Peske F, Urlaub H, Rodnina MV. 2013. EF-P is essential for rapid synthesis of proteins containing consecutive proline residues. *Science* **339:** 85–88.

Drysdale JW, Munro HN. 1965. Failure of actinomycin D to prevent induction of liver apofentin after iron administration. *Biochim Biophys Acta* **103:** 185–188.

* Duchaine TF, Fabian MR. 2018. Mechanistic insights into microRNA-mediated gene silencing. *Cold Spring Harb Perspect Biol* doi: 10.1101/cshperspect.a032771.

Eboué-Bonis D, Chambaut AM, Volfin P, Clauser H. 1963. Action of insulin on the isolated rat diaphragm in the presence of actinomycin D and puromycin. *Nature* **199:** 1183–1184.

Edmonds M, Vaughan MH Jr, Nakazato H. 1971. Polyadenylic acid sequences in the heterogeneous nuclear RNA and rapidly-labeled polyribosomal RNA of HeLa cells: Possible evidence for a precursor relationship. *Proc Natl Acad Sci* **68:** 1336–1340.

Ehrenfeld E, Hunt T. 1971. Double-stranded poliovirus RNA inhibits initiation of protein synthesis by reticulocyte lysates. *Proc Natl Acad Sci* **68:** 1075–1078.

Ehrenfeld E, Lund H. 1977. Untranslated vesicular stomatitis virus messenger RNA after poliovirus infection. *Virology* **80:** 297–308.

Fan H, Penman S. 1970. Regulation of protein synthesis in mammalian cells. II: Inhibition of protein synthesis at the level of initiation during mitosis. *J Mol Biol* **50:** 655–670.

Farrell PJ, Balkow K, Hunt T, Jackson RJ, Trachsel H. 1977. Phosphorylation of initiation factor eIF-2 and the control of reticulocyte protein synthesis. *Cell* **11:** 187–200.

Fonseca BD, Zakaria C, Jia JJ, Graber TE, Svitkin Y, Tahmasebi S, Healy D, Hoang HD, Jensen JM, Diao IT, et al. 2015. La-related protein 1 (LARP1) represses terminal oligopyrimidine (TOP) mRNA translation downstream of mTOR complex 1 (mTORC1). *J Biol Chem* **290:** 15996–16020.

Furuichi Y, Miura K. 1975. A blocked structure at the 5′ terminus of mRNA from cytoplasmic polyhedrosis virus. *Nature* **253:** 374–375.

Furuichi Y, Morgan M, Muthukrishnan S, Shatkin AJ. 1975a. Reovirus messenger RNA contains a methylated, blocked 5′-terminal structure: m^{7}G(5′)ppp(5′)G-MpCp. *Proc Natl Acad Sci* **72:** 362–366.

Furuichi Y, Morgan M, Shatkin AJ, Jelinek W, Salditt-Georgieff M, Darnell JE. 1975b. Methylated, blocked 5 termini in HeLa cell mRNA. *Proc Natl Acad Sci* **72:** 1904–1908.

Garren LD, Howell RR, Tomkins GM, Crocco RM. 1964. A paradoxical effect of actinomycin D: The mechanism of regulation of enzyme synthesis by hydrocortisone. *Proc Natl Acad Sci* **52:** 1121–1129.

Genuth NR, Barna M. 2018. Heterogeneity and specialized functions of translation machinery: From genes to organisms. *Nat Rev Genet* doi: 10.1038/s41576-018-0008-z.

Gierer A. 1963. Function of aggregated reticulocyte ribosomes in protein synthesis. *J Mol Biol* **6:** 148–157.

Gilmore R, Blobel G, Walter P. 1982a. Protein translocation across the endoplasmic reticulum. I: Detection in the microsomal membrane of a receptor for the signal recognition particle. *J Cell Biol* **95:** 463–469.

Gilmore R, Walter P, Blobel G. 1982b. Protein translocation across the endoplasmic reticulum. II: Isolation and characterization of the signal recognition particle receptor. *J Cell Biol* **95:** 470–477.

Goodman HM, Rich A. 1963. Mechanism of polyribosome action during protein synthesis. *Nature* **199:** 318–322.

Gros F, Hiatt H, Gilbert W, Kurland GG, Risebrough RW, Watson JD. 1961. Unstable ribonucleic acid revealed by pulse labelling of *Escherichia coli*. *Nature* **190:** 581–585.

Gross PR, Malkin LI, Moyer WA. 1964. Templates for the first proteins of embryonic development. *Proc Natl Acad Sci* **51:** 407–414.

Hardesty B, Miller R, Schweet R. 1963. Polyribosome breakdown and hemoglobin synthesis. *Proc Natl Acad Sci* **50:** 924–931.

Heitman J, Movva NR, Hall MN. 1991. Targets for cell cycle arrest by the immunosuppressant rapamycin in yeast. *Science* **253:** 905–909.

* Hellen CUT. 2018. Translation termination and ribosome recycling in eukaryotes. *Cold Spring Harb Perspect Biol* doi: 10.1101/cshperspect.a032656.

* Hershey JWB, Sonenberg N, Mathews MB. 2018. Principles of translational control. *Cold Spring Harb Perspect Biol* doi: 10.1101/cshperspect.a032607.

Heywood SM. 1970. Specificity of mRNA binding factor in eukaryotes. *Proc Natl Acad Sci* **67:** 1782–1788.

Hoagland MB, Stephenson ML, Scott JF, Hecht LI, Zamecnik PC. 1958. A soluble ribonucleic acid intermediate in protein synthesis. *J Biol Chem* **231:** 241–257.

Hoagland MB, Zamecnik PC, Stephenson ML. 1959. A hypothesis concerning the roles of particulate and soluble ribonucleic acids in protein synthesis. In *A symposium on molecular biology* (ed. Zirkle RE). University of Chicago, Chicago, IL.

Hodge LD, Robbins E, Scharff MD. 1969. Persistence of messenger RNA through mitosis in HeLa cells. *J Cell Biol* **40:** 497–507.

Holley RW, Apgar J, Everett GA, Madison JT, Marquisee M, Merrill SH, Penswick JR, Zamir A. 1965. Structure of a Ribonucleic Acid. *Science* **147:** 1462–1465.

Hultin T. 1961. Activation of ribosomes in sea urchin eggs in response to fertilization. *Exp Cell Res* **25:** 405–417.

Hultin T, Beskow G. 1956. The incorporation of ^{14}C-L-leucine into rat liver proteins in vitro visualized as a two-step reaction. *Exp Cell Res* **11**: 664–666.

Humphreys T. 1969. Efficiency of translation of messenger-RNA before and after fertilization in sea urchins. *Dev Biol* **20**: 435–458.

Humphreys T. 1971. Measurements of messenger RNA entering polysomes upon fertilization of sea urchin eggs. *Dev Biol* **26**: 201–208.

Ikemura T. 1981. Correlation between the abundance of *Escherichia coli* transfer RNAs and the occurrence of the respective codons in its protein genes: A proposal for a synonymous codon choice that is optimal for the *E. coli* translational system. *J Mol Biol* **151**: 389–409.

Ingolia NT, Ghaemmaghami S, Newman JR, Weissman JS. 2009. Genome-wide analysis in vivo translation with nucleotide resolution using ribosome profiling. *Science* **324**: 218–223.

* Ingolia NT, Hussmann JA, Weissman JS. 2018. Ribosome profiling: Global views of translation. *Cold Spring Harb Perspect Biol* doi: 10.1101/cshperspect.a032698.

Jacob F, Monod J. 1961. Genetic regulatory mechanisms in the synthesis of proteins. *J Mol Biol* **3**: 318–356.

Jang SK, Krausslich HG, Nicklin MJ, Duke GM, Palmenberg AC, Wimmer E. 1988. A segment of the 5′ nontranslated region of encephalomyocarditis virus RNA directs internal entry of ribosomes during in vitro translation. *J Virol* **62**: 2636–2643.

Jefferies HB, Reinhard C, Kozma SC, Thomas G. 1994. Rapamycin selectively represses translation of the "polypyrimidine tract" mRNA family. *Proc Natl Acad Sci* **91**: 4441–4445.

* Jobe A, Liu Z, Gutierrez-Vargas C, Frank J. 2018. New insights into ribosome structure and function. *Cold Spring Harb Perspect Biol* doi: 10.1101/cshperspect.a032615.

Kang HA, Hershey JW. 1994. Effect of initiation factor eIF-5A depletion on protein synthesis and proliferation of *Saccharomyces cerevisiae*. *J Biol Chem* **269**: 3934–3940.

Kearse MG, Wilusz JE. 2017. Non-AUG translation: A new start for protein synthesis in eukaryotes. *Genes Dev* **31**: 1717–1731.

Keller EB, Zamecnik PC. 1956. The effect of guanosine diphosphate and triphosphate on the incorporation of labeled amino acids into proteins. *J Biol Chem* **221**: 45–59.

Khajuria RK, Munschauer M, Ulirsch JC, Fiorini C, Ludwig LS, McFarland SK, Abdulhay NJ, Specht H, Keshishian H, Mani DR, et al. 2018. Ribosome levels selectively regulate translation and lineage commitment in human hematopoiesis. *Cell* **173**: 90–103.e119.

King HA, Cobbold LC, Willis AE. 2010. The role of IRES *trans*-acting factors in regulating translation initiation. *Biochem Soc Trans* **38**: 1581–1586.

Komar AA, Gross SR, Barth-Baus D, Strachan R, Hensold JO, Goss Kinzy T, Merrick WC. 2005. Novel characteristics of the biological properties of the yeast *Saccharomyces cerevisiae* eukaryotic initiation factor 2A. *J Biol Chem* **280**: 15601–15611.

Kondrashov N, Pusic A, Stumpf CR, Shimizu K, Hsieh AC, Ishijima J, Shiroishi T, Barna M. 2011. Ribosome-mediated specificity in Hox mRNA translation and vertebrate tissue patterning. *Cell* **145**: 383–397.

Kosower NS, Vanderhoff GA, Benerofe B, Hunt T, Kosower EM. 1971. Inhibition of protein synthesis by glutathione disulfide in the presence of glutathione. *Biochem Biophys Res Commun* **45**: 816–821.

Kozak M. 1978. How do eukaryotic ribosomes select initiation regions in messenger RNA. *Cell* **15**: 1109–1123.

Kozak M. 1986. Point mutations define a sequence flanking the AUG initiator codon that modulates translation by eukaryotic ribosomes. *Cell* **44**: 283–292.

Kozak M. 1987. An analysis of 5′-noncoding sequences from 699 vertebrate messenger RNAs. *Nucleic Acids Res* **15**: 8125–8148.

Kozak M, Shatkin AJ. 1978. Migration of 40 S ribosomal subunits on messenger RNA in the presence of edeine. *J Biol Chem* **253**: 6568–6577.

Kruh J, Borsook H. 1956. Hemoglobin synthesis in rabbit reticulocytes in vitro. *J Biol Chem* **220**: 905–915.

Kuntzel H, Noll H. 1967. Mitochondrial and cytoplasmic polysomes from *Neurospora crassa*. *Nature* **215**: 1340–1345.

* Kwan T, Thompson SR. 2018. Noncanonical translation initiation in eukaryotes. *Cold Spring Harb Perspect Biol* doi: 10.1101/cshperspect.a032672.

Lahr RM, Fonseca BD, Ciotti GE, Al-Ashtal HA, Jia JJ, Niklaus MR, Blagden SP, Alain T, Berman AJ. 2017. La-related protein 1 (LARP1) binds the mRNA cap, blocking eIF4F assembly on TOP mRNAs. *eLife* **6**: e24146.

Lamfrom H, Knopf PM. 1964. Initiation of haemoglobin synthesis in cell-free systems. *J Mol Biol* **9**: 558–575.

Laycock DG, Hunt JA. 1969. Synthesis of rabbit globin by a bacterial cell free system. *Nature* **221**: 1118–1122.

Lee SY, Mendecki J, Brawerman G. 1971. A polynucleotide segment rich in adenylic acid in the rapidly-labeled polyribosomal RNA component of mouse sarcoma 180 ascites cells. *Proc Natl Acad Sci* **68**: 1331–1335.

Lee KM, Chen CJ, Shih SR. 2017. Regulation mechanisms of viral IRES-driven translation. *Trends Microbiol* **25**: 546–561.

Leibowitz R, Penman S. 1971. Regulation of protein synthesis in HeLa cells. III: Inhibition during poliovirus infection. *J Virol* **8**: 661–668.

Levy S, Avni D, Hariharan N, Perry RP, Meyuhas O. 1991. Oligopyrimidine tract at the 5′ end of mammalian ribosomal protein mRNAs is required for their translational control. *Proc Natl Acad Sci* **88**: 3319–3323.

Lisowska-Bernstein B, Lamm ME, Vassalli P. 1970. Synthesis of immunoglobulin heavy and light chains by the free ribosomes of a mouse plasma cell tumor. *Proc Natl Acad Sci* **66**: 425–432.

Liu ZY, Jia KT, Li C, Weng SP, Guo CJ, He JG. 2013. A truncated Danio rerio PKZ isoform functionally interacts with eIF2α and inhibits protein synthesis. *Gene* **527**: 292–300.

Lockard RE, Lingrel JB. 1969. The synthesis of mouse hemoglobin β-chains in a rabbit reticulocyte cell-free system programmed with mouse reticulocyte 9S RNA. *Biochem Biophys Res Commun* **37**: 204–212.

Lodish HF, Zinder ND. 1966. Mutants of the bacteriophage f2. VIII: Control mechanisms for phage-specific syntheses. *J Mol Biol* **19**: 333–348.

Maggio R, Vittorelli ML, Rinaldi AM, Monroy A. 1964. In vitro incorporation of amino acids into proteins stimulated by RNA from unfertilized sea urchin eggs. *Biochem Biophys Res Commun* **15:** 436–441.

Marks PA, Burka ER, Schlessinger D. 1962. Protein synthesis in erythroid cells. I. Reticulocyte ribosomes active in stimulating amino acid incorporation. *Proc Natl Acad Sci* **48:** 2163–2171.

Martin TE, Young FG. 1965. An in vitro action of human growth hormone in the presence of actinomycin D. *Nature* **208:** 684–685.

Mathews M, Korner A. 1970. Mammalian cell-free protein synthesis directed by viral ribonucleic acid. *Eur J Biochem* **17:** 328–338.

Mathews MB, Hunt T, Brayley A. 1973. Specificity of the control of protein synthesis by haemin. Mammalian messenger RNA. *Nat New Biol* **243:** 230–233.

Mathews MB, Sonenberg N, Hershey JW. 2007. Origins and principles of translational control. In *Translational control in biology and medicine* (ed. Mathews MB, Sonenberg N, Hershey JWB). Cold Spring Harbor Laboratory Press, Cold Spring Harbor, NY.

McCormick W, Penman S. 1969. Regulation of protein synthesis in HeLa cells: Translation at elevated temperatures. *J Mol Biol* **39:** 315–333.

McQuillen K, Roberts RB, Britten RJ. 1959. Synthesis of nascent protein by ribosomes in *Escherichia coli*. *Proc Natl Acad Sci* **45:** 1437–1447.

* Merrick WC, Pavitt GD. 2018. Protein synthesis initiation in eukaryotic cells. *Cold Spring Harb Perspect Biol* doi: 10.1101/cshperspect.a033092.

Meyer DI, Krause E, Dobberstein B. 1982. Secretory protein translocation across membranes-the role of the "docking protein." *Nature* **297:** 647–650.

Meyer KD, Patil DP, Zhou J, Zinoviev A, Skabkin MA, Elemento O, Pestova TV, Qian SB, Jaffrey SR. 2015. 5′ UTR m^{6}A promotes cap-independent translation. *Cell* **163:** 999–1010.

Mills EW, Green R. 2017. Ribosomopathies: There's strength in numbers. *Science* **358:** eaan2755.

Milstein C, Brownlee GG, Harrison TM, Mathews MB. 1972. A possible precursor of immunoglobulin light chains. *Nat New Biol* **239:** 117–120.

Monroy A, Maggio R, Rinaldi AM. 1965. Experimentally induced activation of the ribosomes of the unfertilized sea urchin egg. *Proc Natl Acad Sci* **54:** 107–111.

Nakamoto TC, Conway TW, Allende JE, Spyrides G, Lipmann F. 1963. Formation of peptide bonds. I: Peptide formation from aminoacyl-S-RNA. *Cold Spring Harb Quant Biol* **28:** 227.

Nemer M. 1962. Interrelation of messenger polyribonucleotides and ribosomes in the sea urchin egg during embryonic development. *Biochem Biophys Res Commun* **8:** 511–515.

Nirenberg MW, Matthaei JH. 1961. The dependence of cell-free protein synthesis in *E. coli* upon naturally occurring or synthetic polyribonucleotides. *Proc Natl Acad Sci* **47:** 1588–1602.

Noll HS, Wettstein T. 1963. Ribosomal aggregates engaged in protein synthesis: Ergosome breakdown and messenger ribonucleic acid transport. *Nature* **198:** 632–638.

Nomura MA, Hall BD, Spiegelman BM. 1960. Characterization of RNA synthesized in *Escherichia coli* after bacteriophage T2 infection. *J Mol Biol* **2:** 306–326.

Ohtaka Y, Spiegelman S. 1963. Translational control of protein synthesis in a cell-free system directed by a polycistronic viral RNA. *Science* **142:** 493–497.

Palade GE. 1955. A small particulate component of the cytoplasm. *J Biophys Biochem Cytol* **1:** 59–68.

* Peer Y, Moshitch-Moshkovitz S, Rechavi G, Dominissini D. 2018. The epitranscriptome in translation regulation. *Cold Spring Harb Perspect Biol* doi: 10.1101/cshperspect.a032623.

Pelham HRB, Jackson RJ. 1976. An efficient mRNA-dependent translation system from reticulocyte lysates. *Eur J Biochem* **67:** 247–256.

Pelletier J, Sonenberg N. 1988. Internal initiation of translation of eukaryotic mRNA directed by a sequence derived from poliovirus RNA. *Nature* **334:** 320–325.

Penman S, Summers D. 1965. Effects on host cell metabolism following synchronous infection with poliovirus. *Virol* **27:** 614–620.

Penman S, Scherrer K, Becker Y, Darnell JE. 1963. Polyribosomes in normal and poliovirus-infected HeLa cells and their relationship to messenger RNA. *Proc Natl Acad Sci* **49:** 654–662.

Perry RP, Kelley DE. 1975. Methylated constituents of heterogeneous nuclear RNA: Presence in blocked 5′ terminal structures. *Cell* **6:** 13–19.

* Proud CG. 2018. Phosphorylation and signal transduction pathways in translational control. *Cold Spring Harb Perspect Biol* doi: 10.1101/cshperspect.a033050.

* Robichaud N, Sonenberg N, Ruggero D, Schneider RJ. 2018. Translational control in cancer. *Cold Spring Harb Perspect Biol* doi: 10.1101/cshperspect.a032896.

* Rodnina MV. 2018. Translation in prokaryotes. *Cold Spring Harb Perspect Biol* doi: 10.1101/cshperspect.a032664.

Rosbash M. 1972. Formation of membrane-bound polyribosomes. *J Mol Biol* **65:** 413–422.

Sanger F, Tuppy H. 1951. The amino-acid sequence in the phenylalanyl chain of insulin. 2. The investigation of peptides from enzymic hydrolysates. *Biochem J* **49:** 481–490.

Sauert M, Temmel H, Moll I. 2015. Heterogeneity of the translational machinery: Variations on a common theme. *Biochimie* **114:** 39–47.

Shine J, Dalgarno L. 1975. Determinant of cistron specificity in bacterial ribosomes. *Nature* **254:** 34–38.

Siekevitz P, Zamecnik PC. 1951. In vitro incorporation of 1-^{14}C-DL-alanine into proteins of rat-liver granular fractions. *Fed Proc* **10:** 246–247.

Sonenberg N, Rupprecht KM, Hecht SM, Shatkin AJ. 1979. Eukaryotic mRNA cap binding protein: Purification by affinity chromatography on sepharose-coupled m^{7}GDP. *Proc Natl Acad Sci* **76:** 4345–4349.

Spirin AS, Nemer M. 1965. Messenger RNA in early sea urchin embryos: Cytoplasmic particles. *Science* **150:** 214–217.

Starck SR, Jiang V, Pavon-Eternod M, Prasad S, McCarthy B, Pan T, Shastri N. 2012. Leucine-tRNA initiates at CUG start codons for protein synthesis and presentation by MHC class I. *Science* **336:** 1719–1723.

* Stern-Ginossar N, Thompson SR, Mathews MB, Mohr I. 2018. Translational control in virus-infected cells. *Cold Spring Harb Perspect Biol* doi: 10.1101/cshperspect. a033001.

Steward DL, Shaeffer JR, Humphrey RM. 1968. Breakdown and assembly of polyribosomes in synchronized Chinese hamster cells. *Science* **161**: 791–793.

Summers DF, Maizel JV. 1967. Disaggregation of HeLa cell polysomes after infection with poliovirus. *Virol* **31**: 550–552.

Summers DF, Maizel JV, Darnell JE. 1965. Evidence for virus-specific noncapsid proteins in poliovirus-infected HeLa cells. *Proc Natl Acad Sci* **54**: 505–513.

Taghavi N, Samuel CE. 2013. RNA-dependent protein kinase PKR and the Z-DNA binding orthologue PKZ differ in their capacity to mediate initiation factor eIF2α-dependent inhibition of protein synthesis and virus-induced stress granule formation. *Virology* **443**: 48–58.

Tahmasebi S, Khoutorsky A, Mathews MB, Sonenberg N. 2018. Translation deregulation in human disease. *Nat Rev Mol Cell Biol* doi: 10.1038/s41580-018-0034-x.

* Teixeira FK, Lehmann R. 2018. Translational control during development transitions. *Cold Spring Harb Perspect Biol* doi: 10.1101/cshperspect.a032987.

Terada N, Patel HR, Takase K, Kohno K, Nairn AC, Gelfand EW. 1994. Rapamycin selectively inhibits translation of mRNAs encoding elongation factors and ribosomal proteins. *Proc Natl Acad Sci* **91**: 11477–11481.

Thomas GP, Mathews MB. 1984. Alterations of transcription and translation in HeLa cells exposed to amino acid analogs. *Mol Cell Biol* **4**: 1063–1072.

Tomkins GM, Garren LD, Howell RR, Peterkofsky B. 1965. The regulation of enzyme synthesis by steroid hormones: the role of translation. *J Cell Comp Physiol* **66**: 137–151.

Ude S, Lassak J, Starosta AL, Kraxenberger T, Wilson DN, Jung K. 2013. Translation elongation factor EF-P alleviates ribosome stalling at polyproline stretches. *Science* **339**: 82–85.

Ventoso I, Sanz MA, Molina S, Berlanga JJ, Carrasco L, Esteban M. 2006. Translational resistance of late α virus mRNA to eIF2α phosphorylation: A strategy to overcome the antiviral effect of protein kinase PKR. *Genes Dev* **20**: 87–100.

Volkin E, Astrachan L. 1956. Phosphorus incorporation in *Escherichia coli* ribonucleic acid after infection with bacteriophage T2. *Virology* **2**: 149–161.

Walter P, Blobel G. 1981. Translocation of proteins across the endoplasmic reticulum. II: Signal recognition protein (SRP) mediates the selective binding to microsomal membranes of in-vitro-assembled polysomes synthesizing secretory protein. *J Cell Biol* **91**: 551–556.

Walter P, Ibrahimi I, Blobel G. 1981. Translocation of proteins across the endoplasmic reticulum. I: Signal recognition protein (SRP) binds to in-vitro-assembled polysomes synthesizing secretory protein. *J Cell Biol* **91**: 545–550.

Warner JR, Rich A, Hall CE. 1962. Electron microscope studies of ribosomal clusters synthesizing hemoglobin. *Science* **138**: 1399–1403.

Warner JR, Knopf PM, Rich A. 1963. A multiple ribosomal structure in protein synthesis. *Proc Natl Acad Sci* **49**: 122–129.

Waxman HS, Rabinowitz M. 1966. Control of reticulocyte polyribosome content and hemoglobin synthesis by heme. *Biochim Biophys Acta* **129**: 369–379.

Wei CM, Moss B. 1975. Methylated nucleotides block 5′-terminus of vaccinia virus messenger RNA. *Proc Natl Acad Sci* **72**: 318–322.

* Wek RC. 2018. Role of eIF2α kinases in translational control and adaptation to cellular stress. *Cold Spring Harb Perspect Biol* doi: 10.1101/cshperspect.a032870.

Wettstein FO, Staehelin T, Noll H. 1963. Ribosomal aggregate engaged in protein synthesis: characterization of the ergosome. *Nature* **197**: 430–435.

Wilt FH, Hultin T. 1962. Stimulation of phenylalanine incorporation by polyuridylic acid in homogenates of sea urchin eggs. *Biochem Biophys Res Commun* **9**: 313–317.

Witherell GW, Gott JM, Uhlenbeck OC. 1991. Specific interaction between RNA phage coat proteins and RNA. *Prog Nucleic Acid Res Mol Biol* **40**: 185–220.

Zamecnik PC. 1960. Historical and current aspects of the problem of protein synthesis. *Harvey Lectures* **54**: 256–281.

Zamecnik PC. 1969. An historical account of protein synthesis, with current overtones—A personalized view. *Cold Spring Harb Symp Quant Biol* **34**: 1–16.

Zamecnik PC, Keller EB. 1954. Relation between phosphate energy donors and incorporation of labeled amino acids into proteins. *J Biol Chem* **209**: 337–354.

Zhou J, Wan J, Gao X, Zhang X, Jaffrey SR, Qian SB. 2015. Dynamic m⁶A mRNA methylation directs translational control of heat shock response. *Nature* **526**: 591–594.

Zu T, Gibbens B, Doty NS, Gomes-Pereira M, Huguet A, Stone MD, Margolis J, Peterson M, Markowski TW, Ingram MA, et al. 2011. Non-ATG-initiated translation directed by microsatellite expansions. *Proc Natl Acad Sci* **108**: 260–265.

* Zu T, Pattamatta A, Ranum LPW. 2018. RAN translation in neurological diseases. *Cold Spring Harb Perspect Biol* doi: 10.1101/cshperspect.a033019.

Protein Synthesis Initiation in Eukaryotic Cells

William C. Merrick[1] and Graham D. Pavitt[2]

[1]Department of Biochemistry, School of Medicine, Case Western Reserve University, Cleveland, Ohio 44106

[2]Division of Molecular and Cellular Function, Faculty of Biology Medicine and Health, Manchester Academic Health Science Centre, The University of Manchester, Manchester M13 9PT, United Kingdom

Correspondence: wcm2@case.edu; graham.pavitt@manchester.ac.uk

This review summarizes our current understanding of the major pathway for the initiation phase of protein synthesis in eukaryotic cells, with a focus on recent advances. We describe the major scanning or messenger RNA (mRNA) m^{7}G cap-dependent mechanism, which is a highly coordinated and stepwise regulated process that requires the combined action of at least 12 distinct translation factors with initiator transfer RNA (tRNA), ribosomes, and mRNAs. We limit our review to studies involving either mammalian or budding yeast cells and factors, as these represent the two best-studied experimental systems, and only include a reference to other organisms where particular insight has been gained. We close with a brief description of what we feel are some of the major unknowns in eukaryotic initiation.

We present a summary of the current knowledge of the molecular mechanisms enabling the initiation of protein synthesis in eukaryotic cells. We focus on the major m^{7}G cap-dependent pathway, and only briefly consider alternative initiation routes. We cannot cover all the work that has been performed and so we focus primarily on more recent studies. The evidence for the pathway derives from a wide variety of complementary biochemical, genetic, and structural biology approaches used to mainly study both mammalian translation initiation and that of the yeast *Saccharomyces cerevisiae*. The systems are generally highly similar, but there are some differences and increased complexity is found in mammalian cells. One advantage of the yeast system is its high amenability to genetics that, when combined with other complementary approaches, has provided deep mechanistic insight. We begin with a brief overview of the general initiation pathway, focusing on the roles of the initiation factors and then discuss each step in detail, highlighting key points of mechanistic understanding. The review concludes with a consideration of important questions that remain as yet unanswered.

AN OVERVIEW OF THE INITIATION PATHWAY

The initiation of protein synthesis is the process that results in bringing together an 80S ribosome with a messenger RNA (mRNA) and initiator methionyl-transfer RNA (Met-tRNA$_i$). These three components combine such that Met-tRNA$_i$ makes codon–anticodon base-pair interactions with the correct initiation codon at the start of an open-reading frame (ORF) within the

Cite this article as *Cold Spring Harb Perspect Biol* doi: 10.1101/cshperspect.a033092

mRNA. AUG is the typical initiation codon, although near-cognate codons are used to initiate on some mRNAs (Starck and Shastri 2016; Kearse and Wilusz 2017). The codon–anticodon interaction is confined within the P site of an 80S ribosome. This topology then allows the elongation phase of protein synthesis to begin, by binding an elongator aminoacyl-tRNA to the vacant 80S A site to read the adjacent codon downstream. Although this may appear relatively simple, eukaryotes have evolved a complex translation initiation pathway. There are at least 12 dedicated proteins, termed eukaryotic initiation factors (eIFs), that each play critical roles in the process, and several of these factors are comprised of multisubunit proteins. Why so involved, when prokaryotes initiate translation using just three factors (Rodnina 2018)? Of course, this is a matter of conjecture, but it likely reflects the need to tightly control a process that consumes a large amount of cellular energy in the form of ATP and GTP and can cause serious consequences when out of balance, as evidenced by many disease states caused by aberrant translation (see Stern-Ginossar et al. 2018; Wek 2018). In addition to the discussion below, readers may find it informative to consult previous reviews covering the topics we discuss here (Hinnebusch and Lorsch 2012; Hinnebusch 2014, 2017; Dever et al. 2016).

The initiation pathway is shown schematically in Figure 1 and is envisaged as a series of steps (blue text) in which the eIFs guide the Met-tRNA$_i$ and ribosomal subunits to the mRNA AUG codon. Met-tRNA$_i$ is brought to the ribosome in a complex with eIF2•GTP. eIF2 is a GDP/GTP binding "G" protein and has high affinity for Met-tRNA$_i$ only when GTP-bound (Kapp and Lorsch 2004). The eIF2•GTP•Met-tRNA$_i$ complex is widely known as the ternary complex (TC). This is typically considered the first step in translation initiation. As eIF2•GDP is the stable form of eIF2 (Panniers et al. 1988; Erickson and Hannig 1996), it must first be activated to eIF2•GTP by the guanine nucleotide exchange factor (GEF) eIF2B before the TC can form (Rowlands et al. 1988; Pavitt et al. 1998). Next, the TC binds to the small ribosomal subunit (40S) with eIF5, eIF3, eIF1, and eIF1A to form a larger 43S preinitiation complex (PIC). There may be several routes to PIC formation. eIF1, eIF1A, and eIF3 can dissociate 80S complexes and bind to 40S subunits (Pisarev et al. 2007) and then recruit eIF5/TC. Alternatively, eIF3 and eIF1 can form an independent multifactor complex (MFC) with TC and eIF5 before binding the 40S subunit (Asano et al. 2000; Sokabe et al. 2012).

mRNAs are activated by binding eIF4F at the 5' 7-methylguanosine cap (m^{7}G cap) and the poly(A) binding protein (PABP) at the 3' poly(A) tail (Jackson et al. 2010). eIF4F comprises a complex formed between eIF4E, eIF4G, and eIF4A. The large eIF4G subunit has binding sites for eIF4E, PABP, and RNA and can form a "closed-loop" circularized mRNA/RNA-binding protein (RPB) complex (Fig. 1). eIF4A is an RNA helicase and both eIF4F formation and the binding to eIF4A of an accessory factor eIF4B enhance eIF4A activity. eIF4A unwinds mRNA secondary structures to facilitate 43S PIC recruitment at, or close to, the m^{7}G cap, forming an mRNA•43S complex (Fig. 1) (Kumar et al. 2016). eIFs 1 and 1A bound at the P and A sites of the 40S subunit, respectively, open a cleft in the 40S between its "head" and "body" that facilitates single-stranded mRNA binding and the subsequent scanning step. Further interactions between eIF3 in the 43S PIC and a region of eIF4G on the activated mRNA facilitate forming and stabilizing this intermediate complex (Villa et al. 2013). mRNAs have 5' leaders, typically called 5' untranslated regions (5'UTRs), that span the distance between the m^{7}G cap and the initiation codon for the major ORF. 5'UTRs vary in length, sequence, and structure. Usually the AUG codon closest to the m^{7}G cap is used to initiate protein synthesis, but there are exceptions.

Scanning describes the movement of the 43S PIC along the mRNA from the m^{7}G cap, in a 3' direction, searching for an AUG initiation codon in a suitable context to make a stable mRNA codon–tRNA$_i$ anticodon interaction. This requires energy in the form of ATP. In Figure 1, this is depicted with PIC/m^{7}G cap complex interactions maintained. This "open" scanning form transitions to a "closed" complex on

Cite this article as *Cold Spring Harb Perspect Biol* doi: 10.1101/cshperspect.a033092

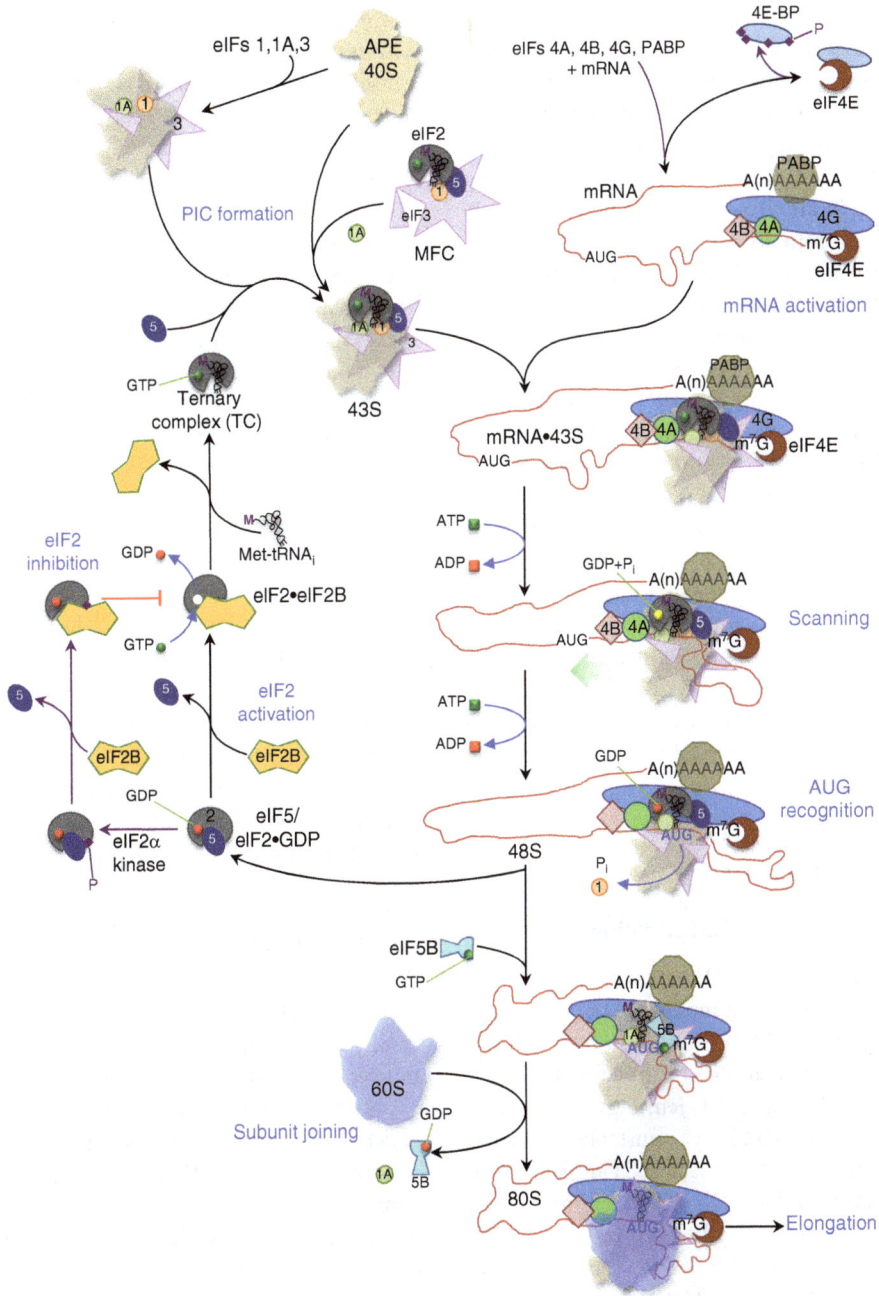

Figure 1. Overview of the general eukaryotic translation initiation pathway. The pathway for recruiting initiator transfer RNA (tRNA) to the messenger RNA (mRNA) AUG codon in the context of an 80S ribosome (*bottom right*) is depicted as a series of major steps, labeled with blue text, linked with black arrows. Individual eukaryotic initiation factor (eIF) cartoons and complexes are labeled with black text and nucleotide hydrolysis/inorganic phosphate release reactions are shown by blue arrows. The broad green arrow indicates the direction of scanning toward the AUG codon. The regulatory reactions leading to eIF2 and eIF4E inhibition are shown with plum and red arrows. All steps are described in the main text, starting with eIF2 activation. The timing of release of some factors from initiating ribosomes/mRNA (eIF4F, eIF4B, or eIF3) is not yet clear, so this is not shown.

AUG recognition, in which base pairing is established between the P-site-bound Met-tRNA$_i$ anticodon and the mRNA AUG codon. Here scanning is halted and initiation complex reorganization is triggered. eIF2-bound GTP can be hydrolyzed to GDP + phosphate (P$_i$) during scanning in an eIF5-dependent reaction, but P$_i$ is released from eIF2 only on AUG recognition (Algire et al. 2005; Majumdar and Maitra 2005). Following AUG recognition, eIF1 relocates, triggering an eIF2 conformational change that likely facilitates P$_i$ release. This significantly lowers the affinity of eIF2 for Met-tRNA$_i$ (Algire et al. 2005). Together these changes facilitate release of eIF2•GDP and eIF5. eIF2 must be reactivated again to form the eIF2•GTP complex to participate in further rounds of initiation. Finally, eIF5B•GTP is recruited and facilitates the joining of the large ribosomal subunit (60S) and reorientation of the initiator tRNA (Yamamoto et al. 2014). Release of eIF5B•GDP and eIF1A allows formation of an 80S ribosome attached to an mRNA with Met-tRNA$_i$ bound to the AUG codon in the P site, poised to begin translation elongation.

43S PIC FORMATION

eIF2 Activation and TC Formation

eIF2 is the main Met-tRNA$_i$ carrier and is a key regulatory switch that modulates the global initiation pathway. Met-tRNA$_i$ has 20- to 50-fold higher affinity for eIF2•GTP than for eIF2•GDP (Kapp and Lorsch 2004; Jennings et al. 2017). However, eIF2•GTP is unstable, whereas eIF2•GDP is relatively stable (Panniers et al. 1988; Kapp and Lorsch 2004; Jennings et al. 2016), necessitating the action of a GEF (eIF2B) to promote GDP dissociation and thereby facilitate GTP and Met-tRNA$_i$ binding to form the TC. During periods of cell stress, eIF2 is subjected to tight control as several protein kinases can phosphorylate the α subunit of eIF2 (eIF2αP) on a conserved serine residue (serine 51 in humans). This converts eIF2 from being a substrate to an inhibitor of eIF2B GEF activity (Fig. 1) and provides a mechanism to impose a global brake on protein synthesis initiation at its outset (Rowlands et al. 1988; Pavitt et al. 1998). This is an important regulatory mechanism, because selective mRNAs are able to escape such global repression. These translational controls are described in greater depth elsewhere (Proud 2018; Wek 2018, but see also Hinnebusch et al. 2016; Young and Wek 2016).

Evidence from yeast suggests that there is little free eIF2•GDP in cells, as eIF5 associates with high affinity forming eIF2•GDP/eIF5 complexes that are produced on release of these factors at the end of each initiation cycle (Algire et al. 2005; Singh et al. 2006). eIF5 can stabilize GDP binding to eIF2•GDP and thereby ensure tight regulation of initiation by preventing any spontaneous release of GDP from eIF2 that might otherwise bypass the eIF2B-centered control mechanism. Thus, eIF5 is a GDP-dissociation inhibitor (GDI) (Jennings and Pavitt 2010). This activity requires the eIF5 carboxy-terminal domain (CTD) and evolutionarily conserved residues within the central linker region, but not the amino-terminal domain (NTD) that is necessary for eIF5's role as a GTPase-activating protein (GAP). The CTD and linker interact with both eIF2γ and β subunits and may constrain eIF2, preventing eIF2B-independent GDP release (Alone and Dever 2006; Jennings and Pavitt 2010; Jennings et al. 2016). Thus, eIF2B displaces eIF5 from eIF2•GDP before performing nucleotide exchange. In yeast, eIF5 displacement requires the eIF2Bγ and ε subunits (Jennings et al. 2013). eIF5 displacement by eIF2B has not yet been studied with mammalian factors.

eIF2B is a large multifunctional protein. Encoded by five genes (subunits α–ε), its α, β, and δ subunits form an eIF2αP-sensing regulatory complex, whereas eIF2Bγ and ε perform eIF2 activation functions (Pavitt 2005; Jennings et al. 2013). eIF2B was recently shown to be a dimer of pentamers (Gordiyenko et al. 2014; Wortham et al. 2014), with the *Schizosaccharomyces pombe* structure revealing that the decamer contains a regulatory (αβδ)$_2$ hexameric "core" with two separate εγ "arms," so it can likely interact with two eIF2 molecules simultaneously. Consistent with prior genetic and biochemical evidence, in vitro cross-linking indicated that the eIF2α domain 1, including serine 51, makes direct contact with an interface composed of all three eIF2B regulatory subunits, whereas eIF2γ (to which GDP/GTP directly binds) interacts with both γ and ε (Kashiwagi et al. 2016). The eIF2Bε carboxy-terminal catalytic domain (Boesen et al. 2004; Mohammad-Qureshi et al. 2007) was not resolved in the eIF2B complex structure, so further studies will be required to elucidate both the mechanism of eIF2B GEF action and its major form of control by eIF2αP.

Following nucleotide exchange, eIF2•GTP binds Met-tRNA$_i$ to form the TC. eIF2 binds with 20-fold enhanced specificity to Met-tRNA$_i$ over the distinct elongator methionyl-tRNA (Met-tRNA$_m$) that decodes internal AUG methionine codons (Kapp et al. 2006). Assessing the function of tRNAs where specific nucleotides have been swapped has identified Met-tRNA$_i$ elements that ensure it binds eIF2 and is excluded from elongation. In Met-tRNA$_i$ these include the A1:U72 base pair that is replaced by a G:C pair in Met-tRNA$_m$. Substitutions here and at bases throughout the tRNA$_i$ contribute to eIF2 binding (Drabkin et al. 1993; Kapp et al. 2006), and make other contacts important for scanning and ensuring the fidelity of start codon recognition (Kolitz and Lorsch 2010; Dong et al. 2014). As eIF2γ is a structural mimic of tRNA-binding elongation factors SelB, EF1A (EF-Tu), and eEF1A, it was proposed that eIF2 would bind tRNA$_i$ in a similar manner to the elongation factors (Schmitt et al. 2010). However, cross-linking and structural studies of both yeast and archaeal eIF2 proteins reveal that the eIF2α and β subunits also contribute to a distinct Met-tRNA$_i$ binding mode where both eIF2γ and eIF2α make extensive contacts with the Met-tRNA$_i$ acceptor stem (Shin et al. 2011; Schmitt et al. 2012; Naveau et al. 2013). As described below, cryoelectron microscopy (cryo-EM) of partial 48S complexes has provided further insight into Met-tRNA$_i$ binding by eIF2 (Hussain et al. 2014; Llacer et al. 2015; Hinnebusch 2017; Jobe et al. 2018).

Although the TC is stable in isolation, it was recently shown that eIF2B can bind and disrupt the TC (Jennings et al. 2017). This unexpected finding suggests that eIF2B can both promote TC formation and destabilize the product of this GEF reaction. Hence, free TC may not be long-lived in cells and may not represent the final product of eIF2 activation. When eIF5 is bound to the TC, it impairs the ability of eIF2B to destabilize the TC, consistent with the idea that TC/eIF5 complexes represent a stable active form of eIF2 that can be recruited to the PIC (Jennings et al. 2017). Therefore, eIF2B can displace eIF5 from eIF2•GDP/eIF5 complexes to perform nucleotide exchange (Jennings et al. 2013), but cannot remove eIF5 from TC/eIF5. Hence, eIF5 binding to TC likely prevents eIF2B from antagonizing the next steps in protein synthesis initiation (Jennings et al. 2017).

eIFs-1, -1A, and -3 Promote TC Binding to 40S

Three additional factors, eIF1, eIF1A, and eIF3, are implicated in recruiting the TC to the 40S ribosomal subunit (Majumdar et al. 2003; Olsen et al. 2003; Kolupaeva et al. 2005; Cheung et al. 2007). Their cooperative binding induces conformational changes that rotate the head and open up a cleft between the head and body of the 40S to facilitate TC binding (Maag et al. 2005; Passmore et al. 2007; Weisser et al. 2013; Sokabe and Fraser 2014). eIF1 is a small protein that binds to the platform of the 40S subunit body close to the P site and mRNA channel (Weisser et al. 2013; Aylett et al. 2015). eIF1A, which is a homolog of bacterial IF1 (Rodnina 2018), sits adjacent to eIF1 in the 40S A site contacting both the head and body. Both eIF1 and eIF1A bind cooperatively to the 40S subunit (Maag et al. 2005; Sokabe and Fraser 2014) above rRNA helix 44, and eIF1A is located at the decoding center (Weisser et al. 2013). The decoding center monitors A-site tRNA-mRNA codon–anticodon pairing during elongation. Both factors play crucial roles in many steps in the initiation pathway and mutations in either factor affect TC recruitment, scanning, and the stringency of AUG recognition (Cheung et al. 2007; Fekete et al. 2007; Martin-Marcos et al. 2013; Hinnebusch 2014). eIF1A is also important for 60S subunit joining. These roles are described below.

Like eIF1 and 1A, eIF3 promotes TC binding to the 43S PIC (Valasek 2012; Sokabe and Fraser

2014; Aitken et al. 2016). The mammalian eIF3 complex is the largest initiation factor comprising 13 subunits (designated a→m) and ~800 kDa. In contrast, yeast eIF3 is smaller, sharing only homologs of the a, b, c, g, i, and j subunits. It has been suggested that the yeast complex represents a minimal core eIF3 (Valasek 2012; Wagner et al. 2016; Valasek et al. 2017). In yeast, an eIF3ac dimer is connected to the eIF3bgij subcomplex via an eIF3a–CTD/3b interaction (Valasek 2012). This eIF3aCTD/bgij complex has been termed the "yeast-like core" of eIF3 (Fig. 2, left) (des Georges et al. 2015) or alternatively the "peripheral" subunits (Simonetti et al. 2016). Six of the additional mammalian subunits (eIF3d–f, h, and k–m) interact with the conserved eIF3ac module to form a large "octamer" structural core of the mammalian complex comprising subunits a, c, e, f, h, k, l, and m that form a five-lobed structure (des Georges et al. 2015) to which eIF3d binds (Fig. 2, right). It is likely that the enhanced complexity of mammalian eIF3 allows for more diverse regulatory inputs to control translation than in yeast.

Structural studies have shown, in progressively increasing detail, that the large eIF3 complex binds and wraps around the solvent exposed (back) surface of the body of the 40S subunit such that it can contact and monitor important events occurring at both the mRNA entry channel/A site as well as the exit channel/E site and factor binding to the intersubunit interface (Hashem et al. 2013b; Erzberger et al. 2014; Aylett et al. 2015; des Georges et al. 2015; Llacer et al. 2015; Simonetti et al. 2016). A cartoon model summarizing some of these factor–40S interactions is shown in Figure 3. The images are a composite derived from both yeast and mammalian studies and show positions for all eIF3 subunits. The structural studies are supported by many functional analyses of eIF3 interactions that indicate a complex network of interactions among the 43S PIC factors that promote its formation and also make critical contributions to later steps in initiation (Valasek 2012; Sokabe and Fraser 2014; Aitken et al. 2016). Specifically, the eIF3 octamer core of the mammalian complex binds around the 40S mRNA exit channel. In yeast, eIF3c binds eIF1 and eIF5 and these subunits make further links to the TC (Valasek et al. 2003, 2004; Yamamoto et al. 2005). In contrast, the eIF3aCTD-3bgij

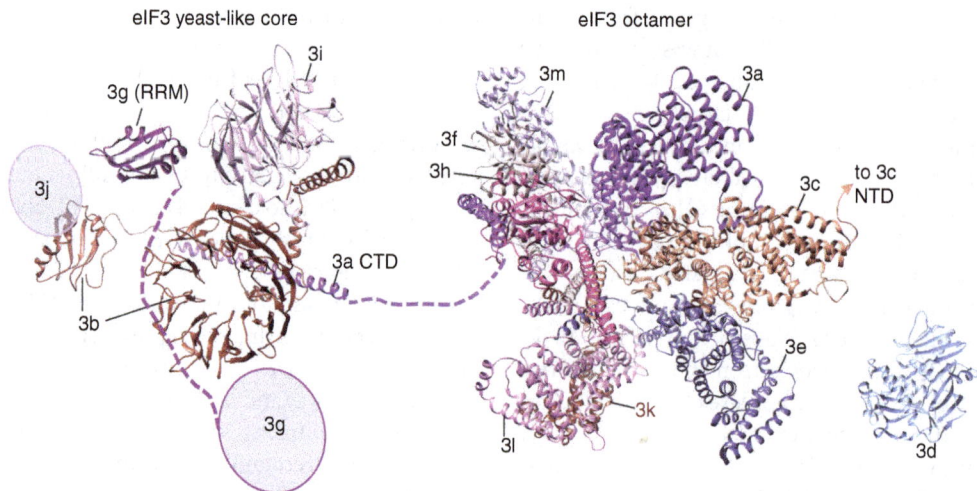

Figure 2. Structural models of eukaryotic initiation factor (eIF)3. A composite model of the structure of eIF3 showing the mammalian octamer (*right*) from PDB 5A5T (des Georges et al. 2015) and the associated eIF3d subunit (PDB 5K4D) (Lee et al. 2016) that is linked via eIF3a to the yeast-like core eIF3bgi complex (PDB 5A5U) (des Georges et al. 2015). The eIF3g is composed of an RNA-recognition motif (RRM) (PDB 2CQ0) and a β-propeller domain similar to eIF3b. eIF3j can interact with the eIF3b RRM. NTD, Amino-terminal domain.

Cite this article as *Cold Spring Harb Perspect Biol* doi: 10.1101/cshperspect.a033092

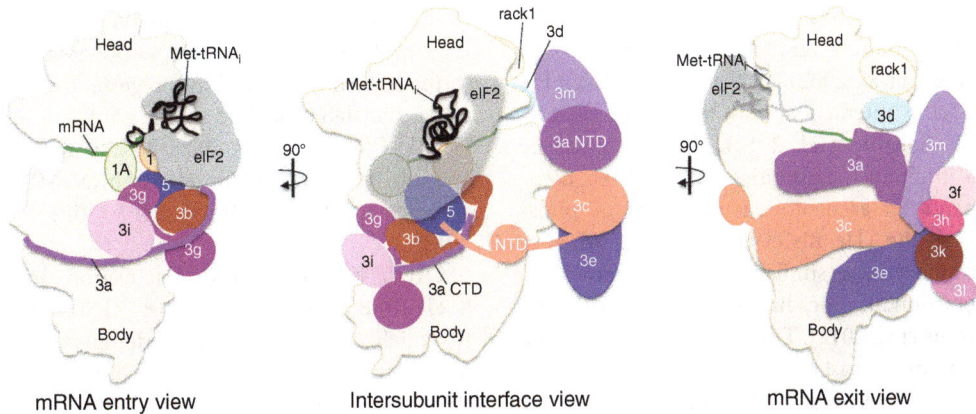

Figure 3. Model for preinitiation complex (PIC) arrangement at AUG recognition. Three 90° rotated cartoon views of an idealized 48S PIC, modeled on recent structural and functional studies. This model was based on the partial yeast 48S closed complex structure (3JAP) (Llacer et al. 2015) and modified to include findings of others (Luna et al. 2012, 2013; Aylett et al. 2015; des Georges et al. 2015; Simonetti et al. 2016; Obayashi et al. 2017). The entry view (*left*) shows the eukaryotic initiation factor (eIF)3 yeast-like core subunits, whereas the exit view (*right*) shows only the mammalian eIF3 octamer complex. eIF2 is shown as semitransparent in the central intersubunit view (*middle*) to indicate the position of factors otherwise hidden below. For further details of eIF2 interactions, see Figure 4. The position of Rack1 on the ribosome head is shown for orientation purposes only and is not discussed in the text. Factor colors correspond to those used in other figures.

yeast-like core complex is located at the mRNA entry channel, where eIF3aCTD binds both eIF1 and eIF2, whereas 3a, 3b, 3g, and 3j contact 18S rRNA and ribosomal proteins (Fig. 3) (Valasek 2012). eIF3j appears more loosely associated with the other eIF3 subunits and has distinct roles suggesting that it should be viewed as an eIF3-associated factor (Valasek et al. 2017). eIF3j enhances the affinity of many interactions among the 43S components when part of eIF3 (Fraser et al. 2007; Sokabe and Fraser 2014). Recent analyses of yeast eIF3 mutants indicate that mutations in eIF3b, 3i, or 3g destabilize TC binding to the PIC and mutations within these subunits or in eIF3a also impair mRNA recruitment (Khoshnevis et al. 2014; Aitken et al. 2016).

To sum up, eIF3 is critical for bringing the TC to the 40S subunit and for stabilizing mRNA interactions. It also plays a role bridging the 43S PIC to the mRNA m^{7}G cap-binding complex as human eIF3c, 3d, and 3e subunits can be cross-linked to a region of eIF4G (Villa et al. 2013). Of these subunits, yeast eIF3 lacks 3d and 3e, suggesting 3c may fulfill this role, although the eIF5-CTD can provide further stabilizing inter-

actions between the 43S PIC and eIF4G in place of the eIF4G/eIF3 contacts found for mammalian eIF3 (Yamamoto et al. 2005; Singh et al. 2012). As described below eIF3 also facilitates later steps in initiation and can play wider roles in translation including the recycling of posttermination ribosomes and in reinitiation (Pisarev et al. 2007; Beznoskova et al. 2015; Mohammad et al. 2017; Valasek et al. 2017; see also Hellen 2018; Wek 2018).

MFC Is an Alternative Route to the 43S PIC

Another route to form the 43S PIC has been proposed following the isolation of a complex containing the TC and eIFs 1, 3, and 5. Originally described in yeast and termed the MFC, it has now also been isolated from both plant and mammalian cells (Asano et al. 2000; Dennis et al. 2009; Sokabe et al. 2012). The MFC likely helps cooperative recruitment of these translation factors to 40S subunits as there are multiple interactions among components. The yeast eIF5-CTD makes independent contacts with eIF3c-NTD, eIF1, and eIF2β (Yamamoto et al. 2005; Luna et al. 2012). Thus, eIF5 bridges an interaction between the TC and eIF3c, in addition to a direct eIF2/eIF3a–CTD interaction (Valasek et al. 2002). In the human MFC, the eIF3/eIF5 interaction may be weaker than in yeast (Sokabe et al. 2012). There is evidence that in plants the affinities of these MFC components for each other is enhanced by phosphorylation (Dennis et al. 2009), but whether phosphorylation similarly affects human MFC interaction affinities has not yet been evaluated (Sokabe et al. 2012). Thus, as indicated in Figure 1, there are at least two routes to the formation of 43S complexes. The relative importance of these 43S PIC formation routes to individual mRNAs in different cells or in response to different stimuli remains to be resolved.

mRNA RECRUITMENT AND SCANNING

Having constructed the 43S PIC, the next step is mRNA recruitment. eIF4A, eIF4B, eIF4E, and eIF4G are all that is required in model in vitro systems. As these proteins have been purified and a range of mutants analyzed, some of the biochemical properties of these proteins are known as independent entities. However, there remains some uncertainty to their sequential utilization in the initiation pathway. eIF4E is the major m^7G cap recognition factor (Gross et al. 2003) and is a site of translational control via a series of eIF4E-binding proteins (4E-BPs) that limit eIF4E access to eIF4G (Richter and Sonenberg 2005). Binding of 4E-BPs to eIF4E is regulated by phosphorylation that alters the structure of the binding protein so that it can no longer bind eIF4E (Bah et al. 2015). When not bound by a 4E-BP, eIF4E can interact with eIF4G (Gross et al. 2003; Gruner et al. 2016; Proud 2018). eIF4G enhances the affinity of eIF4E for the m^7G cap, suggesting the interactions help reinforce each other to promote initiation (Gross et al. 2003; Yanagiya et al. 2009; O'Leary et al. 2013). eIF4E binds to a motif shared among the 4E-BPs and eIF4G: $YX_4L\phi$ (where X and ϕ denote any amino acid and any hydrophobic residue, respectively). There are structures available of eIF4E with short regions of eIF4G, including this motif, which reveal this interaction (Gross et al. 2003; Gruner

et al. 2016). eIF4G is frequently referred to as a large protein scaffold because, as previously described, it contains multiple regions to which many translation factors and mRNA can bind (Fig. 1) (Yanagiya et al. 2009; Park et al. 2011), and can also bridge this eIF4F/mRNA/PABP complex to the 43S PIC via eIF3 (Villa et al. 2013).

At the heart of the mRNA activation process is eIF4A that is one of the best-characterized RNA helicases and one of the original founding members of the DEAD-box RNA helicases (Grifo et al. 1984; Linder et al. 1989). It is the only "traditional" initiation factor that is known to bind ATP and it functions during both mRNA activation and mRNA scanning. Characteristic activities for eIF4A include: ATP-dependent retention of RNA on nitrocellulose filters, RNA-dependent ATP hydrolysis, and RNA duplex unwinding (Merrick 2015). All of these activities are also found in the eIF4F complex where the level of activity is always greater (Abramson et al. 1987; Rozen et al. 1990; Feoktistova et al. 2013; Sokabe and Fraser 2017). Activity enhancements are a reflection of improved binding constants for mRNA as well as enhanced efficiency (as rate). The activities of both eIF4A and eIF4F can be further improved by the presence of eIF4B (Harms et al. 2014). What is unclear currently is whether the ATP-dependent action of eIF4A in the initiation pathway is exclusively through eIF4F or whether there are individual steps that use just eIF4A (Pause et al. 1994).

What is thought to be the most important function of eIF4A is its helicase activity that can be used to reduce or eliminate secondary structure in an mRNA or, by analogy to the DEAD-box RNA helicase NPH-II, to separate protein from RNA (Jankowsky et al. 2001). Both of these elements are important for loading a single strand of RNA into the mRNA channel on the 43S PIC. This conversion of an mRNA (as the mRNA initially exists having exited the nucleus) to an activated mRNA is catalyzed by the sequential recognition by eIF4F of the m^7G cap and the generation of a single strand of RNA (eliminating secondary structure, protein, or both from the 5′ end of the mRNA) that can

Cite this article as *Cold Spring Harb Perspect Biol* doi: 10.1101/cshperspect.a033092

be placed on the 43S PIC. The first step is the recognition of the m^{7}G cap by the eIF4E subunit of eIF4F. This recognition ensures that the bound RNA is an mRNA (has an m^{7}G cap) and that eIF4E is bound to the 5′ end of the mRNA in an ATP-independent manner. The subsequent removal of RNA secondary structure and possible protein is dependent on ATP and is greatly enhanced by eIF4B (Abramson et al. 1987; Rozen et al. 1990). Based on the relatively slow turnover numbers for ATP hydrolysis (Lorsch and Herschlag 1998; Merrick 2015), presumed to be the result of the slow chemical step in the hydrolysis of ATP, the complex of eIF4F and eIF4B remains associated with the mRNA long enough to effect an interaction with eIF3 on the 43S PIC that leads to the binding of the mRNA to the 40S subunit via a protein–protein interaction (eIF4F and eIF3) although it is likely that the 40S subunit also contributes to the mRNA loading process.

A major player in mRNA recruitment is eIF3, both as an essential component in the formation of the 43S PIC as well as in stabilization of the bound mRNA at both the entry and exit channels (Aitken et al. 2016). Of particular importance are the NT and CT domains of eIF3a and the NTD of eIF3c, and this is consistent with the mapping of these portions of eIF3 to various cryo-EM structures of 48S complexes (Aylett et al. 2015; des Georges et al. 2015; Llacer et al. 2015; Jobe et al. 2018). Recently m^{7}G cap-interacting domains were identified within both eIF3d (Lee et al. 2016) and eIF3l (Kumar et al. 2016) that may play roles in transitioning an eIF4E-bound mRNA m^{7}G cap to one where the m^{7}G cap is anchored by eIF3. Such a role for eIF3 would be consistent with the apparent loss of eIF4E when 48S complexes are formed (Kumar et al. 2016). Release of eIF4E would then permit the known regulation by 4E-BP under appropriate circumstances (Merrick 2015; Proud 2018). It should be noted that yeast eIF3 lacks homologs of the d and l subunits, and so further studies will be needed to resolve the relative importance of these m^{7}G cap interactions for general protein synthesis initiation. eIF3j binds to the 40S decoding center and makes contact with eIF1A (Fig. 3) (Fraser et al. 2007;

Aylett et al. 2015). eIF3j binding was shown to impair mRNA recruitment to 40S, except when TC was present, suggesting eIF3j may impair mRNA binding to 40S complexes lacking TC (Fraser et al. 2007; Mitchell et al. 2010). In summary, numerous studies suggest that a network of multiple interactions is established among initiation factors, tRNA, and the 40S ribosome, which act together to facilitate mRNA recruitment to the 48S PIC.

Perhaps the least understood step in the initiation process is scanning. This is in part because of the difficulty in obtaining a 48S PIC with the mRNA in its initial position when first bound to the 40S subunit, as the requirements for mRNA binding and scanning appear to be similar (requiring eIF4A, eIF4B, eIF4F, and ATP). The early work of Kozak established the requirement for ATP in scanning (Kozak 1980). eIF4A uses ATP and, in vitro, is sufficient as either eIF4A alone or within eIF4F to form 48S complexes positioned at the start codon (Pestova et al. 1998; Dmitriev et al. 2003; Aylett et al. 2015; Llacer et al. 2015). The question is how is the "RNA helicase" activity used to drive the scanning process? Some hints at this have been obtained from RNA helicase studies. For yeast eIF4F, there is about a 22-fold preference for a 5′ overhang on a RNA duplex substrate relative to a 3′ one (Rajagopal et al. 2012). This could conceivably orient the mRNA for movement in a 5′ to 3′ direction. However, in contrast, there is no equivalent preference shown by mammalian eIF4F (Rogers et al. 2001). A recent study using a single molecule, optical trapping assay for RNA duplex unwinding was able to show processivity with complexes of human eIF4A, eIF4B, and eIF4G (García-García et al. 2015), although this result has not been reproduced in standard assays of duplex unwinding with these proteins (Rogers et al. 2001; Gao et al. 2016). What is not yet clear in the scanning process is how the TC is able to inspect the mRNA for a start codon while the PIC is traversing the 5′UTR.

Translation of the mRNA is assumed to temporarily remove extensive secondary structure and RNA-associated proteins from both the 5′ leader and ORF. Thus, at termination the release of the 40S and 60S subunits frees them to initiate again (Hellen 2018). Kinetically, reuse of ribosomes to repeatedly translate the same mRNA is likely aided by what is termed the "closed-loop model" wherein PABP and eIF4G interact with each other to circularize the mRNA (Fig. 1) (Wells et al. 1998; Park et al. 2011; Archer et al. 2015; Costello et al. 2015, 2017). Thus, at termination, the ribosomal subunits are likely released in close proximity to the 5′ end of the mRNA and this facilitates subsequent initiation events in a more efficient manner. This may be especially important in cells in log phase growth where the availability of free 40S subunits is limiting (Chu et al. 2014), and is likely more important on some mRNAs than others as, at least in yeast, it is clear that the propensity of some mRNAs to form a closed loop is greater than others (Archer et al. 2015; Costello et al. 2015). Recent computational analyses support the idea that the closed-loop can enhance translation rates, particularly on shorter mRNAs (Rogers et al. 2017).

Some mRNAs have extensive secondary structure or length, or both, in their 5′UTRs and therefore require additional "helicase power," which may come in the form of increased concentrations of eIF4A or eIF4B or of a variety of other RNA helicases, including mammalian Dhx29 and yeast Ded1 (Svitkin et al. 2001; Pisareva et al. 2008; Parsyan et al. 2011; Sen et al. 2015, 2016). At present, it is not clear whether these other helicases function entirely independently or as a partner/subunit of one of the initiation factors, as yeast Ded1 may function as an eIF4F subunit (Gao et al. 2016). The current interpretation is that such helicases regularly participate in many initiation events but are critically important for optimal initiation of mRNAs with highly structured 5′UTRs.

AUG RECOGNITION

AUG Recognition Causes Conformational Changes in the PIC

Typically, the first encountered AUG in an mRNA is used, providing it has a good sequence context. In mammals, the sequence

GCCPuCCAUGG, generally called the Kozak consensus sequence, enhances AUG recognition (Jackson et al. 2010). Within this sequence, the purine (Pu = A or G) nucleotide at −3 and the G at +4 position are most important (both relative to the A of the AUG, designated +1). In contrast, AUG contexts differing significantly from this context can be bypassed, a process called leaky scanning. Leaky scanning is used by specific mRNAs to regulate expression levels and is used as an autoregulatory loop as described for the translation factor eIF1 and eIF5 mRNAs (Ivanov et al. 2010; Martin-Marcos et al. 2011; Loughran et al. 2012; Hinnebusch et al. 2016). Alternatively, leaky scanning may include or exclude an NTD signal sequence that can alter the destination of the final protein. For example, a single mRNA encodes an enzyme with both cytoplasmically and mitochondrially targeted forms where different start codons are used to generate the isoforms (Slusher et al. 1991). Similarly, leaky scanning of an upstream ORF can regulate the flow of ribosomes to the downstream ORF (see Wek 2018). It is also notable that mRNAs with long leaders can be efficiently translated, as can those with very short leaders (Berthelot et al. 2004). How close an AUG codon can be to the m⁷G cap and still be translated has been addressed recently in vitro. It was found that AUG codons immediately adjacent to the m⁷G cap can be used efficiently, providing that eIF1 was excluded from the reactions (Kumar et al. 2016). Hence, it is possible that eIF1 may prevent initiation very close to the m⁷G cap. A potential exception to this is the translation initiator of short 5′UTR (TISU) sequence SAASAUGGCGGC (S = G or C) in mRNAs that appears to be much more permissive for initiation events close to the m⁷G cap (Elfakess et al. 2011; Sinvani et al. 2015; Kwan and Thompson 2018).

To recognize an AUG initiation codon, the Met-tRNA$_i$ anticodon within the scanning PIC must base-pair with the mRNA and signal this interaction to the associated factors in the PIC. These signaling events drive conformational changes in the PIC that switch it from an "open" to a "closed" conformation wherein altered interactions between the components cause release of eIF1, eIF2•GDP, and eIF5 from the PIC (Unbehaun et al. 2004; Cheung et al. 2007). Many of the precise changes that occur are now becoming understood. These ultimately result in a PIC arrangement competent to recruit the 60S subunit. Both biochemistry and yeast genetics implicate eIF1, eIF1A, and eIF5 along with eIF2 in stringent AUG recognition. Highly informative to the mechanism of AUG recognition has been the study of yeast Sui⁻ mutants that enhance inappropriate recognition of a mutated UUG start codon. Compensating Ssu⁻ "suppressor of sui" mutations in these initiation factors have also been described that enhance the stringency of AUG selection, as recently reviewed (Dever et al. 2016).

In the scanning PIC, eIF1 is located close to the P site (Hashem et al. 2013a; Weisser et al. 2013; Hussain et al. 2014). Its location prevents complete Met-tRNAi anticodon–AUG pairing. Recent structural insights revealed that in the open scanning conformation, the eIF2β subunit extends between eIF1 and eIF1A connecting these factors with Met-tRNAi that is bound to the 40S head. It was proposed that contacts observed between eIF1, eIF2β, and the Met-tRNAi

anticodon stem help to stabilize the scanning conformation (Fig. 4, "open") (Llacer et al. 2015). Upon AUG recognition, there are a number of rearrangements that occur following the transition to the intermediate closed conformation that occurs following scanning arrest (Fig. 4, "closed"). These include movements of the 40S that define these conformations (Llacer et al. 2015), as well as changes in the relative positions of the bound factors that are highlighted in Figure 4. Met-tRNAi is repositioned and establishes codon–anticodon pairing with the mRNA. The eIF1A amino-terminal tail (NTT) appears to monitor or stabilize the closed codon–anticodon duplex in agreement with prior biochemical observations (Fekete et al. 2007). Likewise eIF2α contacts the conserved mRNA nucleotide at the −3 position using Arg54, which is close to the regulatory Ser51 residue, in agreement with prior cross-linking experiments (Pisarev et al. 2006). In contrast, eIF2β retracts from both the tRNA acceptor stem and eIF1A. In the closed conformation structure, eIF1 is partially displaced to enable codon–anticodon pairing between Met-tRNAi and the AUG codon of the mRNA.

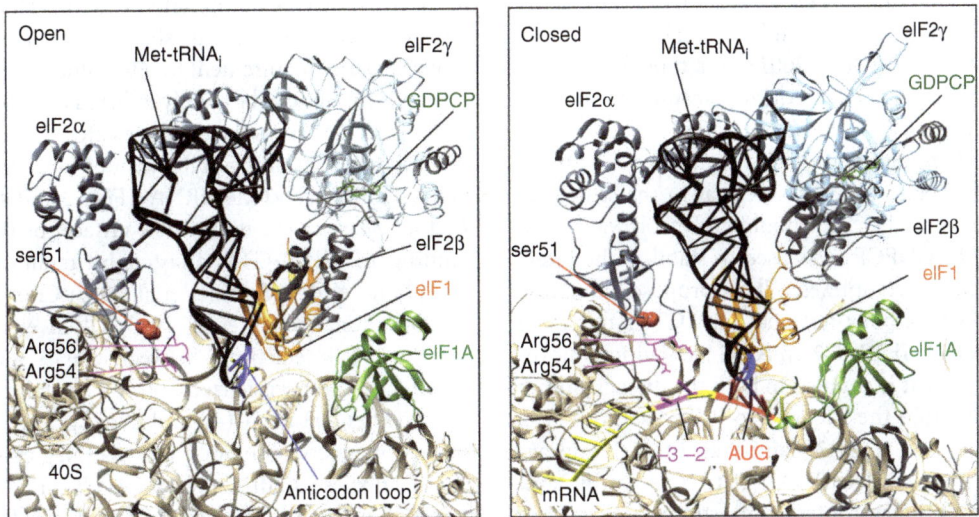

Figure 4. Altered conformation of initiation factors on AUG recognition. Cartoons derived from cryoelectron microscopy (cryo-EM) analyses of partial preinitiation complexes in a scanning mode (open, *left* panel) and AUG recognition mode (closed, *right* panel) (Llacer et al. 2015). Note, in particular, the movement of eukaryotic initiation factor (eIF)1, eIF1A (tail), and eIF2β. eIF3 and part of the 40S head structures were removed for clarity. Images were created with University of California San Francisco (UCSF) Chimera software from PDB files 3JAP and 3JAQ. See text for details.

Although not resolved in these PIC structures, eIF5 plays critical roles in AUG recognition and is thought to move during the PIC rearrangements mediating the open–closed transitions (Aylett et al. 2015; Llacer et al. 2015; Aitken et al. 2016; Obayashi et al. 2017). eIF1A and eIF5 interactions help retain eIF1 in the open scanning PIC (Maag et al. 2005; Luna et al. 2012, 2013) and eIF5–eIF3c interactions contribute to coordinating the rearrangements (Obayashi et al. 2017). It has also been shown that the 40S ribosomal proteins make important contacts with mRNA, Met-tRNA$_i$, and translation factors that assist in determining accuracy of AUG codon recognition. Rps5/uS7 is located at the 40S mRNA exit channel and contacts TC and AUG −3/−4 context nucleotides (Pisarev et al. 2008), and mutations in uS7 were shown to impair AUG recognition analogous to mutations in translation initiation factors (Visweswaraiah et al. 2015) or in Met-tRNA$_i$ itself (Dong et al. 2014).

eIF2•GTP Hydrolysis and Factor Release

Although the eIF5-CTD makes a series of critical contacts during scanning, it is the Arg15 residue within the eIF5-NTD that stimulates hydrolysis of eIF2-bound GTP (Das and Maitra 2001; Paulin et al. 2001). It is thought that hydrolysis can occur during scanning, but that release of P_i to form eIF2•GDP occurs only at AUG recognition (Algire et al. 2005; Majumdar and Maitra 2005). (Note that in the structures shown in Fig. 4, a nonhydrolyzable variant of GTP [GDPCP] was used to stabilize the intermediate complexes.) It is proposed that the eIF1A carboxy-terminal tail (CTT) moves to contact eIF5-NTD on AUG recognition, movement that is coupled to both the dissociation of eIF1 from the PIC and to P_i release from eIF2 (Nanda et al. 2013; Saini et al. 2014). It appears likely that Met-tRNA$_i$–AUG base-pairing and multiple factor interactions stimulate this series of events (Aitken et al. 2016; Obayashi et al. 2017). Although not shown in Figure 4, structural analysis of a late-stage 48S complex reveals that large movements of the eIF3 yeast-like core subunits occur before eIF2•GDP release (Simonetti et al. 2016).

Following GTP hydrolysis, eIF2•GDP has a low affinity for Met-tRNA$_i$ (Kapp and Lorsch 2004; Jennings et al. 2017), enabling eIF2•GDP to be released from the 48S complex. It is proposed that eIF5 leaves with eIF2 (Unbehaun et al. 2004) where it prevents premature release of GDP, as described above (Jennings and Pavitt 2010), before subsequent reactivation for further rounds of translation. In contrast, eIF1A remains ribosome-bound to stimulate 60S joining. The precise timing of when eIF3 and the eIF4 factors dissociate is not yet clear and so the release of these factors is not shown in Figure 1.

60S SUBUNIT JOINING

The ribosomal 40S and 60S subunits naturally bind to each other, especially in the presence of elevated concentrations of Mg^{2+}. Thus, it was not too surprising that one of the early activities ascribed to eIF5B was its subunit-dependent GTPase activity, an activity not dependent on any other initiation factor or Met-tRNA$_i$ (Merrick et al. 1975). Early studies established that the use of GDPNP (or GDPCP) blocked initiation complex formation at the level of a 48S complex (with either artificial or natural mRNAs). Thus, release of eIF2 from the 40S subunit was an established early requirement for 60S joining. The rationale for this can be readily inferred from the recent 48S structures that provide extensive detail into the positioning of the various initiation factors, mRNA, and Met-tRNA$_i$ (Figs. 3 and 4). Factor release opens up the 40S surface for the binding of eIF5B•GTP, assisted by retaining 40S bound eIF1A (Schreier et al. 1977; Choi et al. 2000; Marintchev et al. 2003; Olsen et al. 2003). The interactions between eIF5B and eIF1A have been visualized both by crystallography (Zheng et al. 2014), and dynamically by nuclear magnetic resonance (NMR) (Nag et al. 2016). In these studies, the eIF1A-CTT is observed to interact with domain IV of eIF5B. Together these events are thought to stabilize Met-tRNA$_i$–AUG interactions at the P-site and recruit the 60S subunit.

Uncertain in these events is whether eIF5B drives or facilitates the release of eIF2•GDP/eIF5 or the release of eIF3. Cross-linking studies with an mRNA containing 4-thioU at the –3 position indicate that there is a conformational change (and change in cross-linking to eIF2α) that occurs in the presence of eIF5 and eIF5B (Pisareva and Pisarev 2014). Yet in studies that have examined the formation of 80S complexes on model mRNAs, eIF3 was routinely associated with the 40S subunits, but little or none was found in the 80S complexes (Peterson et al. 1979; Pisareva and Pisarev 2014). This would suggest that there may be other mechanisms to promote the release of eIF3.

When trapped as a GDPNP (or GDPCP) complex, eIF5B has been located on the 80S ribosome (Fernandez et al. 2013; Yamamoto et al. 2014). It is positioned across the interface of the 40S subunit such that contact between eIF5B domain IV places the Met-tRNA$_i$ in the P/I conformation, which with hydrolysis of the GTP would lead the Met-tRNA$_i$ into the P position. There is a large change in the conformation of 80S-bound eIF5B, where domains III and IV are rotated 60°–65° relative to free eIF5B (Fernandez et al. 2013; Yamamoto et al. 2014). A recent "remodeling" performed by Kuhle and Ficner (2014) places the Met-tRNA$_i$ 3′ CCA end and the methionine in domain IV of eIF5B in a position equivalent to that seen with either IF2 or EF1A (EF-Tu) and their aminoacyl-tRNAs. It appears that the hydrolysis of GTP by eIF5B is sufficient to lead to its removal from the 80S ribosome (as is also true for bacterial IF2 (Rodnina 2018). eIF5B•GDP release is coupled to the removal of eIF1A (Fringer et al. 2007). The 80S ribosome is then ready to participate in the elongation phase of protein synthesis, with Met-tRNA$_i$ in the P site and a vacant A site.

ALTERNATIVE INITIATION EVENTS

For cells that are in log phase growth, almost all the translation is thought to proceed as described earlier. However, that does leave a small percentage of mRNAs translated by a different route. Characteristic of the expression of these mRNAs is that they are either cell-type-specific or they are expressed under conditions of stress. As such, these mechanisms usually have limited impact on overall expression because of their

lack of competitiveness for the translational machinery and the consequence is that these systems have been difficult to recapitulate in vitro. Two of the best-studied alternative schemes are the internal ribosome entry site (IRES) promoted initiation and regulated-reinitiation on *GCN4/ATF4* and related mRNAs with multiple upstream ORFs (Pelletier and Sonenberg 1988; Dever et al. 1992). IRESs provide an alternative 40S recruitment strategy, whereas *GCN4/ATF4* and other mRNAs use the scanning mechanism described above (see Kwan and Thompson 2018 and Wek 2018, respectively, for coverage of these examples in greater detail). Further mechanisms include the previously mentioned TISU, an alternate 5′ sequence that promotes efficient initiation at AUG codons close to the 5′ end in the absence of scanning (Sinvani et al. 2015; Haimov et al. 2017; Kwan and Thompson 2018), and m^6 methylation of 5′UTR adenine residues, which can promote cap-independent translation (Meyer et al. 2015; Peer et al. 2018).

The more curious events are those where eIF2 does not seem to be the protein responsible for directing the binding of the initiator Met-tRNA$_i$. The most commonly cited player is eIF2A, which is a single polypeptide of 65 kDa (Merrick and Anderson 1975). eIF2A appears to play a major role in initiating translation for major histocompatibility complex (MHC) class I peptides (Starck and Shastri 2016), in the integrated stress response (Starck et al. 2016), tumor progression (Sendoel et al. 2017), and viral replication (Ventoso et al. 2006; Kim et al. 2011). Other proteins such as eIF2D (also known as ligatin), eIF5B, or MCT-1/DENR have been implicated in binding initiator tRNA to ribosomes (Pestova et al. 2008; Dmitriev et al. 2010; Skabkin et al. 2010; Weisser et al. 2017). Similarly, initiator Met-tRNA$_i$ may not always be used to initiate. Some initiation at CUG leucine codons has been found independent of eIF2 (Schwab et al. 2004). Biochemical analysis revealed that cells use an elongator leucine-bound transfer RNA (Leu-tRNA) to initiate translation at cryptic CUG start codons in an eIF2A-dependent manner (Starck et al. 2012). The ability to use these alternate proteins to initiate protein synthesis is generally assisted by the activation of one of the eIF2 kinases and the subsequent phosphorylation of serine 51 of the α subunit of eIF2, yielding a reduced concentration of ternary complexes (Fig. 1). The definition of the exact pathway of initiation for alternate initiating aminoacyl-tRNA-binding proteins remains to be elucidated.

The strangest and least understood to date is repeat associated non-ATG (RAN) translation, which occurs in microsatellite repeat diseases (Zu et al. 2013) where initiation occurs in all three reading frames. The ability to initiate in all three reading frames would be consistent with the properties of either eIF2D or eEF1A, but currently no mechanism is proposed. RAN translation is described elsewhere (Zu et al. 2018). It is likely that the study of various diseases in the future may yield yet additional mechanisms, although it is anticipated that these, like those above, would be dominated by the most common binder of the initiator tRNA, eIF2.

REGULATION OF THE 80S INITIATION PATHWAY

It is of interest to know the molar concentration of each initiation factor as this is also important for regulation where the alteration of the concentration of active initiation factors can influence both the amount of total protein made and mRNA selectivity. Shown in Table 1 is a subset of the data obtained by Kulak et al. (2014), using a label-free mass spectrometry protocol (numbers rounded off) for exponentially growing yeast and HeLa cell cultures, thereby enabling comparison across species from a single study. These values share consistencies with some earlier evaluations (Duncan and Hershey 1983; von der Haar and McCarthy 2002), although there are differences that may reflect specific difficulties with the estimation of one or more of the individual factors by various methods. If the numbers in Table 1 are roughly correct, then the following conclusions can be drawn:

1. Most initiation factors are less abundant than the number of ribosomes, except for eIF4A, which is roughly equimolar.

Table 1. Number of protein molecules per cell

Protein	Yeast ($\times 10^3$)	HeLa ($\times 10^4$)
Cytoplasmic ribosomes	310	1064
eIF1	41	198
eIF1A	60	101
eIF2	31	210
eIF2B	6	75
eIF2A	11	72
eIF3	29	164
eIF4A	196	801
eIF4B	18	131
eIF4E	36	41
eIF4G (1 + 2)	11	74
eIF4H	-	103
eIF5	48	91
eIF5B	13	76
PABP	96	252
DED1/DDX3	32	138
DHX29	-	12
Caf20	10	-
Eap1	1.6	-
4E-BP1	-	33
4E-BP2	-	6
4E-T	-	0.4
Pdcd4	-	84

These data have been reorganized from mass spectrometry data reported previously (Kulak et al. 2014). For multisubunit factors and ribosomes, the value given corresponds to the most abundant subunit value reported.

2. eIF2B levels are lower than eIF2 and eIF5, consistent with nucleotide exchange being a regulated step in initiation.

3. In yeast, eIF4G is limiting for the formation of eIF4F, whereas in HeLa cells eIF4E is limiting.

4. The concentration of "extra" RNA helicases that might assist in the initiation process is about one-fifth the concentration of eIF4A, consistent with their proposed more specialized role.

5. The level of the 4E-BP regulators is sufficient to completely shut down eIF4F-dependent protein synthesis (Korets et al. 2011), but the level of the eIF4A regulator Pdcd4 in HeLa cells is not (Lankat-Buttgereit and Goke 2009).

Cite this article as *Cold Spring Harb Perspect Biol* doi: 10.1101/cshperspect.a033092

Posttranslational modification of specific initiation factors renders it possible that any given factor might become more or less active. At present, the major global regulatory steps in initiation are understood as the phosphorylation of eIF2 (serine 51 of the α subunit) and the dephosphorylation of 4E-BPs, both of which are inhibitory (Fig. 1). These and other mechanisms are discussed elsewhere in the literature (Proud 2018; Wek 2018).

PERSPECTIVE: WHAT WE MIGHT LIKE TO KNOW

The size and complexity of the initiation factors and the ribosomal subunits have made detailed studies challenging, but methods continue to improve and offer hopes for a more complete understanding of protein synthesis in the near future. One advance would be to detail the precise timing and sequence of the binding, movement, and release events depicted in Figure 1, the 80S initiation pathway. With this, one would hope to define the rate-limiting step(s). From the observation of 43S and 48S complexes by sucrose gradient analysis, it would seem likely that there are at least two slow steps, the binding of the mRNA (hence 43S complexes) and either the scanning of the mRNA or subunit joining (hence the 48S complexes). Kinetic analyses of translation should address this point and recent progress is described elsewhere (Sokabe and Fraser 2018).

For a single complete initiation event, is only one copy of each initiation factor used or do some factors recycle multiple times or does initiation require multiple copies of some factors to be engaged? For the initiation factors eIF1, eIF1A, eIF2, eIF2B, eIF3, eIF5, and eIF5B it would seem likely that only a single copy of each factor is required. However, for eIF4A (and perhaps eIF4F), because multiple rounds of ATP hydrolysis are required, and as eIF4A is significantly more abundant than other initiation factors (Table 1), it is plausible that multiple copies are used, with the number likely dependent on the complexity of the mRNA 5'UTR (e.g., amount of secondary structure and length) (Svitkin et al. 2001). As has been noted for a few

particular mRNAs, either in vitro or in vivo, if increased concentrations or other RNA helicases are required, multiple rounds of use for each of these may also be required (Pause et al. 1994).

During scanning on longer 5' leaders, is contact maintained between eIF4F and the m^7G cap until subunit joining or through multiple rounds of initiation? Energetically, it makes sense for eIF4F to remain bound to the m^7G cap and assist with multiple rounds of initiation before detaching; however, it was recently proposed that eIF4E may detach from the mRNA during scanning (Kumar et al. 2016). Similarly, is contact between eIF4F and the PIC maintained during scanning? This tethered scenario is depicted in Figure 1, with the mRNA threaded though the channel between the head and body of the 40S subunit. However, other models of scanning propose that the PIC can detach from eIF4F (Archer et al. 2016).

There are several well-studied examples of the regulated expression of proteins via reinitiation, including GCN4, ATF4, and the glutamine amidotransferase subunit of Arg-specific carbamoyl phosphate synthetase (Elbarghati et al. 2008; Spevak et al. 2010; Dever et al. 2016). For the regulated expression of these mRNAs, what are the important translation factors or other trans-acting elements and what are the important cis-acting elements (such as upstream ORF [uORF] length or sequence, mRNA sequence, or structure contexts, etc.)? Current understanding of GCN4 translation, the most well-studied reinitiation system, implicates several sequences around its uORFs and eIF3 (Szamecz et al. 2008; Munzarova et al. 2011; Gunisova and Valasek 2014; Mohammad et al. 2017). Given the findings that a high percentage of mRNAs contain either bona fide uORFs or uORFs with non-AUG start codons (Calvo et al. 2009), it is highly probable that both nonstandard and reinitiation mechanisms will turn out to be more common than currently thought (Hinnebusch et al. 2016).

Finally, it is clear that there are rare (tissue- or disease-specific), but important initiation events for which we have little idea as to how they are accomplished: MHC class I peptide synthesis, tumor progression, and RAN translation being examples studied (Malys and McCarthy 2011; Starck et al. 2012; Zu et al. 2013; Sendoel et al. 2017). The ability to better understand each of these processes could enable the development of treatments to defeat these rare events with fewer side effects than drugs that are currently available. There is still a lot to do.

ACKNOWLEDGMENTS

Research in the Pavitt laboratory is funded by United Kingdom Biotechnology and Biological Sciences Research Council Grants BB/L000652/1, BB/L020157/1, and BB/M006565/1. W.C.M. is supported by the School of Medicine at Case Western Reserve University. We declare no competing interests.

REFERENCES

*Reference is also in this collection.

Abramson RD, Dever TE, Lawson TG, Ray BK, Thach RE, Merrick WC. 1987. The ATP-dependent interaction of eukaryotic initiation factors with mRNA. *J Biol Chem* 262: 3826–3832.

Aitken CE, Beznoskova P, Vlckova V, Chiu WL, Zhou F, Valasek LS, Hinnebusch AG, Lorsch JR. 2016. Eukaryotic translation initiation factor 3 plays distinct roles at the mRNA entry and exit channels of the ribosomal preinitiation complex. *eLife* 5: e20934.

Algire MA, Maag D, Lorsch JR. 2005. P$_i$ release from eIF2, not GTP hydrolysis, is the step controlled by start-site selection during eukaryotic translation initiation. *Mol Cell* 20: 251–262.

Alone PV, Dever TE. 2006. Direct binding of translation initiation factor eIF2γ-G domain to its GTPase-activating and GDP-GTP exchange factors eIF5 and eIF2Bε. *J Biol Chem* 281: 12636–12644.

Archer SK, Shirokikh NE, Hallwirth CV, Beilharz TH, Preiss T. 2015. Probing the closed-loop model of mRNA translation in living cells. *RNA Biol* 12: 248–254.

Archer SK, Shirokikh NE, Beilharz TH, Preiss T. 2016. Dynamics of ribosome scanning and recycling revealed by translation complex profiling. *Nature* 535: 570–574.

Asano K, Clayton J, Shalev A, Hinnebusch AG. 2000. A multifactor complex of eukaryotic initiation factors, eIF1, eIF2, eIF3, eIF5, and initiator tRNA(Met) is an important translation initiation intermediate in vivo. *Genes Dev* 14: 2534–2546.

Aylett CH, Boehringer D, Erzberger JP, Schaefer T, Ban N. 2015. Structure of a yeast 40S-eIF1-eIF1A-eIF3-eIF3j initiation complex. *Nat Struct Mol Biol* 22: 269–271.

Bah A, Vernon RM, Siddiqui Z, Krzeminski M, Muhandiram R, Zhao C, Sonenberg N, Kay LE, Forman-Kay JD. 2015. Folding of an intrinsically disordered protein by phosphorylation as a regulatory switch. *Nature* 519: 106–109.

Berthelot K, Muldoon M, Rajkowitsch L, Hughes J, McCarthy JE. 2004. Dynamics and processivity of 40S ribosome scanning on mRNA in yeast. *Mol Microbiol* 51: 987–1001.

Beznoskova P, Wagner S, Jansen ME, von der Haar T, Valasek LS. 2015. Translation initiation factor eIF3 promotes programmed stop codon readthrough. *Nucleic Acids Res* 43: 5099–5111.

Boesen T, Mohammad SS, Pavitt GD, Andersen GR. 2004. Structure of the catalytic fragment of translation initiation factor 2B and identification of a critically important catalytic residue. *J Biol Chem* 279: 10584–10592.

Calvo SE, Pagliarini DJ, Mootha VK. 2009. Upstream open reading frames cause widespread reduction of protein expression and are polymorphic among humans. *Proc Natl Acad Sci* 106: 7507–7512.

Cheung YN, Maag D, Mitchell SF, Fekete CA, Algire MA, Takacs JE, Shirokikh N, Pestova T, Lorsch JR, Hinnebusch AG. 2007. Dissociation of eIF1 from the 40S ribosomal subunit is a key step in start codon selection in vivo. *Genes Dev* 21: 1217–1230.

Choi SK, Olsen DS, Roll-Mecak A, Martung A, Remo KL, Burley SK, Hinnebusch AG, Dever TE. 2000. Physical and functional interaction between the eukaryotic orthologs of prokaryotic translation initiation factors IF1 and IF2. *Mol Cell Biol* 20: 7183–7191.

Chu D, Kazana E, Bellanger N, Singh T, Tuite MF, von der Haar T. 2014. Translation elongation can control translation initiation on eukaryotic mRNAs. *EMBO J* 33: 21–34.

Costello J, Castelli LM, Rowe W, Kershaw CJ, Talavera D, Mohammad-Qureshi SS, Sims PF, Grant CM, Pavitt GD, Hubbard SJ, et al. 2015. Global mRNA selection mechanisms for translation initiation. *Genome Biol* 16: 10.

Costello JL, Kershaw CJ, Castelli LM, Talavera D, Rowe W, Sims PFG, Ashe MP, Grant CM, Hubbard SJ, Pavitt GD. 2017. Dynamic changes in eIF4F-mRNA interactions revealed by global analyses of environmental stress responses. *Genome Biol* 18: 201.

Das S, Maitra U. 2001. Functional significance and mechanism of eIF5-promoted GTP hydrolysis in eukaryotic translation initiation. *Prog Nucleic Acid Res Mol Biol* 70: 207–231.

Dennis MD, Person MD, Browning KS. 2009. Phosphorylation of plant translation initiation factors by CK2 enhances the in vitro interaction of multifactor complex components. *J Biol Chem* 284: 20615–20628.

des Georges A, Dhote V, Kuhn L, Hellen CU, Pestova TV, Frank J, Hashem Y. 2015. Structure of mammalian eIF3 in the context of the 43S preinitiation complex. *Nature* 525: 491–495.

Dever TE, Feng L, Wek RC, Cigan AM, Donahue TF, Hinnebusch AG. 1992. Phosphorylation of initiation factor 2α by protein kinase GCN2 mediates gene-specific translational control of GCN4 in yeast. *Cell* 68: 585–596.

Dever TE, Kinzy TG, Pavitt GD. 2016. Mechanism and regulation of protein synthesis in *Saccharomyces cerevisiae*. *Genetics* 203: 65–107.

Dmitriev SE, Pisarev AV, Rubtsova MP, Dunaevsky YE, Shatsky IN. 2003. Conversion of 48S translation preini-

Cite this article as *Cold Spring Harb Perspect Biol* doi: 10.1101/cshperspect.a033092

tiation complexes into 80S initiation complexes as revealed by toeprinting. *FEBS Lett* **533**: 99–104.

Dmitriev SE, Terenin IM, Andreev DE, Ivanov PA, Dunaevsky JE, Merrick WC, Shatsky IN. 2010. GTP-independent tRNA delivery to the ribosomal P-site by a novel eukaryotic translation factor. *J Biol Chem* **285**: 26779–26787.

Dong J, Munoz A, Kolitz SE, Saini AK, Chiu WL, Rahman H, Lorsch JR, Hinnebusch AG. 2014. Conserved residues in yeast initiator tRNA calibrate initiation accuracy by regulating preinitiation complex stability at the start codon. *Genes Dev* **28**: 502–520.

Drabkin HJ, Helk B, RajBhandary UL. 1993. The role of nucleotides conserved in eukaryotic initiator methionine tRNAs in initiation of protein synthesis. *J Biol Chem* **268**: 25221–25228.

Duncan R, Hershey JW. 1983. Identification and quantitation of levels of protein synthesis initiation factors in crude HeLa cell lysates by two-dimensional polyacrylamide gel electrophoresis. *J Biol Chem* **258**: 7228–7235.

Elbarghati L, Murdoch C, Lewis CE. 2008. Effects of hypoxia on transcription factor expression in human monocytes and macrophages. *Immunobiology* **213**: 899–908.

Elfakess R, Sinvani H, Haimov O, Svitkin Y, Sonenberg N, Dikstein R. 2011. Unique translation initiation of mRNAs-containing TISU element. *Nucleic Acids Res* **39**: 7598–7609.

Erickson FL, Hannig EM. 1996. Ligand interactions with eukaryotic translation initiation factor 2: Role of the γ-subunit. *EMBO J* **15**: 6311–6320.

Erzberger JP, Stengel F, Pellarin R, Zhang S, Schaefer T, Aylett CH, Cimermancic P, Boehringer D, Sali A, Aebersold R, et al. 2014. Molecular architecture of the 40S•eIF1•eIF3 translation initiation complex. *Cell* **158**: 1123–1135.

Fekete CA, Mitchell SF, Cherkasova VA, Applefield D, Algire MA, Maag D, Saini AK, Lorsch JR, Hinnebusch AG. 2007. N- and C-terminal residues of eIF1A have opposing effects on the fidelity of start codon selection. *EMBO J* **26**: 1602–1614.

Feoktistova K, Tuvshintogs E, Do A, Fraser CS. 2013. Human eIF4E promotes mRNA restructuring by stimulating eIF4A helicase activity. *Proc Natl Acad Sci* **110**: 13339–13344.

Fernandez IS, Bai XC, Hussain T, Kelley AC, Lorsch JR, Ramakrishnan V, Scheres SH. 2013. Molecular architecture of a eukaryotic translational initiation complex. *Science* **342**: 1240585.

Fraser CS, Berry KE, Hershey JW, Doudna JA. 2007. eIF3j is located in the decoding center of the human 40S ribosomal subunit. *Mol Cell* **26**: 811–819.

Fringer JM, Acker MG, Fekete CA, Lorsch JR, Dever TE. 2007. Coupled release of eukaryotic translation initiation factors 5B and 1A from 80S ribosomes following subunit joining. *Mol Cell Biol* **27**: 2384–2397.

Gao Z, Putnam AA, Bowers HA, Guenther UP, Ye X, Kindsfather A, Hilliker AK, Jankowsky E. 2016. Coupling between the DEAD-box RNA helicases Ded1p and eIF4A. *eLife* **5**: e16408.

García-García C, Frieda KL, Feoktistova K, Fraser CS, Block SM. 2015. Factor-dependent processivity in human eIF4A DEAD-box helicase. *Science* **348**: 1486–1488.

Gordiyenko Y, Schmidt C, Jennings MD, Matak-Vinkovic D, Pavitt GD, Robinson CV. 2014. eIF2B is a decameric guanine nucleotide exchange factor with a γ2ε2 tetrameric core. *Nat Commun* **5**: 3902.

Grifo JA, Abramson RD, Satler CA, Merrick WC. 1984. RNA-stimulated ATPase activity of eukaryotic initiation factors. *J Biol Chem* **259**: 8648–8654.

Gross JD, Moerke NJ, von der Haar T, Lugovskoy AA, Sachs AB, McCarthy JE, Wagner G. 2003. Ribosome loading onto the mRNA cap is driven by conformational coupling between eIF4G and eIF4E. *Cell* **115**: 739–750.

Gruner S, Peter D, Weber R, Wohlbold L, Chung MY, Weichenrieder O, Valkov E, Igreja C, Izaurralde E. 2016. The structures of eIF4E–eIF4G complexes reveal an extended interface to regulate translation initiation. *Mol Cell* **64**: 467–479.

Gunisova S, Valasek LS. 2014. Fail-safe mechanism of GCN4 translational control—uORF2 promotes reinitiation by analogous mechanism to uORF1 and thus secures its key role in GCN4 expression. *Nucleic Acids Res* **42**: 5880–5893.

Haimov O, Sinvani H, Martin F, Ulitsky I, Emmanuel R, Tamarkin-Ben-Harush A, Vardy A, Dikstein R. 2017. Efficient and accurate translation initiation directed by TISU involves RPS3 and RPS10e binding and differential eukaryotic initiation factor 1A regulation. *Mol Cell Biol* **37**: e00150.

Harms U, Andreou AZ, Gubaev A, Klostermeier D. 2014. eIF4B, eIF4G and RNA regulate eIF4A activity in translation initiation by modulating the eIF4A conformational cycle. *Nucleic Acids Res* **42**: 7911–7922.

Hashem Y, des Georges A, Dhote V, Langlois R, Liao HY, Grassucci RA, Hellen CU, Pestova TV, Frank J. 2013a. Structure of the mammalian ribosomal 43S preinitiation complex bound to the scanning factor DHX29. *Cell* **153**: 1108–1119.

Hashem Y, des Georges A, Dhote V, Langlois R, Liao HY, Grassucci RA, Pestova TV, Hellen CU, Frank J. 2013b. Hepatitis-C-virus-like internal ribosome entry sites displace eIF3 to gain access to the 40S subunit. *Nature* **503**: 539–543.

* Hellen CUT. 2018. Translation termination and ribosome recycling in eukaryotes. *Cold Spring Harb Perspect Biol* doi: 10.1101/cshperspect.a032656.

Hinnebusch AG. 2014. The scanning mechanism of eukaryotic translation initiation. *Annu Rev Biochem* **83**: 779–812.

Hinnebusch AG. 2017. Structural insights into the mechanism of scanning and start codon recognition in eukaryotic translation initiation. *Trends Biochem Sci* **42**: 589–611.

Hinnebusch AG, Lorsch JR. 2012. The mechanism of eukaryotic translation initiation: New insights and challenges. *Cold Spring Harb Perspect Biol* **4**: a011544.

Hinnebusch AG, Ivanov IP, Sonenberg N. 2016. Translational control by 5′-untranslated regions of eukaryotic mRNAs. *Science* **352**: 1413–1416.

Hussain T, Llacer JL, Fernandez IS, Munoz A, Martin-Marcos P, Savva CG, Lorsch JR, Hinnebusch AG, Ramakrishnan V. 2014. Structural changes enable start codon recognition by the eukaryotic translation initiation complex. *Cell* **159**: 597–607.

Ivanov IP, Loughran G, Sachs MS, Atkins JF. 2010. Initiation context modulates autoregulation of eukaryotic translation initiation factor 1 (eIF1). *Proc Natl Acad Sci* **107:** 18056–18060.

Jackson RJ, Hellen CU, Pestova TV. 2010. The mechanism of eukaryotic translation initiation and principles of its regulation. *Nat Rev Mol Cell Biol* **11:** 113–127.

Jankowsky E, Gross CH, Shuman S, Pyle AM. 2001. Active disruption of an RNA-protein interaction by a DExH/D RNA helicase. *Science* **291:** 121–125.

Jennings MD, Pavitt GD. 2010. eIF5 has GDI activity necessary for translational control by eIF2 phosphorylation. *Nature* **465:** 378–381.

Jennings MD, Zhou Y, Mohammad-Qureshi SS, Bennett D, Pavitt GD. 2013. eIF2B promotes eIF5 dissociation from eIF2*GDP to facilitate guanine nucleotide exchange for translation initiation. *Genes Dev* **27:** 2696–2707.

Jennings MD, Kershaw CJ, White C, Hoyle D, Richardson JP, Costello JL, Donaldson IJ, Zhou Y, Pavitt GD. 2016. eIF2β is critical for eIF5-mediated GDP-dissociation inhibitor activity and translational control. *Nucleic Acids Res* **44:** 9698–9709.

Jennings MD, Kershaw CJ, Adomavicius T, Pavitt GD. 2017. Fail-safe control of translation initiation by dissociation of eIF2α phosphorylated ternary complexes. *eLife* **6:** e24542.

* Jobe A, Liu Z, Gutierrez-Vargas C, Frank J. 2018. New insights into ribosome structure and function. *Cold Spring Harb Perspect Biol* doi: 10.1101/cshperspect.a032615.

Kapp LD, Lorsch JR. 2004. GTP-dependent recognition of the methionine moiety on initiator tRNA by translation factor eIF2. *J Mol Biol* **335:** 923–936.

Kapp LD, Kolitz SE, Lorsch JR. 2006. Yeast initiator tRNA identity elements cooperate to influence multiple steps of translation initiation. *RNA* **12:** 751–764.

Kashiwagi K, Takahashi M, Nishimoto M, Hiyama TB, Higo T, Umehara T, Sakamoto K, Ito T, Yokoyama S. 2016. Crystal structure of eukaryotic translation initiation factor 2B. *Nature* **531:** 122–125.

Kearse MG, Wilusz JE. 2017. Non-AUG translation: A new start for protein synthesis in eukaryotes. *Genes Dev* **31:** 1717–1731.

Khoshnevis S, Gunisova S, Vlckova V, Kouba T, Neumann P, Beznoskova P, Ficner R, Valasek LS. 2014. Structural integrity of the PCI domain of eIF3a/TIF32 is required for mRNA recruitment to the 43S pre-initiation complexes. *Nucleic Acids Res* **42:** 4123–4139.

Kim JH, Park SM, Park JH, Keum SJ, Jang SK. 2011. eIF2A mediates translation of hepatitis C viral mRNA under stress conditions. *EMBO J* **30:** 2454–2464.

Kolitz SE, Lorsch JR. 2010. Eukaryotic initiator tRNA: Finely tuned and ready for action. *FEBS Lett* **584:** 396–404.

Kolupaeva VG, Unbehaun A, Lomakin IB, Hellen CU, Pestova TV. 2005. Binding of eukaryotic initiation factor 3 to ribosomal 40S subunits and its role in ribosomal dissociation and anti-association. *RNA* **11:** 470–486.

Korets SB, Czok S, Blank SV, Curtin JP, Schneider RJ. 2011. Targeting the mTOR/4E-BP pathway in endometrial cancer. *Clin Cancer Res* **17:** 7518–7528.

Kozak M. 1980. Role of ATP in binding and migration of 40S ribosomal subunits. *Cell* **22:** 459–467.

Kuhle B, Ficner R. 2014. eIF5B employs a novel domain release mechanism to catalyze ribosomal subunit joining. *EMBO J* **33:** 1177–1191.

Kulak NA, Pichler G, Paron I, Nagaraj N, Mann M. 2014. Minimal, encapsulated proteomic-sample processing applied to copy-number estimation in eukaryotic cells. *Nat Methods* **11:** 319–324.

Kumar P, Hellen CU, Pestova TV. 2016. Toward the mechanism of eIF4F-mediated ribosomal attachment to mammalian capped mRNAs. *Genes Dev* **30:** 1573–1588.

* Kwan T, Thompson SR. 2018. Noncanonical translation initiation in eukaryotes. *Cold Spring Harb Perspect Biol* doi: 10.1101/cshperspect.a032672.

Lankat-Buttgereit B, Goke R. 2009. The tumour suppressor Pdcd4: Recent advances in the elucidation of function and regulation. *Biol Cell* **101:** 309–317.

Lee AS, Kranzusch PJ, Doudna JA, Cate JH. 2016. eIF3d is an mRNA cap-binding protein that is required for specialized translation initiation. *Nature* **536:** 96–99.

Linder P, Lasko PF, Ashburner M, Leroy P, Nielsen PJ, Nishi K, Schnier J, Slonimski PP. 1989. Birth of the D-E-A-D box. *Nature* **337:** 121–122.

Llacer JL, Hussain T, Marler L, Aitken CE, Thakur A, Lorsch JR, Hinnebusch AG, Ramakrishnan V. 2015. Conformational differences between open and closed states of the eukaryotic translation initiation complex. *Mol Cell* **59:** 399–412.

Lorsch JR, Herschlag D. 1998. The DEAD-box protein eIF4A. 1: A minimal kinetic and thermodynamic framework reveals coupled binding of RNA and nucleotide. *Biochemistry* **37:** 2180–2193.

Loughran G, Sachs MS, Atkins JF, Ivanov IP. 2012. Stringency of start codon selection modulates autoregulation of translation initiation factor eIF5. *Nucleic Acids Res* **40:** 2898–2906.

Luna RE, Arthanari H, Hiraishi H, Nanda J, Martin-Marcos P, Markus MA, Akabayov B, Milbradt AG, Luna LE, Seo HC, et al. 2012. The C-terminal domain of eukaryotic initiation factor 5 promotes start codon recognition by its dynamic interplay with eIF1 and eIF2β. *Cell Rep* **1:** 689–702.

Luna RE, Arthanari H, Hiraishi H, Akabayov B, Tang L, Cox C, Markus MA, Luna LE, Ikeda Y, Watanabe R, et al. 2013. The interaction between eukaryotic initiation factor 1A and eIF5 retains eIF1 within scanning preinitiation complexes. *Biochemistry* **52:** 9510–9518.

Maag D, Fekete CA, Gryczynski Z, Lorsch JR. 2005. A conformational change in the eukaryotic translation preinitiation complex and release of eIF1 signal recognition of the start codon. *Mol Cell* **17:** 265–275.

Majumdar R, Maitra U. 2005. Regulation of GTP hydrolysis prior to ribosomal AUG selection during eukaryotic translation initiation. *EMBO J* **24:** 3737–3746.

Majumdar R, Bandyopadhyay A, Maitra U. 2003. Mammalian translation initiation factor eIF1 functions with eIF1A and eIF3 in the formation of a stable 40S preinitiation complex. *J Biol Chem* **278:** 6580–6587.

Malys N, McCarthy JE. 2011. Translation initiation: Variations in the mechanism can be anticipated. *Cell Mol Life Sci* **68:** 991–1003.

Marintchev A, Kolupaeva VG, Pestova TV, Wagner G. 2003. Mapping the binding interface between human eukaryotic initiation factors 1A and 5B: A new interaction between old partners. *Proc Natl Acad Sci* **100**: 1535–1540.

Martin-Marcos P, Cheung YN, Hinnebusch AG. 2011. Functional elements in initiation factors 1, 1A, and 2β discriminate against poor AUG context and non-AUG start codons. *Mol Cell Biol* **31**: 4814–4831.

Martin-Marcos P, Nanda J, Luna RE, Wagner G, Lorsch JR, Hinnebusch AG. 2013. β-Hairpin loop of eukaryotic initiation factor 1 (eIF1) mediates 40S ribosome binding to regulate initiator tRNA(Met) recruitment and accuracy of AUG selection *in vivo*. *J Biol Chem* **288**: 27546–27562.

Merrick WC. 2015. eIF4F: A retrospective. *J Biol Chem* **290**: 24091–24099.

Merrick WC, Anderson WF. 1975. Purification and characterization of homogeneous protein synthesis initiation factor M1 from rabbit reticulocytes. *J Biol Chem* **250**: 1197–1206.

Merrick WC, Kemper WM, Anderson WF. 1975. Purification and characterization of homogeneous initiation factor M2A from rabbit reticulocytes. *J Biol Chem* **250**: 5556–5562.

Meyer KD, Patil DP, Zhou J, Zinoviev A, Skabkin MA, Elemento O, Pestova TV, Qian SB, Jaffrey SR. 2015. 5′UTR m⁶A promotes cap-independent translation. *Cell* **163**: 999–1010.

Mitchell SF, Walker SE, Algire MA, Park EH, Hinnebusch AG, Lorsch JR. 2010. The 5′-7-methylguanosine cap on eukaryotic mRNAs serves both to stimulate canonical translation initiation and to block an alternative pathway. *Mol Cell* **39**: 950–962.

Mohammad MP, Munzarova Pondelickova V, Zeman J, Gunisova S, Valasek LS. 2017. In vivo evidence that eIF3 stays bound to ribosomes elongating and terminating on short upstream ORFs to promote reinitiation. *Nucleic Acids Res* **45**: 2658–2674.

Mohammad-Qureshi SS, Haddad R, Hemingway EJ, Richardson JP, Pavitt GD. 2007. Critical contacts between the eukaryotic initiation factor 2B (eIF2B) catalytic domain and both eIF2β and -2γ mediate guanine nucleotide exchange. *Mol Cell Biol* **27**: 5225–5234.

Munzarova V, Panek J, Gunisova S, Danyi I, Szamecz B, Valasek LS. 2011. Translation reinitiation relies on the interaction between eIF3a/TIF32 and progressively folded *cis*-acting mRNA elements preceding short uORFs. *PLoS Genet* **7**: e1002137.

Nag N, Lin KY, Edmonds KA, Yu J, Nadkarni D, Marintcheva B, Marintchev A. 2016. eIF1A/eIF5B interaction network and its functions in translation initiation complex assembly and remodeling. *Nucleic Acids Res* **44**: 7441–7456.

Nanda JS, Saini AK, Munoz AM, Hinnebusch AG, Lorsch JR. 2013. Coordinated movements of eukaryotic translation initiation factors eIF1, eIF1A, and eIF5 trigger phosphate release from eIF2 in response to start codon recognition by the ribosomal preinitiation complex. *J Biol Chem* **288**: 5316–5329.

Naveau M, Lazennec-Schurdevin C, Panvert M, Dubiez E, Mechulam Y, Schmitt E. 2013. Roles of yeast eIF2α and eIF2β subunits in the binding of the initiator methionyl-tRNA. *Nucleic Acids Res* **41**: 1047–1057.

Obayashi E, Luna RE, Nagata T, Martin-Marcos P, Hiraishi H, Singh CR, Erzberger JP, Zhang F, Arthanari H, Morris J, et al. 2017. Molecular landscape of the ribosome preinitiation complex during mRNA scanning: Structural role for eIF3c and its control by eIF5. *Cell Rep* **18**: 2651–2663.

O'Leary SE, Petrov A, Chen J, Puglisi JD. 2013. Dynamic recognition of the mRNA cap by *Saccharomyces cerevisiae* eIF4E. *Structure* **21**: 2197–2207.

Olsen DS, Savner EM, Mathew A, Zhang F, Krishnamoorthy T, Phan L, Hinnebusch AG. 2003. Domains of eIF1A that mediate binding to eIF2, eIF3, and eIF5B and promote ternary complex recruitment in vivo. *EMBO J* **22**: 193–204.

Panniers R, Rowlands AG, Henshaw EC. 1988. The effect of Mg^{2+} and guanine nucleotide exchange factor on the binding of guanine nucleotides to eukaryotic initiation factor 2. *J Biol Chem* **263**: 5519–5525.

Park EH, Walker SE, Lee JM, Rothenburg S, Lorsch JR, Hinnebusch AG. 2011. Multiple elements in the eIF4G1 N-terminus promote assembly of eIF4G1*PABP mRNPs in vivo. *EMBO J* **30**: 302–316.

Parsyan A, Svitkin Y, Shahbazian D, Gkogkas C, Lasko P, Merrick WC, Sonenberg N. 2011. mRNA helicases: The tacticians of translational control. *Nat Rev Mol Cell Biol* **12**: 235–245.

Passmore LA, Schmeing TM, Maag D, Applefield DJ, Acker MG, Algire MA, Lorsch JR, Ramakrishnan V. 2007. The eukaryotic translation initiation factors eIF1 and eIF1A induce an open conformation of the 40S ribosome. *Mol Cell* **26**: 41–50.

Paulin FE, Campbell LE, O'Brien K, Loughlin J, Proud CG. 2001. Eukaryotic translation initiation factor 5 (eIF5) acts as a classical GTPase-activator protein. *Curr Biol* **11**: 55–59.

Pause A, Methot N, Svitkin Y, Merrick WC, Sonenberg N. 1994. Dominant negative mutants of mammalian translation initiation factor eIF-4A define a critical role for eIF-4F in cap-dependent and cap-independent initiation of translation. *EMBO J* **13**: 1205–1215.

Pavitt GD. 2005. eIF2B, a mediator of general and gene-specific translational control. *Biochem Soc Trans* **33**: 1487–1492.

Pavitt GD, Ramaiah KV, Kimball SR, Hinnebusch AG. 1998. eIF2 independently binds two distinct eIF2B subcomplexes that catalyze and regulate guanine-nucleotide exchange. *Genes Dev* **12**: 514–526.

* Peer E, Moshitch-Moshkovitz S, Rechavi G, Dominissini D. 2018. The epitranscriptome in translation regulation. *Cold Spring Harb Perspect Biol* doi: 10.1101/cshperspect. a032623.

Pelletier J, Sonenberg N. 1988. Internal initiation of translation of eukaryotic mRNA directed by a sequence derived from poliovirus RNA. *Nature* **334**: 320–325.

Pestova TV, Borukhov SI, Hellen CU. 1998. Eukaryotic ribosomes require initiation factors 1 and 1A to locate initiation codons. *Nature* **394**: 854–859.

Pestova TV, de Breyne S, Pisarev AV, Abaeva IS, Hellen CU. 2008. eIF2-dependent and eIF2-independent modes of initiation on the CSFV IRES: A common role of domain II. *EMBO J* **27**: 1060–1072.

Peterson DT, Merrick WC, Safer B. 1979. Binding and release of radiolabeled eukaryotic initiation factors 2 and 3 during 80S initiation complex formation. *J Biol Chem* **254:** 2509–2516.

Pisarev AV, Kolupaeva VG, Pisareva VP, Merrick WC, Hellen CU, Pestova TV. 2006. Specific functional interactions of nucleotides at key −3 and +4 positions flanking the initiation codon with components of the mammalian 48S translation initiation complex. *Genes Dev* **20:** 624–636.

Pisarev AV, Hellen CU, Pestova TV. 2007. Recycling of eukaryotic posttermination ribosomal complexes. *Cell* **131:** 286–299.

Pisarev AV, Kolupaeva VG, Yusupov MM, Hellen CU, Pestova TV. 2008. Ribosomal position and contacts of mRNA in eukaryotic translation initiation complexes. *EMBO J* **27:** 1609–1621.

Pisareva VP, Pisarev AV. 2014. eIF5 and eIF5B together stimulate 48S initiation complex formation during ribosomal scanning. *Nucleic Acids Res* **42:** 12052–12069.

Pisareva VP, Pisarev AV, Komar AA, Hellen CU, Pestova TV. 2008. Translation initiation on mammalian mRNAs with structured 5′UTRs requires DExH-box protein DHX29. *Cell* **135:** 1237–1250.

* Proud CG. 2018. Regulation of translation by cap-binding proteins. *Cold Spring Harb Perspect Biol* doi: 10.1101/cshperspect.a033050.

Rajagopal V, Park EH, Hinnebusch AG, Lorsch JR. 2012. Specific domains in yeast translation initiation factor eIF4G strongly bias RNA unwinding activity of the eIF4F complex toward duplexes with 5′-overhangs. *J Biol Chem* **287:** 20301–20312.

Richter JD, Sonenberg N. 2005. Regulation of cap-dependent translation by eIF4E inhibitory proteins. *Nature* **433:** 477–480.

* Rodnina MV. 2018. Translation in prokaryotes. *Cold Spring Harb Perspect Biol* doi: 10.1101/cshperspect.a032664.

Rogers GW Jr, Richter NJ, Lima WF, Merrick WC. 2001. Modulation of the helicase activity of eIF4A by eIF4B, eIF4H, and eIF4F. *J Biol Chem* **276:** 30914–30922.

Rogers DW, Bottcher MA, Traulsen A, Greig D. 2017. Ribosome reinitiation can explain length-dependent translation of messenger RNA. *PLoS Comput Biol* **13:** e1005592.

Rowlands AG, Panniers R, Henshaw EC. 1988. The catalytic mechanism of guanine nucleotide exchange factor action and competitive inhibition by phosphorylated eukaryotic initiation factor 2. *J Biol Chem* **263:** 5526–5533.

Rozen F, Edery I, Meerovitch K, Dever TE, Merrick WC, Sonenberg N. 1990. Bidirectional RNA helicase activity of eucaryotic translation initiation factors 4A and 4F. *Mol Cell Biol* **10:** 1134–1144.

Saini AK, Nanda JS, Martin-Marcos P, Dong J, Zhang F, Bhardwaj M, Lorsch JR, Hinnebusch AG. 2014. Eukaryotic translation initiation factor eIF5 promotes the accuracy of start codon recognition by regulating P_i release and conformational transitions of the preinitiation complex. *Nucleic Acids Res* **42:** 9623–9640.

Schmitt E, Naveau M, Mechulam Y. 2010. Eukaryotic and archaeal translation initiation factor 2: A heterotrimeric tRNA carrier. *FEBS Lett* **584:** 405–412.

Schmitt E, Panvert M, Lazennec-Schurdevin C, Coureux PD, Perez J, Thompson A, Mechulam Y. 2012. Structure of the ternary initiation complex aIF2-GDPNP-methionylated initiator tRNA. *Nat Struct Mol Biol* **19:** 450–454.

Schreier MH, Erni B, Staehelin T. 1977. Initiation of mammalian protein synthesis. I. Purification and characterization of seven initiation factors. *J Mol Biol* **116:** 727–753.

Schwab SR, Shugart JA, Horng T, Malarkannan S, Shastri N. 2004. Unanticipated antigens: Translation initiation at CUG with leucine. *PLoS Biol* **2:** e366.

Sen ND, Zhou F, Ingolia NT, Hinnebusch AG. 2015. Genome-wide analysis of translational efficiency reveals distinct but overlapping functions of yeast DEAD-box RNA helicases Ded1 and eIF4A. *Genome Res* **25:** 1196–1205.

Sen ND, Zhou F, Harris MS, Ingolia NT, Hinnebusch AG. 2016. eIF4B stimulates translation of long mRNAs with structured 5′UTRs and low closed-loop potential but weak dependence on eIF4G. *Proc Natl Acad Sci* **113:** 10464–10472.

Sendoel A, Dunn JG, Rodriguez EH, Naik S, Gomez NC, Hurwitz B, Levorse J, Dill BD, Schramek D, Molina H, et al. 2017. Translation from unconventional 5′ start sites drives tumour initiation. *Nature* **541:** 494–499.

Shin BS, Kim JR, Walker SE, Dong J, Lorsch JR, Dever TE. 2011. Initiation factor eIF2γ promotes eIF2-GTP-Met-tRNA$_i^{Met}$ ternary complex binding to the 40S ribosome. *Nat Struct Mol Biol* **18:** 1227–1234.

Simonetti A, Brito Querido J, Myasnikov AG, Mancera-Martinez E, Renaud A, Kuhn L, Hashem Y. 2016. eIF3 peripheral subunits rearrangement after mRNA binding and start-codon recognition. *Mol Cell* **63:** 206–217.

Singh CR, Lee B, Udagawa T, Mohammad-Qureshi SS, Yamamoto Y, Pavitt GD, Asano K. 2006. An eIF5/eIF2 complex antagonizes guanine nucleotide exchange by eIF2B during translation initiation. *EMBO J* **25:** 4537–4546.

Singh CR, Watanabe R, Chowdhury W, Hiraishi H, Murai MJ, Yamamoto Y, Miles D, Ikeda Y, Asano M, Asano K. 2012. Sequential eukaryotic translation initiation factor 5 (eIF5) binding to the charged disordered segments of eIF4G and eIF2β stabilizes the 48S preinitiation complex and promotes its shift to the initiation mode. *Mol Cell Biol* **32:** 3978–3989.

Sinvani H, Haimov O, Svitkin Y, Sonenberg N, Tamarkin-Ben-Harush A, Viollet B, Dikstein R. 2015. Translational tolerance of mitochondrial genes to metabolic energy stress involves TISU and eIF1-eIF4GI cooperation in start codon selection. *Cell Metab* **21:** 479–492.

Skabkin MA, Skabkina OV, Dhote V, Komar AA, Hellen CU, Pestova TV. 2010. Activities of Ligatin and MCT-1/DENR in eukaryotic translation initiation and ribosomal recycling. *Genes Dev* **24:** 1787–1801.

Slusher LB, Gillman EC, Martin NC, Hopper AK. 1991. mRNA leader length and initiation codon context determine alternative AUG selection for the yeast gene MOD5. *Proc Natl Acad Sci* **88:** 9789–9793.

Sokabe M, Fraser CS. 2014. Human eukaryotic initiation factor 2 (eIF2)-GTP-Met-tRNA$_i$ ternary complex and eIF3 stabilize the 43S preinitiation complex. *J Biol Chem* **289:** 31827–31836.

Sokabe M, Fraser CS. 2017. A helicase-independent activity of eIF4A in promoting mRNA recruitment to the human ribosome. *Proc Natl Acad Sci* **114:** 6304–6309.

* Sokabe M, Fraser CS. 2018. Toward a kinetic understanding of eukaryotic translation. *Cold Spring Harb Perspect Biol* doi: 10.1101/cshperspect.a032706.

Sokabe M, Fraser CS, Hershey JW. 2012. The human translation initiation multi-factor complex promotes methionyl-tRNA$_i$ binding to the 40S ribosomal subunit. *Nucleic Acids Res* **40:** 905–913.

Spevak CC, Ivanov IP, Sachs MS. 2010. Sequence requirements for ribosome stalling by the arginine attenuator peptide. *J Biol Chem* **285:** 40933–40942.

Starck SR, Shastri N. 2016. Nowhere to hide: Unconventional translation yields cryptic peptides for immune surveillance. *Immunol Rev* **272:** 8–16.

Starck SR, Jiang V, Pavon-Eternod M, Prasad S, McCarthy B, Pan T, Shastri N. 2012. Leucine-tRNA initiates at CUG start codons for protein synthesis and presentation by MHC class I. *Science* **336:** 1719–1723.

Starck SR, Tsai JC, Chen K, Shodiya M, Wang L, Yahiro K, Martins-Green M, Shastri N, Walter P. 2016. Translation from the 5′ untranslated region shapes the integrated stress response. *Science* **351:** aad3867.

* Stern-Ginossar N, Thompson SR, Mathews MB, Mohr I. 2018. Translational control in virus-infected cells. *Cold Spring Harb Perspect Biol* doi: 10.1101/cshperspect.a033001.

Svitkin YV, Pause A, Haghighat A, Pyronnet S, Witherell G, Belsham GJ, Sonenberg N. 2001. The requirement for eukaryotic initiation factor 4A (eIF4A) in translation is in direct proportion to the degree of mRNA 5′ secondary structure. *RNA* **7:** 382–394.

Szamecz B, Rutkai E, Cuchalova L, Munzarova V, Herrmannova A, Nielsen KH, Burela L, Hinnebusch AG, Valasek L. 2008. eIF3a cooperates with sequences 5′ of uORF1 to promote resumption of scanning by post-termination ribosomes for reinitiation on GCN4 mRNA. *Genes Dev* **22:** 2414–2425.

Unbehaun A, Borukhov SI, Hellen CU, Pestova TV. 2004. Release of initiation factors from 48S complexes during ribosomal subunit joining and the link between establishment of codon-anticodon base-pairing and hydrolysis of eIF2-bound GTP. *Genes Dev* **18:** 3078–3093.

Valasek LS. 2012. "Ribozoomin"—Translation initiation from the perspective of the ribosome-bound eukaryotic initiation factors (eIFs). *Curr Protein Pept Sci* **13:** 305–330.

Valasek L, Nielsen KH, Hinnebusch AG. 2002. Direct eIF2-eIF3 contact in the multifactor complex is important for translation initiation in vivo. *EMBO J* **21:** 5886–5898.

Valasek L, Mathew AA, Shin BS, Nielsen KH, Szamecz B, Hinnebusch AG. 2003. The yeast eIF3 subunits TIF32/a, NIP1/c, and eIF5 make critical connections with the 40S ribosome in vivo. *Genes Dev* **17:** 786–799.

Valasek L, Nielsen KH, Zhang F, Fekete CA, Hinnebusch AG. 2004. Interactions of eukaryotic translation initiation factor 3 (eIF3) subunit NIP1/c with eIF1 and eIF5 promote preinitiation complex assembly and regulate start codon selection. *Mol Cell Biol* **24:** 9437–9455.

Valasek LS, Zeman J, Wagner S, Beznoskova P, Pavlikova Z, Mohammad MP, Hronova V, Herrmannova A, Hashem Y, Gunisova S. 2017. Embraced by eIF3: Structural and functional insights into the roles of eIF3 across the translation cycle. *Nucleic Acids Res* **45:** 10948–10968.

Ventoso I, Sanz MA, Molina S, Berlanga JJ, Carrasco L, Esteban M. 2006. Translational resistance of late αvirus mRNA to eIF2α phosphorylation: A strategy to overcome the antiviral effect of protein kinase PKR. *Genes Dev* **20:** 87–100.

Villa N, Do A, Hershey JW, Fraser CS. 2013. Human eukaryotic initiation factor 4G (eIF4G) protein binds to eIF3c, -d, and -e to promote mRNA recruitment to the ribosome. *J Biol Chem* **288:** 32932–32940.

Visweswaraiah J, Pittman Y, Dever TE, Hinnebusch AG. 2015. The β-hairpin of 40S exit channel protein Rps5/uS7 promotes efficient and accurate translation initiation in vivo. *eLife* **4:** e07939.

von der Haar T, McCarthy JE. 2002. Intracellular translation initiation factor levels in *Saccharomyces cerevisiae* and their role in cap-complex function. *Mol Microbiol* **46:** 531–544.

Wagner S, Herrmannova A, Sikrova D, Valasek LS. 2016. Human eIF3b and eIF3a serve as the nucleation core for the assembly of eIF3 into two interconnected modules: The yeast-like core and the octamer. *Nucleic Acids Res* **44:** 10772–10788.

Weisser M, Voigts-Hoffmann F, Rabl J, Leibundgut M, Ban N. 2013. The crystal structure of the eukaryotic 40S ribosomal subunit in complex with eIF1 and eIF1A. *Nat Struct Mol Biol* **20:** 1015–1017.

Weisser M, Schafer T, Leibundgut M, Bohringer D, Aylett CHS, Ban N. 2017. Structural and functional insights into human re-initiation complexes. *Mol Cell* **67:** 447–456. e447.

* Wek RC. 2018. Role of eIF2α kinases in translational control and adaptation to cellular stress. *Cold Spring Harb Perspect Biol* doi: 10.1101/cshperspect.a032870.

Wells SE, Hillner PE, Vale RD, Sachs AB. 1998. Circularization of mRNA by eukaryotic translation initiation factors. *Mol Cell* **2:** 135–140.

Wortham NC, Martinez M, Gordiyenko Y, Robinson CV, Proud CG. 2014. Analysis of the subunit organization of the eIF2B complex reveals new insights into its structure and regulation. *FASEB J* **28:** 2225–2237.

Yamamoto Y, Singh CR, Marintchev A, Hall NS, Hannig EM, Wagner G, Asano K. 2005. The eukaryotic initiation factor (eIF) 5 HEAT domain mediates multifactor assembly and scanning with distinct interfaces to eIF1, eIF2, eIF3, and eIF4G. *Proc Natl Acad Sci* **102:** 16164–16169.

Yamamoto H, Unbehaun A, Loerke J, Behrmann E, Collier M, Burger J, Mielke T, Spahn CM. 2014. Structure of the mammalian 80S initiation complex with initiation factor 5B on HCV-IRES RNA. *Nat Struct Mol Biol* **21:** 721–727.

Yanagiya A, Svitkin YV, Shibata S, Mikami S, Imataka H, Sonenberg N. 2009. Requirement of RNA binding of mammalian eukaryotic translation initiation factor 4GI (eIF4GI) for efficient interaction of eIF4E with the mRNA cap. *Mol Cell Biol* **29:** 1661–1669.

Young SK, Wek RC. 2016. Upstream open reading frames differentially regulate gene-specific translation in the integrated stress response. *J Biol Chem* **291:** 16927–16935.

Zheng A, Yu J, Yamamoto R, Ose T, Tanaka I, Yao M. 2014. X-ray structures of eIF5B and the eIF5B-eIF1A complex: The conformational flexibility of eIF5B is restricted on the ribosome by interaction with eIF1A. *Acta Crystallogr D Biol Crystallogr* **70:** 3090–3098.

Zu T, Liu Y, Banez-Coronel M, Reid T, Pletnikova O, Lewis J, Miller TM, Harms MB, Falchook AE, Subra- mony SH, et al. 2013. RAN proteins and RNA foci from antisense transcripts in C9ORF72 ALS and fron- totemporal dementia. *Proc Natl Acad Sci* **110:** E4968– E4977.

* Zu T, Pattamatta A, Ranum LPW. 2018. RNA translation in neurological diseases. *Cold Spring Harb Perspect Biol* doi: 10.1101/cshperspect.a033019.

Cite this article as *Cold Spring Harb Perspect Biol* doi: 10.1101/cshperspect.a033092

Translation Elongation and Recoding in Eukaryotes

Thomas E. Dever,[1] Jonathan D. Dinman,[2] and Rachel Green[3]

[1]Eunice Kennedy Shriver National Institute of Child Health and Human Development, National Institutes of Health, Bethesda, Maryland 20892

[2]Department of Cell Biology and Molecular Genetics, University of Maryland, College Park, Maryland 20742

[3]Department of Molecular Biology and Genetics, Johns Hopkins University School of Medicine, Baltimore, Maryland 21205

Correspondence: tdever@nih.gov; dinman@umd.edu; ragreen@jhmi.edu

In this review, we highlight the current understanding of translation elongation and recoding in eukaryotes. In addition to providing an overview of the process, recent advances in our understanding of the role of the factor eIF5A in both translation elongation and termination are discussed. We also highlight mechanisms of translation recoding with a focus on ribosomal frameshifting during elongation. We see that the balance between the basic steps in elongation and the less common recoding events is determined by the kinetics of the different processes as well as by specific sequence determinants.

OVERVIEW OF TRANSLATION ELONGATION

The mechanism of translation elongation is conserved in all kingdoms of life. Whereas most of the mechanistic details of the process have been elucidated in studies of bacterial translation (see Rodnina 2018), the key steps are shared between eukaryotes and bacteria. In eukaryotes, translation initiation culminates with formation of an 80S initiation complex in which Met-tRNA$_i^{Met}$ is bound in the P (peptidyl) site of the ribosome. The anticodon of the Met-tRNA$_i^{Met}$ is base-paired with the start codon of the messenger RNA (mRNA), and the second codon of the open reading frame (ORF) is in the A (aminoacyl) site of the ribosome. Elongation commences with delivery of the cognate elongating aminoacyl-tRNA (transfer RNA) to the A site of the ribosome (Fig. 1). The eukaryotic translation elongation factor eEF1A, like its bacterial ortholog EF-Tu, is activated upon binding guanosine triphosphate (GTP) and forms a ternary complex upon binding an aminoacyl-tRNA. The eEF1A•GTP•aminoacyl-tRNA complex binds in the A site. Base-pairing interactions between the anticodon of the aminoacyl-tRNA and the A-site codon trigger GTP hydrolysis by eEF1A. The eEF1A•GDP complex is released and the aminoacyl-tRNA is accommodated into the A site.

High-resolution cryoelectron microscopy (EM) structures of decoding complexes have provided insights into how the bacterial and eukaryotic ribosomes sense proper decoding (Jobe et al. 2018). The 18S ribosomal RNA (rRNA)

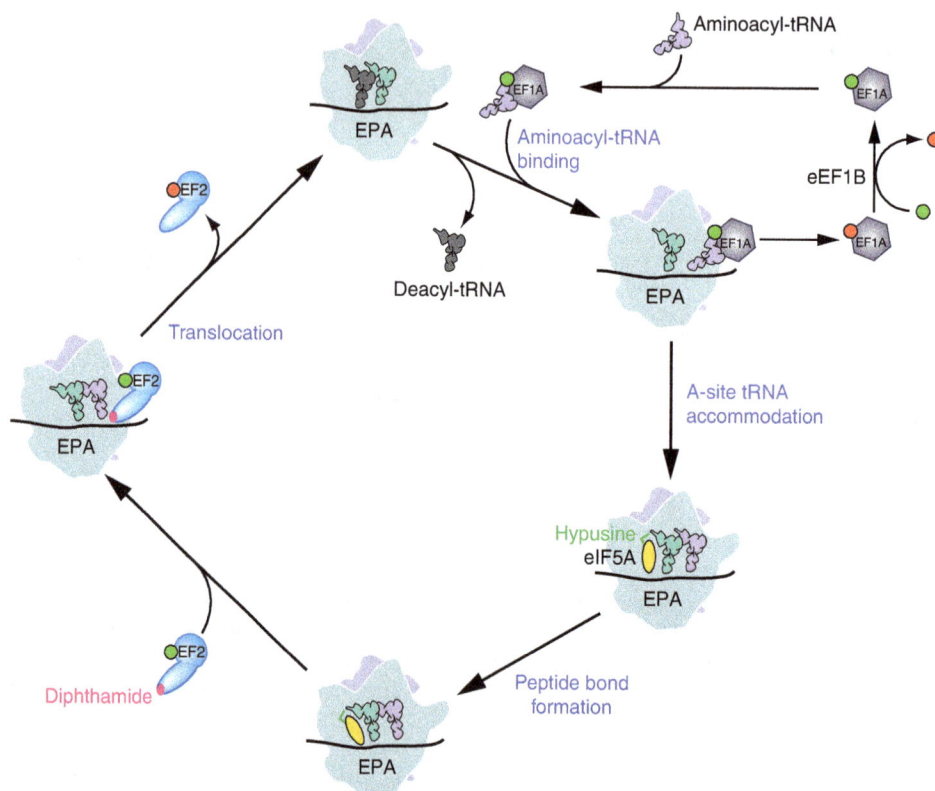

Figure 1. Model of the eukaryotic translation elongation pathway. At the *top*, an eEF1A•GTP•aminoacyl-tRNA (transfer RNA) ternary complex binds to the A (aminoacyl) site of an 80S ribosome with the anticodon loop of the tRNA in contact with the messenger RNA (mRNA). Following GTP hydrolysis and release of an eEF1A•GDP binary complex, the aminoacyl-tRNA is accommodated into the A site, and the eEF1A•GDP is recycled to eEF1A•GTP by the exchange factor eEF1B. During catalysis of peptide bond formation, the A- and P (peptidyl)-site tRNAs shift into hybrid states with the acceptor ends of the tRNAs moving to the P and E sites, respectively. Substrate positioning for peptide bond formation is aided by binding of the factor eIF5A and its hypusine modification (green) in the E site. Following peptide bond formation, the factor eEF2•GTP with its diphthamide modification (magenta) binds in the A site and promotes translocation of the tRNAs into the canonical P and E sites. Following release of the deacylated tRNA from the E site, the next cycle of elongation commences with binding of the appropriate eEF1A•GTP•aminoacyl-tRNA to the A site. Throughout, GTP is depicted as a green ball and GDP as a red ball; also, the large ribosomal subunit (light blue) is displayed transparently to enable visualization of the tRNAs, factors, and mRNA bound to the decoding center at the interface between the large and small subunits and of tRNAs, interacting with the peptidyl transferase center in the large subunit. Note, however, that the positions of the mRNA, tRNAs, and factors are drawn for clarity and are not meant to specify their exact places on the ribosome.

helix h44 residues A1824 and A1825 in mammalian (rabbit) ribosomes (A1755 and A1756 in *Saccharomyces cerevisiae* and A1492 and A1493 in *Escherichia coli*, respectively) as well as the residue G626 in rabbit ribosomes (G577 in *S. cerevisiae* and G530 in *E. coli*) interact with the minor groove of the codon–anticodon helix and stabilize A-site tRNA binding by hydrogen bonding (Ogle et al. 2001; Shao et al. 2016; Loveland et al. 2017). Interestingly, when flipped out of helix 44 (h44), the residues A1824 and A1825 (or A1492 and A1493 in bacteria) interact with the first two codon pairs in the codon–anticodon duplex, enabling the +3 position to participate in wobble interactions related to the degeneracy of the genetic code (Loveland et al. 2017).

As the h44 residues also flip out to interact with mispaired codon–anticodon helices formed with near-cognate tRNAs in the A site (Demeshkina et al. 2012), it has been proposed that the interaction of G626 (G530 in bacteria) may perform a more crucial function as the latching nucleotide that fixes the codon–anticodon helix in the decoding center of the ribosome (Loveland et al. 2017). In addition to providing insights into decoding, the recent structures of eukaryotic ribosomal complexes have provided insights into GTPase activation of eEF1A as well as of eRF3 in termination complexes and of Hbs1 in ribosome rescue complexes (Shao et al. 2016). In these structures, interactions between the sarcin-ricin loop of the large ribosomal subunit and the Switch 2 loop of the GTPase domains helps position the catalytic His residue to promote GTP hydrolysis (Shao et al. 2016). Moreover, the amino terminus of the eukaryote-specific ribosomal protein eS30 becomes ordered upon cognate codon–anticodon interaction in the A site, and a conserved His residue inserts into the decoding center to form potentially stabilizing contacts (Shao et al. 2016). These novel interactions may contribute to the reported enhanced accuracy of eukaryotic versus bacterial elongation (Kramer et al. 2010). Finally, the structural studies of the eukaryotic elongation complex revealed a conserved binding site for inhibitors of eEF1A and EF-Tu. The translational inhibitor didemnin B, which specifically impairs eukaryotic elongation, was found to bind in a cleft between the G domain and domain III of eEF1A in a position that overlaps with the binding site of the structurally unrelated antibiotic kirromycin on EF-Tu (Shao et al. 2016). Like kirromycin, didemnin B and the functionally related eEF1A inhibitor ternatin (Carelli et al. 2015) are thought to prevent the structural rotations in eEF1A required for release of the factor from the ribosome following GTP hydrolysis.

The heart of protein synthesis is peptide bond formation, and the conservation of the ribosome active site structure suggests that the mechanism of peptide bond formation is universally conserved. Following release of eEF1A and accommodation of the aminoacyl-tRNA

into the A site, peptide bond formation with the peptidyl-tRNA in the P site occurs rapidly. Composed of conserved rRNA elements in the large ribosomal subunit, the peptidyl transferase center (PTC) of the ribosome is well conserved between bacteria and eukaryotes (Ben-Shem et al. 2010, 2011; Klinge et al. 2011), and principally functions by positioning the peptidyl- and aminoacyl-tRNAs for catalysis. The factor eIF5A, the ortholog of bacterial elongation factor P (EF-P), binds in the E site, interacts with the acceptor arm of the peptidyl-tRNA, and is thought, like EF-P (Doerfel et al. 2015), to promote peptide bond formation by inducing a favorable positioning of the substrates (Gutierrez et al. 2013; Melnikov et al. 2016a,b; Schmidt et al. 2016; Shin et al. 2017).

During peptide bond formation, the nascent peptide is transferred from the peptidyl-tRNA in the P site to the amino group of the A-site aminoacyl-tRNA (aa-tRNA) to form a new extended peptidyl-tRNA. Peptide bond formation is accompanied by repositioning of the tRNAs into hybrid states (Moazed and Noller 1989) and by subunit rotation. Cryo-EM imaging of eukaryotic elongation complexes has revealed three states of tRNA binding and subunit rotation upon peptide bond formation: in unrotated complexes, the newly formed peptidyl-tRNA is in the A site and deacylated tRNA is in the P site; in rotated-1 complexes, the deacylated tRNA adopts a hybrid P/E state with the anticodon paired with the mRNA in the P site and the acceptor arm of the tRNA in the E site, while the peptidyl-tRNA remains in the classic A site; and in the rotated-2 state, the deacylated tRNA is in the hybrid P/E state and the peptidyl-tRNA is repositioned into a hybrid A/P state with the anticodon paired with mRNA in the A site and the peptide attached to the acceptor arm in the P site (Budkevich et al. 2011; Behrmann et al. 2015). Translocation of the tRNAs to the canonical E and P sites is promoted by the elongation factor eEF2, the eukaryotic ortholog of the bacterial factor EF-G. Structural studies have revealed that eEF2 binds in the A site where it is thought to "unlock" the decoding interaction of the helix h44 nucleotides (A1755 and A1756 in *S. cerevisiae*) with the codon–anticodon duplex in the A site (Spahn et al. 2004; Taylor et al. 2007; Abeyrathne et al. 2016; Murray et al. 2016). Whereas no structures of canonical eukaryotic translocation intermediates (ribosomes with eEF2 and two translocating tRNAs) have been reported, based on similarity with the bacterial system swiveling of the small subunit head and reverse rotation of the small subunit relative to the large subunit are thought to accompany movement of the tRNAs into the canonical P and E sites (P/P and E/E states) (Ratje et al. 2010; Ermolenko and Noller 2011; Ramrath et al. 2013) and to allow for release of eEF2•GDP from the posttranslocation ribosome.

Following translocation, a deacylated tRNA occupies the E site and peptidyl-tRNA is positioned in the P site. It was previously proposed that release of the E-site tRNA from eukaryotic ribosomes is coupled to binding of the eEF1A•GTP•aminoacyl-tRNA ternary complex in the A site (Triana-Alonso et al. 1995; Anand et al. 2003) as also proposed for the bacterial system (Burkhardt et al. 1998). At odds with this proposal, various kinetic analyses revealed that release of the E-site tRNA and binding of the A-site tRNA are not strictly coupled in bacterial systems (Semenkov et al. 1996; Uemura et al. 2010; Chen et al. 2011; Petropoulos and Green 2012). Indeed, in one single molecule kinetic study, it was seen that deacylated tRNA is released slowly from the E site of human ribosomes following translocation, but independent of the binding of A-site aa-tRNA (Ferguson et al. 2015). Higher affinity binding of deacylated tRNA to the E site may be a feature that distinguishes eukaryotic from bacterial translation, and could impose novel or additional requirements on the translation factors that function at the E site.

As stated above, the basic mechanism of translation elongation is conserved between bacteria and eukaryotes. Whereas many of the studies elucidating the ribosomal and translation factor contributions to translation elongation have focused on the bacterial system, we focus this review on eukaryotic translation elongation and the features that distinguish eukaryotic from bacterial translation elongation.

THE EUKARYOTIC TRANSLATION ELONGATION FACTORS

In contrast to the complex factor requirements in translation initiation, elongation is assisted by a minimal set of factors. In addition to the canonical factors eEF1A/EF-Tu and eEF2/EF-G, the elongation factor eIF5A/EF-P is also conserved between eukaryotes and bacteria. In contrast, the ATPase eEF3 appears to be restricted to fungi and perhaps some other single-cell eukaryotes. In this section, we will highlight properties of the eukaryotic translation elongation factors with a focus on the unique features that distinguish the eukaryotic factors from their bacterial counterparts.

eEF1A–eEF1B

The GTPase eEF1A binds aminoacyl-tRNA in a ternary complex with GTP. Following GTP hydrolysis on the ribosome, the eIF1A is released in a binary complex with GDP. As with EF-Tu, the spontaneous rate of GDP dissociation from eEF1A is slow and the guanine nucleotide exchange factor eEF1B is required to recycle inactive eEF1A•GDP to active eEF1A•GTP (Gromadski et al. 2007). Whereas the complementary factor EF-Ts in bacteria is a single polypeptide, eEF1B is composed of two or three subunits (depending on the organism) and destabilizes GDP binding to eEF1A by a mechanism that is distinct from that employed by EF-Ts (Andersen et al. 2001; Rodnina and Wintermeyer 2009). The catalytic eEF1Bα subunit forms a dimeric complex with eEF1Bγ in yeast and forms trimeric complexes in mammals and plants consisting of eEF1Bα, eEF1Bγ, and either eIF1Bβ (mammals) or eIF1Bδ (plants). Overexpression of eIF1A or mutations in eEF1A that lower guanine nucleotide binding affinity bypass the essential requirement for eEF1Bα in yeast (Kinzy and Woolford 1995; Carr-Schmid et al. 1999); however, the suppression is incomplete as the eEF1A mutations increase nonsense suppression and show increased sensitivity to translation elongation inhibitors (Carr-Schmid et al. 1999).

In humans, eEF1A is encoded by two genes: *EEF1A1* and *EEF1A2*. Mutations in *EEF1A2*

Cite this article as *Cold Spring Harb Perspect Biol* doi: 10.1101/cshperspect.a032649

have been linked to a novel intellectual disability and epilepsy syndrome (Nakajima et al. 2015; Inui et al. 2016; Lam et al. 2016), and overexpression of eEF1A2 has been reported in a variety of cancers (Lee and Surh 2009). eEF1A in mammals and yeast is posttranslationally modified on several residues (Dever et al. 1989; Cavallius et al. 1993), most notably by methylation of lysines. Several eEF1A lysine methyltransferases have recently been identified (Lipson et al. 2010; Jakobsson et al. 2015, 2017; Hamey et al. 2016; Malecki et al. 2017), and loss of methylation has been linked to altered translation (Jakobsson et al. 2017; Malecki et al. 2017). These connections to human health are consistent with the expectation that gene expression is quite precisely tuned at the level of translation.

eEF2

Translation elongation factor eEF2, like EF-G, is a structural mimic of the eEF1A•GTP•aminoacyl-tRNA ternary complex (Jorgensen et al. 2003). Domain IV of eEF2, like the anticodon loop of tRNA, binds deep in the A-site decoding center to promote translocation of the tRNAs and mRNA on the ribosome following peptide bond formation. A conserved His residue at the tip of domain IV of eEF2 is posttranslationally modified to diphthamide (2-[3-carboxyamido-3-(trimethylammonio)propyl]histidine); this diphthamide modification is conserved in eukaryotes and archaea (Su et al. 2013; Schaffrath et al. 2014) but is not present on bacterial EF-G. The diphthamide residue is ADP-ribosylated by diphtheria toxin produced by *Corynbacterium diphtheriae*, as well as by exotoxin A from *Pseudomonas aeruginosa* and cholix toxin from *Vibrio cholera* (Schaffrath et al. 2014). The ADP-ribosylation of eEF2 inactivates the factor, blocks protein synthesis, and impairs cell growth; however, the molecular basis for how ADP-ribosylation impairs eEF2 function has not been fully resolved (Davydova and Ovchinnikov 1990; Nygard and Nilsson 1990; Jorgensen et al. 2004; Taylor et al. 2007; Mateyak and Kinzy 2013).

Synthesis of diphthamide requires a four-step pathway and the action of seven gene products (DPH1-7). Lack of the diphthamide modification is lethal in mice as a result of significant developmental defects (Chen and Behringer 2004; Nobukuni et al. 2005; Liu et al. 2006; Webb et al. 2008; Thakur et al. 2012; Yu et al. 2014). Moreover, mutations in DPH1/OVCA1 have been linked to ovarian cancer (Chen and Behringer 2004). Surprisingly, despite its deep conservation, diphthamide is not essential in yeast. Yeast lacking the first enzyme required for diphthamide synthesis and, thus, presenting an unmodified His residue at the tip of domain IV, grow normally, suggesting perhaps that diphthamide plays a role in translational fidelity rather than the fundamental mechanism of protein synthesis. Mice lacking the diphthamide biosynthetic enzymes DPH1, DPH3, or DPH4 exhibit severe developmental defects or embryonic lethality (Chen and Behringer 2004; Liu et al. 2006; Webb et al. 2008), indicating that diphthamide synthesis or perhaps another function of these enzymes is required during development. Despite this critical role of diphthamide in development, mammalian CHO and MCF7 cells lacking the ability to synthesize diphthamide are viable (Liu et al. 2004; Stahl et al. 2015), and the only clear phenotypes in these mutants are their insensitivity to diphtheria toxin and altered nuclear factor (NF)-κB and tumor necrosis factor pathways. However, both yeast (Ortiz et al. 2006) and mammalian (Liu et al. 2012) cells lacking diphthamide show increased levels of programmed −1 ribosomal frameshifting, revealing a positive impact of diphthamide on translational fidelity and suggesting that diphthamide may augment the function of eEF2 in promoting precise ribosomal translocation.

Recent biochemical studies examining eEF2 function in a novel translocation reaction required for translation initiation on the internal ribosome entry sites (IRESs) from the cricket paralysis virus (CrPV) and the Taura syndrome virus (TSV) have provided additional insights into the function of the diphthamide modification on eEF2. Pseudoknot I (PKI) of the IRES binds in the A site and mimics a tRNA bound to its mRNA codon. To enable translation, PKI must be translocated to the P site (reviewed in Butcher and Jan 2016). Whereas eEF2 with or without the diphthamide modification functions equivalently on canonically initiated elongation complexes analyzed in an in vitro peptide synthesis assay, peptide synthesis initiated on the CrPV IRES is impaired when eEF2 lacks the diphthamide modification (Murray et al. 2016). These results likely reflect a heightened requirement for diphthamide to promote high-fidelity translocation by the IRES and are supported by recent cryo-EM structures of translocation complexes that reveal interactions between diphthamide and PKI (Abeyrathne et al. 2016; Murray et al. 2016). Domain IV of eEF2 is inserted into the A site where it stabilizes PKI in a conformation that resembles a hybrid state with the pseudocodon–anticodon interaction in the A site. The diphthamide residue directly interacts with the pseudocodon–anticodon helix of PKI (Abeyrathne et al. 2016) and appears to disrupt the interaction of the ribosome decoding center h44 residues A1753 and A1754 (*Kluyveromyces lactis* 18S rRNA) with the PKI (Abeyrathne et al. 2016; Murray et al. 2016), and thus may directly facilitate translocation. It is tempting to speculate that loss of the interaction between diphthamide and the codon–anticodon helix during canonical elongation might contribute to the increased ribosomal frameshifting observed in cells lacking diphthamide. Taken together, these structural and biochemical studies of the IRES-dependent translation, together with the in vivo experiments revealing heightened ribosomal frameshifting in cells lacking diphthamide, suggest that diphthamide functions to optimize the efficiency and fidelity of ribosomal translocation during translation elongation.

In addition to the diphthamide modification, eEF2 is also modified by phosphorylation. The Ca^{2+}-activated kinase eEF2K phosphorylates eEF2 in metazoans on Thr56 and blocks translation by impairing eEF2 binding to the ribosome (Carlberg et al. 1990). As covered in other reviews (see Proud 2018), the activity of eEF2K is regulated by nutrients via mammalian target of rapamycin complex 1 (mTORC1) and/ or AMP-activated protein kinase (AMPK) (Kenney et al. 2014) and during neuronal signaling (Taha et al. 2013).

eEF3

The translation elongation factor eEF3 is restricted to fungi and appears to be specifically required for protein synthesis with yeast ribosomes. Whereas yeast eEF1 and eEF2 will functionally substitute for their mammalian counterparts to promote translation with mammalian ribosomes in vitro, the mammalian factors eEF1 and eEF2 will only work with yeast ribosomes when eEF3 is added as well (Skogerson and Engelhardt 1977). eEF3 contains two ATP-binding cassettes (ABCs) and possesses ribosome-stimulated ATPase activity. Mutations in a chromodomain insert in the second ABC domain impair general translation and ribosome-stimulated ATPase activity (Sasikumar and Kinzy 2014). A low-resolution cryo-EM structure of eEF3 bound to the ribosome revealed that eEF3 contacts both the central protuberance of the 60S subunit and the head of the 40S subunit (Andersen et al. 2006). In this structure, the chromodomain of eEF3 is located near the E site of the ribosome, consistent with the model that eEF3 may promote release of deacylated tRNA from the E site following translocation (Triana-Alonso et al. 1995; Andersen et al. 2006). It is unclear why yeast ribosomes require eEF3 when similar ATPases are neither required for translation nor are obviously present in the genomes of higher eukaryotes.

eIF5A

In addition to eEF1A/EF-Tu and eEF2/EF-G, a third universally conserved factor, eIF5A/EF-P, also functions in translation elongation. eIF5A and EF-P were originally identified based on their abilities to stimulate the yield of methionyl-puromycin in a model assay of first peptide bond formation (Glick and Ganoza 1975; Kemper et al. 1976), and, so, eIF5A was considered an initiation factor with a critical role in first peptide bond formation. However, it is noteworthy that puromycin is a poor substrate because of unfavorable positioning in the PTC (Youngman et al. 2004; Wohlgemuth et al. 2008). Thus, eIF5A stimulation of the puromycin reaction might reflect an ability of eIF5A to enhance the

 Cite this article as *Cold Spring Harb Perspect Biol* doi: 10.1101/cshperspect.a032649

reactivity of a poor substrate like puromycin rather than a role for eIF5A in first peptide bond synthesis. The results of dipeptide synthesis assays employing canonical aminoacyl-tRNA substrates argue strongly against a critical role for eIF5A in first peptide bond formation. The dipeptides Met-Phe and Met-Pro, or related polypeptides initiating with these residues, were efficiently synthesized in fully reconstituted yeast in in vitro translation assays lacking eIF5A, and addition of the factor resulted in only a modest stimulation in peptide yield (Gutierrez et al. 2013; Schuller et al. 2017; Shin et al. 2017). The absence of a strong eIF5A dependence for synthesis of these peptides indicates that first peptide bond synthesis does not impose a heightened requirement for the factor. An in vivo study supporting a role for eIF5A in first peptide bond synthesis reported that depletion of eIF5A in yeast resulted in reduced levels of large polysomes and accumulation of smaller polysomes and monosomes, suggestive of an initiation defect (Henderson and Hershey 2011). At odds with this finding, a separate study in yeast using a different degron to deplete eIF5A reported the maintenance of polysomes upon depletion of eIF5A, even in the absence of the elongation inhibitor cycloheximide (Saini et al. 2009). Thus, depletion of eIF5A mimicked cycloheximide treatment and was suggestive of impaired translation elongation. Moreover, this latter study also reported that rapid inactivation of temperature-sensitive eIF5A mutants in yeast resulted in the accumulation of polysomes in the absence of cycloheximide (Saini et al. 2009). These findings indicate a rate-limiting role for eIF5A in translation elongation rather than translation initiation or first peptide bond formation. In further support for a role for eIF5A in translation elongation or termination, inactivation of eIF5A resulted in increased ribosomal transit times (Gregio et al. 2009; Saini et al. 2009). Consistent with the findings in these latter in vivo studies, addition of eIF5A stimulated the rate of peptide synthesis in in vitro elongation assays and of release in termination assays in the presence of eRF1 and eRF3 (Saini et al. 2009). Taken together, the findings from the in vitro peptide synthesis assays using authentic amino-acyl-tRNA substrates and from the eIF5A inactivation studies in yeast argue that the factor plays a critical role in translation elongation, but not in translation initiation or first peptide bond formation.

Further studies into the function of EF-P revealed that the factor stimulated the synthesis of proteins containing runs of consecutive proline residues (Doerfel et al. 2013; Ude et al. 2013). Complementary studies in yeast cells revealed that inactivation of eIF5A impaired translation of reporter genes containing runs of polyproline residues and that eIF5A and its hypusine modification are required for the synthesis of polyproline peptides in vitro (Gutierrez et al. 2013). Consistent with these results, the synthesis of native yeast proteins containing polyproline motifs was impaired in eIF5A mutants (Gutierrez et al. 2013; Li et al. 2014a).

Whereas the critical role of eIF5A to promote translation of polyproline motifs is consistent with previous reports suggesting that eIF5A stimulates the translation of only a subset of mRNAs in the cell (Kang and Hershey 1994), polysome profile analyses in cells depleted of eIF5A revealed a pervasive elongation defect affecting a substantial fraction of cellular mRNAs (Saini et al. 2009). Moreover, recent ribosomal profiling analyses revealed that eIF5A functions globally to promote translation elongation and that its function is not restricted to polyproline motifs (Pelechano and Alepuz 2017; Schuller et al. 2017). Importantly, the stimulation of nonpolyproline peptide synthesis by eIF5A was also observed in vitro (Schuller et al. 2017). In addition to detecting ribosomal pausing during translation elongation, the ribosomal profiling studies of cells depleted for eIF5A revealed a pronounced accumulation of ribosomes at stop codons (Schuller et al. 2017). Using in vitro peptide release assays, eIF5A was shown to promote the eRF1 and eRF3-dependent translation termination (Schuller et al. 2017). Taken together, these new findings reveal a genome-wide role for eIF5A in translation elongation and termination. Interestingly, in contrast to eIF5A, ribosomal profiling of bacteria lacking EF-P did not detect any impact on translation termination (Elgamal et al. 2014; Woolstenhulme et al.

2015); the differences between eukaryotic and bacterial translation termination that underlie the stimulatory effect of eIF5A are unclear.

Recent structural and biochemical studies have provided insights into how eIF5A and EF-P may stimulate translation. The amino- and carboxy-terminal domains of eIF5A resemble domains I and II of EF-P; however, the bacterial factor possesses a carboxy-terminal domain III that is not present in eIF5A (Dever et al. 2014; Lassak et al. 2016). Two isoforms of eIF5A are differentially expressed in both yeast and mammalian cells (Dever et al. 2014; Mathews and Hershey 2015). Although no biochemical studies have reported functional differences between eIF5A isoforms, the eIF5A2 isoform has been linked to cancer (Mathews and Hershey 2015). It is noteworthy that the amino acid hypusine is formed posttranslationally on eIF5A by transfer of an N-butylamine group from spermidine to the ε-amino group of a conserved Lys residue forming deoxyhypusine (Park et al. 2010). Hydroxylation of deoxyhypusine completes the modification. The modified Lys residue resides at the tip of a loop in domain I of eIF5A (Kim et al. 1998), and the corresponding Lys or Arg residue in EF-P is modified by the addition of hydroxylated β-lysine, 5-aminopentanol, or the sugar rhamnose in different species (Dever et al. 2014; Lassak et al. 2016). The posttranslational modification of eIF5A and EF-P is required for stimulation of methionyl-puromycin synthesis (Park et al. 1991, 2012), polyproline synthesis (Gutierrez et al. 2013; Ude et al. 2013; Doerfel et al. 2015), and translation termination (Schuller et al. 2017). As in the X-ray structure of EF-P bound to the bacterial ribosome (Blaha et al. 2009), cryo-EM (Schmidt et al. 2016), and X-ray (Melnikov et al. 2016b) structures of eIF5A bound to the yeast ribosome revealed the factor binding in the E site. In these structures, the factor abuts the peptidyl-tRNA in the P site and the hypusine residue interacts with the acceptor arm of the peptidyl-tRNA. Exploiting the heightened requirement for eIF5A/EF-P for polyproline synthesis, biochemical studies using misacylated tRNAs revealed that the imino acid proline rather than tRNA^{Pro} imposes the requirement for eIF5A (Shin et al. 2017), although in bacteria the D-arm of the tRNA^{Pro} contributes to EF-P stimulation of translation (Katoh et al. 2016). Further analysis of bacterial translation revealed that EF-P provides an entropic benefit to peptide synthesis (Doerfel et al. 2015), and studies in yeast revealed that a more flexible proline analog lessened the requirement for eIF5A in peptide synthesis (Shin et al. 2017). These findings support a model in which the hypusine side chain acts sterically to position the acceptor arm of the P-site tRNA for favorable interaction with the A-site substrate in the ribosome PTC. Although this repositioning is likely to assist synthesis of all peptide bonds, some substrates like polyproline may show a greater requirement because of their inherently poor positioning in the PTC.

Assuming that eIF5A and deacyl-tRNA cannot simultaneously occupy the E site, as would be predicted based on the structures of the relevant ribosomal complexes, eIF5A binding to the E site is likely restricted until after dissociation of the deacyl-tRNA. While single-molecule studies in a human eukaryotic translation elongation system indicate that dissociation of the deacyl-tRNA from the E site is slow (Ferguson et al. 2015), it is noted that these studies were performed in the absence of eIF5A and using bacterial tRNAs. Given the high abundance of eIF5A in both yeast and mammalian cells (Duncan and Hershey 1986; Firczuk et al. 2013), more than twofold greater than the concentration of total ribosomes, and its strong affinity for the ribosome (unpublished data), eIF5A is predicted to rapidly fill the E site following deacyl-tRNA release and contribute to each peptide bond and termination reaction. Finally, biochemical studies in both mammalian and yeast systems have revealed an interplay between eIF5A and polyamines. Inclusion of eIF5A lowered the optimum Mg^{2+} concentration for globin mRNA translation in mammalian assays lacking spermidine (Schreier et al. 1977). Moreover, whereas most of the eIF5A activities in yeast assays can be attributed to the hypusine residue, suggesting that the body of eIF5A might function simply as a hypusine delivery agent, it is notable that unhypusinated eIF5A can substitute for polyamines in the stimulation of general translation in vitro (Shin et al. 2017). Whereas

Cite this article as *Cold Spring Harb Perspect Biol* doi: 10.1101/cshperspect.a032649

the function of polyamines in translation elongation is not clear, these results suggest that polyamines, like the body of eIF5A, may interact with the peptidyl-tRNA to facilitate its proper and stable positioning required for peptide bond formation.

TRANSLATIONAL RECODING

Recoding Definition

It is generally assumed that (1) all codons encode identical information in all organisms (with few exceptions [Ling et al. 2015]), and (2) the reading frame is invariant. Beginning in the mid-1970s, mRNA elements were discovered that direct ribosomes to reassign the meanings of codons, induce ribosomes to slip into alternative reading frames (programmed ribosomal frameshifting [PRF]), and even bypass long stretches of mRNA sequence (ribosome shunting). All of these were eventually subsumed under the general heading of "translational recoding," defined as instances in which "...the rules for decoding are temporarily altered through the action of specific signals built into the mRNA sequences" (Gesteland and Atkins 1996).

A Unifying Mechanistic Concept: Recoding Is Driven by Kinetic Traps

At the biophysical level, translational recoding events are driven by *cis*-acting elements on mRNAs that alter the processivity kinetics of elongating ribosomes. These "kinetic traps" alter rates of kinetic partitioning between the normal "forward reaction" (i.e., canonical decoding) and "side reactions" (i.e., recoding events). Typically, these kinetic traps direct ribosomes to pause at a specific location on an mRNA. Most studies suggest that directed pausing is critical; the sequence over which a ribosome is stopped is thought to lower the energy barrier to a particular alternative coding solution. Some *cis*-acting elements can be very simple "flat" sequences (i.e., defined by primary sequence alone and not by the ability to form higher-order structures). More often, recoding elements involve complex mRNA topological features that induce ribosomal pausing. These can be in *cis* (i.e., entirely composed of mRNA), in *trans* (i.e., composed of proteins and/or other RNAs that interact with specific mRNA sequences), or a combination of the two. The combination of the sequence at which the ribosome is paused plus the kinetic substep that is affected determines the functional output of the ribosome (i.e., the nature and extent of the recoding event).

Molecular Mechanisms of Recoding

Recoding Directed by "Flat" cis-Acting Sequence Elements

This term refers to recoding elements in which *cis*-acting mRNA structural elements do contribute to defining the recoding signal. Typically, in these cases, low abundance of a translation factor that would normally be required for the next step of the normal coding program induces ribosome pausing at a specific recoding sequence. For example, +1 PRF by the yeast *Ty1* retrotransposable element is effected by the simple heptameric sequence CUU AGG C (where the incoming 0 frame is indicated by spaces). Here, the kinetic trap is supplied by the rare 0 frame A-site AGG codon, which is decoded by the very low abundance Arg-tRNA^{CCU} tRNA. Ribosomal pausing at this codon allows the P-site tRNA to slip from the 0 frame CUU to the +1 frame UUA (Belcourt and Farabaugh 1990). Evidence of this mechanism was first based on the observation that high-copy episomal expression of this tRNA caused a 50-fold decrease in +1 PRF, while its deletion from the chromosome caused +1 PRF efficiency to approach 100% (Kawakami et al. 1993). Biochemical studies further demonstrated that mutant yeast ribosomes with altered affinities for tRNAs in the P site displayed changes in *Ty1*-mediated +1 PRF (Meskauskas and Dinman 2001; Rhodin and Dinman 2010; Musalgaonkar et al. 2014) and that these effects could be antagonized by sparsomycin, an antibiotic that increases the affinity of ribosomes for peptidyl-tRNA (Meskauskas and Dinman 2001). The +1 frameshifts of *Ty2* and *Ty4*, other members of the *copia* family of retrotransposable elements, as well as the yeast

ABP140 and *EST3* mRNAs are also thought to use this mechanism of tRNA slippage (Morris and Lundblad 1997; Asakura et al. 1998; Farabaugh et al. 2006). Interestingly, the *Ty3* GCG AGU U slippery site appears to function quite differently because the 0 frame P-site tRNA is not able to base pair with the +1 frame codon (Vimaladithan and Farabaugh 1994). Rather, *Ty3*-directed +1 PRF requires skipping the first A of the 0 frame P-site codon and to instead recognize the +1 frame GUU codon. This is dependent on an unknown feature of the Ala-tRNA^{UGC}, which is shared by four additional tRNAs.

Stop codons can also function as kinetic traps to drive recoding. These kinetic traps are driven by release factor abundance. This is particularly well documented in protozoa and mitochondria where loss of genes encoding a stop codon–specific (typically UGA) release factor leads to this codon being decoded by a suppressor tRNA (Alkalaeva and Mikhailova 2017; Lobanov et al. 2017).

Recoding Directed by cis-Acting Topological Features

Most known translational recoding signals include *cis*-acting mRNA structural elements, typically mRNA stem-loops and pseudoknots. In PRF, both classes of elements are thought to direct ribosomes to pause at special slippery sequences, the nature of which allows re-pairing of tRNAs that are already within the ribosome to shift into a different reading frame (reviewed in Dinman 2012b). Shifting thus requires unpairing of the tRNAs from the initial reading frame, an event that inevitably requires energetic input. Contextualization of PRF within the elongation cycle reveals two translation factors capable of providing the needed energy by virtue of their GTP hydrolysis activities: EF-Tu/eEF1A and EF-G/eEF2 (Harger et al. 2002). Computational kinetic modeling of −1 PRF revealed three steps during the elongation cycle at which this may occur: (1) during translocation of the ribosome into the slippery site; (2) during accommodation of tRNA into a ribosome paused at the slippery site; and (3) during translocation out of

the slippery site (Liao et al. 2008). All three of these mechanisms are supported by data generated using molecular genetics, structural biology, and biochemical analyses in multiple different systems (Jacks et al. 1988a,b; Weiss et al. 1989; Yelverton et al. 1994; Plant et al. 2003; Baranov et al. 2004; Namy et al. 2006; Leger et al. 2007; Caliskan et al. 2014). Thus, we suggest that rather than a monolithic molecular mechanism, −1 PRF should be viewed as a functional outcome that can result from at least three different kinetic pathways.

mRNA pseudoknot stimulation of recoding has been proposed to occur via a torsional restraint model, in which supercoiling of stem 2 forces ribosomes to pause over the slippery site (Plant and Dinman 2005). This model can account for how elongating ribosomes are directed to the slippery site but it does not address the actual mechanism of slippage. A mechanistic model of −1 PRF (Plant et al. 2003) is based on observations that the mRNA pseudoknot region is "pulled into" the ribosome by one base during the process of aa-tRNA accommodation (Noller et al. 2002). This movement pulls the entire mRNA in the 5′ direction (i.e., into the ribosome by this distance). This model is supported by ribosome toe printing studies showing that the lengths of reverse transcriptase primer extension products are reduced by one base after aa-tRNA accommodation (Fredrick and Noller 2002) (i.e., the mRNA is pulled into the ribosome by the distance of one base during this event). The 9 Å model of −1 PRF (Plant et al. 2003) posits that the placement of the downstream stimulatory structure in the ribosome's mRNA entry tunnel impedes this one base movement of the mRNA into the ribosome, stretching the segment of mRNA located between the slippery site and the downstream stimulatory structure. The resulting local region of tension in the mRNA can be resolved either by unwinding the stimulatory structure or by −1 slippage. This mechanism also applies to −1 PRF events that occur during translocation (Caliskan et al. 2014; Chen et al. 2014; Kim et al. 2014; Kim and Tinoco 2017). Regardless of whether it occurs during translocation or aa-tRNA accommodation, the active stretching of

Cite this article as *Cold Spring Harb Perspect Biol* doi: 10.1101/cshperspect.a032649

the spacer region between the slippery site and downstream stimulatory element followed by tRNA unpairing and slippage of the mRNA into an alternative reading frame can be subsumed under the heading of the "tension model." Structural and kinetic analyses using purified *E. coli* ribosomes and elongation factors also revealed that the downstream pseudoknot in the mRNA can impair the closing movement of the large subunit head, delaying dissociation of the translocase EF-G and the release of deacylated tRNA. Release of the tension by ribosomal slippage accelerates completion of translocation, providing a lower energy path for the ribosome to continue translation (Caliskan et al. 2014).

Do downstream stimulatory structures play active or passive roles in directing recoding? Numerous studies suggest that dynamic mRNA structural remodeling helps to physically "push" ribosomes to slip (Ritchie et al. 2012, 2014, 2017; Tinoco et al. 2013; Gupta and Bansal 2014; Moomau et al. 2016; Tsai et al. 2016; Zhong et al. 2016; Kendra et al. 2017). Coordination of base triples in both major and minor grooves provide mechanical resistance to pseudoknot unwinding, and stretches of adenosines confined along the minor groove of a helix prevent it from unwinding. Together, these molecular features contribute to ribosome pausing at the slippery site to help stimulate −1 PRF (Chen et al. 2017). Thus, although it was initially thought that downstream stimulatory structures were mere passive "roadblocks," the most recent research suggests that they play active roles in recoding.

Given the existence of *cis*-acting recoding stimulatory elements, it is logical to assume that elements with the opposing activity may also exist. Indeed, *cis*-acting mRNA structural elements that attenuate −1 PRF activity have been described in coronaviruses (Su et al. 2005; Cho et al. 2013). These consist of stem-loop structures located immediately 5′ of the slippery site sequences. These hairpins are first unwound by elongating ribosomes as they approach the frameshift signal. As they enter the slippery site, however, the ribosome moves past the sequence, enabling the stem-loop to re-form. It is reasoned that this structure can then resist

the backward slippage of the ribosome caused by the −1 PRF signal. This regulation of translational recoding via formation of a stem-loop after ribosome clearance is reminiscent of Rho-independent transcription termination in bacteria (reviewed in Henkin and Yanofsky 2002).

Recoding Directed by trans-Acting Factors

Translational recoding is also subject to regulation through the action of *trans*-acting factors. These can be divided into three general classes: small molecules, nucleic acids, and proteins.

Small molecules. Programmed +1 ribosomal frameshifting on mRNAs encoding ornithine decarboxylase antizyme (OAZ) is stimulated by polyamines (Ivanov et al. 2000). OAZ +1 PRF is autoregulated by the availability of small molecules in the form of polyamines (i.e., the products of the synthetic pathway controlled by OAZ). OAZ downregulates polyamine synthesis by stimulating ubiquitin-independent degradation of ornithine decarboxylase (ODC), the enzyme that catalyzes the first step in polyamine biosynthesis. When polyamine levels are low, +1 PRF on the OAZ mRNA is low, thus downregulating OAZ synthesis and resulting in increased levels of polyamines. These levels of polyamines, in turn, feed back to increase +1 PRF and OAZ synthesis, negatively feeding back on polyamine synthesis.

Trans-acting proteins. There is a growing list of *trans*-acting proteins that stimulate translational recoding in all domains of life. Synthesis of bacterial release factor 2 (RF2), one of the two versions of the protein involved in termination codon recognition in bacteria, requires a +1 PRF event (Larsen et al. 1995). RF2 is required for decoding of the UGA stop codon and additionally contributes to the recognition of some fraction of the UAA stop codons. Importantly, the *prfB* genes encoding RF2 in approximately 87% of bacterial species harbor an in-frame UGA codon located approximately 26 codons downstream from the AUG start codon, with the remainder of the protein coding sequence in the +1 frame (Craigen et al. 1985). This allows for an autoregulatory feedback system: when RF2 levels are high, termination is efficient, whereas when RF2 levels are low, termination on the UGA stop codon is inefficient. Thus, low RF2 levels enhance ribosomal pausing at the termination codon, and this kinetic pause enhances formation of a frameshift-inducing SD-like/anti-SD interaction between the ribosome and the mRNA.

Whereas RF2 technically functions in the prfB system, it only does so through its absence. In contrast, selenocysteine recoding actively requires *trans*-acting proteins. SECIS-binding proteins (SBPs in archaea, SBP2 in eukaryotes) interact with a special domain of a specialized elongation factor eEFsec to enhance recruitment of Sec-tRNA$^{(Ser)Sec}$ to the SECIS element and thus to an elongating ribosome. An additional protein, SECp43, methylates the 2′-hydroxylribosyl moiety in the wobble position of the selenocysteyl-tRNA$^{(Ser)Sec}$ to enhance selenoprotein expression (Ding and Grabowski 1999). In porcine reproductive and respiratory syndrome virus, an unusual −1/−2 PRF mechanism is stimulated in the absence of any apparent downstream RNA structural element, by the binding of a *trans*-acting protein complex composed of the virus-encoded nsp1β replicase subunit and the cellular poly(C) binding protein (Fang et al. 2012) to the mRNA sequence CCCANCUCC located 11 nucleotides 3′ of the GGGUUUUU shift site (Li et al. 2014b). Binding of this complex to the target sequence induces ribosome pausing over this −1/−2 slippery site (Napthine et al. 2016). Encephalomyocarditis virus protein 2A similarly functions to direct −1 PRF so as to decrease expression of its nonstructural gene products and up-regulate structural protein production, during the late phase of its replication cycle (Napthine et al. 2017). The possibility that this mechanism may be employed by many picornaviruses (e.g., hepatitis C virus, poliovirus, and rhinoviruses) suggests a novel target for antiviral therapeutic interventions. In each case, a *trans*-acting protein binding to the mRNA impacts the output of gene expression through modulation of a frameshifting event.

Trans-acting nucleic acids. Hybridization of small synthetic nucleic acids to mRNAs 3′ of canonical slippery sites has been demonstrated to *trans*-activate efficient frameshifting in vitro (Aupeix-Scheidler et al. 2000; Howard et al. 2004; Olsthoorn et al. 2004; Henderson et al. 2006; Yu et al. 2010). The spacing between slippery sites and the downstream region of hybridization is important, supporting the idea of mRNA tension as causative (Lin et al. 2012). In live cells, the interaction of a microRNA (miRNA) with a −1 PRF-stimulating mRNA pseudoknot in the human CCR5 mRNA was shown to stimulate −1 PRF (Belew et al. 2014). It is hypothesized that this interaction renders the downstream element even more difficult to resolve, enhancing the probability of kinetic partitioning to the −1 frame at the slippery site. The proposed base-pairing interaction between the −1 PRF signal and miRNA in this case provides the potential for sequence-specific regulation of −1 PRF and, hence, a means to control expression of the CCR5 gene product. Preliminary studies reveal that miRNAs impact frameshifting at other human slippery sites, suggesting that this may be a widely used strategy to regulate gene expression in higher eukaryotes (Belew et al. 2014).

Functional Outcomes of Recoding

Two-for-one. All viruses with positive-sense plus-stranded RNA [(+) ssRNA] genomes, and many with double-stranded RNA genomes, face a common problem: their (+) strands have to serve as both mRNA and as a template for genome replication. Thus, maximization of protein coding information must be achieved in ways that do not alter the genetic information that will be passed to the next generation. Translational recoding is one solution to this problem. The simplest such solution can be thought of as "two-for-one," where PRF or termination suppression mechanisms are used to produce carboxy-terminally extended fusion proteins in addition to the peptides synthesized by canonical translation. Numerous such examples are well documented in many virus families (reviewed in Dinman 2012a; Firth and Brierley 2012). Many studies have shown that viruses have evolved to optimize recoding rates so as to optimize ratios of viral proteins, and that altering recoding efficiency has deleterious effects on viral propagation (reviewed in Dinman 2012b). As such, recoding is a potential target for antiviral therapeutics (Dinman et al. 1998). Efforts targeting the HIV-1 −1 PRF signal in particular have identified promising candidates for therapeutic development (Lonnroth et al. 1988; Dinman et al. 1997; Hung et al. 1998; Aupeix-Scheidler et al. 2000; McNaughton et al. 2007; Dulude et al. 2008; Marcheschi et al. 2009, 2011; Kobayashi et al. 2010; Palde et al. 2010; Ofori et al. 2014; Cardno et al. 2015; Hilimire et al. 2016; Hu et al. 2016). As a note of caution, however, in light of the finding that ∼10% of chromosomally encoded genes harbor potential −1 PRF signals (Belew et al. 2008), and that global dysregulation of −1 PRF has deleterious effects on cell growth and replication (reviewed in Dinman 2012b), drug development efforts must be tailored to specific recoding elements as opposed to a broad targeting of all recoding.

mRNA destabilizing elements. Analysis of −1 PRF signals located in chromosomally encoded mRNAs revealed the counterintuitive finding that >99% of all predicted frameshifts would direct elongating ribosomes to premature termination codons (Jacobs et al. 2007; Belew et al. 2008). This prompted the hypothesis that these elements might serve to limit gene expression through the nonsense-mediated mRNA decay (NMD) pathway. Further, the ability of −1 PRF stimulatory elements to cause ribosomes to pause for relatively long periods of time (Heller et al. 1976; Caliskan et al. 2014) suggested that these elements may also render mRNAs substrates for degradation through the No-Go mRNA decay (NGD) pathway. Both of these mechanisms were validated using endogenous −1 PRF signals in yeast (Belew et al. 2011). Moreover, −1 PRF-directed NMD has been shown to control gene expression in human cells on many genes (Belew et al. 2014). Evidence is emerging that this strategy is used to control telomere maintenance in yeast (Advani et al. 2013), the cell cycle (Belew and Dinman 2015), and many more cellular pathways (Advani and Dinman 2016; Meydan et al. 2017). Further, global dysregulation of −1 PRF may be linked to a wide variety of human diseases (Jack et al. 2011; Hekman et al. 2012; Sulima et al. 2014; Belew and Dinman 2015; De Keersmaecker et al. 2015; Paolini et al. 2017).

CONCLUDING REMARKS

Fulfilling the fundamental role of decoding the genetic code, high-fidelity translation elongation is critical for proper cellular function. Recent studies have provided new insights into the general mechanism of translation elongation, its regulation, and the means to exploit the process for alternative decoding events. As is typical in biology, complex regulation is achieved through the modest manipulation of the core events of the process, not through the acquisition of wholly novel elements that redirect the system. As such, continued progress in obtaining high-resolution structural images of translation elongation intermediates, combined with rigorous biochemical and kinetic dissection of the partial reactions in elongation, and further exploitation of ribosomal profiling strategies to interrogate the translation elongation process, offer the exciting opportunity for even greater insights in the near future.

REFERENCES

*Reference is also in this collection.

Abeyrathne PD, Koh CS, Grant T, Grigorieff N, Korostelev AA. 2016. Ensemble cryo-EM uncovers inchworm-like translocation of a viral IRES through the ribosome. *eLife* **5:** e14874.

Advani VM, Dinman JD. 2016. Reprogramming the genetic code: The emerging role of ribosomal frameshifting in regulating cellular gene expression. *Bioessays* **38:** 21–26.

Advani VM, Belew AT, Dinman JD. 2013. Yeast telomere maintenance is globally controlled by programmed ribosomal frameshifting and the nonsense-mediated mRNA decay pathway. *Translation (Austin)* **1:** e24418.

Alkalaeva E, Mikhailova T. 2017. Reassigning stop codons via translation termination: How a few eukaryotes broke the dogma. *Bioessays* **39:** 1600213.

Anand M, Chakraburtty K, Marton MJ, Hinnebusch AG, Kinzy TG. 2003. Functional interactions between yeast translation eukaryotic elongation factor (eEF) 1A and eEF3. *J Biol Chem* **278:** 6985–6991.

Andersen GR, Valente L, Pedersen L, Kinzy TG, Nyborg J. 2001. Crystal structures of nucleotide exchange intermediates in the eEF1A-eEF1Bα complex. *Nat Struct Biol* **8:** 531–534.

Andersen CB, Becker T, Blau M, Anand M, Halic M, Balar B, Mielke T, Boesen T, Pedersen JS, Spahn CM, et al. 2006. Structure of eEF3 and the mechanism of transfer RNA release from the E-site. *Nature* **443**: 663–668.

Asakura T, Sasaki T, Nagano F, Satoh A, Obaishi H, Nishioka H, Imamura H, Hotta K, Tanaka K, Nakanishi H, et al. 1998. Isolation and characterization of a novel actin filament-binding protein from *Saccharomyces cerevisiae*. *Oncogene* **16**: 121–130.

Aupeix-Scheidler K, Chabas S, Bidou L, Rousset JP, Leng M, Toulme JJ. 2000. Inhibition of in vitro and ex vivo translation by a transplatin-modified oligo(2′-*O*-methylribonucleotide) directed against the HIV-1 gag-pol frameshift signal. *Nucleic Acids Res* **28**: 438–445.

Baranov PV, Gesteland RF, Atkins JF. 2004. P-site tRNA is a crucial initiator of ribosomal frameshifting. *RNA* **10**: 221–230.

Behrmann E, Loerke J, Budkevich TV, Yamamoto K, Schmidt A, Penczek PA, Vos MR, Burger J, Mielke T, Scheerer P, et al. 2015. Structural snapshots of actively translating human ribosomes. *Cell* **161**: 845–857.

Belcourt MF, Farabaugh PJ. 1990. Ribosomal frameshifting in the yeast retrotransposon Ty: tRNAs induce slippage on a 7 nucleotide minimal site. *Cell* **62**: 339–352.

Belew AT, Dinman JD. 2015. Cell cycle control (and more) by programmed −1 ribosomal frameshifting: Implications for disease and therapeutics. *Cell Cycle* **14**: 172–178.

Belew AT, Hepler NL, Jacobs JL, Dinman JD. 2008. PRFdb: A database of computationally predicted eukaryotic programmed −1 ribosomal frameshift signals. *BMC Genomics* **9**: 339.

Belew AT, Advani VM, Dinman JD. 2011. Endogenous ribosomal frameshift signals operate as mRNA destabilizing elements through at least two molecular pathways in yeast. *Nucleic Acids Res* **39**: 2799–2808.

Belew AT, Meskauskas A, Musalgaonkar S, Advani VM, Sulima SO, Kasprzak WK, Shapiro BA, Dinman JD. 2014. Ribosomal frameshifting in the CCR5 mRNA is regulated by miRNAs and the NMD pathway. *Nature* **512**: 265–269.

Ben-Shem A, Jenner L, Yusupova G, Yusupov M. 2010. Crystal structure of the eukaryotic ribosome. *Science* **330**: 1203–1209.

Ben-Shem A, Garreau de Loubresse N, Melnikov S, Jenner L, Yusupova G, Yusupov M. 2011. The structure of the eukaryotic ribosome at 3.0 Å resolution. *Science* **334**: 1524–1529.

Blaha G, Stanley RE, Steitz TA. 2009. Formation of the first peptide bond: The structure of EF-P bound to the 70S ribosome. *Science* **325**: 966–970.

Budkevich T, Giesebrecht J, Altman RB, Munro JB, Mielke T, Nierhaus KH, Blanchard SC, Spahn CM. 2011. Structure and dynamics of the mammalian ribosomal pretranslocation complex. *Mol Cell* **44**: 214–224.

Burkhardt N, Junemann R, Spahn CM, Nierhaus KH. 1998. Ribosomal tRNA binding sites: Three-site models of translation. *Crit Rev Biochem Mol Biol* **33**: 95–149.

Butcher SE, Jan E. 2016. tRNA-mimicry in IRES-mediated translation and recoding. *RNA Biol* **13**: 1068–1074.

Caliskan N, Katunin VI, Belardinelli R, Peske F, Rodnina MV. 2014. Programmed −1 frameshifting by kinetic par-titioning during impeded translocation. *Cell* **157**: 1619–1631.

Cardno TS, Shimaki Y, Sleebs BE, Lackovic K, Parisot JP, Moss RM, Crowe-McAuliffe C, Mathew SF, Edgar CD, Kleffmann T, et al. 2015. HIV-1 and human PEG10 frameshift elements are functionally distinct and distinguished by novel small molecule modulators. *PLoS ONE* **10**: e0139036.

Carelli JD, Sethofer SG, Smith GA, Miller HR, Simard JL, Merrick WC, Jain RK, Ross NT, Taunton J. 2015. Ternatin and improved synthetic variants kill cancer cells by targeting the elongation factor-1A ternary complex. *eLife* **4**: e10222.

Carlberg U, Nilsson A, Nygard O. 1990. Functional properties of phosphorylated elongation factor 2. *Eur J Biochem* **191**: 639–645.

Carr-Schmid A, Durko N, Cavallius J, Merrick WC, Kinzy TG. 1999. Mutations in a GTP-binding motif of eukaryotic elongation factor 1A reduce both translational fidelity and the requirement for nucleotide exchange. *J Biol Chem* **274**: 30297–30302.

Cavallius J, Zoll W, Chakraburtty K, Merrick WC. 1993. Characterization of yeast EF-1α: Non-conservation of post-translational modifications. *Biochim Biophys Acta* **1163**: 75–80.

Chen CM, Behringer RR. 2004. Ovca1 regulates cell proliferation, embryonic development, and tumorigenesis. *Genes Dev* **18**: 320–332.

Chen C, Stevens B, Kaur J, Smilansky Z, Cooperman BS, Goldman YE. 2011. Allosteric vs. spontaneous exit-site (E-site) tRNA dissociation early in protein synthesis. *Proc Natl Acad Sci* **108**: 16980–16985.

Chen J, Petrov A, Johansson M, Tsai A, O'Leary SE, Puglisi JD. 2014. Dynamic pathways of −1 translational frameshifting. *Nature* **512**: 328–332.

Chen YT, Chang KC, Hu HT, Chen YL, Lin YH, Hsu CF, Chang CF, Chang KY, Wen JD. 2017. Coordination among tertiary base pairs results in an efficient frameshift-stimulating RNA pseudoknot. *Nucleic Acids Res* **45**: 6011–6022.

Cho CP, Lin SC, Chou MY, Hsu HT, Chang KY. 2013. Regulation of programmed ribosomal frameshifting by co-translational refolding RNA hairpins. *PLoS ONE* **8**: e62283.

Craigen WJ, Cook RG, Tate WP, Caskey CT. 1985. Bacterial peptide chain release factors: Conserved primary structure and possible frameshift regulation of release factor 2. *Proc Natl Acad Sci* **82**: 3616–3620.

Davydova EK, Ovchinnikov LP. 1990. ADP-ribosylated elongation factor 2 (ADP-ribosyl-EF-2) is unable to promote translocation within the ribosome. *FEBS Lett* **261**: 350–352.

De Keersmaecker K, Sulima SO, Dinman JD. 2015. Ribosomopathies and the paradox of cellular hypo- to hyperproliferation. *Blood* **125**: 1377–1382.

Demeshkina N, Jenner L, Westhof E, Yusupov M, Yusupova G. 2012. A new understanding of the decoding principle on the ribosome. *Nature* **484**: 256–259.

Dever TE, Costello CE, Owens CL, Rosenberry TL, Merrick WC. 1989. Location of seven post-translational modifications in rabbit elongation factor 1α including dimethyllysine, trimethyllysine, and glycerylphosphorylethanolamine. *J Biol Chem* **264:** 20518–20525.

Dever TE, Gutierrez E, Shin BS. 2014. The hypusine-containing translation factor eIF5A. *Crit Rev Biochem Mol Biol* **49:** 413–425.

Ding F, Grabowski PJ. 1999. Identification of a protein component of a mammalian tRNA(Sec) complex implicated in the decoding of UGA as selenocysteine. *RNA* **5:** 1561–1569.

Dinman JD. 2012a. Control of gene expression by translational recoding. *Adv Protein Chem Struct Biol* **86:** 129–149.

Dinman JD. 2012b. Mechanisms and implications of programmed translational frameshifting. *Wiley Interdiscip Rev RNA* **3:** 661–673.

Dinman JD, Ruiz-Echevarria MJ, Czaplinski K, Peltz SW. 1997. Peptidyl-transferase inhibitors have antiviral properties by altering programmed −1 ribosomal frameshifting efficiencies: Development of model systems. *Proc Natl Acad Sci* **94:** 6606–6611.

Dinman JD, Ruiz-Echevarria MJ, Peltz SW. 1998. Translating old drugs into new treatments: ribosomal frameshifting as a target for antiviral agents. *Trends Biotechnol* **16:** 190–196.

Doerfel LK, Wohlgemuth I, Kothe C, Peske F, Urlaub H, Rodnina MV. 2013. EF-P is essential for rapid synthesis of proteins containing consecutive proline residues. *Science* **339:** 85–88.

Doerfel LK, Wohlgemuth I, Kubyshkin V, Starosta AL, Wilson DN, Budisa N, Rodnina MV. 2015. Entropic contribution of elongation factor P to proline positioning at the catalytic center of the ribosome. *J Am Chem Soc* **137:** 12997–13006.

Dulude D, Theberge-Julien G, Brakier-Gingras L, Heveker N. 2008. Selection of peptides interfering with a ribosomal frameshift in the human immunodeficiency virus type 1. *RNA* **14:** 981–991.

Duncan RF, Hershey JW. 1986. Changes in eIF-4D hypusine modification or abundance are not correlated with translational repression in HeLa cells. *J Biol Chem* **261:** 12903–12906.

Elgamal S, Katz A, Hersch SJ, Newsom D, White P, Navarre WW, Ibba M. 2014. EF-P dependent pauses integrate proximal and distal signals during translation. *PLoS Genet* **10:** e1004553.

Ermolenko DN, Noller HF. 2011. mRNA translocation occurs during the second step of ribosomal intersubunit rotation. *Nat Struct Mol Biol* **18:** 457–462.

Fang Y, Treffers EE, Li Y, Tas A, Sun Z, van der Meer Y, de Ru AH, van Veelen PA, Atkins JF, Snijder EJ, et al. 2012. Efficient −2 frameshifting by mammalian ribosomes to synthesize an additional arterivirus protein. *Proc Natl Acad Sci* **109:** E2920–E2928.

Farabaugh PJ, Kramer E, Vallabhaneni H, Raman A. 2006. Evolution of +1 programmed frameshifting signals and frameshift-regulating tRNAs in the order *Saccharomycetales*. *J Mol Evol* **63:** 545–561.

Ferguson A, Wang L, Altman RB, Terry DS, Juette MF, Burnett BJ, Alejo JL, Dass RA, Parks MM, Vincent CT, et al. 2015. Functional rynamics within the human ribosome regulate the rate of active protein synthesis. *Mol Cell* **60:** 475–486.

Firczuk H, Kannambath S, Pahle J, Claydon A, Beynon R, Duncan J, Westerhoff H, Mendes P, McCarthy JE. 2013. An in vivo control map for the eukaryotic mRNA translation machinery. *Mol Syst Biol* **9:** 635.

Firth AE, Brierley I. 2012. Non-canonical translation in RNA viruses. *J Gen Virol* **93:** 1385–1409.

Fredrick K, Noller HF. 2002. Accurate translocation of mRNA by the ribosome requires a peptidyl group or its analog on the tRNA moving into the 30S P site. *Mol Cell* **9:** 1125–1131.

Gesteland RF, Atkins JF. 1996. Recoding: Dynamic reprogramming of translation. *Annu Rev Biochem* **65:** 741–768.

Glick BR, Ganoza MC. 1975. Identification of a soluble protein that stimulates peptide bond synthesis. *Proc Natl Acad Sci* **72:** 4257–4260.

Gregio AP, Cano VP, Avaca JS, Valentini SR, Zanelli CF. 2009. eIF5A has a function in the elongation step of translation in yeast. *Biochem Biophys Res Commun* **380:** 785–790.

Gromadski KB, Schummer T, Stromgaard A, Knudsen CR, Kinzy TG, Rodnina MV. 2007. Kinetics of the interactions between yeast elongation factors 1A and 1Bα, guanine nucleotides, and aminoacyl-tRNA. *J Biol Chem* **282:** 35629–35637.

Gupta A, Bansal M. 2014. Local structural and environmental factors define the efficiency of an RNA pseudoknot involved in programmed ribosomal frameshift process. *J Phys Chem B* **118:** 11905–11920.

Gutierrez E, Shin BS, Woolstenhulme CJ, Kim JR, Saini P, Buskirk AR, Dever TE. 2013. eIF5A promotes translation of polyproline motifs. *Mol Cell* **51:** 35–45.

Hamey JJ, Winter DL, Yagoub D, Overall CM, Hart-Smith G, Wilkins MR. 2016. Novel N-terminal and lysine methyltransferases that target translation elongation factor 1A in yeast and human. *Mol Cell Proteomics* **15:** 164–176.

Harger JW, Meskauskas A, Dinman JD. 2002. An "integrated model" of programmed ribosomal frameshifting. *Trends Biochem Sci* **27:** 448–454.

Hekman KE, Yu GY, Brown CD, Zhu H, Du X, Gervin K, Undlien DE, Peterson A, Stevanin G, Clark HB, et al. 2012. A conserved eEF2 coding variant in SCA26 leads to loss of translational fidelity and increased susceptibility to proteostatic insult. *Hum Mol Genet* **21:** 5472–5483.

Heller JS, Fong WF, Canellakis ES. 1976. Induction of a protein inhibitor to ornithine decarboxylase by the end products of its reaction. *Proc Natl Acad Sci* **73:** 1858–1862.

Henderson A, Hershey JW. 2011. Eukaryotic translation initiation factor (eIF) 5A stimulates protein synthesis in *Saccharomyces cerevisiae*. *Proc Natl Acad Sci* **108:** 6415–6419.

Henderson CM, Anderson CB, Howard MT. 2006. Antisense-induced ribosomal frameshifting. *Nucleic Acids Res* **34:** 4302–4310.

Henkin TM, Yanofsky C. 2002. Regulation by transcription attenuation in bacteria: How RNA provides instructions for transcription termination/antitermination decisions. *Bioessays* **24:** 700–707.

Hilimire TA, Bennett RP, Stewart RA, Garcia-Miranda P, Blume A, Becker J, Sherer N, Helms ED, Butcher SE, Smith HC, et al. 2016. N-methylation as a strategy for enhancing the affinity and selectivity of RNA-binding peptides: Application to the HIV-1 frameshift-stimulating RNA. *ACS Chem Biol* **11:** 88–94.

Howard MT, Gesteland RF, Atkins JF. 2004. Efficient stimulation of site-specific ribosome frameshifting by antisense oligonucleotides. *RNA* **10:** 1653–1661.

Hu HT, Cho CP, Lin YH, Chang KY. 2016. A general strategy to inhibiting viral −1 frameshifting based on upstream attenuation duplex formation. *Nucleic Acids Res* **44:** 256–266.

Hung M, Patel P, Davis S, Green SR. 1998. Importance of ribosomal frameshifting for human immunodeficiency virus type 1 particle assembly and replication. *J Virol* **72:** 4819–4824.

Inui T, Kobayashi S, Ashikari Y, Sato R, Endo W, Uematsu M, Oba H, Saitsu H, Matsumoto N, Kure S, et al. 2016. Two cases of early-onset myoclonic seizures with continuous parietal delta activity caused by EEF1A2 mutations. *Brain Dev* **38:** 520–524.

Ivanov IP, Gesteland RF, Atkins JF. 2000. Antizyme expression: A subversion of triplet decoding, which is remarkably conserved by evolution, is a sensor for an autoregulatory circuit. *Nucleic Acids Res* **28:** 3185–3196.

Jack K, Bellodi C, Landry DM, Niederer RO, Meskauskas A, Musalgaonkar S, Kopmar N, Krasnykh O, Dean AM, Thompson SR, et al. 2011. rRNA pseudouridylation defects affect ribosomal ligand binding and translational fidelity from yeast to human cells. *Mol Cell* **44:** 660–666.

Jacks T, Madhani HD, Masiarz FR, Varmus HE. 1988a. Signals for ribosomal frameshifting in the Rous sarcoma virus gag-pol region. *Cell* **55:** 447–458.

Jacks T, Power MD, Masiarz FR, Luciw PA, Barr PJ, Varmus HE. 1988b. Characterization of ribosomal frameshifting in HIV-1 gag-pol expression. *Nature* **331:** 280–283.

Jacobs JL, Belew AT, Rakauskaite R, Dinman JD. 2007. Identification of functional, endogenous programmed −1 ribosomal frameshift signals in the genome of *Saccharomyces cerevisiae*. *Nucleic Acids Res* **35:** 165–174.

Jakobsson ME, Davydova E, Malecki J, Moen A, Falnes PO. 2015. *Saccharomyces cerevisiae* eukaryotic elongation factor 1A (eEF1A) is methylated at Lys-390 by a METTL21-like methyltransferase. *PLoS ONE* **10:** e0131426.

Jakobsson ME, Malecki J, Nilges BS, Moen A, Leidel SA, Falnes PO. 2017. Methylation of human eukaryotic elongation factor 1 α (eEF1A) by a member of a novel protein lysine methyltransferase family modulates mRNA translation. *Nucleic Acids Res* **45:** 8239–8254.

* Jobe A, Liu Z, Gutierrez-Vargas C, Frank J. 2018. New insights into ribosome structure and function. *Cold Spring Harb Perspect Med* doi: 10.1101/cshperspect.a032615.

Jorgensen R, Ortiz PA, Carr-Schmid A, Nissen P, Kinzy TG, Andersen GR. 2003. Two crystal structures demonstrate large conformational changes in the eukaryotic ribosomal translocase. *Nat Struct Biol* **10:** 379–385.

Jorgensen R, Yates SP, Teal DJ, Nilsson J, Prentice GA, Merrill AR, Andersen GR. 2004. Crystal structure of ADP-ribosylated ribosomal translocase from *Saccharomyces cerevisiae*. *J Biol Chem* **279:** 45919–45925.

Kang HA, Hershey JWB. 1994. Effect of initiation factor eIF-5A depletion on protein synthesis and proliferation of *Saccharomyces cerevisiae*. *J Biol Chem* **269:** 3934–3940.

Katoh T, Wohlgemuth I, Nagano M, Rodnina MV, Suga H. 2016. Essential structural elements in tRNA(Pro) for EF-P-mediated alleviation of translation stalling. *Nat Commun* **7:** 11657.

Kawakami K, Pande S, Faiola B, Moore DP, Boeke JD, Farabaugh PJ, Strathern JN, Nakamura Y, Garfinkel DJ. 1993. A rare tRNA-Arg(CCU) that regulates Ty1 element ribosomal frameshifting is essential for Ty1 retrotransposition in *Saccharomyces cerevisiae*. *Genetics* **135:** 309–320.

Kemper WM, Berry KW, Merrick WC. 1976. Purification and properties of rabbit reticulocyte protein synthesis initiation factors M2Bα and M2Bβ. *J Biol Chem* **251:** 5551–5557.

Kendra JA, de la Fuente C, Brahms A, Woodson C, Bell TM, Chen B, Khan YA, Jacobs JL, Kehn-Hall K, Dinman JD. 2017. Ablation of programmed −1 ribosomal frameshifting in Venezuelan equine encephalitis virus results in attenuated neuropathogenicity. *J Virol* **91:** e01766.

Kenney JW, Moore CE, Wang X, Proud CG. 2014. Eukaryotic elongation factor 2 kinase, an unusual enzyme with multiple roles. *Adv Biol Regul* **55:** 15–27.

Kim HK, Tinoco I Jr. 2017. EF-G catalyzed translocation dynamics in the presence of ribosomal frameshifting stimulatory signals. *Nucleic Acids Res* **45:** 2865–2874.

Kim KK, Hung LW, Yokota H, Kim R, Kim SH. 1998. Crystal structures of eukaryotic translation initiation factor 5A from *Methanococcus jannaschii* at 1.8 Å resolution. *Proc Natl Acad Sci* **95:** 10419–10424.

Kim HK, Liu F, Fei J, Bustamante C, Gonzalez RL Jr, Tinoco I Jr. 2014. A frameshifting stimulatory stem loop destabilizes the hybrid state and impedes ribosomal translocation. *Proc Natl Acad Sci* **111:** 5538–5543.

Kinzy TG, Woolford JL Jr. 1995. Increased expression of *Saccharomyces cerevisiae* translation elongation factor 1α bypasses the lethality of a TEF5 null allele encoding elongation factor 1β. *Genetics* **141:** 481–489.

Klinge S, Voigts-Hoffmann F, Leibundgut M, Arpagaus S, Ban N. 2011. Crystal structure of the eukaryotic 60S ribosomal subunit in complex with initiation factor 6. *Science* **334:** 941–948.

Kobayashi Y, Zhuang J, Peltz S, Dougherty J. 2010. Identification of a cellular factor that modulates HIV-1 programmed ribosomal frameshifting. *J Biol Chem* **285:** 19776–19784.

Kramer EB, Vallabhaneni H, Mayer LM, Farabaugh PJ. 2010. A comprehensive analysis of translational missense errors in the yeast *Saccharomyces cerevisiae*. *RNA* **16:** 1797–1808.

Lam WW, Millichap JJ, Soares DC, Chin R, McLellan A, FitzPatrick DR, Elmslie F, Lees MM, Schaefer GB, Study DDD, et al. 2016. Novel de novo EEF1A2 missense mutations causing epilepsy and intellectual disability. *Mol Genet Genomic Med* **4:** 465–474.

Larsen B, Peden J, Matsufuji S, Matsufuji T, Brady K, Maldonado R, Wills NM, Fayet O, Atkins JF, Gesteland RF. 1995. Upstream stimulators for recoding. *Biochem Cell Biol* **73:** 1123–1129.

Lassak J, Wilson DN, Jung K. 2016. Stall no more at polyproline stretches with the translation elongation factors EF-P and IF-5A. *Mol Microbiol* **99**: 219–235.

Lee MH, Surh YJ. 2009. eEF1A2 as a putative oncogene. *Ann NY Acad Sci* **1171**: 87–93.

Leger M, Dulude D, Steinberg SV, Brakier-Gingras L. 2007. The three transfer RNAs occupying the A, P and E sites on the ribosome are involved in viral programmed −1 ribosomal frameshift. *Nucleic Acids Res* **35**: 5581–5592.

Li T, Belda-Palazon B, Ferrando A, Alepuz P. 2014a. Fertility and polarized cell growth depends on eIF5A for translation of polyproline-rich formins in *Saccharomyces cerevisiae*. *Genetics* **197**: 1191–1200.

Li Y, Treffers EE, Napthine S, Tas A, Zhu L, Sun Z, Bell S, Mark BL, van Veelen PA, van Hemert MJ, et al. 2014b. Transactivation of programmed ribosomal frameshifting by a viral protein. *Proc Natl Acad Sci* **111**: E2172–E2181.

Liao PY, Gupta P, Petrov AN, Dinman JD, Lee KH. 2008. A new kinetic model reveals the synergistic effect of E-, P- and A-sites on +1 ribosomal frameshifting. *Nucleic Acids Res* **36**: 2619–2629.

Lin Z, Gilbert RJ, Brierley I. 2012. Spacer-length dependence of programmed −1 or −2 ribosomal frameshifting on a U6A heptamer supports a role for messenger RNA (mRNA) tension in frameshifting. *Nucleic Acids Res* **40**: 8674–8689.

Ling J, O'Donoghue P, Soll D. 2015. Genetic code flexibility in microorganisms: Novel mechanisms and impact on physiology. *Nat Rev Microbiol* **13**: 707–721.

Lipson RS, Webb KJ, Clarke SG. 2010. Two novel methyltransferases acting upon eukaryotic elongation factor 1A in *Saccharomyces cerevisiae*. *Arch Biochem Biophys* **500**: 137–143.

Liu S, Milne GT, Kuremsky JG, Fink GR, Leppla SH. 2004. Identification of the proteins required for biosynthesis of diphthamide, the target of bacterial ADP-ribosylating toxins on translation elongation factor 2. *Mol Cell Biol* **24**: 9487–9497.

Liu S, Wiggins JF, Sreenath T, Kulkarni AB, Ward JM, Leppla SH. 2006. Dph3, a small protein required for diphthamide biosynthesis, is essential in mouse development. *Mol Cell Biol* **26**: 3835–3841.

Liu S, Bachran C, Gupta P, Miller-Randolph S, Wang H, Crown D, Zhang Y, Wein AN, Singh R, Fattah R, et al. 2012. Diphthamide modification on eukaryotic elongation factor 2 is needed to assure fidelity of mRNA translation and mouse development. *Proc Natl Acad Sci* **109**: 13817–13822.

Lobanov AV, Heaphy SM, Turanov AA, Gerashchenko MV, Pucciarelli S, Devaraj RR, Xie F, Petyuk VA, Smith RD, Klobutcher LA, et al. 2017. Position-dependent termination and widespread obligatory frameshifting in Euplotes translation. *Nat Struct Mol Biol* **24**: 61–68.

Lonnroth P, Appell KC, Wesslau C, Cushman SW, Simpson IA, Smith U. 1988. Insulin-induced subcellular redistribution of insulin-like growth factor II receptors in the rat adipose cell. Counterregulatory effects of isoproterenol, adenosine, and cAMP analogues. *J Biol Chem* **263**: 15386–15391.

Loveland AB, Demo G, Grigorieff N, Korostelev AA. 2017. Ensemble cryo-EM elucidates the mechanism of translation fidelity. *Nature* **546**: 113–117.

Malecki J, Aileni VK, Ho AYY, Schwarz J, Moen A, Sorensen V, Nilges BS, Jakobsson ME, Leidel SA, Falnes PO. 2017. The novel lysine specific methyltransferase METTL21B affects mRNA translation through inducible and dynamic methylation of Lys-165 in human eukaryotic elongation factor 1α (eEF1A). *Nucleic Acids Res* **45**: 4370–4389.

Marcheschi RJ, Mouzakis KD, Butcher SE. 2009. Selection and characterization of small molecules that bind the HIV-1 frameshift site RNA. *ACS Chem Biol* **4**: 844–854.

Marcheschi RJ, Tonelli M, Kumar A, Butcher SE. 2011. Structure of the HIV-1 frameshift site RNA bound to a small molecule inhibitor of viral replication. *ACS Chem Biol* **6**: 857–864.

Mateyak MK, Kinzy TG. 2013. ADP-ribosylation of translation elongation factor 2 by diphtheria toxin in yeast inhibits translation and cell separation. *J Biol Chem* **288**: 24647–24655.

Mathews MB, Hershey JW. 2015. The translation factor eIF5A and human cancer. *Biochim Biophys Acta* **1849**: 836–844.

McNaughton BR, Gareiss PC, Miller BL. 2007. Identification of a selective small-molecule ligand for HIV-1 frameshift-inducing stem-loop RNA from an 11,325 member resin bound dynamic combinatorial library. *J Am Chem Soc* **129**: 11306–11307.

Melnikov S, Mailliot J, Rigger L, Neuner S, Shin BS, Yusupova G, Dever TE, Micura R, Yusupov M. 2016a. Molecular insights into protein synthesis with proline residues. *EMBO Rep* **17**: 1776–1784.

Melnikov S, Mailliot J, Shin BS, Rigger L, Yusupova G, Micura R, Dever TE, Yusupov M. 2016b. Crystal structure of hypusine containing translation factor eIF5A bound to a rotated eukaryotic ribosome. *J Mol Biol* **428**: 3570–3576.

Meskauskas A, Dinman JD. 2001. Ribosomal protein L5 helps anchor peptidyl-tRNA to the P-site in *Saccharomyces cerevisiae*. *RNA* **7**: 1084–1096.

Meydan S, Klepacki D, Karthikeyan S, Margus T, Thomas P, Jones JE, Khan Y, Briggs J, Dinman JD, Vazquez-Laslop N, et al. 2017. Programmed ribosomal frameshifting generates a copper transporter and a copper chaperone from the same gene. *Mol Cell* **65**: 207–219.

Moazed D, Noller HF. 1989. Intermediate states in the movement of transfer RNA in the ribosome. *Nature* **342**: 142–148.

Moomau C, Musalgaonkar S, Khan YA, Jones JE, Dinman JD. 2016. Structural and functional characterization of programmed ribosomal frameshift signals in West Nile virus strains reveals high structural plasticity among *cis*-acting RNA elements. *J Biol Chem* **291**: 15788–15795.

Morris DK, Lundblad V. 1997. Programmed translational frameshifting in a gene required for yeast telomere replication. *Curr Biol* **7**: 969–976.

Murray J, Savva CG, Shin BS, Dever TE, Ramakrishnan V, Fernandez IS. 2016. Structural characterization of ribosome recruitment and translocation by type IV IRES. *eLife* **5**: e13567.

Musalgaonkar S, Moomau CA, Dinman JD. 2014. Ribosomes in the balance: Structural equilibrium ensures translational fidelity and proper gene expression. *Nucleic Acids Res* **42**: 13384–13392.

Nakajima J, Okamoto N, Tohyama J, Kato M, Arai H, Funahashi O, Tsurusaki Y, Nakashima M, Kawashima H, Saitsu H, et al. 2015. De novo *EEF1A2* mutations in patients with characteristic facial features, intellectual disability, autistic behaviors and epilepsy. *Clin Genet* **87**: 356–361.

Namy O, Moran SJ, Stuart DI, Gilbert RJ, Brierley I. 2006. A mechanical explanation of RNA pseudoknot function in programmed ribosomal frameshifting. *Nature* **441**: 244–247.

Napthine S, Treffers EE, Bell S, Goodfellow I, Fang Y, Firth AE, Snijder EJ, Brierley I. 2016. A novel role for poly(C) binding proteins in programmed ribosomal frameshifting. *Nucleic Acids Res* **44**: 5491–5503.

Napthine S, Ling R, Finch LK, Jones JD, Bell S, Brierley I, Firth AE. 2017. Protein-directed ribosomal frameshifting temporally regulates gene expression. *Nat Commun* **8**: 15582.

Nobukuni Y, Kohno K, Miyagawa K. 2005. Gene trap mutagenesis-based forward genetic approach reveals that the tumor suppressor OVCA1 is a component of the biosynthetic pathway of diphthamide on elongation factor 2. *J Biol Chem* **280**: 10572–10577.

Noller HF, Yusupov MM, Yusupova GZ, Baucom A, Cate JH. 2002. Translocation of tRNA during protein synthesis. *FEBS Lett* **514**: 11–16.

Nygard O, Nilsson L. 1990. Kinetic determination of the effects of ADP-ribosylation on the interaction of eukaryotic elongation factor 2 with ribosomes. *J Biol Chem* **265**: 6030–6034.

Ofori LO, Hilimire TA, Bennett RP, Brown NW Jr, Smith HC, Miller BL. 2014. High-affinity recognition of HIV-1 frameshift-stimulating RNA alters frameshifting in vitro and interferes with HIV-1 infectivity. *J Med Chem* **57**: 723–732.

Ogle JM, Brodersen DE, Clemons WM Jr, Tarry MJ, Carter AP, Ramakrishnan V. 2001. Recognition of cognate transfer RNA by the 30S ribosomal subunit. *Science* **292**: 897–902.

Olsthoorn RC, Laurs M, Sohet F, Hilbers CW, Heus HA, Pleij CW. 2004. Novel application of sRNA: Stimulation of ribosomal frameshifting. *RNA* **10**: 1702–1703.

Ortiz PA, Ulloque R, Kihara GK, Zheng H, Kinzy TG. 2006. Translation elongation factor 2 anticodon mimicry domain mutants affect fidelity and diphtheria toxin resistance. *J Biol Chem* **281**: 32639–32648.

Palde PB, Ofori LO, Gareiss PC, Lerea J, Miller BL. 2010. Strategies for recognition of stem-loop RNA structures by synthetic ligands: Application to the HIV-1 frameshift stimulatory sequence. *J Med Chem* **53**: 6018–6027.

Paolini NA, Attwood M, Sondalle SB, Vieira CM, van Adrichem AM, di Summa FM, O'Donohue MF, Gleizes PE, Rachuri S, Briggs JW, et al. 2017. A ribosomopathy reveals decoding defective ribosomes driving human dysmorphism. *Am J Hum Genet* **100**: 506–522.

Park MH, Wolff EC, Smit-McBride Z, Hershey JW, Folk JE. 1991. Comparison of the activities of variant forms of eIF-4D. The requirement for hypusine or deoxyhypusine. *J Biol Chem* **266**: 7988–7994.

Park MH, Nishimura K, Zanelli CF, Valentini SR. 2010. Functional significance of eIF5A and its hypusine modification in eukaryotes. *Amino Acids* **38**: 491–500.

Park JH, Johansson HE, Aoki H, Huang B, Kim HY, Ganoza MC, Park MH. 2012. Post-translational modification by β-lysylation is required for the activity of *E. coli* elongation factor P (EF-P). *J Biol Chem* **287**: 2579–2590.

Pelechano V, Alepuz P. 2017. eIF5A facilitates translation termination globally and promotes the elongation of many non polyproline-specific tripeptide sequences. *Nucleic Acids Res* **45**: 7326–7338.

Petropoulos AD, Green R. 2012. Further in vitro exploration fails to support the allosteric three-site model. *J Biol Chem* **287**: 11642–11648.

Plant EP, Dinman JD. 2005. Torsional restraint: A new twist on frameshifting pseudoknots. *Nucleic Acids Res* **33**: 1825–1833.

Plant EP, Jacobs KL, Harger JW, Meskauskas A, Jacobs JL, Baxter JL, Petrov AN, Dinman JD. 2003. The 9-Å solution: How mRNA pseudoknots promote efficient programmed −1 ribosomal frameshifting. *RNA* **9**: 168–174.

* Proud CG. 2018. Phosphorylation and signal transduction pathways. *Cold Spring Harb Perspect Biol* doi: 10.1101/cshperspect.a033050.

Ramrath DJ, Lancaster L, Sprink T, Mielke T, Loerke J, Noller HF, Spahn CM. 2013. Visualization of two transfer RNAs trapped in transit during elongation factor G-mediated translocation. *Proc Natl Acad Sci* **110**: 20964–20969.

Ratje AH, Loerke J, Mikolajka A, Brunner M, Hildebrand PW, Starosta AL, Donhofer A, Connell SR, Fucini P, Mielke T, et al. 2010. Head swivel on the ribosome facilitates translocation by means of intra-subunit tRNA hybrid sites. *Nature* **468**: 713–716.

Rhodin MH, Dinman JD. 2010. A flexible loop in yeast ribosomal protein L11 coordinates P-site tRNA binding. *Nucleic Acids Res* **38**: 8377–8389.

Ritchie DB, Foster DA, Woodside MT. 2012. Programmed −1 frameshifting efficiency correlates with RNA pseudoknot conformational plasticity, not resistance to mechanical unfolding. *Proc Natl Acad Sci* **109**: 16167–16172.

Ritchie DB, Soong J, Sikkema WK, Woodside MT. 2014. Anti-frameshifting ligand reduces the conformational plasticity of the SARS virus pseudoknot. *J Am Chem Soc* **136**: 2196–2199.

Ritchie DB, Cappellano TR, Tittle C, Rezajooei N, Rouleau L, Sikkema WK, Woodside MT. 2017. Conformational dynamics of the frameshift stimulatory structure in HIV-1. *RNA* **23**: 1376–1384.

* Rodnina MV. 2018. Translation in prokaryotes. *Cold Spring Harb Perspect Biol* doi: 10.1101/cshperspect.a032664.

Rodnina MV, Wintermeyer W. 2009. Recent mechanistic insights into eukaryotic ribosomes. *Curr Opin Cell Biol* **21**: 435–443.

Saini P, Eyler DE, Green R, Dever TE. 2009. Hypusine-containing protein eIF5A promotes translation elongation. *Nature* **459**: 118–121.

Sasikumar AN, Kinzy TG. 2014. Mutations in the chromodomain-like insertion of translation elongation factor 3 compromise protein synthesis through reduced ATPase activity. *J Biol Chem* **289**: 4853–4860.

Schaffrath R, Abdel-Fattah W, Klassen R, Stark MJ. 2014. The diphthamide modification pathway from *Saccharomyces cerevisiae*—Revisited. *Mol Microbiol* **94**: 1213–1226.

Schmidt C, Becker T, Heuer A, Braunger K, Shanmuganathan V, Pech M, Berninghausen O, Wilson DN, Beckmann R. 2016. Structure of the hypusinylated eukaryotic translation factor eIF-5A bound to the ribosome. *Nucleic Acids Res* **44**: 1944–1951.

Schreier MH, Erni B, Staehelin T. 1977. Initiation of mammalian protein synthesis: Purification and characterization of seven initiation factors. *J Mol Biol* **116**: 727–753.

Schuller AP, Wu CC, Dever TE, Buskirk AR, Green R. 2017. eIF5A functions globally in translation elongation and termination. *Mol Cell* **66**: 194–205.

Semenkov YP, Rodnina MV, Wintermeyer W. 1996. The "allosteric three-site model" of elongation cannot be confirmed in a well-defined ribosome system from *Escherichia coli*. *Proc Natl Acad Sci* **93**: 12183–12188.

Shao S, Murray J, Brown A, Taunton J, Ramakrishnan V, Hegde RS. 2016. Decoding mammalian ribosome-mRNA states by translational GTPase complexes. *Cell* **167**: 1229–1240.

Shin BS, Katoh T, Gutierrez E, Kim JR, Suga H, Dever TE. 2017. Amino acid substrates impose polyamine, eIF5A, or hypusine requirement for peptide synthesis. *Nucleic Acids Res* **45**: 8392–8402.

Skogerson L, Engelhardt D. 1977. Dissimilarity in protein chain elongation factor requirements between yeast and rat liver ribosomes. *J Biol Chem* **252**: 1471–1475.

Spahn CM, Gomez-Lorenzo MG, Grassucci RA, Jorgensen R, Andersen GR, Beckmann R, Penczek PA, Ballesta JP, Frank J. 2004. Domain movements of elongation factor eEF2 and the eukaryotic 80S ribosome facilitate tRNA translocation. *EMBO J* **23**: 1008–1019.

Stahl S, da Silva Mateus Seidl AR, Ducret A, Kux van Geijtenbeek S, Michel S, Racek T, Birzele F, Haas AK, Rueger R, Gerg M, et al. 2015. Loss of diphthamide pre-activates NF-κB and death receptor pathways and renders MCF7 cells hypersensitive to tumor necrosis factor. *Proc Natl Acad Sci* **112**: 10732–10737.

Su MC, Chang CT, Chu CH, Tsai CH, Chang KY. 2005. An atypical RNA pseudoknot stimulator and an upstream attenuation signal for −1 ribosomal frameshifting of SARS coronavirus. *Nucleic Acids Res* **33**: 4265–4275.

Su X, Lin Z, Lin H. 2013. The biosynthesis and biological function of diphthamide. *Crit Rev Biochem Mol Biol* **48**: 515–521.

Sulima SO, Patchett S, Advani VM, De Keersmaecker K, Johnson AW, Dinman JD. 2014. Bypass of the pre-60S ribosomal quality control as a pathway to oncogenesis. *Proc Natl Acad Sci* **111**: 5640–5645.

Taha E, Gildish I, Gal-Ben-Ari S, Rosenblum K. 2013. The role of eEF2 pathway in learning and synaptic plasticity. *Neurobiol Learn Mem* **105**: 100–106.

Taylor DJ, Nilsson J, Merrill AR, Andersen GR, Nissen P, Frank J. 2007. Structures of modified eEF2 80S ribosome complexes reveal the role of GTP hydrolysis in translocation. *EMBO J* **26**: 2421–2431.

Thakur A, Chitoor B, Goswami AV, Pareek G, Atreya HS, D'Silva P. 2012. Structure and mechanistic insights into novel iron-mediated moonlighting functions of human J-

protein cochaperone, Dph4. *J Biol Chem* **287**: 13194–13205.

Tinoco I Jr, Kim HK, Yan S. 2013. Frameshifting dynamics. *Biopolymers* **99**: 1147–1166.

Triana-Alonso FJ, Chakraburtty K, Nierhaus KH. 1995. The elongation factor unique in higher fungi and essential for protein biosynthesis is an E site factor. *J Biol Chem* **270**: 20473–20478.

Tsai A, Puglisi JD, Uemura S. 2016. Probing the translation dynamics of ribosomes using zero-mode waveguides. *Prog Mol Biol Transl Sci* **139**: 1–43.

Ude S, Lassak J, Starosta AL, Kraxenberger T, Wilson DN, Jung K. 2013. Translation elongation factor EF-P alleviates ribosome stalling at polyproline stretches. *Science* **339**: 82–85.

Uemura S, Aitken CE, Korlach J, Flusberg BA, Turner SW, Puglisi JD. 2010. Real-time tRNA transit on single translating ribosomes at codon resolution. *Nature* **464**: 1012–1017.

Vimaladithan A, Farabaugh PJ. 1994. Special peptidyl-tRNA molecules can promote translational frameshifting without slippage. *Mol Cell Biol* **14**: 8107–8116.

Webb TR, Cross SH, McKie L, Edgar R, Vizor L, Harrison J, Peters J, Jackson IJ. 2008. Diphthamide modification of eEF2 requires a J-domain protein and is essential for normal development. *J Cell Sci* **121**: 3140–3145.

Weiss RB, Dunn DM, Shuh M, Atkins JF, Gesteland RF. 1989. *E. coli* ribosomes re-phase on retroviral frameshift signals at rates ranging from 2 to 50 percent. *New Biol* **1**: 159–169.

Wohlgemuth I, Brenner S, Beringer M, Rodnina MV. 2008. Modulation of the rate of peptidyl transfer on the ribosome by the nature of substrates. *J Biol Chem* **283**: 32229–32235.

Woolstenhulme CJ, Guydosh NR, Green R, Buskirk AR. 2015. High-precision analysis of translational pausing by ribosome profiling in bacteria lacking EFP. *Cell Rep* **11**: 13–21.

Yelverton E, Lindsley D, Yamauchi P, Gallant JA. 1994. The function of a ribosomal frameshifting signal from human immunodeficiency virus-1 in *Escherichia coli*. *Mol Microbiol* **11**: 303–313.

Youngman EM, Brunelle JL, Kochaniak AB, Green R. 2004. The active site of the ribosome is composed of two layers of conserved nucleotides with distinct roles in peptide bond formation and peptide release. *Cell* **117**: 589–599.

Yu CH, Noteborn MH, Olsthoorn RC. 2010. Stimulation of ribosomal frameshifting by antisense LNA. *Nucleic Acids Res* **38**: 8277–8283.

Yu YR, You LR, Yan YT, Chen CM. 2014. Role of OVCA1/DPH1 in craniofacial abnormalities of Miller–Dieker syndrome. *Hum Mol Genet* **23**: 5579–5596.

Zhong Z, Yang L, Zhang H, Shi J, Vandana JJ, Lam DT, Olsthoorn RC, Lu L, Chen G. 2016. Mechanical unfolding kinetics of the SRV-1 gag-pro mRNA pseudoknot: Possible implications for −1 ribosomal frameshifting stimulation. *Sci Rep* **6**: 39549.

Translation Termination and Ribosome Recycling in Eukaryotes

Christopher U.T. Hellen

Department of Cell Biology, State University of New York, Downstate Medical Center, New York, New York 11203

Correspondence: christopher.hellen@downstate.edu

Termination of mRNA translation occurs when a stop codon enters the A site of the ribosome, and in eukaryotes is mediated by release factors eRF1 and eRF3, which form a ternary eRF1/eRF3–guanosine triphosphate (GTP) complex. eRF1 recognizes the stop codon, and after hydrolysis of GTP by eRF3, mediates release of the nascent peptide. The posttermination complex is then disassembled, enabling its constituents to participate in further rounds of translation. Ribosome recycling involves splitting of the 80S ribosome by the ATP-binding cassette protein ABCE1 to release the 60S subunit. Subsequent dissociation of deacylated transfer RNA (tRNA) and messenger RNA (mRNA) from the 40S subunit may be mediated by initiation factors (priming the 40S subunit for initiation), by ligatin (eIF2D) or by density-regulated protein (DENR) and multiple copies in T-cell lymphoma-1 (MCT1). These events may be subverted by suppression of termination (yielding carboxy-terminally extended readthrough polypeptides) or by interruption of recycling, leading to reinitiation of translation near the stop codon.

OVERVIEW OF TRANSLATION TERMINATION AND RECYCLING

Translation is a cyclical process that comprises initiation, elongation, termination, and ribosome recycling stages (Jackson et al. 2010). Termination is triggered when a stop codon enters the A site of the ribosome, and is mediated by the release factors eRF1 and eRF3 (Dever and Green 2012; Jackson et al. 2012). eRF1 is omnipotent, that is, it is responsible for recognition of all three stop codons, and induces release of the nascent polypeptide from the P-site peptidyl-transfer RNA (tRNA), whereas eRF3 is a GTPase that enhances polypeptide release. The resulting posttermination complex (post-TC) is recycled by splitting of the ribosome, which is mediated by ABCE1. This step is followed by release of deacylated tRNA and messenger RNA (mRNA) from the 40S subunit via redundant pathways involving initiation factors, ligatin (eIF2D) or density-regulated protein (DENR), and multiple copies in T-cell lymphoma-1 (MCT1). Recycling enables ribosomes and mRNAs to participate in multiple rounds of translation.

Here, I focus on recent advances in understanding of mechanisms of termination and ABCE1-mediated ribosomal splitting, discuss unresolved aspects of recycling, and consider how termination or recycling is subverted to

65

allow translation of carboxy-terminally extended "readthrough" polypeptides or of downstream open reading frames (ORFs).

TERMINATION

eRF1–eRF3 Interactions and Ternary Complex Formation

Termination is mediated by eRF1 and eRF3 (Alkalaeva et al. 2006). eRF1 has an amino-terminal domain (N) that is responsible for recognition of the stop codon in the A site (Bertram et al. 2000); a middle domain (M) containing a universally conserved apical GGQ motif that induces release of the nascent polypeptide from peptidyl-tRNA in the ribosomal P site (Frolova et al. 1999); and a carboxy-terminal domain (C) that binds to eRF3 and ABCE1, and contains a mini-domain that affects stop codon specificity (Fig. 1A) (Song et al. 2000; Mantsyzov et al. 2010). eRF3 consists of (1) a nonconserved amino-terminal domain that is not required for eRF3's function in termination, but binds the poly(A)-binding protein (PABP) (Hoshino et al. 1999; Kozlov and Gehring 2010)

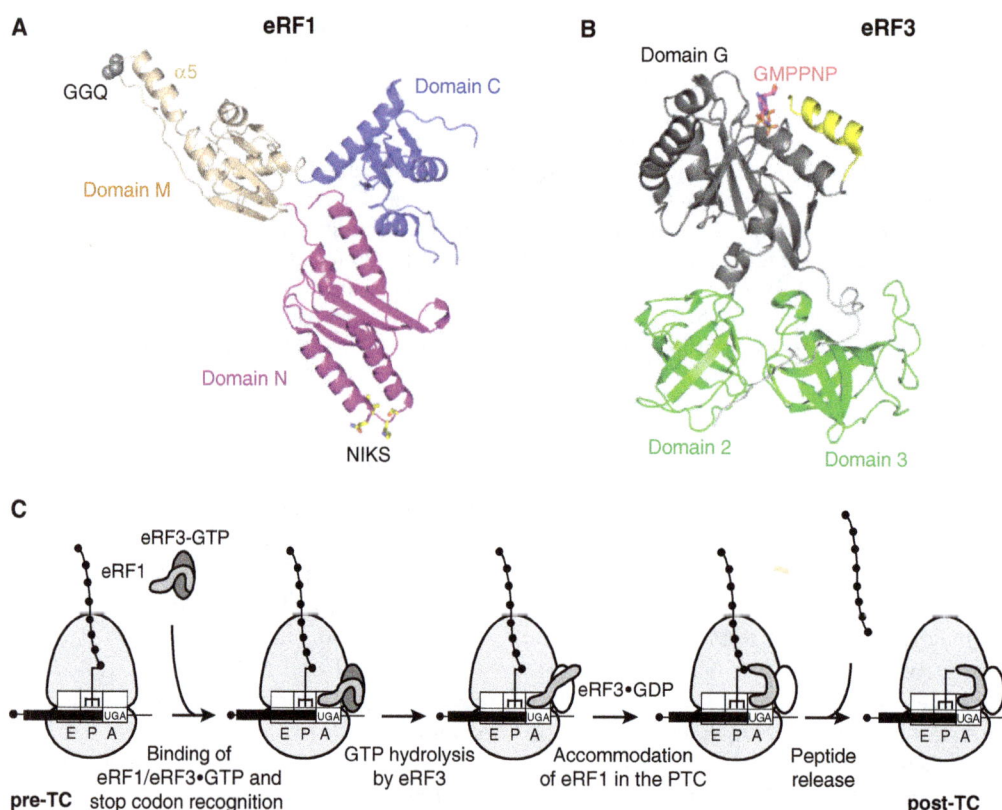

Figure 1. Translation termination in eukaryotes. Ribbon representations of (A) human eRF1 (protein data bank [PDB]: 1DT9), with the Cα atoms of the GGQ motif in domain M and the NIKS motif in domain N shown in CPK and stick models, respectively, and (B) *Schizosaccharomyces pombe* eRF3 (amino acids 215–662, and thus lacking the nonconserved amino-terminal domain) (PDB: 1R5B), with bound GMPPNP shown in a stick model. (C) Outline of the termination process. The pretermination complex (pre-TC) contains peptidyl-transfer RNA (tRNA) in the P site. The eRF1/eRF3–guanosine triphosphate (GTP) complex binds to the A site of the pre-TC, and eRF1 recognizes the stop codon, which, with the +4 nucleotide, binds in a pocket formed by eRF1 and the 40S subunit. eRF1's M domain dissociates from eRF3's Switch I/Switch II elements, and after GTP hydrolysis by eRF3, accommodates in the peptidyl-transferase center (PTC), inducing peptide release. eRF1 and possibly eRF3-guanosine diphosphate (GDP) remain associated with posttermination complexes (post-TCs).

and the nonsense-mediated decay (NMD) factor UPF3b (Neu-Yilik et al. 2017), and (2) a canonical guanosine triphosphate (GTP)-binding domain (G) and two β-barrel domains (2 and 3), which are homologous to GTP-binding translation factors such as EF-Tu, eEF1A, and the carboxy-terminal region of the ribosome rescue factor Hbs1 (Fig. 1B) (Kong et al. 2004; Atkinson et al. 2008; van den Elzen et al. 2010). As in all GTPases, the G domain of eRF3 contains "Switch I" and "Switch II" elements that are essential for binding and hydrolysis of GTP and that regulate the nucleotide-dependent conformational status of the factor. Switch I and Switch II are disordered in free eRF3 (Kong et al. 2004) but become ordered on binding to the γ-phosphate of GTP in the presence of eRF1, with which they also interact (Cheng et al. 2009; des Georges et al. 2014; Preis et al. 2014; Shao et al. 2016). There are two isoforms of eRF3 that are encoded by different genes and have different amino-terminal domains; both bind eRF1 and are functional termination factors (Hoshino et al. 1998; Chauvin et al. 2005). eRF3b is predominantly expressed in brain tissue, whereas eRF3a is ubiquitously expressed (Hoshino et al. 1998; Chauvin et al. 2005).

eRF1 enhances binding of GTP to eRF3 by acting as a GTP dissociation inhibitor, promoting formation of a stable eRF1/eRF3•GTP complex (Fig. 1C) (Hauryliuk et al. 2006; Mitkevich et al. 2006; Pisareva et al. 2006). eRF1 and eRF3 interact extensively with each other on and off the ribosome (Cheng et al. 2009; Taylor et al. 2012; des Georges et al. 2014; Preis et al. 2014; Shao et al. 2016). In pretermination TCs (pre-TCs) containing eRF1/eRF3 before GTP hydrolysis, eRF1's M domain packs against eRF3, inserting into a cleft between domain 2 and the G domain such that the GGQ motif is fixed close to Switch I. The M domain may engage with Switch II (Cheng et al. 2009; des Georges et al. 2014; Preis et al. 2014; Shao et al. 2016). The "preaccommodation" conformation of the M domain before GTP hydrolysis is incompatible with peptide-release activity because the catalytic GGQ motif is sequestered >80 Å from the P-site peptidyl-tRNA's ester bond in the peptidyl-transferase center (PTC) of the 60S subunit.

eRF1's N domain extends into the 40S subunit's decoding center (see below), while domain C interacts with the stalk base of the 60S subunit and, via its mini-domain, with the 40S subunit beak. Like other translational GTPases, eRF3 binds to the GTPase-associated center (GAC), between the sarcin–ricin loop (SRL) of the 60S subunit and helices h5 and h14 of 18S ribosomal RNA (rRNA) on the 40S subunit (des Georges et al. 2014; Preis et al. 2014; Shao et al. 2016).

Stop Codon Recognition

The canonical genetic code has three stop codons (UAA, UAG, and UGA), and the efficiency of termination is enhanced by a purine residue in the +4 position and by a +5 purine if the +4 residue is a pyrimidine (McCaughan et al. 1995). In a few organisms, including some ciliate protists, green algae, and diplomonads, UGA is reassigned as a sense codon and UAA and UAG are retained as stop codons, UAA and UAG are reassigned as sense codons with UGA as the sole termination codon, or UAG is reassigned as a sense codon and UAA and UGA function as stop codons (Keeling 2016; Pánek et al. 2017). In extreme examples, such as the ciliate protist *Condylostoma magnum* and a member of the proposed *Blastocrithidia* clade of trypanosomatids (Heaphy et al. 2016; Swart et al. 2016; Záhonová et al. 2016), all three stop codons serve as sense codons at internal positions but specify termination when located close to the 3' end of mRNA. A hypothesis concerning position-dependent termination is discussed below. Stop codon reassignment depends in part on alterations in eRF1's sequence, and mutational studies coupled with analysis of eRF1 from "variant code" organisms identified highly conserved motifs in the N domain that influence stop codon recognition (Jackson et al. 2010; Blanchet et al. 2015). These motifs include GTS_{31-33}, E_{55}, $TASNIKS_{58-64}$, and $YxCxxxF_{125-130}$ (numbering for human eRF1). Recent cryoelectron microscopy (cryo-EM) studies (Brown et al. 2015; Matheisl et al. 2015; Shao et al. 2016) have established a molecular framework for this process.

In eRF1-bound ribosomal complexes, the N domain reaches into the A site, forming a pocket that accommodates the stop codon and the +4 nucleotide in a compacted conformation (Fig. 2). This compacted state and eRF1's interactions with the stop codon are likely maintained throughout the termination process until eRF1 dissociates (Shao et al. 2016). Stabilization of the compacted state requires a +1U, which is thus a determinant of stop codon recognition, as are the stacking interactions of the +2, +4, and +5 nucleotides with A_{1825}, G_{626}, and C_{1698} of 18S rRNA, respectively, as well as the stop codon's multiple interactions with eRF1. The stabilizing interaction of the +1 uridine with N_{61} and K_{63} of the TASNIKS motif would not be possible for cytidine at this position, and steric hindrance discriminates against +1 purines. Interactions of the YxCxxxF motif and E_{55} with +2 and +3 nucleotides are possible only with purines, and T_{32} of the GTS motif can hydrogen bond with the +3 nucleotide of UAG but not UGA or UGG codons. The specificity of the flexible GTS motif, and the mutual repulsion of G residues at +2 and +3 positions, account for eRF1's discrimination against UGG codons. These observations are consistent with site-directed cross-linking anal-

ysis of stop codon/eRF1 interactions (Chavatte et al. 2002; Bulygin et al. 2010) and mutational analyses of stop codon recognition and specificity (Cheng et al. 2009; Jackson et al. 2010; Conard et al. 2012; Blanchet et al. 2015). Recognition of stop codons, particularly UGA, likely occurs via multiple steps involving RNA compaction in the A site, conformational changes in TASNIKS, and GTS motifs in eRF1 and other localized eRF3-induced changes (Fan-Minogue et al. 2008; Wong et al. 2012; Kryuchkova et al. 2013).

The Mechanism of GTP Hydrolysis by eRF3

eRF3 belongs to a family of translational GTPases, including eEF1A, EF-Tu, and Hbs1, which deliver aminoacyl-tRNA, eRF1 or the related protein Pelota to the A site (Atkinson et al. 2008; Pisareva et al. 2011). In the case of ribosome-bound eEF1A/EF-Tu, establishment of complementarity between cognate aminoacyl-tRNA and the A-site codon induces GTP hydrolysis and is required for release of the tRNA "cargo" so that it can accommodate in

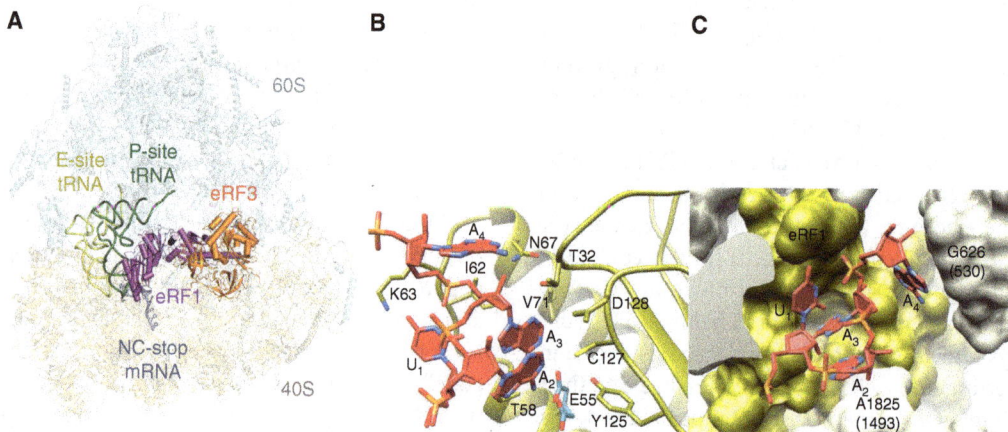

Figure 2. Structure of the mammalian termination complex. (A) The ribosomal termination complex assembled with eRF1 (purple) and eRF3 (orange). (Panel A from Shao et al. 2016; reprinted, with permission, under the terms of Creative Commons Attribution License CC-BY). (B) Close-up view of the stop codon binding pocket of the eRF1 N domain bound to a UAA(A) stop codon (dark red), showing residues important for stop codon recognition, including GTS(31–33), E55, TASNIKS(58–64), N67, V71, and YxCxxxF(125–131). (C) Surface representation of the UAA(A)-binding cavity formed by the eRF1 N domain and 18S ribosomal RNA (rRNA) in the decoding center of the 40S ribosomal subunit, showing A1825 and G626, which engage in stacking interactions with the +2 and +4 nucleotides of the stop codon. (Panel C from Matheisl et al. 2015; reprinted, with permission, from Oxford University Press © 2015.)

Cite this article as *Cold Spring Harb Perspect Biol* doi: 10.1101/cshperspect.a032656

the PTC. The decoding center on the 40S subunit is >70 Å from the active site of these GTPases, so that activation of these factors requires long-range signaling. Binding of cognate aminoacyl-tRNA to the A-site codon leads to domain closure in the small ribosomal subunit, which moves EF-Tu's GTPase domain so that it binds to the SRL (Loveland et al. 2017). This is an obligatory step in the activation of EF-Tu and its hydrolysis of GTP (Maracci and Rodnina 2016). Binding of cognate aminoacyl-tRNA to the A site of eukaryotic ribosomal complexes leads to a similar domain closure and activation of eEF1A (Shao et al. 2016).

The mechanism of activation of eRF3's GTPase activity is somewhat different. eRF3's GTPase activity requires eRF1's M and C domains and is ribosome-dependent, and although eRF1 can stimulate this activity in the absence of its N domain and an A-site termination codon (Frolova et al. 1996; Kononenko et al. 2008), stop codon recognition by eRF1 is thought to accelerate GTP hydrolysis by eRF3 (e.g., Wada and Ito 2014). eRF3, eEF1A, and EF-Tu all engage in similar interactions with the shoulder of the small subunit (Voorhees et al. 2010; des Georges et al. 2014; Shao et al. 2016). Termination complexes containing eRF1 and eRF3 at a stage just before hydrolysis of eRF3-bound GTP have not been visualized. However, several observations suggest that the SRL is required for activation of eRF3, as for other translational GTPases. Thus, eRF3's Switch I interacts with G_{4600} of the SRL (Shao et al. 2016), and termination defects are caused by a substitution in the *Saccharomyces cerevisiae* SRL (Liu and Liebman 1996) and by substitutions in eRF3 that are close to the SRL in pre-TCs (Wada and Ito 2014). Domain closure in the 40S subunit is not induced by binding of eRF1 or Pelota to the A site (Hilal et al. 2016; Shao et al. 2016), however, indicating that a different mechanism must be responsible for the relative repositioning of eRF3 so that Switch I and Switch II can interact with the SRL. This necessary dissociation of eRF1's M domain from Switch I/Switch II to remove a steric block to GTP hydrolysis might be induced by the establishment of interactions between h14 of the 40S subunit and the conserved R_{192} in eRF1's M

domain, and of h5 with eRF3 domain 2 following binding of eRF1/eRF3•GTP to the pre-TC (des Georges et al. 2014). These interactions are important for eRF1-mediated stimulation of eRF3's GTP hydrolysis activity (Cheng et al. 2009) and mimic a step in EF-Tu activation that leads to displacement of the 3′ end of tRNA from the Switch I loop (Voorhees et al. 2010), respectively.

Conformational Rearrangements following Hydrolysis of eRF3-Bound GTP

The timing of events after hydrolysis of eRF3-bound GTP has not been established, but by analogy with EF-Tu (Pape et al. 1998), rapid GTP hydrolysis would be followed by slower accommodation of eRF1 in the PTC and dissociation of eRF3. eRF3-guanosine diphosphate (GDP) appears to remain on the ribosome for an extended period (Bulygin et al. 2016).

After hydrolysis of eRF3-bound GTP, eRF1 adopts an extended conformation that allows the catalytic GGQ motif at the tip of domain M to accommodate in the PTC (Matheisl et al. 2015; Muhs et al. 2015; Shao et al. 2016). This transition results from domain M undergoing a 140° rotation relative to domain N, which remains bound to the stop codon in the A site, as well as rotation of domain C. The motive force for the reorientation of M and C domains may be relaxation of an eRF3-enforced kink between α-helix 8 (in domain M) and α-helix 9 (in domain C), enabling them to form a single continuous α-helix (Shao et al. 2016). The inhibition of eRF1-mediated peptide release by eRF3•GMPPNP (Alkalaeva et al. 2006; Fan-Minogue et al. 2008) is likely because it "locks" eRF1 so that it cannot switch from the compact to the extended conformation. Conformational changes in eRF1 domain C and ribosomal protein uL11 disrupt the interaction of the eRF1 mini-domain with the head of the 40S subunit and lead to establishment of new interactions with uL11 and the L7/L12 stalk base that may stabilize binding of domain C. Consistently, termination is impaired in an *S. cerevisiae* uL11 deletion strain (Salas-Marco and Bedwell 2005).

Although eRF1 alone can induce peptide release, this activity is strongly increased by eRF3 (Alkalaeva et al. 2006), confirming suggestions that eRF3's GTPase activity couples stop codon recognition and peptidyl-tRNA hydrolysis by eRF1 (Salas-Marco and Bedwell 2004). This effect is apparent even when recycling of eRF1 is not required, and it can therefore not be solely attributed to enhanced release of eRF1 from post-TCs or from ribosomal subunits after dissociation of post-TCs (e.g., Eyler et al. 2013). Enhancement by eRF3 of peptide release could be the result of increased eRF1 recruitment to pre-TCs (by eRF3 escorting it to the A site in an EF-Tu-like manner) or by augmentation of the catalytic rate of peptidyl-tRNA hydrolysis. The characteristic +2 nt toeprint shift induced by binding of eRF1 to pre-TCs (Alkalaeva et al. 2006) correlates with the compaction of the extended stop codon sequence (Brown et al. 2015; Matheisl et al. 2015). eRF3 enhances the shift induced by wild-type eRF1 and leads to its appearance in the presence of some eRF1 mutants (Kryuchkova et al. 2013), suggesting that eRF3 promotes stop codon recognition or stabilizes the resulting pre-TC. As well as enhancing the rate of termination, eRF3 could potentially increase its fidelity by kinetic proofreading, by introducing an irreversible GTP hydrolysis step between stop codon recognition and peptidyl-tRNA hydrolysis.

The Mechanism of Peptide Release

In the extended conformation of eRF1, domain M positions the GGQ motif in the PTC close to the CCA end of the P-site peptidyl-tRNA, with Q_{135} adjacent to the ester bond that links the tRNA to the nascent polypeptide (Preis et al. 2014; Brown et al. 2015; Muhs et al. 2015). The similar conformations of the GGQ motif of eRF1 and the GGQ motifs of the otherwise unrelated bacterial RF1/RF2 release factors, and the interactions of the eRF1 M domain and domain 3 of bacterial RF1/RF2 with conserved elements of the ribosome (e.g., Laurberg et al. 2008; Jin et al. 2010) suggest that they function analogously to promote peptidyl-tRNA hydrolysis. Placement of the bacterial GGQ motif in the PTC induces conformational changes that expose the peptidyl-tRNA ester bond to nucle-

ophilic attack by water and stabilize the tetrahedral transition state, promoting cleavage and releasing the nascent polypeptide (Jin et al. 2010; see the discussion of models for this process in Rodnina 2018).

RECYCLING

Recycling of posttermination ribosomes (Fig. 3) is initiated by the highly conserved, essential protein ABCE1, which also recycles vacant 80S ribosomes and stalled ribosomal elongation complexes after their recognition by Hbs1/Pelota (Pisarev et al. 2010, 2011; Franckenberg et al. 2012; Jackson et al. 2012). ABCE1-mediated recycling of post-TCs depends on the presence of eRF1 in the A site (Pisarev et al. 2010). eRF1 therefore participates in two successive stages of protein synthesis: termination and recycling.

The Structure of ABCE1 and Its Interactions with eRF1

ABCE1 has twin nucleotide-binding domains (NBDs) with two composite nucleotide-binding sites formed by motifs from both domains, a helix–loop–helix (HLH) motif in NBD1 and a unique amino-terminal FeS domain containing two $[4Fe-4S]^{2+}$ clusters, which is connected to the NBD core by a hinged cantilever arm (Fig. 3A,B) (Barthelme et al. 2007, 2011; Karcher et al. 2008). The NBDs adopt an "open" state when bound to ADP or nucleotide-free (Karcher et al. 2008; Barthelme et al. 2011), but transition to a closed form with an extensive interface between NBDs on binding ATP (Heuer et al. 2017). ABCE1 has been observed on the 80S ribosome in an intermediate semi-closed state (Becker et al. 2012; Preis et al. 2014; Brown et al. 2015), but must undergo domain closure to manifest its ATPase activity and functionality in recycling.

ABCE1 binds in the ribosomal intersubunit space, interacting extensively via the HLH and hinge elements with sites on the 40S subunit (h5–h15 and h8–h14), which constitute contact points for ribosome-bound translational GTPases (such as eRF3) and via NBD2 at a single site (rpL9) on the 60S subunit. The amino-terminal FeS domain binds NBD2 and eRF1's C

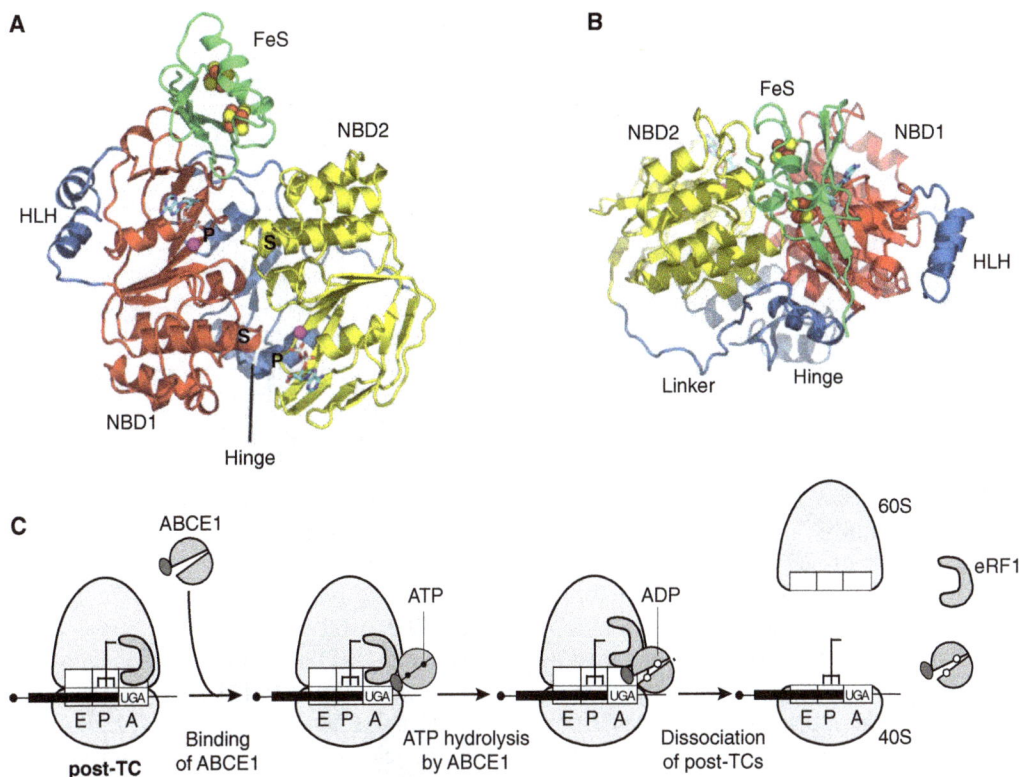

Figure 3. Ribosome recycling in eukaryotes. (*A,B*) Ribbon representations of ABCE1 from *Pyrococcus abyssi* bound to ADP (protein data bank [PDB]: 3BK7). *Top* (*A*) and *front* (*B*) views, showing nucleotide-binding domain NBD1 (red), with its helix–loop–helix insertion (HLH), linked to NBD2 (yellow) by the hinge domain. The FeS domain includes two [4Fe-S] clusters (red and yellow spheres), and binds to the lateral opening of the nucleotide-binding cleft. Nucleotides bind to the two composite binding sites formed by P-loop/Walker A ("P") and signature (S) motifs. (From Karcher et al. 2008; reprinted, with permission, from The American Society for Biochemistry and Molecular Biology © 2008.) (*C*) Model for ribosome recycling (Pisarev et al. 2010), in which ABCE1 binds to eRF1 on the posttermination complex (post-TC), and ATP hydrolysis leads to a power stroke that splits the posttermination ribosome, yielding a 60S subunit, eRF1, ABCE1, and a 40S subunit still bound to messenger RNA (mRNA) and deacylated transfer RNA (tRNA).

domain. Given that ABCE1 binds sites on eRF1 (Preis et al. 2014; Shao et al. 2016) and the 40S subunit that interact with eRF3, ABCE1's association with post-TCs requires prior dissociation of eRF3.

The Mechanism of ABCE1-Mediated Ribosomal Splitting

Cycles of binding of ATP to the twin nucleotide-binding sites in ABC proteins (yielding a "closed" state), hydrolysis of ATP and release of ADP (leading to an "open" state) induce conformational changes in these proteins that are

thought to generate power strokes that drive structural changes in associated domains or macromolecules (Rees et al. 2009). In ribosome-bound complexes, ABCE1 is in an intermediate semi-closed state, in which NBD2 has rotated, leading to repositioning of the FeS domain (Becker et al. 2012; Franckenberg et al. 2012). These changes are accompanied by movement of the eRF1 C domain, because of its close association with the FeS domain, and possibly result in further conformational shifts in eRF1. Subsequent conformational changes in ABCE1 could destabilize intersubunit bridges, leading to ribosomal splitting (Fig. 3C). Movement of the FeS domain has been observed, albeit in complexes that had been assembled in vitro from archaeal 30S subunits or yeast 40S subunits and ABCE1•AMP-PNP without eRF1 (Kiosze-Becker et al. 2016; Heuer et al. 2017). However, a complete structural outline of the mechanism of recycling, linking specific conformational changes in ABCE1, eRF1, and the ribosome to binding and hydrolysis of ATP in the twin nucleotide-binding sites of ABCE1, remains to be established.

ATP hydrolysis by ABCE1 is required for it to split eRF1-bound post-TCs and Pelota-associated ribosomal complexes (Pisarev et al. 2010, 2011; Shoemaker and Green 2011; Becker et al. 2012). In one report, splitting of 80S ribosomes lacking bound mRNA and P-site peptidyl-tRNA was dependent on ATP binding but not its hydrolysis, which was, however, required for release of ABCE1 from the small ribosomal subunit (Barthelme et al. 2011). The basis for this discrepancy remains unclear. This latter alternative model proposes that retention of ABCE1 on the small subunit primes the next round of initiation by enhancing recruitment of initiation factors (Heuer et al. 2017), but neither the stage at which ABCE1 is released in this hypothetical process, nor the trigger for ATP hydrolysis by 40S-bound ABCE1 that would induce release have been determined. Whereas ABCE1's intrinsic ATPase activity is strongly activated by eRF1-bound ribosomes, stimulation by 40S subunits is weak, even though they bind avidly to ABCE1 (Pisarev et al. 2010).

Release of mRNA and Deacylated tRNA from Posttermination Ribosomal Complexes

Deacylated tRNA and mRNA remain bound to the 40S subunit after release of the 60S subunit, and can be released by eIF1, eIF1A, eIF3 and its weakly associated eIF3j subunit (Pisarev et al. 2007, 2010). The multi-subunit eIF3 binds to the 40S subunit's solvent side, extending from the mRNA entrance to the exit (des Georges et al. 2015), interacting with eIF3j at the former site and with eIF1 at the latter. eIF1 and eIF1A bind to the subunit's interface, on the platform

adjacent to P-site initiator tRNA and in the A site, respectively. eIF1 discriminates against noninitiator tRNAs in the P site (Lomakin et al. 2006) and destabilizes their binding in a manner that is augmented by eIF1A and particularly eIF3 (Pisarev et al. 2010). Release of P-site tRNA leads to dissociation of mRNA in the absence of eIF3 and its partial retention in its presence. eIF3j enhances release of mRNA. At concentrations below 1 mM free Mg^{2+}, these factors can recycle post-TCs, eIF3 being primarily responsible for ribosome splitting (Pisarev et al. 2007). Interestingly, eIF3-mediated splitting is inhibited by eIF4F (Skabkin et al. 2013).

There is significant redundancy in the process of tRNA/mRNA release from 40S subunits after ABCE1-mediated dissociation of post-TCs (Fig. 4A). This process can also be mediated by ligatin (eIF2D) and, less effectively, by MCT-1 and DENR, which are interacting proteins that are homologs of the amino- and carboxy-terminal regions of ligatin (eIF2D) (Fig. 4B) (Skabkin et al. 2010). MCT-1 and ligatin's (eIF2D) amino-terminal region contains DUF1947 and PUA domains (Tempel et al. 2013), whereas DENR and ligatin's (eIF2D) carboxy-terminal region contains SWIB/MDM2 and SUI1/eIF1 domains (Fig. 4C) (Vaidya et al. 2017; Weisser et al. 2017). Ligatin (eIF2D) also contains a central winged-helix domain (WHD). The location of the MCT1-like and WHD domains on the interface surface of the 40S subunit may be incompatible with binding of peripheral elements of eIF3a and eIF3c or eIF3b (Lomakin et al. 2017; Weisser et al. 2017), and the SUI1/eIF1 domains of DENR and ligatin (eIF2D) could clash with the anticodon stem-loop of deacylated P-site tRNA and eject it.

REGULATION OF TERMINATION

Various *trans*-acting factors regulate termination and integrate it with processes such as NMD (see Karousis and Mühlemann 2018). Termination may be regulated to arrest ribosomes at the stop codon of an upstream ORF, thereby reducing translation of downstream ORFs, or to modulate readthrough, in which a near-cognate or natural suppressor tRNA decodes the stop codon, allowing translation to

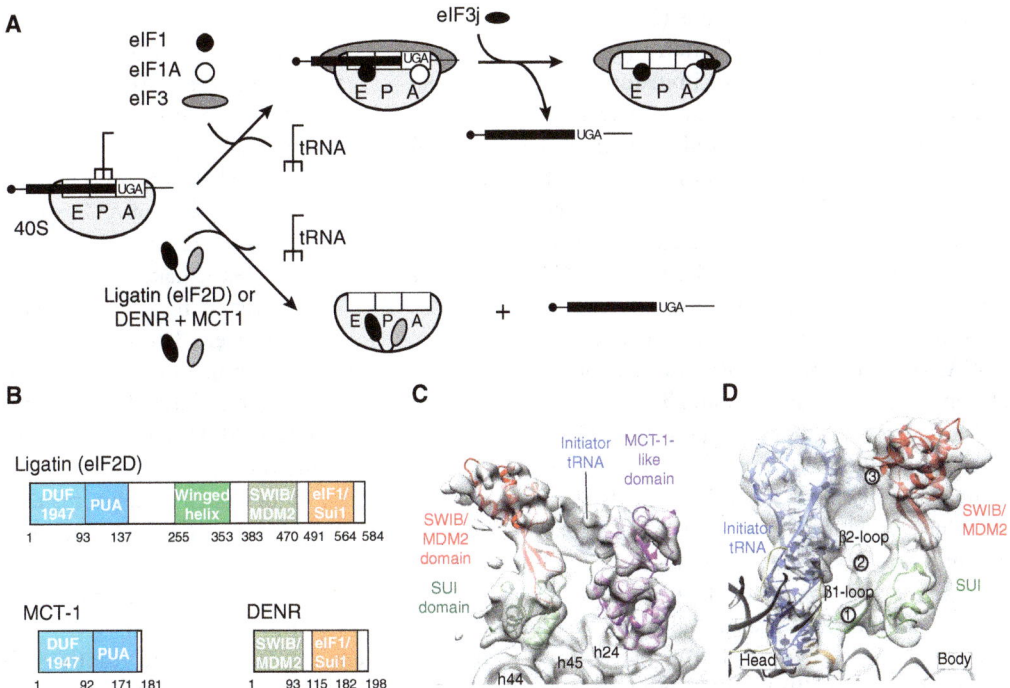

Figure 4. Release of messenger RNA (mRNA) and deacylated transfer RNA (tRNA) from recycled 40S subunits. (*A*) Outline of the release process mediated by initiation factors 1, 1A, 3, and 3j, by ligatin (eIF2D), or by multiple copies in T-cell lymphoma-1 (MCT1) and density-regulated protein (DENR). (*B*) Domain organization of human ligatin (eIF2D), MCT-1, and DENR. (*C*) The SUI domain (green) and the MCT-1-like domain (purple) of ligatin (eIF2D) bound to the 40S subunit (gray) and interacting with P-site tRNA, seen from the foot of the 40S subunit. (*D*) A model derived by docking the crystal structure of the SWIB/MDM2-SUI domain of ligatin (eIF2D) into the cryo-EM map of a ligatin (eIF2D)-bound ribosomal complex containing P-site tRNA and seen from the A site, showing interactions of the SUI loop b1 with the codon–anticodon duplex (1), loop b2 with C11 and G12 in the D-loop (2), and the SWIB/MDM2 domain with C66, G67, and G68 of the acceptor stem (3). (Panels *B–D* from Weisser et al. 2017; reprinted, with permission, from Elsevier © 2017.)

proceed in the same reading frame (e.g., Beier and Grimm 2001). Stop codon suppression allows the expression of carboxy-terminally extended forms of a protein, potentially in a tissue-specific or developmentally regulated manner, and is used by some viruses to modulate the relative levels of expression of structural and nonstructural polyproteins (Firth and Brierley 2012). In yeast, insertion of near-cognate tRNA at premature stop codons is determined by the ability of the ribosomal A site to accommodate mispairing, leading to insertion of Trp, Arg, and Cys at UGA codons and Gln, Tyr, and Lys at UAA and UAG codons (Blanchet et al. 2014; Roy et al. 2015). Flanking residues, particularly the +4 nucleotide, influence selection of

near-cognate aminoacyl-tRNAs in yeast and humans (Beznosková et al. 2016; Xue et al. 2017).

Regulation of Termination by *trans*-Acting Factors

mRNAs harboring premature termination codons are targeted for destruction by NMD, which is thought to be activated because termination at such codons is slower and less efficient than at "normal" stop codons (reviewed in He and Jacobson 2015). Termination at premature termination codons is thought to be impaired both by attenuation of the activity of termination enhancers and by negative regulator(s). PABP binds to eRF3 (Hoshino et al. 1999) and promotes termination in humans (Ivanov et al. 2008, 2016). Impaired termination might thus reflect attenuation of PABP's stimulatory influence caused by the greater spacing between the PABP-bound 3′-poly(A) tail and premature vs. "normal" stop codons. A similar PABP-mediated position-dependent effect may underlie termination on mRNAs lacking dedicated termination codons in which UAA, UAG, and UGA codons at internal positions are decoded as sense codons, whereas those near the 3′ end of mRNAs are recognized by eRF1 as stop codons (Heaphy et al. 2016; Swart et al. 2016).

Various observations suggest that NMD factors also influence termination, possibly in an organism-specific manner. UPF1, UPF2, and UPF3 interact with eRF1 and eRF3 at the A-site of terminating ribosomes (Kashima et al. 2006), but whereas deletion of UPF genes in yeast increased stop codon readthrough (e.g., Wang et al. 2001), small-interfering RNA (siRNA)-mediated depletion of UPF1, UPF2, and UPF3B in human cells had the opposite effect (Ivanov et al. 2008; Jia et al. 2017). ATP hydrolysis by UPF1 is required for efficient termination at PTCs in yeast (Serdar et al. 2016), whereas human UPF1 had no influence on termination in a fully reconstituted translation termination system (Neu-Yilik et al. 2017). However, UPF3B, which binds directly to eRF3 and eRF1, impaired stop codon recognition and peptide release in this system, and dissociated posttermination complexes, by as-yet-undetermined mechanisms.

eIF5A promotes elongation, particularly on polyproline-encoding sequences (Saini et al. 2009). Ribosomal profiling determined that termination is also significantly impaired in eIF5A-depleted S. cerevisiae (Pelechano and Alepuz 2017; Schuller et al. 2017). Estimates of eIF5A's stimulation of the termination rate range from 2- to 17-fold (Saini et al. 2009; Schuller et al. 2017). eIF5A binds between the ribosomal P and E sites, and interacts with the CCA end of P-site tRNA, potentially stabilizing it in an orientation that favors peptidyl-tRNA hydrolysis (Melnikov et al. 2016; Schmidt et al. 2016). Depletion of eIF5A in HeLa cells (Hoque et al. 2017) and a temperature-sensitive mutation in yeast eIF5A (Schrader et al. 2006) both attenu-

ated NMD, hinting that some termination complexes may interact with eIF5A in a manner that targets them for NMD.

ABCE1 has a critical role in posttermination ribosome recycling, but in S. cerevisiae also promotes eRF1-mediated peptide release in a manner that is independent of ATP hydrolysis by ABCE1 and that is enhanced by but not dependent on eRF3 (Shoemaker and Green 2011). ABCE1's activity in recycling depends on prior dissociation of eRF3 from eRF1, but its enhancement of termination is not caused by stabilization of binding of eRF1 to ribosomes (Shoemaker and Green 2011). The basis for this activity remains unknown.

Trans-acting factors may also down-regulate termination by sequestering release factors. The Moloney murine leukemia virus (MuLV) reverse transcriptase (RT) binds to the carboxy-terminal domain of eRF1, outcompeting eRF3 and thereby promoting stop codon readthrough (Tang et al. 2016). Suppression of the *gag* UAG stop codon permits readthrough to the in-frame *pol* gene and synthesis of replicase proteins (including RT) as a Gag-Pol fusion protein.

Regulation of Termination by *cis*-Acting RNA Elements

In MuLV, readthrough is also promoted by a pseudoknot downstream from the *gag* stop codon (Houck-Loomis et al. 2011) by an unknown mechanism that might involve impairment of ribosome function or release factor binding. Numerous viruses use readthrough, in many instances promoted by secondary structures immediately downstream from the stop codon (Firth et al. 2011; Napthine et al. 2012). Until a decade ago, readthrough was thought to be exceedingly rare in nonviral mRNAs, but is now known to be pervasive in many eukaryotes (Jungreis et al. 2011, 2016; Dunn et al. 2013; Eswarappa et al. 2014; Loughran et al. 2014). Several suppressible stop codons are associated with downstream structural elements, which in the case of the *Drosophila hdc* stop codon, is a hairpin that can function in heterologous mRNAs (Steneberg and Samakovlis 2001; Jungreis et al. 2011). However, other mRNAs lack such ele-

Cite this article as *Cold Spring Harb Perspect Biol* doi: 10.1101/cshperspect.a032656

ments, and readthrough is instead promoted by a downstream element that binds hnRNP A2/B1 (Eswarappa et al. 2014), by a downstream GUAC motif (Loughran et al. 2014) or by as-yet-unidentified effectors, in all instances by unknown mechanisms.

Regulation of Termination by Posttranslational Modification

Components of the translation apparatus, including eRF1, eRF3, and the ribosome, are posttranslationally modified, but the functional consequences of these modifications remain largely uncharacterized. Hydroxylation of mammalian eRF1 at K_{63} in the TASNIKS motif enhances termination efficiency (Feng et al. 2014), possibly by establishing an additional hydrogen-bonding interaction with mRNA (Brown et al. 2015). OGFOD1 catalyzes hydroxylation of a conserved prolyl residue in rps23 at a site that projects into the decoding center. This modification has varying effects on termination in yeast and human cells, and at stop codons with different contexts (e.g., Keeling et al. 2006; Loenarz et al. 2014; Singleton et al. 2014). It could potentially regulate readthrough on specific mRNAs in specific circumstances. The Gln of the GGQ motif of yeast eRF1 is methylated by a methyltransferase that consists of the catalytic Mtq2 subunit and the zinc finger protein Trm112 (Graille et al. 2012). It is not known whether this modification enhances eRF1's function in promoting peptide release, as has been reported in bacteria for RF1 and RF2 (Pierson et al. 2016).

Regulation of Termination by Nascent Peptides

Translation of a subset of mRNAs is regulated by the peptide encoded by an upstream ORF (uORF) in the 5′-leader region, causing ribosomes to stall at the uORF stop codon. The human cytomegalovirus uORF2-encoded peptide impairs translation of the gp48 gene (Janzen et al. 2002), and, in fungi, the uORF-encoded arginine attenuator peptide stalls ribosomes at the uORF termination codon of the arg-2 gene when arginine levels are high, resulting in feedback inhi-

bition of synthesis of the first enzyme specific for arginine biosynthesis (Wang and Sachs 1997). These regulatory nascent peptides bind to the ribosomal tunnel and perturb the PTC so that although the eRF1 GGQ-loop is appropriately positioned, it cannot promote peptide release (Bhushan et al. 2010; Matheisl et al. 2015). Systematic analysis of uORFs in *Arabidopsis thaliana* suggests that this mechanism is prevalent in all eukaryotes (Ebina et al. 2015). Stalling in *A. thaliana* is responsive to small molecules such as boric acid (Tanaka et al. 2016) and sucrose (Yamashita et al. 2017), allowing for feedback metabolic control of gene expression.

Therapeutic Enhancement of Stop Codon Readthrough

Nonsense mutations change sense codons to premature termination codons, leading to a loss of function because of the synthesis of defective truncated proteins and/or decreased mRNA stability (Keeling et al. 2014). They account for ~11% of the mutations that cause inherited human diseases (Mort et al. 2008), and PTC readthrough therefore has the potential to ameliorate numerous genetic disorders. Termination suppression occurs at ~10-fold higher levels at PTCs than at naturally occurring stop codons and is augmented by aminoglycoside antibiotics, which bind to the ribosomal decoding center, promoting misincorporation of near-cognate aminoacyl-tRNAs at premature termination codons without significantly affecting the fidelity of elongation (Keeling et al. 2014). The small molecule drug ataluren (PTC124), which is being developed to ameliorate diseases caused by nonsense mutations, has similarly been reported to enhance insertion of near-cognate aminoacyl-tRNAs (Roy et al. 2016), although doubts about its efficacy have been raised (McElroy et al. 2013).

REINITIATION OF TRANSLATION FOLLOWING INTERRUPTED RIBOSOME RECYCLING

Reinitiation after translation of a uORF, first reported more than 30 years ago (Kozak 1984, 1987), is now recognized as a key regulatory process in posttranscriptional control of eukaryotic gene expression. Genome-wide ribosomal profiling, bioinformatics, and proteomic analyses indicate that ~50% of mammalian mRNAs contain a uORF, many of which are translated (Calvo et al. 2009; Ingolia et al. 2011; Plaza et al. 2017). Only 13% of yeast mRNAs contain uORFs (Lawless et al. 2009). Reinitiation of translation can occur by different mechanisms depending on how far the recycling process has progressed.

Reinitiation of Translation Mediated by Canonical Initiation Factors

Reinitiation after ABCE1-mediated splitting of post-TCs occurs in circumstances that prevent dissociation of mRNA from 40S subunits. Efficient reinitiation usually occurs only after translation of short ORFs, and the level of reinitiation drops with uORF length (Luukkonen et al. 1995; Kozak 2001). The realization that reinitiation efficiency is determined by the time taken to translate a uORF rather than by its length led to the hypothesis that reinitiation depends on ribosomal retention of a critical factor during elongation and termination (Kozak 2001). This factor would dissociate stochastically, and only those ribosomes that retained it would be reinitiation-competent. Notably, initiation factors that bind to the 40S subunit's interface surface are displaced during subunit joining (Unbehaun et al. 2004). Analysis of reinitiation on mRNAs that use different sets of factors for initiation indicated that reinitiation depends on the initial involvement of eIF4F, presumably retained on ribosomes via its interaction with eIF3 (Pöyry et al. 2004). Consistently, eIF3 binds primarily on the solvent surface of the 40S subunit (des Georges et al. 2015); cross-linking showed that eIF3 is retained on ribosomes in yeast during elongation on uORFs (Mohammad et al. 2017) and, in the absence of eIF3j, eIF3 impairs release of mRNA from recycled post-TCs (Pisarev et al. 2007), reflecting its activity in stabilizing 40S/mRNA association (Kolupaeva et al. 2005). In vitro reconstitution experiments (Skabkin et al. 2013) yielded a more complete overview of the process, showing that

eIF1, eIF1A, eIF2, eIF3, and Met-tRNA$_i^{Met}$ must be present for reinitiation to occur; eIF2•GTP/Met-tRNA$_i^{Met}$ can be reacquired during scanning (Hinnebusch 2005). eIF4F imposes 5'-3' directionality on scanning, but in its absence 40S subunits move bidirectionally on unstructured regions of mRNA and initiate upstream or downstream from the stop codon.

Reinitiation of Translation Mediated by Reinitiation Factors

In addition to their involvement in recycling, ligatin (eIF2D) and MCT1/DENR promote reinitiation downstream from uORFs in in vitro reconstituted reactions (Skabkin et al. 2013) in *Drosophila* and in human cells (Schleich et al. 2014, 2017). These in vivo reports indicated an extreme dependence on uORF length, with reinitiation mediated by MCT1/DENR occurring in human cells only after translation of uORFs containing one or two sense codons. However, in vitro, ligatin (eIF2D) promoted reinitiation after translation of longer ORFs (Skabkin et al. 2013; Zinoviev et al. 2015) and in NIH3T3 cells, several mRNAs containing significantly longer uORFs are translated in a DENR-dependent manner (Janich et al. 2015). They encode proteins involved in establishing circadian rhythms and, consistently, silencing of DENR led to shortened circadian periods (Janich et al. 2015).

Structural studies indicated that rather than delivering Met-tRNA$_i^{Met}$ to 40S subunits, ligatin (eIF2D) and MCT1/DENR engage in multiple interactions with P-site tRNA that could anchor it on the 40S subunit (Fig. 4D), showing that the functionally important phosphorylation of MCT1 (Schleich et al. 2014) may modulate positioning of the Met-tRNA$_i^{Met}$ acceptor stem, and suggesting that these factors could sterically prevent binding of multiple canonical eIFs to the 40S subunit interface surface (Lomakin et al. 2017; Vaidya et al. 2017; Weisser et al. 2017). Many aspects of the mechanism of reinitiation promoted by these noncanonical factors remain to be established, including what determines whether they act to promote recycling or reinitiation, to what degree their functions are compatible with scanning, whether they can promote reinitiation with or discriminate against specific elongator tRNAs, whether and how they may function in conjunction with other factors such as eIF3, and how ligatin (eIF2D)/MCT-1 dissociate from the aminoacylated-CCA end of Met-tRNA$_i^{Met}$ before its accommodation into the P site of the 60S subunit.

Mobility of Posttermination 80S Ribosomes and Reinitiation of Translation

Posttermination ribosomes are usually weakly anchored to mRNA, and if they are not split by ABCE1 in in vitro reconstituted mammalian translation reactions (Skabkin et al. 2013; Zinoviev et al. 2015), in yeast cells in which it has been depleted (Young et al. 2015) or in primary human platelets and reticulocytes in which ABCE1 has been degraded during terminal differentiation (Mills et al. 2016), can migrate by sliding upstream and downstream from the stop codon. They then stop at triplets that are cognate to the P-site deacylated tRNA and reinitiate translation. In a model (Skabkin et al. 2013) that accounts for these phenomena, dissociation of eRF1 from post-TCs allows P-site tRNA to adopt the P/E hybrid state, disrupting P-site codon–anticodon base pairing, which enables the ribosomes to migrate bidirectionally by sliding in a manner that may be biased by local mRNA structure. Binding of cognate eIF1A•GTP/aa-tRNA to the A-site sense codon is thought to be followed by pseudo-translocation, leading to resumption of translation and accounting for reinitiation occurring in vivo without codon preference. This mechanism is not known to be used for the synthesis of functional gene products, likely because of the semirandom nature of start codon selection, but it may account for the translation of rare peptides from 3'UTR ORFs lacking AUG codons that are presented to major histocompatibility complex (MHC) class I molecules for immune surveillance (Schwab et al. 2003).

There are circumstances in which post-TCs become tethered to mRNA in a manner that promotes reinitiation near the stop codon. For example, the bicistronic subgenomic mRNAs of caliciviruses encode major and minor capsid proteins, and translation of ORF2 occurs by reinitiation. Although reinitiation generally does not occur after long ORFs, likely caused by dissociation of factors (eIF3, eIF4F) that bind to the ribosome's solvent surface, calicivirus mRNAs can support this process because they contain an essential 40- to 80-nt-long structured "termination upstream ribosome-binding site" (TURBS) upstream of the restart AUG. The TURBS binds to eIF3 and base pairs with the apical loop of h26 of 18S rRNA, likely to trap 80S posttermination ribosomes, and this leads to reinitiation being favored over ribosomal dissociation (Pöyry et al. 2007; Luttermann and Meyers 2009; Zinoviev et al. 2015). The TURBS also supports reinitiation by eIFs or ligatin (eIF2D) with 40S subunits generated by ABCE1-mediated splitting of post-TCs; by arresting the ribosome, it also promotes initiation at near-cognate and even at noncognate codons, in the latter case, either by posttermination 80S ribosomes or by 40S subunits and ligatin (eIF2D) (Zinoviev et al. 2015).

In conclusion, recycling can be interrupted at distinct steps to allow posttermination ribosomes to reinitiate translation by diverse mechanisms. This allows access of ribosomes to the principal ORF in an mRNA to be regulated, but also enables the coding capacity of viral mRNAs to be maximized and, as suggested previously (Skabkin et al. 2013), could lead to the translation of sequences that could ultimately evolve into novel genes by permitting ribosomal access to 3'UTR sequences or to alternative ORFs overlapping the stop codon (Carvunis et al. 2012).

ACKNOWLEDGMENTS

We thank M. Weisser and N. Ban for the figures. Research in the author's laboratory is supported by National Institutes of Health (NIH) Grant AI123406.

REFERENCES

*Reference is also in this collection.

Alkalaeva EZ, Pisarev AV, Frolova LY, Kisselev LL, Pestova TV. 2006. In vitro reconstitution of eukaryotic translation reveals cooperativity between release factors eRF1 and eRF3. *Cell* 125: 1125–1136.

Atkinson GC, Baldauf SL, Hauryliuk V. 2008. Evolution of nonstop, no-go and nonsense-mediated mRNA decay and their termination factor-derived components. *BMC Evol Biol* 8: 290.

Barthelme D, Scheele U, Dinkelaker S, Janoschka A, Macmillan F, Albers SV, Driessen AJ, Stagni MS, Bill E, Meyer-Klaucke W, et al. 2007. Structural organization of essential iron-sulfur clusters in the evolutionarily highly conserved ATP-binding cassette protein ABCE1. *J Biol Chem* 282: 14598–14607.

Barthelme D, Dinkelaker S, Albers SV, Londei P, Ermler U, Tampé R. 2011. Ribosome recycling depends on a mechanistic link between the FeS cluster domain and a conformational switch of the twin-ATPase ABCE1. *Proc Natl Acad Sci* 108: 3228–3233.

Becker T, Franckenberg S, Wickles S, Shoemaker CJ, Anger AM, Armache JP, Sieber H, Ungewickell C, Berninghausen O, Daberkow I, et al. 2012. Structural basis of highly conserved ribosome recycling in eukaryotes and archaea. *Nature* 482: 501–506.

Beier H, Grimm M. 2001. Misreading of termination codons in eukaryotes by natural nonsense suppressor tRNAs. *Nucleic Acids Res* 29: 4767–4782.

Bertram G, Bell HA, Ritchie DW, Fullerton G, Stansfield I. 2000. Terminating eukaryote translation: Domain 1 of release factor eRF1 functions in stop codon recognition. *RNA* 6: 1236–1247.

Beznosková P, Gunišová S, Valášek LS. 2016. Rules of UGA-N decoding by near-cognate tRNAs and analysis of readthrough on short uORFs in yeast. *RNA* 22: 456–466.

Bhushan S, Meyer H, Starosta AL, Becker T, Mielke T, Berninghausen O, Sattler M, Wilson DN, Beckmann R. 2010. Structural basis for translational stalling by human cytomegalovirus and fungal arginine attenuator peptide. *Mol Cell* 40: 138–146.

Blanchet S, Cornu D, Argentini M, Namy O. 2014. New insights into the incorporation of natural suppressor tRNAs at stop codons in *Saccharomyces cerevisiae*. *Nucleic Acids Res* 42: 10061–10072.

Blanchet S, Rowe M, Von der Haar T, Fabret C, Demais S, Howard MJ, Namy O. 2015. New insights into stop codon recognition by eRF1. *Nucleic Acids Res* 43: 3298–3308.

Brown A, Shao S, Murray J, Hegde RS, Ramakrishnan V. 2015. Structural basis for stop codon recognition in eukaryotes. *Nature* 524: 493–496.

Bulygin KN, Khairulina YS, Kolosov PM, Ven'yaminova AG, Graifer DM, Vorobjev YN, Frolova LY, Kisselev LL, Karpova GG. 2010. Three distinct peptides from the N domain of translation termination factor eRF1 surround stop codon in the ribosome. *RNA* 16: 1902–1914.

Bulygin KN, Bartuli YS, Malygin AA, Graifer DM, Frolova LY, Karpova GG. 2016. Chemical footprinting reveals conformational changes of 18S and 28S rRNAs at different steps of translation termination on the human ribosome. *RNA* 22: 278–289.

Calvo SE, Pagliarini DJ, Mootha VK. 2009. Upstream open reading frames cause widespread reduction of protein expression and are polymorphic among humans. *Proc Natl Acad Sci* 106: 7507–7512.

Carvunis AR, Rolland T, Wapinski I, Calderwood MA, Yildirim MA, Simonis N, Charloteaux B, Hidalgo CA, Barbette J, Santhanam B, et al. 2012. Proto-genes and de novo gene birth. *Nature* 487: 370–374.

Chauvin C, Salhi S, Le Goff C, Viranaicken W, Diop D, Jean-Jean O. 2005. Involvement of human release factors eRF3a and eRF3b in translation termination and regulation of the termination complex formation. *Mol Cell Biol* 25: 5801–5811.

Chavatte L, Seit-Nebi A, Dubovaya V, Favre A. 2002. The invariant uridine of stop codons contacts the conserved NIKSR loop of human eRF1 in the ribosome. *EMBO J* 21: 5302–5311.

Cheng Z, Saito K, Pisarev AV, Wada M, Pisareva VP, Pestova TV, Gajda M, Round A, Kong C, Lim M, et al. 2009. Structural insights into eRF3 and stop codon recognition by eRF1. *Genes Dev* 23: 1106–1118.

Conard SE, Buckley J, Dang M, Bedwell GJ, Carter RL, Khass M, Bedwell DM. 2012. Identification of eRF1 residues that play critical and complementary roles in stop codon recognition. *RNA* 18: 1210–1221.

des Georges A, Hashem Y, Unbehaun A, Grassucci RA, Taylor D, Hellen CU, Pestova TV, Frank J. 2014. Structure of the mammalian ribosomal pre-termination complex associated with eRF1•eRF3•GDPNP. *Nucleic Acids Res* 42: 3409–3418.

des Georges A, Dhote V, Kuhn L, Hellen CU, Pestova TV, Frank J, Hashem Y. 2015. Structure of mammalian eIF3 in the context of the 43S preinitiation complex. *Nature* 525: 491–495.

Dever TE, Green R. 2012. The elongation, termination, and recycling phases of translation in eukaryotes. *Cold Spring Harb Perspect Biol* 4: a013706.

Dunn JG, Foo CK, Belletier NG, Gavis ER, Weissman JS. 2013. Ribosome profiling reveals pervasive and regulated stop codon readthrough in *Drosophila melanogaster*. *eLife* 2: e01179.

Ebina I, Takemoto-Tsutsumi M, Watanabe S, Koyama H, Endo Y, Kimata K, Igarashi T, Murakami K, Kudo R, Ohsumi A, et al. 2015. Identification of novel *Arabidopsis thaliana* upstream open reading frames that control expression of the main coding sequences in a peptide sequence-dependent manner. *Nucleic Acids Res* 43: 1562–1576.

Eswarappa SM, Potdar AA, Koch WJ, Fan Y, Vasu K, Lindner D, Willard B, Graham LM, DiCorleto PE, Fox PL. 2014. Programmed translational readthrough generates antiangiogenic VEGF-Ax. *Cell* 157: 1605–1618.

Eyler DE, Wehner KA, Green R. 2013. Eukaryotic release factor 3 is required for multiple turnovers of peptide release catalysis by eukaryotic release factor 1. *J Biol Chem* 288: 29530–29538.

Fan-Minogue H, Du M, Pisarev AV, Kallmeyer AK, Salas-Marco J, Keeling KM, Thompson SR, Pestova TV, Bedwell DM. 2008. Distinct eRF3 requirements suggest alternate eRF3 conformations mediate peptide release during eukaryotic translation termination. *Mol Cell* 30: 599–609.

Feng T, Yamamoto A, Wilkins SE, Sokolova E, Yates LA, Münzel M, Singh P, Hopkinson RJ, Fischer R, Cockman ME, et al. 2014. Optimal translational termination requires C4 lysyl hydroxylation of eRF1. *Mol Cell* 53: 645–654.

Cite this article as *Cold Spring Harb Perspect Biol* doi: 10.1101/cshperspect.a032656

Firth AE, Brierley I. 2012. Non-canonical translation in RNA viruses. *J Gen Virol* **93**: 1385–1409.

Firth AE, Wills NM, Gesteland RF, Atkins JF. 2011. Stimulation of stop codon readthrough: Frequent presence of an extended 3′ RNA structural element. *Nucleic Acids Res* **39**: 6679–6691.

Franckenberg S, Becker T, Beckmann R. 2012. Structural view on recycling of archaeal and eukaryotic ribosomes after canonical termination and ribosome rescue. *Curr Opin Struct Biol* **22**: 786–796.

Frolova L, Le Goff X, Zhouravleva G, Davydova E, Philippe M, Kisselev L. 1996. Eukaryotic polypeptide chain release factor eRF3 is an eRF1- and ribosome-dependent guanosine triphosphatase. *RNA* **2**: 334–341.

Frolova LY, Tsivkovskii RY, Sivolobova GF, Oparina NY, Serpinsky OI, Blinov VM, Tatkov SI, Kisselev LL. 1999. Mutations in the highly conserved GGQ motif of class 1 polypeptide release factors abolish ability of human eRF1 to trigger peptidyl-tRNA hydrolysis. *RNA* **5**: 1014–1020.

Graille M, Figaro S, Kervestin S, Buckingham RH, Liger D, Heurgué-Hamard V. 2012. Methylation of class I translation termination factors: Structural and functional aspects. *Biochimie* **94**: 1533–1543.

Hauryliuk V, Zavialov A, Kisselev L, Ehrenberg M. 2006. Class-1 release factor eRF1 promotes GTP binding by class-2 release factor eRF3. *Biochimie* **88**: 747–757.

He F, Jacobson A. 2015. Nonsense-mediated mRNA decay: Degradation of defective transcripts is only part of the story. *Annu Rev Genet* **49**: 339–366.

Heaphy SM, Mariotti M, Gladyshev VN, Atkins JF, Baranov PV. 2016. Novel ciliate genetic code variants including the reassignment of all three stop codons to sense codons in *Condylostoma magnum*. *Mol Biol Evol* **33**: 2885–2889.

Heuer A, Gerovac M, Schmidt C, Trowitzsch S, Preis A, Kötter P, Berninghausen O, Becker T, Beckmann R, Tampé R. 2017. Structure of the 40S-ABCE1 post-splitting complex in ribosome recycling and translation initiation. *Nat Struct Mol Biol* **24**: 453–460.

Hilal T, Yamamoto H, Loerke J, Bürger J, Mielke T, Spahn CM. 2016. Structural insights into ribosomal rescue by Dom34 and Hbs1 at near-atomic resolution. *Nat Commun* **7**: 13521.

Hinnebusch AG. 2005. Translational regulation of GCN4 and the general amino acid control of yeast. *Annu Rev Microbiol* **59**: 407–450.

Hoque M, Park JY, Chang YJ, Luchessi AD, Cambiaghi TD, Shamanna R, Hanauske-Abel HM, Holland B, Pe'ery T, Tian B, et al. 2017. Regulation of gene expression by translation factor eIF5A: Hypusine-modified eIF5A enhances nonsense-mediated mRNA decay in human cells. *Translation* **5**: e1366294.

Hoshino S, Imai M, Mizutani M, Kikuchi Y, Hanaoka F, Ui M, Katada T. 1998. Molecular cloning of a novel member of the eukaryotic polypeptide chain-releasing factors (eRF). Its identification as eRF3 interacting with eRF1. *J Biol Chem* **273**: 22254–22259.

Hoshino S, Imai M, Kobayashi T, Uchida N, Katada T. 1999. The eukaryotic polypeptide chain releasing factor (eRF3/GSPT) carrying the translation termination signal to the 3′-Poly(A) tail of mRNA. Direct association of eRF3/GSPT with polyadenylate-binding protein. *J Biol Chem* **274**: 16677–16680.

Houck-Loomis B, Durney MA, Salguero C, Shankar N, Nagle JM, Goff SP, D'Souza VM. 2011. An equilibrium-dependent retroviral mRNA switch regulates translational recoding. *Nature* **480**: 561–564.

Ingolia NT, Lareau LF, Weissman JS. 2011. Ribosome profiling of mouse embryonic stem cells reveals the complexity and dynamics of mammalian proteomes. *Cell* **147**: 789–802.

Ivanov PV, Gehring NH, Kunz JB, Hentze MW, Kulozik AE. 2008. Interactions between UPF1, eRFs, PABP and the exon junction complex suggest an integrated model for mammalian NMD pathways. *EMBO J* **27**: 736–747.

Ivanov A, Mikhailova T, Eliseev B, Yeramala L, Sokolova E, Susorov D, Shuvalov A, Schaffitzel C, Alkalaeva E. 2016. PABP enhances release factor recruitment and stop codon recognition during translation termination. *Nucleic Acids Res* **44**: 7766–7776.

Jackson RJ, Hellen CU, Pestova TV. 2010. The mechanism of eukaryotic translation initiation and principles of its regulation. *Nat Rev Mol Cell Biol* **11**: 113–127.

Jackson RJ, Hellen CU, Pestova TV. 2012. Termination and post-termination events in eukaryotic translation. *Adv Protein Chem Struct Biol* **86**: 45–93.

Janich P, Arpat AB, Castelo-Szekely V, Lopes M, Gatfield D. 2015. Ribosome profiling reveals the rhythmic liver translatome and circadian clock regulation by upstream open reading frames. *Genome Res* **25**: 1848–1859.

Janzen DM, Frolova L, Geballe AP. 2002. Inhibition of translation termination mediated by an interaction of eukaryotic release factor 1 with a nascent peptidyl-tRNA. *Mol Cell Biol* **22**: 8562–8570.

Jia J, Werkmeister E, Gonzalez-Hilarion S, Leroy C, Gruenert DC, Lafont F, Tulasne D, Lejeune F. 2017. Premature termination codon readthrough in human cells occurs in novel cytoplasmic foci and requires UPF proteins. *J Cell Sci* **130**: 3009–3022.

Jin H, Kelley AC, Loakes D, Ramakrishnan V. 2010. Structure of the 70S ribosome bound to release factor 2 and a substrate analog provides insights into catalysis of peptide release. *Proc Natl Acad Sci* **107**: 8593–8598.

Jungreis I, Lin MF, Spokony R, Chan CS, Negre N, Victorsen A, White KP, Kellis M. 2011. Evidence of abundant stop codon readthrough in *Drosophila* and other metazoa. *Genome Res* **21**: 2096–2113.

Jungreis I, Chan CS, Waterhouse RM, Fields G, Lin MF, Kellis M. 2016. Evolutionary dynamics of abundant stop codon readthrough. *Mol Biol Evol* **33**: 3108–3132.

Karcher A, Schele A, Hopfner KP. 2008. X-ray structure of the complete ABC enzyme ABCE1 from *Pyrococcus abyssi*. *J Biol Chem* **283**: 7962–7971.

* Karousis ED, Mühlemann O. 2018. Nonsense-mediated mRNA decay begins where translation ends. *Cold Spring Harb Perspect Biol* doi: 10.1101/cshperspect.a032862.

Kashima I, Yamashita A, Izumi N, Kataoka N, Morishita R, Hoshino S, Ohno M, Dreyfuss G, Ohno S. 2006. Binding of a novel SMG-1-Upf1-eRF1-eRF3 complex (SURF) to the exon junction complex triggers Upf1 phosphorylation and nonsense-mediated mRNA decay. *Genes Dev* **20**: 355–367.

Keeling PJ. 2016. Genomics: Evolution of the genetic code. *Curr Biol* **26**: R851–R853.

Keeling KM, Salas-Marco J, Osherovich LZ, Bedwell DM. 2006. Tpa1p is part of an mRNP complex that influences translation termination, mRNA deadenylation, and mRNA turnover in *Saccharomyces cerevisiae*. *Mol Cell Biol* **26**: 5237–5248.

Keeling KM, Xue X, Gunn G, Bedwell DM. 2014. Therapeutics based on stop codon readthrough. *Annu Rev Genomics Hum Genet* **15**: 371–394.

Kiosze-Becker K, Ori A, Gerovac M, Heuer A, Nürenberg-Goloub E, Rashid UJ, Becker T, Beckmann R, Beck M, Tampé R. 2016. Structure of the ribosome post-recycling complex probed by chemical cross-linking and mass spectrometry. *Nat Commun* **7**: 13248.

Kolupaeva VG, Unbehaun A, Lomakin IB, Hellen CU, Pestova TV. 2005. Binding of eukaryotic initiation factor 3 to ribosomal 40S subunits and its role in ribosomal dissociation and anti-association. *RNA* **11**: 470–486.

Kong C, Ito K, Walsh MA, Wada M, Liu Y, Kumar S, Barford D, Nakamura Y, Song H. 2004. Crystal structure and functional analysis of the eukaryotic class II release factor eRF3 from *S. pombe*. *Mol Cell* **14**: 233–245.

Kononenko AV, Mitkevich VA, Dubovaya VI, Kolosov PM, Makarov AA, Kisselev LL. 2008. Role of the individual domains of translation termination factor eRF1 in GTP binding to eRF3. *Proteins* **70**: 388–393.

Kozak M. 1984. Selection of initiation sites by eucaryotic ribosomes: Effect of inserting AUG triplets upstream from the coding sequence for preproinsulin. *Nucleic Acids Res* **12**: 3873–3893.

Kozak M. 1987. Effects of intercistronic length on the efficiency of reinitiation by eucaryotic ribosomes. *Mol Cell Biol* **7**: 3438–3445.

Kozak M. 2001. Constraints on reinitiation of translation in mammals. *Nucleic Acids Res* **29**: 5226–5232.

Kozlov G, Gehring K. 2010. Molecular basis of eRF3 recognition by the MLLE domain of poly(A)-binding protein. *PLoS ONE* **5**: e10169.

Kryuchkova P, Grishin A, Eliseev B, Karyagina A, Frolova L, Alkalaeva E. 2013. Two-step model of stop codon recognition by eukaryotic release factor eRF1. *Nucleic Acids Res* **41**: 4573–4586.

Laurberg M, Asahara H, Korostelev A, Zhu J, Trakhanov S, Noller HF. 2008. Structural basis for translation termination on the 70S ribosome. *Nature* **454**: 852–857.

Lawless C, Pearson RD, Selley JN, Smirnova JB, Grant CM, Ashe MP, Pavitt GD, Hubbard SJ. 2009. Upstream sequence elements direct post-transcriptional regulation of gene expression under stress conditions in yeast. *BMC Genomics* **10**: 7.

Liu R, Liebman SW. 1996. A translational fidelity mutation in the universally conserved sarcin/ricin domain of 25S yeast ribosomal RNA. *RNA* **2**: 254–263.

Loenarz C, Sekirnik R, Thalhammer A, Ge W, Spivakovsky E, Mackeen MM, McDonough MA, Cockman ME, Kessler BM, Ratcliffe PJ, et al. 2014. Hydroxylation of the eukaryotic ribosomal decoding center affects translational accuracy. *Proc Natl Acad Sci* **111**: 4019–4024.

Lomakin IB, Shirokikh NE, Yusupov MM, Hellen CU, Pestova TV. 2006. The fidelity of translation initiation: Reciprocal activities of eIF1, IF3 and YciH. *EMBO J* **25**: 196–210.

Lomakin IB, Stolboushkina EA, Vaidya AT, Zhao C, Garber MB, Dmitriev SE, Steitz TA. 2017. Crystal structure of the human ribosome in complex with DENR-MCT-1. *Cell Rep* **20**: 521–528.

Loughran G, Chou MY, Ivanov IP, Jungreis I, Kellis M, Kiran AM, Baranov PV, Atkins JF. 2014. Evidence of efficient stop codon readthrough in four mammalian genes. *Nucleic Acids Res* **42**: 8928–8938.

Loveland AB, Demo G, Grigorieff N, Korostelev AA. 2017. Ensemble cryo-EM elucidates the mechanism of translation fidelity. *Nature* **546**: 113–117.

Luttermann C, Meyers G. 2009. The importance of inter- and intramolecular base pairing for translation reinitiation on a eukaryotic bicistronic mRNA. *Genes Dev* **23**: 331–344.

Luukkonen BG, Tan W, Schwartz S. 1995. Efficiency of reinitiation of translation on human immunodeficiency virus type 1 mRNAs is determined by the length of the upstream open reading frame and by intercistronic distance. *J Virol* **69**: 4086–4094.

Mantsyzov AB, Ivanova EV, Birdsall B, Alkalaeva EZ, Kryuchkova PN, Kelly G, Frolova LY, Polshakov VI. 2010. NMR solution structure and function of the C-terminal domain of eukaryotic class 1 polypeptide chain release factor. *FEBS J* **277**: 2611–2627.

Maracci C, Rodnina MV. 2016. Review: Translational GTPases. *Biopolymers* **105**: 463–475.

Matheisl S, Berninghausen O, Becker T, Beckmann R. 2015. Structure of a human translation termination complex. *Nucleic Acids Res* **43**: 8615–8626.

McCaughan KK, Brown CM, Dalphin ME, Berry MJ, Tate WP. 1995. Translational termination efficiency in mammals is influenced by the base following the stop codon. *Proc Natl Acad Sci* **92**: 5431–5435.

McElroy SP, Nomura T, Torrie LS, Warbrick E, Gartner U, Wood G, McLean WH. 2013. A lack of premature termination codon read-through efficacy of PTC124 (Ataluren) in a diverse array of reporter assays. *PLoS Biol* **11**: e1001593.

Melnikov S, Mailliot J, Shin BS, Rigger L, Yusupova G, Micura R, Dever TE, Yusupov M. 2016. Crystal structure of hypusine-containing translation factor eIF5A bound to a rotated eukaryotic ribosome. *J Mol Biol* **428**: 3570–3576.

Mills EW, Wangen J, Green R, Ingolia NT. 2016. Dynamic regulation of a ribosome rescue pathway in erythroid cells and platelets. *Cell Rep* **17**: 1–10.

Mitkevich VA, Kononenko AV, Petrushanko IY, Yanvarev DV, Makarov AA, Kisselev LL. 2006. Termination of translation in eukaryotes is mediated by the quaternary eRF1•eRF3•GTP•Mg^{2+} complex. The biological roles of eRF3 and prokaryotic RF3 are profoundly distinct. *Nucleic Acids Res* **34**: 3947–3954.

Mohammad MP, Munzarová Pondelíčková V, Zeman J, Gunišová S, Valášek LS. 2017. In vivo evidence that eIF3 stays bound to ribosomes elongating and terminating on short upstream ORFs to promote reinitiation. *Nucleic Acids Res* **45**: 2658–2674.

Mort M, Ivanov D, Cooper DN, Chuzhanova NA. 2008. A meta-analysis of nonsense mutations causing human genetic disease. *Hum Mutat* **29**: 1037–1047.

Muhs M, Hilal T, Mielke T, Skabkin MA, Sanbonmatsu KY, Pestova TV, Spahn CM. 2015. Cryo-EM of ribosomal 80S complexes with termination factors reveals the translocated cricket paralysis virus IRES. *Mol Cell* **57:** 422–432.

Napthine S, Yek C, Powell ML, Brown TD, Brierley I. 2012. Characterization of the stop codon readthrough signal of Colorado tick fever virus segment 9 RNA. *RNA* **18:** 241–252.

Neu-Yilik G, Raimondeau E, Eliseev B, Yeramala L, Amthor B, Deniaud A, Huard K, Kerschgens K, Hentze MW, Schaffitzel C, et al. 2017. Dual function of UPF3B in early and late translation termination. *EMBO J* **36:** 2968–2986.

Pánek T, Žihala D, Sokol M, Derelle R, Klimeš V, Hradilová M, Zadrobílková E, Susko E, Roger AJ, Čepička I, et al. 2017. Nuclear genetic codes with a different meaning of the UAG and the UAA codon. *BMC Biol* **15:** 8.

Pape T, Wintermeyer W, Rodnina MV. 1998. Complete kinetic mechanism of elongation factor Tu-dependent binding of aminoacyl-tRNA to the A site of the *E. coli* ribosome. *EMBO J* **17:** 7490–7497.

Pelechano V, Alepuz P. 2017. eIF5A facilitates translation termination globally and promotes the elongation of many non polyproline-specific tripeptide sequences. *Nucleic Acids Res* **45:** 7326–7338.

Pierson WE, Hoffer ED, Keedy HE, Simms CL, Dunham CM, Zaher HS. 2016. Uniformity of peptide release is maintained by methylation of release factors. *Cell Rep* **17:** 11–18.

Pisarev AV, Hellen CU, Pestova TV. 2007. Recycling of eukaryotic posttermination ribosomal complexes. *Cell* **131:** 286–299.

Pisarev AV, Skabkin MA, Pisareva VP, Skabkina OV, Rakotondrafara AM, Hentze MW, Hellen CU, Pestova TV. 2010. The role of ABCE1 in eukaryotic posttermination ribosomal recycling. *Mol Cell* **37:** 196–210.

Pisareva VP, Pisarev AV, Hellen CU, Rodnina MV, Pestova TV. 2006. Kinetic analysis of interaction of eukaryotic release factor 3 with guanine nucleotides. *J Biol Chem* **281:** 40224–40235.

Pisareva VP, Skabkin MA, Hellen CU, Pestova TV, Pisarev AV. 2011. Dissociation by Pelota, Hbs1 and ABCE1 of mammalian vacant 80S ribosomes and stalled elongation complexes. *EMBO J* **30:** 1804–1817.

Plaza S, Menschaert G, Payre F. 2017. In search of lost small peptides. *Annu Rev Cell Dev Biol* **33:** 391–416.

Pöyry TA, Kaminski A, Jackson RJ. 2004. What determines whether mammalian ribosomes resume scanning after translation of a short upstream open reading frame? *Genes Dev* **18:** 62–75.

Pöyry TA, Kaminski A, Connell EJ, Fraser CS, Jackson RJ. 2007. The mechanism of an exceptional case of reinitiation after translation of a long ORF reveals why such events do not generally occur in mammalian mRNA translation. *Genes Dev* **21:** 3149–3162.

Preis A, Heuer A, Barrio-Garcia C, Hauser A, Eyler DE, Berninghausen O, Green R, Becker T, Beckmann R. 2014. Cryoelectron microscopic structures of eukaryotic translation termination complexes containing eRF1-eRF3 or eRF1-ABCE1. *Cell Rep* **8:** 59–65.

Rees DC, Johnson E, Lewinson O. 2009. ABC transporters: The power to change. *Nat Rev Mol Cell Biol* **10:** 218–227.

* Rodnina MV. 2018. Translation in prokaryotes. *Cold Spring Harb Perspect Biol* doi: 10.1101/cshperspect.a032664.

Roy B, Leszyk JD, Mangus DA, Jacobson A. 2015. Nonsense suppression by near-cognate tRNAs employs alternative base pairing at codon positions 1 and 3. *Proc Natl Acad Sci* **112:** 3038–3043.

Roy B, Friesen WJ, Tomizawa Y, Leszyk JD, Zhuo J, Johnson B, Dakka J, Trotta CR, Xue X, Mutyam V, et al. 2016. Ataluren stimulates ribosomal selection of near-cognate tRNAs to promote nonsense suppression. *Proc Natl Acad Sci* **113:** 12508–12513.

Saini P, Eyler DE, Green R, Dever TE. 2009. Hypusine-containing protein eIF5A promotes translation elongation. *Nature* **459:** 118–121.

Salas-Marco J, Bedwell DM. 2004. GTP hydrolysis by eRF3 facilitates stop codon decoding during eukaryotic translation termination. *Mol Cell Biol* **24:** 7769–7778.

Salas-Marco J, Bedwell DM. 2005. Discrimination between defects in elongation fidelity and termination efficiency provides mechanistic insights into translational readthrough. *J Mol Biol* **348:** 801–815.

Schleich S, Strassburger K, Janiesch PC, Koledachkina T, Miller KK, Haneke K, Cheng YS, Kuechler K, Stoecklin G, Duncan KE, et al. 2014. DENR–MCT-1 promotes translation re-initiation downstream of uORFs to control tissue growth. *Nature* **512:** 208–212.

Schleich S, Acevedo JM, Clemm von Hohenberg K, Teleman AA. 2017. Identification of transcripts with short stuORFs as targets for DENR•MCTS1-dependent translation in human cells. *Sci Rep* **7:** 3722.

Schmidt C, Becker T, Heuer A, Braunger K, Shanmuganathan V, Pech M, Berninghausen O, Wilson DN, Beckmann R. 2016. Structure of the hypusinylated eukaryotic translation factor eIF-5A bound to the ribosome. *Nucleic Acids Res* **44:** 1944–1951.

Schrader R, Young C, Kozian D, Hoffmann R, Lottspeich F. 2006. Temperature-sensitive eIF5A mutant accumulates transcripts targeted to the nonsense-mediated decay pathway. *J Biol Chem* **281:** 35336–35346.

Schuller AP, Wu CC, Dever TE, Buskirk AR, Green R. 2017. eIF5A functions globally in translation elongation and termination. *Mol Cell* **66:** 194–205.e5.

Schwab SR, Li KC, Kang C, Shastri N. 2003. Constitutive display of cryptic translation products by MHC class I molecules. *Science* **301:** 1367–1371.

Serdar LD, Whiteside DL, Baker KE. 2016. ATP hydrolysis by UPF1 is required for efficient translation termination at premature stop codons. *Nat Commun* **7:** 14021.

Shao S, Murray J, Brown A, Taunton J, Ramakrishnan V, Hegde RS. 2016. Decoding mammalian ribosome-mRNA states by translational GTPase complexes. *Cell* **167:** 1229–1240.e15.

Shoemaker CJ, Green R. 2011. Kinetic analysis reveals the ordered coupling of translation termination and ribosome recycling in yeast. *Proc Natl Acad Sci* **108:** E1392–E1398.

Singleton RS, Liu-Yi P, Formenti F, Ge W, Sekirnik R, Fischer R, Adam J, Pollard PJ, Wolf A, Thalhammer A, et al. 2014. OGFOD1 catalyzes prolyl hydroxylation of RPS23 and is involved in translation control and stress granule formation. *Proc Natl Acad Sci* **111:** 4031–4036.

Skabkin MA, Skabkina OV, Dhote V, Komar AA, Hellen CU, Pestova TV. 2010. Activities of ligatin and MCT-1/DENR in eukaryotic translation initiation and ribosomal recycling. *Genes Dev* **24:** 1787–1801.

Skabkin MA, Skabkina OV, Hellen CUT, Pestova TV. 2013. Reinitiation and other unconventional post-termination events during eukaryotic translation. *Mol Cell* **51:** 249–264.

Song H, Mugnier P, Das AK, Webb HM, Evans DR, Tuite MF, Hemmings BA, Barford D. 2000. The crystal structure of human eukaryotic release factor eRF1—Mechanism of stop codon recognition and peptidyl-tRNA hydrolysis. *Cell* **100:** 311–321.

Steneberg P, Samakovlis C. 2001. A novel stop codon readthrough mechanism produces functional headcase protein in *Drosophila* trachea. *EMBO Rep* **2:** 593–597.

Swart EC, Serra V, Petroni G, Nowacki M. 2016. Genetic codes with no dedicated stop codon: Context-dependent translation termination. *Cell* **166:** 691–702.

Tanaka M, Sotta N, Yamazumi Y, Yamashita Y, Miwa K, Murota K, Chiba Y, Hirai MY, Akiyama T, Onouchi H, et al. 2016. The minimum open reading frame, AUG-Stop, induces boron-dependent ribosome stalling and mRNA degradation. *Plant Cell* **28:** 2830–2849.

Tang X, Zhu Y, Baker SL, Bowler MW, Chen BJ, Chen C, Hogg JR, Goff SP, Song H. 2016. Structural basis of suppression of host translation termination by Moloney murine leukemia virus. *Nat Commun* **7:** 12070.

Taylor D, Unbehaun A, Li W, Das S, Lei J, Liao HY, Grassucci RA, Pestova TV, Frank J. 2012. Cryo-EM structure of the mammalian eukaryotic release factor eRF1–eRF3-associated termination complex. *Proc Natl Acad Sci* **109:** 18413–18418.

Tempel W, Dimov S, Tong Y, Park HW, Hong BS. 2013. Crystal structure of human multiple copies in T-cell lymphoma-1 oncoprotein. *Proteins* **81:** 519–525.

Unbehaun A, Borukhov SI, Hellen CU, Pestova TV. 2004. Release of initiation factors from 48S complexes during ribosomal subunit joining and the link between establishment of codon-anticodon base-pairing and hydrolysis of eIF2-bound GTP. *Genes Dev* **18:** 3078–3093.

Vaidya AT, Lomakin IB, Joseph NN, Dmitriev SE, Steitz TA. 2017. Crystal structure of the C-terminal domain of human eIF2D and its implications on eukaryotic translation initiation. *J Mol Biol* **429:** 521–528.

van den Elzen AM, Henri J, Lazar N, Gas ME, Durand D, Lacroute F, Nicaise M, van Tilbeurgh H, Séraphin B, Graille M. 2010. Dissection of Dom34–Hbs1 reveals independent functions in two RNA quality control pathways. *Nat Struct Mol Biol* **17:** 1446–1452.

Voorhees RM, Schmeing TM, Kelley AC, Ramakrishnan V. 2010. The mechanism for activation of GTP hydrolysis on the ribosome. *Science* **330:** 835–838.

Wada M, Ito K. 2014. A genetic approach for analyzing the co-operative function of the tRNA mimicry complex, eRF1/eRF3, in translation termination on the ribosome. *Nucleic Acids Res* **42:** 7851–7866.

Wang Z, Sachs MS. 1997. Ribosome stalling is responsible for arginine specific translational attenuation in *Neurospora crassa*. *Mol Cell Biol* **17:** 4904–4913.

Wang W, Czaplinski K, Rao Y, Peltz SW. 2001. The role of Upf proteins in modulating the translation read-through of nonsense-containing transcripts. *EMBO J* **20:** 880–890.

Weisser M, Schäfer T, Leibundgut M, Böhringer D, Aylett CHS, Ban N. 2017. Structural and functional insights into human re-initiation complexes. *Mol Cell* **67:** 447–456.e7.

Wong LE, Li Y, Pillay S, Frolova L, Pervushin K. 2012. Selectivity of stop codon recognition in translation termination is modulated by multiple conformations of GTS loop in eRF1. *Nucleic Acids Res* **40:** 5751–5765.

Xue X, Mutyam V, Thakerar A, Mobley J, Bridges RJ, Rowe SM, Keeling KM, Bedwell DM. 2017. Identification of the amino acids inserted during suppression of CFTR nonsense mutations and determination of their functional consequences. *Hum Mol Genet* **26:** 3116–3129.

Yamashita Y, Takamatsu S, Glasbrenner M, Becker T, Naito S, Beckmann R. 2017. Sucrose sensing through nascent peptide-meditated ribosome stalling at the stop codon of *Arabidopsis bZIP11* uORF2. *FEBS Lett* **591:** 1266–1277.

Young DJ, Guydosh NR, Zhang F, Hinnebusch AG, Green R. 2015. Rli1/ABCE1 recycles terminating ribosomes and controls translation reinitiation in 3′UTRs in vivo. *Cell* **162:** 872–884.

Záhonová K, Kostygov AY, Ševčíková T, Yurchenko V, Eliáš M. 2016. An unprecedented non-canonical nuclear genetic code with all three termination codons reassigned as sense codons. *Curr Biol* **26:** 2364–2369.

Zinoviev A, Hellen CU, Pestova TV. 2015. Multiple mechanisms of reinitiation on bicistronic calicivirus mRNAs. *Mol Cell* **57:** 1059–1073.

Cite this article as *Cold Spring Harb Perspect Biol* doi: 10.1101/cshperspect.a032656

Translation in Prokaryotes

Marina V. Rodnina

Department of Physical Biochemistry, Max Planck Institute for Biophysical Chemistry, Goettingen 37077, Germany

Correspondence: rodnina@mpibpc.mpg.de

This review summarizes our current understanding of translation in prokaryotes, focusing on the mechanistic and structural aspects of each phase of translation: initiation, elongation, termination, and ribosome recycling. The assembly of the initiation complex provides multiple checkpoints for messenger RNA (mRNA) and start-site selection. Correct codon–anticodon interaction during the decoding phase of elongation results in major conformational changes of the small ribosomal subunit and shapes the reaction pathway of guanosine triphosphate (GTP) hydrolysis. The ribosome orchestrates proton transfer during peptide bond formation, but requires the help of elongation factor P (EF-P) when two or more consecutive Pro residues are to be incorporated. Understanding the choreography of transfer RNA (tRNA) and mRNA movements during translocation helps to place the available structures of translocation intermediates onto the time axis of the reaction pathway. The nascent protein begins to fold cotranslationally, in the constrained space of the polypeptide exit tunnel of the ribosome. When a stop codon is reached at the end of the coding sequence, the ribosome, assisted by termination factors, hydrolyzes the ester bond of the peptidyl-tRNA, thereby releasing the nascent protein. Following termination, the ribosome is dissociated into subunits and recycled into another round of initiation. At each step of translation, the ribosome undergoes dynamic fluctuations between different conformation states. The aim of this article is to show the link between ribosome structure, dynamics, and function.

Translation is the last step in gene expression, during which the coding sequence of mRNA is translated into the amino-acid sequence of a protein. Translation is a highly dynamic process that entails four major phases: initiation, elongation, termination, and ribosome recycling. During each phase, ribosomes form transient complexes with auxiliary translation factors that facilitate protein synthesis. In addition to the compositional dynamics of translating ribosome complexes, conformational fluctuations of the ribosome and the translation factors play an important role in promoting the directionality of the process. A major challenge is to understand how the loosely coupled motions of the translational components lead to rapid and accurate protein production. Here, we summarize recent results of biochemical, biophysical, and structural work that dissected the order of events at each step of translation in bacteria, identified dynamic components, and captured structures of individual intermediates. Understanding the dynamics of ribosome complexes during translation could ultimately reveal how macromolec-

Cite this article as *Cold Spring Harb Perspect Biol* doi: 10.1101/cshperspect.a032664

ular machines navigate through the available conformational space and how their dynamics translates into function.

INITIATION

During translation initiation, the ribosome recruits an mRNA and selects the start codon of the open reading frame (ORF) (for recent reviews, see Milon and Rodnina 2012; Duval et al. 2015; Gualerzi and Pon 2015; see also Merrick and Pavitt 2018). In bacteria, translation initiation occurs cotranscriptionally, with the RNA polymerase (RNAP) and the ribosome physically interacting with each other (Kohler et al. 2017). The ribosome binds to the ribosome binding site (RBS) of the mRNA as soon as it emerges from the RNAP. Inhibition of translation leads to increased RNAP pausing, suggesting that transcription and translation are kinetically coupled (Landick et al. 1985; Proshkin et al. 2010). So far, almost nothing is known about the mechanism of initiation in the transcription–translation complex, a molecular machine denoted as the expressome (Kohler et al. 2017). Similarly, very little is known about initiation on mRNAs that are engaged in polysomes (Mitarai et al. 2008; Espah Borujeni and Salis 2016), as most of the mechanistic knowledge comes from studies that used free mRNAs not attached to the RNAP or to a preceding ribosome. Further studies are needed to determine whether initiation in expressomes or polysomes follows the same mechanism as initiation by the pioneering ribosome on free mRNA.

Among the different types of mRNAs found in prokaryotes, mRNAs containing the Shine–Dalgarno (SD) sequence are particularly well studied. They usually have an extended 5' untranslated region (5'UTR) and an SD sequence located 8–10 nt upstream of the start codon (usually AUG). During SD-led initiation, the small subunit ([SSU], 30S in bacteria) is recruited to the RBS through interactions between the SD sequence and the complementary anti-SD (aSD) sequence in 16S ribosomal RNA (rRNA). Initiation on SD-led mRNAs is promoted by initiation factors IF1, IF2, and IF3.

These bacterial factors display activities that resemble those of eIF1A, eIF2, and eIF1 in eukaryotes, respectively, but there is very little sequence homology between these prokaryotic and eukaryotic initiation factors. IF2 is homologous with eukaryotic initiation factor eIF5B. IF1 enhances the activities of IF2 and IF3. IF2 is a GTPase that recruits the initiator fMet-tRNA^{fMet}. IF3 interferes with subunit association, ensures the fidelity of fMet-tRNA^{fMet} selection over the elongator aminoacyl-tRNAs (aa-tRNAs), and helps to discriminate against mRNAs with unfavorable translation initiation regions (TIRs) (Milon and Rodnina 2012; Duval et al. 2015; Gualerzi and Pon 2015, and references therein).

However, not all mRNAs have an SD sequence. mRNAs lacking the SD sequence exist in most bacteria and archaea (Tolstrup et al. 2000; Weiner et al. 2000; Ma et al. 2002; Chang et al. 2006). The number of SD-led genes among 162 completed prokaryotic genomes varies from ~12% to 90%, suggesting a significant number of non-SD-led or leaderless mRNAs (Chang et al. 2006). Very little is known about initiation on non-SD-led mRNAs except that the 5'UTR is usually unfolded and the AUG start codon resides in a single-stranded mRNA region (Scharff et al. 2011). In archaea and some bacteria, internal ORFs of multicistronic mRNAs are more likely to have an SD sequence than the leading ORF; genes with an AUG start codon are more likely to have an SD sequence than those with GUG or UUG start codons (Ma et al. 2002; Chang et al. 2006).

Another group of mRNAs comprises leaderless mRNAs that lack a 5'UTR. Such mRNAs are widespread in a variety of bacteria (Zheng et al. 2011) and may play an important role in regulating the stress response (Grill et al. 2000; Vesper et al. 2011). A major determinant for leaderless initiation is the presence of an AUG start codon close to the 5' end of the mRNA (Krishnan et al. 2010). Leaderless mRNAs bind to 70S ribosomes directly; recruitment of fMet-tRNA^{fMet} is facilitated by IF2 and IF3 (Grill et al. 2000; Yamamoto et al. 2016). Whereas IF2 can bind in a similar way to either the 30S subunit or 70S ribosome (Goyal et al. 2015), IF3

Cite this article as *Cold Spring Harb Perspect Biol* doi: 10.1101/cshperspect.a032664

must move from its binding site on the 30S subunit on 50S subunit joining. Binding of IF3 to 70S ribosomes promotes their dissociation into subunits. This raises the question how IF3 can promote initiation on 70S ribosomes without splitting them into subunits. Recent results suggest that after dissociating from its 30S site on 50S subunit joining, IF3 may remain bound at the noncanonical binding site on the 50S subunit, which would allow the factor to act in leaderless initiation without promoting the dissociation of the 70S ribosome into subunits (Goyal et al. 2017). After translating the first ORF of a polycistronic mRNA, the ribosome can also reinitiate downstream at a second ORF using a 70S-scanning mechanism that requires fMet-tRNA^{fMet} and IF3 (Yamamoto et al. 2016).

Translation initiation on SD-led mRNAs in *Escherichia coli* proceeds through three main assembly intermediates (Fig. 1) (Milon and Rodnina 2012; Duval et al. 2015; Gualerzi and Pon 2015). The SSU, IF1, IF2, IF3, and fMet-tRNA^{fMet} form a labile 30S preinitiation complex (30S PIC). As soon as mRNA is recruited, start codon recognition converts the 30S PIC into the stable 30S initiation complex (30S IC). Joining of the large subunit ([LSU], 50S in bacteria) triggers the dissociation of the initiation factors, the accommodation of fMet-tRNA^{fMet} in the P site, and the formation of the mature

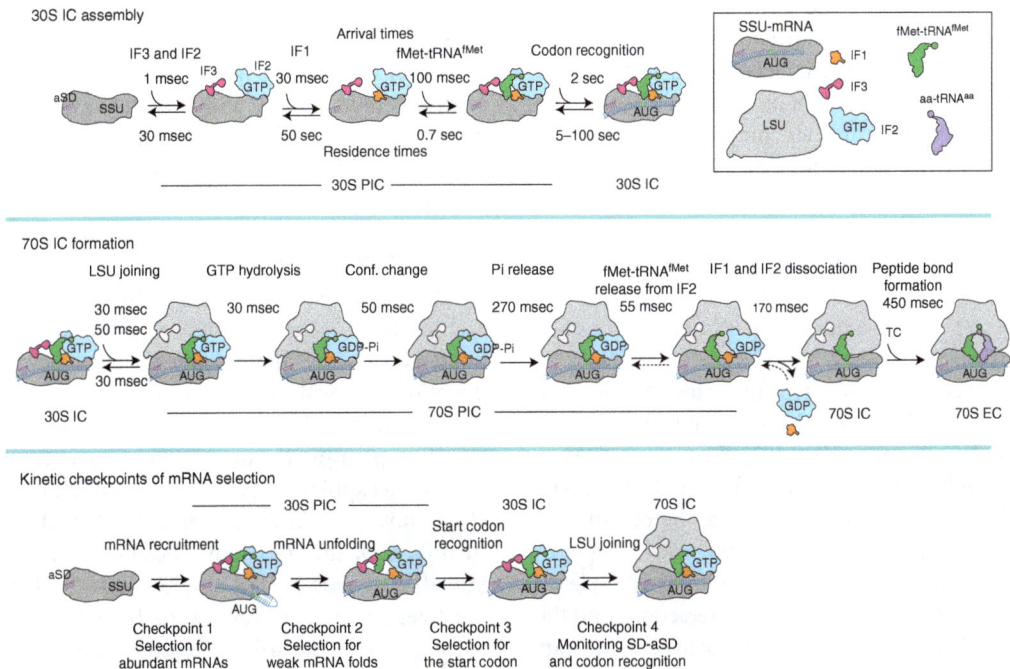

Figure 1. Kinetic model of translation initiation. (*Top*) Assembly of the 30S preinitiation complex (PIC) and 30S initiation complex (IC). Arrival times are calculated using experimentally measured bimolecular association rate constants and the in vivo concentrations of initiation factors in *E. coli*. Residence times are calculated from the measured dissociation rate constants of the individual components; mRNA binding is shown as a last step, but can occur at any step of the assembly pathway, independent of the presence of initiation factors or fMet-tRNA^{fMet}. Recognition of the start codon signifies the transition to the 30S IC (based on data in Milon et al. 2012). (*Middle*) Formation and maturation of the 70S IC. After subunit joining, IF3 may remain loosely bound to a site on the large subunit (LSU) (based on data in Goyal et al. 2017). (*Bottom*) Checkpoints of mRNA selection. From an mRNA-centric point of view, structured mRNAs can be recruited to the platform of the small subunit (SSU), unfold, and then accommodate in the mRNA-binding channel of the SSU (based on data in Milon et al. 2008 and Milon and Rodnina 2012).

70S IC, which is ready for translation elongation. The assembly pathway of the 30S PIC does not follow a strict order of factor addition. The factors can bind to the SSU independently of each other. However, there is a kinetically preferred sequence of factor association in the order IF3 and IF2, then IF1, followed by the recruitment of fMet-tRNA^{fMet} through IF2 (Fig. 1) (Milon et al. 2012). Occasionally, fMet-tRNA^{fMet} can form an IF2•GTP/fMet-tRNA^{fMet} complex (Tsai et al. 2012), but this complex does not constitute an obligatory delivery pathway for fMet-tRNA^{fMet} (Milon et al. 2010). The mRNA can bind to the SSU at any time, independent of the presence of the initiation factors (Studer and Joseph 2006; Milon et al. 2012). The association rate depends on the properties of the mRNA, such as the presence of secondary structures in the RBS, as well as the mRNA concentration (Studer and Joseph 2006). Codon recognition changes the conformation of the complex (Milon et al. 2008, 2012; Simonetti et al. 2008; Julian et al. 2011), stabilizes tRNA binding and destabilizes IF3 binding (Milon et al. 2012; Qin et al. 2012; Elvekrog and Gonzalez 2013; Hussain et al. 2016). IF3 changes its position on the ribosome in response to codon recognition (Hussain et al. 2016).

The next major step entails the LSU docking onto the 30S IC (Fig. 1). Rapid docking depends on the presence of IF1, IF3, IF2•GTP, and fMet-tRNA^{fMet} (Antoun et al. 2006; Milon et al. 2008; Goyal et al. 2015). In addition, the rate of subunit joining is attenuated by the mRNA depending on the sequence of the RBS, for example on the strength of the SD–aSD interactions and the length of the spacer between the SD and the start codon (Milon et al. 2008). After GTP hydrolysis by IF2, fMet-tRNA^{fMet} accommodates in the P site (Grigoriadou et al. 2007; Milon et al. 2008; Goyal et al. 2015). Displacement of IF3 from its 30S binding site and dissociation of IF1 and IF2 from the complex allows the ribosome to make intersubunit bridges and leads to formation of the mature 70S IC (Fig. 1) (Fabbretti et al. 2007; Chen et al. 2015; Goyal et al. 2015, 2017; Liu and Fredrick 2015; MacDougall and Gonzalez 2015). The irreversible steps of start-codon rec-

ognition and GTP hydrolysis promote conformational changes of the 30S subunit and induce rotation of the two subunits relative to each other (Allen et al. 2005; Myasnikov et al. 2005; Marshall et al. 2009; Julian et al. 2011; Coureux et al. 2016; Sprink et al. 2016).

A key question is which features of the mRNA determine its translational efficiency. In bacteria, the RBS spans nucleotides −20 to +15 around the translation start codon. Translational efficiency is modulated by the nature of the codon used for initiation (AUG, GUG, or UUG), the SD sequence and the spacer between the SD sequence and the start codon, the mRNA secondary structure near the start site, and A/U-rich elements in the mRNA that are recognized by the SSU protein bS1. bS1, which is the largest and most acidic ribosomal protein, is required for the binding and unfolding of structured mRNAs (Duval et al. 2013; Byrgazov et al. 2015). The relative contribution of each specific element is not clear. The available on-line tools used to estimate translational efficiency from the thermodynamic properties of the RBS yield predictions that are quite good for engineered mRNAs (Salis et al. 2009; Kosuri et al. 2013; Reeve et al. 2014; Bonde et al. 2016). However, in natural mRNAs, each element of the RBS alone appears to have limited effect and can modulate the efficiency of initiation only within a certain context.

A more holistic approach conceptualizes the initiation pathway as comprising a sequence of kinetic checkpoints (Fig. 1) (Milon and Rodnina 2012; Duval et al. 2015; Gualerzi and Pon 2015). In this view, the initiation efficiency is determined by kinetic partitioning between the forward steps on the pathway toward the mature 70S IC, and the backward or rejection steps. The structure and thermodynamic stability of the RBS affect the association (step 1) and unfolding (step 2) of the mRNA. The identity of the start codon determines the stability of the codon–anticodon complex (step 3). Finally, the overall conformation of the 30S IC, which is modulated by the sequence context of the RBS, defines the rate of LSU joining (step 4). The kinetic model can explain any variations in the translational efficiency of different mRNAs. If the rate constants of the elemental steps are known, the

Cite this article as *Cold Spring Harb Perspect Biol* doi: 10.1101/cshperspect.a032664

translational efficiency can be predicted. In the few cases where such measurements were possible, the calculated value matched well with the directly measured translational efficiency (Milon et al. 2008, 2012). However, for most mRNAs the elemental rate constants are unknown, which hinders the use of the kinetic parameters as descriptors in global bioinformatics analysis. Although the mechanism of translation initiation is generally quite different in pro- and eukaryotes, the principles of kinetic partitioning most likely play a major role in start-site selection in eukaryotes as well (see Sokabe and Fraser 2018).

ELONGATION

Elongation entails repetitive cycles of decoding, peptide bond formation, and translocation. Elongation begins as soon as the second codon of the ORF becomes accessible for reading by elongator aa-tRNAs and ends when the ribosome arrives at the stop codon. The basic mechanism of elongation is very similar in prokaryotes and eukaryotes (see Dever et al. 2018) and is facilitated by homologous translation factors (EF-Tu/eEF1A, EF-G/eEF2, EF-P/eIF5A, SelB/ EFsec), with some notable additions, such as eEF3, which is found in fungi.

Decoding

During decoding, the ribosome translates the sequence of codons in an mRNA into the amino acid sequence of a protein. A codon exposed in the A site is recognized by aa-tRNAs, which in bacteria are delivered to the ribosome in a ternary complex with EF-Tu and GTP. The initial recruitment of the EF-Tu•GTP/aa-tRNA complex occurs through the interactions with the bL12 stalk of the ribosome (Kothe et al. 2004; Diaconu et al. 2005). Interaction of the aa-tRNA anticodon with the mRNA codon in the decoding site of the SSU triggers GTP hydrolysis by EF-Tu. After Pi release, EF-Tu rearranges into the guanosine diphosphate (GDP)-bound form and releases the aa-tRNA. The aa-tRNA accommodates in the A site of the LSU, while EF-Tu•GDP dissociates from the ribosome (Rodnina et al. 2017).

The structures of several key intermediates of decoding, which were initially identified by biochemical and biophysical studies, have been determined by cryoelectron microscopy (cryo-EM). Currently, a sequence of snapshots of cognate decoding is available for EF-Tu•GTP/ Phe-tRNA^{Phe} (Loveland et al. 2017) and SelB•GTP/Sec-tRNA^{Sec} (Fischer et al. 2016). In contrast to EF-Tu, which is a general translation factor that directs every elongator aa-tRNA to the A site, SelB is a specialized elongation factor that is responsible for the delivery of the 21st natural proteinogenic amino acid, selenocysteine (Forchhammer et al. 1989). Whereas the two cryo-EM structures capture somewhat different intermediates on the decoding pathway, and some details appear specific for EF-Tu or SelB, the overall sequence of rearrangements is remarkably similar. Both reports (Fischer et al. 2016; Loveland et al. 2017) identify an early decoding intermediate where the ternary complex is bound to the SSU, but the anticodon does not yet base-pair with the codon (Fig. 2). These structures have an open SSU-domain conformation similar to or even more open than in ribosomes with a vacant A site (Fischer et al. 2016; Loveland et al. 2017). The 16S rRNA residues forming the local core in the SSU decoding center point away from the codon–anticodon complex, although there seem to be subtle differences in the orientation of different nucleotides between the EF-Tu and SelB structures. In both structures, the key residue A1492 is in the "flipped-out" (Ogle et al. 2001) open form oriented away from the codon–anticodon complex. A1493 is in a semi-open orientation in the ribosome–EF-Tu complex, but in an open orientation in the SelB complex; also the mobility of G530 of 26S rRNA appears to differ between the two complexes (Fischer et al. 2016; Loveland et al. 2017).

The second intermediate of the SelB complex captures an early decoding state in which only a single potential base pair is formed between codon and anticodon, whereas in the EF-Tu-bound complex the anticodon is fully base-paired with the codon; in both cases, the ribosome remains in an open conformation. In the EF-Tu complex, the GTP-binding pocket remains distant from the LSU (Loveland et al.

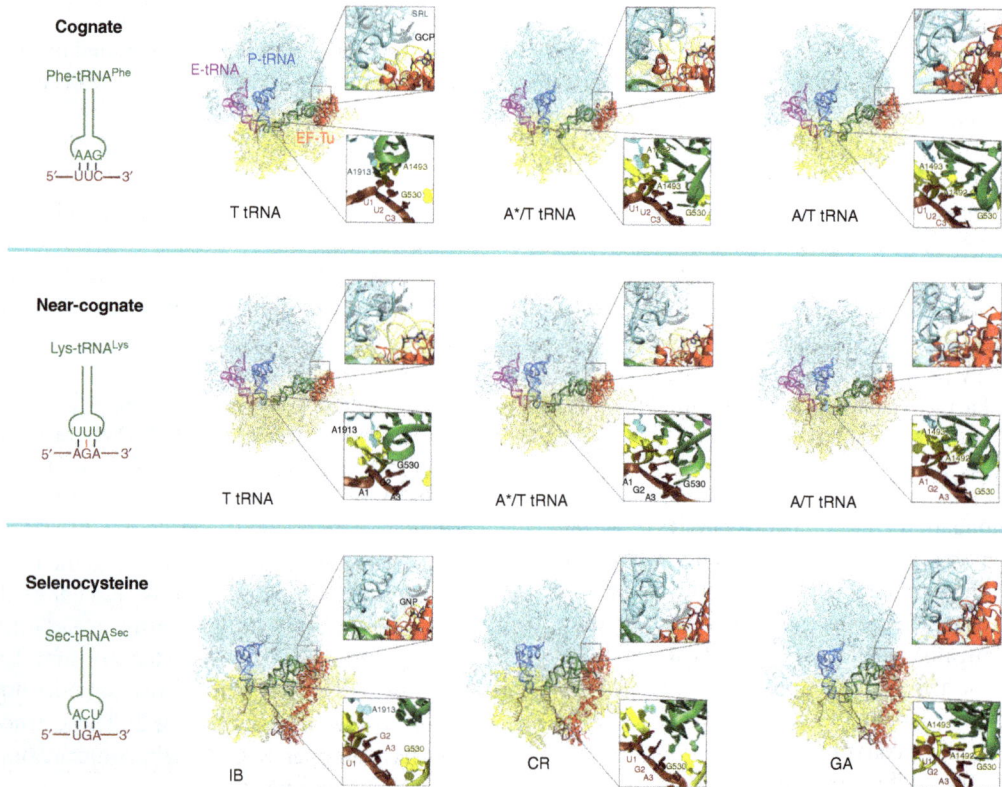

Figure 2. Structural mechanism of decoding as visualized by cryoelectron microscopy (cryo-EM). (*Top*) Intermediates of cognate decoding by elongation factor (EF)-Tu•GDP–Phe-tRNA[Phe]. (*Left*) Schematic of the cognate codon–anticodon interaction between the UUC mRNA codon and the AAG anticodon of Phe-tRNA[Phe]. Other panels show decoding intermediates from the T state prior to codon reading, A*/T, where the codon has been recognized but EF-Tu did not move onto the sarcin-ricin loop (SRL) of the SSU and A/T state with the correct codon–anticodon interaction and EF-Tu docked on the SRL. *Insets* on *top* show the orientation of the G-domain of EF-Tu relative to the SRL. GCP, nonhydrolyzable GTP analog GDPCP. *Insets* at the *bottom* show the codon–anticodon complex and the key residues of 16S ribosomal RNA (rRNA) interacting with it. (*Middle*) Same as above for a near-cognate pair with a single G U mismatch in the second position of the codon anticodon complex. (*Bottom*) Intermediates of cognate decoding by SelB•GDPNP/Sec-tRNA[Sec]. GNP, nonhydrolyzable GTP analog GDPNP; IB, initial binding prior to codon reading; CR, codon reading complex in which the anticodon of the tRNA comes into the proximity of the codon, but prior to base pairing; GA, GTPase-activated complex analogous to the A/T state. (Figure was prepared using structure coordinates from Fischer et al. 2016 and Loveland et al. 2017, PDB 5UYK, 5UYL, 5UYM, 5UYN, 5UYP, 5UYQ, 5LZB, 5LZC and 5LZD.)

2017), and thus the GTPase activity of EF-Tu, which requires an interaction of EF-Tu with the sarcin-ricin loop (SRL) of the 23S rRNA as a GTPase-activating element, remains low (Maracci et al. 2014). In contrast to EF-Tu, SelB interacts with the LSU in both intermediates prior to codon recognition, but the contact with the SRL is blocked by Sec-tRNA[Sec] (Fischer et al. 2016). Interestingly, a similar protective

interaction of aa-tRNA with the SRL was identified among decoding intermediates of eukaryotic translation, where it blocks the access of eEF1A to the GTPase-activating center (Budkevich et al. 2014).

After the codon–anticodon interaction has been established, the 16S rRNA residues at the decoding center change their orientation to the "flipped-in" (Ogle et al. 2001) conformation and

Cite this article as *Cold Spring Harb Perspect Biol* doi: 10.1101/cshperspect.a032664

close on the codon–anticodon complex (Fischer et al. 2016; Loveland et al. 2017), consistent with the local rearrangements inferred from comparisons of SSU structures with or without A-site tRNA anticodon stem-loops (ASLs) (Ogle et al. 2001). In the EF-Tu complex, G530 appears to act as a latch that fastens the codon–anticodon helix into the decoding center. The local closure of the decoding center coincides with the SSU domain closure, which drags the tRNA and EF-Tu toward the SRL. The magnitude of the conformational changes that take place on domain closure (Fischer et al. 2016) appears to be even larger than those previously reported for the SSU–ASL complex (Ogle et al. 2001). The aa-tRNA becomes distorted on codon recognition (Valle et al. 2003; Schmeing et al. 2009; Schmeing et al. 2011; Fischer et al. 2016; Loveland et al. 2017). In solution, tRNA can adopt such distorted tRNA conformations spontaneously within less than a microsecond (Fischer et al. 2016). The ribosome seems to stabilize specific subsets of conformations in a given state, depending on the interactions at the decoding center. Docking of the G-domain of EF-Tu activates GTP hydrolysis, the irreversible step that separates initial selection from the subsequent proofreading step.

GTP hydrolysis in EF-Tu and other translational GTPases relies on the universally conserved His residue (His84 in *E. coli* EF-Tu) in the Switch II region and on the conserved Asp residue (Asp21 in EF-Tu) in the P loop (Maracci et al. 2014). The SRL of the LSU acts as GTPase activator (Wool et al. 1992; Schmeing et al. 2009). The interaction of the SRL with His84 shifts the pK_a value of His84 in such a way that the side chain becomes positively charged at neutral pH and positions the nucleophilic water molecule for the attack on the γ-phosphate of GTP (Adamczyk and Warshel 2011; Wallin et al. 2013; Aqvist and Kamerlin 2015). Computer simulations favor the reaction mechanism with an early proton transfer from water to the γ-phosphate, followed by nucleophilic attack by a hydroxide ion, a scenario that appears to be consistent with the lack of pH-dependence of GTP hydrolysis at near-neutral pH and a negligible kinetic solvent isotope effect (Maracci et al. 2014). The conserved Asp21 complexed to Mg^{2+}

may contribute to the acceleration of GTP hydrolysis by "pushing" the negative charge toward His84 (Aqvist and Kamerlin 2015). Thus, GTP hydrolysis is primarily governed by the electrostatics of the reaction center (Adamczyk and Warshel 2011; Prasad et al. 2013; Maracci et al. 2014; Aqvist and Kamerlin 2015). Also, ribosomal protein bL12 contributes to the GTPase activation through an as-yet-undetermined mechanism (Mohr et al. 2002; Diaconu et al. 2005). The mechanism of GTP hydrolysis is likely to be conserved in all translational GTPases, such as EF-Tu, EF-G, SelB, IF2, and RF3 and their eukaryotic homologs.

During decoding, the ribosome has to select an aa-tRNA cognate to the given codon from the pool of different aa-tRNAs. The fidelity of aa-tRNA selection is high, with error frequencies of 10^{-3} or less (Drummond and Wilke 2009). One important unresolved question is how the ribosome responds to codon–anticodon mismatches. The early model based on the comparison of the 30S–mRNA complexes with cognate or near-cognate ASLs suggested that mismatches disturb SSU domain closure (Ogle et al. 2002). However, high-resolution crystal structures of non- and near-cognate ribosome–mRNA–tRNA complexes all showed an identical local and global arrangement of the SSU (Demeshkina et al. 2012; Rozov et al. 2015, 2016a,b). Similarly, in the cryo-EM structure, near-cognate EF-Tu•GTP/Lys-tRNA (anticodon UUU, which makes a second position mismatch to the AGA codon) induced the same local and global conformational changes as EF-Tu•GTP/Phe-tRNA^{Phe} on the cognate UUC codon (Loveland et al. 2017). However, only a minor portion of the near-cognate ternary complexes adopts the GTPase-activated state, whereas the majority is rejected at the initial selection stage prior to GTP hydrolysis (Rodnina et al. 2017). Thus, it is possible that the available structures of the near-cognate complexes provide snapshots of translational errors, rather than of regular decoding intermediates. The visualized complexes represent those that passed the selection screens of the ribosome and will result in the incorporation of an incorrect amino acid into the protein. In contrast, the structures of the rejected majority of near-cognate complexes may be similar to that of the complexes prior to codon–anticodon pairing. Notably, such complexes must be even less stable than the pre-codon-recognition intermediates of the cognate complexes, as they appear too transient to be captured by crystallography or cryo-EM. It should be emphasized that the structures of near-cognate complexes are extremely valuable, as they show how tautomerization or the presence of tRNA modifications help the mismatched complexes to adopt the geometry that the ribosome recognizes as correct.

After GTP hydrolysis and a slightly delayed Pi release (Kothe and Rodnina 2006), EF-Tu rearranges into the GDP-bound conformation. Molecular-dynamic simulations suggested that this may propel the $3'$ end of aa-tRNA toward the A site on the LSU into a position where the tRNA elbow interacts with H89 of the LSU (Noel and Whitford 2016). Aa-tRNA may fluctuate between the two conformational states until EF-Tu is released. The accommodation of the tRNA $3'$ end in the A site constitutes a separate step (Geggier et al. 2010). This notion is supported by recent biochemical data (Ieong et al. 2016; Ranjan and Rodnina 2017), although the two steps are not kinetically resolved for every tRNA. The accommodation is slower for near-cognate, compared to cognate aa-tRNA, which provides a second chance to reject a tRNA with mismatches in the codon–anticodon complex in the proofreading stage (Rodnina et al. 2017).

Peptide Bond Formation

In the peptidyl transferase center of the ribosome, peptidyl-tRNA in the P site and aa-tRNA in the A site react to form a peptide bond. In comparison with the reaction between model substrates in solution, the reaction on the ribosome is accelerated about 10^7-fold (Sievers et al. 2004). The ribosome's active site is comprised of rRNA (Ban et al. 2000; Polikanov et al. 2014), and thus the ribosome is the largest known RNA-catalyst and the only known natural ribozyme that has polymerase activity. Unlike protein enzymes, the ribosome does not provide catalytic groups with pK_a values at neutral pH (Youngman et al. 2004; Bieling et al. 2006).

The catalysis is mainly entropic (Sievers et al. 2004). The ribosome facilitates the reaction by ordering water molecules, positioning of rRNA and tRNA residues, and electrostatic shielding (Sharma et al. 2005; Wallin and Aqvist 2010).

Peptide bond formation proceeds by nucleophilic attack of the amino group of aa-tRNA on the carbonyl carbon of the ester bond in peptidyl-tRNA. In solution, the reaction is expected to have two intermediates, a zwitterionic tetrahedral intermediate ($T^{\pm}$), that is deprotonated and forms a second intermediate (T^{-}), and then decomposes to form the reaction products (Satterthwait and Jencks 1974). A comprehensive analysis of heavy-atom kinetic isotope effects indicated that the ribosome alters the reaction pathway in such a way that the $T^{\pm}$ intermediate no longer accumulates. Proton transfer from the attacking nitrogen and formation of the tetrahedral intermediate take place during the rate-limiting step, while tetrahedral intermediate breakdown happens in a separate rapid step (Hiller et al. 2011). Analysis of kinetic solvent isotope effects showed that in the rate-limiting transition state three protons move in a concerted manner (Kuhlenkoetter et al. 2011). Notably, there are several water molecules within the peptidyl transferase center that can exchange protons (Polikanov et al. 2014). Also the 2′OH group of A76 of the P-site tRNA (Zaher et al. 2011) and the 2′OH of A2451 of 23S rRNA (Erlacher et al. 2006) contribute to proton transfer and/or stabilize the charges developing as the reaction proceeds.

Currently there are two models that account for the movement of protons in the active site during peptide bond formation (Fig. 3). One model, referred to as the eight-membered proton shuttle (Wallin and Aqvist 2010), suggests that the attack of the α-amino group on the ester carbonyl carbon results in an eight-membered transition state in which a proton from the α-amino group is received by the 2′OH group of A76, which at the same time donates its proton to the carbonyl oxygen by way of an adjacent water molecule (Kuhlenkoetter et al. 2011). Protonation of the 3′OH is an independent rapid step. The alternative model, referred to as the "proton wire" model, suggests that one of the

Eight-membered proton shuttle

Proton wire

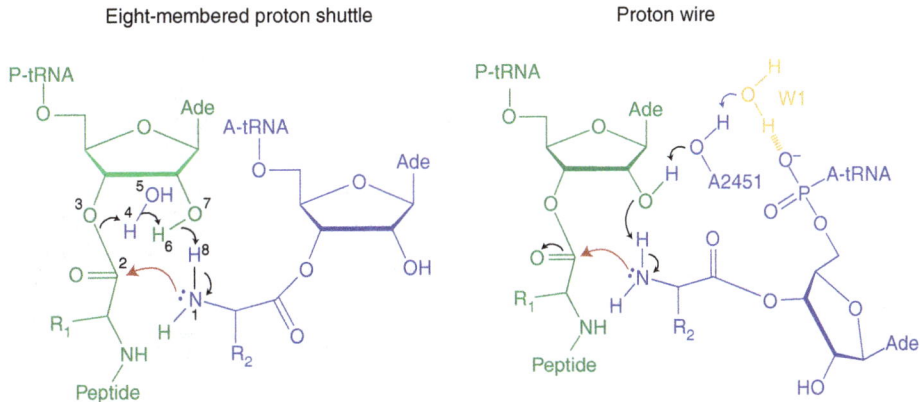

Figure 3. Models for proton transfer during peptide bond formation. Reaction schemes are shown for the eight-membered proton shuttle and proton wire mechanisms. P-site tRNA is shown in green, A-site tRNA in blue, and 2′OH of A2451 in black. The nucleophilic attack is depicted by red arrows. In the eight-membered proton shuttle, the attack of the α-amino group on the ester carbonyl carbon results in an eight-membered rate-limiting transition state in which a proton from the α-amino group is received by the 2′OH group of A76, which at the same time donates its proton to the carbonyl oxygen via an adjacent water molecule (Kuhlenkoetter et al. 2011). In the proton wire model, the proton from the α-amino group is received by the 2′OH group of A76, which in turn donates a proton to the 2′OH of A2451, and then to a water molecule (W1), which is partially negatively charged (Polikanov et al. 2014). Both models account for the concerted movement of three protons in the rate-limiting transition state (Kuhlenkoetter et al. 2011). (From Polikanov et al. 2014; adapted, with permission, from Nature Publishing Group under a Creative Commons license.)

water molecules residing in the peptidyl transferase center is partially negatively charged by the vicinity of the deprotonated amino-terminal α-amino group of ribosomal protein bL27 and of the negatively charged 5′-phosphate oxygen of the A-site A76 (Polikanov et al. 2014). Upon formation of a tetrahedral intermediate, the emerging positive and negative charges may become separated in space and delocalized over the pockets containing water molecules. The two models account for the three protons moving in a concerted manner in the rate-limiting transition state, but disagree on the exact proton transfer pathway. One argument against the proton wire model (Polikanov et al. 2014) is that deletion of bL27 has no effect on peptide bond formation (Maracci et al. 2015). Furthermore, the role of the 2′OH of A2451 of the 23S rRNA, which plays an essential role in the proton wire model, has not been tested at conditions of rapid peptide bond formation (Erlacher et al. 2006). On the other hand, the proton shuttle model has a less optimal stereochemistry than the proton wire model (Polikanov et al. 2014).

The reactivities of natural amino acids in the peptidyl transferase reaction differ substantially (Wohlgemuth et al. 2008). Nevertheless, the ribosome can make peptides with most amino acid combinations without the help of any additional auxiliary factors. One notable exception is the synthesis of poly-Pro stretches with three or more consecutive prolines or of distinct XPPX sequences with two prolines flanked by specific amino acids (Hersch et al. 2013; Peil et al. 2013; Woolstenhulme et al. 2013). During synthesis of such peptides the ribosome stalls because of a low rate of peptide bond formation (e.g., for the PPP motif the ribosome is stalled after incorporation of the second Pro) (Ude et al. 2013). The stalling is alleviated by EF-P (Doerfel et al. 2013; Ude et al. 2013), a specialized translation factor that enters the E site of the ribosome and acts by entropic steering of the P- and A-site substrates toward their catalytically productive orientation in the peptidyl transferase center (Fig. 4) (Doerfel et al. 2015; Huter et al. 2017). The eukaryotic homolog of EF-P, eIF5A, also accelerates the formation of poly-Pro chains

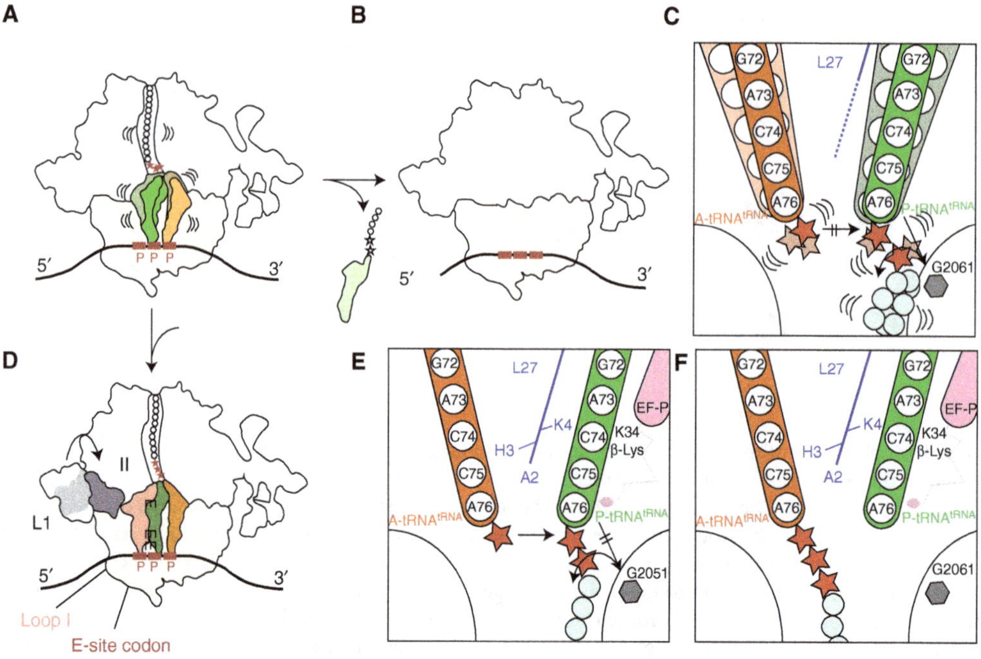

Figure 4. Action of elongation factor P (EF-P) on ribosomes stalled at polyproline stretches. (*A*) Ribosomes stall during translation of proteins containing three consecutive prolines (red stars). Binding of the peptidyl-tRNA (green) to the P site is destabilized, which (*B*) can lead to peptidyl-tRNA drop-off. (*C*) The all-*trans* or all-*cis* conformations of polyprolines of the nascent chain are not possible because of a steric clash with G2061 (gray) within the tunnel wall. Peptidyl-tRNA is destabilized and prevents accommodation of the A-site tRNA (orange) and peptide bond formation. (*D*) Ribosomes stalled on polyproline stretches are recognized by EF-P (pink), which binds within the E-site region and stabilizes the peptidyl-tRNA. EF-P binding is facilitated via contacts with the L1 stalk and the P-site tRNA as well as E-site codon. (*E*) Interaction of the ε(R)-β-lysyl-hydroxylysine with the CCA end of P-site tRNA^{Pro} stabilizes the P-site tRNA as well as the nascent chain, by forcing the prolines to adopt an alternative conformation that passes into the ribosomal exit tunnel. (*F*) Thus, an optimal geometry between the nascent chain and the aminoacyl-tRNA in the A site is achieved and peptide bond formation can occur. (From Huter et al. 2017; adapted, with permission, from Elsevier © 2017.)

(Gutierrez et al. 2013). eIF5A appears to have additional functions in elongation and termination (Schuller et al. 2017; also see Dever et al. 2018). EF-P does not have such a broad functionality, because EF-P recognizes the tRNA in the P site and favors interactions with tRNA^{Pro} and tRNA^{fMet}, but is considerably less active with other tRNAs (Katoh et al. 2016).

EF-P and eIF5A are posttranslationally modified (for review see Lassak et al. 2016). The modification is essential for function (Doerfel et al. 2013; Gutierrez et al. 2013; Ude et al. 2013), but varies between different groups of bacteria and eukaryotes. In *E. coli*, Lys34 of EF-P is posttranslationally modified to a lysyl-lysine (Yanagisawa et al. 2010). In *Pseudomonas aeruginosa* and *Shewanella oneidensis* Arg32 (which occupies the position of *E. coli* Lys34) is rhamnosylated (Lassak et al. 2015), whereas EF-P from *Bacillus subtilis* bears a 5-aminopentanol moiety attached to Lys32 (Rajkovic et al. 2016). In eukaryotes, the conserved lysine residue of eIF5A is modified to hypusine (Cooper et al. 1983).

Translocation

After peptide bond formation, the ribosomal subunits move relative to each other, from the nonrotated (N) state with the two tRNAs bound to the P and A sites both on the SSU and LSU, to the rotated (R) state with the tRNAs bound in hybrid P/E and A/P states. At the same time, the uL1 stalk moves from an open to a closed conformation toward the P-site tRNA (for review, see Frank and Gonzalez 2010). Ribosomes in the pretranslocation state (PRE) fluctuate between N and R states. Molecular dynamics simulations suggest that ribosome rotation and the concomitant internal movement of the SSU head and body domains (referred to as head swiveling) are intrinsically very rapid and can occur in the microsecond time scale (Bock et al. 2013). However, tRNA translocation does not occur spontaneously because the interactions of the two tRNAs with the ribosome restrict the movement. In the posttranslocation (POST) state the ribosomes are predominantly in the N state with tRNAs in their classical positions.

Translocation is promoted by EF-G at the cost of GTP hydrolysis. Movement of the tRNAs and the mRNA during translocation is a multistep process (Fig. 5). Current models distinguish up to eight discrete steps based on structural information as well as ensemble and single-molecule kinetic studies using a large variety of fluorescence reporters and fluorescence resonance energy transfer (FRET) pairs (Guo and Noller 2012; Adio et al. 2015; Belardinelli et al. 2016a; Sharma et al. 2016b; Wasserman et al. 2016). In

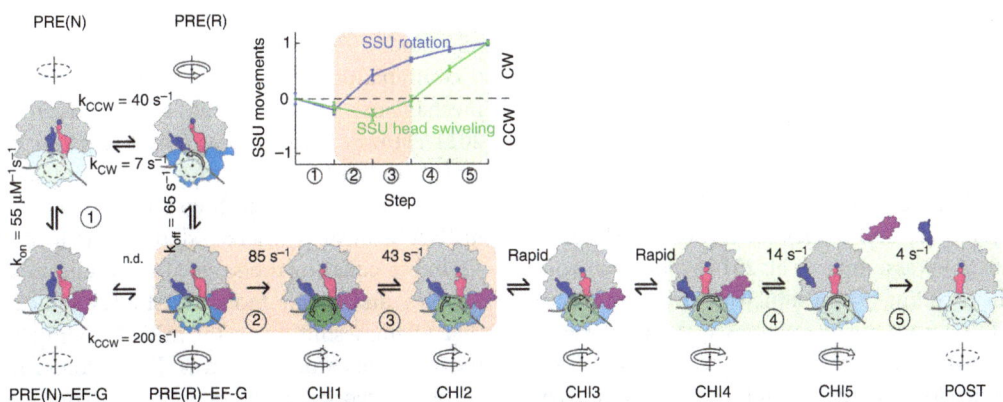

Figure 5. Kinetic model of translocation. The rotation states of the small subunit (SSU) relative to the large subunit (LSU) (gray) are indicated by color intensity of the SSU body (light blue for nonrotated, N; dark blue for rotated, R). The swiveling motions of the SSU head are depicted by a color gradient from light green (classical nonswiveled SSU head position) to forest green (maximum degree of SSU head domain swiveling relative to the SSU body domain) (Belardinelli et al. 2016a). Peptidyl- and deacylated tRNA in the pretranslocation state (PRE) complex are shown in magenta and blue, respectively. EF-G (purple) is shown in two conformations, a compact (Lin et al. 2015) and an extended one after engagement with the ribosome (Ramrath et al. 2013; Zhou et al. 2014). The rates of transitions between PRE(N) and PRE(R) and PRE(N)–EF-G and PRE(R)–EF-G (k_{CW} and k_{CCW} at 37°C) are modified from data in Sharma et al. (2016b). The rate constants for the kinetically defined steps 1, 2, 3, 4, and 5 are from ensemble kinetics studies at 37°C (Belardinelli et al. 2016a). Step 1, EF-G binding. Step 2, GTP hydrolysis (Rodnina et al. 1997; Savelsbergh et al. 2003) and opening of the SSU because of the opposite movements of the SSU head and body domains (see *inset*) (Belardinelli et al. 2016a). Step 3, unlocking of the tRNA–mRNA complex on the SSU, which is rate-limiting for tRNA movement and Pi release from EF-G (Savelsbergh et al. 2003). The existence of rapid transitions between steps 3 and 4 was shown using stalled translocation intermediates (Savelsbergh et al. 2003; Holtkamp et al. 2014). Translocation intermediates (CHI1 to CHI4) are adopted from smFRET data (Adio et al. 2015). The posttranslocation (POST) state may entail further conformational substates (Wasserman et al. 2016). The light red background indicates complexes undergoing unlocking; the light green background shows complexes that move toward relocking. (*Inset*) Distinct timing of counterclockwise (CCW) and clockwise (CW) movements of the SSU body relative to LSU (blue symbols) and of the SSU head (green symbols) (Belardinelli et al. 2016a). (From Belardinelli et al. 2016b; with permission from Taylor & Francis and Creative Commons Public Domain licensing.)

addition to PRE and POST states, the ribosome can adopt several intermediate conformations that differ in the positions of the tRNAs with respect to the SSU head and body domains and the A- and P-site loops on the LSU. The tRNA positions correlate with the degree of subunit rotation and the SSU head domain swiveling. These states are referred to as chimeric (CHI) states (e.g., ap/P and pe/E [Ramrath et al. 2013]; ap/ap [Zhou et al. 2014]; or noncanonical states identified by smFRET [Chen et al. 2011; Adio et al. 2015; Wasserman et al. 2016]).

After EF-G binding to the ribosome, the SSU head and body domains move in the same counterclockwise (CCW) direction relative to the LSU, which is referred to as forward, because it corresponds with the direction of translocation (Fig. 5) (Guo and Noller 2012; Belardinelli et al. 2016b; Wasserman et al. 2016). EF-G hydrolyzes GTP, but retains the Pi (Savelsbergh et al. 2003, 2005). Then, the SSU body begins moving backward in the clockwise (CW) direction, whereas the SSU head remains in the forward-swiveled state (Guo and Noller 2012; Belardinelli et al. 2016b; Wasserman et al. 2016). This may open the decoding region sufficiently to uncouple the tRNAs from the interactions with the ribosome elements that hold the mRNA and the tRNA anticodons in the A and P site, respectively. This explains how CHI states are formed: while the tRNA positions on the SSU head domain are retained, the SSU body moves, leading to the displacement of the tRNAs to the CHI states.

After the unlocking of the codon–anticodon complexes from the SSU, the SSU head domain starts to move backward (Guo and Noller 2012; Belardinelli et al. 2016b; Wasserman et al. 2016). The tRNAs adopt their canonical POST positions in P and E sites and EF-G releases Pi (Savelsbergh et al. 2003). Next, the E-site tRNA moves further away from the P-site tRNA, which is accompanied by the loss of the E-site codon–anticodon interaction, while the SSU head moves further backward (Adio et al. 2015; Belardinelli et al. 2016a). Finally, the dissociation of E-site tRNA and EF-G restores the N state with a classical P/P position of the peptidyl-tRNA. The role of EF-G is to accelerate and stabilize the R state of the ribosome, to unlock the tRNA–mRNA complex from the SSU, and to ensure the forward movement of the tRNAs by inserting EF-G domain 4 into the A site on the SSU. In other words, EF-G acts as a ratchet to rectify the Brownian motions of the ribosome into directed movement and promotes a key conformational rearrangement of the SSU that disrupts its interactions with the tRNAs, thereby allowing for rapid tRNA translocation (for review see Belardinelli et al. 2016b).

Translocation and Recoding

While moving along the mRNA the ribosome may encounter structures such as stem-loops or pseudoknots. In most cases, the ribosome helicase composed of ribosomal proteins uS3, uS4, and uS5 ensures mRNA unwinding by a combination of active and passive helicase mechanisms (Takyar et al. 2005; Qu et al. 2011). However, mRNA secondary structure elements in combination with slippery sequences can lead to −1 frameshifting. Slippery sites are mRNA sequences where codons in the 0- and −1-frame code for the same tRNAs. Ensemble kinetics and single-molecule studies indicate that −1 frameshifting on the mRNA coding for the infectious bronchitis virus (IBV) protein 1a/1b (which works in vivo in both eukaryotic and bacterial systems [Napthine et al. 2003]) or *E. coli dnaX* occurs during translocation of two tRNAs attached to the slippery site codons (Caliskan et al. 2014; Chen et al. 2014; Kim et al. 2014; Yan et al. 2015). Recoding in eukaryotes is described in Dever et al. (2018). The secondary structure element disturbs the ribosome dynamics during translocation. While early steps of translocation proceed with unperturbed efficiency, the backward swiveling of the SSU head domain is impeded by the structure obstacle in the mRNA (Caliskan et al. 2014; Kim et al. 2014). Ribosomes that are stalled in an unlocked state are apparently less stringent in maintaining the reading frame and can slip in the −1 direction.

Cotranslational Protein Folding

During elongation, the nascent peptide moves through the polypeptide exit tunnel of the ribo-

Cite this article as *Cold Spring Harb Perspect Biol* doi: 10.1101/cshperspect.a032664

some, which spans from the peptidyl transferase center to the cytoplasmic surface of the LSU (Fig. 6). The exit tunnel is about 100 Å long and can accommodate 30 or more amino acids, depending on the structure of the peptide, before the nascent chain emerges from the ribosome. Emerging nascent proteins can start folding within the tunnel, which can accommodate secondary structure elements or even small domains (for reviews see Balchin et al. 2016; Rodnina 2016; Thommen et al. 2017). While traveling through the exit tunnel, the peptide can adopt a nonnative compact structure, which is rearranged to a native form when the whole domain emerges from the exit port (Holtkamp et al. 2015). The ribosome affects protein folding by imposing a vectorial folding pathway and by coupling folding to the pace of translation. Initial folding events are restricted by the confined space within the exit tunnel and are further modulated by interactions between the peptide emerging from the exit tunnel and the surface of the LSU. The relationship between the rate of translation and the direction of protein folding is quite complex. The presence of rare codons in the mRNA slows down translation, alters the kinetics of cotranslational folding, and changes the distribution of protein conformations in the resulting mature protein pool (Clarke and Clark 2008; Tsai et al. 2008; Zhang et al. 2009; Siller et al. 2010; Spencer et al. 2012; Yu et al. 2015; Buhr et al. 2016; Sharma et al. 2016a). Computer

modeling suggests that changes of the translation rate can coordinate local folding rates and induce or prevent misfolding (O'Brien et al. 2014; Trovato and O'Brien 2017). Cotranslational protein folding is also affected by the proteins that bind in the vicinity of the exit port, such as the chaperone trigger factor (Gloge et al. 2014; Balchin et al. 2016).

TERMINATION

Termination occurs when the ribosome encounters a stop codon in the mRNA. In bacteria, stop codons are recognized by the termination (or release) factors RF1 and RF2, which read the codons UAG/UAA and UGA/UAA, respectively. Another termination factor, RF3, facilitates turnover of RF1 and RF2 but is not required for peptidyl-tRNA hydrolysis. Crystal and cryo-EM structures as well as smFRET studies show that RF1 and RF2 stabilize the ribosome in the N state (Rawat et al. 2003, 2006; Korostelev et al. 2008, 2010; Weixlbaumer et al. 2008), whereas RF3 alone stabilizes the R state (Gao et al. 2007; Jin et al. 2011). When RF3 binds to the RF1–ribosome complex, RF1 changes its position and RF3 adopts a different conformation than in complexes with just one factor bound (Pallesen et al. 2013).

The mechanism of termination entails three steps: recognition of the stop codon, hydrolysis of the ester bond of the peptidyl-tRNA (these

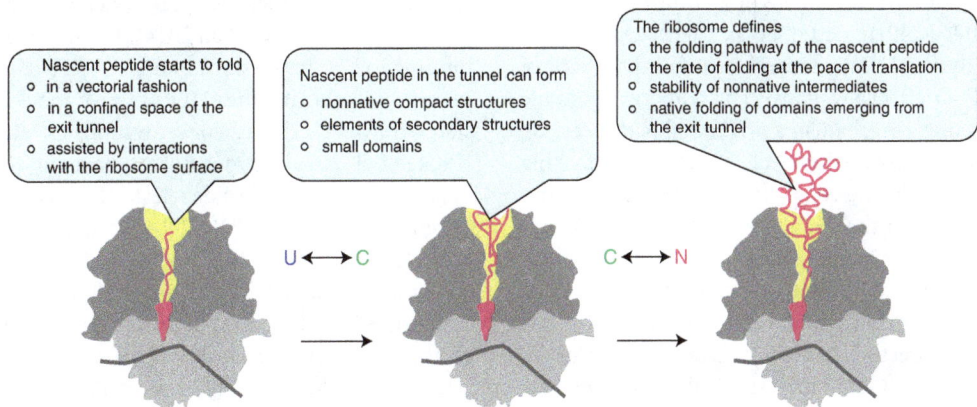

Figure 6. Cotranslational protein folding. Callouts summarize the potential effects at each step (Rodnina 2016). U, unfolded state; C, compact transient state or folding intermediate; N, native fold.

two steps are accomplished by RF1 or RF2) and dissociation of RF1/RF2 with the help of RF3. RF1 and RF2 select the respective stop codons by conserved recognition motifs, PVT in RF1 or SPF in RF2. Crystal structures identified the principal elements involved in the recognition (Korostelev et al. 2008, 2010; Laurberg et al. 2008; Weixlbaumer et al. 2008; Santos et al. 2013). The uracil in the first position of all three stop codons is recognized by the amino terminus of helix α5 of either RF1 or RF2. In RF1, interactions of the Thr residue in the PVT motif restrict reading to an A at the second position of the codon. In contrast, in RF2 the Ser residue in the SPV motif can bind to both A and G. The ability of RF1 to read both A and G in the third codon position is explained by a rotation of the side-chain amide of a Gln residue that is conserved in RF1, but not in RF2. RF2 has a hydrophobic Val residue at the homologous position and is restricted to an A in the third codon position.

Peptidyl-tRNA hydrolysis is catalyzed by the peptidyl transferase center of the ribosome with the help of the GGQ motif that is conserved in RF1 and RF2. The Gln residue of the GGQ motif is methylated; the modification increases the rate of peptide release, in particular for RF2. The reaction is expected to move through a tetrahedral intermediate, which breaks down and forms free peptide and deacylated tRNA (Jin et al. 2010). Proton inventories and computer simulations indicate that the transition state of the RF2-dependent hydrolysis reaction involves only one proton in flight (Trobro and Aqvist 2009; Kuhlenkoetter et al. 2011). Mutational analysis, pH/rate dependencies, and the lack of a D_2O effect are consistent with two possible reaction mechanisms (Kuhlenkoetter et al. 2011; Indrisiunaite et al. 2015). The attacking water molecule could donate a proton to a hydroxide ion that facilitates both proton transfer and nucleophilic attack. Alternatively, a hydroxide ion could act as an attacking group in which the proton in flight could be transferred to the 2′OH directly or through another water molecule. The 2′OH of the P-site substrate is vital for orienting the nucleophile in a hydrogen-bonding network productive for catalysis (for review, see Rodnina 2013).

RF3 is required to release RF1/RF2 from the ribosome, but the mechanism of RF3 action is controversial (Zavialov et al. 2001, 2002; Pallesen et al. 2013; Koutmou et al. 2014; Peske et al. 2014; Shi and Joseph 2016). RF3 has a higher binding affinity for GDP than for GTP and can be activated by the ribosome through accelerating the nucleotide exchange. The original model for the RF3 mechanism relied on measurements of the affinity of RF3 for GTP and GDP, which suggested that at cellular concentrations of GTP/GDP most RF3 should be in the GDP-bound form (Zavialov et al. 2001). This model suggested that RF3•GDP can bind to the ribosome in a complex with RF1 or RF2 only after the peptide is released, thereby forming an unstable encounter complex. Subsequent dissociation of GDP would lead to a stable high-affinity complex with RF3 in the nucleotide-free state. The subsequent binding of GTP to RF3 would promote RF1/RF2 dissociation. In the final step of the model, RF3 hydrolyzes GTP and dissociates from the ribosome (Zavialov et al. 2001, 2002). However, more direct measurements of the RF3 affinity for GTP and GDP did not confirm a large preference of RF3 for GDP (Koutmou et al. 2014; Peske et al. 2014). Whereas GDP binding is favored, the affinity difference to GTP is only 10-fold, which suggests that at cellular conditions, where GTP is present in large excess over GDP, RF3 adopts the GTP-bound form. Furthermore, RF3•GTP binding is independent of peptide release, because a catalytically inactive RF2 mutant activates nucleotide exchange in RF3 (Peske et al. 2014). Because GTP binding to RF3 in the ribosome–RF2–RF3 complex is rapid, the lifetime of the putative apo-RF3 state (5 ms at cellular concentrations of RF3 [Peske et al. 2014]) is too short to have physiological significance. Peptide release results in the stabilization of the RF3•GTP–ribosome complex, thereby promoting the dissociation of RF1/2, followed by GTP hydrolysis and dissociation of RF3•GDP from the ribosome (Peske et al. 2014). Further work will be required to clarify the differences and come to a unifying model of translation termination.

Interestingly, the mechanism of translation termination appears different in prokaryotes

Cite this article as *Cold Spring Harb Perspect Biol* doi: 10.1101/cshperspect.a032664

and eukaryotes, where only two factors, eRF1 and eRF3, are responsible for termination on all three codons. In eukaryotes, eRF1 and eRF3 form a stable complex, which is recruited to the stop codon. Peptide release requires GTP hydrolysis by eRF3. Further steps of termination and ribosome recycling in eukaryotes require factors that do not exist in prokaryotes, such as ABCE1 (Rli1) (for reviews, see Dever and Green 2012; Jackson et al. 2012; see also Hellen 2018). The functional significance of these differences between prokaryotes and eukaryotes remains unclear.

RIBOSOME RECYCLING

After termination, the ribosomes still contain mRNA and tRNA, which have to be released to allow for the reuse of ribosomal subunits in the next round of initiation. In bacteria, subunit splitting is catalyzed by the ribosome recycling factor (RRF) and EF-G. RRF binds to the A site of the ribosome (Gao et al. 2005) and stabilizes a fully rotated state of the ribosome, with the P-site tRNA in the hybrid P/E binding state (Dunkle et al. 2011). GTP hydrolysis by EF-G promotes the push of RRF against the key intersubunit bridge, thereby promoting subunit splitting (Gao et al. 2005). An earlier suggestion that EF-G facilitates a translocation-like movement of RRF that acts as a tRNA mimic has been refuted by biochemical (Peske et al. 2005) and structural studies (see Fu et al. 2016 for references).

Efficient ribosome recycling occurs when RRF binds to the posttermination complex before EF-G is recruited (Borg et al. 2016). In principle, RRF and EF-G can bind independently of each other, but binding of EF-G•GTP in the absence of RRF leads to futile GTP hydrolysis without ribosome splitting (Seo et al. 2004; Savelsbergh et al. 2009; Borg et al. 2016). The sequence of subsequent events is currently a matter of debate. One model suggests that GTP hydrolysis by EF-G and the subsequent delayed Pi release result in the ribosome splitting into subunits. The resulting SSU still carries the mRNA and tRNA. Dissociation of tRNA is promoted by IF3, while mRNA is exchanged spontaneously. This sequence of events is supported by extensive kinetic experiments from different groups (Seo et al. 2004; Peske et al. 2005; Savelsbergh et al. 2009; Borg et al. 2016), smFRET study (Prabhakar et al. 2017), and time-resolved cryo-EM (Fu et al. 2016). However, in all these experiments, model mRNAs were used that contained an SD sequence that could stabilize the mRNA binding. An alternative pathway was recently suggested for mRNAs that do not contain an SD sequence in the vicinity of the stop codon (which is a normal case after termination) (Chen et al. 2017). In the latter case, GTP hydrolysis by EF-G was suggested to promote mRNA release, which would be a novel, unexpected function for EF-G. This is followed by tRNA dissociation and finally subunit splitting (Chen et al. 2017). Future experiments will show which order of events is correct or whether multiple different dissociation pathways are possible.

CONCLUSIONS AND PERSPECTIVES

The most important result of the last decade of studies on prokaryotic translation mechanisms is the view of the ribosome as a dynamic molecular machine, which changes its conformations at each step of translation. The movements of ribosome elements, including subunit rotations, flexing of SSU domains, or large movements of the S1 and L12 stalks, are inherently rapid and are gated by its ligands, such as tRNAs and translation factors. We can now describe the choreography of many steps of translation using a combination of ensemble kinetics and single-molecule techniques. Once the order of events is established, crystallography and cryo-EM can provide structures of intermediates. In the future, it would be exciting to obtain structural studies in a time-resolved fashion without the use of antibiotics or nonhydrolyzable GTP analogs that are now used to block intermediates (e.g., by performing time-resolved cryo-EM or XFEL [X-ray free-electron laser] crystallography). An improvement of the time resolution of single-molecule techniques may yield additional, yet uncharacterized transient intermediates. Tracking translation in living cells is another emerging direction. Thus, it is likely that soon we will have a near-complete 3D picture of prokaryotic translation resolved in time and space and possibly in living cells. The major challenge for years to come is to understand how eukaryotic ribosomes work. Whereas, in principle, the same experimental approaches can be used to study translation in bacteria and eukaryotes, the technical hurdles of reconstructing functional pathways for each step of translation are extremely high and amplified by a larger number of accessory proteins, extensive heterogeneity of components and complexes, and probably a larger degree of conformational mobility of translational components in eukaryotes. Solving these technical challenges and dissecting the detailed structural and functional mechanisms of the eukaryotic ribosome will constitute a milestone toward understanding translation and its regulation in health and disease.

ACKNOWLEDGMENTS

I apologize to those authors whose work was not cited because of space limitations. I am indebted to Cristina Maracci for the help with figure preparation. I thank Akanksha Goyal, Cristina Maracci, and Wolfgang Wintermeyer for critical reading of the manuscript. The work in my laboratory is supported by the Max Planck Society and Grants of the Deutsche Forschungsgemeinschaft (SFB 860 and FOR 1805).

REFERENCES

*Reference is also in this collection.

Adamczyk AJ, Warshel A. 2011. Converting structural information into an allosteric-energy-based picture for elongation factor Tu activation by the ribosome. *Proc Natl Acad Sci* **108:** 9827–9832.

Adio S, Senyushkina T, Peske F, Fischer N, Wintermeyer W, Rodnina MV. 2015. Fluctuations between multiple EF-G-induced chimeric tRNA states during translocation on the ribosome. *Nat Commun* **6:** 7442.

Allen GS, Zavialov A, Gursky R, Ehrenberg M, Frank J. 2005. The cryo-EM structure of a translation initiation complex from *Escherichia coli*. *Cell* **121:** 703–712.

Antoun A, Pavlov MY, Lovmar M, Ehrenberg M. 2006. How initiation factors tune the rate of initiation of protein synthesis in bacteria. *EMBO J* **25:** 2539–2550.

Aqvist J, Kamerlin SC. 2015. The conformation of a catalytic loop is central to GTPase activity on the ribosome. *Biochemistry* **54:** 546–556.

Balchin D, Hayer-Hartl M, Hartl FU. 2016. In vivo aspects of protein folding and quality control. *Science* **353:** aac4354.

Ban N, Nissen P, Hansen J, Moore PB, Steitz TA. 2000. The complete atomic structure of the large ribosomal subunit at 2.4 Å resolution. *Science* **289:** 905–920.

Belardinelli R, Sharma H, Caliskan N, Cunha CE, Peske F, Wintermeyer W, Rodnina MV. 2016a. Choreography of molecular movements during ribosome progression along mRNA. *Nat Struct Mol Biol* **23:** 342–348.

Belardinelli R, Sharma H, Peske F, Wintermeyer W, Rodnina MV. 2016b. Translocation as continuous movement through the ribosome. *RNA Biol* **13:** 1197–1203.

Bieling P, Beringer M, Adio S, Rodnina MV. 2006. Peptide bond formation does not involve acid-base catalysis by ribosomal residues. *Nat Struct Mol Biol* **13:** 423–428.

Bock LV, Blau C, Schroder GF, Davydov II, Fischer N, Stark H, Rodnina MV, Vaiana AC, Grubmuller H. 2013. Energy barriers and driving forces in tRNA translocation through the ribosome. *Nat Struct Mol Biol* **20:** 1390–1396.

Bonde MT, Pedersen M, Klausen MS, Jensen SI, Wulff T, Harrison S, Nielsen AT, Herrgard MJ, Sommer MO. 2016. Predictable tuning of protein expression in bacteria. *Nat Methods* **13:** 233–236.

Borg A, Pavlov M, Ehrenberg M. 2016. Complete kinetic mechanism for recycling of the bacterial ribosome. *RNA* **22:** 10–21.

Budkevich TV, Giesebrecht J, Behrmann E, Loerke J, Ramrath DJ, Mielke T, Ismer J, Hildebrand PW, Tung CS, Nierhaus KH, et al. 2014. Regulation of the mammalian elongation cycle by subunit rolling: A eukaryotic-specific ribosome rearrangement. *Cell* **158:** 121–131.

Buhr F, Jha S, Thommen M, Mittelstaet J, Kutz F, Schwalbe H, Rodnina MV, Komar AA. 2016. Synonymous codons direct cotranslational folding toward different protein conformations. *Mol Cell* **61:** 341–351.

Byrgazov K, Grishkovskaya I, Arenz S, Coudevylle N, Temmel H, Wilson DN, Djinovic-Carugo K, Moll I. 2015. Structural basis for the interaction of protein S1 with the *Escherichia coli* ribosome. *Nucleic Acids Res* **43:** 661–673.

Caliskan N, Katunin VI, Belardinelli R, Peske F, Rodnina MV. 2014. Programmed -1 frameshifting by kinetic partitioning during impeded translocation. *Cell* **157:** 1619–1631.

Chang B, Halgamuge S, Tang SL. 2006. Analysis of SD sequences in completed microbial genomes: Non-SD-led genes are as common as SD-led genes. *Gene* **373:** 90–99.

Chen C, Stevens B, Kaur J, Cabral D, Liu H, Wang Y, Zhang H, Rosenblum G, Smilansky Z, Goldman YE, et al. 2011. Single-molecule fluorescence measurements of ribosomal translocation dynamics. *Mol Cell* **42:** 367–377.

Chen J, Petrov A, Johansson M, Tsai A, O'Leary SE, Puglisi JD. 2014. Dynamic pathways of −1 translational frameshifting. *Nature* **512:** 328–332.

Chen B, Kaledhonkar S, Sun M, Shen B, Lu Z, Barnard D, Lu TM, Gonzalez RL Jr, Frank J. 2015. Structural dynamics of ribosome subunit association studied by mixing-spraying time-resolved cryogenic electron microscopy. *Structure* **23:** 1097–1105.

Chen Y, Kaji A, Kaji H, Cooperman BS. 2017. The kinetic mechanism of bacterial ribosome recycling. *Nucleic Acids Res* **45:** 10168–10177.

Clarke TFt, Clark PL. 2008. Rare codons cluster. *PLoS ONE* **3:** e3412.

Cooper HL, Park MH, Folk JE, Safer B, Braverman R. 1983. Identification of the hypusine-containing protein hy+ as translation initiation factor eIF-4D. *Proc Natl Acad Sci* **80:** 1854–1857.

Coureux PD, Lazennec-Schurdevin C, Monestier A, Larquet E, Cladiere L, Klaholz BP, Schmitt E, Mechulam Y. 2016. Cryo-EM study of start codon selection during archaeal translation initiation. *Nat Commun* **7:** 13366.

Demeshkina N, Jenner L, Westhof E, Yusupov M, Yusupova G. 2012. A new understanding of the decoding principle on the ribosome. *Nature* **484:** 256–259.

Dever TE, Green R. 2012. The elongation, termination, and recycling phases of translation in eukaryotes. *Cold Spring Harb Perspect Biol* **4:** a013706.

* Dever TE, Dinman JD, Green R. 2018. Translation elongation and recoding in eukaryotes. *Cold Spring Harb Perspect Biol* doi: 10.1101/cshperspect.a032649.

Diaconu M, Kothe U, Schlunzen F, Fischer N, Harms JM, Tonevitsky AG, Stark H, Rodnina MV, Wahl MC. 2005. Structural basis for the function of the ribosomal L7/12 stalk in factor binding and GTPase activation. *Cell* **121:** 991–1004.

Doerfel LK, Wohlgemuth I, Kothe C, Peske F, Urlaub H, Rodnina MV. 2013. EF-P is essential for rapid synthesis of proteins containing consecutive proline residues. *Science* **339:** 85–88.

Doerfel LK, Wohlgemuth I, Kubyshkin V, Starosta AL, Wilson DN, Budisa N, Rodnina MV. 2015. Entropic contribution of elongation factor P to proline positioning at the catalytic center of the ribosome. *J Am Chem Soc* **137:** 12997–13006.

Drummond DA, Wilke CO. 2009. The evolutionary consequences of erroneous protein synthesis. *Nat Rev Genet* **10:** 715–724.

Dunkle JA, Wang L, Feldman MB, Pulk A, Chen VB, Kapral GJ, Noeske J, Richardson JS, Blanchard SC, Cate JH. 2011. Structures of the bacterial ribosome in classical and hybrid states of tRNA binding. *Science* **332:** 981–984.

Duval M, Korepanov A, Fuchsbauer O, Fechter P, Haller A, Fabbretti A, Choulier L, Micura R, Klaholz BP, Romby P, et al. 2013. *Escherichia coli* ribosomal protein S1 unfolds structured mRNAs onto the ribosome for active translation initiation. *PLoS Biol* **11:** e1001731.

Duval M, Simonetti A, Caldelari I, Marzi S. 2015. Multiple ways to regulate translation initiation in bacteria: Mechanisms, regulatory circuits, dynamics. *Biochimie* **114:** 18–29.

Elvekrog MM, Gonzalez RL Jr. 2013. Conformational selection of translation initiation factor 3 signals proper substrate selection. *Nat Struct Mol Biol* **20:** 628–633.

Erlacher MD, Lang K, Wotzel B, Rieder R, Micura R, Polacek N. 2006. Efficient ribosomal peptidyl transfer critically relies on the presence of the ribose 2′-OH at A2451 of 23S rRNA. *J Am Chem Soc* **128:** 4453–4459.

Espah Borujeni A, Salis HM. 2016. Translation initiation is controlled by RNA folding kinetics via a ribosome drafting mechanism. *J Am Chem Soc* **138:** 7016–7023.

Fabbretti A, Pon CL, Hennelly SP, Hill WE, Lodmell JS, Gualerzi CO. 2007. The real-time path of translation factor IF3 onto and off the ribosome. *Mol Cell* **25:** 285–296.

Fischer N, Neumann P, Bock LV, Maracci C, Wang Z, Paleskava A, Konevega AL, Schroder GF, Grubmuller H, Ficner R, et al. 2016. The pathway to GTPase activation of elongation factor SelB on the ribosome. *Nature* **540:** 80–85.

Forchhammer K, Leinfelder W, Bock A. 1989. Identification of a novel translation factor necessary for the incorporation of selenocysteine into protein. *Nature* **342:** 453–456.

Frank J, Gonzalez RL Jr. 2010. Structure and dynamics of a processive Brownian motor: The translating ribosome. *Annu Rev Biochem* **79:** 381–412.

Fu Z, Kaledhonkar S, Borg A, Sun M, Chen B, Grassucci RA, Ehrenberg M, Frank J. 2016. Key intermediates in ribosome recycling visualized by time-resolved cryoelectron microscopy. *Structure* **24:** 2092–2101.

Gao N, Zavialov AV, Li W, Sengupta J, Valle M, Gursky RP, Ehrenberg M, Frank J. 2005. Mechanism for the disassembly of the posttermination complex inferred from cryo-EM studies. *Mol Cell* **18:** 663–674.

Gao H, Zhou Z, Rawat U, Huang C, Bouakaz L, Wang C, Cheng Z, Liu Y, Zavialov A, Gursky R, et al. 2007. RF3 induces ribosomal conformational changes responsible for dissociation of class I release factors. *Cell* **129:** 929–941.

Geggier P, Dave R, Feldman MB, Terry DS, Altman RB, Munro JB, Blanchard SC. 2010. Conformational sampling of aminoacyl-tRNA during selection on the bacterial ribosome. *J Mol Biol* **399:** 576–595.

Gloge F, Becker AH, Kramer G, Bukau B. 2014. Co-translational mechanisms of protein maturation. *Curr Opin Struct Biol* **24:** 24–33.

Goyal A, Belardinelli R, Maracci C, Milon P, Rodnina MV. 2015. Directional transition from initiation to elongation in bacterial translation. *Nucleic Acids Res* **43:** 10700–10712.

Goyal A, Belardinelli R, Rodnina MV. 2017. Non-canonical binding site for bacterial initiation factor 3 on the large ribosomal subunit. *Cell Rep* **20:** 3113–3122.

Grigoriadou C, Marzi S, Kirillov S, Gualerzi CO, Cooperman BS. 2007. A quantitative kinetic scheme for 70 S translation initiation complex formation. *J Mol Biol* **373:** 562–572.

Grill S, Gualerzi CO, Londei P, Blasi U. 2000. Selective stimulation of translation of leaderless mRNA by initiation factor 2: Evolutionary implications for translation. *EMBO J* **19:** 4101–4110.

Gualerzi CO, Pon CL. 2015. Initiation of mRNA translation in bacteria: Structural and dynamic aspects. *Cell Mol Life Sci* **72:** 4341–4367.

Guo Z, Noller HF. 2012. Rotation of the head of the 30S ribosomal subunit during mRNA translocation. *Proc Natl Acad Sci* **109:** 20391–20394.

Gutierrez E, Shin BS, Woolstenhulme CJ, Kim JR, Saini P, Buskirk AR, Dever TE. 2013. eIF5A promotes translation of polyproline motifs. *Mol Cell* **51:** 35–45.

* Hellen CUT. 2018. Translation termination and ribosome recycling in eukaryotes. *Cold Spring Harb Perspect Biol* doi: 10.1101/cshperspect.a032656.

Hersch SJ, Wang M, Zou SB, Moon KM, Foster LJ, Ibba M, Navarre WW. 2013. Divergent protein motifs direct elongation factor P-mediated translational regulation in *Salmonella enterica* and *Escherichia coli*. *MBio* **4**: e00180-13.

Hiller DA, Singh V, Zhong M, Strobel SA. 2011. A two-step chemical mechanism for ribosome-catalysed peptide bond formation. *Nature* **476**: 236–239.

Holtkamp W, Cunha CE, Peske F, Konevega AL, Wintermeyer W, Rodnina MV. 2014. GTP hydrolysis by EF-G synchronizes tRNA movement on small and large ribosomal subunits. *EMBO J* **33**: 1073–1085.

Holtkamp W, Kokic G, Jager M, Mittelstaet J, Komar AA, Rodnina MV. 2015. Cotranslational protein folding on the ribosome monitored in real time. *Science* **350**: 1104–1107.

Hussain T, Llacer JL, Wimberly BT, Kieft JS, Ramakrishnan V. 2016. Large-scale movements of IF3 and tRNA during bacterial translation initiation. *Cell* **167**: 133–144.e113.

Huter P, Arenz S, Bock LV, Graf M, Frister JO, Heuer A, Peil L, Starosta AL, Wohlgemuth I, Peske F, et al. 2017. Structural basis for polyproline-mediated ribosome stalling and rescue by the translation elongation factor EF-P. *Mol Cell* **68**: 515–527.e516.

Ieong KW, Uzun U, Selmer M, Ehrenberg M. 2016. Two proofreading steps amplify the accuracy of genetic code translation. *Proc Natl Acad Sci* **113**: 13744–13749.

Indrisiunaite G, Pavlov MY, Heurgue-Hamard V, Ehrenberg M. 2015. On the pH dependence of class-1 RF-dependent termination of mRNA translation. *J Mol Biol* **427**: 1848–1860.

Jackson RJ, Hellen CU, Pestova TV. 2012. Termination and post-termination events in eukaryotic translation. *Adv Protein Chem Struct Biol* **86**: 45–93.

Jin H, Kelley AC, Loakes D, Ramakrishnan V. 2010. Structure of the 70S ribosome bound to release factor 2 and a substrate analog provides insights into catalysis of peptide release. *Proc Natl Acad Sci* **107**: 8593–8598.

Jin H, Kelley AC, Ramakrishnan V. 2011. Crystal structure of the hybrid state of ribosome in complex with the guanosine triphosphatase release factor 3. *Proc Natl Acad Sci* **108**: 15798–15803.

Julian P, Milon P, Agirrezabala X, Lasso G, Gil D, Rodnina MV, Valle M. 2011. The cryo-EM structure of a complete 30S translation initiation complex from *Escherichia coli*. *PLoS Biol* **9**: e1001095.

Katoh T, Wohlgemuth I, Nagano M, Rodnina MV, Suga H. 2016. Essential structural elements in tRNA(Pro) for EF-P-mediated alleviation of translation stalling. *Nat Commun* **7**: 11657.

Kim HK, Liu F, Fei J, Bustamante C, Gonzalez RL Jr, Tinoco I Jr. 2014. A frameshifting stimulatory stem loop destabilizes the hybrid state and impedes ribosomal translocation. *Proc Natl Acad Sci* **111**: 5538–5543.

Kohler R, Mooney RA, Mills DJ, Landick R, Cramer P. 2017. Architecture of a transcribing-translating expressome. *Science* **356**: 194–197.

Korostelev A, Asahara H, Lancaster L, Laurberg M, Hirschi A, Zhu J, Trakhanov S, Scott WG, Noller HF. 2008. Crystal structure of a translation termination complex formed with release factor RF2. *Proc Natl Acad Sci* **105**: 19684–19689.

Korostelev A, Zhu J, Asahara H, Noller HF. 2010. Recognition of the amber UAG stop codon by release factor RF1. *EMBO J* **29**: 2577–2585.

Kosuri S, Goodman DB, Cambray G, Mutalik VK, Gao Y, Arkin AP, Endy D, Church GM. 2013. Composability of regulatory sequences controlling transcription and translation in *Escherichia coli*. *Proc Natl Acad Sci* **110**: 14024–14029.

Kothe U, Rodnina MV. 2006. Delayed release of inorganic phosphate from elongation factor Tu following GTP hydrolysis on the ribosome. *Biochemistry* **45**: 12767–12774.

Kothe U, Wieden HJ, Mohr D, Rodnina MV. 2004. Interaction of helix D of elongation factor Tu with helices 4 and 5 of protein L7/12 on the ribosome. *J Mol Biol* **336**: 1011–1021.

Koutmou KS, McDonald ME, Brunelle JL, Green R. 2014. RF3:GTP promotes rapid dissociation of the class 1 termination factor. *RNA* **20**: 609–620.

Krishnan KM, Van Etten WJ III, Janssen GR. 2010. Proximity of the start codon to a leaderless mRNA's 5′ terminus is a strong positive determinant of ribosome binding and expression in *Escherichia coli*. *J Bacteriol* **192**: 6482–6485.

Kuhlenkoetter S, Wintermeyer W, Rodnina MV. 2011. Different substrate-dependent transition states in the active site of the ribosome. *Nature* **476**: 351–354.

Landick R, Carey J, Yanofsky C. 1985. Translation activates the paused transcription complex and restores transcription of the trp operon leader region. *Proc Natl Acad Sci* **82**: 4663–4667.

Lassak J, Keilhauer EC, Furst M, Wuichet K, Godeke J, Starosta AL, Chen JM, Sogaard-Andersen L, Rohr J, Wilson DN, et al. 2015. Arginine-rhamnosylation as new strategy to activate translation elongation factor P. *Nat Chem Biol* **11**: 266–270.

Lassak J, Wilson DN, Jung K. 2016. Stall no more at polyproline stretches with the translation elongation factors EF-P and IF-5A. *Mol Microbiol* **99**: 219–235.

Laurberg M, Asahara H, Korostelev A, Zhu J, Trakhanov S, Noller HF. 2008. Structural basis for translation termination on the 70S ribosome. *Nature* **454**: 852–857.

Lin J, Gagnon MG, Bulkley D, Steitz TA. 2015. Conformational changes of elongation factor G on the ribosome during tRNA translocation. *Cell* **160**: 219–227.

Liu Q, Fredrick K. 2015. Roles of helix H69 of 23S rRNA in translation initiation. *Proc Natl Acad Sci* **112**: 11559–11564.

Loveland AB, Demo G, Grigorieff N, Korostelev AA. 2017. Ensemble cryo-EM elucidates the mechanism of translation fidelity. *Nature* **546**: 113–117.

Ma J, Campbell A, Karlin S. 2002. Correlations between Shine–Dalgarno sequences and gene features such as predicted expression levels and operon structures. *J Bacteriol* **184**: 5733–5745.

MacDougall DD, Gonzalez RL Jr. 2015. Translation initiation factor 3 regulates switching between different modes of ribosomal subunit joining. *J Mol Biol* **427**: 1801–1818.

Maracci C, Peske F, Dannies E, Pohl C, Rodnina MV. 2014. Ribosome-induced tuning of GTP hydrolysis by a translational GTPase. *Proc Natl Acad Sci* **111:** 14418–14423.

Maracci C, Wohlgemuth I, Rodnina MV. 2015. Activities of the peptidyl transferase center of ribosomes lacking protein L27. *RNA* **21:** 2047–2052.

Marshall RA, Aitken CE, Puglisi JD. 2009. GTP hydrolysis by IF2 guides progression of the ribosome into elongation. *Mol Cell* **35:** 37–47.

* Merrick WC, Pavitt GD. 2018. Protein synthesis initiation in eukaryotic cells. *Cold Spring Harb Perspect Biol* doi: 10.1101/cshperspect.a033092.

Milon P, Rodnina MV. 2012. Kinetic control of translation initiation in bacteria. *Crit Rev Biochem Mol Biol* **47:** 334–348.

Milon P, Konevega AL, Gualerzi CO, Rodnina MV. 2008. Kinetic checkpoint at a late step in translation initiation. *Mol Cell* **30:** 712–720.

Milon P, Carotti M, Konevega AL, Wintermeyer W, Rodnina MV, Gualerzi CO. 2010. The ribosome-bound initiation factor 2 recruits initiator tRNA to the 30S initiation complex. *EMBO Rep* **11:** 312–316.

Milon P, Maracci C, Filonava L, Gualerzi CO, Rodnina MV. 2012. Real-time assembly landscape of bacterial 30S translation initiation complex. *Nat Struct Mol Biol* **19:** 609–615.

Mitarai N, Sneppen K, Pedersen S. 2008. Ribosome collisions and translation efficiency: Optimization by codon usage and mRNA destabilization. *J Mol Biol* **382:** 236–245.

Mohr D, Wintermeyer W, Rodnina MV. 2002. GTPase activation of elongation factors Tu and G on the ribosome. *Biochemistry* **41:** 12520–12528.

Myasnikov AG, Marzi S, Simonetti A, Giuliodori AM, Gualerzi CO, Yusupova G, Yusupov M, Klaholz BP. 2005. Conformational transition of initiation factor 2 from the GTP- to GDP-bound state visualized on the ribosome. *Nat Struct Mol Biol* **12:** 1145–1149.

Napthine S, Vidakovic M, Girnary R, Namy O, Brierley I. 2003. Prokaryotic-style frameshifting in a plant translation system: Conservation of an unusual single-tRNA slippage event. *EMBO J* **22:** 3941–3950.

Noel JK, Whitford PC. 2016. How EF-Tu can contribute to efficient proofreading of aa-tRNA by the ribosome. *Nat Commun* **7:** 13314.

O'Brien EP, Ciryam P, Vendruscolo M, Dobson CM. 2014. Understanding the influence of codon translation rates on cotranslational protein folding. *Acc Chem Res* **47:** 1536–1544.

Ogle JM, Brodersen DE, Clemons WM Jr, Tarry MJ, Carter AP, Ramakrishnan V. 2001. Recognition of cognate transfer RNA by the 30S ribosomal subunit. *Science* **292:** 897–902.

Ogle JM, Murphy FV, Tarry MJ, Ramakrishnan V. 2002. Selection of tRNA by the ribosome requires a transition from an open to a closed form. *Cell* **111:** 721–732.

Pallesen J, Hashem Y, Korkmaz G, Koripella RK, Huang C, Ehrenberg M, Sanyal S, Frank J. 2013. Cryo-EM visualization of the ribosome in termination complex with apo-RF3 and RF1. *eLife* **2:** e00411.

Peil L, Starosta AL, Lassak J, Atkinson GC, Virumae K, Spitzer M, Tenson T, Jung K, Remme J, Wilson DN. 2013. Distinct XPPX sequence motifs induce ribosome stalling, which is rescued by the translation elongation factor EF-P. *Proc Natl Acad Sci* **110:** 15265–15270.

Peske F, Rodnina MV, Wintermeyer W. 2005. Sequence of steps in ribosome recycling as defined by kinetic analysis. *Mol Cell* **18:** 403–412.

Peske F, Kuhlenkoetter S, Rodnina MV, Wintermeyer W. 2014. Timing of GTP binding and hydrolysis by translation termination factor RF3. *Nucleic Acids Res* **42:** 1812–1820.

Polikanov YS, Steitz TA, Innis CA. 2014. A proton wire to couple aminoacyl-tRNA accommodation and peptide-bond formation on the ribosome. *Nat Struct Mol Biol* **21:** 787–793.

Prabhakar A, Capece MC, Petrov A, Choi J, Puglisi JD. 2017. Post-termination ribosome intermediate acts as the gateway to ribosome recycling. *Cell Rep* **20:** 161–172.

Prasad RB, Plotnikov NV, Lameira J, Warshel A. 2013. Quantitative exploration of the molecular origin of the activation of GTPase. *Proc Natl Acad Sci* **110:** 20509–20514.

Proshkin S, Rahmouni AR, Mironov A, Nudler E. 2010. Cooperation between translating ribosomes and RNA polymerase in transcription elongation. *Science* **328:** 504–508.

Qin D, Liu Q, Devaraj A, Fredrick K. 2012. Role of helix 44 of 16S rRNA in the fidelity of translation initiation. *RNA* **18:** 485–495.

Qu X, Wen JD, Lancaster L, Noller HF, Bustamante C, Tinoco I Jr. 2011. The ribosome uses two active mechanisms to unwind messenger RNA during translation. *Nature* **475:** 118–121.

Rajkovic A, Hummels KR, Witzky A, Erickson S, Gafken PR, Whitelegge JP, Faull KF, Kearns DB, Ibba M. 2016. Translation control of swarming proficiency in *Bacillus subtilis* by 5-amino-pentanolylated elongation factor P. *J Biol Chem* **291:** 10976–10985.

Ramrath DJ, Lancaster L, Sprink T, Mielke T, Loerke J, Noller HF, Spahn CM. 2013. Visualization of two transfer RNAs trapped in transit during elongation factor G-mediated translocation. *Proc Natl Acad Sci* **110:** 20964–20969.

Ranjan N, Rodnina MV. 2017. Thio-modification of tRNA at the wobble position as regulator of the kinetics of decoding and translocation on the ribosome. *J Am Chem Soc* **139:** 5857–5864.

Rawat UB, Zavialov AV, Sengupta J, Valle M, Grassucci RA, Linde J, Vestergaard B, Ehrenberg M, Frank J. 2003. A cryo-electron microscopic study of ribosome-bound termination factor RF2. *Nature* **421:** 87–90.

Rawat U, Gao H, Zavialov A, Gursky R, Ehrenberg M, Frank J. 2006. Interactions of the release factor RF1 with the ribosome as revealed by cryo-EM. *J Mol Biol* **357:** 1144–1153.

Reeve B, Hargest T, Gilbert C, Ellis T. 2014. Predicting translation initiation rates for designing synthetic biology. *Front Bioeng Biotechnol* **2:** 1.

Rodnina MV. 2013. The ribosome as a versatile catalyst: Reactions at the peptidyl transferase center. *Curr Opin Struct Biol* **23:** 595–602.

Rodnina MV. 2016. The ribosome in action: Tuning of translational efficiency and protein folding. *Protein Sci* **25**: 1390–1406.

Rodnina MV, Savelsbergh A, Katunin VI, Wintermeyer W. 1997. Hydrolysis of GTP by elongation factor G drives tRNA movement on the ribosome. *Nature* **385**: 37–41.

Rodnina MV, Fischer N, Maracci C, Stark H. 2017. Ribosome dynamics during decoding. *Philos Trans R Soc Lond B Biol Sci* **372**.

Rozov A, Demeshkina N, Westhof E, Yusupov M, Yusupova G. 2015. Structural insights into the translational infidelity mechanism. *Nat Commun* **6**: 7251.

Rozov A, Demeshkina N, Khusainov I, Westhof E, Yusupov M, Yusupova G. 2016a. Novel base-pairing interactions at the tRNA wobble position crucial for accurate reading of the genetic code. *Nat Commun* **7**: 10457.

Rozov A, Westhof E, Yusupov M, Yusupova G. 2016b. The ribosome prohibits the G*U wobble geometry at the first position of the codon-anticodon helix. *Nucleic Acids Res* **44**: 6434–6441.

Salis HM, Mirsky EA, Voigt CA. 2009. Automated design of synthetic ribosome binding sites to control protein expression. *Nat Biotechnol* **27**: 946–950.

Santos N, Zhu J, Donohue JP, Korostelev AA, Noller HF. 2013. Crystal structure of the 70S ribosome bound with the Q253P mutant form of release factor RF2. *Structure* **21**: 1258–1263.

Satterthwait AC, Jencks WP. 1974. The mechanism of the aminolysis of acetate esters. *J Am Chem Soc* **96**: 7018–7031.

Savelsbergh A, Katunin VI, Mohr D, Peske F, Rodnina MV, Wintermeyer W. 2003. An elongation factor G-induced ribosome rearrangement precedes tRNA-mRNA translocation. *Mol Cell* **11**: 1517–1523.

Savelsbergh A, Mohr D, Kothe U, Wintermeyer W, Rodnina MV. 2005. Control of phosphate release from elongation factor G by ribosomal protein L7/12. *EMBO J* **24**: 4316–4323.

Savelsbergh A, Rodnina MV, Wintermeyer W. 2009. Distinct functions of elongation factor G in ribosome recycling and translocation. *RNA* **15**: 772–780.

Scharff LB, Childs L, Walther D, Bock R. 2011. Local absence of secondary structure permits translation of mRNAs that lack ribosome-binding sites. *PLoS Genet* **7**: e1002155.

Schmeing TM, Voorhees RM, Kelley AC, Gao YG, Murphy FVt, Weir JR, Ramakrishnan V. 2009. The crystal structure of the ribosome bound to EF-Tu and aminoacyl-tRNA. *Science* **326**: 688–694.

Schmeing TM, Voorhees RM, Kelley AC, Ramakrishnan V. 2011. How mutations in tRNA distant from the anticodon affect the fidelity of decoding. *Nat Struct Mol Biol* **18**: 432–436.

Schuller AP, Wu CC, Dever TE, Buskirk AR, Green R. 2017. eIF5A functions globally in translation elongation and termination. *Mol Cell* **66**: 194–205 e195.

Seo HS, Kiel M, Pan D, Raj VS, Kaji A, Cooperman BS. 2004. Kinetics and thermodynamics of RRF, EF-G, and thiostrepton interaction on the *Escherichia coli* ribosome. *Biochemistry* **43**: 12728–12740.

Sharma PK, Xiang Y, Kato M, Warshel A. 2005. What are the roles of substrate-assisted catalysis and proximity effects in peptide bond formation by the ribosome? *Biochemistry* **44**: 11307–11314.

Sharma AK, Bukau B, O'Brien EP. 2016a. Physical origins of codon positions that strongly influence cotranslational folding: A framework for controlling nascent-protein folding. *J Am Chem Soc* **138**: 1180–1195.

Sharma H, Adio S, Senyushkina T, Belardinelli R, Peske F, Rodnina MV. 2016b. Kinetics of spontaneous and EF-G-accelerated rotation of ribosomal subunits. *Cell Rep* **16**: 2187–2196.

Shi X, Joseph S. 2016. Mechanism of translation termination: RF1 dissociation follows RF3 dissociation from the ribosome. *Biochemistry* **55**: 6344–6354.

Sievers A, Beringer M, Rodnina MV, Wolfenden R. 2004. The ribosome as an entropy trap. *Proc Natl Acad Sci* **101**: 7897–7901.

Siller E, DeZwaan DC, Anderson JF, Freeman BC, Barral JM. 2010. Slowing bacterial translation speed enhances eukaryotic protein folding efficiency. *J Mol Biol* **396**: 1310–1318.

Simonetti A, Marzi S, Myasnikov AG, Fabbretti A, Yusupov M, Gualerzi CO, Klaholz BP. 2008. Structure of the 30S translation initiation complex. *Nature* **455**: 416–420.

* Sokabe M, Fraser CS. 2018. Toward a kinetic understanding of eukaryotic translation. *Cold Spring Harb Perspect Biol* doi: 10.1101/cshperspect.a032706.

Spencer PS, Siller E, Anderson JF, Barral JM. 2012. Silent substitutions predictably alter translation elongation rates and protein folding efficiencies. *J Mol Biol* **422**: 328–335.

Sprink T, Ramrath DJ, Yamamoto H, Yamamoto K, Loerke J, Ismer J, Hildebrand PW, Scheerer P, Burger J, Mielke T, et al. 2016. Structures of ribosome-bound initiation factor 2 reveal the mechanism of subunit association. *Sci Adv* **2**: e1501502.

Studer SM, Joseph S. 2006. Unfolding of mRNA secondary structure by the bacterial translation initiation complex. *Mol Cell* **22**: 105–115.

Takyar S, Hickerson RP, Noller HF. 2005. mRNA helicase activity of the ribosome. *Cell* **120**: 49–58.

Thommen M, Holtkamp W, Rodnina MV. 2017. Co-translational protein folding: Progress and methods. *Curr Opin Struct Biol* **42**: 83–89.

Tolstrup N, Sensen CW, Garrett RA, Clausen IG. 2000. Two different and highly organized mechanisms of translation initiation in the archaeon *Sulfolobus solfataricus*. *Extremophiles* **4**: 175–179.

Trobro S, Aqvist J. 2009. Mechanism of the translation termination reaction on the ribosome. *Biochemistry* **48**: 11296–11303.

Trovato F, O'Brien EP. 2017. Fast protein translation can promote co- and posttranslational folding of misfolding-prone proteins. *Biophys J* **112**: 1807–1819.

Tsai CJ, Sauna ZE, Kimchi-Sarfaty C, Ambudkar SV, Gottesman MM, Nussinov R. 2008. Synonymous mutations and ribosome stalling can lead to altered folding pathways and distinct minima. *J Mol Biol* **383**: 281–291.

Tsai A, Petrov A, Marshall RA, Korlach J, Uemura S, Puglisi JD. 2012. Heterogeneous pathways and timing of factor departure during translation initiation. *Nature* **487**: 390–393.

Ude S, Lassak J, Starosta AL, Kraxenberger T, Wilson DN, Jung K. 2013. Translation elongation factor EF-P alleviates ribosome stalling at polyproline stretches. *Science* **339:** 82–85.

Valle M, Zavialov A, Li W, Stagg SM, Sengupta J, Nielsen RC, Nissen P, Harvey SC, Ehrenberg M, Frank J. 2003. Incorporation of aminoacyl-tRNA into the ribosome as seen by cryo-electron microscopy. *Nat Struct Biol* **10:** 899–906.

Vesper O, Amitai S, Belitsky M, Byrgazov K, Kaberdina AC, Engelberg-Kulka H, Moll I. 2011. Selective translation of leaderless mRNAs by specialized ribosomes generated by MazF in *Escherichia coli*. *Cell* **147:** 147–157.

Wallin G, Aqvist J. 2010. The transition state for peptide bond formation reveals the ribosome as a water trap. *Proc Natl Acad Sci* **107:** 1888–1893.

Wallin G, Kamerlin SC, Aqvist J. 2013. Energetics of activation of GTP hydrolysis on the ribosome. *Nat Commun* **4:** 1733.

Wasserman MR, Alejo JL, Altman RB, Blanchard SC. 2016. Multiperspective smFRET reveals rate-determining late intermediates of ribosomal translocation. *Nat Struct Mol Biol* **23:** 333–341.

Weiner J III, Herrmann R, Browning GF. 2000. Transcription in *Mycoplasma pneumoniae*. *Nucl Acids Res* **28:** 4488–4496.

Weixlbaumer A, Jin H, Neubauer C, Voorhees RM, Petry S, Kelley AC, Ramakrishnan V. 2008. Insights into translational termination from the structure of RF2 bound to the ribosome. *Science* **322:** 953–956.

Wohlgemuth I, Brenner S, Beringer M, Rodnina MV. 2008. Modulation of the rate of peptidyl transfer on the ribosome by the nature of substrates. *J Biol Chem* **283:** 32229–32235.

Wool IG, Gluck A, Endo Y. 1992. Ribotoxin recognition of ribosomal RNA and a proposal for the mechanism of translocation. *Trends Biochem Sci* **17:** 266–269.

Woolstenhulme CJ, Parajuli S, Healey DW, Valverde DP, Petersen EN, Starosta AL, Guydosh NR, Johnson WE, Wilson DN, Buskirk AR. 2013. Nascent peptides that block protein synthesis in bacteria. *Proc Natl Acad Sci* **110:** E878–E887.

Yamamoto H, Wittek D, Gupta R, Qin B, Ueda T, Krause R, Yamamoto K, Albrecht R, Pech M, Nierhaus KH. 2016. 70S-scanning initiation is a novel and frequent initiation mode of ribosomal translation in bacteria. *Proc Natl Acad Sci* **113:** E1180–E1189.

Yan S, Wen JD, Bustamante C, Tinoco I Jr. 2015. Ribosome excursions during mRNA translocation mediate broad branching of frameshift pathways. *Cell* **160:** 870–881.

Yanagisawa T, Sumida T, Ishii R, Takemoto C, Yokoyama S. 2010. A paralog of lysyl-tRNA synthetase aminoacylates a conserved lysine residue in translation elongation factor P. *Nat Struct Mol Biol* **17:** 1136–1143.

Youngman EM, Brunelle JL, Kochaniak AB, Green R. 2004. The active site of the ribosome is composed of two layers of conserved nucleotides with distinct roles in peptide bond formation and peptide release. *Cell* **117:** 589–599.

Yu CH, Dang Y, Zhou Z, Wu C, Zhao F, Sachs MS, Liu Y. 2015. Codon usage influences the local rate of translation elongation to regulate co-translational protein folding. *Mol Cell* **59:** 744–754.

Zaher HS, Shaw JJ, Strobel SA, Green R. 2011. The 2′-OH group of the peptidyl-tRNA stabilizes an active conformation of the ribosomal PTC. *EMBO J* **30:** 2445–2453.

Zavialov AV, Buckingham RH, Ehrenberg M. 2001. A post-termination ribosomal complex is the guanine nucleotide exchange factor for peptide release factor RF3. *Cell* **107:** 115–124.

Zavialov AV, Mora L, Buckingham RH, Ehrenberg M. 2002. Release of peptide promoted by the GGQ motif of class 1 release factors regulates the GTPase activity of RF3. *Mol Cell* **10:** 789–798.

Zhang G, Hubalewska M, Ignatova Z. 2009. Transient ribosomal attenuation coordinates protein synthesis and co-translational folding. *Nat Struct Mol Biol* **16:** 274–280.

Zheng X, Hu GQ, She ZS, Zhu H. 2011. Leaderless genes in bacteria: Clue to the evolution of translation initiation mechanisms in prokaryotes. *BMC Genomics* **12:** 361.

Zhou J, Lancaster L, Donohue JP, Noller HF. 2014. How the ribosome hands the A-site tRNA to the P site during EF-G-catalyzed translocation. *Science* **345:** 1188–1191.

New Insights into Ribosome Structure and Function

Amy Jobe,[1] Zheng Liu,[1] Cristina Gutierrez-Vargas,[2] and Joachim Frank[1,2]

[1]Department of Biochemistry and Molecular Biophysics, Columbia University, New York, New York 10032

[2]Department of Biological Sciences, Columbia University, New York, New York 10032

Correspondence: jf2192@cumc.columbia.edu

In the past 4 years, because of the advent of new cameras, many ribosome structures have been solved by cryoelectron microscopy (cryo-EM) at high, often near-atomic resolution, bringing new mechanistic insights into the processes of translation initiation, peptide elongation, termination, and recycling. Thus, cryo-EM has joined X-ray crystallography as a powerful technique in structural studies of translation. The significance of this new development is that structures of ribosomes in complex with their functional binding partners can now be determined to high resolution in multiple states as they perform their work. The aim of this article is to provide an overview of these new studies and assess the contributions they have made toward an understanding of translation and translational control.

To understand translation and translational control, an understanding of structure as well as dynamics of the various processes is required. The ability to see interactions at the functional sites in atomic detail, facilitated by much of the recent work, has made a large difference in this respect. One mechanism can be singled out that is ubiquitous, with different ramifications in initiation, translocation, decoding, termination, and recycling; in each of these steps of translation, the action of a GTPase is required, and typically the engagement of a factor creates an atomic constellation that triggers guanosine triphosphate (GTP) hydrolysis, followed by changes in both the factor and the ribosome.

There are tantalizing questions on how this action unfolds in each case, and the reason so much is still unanswered is that it occurs quite rapidly, and that its structural characterization therefore requires some kind of interference, such as the use of a nonhydrolyzable GTP analog or mutation at a key site, each time invoking questions about the authenticity of the state trapped. Thus, the nature of the decisive power-stroke that brings us from point A to point B in the processive machinery is uncertain.

Among all steps of translation, eukaryotic initiation is probably the least understood because of the multitude of steps and the complexity of the scanning mechanism leading to the placement of the AUG codon at the P site of the small subunit. Next there come the variants of translation initiation through the viral internal ribosome entry site (IRES) mechanism; here it is necessary to understand how the IRES is able to "push the right buttons" on the ribosome to take over control. Another highlight of this review is the recent progress in the structural characterization of ribosomes of parasitic protozoans; here, the reason little is known thus far is related to the prevalence of the diseases that they cause in underdeveloped countries and the politics of science funding.

As translation of the genetic code into polypeptides proceeds through its main phases of initiation, peptide elongation, termination, and recycling, the ribosome undergoes numerous conformational changes and ligand binding/unbinding events. For a long time, steps toward the goal of obtaining information on the structural basis of these events, critical for understanding the mechanistic basis of biological function, have been handicapped by shortcomings of the two main structural visualization techniques, X-ray crystallography and single-particle cryo-electron microscopy (cryo-EM). Single-particle cryo-EM, recently highlighted by the 2017 Nobel Prize in Chemistry, is the visualization of molecules in a thin layer of vitreous ice following rapid plunge-freezing of the sample into a cryogen, or a freezing agent, at liquid-nitrogen temperature (McDowall et al. 1983). Raw data collected on the electron microscope comes in the form of two-dimensional "snapshots" called micrographs, each of which include a few to over 100 copies of a biological molecule. Because molecules are randomly oriented, all view directions are sampled when a sufficient number of micrographs are taken. As the electron dose is spread over a large number of molecules, radiation damage is avoided or strongly reduced. Raw data are subjected to a specialized series of computational techniques, during which individual copies of the specimen are identified and the two-dimensional view represented by each copy is assigned a viewing angle, and a three-dimensional reconstruction of the Coulomb potential distribution (colloquially called "density map") can ultimately be computed.

While X-ray crystallography, starting with publications in 2000 (Ban et al. 2000; Schluenzen et al. 2000; Wimberly et al. 2000), has been the source of many high-resolution atomic models of the ribosome structure, it is intrinsically limited as a technique by the conformational selection taking place during the formation of the crystal, such that many interesting functional states will escape visualization. Conversely, states that are only transiently visited may receive inordinate attention because they happen to be captured in crystals diffracting to high resolution. Cryo-EM, on the other hand, apart from the fact that it does not require crystals, has the advantage that it is capable of displaying all states coexisting in the sample without restriction. The much lower sample volume required is another benefit. However, until recently, this technique has been limited in resolution to 5–6 Ångströms (Å, 10^{-10} meters) for asymmetric molecules (e.g., Hashem et al. 2013a) because of the low quality of the recording medium. This situation has been radically changed with the commercial introduction, in 2012, of direct electron detecting cameras (see McMullan et al. 2009), which allow near-atomic resolution to be reached. For the ribosome field, this has been a game changer as it is now possible to study ligand-binding events (e.g., with transfer RNA [tRNA] or various translation factors) in minute detail. Even ab initio structural modeling is now possible over large stretches of the reconstructed density map; magnesium ions can be located; and ribosomal RNA (rRNA) modifications can be directly identified by the positions of extra density in the map (see Fischer et al. 2015; Liu et al. 2016). This greatly expanded potential of cryo-EM has led an increasing number of crystallographers to adopt the new technique, helped by the fact that both techniques of structure research share many mathematical concepts and modeling tools.

The aim of this article is to give an overview of the recent gain in fundamental knowledge on ribosome structure and function, now that cryo-EM has joined X-ray crystallography as a powerful alternative technique of high-resolution structure research. However, the ensuing large proliferation of structures precludes an exhaustive coverage, and some areas receive particular attention.

Readers unfamiliar with the ribosome may benefit from a brief introduction to its structure (Fig. 1). In all organisms, the ribosome is a 2- to 4.5-megadalton (MDa) structure made up of a small subunit (SSU) and a large subunit (LSU). Both subunits are composed of rRNA surround-

 Cite this article as *Cold Spring Harb Perspect Biol* doi: 10.1101/cshperspect.a032615

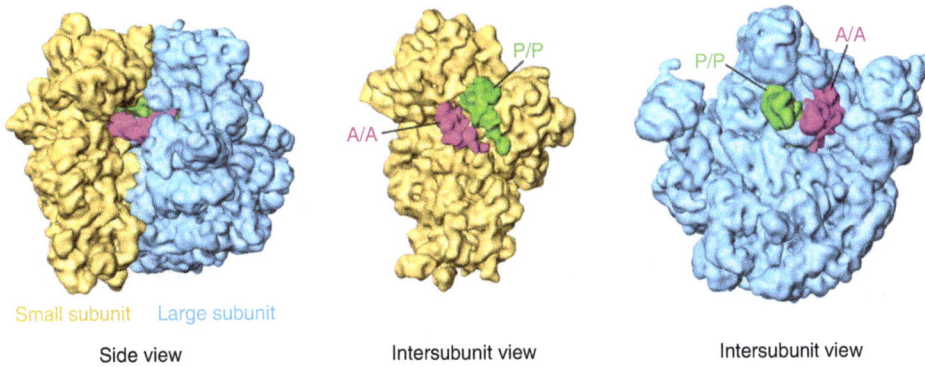

Side view   Intersubunit view   Intersubunit view

Figure 1. General features of ribosome structure and positions of transfer RNAs (tRNAs). The large subunit appears in blue and the small subunit in yellow. A-site (A/A) tRNA is pink, and P-site (P/P) tRNA is green. Note that the E site is to the *right* of the P site as shown from the intersubunit view of the small subunit and simultaneously to the *left* of the P site as shown from the view of the large subunit. A hybrid-state A/P tRNA would appear as an occupied A site on the small subunit and an occupied P site on the large subunit; in the same way, a P/E tRNA occupies the P site on the small subunit and the E site on the large subunit. (From Agirrezabala et al. 2008; adapted, with permission, from Elsevier © 2008.)

ed by many smaller ribosomal proteins (r-proteins). From a side view, the solvent side of the SSU appears relatively flat, whereas that of the LSU is rounded. A view of the SSU's intersubunit side displays its head region, complete with a leftward-facing beak, above a thin neck region that separates the head from the body of the subunit. The body features a large platform on the right and a smaller shoulder on the left, narrowing somewhat toward a pointed left foot and right foot in eukaryotes, or to a single spur in prokaryotes. Meanwhile, the solvent side of the large subunit appears roughly round with the central protuberance at the top. The messenger RNA (mRNA) passes through the mRNA channel created by the groove between the head and the body of the SSU at its intersubunit face. The mRNA and tRNAs both pass through the three adjacent A (aminoacyl), P (peptidyl), and E (exit) sites, which are located on both the SSU and LSU intersubunit faces.

EUBACTERIAL TRANSLATION AND TRANSLATIONAL CONTROL

In previous studies, before 2013, cryo-EM reconstructions could only be used to study large conformational changes such as intersubunit rotation or subunit head swivel, at intermediate

($\sim$5–9 Å) resolutions. At high (2.5–3.5 Å) resolutions readily attainable now, where domain motions and molecular contacts are much better defined, insights have been gained into key events during translation, and into diverse mechanisms of translational control. Additionally, the capability of single-particle cryo-EM to visualize multiple states in a single sample has particularly benefited the understanding of multistep processes such as mRNA-tRNA translocation.

Detailed structural studies have focused on two pivotal stages of the polypeptide elongation cycle (see reviews by Voorhees and Ramakrishnan 2013; Rodnina and Wintermeyer 2016): elongation factor G (EF-G)-assisted mRNA-tRNA translocation and tRNA selection/decoding. As fast and highly dynamic events involving a multitude of domain motions, these are very hard to capture by cryo-EM because of the high degree of ensuing heterogeneity. Particular difficulties are posed by structural states that differ in structural constellations (e.g., base flipping) in a local region only, whereas remaining virtually unchanged elsewhere. In these cases, a technique called local classification can still be used, by singling out the region of interest through imposition of an appropriate mask in the application of the classification algorithm (Penczek et al. 2006). In a variant of the technique introduced by Scheres (2016), the projected average density is subtracted from the data, thereby increasing the sensitivity of the local classification.

Research on the structural basis of mRNA-tRNA translocation was recently reviewed by Ling and Ermolenko (2016) and Frank (2017). Despite a long history of structural studies, the role of EF-G as it interacts with the ribosome during mRNA-tRNA translocation is still not well understood, even though there is basic agreement on the factor's catalytic action leading to a large (10^4- to 10^6-fold) acceleration of the translocation rate (Katunin et al. 2002). Brownian intersubunit motion of the ribosome has been observed in the absence of factors (Agirrezabala et al. 2008, 2012; Cornish et al. 2008; Julián et al. 2008), and the role of EF-G as providing a "pawl" in the promotion of translocation is generally accepted, implying the existence of some type of power stroke. Specifically, this pawl is manifest in the tip of EF-G's domain IV, which disrupts the codon–anticodon interaction (Frank et al. 2007; Taylor et al. 2007), a notion that has received renewed support in a recent mutation study (Liu et al. 2014).

Intermediate states in factor-free intersubunit rotation (Fischer et al. 2010; Agirrezabala et al. 2012), also observed by X-ray crystallography (Zhang et al. 2009; Tourigny et al. 2013; Zhou et al. 2014), reflect a trajectory across the ribosome's "metastable" free-energy landscape accessible through ambient thermal energy (Munro et al. 2009). Brilot et al. (2013) used cryo-EM to determine the structure of the 70S ribosome bound with EF-G, trapped in the pretranslocation state using the antibiotic viomycin. The structure shows a normally transient state where the A-site tRNA is still bound to the ribosome as well. A surprising discovery of the Steitz group made by X-ray crystallography (Lin et al. 2015) is a large conformational change of EF-G that apparently occurs on the binding of this factor to the ribosome. It is still unclear how this result fits into the overall picture of translocation. In their cryo-EM study, Li et al. (2015) were able to trap EF-G on the pretranslocational ribosome in two different binding configurations, one bound to the rotated and the other

to the unrotated ribosome, making use of the mutation H94A. The results of the study allowed the investigators to formulate necessary and sufficient conditions for GTPase activation. Accordingly, stabilization of the factor by contacts with protein S12 and the L11-lobe is required, a condition that is only fulfilled in the rotated configuration.

In the area of tRNA selection, the most interesting question is the discrimination between cognate and near-cognate codon–anticodon pairing, essential for the fidelity of translation. Here it is clear from previous work (Ogle et al. 2001) that discrimination is achieved as a result of at least two successive steps of probing the short codon–anticodon helix with the help of three bases: A1492, A1493, and G650. It has been a challenge to image the ribosome-A/T-tRNA complex for a programmed ribosome and bring order into the sequence of events. Here a comparison between two cryo-EM studies, Agirrezabala et al. (2011) and Loveland et al. (2017) shows the progress recently achieved in the examination of the near-cognate case in a striking way, as the former did not achieve resolution sufficient to delineate the positions of the bases with certainty, whereas the latter shows the complex at close to 3 Å resolution in multiple states of intermediate engagement. In passing, it should also be noted that Fischer et al. (2015), working with cryo-EM, achieved 2.9 Å resolution with the cognate aminoacylated tRNA-EF-Tu•GDP-ribosome complex stabilized with kirromycin, a big step forward from the previous reconstruction of the same complex by Villa et al. (2009), while shedding no new light upon dynamics of tRNA selection. The main gain of the study was the identification and enumeration of rRNA modifications.

It is increasingly recognized that structural snapshots are not sufficient to understand translocation (or any of the steps of translation, for that matter) and that information on real-time dynamics is required. A whole series of recent studies uses single-molecule fluorescence resonance energy transfer (FRET) to study the timing of binding, domain motions on EF-G, intersubunit rotation, small subunit head swivel, and mRNA-tRNA translocation with respect to the

Cite this article as *Cold Spring Harb Perspect Biol* doi: 10.1101/cshperspect.a032615

small subunit (Chen et al. 2011, 2014, 2016; Adio et al. 2015; Salsi et al. 2015; Sharma et al. 2016; Wasserman et al. 2016; Kim and Tinoco 2017). A detailed model for the sequence of events inferred from structural and single-molecule FRET (smFRET) studies is described by Wasserman et al. (2016). A new method called single-molecule polarized total internal reflection fluorescence (polTIRF) microscopy provides information not just on distance but on relative orientation between domains. Application of this method to the ribosome indeed allowed the investigators to observe a power stroke of domain IV of EF-G during translocation (Chen et al. 2016).

Important inroads have also been made into the understanding of translational control in eubacteria involving the binding of specific protein factors, an advance helped by the improvement in spatial resolution. Here the emerging common theme is the existence of very specific mechanisms for structural recognition of an abnormal state of the translation process, often going hand-in-hand with stress conditions in the cell, which requires intervention and rescue. EttA is a newly found ATP-binding cassette (ABC) trans-

porter protein, which is found to bind at the ribosomal E site and suppresses translation when the energy supply in the cell falls below a critical level (Boel et al. 2014; Chen et al. 2014). EttA has a "feeler" domain that is apparently able to sense the presence or absence of a peptide on the P-site tRNA and controls the action of the factor through a mechanism of conformational signaling accordingly. RelA is a protein factor active in the stringent response, a response to conditions of amino acid starvation. In *Escherichia coli*, RelA synthesizes alarmones on binding to the ribosome and encountering a deacylated tRNA at the A site (Agirrezabala et al. 2013). Recent cryo-EM studies have uncovered the detailed molecular mechanism, which involves sensing the absence of the aminoacyl group on the CCA end of the strongly distorted, A/T-shaped A-site tRNA (Fig. 2) (Brown et al. 2016; Loveland et al. 2016). BipA is a GTPase closely related to EF-G and engaged at the same factor-binding site of the ribosome, but only if a tRNA is bound at the A site. Curiously, the ribosome was found in the rotated position (Kumar et al. 2015), again implying specific recognition of a local conformational anomaly.

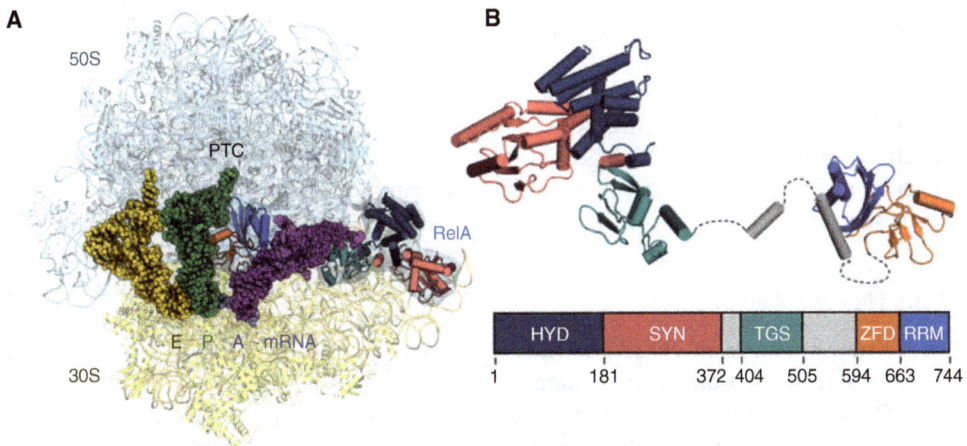

Figure 2. RelA in action on the ribosome, as revealed by cryoelectron microscopy (cryo-EM). (*A*) RelA monitors the CCA end of an incoming putative aminoacyl-transfer RNA (tRNA) and binds to the ribosome on recognizing the absence of an amino acid. (*B*) Model of ribosome-bound form of RelA oriented from amino to carboxyl terminus, with the domain organization shown below; hydrolase (HYD), synthetase (SYN), TGS (threonyl-tRNA synthetase, GTPase and SpoT), Zinc-finger (ZFD) and RNA recognition motif (RRM) domains. Flexible elements between ordered RelA domains are shown in dashed lines. (From Brown et al. 2016; adapted, with permission, from Springer Nature © 2016.)

BipA is known to be essential to survival at low temperature, nutrient depletion, and various other stress conditions, but the molecular mechanism of its action is as yet unknown. Finally, ArfA is a factor that recruits release factor RF2 on encountering a ribosome containing truncated mRNA, a condition characterized by the presence of a peptidyl-tRNA in the P site and an empty A site. Thus ArfA rescues translation in bacteria lacking the trans-translation mechanism. In their recent work, Demo et al. (2017) were able to elucidate the structure of the ArfA-RF2-bound ribosome and describe its likely mechanism of action. Interestingly, this capability adds to the versatility of RF2, the protein factor already known from its classical role as a class I release factor in stop codon recognition and, more recently, in translation quality control (Zaher and Green 2009).

EUKARYOTIC RIBOSOME STRUCTURE AND FUNCTION

In the past few years, studies of ribosome structure for eukaryotes have led to a better appreciation of their diversity, particularly because of the great variation in rRNA expansion segments and the addition of a variety of eukaryotic-specific proteins. As we describe below, special cases are the ribosomes of parasitic protozoans, which for the first time reached resolutions that bring them within reach of drug design, as resolutions in the 2.5 to 3.5 Å range allow bound drug molecules to be directly located and visualized.

Yeast and Higher Eukaryotes

The first X-ray structures of eukaryotic ribosomes, denoted as 80S ribosomes, each comprising one 40S small subunit and one 60S large subunit, were from *Tetrahymena thermophila* (a 40S•eIF1 complex [Rabl et al. 2011] and a 60S•eIF6 complex [Klinge et al. 2011]) and yeast (80S [Ben-Shem et al. 2011]), and were reviewed by Wilson and Doudna Cate (2012) and Klinge et al. (2012). Specific progress with the ribosome structure of yeast was discussed by Yusupova and Yusupov (2014). Yusupov's group also pub-

lished 16 X-ray structures of the drug-bound yeast ribosome (de Loubresse et al. 2014; Yusupova and Yusupov 2017). 80S ribosomes from a growing number of species were solved by cryo-EM in various states of translation, at close to atomic resolutions: canine (Voorhees and Hegde 2016), porcine (Voorhees et al. 2013, 2014), and human (Khatter et al. 2015).

A number of studies focused on translation mediated by IRESs from viruses on ribosomes from yeast (Fernández et al. 2014; Abeyrathne et al. 2016; Murray et al. 2016), rabbit (Hashem et al. 2013c; Yamamoto et al. 2014, 2015; Muhs et al. 2015), and human (Quade et al. 2015). Besides adding to our rapidly growing structural knowledge base, these reconstructions offer novel insights into the way the eukaryotic ribosome is hijacked by viral mRNA. These works are reviewed in further detail below.

Parasitic Protozoans

In recent cryo-EM studies, several structures of ribosomes from a number of protozoan parasites have been solved at near-atomic resolutions (2.5 to 3 Å), elucidating their unique features, which are distinct from those of other eukaryotes. The resulting structures provide a basis for future functional studies of the translational machinery of protozoan parasites and have established the ribosome as a promising drug target against these human pathogens. These include a number of trypanosomatids and *Plasmodium falciparum*, all causing debilitating and often-fatal diseases. Current treatments against these parasites have poor efficacy, high toxicity, and increasing levels of drug resistance (Andrews et al. 2014). The importance of these parasites in epidemic diseases warrants a special highlight in our review.

Among pathogenic protozoans, the trypanosomatids, *Trypanosoma cruzi*, *Trypanosoma brucei*, and *Leishmania* spp., are a unique family causing insect-borne diseases: Chagas disease, African trypanosomiasis (sleeping sickness), and Leishmaniasis, respectively. Altogether, an estimated 37 million people are infected with these parasites worldwide. The first cryo-EM structures of ribosomes from this family (for

T. cruzi at 12 Å [Gao et al. 2005] and for *T. brucei* at 5.5 Å [Hashem et al. 2013a]) provided a first glimpse and overview of their architecture. More detailed insights have only recently emerged with the advent of the new direct-detector technology. Cryo-EM reconstructions of native 60S subunits from *T. cruzi* at 2.5 Å (Liu et al. 2016, 2017) and *Leishmania donovani* at 2.8 Å (Shalev-Benami et al. 2016), as well as 80S ribosomes of *L. donovani* at 2.9 Å (Zhang et al. 2016) have revealed the ribosome structure in unprecedented detail, making it possible to discern rRNA nucleotides, amino acid side chains, rRNA modifications, solvent molecules, and metal ions. As a result, atomic models are now available for the ribosomes of all three species.

A very striking feature of trypanosomatid ribosomes is the fragmentation of the 28S rRNA in the large subunit, discovered by use of nucleic acid electrophoresis (Campbell et al. 1987). Following the terminology of Hashem et al. (2013a), we refer to the six fragments as LSU-α, LSU-β, and srRNA1 through srRNA4. The two largest pieces, LSU-α and LSU-β, are located on the solvent and interface side, respectively, and provide the scaffold for the large subunit. Locations of these rRNA components and the unique ways they are knitted together in the mature ribosome were described in various depths (Fig. 3) (Hashem et al. 2013a; Liu et al. 2016; Shalev-Benami et al. 2016; Zhang et al. 2016). The srRNA1 piece is at the bottom of the large subunit, as viewed from the solvent side, opposite the 5S rRNA. It lies close to the 5.8S rRNA, without being associated with it. The remaining three pieces srRNA2-4 are in close contact with one another and situated mainly on the left side of the large subunit under the P

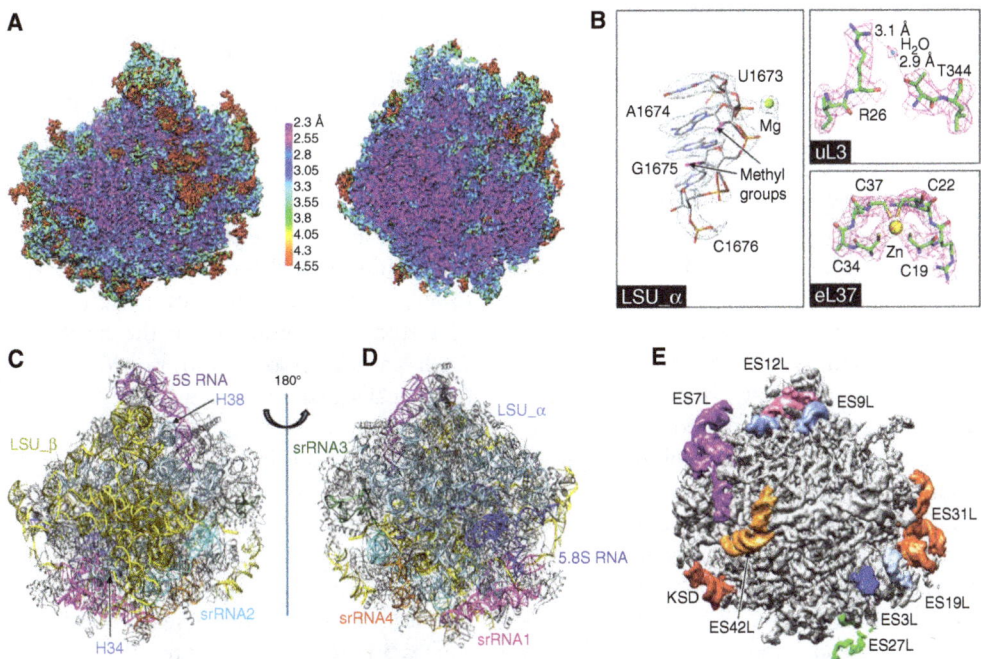

Figure 3. Structure of the large ribosomal subunit of *Trypanosoma cruzi*. (*A*) Sharpened cryoelectron microscopy (cryo-EM) map of the 60S subunit, colored by local resolution and viewed from the subunit interface. (*Left*) Surface view. (*Right*) Central cutaway view. (*B*) Selected views of density for ribosomal RNA (rRNA) and proteins, with associated ions and water molecule. (*C,D*) rRNA architecture of the large subunit: interface (*C*) and solvent (*D*) view. (*E*) Some rRNA expansion segments in the unsharpened map of a large subunit, viewed from the solvent side. (From Liu et al. 2016; adapted, with permission, from the National Academy of Sciences in conjunction with Creative Commons licensing.)

stalk (solvent-side view). Strikingly, despite the fragmented nature of the rRNA, the ribosomal functional cores adopt the complete 3D architecture that is common to other eukaryotic ribosomes. Although the exact mechanism of rRNA fragmentation in trypanosomatids is unknown, the high-resolution structures of the *T. cruzi* and *L. donovani* 60S large ribosomal subunit suggest that cleavage occurs during or postassembly (Liu et al. 2016; Shalev-Benami et al. 2016). Further, these structures contain clues on how these rRNA fragments are most likely assembled (Liu et al. 2016). In fact, this "archeological" approach to the interpretation of structure has resulted in a proposal for the chronology of biosynthesis of trypanosomatid ribosomes (Liu et al. 2016).

Compared with other eukaryotic ribosomes, in particular with the yeast ribosome, the cytosolic ribosome in trypanosomatids shows a set of trypanosome-specific components both in RNAs and in proteins. In the large subunit, a number of rRNA kinetoplastid-specific expansion segments (ESs) are located in the periphery and are unusually large. Many r-proteins have carboxy-terminal or amino-terminal extensions. Most remarkably, the r-protein eL19 contains an extremely long carboxy-terminal extension, doubling its size compared to its yeast counterpart. In addition, kinetoplastid ribosomes lack the r-protein eL41, resulting in the absence of the eukaryote/archaea-specific intersubunit bridges eB12 and eB14. In contrast to the large differences seen in the large subunit rRNA, the small subunit is more conserved.

P. falciparum, an apicomplexan parasite transmitted by the bite of an infected female mosquito (*Anopheles* spp.), is the causative agent of the most severe form of human malaria. Malaria constitutes an immense global public health burden; in 2015, it caused an estimated 214 million clinical episodes and 438,000 deaths (www .who.int/mediacentre/factsheets/fs094/en). Using cryo-EM, the parasite-specific features and structural dynamics of the *P. falciparum* 80S (Pf80S) ribosome were recently characterized (Wong et al. 2014, 2017; Sun et al. 2015). In the first of these studies, Wong et al. (2014) solved the structure of the Pf80S ribosome bound to emetine, a eukaryotic protein synthesis inhibitor, at a resolution of 3.2 Å. The Pf80S ribosome displays extensive differences in the lengths of its rRNA expansion segments and protein extensions in comparison to the human ribosome.

As for the r-proteins, 15 of these are extended in *P. falciparum* when compared to yeast ribosomes, which are commonly used as a reference against other eukaryotic ribosomes. Importantly, these extensions result in unique interactions not observed in the ribosomes of the human host (Wong et al. 2014; Sun et al. 2015). For instance, r-protein eL41 has a 14-residue amino-terminal extension that reaches into a pocket formed by 18S rRNA, bridging the large and small subunits. An additional small bridge is formed between the platform of the small subunit and the region around the L1 stalk via interactions between the carboxy-terminal helix extension of eL8 and the carboxy-terminal helix eS1. Stabilizing interactions are also observed near the L1 and P stalks, likely with functional implications on translation in *P. falciparum*. Sun et al. (2015), with a sample purified from a schizont-stage *P. falciparum* cell extract, obtained five reconstructions showing the ribosome either empty or in four different tRNA-binding states, two of which (an 80S with tRNAs in the P/P and E/E sites, and a rotated 80S bound by hybrid-state A/P and P/E tRNAs) are recognizable as part of the canonical elongation cycle. This study is an example for the capability of single-particle cryo-EM to provide multiple functionally relevant structures from the same sample.

Altogether, the newly determined structures of ribosomes from human parasites provide a platform for functional studies of parasite-specific features (rRNA expansion segments, r-protein extensions, dynamics, assembly, etc.) and the development of new therapeutics. Future studies are expected to explore the binding of translation initiation, elongation, and release factors in these parasites, as they are incompletely understood. Thus, there is vast potential for high-resolution cryo-EM to characterize the functions of parasite-specific elements as well as complexes to guide the design of drugs targeting their translational machinery.

Cite this article as *Cold Spring Harb Perspect Biol* doi: 10.1101/cshperspect.a032615

Canonical Eukaryotic Translation Initiation

Eukaryotic translation initiation occurs in four stages: 43S preinitiation complex formation, mRNA attachment, scanning and start codon recognition, and large ribosomal subunit joining. This process requires the concerted action of over a dozen protein factors at different times, with concomitant conformational changes in those factors, the ribosome, tRNA, and mRNA (Fig. 4).

In the field of eukaryotic initiation, structural knowledge has lagged far behind the knowledge in other steps of translation. This has been the result mainly of the low stability and structural variability of the various complexes engaged in preinitiation and their pathway toward assembly. The resulting compositional and conformational heterogeneity of such complexes poses challenges in visualization. A breakthrough was achieved with the dis-

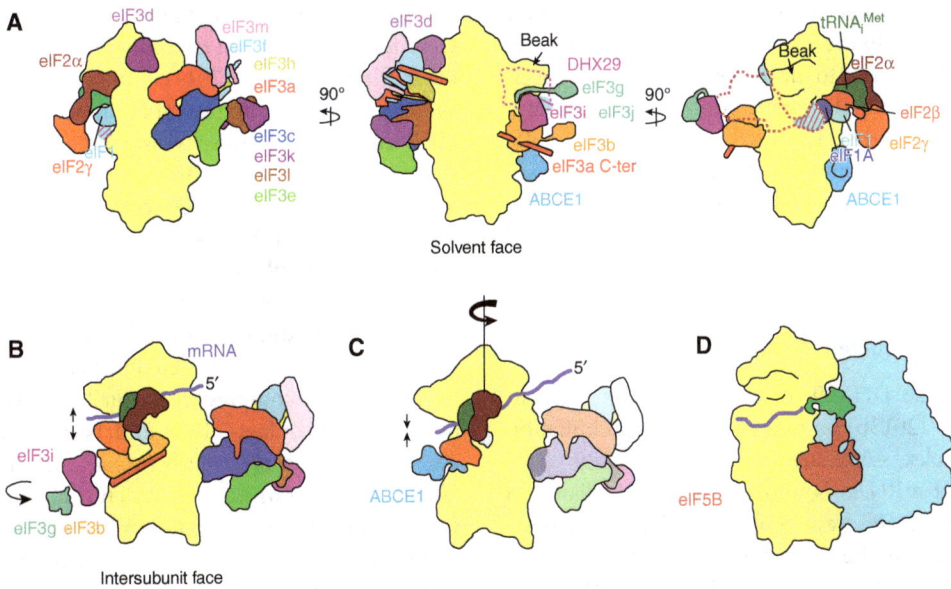

Figure 4. A model of eukaryotic translation initiation. (Note that eukaryotic initiation factor [eIF]4 subunits and eIF5 are not shown.) (A) 43S complex. Factors eIF1, eIF1A, ternary complex, and eIF3 bind to the 40S subunit. DHX29 is required only for initiation on highly structured messenger RNAs (mRNAs) in mammals, and is shown as a dotted outline. The binding site of the substoichiometric subunit eIF3j overlaps with part of DHX29's binding site; the area of overlap is indicated with diagonal lines. The mammalian homolog of factor eIF3 is shown; yeast eIF3 comprises only core subunits a and c and peripheral subunits b, i, g, and j. Factor ABCE1 likely binds to the 40S subunit interface, at a conserved GTPase binding site. (B) 48S-open complex. mRNA is delivered to the complex, and start codon scanning begins. The latch between rRNA helices 18 and 24 on the 40S subunit opens to allow mRNA entry. Subunits eIF3b, i, and g move from the solvent face to the intersubunit face of the 40S subunit, likely on mRNA binding. (C) 48S-closed complex. Scanning concludes on start codon recognition in the 40S subunit P site. The 40S subunit head rotates toward the solvent side of the subunit, aiding the closure of the latch. Subunits eIF3b, i, and g have dissociated from the 40S subunit. Factor ABCE1 binds the 40S subunit interface. Initiator tRNA is accommodated fully in the P site, caused, in part, by eIF2β's releasing its contacts with eIF1 and eIF1A in favor of contacting the 40S subunit head, moving the ternary complex toward the 40S subunit head away from the body, whereas the anticodon remains securely in the P site. Factors eIF1 (shown transparently) and the carboxy-terminal tail (CTT) of eIF1A (not shown) begin to dissociate. Factor eIF5 (not shown) catalyzes GTP hydrolysis on eIF2, after which both factors will dissociate. (D) 80S complex. Factor eIF5B catalyzes 60S subunit joining. Upon subunit joining, eIF5B will dissociate with eIF1A (not shown), leaving an elongation-competent 80S complex. (Figure based on data in des Georges et al. 2015 and Simonetti et al. 2016.)

covery that DHX29, a newly discovered initiation factor (Pisareva et al. 2008; Parsyan et al. 2009), stabilizes the preinitiation complex to some extent, sufficient for visualization by cryo-EM (Hashem et al. 2013b). It is worth noting, however, that only 4% of the data showed the complete complex (with the exception of eIF1), and that all other classes presented complexes where some of the factors were missing. It is only because of the power of today's classification methods that such a small portion of the data can be identified. The reason for the low yield of intact complexes is unknown, but it may be related to the low binding affinity of some of the factors and to the forces that act on the specimen during the preparation of the EM grid (Glaeser 2016). This low yield explains the relatively low resolution of the reconstruction. Nevertheless, in this work, the structure and position of eIF3 on the ribosome were well defined for the first time.

43S Preinitiation Complex

Helicase DHX29, first visualized by Hashem et al. (2013b) as part of the 43S preinitiation complex, is required to translate highly structured mRNAs bound to the mammalian 43S complex (Pisareva et al. 2008). DHX29 is positioned on the 40S subunit shoulder near the mRNA entry channel latch, consistent with a role in unwinding double-stranded stems of incoming mRNA. Additionally, the resolution of this 43S complex, 11.6Å, was sufficient to clarify the orientation of the eIF3 core octamer bound to the 40S subunit, which differed significantly from the previous understanding of its orientation (Siridechadilok et al. 2005); in its actual orientation, the eIF3 core contacts much less of the 40S subunit than previously thought.

des Georges et al. (2015) presented the first set of cryo-EM reconstructions that collectively show all but eIF3j of the 13 subunits of mammalian eIF3 in the context of the 43S complex. The PCI/MPN octameric core is attached to the platform of the 40S subunit via subunits eIF3a and eIF3c; the eIF3 core in mammals differs substantially from that in yeast, which comprises only subunits eIF3a and eIF3c. In a separate reconstruction, the investigators resolved peripheral subunits eIF3bgi on the solvent side of the 40S subunit near the shoulder. In a third reconstruction, a density seen previously (Hashem et al. 2013b) at low resolution is now visible in des Georges et al. (2015) at intermediate resolution and tentatively assigned to eIF3d, the first time a cryo-EM density is attributed to this peripheral subunit.

Subunit eIF3j was finally visualized by Aylett et al. (2015) as part of a yeast 40S•eIF1•eIF1A•eIF3 complex. These investigators enriched for complete complexes using a lysine cross-linking technique that tethered components to one another (Erzberger et al. 2014). Subunit eIF3j is a substoichiometric, nonessential component of eIF3, conserved between yeast and mammals (Valásek et al. 2001; see Browning et al. 2001); in the 40S•eIF1•eIF1A•eIF3 complex, eIF3j is situated at the mRNA entry channel on the 40S subunit shoulder at the intersubunit face, near the mRNA entry channel (Fraser et al. 2007). This binding site overlaps with that of DHX29, indicating that their binding associations in the 43S complex are mutually exclusive. Subunit eIF3j contacts eIF1A and, because eIF1 and eIF1A bind cooperatively (Maag and Lorsch 2003), the cryo-EM study by Aylett et al. (2015) is the first evidence that links binding of eIF3 to that of eIF1 and eIF1A. eIF3 makes relatively few contacts with the 40S subunit, so its interactions with eIF1 and eIF1A are particularly important.

Also visible for the first time in a cryo-EM reconstruction is the helical, spectrin-like carboxy-terminal domain of subunit eIF3a, which extends along the solvent side of the 40S subunit, from the eIF3ac subunit core situated on the platform to the peripheral eIF3bgi subunit module near the shoulder.

48S Translation Initiation Complex

For some time, it was difficult to directly assess structural differences between the stages of scanning and start codon recognition during initiation, because the two were thought to differ only by relatively small conformational changes and by start codon placement. Hussain et al. (2014) broke ground in this area, using a tRNA variant to stabilize start codon recognition to yield a reconstruction of a partial yeast 48S complex (py48S), containing the 40S subunit, eIF1, eIF1A, ternary complex (i.e., eIF2•GTP-Met-tRNA$_i^{Met}$ or nonhydrolyzable GTP analog), and mRNA, with partial putative eIF5 occupancy. The 4 Å structure, which represents an initiation complex just after start codon recognition, presents initiator tRNA, eIF1, and the amino-terminal tail (NTT) of eIF1A together for the first time. According to this structural mapping, the initiator tRNA interacts with the mRNA start codon in the P site; the eIF1A NTT is ordered and stabilizes the codon–anticodon duplex; and eIF1 is seen in a novel conformation in which its two β-hairpins have shifted to avoid a clash with the tRNA in its current close proximity to the mRNA. Additionally, the mRNA entry channel latch between h18 on the 40S subunit head and h34 on its body (Spahn et al. 2001) is in a closed state, as expected on start codon recognition. Furthermore, the reconstruction reveals a 13° clockwise swivel of the 40S subunit head, with the beak having moved toward the solvent face of the subunit, about the axis of h28, the "neck" between the 40S subunit head and body. This swivel motion brings h29 of the 18S rRNA from the 40S subunit head into contact with the anticodon stem-loop (ASL) of initiator tRNA, thereby stabilizing the tRNA in its start codon-bound state.

Llácer et al. (2015) built on this result by introducing at 6 Å the first open-state py48S complex, omitting eIF5 and using an mRNA variant to stabilize a scanning complex (with 40S subunit, eIF1, eIF1A, ternary complex, mRNA, and parts of eIF3 visible) in addition to another py48S-closed complex at 4.9 Å similar to the one in Hussain et al. (2014) (with the same components as the open complex in addition to eIF5). Here "open" and "closed" refer to the state of the mRNA entry channel latch. The 40S head tilts away from the body to open the latch during scanning, which permits mRNA movement through the now widened channel between the 40S subunit head and neck; the latch closes after start codon recognition, to "lock" the mRNA in position.

Specifically, Llácer et al. (2015) showed that on start codon recognition, h28 is compressed,

bringing the 40S subunit head closer to the body to close the latch. The 40S subunit body component of the P site is brought into contact with the initiator tRNA's ASL, helping to accommodate tRNA$_i$ fully in the P site. Concomitantly, subunit eIF2β releases its contacts with eIF1 and eIF1A in favor of contacting the 40S subunit head. As a result, the bulk of the ternary complex moves toward the 40S subunit head and away from the body, whereas the ASL is stationary in the 40S subunit P site (Llácer et al. 2015; Simonetti et al. 2016). In conjunction with rearrangements of other initiation factors, including eIF1, eIF1A, and eIF2α, the 40S subunit head positions the initiator tRNA in the P site in a closed, scanning-arrested state.

Simonetti et al. (2016), via a late-stage mammalian 48S subunit reconstruction (containing the 40S subunit, ternary complex, mRNA, eIF3i, and eIF3g), reinterpreted Llácer et al.'s results by swapping the attributions of eIF3b and eIF3i, both of which bear WD40 domains of similar size. The reassigned factor eIF3b appears on the intersubunit face of the 40S subunit in both open and closed complexes, which suggests that it has moved from the 40S subunit solvent face before scanning and start codon recognition, and likely on mRNA attachment. The relocated eIF3b contacts eIF2γ, maintaining the ternary complex in a scanning-competent conformation, as opposed to the 40S head-proximal conformation mentioned above on start codon recognition. A follow-up analysis of the same 48S subunit reconstruction (Mancera-Martínez et al. 2017) indicates that eIF3i and eIF3g likely relocate with eIF3b from the 40S subunit solvent face to the intersubunit face as a module, although whether eIF3i+g remain bound to the 40S subunit is unclear. Importantly, the investigators additionally reassign a density on the 40S subunit intersubunit face to the NTPase ABCE1, positing that ABCE1 binds the interface on start codon recognition, after dissociation of the eIF3b (along with i+g). Moreover, ABCE1 may cycle on and off the 40S subunit throughout ribosome recycling and subsequent translation initiation, possibly acting as an anti-association factor and a gatekeeper to multiple steps of translation initiation.

Fernández et al. (2013) visualized a complex representing the result of the final step of eukaryotic initiation, eIF5B-mediated subunit joining. Their 6.6 Å structure of factor eIF5B in complex with the yeast 80S ribosome and the initiator aminoacylated tRNA in the P/I state—a conformation in the P site with the large-subunit end bound to an initiation factor—was achieved by optimizing the sample in a stepwise fashion in response to shortcomings of early reconstructions of the complex, because of partial occupancy of some components and tRNA heterogeneity. The resultant reconstruction shows that eIF5B binds tightly to ribosomal protein eL40 at the base of the P stalk, such that there would be a steric clash if eL40 were ubiquitinated. In fact, as eL40 is ubiquitinated through the late stages of 60S subunit maturation (Fernández-Pevida et al. 2012), eIF5B promotes subunit joining strictly with mature 60S subunits. Furthermore, some aspects of this complex that are conserved in bacteria—the P/I conformation of the initiator tRNA, salient intersubunit rotation, and the general conformation of ribosome-bound eIF5B—corroborate an earlier low-resolution cryo-EM reconstruction in bacteria (Allen et al. 2005) that featured eIF5B's prokaryotic homolog IF2. This suggests that eIF5B's role in initiation is universally conserved, which would explain why eIF5B is required even when most initiation factors are not during many types of noncanonical 5′-cap-independent initiation.

With all these pieces in place, it is possible now to put a composite model together for the stepwise progression of canonical translation initiation in eukaryotes (Fig. 4).

IRES-Mediated Initiation and Translocation

Cryo-EM has advanced the understanding of noncanonical translation initiation via IRESs. In viruses, IRESs are classified into four types, according in part to the subset of initiation factors that they require. Type 3 IRESs, which require very few initiation factors, and type 4 IRESs, which require none, depart the most from canonical requirements, and recent cryo-EM studies have helped to elucidate the mechanisms of types 3 and 4 IRES function (Kwan

and Thompson 2018) and of IRES-bound ribosome behavior.

Type 3 IRESs, such as the classical swine fever virus (CSFV) IRES, bind to the 40S subunit in a way that clashes with the mammalian eIF3 core's ribosomal contacts, eS1, eS26, and eS27. Hashem et al. (2013c) observed that the IRES resolves the clash by displacing the eIF3 core entirely from the ribosome and instead binding eIF3 via the IRES's own domain III, holding eIF3 in a position 55 Å shifted and 60° rotated from its canonical position on the 40S subunit platform.

Initiation with the type 3 IRES appears to rely on 40S subunit *rolling*, a movement usually confined to the elongation stage during canonical translation. Specific to eukaryotes, rolling is a ~6° rotation about the longest axis of the 40S subunit, orthogonal to intersubunit rotation (Budkevich et al. 2011) with the beak moving closer to the 60S subunit. Canonically, rolling occurs in the early pretranslocation state where it serves to shift an incoming tRNA in the A/T conformation, in which the 3′ acceptor end of the tRNA is not yet situated in the 60S A site, into the A/A conformation, which is fully accommodated in the A site. Reverse rolling takes place concurrently with intersubunit reverse rotation to yield the posttranslocation complex (Budkevich et al. 2014). However, Yamamoto et al. (2014) showed that rolling occurs in the final steps of hepatitis C virus (HCV) IRES-mediated initiation. The subunit-joining GTPase eIF5B recruits the 60S subunit to a 40S•HCV-IRES•Met-tRNA$_i^{Met}$ complex in the rolled state, yielding a pretranslocation-like 80S complex presented at 8.2 Å. eIF5B•GTP hydrolysis induces reverse rolling as eIF5B begins to disengage from the complex; this posttranslocation-like state is captured in an 8.6 Å reconstruction. Yamamoto et al. propose that subsequent eIF5B dissociation yields an elongation-competent complex. Thus, the HCV IRES uses canonical-elongation-like rolling to complete initiation.

Fernández et al. (2014) obtained two reconstructions of a model type 4 IRES, the cricket paralysis virus (CrPV) intergenic IRES, bound to a yeast 80S ribosome in rotated and nonrotated states, at sufficiently high resolution (3.7 Å and 3.8 Å, respectively) to build an atomic model of the full-length IRES. They took advantage of the fact that the 80S ribosome from the yeast *Kluyveromyces lactis* is stable (Rodicio and Heinisch 2013) and does not aggregate at pH 6, and used it to build complexes with the CrPV IRES RNA, as RNA and especially RNA pseudoknots, which are found in the CrPV IRES, are stable at pH 6 (Li and Breaker 1999; Nixon and Giedroc 2000). Importantly, the resultant high-resolution structures revealed that the CrPV IRES binds the ribosome with its start codon-initiator tRNA mimic (Costantino et al. 2008), called pseudoknot I (PKI), in the A site rather than in the P site as the canonical start codon-initiator tRNA complex does; the 80S ribosome•CrPV IRES assembly therefore requires translocation to occur before elongation proceeds, and represents a pretranslocation complex rather than an initiation complex. In fact, the CrPV IRES-bound 80S ribosome is in equilibrium between rotated and nonrotated states (Fernández et al. 2014).

Type 3 and 4 IRESs both bind between the head and body of the 40S subunit in a way that constrains the flexibility of the 40S head position, tilting the head away from the body and thereby forcing the latch into its open position (Quade et al. 2015; Yamamoto et al. 2015; Murray et al. 2016). In this way, the IRESs mimic the induction of head tilt and latch opening by factors eIF1, eIF1A, and eIF3 during canonical initiation (Aylett et al. 2015; Llácer et al. 2015). In the case of the HCV IRES, the model type 3 IRES head tilt of about 17° persists until P-site tRNA binding occurs, which takes place after 60S subunit joining in HCV IRES-driven initiation (Yamamoto et al. 2015). The incoming tRNA displaces the HCV IRES domain II from its position as a wedge between the 40S subunit head and body, and domain II undergoes a 55° rotation resulting in contact with the 28S rRNA on the 60S subunit. In type 4 IRES-driven initiation, stem loops IV and V of the IRES insert themselves between the head and body of the 40S subunit (Fernández et al. 2014; Koh et al. 2014).

During eEF2-dependent translocation, the type 4 IRES must shift through the intersubunit space as a single flexible body. Two groups (Muhs et al. 2015; Murray et al. 2016) elucidated individual mid- and posttranslocation states, respectively, of ribosome-bound type 4 CrPV IRES. In a tour-de-force of cryo-EM data collection and classification, Abeyrathne et al. (2016) captured the entire arc of IRES translocational motion by five high-resolution (3.5 Å to 4.2 Å) cryo-EM reconstructions of the type 4 IRES from Taura syndrome virus (TSV) bound to the 80S ribosome, which was bound in turn by eEF2. The TSV IRES, in an extended conformation before translocation, compresses and then extends again in a stepwise fashion, whereas its PK1 moves from the A site to the P site. Concerted intersubunit rotation and 40S head swivel accompany the movement of the IRES. As the ribosome proceeds from the rotated state (3° more rotated than when eEF2 is absent [Murray et al. 2016]) to the nonrotated state, 40S subunit head swivel progresses from moderate (12°) to high (17°) to minimal (1°). The swiveling action toward the 60S subunit appears to align the 40S subunit head A-site region with the body P-site region to permit the IRES PK1 to advance from the A site to the P site (Abeyrathne et al. 2016) in addition to a previously known role in which swiveling permits P-site tRNA's shift to the E site (Spahn et al. 2004; Schuwirth et al. 2005).

CONCLUSION

When compared to the knowledge of only 4 years ago, the structural information available on ribosomes from eubacteria and eukaryotes is now much richer and more detailed. Especially, the addition of information on multiple states and dynamics obtained by single-particle cryo-EM has helped to broaden the functional knowledge base in eubacterial and eukaryotic translation. As we have shown, there are beautiful examples for the elucidation of detailed mechanisms in the recent structural literature on translational control and tRNA selection in eukaryotes, and IRES-mediated initiation and translocation in eukaryotes. In all, we can look forward to a much-enriched integration of biochemical, structural, and single-molecule

FRET data relating to translation than previously thought possible.

REFERENCES

*Reference is also in this collection.

Abeyrathne PD, Koh CS, Grant T, Grigorieff N, Korostelev AA. 2016. Ensemble cryo-EM uncovers inchworm-like translocation of a viral IRES through the ribosome. *eLife* **5:** e14874.

Adio S, Senyushkina T, Peske F, Fischer N, Wintermeyer W, Rodnina MV. 2015. Fluctuations between multiple EF-G-induced chimeric tRNA states during translocation on the ribosome. *Nat Commun* **6:** 7442.

Agirrezabala X, Lei J, Brunelle JL, Ortiz-meoz RF, Green R, Frank J. 2008. Visualization of the hybrid state of tRNA binding promoted by spontaneous ratcheting of the ribosome. *Mol Cell* **32:** 190–197.

Agirrezabala X, Schreiner E, Trabuco LG, Lei J, Ortiz-Meoz RF, Schulten K, Green R, Frank J. 2011. Structural insights into cognate vs. near-cognate discrimination during decoding. *EMBO J* **30:** 1497–1507.

Agirrezabala X, Liao HY, Schreiner E, Fu J, Ortiz-Meoz RF, Schulten K, Green R, Frank J. 2012. Structural characterization of mRNA-tRNA translocation intermediates. *Proc Natl Acad Sci* **109:** 6094–6099.

Agirrezabala X, Fernandez IS, Kelley AC, Carton DG, Ramakrishnan V, Valle M. 2013. The ribosome triggers the stringent response by RelA via a highly distorted tRNA. *EMBO Rep* **14:** 811–816.

Allen GS, Zavialov A, Gursky R, Ehrenberg M, Frank J. 2005. The cryo-EM structure of a translation initiation complex from *Escherichia coli*. *Cell* **121:** 703–712.

Andrews KT, Fisher G, Skinner-Adams TS. 2014. Drug repurposing and human parasitic protozoan diseases. *Int J Parasitol Drugs Drug Resist* **4:** 95–111.

Aylett CHS, Boehringer D, Erzberger JP, Schaefer T, Nenad Ban N. 2015. Structure of a Yeast 40S–eIF1–eIF1A–eIF3–eIF3j initiation complex. *Nat Struct Mol Biol* **22:** 269–271.

Ban N, Nissen P, Hansen JJ, Moore PB, Steitz TA. 2000. The complete atomic structure of the large ribosomal subunit at 2.4 Å resolution. *Science* **289:** 905–920.

Ben-Shem A, de Loubresse NG, Melnikov S, Jenner L, Yusupova G, Yusupov M. 2011. The structure of the Eukaryotic ribosome at 3.0 Å resolution. *Science* **334:** 1524–1529.

Boel G, Smith PC, Ning W, Englander MT, Chen B, Hashem Y, Testa AJ, Fischer JJ, Wieden HJ, Frank J, et al. 2014. The ABC-F protein EttA gates ribosome entry into the translation elongation cycle. *Nat Struct Mol Biol* **21:** 143–151.

Brilot AF, Korostelev AA, Ermolenko DN, Grigorieff N. 2013. Structure of the ribosome with elongation factor G trapped in the pretranslocation state. *Proc Natl Acad Sci* **110:** 20994–20999.

Brown A, Fernández IS, Gordiyenko Y, Ramakrishnan V. 2016. Ribosome-dependent activation of stringent control. *Nature* **534:** 277–280.

Browning KS, Gallie DR, Hershey JW, Hinnebusch AG, Maitra U, Merrick WC, Norbury C. 2001. Unified nomencla-

ture for the subunits of eukaryotic initiation factor 3. *Trends Biochem Sci* **26:** 284.

Budkevich T, Giesebrecht J, Altman RB, Munro JB, Mielke T, Nierhaus KH, Blanchard SC, Spahn CMT. 2011. Structure and dynamics of the mammalian ribosomal pre-translocation complex. *Mol Cell* **44:** 214–224.

Budkevich TV, Giesebrecht J, Behrmann E, Loerke J, Ramrath DJF, Mielke T, Ismer J, Hildebrand PW, Tung C-S, Nierhaus KH, et al. 2014. Regulation of the mammalian elongation cycle by subunit rolling: A eukaryotic-specific ribosome rearrangement. *Cell* **158:** 121–131.

Campbell DA, Kubo K, Graham Clark C, Boothroyd JC. 1987. Precise identification of cleavage sites involved in the unusual processing of trypanosome ribosomal RNA. *J Mol Biol* **196:** 113–124.

Chen C, Stevens B, Kaur J, Cabral D, Liu H, Wang Y, Zhang H, Rosenblum G, Smilansky Z, Goldman YE, et al. 2011. Single-molecule fluorescence measurements of ribosomal translocation dynamics. *Mol Cell* **42:** 367–377.

Chen B, Boel G, Hashem Y, Ning W, Fei J, Wang C, Gonzalez RL Jr, Hunt JF, Frank J. 2014. EttA regulates translation by binding the ribosomal E site and restricting ribosome-tRNA dynamics. *Nat Struct Mol Biol* **21:** 152–159.

Chen C, Cui X, Beausang JF, Zhang H, Farrell I, Cooperman BS, Goldman YE. 2016. Elongation factor G initiates translocation through a power stroke. *Proc Natl Acad* **113:** 7515–7520.

Cornish PV, Ermolenko DN, Noller HF, Ha T. 2008. Spontaneous intersubunit rotation in single ribosomes. *Mol Cell* **30:** 578–588.

Costantino DA, Pfingsten JS, Rambo RP, Kieft JS. 2008. tRNA-mRNA mimicry drives translation initiation from a viral IRES. *Nat Struct Mol Biol* **15:** 57–64.

De Loubresse NG, Prokhorova I, Holtkamp W, Rodnina MV, Yusupova G, Yusupov M. 2014. Structural basis for the inhibition of the eukaryotic ribosome. *Nature* **513:** 517–522.

Demo G, Svidritskiy E, Madireddy R, Diaz-Avalos R, Grant T, Grigorieff N, Sousa D, Korostelev AA. 2017. Mechanism of ribosome rescue by ArfA and RF2. *eLife* doi: 10.7554/eLife.23687.

des Georges A, Dhote V, Kuhn L, Hellen CU, Pestova TV, Frank J, Hashem Y. 2015. Structure of mammalian eIF3 in the context of the 43S preinitiation complex. *Nature* **525:** 491–495.

Erzberger JP, Stengel F, Pellarin R, Zhang S, Schaefer T, Aylett CHS, Cimermančič P, Boehringer D, Sali A, Aebersold R, et al. 2014. Molecular architecture of the 40S•eIF1•eIF3 translation initiation complex. *Cell* **158:** 1123–1135.

Fernández IS, Bai XC, Hussain T, Kelley AC, Lorsch JR, Ramakrishnan V, Scheres SH. 2013. Molecular architecture of a eukaryotic translational initiation complex. *Science* **342:** 1240585.

Fernández IS, Bai XC, Murshudov G, Scheres SH, Ramakrishnan V. 2014. Initiation of translation by cricket paralysis virus IRES requires its translocation in the ribosome. *Cell* **157:** 823–831.

Fernández-Pevida A, Rodríguez-Galán O, Díaz-Quintana A, Kressler D, de la Cruz J. 2012. Yeast ribosomal protein L40 assembles late into precursor 60S ribosomes and is required for their cytoplasmic maturation. *J Biol Chem* **287:** 38390–383407.

Fischer N, Konevega AL, Wintermeyer W, Rodnina MV, Stark H. 2010. Ribosome dynamics and tRNA movement by time-resolved electron cryomicroscopy. *Nature* **466:** 329–333.

Fischer N, Neumann P, Konevega AL, Bock LV, Ficner R, Rodnina MV, Stark H. 2015. Structure of the *E. coli* ribosome–EF-Tu complex at <3 Å resolution by Cs-corrected cryo-EM. *Nature* **520:** 567–570.

Frank J. 2017. The translation elongation cycle—Capturing multiple states by cryo-electron microscopy. *Philos Trans R Soc Lond B Biol Sci* **372:** 20160180.

Frank J, Gao H, Sengupta J, Gao N, Taylor DJ. 2007. The process of mRNA-tRNA translocation. *Proc Natl Acad Sci* **104:** 19671–19678.

Fraser CS, Berry KE, Hershey JW, Doudna JA. 2007. eIF3j is located in the decoding center of the human 40S ribosomal subunit. *Mol Cell* **26:** 811–819.

Gao H, Ayub MJ, Levin MJ, Frank J. 2005. The structure of the 80S ribosome from *Trypanosoma cruzi* reveals unique rRNA components. *Proc Natl Acad Sci* **102:** 10206–10211.

Glaeser RM. 2016. How good can cryo-EM become? *Nat Methods* **13:** 28–32.

Hashem Y, des Georges A, Fu J, Buss SN, Jossinet F, Jobe A, Zhang Q, Liao HY, Grassucci RA, Bajaj C, et al. 2013a. High-resolution cryo-electron microscopy structure of the *Trypanosoma brucei* ribosome. *Nature* **494:** 385–389.

Hashem Y, des Georges A, Dhote V, Langlois R, Liao HL, Grassucci RA, Hellen CUT, Pestova TV, Frank J. 2013b. Structure of the mammalian ribosomal 43S preinitiation complex bound to the scanning factor DHX29. *Cell* **153:** 1108–1119.

Hashem Y, des Georges A, Dhote V, Langlois R, Liao HY, Grassucci RA, Pestova TV, Hellen CU, Frank J. 2013c. Hepatitis-C-virus-like internal ribosome entry sites displace eIF3 to gain access to the 40S subunit. *Nature* **503:** 539–543.

Hussain T, Llácer JL, Fernández IS, Munoz A, Martin-Marcos P, Savva CG, Lorsch JR, Hinnebusch AG, Ramakrishnan V. 2014. Structural changes enable start codon recognition by the eukaryotic translation initiation complex. *Cell* **159:** 597–607.

Julián P, Koneveg AL, Scheres SHW, Lázaro M, Gil D, Wintermeyer W, Rodnina MV, Valle M. 2008. Structure of ratcheted ribosomes with tRNAs in hybrid states. *Proc Natl Acad Sci* **105:** 16924–16927.

Katunin VI, Savelsbergh A, Rodnina MV, Wintermeyer W. 2002. Coupling of GTP hydrolysis by elongation factor G to translocation and factor recycling on the ribosome. *Biochemistry* **41:** 12806–12812.

Khatter H, Myasnikov AG, Natchiar SK. 2015. Structure of the human 80S ribosome. *Nature* **520:** 640–645.

Kim HK, Tinoco I Jr. 2017. EF-G catalyzed translocation dynamics in the presence of ribosomal frameshifting stimulatory signals. *Nucl Acids Res* **45:** 2865–2874.

Klinge S, Voigts-Hoffmann F, Leibundgut M, Arpagaus S, Ban N. 2011. Crystal structure of the eukaryotic 60S ribosomal subunit in complex with initiation factor 6. *Science* **334:** 941–948.

Klinge S, Voigts-Hoffmann M, Leibundgut M, Ban N. 2012. Atomic structures of the eukaryotic ribosome. *Trends Biochem Sci* **37:** 189–198.

Koh CS, Brilot AF, Grigorieff N, Korostelev AA. 2014. Taura syndrome virus IRES initiates translation by binding its tRNA-mRNA-like structural element in the ribosomal decoding center. *Proc Natl Acad Sci* **111:** 9139–9144.

Kumar V, Chen Y, Ero R, Ahmed T, Tan J, Li Z, Wong ASW, Bhushan S, Gao YG. 2015. Structure of BipA in GTP form bound to the ratcheted ribosome. *Proc Natl Acad Sci* **112:** 10944–10949.

* Kwan T, Thompson SR. 2018. Noncanonical translation initiation in eukaryotes. *Cold Spring Harb Perspect Biol* doi: 10.1101/cshperspect.a032672.

Li Y, Breaker RR. 1999. Kinetics of RNA degradation by specific base catalysis of transesterification involving the 2′-hydroxyl group. *J Am Chem Soc* **121:** 5364–5372.

Li W, Liu Z, Koripella RK, Langlois R, Sanyal S, Frank J. 2015. Activation of GTP hydrolysis in mRNA-tRNA translocation by elongation factor G. *Sci Adv* **1:** e1500169.

Lin J, Gagnon MG, Bulkley D, Steitz TA. 2015. Conformational changes of elongation factor G on the ribosome during tRNA translocation. *Cell* **160:** 219–227.

Ling C, Ermolenko DN. 2016. Structural insights into ribosome translocation. *WIREs RNA* **7:** 620–636.

Liu G, Song G, Zhang D, Zhang D, Li Z, Lyu Z, Dong J, Achenbach J, Gong W, Zhao XS, et al. 2014. EF-G catalyzes tRNA translocation by disrupting interactions between decoding center and codon–anticodon duplex. *Nat Struct Mol Biol* **21:** 817–824.

Liu Z, Gutierrez-Vargas C, Wei J, Grassucci RA, Ramesh M, Espina N, Sun M, Tutuncuoglu B, Madison-Antenucci S, Woolford JL, et al. 2016. Structure and assembly model for the *Trypanosoma cruzi* 60S ribosomal subunit. *Proc Natl Acad Sci* **113:** 12174–12179.

Liu Z, Gutierrez-Vargas C, Wei J, Grassucci RA, Sun M, Espina N, Madison-Antenucci S, Tong L, Frank J. 2017. Determination of the ribosome structure to a resolution of 2.5 Å by single-particle cryo-EM. *Prot Sci* **26:** 82–92.

Llácer JL, Hussain T, Marler L, Aitken CE, Thakur A, Lorsch JR, Hinnebusch AG, Ramakrishnan V. 2015. Conformational differences between open and closed states of the eukaryotic translation initiation complex. *Mol Cell* **59:** 399–412.

Loveland AB, Bah E, Madireddy R, Zhang Y, Brilot AF, Grigorieff N, Korostelev AA. 2016. Ribosome•RelA structures reveal the mechanism of stringent response activation. *eLife* doi: 10.7554/eLife.17029.

Loveland AB, Demo G, Grigorieff N, Korostelev AA. 2017. Ensemble cryo-EM elucidates the mechanism of translation fidelity. *Nature* **546:** 113–117.

Maag D, Lorsch JR. 2003. Communication between eukaryotic translation initiation factors 1 and 1A on the yeast small ribosomal subunit. *J Mol Biol* **330:** 917–924.

Mancera-Martínez E, Querido J, Valasek LS, Simonetti A, Hashem Y. 2017. ABCE1: A special factor that orchestrates translation at the crossroad between recycling and initiation. *RNA Biol* doi: 10.1080/15476286.2016.1269993.

McDowall AW, Chang JJ, Freeman R, Lepault J, Walter CA, Dubochet J. 1983. Electron microscopy of frozen hydrated sections of vitreous ice and vitrified biological samples. *J Microsc* **131**: 1–9.

McMullan G, Chen S, Henderson R, Faruqi A. 2009. Detective quantum efficiency of electron area detectors in electron microscopy. *Ultramicroscopy* **109**: 1126–1143.

Muhs M, Hilal T, Mielke T, Skabkin MA, Sanbonmatsu KY, Pestova TV, Spahn CMT. 2015. Cryo-EM of ribosomal 80S complexes with termination factors reveals the translocated cricket paralysis virus IRES. *Mol Cell* **57**: 422–432.

Munro JB, Sanbonmatsu KY, Spahn CMT, Blanchard SC. 2009. Navigating the ribosome's metastable energy landscape. *Trends Biochem Sci* **34**: 390–400.

Murray J, Savva CG, Shin BS, Dever TE, Ramakrishnan V, Fernández IS. 2016. Structural characterization of ribosome recruitment and translocation by type IV IRES. *eLife* doi: 10.7554/eLife.13567.001.

Nixon PL, Giedroc DP. 2000. Energetics of a strongly pH dependent RNA tertiary structure in a frameshifting pseudoknot. *J Mol Biol* **296**: 659–671.

Ogle JM, Brodersen DE, Clemons WM Jr, Tarry MJ, Carter AP, Ramakrishnan V. 2001. Recognition of cognate transfer RNA by the 30S ribosomal subunit. *Science* **292**: 897–902.

Parsyan A, Shahbaziana D, Martineaua Y, Petroulakisa E, Alaina T, Larssona O, Mathonneta G, Tettweilera G, Hellen CU, Pestova TV, et al. 2009. The helicase protein DHX29 promotes translation initiation, cell proliferation, and tumorigenesis. *Proc Natl Acad Sci* **106**: 22217–22222.

Penczek PA, Frank J, Spahn CM. 2006. A method of focused classification, based on the bootstrap 3D variance analysis, and its application to EF-G-dependent translocation. *J Struct Biol* **154**: 184–194.

Pisareva VP, Pisarev AV, Komar AA, Hellen CU, Pestova TV. 2008. Translation initiation on mammalian mRNAs with structured 5′UTRs requires DExH-box protein DHX29. *Cell* **135**: 1237–1250.

Quade N, Boehringer D, Leibundgut M, van den Heuvel J, Ban N. 2015. Cryo-EM structure of Hepatitis C virus IRES bound to the human ribosome at 3.9 Å resolution. *Nat Commun* **6**: 8646.

Rabl J, Leibundgut M, Ataide SF, Haag A, Ban N. 2011. Crystal structure of the eukaryotic 40S ribosomal subunit in complex with initiation factor 1. *Science* **331**: 730–736.

Rodicio R, Heinisch JJ. 2013. Yeast on the milky way: Genetics, physiology and biotechnology of *Kluyveromyces lactis*. *Yeast* **30**: 165–177.

Rodnina MV, Wintermeyer W. 2016. Protein elongation, co-translational folding and targeting. *J Mol Biol* **428**: 2165–2185.

Salsi E, Farah E, Netter Z, Dann J, Ermolenko DN. 2015. Movement of elongation factor G between compact and extended conformations. *J Mol Biol* **427**: 454–467.

Scheres SH. 2016. Processing of structurally heterogeneous cryo-EM data in RELION. *Methods Enzymol* **579**: 125–157.

Schluenzen F, Tocilj A, Zarivach R, Harms J, Gluehmann M, Janell D, Bashan A, Bartels H, Agmon H, Franceschi F, et al. 2000. Structure of functionally activated small ribosomal subunit at 3.3 Å resolution. *Cell* **102**: 615–623.

Schuwirth BS, Borovinskaya MA, Hau CW, Zhang W, Vila-Sanjurjo A, Holton JM, Cate JH. 2005. Structures of the

bacterial ribosome at 3.5 Å resolution. *Science* **310**: 827–834.

Simonetti A, Brito Querido J, Myasnikov AG, Mancera-Martínez E, Renaud A, Kuhn L, Hashem Y. 2016. eIF3 peripheral subunits rearrangement after mRNA binding and start-codon recognition. *Mol Cell* **63**: 206–217.

Siridechadilok B, Fraser CS, Hall RJ, Doudna JA, Nogales E. 2005. Structural roles for human translation factor eIF3 in initiation of protein synthesis. *Science* **310**: 1513–1515.

Shalev-Benami M, Zhang Y, Matzov D, Halfon Y, Zackay A, Rozenberg H, Zimmerman E, Bashan A, Jaffe CL, Yonath A, et al. 2016. 2.8 Å cryo-EM structure of the large ribosomal subunit from the eukaryotic parasite *Leishmania*. *Cell Rep* **16**: 288–294.

Sharma H, Adio S, Senyushkina T, Belardinelli R, Peske F, Rodnina MV. 2016. Kinetics of spontaneous and EF-G-accelerated rotation of ribosomal subunits. *Cell Rep* **16**: 2187–2196.

Spahn CM, Kieft JS, Grassucci RA, Penczek PA, Zhou K, Doudna JA, Frank J. 2001. Hepatitis C virus IRES RNA-induced changes in the conformation of the 40s ribosomal subunit. *Science* **291**: 1959–1962.

Spahn CM, Gomez-Lorenzo MG, Grassucci RA, Jørgensen R, Andersen GR, Beckmann R, Penczek PA, Ballesta JPG, Frank J. 2004. Domain movements of elongation factor eEF2 and the eukaryotic 80S ribosome facilitate tRNA translocation. *EMBO J* **23**: 1008–1019.

Sun M, Li W, Blomqvist K, Das S, Hashem Y, Dvorin JD, Frank J. 2015. Dynamical features of the *Plasmodium falciparum* ribosome during translation. *Nucl Acids Res* **43**: 10515–10524.

Taylor DJ, Nilsson J, Merrill AR, Andersen GR, Nissen P, Frank J. 2007. Structures of modified eEF2 80S ribosome complexes reveal the role of GTP hydrolysis in translocation. *EMBO J* **26**: 2421–2431.

Tourigny DS, Fernandez IS, Kelley AC, Ramakrishnan V. 2013. Elongation factor G bound to the ribosome in an intermediate state of translocation. *Science* **340**: 1235490.

Valásek L, Phan L, Schoenfeld LW, Valásková V, Hinnebusch AG. 2001. Related eIF3 subunits TIF32 and HCR1 interact with an RNA recognition motif in PRT1 required for eIF3 integrity and ribosome binding. *EMBO J* **20**: 891–904.

Villa E, Sengupta J, Trabuco LG, LeBarron J, Baxter WT, Shaikh TR, Grassucci RA, Nissen P, Ehrenberg M, Schulten K, Frank J. 2009. Ribosome-induced changes in elongation factor Tu conformation control GTP hydrolysis. *Proc Natl Acad Sci* **106**: 1063–1068.

Voorhees RM, Hegde RS. 2016. Structure of the Sec61 channel opened by a signal sequence. *Science* **351**: 88–91.

Voorhees RM, Ramakrishnan V. 2013. Structural basis of the translational elongation cycle. *Annu Rev Biochem* **82**: 203–236.

Voorhees RM, Fernández IS, Scheres SHW, Hegde RS. 2013. Structure of the mammalian ribosome-Sec61 complex to 3.4 Å resolution. *Cell* **157**: 1632–1643.

Voorhees RM, Fernandez IS, Scheres SHW, Hegde RS. 2014. Structure of the mammalian ribosome-Sec61 complex to 3.4 Å resolution. *Cell* **157**: 1632–1643.

Wasserman MR, Alejo, JL, Altman, RB, Blanchard, SC. 2016. Multiperspective smFRET reveals rate-determining late intermediates of ribosomal translocation. *Nat Struct Mol Biol* **23:** 333–341.

Wilson DN, Doudna Cate JH. 2012. The structure and function of the Eukaryotic ribosome. *Cold Spring Harb Perspect Biol* doi: 10.1101/cshperspect.a011536.

Wimberly BT, Brodersen DE, Clemons WM, Morgan-Warren RJ, Carter AP, Vonrhein C, Hartsch T, Ramakrishnan V. 2000. Structure of the 30S ribosomal subunit. *Nature* **407:** 327–339.

Wong W, Bai X, Brown A, Fernandez IS, Hanssen E, Condron M, Tan YH, Baum J, Scheres SHW. 2014. Cryo-EM structure of the *Plasmodium falciparum* 80S ribosome bound to the anti-protozoan drug emetine. *eLife* **3:** e03080.

Wong W, Bai XC, Sleebs BE, Triglia T, Brown A, Thompson JK, Jackson KE, Hanssen E, Marapana DS, Fernandez IS, et al. 2017. Mefloquine targets the *Plasmodium falciparum* 80S ribosome to inhibit protein synthesis. *Nat Microbiol* **2:** 17031.

Yamamoto H, Unbehaun A, Loerke J, Behrmann E, Collier M, Bürger J, Mielke T, Spahn CM. 2014. Structure of the mammalian 80S initiation complex with initiation factor 5B on HCV-IRES RNA. *Nat Struct Mol Biol* **21:** 721–727.

Yamamoto H, Collier M, Loerke J, Ismer J, Schmidt A, Hilal T, Sprink T, Yamamoto K, Mielke T, Bürger J, et al. 2015. Molecular architecture of the ribosome-bound hepatitis C virus internal ribosomal entry site RNA. *EMBO J* **34:** 3042–3058.

Yusupova G, Yusupov M. 2014. High-resolution structure of the eukaryotic 80S ribosome. *Ann Rev Biochem* **83:** 467–486.

Yusupova G, Yusupov M. 2017. Crystal structure of eukaryotic ribosome and its complexes with inhibitors. *Philos Trans R Soc Lond B Biol Sci* **372:** 20160184.

Zaher HS, Green R. 2009. Quality control by the ribosome following peptide bond formation. *Nature* **457:** 161–166.

Zhang W, Dunkle J, Cate JHD. 2009. Structures of the ribosome in intermediate states of ratcheting. *Science* **325:** 1014–1017.

Zhang X, Lai M, Chang W, Yu I, Ding K, Mrazek J, Ng HL, Yang OO, Maslov DA, Zhou ZH. 2016. Structures and stabilization of kinetoplastid-specific split rRNAs revealed by comparing leishmanial and human ribosomes. *Nat Commun* **7:** 13223.

Zhou J, Lancaster L, Donohue JP, Noller HF. 2014. How the ribosome hands the A-site tRNA to the P site during EF-G-catalyzed translocation. *Science* **345:** 1188–1191.

Ribosome Profiling: Global Views of Translation

Nicholas T. Ingolia,[1] Jeffrey A. Hussmann,[2,3] and Jonathan S. Weissman[2,3]

[1]Department of Molecular and Cell Biology, University of California, Berkeley, California 94720

[2]Department of Cellular and Molecular Pharmacology, University of California, San Francisco, California 94158

[3]Howard Hughes Medical Institute, San Francisco, California 94158

Correspondence: ingolia@berkeley.edu; jonathan.weissman@ucsf.edu

The translation of messenger RNA (mRNA) into protein and the folding of the resulting protein into an active form are prerequisites for virtually every cellular process and represent the single largest investment of energy by cells. Ribosome profiling-based approaches have revolutionized our ability to monitor every step of protein synthesis in vivo, allowing one to measure the rate of protein synthesis across the proteome, annotate the protein coding capacity of genomes, monitor localized protein synthesis, and explore cotranslational folding and targeting. The rich and quantitative nature of ribosome profiling data provides an unprecedented opportunity to explore and model complex cellular processes. New analytical techniques and improved experimental protocols will provide a deeper understanding of the factors controlling translation speed and its impact on protein function and cell physiology as well as the role of ribosomal RNA and mRNA modifications in regulating translation.

The translation of messenger RNA (mRNA) into protein and the folding of the resulting polypeptide into an active form connect genetic information to functional proteins—a prerequisite for virtually every cellular process. Translation is also a costly biosynthetic process that often comprises the single largest investment of energy by cells (Verduyn et al. 1991; Russell and Cook 1995). Translation is thus highly regulated to ensure that the right proteins are made in the right places within the cell. Ensuring that newly made proteins fold, assemble, and function properly is also a major challenge to the cell (Gloge et al. 2014). The biogenesis of functional proteins depends on cotranslational folding chaperones as well as the speed of translation itself, and quality control of aberrant translation suppresses the harmful effects of mutations and errors (Brandman and Hegde 2016).

The central role played by translation has motivated the development of experimental approaches to analyze both the proteins produced by the cell and the process of their synthesis. In particular, the advent of genomics and gene expression profiling has driven interest in extending such global analyses to the study of translation (Vogel and Marcotte 2012). Beyond taking a complete and quantitative inventory of proteins produced in the cell, there is also great interest in the dynamic, multistep process of synthesizing these proteins. The ribosome incorporates amino acids with varying chemical properties by translating mRNAs that show biased use of synonymous codons, show secondary structures, and are decorated with chemical modifications. It seems natural to ask how these features affect the speed of the ribosome and what consequences result from varying the speed of elongation (Plotkin and Kudla 2011). These variations, encoded within the mRNA sequence, impact mRNA stability (Presnyak et al. 2015; Bazzini et al. 2016; Chan et al. 2017) and the rate of protein synthesis (Gingold et al. 2014), as well as the identity (Kawakami et al. 1993), folding (Kimchi-Sarfaty et al. 2007; Zhang et al. 2009), and localization (Pechmann et al. 2014) of the resulting protein.

Ribosome profiling, in which next-generation sequencing is used to identify ribosome-protected mRNA fragments, thereby revealing the positions of the full set of ribosomes engaged in translation, has emerged as a transformative technique for enabling global analyses of in vivo translation and coupled, cotranslational events. Historically, it has been challenging to measure these even in vitro; now, ribosome profiling provides a comprehensive view in living cells. It has been applied to address a remarkably broad diversity of mechanistic and physiological questions (Fig. 1). In this review, we present the historical context for ribosome profiling and highlight studies that exemplify the insights it can provide. We cannot at this point hope to cover all uses of this technique. It has been reviewed extensively (Michel and Baranov 2013; Ingolia 2014, 2016; Brar and Weissman 2015), and so we place emphasis here on recent developments.

RIBOSOME FOOTPRINT PROFILING OF IN VIVO TRANSLATION

Ribosome profiling relies on deep sequencing of ribosome footprints—the short (typically, ~30 nucleotide [nt]) fragments of mRNA that are physically enclosed by the ribosome and shielded from nuclease digestion (Fig. 2). These footprints are converted into a library of DNA fragments and analyzed by next-generation sequencing (Ingolia et al. 2012; McGlincy and Ingolia 2017). Each sequenced footprint reports on the position of one ribosome, revealing what transcript that ribosome was translating and where along the coding sequence it was captured during cell lysis, often with single-nucleotide resolution. Current deep-sequencing technologies analyze hundreds of millions of individual short reads in one experiment. When applied to libraries of ribosome footprints, this sequencing yields a comprehensive view of the translational landscape that can address many fundamental questions about translation (Fig. 1). Whereas there remain important experimental and analytical challenges to fully exploiting these data, especially in regard to the analysis of ribosome pause sites as discussed below, the presence of ribosome footprints indicates which sequences are being translated in the cell, and thus what protein is being produced. The overall density of ribosome footprints reflects the rate of translation occurring on different transcripts, allowing a direct and quantitative measure of how rapidly a cell is producing each of its proteins. These densities must be corrected for differences in the average elongation rate of each mRNA, which, when explored, have been

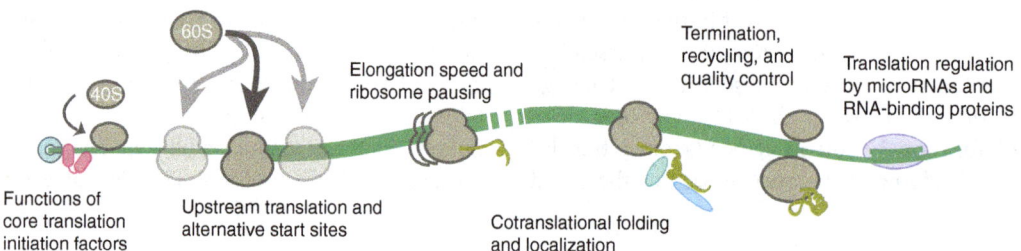

Figure 1. Insights from ribosome profiling. Ribosome profiling experiments have addressed many aspects of the mechanisms of protein synthesis and its regulation in the cell as well as related, cotranslational processes.

Cite this article as *Cold Spring Harb Perspect Biol* doi: 10.1101/cshperspect.a032698

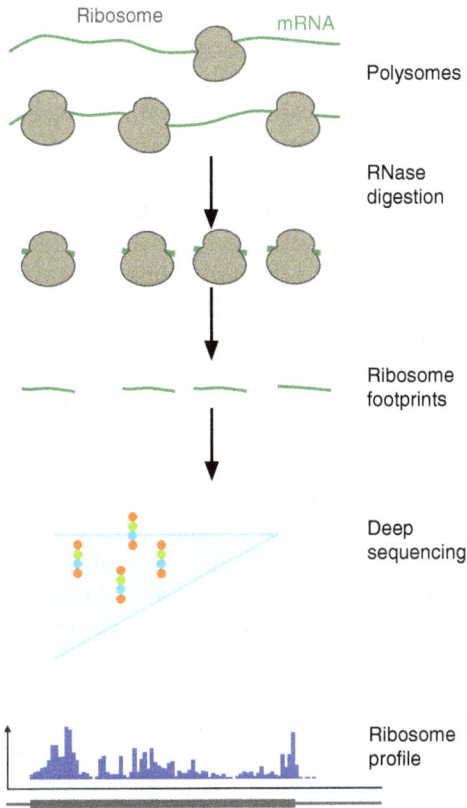

Figure 2. Ribosome footprint profiling. Steps in a typical ribosome profiling experiment are shown. Polysomes reflecting in vivo translation are isolated from cells and subjected to RNase digestion, which degrades unprotected messenger RNA (mRNA). The ribosome-protected footprints are analyzed by deep sequencing, schematized by the flowcell (light blue) with clusters of fluorescently labeled DNA attached to it (colored dots). Aligning these footprint sequences back to the transcriptome produces a quantitative profile of ribosome occupancy.

found to be relatively modest (Li et al. 2014). The detailed pattern of ribosome footprints within a coding sequence varies substantially, however, and reveals the relative speed of the ribosome along the transcript. As described below, these rich data can be augmented further, using drugs that modulate translation or targeted purification of interesting ribosomal subpopulations to address a wide array of biological questions.

QUANTIFYING GENE-SPECIFIC TRANSLATION

An important antecedent to the ribosome profiling approach was a study from Joan Steitz (1969), who mapped the sites of translation initiation in bacteriophage RNA by analyzing the RNase-resistant mRNA fragments protected by initiating ribosomes assembled in vitro. Subsequently, Wolin and Walter (1988, 1989) identified sites of in vitro translational pausing by mapping the relative density of ribosome-protected fragments using a primer extension assay. These studies made fundamental contributions to our understanding of translation through the analysis of ribosome footprints, but they were limited to the study of individual mRNAs translated in vitro.

Historically, studies of in vivo translation relied on analyzing polysomes recovered from cells and tissues (Mathews et al. 2000). These polysomes comprise multiple ribosomes translating a single mRNA template, and they can be fractionated according to the number of ribosomes they contain by ultracentrifugation through a sucrose density gradient. The distribution of an mRNA across these fractions reflects its translational status. Gene-specific and global polysome analyses (Arava et al. 2003) have provided a wealth of information about translation, but their quantitative resolution is limited by the poor separation of heavier polysomes and the difficulty of distinguishing ribosomes on different open reading frames (ORFs) in polycistronic mRNAs and transcripts with regulatory upstream translation. This challenge is exacerbated when comparing translation between different genes, as the number of ribosomes on a transcript scales with the length of the coding sequence as well as the translation level.

Ribosome profiling circumvents these limitations and precisely measures translation levels by counting discrete ribosome footprints (Ingolia et al. 2009). This quantitative precision revealed a principle of "proportional synthesis" that holds in bacterial and eukaryotic cells: protein subunits of multimeric complexes are synthesized in proportion to their stoichiometry in the assemblies (Li et al. 2014) thus reducing waste of producing unneeded subunits and eliminating the need to dispose of such uncomplexed species (Fig. 3). In bacteria, proteins with differing stoichiometry are often translated from a single polycistronic transcript, and so these differences likely reflect differential translation of the individual reading frames within that RNA. Translation-driven proportional synthesis can even be seen in chloroplasts, in which plastid ribosome profiling revealed that this organelle produces photosystem components in a tight stoichiometric ratio not seen in mRNA abundance (Chotewutmontri and Barkan 2016). More broadly, the fine-tuning seen in proportional synthesis highlights the accuracy and precision of ribosome profiling measurements.

This fine-tuning of stoichiometry also emphasizes how protein abundance is typically the functionally relevant output of gene expression in the cell, and selective constraints arise on protein levels rather than mRNA levels. Protein abundance correlates better with ribosome profiling measurements than with mRNA levels (Liu et al. 2017b; Cheng et al. 2018), and so experiments capture additional, biologically relevant information about gene expression by incorporating profiling data. Combining ribosome profiling with transcriptomic and proteomic data promises the opportunity to learn more about the genetic determinants of protein levels that impact translation as well as transcription.

Ribosome profiling has enabled the study of interallelic (Muzzey et al. 2014), interstrain (Albert et al. 2014), and interspecies (Artieri and Fraser 2014b; McManus et al. 2014) translational differences in yeast. It has also been applied to study gene expression variation between individual humans (Battle et al. 2015; Cenik et al. 2015). In some cases, transcriptional divergence is buffered by translational changes to conserve protein levels, whereas in other cases transcriptional and translational changes reinforce each other. Future ribosome profiling studies promise further insight into the roles of translational changes as well as the nature of the polymorphisms that drive these changes.

During dynamic remodeling of cell physiology, global expression profiling has shown how genes are induced just in time to fulfill their functional roles (Brown and Botstein 1999). Ribosome profiling in meiotic yeast has revealed how this principle of just-in-time regulation extends to the translational as well as transcriptional control of the proteins synthesized for each stage of this highly ordered process (Brar et al. 2012). Ribosome profiling in mitosis, too, points to translational control of protein production in concert with cell cycle progression (Stumpf et al. 2013; Tanenbaum et al. 2015).

Subsequently, concerted programs of translational regulation have emerged in many other models, ranging from basal eukaryotic parasites

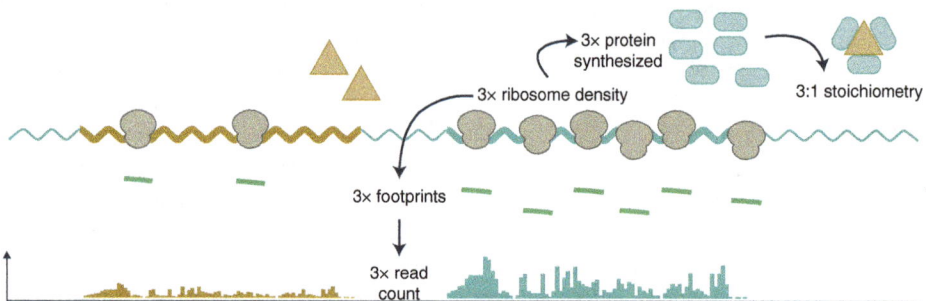

Figure 3. Quantifying protein synthesis. The number of ribosomes translating a reading frame determines the number of footprints generated in a profiling experiment, and so counting the footprint sequences derived from a reading frame indicates the amount of the encoded protein that is being synthesized. An exemplary polycistronic bacterial transcript is shown, with two open reading frames ([ORFs] A and B) encoding a pair of proteins that assemble with a 1:3 stoichiometric ratio. To achieve this stoichiometry, ORF B is translated threefold more heavily than ORF A, leading to threefold higher ribosome density and threefold more ribosome footprints.

Cite this article as *Cold Spring Harb Perspect Biol* doi: 10.1101/cshperspect.a032698

Trypanosoma (Jensen et al. 2014; Vasquez et al. 2014) and *Plasmodium* (Caro et al. 2014) to circadian cycles in mammalian tissue (Janich et al. 2015). This principle even extends to the distinct mitochondrial translational apparatus, in which profiling of mitoribosomes points to coordinated synthesis of oxidative phosphorylation components translated in the mitochondrion with those produced in the cytosol (Couvillion et al. 2016).

One common theme arising in many studies is the coordinated regulation of ribosomal proteins and ribosome biogenesis factors linked to the rate of cell growth. In metazoa, ribosome production and protein synthesis levels are regulated in part by the protein kinase, mammalian target of rapamycin (mTOR), which serves as a master regulator of growth (Hindupur et al. 2015), in part through controlling the translation of mRNAs encoding ribosomal proteins, but affecting many other transcripts as well (Proud 2018). This mTOR-driven translation supports proliferating cells during normal development and malignant growth in cancer (Dowling et al. 2010; Alain et al. 2012; Robichaud et al. 2018). As mTOR activity drives cell growth downstream from well-known oncogenic signaling pathways, there is great interest in developing active-site mTOR inhibitors as clinically useful anticancer therapies (Bhat et al. 2015). Ribosome profiling studies of cells treated with these inhibitors has revealed a broad range of target transcripts beyond ribosomal proteins that seem to support the cancer cell phenotype (Hsieh et al. 2012; Thoreen et al. 2012). Intriguingly, many of the same genes are translationally repressed in mouse embryonic stem cells induced to differentiate into embryoid bodies rather than continue rapid proliferation (Ingolia et al. 2011).

MOLECULAR MECHANISMS OF TRANSLATIONAL CONTROL

Ribosome profiling has revealed the mechanism underlying the action of other anticancer drugs that target the translational apparatus (Chu and Pelletier 2018). Rocaglate drugs are a class of natural products that target eukaryotic initiation factor 4A (eIF4A), a prototypical DEAD-box RNA helicase, selectively killing cancer cells (Santagata et al. 2013). Ribosome profiling revealed that rocaglates inhibit translation of specific mRNAs, suggesting that this targeted inhibition could explain their selectivity for cancer cells (Wolfe et al. 2014). By measuring the relative sensitivity of different transcripts to rocaglate treatment—which differed greatly from their sensitivity to hippuristanol, a more conventional eIF4A inhibitor—ribosome profiling further elucidated the unique repressive mechanism of these drugs. Rather than mimicking the loss of eIF4A function, rocaglate drugs clamp eIF4A onto certain polypurine RNA sequences, where it serves as a roadblock to translation initiation (Iwasaki et al. 2016). Transcripts with polypurine-rich transcript leaders are thus particularly sensitive to rocaglate treatment.

Similar correspondences between translational changes measured by ribosome profiling and transcript features have provided new insights into the normal function of eIF4A as well as other translation initiation factors, often challenging our current understanding of their roles. Ribosome profiling measurements conducted after inactivation of eIF4A revealed profound but fairly uniform reduction in translation (Sen et al. 2015). Conditional inactivation of yeast Ded1, another DEAD-box translation initiation factor, argued that it is particularly important for translating mRNAs with longer and more structured $5'$ untranslated regions (UTRs) (Sen et al. 2015). The scaffolding protein eIF4G recruits eIF4A and stimulates its ATPase activity (Merrick and Pavitt 2018; Sokabe and Fraser 2018). Although eIF4G is typically thought to be recruited to mRNAs through its interactions with the cap-binding protein eIF4E and poly(A)-binding protein, recent studies suggest that in yeast it preferentially binds and promotes the translation of mRNAs with oligo (U) tracts in their $5'$UTRs (Zinshteyn et al. 2017). Transcripts that depend on eIF4G also show reduced translation in the absence of the ribosome-associated factor Asc1/RACK1 (Thompson et al. 2016). In plants, poly(A)-binding proteins interact with A-rich motifs directly in the 5'UTR, in which they promote translation in pattern-triggered immune response (Xu et al. 2017).

Global ribosome profiling, combined with mRNA and protein abundance measurements, has also provided critical insights into the mechanism of microRNA-mediated repression (Duchaine and Fabian 2018), making it possible to disentangle the effects of mRNA destabilization from reduced translation. Early ribosome profiling studies measured concordant effects at both stages of expression, with translational repression contributing ~1/6th of the total reduction in protein abundance at steady state (Guo et al. 2010). Later work in developing zebrafish embryos showed that strong translational repression precedes mRNA decay during the activation of miR-430 in the maternal-to-zygotic transition (Bazzini et al. 2012). It remains unclear whether this reflects a general kinetic pathway for microRNA-mediated repression, or a shift in the mode of action between earlier or later stages of embryogenesis (Subtelny et al. 2014). Similarly, ribosome profiling has distinguished the translational effects of RNA modifications such as adenosine N^6 methylation, which seem to influence both translation and mRNA stability (Wang et al. 2015; Zhou et al. 2015; Coots et al. 2017; Slobodin et al. 2017; Vu et al. 2017; Peer et al. 2018).

Quantitative and comprehensive ribosome profiling measurements broadly offer the insights available from transcriptome profiling, augmented with information about translational regulation. Using these comprehensive measurements as input, modeling approaches have been used to systematically quantify the roles of various transcript features in determining the translational output of an mRNA (Weinberg et al. 2016; Hockenberry et al. 2017). This approach can also be applied to learn about cellular physiology by profiling natural biological processes and to learn about regulatory mechanisms by observing the effects of targeted molecular disruptions, as shown by the examples above. Ribosome profiling contains information about the exact positions of ribosomes as well; however, that goes beyond expression profiling.

DISCOVERY OF NONCANONICAL TRANSLATION EVENTS

The prevalence of ribosome footprints in 5'UTRs was one of the most striking features we observed in the first ribosome profiling data from yeast and mammalian cells (Fig. 4A) (Ingolia et al. 2009, 2011). Upstream translation itself can serve to repress expression of downstream protein-coding genes when scanning ribosomes initiate at upstream ORFs (uORFs) and

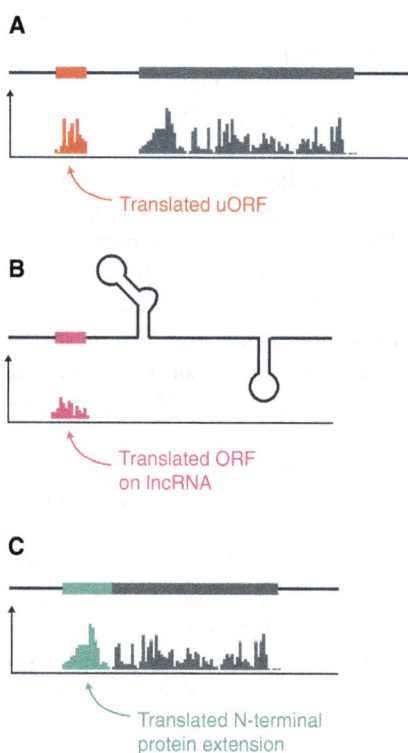

Figure 4. Annotating the proteome with ribosome profiling. The figure diagrams mRNAs (*top*) showing the frequency of ribosome footprints along them (*below*). (*A*) Ribosome footprint sequences mapping to the 5' leader of a transcript indicates the translation of an upstream open reading frame (uORF, red segment) and downstream ORF (gray segment). (*B*) Likewise, ribosome footprint sequences on a noncoding RNA indicate the presence of a translated region, typically near the 5' end of the transcript. (*C*) Alternative protein isoforms translated in addition to or in place of annotated reading frames also appear in ribosome profiling data.

 Cite this article as *Cold Spring Harb Perspect Biol* doi: 10.1101/cshperspect.a032698

translate them instead of proceeding on to the main reading frame (Sonenberg and Hinnebusch 2009). Ribosome profiling data are critical for understanding this mode of regulation, as not all uORFs are translated, and their repressive effects can be modulated by genetic variation between individuals, including polymorphisms that create or destroy uORFs (Calvo et al. 2009; Cenik et al. 2015). Indeed, in some cases, uORFs can promote the translation of the downstream coding sequence (Sonenberg and Hinnebusch 2009). Alternative transcript isoforms can also include or exclude upstream translated regions, thereby modulating translation. In the most extreme cases, totally unproductive transcript isoforms may result from the inclusion of long upstream regions replete with uORFs (Brar et al. 2012; Chen et al. 2017; Cheng et al. 2018).

Regulatory uORFs control the paradoxical induction of specific mRNAs such as *ATF4* and *CHOP* during stress-induced translational shutoff, mediated by the phosphorylation of the α subunit of eukaryotic initiation factor 2 (eIF2α) (Sonenberg and Hinnebusch 2009; Wek 2018). Ribosome profiling has now provided a comprehensive list of dozens of genes that show similar up-regulation (Andreev et al. 2015; Sidrauski et al. 2015). Surprisingly, although ribosome profiling confirms the translation of *ATF4* and *CHOP* uORFs, many other targets lack ribosome occupancy in their 5′UTRs, raising the question of how their translation is controlled. Furthermore, translated uORFs appear to be widespread across the transcriptome (Johnstone et al. 2016), not restricted to the small number of phospho-eIF2α-induced genes (Andreev et al. 2015; Sidrauski et al. 2015).

Much of the upstream translation seen in ribosome profiling cannot be attributed to AUG codons, and instead appears to initiate at a near-cognate, non-AUG codon. As the most common non-AUG initiation sites occur at codons that are quite similar to AUG (e.g., CUG), some of this translation results from mispairing of the initiator transfer RNA (tRNA) with the noncanonical start codon. There is also evidence for uORF peptide products that begin with amino acids other than

methionine. These alternative start codon products are induced when eIF2 is inhibited by phosphorylation, and depend on the poorly understood initiation factor eIF2A, which is unrelated to the canonical eIF2 complex (Starck et al. 2016; Merrick and Pavitt 2018). Ribosome profiling recently uncovered a shift toward *EIF2A*-dependent, non-AUG initiation in a *SOX2*-driven mouse model of squamous cell carcinoma (Sendoel et al. 2017). Many oncogenic transcripts were induced in this translational program, and tumor progression depended on *EIF2A*, pointing toward a causal link between unconventional 5′UTR translation and cancer. The recent development of translation complex profile sequencing (TCP-Seq) promises a more direct view of the mechanism of translation initiation. TCP-Seq augments the standard ribosome profiling approach with formaldehyde cross-linking that stabilizes preinitiation complexes on mRNAs to profile 40S subunits scanning through 5′UTRs, providing insights into the molecular choreography of the scanning process (Archer et al. 2016).

Ribosome footprints were seen on many presumptively noncoding RNAs in addition to the 5′UTRs of coding genes (Fig. 4B) (Ingolia et al. 2011, 2014; Chew et al. 2013; Ji et al. 2015). The patterns of footprints on long noncoding RNAs (lncRNAs) (Chekulaeva and Rajewsky 2018) matched our expectations for those of translating ribosomes; they fell in AUG-initiated reading frames near the 5′ ends of transcripts and in aggregate showed three-nucleotide periodicity (Ingolia et al. 2011; Calviello et al. 2016). The lack of canonical features of protein-coding sequences in these translated regions motivated a variety of experiments aimed at validating the profiling results. lncRNA footprints copurified with affinity-tagged ribosomes and responded to drugs that targeted the ribosome, and thus represented ribosome-protected footprints rather than nonribosomal background (Ingolia et al. 2014). We have also reported evidence for protein products derived from lncRNA translation (see Chekulaeva and Rajewsky 2018). Viral infection leads to an immunological memory of epitopes derived from lncRNA translation (Stern-Ginossar et al. 2012), similar to the epitopes produced by noncanonical upstream translation (Starck et al. 2012). Certain peptides have also been detected directly (Calviello et al. 2016), although in general these short products are challenging targets for proteomics, and ribosome profiling predictions can greatly aid in finding them (Menschaert et al. 2013).

ANNOTATING THE EXPANDED PROTEOME

The functional impact of pervasive alternative translation remains an important question. In a few cases, ribosome profiling has directly identified short, functional proteins such as Toddler (Pauli et al. 2014). Ribosome profiling points to a remarkable breadth of short, translated ORFs supported by conservation analysis (Menschaert et al. 2013; Aspden et al. 2014; Bazzini et al. 2014; Fields et al. 2015; Calviello et al. 2016) that could add to our growing catalog of functional micropeptides (Anderson et al. 2015; D'Lima et al. 2017). In many cases, however, this translation may be adventitious and subject principally to negative selection to avoid harmful effects from protein products or RNA destabilization (Ulitsky and Bartel 2013). Even nonfunctional proteins can serve as antigens, however (Ingolia 2014), and understanding this cryptic source of antigens (Starck et al. 2012) has implications for cancer immunotherapy (Schumacher and Schreiber 2015) and immunobiology more generally.

Ribosome profiling has also revealed alternative translation that extends or truncates classical protein-coding genes (Fig. 4C). This variation can add or remove entire domains, changing or reversing the function of the protein product. For instance, a truncated form of the innate immune signaling protein, mitochondrial antiviral signaling protein (MAVS), called miniMAVS, seems to antagonize the function of full-length MAVS in antiviral gene expression (Brubaker et al. 2014). Our data suggest that such alternative isoforms are widespread (Ingolia et al. 2011; Fields et al. 2015).

The diversity of translation products has spurred adaptations of ribosome profiling optimized for identifying translated regions of the transcriptome. We reported on the use of har-

ringtonine, a drug that immobilizes initiating ribosomes, producing footprints that mark sites of translation initiation (Ingolia et al. 2011). Others showed, independently, that lactimidomycin or pateamine A could likewise trap initiating ribosome footprints (Lee et al. 2012; Gao et al. 2015; Popa et al. 2016), while high doses of the drug puromycin could drive rapid premature termination to produce a similar initiation-specific footprint profile (Fritsch et al. 2012). Joint analysis of lactimidomycin- and harringtonine-treated profiling data promises more robust identification of translational start sites appearing in both data sets by excluding possible artifacts resulting from either of these mechanistically distinct drugs (Stern-Ginossar et al. 2012; Arias et al. 2014). This combined analysis was used to define translated reading frames in human cytomegalovirus, a large herpesvirus with a complex life cycle that expresses a variety of alternative translation products, some of which seem to display specific molecular function (Stern-Ginossar et al. 2012). Novel translated ORFs upstream of, or within, known ORFs have also been detected in other viruses (Stern-Ginossar et al. 2018). More recently, a systematic, regression-based combination of ribosome profiling data generated with these different drugs revealed hundreds of novel coding sequences in mammalian cells along with a wealth of shorter, translated reading frames (Fields et al. 2015).

To draw accurate inferences about in vivo translation from deep-sequencing data, it is essential to know that the RNA fragments being sequenced are ribosome-protected footprints. We have provided several lines of evidence showing that this is true in general (Ingolia et al. 2014), and we and others have shown how straightforward computational approaches can distinguish signatures of translation from footprints left by nonribosomal RNA–protein interactions (Ingolia et al. 2014; Ji et al. 2016). The bulk of these nonribosomal reads derive from abundant structural RNAs, including tRNAs, spliceosomal small nuclear RNAs (snRNAs), and small nucleolar RNAs (snoRNAs) (Ji et al. 2016). Fragments of these RNAs, along with other nonribosomal background, can be identified and excluded from ribosome footprint analysis because they differ in length from ribosome footprints and lack triplet periodicity (Ingolia et al. 2014; Ji et al. 2016).

To equate ribosome footprint density with protein production, it is also important to know that these ribosomes are translating productively. We have followed run-off elongation after blocking new initiation with drugs and shown that coding sequences are quickly depleted of ribosomes (Ingolia et al. 2011), except under conditions in which elongation is arrested (Barry et al. 2017). These basic features seem to hold in most systems, although it does not obviate the need to evaluate them in unusual biological contexts.

TRACKING THE FOOTPRINTS OF TRANSLATION ELONGATION

Ribosome profiling has broad applications in annotating genes as well as measuring expression, but its most distinctive contributions may stem from insights into the activities of ribosomes in vivo. We know that the speed of translation elongation can vary across a coding sequence, presumably as a result of variations in the mRNA template and the protein product. Ribosomes will spend more time at positions of slow elongation, and so we will observe a higher density of footprints at these sites (Fig. 5).

Indeed, footprint counts vary substantially across genes and accumulate at specific "pause" sites. There is great interest in understanding the features that correlate with the speed of translation elongation and deconvolving this from biases in capturing and sequencing footprints (Stadler and Fire 2011; Dana and Tuller 2012; Qian et al. 2012; Charneski and Hurst 2013; Lareau et al. 2014; Pop et al. 2014; Liu and Song 2016; O'Connor et al. 2016; Weinberg et al. 2016; Dao Duc et al. 2017). Likewise, there is interest in understanding the factors that drive dramatic ribosome pausing at specific locations, which may reflect a qualitatively different process than the variation in translation speed seen across typical codons (Han et al. 2014; Li et al. 2014; Martens et al. 2015; Mohammad et al. 2016; Zhang et al. 2017). Various measures of codon usage bias correlate with footprint occupancy, suggesting that favored codons are decoded more quickly. However, consensus has not emerged on the exact basis of this effect, which seems to extend beyond the time required for decoding and tRNA recruitment. Elongation rates learned from ribosome profiling nonetheless provide an empirical basis for tuning the translation of a coding sequence, thereby controlling its expression (Tunney et al. 2017).

Translation of even a single codon is a complicated, multistep process (Dever et al. 2018; Rodnina 2018), and ribosome profiling has opened a new window into the operation

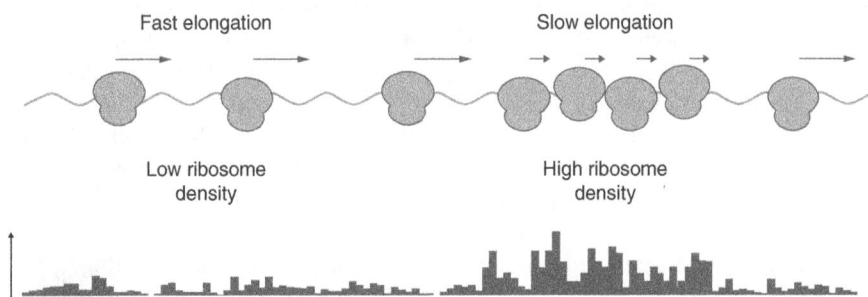

Figure 5. Inferring elongation speed from variations in ribosome footprint density. The *lower* part of the figure reports the frequency of ribosomal footprints along the mRNA. Regions of slow elongation will accumulate higher ribosome occupancy than regions of faster elongation on the same transcript. These differences in ribosome density are visible in profiling data, and they can be used to infer how codon usage, peptide sequence, and other features control the speed of translation.

of the translational machinery by reporting on normal elongation and on the effects of mutations. tRNAs in particular are heavily modified, and disrupting these modifications can change the speed of translation (Zinshteyn and Gilbert 2013) and thereby disrupt protein folding (Nedialkova and Leidel 2015). A broad survey of tRNA modifications revealed diverse effects on specific codons and on gene expression (Chou et al. 2017). In a similar fashion, base modifications on mRNA can affect decoding (Choi et al. 2016; Li et al. 2017), representing another factor that can contribute. In yeast, ribosome profiling provided in vivo confirmation that certain pairs of synonymous codons induce major ribosome pausing only when adjacent and in a particular order, suggesting structural cross talk between tRNAs in the A and P sites (Gamble et al. 2016). Ribosome profiling can distinguish between different phases of the translation elongation cycle, as different ribosome conformations protect footprints of differing length (Lareau et al. 2014). The longer (~28 nt) footprints, captured in most ribosome profiling experiments, probably reflect unrotated ribosomes. Cycloheximide treatment traps ribosomes in this long-footprint conformation, and yeast ribosome profiling performed without cycloheximide revealed a population of shorter (~21 nt) footprints that are attributed to rotated ribosomes. Although the abundance of long ribosome footprints correlates with codon usage and tRNA availability, short footprint density correlates with physicochemical amino acid properties instead, likely reflecting effects on translocation rather than decoding. Short footprints accumulate at certain tRNA-dependent stalls (Matsuo et al. 2017), suggesting that this cross talk affects translocation rather than tRNA recruitment (Lareau et al. 2014). Notably, these short footprints differ from the ~16 nt footprints reflecting a ribosome stalled at the end of a broken mRNA (Guydosh and Green 2014). More generally, these examples show the importance of identifying and quantifying all ribosome footprints regardless of their length (Mohammad et al. 2016).

Translation elongation also slows in response to amino acid limitation, leading to ribosome footprint accumulation on codons encoding the affected amino acids. Footprint density peaks induced by histidine deprivation were used to generate fiduciary marks in ribosome profiling data (Guydosh and Green 2014; Lareau et al. 2014), and inadvertent serine restriction likewise caused a buildup of footprints on serine codons in bacteria (Li et al. 2014). Systematic amino acid starvation coupled with ribosome profiling provided a spectrum of perturbed ribosome occupancy profiles that inform biophysical models of bacterial translation (Subramaniam et al. 2014). Remarkably, ribosome profiling likewise uncovered proline limitations in certain human tumors, based on slowed elongation when decoding proline codons (Loayza-Puch et al. 2016), and may serve more generally to probe for metabolic disruptions in cancer and other diseases.

Ribosome profiling has also facilitated the study of peptide-mediated translational pausing that occurs naturally in cells (Nakatogawa and Ito 2002). We observed ribosome footprint accumulation at certain tandem proline codons in mammalian cells (Ingolia et al. 2011), consistent with the slowed rate of elongation at these sites in vitro and the unfavorable conformation of polyprolyl nascent chains in the ribosome (Huter et al. 2017). The universally conserved elongation factor EF-P/eIF5A is implicated in translation of polyproline peptides (Doerfel et al. 2013; Gutierrez et al. 2013; Ude et al. 2013; Dever et al. 2018; Rodnina 2018) but ribosome profiling after eIF5A depletion reveals broader perturbation of ribosome footprint profiles, supporting a wider role for eIF5A in elongation through many unfavorable peptide sequences and in efficient translation termination (Schuller et al. 2017). Ribosome footprinting of these eIF5A stalls agreed with stalling sites identified by 5PSeq (Pelechano and Alepuz 2017), which focuses on in vivo RNA degradation intermediates whose 5′ terminus marks the trailing edge of the last translating ribosome (Pelechano et al. 2015). In contrast, ribosome profiling showed that *Legionella* toxins targeting elongation factor 1A (eEF1A) show no such specificity (Barry et al. 2017). Many peptide sequences can block bacterial translation (Woolstenhulme et al. 2013), and this effect is often

 Cite this article as *Cold Spring Harb Perspect Biol* doi: 10.1101/cshperspect.a032698

exploited for biological regulation. Bacterial ribosome profiling has also defined peptide-specific arrest caused by antibiotics targeting the translational machinery (Kannan et al. 2014; Marks et al. 2016). Stalling also occurs at programmed ribosomal frameshifting, which can stand out dramatically in footprint profiles (Michel et al. 2012; Napthine et al. 2017).

Detailed analyses of ribosome footprint occupancy patterns on individual mRNAs are particularly impacted by technical challenges. Studies in prokaryotes face a unique obstacle: ribonucleases do not degrade unprotected RNA precisely to the edge of the prokaryotic ribosome (Oh et al. 2011), making it challenging to precisely identify functionally relevant positions within each footprint (Woolstenhulme et al. 2013). More universally, all methods for converting ribosome footprints into a deep-sequencing library display biases that over- or underrepresent certain footprints, thereby distorting the apparent ribosome occupancy observed after sequencing (Artieri and Fraser 2014a; Bartholomaus et al. 2016; Lecanda et al. 2016; Tunney et al. 2017). Translation inhibitors used before cell lysis can distort ribosome occupancy profiles more directly (Gerashchenko and Gladyshev 2014; Hussmann et al. 2015). Eukaryotic cells are often treated with cycloheximide before ribosome profiling to immobilize and stabilize ribosomes. In our early studies, we reported that this treatment did not affect overall ribosome occupancy across a coding sequence but did change the pattern of footprints within that sequence (Ingolia et al. 2011). Subsequently, use of cycloheximide varied between studies. Later analysis showed that peaks of ribosome density appear to shift downstream in cycloheximide-treated samples relative to untreated ones (Gerashchenko and Gladyshev 2014; Hussmann et al. 2015).

Interest in a quantitative understanding of elongation has been heightened by recent studies that identified a potential role for elongation rates in dictating mRNA half-lives (Presnyak et al. 2015; Chan et al. 2017), either through direct surveillance of ribosome speed by mRNA decay machinery (Radhakrishnan et al. 2016) or indirectly by inducing ribosome collisions that then trigger decay pathways (Ferrin and Subramaniam 2017; Simms et al. 2017). Ribosome profiling will undoubtedly play a key role in unraveling the molecular mechanisms connecting elongation to decay and in quantifying the role of elongation in determining steady-state mRNA levels.

TRANSLATION TERMINATION AND BEYOND

In most ribosome profiling studies, 3′UTRs are devoid of footprints, in contrast to the surprising abundance of upstream initiation. Stop codon readthrough causes a specific accumulation of in-frame ribosome footprints in 3′UTRs, which are particularly prominent in *Drosophila* (Dunn et al. 2013). Defects in posttermination ribosome recycling (Hellen 2018) allow unrecycled ribosomes to enter 3′UTRs with no particular reading frame, and then reinitiate translation in some different reading frame, producing 3′UTR footprints out of frame from the coding DNA sequence (CDS) (Young et al. 2015). It appears that the ribosome rescue factors Dom34/Pelota and Hbs1 typically rescue many posttermination, unrecycled ribosomes, as the loss of these factors also causes an accumulation of vacant ribosomes past the stop codon (Guydosh and Green 2014). This distinctive accumulation of 3′UTR footprint patterns arose in reticulocytes and platelets, bringing to light a depletion in normal recycling factors and a disruption of ribosome homeostasis in both of these anucleate blood lineages (Mills et al. 2016). Indeed, translational regulation is pervasive in hematopoiesis (Alvarez-Dominguez et al. 2017), and altered ribosome recycling may underlie the particular sensitivity of red blood cells to defects in the translational machinery (Mills and Green 2017). It seems that the loss of ribosome rescue may serve a positive role in platelets, however. The rescue of ribosomes is linked to quality control processes that degrade aberrant protein products and mRNA templates (Brandman and Hegde 2016) and the loss of ribosome rescue factors seems to stabilize mRNAs that cannot be replaced by transcription (Mills et al. 2017).

PICKING THE RIGHT FOOTPRINTS

Footprinting of purified ribosome subpopulations enables profiling of cotranslational processes that act on proteins but, using a sequencing-based assay. Selective ribosome profiling by purifying ribosomes that are engaged by specific chaperones or targeting factors has revealed the in vivo substrates and engagement patterns in bacteria (Oh et al. 2011) and eukaryotes (Döring et al. 2017). Likewise, profiling of ribosome footprints engaged with the signal recognition particle (SRP) monitors cotranslational secretion and suggests that determinants beyond the classic signal sequence may aid SRP targeting (Chartron et al. 2016). Profiling has even been adapted to profile the folding state of nascent protein chains directly (Han et al. 2014), highlighting the fact that protein folding can be coupled directly to translation (Gloge et al. 2014). Indeed, selective ribosome profiling of ribosomes associated with different members of a multiprotein complex has suggested that complex assembly can begin cotranslationally (Shieh et al. 2015). Such cotranslational assembly could couple with the degradation of monomers that lack partner proteins for complex formation (Ishikawa et al. 2017) to complement proportional synthesis (Li et al. 2014) in maintaining proteome stoichiometry. This selective profiling strategy has been extended to study ribosomal subpopulations with varying composition. After identifying proteins RPL10A and RPL38 as substoichiometric in ribosomes, selective ribosome profiling of only those ribosomes containing these proteins revealed a potential role for heterogeneity between ribosomes in shaping overall translational output (Shi et al. 2017).

Recently, an approach termed proximity-specific ribosome profiling has enabled selective profiling of ribosomes at specific subcellular locations. Subcellular organization is inevitably disrupted by lysis and homogenization, but proximity labeling with a localized biotin ligase can mark ribosomes according to their in vivo localization for subsequent purification and footprinting. This approach was used first to identify the ribosomes that localize near the endoplasmic reticulum and the mitochondria in

yeast, providing further insight into protein targeting (Jan et al. 2014; Williams et al. 2014). We have recently combined this method with rapid and specific depletion of SRP to comprehensively characterize the role of SRP in cotranslational localization. This approach uncovered an unexpected class of mRNAs encoding proteins that are normally secreted but become mistargeted to mitochondria in the absence of SRP (Costa et al. 2018). Recent results suggest that subcellular organization of protein synthesis may be widespread, especially in tissues in which cells polarize and form three-dimensional structures (Moor et al. 2017). Localized translational control is particularly prominent in neurons, in which it is implicated in fundamental neural processes such as long-term potentiation and depression (Glock et al. 2017; Biswas et al. 2018; Sossin and Costa-Mattioli 2018).

Ribosome affinity purification has also emerged as a tool for cell type–specific translational profiling in animals, by using translating ribosome affinity purification (TRAP) (Doyle et al. 2008; Heiman et al. 2008) and RiboTag (Sanz et al. 2009). These approaches seem quite complementary to ribosome profiling, and indeed, tissue-specific ribosome profiling was recently shown in *Drosophila* (Chen and Dickman 2017). TRAP has seen its broadest application in the nervous system, which is characterized by extreme cell type diversity as well as a prominent role for translational control in synaptic plasticity (Sossin and Costa-Mattioli 2018). Further integration of cell type–specific ribosome profiling seems particularly promising in understanding the molecular basis of neuronal functions.

PERSPECTIVE

The translation of mRNA into protein and the folding of the resulting protein into an active form are prerequisites for virtually every cellular process and represent the single largest investment of energy by cells. Ribosome profiling-based approaches have revolutionized our ability to monitor protein synthesis in vivo, making it possible to determine the start, stop, reading frame, chaperone engagement, subcellular tar-

Cite this article as *Cold Spring Harb Perspect Biol* doi: 10.1101/cshperspect.a032698

geting, and rate of translation for virtually every mRNA and protein encoded in a cell. The rich and quantitative nature of ribosome profiling data provides an unprecedented opportunity to explore and model complex cellular processes. Finally, by virtue of the precise genomic positional information obtained by ribosome profiling, the protein coding capacity of genomes can now be explored experimentally.

Nonetheless, important technical and conceptual questions remain. For example, the function of the many novel, short, and alternate translated regions identified thus far by ribosome profiling remains an intriguing and largely open question and one whose answer could fundamentally change the way that we believe about information encoding in genomes. Newly available CRISPR-based methods now make it possible to shut down the expression of any transcript (Gilbert et al. 2013, 2014; Liu et al. 2017a) or introduce nonsense mutations into any ORF (Hess et al. 2017). These approaches provide a central tool for efforts to define the functional roles for this broad array of newly identified translation products.

We have already seen demonstrations of specialized alterations to ribosome profiling that will advance its utility in complex systems. These developments include the analysis of molecularly defined subsets of ribosomes, either associated with specific factors or protein modifications, or even specialized ribosomes missing a core ribosomal protein entirely. Similar approaches allow the analysis of localized ribosomes within increasingly specific cell types or subcellular locations. We also know little about how ribosomes are distributed across individual transcripts of the same gene: Is the spacing between ribosomes purely stochastic, or are initiation and elongation "metered" to shape the traffic of ribosomes and minimize collisions? Along this line, understanding the biological roles for the use of synonymous codons remains one of the oldest outstanding questions in the field. Ribosome profiling provides an unprecedented view of their impact by yielding position-specific densities of ribosomes along a message. However, better protocols are needed to ensure that in vivo ribosome positions are captured

faithfully and turned into sequencing libraries free of biases or distortions. Finally, transformative advances are likely to emerge from progressively more sophisticated and creative analysis of the rich data sets generated from ribosome profiling experiments, enabling major surprises to be revealed, even in systems that were thought to be well characterized.

REFERENCES

*Reference is also in this collection.

Alain T, Morita M, Fonseca BD, Yanagiya A, Siddiqui N, Bhat M, Zammit D, Marcus V, Metrakos P, Voyer LA, et al. 2012. eIF4E/4E-BP ratio predicts the efficacy of mTOR targeted therapies. *Cancer Res* **72:** 6468–6476.

Albert FW, Muzzey D, Weissman JS, Kruglyak L. 2014. Genetic influences on translation in yeast. *PLoS Genet* **10:** e1004692.

Alvarez-Dominguez JR, Zhang X, Hu W. 2017. Widespread and dynamic translational control of red blood cell development. *Blood* **129:** 619–629.

Anderson DM, Anderson KM, Chang CL, Makarewich CA, Nelson BR, McAnally JR, Kasaragod P, Shelton JM, Liou J, Bassel-Duby R, et al. 2015. A micropeptide encoded by a putative long noncoding RNA regulates muscle performance. *Cell* **160:** 595–606.

Andreev DE, O'Connor PB, Fahey C, Kenny EM, Terenin IM, Dmitriev SE, Cormican P, Morris DW, Shatsky IN, Baranov PV. 2015. Translation of 5′ leaders is pervasive in genes resistant to eIF2 repression. *eLife* **4:** e03971.

Arava Y, Wang Y, Storey JD, Liu CL, Brown PO, Herschlag D. 2003. Genome-wide analysis of mRNA translation profiles in Saccharomyces cerevisiae. *Proc Natl Acad Sci* **100:** 3889–3894.

Archer SK, Shirokikh NE, Beilharz TH, Preiss T. 2016. Dynamics of ribosome scanning and recycling revealed by translation complex profiling. *Nature* **535:** 570–574.

Arias C, Weisburd B, Stern-Ginossar N, Mercier A, Madrid AS, Bellare P, Holdorf M, Weissman JS, Ganem D. 2014. KSHV 2.0: A comprehensive annotation of the Kaposi's sarcoma-associated herpesvirus genome using next-generation sequencing reveals novel genomic and functional features. *PLoS Pathog* **10:** e1003847.

Artieri CG, Fraser HB. 2014a. Accounting for biases in riboprofiling data indicates a major role for proline in stalling translation. *Genome Res* **24:** 2011–2021.

Artieri CG, Fraser HB. 2014b. Evolution at two levels of gene expression in yeast. *Genome Res* **24:** 411–421.

Aspden JL, Eyre-Walker YC, Phillips RJ, Amin U, Mumtaz MA, Brocard M, Couso JP. 2014. Extensive translation of small open reading frames revealed by Poly-Ribo-Seq. *eLife* **3:** e03528.

Barry KC, Ingolia NT, Vance RE. 2017. Global analysis of gene expression reveals mRNA superinduction is required for the inducible immune response to a bacterial pathogen. *eLife* **6:** e22707.

Bartholomaus A, Del Campo C, Ignatova Z. 2016. Mapping the non-standardized biases of ribosome profiling. *Biol Chem* **397:** 23–35.

Battle A, Khan Z, Wang SH, Mitrano A, Ford MJ, Pritchard JK, Gilad Y. 2015. Genomic variation. Impact of regulatory variation from RNA to protein. *Science* **347:** 664–667.

Bazzini AA, Lee MT, Giraldez AJ. 2012. Ribosome profiling shows that miR-430 reduces translation before causing mRNA decay in zebrafish. *Science* **336:** 233–237.

Bazzini AA, Johnstone TG, Christiano R, Mackowiak SD, Obermayer B, Fleming ES, Vejnar CE, Lee MT, Rajewsky N, Walther TC, et al. 2014. Identification of small ORFs in vertebrates using ribosome footprinting and evolutionary conservation. *EMBO J* **33:** 981–993.

Bazzini AA, Del Viso F, Moreno-Mateos MA, Johnstone TG, Vejnar CE, Qin Y, Yao J, Khokha MK, Giraldez AJ. 2016. Codon identity regulates mRNA stability and translation efficiency during the maternal-to-zygotic transition. *EMBO J* **35:** 2087–2103.

Bhat M, Robichaud N, Hulea L, Sonenberg N, Pelletier J, Topisirovic I. 2015. Targeting the translation machinery in cancer. *Nat Rev Drug Discov* **14:** 261–278.

* Biswas J, Liu Y, Singer RH, Wu B. 2018. Fluorescence imaging methods to investigate translation in single cells. *Cold Spring Harb Perspect Biol* doi: 10.1101/cshperspect.a032722.

Brandman O, Hegde RS. 2016. Ribosome-associated protein quality control. *Nat Struct Mol Biol* **23:** 7–15.

Brar GA, Weissman JS. 2015. Ribosome profiling reveals the what, when and how of protein synthesis. *Nat Rev Mol Cell Biol* **16:** 651–664.

Brar GA, Yassour M, Friedman N, Regev A, Ingolia NT, Weissman JS. 2012. High-resolution view of the yeast meiotic program revealed by ribosome profiling. *Science* **335:** 552–557.

Brown PO, Botstein D. 1999. Exploring the new world of the genome with DNA microarrays. *Nat Genet* **21:** 33–37.

Brubaker SW, Gauthier AE, Mills EW, Ingolia NT, Kagan JC. 2014. A bicistronic MAVS transcript highlights a class of truncated variants in antiviral immunity. *Cell* **156:** 800–811.

Calviello L, Mukherjee N, Wyler E, Zauber H, Hirsekorn A, Selbach M, Landthaler M, Obermayer B, Ohler U. 2016. Detecting actively translated open reading frames in ribosome profiling data. *Nat Methods* **13:** 165–170.

Calvo SE, Pagliarini DJ, Mootha VK. 2009. Upstream open reading frames cause widespread reduction of protein expression and are polymorphic among humans. *Proc Natl Acad Sci* **106:** 7507–7512.

Caro F, Ahyong V, Betegon M, DeRisi JL. 2014. Genome-wide regulatory dynamics of translation in the Plasmodium falciparum asexual blood stages. *eLife* **3:** e04106.

Cenik C, Cenik ES, Byeon GW, Grubert F, Candille SI, Spacek D, Alsallakh B, Tilgner H, Araya CL, Tang H, et al. 2015. Integrative analysis of RNA, translation, and protein levels reveals distinct regulatory variation across humans. *Genome Res* **25:** 1610–1621.

Chan LY, Mugler CF, Heinrich S, Vallotton P, Weis K. 2017. Non-invasive measurement of mRNA decay reveals translation initiation as the major determinant of mRNA stability. *bioRxiv* doi: 10.1101/214775.

Charneski CA, Hurst LD. 2013. Positively charged residues are the major determinants of ribosomal velocity. *PLoS Biol* **11:** e1001508.

Chartron JW, Hunt KC, Frydman J. 2016. Cotranslational signal-independent SRP preloading during membrane targeting. *Nature* **536:** 224–228.

* Chekulaeva M, Rajewsky N. 2018. Roles of long noncoding RNAs and circular RNAs in translation. *Cold Spring Harb Perspect Biol* doi: 10.1101/cshperspect.a032680.

Chen X, Dickman D. 2017. Development of a tissue-specific ribosome profiling approach in *Drosophila* enables genome-wide evaluation of translational adaptations. *PLoS Genet* **13:** e1007117.

Chen J, Tresenrider A, Chia M, McSwiggen DT, Spedale G, Jorgensen V, Liao H, van Werven FJ, Unal E. 2017. Kinetochore inactivation by expression of a repressive mRNA. *eLife* **6:** e27417.

Cheng Z, Otto GM, Powers E, Keskin A, Mertins P, Carr S, Jovanovic M, Brar GA. 2018. Pervasive, coordinated protein level changes driven by transcript isoform switching during meiosis. *Cell* **172:** 910–923.

Chew GL, Pauli A, Rinn JL, Regev A, Schier AF, Valen E. 2013. Ribosome profiling reveals resemblance between long non-coding RNAs and 5′ leaders of coding RNAs. *Development* **140:** 2828–2834.

Choi J, Ieong KW, Demirci H, Chen J, Petrov A, Prabhakar A, O'Leary SE, Dominissini D, Rechavi G, Soltis SM, et al. 2016. N^6-methyladenosine in mRNA disrupts tRNA selection and translation-elongation dynamics. *Nat Struct Mol Biol* **23:** 110–115.

Chotewutmontri P, Barkan A. 2016. Dynamics of chloroplast translation during chloroplast differentiation in maize. *PLoS Genet* **12:** e1006106.

Chou HJ, Donnard E, Gustafsson HT, Garber M, Rando OJ. 2017. Transcriptome-wide analysis of roles for tRNA modifications in translational regulation. *Mol Cell* **68:** 978–992 e974.

* Chu J, Pelletier J. 2018. Translating therapeutics. *Cold Spring Harb Perspect Biol* doi: 10.1101/cshperspect.a032995.

Coots RA, Liu XM, Mao Y, Dong L, Zhou J, Wan J, Zhang X, Qian SB. 2017. m^6A facilitates eIF4F-independent mRNA translation. *Mol Cell* doi: 10.1016/j.molcel.2017.10.002.

Costa EA, Subramanian K, Nunnari J, Weissman JS. 2018. Defining the physiological role of SRP in protein targeting efficiency and specificity. *Science* **359:** 689–692.

Couvillion MT, Soto IC, Shipkovenska G, Churchman LS. 2016. Synchronized mitochondrial and cytosolic translation programs. *Nature* **533:** 499–503.

Dana A, Tuller T. 2012. Determinants of translation elongation speed and ribosomal profiling biases in mouse embryonic stem cells. *PLoS Comput Biol* **8:** e1002755.

Dao Duc K, Saleem ZH, Song YS. 2018. Theoretical analysis of the distribution of isolated particles in the TASEP: Application to mRNA translation rate estimation. *Phys Rev E* **97:** 012106.

* Dever TE, Dinman JD, Green R. 2018. Translation elongation and recoding in eukaryotes. *Cold Spring Harb Perspect Biol* doi: 10.1101/cshperspect.a032649.

Cite this article as *Cold Spring Harb Perspect Biol* doi: 10.1101/cshperspect.a032698

D'Lima NG, Ma J, Winkler L, Chu Q, Loh KH, Corpuz EO, Budnik BA, Lykke-Andersen J, Saghatelian A, Slavoff SA. 2017. A human microprotein that interacts with the mRNA decapping complex. *Nat Chem Biol* **13:** 174–180.

Doerfel LK, Wohlgemuth I, Kothe C, Peske F, Urlaub H, Rodnina MV. 2013. EF-P is essential for rapid synthesis of proteins containing consecutive proline residues. *Science* **339:** 85–88.

Döring K, Ahmed N, Riemer T, Suresh HG, Vainshtein Y, Habich M, Riemer J, Mayer MP, O'Brien EP, Kramer G, et al. 2017. Profiling Ssb-nascent chain interactions reveals principles of Hsp70-assisted folding. *Cell* **170:** 298–311e220.

Dowling RJ, Topisirovic I, Alain T, Bidinosti M, Fonseca BD, Petroulakis E, Wang X, Larsson O, Selvaraj A, Liu Y, et al. 2010. mTORC1-mediated cell proliferation, but not cell growth, controlled by the 4E-BPs. *Science* **328:** 1172–1176.

Doyle JP, Dougherty JD, Heiman M, Schmidt EF, Stevens TR, Ma G, Bupp S, Shrestha P, Shah RD, Doughty ML, et al. 2008. Application of a translational profiling approach for the comparative analysis of CNS cell types. *Cell* **135:** 749–762.

* Duchaine TF, Fabian MR. 2018. Mechanistic insights into microRNA-mediated gene silencing. *Cold Spring Harb Perspect Biol* doi: 10.1101/cshperspect.a032771.

Dunn JG, Foo CK, Belletier NG, Gavis ER, Weissman JS. 2013. Ribosome profiling reveals pervasive and regulated stop codon readthrough in *Drosophila melanogaster*. *eLife* **2:** e01179.

Ferrin MA, Subramaniam AR. 2017. Kinetic modeling predicts a stimulatory role for ribosome collisions at elongation stall sites in bacteria. *eLife* **6:** e23629.

Fields AP, Rodriguez EH, Jovanovic M, Stern-Ginossar N, Haas BJ, Mertins P, Raychowdhury R, Hacohen N, Carr SA, Ingolia NT, et al. 2015. A regression-based analysis of ribosome-profiling data reveals a conserved complexity to mammalian translation. *Mol Cell* **60:** 816–827.

Fritsch C, Herrmann A, Nothnagel M, Szafranski K, Huse K, Schumann F, Schreiber S, Platzer M, Krawczak M, Hampe J, et al. 2012. Genome-wide search for novel human uORFs and N-terminal protein extensions using ribosomal footprinting. *Genome Res* **22:** 2208–2218.

Gamble CE, Brule CE, Dean KM, Fields S, Grayhack EJ. 2016. Adjacent codons act in concert to modulate translation efficiency in yeast. *Cell* **166:** 679–690.

Gao X, Wan J, Liu B, Ma M, Shen B, Qian SB. 2015. Quantitative profiling of initiating ribosomes in vivo. *Nat Methods* **12:** 147–153.

Gerashchenko MV, Gladyshev VN. 2014. Translation inhibitors cause abnormalities in ribosome profiling experiments. *Nucleic Acids Res* **42:** e134.

Gilbert LA, Larson MH, Morsut L, Liu Z, Brar GA, Torres SE, Stern-Ginossar N, Brandman O, Whitehead EH, Doudna JA, et al. 2013. CRISPR-mediated modular RNA-guided regulation of transcription in eukaryotes. *Cell* **154:** 442–451.

Gilbert LA, Horlbeck MA, Adamson B, Villalta JE, Chen Y, Whitehead EH, Guimaraes C, Panning B, Ploegh HL, Bassik MC, et al. 2014. Genome-scale CRISPR-mediated control of gene repression and activation. *Cell* **159:** 647–661.

Gingold H, Tehler D, Christoffersen NR, Nielsen MM, Asmar F, Kooistra SM, Christophersen NS, Christensen LL, Borre M, Sorensen KD, et al. 2014. A dual program for translation regulation in cellular proliferation and differentiation. *Cell* **158:** 1281–1292.

Glock C, Heumuller M, Schuman EM. 2017. mRNA transport and local translation in neurons. *Curr Opin Neurobiol* **45:** 169–177.

Gloge F, Becker AH, Kramer G, Bukau B. 2014. Co-translational mechanisms of protein maturation. *Curr Opin Struct Biol* **24:** 24–33.

Guo H, Ingolia NT, Weissman JS, Bartel DP. 2010. Mammalian microRNAs predominantly act to decrease target mRNA levels. *Nature* **466:** 835–840.

Gutierrez E, Shin BS, Woolstenhulme CJ, Kim JR, Saini P, Buskirk AR, Dever TE. 2013. eIF5A promotes translation of polyproline motifs. *Mol Cell* **51:** 35–45.

Guydosh NR, Green R. 2014. Dom34 rescues ribosomes in 3′ untranslated regions. *Cell* **156:** 950–962.

Han Y, Gao X, Liu B, Wan J, Zhang X, Qian SB. 2014. Ribosome profiling reveals sequence-independent post-initiation pausing as a signature of translation. *Cell Res* **24:** 842–851.

Heiman M, Schaefer A, Gong S, Peterson JD, Day M, Ramsey KE, Suarez-Farinas M, Schwarz C, Stephan DA, Surmeier DJ, et al. 2008. A translational profiling approach for the molecular characterization of CNS cell types. *Cell* **135:** 738–748.

* Hellen CUT. 2018. Translation termination and ribosome recycling in eukaryotes. *Cold Spring Harb Perspect Biol* doi: 10.1101/cshperspect.a032656.

Hess GT, Tycko J, Yao D, Bassik MC. 2017. Methods and applications of CRISPR-mediated base editing in eukaryotic genomes. *Mol Cell* **68:** 26–43.

Hindupur SK, Gonzalez A, Hall MN. 2015. The opposing actions of target of rapamycin and AMP-activated protein kinase in cell growth control. *Cold Spring Harb Perspect Biol* **7:** a019141.

Hockenberry AJ, Pah AR, Jewett MC, Amaral LA. 2017. Leveraging genome-wide datasets to quantify the functional role of the anti-Shine–Dalgarno sequence in regulating translation efficiency. *Open Biol* **7:** 160239.

Hsieh AC, Liu Y, Edlind MP, Ingolia NT, Janes MR, Sher A, Shi EY, Stumpf CR, Christensen C, Bonham MJ, et al. 2012. The translational landscape of mTOR signalling steers cancer initiation and metastasis. *Nature* **485:** 55–61.

Hussmann JA, Patchett S, Johnson A, Sawyer S, Press WH. 2015. Understanding biases in ribosome profiling experiments reveals signatures of translation dynamics in yeast. *PLoS Genet* **11:** e1005732.

Huter P, Arenz S, Bock LV, Graf M, Frister JO, Heuer A, Peil L, Starosta AL, Wohlgemuth I, Peske F, et al. 2017. Structural basis for polyproline-mediated ribosome stalling and rescue by the translation elongation factor EF-P. *Mol Cell* **68:** 515–527.e516.

Ingolia NT. 2014. Ribosome profiling: New views of translation, from single codons to genome scale. *Nat Rev Genet* **15:** 205–213.

Ingolia NT. 2016. Ribosome footprint profiling of translation throughout the genome. *Cell* **165:** 22–33.

Ingolia NT, Ghaemmaghami S, Newman JR, Weissman JS. 2009. Genome-wide analysis in vivo of translation with nucleotide resolution using ribosome profiling. *Science* **324:** 218–223.

Ingolia NT, Lareau LF, Weissman JS. 2011. Ribosome profiling of mouse embryonic stem cells reveals the complexity and dynamics of mammalian proteomes. *Cell* **147:** 789–802.

Ingolia NT, Brar GA, Rouskin S, McGeachy AM, Weissman JS. 2012. The ribosome profiling strategy for monitoring translation in vivo by deep sequencing of ribosome-protected mRNA fragments. *Nat Protoc* **7:** 1534–1550.

Ingolia NT, Brar GA, Stern-Ginossar N, Harris MS, Talhouarne GJ, Jackson SE, Wills MR, Weissman JS. 2014. Ribosome profiling reveals pervasive translation outside of annotated protein-coding genes. *Cell Rep* **8:** 1365–1379.

Ishikawa K, Makanae K, Iwasaki S, Ingolia NT, Moriya H. 2017. Post-translational dosage compensation buffers genetic perturbations to stoichiometry of protein complexes. *PLoS Genet* **13:** e1006554.

Iwasaki S, Floor SN, Ingolia NT. 2016. Rocaglates convert DEAD-box protein eIF4A into a sequence-selective translational repressor. *Nature* **534:** 558–561.

Jan CH, Williams CC, Weissman JS. 2014. Principles of ER cotranslational translocation revealed by proximity-specific ribosome profiling. *Science* **346:** 1257521.

Janich P, Arpat AB, Castelo-Szekely V, Lopes M, Gatfield D. 2015. Ribosome profiling reveals the rhythmic liver translatome and circadian clock regulation by upstream open reading frames. *Genome Res* **25:** 1848–1859.

Jensen BC, Ramasamy G, Vasconcelos EJ, Ingolia NT, Myler PJ, Parsons M. 2014. Extensive stage-regulation of translation revealed by ribosome profiling of *Trypanosoma brucei*. *BMC Genomics* **15:** 911.

Ji Z, Song R, Regev A, Struhl K. 2015. Many lncRNAs, 5′UTRs, and pseudogenes are translated and some are likely to express functional proteins. *eLife* **4:** e08890.

Ji Z, Song R, Huang H, Regev A, Struhl K. 2016. Transcriptome-scale RNase-footprinting of RNA–protein complexes. *Nat Biotechnol* **34:** 410–413.

Johnstone TG, Bazzini AA, Giraldez AJ. 2016. Upstream ORFs are prevalent translational repressors in vertebrates. *EMBO J* **35:** 706–723.

Kannan K, Kanabar P, Schryer D, Florin T, Oh E, Bahroos N, Tenson T, Weissman JS, Mankin AS. 2014. The general mode of translation inhibition by macrolide antibiotics. *Proc Natl Acad Sci* **111:** 15958–15963.

Kawakami K, Pande S, Faiola B, Moore DP, Boeke JD, Farabaugh PJ, Strathern JN, Nakamura Y, Garfinkel DJ. 1993. A rare tRNA-Arg(ccu) that regulates Ty1 element ribosomal frameshifting is essential for Ty1 retrotransposition in *Saccharomyces cerevisiae*. *Genetics* **135:** 309–320.

Kimchi-Sarfaty C, Oh JM, Kim IW, Sauna ZE, Calcagno AM, Ambudkar SV, Gottesman MM. 2007. A "silent" polymorphism in the *MDR1* gene changes substrate specificity. *Science* **315:** 525–528.

Lareau LF, Hite DH, Hogan GJ, Brown PO. 2014. Distinct stages of the translation elongation cycle revealed by sequencing ribosome-protected mRNA fragments. *eLife* **3:** e01257.

Lecanda A, Nilges BS, Sharma P, Nedialkova DD, Schwarz J, Vaquerizas JM, Leidel SA. 2016. Dual randomization of oligonucleotides to reduce the bias in ribosome-profiling libraries. *Methods* **107:** 89–97.

Lee S, Liu B, Lee S, Huang SX, Shen B, Qian SB. 2012. Global mapping of translation initiation sites in mammalian cells at single-nucleotide resolution. *Proc Natl Acad Sci* **109:** E2424–E2432.

Li GW, Burkhardt D, Gross C, Weissman JS. 2014. Quantifying absolute protein synthesis rates reveals principles underlying allocation of cellular resources. *Cell* **157:** 624–635.

Li X, Xiong X, Zhang M, Wang K, Chen Y, Zhou J, Mao Y, Lv J, Yi D, Chen XW, et al. 2017. Base-resolution mapping reveals distinct m^1A methylome in nuclear- and mitochondrial-encoded transcripts. *Mol Cell* **68:** 993–1005. e1009.

Liu TY, Song YS. 2016. Prediction of ribosome footprint profile shapes from transcript sequences. *Bioinformatics* **32:** i183–i191.

Liu SJ, Horlbeck MA, Cho SW, Birk HS, Malatesta M, He D, Attenello FJ, Villalta JE, Cho MY, Chen Y, et al. 2017a. CRISPRi-based genome-scale identification of functional long noncoding RNA loci in human cells. *Science* **355:** aah7111.

Liu TY, Huang HH, Wheeler D, Xu Y, Wells JA, Song YS, Wiita AP. 2017b. Time-resolved proteomics extends ribosome profiling-based measurements of protein synthesis dynamics. *Cell Syst* **4:** 636–644.e639.

Loayza-Puch F, Rooijers K, Buil LC, Zijlstra J, Oude Vrielink JF, Lopes R, Ugalde AP, van Breugel P, Hofland I, Wesseling J, et al. 2016. Tumour-specific proline vulnerability uncovered by differential ribosome codon reading. *Nature* **530:** 490–494.

Marks J, Kannan K, Roncase EJ, Klepacki D, Kefi A, Orelle C, Vazquez-Laslop N, Mankin AS. 2016. Context-specific inhibition of translation by ribosomal antibiotics targeting the peptidyl transferase center. *Proc Natl Acad Sci* **113:** 12150–12155.

Martens AT, Taylor J, Hilser VJ. 2015. Ribosome A and P sites revealed by length analysis of ribosome profiling data. *Nucleic Acids Res* **43:** 3680–3687.

Mathews MB, Nahum S, John WBH. 2000. Origins and principles of translational control. In *Cold Spring Harbor monograph archive; Volume 39: Translational control of gene expression*. Cold Spring Harbor Laboratory Press, Cold Spring Harbor, NY.

Matsuo Y, Ikeuchi K, Saeki Y, Iwasaki S, Schmidt C, Udagawa T, Sato F, Tsuchiya H, Becker T, Tanaka K, et al. 2017. Ubiquitination of stalled ribosome triggers ribosome-associated quality control. *Nat Commun* **8:** 159.

McGlincy NJ, Ingolia NT. 2017. Transcriptome-wide measurement of translation by ribosome profiling. *Methods* **126:** 112–129.

McManus CJ, May GE, Spealman P, Shteyman A. 2014. Ribosome profiling reveals post-transcriptional buffering of divergent gene expression in yeast. *Genome Res* **24:** 422–430.

Menschaert G, Van Criekinge W, Notelaers T, Koch A, Crappe J, Gevaert K, Van Damme P. 2013. Deep proteome coverage based on ribosome profiling aids mass spectrometry-based protein and peptide discovery and

Cite this article as *Cold Spring Harb Perspect Biol* doi: 10.1101/cshperspect.a032698

provides evidence of alternative translation products and near-cognate translation initiation events. *Mol Cell Proteomics* **12**: 1780–1790.

* Merrick WC, Pavitt GD. 2018. Protein synthesis initiation in eukaryotic cells. *Cold Spring Harb Perspect Biol* doi: 10.1101/cshperspect.a033092.

Michel AM, Baranov PV. 2013. Ribosome profiling: A hi-def monitor for protein synthesis at the genome-wide scale. *Wiley Interdiscip Rev RNA* **4**: 473–490.

Michel AM, Choudhury KR, Firth AE, Ingolia NT, Atkins JF, Baranov PV. 2012. Observation of dually decoded regions of the human genome using ribosome profiling data. *Genome Res* **22**: 2219–2229.

Mills EW, Green R. 2017. Ribosomopathies: There's strength in numbers. *Science* **358**: eaan2755.

Mills EW, Wangen J, Green R, Ingolia NT. 2016. Dynamic regulation of a ribosome rescue pathway in erythroid cells and platelets. *Cell Rep* **17**: 1–10.

Mills EW, Green R, Ingolia NT. 2017. Slowed decay of mRNAs enhances platelet specific translation. *Blood* **129**: e38–e48.

Mohammad F, Woolstenhulme CJ, Green R, Buskirk AR. 2016. Clarifying the translational pausing landscape in bacteria by ribosome profiling. *Cell Rep* **14**: 686–694.

Moor AE, Golan M, Massasa EE, Lemze D, Weizman T, Shenhav R, Baydatch S, Mizrahi O, Winkler R, Golani O, et al. 2017. Global mRNA polarization regulates translation efficiency in the intestinal epithelium. *Science* **357**: 1299–1303.

Muzzey D, Sherlock G, Weissman JS. 2014. Extensive and coordinated control of allele-specific expression by both transcription and translation in *Candida albicans*. *Genome Res* **24**: 963–973.

Nakatogawa H, Ito K. 2002. The ribosomal exit tunnel functions as a discriminating gate. *Cell* **108**: 629–636.

Napthine S, Ling R, Finch LK, Jones JD, Bell S, Brierley I, Firth AE. 2017. Protein-directed ribosomal frameshifting temporally regulates gene expression. *Nat Commun* **8**: 15582.

Nedialkova DD, Leidel SA. 2015. Optimization of codon translation rates via tRNA modifications maintains proteome integrity. *Cell* **161**: 1606–1618.

O'Connor PB, Andreev DE, Baranov PV. 2016. Comparative survey of the relative impact of mRNA features on local ribosome profiling read density. *Nat Commun* **7**: 12915.

Oh E, Becker AH, Sandikci A, Huber D, Chaba R, Gloge F, Nichols RJ, Typas A, Gross CA, Kramer G, et al. 2011. Selective ribosome profiling reveals the cotranslational chaperone action of trigger factor in vivo. *Cell* **147**: 1295–1308.

Pauli A, Norris ML, Valen E, Chew GL, Gagnon JA, Zimmerman S, Mitchell A, Ma J, Dubrulle J, Reyon D, et al. 2014. Toddler: An embryonic signal that promotes cell movement via Apelin receptors. *Science* **343**: 1248636.

Pechmann S, Chartron JW, Frydman J. 2014. Local slowdown of translation by nonoptimal codons promotes nascent-chain recognition by SRP in vivo. *Nat Struct Mol Biol* **21**: 1100–1105.

* Peer E, Moshitch-Moshkovitz S, Rechavi G, Dominissini D. 2018. The epitranscriptome in translation regulation.

Cold Spring Harb Perspect Biol doi: 10.110l/cshperspect.a032623.

Pelechano V, Alepuz P. 2017. eIF5A facilitates translation termination globally and promotes the elongation of many non polyproline-specific tripeptide sequences. *Nucleic Acids Res* **45**: 7326–7338.

Pelechano V, Wei W, Steinmetz LM. 2015. Widespread cotranslational RNA decay reveals ribosome dynamics. *Cell* **161**: 1400–1412.

Plotkin JB, Kudla G. 2011. Synonymous but not the same: The causes and consequences of codon bias. *Nat Rev Genet* **12**: 32–42.

Pop C, Rouskin S, Ingolia NT, Han L, Phizicky EM, Weissman JS, Koller D. 2014. Causal signals between codon bias, mRNA structure, and the efficiency of translation and elongation. *Mol Syst Biol* **10**: 770.

Popa A, Lebrigand K, Barbry P, Waldmann R. 2016. Pateamine A-sensitive ribosome profiling reveals the scope of translation in mouse embryonic stem cells. *BMC Genomics* **17**: 52.

Presnyak V, Alhusaini N, Chen YH, Martin S, Morris N, Kline N, Olson S, Weinberg D, Baker KE, Graveley BR, et al. 2015. Codon optimality is a major determinant of mRNA stability. *Cell* **160**: 1111–1124.

* Proud CG. 2018. Phosphorylation and signal transduction pathways in translational control. *Cold Spring Harb Perspect Biol* doi: 10.1101/cshperspect.a033050.

Qian W, Yang JR, Pearson NM, Maclean C, Zhang J. 2012. Balanced codon usage optimizes eukaryotic translational efficiency. *PLoS Genet* **8**: e1002603.

Radhakrishnan A, Chen YH, Martin S, Alhusaini N, Green R, Coller J. 2016. The DEAD-Box protein Dhh1p couples mRNA decay and translation by monitoring codon optimality. *Cell* **167**: 122–132 e129.

* Robichaud N, Sonenberg N, Ruggero D, Schneider RJ. 2018. Translational control in cancer. *Cold Spring Harb Perspect Biol* doi: 10.1101/cshperspect.a032896.

* Rodnina MV. 2018. Translation in prokaryotes. *Cold Spring Harb Perspect Biol* doi: 10.1101/cshperspect.a032664.

Russell JB, Cook GM. 1995. Energetics of bacterial growth: Balance of anabolic and catabolic reactions. *Microbiol Rev* **59**: 48–62.

Santagata S, Mendillo ML, Tang YC, Subramanian A, Perley CC, Roche SP, Wong B, Narayan R, Kwon H, Koeva M, et al. 2013. Tight coordination of protein translation and HSF1 activation supports the anabolic malignant state. *Science* **341**: 1238303.

Sanz E, Yang L, Su T, Morris DR, McKnight GS, Amieux PS. 2009. Cell-type-specific isolation of ribosome-associated mRNA from complex tissues. *Proc Natl Acad Sci* **106**: 13939–13944.

Schuller AP, Wu CC, Dever TE, Buskirk AR, Green R. 2017. eIF5A functions globally in translation elongation and termination. *Mol Cell* **66**: 194–205 e195.

Schumacher TN, Schreiber RD. 2015. Neoantigens in cancer immunotherapy. *Science* **348**: 69–74.

Sen ND, Zhou F, Ingolia NT, Hinnebusch AG. 2015. Genome-wide analysis of translational efficiency reveals distinct but overlapping functions of yeast DEAD-box RNA helicases Ded1 and eIF4A. *Genome Res* **25**: 1196–1205.

Sendoel A, Dunn JG, Rodriguez EH, Naik S, Gomez NC, Hurwitz B, Levorse J, Dill BD, Schramek D, Molina H, et al. 2017. Translation from unconventional 5′ start sites drives tumour initiation. *Nature* 541: 494–499.

Shi Z, Fujii K, Kovary KM, Genuth NR, Rost HL, Teruel MN, Barna M. 2017. Heterogeneous ribosomes preferentially translate distinct subpools of mRNAs genome-wide. *Mol Cell* 67: 71–83.e7.

Shieh YW, Minguez P, Bork P, Auburger JJ, Guilbride DL, Kramer G, Bukau B. 2015. Operon structure and cotranslational subunit association direct protein assembly in bacteria. *Science* 350: 678–680.

Sidrauski C, McGeachy AM, Ingolia NT, Walter P. 2015. The small molecule ISRIB reverses the effects of eIF2α phosphorylation on translation and stress granule assembly. *eLife* 4: 05033.

Simms CL, Yan LL, Zaher HS. 2017. Ribosome collision is critical for quality control during no-go decay. *Mol Cell* 68: 361–373.e365.

Slobodin B, Han R, Calderone V, Vrielink J, Loayza-Puch F, Elkon R, Agami R. 2017. Transcription impacts the efficiency of mRNA translation via co-transcriptional N^6-adenosine methylation. *Cell* 169: 326–337.e312.

* Sokabe M, Fraser CS. 2018. Toward a kinetic understanding of eukaryotic translation. *Cold Spring Harb Perspect Biol* doi: 10.1101/cshperspect.a032706.

Sonenberg N, Hinnebusch AG. 2009. Regulation of translation initiation in eukaryotes: Mechanisms and biological targets. *Cell* 136: 731–745.

* Sossin WS, Costa-Mattioli M. 2018. Translational control in the brain in health and disease. *Cold Spring Harb Perspect Biol* doi: 10.1101/cshperspect.a032912.

Stadler M, Fire A. 2011. Wobble base-pairing slows in vivo translation elongation in metazoans. *RNA* 17: 2063–2073.

Starck SR, Jiang V, Pavon-Eternod M, Prasad S, McCarthy B, Pan T, Shastri N. 2012. Leucine-tRNA initiates at CUG start codons for protein synthesis and presentation by MHC class I. *Science* 336: 1719–1723.

Starck SR, Tsai JC, Chen K, Shodiya M, Wang L, Yahiro K, Martins-Green M, Shastri N, Walter P. 2016. Translation from the 5′ untranslated region shapes the integrated stress response. *Science* 351: aad3867.

Steitz JA. 1969. Nucleotide sequences of the ribosomal binding sites of bacteriophage R17 RNA. *Cold Spring Harb Symp Quant Biol* 34: 621–630.

Stern-Ginossar N, Weisburd B, Michalski A, Le VT, Hein MY, Huang SX, Ma M, Shen B, Qian SB, Hengel H, et al. 2012. Decoding human cytomegalovirus. *Science* 338: 1088–1093.

* Stern-Ginossar N, Thompson SR, Mathews MB, Mohr I. 2018. Translational control in virus-infected cells. *Cold Spring Harb Perspect Biol* doi: 10.1101/cshperspect.a033001.

Stumpf CR, Moreno MV, Olshen AB, Taylor BS, Ruggero D. 2013. The translational landscape of the mammalian cell cycle. *Mol Cell* 52: 574–582.

Subramaniam AR, Zid BM, O'Shea EK. 2014. An integrated approach reveals regulatory controls on bacterial translation elongation. *Cell* 159: 1200–1211.

Subtelny AO, Eichhorn SW, Chen GR, Sive H, Bartel DP. 2014. Poly(A)-tail profiling reveals an embryonic switch in translational control. *Nature* 508: 66–71.

Tanenbaum ME, Stern-Ginossar N, Weissman JS, Vale RD. 2015. Regulation of mRNA translation during mitosis. *eLife* 4: 07957.

Thompson MK, Rojas-Duran MF, Gangaramani P, Gilbert WV. 2016. The ribosomal protein Asc1/RACK1 is required for efficient translation of short mRNAs. *eLife* 5: e11154.

Thoreen CC, Chantranupong L, Keys HR, Wang T, Gray NS, Sabatini DM. 2012. A unifying model for mTORC1-mediated regulation of mRNA translation. *Nature* 485: 109–113.

Tunney RJ, McGlincy NJ, Graham ME, Naddaf N, Pachter L, Lareau L. 2017. Accurate design of translational output by a neural network model of ribosome distribution. *bioRxiv* doi: 10.1101/201517.

Ude S, Lassak J, Starosta AL, Kraxenberger T, Wilson DN, Jung K. 2013. Translation elongation factor EF-P alleviates ribosome stalling at polyproline stretches. *Science* 339: 82–85.

Ulitsky I, Bartel DP. 2013. lincRNAs: Genomics, evolution, and mechanisms. *Cell* 154: 26–46.

Vasquez JJ, Hon CC, Vanselow JT, Schlosser A, Siegel TN. 2014. Comparative ribosome profiling reveals extensive translational complexity in different *Trypanosoma brucei* life cycle stages. *Nucleic Acids Res* 42: 3623–3637.

Verduyn C, Stouthamer AH, Scheffers WA, van Dijken JP. 1991. A theoretical evaluation of growth yields of yeasts. *Antonie Van Leeuwenhoek* 59: 49–63.

Vogel C, Marcotte EM. 2012. Insights into the regulation of protein abundance from proteomic and transcriptomic analyses. *Nat Rev Genet* 13: 227–232.

Vu LP, Pickering BF, Cheng Y, Zaccara S, Nguyen D, Minuesa G, Chou T, Chow A, Saletore Y, MacKay M, et al. 2017. The N^6-methyladenosine (m^{6}A)-forming enzyme METTL3 controls myeloid differentiation of normal hematopoietic and leukemia cells. *Nat Med* 23: 1369–1376.

Wang X, Zhao BS, Roundtree IA, Lu Z, Han D, Ma H, Weng X, Chen K, Shi H, He C. 2015. N^6-methyladenosine modulates messenger RNA translation efficiency. *Cell* 161: 1388–1399.

Weinberg DE, Shah P, Eichhorn SW, Hussmann JA, Plotkin JB, Bartel DP. 2016. Improved ribosome-footprint and mRNA measurements provide insights into dynamics and regulation of yeast translation. *Cell Rep* 14: 1787–1799.

* Wek RC. 2018. Role of eIF2α kinases in translational control and adaptation to cellular stress. *Cold Spring Harb Perspect Biol* doi: 10.1101/cshperspect.a032870.

Williams CC, Jan CH, Weissman JS. 2014. Targeting and plasticity of mitochondrial proteins revealed by proximity-specific ribosome profiling. *Science* 346: 748–751.

Wolfe AL, Singh K, Zhong Y, Drewe P, Rajasekhar VK, Sanghvi VR, Mavrakis KJ, Jiang M, Roderick JE, Van der Meulen J, et al. 2014. RNA G-quadruplexes cause eIF4A-dependent oncogene translation in cancer. *Nature* 513: 65–70.

Wolin SL, Walter P. 1988. Ribosome pausing and stacking during translation of a eukaryotic mRNA. *EMBO J* **7**: 3559–3569.

Wolin SL, Walter P. 1989. Signal recognition particle mediates a transient elongation arrest of preprolactin in reticulocyte lysate. *J Cell Biol* **109**: 2617–2622.

Woolstenhulme CJ, Parajuli S, Healey DW, Valverde DP, Petersen EN, Starosta AL, Guydosh NR, Johnson WE, Wilson DN, Buskirk AR. 2013. Nascent peptides that block protein synthesis in bacteria. *Proc Natl Acad Sci* **110**: E878–E887.

Xu G, Greene GH, Yoo H, Liu L, Marques J, Motley J, Dong X. 2017. Global translational reprogramming is a fundamental layer of immune regulation in plants. *Nature* **545**: 487–490.

Young DJ, Guydosh NR, Zhang F, Hinnebusch AG, Green R. 2015. Rli1/ABCE1 recycles terminating ribosomes and controls translation reinitiation in 3′UTRs in vivo. *Cell* **162**: 872–884.

Zhang G, Hubalewska M, Ignatova Z. 2009. Transient ribosomal attenuation coordinates protein synthesis and co-translational folding. *Nat Struct Mol Biol* **16**: 274–280.

Zhang S, Hu H, Zhou J, He X, Jiang T, Zeng J. 2017. Analysis of ribosome stalling and translation elongation dynamics by deep learning. *Cell Syst* **5**: 212–220. e216.

Zhou J, Wan J, Gao X, Zhang X, Jaffrey SR, Qian SB. 2015. Dynamic m^6A mRNA methylation directs translational control of heat shock response. *Nature* **526**: 591–594.

Zinshteyn B, Gilbert WV. 2013. Loss of a conserved tRNA anticodon modification perturbs cellular signaling. *PLoS Genet* **9**: e1003675.

Zinshteyn B, Rojas-Duran MF, Gilbert WV. 2017. Translation initiation factor eIF4G1 preferentially binds yeast transcript leaders containing conserved oligo-uridine motifs. *RNA* **23**: 1365–1375.

Toward a Kinetic Understanding of Eukaryotic Translation

Masaaki Sokabe and Christopher S. Fraser

Department of Molecular and Cellular Biology, College of Biological Sciences, University of California, Davis, California 95616

Correspondence: csfraser@ucdavis.edu

The eukaryotic translation pathway has been studied for more than four decades, but the molecular mechanisms that regulate each stage of the pathway are not completely defined. This is in part because we have very little understanding of the kinetic framework for the assembly and disassembly of pathway intermediates. Steps of the pathway are thought to occur in the subsecond to second time frame, but most assays to monitor these events require minutes to hours to complete. Understanding translational control in sufficient detail will therefore require the development of assays that can precisely monitor the kinetics of the translation pathway in real time. Here, we describe the translation pathway from the perspective of its kinetic parameters, discuss advances that are helping us move toward the goal of a rigorous kinetic understanding, and highlight some of the challenges that remain.

Determining how the process of translation is regulated in eukaryotic cells remains a central challenge in biology. Translation can be studied in vitro and in vivo using techniques that range from single-molecule analysis to genome-wide measurements. The latter are playing a key part in revealing the extent to which translational control plays a central role in regulating many aspects of cell physiology. The canonical pathway of translation can be depicted as four main stages (Fig. 1). The initiation stage determines which messenger RNA (mRNA) is recruited to the ribosome and which start codon is selected. During elongation, the ribosome translates the open reading frame (ORF) into a polypeptide chain. Termination occurs when the termination codon enters the aminoacyl (A) site of the ribosome, leading to the release of the polypep-

tide chain. Finally, recycling of the ribosome takes place so that it can enter another cycle of translation.

Translational efficiency (TE) relates the abundance of an mRNA to its rate of translation into protein. In yeast and mammals, the TE can vary between mRNAs by at least 20-fold, which provides a considerable range of translational control (Ingolia et al. 2009; Weinberg et al. 2016). Genome-wide analysis of the translatome indicates that the TE of a large number of different mRNAs is altered in response to changes in environmental conditions and events such as progression through stages of the cell cycle (Ingolia et al. 2011; Brar et al. 2012; Stumpf et al. 2013; Tanenbaum et al. 2015). A major challenge is to elucidate the molecular mechanisms that reprogram the translational machinery to determine:

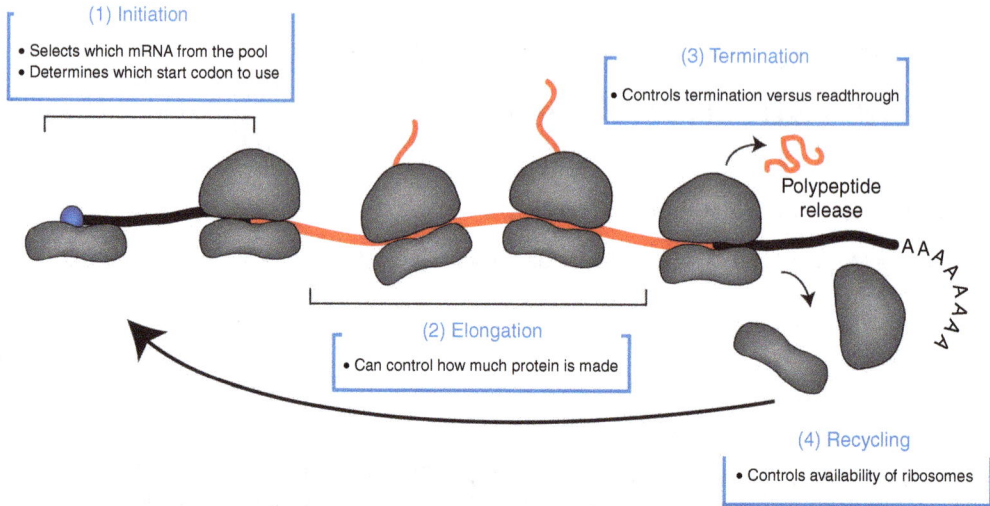

Figure 1. Pathway of eukaryotic translation. The model depicts four main stages of translation. (*1*) The initiation stage includes the recruitment of the 40S subunit to the m^{7}G cap structure, followed by its migration in a 3′ direction. Following start codon selection, the 60S subunit joins to form the 80S ribosome. This stage determines which messenger (mRNA) is selected from the pool and which start codon is chosen to initiate translation. (*2*) The elongation stage involves decoding the open reading frame (ORF) into a polypeptide chain. This stage can determine the rate at which protein is synthesized when initiation is very rapid (see main text). (*3*) Translation termination occurs when a termination codon is positioned in the A site of the ribosome, which results in the release of the polypeptide chain. The regulation of termination versus readthrough occurs at this step. (*4*) Following termination, the 40S and 60S subunits are recycled so that they can take part in another round of translation. For simplicity, the model depicts the recycling of the 40S subunit back onto the same mRNA. The rate of recycling can regulate the availability of ribosomal subunits for translation.

(1) what mRNA is selected for translation from the pool of mRNAs in the cell; (2) which start codon is used to specify the ORF that is translated; (3) how much protein is made from the selected mRNA; and (4) which termination codon is used.

As we discuss below, each stage of translation is generally believed to occur as a series of intermediate sequential substeps. Pathway intermediates are typically monitored at steady-state, which provides only limited information about how the pathway is regulated. To fully understand translational control, it will be essential to determine the rate at which intermediates are formed. This will require precise monitoring of the dynamics of the translational machinery in real time throughout the translation pathway in vitro and in vivo. In this review, we discuss the progress being made toward this goal and how we may achieve this with continued technological progress.

INITIATION

During initiation, an mRNA is recruited to the ribosome and a start codon is selected (for details of the initiation pathway, see Merrick and Pavitt 2018). Of all the stages of translation, initiation is the most different between the domains of life. This is highlighted by the fact that the combined molecular weight of the initiation factors is roughly an order of magnitude more in eukaryotes as compared to bacteria. Initiation can be separated into substeps, which are depicted in Figure 2. It is important to note, however, that the proposed substeps have generally been detected and characterized by virtue of their being relatively long-lived thermodynamically stable complexes. Being that many intermediates in the pathway are likely to exist on the subsecond or second timescale, new approaches for rapidly detecting pathway intermediates will be needed to verify the importance of these

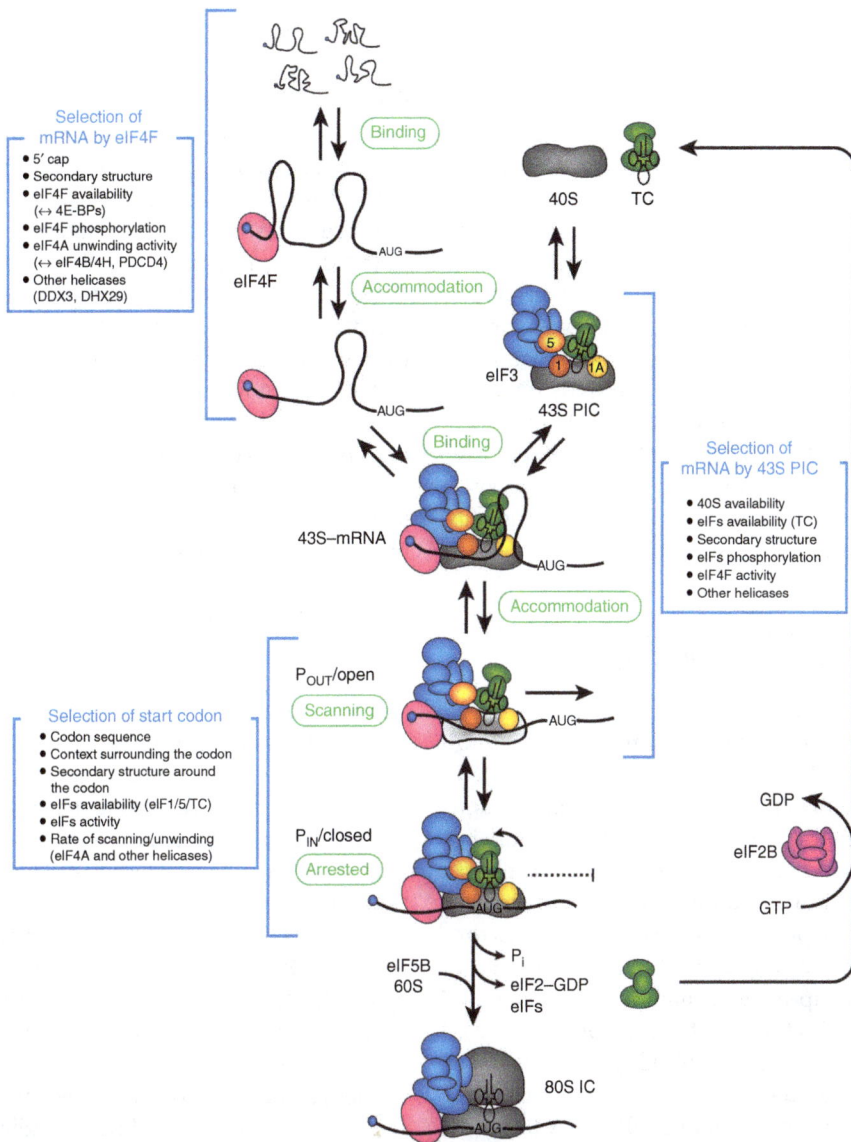

Figure 2. Cap-dependent translation initiation. The eukaryotic translation initiation pathway is depicted as a series of substeps, many of which are likely to be reversible (see main text for details). Potential sources of regulation for each substep are summarized in boxes and described in more detail in the main text. Selection of messenger RNA (mRNA) by eukaryotic initiation factor (eIF)4F: the mRNA is selected for translation by the binding and accommodation of the eIF4F complex. The accommodation step requires secondary structure in the mRNA to be unwound by the ATP-dependent helicase activity of eIF4A (secondary structure is depicted as "loops" in the 5′UTR). Selection of an mRNA by the 43S PIC: the 43S PIC (40S subunit, eIF1, eIF1A, ternary complex [TC], eIF3, and eIF5) is recruited to the 5′ end of the mRNA through its interaction with eIF4F. Productive recruitment occurs through initial binding and accommodation steps. The accommodation step requires a conformational change in the 40S subunit to a P_{OUT}/open conformation (described in the main text and depicted as a light gray-colored 40S subunit). Selection of the start codon: the base-pairing between the methionyl-transfer RNA (Met-tRNA$_i$) and the start codon generates a scanning arrested 43S preinitiation complex (PIC). The arrested complex possesses a P_{IN}/closed conformation. The final step of the pathway involves the recruitment of the 60S subunit to form the 80S initiation complex (IC). This step is stimulated by the GTPase activity of eIF5B. Following its release from the 40S subunit, the eIF2•GDP complex is recycled to eIF2•GTP by the guanine nucleotide exchange factor (GEF) activity of eIF2B. It should be noted that the possible function of poly(A)-binding protein (PABP) in circularization of the mRNA is not shown for clarity.

proposed intermediates. Moreover, whereas the initiation pathway is presented as a sequence of ordered steps, the order of binding and release of components during initiation is not known. It is also not clear whether different initiation factors exist in a free form, or are bound together in larger multifactor complexes. Two unanswered fundamental questions about the regulation of initiation are (1) how does the translation machinery discriminate between mRNAs for their recruitment to the ribosome; and (2) how is the selection of the ORF regulated? Here, we discuss how we can move toward a kinetic framework for the initiation stage and explain why this will be needed to fully answer these two questions.

RECRUITMENT OF mRNA TO THE 43S PREINITIATION COMPLEX (PIC)

In response to various physiological stimuli, the translation machinery is reprogrammed to change which mRNAs are preferentially selected for recruitment to the ribosome. A large amount of data indicate that reprogramming occurs in response to changes in the availability of at least two canonical initiation factors. These are the cap-binding protein, eukaryotic initiation factor (eIF)4E, which together with eIF4A and eIF4G forms the cap-binding complex, eIF4F; and the eIF2•GTP•Met-tRNA$_i$ ternary complex (TC). An increase in the availability of eIF4E (see Proud 2018) correlates with increased recruitment of growth-promoting mRNAs, whereas a decrease in its availability correlates with reduced recruitment of these mRNAs. In contrast, reduced availability of the TC more generally lowers global rates of translation, although the translation of some mRNAs containing upstream ORFs (uORFs) can be increased in response to reduced TC availability (see Wek 2018).

Regulation of 43S PIC Availability

The 43S PIC can be purified from eukaryotic cells as a stable complex containing the 40S subunit, eIF1, eIF1A, TC, eIF3, and eIF5. Thermodynamic frameworks for the yeast and human 43S PIC in the absence of mRNA have revealed

that the stability of this complex is governed by a network of direct and indirect interactions among its components (reviewed in Fraser 2015). This complex can be formed on free ribosomal subunits, or on 40S subunits that are generated during ribosome recycling (see Hellen 2018). A kinetic framework for association/dissociation rates of eIF1, eIF1A, TC, eIF3, and eIF5 from the 40S subunit has not been determined and it is not known whether different kinetically favored complexes regulate the subsequent recruitment of specific mRNAs to the 43S PIC.

The availability of TC is one 43S PIC component that is regulated in response to changes in environmental conditions. In response to different cellular stresses, eIF2 is phosphorylated, which results in reduced TC availability for 40S binding by virtue of an altered interaction with its guanine nucleotide exchange factor, eIF2B (see Merrick and Pavitt 2018; Wek 2018). Initiation pathway models imply that binding of the TC to the 40S subunit is required before mRNA recruitment, suggesting that reduced TC levels would globally reduce translation. However, sucrose gradient, gel shift, and toeprinting assays that are generally used to monitor the mRNA•eIF4F•43S PIC complex (often described as the 48S PIC) rely on start codon recognition by the TC to stabilize the complex. Thus, it is unclear whether prior binding of the TC is required before mRNA recruitment to the 40S ribosomal subunit. To investigate this, we recently generated an equilibrium-binding assay that revealed a 10-fold reduction in the affinity of eIF3j for the eIF4F•43S PIC on mRNA accommodation into the mRNA entry channel (Sokabe and Fraser 2017). The TC was found to be absolutely required for the reduced eIF3j affinity in an mRNA-dependent, but AUG codon-independent manner. This is consistent with the recruitment of the TC being needed for mRNA to fully accommodate into the mRNA entry channel of the 40S subunit. What is currently not known is whether mRNA accommodation into the entry channel is absolutely required for the scanning stage of initiation, or whether scanning can occur on an mRNA that is partially accommodated into the entry channel of the 40S subunit. An m^{7}G cap-proximal AUG codon is more frequently bypassed when TC availability is reduced in cells in response to increased eIF2 phosphorylation (Palam et al. 2011). A similar observation was made in cell-free extracts when mRNA concentration was increased (Dasso et al. 1990). These findings are consistent with TC not being absolutely required for mRNA recruitment to and accommodation into the decoding site of the 40S subunit, or the subsequent translocation of the 40S subunit. However, TC is of course absolutely necessary for the recognition of the initiation codon. More investigation will be needed to establish the precise role of TC in mRNA recruitment to the 43S PIC and whether its availability can influence mRNA selection in addition to start site selection. In addition to TC availability, it is possible that the amount of free 43S PICs for mRNA recruitment becomes limiting for translation when the majority of ribosomes are engaged in translation in actively growing cells. Under such conditions, the rate of ribosome recycling after termination would be expected to play a significant role in controlling translation (see below).

The Role of the eIF4F Complex in mRNA Selection

The molecular details of how eIF4F regulates the selective recruitment of different messages to the ribosome are complex. eIF4F comprises eIF4E, the m^7G-cap-binding subunit; eIF4A, an RNA helicase protein; and eIF4G, a large scaffolding protein. It therefore is capable of numerous activities: binding to the m^7G cap; melting of RNA secondary structure; and interacting through eIF4G with other proteins such as the poly(A)-binding protein (PABP), eIF3, DDX3/ded1, and the protein kinase MNK1 (Lamphear et al. 1995; Imataka et al. 1998; Pyronnet et al. 1999; Korneeva et al. 2000; Hilliker et al. 2011; Soto-Rifo et al. 2012). Human eIF4F can be purified from cells as a stable trimeric complex, but in cells may also be present as a mixture of larger complexes containing the eIF4G-binding proteins and possessing different mRNA-binding properties (Merrick 2015). Given the dynamic aspects of eIF4F and its complexes, we

will evaluate how the different substeps of mRNA selection are accomplished.

The first step in the interaction of eIF4F with mRNA is its binding to the m^7G cap through its eIF4E subunit. The rate of cap binding can be reduced if secondary structure in the mRNA "hides" the cap, sterically hindering its interaction with eIF4E (reviewed in Kozak 2005; Livingstone et al. 2010; Fraser 2015). Such weak RNA secondary structures are likely dynamic, being partially or completely unwound through thermal energy in the cell. When the m^7G cap is sufficiently exposed, eIF4E binds and forms an eIF4F•mRNA complex called the preaccommodated state. This complex is not thought to be very stable, although its rates of binding and dissociation are not well elucidated. It seems likely that mRNAs differ in how exposed their cap structures are. Under conditions of limiting eIF4F, the mRNAs must compete for binding to the factor. Those mRNAs that are best able to bind with well-exposed 5′ caps are often called "strong" mRNAs (reviewed in Gingras et al. 1999).

Thermodynamic frameworks have shown that eIF4E can bind to the m^7G cap in its free form, bound to eIF4G (and eIF4F associated factors), or associated with 4E-BPs. The binding of eIF4F to the cap is thermodynamically favored over that of uncomplexed eIF4E binding (Niedzwiecka et al. 2002), but this does not inform about the order in which these components bind to the cap structure. Heterogeneity in the order of binding is likely to occur, but whether this contributes to regulating specific mRNA recruitment is not known. The use of single-molecule analysis similar to that recently used to study the order of binding and release of components during bacterial initiation will likely help solve this question (Tsai et al. 2012; see Prabhakar et al. 2018).

The next step is the formation of an additional interaction involving the RNA-binding region of eIF4G and a cap-proximal portion of the mRNA that is unstructured. The resulting eIF4F•mRNA complex, called the accommodated state, is more stable, although again its rate of formation and dissociation are not known. Based on cross-linking experiments, it was shown that formation of the accommodated state is inhibited by cap-proximal secondary structure (Pelletier and Sonenberg 1985; Lawson et al. 1986). DEAD-box helicases, such as eIF4A, unwind duplex regions by local strand separation, whereby they can bind directly to the duplex region and destabilize it (Yang et al. 2007). Whether the opening of cap-proximal structures by thermal energy to allow eIF4A access to a duplex contributes to eIF4F accommodation is unknown. It is thought that eIF4F accommodation could function as a kinetic checkpoint for mRNA selection, as the accommodated state is more stable than the preaccommodated state and therefore more likely to progress down the initiation pathway. Should the rate of accommodation of eIF4F not be suitably fast, it is possible that the entire eIF4F complex will dissociate from the mRNA, preventing its selection for translation. The accommodation model will need to be experimentally tested by determining detailed kinetic frameworks of eIF4F binding to different mRNAs. Obtaining high-resolution structures of mRNAs in the absence and presence of eIF4F (or other RNA-binding proteins) will provide the necessary structural information to explain how mRNAs are selected by eIF4F.

How can the mechanism of mRNA binding by eIF4F be regulated so that mRNA recruitment can be reprogrammed? It has long been proposed that mRNAs compete for limiting amounts of eIF4F in the cell (Gingras et al. 1999). The regulation of eIF4F availability is controlled by eIF4E-binding proteins (4E-BPs) acting as competitive inhibitors to prevent eIF4E binding to eIF4G (Richter and Sonenberg 2005). It is generally assumed that eIF4G (and associated proteins) cannot productively bind to the 5′ end of an mRNA and successfully recruit it to the ribosome in the absence of eIF4E. However, this can be accomplished in vitro by an eIF4G truncation lacking its eIF4E-binding domain (eIF4G$_{682-1599}$ [Ali et al. 2001]). Whether full-length eIF4G can accomplish this has not been rigorously tested.

Following formation of the eIF4F•mRNA accommodated state, binding of the 43S PIC to the 5′ end of the mRNA occurs through an interaction between eIF4G and eIF3 (Lamphear et al. 1995; Korneeva et al. 2000; Morino et al. 2000; Villa et al. 2013). Binding and mRNA entry into the mRNA channel of the 40S ribosomal subunit requires an unstructured portion of the mRNA. To enable 43S PIC binding, a secondary structure near the cap-bound eIF4F complex is unwound by the helicase activity of its eIF4A subunit, together with eIF4B (or eIF4H) (Rozen et al. 1990; Rogers et al. 2001; Nielsen et al. 2011; Özeş et al. 2011; Harms et al. 2014; see Merrick and Pavitt 2018). The ability of eIF4F to unwind hairpins in vitro is thought to be limited to those with modest stability ($\sim\Delta G$ −30 kcal/mol [Linder and Jankowsky 2011]). A limited unwinding potential of eIF4F may be caused by the fact that eIF4A, like other DEAD-box helicase proteins, functions as a nonprocessive helicase in the absence of eIF4G (Linder and Jankowsky 2011; García-García et al. 2015). However, using a force-generated, single-molecule assay, eIF4A was shown to behave as a processive helicase protein in the presence of eIF4G and eIF4B/4H (García-García et al. 2015). Single-molecule fluorescence resonance energy transfer (smFRET) has shown that eIF4B, eIF4H, and eIF4G promote unwinding by increasing the rate of cycling between "open" and "closed" conformations of eIF4A (Sun et al. 2012; Harms et al. 2014). It should be noted that the processivity mechanism used by eIF4A is still not clear. It is possible that, whereas eIF4A itself may not remain stably bound to an mRNA following ATP hydrolysis, cooperative interactions between eIF4A, eIF4G, eIF4B, and RNA may prevent eIF4A from dissociating from the eIF4F•eIF4B complex (thereby enabling eIF4A to take part in another round of unwinding). Monitoring rapid kinetics of the association and dissociation rates between eIF4F components during this process will explain this in greater detail. The interaction of Pdcd4 with eIF4A may function to inhibit the availability and/or unwinding activity of eIF4A, thereby decreasing 43S PIC binding on structured mRNAs (Yang et al. 2003). Finally, the recruitment of additional helicase proteins, such as DDX3/ded1p, to the eIF4F complex may regulate its ability to unwind secondary structure (Gao et al. 2016).

Accommodation of the 43S PIC onto mRNA

It is possible that the final stage of mRNA selection for translation is completed when the mRNA fully accommodates into the mRNA entry channel and decoding site of the 43S PIC (Fig. 2). However, it should be noted that the dissociation rate of a decoding site–accommodated mRNA has not been measured, so it is entirely possible that start codon recognition could serve as the final stage of mRNA selection. For mRNA accommodation into the entry channel and decoding site of the 40S subunit, the 43S PIC must adopt an open conformation of the decoding site. The binding of eIF1 and eIF1A to the 40S subunit generates this conformation (Pestova et al. 1998; Passmore et al. 2007; Hussain et al. 2014). However, other initiation components may contribute to stabilizing this conformation because structures of 40S subunit complexes that include eIF1 and eIF1A adopt a closed conformation (Hashem et al. 2013; Lomakin and Steitz 2013; Weisser et al. 2013). All structures to date have been determined in the absence of the eIF4F complex. It is therefore unknown whether this complex affects the conformation of the 40S subunit decoding site. In support of this possibility, eIF4F and eIF4B were recently shown to lower the affinity of eIF3j for the 43S PIC, but only in the presence of mRNA (Sokabe and Fraser 2017). This is likely caused by the fact that eIF3j binds to the mRNA entry channel and A site of the 43S PIC with a reduced affinity when mRNA is stabilized in the mRNA entry channel (Unbehaun et al. 2004; Fraser et al. 2007; Aylett et al. 2015; Sharifulin et al. 2016). Notably, an unstructured mRNA is required for the reduction of the affinity of eIF3j for the 43S PIC. This is consistent with the observation that secondary structure serves as a barrier to productive 48S PIC formation on mRNAs containing the iron responsive element ([IRE]; Muckenthaler et al. 1998). Thus, the 43S PIC likely binds to an mRNA first in a preaccommodated state (with the mRNA not stably bound to the mRNA entry channel) and then in an accommodated state (with the mRNA stably bound to the mRNA entry channel). It is anticipated that high-resolution structures of eIF4F

bound to the 43S PIC will help determine whether this affects the conformation of the 40S subunit decoding site to enable mRNA entry. Equally important is the need to generate kinetic assays to monitor the rate by which prebound mRNA accommodates into the decoding site of the 43S PIC or dissociates from the ribosome.

Generating a kinetic assay to monitor the rate of mRNA accommodation into the decoding site of the 43S PIC will help establish the extent to which the mRNA accommodation step can serve as a kinetic checkpoint to regulate mRNA selection. Reduced availability of the 43S PIC would be expected to regulate the preaccommodation step if the 43S PIC becomes limiting for translation. This would also be expected to be the case when TC is limiting (when eIF2 is phosphorylated). Comparable to the regulation of eIF4F accommodation onto an mRNA, the rate at which secondary structure is unwound by eIF4A and other helicases could play a critical role in controlling such a checkpoint. In mammals, a direct interaction between eIF4G and eIF3 plays an important role in stabilizing the 43S PIC on an mRNA (Hinton et al. 2007; Villa et al. 2013). Immunoprecipitation assays have indicated that the affinity of eIF4G binding to eIF3 is enhanced by TORC1 activation by a yet-to-be-determined mechanism (Harris et al. 2006; Thoreen et al. 2012). Altering the lifetime of this interaction could regulate 48S PIC formation on different mRNAs. Should the rate of accommodation of mRNA into the decoding site of the 43S PIC not be suitably fast, it is entirely possible that the entire 43S PIC dissociates from the mRNA, preventing its selection for translation.

SCANNING AND START CODON SELECTION

Following accommodation of the mRNA into the entry channel and decoding site of the 43S PIC, the complex migrates in a 3' direction until it selects a suitable codon with which to initiate protein synthesis. The scanning mechanism was first proposed by Kozak and Shatkin and has remained the primary mechanism to explain the process of start codon selection on the majority of eukaryotic cellular mRNAs (Kozak and Shatkin 1978; Kozak 1980). Progress toward elucidating the thermodynamic and kinetic frameworks of start site selection has been made, but how the 43S PIC scans/migrates in a 5′ to 3′ direction is less understood. This is largely caused by the fact that kinetic assays to monitor 43S PIC migration have not yet been developed. Thus, little is known about the extent to which the rate of scanning can contribute to the regulation of translation or selection of mRNA.

The Scanning Mechanism

The ability of the eIF4F•4B complex to migrate along an mRNA in a 5′ to 3′ processive manner is consistent with eIF4A serving as the translocation motor for the scanning 43S PIC. The unwinding step size of eIF4A is 11 ± 2 base pairs, which is the same as the major steps taken by the hepatitis C virus NS3 DEx(H/D) helicase (Dumont et al. 2006; García-García et al. 2015). This relatively large step size is remarkable given that the 43S PIC must inspect each nucleotide during the scanning process. Generating single-molecule assays that are able to monitor scanning of the 43S PIC in real time at single-nucleotide resolution will likely play an important role in determining how the step size of eIF4A helicase activity relates to the migration of the 43S PIC. In addition, the emergence of high-resolution structures of scanning intermediates will help provide a structural basis for scanning. The relationship between scanning and start site selection is discussed in more detail in the following section.

Kinetics of Start Codon Recognition

The fidelity of start site selection is primarily governed by mRNA cis-elements, such as the codon sequence, the context surrounding the codon (Kozak sequence), and the position and stability of secondary structure. An ORF starting with the AUG codon is >50-fold better translated than one with a near-AUG codon, such as UUG (Fekete et al. 2005). Similarly, an AUG in the optimal context can be >20-fold more efficiently selected than one in the weakest context

(Kozak 1986). Start codon selection is one of the most well-characterized steps in the eukaryotic translation pathway (Lorsch and Dever 2010; Hinnebusch 2014, 2017). Recognition of the start codon requires concerted structural rearrangements of initiation factors, initiator Met-tRNA$_i$, and the 40S subunit. This results in a stable, scanning-arrested, 43S PIC harboring Met-tRNA$_i$ base-paired with the start codon at the P site (see Jobe et al. 2018 and Merrick and Pavitt 2018 for details). Structurally, codon–anticodon base-pairing results in transition of the 40S subunit from an open to a closed conformation together with a movement of the Met-tRNA$_i$ from an unaccommodated (P_{OUT}) to an accommodated (P_{IN}) state (Llácer et al. 2015; Hinnebusch 2017). The release of the γ phosphate from the hydrolyzed TC-bound GTP ensures that the process is irreversible. The kinetics of these events have been primarily identified and characterized using a purified reconstituted yeast system in conjunction with structural and genetic studies (reviewed in Hinnebusch 2017).

The Lorsch laboratory has pioneered a reconstituted purified yeast system to monitor start site selection using fluorescently tagged components. What has emerged from this work is a model that describes a complex set of dynamic interactions between initiation components and the decoding site of the 40S subunit to ensure the fidelity of start site selection. Fast kinetic monitoring of a purified 40S subunit together with eIF1, eIF1A, and TC (40S•1•1A•TC) revealed that the rate of eIF5-dependent GTP hydrolysis is rapid (within a second) and somewhat insensitive to the presence of an mRNA (only a ~1.5-fold stimulation was observed by adding an mRNA with an AUG codon) (Algire et al. 2005). In contrast to GTP hydrolysis, the rate of P_i release from 40S•1•1A•TC is sensitive to the presence of a prebound mRNA, but only if it possesses an AUG codon. The rate of P_i release was 0.04–0.06/sec (taking >40 sec to completion) in the absence of an AUG codon and was >100-fold accelerated in the presence of an mRNA with an AUG codon (6.7/sec). The rates of GTP hydrolysis and P_i release are similar in the presence of an mRNA containing an AUG codon, which is consistent with a model where-

Cite this article as Cold Spring Harb Perspect Biol doi: 10.1101/cshperspect.a032706

by GTP and GDP•P_i (possibly a hydrolysis transition state) are in rapid equilibrium within the 40S•1•1A•TC in the presence of eIF5 and independent of the codon in the P site. Upon start site selection, P_i release is the commitment step that makes GTP hydrolysis irreversible. The Lorsch laboratory also found that P_i release from 40S•1•1A•TC is >10-fold slower (0.54/sec) when saturating mRNA with an AUG codon was added together with eIF5 to trigger the reaction, instead of the mRNA being prebound to 40S•1•1A•TC (Algire et al. 2005). Because the rate of mRNA binding was not limiting for a saturated concentration, this 10-fold reduced rate is entirely consistent with P_i release being limited by a conformational change that occurs on placement of the AUG codon of an mRNA into the P site.

To characterize the conformational change of the 40S subunit on start site selection, a kinetic analysis was undertaken using a FRET signal between eIF1A and eIF1 to monitor the dissociation rate of eIF1 from 40S•1•1A•TC (Maag et al. 2005). The dissociation rate of eIF1 was accelerated 40-fold by the mRNA with an AUG codon, resulting in a dissociation rate (0.6/sec) that is the same as that of P_i release triggered by the same mRNA and eIF5 (Maag et al. 2005). This is consistent with eIF1 dissociation from the 40S•1•1A•TC complex acting as the rate-limiting step for P_i release. To verify this, when slowly or rapidly dissociating mutants of eIF1 (G107K or G107E) were used in this assay, the P_i release rate was appropriately reduced or increased (Algire et al. 2005; Nanda et al. 2013). It is worth noting that the rate of decrease in eIF1-eIF1A FRET in response to an mRNA containing an AUG codon would also be consistent with a rapid conformation change of eIF1 preceding its dissociation (with the rate of 9/sec). Such a conformational change may be similar to the recent structural model in which eIF1 adopts an altered conformation when Met-$tRNA_i$ adopts the accommodated conformation (P_{IN} state described above) (Hussain et al. 2014; Llácer et al. 2015). It is important to note that the kinetic experiments were performed in the absence of eIF3 and eIF4F, so possible contributions of these components in this process could

not be determined. Using a mammalian reconstituted system, both eIF3 and eIF4F (bound to the m^7G cap) have been shown to inhibit eIF5-induced GTP hydrolysis in the absence of the AUG codon (Majumdar and Maitra 2005). This indicates that GTP hydrolysis is regulated by these initiation components, which were not present in the yeast work. It will be important in the future to rigorously explore these regulatory roles of eIF4F and eIF3 on GTP hydrolysis and P_i release using kinetic assays.

Upon start site selection, a conformational change occurs between the carboxy-terminal tail of eIF1A (1A-CTT) and the amino terminus of eIF5 (5-amino-terminal domain [NTD]). Specifically, the placement of the AUG codon at the P site induces a biphasic increase in a FRET signal between 1A-CTT and 5-NTD (24/sec and 0.4/sec in rates) (Nanda et al. 2013). This is consistent with a two-step conformational change of these proteins during start site selection. The rate of the slow phase in this process is once again very similar to the rate of AUG-dependent eIF1 dissociation. As expected, the slow kinetic phase can be reduced further, or increased, when more slowly or rapidly dissociating mutants of eIF1 were used (Nanda et al. 2013). Mutations in the 1A-CTT that reduce this slow phase (to 0.01–0.02/sec) further reduced the P_i release rate but did not alter the dissociation rate of eIF1. This implies that eIF1 dissociation is not in fact sufficient to induce P_i release, but is likely a key mediatory event in start site selection followed by the altered binding of eIF1A and eIF5.

How is the conformation of the 43S PIC altered on start codon recognition? Upon start codon recognition, the 43S PIC adopts a closed/P_{IN} conformation, in which eIF1A binding is stabilized. This is shown in recent cryoelectron microscopy (cryo-EM) structures where eIF1A bridges the narrowed gap between head and body domains with its NTD interacting with the codon–anticodon duplex and +4 to +6 nucleotides of mRNA (Hussain et al. 2014; Llácer et al. 2015). The dissociation rate of fluorescently labeled eIF1A shows a biphasic pattern (with fast and slow phases) in the absence of an AUG codon (Maag et al. 2006). This is consistent with two distinct populations, each possessing a defined conformation of the 40S subunit. Importantly, AUG recognition increases the occupancy of the slow phase population from ~30% to ~90%. The 43S PIC may therefore be spontaneously switching back and forth between open and closed states even in the absence of mRNA, with AUG recognition shifting the equilibrium between these conformations toward the closed state. Thus, rather than a static model in which an open conformation 40S subunit slides along the mRNA during scanning, the scanning complex may dynamically sample the codon–anticodon interaction, whereas initiation factors function to help stabilize the conformation with stable base-pairing.

Start Codon Recognition during Scanning

The above kinetic studies provide precise information about the molecular steps and time frame for start codon recognition, but one wonders if the observed ~twofold difference of the P_i release rates at AUG and near-AUG (UUG/AUU) codons is sufficient to enable a >50-fold difference in their translation efficiencies (Cheung et al. 2007; Nanda et al. 2009). As noted above, it is possible that the partially reconstituted system used in these studies (minus eIF4F and eIF3) possesses altered kinetics and/or reduced fidelity. This partial system allows for direct loading of an unstructured short mRNA to the 43S PIC without imposing migration of the 43S PIC driven by ATP hydrolysis, which is mediated by eIF4F. Lack of scanning may enable the prolonged lifetime (or iterative repositioning) of a near-cognate codon in the P site compared to a 43S PIC that migrates at a certain rate using an energy-dependent translocase (eIF4A). It will therefore be important to determine the rate of P_i release (i.e., fidelity) of the 43S PIC under more stringent physiologically relevant scanning conditions. Interestingly, placing a moderately stable hairpin (19 kcal/mol) ~14 nt downstream of a weak AUG codon (near the leading edge of the 43S PIC) dramatically increases the likelihood of initiating translation at that codon (Kozak 1990). This is entirely con-

sistent with improved recognition of a weak codon when the rate of scanning is reduced.

The rate of 43S PIC migration is generally believed not to be limiting for the overall rate of translation, as long as the 5′UTR is unstructured. This has been inferred from experiments that have shown that the extension of a 5′UTR does not result in decreased translation, and may even lead to increased translation (possibly caused by the increased capacity for loading multiple 43S PICs) (Kozak 1991). Nevertheless, secondary structure in the mRNA could reduce the kinetics of 43S PIC migration to a rate comparable to that needed to select a weak AUG codon. As mentioned earlier, the mechanism of codon recognition requires a dynamic set of interactions and conformational changes that likely reduces the rate of scanning. These include interactions of the Kozak sequence with eIFs and ribosomal proteins, codon–anticodon base-pairing, a closing of the mRNA decoding site conformation, and the generation of a P_{IN} conformation. Consistent with this, in a cell-free extract, the closed/P_{IN} scanning 43S PIC located at an optimum AUG (formed with nonhydrolyzable GTP) is stable at least for an hour, with marginal "sliding" to downstream start codons (Terenin et al. 2015). Presumably, the lifetime of the closed/P_{IN} state formed at a weak/noncognate codon is much shorter to allow 43S PIC moving downstream more occasionally before P_i release.

Regulation of Start Site Selection

Early studies identified eIF1, eIF2, and eIF5 as the primary components that maintain the fidelity of start site selection in yeast (Donahue et al. 1988; Cigan et al. 1989; Yoon and Donahue 1992; Huang et al. 1997). Subsequent work characterizing additional yeast mutants has revealed that the fidelity of start site selection is maintained by coordinated interactions between the mRNA, the 40S subunit, and all initiation factors that make up the 43S PIC (eIF1, eIF1A, eIF2, eIF3, and eIF5 [reviewed in Hinnebusch 2014, 2017]). A conserved autoregulatory translation mechanism maintains the cellular concentrations of both eIF1 and eIF5, underscoring the importance of these components in maintaining fidelity (Ivanov et al. 2010; Loughran et al. 2012). What appears to be missing is a regulatory mechanism to control the fidelity of start site selection to reprogram the proteome in response to different cell stimuli. An attractive mechanism would be provided by regulated phosphorylation of fidelity-controlling components. Interestingly, Thr72 in eIF1 can be phosphorylated in response to arsenite stress, and this correlates with the upregulation of arsenite-inducible regulatory particle-associated protein (AIRAP) synthesis (Zach et al. 2014). This has been proposed to occur by increased leaky scanning of an inhibitory uORF that possesses a suboptimal AUG codon (Zach et al. 2014). The phosphorylation of eIF5 occurs at several sites in the carboxy-terminal domain in a cell-cycle-dependent manner (Homma et al. 2005). These phosphorylation sites are correlated with increased formation of the 43S PIC and cyclin B1 translation, although it is not clear whether they lead to selective or general translation activation. One anticipates that regulatory mechanisms will be identified that are able to precisely control start site selection. The ability to monitor the kinetics of start site selection using in vitro assays will undoubtedly help provide the necessary molecular insight into these mechanisms as they are found.

ELONGATION

During elongation, the 80S ribosome decodes the ORF of the mRNA into protein (for a review of the elongation pathway in eukaryotes, see Dever et al. 2018). The process of elongation can be regulated at all three substeps: (1) decoding of the codon, (2) peptide bond formation, and (3) translocation of the reading frame (Fig. 3). The rate-limiting step of the elongation process is almost always the decoding of the codon, which is governed by the rate of delivery and accommodation of the cognate aminoacyl transfer RNA (aa-tRNA) to the A site of the ribosome (reviewed in Wohlgemuth et al. 2011). Importantly, the peptidyl-transferase center (PTC) of the ribosome is one of its most highly conserved regions, highlighting the fact that the fundamental mechanism of peptide bond formation is con-

served across all organisms (reviewed in Rodnina and Wintermeyer 2009; Simonovic and Steitz 2009). Maintaining the speed and accuracy of translation requires precise coordination between the ribosome, mRNA, tRNA, and elongation factors (reviewed in Rodnina et al. 2017). The rate of elongation can in some instances control protein abundance, which implies that it can serve as the rate-limiting step of translation. However, this almost certainly depends on the mRNA that is being studied and the growth state of the cell. Recent analysis of yeast ribosome profiling data together with methods to monitor translation in real time have indicated that initiation rates in yeast likely vary between 4 and 240 sec (~60-fold) on different mRNAs, whereas elongation rates appear to vary between three and five amino acids per second (~twofold) (Shah et al. 2013; Wang et al. 2016; Yan et al. 2016). However, direct measurements of translation on mRNAs containing different rare codons in vitro have indicated that elongation rates can in fact vary by as much as ~20-fold (reviewed in Rodnina 2016). This suggests that elongation can provide a mechanism with considerable potential to regulate protein synthesis, especially on mRNAs for which initiation is not rate limiting. It is anticipated that more precise methods to measure initiation and elongation in real time will help to fully understand the precise contributions that initiation and elongation stages can have on regulating translation on different mRNAs.

Regulation of Elongation by Codon Optimality

The rate of elongation in all organisms can be regulated by the degree of codon optimality of an mRNA. Optimal codons correspond to tRNA species that are abundant and rapidly accommodated into the ribosome during elongation, whereas nonoptimal codons (rare codons) interact with tRNAs that are low in abundance and therefore more slowly accommodated into the A site of the ribosome during elongation (reviewed in Chamary et al. 2006; Plotkin and Kudla 2011; Koutmou et al. 2015; Rodnina 2016; Hanson and Coller 2018). The introduction of

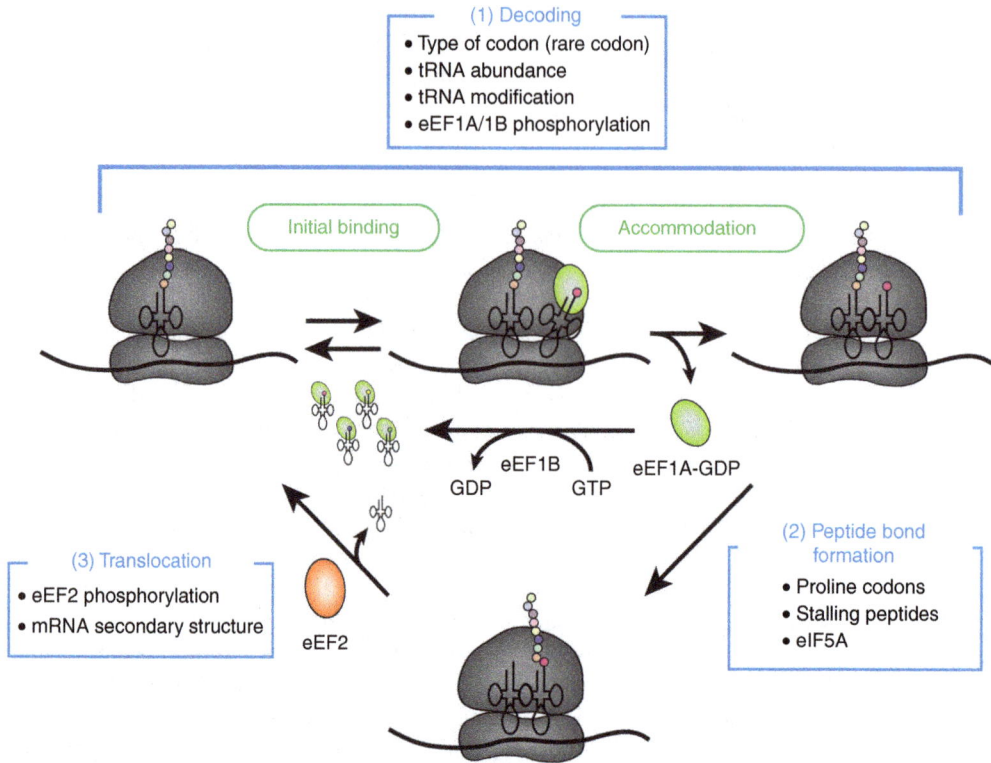

Figure 3. Model of eukaryotic elongation. Elongation is depicted as three main stages: (*1*) During the decoding step, eEF1A•GTP recruits the appropriate aminoacyl-transfer RNA (aa-tRNA) to the A site of the ribosome. Productive recruitment occurs through initial binding and accommodation steps. The accommodation step requires the hydrolysis of eEF1A-bound GTP, followed by release of eEF1A•GDP. The rate of decoding can be reduced by the presence of rare codons. Following its release, eEF1A•GDP is recycled by the guanine nucleotide exchange factor (GEF) activity of eEF1B. (*2*) Peptide bond formation between the aa-tRNA in the A site and the peptidyl-tRNA in the P site is catalyzed by the ribosomal RNA (rRNA) through different substeps (not shown). The rate of peptide bond formation is affected by recruitment of eIF5A. (*3*) The translocation step moves the mRNA relative to the ribosome by one codon. This step requires the recruitment of eEF2•GTP, followed by GTP hydrolysis and release of eEF2•GDP. Potential sources of regulation for each step are summarized in boxes and described in more detail in the main text.

ribosome profiling (see Ingolia et al. 2018) and other techniques to monitor translation in real time in living cells is helping to reveal the extent to which codon optimality contributes to the regulation of translation by influencing ribosome pausing at these codons (Gardin et al. 2014; Lareau et al. 2014; Weinberg et al. 2016; Yan et al. 2016). It has been estimated that codon optimality can account for about 30% of the variation that exists between mRNA abundance and protein abundance in human cells (Vogel et al. 2010; Hanson and Coller 2018). A region of mRNA where rare codons

have a particularly strong effect on translation is that directly following the start codon. This is likely caused by the fact that slowing ribosomes in this region of the mRNA further decreases the rate of initiation by slowing the rate that ribosomes clear the start codon (Mitarai et al. 2008; Potapov et al. 2012; Chu et al. 2014). How can the regulation of elongation through rare codons be regulated in response to changes in cell stimuli? One solution would be to alter the concentration of tRNAs that decode rare codons. Accordingly, tRNA concentrations can be increased or decreased in cells by altering both transcription and turnover rates in response to different cell stimuli (Kirchner and Ignatova 2015; Wilusz 2015). This would enable control of the translation of mRNAs that contain a disproportionate number of rare codons in their ORF.

An additional function of rare codons is to slow down ribosomes in mRNA sequences that encode low complexity regions found in between folded domains. This ensures that complex domains are given suitable time to correctly fold (reviewed in Komar 2009). This function can be augmented by structural features in an mRNA that slow the rate of elongation at the translocation step (Fig. 3) (Nackley et al. 2006; Kudla et al. 2009). Importantly, in addition to regulating the rate at which a protein is synthesized from its mRNA, the regulation of elongation by rare codons also contributes to the regulation of translation fidelity, protein folding, and mRNA stability (reviewed in Hanson and Coller 2018).

Regulation of Translation by Elongation Factors

The elongation step of translation in all organisms requires two GTP-binding factors (reviewed in Dever et al. 2018). In eukaryotes, eEF1A binds to an aa-tRNA and recruits it to the ribosome in a GTP-dependent manner. Following the accommodation of the aa-tRNA in the A site of the ribosome, the GTP is hydrolyzed and eEF1A•GDP is released. The GDP-bound form of eEF1A is then recycled back to eEF1A•GTP by the guanine nucleotide exchange factor (GEF), eEF1B. Both eEF1A and eEF1B can be differentially phosphorylated in cells, with evidence suggesting that phosphorylation enhances the rate of GEF activity (reviewed in Traugh 2001; Browne and Proud 2002). This could increase the rate at which eEF1A can rebind to an aa-tRNA following each round of elongation and therefore increase the overall rate of aa-tRNA recruitment and translation.

More attention has been given to the regulated phosphorylation of eEF2 in cells, in part because the eEF2 kinase (eEF2K) is an attractive drug target in cancer (reviewed in Browne and

Proud 2002; Liu and Proud 2016). Similar to EF-G in bacteria, eEF2 promotes the translocation step of eukaryotic elongation. Importantly, the phosphorylation of eEF2 inhibits its activity by reducing its affinity for the ribosome by roughly an order of magnitude (Carlberg et al. 1990). Inhibiting elongation (and therefore translation) by eEF2 phosphorylation therefore enables cells to survive nutrient deprivation (Leprivier et al. 2013). Although it has not been rigorously explored, one would expect that stalling translation by eEF2 phosphorylation would lead to mRNA decay in a similar way to that observed for other mechanisms of elongation stalling (reviewed in Hanson and Coller 2018; Heck and Wilusz 2018). It will be interesting in the future to determine on a genome-wide basis whether this kind of regulation of elongation functions to preferentially target specific mRNAs. Generating this information will help design additional experiments on an mRNA-by-mRNA basis to fully understand how this mechanism of elongation regulation functions to control translation in cells.

Following selection and accommodation of aa-tRNA into the A site of the ribosome, the intrinsic rate of peptide bond formation is generally not limiting for elongation (Wohlgemuth et al. 2010). However, this can in fact be limiting for certain combinations of peptidyl-tRNAs located in the P site and aa-tRNAs in the A site. In particular, the amino acid proline (Pro) is a poor acceptor and donor for peptide bond formation in bacteria (Pavlov et al. 2009). Consistent with this, stretches of Pro codons in bacteria and eukaryotes appear to stall the elongating ribosome by virtue of slowing down the rate of peptidyl transfer. This stalling can be overcome by the addition of EF-P and eIF5A, respectively, in bacterial and eukaryotic translation assays (Doerfel et al. 2013; Gutierrez et al. 2013; Ude et al. 2013; Schuller et al. 2017; Dever et al. 2018; Rodnina 2018). Further characterization of eIF5A has indicated that it may have a more general role in accelerating the rate of peptide bond formation during every round of elongation no matter what the sequence (Schuller et al. 2017). Importantly, the ability of EF-P to limit the rate of translation depends on the rate of initiation being fast enough to enable elongation to act as the rate-limiting step (Hersch et al. 2014). Consistent with this, reporters used to study the regulation of elongation by eIF5A using in vitro and in vivo assays have generally contained very short unstructured 5′UTRs to avoid initiation becoming rate limiting for translation. No mechanism has yet been identified that naturally regulates the availability and/or activity of eIF5A, although eIF5A abundance is reduced in response to drugs that inhibit its modification (reviewed in Mathews and Hershey 2015). While eIF5A is phosphorylated in yeast (Kang et al. 1993), this modification does not appear to regulate its function. It will therefore be important in the future to determine whether the function of this important protein is regulated in response to different cell stimuli.

The Regulation of Elongation

A detailed kinetic framework for the process of elongation has been made by using a reconstituted purified system from bacteria (reviewed in Wohlgemuth et al. 2011; Rodnina 2016). Similar work is only just beginning to use purified eukaryotic systems and it is anticipated that equally detailed kinetic frameworks will eventually be generated. It has become clear that the regulation of elongation is complex and can have profound outcomes affecting the overall rate of translation, frameshifting, protein folding, and mRNA stability. Nevertheless, separating these outcomes can be difficult, especially in intact cells. A challenge for the future will therefore be to generate assays that can precisely monitor all these events in real time so that a more complete understanding of how regulation of the rate of elongation controls translation and the proteome.

TERMINATION AND RECYCLING

The stages of termination and recycling in eukaryotes (see Hellen 2018) appear to be rather different from those described in bacteria. A fundamental difference between the systems is that the steps occur independently in bacteria, but are tightly coupled in eukaryotic cells. This

was only possible to determine through the use of reconstituted purified systems and a careful kinetic analysis of the process (reviewed in Dever and Green 2012; Jackson et al. 2012). A kinetic checkpoint exists at the point when a termination codon enters the A site of the ribosome (Fig. 4). At this checkpoint, one of two outcomes is possible: (1) the eRF1/eRF3•GTP complex can be recruited, resulting in the termination of polypeptide synthesis; or (2) a noncognate aa-tRNA can be recruited to enable the process of elongation to continue. A precise kinetic framework to explain how this decision is made is not yet clear, but some mechanistic details have emerged as to the steps that are taken during each of these two alternative outcomes. Understanding how this checkpoint works is a major goal for translation regulation, in part because of the fact that many genetic diseases can be attributed to the generation of a premature termination codon being introduced into the ORF of various genes (Mort et al. 2008; Keeling et al. 2014).

Termination Codon Recognition

A cartoon that depicts the generally accepted termination pathway is shown in Figure 4 (see also Hellen 2018). In the first step of termination, a pretermination complex (pre-TC) is generated with a termination codon in the A site and a peptidyl-tRNA in the P site of the ribosome. The empty A site of the ribosome allows for a trimeric complex (eRF1/eRF3•GTP) to be stably recruited. Recruitment to the ribosome stimulates the hydrolysis rate of GTP bound to eRF3, promoting its release from the ribosome (Frolova et al. 1996). The release of eRF3 enables eRF1 to correctly position its GGQ loop into the PTC of the 60S subunit to help promote peptide release (Alkalaeva et al. 2006; Shoemaker and Green 2011). Unexpectedly, it was discovered that the binding of the ATP-binding cassette protein E1 (ABCE1 in mammals and Rli1 in *Saccharomyces cerevisiae*) together with eRF3 synergistically accelerates peptide release (Shoemaker and Green 2011). The important thing to note in making this discovery is that it was only possible through

Cite this article as *Cold Spring Harb Perspect Biol* doi: 10.1101/cshperspect.a032706

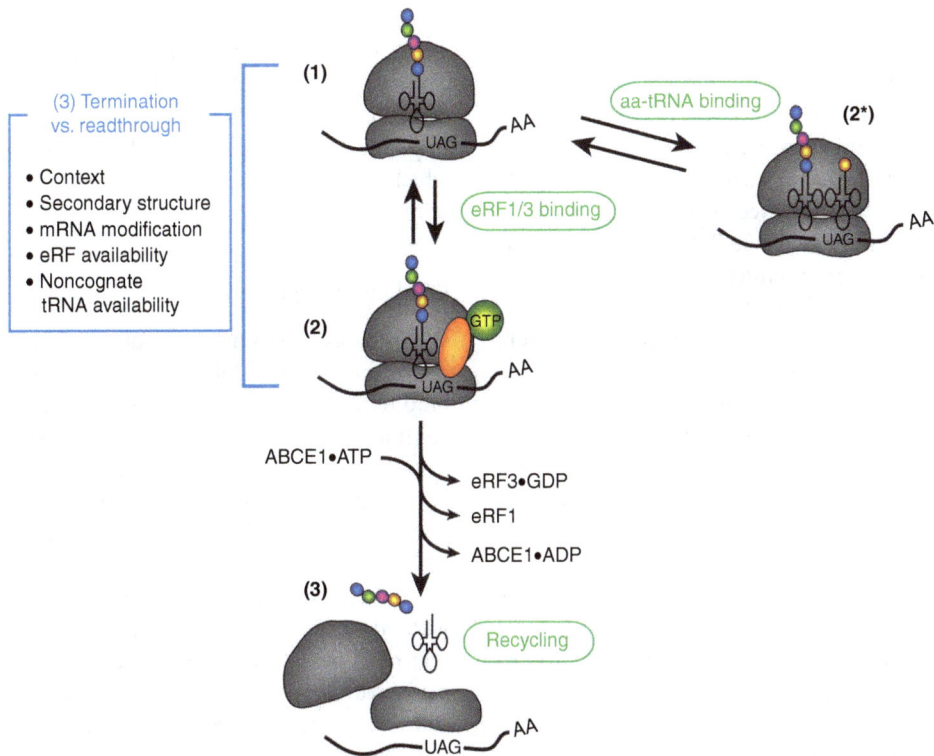

Figure 4. Translation termination and ribosome recycling. (*1*) A pretermination complex is shown to contain a stop codon in the A site of the ribosome. Two possible pathways for subsequent steps are depicted. (*2*) The recruitment of an eRF1/eRF3•GTP complex to the ribosome in the first step of the termination pathway. (*2**) An alternative possibility is that a noncognate aminoacyl transfer RNA (aa-tRNA) is recruited to the A site of the ribosome to continue the elongation stage of translation (stop codon readthrough). Potential sources of regulation for determining which pathway is followed are summarized in the box and described in more detail in the main text. (*3*) Following the recruitment of eRF1/eRF3•GTP to the ribosome, a number of steps are undertaken to complete translation termination and are described in detail in the main text. Briefly, these steps involve the release of eRF3•GDP, recruitment of ABCE1•ATP, and the release of the polypeptide chain. Finally, the 60S ribosomal subunit is dissociated by the hydrolysis of the ATP bound to ABCE1. As mentioned in the main text, the precise pathway of 40S subunit recycling following termination is poorly understood.

the kinetic analysis of the termination process. This kinetic assay further revealed that eIF5A binding to the ribosome further stimulates the rate of peptide release by ~17-fold (Schuller et al. 2017). This is likely caused by eIF5A binding to the exit site (E site) of the ribosome following release of the deacylated tRNA, where it stabilizes the peptidyl-tRNA. While the ATPase activity of ABCE1/Rli1 is not required for promoting peptide release, it is required for promoting the subsequent release of the 60S subunit (Pisarev et al. 2010; Barthelme et al. 2011; Shoemaker and Green 2011). Importantly,

this function of ABCE1/Rli1 was subsequently verified in live cells by using ribosome profiling to monitor the presence of ribosomes around termination codons and the 3′UTR of mRNAs (Young et al. 2015).

Ribosome Recycling

Following the release of the 60S subunit, the separated 40S ribosomal subunits are bound by initiation factors so that they can take part in another round of translation (Pisarev et al. 2007). The binding of initiation factors to posttermination complexes is often depicted as an ordered process, but it should be noted that the order of release and binding of components from the separated ribosomal subunits is still not known. This is because these events have been monitored exclusively by using sucrose gradients, which do not allow for the order of binding to be determined (Pisarev et al. 2007). Importantly, the expected time for the entire process of termination and recycling is likely to be on the order of seconds. It will therefore be necessary in the future to generate a kinetic assay that can monitor the process of termination and ribosome recycling in real time to solve this question. Recent studies have suggested that eIF3 can play an active role in the termination process in vivo (Beznoskova et al. 2013, 2015). However, these studies were limited by the fact that cross-linking agents were necessary to stabilize complexes before sucrose gradient analysis. A kinetic assay will therefore be needed to verify whether eIF3 plays a role in termination, or whether it functions only in the recycling of the 40S subunit (as suggested from reconstitution assays).

It is likely that some initiation factors can bind to a 40S subunit before the release of mRNA. This is implied by the fact that a post-termination 40S subunit can resume scanning in a 3' direction following the translation of a short uORF. The scanning 40S subunit must recruit a new TC before it can select a start codon for another round of translation on the same mRNA. This process is termed reinitiation and has been reviewed in detail elsewhere (Hinnebusch 2011; Jackson et al. 2012). The fact that the TC is not required to resume scanning supports a mechanism by which this complex is not needed for stabilizing mRNA in the decoding site of the 40S subunit. However, this may only be the case for a 40S subunit that is already bound to an mRNA following the termination event. Importantly, the resumption of scanning following termination is restricted to uORFs that are translated in a short amount of time (Kozak 2001). It has been suggested that this limitation is caused by the dissociation rate of eIF4G from eIF3, or possibly the dissociation rate of eIF3 from the 40S subunit (Pöyry et al. 2004; Mohammad et al. 2017). As mentioned above, the

affinity of eIF4G for eIF3 is regulated in a TORC1-dependent manner (Harris et al. 2006; Thoreen et al. 2012). It will be interesting to determine whether this change in affinity corresponds to a change in the dissociation rate, which could conceivably alter the efficiency of reinitiation.

Reading through Termination Codons

When paused with a termination codon in its A site, there is a possibility that an aa-tRNA will bind to the ribosome to decode the codon and continue the elongation cycle (a process called stop codon readthrough). Because the fidelity of termination is high, the competition between noncognate, or cognate, aa-tRNA binding and eRF1/eRF3•GTP binding must favor eRF1/eRF3•GTP binding. However, recent genome-wide data sets have unexpectedly indicated that readthrough can be regulated and is more pervasive than previously thought (although still comparatively rare) (Dunn et al. 2013). A number of factors have been shown to alter the likelihood of readthrough in cells. These include the sequence of the termination codon, the context surrounding the codon, secondary structure in the mRNA, mRNA modifications, and the concentration of endogenous suppressor aa-tRNAs (Cassan and Rousset 2001; Chao et al. 2003; Firth et al. 2011; Karijolich and Yu 2011). Often, various combinations of these and other factors regulate readthrough, making it difficult to precisely understand the mechanism(s) by which readthrough is regulated. Some organisms even use cognate aa-tRNAs to decode termination codons by default, and terminate at these codons only if a poly(A) tail is in close proximity (Heaphy et al. 2016; Swart et al. 2016; Zahonova et al. 2016). Context-dependent readthrough regulation is also apparent in the process of nonsense-mediated mRNA decay (NMD) in eukaryotes (see Karousis and Mühlemann 2018). For NMD, an mRNA is degraded when a termination codon is present within the ORF of an mRNA, but not when it is in the correct position (reviewed in Popp and Maquat 2013; Kurosaki and Maquat 2016; Celik et al. 2017). The ability of cells to differentiate between termination codons in a context-dependent manner raises a fundamental question of how termination codons are recognized by the translation machinery. As mentioned above, the kinetics of termination has begun to be elucidated, but the mechanism of regulated readthrough is still poorly defined. To address this, it is anticipated that genome-wide data sets from cells that show regulated readthrough will continue to provide a list of mRNAs for further study. Biochemical approaches will then be needed to determine the mechanism by which the decision is made to read through versus terminate translation. One goal will be to reconstitute the process of readthrough using a purified system. Although this is an ambitious challenge, the ability to monitor this checkpoint in real time will enable a complete kinetic framework to be developed. Both bulk assays and single-molecule approaches will be needed for developing these frameworks. This information will ultimately help toward development of small-molecule drugs that can precisely alter this process to promote readthrough of premature termination codons that cause a number of genetic diseases.

FUTURE DIRECTIONS

A goal of future work will be to precisely determine how the translation pathway can be reprogrammed to control what mRNA is selected for translation, which initiation/termination codons are used, and how much protein is synthesized from individual mRNAs. This will require the identification of where specific regulation checkpoints in the pathway are located. Using reconstituted purified systems for real-time bulk and single-molecule assays will establish a kinetic framework for the pathway. Using these assays, one can precisely determine how variables such as the concentration of components can alter progress of different mRNAs through the pathway. Nevertheless, reconstituted systems also suffer from limitations that must be considered. For example, the kinetics of these systems is unlikely to approach that found in intact cells, which may mean that the rate-limiting step of the pathway in vitro is different from that found in vivo. Reconstituted systems typically use in

vitro transcribed mRNAs, so possible contributions of other mRNA-binding proteins that are present in cells (forming various mRNPs) and RNA modifications are not easy to test. Therefore, whereas reconstituted systems will be important in generating in vitro models for the pathway of translation, it is essential that these models be tested in intact cells. The power of this combination has been highlighted by the successful test of start site selection models in yeast (reviewed in Hinnebusch and Lorsch 2012; Hinnebusch 2014, 2017). It is anticipated that the use of new genome engineering methods will further enhance the ability to test in vitro models for the mammalian translation pathway. Continued improvements to imaging techniques will also make it increasingly possible to precisely monitor translation regulation on single mRNAs in live cells (see Biswas et al. 2018). It is hoped that once the regulation of translation is understood in sufficient detail, it will be possible to understand how the dysregulation of translation can lead to the generation and maintenance of different disease states.

ACKNOWLEDGMENTS

We thank John Hershey for many thought-provoking conversations about the regulation of eukaryotic translation. We are grateful to the Fraser laboratory for stimulating discussion and comments on the manuscript. We gratefully acknowledge support for this work from National Institutes of Health (NIH) Grant R01 GM092927.

REFERENCES

*Reference is also in this collection.

Ali IK, McKendrick L, Morley SJ, Jackson RJ. 2001. Truncated initiation factor eIF4G lacking an eIF4E binding site can support capped mRNA translation. *EMBO J* **20:** 4233–4242.

Algire MA, Maag D, Lorsch JR. 2005. P_i release from eIF2, not GTP hydrolysis, is the step controlled by start-site selection during eukaryotic translation initiation. *Mol Cell* **20:** 251–262.

Alkalaeva EZ, Pisarev AV, Frolova LY, Kisselev LL, Pestova TV. 2006. In vitro reconstitution of eukaryotic translation reveals cooperativity between release factors eRF1 and eRF3. *Cell* **125:** 1125–1136.

Aylett CH, Boehringer D, Erzberger JP, Schaefer T, Ban N. 2015. Structure of a yeast 40S-eIF1-eIF1A-eIF3-eIF3j initiation complex. *Nat Struct Mol Biol* **22:** 269–271.

Barthelme D, Dinkelaker S, Albers SV, Londei P, Ermler U, Tampe R. 2011. Ribosome recycling depends on a mechanistic link between the FeS cluster domain and a conformational switch of the twin-ATPase ABCE1. *Proc Natl Acad Sci* **108:** 3228–3233.

Beznoskova P, Cuchalova L, Wagner S, Shoemaker CJ, Gunisova S, von der Haar T, Valasek LS. 2013. Translation initiation factors eIF3 and HCR1 control translation termination and stop codon read-through in yeast cells. *PLoS Genet* **9:** e1003962.

Beznoskova P, Wagner S, Jansen ME, von der Haar T, Valasek LS. 2015. Translation initiation factor eIF3 promotes programmed stop codon readthrough. *Nucleic Acids Res* **43:** 5099–5111.

* Biswas J, Liu Y, Singer RH, Wu B. 2018. Fluorescence imaging methods to investigate translation in single cells. *Cold Spring Harb Perspect Biol* doi: 10.1101/cshperspect.a032722.

Brar GA, Yassour M, Friedman N, Regev A, Ingolia NT, Weissman JS. 2012. High-resolution view of the yeast meiotic program revealed by ribosome profiling. *Science* **335:** 552–557.

Browne GJ, Proud CG. 2002. Regulation of peptide-chain elongation in mammalian cells. *Eur J Biochem* **269:** 5360–5368.

Carlberg U, Nilsson A, Nygard O. 1990. Functional properties of phosphorylated elongation factor 2. *Eur J Biochem* **191:** 639–645.

Cassan M, Rousset JP. 2001. UAG read through in mammalian cells: Effect of upstream and downstream stop codon contexts reveal different signals. *BMC Mol Biol* **2:** 3.

Celik A, He F, Jacobson A. 2017. NMD monitors translational fidelity 24/7. *Curr Genet* **63:** 1007–1010.

Chamary JV, Parmley JL, Hurst LD. 2006. Hearing silence: Non-neutral evolution at synonymous sites in mammals. *Nat Rev Genet* **7:** 98–108.

Chao AT, Dierick HA, Addy TM, Bejsovec A. 2003. Mutations in eukaryotic release factors 1 and 3 act as general nonsense suppressors in *Drosophila*. *Genetics* **165:** 601–612.

Cheung YN, Maag D, Mitchell SF, Fekete CA, Algire MA, Takacs JE, Shirokikh N, Pestova T, Lorsch JR, Hinnebusch AG. 2007. Dissociation of eIF1 from the 40S ribosomal subunit is a key step in start codon selection in vivo. *Genes Dev* **21:** 1217–1230.

Chu D, Kazana E, Bellanger N, Singh T, Tuite MF, von der Haar T. 2014. Translation elongation can control translation initiation on eukaryotic mRNAs. *EMBO J* **33:** 21–34.

Cigan AM, Pabich EK, Feng L, Donahue TF. 1989. Yeast translation initiation suppressor sui2 encodes the α subunit of eukaryotic initiation factor 2 and shares sequence identity with the human α subunit. *Proc Natl Acad Sci* **86:** 2784–2788.

Dasso MC, Milburn SC, Hershey JW, Jackson RJ. 1990. Selection of the 5′-proximal translation initiation site is influenced by mRNA and eIF-2 concentrations. *Eur J Biochem* **187:** 361–371.

Dever TE, Green R. 2012. The elongation, termination, and recycling phases of translation in eukaryotes. *Cold Spring Harb Perspect Biol* **4:** a013706.

* Dever TE, Dinman JD, Green R. 2018. Translation elongation and recoding in eukaryotes. *Cold Spring Harb Perspect Biol* doi: 10.1101/cshperspect.a032649.

Doerfel LK, Wohlgemuth I, Kothe C, Peske F, Urlaub H, Rodnina MV. 2013. EF-P is essential for rapid synthesis of proteins containing consecutive proline residues. *Science* **339:** 85–88.

Donahue TF, Cigan AM, Pabich EK, Valavicius BC. 1988. Mutations at a Zn(II) finger motif in the yeast eIF-2 β gene alter ribosomal start-site selection during the scanning process. *Cell* **54:** 621–632.

Dumont S, Cheng W, Serebrov V, Beran RK, Tinoco I Jr, Pyle AM, Bustamante C. 2006. RNA translocation and unwinding mechanism of HCV NS3 helicase and its coordination by ATP. *Nature* **439:** 105–108.

Dunn JG, Foo CK, Belletier NG, Gavis ER, Weissman JS. 2013. Ribosome profiling reveals pervasive and regulated stop codon readthrough in *Drosophila melanogaster*. *eLife* **2:** e01179.

Fekete CA, Applefield DJ, Blakely SA, Shirokikh N, Pestova T, Lorsch JR, Hinnebusch AG. 2005. The eIF1A C-terminal domain promotes initiation complex assembly, scanning and AUG selection in vivo. *EMBO J* **24:** 3588–3601.

Firth AE, Wills NM, Gesteland RF, Atkins JF. 2011. Stimulation of stop codon readthrough: frequent presence of an extended 3′ RNA structural element. *Nucleic Acids Res* **39:** 6679–6691.

Fraser CS. 2015. Quantitative studies of mRNA recruitment to the eukaryotic ribosome. *Biochimie* **114:** 58–71.

Fraser CS, Berry KE, Hershey JW, Doudna JA. 2007. eIF3j is located in the decoding center of the human 40S ribosomal subunit. *Mol Cell* **26:** 811–819.

Frolova L, Le Goff X, Zhouravleva G, Davydova E, Philippe M, Kisselev L. 1996. Eukaryotic polypeptide chain release factor eRF3 is an eRF1- and ribosome-dependent guanosine triphosphatase. *RNA* **2:** 334–341.

Gao Z, Putnam AA, Bowers HA, Guenther UP, Ye X, Kindsfather A, Hilliker AK, Jankowsky E. 2016. Coupling between the DEAD-box RNA helicases Ded1p and eIF4A. *eLife* **5:** e16408.

García-García C, Frieda KL, Feoktistova K, Fraser CS, Block SM. 2015. RNA BIOCHEMISTRY. Factor-dependent processivity in human eIF4A DEAD-box helicase. *Science* **348:** 1486–1488.

Gardin J, Yeasmin R, Yurovsky A, Cai Y, Skiena S, Futcher B. 2014. Measurement of average decoding rates of the 61 sense codons in vivo. *eLife* **3:** e03735.

Gingras AC, Raught B, Sonenberg N. 1999. eIF4 initiation factors: Effectors of mRNA recruitment to ribosomes and regulators of translation. *Annu Rev Biochem* **68:** 913–963.

Gutierrez E, Shin BS, Woolstenhulme CJ, Kim JR, Saini P, Buskirk AR, Dever TE. 2013. eIF5A promotes translation of polyproline motifs. *Mol Cell* **51:** 35–45.

Hanson G, Coller J. 2018. Codon optimality, bias and usage in translation and mRNA decay. *Nat Rev Mol Cell Biol* **19:** 20–30.

Harms U, Andreou AZ, Gubaev A, Klostermeier D. 2014. eIF4B, eIF4G and RNA regulate eIF4A activity in trans-

Cite this article as *Cold Spring Harb Perspect Biol* doi: 10.1101/cshperspect.a032706

lation initiation by modulating the eIF4A conformational cycle. *Nucleic Acids Res* **42:** 7911–7922.

Harris TE, Chi A, Shabanowitz J, Hunt DF, Rhoads RE, Lawrence JC Jr. 2006. mTOR-dependent stimulation of the association of eIF4G and eIF3 by insulin. *EMBO J* **25:** 1659–1668.

Hashem Y, des Georges A, Dhote V, Langlois R, Liao HY, Grassucci RA, Hellen CU, Pestova TV, Frank J. 2013. Structure of the mammalian ribosomal 43S preinitiation complex bound to the scanning factor DHX29. *Cell* **153:** 1108–1119.

Heaphy SM, Mariotti M, Gladyshev VN, Atkins JF, Baranov PV. 2016. Novel ciliate genetic code variants including the reassignment of all three stop codons to sense codons in Condylostoma magnum. *Mol Biol Evol* **33:** 2885–2889.

* Heck AM, Wilusz J. 2018. The interplay between the RNA decay and translation machinery in eukaryotes. *Cold Spring Harb Perspect Biol* doi: 10.1101/cshperspect.a032839.

* Hellen CUT. 2018. Translation termination and ribosome recycling in eukaryotes. *Cold Spring Harb Perspect Biol* doi: 10.1101/cshperspect.a032656.

Hersch SJ, Elgamal S, Katz A, Ibba M, Navarre WW. 2014. Translation initiation rate determines the impact of ribosome stalling on bacterial protein synthesis. *J Biol Chem* **289:** 28160–28171.

Hilliker A, Gao Z, Jankowsky E, Parker R. 2011. The DEAD-box protein Ded1 modulates translation by the formation and resolution of an eIF4F-mRNA complex. *Mol Cell* **43:** 962–972.

Hinnebusch AG. 2011. Molecular mechanism of scanning and start codon selection in eukaryotes. *Microbiol Mol Biol Rev* **75.**

Hinnebusch AG. 2014. The scanning mechanism of eukaryotic translation initiation. *Annu Rev Biochem* **83:** 779–812.

Hinnebusch AG. 2017. Structural insights into the mechanism of scanning and start codon recognition in eukaryotic translation initiation. *Trends Biochem Sci* **42:** 589–611.

Hinnebusch AG, Lorsch JR. 2012. The mechanism of eukaryotic translation initiation: New insights and challenges. *Cold Spring Harb Perspect Biol* **4:** a011544.

Hinton TM, Coldwell MJ, Carpenter GA, Morley SJ, Pain VM. 2007. Functional analysis of individual binding activities of the scaffold protein eIF4G. *J Biol Chem* **282:** 1695–1708.

Homma MK, Wada I, Suzuki T, Yamaki J, Krebs EG, Homma Y. 2005. CK2 phosphorylation of eukaryotic translation initiation factor 5 potentiates cell cycle progression. *Proc Natl Acad Sci* **102:** 15688–15693.

Huang HK, Yoon H, Hannig EM, Donahue TF. 1997. GTP hydrolysis controls stringent selection of the AUG start codon during translation initiation in *Saccharomyces cerevisiae*. *Genes Dev* **11:** 2396–2413.

Hussain T, Llácer JL, Fernández IS, Munoz A, Martin-Marcos P, Savva Christos G, Lorsch JR, Hinnebusch AG, Ramakrishnan V. 2014. Structural changes enable start codon recognition by the eukaryotic translation initiation complex. *Cell* **159:** 597–607.

Imataka H, Gradi A, Sonenberg N. 1998. A newly identified N-terminal amino acid sequence of human eIF4G binds poly(A)-binding protein and functions in poly(A)-dependent translation. *EMBO J* **17:** 7480–7489.

Ingolia NT, Ghaemmaghami S, Newman JR, Weissman JS. 2009. Genome-wide analysis in vivo of translation with nucleotide resolution using ribosome profiling. *Science* **324:** 218–223.

Ingolia NT, Lareau LF, Weissman JS. 2011. Ribosome profiling of mouse embryonic stem cells reveals the complexity and dynamics of mammalian proteomes. *Cell* **147:** 789–802.

* Ingolia NT, Hussmann JA, Weissman JS. 2018. Ribosome profiling: Global views of translation. *Cold Spring Harb Perspect Biol* doi: 10.1101/cshperspect.a032698.

Ivanov IP, Loughran G, Sachs MS, Atkins JF. 2010. Initiation context modulates autoregulation of eukaryotic translation initiation factor 1 (eIF1). *Proc Natl Acad Sci* **107:** 18056–18060.

Jackson RJ, Hellen CU, Pestova TV. 2012. Termination and post-termination events in eukaryotic translation. *Adv Protein Chem Struct Biol* **86:** 45–93.

* Jobe A, Liu Z, Gutierrez-Vargas C, Frank J. 2018. New insights into ribosome structure and function. *Cold Spring Harb Perspect Biol* doi: 10.1101/cshperspect.a032615.

Kang HA, Schwelberger HG, Hershey JW. 1993. Translation initiation factor eIF-5A, the hypusine-containing protein, is phosphorylated on serine in *Saccharomyces cerevisiae*. *J Biol Chem* **268:** 14750–14756.

Karijolich J, Yu YT. 2011. Converting nonsense codons into sense codons by targeted pseudouridylation. *Nature* **474:** 395–398.

* Karousis ED, Mühlemann O. 2018. Nonsense-mediated mRNA decay begins where translation ends. *Cold Spring Harb Perspect Biol* doi: 10.1101/cshperspect.a032862.

Keeling KM, Xue X, Gunn G, Bedwell DM. 2014. Therapeutics based on stop codon readthrough. *Annu Rev Genomics Hum Genet* **15:** 371–394.

Kirchner S, Ignatova Z. 2015. Emerging roles of tRNA in adaptive translation, signalling dynamics and disease. *Nat Rev Genet* **16:** 98–112.

Komar AA. 2009. A pause for thought along the co-translational folding pathway. *Trends Biochem Sci* **34:** 16–24.

Korneeva NL, Lamphear BJ, Hennigan FL, Rhoads RE. 2000. Mutually cooperative binding of eukaryotic translation initiation factor (eIF)3 and eIF4A to human eIF4G-1. *J Biol Chem* **275:** 41369–41376.

Koutmou KS, Radhakrishnan A, Green R. 2015. Synthesis at the speed of codons. *Trends Biochem Sci* **40:** 717–718.

Kozak M. 1980. Evaluation of the "scanning model" for initiation of protein synthesis in eukaryotes. *Cell* **22:** 7–8.

Kozak M. 1986. Point mutations define a sequence flanking the AUG initiator codon that modulates translation by eukaryotic ribosomes. *Cell* **44:** 283–292.

Kozak M. 1991. Effects of long 5′ leader sequences on initiation by eukaryotic ribosomes in vitro. *Gene Expr* **1:** 117–125.

Kozak M. 2001. Constraints on reinitiation of translation in mammals. *Nucleic Acids Res* **29:** 5226–5232.

Kozak M. 2005. Regulation of translation via mRNA structure in prokaryotes and eukaryotes. *Gene* **361**: 13–37.

Kozak M, Shatkin AJ. 1978. Migration of 40 S ribosomal subunits on messenger RNA in the presence of edeine. *J Biol Chem* **253**: 6568–6577.

Kudla G, Murray AW, Tollervey D, Plotkin JB. 2009. Coding-sequence determinants of gene expression in *Escherichia coli. Science* **324**: 255–258.

Kurosaki T, Maquat LE. 2016. Nonsense-mediated mRNA decay in humans at a glance. *J Cell Sci* **129**: 461–467.

Lamphear BJ, Kirchweger R, Skern T, Rhoads RE. 1995. Mapping of functional domains in eukaryotic protein synthesis initiation factor 4G (eIF4G) with picornaviral proteases. Implications for cap-dependent and cap-independent translational initiation. *J Biol Chem* **270**: 21975–21983.

Lareau LF, Hite DH, Hogan GJ, Brown PO. 2014. Distinct stages of the translation elongation cycle revealed by sequencing ribosome-protected mRNA fragments. *eLife* **3**: e01257.

Lawson TG, Ray BK, Dodds JT, Grifo JA, Abramson RD, Merrick WC, Betsch DF, Weith HL, Thach RE. 1986. Influence of 5′ proximal secondary structure on the translational efficiency of eukaryotic mRNAs and on their interaction with initiation factors. *J Biol Chem* **261**: 13979–13989.

Leprivier G, Remke M, Rotblat B, Dubuc A, Mateo AR, Kool M, Agnihotri S, El-Naggar A, Yu B, Somasekharan SP, et al. 2013. The eEF2 kinase confers resistance to nutrient deprivation by blocking translation elongation. *Cell* **153**: 1064–1079.

Linder P, Jankowsky E. 2011. From unwinding to clamping—The DEAD box RNA helicase family. *Nat Rev Mol Cell Biol* **12**: 505–516.

Liu R, Proud CG. 2016. Eukaryotic elongation factor 2 kinase as a drug target in cancer, and in cardiovascular and neurodegenerative diseases. *Acta Pharmacol Sin* **37**: 285–294.

Livingstone M, Atas E, Meller A, Sonenberg N. 2010. Mechanisms governing the control of mRNA translation. *Phys Biol* **7**: 021001.

Llácer JL, Hussain T, Marler L, Aitken CE, Thakur A, Lorsch JR, Hinnebusch AG, Ramakrishnan V. 2015. Conformational differences between open and closed states of the eukaryotic translation initiation complex. *Mol Cell* **59**: 399–412.

Lomakin IB, Steitz TA. 2013. The initiation of mammalian protein synthesis and mRNA scanning mechanism. *Nature* **500**: 307–311.

Lorsch JR, Dever TE. 2010. Molecular view of 43 S complex formation and start site selection in eukaryotic translation initiation. *J Biol Chem* **285**: 21203–21207.

Loughran G, Sachs MS, Atkins JF, Ivanov IP. 2012. Stringency of start codon selection modulates autoregulation of translation initiation factor eIF5. *Nucleic Acids Res* **40**: 2898–2906.

Maag D, Fekete CA, Gryczynski Z, Lorsch JR. 2005. A conformational change in the eukaryotic translation preinitiation complex and release of eIF1 signal recognition of the start codon. *Mol Cell* **17**: 265–275.

Maag D, Algire MA, Lorsch JR. 2006. Communication between eukaryotic translation initiation factors 5 and 1A within the ribosomal pre-initiation complex plays a role in start site selection. *J Mol Biol* **356**: 724–737.

Majumdar R, Maitra U. 2005. Regulation of GTP hydrolysis prior to ribosomal AUG selection during eukaryotic translation initiation. *EMBO J* **24**: 3737–3746.

Mathews MB, Hershey JW. 2015. The translation factor eIF5A and human cancer. *Biochim Biophys Acta* **1849**: 836–844.

Merrick WC. 2015. eIF4F: A retrospective. *J Biol Chem* **290**: 24091–24099.

* Merrick WC, Pavitt GD. 2018. Protein synthesis initiation in eukaryotic cells. *Cold Spring Harb Perspect Biol* doi: 10.1101/cshperspect.a033092.

Mitarai N, Sneppen K, Pedersen S. 2008. Ribosome collisions and translation efficiency: Optimization by codon usage and mRNA destabilization. *J Mol Biol* **382**: 236–245.

Mohammad MP, Munzarova Pondelickova V, Zeman J, Gunisova S, Valasek LS. 2017. In vivo evidence that eIF3 stays bound to ribosomes elongating and terminating on short upstream ORFs to promote reinitiation. *Nucleic Acids Res* **45**: 2658–2674.

Morino S, Imataka H, Svitkin YV, Pestova TV, Sonenberg N. 2000. Eukaryotic translation initiation factor 4E (eIF4E) binding site and the middle one-third of eIF4GI constitute the core domain for cap-dependent translation, and the C-terminal one-third functions as a modulatory region. *Mol Cell Biol* **20**: 468–477.

Mort M, Ivanov D, Cooper DN, Chuzhanova NA. 2008. A meta-analysis of nonsense mutations causing human genetic disease. *Hum Mutat* **29**: 1037–1047.

Muckenthaler M, Gray NK, Hentze MW. 1998. IRP-1 binding to ferritin mRNA prevents the recruitment of the small ribosomal subunit by the cap-binding complex eIF4F. *Mol Cell* **2**: 383–388.

Nackley AG, Shabalina SA, Tchivileva IE, Satterfield K, Korchynskyi O, Makarov SS, Maixner W, Diatchenko L. 2006. Human catechol-*O*-methyltransferase haplotypes modulate protein expression by altering mRNA secondary structure. *Science* **314**: 1930–1933.

Nanda JS, Cheung YN, Takacs JE, Martin-Marcos P, Saini AK, Hinnebusch AG, Lorsch JR. 2009. eIF1 controls multiple steps in start codon recognition during eukaryotic translation initiation. *J Mol Biol* **394**: 268–285.

Nanda JS, Saini AK, Munoz AM, Hinnebusch AG, Lorsch JR. 2013. Coordinated movements of eukaryotic translation initiation factors eIF1, eIF1A, and eIF5 trigger phosphate release from eIF2 in response to start codon recognition by the ribosomal preinitiation complex. *J Biol Chem* **288**: 5316–5329.

Niedzwiecka A, Marcotrigiano J, Stepinski J, Jankowska-Anyszka M, Wyslouch-Cieszynska A, Dadlez M, Gingras AC, Mak P, Darzynkiewicz E, Sonenberg N, et al. 2002. Biophysical studies of eIF4E cap-binding protein: Recognition of mRNA 5′ cap structure and synthetic fragments of eIF4G and 4E-BP1 proteins. *J Mol Biol* **319**: 615–635.

Nielsen KH, Behrens MA, He Y, Oliveira CL, Jensen LS, Hoffmann SV, Pedersen JS, Andersen GR. 2011. Synergistic activation of eIF4A by eIF4B and eIF4G. *Nucleic Acids Res* **39**: 2678–2689.

Cite this article as *Cold Spring Harb Perspect Biol* doi: 10.1101/cshperspect.a032706

Özeş AR, Feoktistova K, Avanzino BC, Fraser CS. 2011. Duplex unwinding and ATPase activities of the DEAD-box helicase eIF4A are coupled by eIF4G and eIF4B. *J Mol Biol* **412:** 674–687.

Palam LR, Baird TD, Wek RC. 2011. Phosphorylation of eIF2 facilitates ribosomal bypass of an inhibitory upstream ORF to enhance CHOP translation. *J Biol Chem* **286:** 10939–10949.

Passmore LA, Schmeing TM, Maag D, Applefield DJ, Acker MG, Algire MA, Lorsch JR, Ramakrishnan V. 2007. The eukaryotic translation initiation factors eIF1 and eIF1A induce an open conformation of the 40S ribosome. *Mol Cell* **26:** 41–50.

Pavlov MY, Watts RE, Tan Z, Cornish VW, Ehrenberg M, Forster AC. 2009. Slow peptide bond formation by proline and other *N*-alkylamino acids in translation. *Proc Natl Acad Sci* **106:** 50–54.

Pelletier J, Sonenberg N. 1985. Photochemical cross-linking of cap binding proteins to eukaryotic mRNAs: Effect of mRNA 5′ secondary structure. *Mol Cell Biol* **5:** 3222–3230.

Pestova TV, Borukhov SI, Hellen CU. 1998. Eukaryotic ribosomes require initiation factors 1 and 1A to locate initiation codons. *Nature* **394:** 854–859.

Pisarev AV, Hellen CU, Pestova TV. 2007. Recycling of eukaryotic posttermination ribosomal complexes. *Cell* **131:** 286–299.

Pisarev AV, Skabkin MA, Pisareva VP, Skabkina OV, Rakotondrafara AM, Hentze MW, Hellen CU, Pestova TV. 2010. The role of ABCE1 in eukaryotic posttermination ribosomal recycling. *Mol Cell* **37:** 196–210.

Plotkin JB, Kudla G. 2011. Synonymous but not the same: The causes and consequences of codon bias. *Nat Rev Genet* **12:** 32–42.

Popp MW, Maquat LE. 2013. Organizing principles of mammalian nonsense-mediated mRNA decay. *Annu Rev Genet* **47:** 139–165.

Potapov I, Makela J, Yli-Harja O, Ribeiro AS. 2012. Effects of codon sequence on the dynamics of genetic networks. *J Theor Biol* **315:** 17–25.

Pöyry TA, Kaminski A, Jackson RJ. 2004. What determines whether mammalian ribosomes resume scanning after translation of a short upstream open reading frame? *Genes Dev* **18:** 62–75.

* Prabhakar A, Puglisi EV, Puglisi JD. 2018. Single-molecule fluorescence applied to translation. *Cold Spring Harb Perspect Biol* doi: 10.1101/cshperspect.a032714.

* Proud CG. 2018. Regulation of translation by cap-binding proteins. *Cold Spring Harb Perspect Biol* doi: 10.1101/cshperspect.a033050.

Pyronnet S, Imataka H, Gingras AC, Fukunaga R, Hunter T, Sonenberg N. 1999. Human eukaryotic translation initiation factor 4G (eIF4G) recruits mnk1 to phosphorylate eIF4E. *EMBO J* **18:** 270–279.

Richter JD, Sonenberg N. 2005. Regulation of cap-dependent translation by eIF4E inhibitory proteins. *Nature* **433:** 477–480.

Rodnina MV. 2016. The ribosome in action: Tuning of translational efficiency and protein folding. *Protein Sci* **25:** 1390–1406.

* Rodnina MV. 2018. Translation in prokaryotes. *Cold Spring Harb Perspect Biol* doi: 10.1101/cshperspect.a032664.

Rodnina MV, Wintermeyer W. 2009. Recent mechanistic insights into eukaryotic ribosomes. *Curr Opin Cell Biol* **21:** 435–443.

Rodnina MV, Fischer N, Maracci C, Stark H. 2017. Ribosome dynamics during decoding. *Philos Trans R Soc Lond B Biol Sci* **372**.

Rogers GW Jr, Richter NJ, Lima WF, Merrick WC. 2001. Modulation of the helicase activity of eIF4A by eIF4B, eIF4H, and eIF4F. *J Biol Chem* **276:** 30914–30922.

Rozen F, Edery I, Meerovitch K, Dever TE, Merrick WC, Sonenberg N. 1990. Bidirectional RNA helicase activity of eukaryotic translation initiation factors 4A and 4F. *Mol Cell Biol* **10:** 1134–1144.

Schuller AP, Wu CC, Dever TE, Buskirk AR, Green R. 2017. eIF5A functions globally in translation elongation and termination. *Mol Cell* **66:** 194–205.e5.

Shah P, Ding Y, Niemczyk M, Kudla G, Plotkin JB. 2013. Rate-limiting steps in yeast protein translation. *Cell* **153:** 1589–1601.

Sharifulin DE, Bartuli YS, Meschaninova MI, Ven'yaminova AG, Graifer DM, Karpova GG. 2016. Exploring accessibility of structural elements of the mammalian 40S ribosomal mRNA entry channel at various steps of translation initiation. *Biochim Biophys Acta* **1864:** 1328–1338.

Shoemaker CJ, Green R. 2011. Kinetic analysis reveals the ordered coupling of translation termination and ribosome recycling in yeast. *Proc Natl Acad Sci* **108:** E1392–E1398.

Simonovic M, Steitz TA. 2009. A structural view on the mechanism of the ribosome-catalyzed peptide bond formation. *Biochim Biophys Acta* **1789:** 612–623.

Sokabe M, Fraser CS. 2017. A helicase-independent activity of eIF4A in promoting mRNA recruitment to the human ribosome. *Proc Natl Acad Sci* **114:** 6304–6309.

Soto-Rifo R, Rubilar PS, Limousin T, de Breyne S, Decimo D, Ohlmann T. 2012. DEAD-box protein DDX3 associates with eIF4F to promote translation of selected mRNAs. *EMBO J* **31:** 3745–3756.

Stumpf CR, Moreno MV, Olshen AB, Taylor BS, Ruggero D. 2013. The translational landscape of the mammalian cell cycle. *Mol Cell* **52:** 574–582.

Sun Y, Atas E, Lindqvist L, Sonenberg N, Pelletier J, Meller A. 2012. The eukaryotic initiation factor eIF4H facilitates loop-binding, repetitive RNA unwinding by the eIF4A DEAD-box helicase. *Nucleic Acids Res* **40:** 6199–6207.

Swart EC, Serra V, Petroni G, Nowacki M. 2016. Genetic codes with no dedicated stop codon: Context-dependent translation termination. *Cell* **166:** 691–702.

Tanenbaum ME, Stern-Ginossar N, Weissman JS, Vale RD. 2015. Regulation of mRNA translation during mitosis. *eLife* **4:** e07957.

Terenin IM, Akulich KA, Andreev DE, Polyanskaya SA, Shatsky IN, Dmitriev SE. 2016. Sliding of a 43S ribosomal complex from the recognized AUG codon triggered by a delay in eIF2-bound GTP hydrolysis. *Nucleic Acids Res* **44:** 1882–1893.

Thoreen CC, Chantranupong L, Keys HR, Wang T, Gray NS, Sabatini DM. 2012. A unifying model for mTORC1-

mediated regulation of mRNA translation. *Nature* **485:** 109–113.

Traugh JA. 2001. Insulin, phorbol ester and serum regulate the elongation phase of protein synthesis. *Prog Mol Subcell Biol* **26:** 33–48.

Tsai A, Petrov A, Marshall RA, Korlach J, Uemura S, Puglisi JD. 2012. Heterogeneous pathways and timing of factor departure during translation initiation. *Nature* **487:** 390–393.

Ude S, Lassak J, Starosta AL, Kraxenberger T, Wilson DN, Jung K. 2013. Translation elongation factor EF-P alleviates ribosome stalling at polyproline stretches. *Science* **339:** 82–85.

Unbehaun A, Borukhov SI, Hellen CU, Pestova TV. 2004. Release of initiation factors from 48S complexes during ribosomal subunit joining and the link between establishment of codon–anticodon base-pairing and hydrolysis of eIF2-bound GTP. *Genes Dev* **18:** 3078–3093.

Villa N, Do A, Hershey JW, Fraser CS. 2013. Human eukaryotic initiation factor 4G (eIF4G) protein binds to eIF3c, -d, and -e to promote mRNA recruitment to the ribosome. *J Biol Chem* **288:** 32932–32940.

Vogel C, Abreu Rde S, Ko D, Le SY, Shapiro BA, Burns SC, Sandhu D, Boutz DR, Marcotte EM, Penalva LO. 2010. Sequence signatures and mRNA concentration can explain two-thirds of protein abundance variation in a human cell line. *Mol Syst Biol* **6:** 400.

Wang C, Han B, Zhou R, Zhuang X. 2016. Real-time imaging of translation on single mRNA transcripts in live cells. *Cell* **165:** 990–1001.

Weinberg DE, Shah P, Eichhorn SW, Hussmann JA, Plotkin JB, Bartel DP. 2016. Improved ribosome-footprint and mRNA measurements provide insights into dynamics and regulation of yeast translation. *Cell Rep* **14:** 1787–1799.

Weisser M, Voigts-Hoffmann F, Rabl J, Leibundgut M, Ban N. 2013. The crystal structure of the eukaryotic 40S ribosomal subunit in complex with eIF1 and eIF1A. *Nat Struct Mol Biol* **20:** 1015–1017.

* Wek RC. 2018. Role of eIF2α kinases in translational control and adaptation to cellular stress. *Cold Spring Harb Perspect Biol* doi: 10.1101/cshperspect.a032870.

Wilusz JE. 2015. Controlling translation via modulation of tRNA levels. *Wiley Interdiscip Rev RNA* **6:** 453–470.

Wohlgemuth I, Pohl C, Rodnina MV. 2010. Optimization of speed and accuracy of decoding in translation. *EMBO J* **29:** 3701–3709.

Wohlgemuth I, Pohl C, Mittelstaet J, Konevega AL, Rodnina MV. 2011. Evolutionary optimization of speed and accuracy of decoding on the ribosome. *Philos Trans R Soc Lond B Biol Sci* **366:** 2979–2986.

Yan X, Hoek TA, Vale RD, Tanenbaum ME. 2016. Dynamics of translation of single mRNA molecules in vivo. *Cell* **165:** 976–989.

Yang HS, Jansen AP, Komar AA, Zheng X, Merrick WC, Costes S, Lockett SJ, Sonenberg N, Colburn NH. 2003. The transformation suppressor Pdcd4 is a novel eukaryotic translation initiation factor 4A binding protein that inhibits translation. *Mol Cell Biol* **23:** 26–37.

Yang Q, Del Campo M, Lambowitz AM, Jankowsky E. 2007. DEAD-box proteins unwind duplexes by local strand separation. *Mol Cell* **28:** 253–263.

Yoon HJ, Donahue TF. 1992. The suiI suppressor locus in *Saccharomyces cerevisiae* encodes a translation factor that functions during tRNA(iMet) recognition of the start codon. *Mol Cell Biol* **12:** 248–260.

Young DJ, Guydosh NR, Zhang F, Hinnebusch AG, Green R. 2015. Rli1/ABCE1 recycles terminating ribosomes and controls translation reinitiation in 3′UTRs in vivo. *Cell* **162:** 872–884.

Zach L, Braunstein I, Stanhill A. 2014. Stress-induced start codon fidelity regulates arsenite-inducible regulatory particle-associated protein (AIRAP) translation. *J Biol Chem* **289:** 20706–20716.

Zahonova K, Kostygov AY, Sevcikova T, Yurchenko V, Elias M. 2016. An unprecedented non-canonical nuclear genetic code with all three termination codons reassigned as sense codons. *Curr Biol* **26:** 2364–2369.

Cite this article as *Cold Spring Harb Perspect Biol* doi: 10.1101/cshperspect.a032706

Single-Molecule Fluorescence Applied to Translation

Arjun Prabhakar,[1,2] Elisabetta Viani Puglisi,[1] and Joseph D. Puglisi[1]

[1]Department of Structural Biology, Stanford University School of Medicine, Stanford, California 94305

[2]Program in Biophysics, Stanford University, Stanford, California 94305

Correspondence: puglisi@stanford.edu

Single-molecule fluorescence methods have illuminated the dynamics of the translational machinery. Structural and bulk biochemical experiments have provided detailed atomic and global mechanistic views of translation, respectively. Single-molecule studies of translation have bridged these views by temporally connecting the conformational and compositional states defined from structural data within the mechanistic framework of translation produced from biochemical studies. Here, we discuss the context for applying different single-molecule fluorescence experiments, and present recent applications to studying prokaryotic and eukaryotic translation. We underscore the power of observing single translating ribosomes to delineate and sort complex mechanistic pathways during initiation and elongation, and discuss future applications of current and improved technologies.

Proteins must be translated rapidly, with high fidelity and quality by the ribosome. Translation is thus a highly dynamic process with temporal changes in ligand stoichiometry on the ribosome, and conformational changes within ribosomes and ligands (Noller et al. 2017a,b; Rodnina et al. 2017). During initiation in all organisms (Merrick and Pavitt 2018; Rodnina 2018), ribosomal subunits must assemble at the correct start site on the messenger RNA (mRNA) defined by specific mRNA–ribosome interactions; this establishes the reading frame for translation of the correct protein sequence. On proper initiation, the assembled 70S or 80S ribosome with initiator transfer RNA (tRNA) base-paired with the start codon positioned in the peptidyl-tRNA site (P site) is competent for elongation of the polypeptide. During elonga-

tion (Dever et al. 2018; Rodnina 2018), the genetic code is read by cognate aminoacyl-tRNAs delivered to the aminoacyl-tRNA site (A site) as a complex with a GTPase elongation factor (EF-Tu in bacteria). This dynamic process of tRNA selection occurs with high fidelity (one wrong amino acid incorporated per 3000) (Loftfield 1963), aided by the free energy of GTP hydrolysis, ribosome and factor conformational changes, and induced fit with kinetic proofreading. Once a correct tRNA is accommodated into the peptidyl transferase center (PTC) of the ribosome, peptide bond formation occurs with subsequent conformational changes in the ribosome (intersubunit rotation) and tRNAs (hybrid state formation). The codon–anticodon pairs on the ribosome must be precisely and rapidly moved from the P and A sites to the exit (E site) and P site during the process of translocation, catalyzed by another GTPase elongation factor (EF-G in bacteria), which also resets the conformation of the ribosomal subunit and places the next codon in the A site. This elongation cycle repeats until a stop codon enters the A site, which signals release factors (RFs) to catalyze release of the polypeptide from the P-site tRNA during termination, and then prompts the ribosomal subunits to be split during recycling (Hellen 2018).

PROBING THE DYNAMICS OF TRANSLATION

The basal process of translation is thus highly directional and dynamic. Even after the process of ribosome biogenesis, whereby ribosomal RNAs are transcribed, processed, and assembled with ribosomal proteins (Shajani et al. 2011), the ribosome undergoes numerous changes in conformation and composition during translation (Fig. 1). Ribosomal subunits assemble during initiation; tRNAs flux through the ribosome as it traverses a messenger RNA (mRNA) (Uemura et al. 2010), unfolding potential RNA structures (Wen et al. 2008), while spooling out a growing nascent polypeptide (which also may fold cotranslationally) during elongation (Thommen et al. 2017); and the disassembly of this transla-

tion complex demands coordinated binding and catalysis of many protein factors triggered by the stop codon signal (Petry et al. 2008; Sternberg et al. 2009; Dever and Green 2012; Koutmou et al. 2014; Peske et al. 2014; Prabhakar et al. 2017a). Underlying the process of translation are conformational dynamics of the ribosome and macromolecules that guide translation (Prabhakar et al. 2017b). Layered onto the basal process are regulatory events that modulate temporally the output of translation—the nature, rates, and efficiency of production of the protein product. An understanding of dynamics is thus required to delineate the mechanism of translation.

Traditional study of dynamics involves measurements in bulk on large numbers of molecules in solution (Wintermeyer et al. 2004; Johansson et al. 2008). Chemical and biochemical kinetics methods have provided the lion's share of information on the time evolution of translation. Reactions are initiated by rapid mixing of the components, and a time-dependent signal is measured in a stopped-flow system either (1) in real time (e.g., fluorescence) to report compositional and conformational changes (Rodnina et al. 1994; Studer et al. 2003; Koutmou et al. 2014), or (2) using quenched-flow methods recording a chemical signal as a function of time to report chemical changes (peptide bond formation, GTP or ATP hydrolysis) (Bilgin et al. 1992;

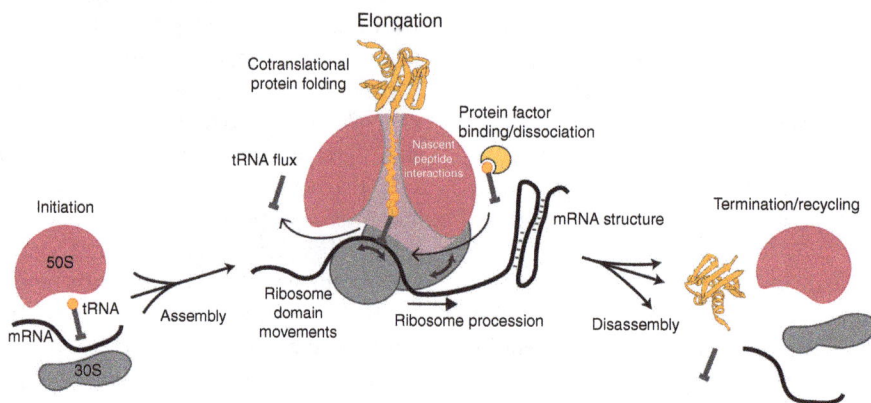

Figure 1. Dynamic picture of translation. Translation is regulated through a complex network of interactions and movements of the ribosomal subunits, the translated messenger RNA (mRNA), the adaptor transfer RNAs (tRNAs), protein translation factors, and the growing nascent peptide chain. Key dynamic features of translation are highlighted.

Cite this article as *Cold Spring Harb Perspect Biol* doi: 10.1101/cshperspect.a032714

Pape et al. 1998). By using these approaches, the basic kinetic mechanisms of initiation (Antoun et al. 2003; Milon et al. 2012), elongation (Gromadski et al. 2002; Johansson et al. 2008), and termination/recycling (Karimi et al. 1999; Zavialov et al. 2001; Peske et al. 2005, 2014; Koutmou et al. 2014) in bacteria have been determined, and progress in understanding the special features of eukaryotic translation made (Shoemaker and Green 2011). More recently, ribosome profiling has revolutionized the study of translation in vivo, by quantifying the occupancy of ribosomes at specific positions on a given mRNA (Ingolia et al. 2009, 2018). These data, derived by deep sequencing of mRNA fragments protected by ribosomes from nuclease treatment, can be quantitatively treated to delineate translation rates and potential pausing sites during elongation (Brar and Weissman 2015).

From these data, the timescales of translation can be determined. Translation initiation occurs on a seconds-to-minutes timescale, whereas elongation rates are 1–20 amino acids per second (depending on in vivo or in vitro systems) (Young and Bremer 1976; Pedersen 1984; Pavlov and Ehrenberg 1996), and termination/recycling also takes seconds (Pavlov et al. 1997). This defines a key window of dynamic processes in the millisecond-to-second range to embrace translation and suggests approaches with corresponding time resolutions to investigate dynamics.

Structural biology has revolutionized the study of translation over the past two decades. Crystal structures of individual subunits and 70S bacterial ribosomes first revealed the architecture of the RNA–protein assemblies (Ban et al. 2000; Schluenzen et al. 2000; Wimberly et al. 2000). Subsequent crystallographic investigations revealed the basis for tRNA–mRNA decoding (Ogle and Ramakrishnan 2005), factor interaction sites (Gao et al. 2009), and GTPase activation of factors (Voorhees et al. 2010), and suggested mechanisms of initiation, translocation, and termination/recycling (Schmeing and Ramakrishnan 2009). In parallel, cryoelectron microscopy (cryo-EM) has contributed enormously to our understanding of structure, in early years at low resolution, and more recently at atomic resolution (Jobe et al. 2018). Cryo-EM has allowed investigation of eukaryotic ribosomal complexes (yeast, human) involved in initiation (Passmore et al. 2007; Fernandez et al. 2013), elongation (Murray et al. 2016), termination (Taylor et al. 2012; Brown et al. 2015; Matheisl et al. 2015), and recycling (Becker et al. 2012). In addition, single-particle cryo-EM methods (including parallel improvements in computation) have revealed conformational and compositional heterogeneity by sorting distinct classes of translational complexes (Llacer et al. 2015; Hussain et al. 2016). These structural studies highlight static conformational flexibility but lack a temporal dimension.

A dream of structural biology is real-time observation of conformational states. Nuclear magnetic resonance (NMR) spectroscopy probes the dynamics of biological systems directly, either at fast (10^{-9} sec), intermediate (10^{-6}–10^{-3} sec), or slow (seconds-to-minutes) timescales (Kovermann et al. 2016). Although NMR provides residue-level information on dynamic movements, correlated domain motion is more difficult to observe. Most critically, solution NMR is limited by size constraints to systems smaller than the ribosome (Huang and Kalodimos 2017). These size limitations can be overcome by application of challenging solid-state NMR measurements (Barbet-Massin et al. 2015; Kurauskas et al. 2016), but, unfortunately, NMR has not been a leading tool in delineating translational dynamics. Time-resolved crystallographic approaches are feasible, but require coordinated changes occurring within a crystalline lattice, often triggered by light (and limited to rather small conformational changes). Recent advances in rapid mixing have allowed a merging of cryo-EM and rapid-mixing kinetics to observe the structural intermediates involved in ribosome recycling (Fu et al. 2016). Further development of time-resolved tools will remain a frontier of structural biology.

STUDYING TRANSLATION WITH SINGLE-MOLECULE FLUORESCENCE

Fluorescence spectroscopy provides a highly sensitive probe of biological dynamics. Using modern dyes, fluorescence from single dyes can be readily detected using standard optics and solid-state cameras (Betzig and Chichester 1993). Stable, bright fluorophores (Cy dyes, Alexa dyes) have been developed that are excited by standard lasers, with fluorescence emission at different visible wavelengths; these bright dyes emit millions of photons as they go through their photocycles before irreversible chemical reactions (photobleaching) that terminate fluorescence (Ha and Tinnefeld 2012). These properties have allowed application of single-molecule approaches over the past two decades (Michalet et al. 2003).

To use single-molecule fluorescence approaches, biological molecules must be fluorescently labeled with bright extrinsic fluorophores. These are usually the organic dyes discussed in the previous paragraph (Ha and Tinnefeld 2012), as fluorescent proteins tend to be dimmer with poorer photophysical properties for single-molecule dynamics studies. For nucleic acids, labeling can be achieved through chemical synthesis of oligonucleotides, hybridization of a labeled DNA or RNA to a target nucleic acid, or modification of thiols or primary amines placed within or at termini of an RNA or DNA (Solomatin and Herschlag 2009). For proteins, dyes can be attached to engineered cysteines using maleimide chemistry, often by mutagenesis to create a single-cysteine version (Nanda and Lorsch 2014). Alternatively, nonnatural amino acids can be incorporated, allowing use of orthogonal chemistries (e.g., CLICK). A number of additional tagging methods have been developed, which include a full protein or peptide domain that can be dye labeled (SNAP, Halo-Tag, ybbR) (Keppler et al. 2003; Yin et al. 2006; Los et al. 2008). For all these labeling approaches, the function of the modified protein must be tested, as modification of the protein (either the sequence, length, or dye labeling) may greatly perturb function. Such biochemical controls are absolutely essential for subsequent valid biophysical investigation.

By observing the behavior of single molecules, complex dynamic behavior of biomolecular systems can be determined and sorted, free from ensemble averaging. Single-molecule fluorescence experiments are often performed with systems immobilized on surfaces, such that they can be observed for long time periods (seconds-to-hours) without diffusing away (Fig. 2A). These experiments harness surface chemistry to make the (usually) SiO_2 surface inert to nonspecific binding of biomolecules, often using polyethylene glycol (PEG) derivatization of the surface, while using biotin–streptavidin interactions to provide specific, stable tethering of biomolecules to a surface (Fig 2A) (Ha 2001). The simplest single-molecule experiment monitors ligand binding and dissociation (Fig. 2B). A fluorescent ligand (labeled protein, DNA/RNA) is present in solution that can bind to a surface-immobilized biomolecule; fluorescence is normally excited using total internal reflectance (TIR), whereby a laser illuminates a quartz surface at a certain angle (dependent on the refractive index of the surface) to create an evanescent field that extends ~100 nm above the surface, leading to wide spatial excitation. Binding of the fluorescent ligand to the surface-immobilized biomolecule leads to a burst in fluorescence, as the fluorophore is localized for a long period to emit photons. The free fluorescent molecule in solution is also excited but if the concentration is kept below ~20–50 nM, the average occupancy of the fluorophore in this "excitation" volume caused by diffusion is less than one, leading to baseline fluorescence lower than the fluorescent burst. Using this experiment, the kinetics of ligand binding and dissociation can be measured at the single-molecule level by analyzing the statistical distribution of binding event lifetimes (dissociation rate), or times between events (association rate). Thus, simple compositional dynamics (ligand occupancy as a function of time) can be measured for biomolecular systems.

Multiple ligands can be tracked at higher free ligand concentration using different single-molecule configurations. We have applied nanophotonic devices called "zero-mode waveguides" (ZMWs) that reduce the excitation volume by orders of magnitude, allowing use of free ligand concentrations up to μM (Levene et al. 2003). This is critical in studying translation, in which many translational components

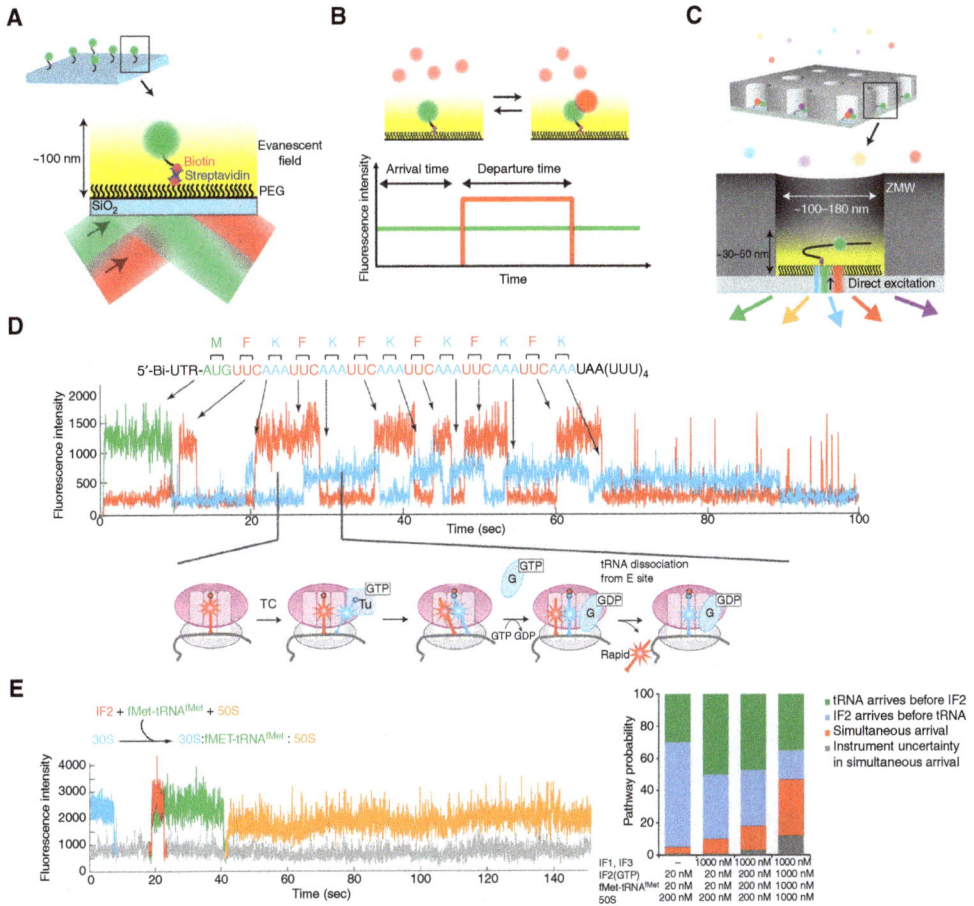

Figure 2. Compositional studies of translation using single-molecule fluorescence. (*A*) Fluorescently labeled biomolecule (green) functionalized with a biotin is immobilized to the polyethylene glycol (PEG)-derivatized SiO_2 surface through biotin–streptavidin interactions. Total internal reflectance (TIR) shown here produces an evanescent field, exciting only fluorophores within 100 nm from the surface. (*B*) Binding of a second labeled biomolecule (red) in solution to the immobilized biomolecule results in detection of red fluorescence burst, from which arrival and departure times can be measured. (*C*) Zero-mode waveguides (ZMWs) allow delivery of up to four different fluorescently labeled ligands at micromolar concentrations caused by its shorter excitation lengths with direct illumination by green (532 nm), red (642 nm), and cyan (488 nm) lasers. (*D*) 70S ribosomes with fMet-Cy3-tRNA^{fMet} (green) were immobilized by the biotinylated mRNA and real-time transit of Phe-Cy5-tRNA^{Phe} (red) and Lys-Cy2-tRNA^{Lys} (cyan) ternary complexes (TCs) was monitored during translation (Uemura et al. 2010). (*E*) Tracking the binding events of Cy5-IF2 (red), fMet-Cy3-tRNA^{fMet} (green), and Cy3.5-50S subunit (yellow) to the Alexa 488-30S-mRNA complex (cyan) (Tsai et al. 2012) revealed multiple pathways of translation initiation in bacteria (see text for details).

are at the micromolar range to drive rapid association of ligands and efficient translation. The ZMWs are roughly cylindrical chambers 100–180 nm in diameter and ~200 nm deep, with transparent quartz (with biotin-PEG) bottoms to immobilize and excite biomolecular and aluminum (or other conductor) side walls (Fig. 2C).

Propagating laser excitation is quenched as it moves up the Z-axis of the ZMW, leading to high excitation only near the surface. We have applied a sophisticated optical system that allows detection of fluorescence with two laser excitation and four different emission channels (Chen et al. 2014). We can thus monitor, in

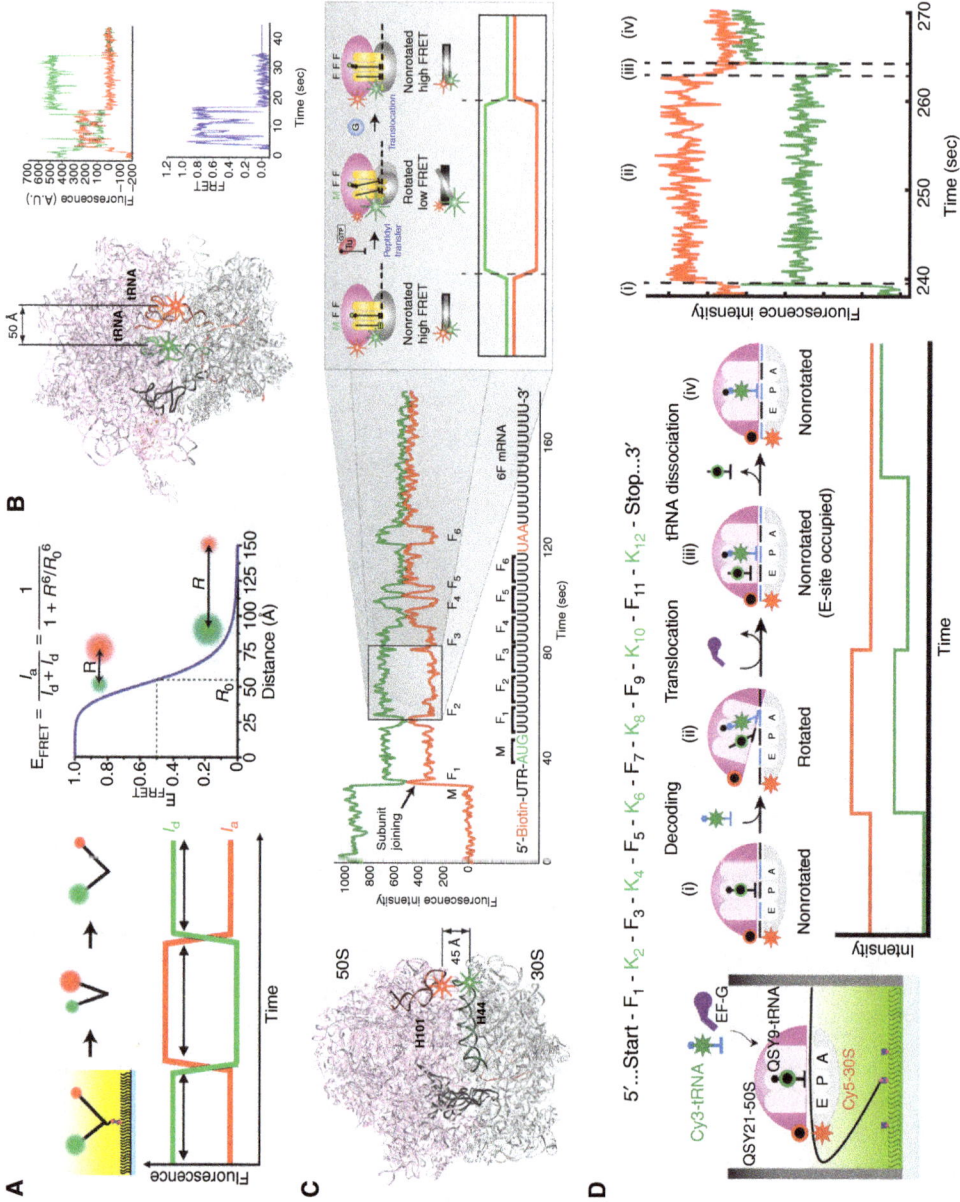

Figure 3. (*See legend on following page.*)

Cite this article as *Cold Spring Harb Perspect Biol* doi: 10.1101/cshperspect.a032714

real time, the ligand occupancy on a translation complex of up to four different ligands simultaneously. This allows determination of the relative order or coordination of ligand arrival and departures in translation, which we outline further in the next section.

EARLY COMPOSITIONAL STUDIES OF TRANSLATION USING OUR ZMW APPROACH

The power of this real-time compositional approach was first shown using labeled tRNAs during bacterial translation initiation and elongation. Using dye-labeled tRNAs (with dyes placed at the elbow position of the L-shaped tRNA, which does not interfere with function), we could track the arrival of tRNAs to an A-site codon, occupancy on the ribosome as the tRNA moves from A to P to E sites, and subsequent departure from the E site at each codon (Fig. 2D) (Uemura et al. 2010). The tRNA-binding events overlap, during two-tRNA occupancy, and from these data a codon-by-codon translation rate could be calculated. Importantly, these data showed a lack of coupling of A-site tRNA arrival with E-site tRNA departure.

We have applied similar approaches to determine the ligand-binding dynamics and pathways of translation initiation. We first investigated prokaryotic translation initiation, using labeled initiator tRNA, subunits, and initiation factor IF2 (Tsai et al. 2012). Using different color

dyes on the different components, we could track the arrival of initiator tRNA, IF2, and 50S subunit simultaneously in real time (Fig. 2E). These experiments showed heterogeneous pathways for fMet-tRNA^{fMet} and IF2 arrival, with the two components (which interact) arriving either sequentially or simultaneously on the 30S subunit. The simultaneous pathway occurs more frequently at higher concentrations of IF2 and tRNA. Both sequential and simultaneous pathways lead to efficient 50S subunit joining, with IF2 departing rapidly from the assembled 70S subunit. These experiments showed the power of tracking translational pathways with single-molecule approaches.

REAL-TIME TRACKING OF CONFORMATION DURING PROKARYOTIC TRANSLATION

A powerful application of single-molecule fluorescence uses Förster resonance energy transfer (FRET) to probe conformational dynamics (Selvin 2000; Ha 2001). FRET relies on through-space, dipolar interactions between donor and acceptor dyes to allow energy transfer between the low-wavelength donor to higher-wavelength acceptor (Fig. 3A). The efficiency of the energy transfer depends on multiple factors (quantum yield of the donor, overlap of the emission spectrum of the donor with the absorbance spectrum of the acceptor, refractive index of the medium, dynamic orientations of the two

Figure 3. Conformational studies of translation using single-molecule Förster resonance energy transfer (FRET). (A) With single-laser excitation of the donor dye, fluorescence intensities of donor (green, I_d) and acceptor (red, I_a) dyes attached to a surface-immobilized biomolecular complex are measured to calculate the FRET efficiency (E_{FRET}), which reports distance between the fluorophores (R) used to define conformational states of a biological process. (B) Cy3 and Cy5 labeling of P-site and A-site transfer RNAs (tRNAs), respectively, was used to follow the tRNA conformations during tRNA selection stage of elongation (Blanchard et al. 2004a). (C) (Left) Helix44 of 30S subunit was labeled with Cy3 and Helix101 of 50S subunit was labeled with Cy5, bringing the dyes within FRET distance to report intersubunit conformational changes (Marshall et al. 2008b). (Right) With immobilization of Cy3-30S (green) as a preinitiation complex and delivery of Cy5-50S (red) along with Phe-tRNA^{Phe} and elongation factors, subunit joining is signaled by the simultaneous burst of Cy5 fluorescence and dip in Cy3 fluorescence at the beginning, and anticorrelated changes in the Cy3 and Cy5 fluorescence intensities correspond to transitions between the nonrotated and rotated conformational states evolved from cycles of peptidyl transfer and translocation during elongation. (D) Replacing the acceptor dyes in tRNA and ribosome labels in B and C with nonfluorescent quenchers (QSY9, QSY21, respectively) allowed simultaneous detection of ribosome intersubunit and tRNA conformations with Cy3 labeling on Lys-tRNA^{Lys} and Cy5 labeling on the 30S subunit (Choi and Puglisi 2017).

dyes) but most critically the interdye distance, R, with $1/R^6$ dependence. For a given dye pair, this efficiency is given by the equation in Figure 3A, in which R_0 is a characteristic of the dye pair (incorporating the parameters discussed above) and represents the distance separation for 50% energy transfer from donor to acceptor. With R_0 values of typical FRET dye pairs (e.g., Cy3 and Cy5) at 40–50 Å, FRET efficiencies vary from 100% at interdye distances below 30 Å to zero beyond 80 Å; in between, FRET depends steeply on interdye distance, making FRET a powerful probe in this range. As such, these FRET measurements are best applied to larger systems with appropriate interdye distances. The ribosome, with its large geometry (250 Å sphere) is an outstanding system to apply FRET.

FRET energy transfer rates are rapid (nanosecond timescale), so FRET provides a powerful probe of conformational dynamics in translation. Changes in interdye distances from a conformational change in a two-state process appear as abrupt changes in single-molecule FRET trajectories that are readily analyzed. We initially characterized tRNA dynamics on the bacterial ribosome using FRET between A- and P-site tRNAs (Fig. 3B) (Blanchard et al. 2004b), as tRNAs were thought to change configurations on the ribosome during translation. Using Cy3 donor dye on the P-site tRNA and Cy5 acceptor dye on the A-site tRNA, FRET monitored the arrival of Cy5-(acceptor) labeled tRNA as a ternary complex with EF-Tu in the A site, and subsequent steps of GTP hydrolysis by EF-Tu, accommodation and conformational changes post-peptide bond formation. Using these experiments, we observed the dynamic pathway of tRNA selection, how near cognate tRNAs are rejected, and how drugs perturb the conformational dynamics of tRNA selection (Blanchard et al. 2004a). Cognate aminoacyl-tRNAs rapidly proceed through an initial low FRET through GTPase activation (medium FRET) to accommodated states (high FRET), whereas near-cognate tRNAs are rejected at earlier steps and rarely proceed to accommodation. Drugs, such as tetracycline, disrupt these processes; with tetracycline blocking the final position of the codon–anticodon on the small subunit, only

brief, low FRET-binding events of cognate ternary complexes were observed. Small molecule drugs are powerful disrupters of ribosome dynamics.

FRET has been used to observe many features of ribosome function. With ribosome structures as blueprints, FRET probes can be placed at appropriate points in the ribosome. Ribosomal proteins can be labeled using single-cysteine mutants or more sophisticated tagging methods (peptide tags, nonnatural amino acid incorporation), and incorporated into ribosomes during biogenesis. The labeled ribosomal protein can be incorporated into ribosomal particles that have single protein deletions in certain cases (Ermolenko et al. 2007) or by in vitro reconstitution (Hickerson et al. 2005). Ribosomal RNA has been labeled using mutations of ribosomal RNA elements to allow hybridization of labeled oligonucleotides (Dorywalska et al. 2005). In all cases, the function of the labeled ribosomes must be tested.

Using labeled bacterial ribosomes, we and others have investigated intersubunit conformation (Ermolenko et al. 2007; Marshall et al. 2008b), head movement of the ribosome (Hickerson et al. 2005), and dynamics of the L1 stalk region (Fei et al. 2008; Cornish et al. 2009). By using our labeling scheme (helix44 on the small subunit, helix101 on the large), the intersubunit rotational state of the ribosome as it passes from initiation, to elongation, to termination, and then to recycling, has been tracked (Marshall et al. 2008b, 2009; Prabhakar et al. 2017a). The two most populated intersubunit states involve the nonrotated and rotated conformations distinguished by rotation of 3°–10° (Valle et al. 2003). The two states interconvert on peptide bond formation or translocation driven by EF-G (or the action of IF2-GTP or RF3-GTP); the nature of the barrier height of this conformational change remains to be determined. However, this conformational signal allowed tracking of translation. Initiation is signaled by the arrival of FRET (subunits join) and elongation by cycles of high to lower to high FRET, representing one round of tRNA selection, peptide bond formation, and translocation (Fig. 3C). We have used these signals to watch translation in real time (Aitken and Puglisi 2010).

These FRET measurements provide time-resolved conformational information on the ribosome during translation. However, all FRET measurements suffer from the sensitivity of FRET to parameters beyond interdye distance (dye mobility, photophysics of the dyes, the interdye medium). Thus, conformational information should be obtained from multiple perspectives using multiple labeling schemes, with each FRET measurement at best representing a single distance within the complexity of the translational apparatus. In addition, conformational changes can be correlated to ligand arrival and departure, but this is challenging with normal FRET as a result of cross talk of the individual dyes required to measure FRET and ligand arrival. By swapping the acceptor dye with a nonfluorescent acceptor that can absorb the energy transfer via FRET but release it as heat and not light, the normal acceptor dye (e.g., Cy5) can be used to label the ligand. We first showed this quencher approach in tracking the ribosome intersubunit conformations during translation (Chen et al. 2012). Using this "one-color" FRET, correlated conformational changes with ligand arrival and departure have been measured, and reviewed extensively elsewhere (see Fig. 4) (Chen et al. 2013; Prabhakar et al. 2017b).

Recently, we expanded this quencher approach to monitor two conformational changes simultaneously (Choi and Puglisi 2017). We used two donor–quencher dye pairs, and designed the separation of the pairs to be greater than 100 Å to avoid cross talk. As such, we could monitor intersubunit conformation directly correlated to tRNA–tRNA conformation during translation. Our results allowed us to show that, as expected from structural data, these conformational fluctuations were correlated. Using this approach, we identified a novel intermediate intersubunit rotational state when three tRNAs reside in the A, P, and E sites. This three-tRNA state slows elongation until E-site tRNA is spontaneously released, allowing EF-G to drive translocation. Further structural characterization is needed to confirm this state, but these results highlight the future interplay of structural and single-molecule approaches.

Biological machines generate force during their function. Using optical trapping methods, the forces generated by these machines can be measured or perturbed in real time, providing a direct view of their mechanochemical cycle. In an optical trapping experiment, biological systems are tethered to large beads, which can be manipulated and trapped by a strong laser beam. As forces are generated, the bead is pulled out of the beam, allowing measurement of the force generated by, or application of force upon, the system. Traditionally, optically trapping has been applied to protein motors (myosins, kinesins, gyrases) and polymerases, because the rigidity of nucleic acids allows more facile trapping and monitoring of forces. It is far more challenging to apply these approaches to translation, which occurs on a flexible single-stranded RNA. By circumventing these difficulties through clever experimental and instrument designs, Bustamante, Tinoco, and colleagues have elegantly explored the mechanisms of ribosome movement during translocation (Wen et al. 2008) and protein folding (Kaiser et al. 2011) on the ribosome. These experiments showed how ribosomes efficiently unfold mRNA structures, and revealed the forces generated by both the ribosome and protein folding within the ribosomal peptide exit tunnel. These force measurements can be combined with fluorescence and/or FRET to provide combined force and conformation/composition information. Such experiments will be powerful probes of translational processes.

RECENT WORK ON BACTERIAL TRANSLATION

To understand the applications of single-molecule fluorescence methods to translation, we highlight below some recent examples from our own work; this, of course, is completely arbitrary, and ignores the outstanding work performed by many other groups through the years (Gonzalez, Blanchard, Cooperman, Goldman, and others). Our goal is not to be comprehensive—many recent reviews have covered the field extensively—but to give examples of how these methods can provide insights into aspects of translation (Mar-

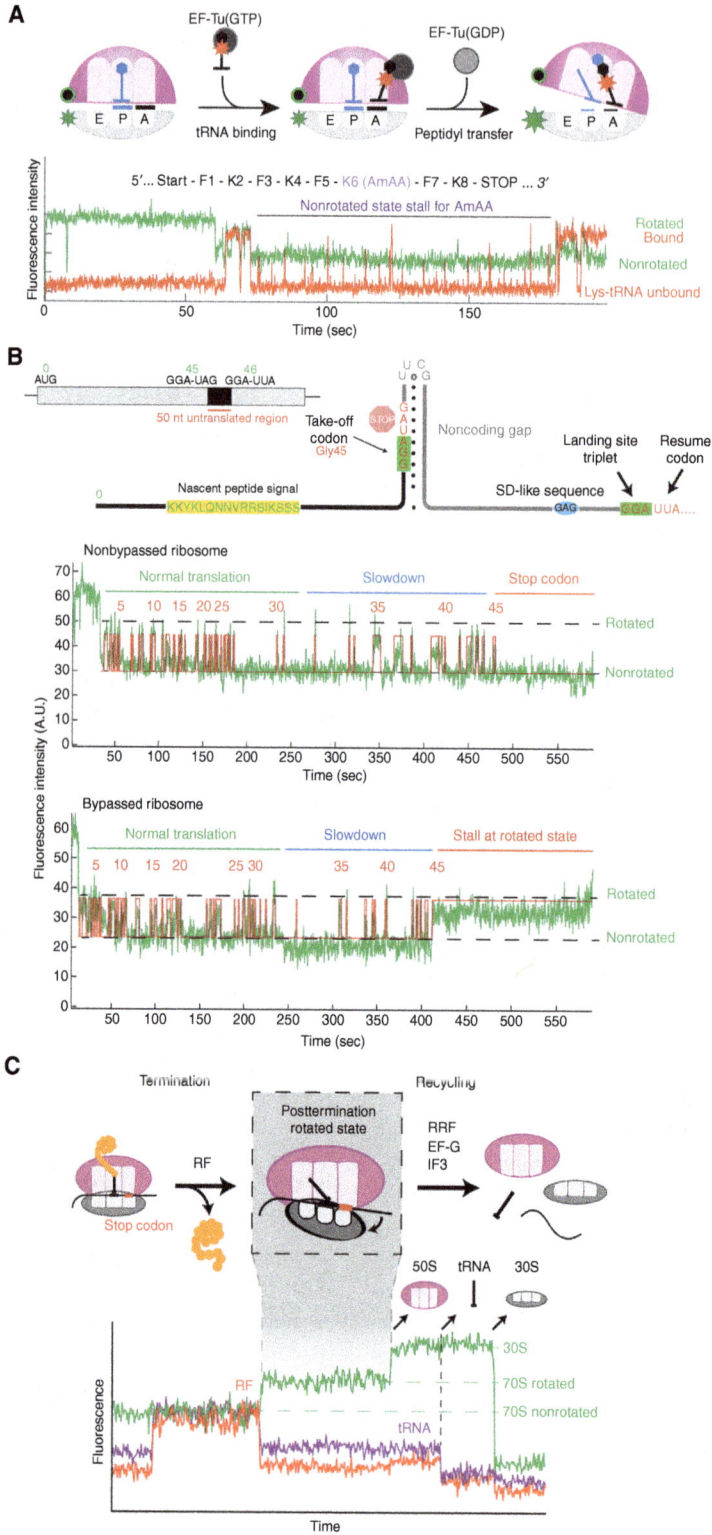

Figure 4. (*See legend on following page.*)

Cite this article as *Cold Spring Harb Perspect Biol* doi: 10.1101/cshperspect.a032714

shall et al. 2008a; Blanchard 2009; Perez and Gonzalez 2011; Petrov et al. 2012).

Effect of mRNA Chemical Modifications on Elongation Dynamics

Recent transcriptome-wide mapping revealed the prevalence of different RNA chemical modifications within the coding sequences of cellular mRNAs (Roundtree et al. 2017). These findings suggested that these modifications may modulate the decoding steps of elongation. We investigated this phenomenon by testing the effects of two frequently occurring modifications, N^6-methyladenosine (m^{6}A) and 2'-O-methylation (Nm), on the dynamics of elongation. By simultaneously tracking binding of labeled tRNA (Cy5) and ribosome conformation (Cy3B-quencher) in our single-molecule translation system (Fig. 4A), combined with bulk biochemical and structural studies, we discovered that the two mRNA modifications perturb different steps of the tRNA-decoding mechanism of elongation. The methylated adenosine of the m^{6}A-containing codon destabilizes the codon–anticodon interactions critical in the initial tRNA selection step, thus increasing the probability of tRNA rejection in the early stage of decoding (Choi et al. 2016). Methylation of the 2'OH of the ribose sugar in Nm, however, does not disrupt codon–anticodon base-pairing; instead, it sterically blocks interactions that allow formation of the correct geometry of ribosomal monitoring bases required to activate the GTPase of EF-Tu, a requisite for successful tRNA accommodation (Choi et al. 2018). In addition to mechanistic differences, Nm results in an enormous slowdown in translation of the codon (1500-fold) in contrast to the modest effects of m^{6}A (18-fold). This disparity may signify diversity in regulatory elements to tune the local translation elongation rates, which could in turn modulate coupled cotranslational processes like protein folding (Kim et al. 2015).

Ribosome and mRNA Structural Rearrangements during Bypassing

Programmed translational bypassing, in which a ribosome hops over a long stretch of mRNA nucleotides during elongation, remains one of the most unusual and understudied translational phenomena. The best-documented example of this occurs in bacteria when the *Escherichia coli* ribosome translates and bypasses a 50-nt stretch of the T4 bacteriophage gene 60 mRNA from the Gly-45 (take-off) codon to another Gly (landing) codon. Two key stimulatory elements that initiate this process are an upstream nascent peptide signal and an mRNA hairpin structure at the take-off site with a UUCG tetraloop (Fig. 4B). The mechanism of stimulation by these elements as well as correct ribosome landing remained unclear. Interested in how these elements changed the translation elongation dynamics, we tracked gene 60 translation using our single-molecule intersubunit dye-quencher system (Chen et al. 2015). Measuring local elongation rates, we observed a gradual slowdown in elongation as the ribosome approached the bypass site. After incorporation of the Gly-tRNA^{Gly} at the take-off codon, 35% of the ribosomes experienced a 10- to 20-fold longer pause in the rotated state, while the rest translocated to

---

Figure 4. Recent work on bacterial translation involved use of intersubunit dye–quencher labeling (Cy3B-30S and BHQ-2-50S; see Fig. 3C for labeling positions) to track ribosome conformation (green signal). (*A*) Monitoring Lys-Cy5-tRNA^{Lys}-binding dynamics (red) on ribosomes with Nm-modified mRNAs (Choi et al. 2018) shows multiple short Cy5 pulses representing tRNA rejection events, explaining the long ribosome stall in nonrotated state. (*B*) Tracking translation of gene *60* mRNA revealed that the subset of ribosomes that bypassed stalled at the take-off Gly-45 codon in the rotated state. (*C*) Simultaneously tracking the ribosome intersubunit conformation (green), Cy5-RF binding (red), and P-site tRNA occupancy (Phe-Cy5.5-tRNA^{Phe}, violet) during translation termination and recycling (Prabhakar et al. 2017a) helped resolve the posttermination rotated state, a key intermediate preceding subunit splitting step (catalyzed by RRF and EF-G) and subsequent 30S complex disassembly (performed by IF3).

the adjacent stop-46 codon and stopped translating. For the 35% that paused at codon Gly-45, elongation continued afterward, representing the subset of ribosomes that successfully bypassed. This long, noncanonical rotated-state pause became the centerpiece of an updated mechanistic model. The model describes the initial slowdown at codons 40–45 to be caused by the peptide interacting with and/or folding within the ribosome peptide exit tunnel. This results in buildup of tension until the rotated state pause at Gly-45 occurs, producing weakened ribosome–tRNA–mRNA interactions key to promoting bypassing. At this point, the take-off hairpin is partially melted within the mRNA channel, so EF-G-catalyzed translocation then leads to refolding of the UUCG tetraloop that then threads mRNA in the 5′ direction, initiating bypassing. Tracking the ribosome position relative to the landing site with ribosome–mRNA FRET showed that the ribosome reaches near the landing site rapidly but spends most of the long rotated-state pause trekking a short distance to find the appropriate landing site. Through this work, we laid a mechanistic framework that places the newly discovered stalled translational state as the focal point and highlights the important coupling of ribosome and mRNA structural rearrangements to successful bypassing.

Dynamic Coupling of Prokaryotic Translation Termination and Ribosome Recycling

The ribosome undergoes multiple cycles of elongation as it decodes every codon with the exception of the three stop codons (UAG, UGA, UAA), which signal the closing events of nascent peptide release (termination) and ribosome disassembly (recycling) (see Rodnina 2018). Termination begins with the recruitment of two classes of RFs. The class I RF (RF1 and RF2 in bacteria) recognizes the stop codon and catalyzes the release of the peptide. Bacterial class II RF, or RF3, is a nonessential (Grentzmann et al. 1994; O'Connor 2015) GTPase factor that accelerates the dissociation of class I RF from the ribosome (Freistroffer et al. 1997), but its general role in translation is still uncer-

tain. Thus, the molecular events that occur after peptide release are still unclear. Eventually, the posttermination ribosome complex is disassembled through a recycling mechanism involving ribosome recycling factor (RRF), elongation factor EF-G, and initiation factor IF3. This mechanism was still unclear owing to the difficulty in resolving transient intermediates of the process.

Using single-molecule real-time tracking of all four stages of translation, we resolved a key intermediate at the transition from termination to recycling and delineated the timings of events during recycling (Prabhakar et al. 2017a). As our intersubunit dye–quencher system monitored initiation and elongation, we tracked binding of Cy5-labeled class I RF during termination (Fig. 4C). We discovered that after successful peptide release, class I RF dissociation triggers an intersubunit rotation from the nonrotated to the rotated state. This posttermination rotated state represents an unstable intermediate that either undergoes uncatalyzed ribosomal subunit splitting or RF rebinding. We found that RRF binds to this intermediate to block RF from rebinding and forms the ribosomal substrate for EF-G, to catalyze subunit disassembly through GTP hydrolysis. IF3 binds to the remaining 30S•tRNA•mRNA complex to induce dissociation of the deacylated tRNA, followed by rapid dissociation of the 30S subunit from the mRNA. By simultaneously tracking ribosome conformation and occupancy of tRNA and protein factors, most of the events of termination and recycling were temporally resolved to bring light to the dynamic coupling between these processes. This method will be powerful in answering further questions surrounding termination and recycling, such as the role of RF3 in this mechanism.

RECENT WORK ON EUKARYOTIC TRANSLATION

The mechanistic pathways by which eukaryotic translation initiation occurs remain unclear and are an outstanding target for single-molecule analyses. However, single-molecule experiments on eukaryotic translation are challenged

 Cite this article as *Cold Spring Harb Perspect Biol* doi: 10.1101/cshperspect.a032714

by the complexity of the process, and the practicality of labeling factors, ligands, and ribosomes for tracking. We have applied a variety of approaches to obtain labeled versions of initiation factors and ribosomes in yeast and human translation systems. These include yeast genetic approaches (Petrov and Puglisi 2010) or CRISPR knockouts (Fuchs et al. 2015) to obtain pure populations of mutant ribosomes, which can be labeled using nucleic acids or labeled proteins added back to the mutants. We have also developed genetic systems for labeling critical initiation factors (such as eIF3) in humans and have shown the power of using these labeled reagents to track translation initiation processes directly (Johnson et al. 2017).

We first applied single-molecule approaches to the much simpler internal ribosome entry site (IRES)-mediated mechanisms, which require a small subset of factors. We showed that human 40S subunits can form dynamic complexes with the hepatitis C virus (HCV) IRES by using FRET between the IRES RNA and labeled ribosomal

protein eS25 on the 40S subunit (Fig. 5A) (Fuchs et al. 2015). We determined the pathways by which the cricket paralysis virus (CrPV) IRES initiates translation directly without initiation factors or initiator tRNA (Petrov et al. 2016). Using direct tracking of 40S and 60S subunits and tRNAs, we could determine the distinct pathways by which the IRES initiates (Fig. 5B). In addition to an expected sequential pathway in which 40S subunits bind to the IRES RNA, followed by 60S subunit binding, we observed a significant population of direct 80S ribosome arrival (simultaneous 40S and 60S binding) to an IRES. Both pathways were equally functional in subsequent elongation, which required the continued presence of eEF2•GTP to allow the first elongator tRNA to bind. We also uncovered heterogeneity in reading frame establishment, with ribosomes initiating on the IRES in both the 0 and +1 frame. Even for a simple system like the CrPV IRES, single-molecule approaches revealed the complex pathways by which initiation occurs (Fig. 5C).

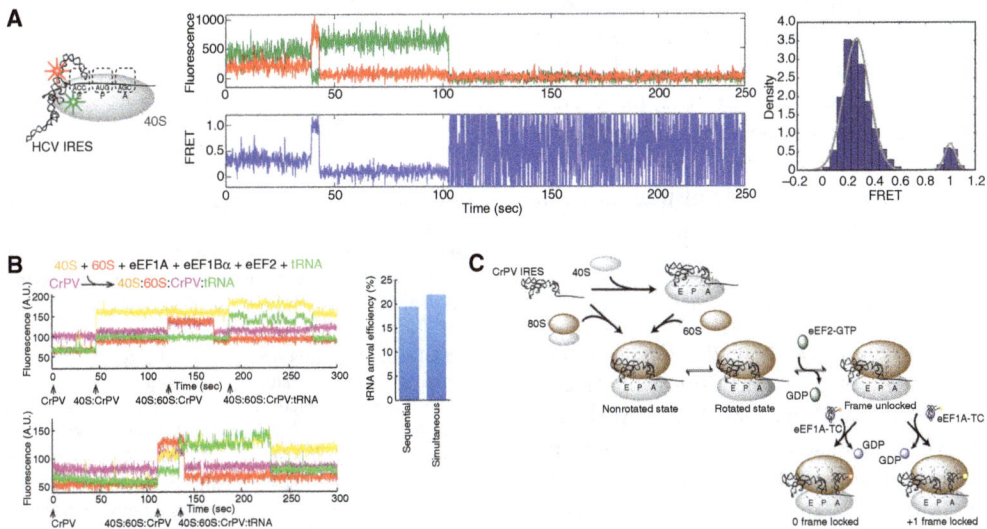

Figure 5. Recent work on eukaryotic translation. (*A*) Conformational dynamics of hepatitis C virus (HCV) internal ribosome entry site (IRES) bound to human 40S subunit was monitored using Cy3-Cy5 Förster resonance energy transfer (FRET) (Fuchs et al. 2015), detecting two different conformational states important in the pathway of HCV IRES initiation. (*B*) Tracking real-time assembly of yeast Cy3.5-40S (yellow), yeast Cy5-60S (red), and Phe-Cy3-tRNA^{Phe} (green) on immobilized Cy5.5-CrPV IRES (violet) (Petrov et al. 2016) led to discovery of parallel pathways of CrPV IRES initiation. (*C*) Petrov et al. (2016) also found that ribosomes initiate on the CrPV IRES in both 0 and +1 reading frames, another pathway branchpoint in the mechanism of CrPV IRES initiation.

CONCLUDING REMARKS

Single-molecule approaches have provided a structurally dynamic view of translation, complementing biochemical and structural experiments. Single-molecule measurements have connected structural states, determined by X-ray crystallography and cryo-EM, into a more coherent pathway of states during normal initiation, elongation, and termination. In short, these methods have animated the static pictures from structural biology and identified transient states that are difficult to capture by these methods. By tracking single ribosomes, the simple, linear view of translation has been adjusted. These methods have parsed distinct, parallel pathways in noncanonical translation during initiation, frameshifting, and bypassing, which directly determine the timing and outcome of protein synthesis. Despite these advances, many open questions remain. In bacterial translation, the interplay of the nascent protein chain and its folding with translational dynamics is a fascinating frontier of investigation. The dynamics of multiple ribosomes (polysomes) on an mRNA, and whether there is communication between ribosomes, should be probed. Eukaryotic translation and its regulation remain a challenging and central topic that has only recently been explored. New methods that access faster timescale dynamics and embrace greater complexity will be needed to elucidate the dynamic role of translation in human health and disease.

ACKNOWLEDGMENTS

Single-molecule research in the Puglisi group is funded by National Institutes of Health (NIH) Grants GM51266, GM113078, AI047365, AI099506, and GM082545. A.P. is funded by NIH Molecular Biophysics Training Grant (T32-GM008294) and a Stanford Interdisciplinary Graduate Fellowship.

REFERENCES

*Reference is also in this collection.

Aitken CE, Puglisi JD. 2010. Following the intersubunit conformation of the ribosome during translation in real time. *Nat Struct Mol Biol* 17: 793–800.

Antoun A, Pavlov MY, Andersson K, Tenson T, Ehrenberg M. 2003. The roles of initiation factor 2 and guanosine triphosphate in initiation of protein synthesis. *EMBO J* 22: 5593–5601.

Ban N, Nissen P, Hansen J, Moore PB, Steitz TA. 2000. The complete atomic structure of the large ribosomal subunit at 2.4 Å resolution. *Science* 289: 905–920.

Barbet-Massin E, Huang CT, Daebel V, Hsu ST, Reif B. 2015. Site-specific solid-state NMR studies of "trigger factor" in complex with the large ribosomal subunit 50S. *Angew Chem Int Ed Engl* 54: 4367–4369.

Becker T, Franckenberg S, Wickles S, Shoemaker CJ, Anger AM, Armache JP, Sieber H, Ungewickell C, Berninghausen O, Daberkow I, et al. 2012. Structural basis of highly conserved ribosome recycling in eukaryotes and archaea. *Nature* 482: 501–506.

Betzig E, Chichester RJ. 1993. Single molecules observed by near-field scanning optical microscopy. *Science* 262: 1422–1425.

Bilgin N, Claesens F, Pahverk H, Ehrenberg M. 1992. Kinetic properties of *Escherichia coli* ribosomes with altered forms of S12. *J Mol Biol* 224: 1011–1027.

Blanchard SC. 2009. Single-molecule observations of ribosome function. *Curr Opin Struct Biol* 19: 103–109.

Blanchard SC, Gonzalez RL, Kim HD, Chu S, Puglisi JD. 2004a. tRNA selection and kinetic proofreading in translation. *Nat Struct Mol Biol* 11: 1008–1014.

Blanchard SC, Kim HD, Gonzalez RL Jr, Puglisi JD, Chu S. 2004b. tRNA dynamics on the ribosome during translation. *Proc Natl Acad Sci* 101: 12893–12898.

Brar GA, Weissman JS. 2015. Ribosome profiling reveals the what, when, where and how of protein synthesis. *Nat Rev Mol Cell Biol* 16: 651–664.

Brown A, Shao S, Murray J, Hegde RS, Ramakrishnan V. 2015. Structural basis for stop codon recognition in eukaryotes. *Nature* 524: 493–496.

Chen J, Tsai A, Petrov A, Puglisi JD. 2012. Nonfluorescent quenchers to correlate single-molecule conformational and compositional dynamics. *J Am Chem Soc* 134: 5734–5737.

Chen J, Petrov A, Tsai A, O'Leary SE, Puglisi JD. 2013. Coordinated conformational and compositional dynamics drive ribosome translocation. *Nat Struct Mol Biol* 20: 718–727.

Chen J, Dalal RV, Petrov AN, Tsai A, O'Leary SE, Chapin K, Cheng J, Ewan M, Hsiung PL, Lundquist P, et al. 2014. High-throughput platform for real-time monitoring of biological processes by multicolor single-molecule fluorescence. *Proc Natl Acad Sci* 111: 664–669.

Chen J, Coakley A, O'Connor M, Petrov A, O'Leary SE, Atkins JF, Puglisi JD. 2015. Coupling of mRNA structure rearrangement to ribosome movement during bypassing of non-coding regions. *Cell* 163: 1267–1280.

Choi J, Puglisi JD. 2017. Three tRNAs on the ribosome slow translation elongation. *Proc Natl Acad Sci* 114: 13691–13696.

Choi J, Ieong KW, Demirci H, Chen J, Petrov A, Prabhakar A, O'Leary SE, Dominissini D, Rechavi G, Soltis SM, et al. 2016. N^6-methyladenosine in mRNA disrupts tRNA selection and translation-elongation dynamics. *Nat Struct Mol Biol* 23: 110–115.

Choi J, Indrisiunaite G, DeMirci H, Ieong KW, Wang J, Petrov A, Prabhakar A, Rechavi G, Dominissini D, He C, et al. 2018. 2′-O-methylation in mRNA disrupts tRNA decoding during translation elongation. *Nat Struct Mol Biol* **25**: 208–216.

Cornish PV, Ermolenko DN, Staple DW, Hoang L, Hickerson RP, Noller HF, Ha T. 2009. Following movement of the L1 stalk between three functional states in single ribosomes. *Proc Natl Acad Sci* **106**: 2571–2576.

Dever TE, Green R. 2012. The elongation, termination, and recycling phases of translation in eukaryotes. *Cold Spring Harb Perspect Biol* **4**: a013706.

* Dever TE, Dinman JD, Green R. 2018. Translation elongation and recoding in eukaryotes. *Cold Spring Harb Perspect Biol* doi: 10.1101/cshperspect.a032649.

Dorywalska M, Blanchard SC, Gonzalez RL, Kim HD, Chu S, Puglisi JD. 2005. Site-specific labeling of the ribosome for single-molecule spectroscopy. *Nucleic Acids Res* **33**: 182–189.

Ermolenko DN, Majumdar ZK, Hickerson RP, Spiegel PC, Clegg RM, Noller HF. 2007. Observation of intersubunit movement of the ribosome in solution using FRET. *J Mol Biol* **370**: 530–540.

Fei J, Kosuri P, MacDougall DD, Gonzalez RL Jr. 2008. Coupling of ribosomal L1 stalk and tRNA dynamics during translation elongation. *Mol Cell* **30**: 348–359.

Fernandez IS, Bai XC, Hussain T, Kelley AC, Lorsch JR, Ramakrishnan V, Scheres SHW. 2013. Molecular architecture of a eukaryotic translational initiation complex. *Science* **342**: 1240585.

Freistroffer DV, Pavlov MY, MacDougall J, Buckingham RH, Ehrenberg M. 1997. Release factor RF3 in *E. coli* accelerates the dissociation of release factors RF1 and RF2 from the ribosome in a GTP-dependent manner. *EMBO J* **16**: 4126–4133.

Fu Z, Kaledhonkar S, Borg A, Sun M, Chen B, Grassucci RA, Ehrenberg M, Frank J. 2016. Key intermediates in ribosome recycling visualized by time-resolved cryoelectron microscopy. *Structure* **24**: 2092–2101.

Fuchs G, Petrov AN, Marceau CD, Popov LM, Chen J, O'Leary SE, Wang R, Carette JE, Sarnow P, Puglisi JD. 2015. Kinetic pathway of 40S ribosomal subunit recruitment to hepatitis C virus internal ribosome entry site. *Proc Natl Acad Sci* **112**: 319–325.

Gao YG, Selmer M, Dunham CM, Weixlbaumer A, Kelley AC, Ramakrishnan V. 2009. The structure of the ribosome with elongation factor G trapped in the posttranslocational state. *Science* **326**: 694–699.

Grentzmann G, Brechemier-Baey D, Heurgue V, Mora L, Buckingham RH. 1994. Localization and characterization of the gene encoding release factor RF3 in *Escherichia coli*. *Proc Natl Acad Sci* **91**: 5848–5852.

Gromadski KB, Wieden HJ, Rodnina MV. 2002. Kinetic mechanism of elongation factor Ts-catalyzed nucleotide exchange in elongation factor Tu. *Biochemistry* **41**: 162–169.

Ha T. 2001. Single-molecule fluorescence resonance energy transfer. *Methods* **25**: 78–86.

Ha T, Tinnefeld P. 2012. Photophysics of fluorescent probes for single-molecule biophysics and super-resolution imaging. *Annu Rev Phys Chem* **63**: 595–617.

* Hellen CUT. 2018. Translation termination and ribosome recycling in eukaryotes. *Cold Spring Harb Perspect Biol* doi: 10.1101/cshperspect.a032656.

Hickerson R, Majumdar ZK, Baucom A, Clegg RM, Noller HF. 2005. Measurement of internal movements within the 30S ribosomal subunit using Forster resonance energy transfer. *J Mol Biol* **354**: 459–472.

Huang C, Kalodimos CG. 2017. Structures of large protein complexes determined by nuclear magnetic resonance spectroscopy. *Annu Rev Biophys* **46**: 317–336.

Hussain T, Llacer JL, Wimberly BT, Kieft JS, Ramakrishnan V. 2016. Large-scale movements of IF3 and tRNA during bacterial translation initiation. *Cell* **167**: 133–144.e113.

Ingolia NT, Ghaemmaghami S, Newman JR, Weissman JS. 2009. Genome-wide analysis in vivo of translation with nucleotide resolution using ribosome profiling. *Science* **324**: 218–223.

* Ingolia NT, Hussmann JA, Weissman JS. 2018. Ribosome profiling: Global views of translation. *Cold Spring Harb Perspect Biol* doi: 10.1101/cshperspect.a032698.

* Jobe A, Liu Z, Gutierrez-Vargas C, Frank J. 2018. New insights into ribosome structure and function. *Cold Spring Harb Perspect Biol* doi: 10.1101/cshperspect.a032615.

Johansson M, Lovmar M, Ehrenberg M. 2008. Rate and accuracy of bacterial protein synthesis revisited. *Curr Opin Microbiol* **11**: 141–147.

Johnson AG, Petrov AN, Fuchs G, Majzoub K, Grosely R, Choi J, Puglisi JD. 2017. Fluorescently-tagged human eIF3 for single-molecule spectroscopy. *Nucleic Acids Res* **46**: e8.

Kaiser CM, Goldman DH, Chodera JD, Tinoco I Jr, Bustamante C. 2011. The ribosome modulates nascent protein folding. *Science* **334**: 1723–1727.

Karimi R, Pavlov MY, Buckingham RH, Ehrenberg M. 1999. Novel roles for classical factors at the interface between translation termination and initiation. *Mol Cell* **3**: 601–609.

Keppler A, Gendreizig S, Gronemeyer T, Pick H, Vogel H, Johnsson K. 2003. A general method for the covalent labeling of fusion proteins with small molecules in vivo. *Nat Biotechnol* **21**: 86–89.

Kim SJ, Yoon JS, Shishido H, Yang Z, Rooney LA, Barral JM, Skach WR. 2015. Protein folding. Translational tuning optimizes nascent protein folding in cells. *Science* **348**: 444–448.

Koutmou KS, McDonald ME, Brunelle JL, Green R. 2014. RF3:GTP promotes rapid dissociation of the class 1 termination factor. *RNA* **20**: 609–620.

Kovermann M, Rogne P, Wolf-Watz M. 2016. Protein dynamics and function from solution state NMR spectroscopy. *Q Rev Biophys* **49**: e6.

Kurauskas V, Crublet E, Macek P, Kerfah R, Gauto DF, Boisbouvier J, Schanda P. 2016. Sensitive proton-detected solid-state NMR spectroscopy of large proteins with selective CH3 labelling: Application to the 50S ribosome subunit. *Chem Commun (Camb)* **52**: 9558–9561.

Levene MJ, Korlach J, Turner SW, Foquet M, Craighead HG, Webb WW. 2003. Zero-mode waveguides for single-molecule analysis at high concentrations. *Science* **299**: 682–686.

Llacer JL, Hussain T, Marler L, Aitken CE, Thakur A, Lorsch JR, Hinnebusch AG, Ramakrishnan V. 2015. Conformational differences between open and closed states of the eukaryotic translation initiation complex. *Mol Cell* **59**: 399–412.

Loftfield RB. 1963. The frequency of errors in protein biosynthesis. *Biochem J* **89**: 82–92.

Los GV, Encell LP, McDougall MG, Hartzell DD, Karassina N, Zimprich C, Wood MG, Learish R, Ohana RF, Urh M, et al. 2008. HaloTag: A novel protein labeling technology for cell imaging and protein analysis. *ACS Chem Biol* **3**: 373–382.

Marshall RA, Aitken CE, Dorywalska M, Puglisi JD. 2008a. Translation at the single-molecule level. *Annu Rev Biochem* **77**: 177–203.

Marshall RA, Dorywalska M, Puglisi JD. 2008b. Irreversible chemical steps control intersubunit dynamics during translation. *Proc Natl Acad Sci* **105**: 15364–15369.

Marshall RA, Aitken CE, Puglisi JD. 2009. GTP hydrolysis by IF2 guides progression of the ribosome into elongation. *Mol Cell* **35**: 37–47.

Matheisl S, Berninghausen O, Becker T, Beckmann R. 2015. Structure of a human translation termination complex. *Nucleic Acids Res* **43**: 8615–8626.

* Merrick WC, Pavitt GD. 2018. Protein synthesis in eukaryotic cells. *Cold Spring Harb Perspect Biol* doi: 10.1101/cshperspect.a32714.

Michalet X, Kapanidis AN, Laurence T, Pinaud F, Doose S, Pflughoefft M, Weiss S. 2003. The power and prospects of fluorescence microscopies and spectroscopies. *Annu Rev Biophys Biomol Struct* **32**: 161–182.

Milon P, Maracci C, Filonava L, Gualerzi CO, Rodnina MV. 2012. Real-time assembly landscape of bacterial 30S translation initiation complex. *Nat Struct Mol Biol* **19**: 609–615.

Murray J, Savva CG, Shin BS, Dever TE, Ramakrishnan V, Fernandez IS. 2016. Structural characterization of ribosome recruitment and translocation by type IV IRES. *eLife* **5**: e13567.

Nanda JS, Lorsch JR. 2014. Labeling of a protein with fluorophores using maleimide derivitization. *Methods Enzymol* **536**: 79–86.

Noller HF, Lancaster L, Mohan S, Zhou J. 2017a. Ribosome structural dynamics in translocation: Yet another functional role for ribosomal RNA. *Q Rev Biophys* **50**: e12.

Noller HF, Lancaster L, Zhou J, Mohan S. 2017b. The ribosome moves: RNA mechanics and translocation. *Nat Struct Mol Biol* **24**: 1021–1027.

O'Connor M. 2015. Interactions of release factor RF3 with the translation machinery. *Mol Genet Genomics* **290**: 1335–1344.

Ogle JM, Ramakrishnan V. 2005. Structural insights into translational fidelity. *Annu Rev Biochem* **74**: 129–177.

Pape T, Wintermeyer W, Rodnina MV. 1998. Complete kinetic mechanism of elongation factor Tu-dependent binding of aminoacyl-tRNA to the A site of the *E. coli* ribosome. *EMBO J* **17**: 7490–7497.

Passmore LA, Schmeing TM, Maag D, Applefield DJ, Acker MG, Algire MA, Lorsch JR, Ramakrishnan V. 2007. The eukaryotic translation initiation factors eIF1 and eIF1A induce an open conformation of the 40S ribosome. *Mol Cell* **26**: 41–50.

Pavlov MY, Ehrenberg M. 1996. Rate of translation of natural mRNAs in an optimized in vitro system. *Arch Biochem Biophys* **328**: 9–16.

Pavlov MY, Freistroffer DV, MacDougall J, Buckingham RH, Ehrenberg M. 1997. Fast recycling of *Escherichia coli* ribosomes requires both ribosome recycling factor (RRF) and release factor RF3. *EMBO J* **16**: 4134–4141.

Pedersen S. 1984. *Escherichia coli* ribosomes translate in vivo with variable rate. *EMBO J* **3**: 2895–2898.

Perez CE, Gonzalez RL Jr. 2011. In vitro and in vivo single-molecule fluorescence imaging of ribosome-catalyzed protein synthesis. *Curr Opin Chem Biol* **15**: 853–863.

Peske F, Rodnina MV, Wintermeyer W. 2005. Sequence of steps in ribosome recycling as defined by kinetic analysis. *Mol Cell* **18**: 403–412.

Peske F, Kuhlenkoetter S, Rodnina MV, Wintermeyer W. 2014. Timing of GTP binding and hydrolysis by translation termination factor RF3. *Nucleic Acids Res* **42**: 1812–1820.

Petrov A, Puglisi JD. 2010. Site-specific labeling of *Saccharomyces cerevisiae* ribosomes for single-molecule manipulations. *Nucleic Acids Res* **38**: e143.

Petrov A, Chen J, O'Leary S, Tsai A, Puglisi JD. 2012. Single-molecule analysis of translational dynamics. *Cold Spring Harb Perspect Biol* **4**: a011551.

Petrov A, Grosely R, Chen J, O'Leary SE, Puglisi JD. 2016. Multiple parallel pathways of translation initiation on the CrPV IRES. *Mol Cell* **62**: 92–103.

Petry S, Weixlbaumer A, Ramakrishnan V. 2008. The termination of translation. *Curr Opin Struct Biol* **18**: 70–77.

Prabhakar A, Capece MC, Petrov A, Choi J, Puglisi JD. 2017a. Post-termination ribosome intermediate acts as the gateway to ribosome recycling. *Cell Rep* **20**: 161–172.

Prabhakar A, Choi J, Wang J, Petrov A, Puglisi JD. 2017b. Dynamic basis of fidelity and speed in translation: Coordinated multistep mechanisms of elongation and termination. *Protein Sci* **26**: 1352–1362.

* Rodnina MV. 2018. Translation in prokaryotes. *Cold Spring Harb Perspect Biol* doi: 10.1101/cshperspect.a032664.

Rodnina MV, Fricke R, Wintermeyer W. 1994. Transient conformational states of aminoacyl-tRNA during ribosome binding catalyzed by elongation factor Tu. *Biochemistry* **33**: 12267–12275.

Rodnina MV, Fischer N, Maracci C, Stark H. 2017. Ribosome dynamics during decoding. *Philos Trans R Soc Lond B Biol Sci* **372**: 20160182.

Roundtree IA, Evans ME, Pan T, He C. 2017. Dynamic RNA modifications in gene expression regulation. *Cell* **169**: 1187–1200.

Schluenzen F, Tocilj A, Zarivach R, Harms J, Gluehmann M, Janell D, Bashan A, Bartels H, Agmon I, Franceschi F, et al. 2000. Structure of functionally activated small ribosomal subunit at 3.3 angstroms resolution. *Cell* **102**: 615–623.

Schmeing TM, Ramakrishnan V. 2009. What recent ribosome structures have revealed about the mechanism of translation. *Nature* **461**: 1234–1242.

Selvin PR. 2000. The renaissance of fluorescence resonance energy transfer. *Nat Struct Biol* **7**: 730–734.

Shajani Z, Sykes MT, Williamson JR. 2011. Assembly of bacterial ribosomes. *Annu Rev Biochem* **80:** 501–526.

Shoemaker CJ, Green R. 2011. Kinetic analysis reveals the ordered coupling of translation termination and ribosome recycling in yeast. *Proc Natl Acad Sci* **108:** E1392–E1398.

Solomatin S, Herschlag D. 2009. Methods of site-specific: Labeling of RNA with fluorescent dyes. *Methods Enzymol* **469:** 47–68.

Sternberg SH, Fei J, Prywes N, McGrath KA, Gonzalez RL Jr. 2009. Translation factors direct intrinsic ribosome dynamics during translation termination and ribosome recycling. *Nat Struct Mol Biol* **16:** 861–868.

Studer SM, Feinberg JS, Joseph S. 2003. Rapid kinetic analysis of EF-G-dependent mRNA translocation in the ribosome. *J Mol Biol* **327:** 369–381.

Taylor D, Unbehaun A, Li W, Das S, Lei J, Liao HY, Grassucci RA, Pestova TV, Frank J. 2012. Cryo-EM structure of the mammalian eukaryotic release factor eRF1-eRF3-associated termination complex. *Proc Natl Acad Sci* **109:** 18413–18418.

Thommen M, Holtkamp W, Rodnina MV. 2017. Co-translational protein folding: Progress and methods. *Curr Opin Struct Biol* **42:** 83–89.

Tsai A, Petrov A, Marshall RA, Korlach J, Uemura S, Puglisi JD. 2012. Heterogeneous pathways and timing of factor departure during translation initiation. *Nature* **487:** 390–393.

Uemura S, Aitken CE, Korlach J, Flusberg BA, Turner SW, Puglisi JD. 2010. Real-time tRNA transit on single translating ribosomes at codon resolution. *Nature* **464:** 1012–1017.

Valle M, Zavialov A, Sengupta J, Rawat U, Ehrenberg M, Frank J. 2003. Locking and unlocking of ribosomal motions. *Cell* **114:** 123–134.

Voorhees RM, Schmeing TM, Kelley AC, Ramakrishnan V. 2010. The mechanism for activation of GTP hydrolysis on the ribosome. *Science* **330:** 835–838.

Wen JD, Lancaster L, Hodges C, Zeri AC, Yoshimura SH, Noller HF, Bustamante C, Tinoco I. 2008. Following translation by single ribosomes one codon at a time. *Nature* **452:** 598–603.

Wimberly BT, Brodersen DE, Clemons WM Jr, Morgan-Warren RJ, Carter AP, Vonrhein C, Hartsch T, Ramakrishnan V. 2000. Structure of the 30S ribosomal subunit. *Nature* **407:** 327–339.

Wintermeyer W, Peske F, Beringer M, Gromadski KB, Savelsbergh A, Rodnina MV. 2004. Mechanisms of elongation on the ribosome: Dynamics of a macromolecular machine. *Biochem Soc Trans* **32:** 733–737.

Yin J, Lin AJ, Golan DE, Walsh CT. 2006. Site-specific protein labeling by Sfp phosphopantetheinyl transferase. *Nat Protoc* **1:** 280–285.

Young R, Bremer H. 1976. Polypeptide-chain-elongation rate in *Escherichia coli* B/r as a function of growth rate. *Biochem J* **160:** 185–194.

Zavialov AV, Buckingham RH, Ehrenberg M. 2001. A post-termination ribosomal complex is the guanine nucleotide exchange factor for peptide release factor RF3. *Cell* **107:** 115–124.

Fluorescence Imaging Methods to Investigate Translation in Single Cells

Jeetayu Biswas,[1,5] Yang Liu,[2,5] Robert H. Singer,[1,3,4] and Bin Wu[2]

[1]Department of Anatomy and Structural Biology, Albert Einstein College of Medicine, Bronx, New York 10461

[2]Department of Biophysics and Biophysical Chemistry, Center for Cell Dynamics, Department of Neuroscience, Johns Hopkins School of Medicine, Baltimore, Maryland 21218

[3]Gruss Lipper Biophotonics Center, Albert Einstein College of Medicine, Bronx, New York 10461

[4]Howard Hughes Medical Institute, Janelia Research Campus, Ashburn, Virginia 20147

Correspondence: bwu20@jhmi.edu

Translation is the fundamental biological process that converts the genetic information in messenger RNAs (mRNAs) into functional proteins. Translation regulation allows cells to control when, where, and how many proteins are synthesized. Much of what we know about translation comes from ensemble approaches that measure the average of many cells. The cellular and molecular heterogeneity in the regulation of translation remains largely elusive. Fluorescence microscopy allows interrogation of biological problems with single-molecule, single-cell sensitivity. In recent years, improved design of reagents and microscopy tools has led to improved spatial and temporal resolution of translation imaging. It is now possible to track global translation in specific subcellular compartments and follow the translation dynamics of single transcripts. Highlighted here is the recent progress in translation imaging with emphasis on in vivo translation dynamics. These tools will be invaluable to the study of translation regulation.

Translation of messenger RNAs (mRNAs) into proteins is precisely controlled in space and time. During development, many mRNAs localize to specific subcellular regions. During *Drosophila* embryogenesis, for instance, over 70% of tested mRNAs are compartmentalized (Lecuyer et al. 2007). To maintain proteome homeostasis, cells respond to environmental cues by translating existing mRNAs without drastically altering transcription rates. For example, mRNAs encoding ribosomal proteins are highly enriched at the apical surface of fasted intestinal epithelium cells and feeding induces a global basal-to-apical shift of mRNAs and increased translation efficiency (Moor et al. 2017). In neurons, high-resolution fluorescence in situ hybridization (FISH) data show a tight correlation between the spatial distribution of mRNAs and their final protein products, suggesting a significant role of mRNAs in protein localization. Deep RNA sequencing (RNA-Seq) has identified more than 2500 mRNAs localized in neuronal dendrites and axons (Cajigas et al. 2012) and, when combined with proteomics, has

[5]These authors contributed equally to this work.

shown that neurite-targeted RNAs encode approximately half of the neurite-localized proteome (Zappulo et al. 2017). Local translation regulates many neuronal processes such as growth cone motility and synaptic plasticity (Holt and Schuman 2013). For instance, treatment with nerve growth factor (NGF) stimulates axonal elongation and pathfinding by trafficking and locally translating Par3 mRNAs in developing axons (Hengst et al. 2009). To understand these molecular processes and biological functions, it is necessary to track when and where specific mRNAs are translated and how newly synthesized proteins are distributed.

Much of our knowledge about the mechanisms underlying translation derives from standard and sophisticated biochemical and molecular biology methods. Various proteins and nucleic acids involved in translation have been purified and reconstituted for in vitro mechanistic studies (Chen et al. 2016). Recently, high-throughput assays such as ribosome profiling have revolutionized our understanding of translation by providing the global translation status of mRNAs (Brar and Weissman 2015; Ingolia et al. 2018). In addition, stable isotope labeling with amino acids in cell culture (SILAC) allows newly synthesized proteins to be quantified genome-wide by quantitative mass spectrometry (Schwanhausser et al. 2011). Although powerful and successful, these ensemble measurements detect overall changes in protein synthesis across millions of cells. Translational information regarding spatiotemporal kinetics, especially cell-to-cell and molecule-to-molecule heterogeneity, have remained largely elusive because of the lack of molecular tools with sufficient resolution and sensitivity.

Fluorescence microscopy has emerged as an ideal tool to complement the aforementioned ensemble approaches. Instead of taking a snapshot of a biological event, imaging studies the translation process in real time with a temporal resolution of milliseconds to hours. The entire process of translation—initiation, elongation, pausing, and eventual termination—can be examined with imaging. Fluorescence imaging records the location of translation with high spatial resolution comparable to the size of a ribo-

some (diameter 25–30 nm). Furthermore, the distribution of newly synthesized protein can be followed in vivo, which offers significant insight into the molecular mechanism. Recent technological advances in fluorescence microscopy have made it possible to visualize single molecules. Fluorescently tagged ribosomes, mRNAs, and newly synthesized proteins have been individually labeled and tracked with single-molecule sensitivity.

In this review, we highlight recently developed microscopy techniques for translation imaging in single cells. These techniques are classified as either pulse-chase approaches or genetically encoded approaches (as summarized in Table 1). We compare these methods and provide a brief outlook regarding possible improvements and new insights that may arise from them.

PULSE-CHASE-BASED APPROACHES

To study the synthesis of protein products, one has to distinguish newly synthesized proteins from preexisting ones. A classical method for achieving this is through the pulse-chase approach, in which newly synthesized proteins are specifically tagged. With this approach, the localization of newly synthesized proteins and the distribution of translation-rich regions (hot spots) can be determined. We broadly divide these techniques into genetically encoded and nongenetically encoded groups (Fig. 1A,B).

Nongenetically Encoded Approaches

The nongenetic imaging approaches (Fig. 1A) allow visualization of the translation of endogenous mRNAs without any genetic modification. These approaches are extensively used to study the spatial distribution of the translation sites and newly produced proteins.

Classically, isotopically labeled amino acids have been powerful tools to study global protein synthesis. The highly specific vibrational energy of a carbon–deuterium bond is amenable to measurement by stimulated Raman scattering (SRS) microscopy (Wei et al. 2013) and pulsed metabolic incorporation of deuterated amino

Table 1. Translation imaging techniques

Techniques	Fixed	Live	Endogenous	Exogenous reporter	Perturbations	Features	References
FUNCAT/PLAs	X		X		Orthologous tRNA system or puromycin	Measure the co-association of endogenous transcripts with newly synthesized proteins; commercial kits available for visualization	tom Dieck et al. 2015
TimeSTAMP 2		X		X	Bicuculline, BDNF, DHPG	Track production of new protein reporter over time	Butko et al. 2012
HaloTag	X	X		X	Photocaged glutamate	Single-molecule protein tracking with the ability to pulse-chase and visualize newly synthesized proteins	Grimm et al. 2015, 2016; Yoon et al. 2016
Ribosome granule imaging	X		X		cLTP	Allows one to determine the masking efficiency (i.e., how much do ribosomes mask specific transcripts); complementary to techniques such as ribosome profiling	Buxbaum et al. 2014
FLAIRM	X		X		Iron supplementation	Assess transcript abundance and association with ribosomes in the same FISH reaction	Burke et al. 2017
Gaussia luciferase		X		X	mGluR and AMPA antagonists	Live-cell visualization of new translation as bioluminescence signal; can then be overlaid onto fluorescence with a modified microscope	Na et al. 2016
Ribosome cotracking		X		X	Hippuristanol, puromycin	Visualize changes in ribosome comovement with RNA	Katz et al. 2016
FCS heterospecies analysis		X		X	Hippuristanol	Interaction between ribosomes and mRNAs at the single-molecule level using two-photon microscopy; allows for single-molecule counting	Wu et al. 2010, 2015
TRICK		X		X	Puromycin, cycloheximide, arsenite	Visualize the translational status of a single mRNA whether or not it has undergone the pioneer round	Halstead et al. 2015
SINAPS and NCT	X	X	X	X	Puromycin, cycloheximide, harringtonine, arsenite	Direct imaging of translation products; allows measurement of nascent peptide dynamics (initiation, elongation, and stalling rates)	Morisaki et al. 2016; Pichon et al. 2016; Wang et al. 2016; Wu et al. 2016; Yan et al. 2016

FUNCAT, Fluorescent noncanonical amino acid tagging; PLAs, pulse-chase approaches; tRNA, transfer RNA; BDNF, brain-derived neurotrophic factor; cLTP, chemical long-term potentiation; FLAIRM, fluorescence assay to detect ribosome interactions with mRNA; FISH, fluorescence in situ hybridization; FCS, fluorescence correlation spectroscopy; mRNA, messenger RNA; TRICK, translating RNA imaging by coat protein knock-off; SINAPS, single-molecule imaging of nascent peptides; NCT, nascent chain tracking.

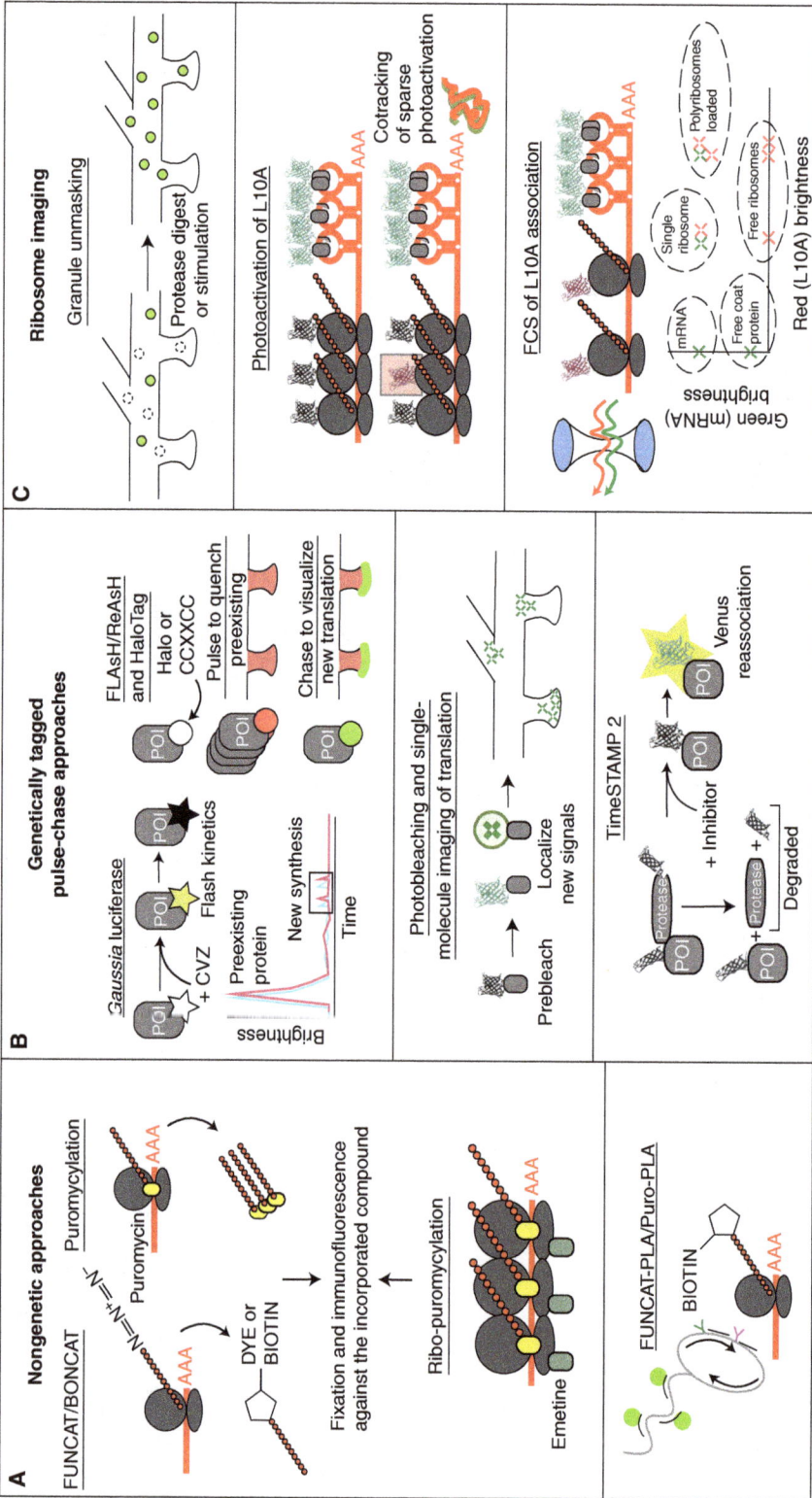

Figure 1. (*See legend on following page.*)

Cite this article as *Cold Spring Harb Perspect Biol* doi: 10.1101/cshperspect.a032722

acid allows nascent protein synthesis to be visualized in live cells. Newly synthesized proteins are highly concentrated in nucleoli, the sites of ribosome biogenesis. The production of ribosomes is a major task in rapidly growing cells and the majority of the newly synthesized nucleolar-localized proteins are presumably ribosomal proteins that are trafficked from the cytoplasm, in agreement with previous observations of fast proteomic turnover in nucleoli (Beatty et al. 2006; Boisvert et al. 2012). The technique has also been generalized to other vibrational tags such as alkynes, allowing protein, RNA, and DNA synthesis to be tracked simultaneously in live cells (Wei et al. 2014).

In fluorescent noncanonical amino acid tagging (FUNCAT; Fig. 1A), azidohomoalanine (AHA), a methionine analog, is incorporated into all newly synthesized proteins in methionine-starved cells. Subsequent fluorescent labeling by a copper-catalyzed click chemical reaction or immunofluorescence allows visualization of the newly synthesized proteins (Dieterich et al. 2010). Puromycylation uses a similar conceptual approach (Fig. 1A). Instead of using methionine-specific labeling, nascent protein synthesis is truncated by low-dose treatment with puromycin, which enters the ribosomal A site and forms a peptide bond with the nascent chain (Starck et al. 2004). The incorporation of puromycin into the newly synthesized nascent proteins is then visualized with fluorescent anti-puromycin antibodies (Schmidt et al. 2009). Alternatively, an alkyne analog of puromycin, O-propargyl-puromycin (OP-puro) can form covalent conjugates with nascent peptides, which can be visualized by an azide-labeled fluorophore via click chemistry (Liu et al. 2012).

FUNCAT has been used to observe local translation in different neuronal compartments. Upon stimulation of brain-derived neurotrophic factor, local production of proteins, such as CamKIIα, increases in postsynaptic dendrites (Dieterich et al. 2010). By combining FUNCAT with a microfluidic device that allows axons to be physically isolated from other neuronal compartments, stimulation-dependent local translation was observed in both axons and dendrites

Figure 1. Techniques to visualize translation at the single-cell and single-molecule level. (A) Both puromycylation and noncanonical amino acid tagging (NCAT) techniques incorporate a small molecule ligand into the protein of interest. For puromycylation, this is a carboxy-terminal puromycin (yellow oval), and for NCAT techniques, this is an amino-terminal azide group that is incorporated into the nascent chain and then can be conjugated to a dye or biotin moiety. To visualize the location of translating polysomes, emetine can be added before fixation (green oval) to allow for puromycin incorporation but prevent ribosome dissociation from the transcript. Conversely, one can use two oligonucleotide-conjugated antibodies to create a bridge for rolling circle amplification and the coincident locations can then be identified by fluorescent dye (green circles)-labeled fluorescence in situ hybridization (FISH) probes. (B) Small molecule dyes or substrates can be used to quench the previously synthesized proteins of interest. For HaloTag and FLAsH ReAsH, the second optically separated dye (green circle) identifies the newly translated proteins. For *Gaussia* luciferase, the substrate (yellow star) is rapidly inactivated (black star) leading to previously synthesized proteins being irreversibly in the dark state and newly synthesized proteins being observed as small flashes of luminescence (small peaks on graph). One can also use continuous photobleaching of fluorescent proteins to isolate only newly synthesized signal. By repeatedly localizing the proteins of interest, it is possible to create maps depicting the translation hot spots (be they at the synapse, neck of the spine, or at dendritic branch points). A similar dark-to-bright approach can be used when a protease site is placed between the two halves of Venus. Normally, continuous protease degradation prevents the two halves of Venus from associating. Upon addition of an inhibitor, one can visualize newly formed proteins by fluorescence rescue (yellow star). (C) Direct and indirect imaging of ribosomes. FISH probes have limited permeability because of the density of granule components in neuronal dendrites, and after chemical long-term potentiation (LTP) or protease digestion granule unmasking exposes an increased number of messenger RNAs (mRNAs) (green circles). Sparse photoactivation of tagged ribosomes can allow for multicolor cotracking of ribosomes associated with single mRNA molecules. Fluorescence correlation spectroscopy (FCS) allows accurate counting of individual molecules as they traverse the femtoliter excitation volume. Heterospecies partition analysis allows one to determine the percentages of transcripts that are associated with ribosomes as well as count the number of ribosomes associated with each transcript.

(Kos et al. 2016). Others have applied FUNCAT to measure translation in brain slices, showing that it occurs in both pre- and postsynaptic compartments. Using this approach, it was found that presynaptic, cap-dependent protein synthesis is required for long-term but not short-term plasticity of a specific population of GABA-releasing axons (Younts et al. 2016). Injection of OP-puro into mice allows protein synthesis to be tracked in whole organisms (Liu et al. 2012). In the mouse intestine, OP-puro strongly stained the cells in the crypts and at the base of the villi and was strongly incorporated into striated muscle fibers.

To observe translating ribosomes, the translation elongation inhibitor emetine was added in combination with puromycin to prevent release of the nascent chain (David et al. 2012; Graber et al. 2013). Using this combined inhibition, dubbed "ribopuromycylation" (Fig. 1A), David et al. (2012) observed a nuclear translation signal. By studying the process in intact cells and avoiding possible contamination from endoplasmic reticulum (ER)-associated ribosomes, the investigators proposed the existence of nuclear translation, a controversial observation needing further independent validation.

Both FUNCAT and puromycylation measure the bulk translation status of the entire proteome rather than the production of specific proteins. Recently, a proximity ligation assay was combined with pulse-chase approaches (FUNCAT-PLA or Puro-PLA) to overcome this disadvantage (tom Dieck et al. 2015). In this work, cells are fixed and incubated with two different primary antibodies. The first antibody detects a protein of interest and the second detects biotin or puromycin, which is followed by the addition of two secondary antibodies conjugated with corresponding DNA oligonucleotides. Only when both secondary antibodies are bound to the same molecule (in close proximity, 10–20 nm), is an enzymatic reaction named rolling circle amplification (RCA) initiated to produce thousands of identical copies of DNA sequences (Fig. 1A, Table 1). Through this strategy, newly synthesized proteins are visualized by fluorescently labeled probes binding to RCA products.

The synthesis of new proteins is required for stabilizing synapses during long-term potentiation and long-term depression. FUNCAT-PLA has been used to visualize local translation of proteins in the soma as well as the neurites. The presynaptically localized protein Bassoon was thought to be translated in the soma and posttranslationally localized to presynapses. FUNCAT-PLA showed that newly synthesized Bassoon was locally made in the presynaptic bouton (tom Dieck et al. 2015). The ability to pulse-chase at different time points has allowed visualization of proteins that are locally translated in the soma and then subsequently localized to the neurites, such as the protein Barentsz, which is a component of the RNA localization machinery (Zappulo et al. 2017).

Genetically Tagged Pulse-Chase

Genetic modification allows incorporation of fluorescent tags into target genes, which gives rise to highly specific fluorescence signals (Fig. 1B). Recently, genetically encoded protein motifs have been designed to bind different ligands. Researchers use this property to distinguish newly synthesized proteins from preexisting ones by pulse-chase labeling with different fluorophores. An example is the tetracysteine motif CCXXCC (where X represents any amino acid), which binds biarsenical dyes with picomolar affinity and converts them from a nonfluorescent to a fluorescent state (Griffin et al. 1998). In a FlAsH-ReAsH system, green (FlAsH) and red (ReAsH) dyes were used in a pulse-chase fashion to determine that β-actin translation sites are more frequently located at cell contacts (Fig. 1B) (Rodriguez et al. 2006). The tiny tetracysteine tag causes minimal perturbation to the function of proteins of interest. However, the arsenical dyes suffer from low brightness and are toxic to cells and can only be used for short-term experiments with relative low resolution. More recently, covalent labeling techniques such as HaloTag (as well as SNAP-tag, CLIP, and TMP) (Gautier et al. 2008; Los et al. 2008; Gallagher et al. 2009) allows irreversible conjugation of bright organic fluorophores such as the Janelia Fluors (JF dyes) (Grimm et al.

2015, 2016). These dyes are 10× brighter than green fluorescent protein (GFP) and allow single-molecule detection in live cells. Based on pulse-chase labeling of HaloTag ligands, translation of newly synthesized proteins has been visualized during synaptic stimulation (Yoon et al. 2016). Caged glutamate molecules (4-methoxy-7-nitroindolinyl-moiety) were photolyzed with a 405 nm laser to stimulate single dendritic spines, which induces rapid localization of endogenous β-actin mRNAs. After preexisting β-actin-HaloTag reporter proteins were covalently labeled to saturation with the far-red dye JF-646, a second HaloTag-ligand (the near-red dye JF-549) was used to detect newly synthesized proteins (Fig. 1B, Table 1) (Grimm et al. 2015). The investigators showed that synaptic activity induces local translation of β-actin, which plays a critical role in stabilizing the growth of dendritic spines.

In addition to traditional pulse-chase approaches that use small molecules as ligands, fluorescent proteins have been used to track newly synthesized proteins. An early approach to visualize translation observed changes in fluorescence intensity over time (Aakalu et al. 2001), which has limited capability to distinguish old proteins from newly synthesized ones. To overcome this limitation, different strategies such as photoconversion or photobleaching have been used. These light-up tags include photoconvertible proteins, fluorescent timers, and Time-STAMP (discussed briefly here as well as in many other reviews [Wu et al. 2011; Zhou and Lin 2013]). To visualize translation of the mRNA encoding the peptide neurotransmitter Sensorin, its 5′ and 3′ untranslated regions (UTRs) were fused to the coding region of Dendra2 (a photoconvertible protein). Conversion of the preexisting proteins (from green to red) with a pulse of ultraviolet (UV) illumination allowed newly synthesized proteins to be visualized in *Aplysia* motor neurons. Translation of this reporter was found to be highly regulated, requiring calcium in the postsynaptic motor neurons, and to occur during long-term facilitation but not during long-term depression (Wang et al. 2009). More recently, the rapid maturation of a yellow fluorescent protein mutant

(named Venus) has been exploited to visualize translation at the single-molecule level. By photobleaching the preexisting Venus fusion protein, it has been possible to localize and observe short bursts of fluorescence from newly synthesized reporter proteins (Yu et al. 2006). By mapping spot localizations over time, translation maps, which mark sites of new protein synthesis, have been made for Arc (activity-related cytoskeleton-associated protein), fragile X mental retardation protein (FMRP), postsynaptic density protein 95 (PSD-95) and β-actin (Fig. 1B). The clustering of translation events was found, either near granules for FMRP and Arc (Tatavarty et al. 2012), synapses for PSD-95 (Ifrim et al. 2015), or dendritic branch points for β-actin (Wong et al. 2017). When using this technique, one must carefully control the cyclical "blinking" of fluorescent proteins.

In addition to reducing preexisting protein signals by photobleaching (or photoconversion), small-molecule-controlled degradation has been adopted as another strategy to remove previously translated proteins. In TimeSTAMP, the nonstructural protease 3 (NS3) of hepatitis C virus is fused to a protein of interest to degrade it by default. Addition of a protease inhibitor stabilizes the target protein and makes it detectable (Lin et al. 2008). TimeSTAMP 2 uses a split fluorescent protein Venus and irreversible bimolecular fluorescence complementation for live cell translation imaging (Butko et al. 2012). With the addition of the protease inhibitor, the two halves of Venus have enough time to come together and the appearance of the fluorescent signal can be used to track newly translated proteins (Fig. 1B, Table 1).

GENETICALLY ENCODED REPORTERS TO VISUALIZE RNAs BEFORE, AFTER, AND DURING TRANSLATION

Fluorescence techniques have been developed to visualize ribosomes and mRNAs at the single-molecule level. FISH is able to detect single endogenous mRNAs or ribosomes in a fixed sample using multiple fluorescently labeled oligonucleotides directed against target transcripts or ribosomal RNAs (rRNAs) (Buxbaum et al.

2014). To image ribosomes or RNA in live cells, genetically encoded fluorescent tags are used. To visualize single mRNAs, MS2-binding sites (MBS) derived from bacteriophage MS2, are inserted as an array of 24 repeats into the 3′UTR of the target gene (Bertrand et al. 1998). Multiple GFP-fused MS2 coat proteins (MCPs) bind to MBS and allow single-mRNA detection. Recently, modified MS2 stem-loops have been developed to follow unstable mRNAs from transcription to degradation (Tutucci et al. 2017). In addition to MS2, other RNA-binding proteins (Daigle and Ellenberg 2007; Chao et al. 2008) are also used for single-molecule tracking. In the next section, we discuss the application of these fluorescence labeling technologies in translation imaging.

IMAGING RIBOSOMES AS A READOUT FOR TRANSLATION

RNAs are complexed with proteins and form cellular particles, ribonucleoproteins (RNPs), and neuronal RNA granules. Granules contain ribosomal subunits but are devoid of initiation factors and tRNAs necessary for active translation (Krichevsky and Kosik 2001). By sequestering RNAs and the required translational machinery into a repressed transporting granule, translational control is achieved both in time and space by relieving the repression (see also Ivanov et al. 2018). It is still unclear how ribosomes interact with the RNA inside the granule. Possibly, the granule represents the accumulation of previously initiated but stalled ribosomes (Graber et al. 2013). Masking of RNA transcripts has long been thought to control translation (Anderson and Kedersha 2006; Kiebler and Bassell 2006; Ivanov et al. 2018). Chemical stimulation rapidly and transiently increases the number of β-actin mRNAs in the neuronal dendrite observable by FISH (Fig. 1C, Table 1). FISH against rRNA showed that β-actin mRNA was indeed present in a masked state and that stimulation lowered the rRNA density around the mRNA (granule unmasking) (Buxbaum et al. 2014).

Techniques that visualize the interaction of specific mRNAs with ribosomes have been de-

veloped. Fluorescence assay to detect ribosome interactions with mRNA (FLAIRM) uses FISH to detect ribosomes in close proximity with the target mRNAs (Burke et al. 2017). A linker oligonucleotide bridges FISH probes targeted to ribosomes and to mRNAs. Signal amplification is achieved by the hybridization chain reaction only when the ribosomes and mRNAs are close to each other. The investigators used this technique to simultaneously track changes in the abundance of the endogenous ferritin heavy chain transcript as well as its ribosome association. They found that, on addition of exogenous iron, transcript abundance increased only modestly, but the mRNA–ribosome interaction increased significantly, indicating enhanced translation.

In live cells, the association between ribosome and mRNA can be inferred if they move together. Abundant ribosomal signals preclude tracking of a single ribosome. One approach is to label ribosomal protein with a photoactivatable fluorescent tag. After sparse photoactivation, single ribosomes can be localized with ∼20 nm precision (Fig. 1C) and the coordinate movements of photoactivated ribosomes and mRNAs tracked. This allowed Katz et al. (2016) to infer when and where translation occurred. The slower, corralled movements of the mRNA-ribosome complex correlate with more active translation of β-actin mRNA around the focal adhesion complex. The association between ribosomes and mRNAs has also been studied with fluorescence fluctuation spectroscopy (FFS). A focused two-photon laser captures bursts of fluorescence when single molecules diffuse in and out of the femtoliter excitation volume. When two interacting molecules pass through the observation volume, their fluorescence signals are detected simultaneously on different detectors (Fig. 1C, Table 1). Statistical analysis is applied to extract interaction information between molecules. By labeling β-actin mRNA in one color and ribosomes in another, and performing subsequent heterospecies partition analysis (Wu et al. 2010), the average numbers of ribosomes associated with mRNAs can be measured (Wu et al. 2015). More ribosomes associate with β-actin mRNA at the periphery of the cell than

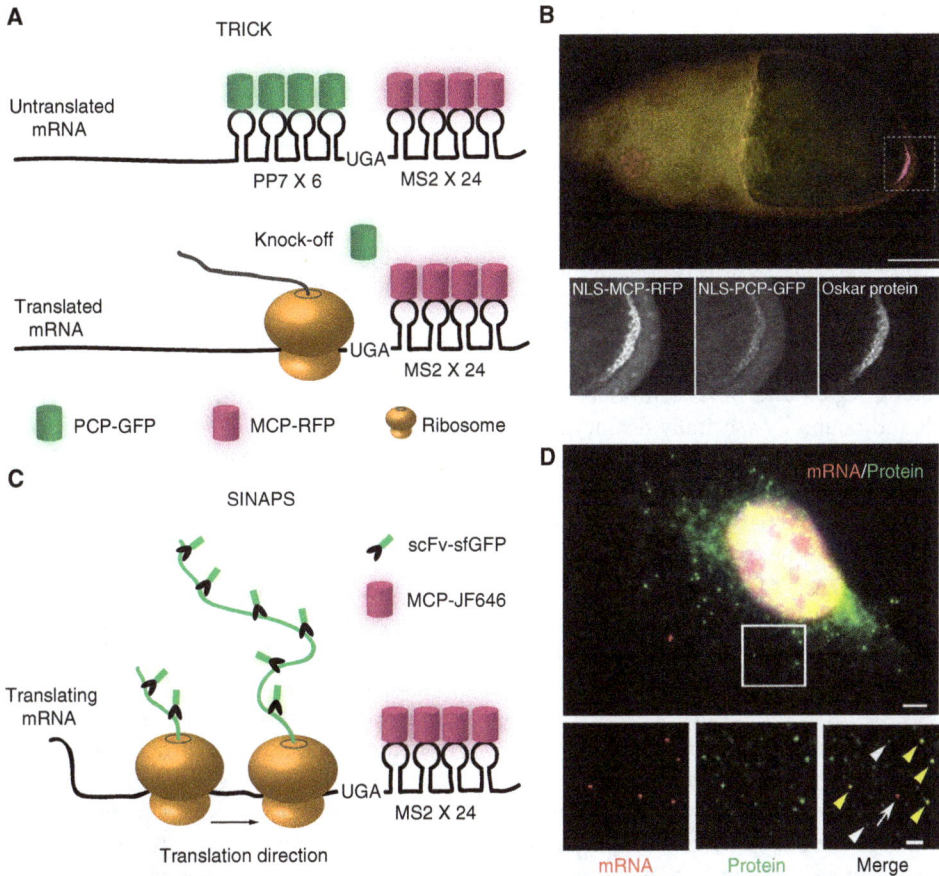

Figure 2. Real-time translation imaging using fluorescent reporter systems. (*A*) Working principle of translating RNA imaging by coat protein knock-off (TRICK). Messenger RNAs (mRNAs) are fused with PP7 bacteriophage stem-loops in the coding region and MS2 stem-loops in the 3′ untranslated region (UTR), to which PP7 coat proteins (PCPs)-GFP (green) and MS2 coat proteins (MCPs)-RFP (red) are bound, respectively. During the pioneer round of translation, ribosomes displace the PCP-GFP and, as a result, a subpopulation of mRNAs lose their green fluorescent signal and retain only their red fluorescent signal, indicating that they had been translated at least once. (*B*) TRICK imaging of Oskar protein expression at the pole of stage 10 *Drosophila* oocyte. Scale bar, 50 μm. (Panel *B* from Halstead et al. 2015; adapted, with permission, from HHS Public Access and *Science* © 2015.) (*C*) Working principle of single-molecule imaging of nascent peptides (SINAPS). Genes of interest are fused with multiple peptidyl epitopes recognized by specific GFP-tagged antibodies. When the nascent peptides emerge from ribosomes, coexpressed fluorescent antibodies (green) spontaneously bind to these tandem epitope repeats, yielding a bright fluorescent spot colocalizing with the MS2/PP7 tagged mRNAs (red). (*D*) SINAPS imaging of translation heterogeneity in living U2OS cells. Translating mRNAs appear in both green (scFv-GFP) and red (MCP-RFP) channels, whereas nontranslating mRNAs appear only in red because of the absence of antibodies. Scale bars, 5 μm (*top*); 2 μm (*lower right*). (Panel *D* from Wu et al. (2016); adapted, with permission, from HHS Public Access and *Science* © 2016.)

in the perinuclear region. This is anticorrelated with the mRNA binding to a well-known translational repressor, the RNA-binding protein, zipcode-binding protein ZBP1, showing how a *trans*-acting factor can regulate the subcellular location of translation.

DYNAMIC APPROACHES TO MEASURE TRANSLATION KINETICS IN LIVE CELLS

Recent technological advances make it possible to measure translation dynamics of single mRNAs in live cells. Halstead et al. (2015) developed a biosensor to directly visualize the first round of translation based on multicolor RNA labeling. The first round of translation, which could happen on both the cap-binding complex (CBC)- and eIF4E-bound mRNAs (Rufener and Mühlemann 2013), is critical for short-lived and responsive proteins (such as those of the cellular immediate early genes) and has been ascribed a role in mRNA quality control through nonsense-mediated mRNA decay (reviewed by Karousis and Mühlemann 2018). Specific mRNAs were tagged with bacteriophage PP7 stem-loops in the coding region and MS2 stem-loops in the 3′UTR and bound by spectrally distinct capsid proteins (green and red, respectively). Nuclear mRNAs and untranslated cytoplasmic mRNAs have both green and red colors. During the first round of translation, ribosomes displace the GFP-fused PP7 capsid protein (PCP-GFP). As a result, mRNAs that have been translated at least once will carry a single color (red, MCP-red fluo-rescent protein [RFP]). Interestingly, spatial profiling of these transcripts suggested many mRNAs underwent initial translation away from the nucleus. In contrast, when translation inhibitors (cycloheximide or puromycin) were preincubated with the cells, over 90% of cytoplasmic mRNAs retained both capsid proteins and were fluorescent in both colors, indicating they remained untranslated. The translation of mRNAs in special cell compartments such as stress granules can be seen using this biosensor, a technique termed translating RNA imaging by coat protein knock-off (Halstead et al. 2015) (TRICK; Fig. 2A). When the TRICK biosensor was constituted with a stress-responsive 5′ terminal oligopyrimidine (TOP) motif and expressed in living cells, greater than 10% of TOP mRNA transcripts were found strongly associated with processing bodies (P-bodies) and translationally repressed on arsenite treatment. More surprisingly, approximately 80% of these associated transcripts remained untranslated even after arsenite relief, suggesting a granule-based mechanism for translation regulation. The investigators further engineered an *oskar*-TRICK reporter to monitor the spatiotemporal translation regulation of Oskar protein during *Drosophila* development. It appeared that the initial round of translation only occurred lo-cally after mRNAs trafficked to the posterior pole area of a stage 10 oocyte (Fig. 2B), consistent with the detection of initial Oskar proteins. This method provides a general framework to visualize the spatial distribution and the kinetics of regulatory processes that are related with the first round of translation.

Dynamic Translation Imaging Using Bioluminescent and Fluorescent Reporter Systems

Continuous monitoring of translation is necessary to follow protein production past the pioneer round. One approach to extend the observation time of translation used the ultrabright bioluminescence flashes generated by *Gaussia* luciferase (Fig. 1B, Table 1). When *Gaussia* luciferase oxidizes the small molecule substrate coelenterazine (CVZ), the luciferase is rapidly inactivated, thus attenuating the bright emission. Upon addition of excess CVZ, a bright flash of luminescence is generated by previously synthesized luciferase, which is then quickly inactivated. Subsequently, small quantal bursts of luminescence are observed, attributed to production of new proteins. Na et al. (2016) used this technique to study the translation of the immediate early protein Arc. Previous literature suggested that Arc transcription and translation are induced by neuronal stimulation. In agreement, the investigators found that new Arc synthesis was enhanced by glutamate stimulation and new translation often occurred repeatedly at the same sites, lending support to the idea of translation hot spots. Because these bursts of Arc translation were observed immediately after glutamate stimulation, the investigators concluded that translation of Arc is caused by previously stalled ribosomes that were poised and waiting for stimulation.

An alternative approach is to directly observe the synthesis of nascent proteins with multivalent fluorescent tags. This can be achieved either by immunofluorescence (IF), where multiple secondary antibodies recognize a primary antibody, or by introducing multiple epitopes to the amino terminus of a protein. One can confidently identify translation sites by colocalizing the IF and FISH signals. Chang et al. 2006 used FISH-IF

Cite this article as *Cold Spring Harb Perspect Biol* doi: 10.1101/cshperspect.a032722

as well as live-cell imaging to conclude that 30% of bright peripherin RNPs were sites of de novo peripherin synthesis before the protein was incorporated into intermediate filaments. Newer technologies have used the same conceptual approach to observe protein synthesis at the single-molecule level.

In 2016, several laboratories simultaneously reported a fluorescence-based imaging technique to visualize translation dynamics of mRNAs in living cells, termed as either single-molecule imaging of nascent peptides (SINAPS) or nascent chain tracking (NCT) (Morisaki et al. 2016; Pichon et al. 2016; Wang et al. 2016; Wu et al. 2016; Yan et al. 2016). Genes of interests are fused with multiple peptidyl epitopes recognized by specific antibodies. When fluorescently labeled antibodies and epitope-tagged genes are coexpressed in living cells, the antibodies spontaneously bind to the nascent peptides emerging from ribosomes. Multiple ribosomes on a single mRNA lead to bright translation sites. In SINAPS, three laboratories used an epitope that is targeted by a single-chain variable fragment (scFv) fused with superfolder GFP (scFv-GCN4-sfGFP), which was previously termed as SUperNova or SunTag (Tanenbaum et al. 2014). This scFv recognizes the GCN4 epitope so as newly synthesized peptides exit the ribosome, multiple scFv-GCN4-sfGFP bind to the epitopes, yielding a bright fluorescent spot colocalizing with the MS2/PP7-tagged mRNAs (Fig. 2C, D). In NCT, Morisaki et al. constructed multimerized FLAG or HA peptides and imaged them with fluorescently labeled anti-FLAG or anti-HA Fab antibody fragments loaded into the cells (Viswanathan et al. 2015; Morisaki et al. 2016). This strategy allowed the investigators to perform multicolor imaging of translation.

To observe signals exclusively from nascent peptides, strategies have been devised to remove undesirable background signals from mature proteins. By adding an auxin-induced degron (AID) to the carboxyl terminus of the translation reporter, treatment with auxin causes the AID-containing mature protein to be rapidly degraded by the proteasome (Wu et al. 2016). Other researchers directly imaged the translation of a protein (ornithine decarboxylase)

with a short lifetime (Wang et al. 2016). Both strategies achieved low backgrounds in the cytoplasm and reduced the sequestration of fluorescent antibodies to mature proteins. A third strategy immobilized PP7-tagged mRNAs onto the cell membrane through a CAAX palmitoylation motif. Because mature peptides diffused away from the cell membrane, the investigators also successfully imaged nascent peptide synthesis without any additional degradation strategies (Yan et al. 2016).

Compared to RNA-Seq and proteomic studies that assess an ensemble of RNAs and proteins, these fluorescence reporters display unique advantages. First, because of the single-molecule sensitivity and resolution, individual mRNAs can be tracked for tens of minutes for the study of the spatial distribution of local translation in polarized cell types such as neurons. Wu et al. (2016) observed that the fraction of translating mRNAs is higher in proximal than distal dendrites. Second, fluorescence reporters provide unprecedented temporal resolution for tracking translation (~10–30 msec). The initiation and elongation rates of translation can be measured based on the fluctuation of fluorescence intensity. For example, these experiments measured ribosome elongation speed at around five amino acids per second and initiation rates in the range of once every 1–4 minutes. Third, translation can be recorded with single-molecule specificity. In many biological processes, it is only a subset of mRNAs that are translationally active at a given time, commonly resulting in a diminished correlation between mRNA and protein levels. In contrast, active and inactive mRNAs are distinguishable using these translation reporters. Therefore, these techniques are well suited for the study of events where translation of only one or a few mRNAs leads to a consequential biological phenotype.

In addition to the spatiotemporal resolution and molecular specificity, fluorescent reporters are especially powerful for studying translation mechanisms. One can investigate the mechanistic roles in translational control of different molecular elements (e.g., UTRs, *cis* elements) at the single-molecule level. For example, Wu et al. (2016) incorporated the β-actin 3'UTR into a translation reporter and investigated localized translation of reporter mRNAs in neuronal dendrites. Yan et al. (2016) examined alternatively spliced forms of the 5′UTR in Emi1 gene transcripts and found that for mRNAs with a long 5′UTR, only 2% were translating and the rest were repressed. To understand how *cis* elements influence translation, Wang et al. (2016) engineered the stress-responsive *cis* element ATF4 to their reporter system. After DTT and arsenite treatment of the cell samples, they observed distinct translation dynamics of the mRNAs. Furthermore, two groups evaluated translation rates after adding regulatory sequences such as ER-targeting sequences and ribosome stalling sequences to the reporter systems (Wang et al. 2016).

Although the aforementioned reporter plasmids were transiently or stably transfected into cells, later work investigated translation of endogenous genes by inserting a SunTag-based translation reporter into a specific gene locus (Pichon et al. 2016). This capability will significantly expand the application of this technique to advance understanding of the role of translation in physiologically relevant events such as neurodegenerative diseases.

CONCLUDING REMARKS/FUTURE DIRECTIONS

Rapid technological advances in translation imaging have occurred because of a revolution in fluorescent reagents and microscopy. By tagging proteins of interest, researchers can track translation products and machineries. Through pulse-chase experiments, one can determine when and where new proteins are synthesized and how they are distributed. Multicolor RNA labeling allows visualization of the pioneering round of translation. Direct labeling of nascent multivalent peptides provides translation dynamics of single mRNAs in live cells. These powerful tools offer exciting opportunities to address questions about the translation dynamics and regulation. For example, how many times is an mRNA translated before it is degraded? The cytoplasmic fate of mRNA is closely related to its history in the nucleus, and researchers can now

determine how different promoters, different splicing conditions, or covalent RNA modifications influence translation at the single-molecule level.

Scientists are continuously pushing the boundary of translation imaging. There are a few directions that require further developments. First, translation of single transcripts needs to be imaged continuously from their birth to death (Tutucci et al. 2018). A key parameter in biology is the number of proteins produced from a single mRNA. Various models and indirect measurements have given an estimation of this parameter. Single-molecule translation assays will provide definitive insight. Second, a quantitative model is necessary to integrate various aspects of an experiment. Totally asymmetric exclusion process (TASEP) has been used to model translation since the 1960s (Zia et al. 2011). It has been used to account for the size of ribosomes, rare codons, circulation of mRNA, finite resources of ribosomes, and tRNAs. Combining TASEP modeling with modern single-molecule translation assays would yield biological insights into the translation mechanism in vivo. Third, a method is needed to image the translation of endogenous transcripts. It has been observed that the cytoplasmic fate of an mRNA is intimately related to its nuclear history (Trcek et al. 2011). Translation may also depend on how the mRNA is produced, for example, transcribed from an endogenous promoter, spliced, edited, capped, and polyadenylated (Zid and O'Shea 2014). Ultimately, the goal is to image translation in live animals to study how endogenous mRNAs are regulated, for instance, during behavior. Local translation has been suggested to play an important role in neuronal plasticity, memory, and learning. It would be ideal to show the synthesis of certain key plasticity-related proteins after the animal is subjected to cognitive challenges. Further development in reagents and microscopy tools will provide increasingly powerful new approaches to these questions.

ACKNOWLEDGMENTS

This publication is supported by National Institutes of Health (NIH) grants NS 83085 and GM

57071 to R.H.S. J.B. is supported by NIH Medical Scientist Training Grant T32-GM007288 and Predoctoral Fellowship F30-CA214009. B.W. is supported by the Pew Biomedical Scholar program from Pew Charitable Trust. The authors also thank Dr. Carolina Eliscovich for discussion and suggestions.

REFERENCES

*Reference is also in this collection.

Aakalu G, Smith WB, Nguyen N, Jiang C, Schuman EM. 2001. Dynamic visualization of local protein synthesis in hippocampal neurons. *Neuron* **30:** 489–502.

Anderson P, Kedersha N. 2006. RNA granules. *J Cell Biol* **172:** 803–808.

Beatty KE, Liu JC, Xie F, Dieterich DC, Schuman EM, Wang Q, Tirrell DA. 2006. Fluorescence visualization of newly synthesized proteins in mammalian cells. *Angew Chem Int Ed Engl* **45:** 7364–7367.

Bertrand E, Chartrand P, Schaefer M, Shenoy SM, Singer RH, Long RM. 1998. Localization of ASH1 mRNA particles in living yeast. *Mol Cell* **2:** 437–445.

Boisvert FM, Ahmad Y, Gierlinski M, Charriere F, Lamont D, Scott M, Barton G, Lamond AI. 2012. A quantitative spatial proteomics analysis of proteome turnover in human cells. *Mol Cell Proteomics* **11:** M111.011429.

Brar GA, Weissman JS. 2015. Ribosome profiling reveals the what, when, where and how of protein synthesis. *Nat Rev Mol Cell Biol* **16:** 651–664.

Burke KS, Antilla KA, Tirrell DA. 2017. A fluorescence in situ hybridization method to quantify mRNA translation by visualizing ribosome-mRNA interactions in single cells. *ACS Cent Sci* **3:** 425–433.

Butko MT, Yang J, Geng Y, Kim HJ, Jeon NL, Shu X, Mackey MR, Ellisman MH, Tsien RY, Lin MZ. 2012. Fluorescent and photo-oxidizing TimeSTAMP tags track protein fates in light and electron microscopy. *Nat Neurosci* **15:** 1742–1751.

Buxbaum AR, Wu B, Singer RH. 2014. Single β-actin mRNA detection in neurons reveals a mechanism for regulating its translatability. *Science* **343:** 419–422.

Cajigas IJ, Tushev G, Will TJ, tom Dieck S, Fuerst N, Schuman EM. 2012. The local transcriptome in the synaptic neuropil revealed by deep sequencing and high-resolution imaging. *Neuron* **74:** 453–466.

Chang L, Shav-Tal Y, Trcek T, Singer RH, Goldman RD. 2006. Assembling an intermediate filament network by dynamic cotranslation. *J Cell Biol* **172:** 747–758.

Chao JA, Patskovsky Y, Almo SC, Singer RH. 2008. Structural basis for the coevolution of a viral RNA–protein complex. *Nat Struct Mol Biol* **15:** 103–105.

Chen J, Choi J, O'Leary SE, Prabhakar A, Petrov A, Grosely R, Puglisi EV, Puglisi JD. 2016. The molecular choreography of protein synthesis: Translational control, regulation, and pathways. *Q Rev Biophys* **49:** e11.

Daigle N, Ellenberg J. 2007. λN-GFP: An RNA reporter system for live-cell imaging. *Nat Methods* **4:** 633–636.

David A, Dolan BP, Hickman HD, Knowlton JJ, Clavarino G, Pierre P, Bennink JR, Yewdell JW. 2012. Nuclear translation visualized by ribosome-bound nascent chain puromycylation. *J Cell Biol* **197:** 45–57.

Dieterich DC, Hodas JJ, Gouzer G, Shadrin IY, Ngo JT, Triller A, Tirrell DA, Schuman EM. 2010. In situ visualization and dynamics of newly synthesized proteins in rat hippocampal neurons. *Nat Neurosci* **13:** 897–905.

Gallagher SS, Sable JE, Sheetz MP, Cornish VW. 2009. An in vivo covalent TMP-tag based on proximity-induced reactivity. *ACS Chem Biol* **4:** 547–556.

Gautier A, Juillerat A, Heinis C, Correa IR Jr, Kindermann M, Beaufils F, Johnsson K. 2008. An engineered protein tag for multiprotein labeling in living cells. *Chem Biol* **15:** 128–136.

Graber TE, Hebert-Seropian S, Khoutorsky A, David A, Yewdell JW, Lacaille JC, Sossin WS. 2013. Reactivation of stalled polyribosomes in synaptic plasticity. *Proc Natl Acad Sci* **110:** 16205–16210.

Griffin BA, Adams SR, Tsien RY. 1998. Specific covalent labeling of recombinant protein molecules inside live cells. *Science* **281:** 269–272.

Grimm JB, English BP, Chen J, Slaughter JP, Zhang Z, Revyakin A, Patel R, Macklin JJ, Normanno D, Singer RH, et al. 2015. A general method to improve fluorophores for live-cell and single-molecule microscopy. *Nat Methods* **12:** 244–250.

Grimm JB, English BP, Choi H, Muthusamy AK, Mehl BP, Dong P, Brown TA, Lippincott-Schwartz J, Liu Z, Lionnet T, et al. 2016. Bright photoactivatable fluorophores for single-molecule imaging. *Nat Methods* **13:** 985–988.

Halstead JM, Lionnet T, Wilbertz JH, Wippich F, Ephrussi A, Singer RH, Chao JA. 2015. Translation. An RNA biosensor for imaging the first round of translation from single cells to living animals. *Science* **347:** 1367–1671.

Hengst U, Deglincerti A, Kim HJ, Jeon NL, Jaffrey SR. 2009. Axonal elongation triggered by stimulus-induced local translation of a polarity complex protein. *Nat Cell Biol* **11:** 1024–1030.

Holt CE, Schuman EM. 2013. The central dogma decentralized: New perspectives on RNA function and local translation in neurons. *Neuron* **80:** 648–657.

Ifrim MF, Williams KR, Bassell GJ. 2015. Single-molecule imaging of PSD-95 mRNA translation in dendrites and its dysregulation in a mouse model of fragile X syndrome. *J Neurosci* **35:** 7116–7130.

* Ingolia NT, Hussmann JA, Weissman JS. 2018. Ribosome profiling: Global views of translation. *Cold Spring Harb Perspect Biol* doi: 10.1101/cshperspect.a032698.

* Ivanov P, Kedersha N, Anderson P. 2018. Stress granules and processing bodies in translational control. *Cold Spring Harb Perspect Biol* doi: 10.1101/cshperspect.a032813.

* Karousis ED, Mühlemann O. 2018. Nonsense-mediated mRNA decay begins where translation ends. *Cold Spring Harb Perspect Biol* doi: 10.1101/cshperspect.a032862.

Katz ZB, English BP, Lionnet T, Yoon YJ, Monnier N, Ovryn B, Bathe M, Singer RH. 2016. Mapping translation "hotspots" in live cells by tracking single molecules of mRNA and ribosomes. *eLife* **5:** e10415.

Kiebler MA, Bassell GJ. 2006. Neuronal RNA granules: Movers and makers. *Neuron* **51**: 685–690.

Kos A, Wanke KA, Gioio A, Martens GJ, Kaplan BB, Aschrafi A. 2016. Monitoring mRNA translation in neuronal processes using fluorescent noncanonical amino acid tagging. *J Histochem Cytochem* **64**: 323–333.

Krichevsky AM, Kosik KS. 2001. Neuronal RNA granules: A link between RNA localization and stimulation-dependent translation. *Neuron* **32**: 683–696.

Lecuyer E, Yoshida H, Parthasarathy N, Alm C, Babak T, Cerovina T, Hughes TR, Tomancak P, Krause HM. 2007. Global analysis of mRNA localization reveals a prominent role in organizing cellular architecture and function. *Cell* **131**: 174–187.

Lin MZ, Glenn JS, Tsien RY. 2008. A drug-controllable tag for visualizing newly synthesized proteins in cells and whole animals. *Proc Natl Acad Sci* **105**: 7744–7749.

Liu J, Xu Y, Stoleru D, Salic A. 2012. Imaging protein synthesis in cells and tissues with an alkyne analog of puromycin. *Proc Natl Acad Sci* **109**: 413–418.

Los GV, Encell LP, McDougall MG, Hartzell DD, Karassina N, Zimprich C, Wood MG, Learish R, Ohana RF, Urh M, et al. 2008. HaloTag: A novel protein labeling technology for cell imaging and protein analysis. *ACS Chem Biol* **3**: 373–382.

Moor AE, Golan M, Massasa EE, Lemze D, Weizman T, Shenhav R, Baydatch S, Mizrahi O, Winkler R, Golani O, et al. 2017. Global mRNA polarization regulates translation efficiency in the intestinal epithelium. *Science* **357**: 1299–1303.

Morisaki T, Lyon K, DeLuca KF, DeLuca JG, English BP, Zhang Z, Lavis LD, Grimm JB, Viswanathan S, Looger LL, et al. 2016. Real-time quantification of single RNA translation dynamics in living cells. *Science* **352**: 1425–1429.

Na Y, Park S, Lee C, Kim DK, Park JM, Sockanathan S, Huganir RL, Worley PF. 2016. Real-time imaging reveals properties of glutamate-induced Arc/Arg 3.1 translation in neuronal dendrites. *Neuron* **91**: 561–573.

Pichon X, Bastide A, Safieddine A, Chouaib R, Samacoits A, Basyuk E, Peter M, Mueller F, Bertrand E. 2016. Visualization of single endogenous polysomes reveals the dynamics of translation in live human cells. *J Cell Biol* **214**: 769–781.

Rodriguez AJ, Shenoy SM, Singer RH, Condeelis J. 2006. Visualization of mRNA translation in living cells. *J Cell Biol* **175**: 67–76.

Rufener SC, Mühlemann O. 2013. eIF4E-bound mRNPs are substrates for nonsense-mediated mRNA decay in mammalian cells. *Nat Struct Mol Biol* **20**: 710–717.

Schmidt EK, Clavarino G, Ceppi M, Pierre P. 2009. SUnSET, a nonradioactive method to monitor protein synthesis. *Nat Methods* **6**: 275–277.

Schwanhausser B, Busse D, Li N, Dittmar G, Schuchhardt J, Wolf J, Chen W, Selbach M. 2011. Global quantification of mammalian gene expression control. *Nature* **473**: 337–342.

Starck SR, Green HM, Alberola-Ila J, Roberts RW. 2004. A general approach to detect protein expression in vivo using fluorescent puromycin conjugates. *Chem Biol* **11**: 999–1008.

Tanenbaum ME, Gilbert LA, Qi LS, Weissman JS, Vale RD. 2014. A protein-tagging system for signal amplification in gene expression and fluorescence imaging. *Cell* **159**: 635–646.

Tatavarty V, Ifrim MF, Levin M, Korza G, Barbarese E, Yu J, Carson JH. 2012. Single-molecule imaging of translational output from individual RNA granules in neurons. *Mol Biol Cell* **23**: 918–929.

tom Dieck S, Kochen L, Hanus C, Heumuller M, Bartnik I, Nassim-Assir B, Merk K, Mosler T, Garg S, Bunse S, et al. 2015. Direct visualization of newly synthesized target proteins in situ. *Nat Methods* **12**: 411–414.

Trcek T, Larson DR, Moldon A, Query CC, Singer RH. 2011. Single-molecule mRNA decay measurements reveal promoter-regulated mRNA stability in yeast. *Cell* **147**: 1484–1497.

Tutucci E, Vera M, Biswas J, Garcia J, Parker R, Singer RH. 2017. An improved MS2 system for accurate reporting of the mRNA life cycle. *Nat Methods* **15**: 81–89.

Tutucci E, Livingston NM, Singer RH, Wu B. 2018. Imaging mRNA in vivo, from birth to death. *Annu Rev Biophys* doi: 10.1146/annurev-biophys-070317-033037.

Viswanathan S, Williams ME, Bloss EB, Stasevich TJ, Speer CM, Nern A, Pfeiffer BD, Hooks BM, Li WP, English BP, et al. 2015. High-performance probes for light and electron microscopy. *Nat Methods* **12**: 568–576.

Wang DO, Kim SM, Zhao Y, Hwang H, Miura SK, Sossin WS, Martin KC. 2009. Synapse- and stimulus-specific local translation during long-term neuronal plasticity. *Science* **324**: 1536–1540.

Wang C, Han B, Zhou R, Zhuang X. 2016. Real-time imaging of translation on single mRNA transcripts in live cells. *Cell* **165**: 990–1001.

Wei L, Yu Y, Shen Y, Wang MC, Min W. 2013. Vibrational imaging of newly synthesized proteins in live cells by stimulated Raman scattering microscopy. *Proc Natl Acad Sci* **110**: 11226–11231.

Wei L, Hu F, Shen Y, Chen Z, Yu Y, Lin CC, Wang MC, Min W. 2014. Live-cell imaging of alkyne-tagged small biomolecules by stimulated Raman scattering. *Nat Methods* **11**: 410–412.

Wong HH, Lin JQ, Strohl F, Roque CG, Cioni JM, Cagnetta R, Turner-Bridger B, Laine RF, Harris WA, Kaminski CF, et al. 2017. RNA docking and local translation regulate site-specific axon remodeling in vivo. *Neuron* **95**: 852–868.e858.

Wu B, Chen Y, Muller JD. 2010. Heterospecies partition analysis reveals binding curve and stoichiometry of protein interactions in living cells. *Proc Natl Acad Sci* **107**: 4117–4122.

Wu B, Piatkevich KD, Lionnet T, Singer RH, Verkhusha VV. 2011. Modern fluorescent proteins and imaging technologies to study gene expression, nuclear localization, and dynamics. *Curr Opin Cell Biol* **23**: 310–317.

Wu B, Buxbaum AR, Katz ZB, Yoon YJ, Singer RH. 2015. Quantifying protein-mRNA interactions in single live cells. *Cell* **162**: 211–220.

Wu B, Eliscovich C, Yoon YJ, Singer RH. 2016. Translation dynamics of single mRNAs in live cells and neurons. *Science* **352**: 1430–1435.

Yan X, Hoek TA, Vale RD, Tanenbaum ME. 2016. Dynamics of translation of single mRNA molecules in vivo. *Cell* **165**: 976–989.

Yoon YJ, Wu B, Buxbaum AR, Das S, Tsai A, English BP, Grimm JB, Lavis LD, Singer RH. 2016. Glutamate-induced RNA localization and translation in neurons. *Proc Natl Acad Sci* **113**: E6877–E6886.

Younts TJ, Monday HR, Dudok B, Klein ME, Jordan BA, Katona I, Castillo PE. 2016. Presynaptic protein synthesis is required for long-term plasticity of GABA release. *Neuron* **92**: 479–492.

Yu J, Xiao J, Ren X, Lao K, Xie XS. 2006. Probing gene expression in live cells, one protein molecule at a time. *Science* **311**: 1600–1603.

Zappulo A, van den Bruck D, Ciolli Mattioli C, Franke V, Imami K, McShane E, Moreno-Estelles M, Calviello L, Filipchyk A, Peguero-Sanchez E, et al. 2017. RNA localization is a key determinant of neurite-enriched proteome. *Nat Commun* **8**: 583.

Zhou XX, Lin MZ. 2013. Photoswitchable fluorescent proteins: Ten years of colorful chemistry and exciting applications. *Curr Opin Chem Biol* **17**: 682–690.

Zia RKP, Dong JJ, Schmittmann B. 2011. Modeling translation in protein synthesis with TASEP: A tutorial and recent developments. *J Stat Phys* **144**: 405–428.

Zid BM, O'Shea EK. 2014. Promoter sequences direct cytoplasmic localization and translation of mRNAs during starvation in yeast. *Nature* **514**: 117–121.

The Epitranscriptome in Translation Regulation

Eyal Peer, Sharon Moshitch-Moshkovitz, Gideon Rechavi, and Dan Dominissini

Sackler School of Medicine, Tel Aviv University, Tel Aviv 6997801, Israel; Cancer Research Center and Wohl Centre for Translational Medicine, Chaim Sheba Medical Center, Tel-Hashomer 5262160, Israel

Correspondence: dan.dominissini@sheba.health.gov.il

The cellular proteome reflects the total outcome of many regulatory mechanisms that affect the metabolism of messenger RNA (mRNA) along its pathway from synthesis to degradation. Accumulating evidence in recent years has uncovered the roles of a growing number of mRNA modifications in every step along this pathway, shaping translational output. mRNA modifications affect the translation machinery directly, by influencing translation initiation, elongation and termination, or by altering mRNA levels and subcellular localization. Features of modification-related translational control are described, charting a new and complex layer of translational regulation.

Cellular homeostasis requires regulation of protein levels, as impaired regulation can result in cellular death or dysregulated proliferation (Hershey et al. 2012). Protein levels are a consequence of three fundamental factors: the abundance of messenger RNA (mRNA), the efficiency of translation, and protein stability. First isolated from ribosomal RNA (rRNA) almost seven decades ago (Cohn 1951), modified RNA bases were later also discovered to be abundant constituents of mRNA, with N^6-methyladenosine (m^{6}A) as the most prevalent internal mRNA modification (Desrosiers et al. 1974). To date, 170 different modified nucleotides have been identified in RNA from all types and species (Boccaletto et al. 2018), expanding the RNA alphabet to embed transcripts with additional information. The introduction of transcriptome-wide mapping methods and the discovery of enzymes capable of removing m^{6}A and of dedicated m^{6}A-binding proteins have

introduced the novel notion that internal chemical modifications of mRNA and long noncoding RNA (lncRNA) are potentially dynamic and sometimes reversible events, which constitute essential regulatory elements in basic RNA-processing steps such as splicing, transport, translation, and decay. The epitranscriptome, as this ensemble is now known, comprises a growing number of chemical adducts: m^{6}A, inosine (I), 5-methylcytidine (5mC), 5-hydroxymethylcytidine (5hmC), pseudouridine (Ψ), 2′-O-methylation (Nm), and N^1-methyladenosine (m^{1}A). This review focuses on m^{6}A as well as on more recently characterized mRNA modifications and their effects on translation.

N^6-METHYLADENOSINE (m^{6}A)

m^{6}A is the most common internal (noncap) mRNA modification found in eukaryotic organisms as well as in RNA of nuclear-replicating viruses (reviewed in Fu et al. 2014). m^6A constitutes 0.1%–0.5% of adenosine residues in mRNA, corresponding to ~3–5 m^6A modifications per transcript. Although m^6A was discovered in mRNA decades ago (Desrosiers et al. 1974), its study was jump-started in recent years with the identification of the m^6A-specific demethylases: the fat mass and obesity-associated protein (FTO) and AlkB homolog 5 (ALKBH5) (Jia et al. 2011; Zheng et al. 2013), followed by the development of high-throughput m^6A mapping tools, m^6A-seq (Dominissini et al. 2012) and methylated RNA immunoprecipitation and sequencing (MeRIP-Seq) (Meyer et al. 2012). Transcriptome-wide mapping of m^6A revealed an evolutionarily conserved, nonrandom distribution (Dominissini et al. 2012; Meyer et al. 2012); m^6A preferentially decorates 5′ untranslated regions (UTRs), long internal exons, and the vicinity of stop codons. Sites preferentially appear within the consensus sequence RRACU (R = A or G) (Dominissini et al. 2012).

The deposition of m^6A in mRNA is performed by a multicomponent methyltransferase complex of which several components have been identified: methyltransferase-like 3 (METTL3), the catalytic component; methyltransferase-like 14 (METTL14), an RNA adaptor needed for METTL3 activity; Wilms Tumor 1–associating protein (WTAP), a regulatory factor that is also responsible for nuclear speckle localization; RNA-binding motif (RBM) proteins 15/15B (RBM15/15B), mediators of methylation specificity by directing the complex to specific transcripts; and KIAA1429/Virilizer, whose function is still unclear (reviewed in Meyer and Jaffrey 2017).

The balance between writing (adding) and erasing m^6A from transcripts confers upon this modification a dynamic nature, which is evident under stress conditions (Dominissini et al. 2012), supporting a role in regulation of gene expression. Indeed, m^6A plays a role in many aspects of gene expression, affecting RNA splicing, nuclear export and retention, translation, and turnover. The major mechanism by which m^6A exerts its effects is by recruiting m^6A reader proteins: fragile X mental retardation 1 (FMR1) (Edupuganti et al. 2017); YTH domain-containing family protein 1 (YTHDF1) and YTH domain-containing family protein 3 (YTHDF3) (regulation of translation); YTH domain-containing family protein 2 (YTHDF2) (regulation of RNA turnover); and heterogeneous nuclear ribonucleoprotein A2/B1 (HNRNPA2B1) and YTH domain-containing 1 (YTHDC1) (processing of primary microRNAs and alternative splicing) (Meyer and Jaffrey 2017). These discoveries opened up the field of epitranscriptomics for accelerated investigation.

PSEUDOURIDINE (Ψ)

Ψ was the first posttranscriptional modification to be identified and, taking all classes of RNA into account, is also the most abundant. Recently it was shown to be more abundant in mRNA than previously believed, with a Ψ/U ratio of 0.2%–0.6% in mammalian mRNA (Li et al. 2015). It is formed by isomerization of uridine in which the base is rotated 180° along the N3-C6 axis. Although the modified base has unaltered Watson–Crick base-pairing properties, it gains an additional hydrogen-bond donor at its non-Watson–Crick edge, thus endowing it with distinct chemical properties (Ge and Yu 2013). Ψ plays roles in the biogenesis and function of spliceosomal small nuclear RNAs (snRNAs) and rRNA. Methods for transcriptome-wide mapping of Ψ using Ψ-specific chemical labeling by CMC [N-cyclohexyl-N′-(2-morpholinoethyl) carbodiimide metho-p-toluenesulfonate] identified hundreds of Ψ sites in yeast and human mRNAs (Carlile et al. 2014; Schwartz et al. 2014; Li et al. 2015). No preferred localization of Ψ to specific regions of mRNA was shown.

Installation of Ψ is catalyzed by pseudouridine synthases (PUSs) and can be achieved through two distinct mechanisms (Li et al. 2016a): (1) RNA-independent pseudouridylation that is catalyzed by a single PUS enzyme responsible for both substrate recognition and catalysis; or (2) RNA-dependent pseudouridylation that relies on RNA–protein complexes consisting of box H/ACA noncoding RNAs in conjunction with the centromere-binding factor 5 (Cbf5)/Dyskerin protein complex. Although mRNA Ψ is dynamically regulated in response to environmental signals (Carlile et al. 2014), pseudouridylation has not been shown to be reversible.

N^1-METHYLADENOSINE (m^{1}A)

Using mass spectrometry and a mapping methodology based on immunocapture and massively parallel sequencing, m^{1}A was recently identified as a new epitranscriptome marker that occurs on thousands of human and mouse mRNAs (Dominissini et al. 2016; Li et al. 2016b). m^{1}A was known to exist in transfer RNA (tRNA) and rRNA and is unique in that it has both a methyl group and a positive charge under physiological conditions, which could markedly alter RNA structure and protein–RNA interactions. Transcriptome-wide mapping of m^{1}A revealed unique features of its distribution that strongly indicate functional roles: (1) m^{1}A is strongly enriched in mRNA 5′UTRs, close to translation initiation sites; (2) m^{1}A is present in highly structured regions of 5′UTRs; (3) m^{1}A positively correlates with translation efficiency and protein levels; (4) the distribution of m^{1}A is highly conserved in all mouse and human cell types examined; and (5) m^{1}A is dynamic in response to stress and physiological signals, and its level varies across tissues (Dominissini et al. 2016; Li et al. 2016b). Together, these attributes suggest a positive and dynamic role for m^{1}A in translation initiation in mammalian cells.

A recently improved detection method identified hundreds of m^{1}A sites with single-nucleotide resolution; most of them are in the mRNA 5′UTR (including a minor fraction in the first transcribed nucleotide, cap + 1), validating and refining earlier studies. Sites fall into three subsets defined by their location, the identity of the writer enzyme, and sequence-structure features: TRMT6/61A-independent and -dependent m^{1}A sites in nuclear-encoded mRNA; and tRNA methyltransferase 61B (TRMT61B)-dependent m^{1}A sites in mitochondrial-encoded mRNA. tRNA methyltransferase 6 (TRMT6)/ tRNA methyltransferase 61A (TRMT61A)-independent m^{1}A sites constitute the largest subset and are strongly enriched in the 5′UTR. TRMT6/61A-dependent m^{1}A sites conform to a GUUCRA tRNA-like motif (R = A or G) and have T-loop-like structures, are evenly distributed along transcript segments, and constitute roughly 10% of all identified sites. Last, TRMT61B-dependent m^{1}A sites are primarily located within the coding region of mitochondrial mRNA, where they inhibit translation (Li et al. 2017b).

2′-O-METHYLATION (Nm)

The ribose ring can be methylated at the 2′ position to form Nm (Boccaletto et al. 2018), an abundant modification present in mRNA, rRNA, tRNA, snRNA, and microRNA, which is essential for their biogenesis, metabolism, and function (Liang et al. 2009; Daffis et al. 2010; Lin et al. 2011; Züst et al. 2011; Deryusheva et al. 2012; Jöckel et al. 2012; Somme et al. 2014). Nm modification is catalyzed either by stand-alone methyltransferases that recognize their targets according to sequence and structure (Somme et al. 2014), or by the enzyme fibrillarin, part of a ribonucleoprotein complex that is guided to its targets by different C/D-box small nucleolar RNAs (snoRNAs) (Shubina et al. 2016). Nm also occurs at the 5′ cap (Byszewska et al. 2014) and internal positions of non-rRNA transcripts (Lacoux et al. 2012). 2′-O-methylation endows nucleotides with greater hydrophobicity, protects against nucleolytic attack, and stabilizes RNA helices (Kumar et al. 2014; Yildirim et al. 2014).

In higher eukaryotes, the 5′ penultimate (the first transcribed) and antepenultimate (the second transcribed) nucleotides in mRNA (m^{7}GpppNmNm) may be 2′-O-methylated by stand-alone enzymes that recognize the cap. Aside from these Nm sites, accumulating evidence suggests that internal positions in mRNA can also be 2′-O-methylated (Gumienny et al. 2016). This possibility is especially intriguing in light of the consequences that such modification could have when present in the coding sequence (CDS) (Hoernes et al. 2016).

A recently developed Nm mapping method, Nm-Seq, uncovered thousands of Nm sites in human mRNA, with Um being the dominant species (Dai et al. 2017). Nm sites are enriched in the 3'UTR and around internal splice sites. In the CDS, the sites are nonrandomly distributed with 60.4% of them occurring in only six codons, suggesting functional roles.

$N^6,2'$-O-DIMETHYLADENOSINE (m^{6}Am)

The 5' end of mRNA transcripts is modified by an N^7-methylguanosine (m^{7}G) cap and 2'-O methylation (Nm) of the ribose sugar of the first, and sometimes the second, transcribed nucleotides (Keith et al. 1978) as mentioned above. Cap-associated modifications recruit translation initiation factors to mRNA and allow the cell to discriminate host from viral mRNA, 2'-O-methylation being especially important for the latter (Daffis et al. 2010). When the first transcribed nucleotide (the 5' penultimate base) is Am, it can be further methylated at the N^6 position by an unidentified methyltransferase to form m^{6}Am (Keith et al. 1978). Taking advantage of the affinity for m^{6}Am of an antibody raised against m^{6}A, hundreds of m^{6}Am sites in mRNAs have recently been mapped (Lacoux et al. 2012). m^{6}Am in the 5' cap is dynamic and reversible; it can be demethylated to Am by FTO (Mauer et al. 2017) to affect transcript decapping and thereby mRNA stability.

5-METHYLCYTIDINE (5mC) AND 5-HYDROXYMETHYLCYTIDINE (5hmC)

Methylation of cytosine at the fifth position was identified in mRNA more than 40 years ago (Dubin and Taylor 1975; Dubin et al. 1977) but other studies failed to recapitulate the finding (Desrosiers et al. 1975; Wei et al. 1976) and some have hypothesized that the detection of 5mC in mRNA was merely because of contamination from other RNA species (Bokar 2005). However, appropriating bisulfite sequencing—a method used in mapping 5mC in DNA—for transcriptome-wide mapping of 5mC in RNA confirmed 5mC as an mRNA modification and unveiled thousands of sites with nonrandom distribution (Squires et al. 2012; Amort et al. 2017). NOP2/Sun RNA methyltransferase family member 2 (NSUN2), a methyltransferase that installs 5mC in tRNA, was identified as re-sponsible for 5mC in some mRNAs as well (Hussain et al. 2013; Khoddami and Cairns 2013).

Recently, ALYREF, a nuclear export factor, has been identified as a reader of 5mC in mRNA (Yang et al. 2017a). Although it is not known to be completely reversible, it was discovered that 5mC in RNA undergoes oxidation by the ten-eleven translocation (Tet) protein family (TET1,2,3) to produce 5hmC (Fu et al. 2014). Transcriptome-wide mapping of 5hmC in *Drosophila melanogaster* was performed by adapting MeRIP-seq to 5hmC (hMeRIP-seq), and yielded over 3000 putative 5hmC sites in mRNA. Although the function of 5hmC is unclear, ribosomal profiling (see Ingolia et al. 2018) revealed that 5hmC is highly enriched in actively translated transcripts, suggesting that 5hmC may facilitate mRNA translation (Delatte et al. 2016).

READING THE EPITRANSCRIPTOME

RNA modifications exert their effects, both directly and indirectly, by altering the chemistry of RNA nucleotides. A modified RNA nucleotide can be specifically recognized by certain RNA-binding proteins, which then act as readers of the modification and determine the fate of the modified mRNA. When the effect is mediated by binding proteins, it is termed indirect. It can also stabilize or destabilize the structure of an RNA molecule, leading to a direct change in level, function, or even translation dynamics. The effect can also be mediated by a combination of direct and indirect mechanisms, where a change in RNA structure improves the accessibility for RNA-binding proteins. In addition, hydrophobic modifications incur a solvation penalty in water, which can be reduced by interaction with protein residues containing hydrophobic side-chains (Noeske et al. 2015; Roundtree et al. 2017).

Each modification carries distinct chemical implications. In m^{6}A, methylation at the sixth position of the base does not alter hydrogen-bonding donors and acceptors and does not appear to affect translational fidelity (You et al. 2017), but the added methyl group creates steric hindrance, thereby changing the energetics of the AU pair and destabilizing the pairing with uracil (Roost et al. 2015). In m^1A, methylation at the N^1 position results in a positive charge, potentially leading to strong electrostatic interactions (He et al. 2005). Additionally, and in contrast to m^6A, the added methyl group protrudes from the Watson–Crick hydrogen-bonding face of adenine, causing the nucleotide to remain unpaired (Lu et al. 2010). Also in contrast with m^6A, m^1A appears to have the potential to block the translation machinery (You et al. 2017). In Ψ, although the isomerization does not change the Watson–Crick base-pairing, it does add another potential hydrogen-bond donor to the base (Ge and Yu 2013). Methylation of the ribose in Nm stabilizes RNA helices and augments the nucleotide's hydrophobicity, protecting it from nucleolytic attack (Kumar et al. 2014; Yildirim et al. 2014).

The direct and indirect mechanisms by which modifications affect mRNA metabolism and translation have been explored mainly for m^6A. m^6A can restructure RNA to control accessibility of sequence motifs to RNA-binding proteins. This dynamic interdependence between RNA structure and modification, termed the m^6A-switch, has functional consequences. For example, heterogeneous nuclear ribonucleoprotein (hnRNP) C does not bind m^6A but rather a U-tract opposing the m^6A site, made accessible by m^6A. Global reduction of m^6A levels masks a significant subset of hnRNP C binding sites, thereby affecting hnRNP C's effect on alternative splicing (Liu et al. 2015).

Genuine direct m^6A readers have been discovered and studied, starting with identification of YTH domain family proteins as bona fide m^6A readers (Dominissini et al. 2012). The YTH domain is a highly conserved RNA-binding domain, identified in over 170 family members in a wide range of eukaryotes, from yeast and plants to vertebrates, appearing in five human proteins: YTHDF1-3 and YTHDC1-2 (Stoilov et al. 2002; Zhang et al. 2010). Initially characterized at the turn of the century (Imai et al. 1998), YTH was recently validated as an m^6A-binding domain (Wang et al. 2014) and YTH domain family proteins were shown to mediate

the effect of m^6A on mRNA stability (Wang et al. 2014; Du et al. 2016; Shi et al. 2017), translation (Wang et al. 2015; Li et al. 2017a; Shi et al. 2017), and processing (Xiao et al. 2016), as well as non-mRNA-related roles such as lncRNA-mediated transcriptional repression (Patil et al. 2016) and translation of circular RNA (circRNA) (Yang et al. 2017b; Chekulaeva and Rajewsky 2018). Although only the YTH domain has been structurally characterized in complex with m^6A (Li et al. 2014; Luo and Tong 2014; Theler et al. 2014; Xu et al. 2014; Zhu et al. 2014), other m^6A readers have been suggested (Edupuganti et al. 2017; Huang et al. 2018).

INDIRECT AND DIRECT EFFECTS OF mRNA MODIFICATIONS ON TRANSLATION

The translation rate of a protein is proportional to the concentration and translational efficiency of its mRNA (Hershey et al. 2012). mRNA modifications define the proteome by influencing these two factors. They play a direct role by attracting translation initiation factors, influencing translation elongation and termination, and potentially recoding the genetic code, and an indirect role by altering mRNA stability, splicing, nuclear export, and subcellular localization.

Indirect Effects

The levels of mRNAs can be modulated by a change in their stability conferred by several mRNA modifications; the half-life of m^6A-methylated transcripts is on average shorter than nonmethylated ones. Degradation of these transcripts is mediated by YTHDF2 that preferentially binds m^6A-containing transcripts and moves them from translatable pools to P-bodies or stress granules (Wang et al. 2014). YTHDF2 recruits the CCR4–NOT deadenylase complex by directly interacting with CNOT1 to initiate deadenylation and mRNA degradation (Du et al. 2016; Heck and Wilusz 2018).

In contrast to m^6A, the presence of m^6Am (Cap1) in mRNA increases the stability of transcripts by conferring resistance to the mRNA decapping enzyme, DCP2. FTO is able to remove the methyl group from the N^6 position, thereby facilitating decapping and degradation (Mauer et al. 2017).

Ψ also affects mRNA stability. In yeast, Pus7p pseudouridylates mRNA in response to heat shock. Depletion of Pus7p results in decreased stability of theses transcripts, supporting a role for Ψ in mRNA degradation (Schwartz et al. 2014).

Translation is also affected by the availability of transcripts for ribosome binding. Thus, subcellular compartmentalization such as nuclear retention controls transcript availability. The effect of m^6A on the nuclear export of transcripts was shown by Fustin et al. (2013) who showed that reduced m^6A levels resulted in nuclear accumulation of otherwise methylated transcripts.

Axonal mRNA can be locally translated in response to specific signals, providing an effective mechanism for rapid protein synthesis. The nonnuclear pool of FTO in the axon regulates local translation of axonal mRNA by m^6A demethylation to increase transcript expression (Yu et al. 2017).

5mC can also affect nuclear export of mRNA transcripts. Depletion of 5mC inhibits recognition of transcripts by the export adaptor ALYREF, an RNA-binding protein that preferentially binds 5mC-modified RNA, resulting in dysregulated nuclear export (Yang et al. 2017a).

Direct Effects

The effects of mRNA modifications on translation can take place at every stage of the translation process. These modifications can act by either altering base-pairing, inducing conformational changes, or modulating the recognition of RNA-binding proteins. Most of the data regarding mRNA modification emerge from m^6A studies. Although it does not affect hydrogen bonding or translational fidelity (You et al. 2017), accumulating evidence suggests that m^6A is involved in translation regulation through several mechanisms.

Transcriptome-wide mapping of m^6A identified a subset of m^6A sites that are located in the 5′UTR of mRNA (Dominissini et al. 2012; Meyer et al. 2012). The 5′UTR is critical for ribo-

some recruitment to mRNA. Typically, translation begins when the 43S ribosomal complex is recruited to the 5′ 7-methylguanosine (m^7G) cap of mRNA transcripts through the cap-binding complex eukaryotic translation initiation factor 4F (eIF4F). Under stress conditions, a switch from canonical eukaryotic translation initiation factor 4E (eIF4E)-dependent to eIF4E-independent translation initiation can take place (see Kwan and Thompson 2018; Merrick and Pavitt 2018; Robichaud et al. 2018). Toeprinting experiments revealed that m^6A at the 5′UTR of mRNA transcripts can allow bypass of eIF4E-dependency through the direct binding of eIF3, which is sufficient to recruit the 43S complex to initiate translation (Meyer et al. 2015). These experiments showed that eIF3 preferentially binds mRNA transcripts that carry m^6A in the 5′UTR, and that m^6A-induced translation initiation occurs only in transcripts containing m^6A in the 5′UTR and not elsewhere within the transcript (including the cap-associated m^6Am).

Another line of evidence for the role of m^6A in translation initiation came from a study that showed direct binding of METTL3 to eIF3 (Lin et al. 2016). METTL3, a predominantly nuclear protein, is also present in the cytoplasm and recruits the ribosome through association with eIF3. This association is dependent on m^6A methylation and independent of METTL3's methyltransferase activity, METTL14, WTAP, and other m^6A reader proteins, including YTHDF1 and YTHDF2. Reduced m^6A levels lead to decreased METTL3 binding to transcripts in the cytoplasm and consequently reduced protein levels (Lin et al. 2016). It is unclear whether METTL3 plays a role in cap-dependent or cap-independent translation.

The effect of m^6A in the 5′UTR on translation initiation was further established in a recent study (Coots et al. 2017) revealing that when eIF4F-dependent translation is impaired, cells use a different mode of translation that is neither cap- nor internal ribosome entry site (IRES)-dependent, but rather m^6A-dependent. This study identified ATP-binding cassette subfamily F member 1 (ABCF1) as a critical mediator of m^6A-dependent, eIF4E-independent translation. ABCF1 serves as an alternative recruiter for the ternary complex during noncanonical translation by interacting with eIF2, which controls ternary complex formation (a rate-limiting step to the overall translational output) and ribosomes, proving its critical role for mRNA translation under stress (see Merrick and Pavitt 2018; Wek 2018). Interestingly, the synthesis of METTL3, required for installing m^6A in mRNA transcripts, is m^6A- and ABCF1-dependent, generating a positive feedback loop and providing a mechanism by which cells activate m^6A-mediated translation on inhibition of cap-dependent translation.

The effect of m^6A on translation initiation involves additional m^6A-binding proteins. Wang et al. (2015) have shown that a member of the YTH domain family is involved in m^6A-mediated regulation of translation initiation. YTHDF1 binds m^6A-modified transcripts and increases translation through interaction with eIF3 (Wang et al. 2015).

Although the effect of m^6A on translation initiation is mediated through its binding by different proteins, other modifications may act by affecting the structure of the 5′UTR. A positively charged m^1A may alter the secondary/tertiary structure of mRNA around translation initiation sites by blocking Watson–Crick base-pairing or introducing charge–charge interactions. Alternatively, potential binding proteins may specifically recognize m^1A and facilitate translation initiation of methylated transcripts in a way analogous to the role of m^6A readers of the YTH domain family. Ribosome profiling and proteomic analyses revealed that m^1A sites in the 5′UTR, but not those in the CDS or 3′UTR, correlated with higher translation efficiency. The mechanisms by which 5′UTR m^1A sites affect translation efficiency are still unknown (Dominissini et al. 2016; Lin et al. 2016; Li et al. 2017b).

During translation elongation, m^6A in mRNA can act as a barrier to tRNA accommodation. The presence of an m^6A within a codon disrupts cognate tRNA selection, with the greatest effect at the most thermodynamically unstable steps of initial selection (Choi et al. 2016). 5mC and Ψ also affect elongation rates in vitro, with in vitro transcribed mRNAs containing both 5mC

and Ψ, requiring a longer time for synthesis of full-length proteins compared to unmodified mRNA (Svitkin et al. 2017).

Translation terminates when the ribosome reaches a nonsense codon (see Hellen 2018; Rodnina 2018). All three possible nonsense codons contain a uridine at the first position, and Karijolich and Yu (2011) have shown that pseudouridylation of this residue suppresses translation termination in vitro and in vivo. An in-depth structural study of the bacterial 30S ribosomal subunit in complex with a pseudouridylated nonsense codon indicates formation of noncanonical base-pairing in the second and third positions (Fernandez et al. 2013). Although it is unclear whether this readthrough mechanism is used naturally, it could be of therapeutic importance, given that many genetic diseases can be attributed to premature translation termination yielding a dysfunctional protein (Karijolich and Yu 2011).

MODIFICATIONS AND INNATE IMMUNITY: GLOBAL EFFECTS ON THE TRANSLATION MACHINERY

Innate immunity, the initial immune response to pathogenic invasion, involves the activation of proteins that distinguish self from nonself by identification of pathogen-associated molecular patterns (PAMPs). DNA and RNA stimulate the mammalian innate immune system by activation of Toll-like receptors (TLRs), retinoic acid-inducible gene I (RIG-I), and the RNA-activated protein kinase R (PKR) (Nallagatla et al. 2011).

Similar to DNA-containing methylated CpG motifs, which do not stimulate TLRs, the modified nucleosides 5mC, m^6A, and Ψ also ablate TLR activity (Kariko et al. 2005). Although 5′-triphosphate RNA is a ligand for RIG-I, 5′ capped or pseudouridylated RNA is not (Hornung et al. 2006; Wang et al. 2010). Consequently, pseudouridylated in vitro transcribed RNA shows significantly enhanced translation compared to nonmodified RNA, whereas a modest effect was also observed when incorporating 5mC (Kariko et al. 2008). Interestingly, enhanced translation was observed in rabbit reticulocyte lysates, which contain the RNA-dependent PKR, but not in wheat germ extracts, which do not (Davis and Watson 1996).

The improved translation of pseudouridylated mRNA is independent of RIG-I activity and further research showed that in vitro transcribed RNA-containing uridine activates PKR, which phosphorylates the α subunit of eIF2, resulting in inhibition of translation (Dever et al. 1992). Replacing uridine with Ψ diminishes PKR activation (Anderson et al. 2010). Curiously, in the same setting, incorporation of m^6A instead of adenosine in mRNAs rendered them untranslatable (Kariko et al. 2008). Recently, Svitkin et al. (2017) showed that N^1-methyl-Ψ outperforms Ψ and several other nucleoside modifications in translation by combining reduced immunogenicity with altered dynamics of the translation process. Collectively, these findings reveal a complex interplay between mRNA modifications, innate immunity, and translation.

DIFFERENTIAL mRNA MODIFICATION AFFECTS TRANSLATION IN COMPLEX BIOLOGICAL PROCESSES

Prior to the emergence of high-throughput methods, studies of the m^6A methyltransferase METTL3 showed the modification to be necessary for elaborate biological processes requiring differentiation, such as embryogenesis in *Arabidopsis thaliana* (Zhong et al. 2008) and oogenesis in *Drosophila melanogaster* (Hongay and Orr-Weaver 2011). The involvement of m^6A in fertility was substantiated when a study of the m^6A demethylase ALKBH5 showed that dysregulated m^6A levels in mice, as a result of ALKBH5 knockout, led to defects in spermatogenesis, possibly because of dysregulated splicing and mRNA expression (Zheng et al. 2013; Tang et al. 2017).

Following the mapping of m^6A in mammalian cells, which revealed the extent and conservation of the modification, its role in differentiation was explored further. Although the first study of m^6A in mouse embryonic stem cells (mESCs) found, by knockdown of the methyltransferases Mettl3 and Mettl14, that m^6A is required for self-renewal of mESCs (Zhu et al.

2014), we and others found that complete knockout of Mettl3 in mESCs abrogates the ability of mESCs to undergo differentiation (Batista et al. 2014; Geula et al. 2015). Differentiation is a cellular process that is hypothesized to require concerted and timely degradation of pluripotency factors, and depletion of m^6A leads to continued expression of pluripotency factors, such as Nanog and Oct4, even as differentiation is triggered. The continued expression of the more stable pluripotent mRNAs, because of their reduced m^6A levels, results in their higher protein levels (Geula et al. 2015).

Another tightly controlled process, early embryogenesis, requires clearance of maternal mRNAs in maternal-to-zygotic transition (MZT) (Lee et al. 2014). Zhao et al. (2017) neatly showed the importance of YTHDF2-mediated removal of m^6A-methylated maternal transcripts for MZT in zebrafish.

The cellular response to stress presents another system that relies on prompt shifts in expression patterns. When we mapped the m^6A methylome, we noticed a dynamic response to various stimuli, such as ultraviolet irradiation and interferon exposure (Dominissini et al. 2012). Recent studies have examined in greater depth the role that m^6A plays in the cellular response to stress. In response to heat shock stress, the m^6A reader YTHDF2 undergoes stress-induced localization to the nucleus, where it protects 5′UTR methylation of pertinent transcripts from demethylation by FTO (Zhou et al. 2015). The protected m^6A residues promote cap-independent translation of heat shock protein 70 (Hsp70) mRNA, providing a mechanism for selective translation under heat shock stress. Knockdown of YTHDF2 disrupts this mechanism, leading to reduced synthesis of Hsp70 after exposure to heat shock stress (Zhou et al. 2015). Further evidence of this mechanism is found in the work of Meyer et al. (2015), which describes transcriptome-wide redistribution of m^6A in response to diverse cellular stresses, resulting in an increased number of mRNAs with 5′UTR methylation capable of promoting cap-independent translation, including similar results regarding the translation of Hsp70 following heat stress.

    Cite this article as *Cold Spring Harb Perspect Biol* doi: 10.1101/cshperspect.a032623

More evidence for the role of m⁶A in regulating translation in response to stress came from researching the role of YTHDF1, where Wang et al. (2015) showed that tethering the noncatalytic amino terminus of YTHDF1 to a transcript augments its translation during recovery from arsenic stress.

Involvement of m⁶A in response to stress also appears to play a part in carcinogenesis. Zhang et al. (2016a,b) found that exposure of breast cancer cells to hypoxia results in reduced m⁶A levels in pluripotency factor NANOG and Kruppel-like factor 4 (KLF4) mRNAs because of both inhibited zinc finger protein 217 (ZNF217)-dependent methylation and increased ALKBH5-dependent demethylation, thus increasing their protein levels and, subsequently, the breast cancer stem-cell phenotype. In contrast to the decrease in m⁶A levels in pluripotency factors in breast cancer cells, examination of the global response of HEK293T cells to hypoxia revealed a general m⁶A increase in poly(A) RNAs, conferring increased mRNA stability and improved recovery of translational efficiency (Fry et al. 2017).

Reduction in m⁶A levels has also been found to promote the tumorigenicity and self-renewal of glioblastoma stem-like cells (GSCs) because of altered mRNA expression of key genes (Cui et al. 2017). In addition, increased ALKBH5 expression is seen in GSCs, where it demethylates transcripts of the transcription factor Forkhead Box M1 (FOXM1), thereby enhancing its expression, thus revealing a new pathway for GSC proliferation and tumorigenesis (Zhang et al. 2017).

In acute myeloid leukemia (AML) a different m⁶A demethylase plays an oncogenic role, although interestingly, by negatively regulating genes important for normal hematopoiesis and differentiation. Unlike the role of ALKBH5 in breast cancer or glioblastoma, FTO promotes leukemogenesis by demethylating target genes, leading to decreased protein levels (Li et al. 2017c). In line with this result, METTL3 plays an oncogenic role in AML by methylating coding regions of genes necessary for leukemogenesis, including the oncogene SP1. This methylation relieves ribosome stalling and leads to increased translation. Disruption of this mechanism resulted in cell-cycle arrest and differentiation of AML cells (Barbieri et al. 2017).

As discussed earlier, m⁶A also positively regulates expression through the activity of METTL3 in its capacity as an m⁶A reader. Increased expression of METTL3 in lung adenocarcinoma promotes growth, survival, and invasion of tumor cells (Lin et al. 2016). Some observations of m⁶A affecting translation are still without mechanistic insight. For example, it has been found that m⁶A allows a flow of information from transcription to translation, as inefficient transcription results in increased m⁶A deposition, causing decreased translation efficiency (Slobodin et al. 2017).

Overall, these findings show that differential methylation, or reading thereof, affects translation in complex biological processes, and as a result plays a key role in determining cell fate. Other recently identified mRNA modifications, such as m¹A (Dominissini et al. 2016) and Ψ (Schwartz et al. 2014), have been reported to be dynamic in response to stress. Deciphering the roles that they play in translational regulation promises to be an interesting avenue of research.

CONCLUDING REMARKS

Research on the epitranscriptome has quickly moved from one of identification and mapping to the investigation of function and of the biological implications of mRNA modifications. Studies of the role that these modifications play in translation paint a complex picture. When examining direct regulation of translation, it appears that the same modification can either promote or repress translation of an mRNA, depending on the location of the modification or the biological system studied. Indirectly, the same modification can either increase mRNA stability, as in the case of hypoxia-induced stabilization of mRNA (Fry et al. 2017), or decrease it, as exemplified in the widely studied YTHDF2-mediated decay of m⁶A-methylated mRNA (Wang et al. 2014; Du et al. 2016). These contrasting effects ultimately have different consequences for translation output. With m⁶A, the most studied modification, we are now beginning to understand that the delicate balance of methylation and demethylation is involved in complex biological processes such as differentiation (Geula et al. 2015) and the stress response (Zhou et al. 2015). Alterations to that balance contribute to different pathologies (Li et al. 2017c; Zhang et al. 2017). The epitranscriptome, consisting of various RNA modifications, is thus beginning to be unraveled as a complex layer of information with major implications for the regulation of translation in healthy and disease states. The continued discovery of more modifications, writers, readers, and erasers will help shape our understanding of how RNA modifications participate in biological and pathological processes, and perhaps design effective pharmacological interventions to disease states.

REFERENCES

*Reference is also in this collection.

Amort T, Rieder D, Wille A, Khokhlova-Cubberley D, Riml C, Trixl L, Jia XY, Micura R, Lusser A. 2017. Distinct 5-methylcytosine profiles in poly(A) RNA from mouse embryonic stem cells and brain. *Genome Biol* **18:** 1.

Anderson BR, Muramatsu H, Nallagatla SR, Bevilacqua PC, Sansing LH, Weissman D, Kariko K. 2010. Incorporation of pseudouridine into mRNA enhances translation by diminishing PKR activation. *Nucleic Acids Res* **38:** 5884–5892.

Barbieri I, Tzelepis K, Pandolfini L, Shi J, Millan-Zambrano G, Robson SC, Aspris D, Migliori V, Bannister AJ, Han N, et al. 2017. Promoter-bound METTL3 maintains myeloid leukaemia by m6A-dependent translation control. *Nature* **552:** 126–131.

Batista PJ, Molinie B, Wang J, Qu K, Zhang J, Li L, Bouley DM, Lujan E, Haddad B, Daneshvar K, et al. 2014. m6A RNA modification controls cell fate transition in mammalian embryonic stem cells. *Cell Stem Cell* **15:** 707–719.

Boccaletto P, Machnicka MA, Purta E, Piątkowski P, Bagiński B, Wirecki TK, de Crécy-Lagard V, Ross R, Limbach PA, Kotter A. 2018. MODOMICS: A database of RNA modification pathways. 2017 update. *Nucleic Acids Res* **46:** D303–D307.

Bokar JA. 2005. The biosynthesis and functional roles of methylated nucleosides in eukaryotic mRNA. In *Fine-tuning of RNA functions by modification and editing* (ed. Grosjean H), pp. 141–177. Springer, Berlin.

Byszewska M, Śmietański M, Purta E, Bujnicki JM. 2014. RNA methyltransferases involved in 5′ cap biosynthesis. *RNA Biol* **11:** 1597–1607.

Carlile TM, Rojas-Duran MF, Zinshteyn B, Shin H, Bartoli KM, Gilbert WV. 2014. Pseudouridine profiling reveals regulated mRNA pseudouridylation in yeast and human cells. *Nature* **515:** 143–146.

* Chekulaeva M, Rajewsky N. 2018. Roles of long noncoding RNAs and circular RNAs in translation. *Cold Spring Harb Perspect Biol* doi: 10.1101/cshperspect.a032680.

Choi J, Ieong KW, Demirci H, Chen J, Petrov A, Prabhakar A, O'Leary SE, Dominissini D, Rechavi G, Soltis SM, et al. 2016. N6-methyladenosine in mRNA disrupts tRNA selection and translation-elongation dynamics. *Nat Struct Mol Biol* **23:** 110–115.

Cohn WE. 1951. Some results of the applications of ion-exchange chromatography to nucleic acid chemistry. *J Cell Comp Physiol* **38:** 21–40.

Coots RA, Liu XM, Mao Y, Dong L, Zhou J, Wan J, Zhang X, Qian SB. 2017. m6A facilitates eIF4F-independent mRNA translation. *Mol Cell* doi: 10.1016/j.molcel.2017.10.001.

Cui Q, Shi H, Ye P, Li L, Qu Q, Sun G, Sun G, Lu Z, Huang Y, Yang CG, et al. 2017. m6A RNA methylation regulates the self-renewal and tumorigenesis of glioblastoma stem cells. *Cell Rep* **18:** 2622–2634.

Daffis S, Szretter KJ, Schriewer J, Li J, Youn S, Errett J, Lin T-Y, Schneller S, Zust R, Dong H. 2010. 2′-O methylation of the viral mRNA cap evades host restriction by IFIT family members. *Nature* **468:** 452.

Dai Q, Moshitch-Moshkovitz S, Han D, Kol N, Amariglio N, Rechavi G, Dominissini D, He C. 2017. Nm-seq maps 2′-O-methylation sites in human mRNA with base precision. *Nat Methods* **14:** 695–698.

Davis S, Watson JC. 1996. In vitro activation of the interferon-induced, double-stranded RNA-dependent protein kinase PKR by RNA from the 3′ untranslated regions of human α-tropomyosin. *Proc Natl Acad Sci* **93:** 508–513.

Delatte B, Wang F, Ngoc LV, Collignon E, Bonvin E, Deplus R, Calonne E, Hassabi B, Putmans P, Awe S, et al. 2016. RNA biochemistry. Transcriptome-wide distribution and function of RNA hydroxymethylcytosine. *Science* **351:** 282–285.

Deryusheva S, Choleza M, Barbarossa A, Gall JG, Bordonné R. 2012. Post-transcriptional modification of spliceosomal RNAs is normal in SMN-deficient cells. *RNA* **18:** 31–36.

Desrosiers R, Friderici K, Rottman F. 1974. Identification of methylated nucleosides in messenger RNA from Novikoff hepatoma cells. *Proc Natl Acad Sci* **71:** 3971–3975.

Desrosiers RC, Friderici KH, Rottman FM. 1975. Characterization of Novikoff hepatoma mRNA methylation and heterogeneity in the methylated 5′ terminus. *Biochemistry* **14:** 4367–4374.

Dever TE, Feng L, Wek RC, Cigan AM, Donahue TF, Hinnebusch AG. 1992. Phosphorylation of initiation factor 2α by protein kinase GCN2 mediates gene-specific translational control of GCN4 in yeast. *Cell* **68:** 585–596.

Dominissini D, Moshitch-Moshkovitz S, Schwartz S, Salmon-Divon M, Ungar L, Osenberg S, Cesarkas K, Jacob-Hirsch J, Amariglio N, Kupiec M, et al. 2012. Topology of the human and mouse m6A RNA methylomes revealed by m6A-seq. *Nature* **485:** 201–206.

Dominissini D, Nachtergaele S, Moshitch-Moshkovitz S, Peer E, Kol N, Ben-Haim MS, Dai Q, Di Segni A, Salmon-Divon M, Clark WC, et al. 2016. The dynamic N1-methyladenosine methylome in eukaryotic messenger RNA. *Nature* **530:** 441–446.

Du H, Zhao Y, He J, Zhang Y, Xi H, Liu M, Ma J, Wu L. 2016. YTHDF2 destabilizes m^{6}A-containing RNA through direct recruitment of the CCR4-NOT deadenylase complex. *Nat Commun* 7: 12626.

Dubin DT, Taylor RH. 1975. The methylation state of poly A-containing messenger RNA from cultured hamster cells. *Nucleic Acids Res* 2: 1653–1668.

Dubin DT, Stollar V, Hsuchen CC, Timko K, Guild GM. 1977. Sindbis virus messenger RNA: The 5′-termini and methylated residues of 26 and 42 S RNA. *Virology* 77: 457–470.

Edupuganti RR, Geiger S, Lindeboom RGH, Shi H, Hsu PJ, Lu Z, Wang S-Y, Baltissen MPA, Jansen PWTC, Rossa M, et al. 2017. N^6-methyladenosine (m^{6}A) recruits and repels proteins to regulate mRNA homeostasis. *Nat Struct Mol Biol* 24: 870–878.

Fernandez IS, Ng CL, Kelley AC, Wu G, Yu YT, Ramakrishnan V. 2013. Unusual base pairing during the decoding of a stop codon by the ribosome. *Nature* 500: 107–110.

Fry NJ, Law BA, Ilkayeva OR, Holley CL, Mansfield KD. 2017. N^6-methyladenosine is required for the hypoxic stabilization of specific mRNAs. *RNA* 23: 1444–1455.

Fu Y, Dominissini D, Rechavi G, He C. 2014. Gene expression regulation mediated through reversible m^{6}A RNA methylation. *Nat Rev Genet* 15: 293–306.

Fustin JM, Doi M, Yamaguchi Y, Hida H, Nishimura S, Yoshida M, Isagawa T, Morioka MS, Kakeya H, Manabe I, et al. 2013. RNA-methylation-dependent RNA processing controls the speed of the circadian clock. *Cell* 155: 793–806.

Ge J, Yu YT. 2013. RNA pseudouridylation: New insights into an old modification. *Trends Biochem Sci* 38: 210–218.

Geula S, Moshitch-Moshkovitz S, Dominissini D, Mansour AA, Kol N, Salmon-Divon M, Hershkovitz V, Peer E, Mor N, Manor YS, et al. 2015. Stem cells. m^{6}A mRNA methylation facilitates resolution of naïve pluripotency toward differentiation. *Science* 347: 1002–1006.

Gumienny R, Jedlinski D, Martin G, Vina-Villaseca A, Zavolan M. 2016. High-throughput identification of C/D box snoRNA targets with CLIP and RiboMeth-seq. *bioRxiv* doi: 10.1101/037259.

He C, Hus JC, Sun LJ, Zhou P, Norman DP, Dotsch V, Wei H, Gross JD, Lane WS, Wagner G, et al. 2005. A methylation-dependent electrostatic switch controls DNA repair and transcriptional activation by *E. coli ada*. *Mol Cell* 20: 117–129.

* Heck AM, Wilusz J. 2018. The interplay between the RNA decay and translation machinery in eukaryotes. *Cold Spring Harb Perspect Biol* doi: 10.1101/cshperspect.a032839.

* Hellen CUT. 2018. Translation termination and ribosome recycling in eukaryotes. *Cold Spring Harb Perspect Biol* doi: 10.1101/cshperspect.a032656.

Hershey JW, Sonenberg N, Mathews MB. 2012. Principles of translational control: An overview. *Cold Spring Harb Perspect Biol* 4: a011528.

Hoernes TP, Clementi N, Faserl K, Glasner H, Breuker K, Lindner H, Hüttenhofer A, Erlacher MD. 2016. Nucleotide modifications within bacterial messenger RNAs regulate their translation and are able to rewire the genetic code. *Nucleic Acids Res* 44: 852–862.

Hongay CF, Orr-Weaver TL. 2011. Drosophila inducer of MEiosis 4 (IME4) is required for Notch signaling during oogenesis. *Proc Natl Acad Sci* 108: 14855–14860.

Hornung V, Ellegast J, Kim S, Brzozka K, Jung A, Kato H, Poeck H, Akira S, Conzelmann KK, Schlee M, et al. 2006. 5′-triphosphate RNA is the ligand for RIG-I. *Science* 314: 994–997.

Huang H, Weng H, Sun W, Qin X, Shi H, Wu H, Zhao BS, Mesquita A, Liu C, Yuan CL, et al. 2018. Recognition of RNA N^6-methyladenosine by IGF2BP proteins enhances mRNA stability and translation. *Nat Cell Biol* 20: 285–295.

Hussain S, Sajini AA, Blanco S, Dietmann S, Lombard P, Sugimoto Y, Paramor M, Gleeson JG, Odom DT, Ule J, et al. 2013. NSun2-mediated cytosine-5 methylation of vault noncoding RNA determines its processing into regulatory small RNAs. *Cell Rep* 4: 255–261.

Imai Y, Matsuo N, Ogawa S, Tohyama M, Takagi T. 1998. Cloning of a gene, YT521, for a novel RNA splicing-related protein induced by hypoxia/reoxygenation. *Brain Res Mol Brain Res* 53: 33–40.

* Ingolia NT, Hussmann JA, Weissman JS. 2018. Ribosome profiling: Global views of translation. *Cold Spring Harb Perspect Biol* doi: 10.1101/cshperspect.a032698.

Jia G, Fu Y, Zhao X, Dai Q, Zheng G, Yang Y, Yi C, Lindahl T, Pan T, Yang YG, et al. 2011. N^6-methyladenosine in nuclear RNA is a major substrate of the obesity-associated FTO. *Nat Chem Biol* 7: 885–887.

Jöckel S, Nees G, Sommer R, Zhao Y, Cherkasov D, Hori H, Ehm G, Schnare M, Nain M, Kaufmann A. 2012. The 2′-O-methylation status of a single guanosine controls transfer RNA–mediated Toll-like receptor 7 activation or inhibition. *J Exp Med* 209: 235–241.

Karijolich J, Yu YT. 2011. Converting nonsense codons into sense codons by targeted pseudouridylation. *Nature* 474: 395–398.

Kariko K, Buckstein M, Ni H, Weissman D. 2005. Suppression of RNA recognition by Toll-like receptors: The impact of nucleoside modification and the evolutionary origin of RNA. *Immunity* 23: 165–175.

Kariko K, Muramatsu H, Welsh FA, Ludwig J, Kato H, Akira S, Weissman D. 2008. Incorporation of pseudouridine into mRNA yields superior nonimmunogenic vector with increased translational capacity and biological stability. *Mol Ther* 16: 1833–1840.

Keith JM, Ensinger MJ, Mose B. 1978. HeLa cell RNA (2′-O-methyladenosine-N^6-)-methyltransferase specific for the capped 5′-end of messenger RNA. *J Biol Chem* 253: 5033–5039.

Khoddami V, Cairns BR. 2013. Identification of direct targets and modified bases of RNA cytosine methyltransferases. *Nat Biotechnol* 31: 458–464.

Kumar S, Mapa K, Maiti S. 2014. Understanding the effect of locked nucleic acid and 2′-O-methyl modification on the hybridization thermodynamics of a miRNA–mRNA pair in the presence and absence of AfPiwi protein. *Biochemistry* 53: 1607–1615.

* Kwan T, Thompson SR. 2018. Noncanonical translation in eukaryotes. *Cold Spring Harb Perspect Biol* doi: 10.1101/cshperspect.a032672.

Lacoux C, Di Marino D, Boyl PP, Zalfa F, Yan B, Ciotti MT, Falconi M, Urlaub H, Achsel T, Mougin A, et al. 2012. BC1-FMRP interaction is modulated by 2′-O-methylation: RNA-binding activity of the tudor domain and translational regulation at synapses. *Nucleic Acids Res* **40:** 4086–4096.

Lee MT, Bonneau AR, Giraldez AJ. 2014. Zygotic genome activation during the maternal-to-zygotic transition. *Annu Rev Cell Dev Biol* **30:** 581–613.

Li F, Zhao D, Wu J, Shi Y. 2014. Structure of the YTH domain of human YTHDF2 in complex with an m6A mononucleotide reveals an aromatic cage for m6A recognition. *Cell Res* **24:** 1490–1492.

Li X, Zhu P, Ma S, Song J, Bai J, Sun F, Yi C. 2015. Chemical pulldown reveals dynamic pseudouridylation of the mammalian transcriptome. *Nat Chem Biol* **11:** 592–597.

Li X, Ma S, Yi C. 2016a. Pseudouridine: The fifth RNA nucleotide with renewed interests. *Curr Opin Chem Biol* **33:** 108–116.

Li X, Xiong X, Wang K, Wang L, Shu X, Ma S, Yi C. 2016b. Transcriptome-wide mapping reveals reversible and dynamic N^1-methyladenosine methylome. *Nat Chem Biol* **12:** 311–316.

Li A, Chen YS, Ping XL, Yang X, Xiao W, Yang Y, Sun HY, Zhu Q, Baidya P, Wang X, et al. 2017a. Cytoplasmic m6A reader YTHDF3 promotes mRNA translation. *Cell Res* **27:** 444–447.

Li X, Xiong X, Zhang M, Wang K, Chen Y, Zhou J, Mao Y, Lv J, Yi D, Chen XW. 2017b. Base-resolution mapping reveals distinct m1A methylome in nuclear-and mitochondrial-encoded transcripts. *Mol Cell* **68:** 993–1005.

Li Z, Weng H, Su R, Weng X, Zuo Z, Li C, Huang H, Nachtergaele S, Dong L, Hu C, et al. 2017c. FTO plays an oncogenic role in acute myeloid leukemia as a N^6-methyladenosine RNA demethylase. *Cancer Cell* **31:** 127–141.

Liang X-h, Liu Q, Fournier MJ. 2009. Loss of rRNA modifications in the decoding center of the ribosome impairs translation and strongly delays pre-rRNA processing. *RNA* **15:** 1716–1728.

Lin J, Lai S, Jia R, Xu A, Zhang L, Lu J, Ye K. 2011. Structural basis for site-specific ribose methylation by box C/D RNA protein complexes. *Nature* **469:** 559–563.

Lin S, Choe J, Du P, Triboulet R, Gregory RI. 2016. The m6A methyltransferase METTL3 promotes translation in human cancer cells. *Mol Cell* **62:** 335–345.

Liu N, Dai Q, Zheng G, He C, Parisien M, Pan T. 2015. N^6-methyladenosine-dependent RNA structural switches regulate RNA-protein interactions. *Nature* **518:** 560–564.

Lu L, Yi C, Jian X, Zheng G, He C. 2010. Structure determination of DNA methylation lesions N1-meA and N3-meC in duplex DNA using a cross-linked protein-DNA system. *Nucleic Acids Res* **38:** 4415–4425.

Luo S, Tong L. 2014. Molecular basis for the recognition of methylated adenines in RNA by the eukaryotic YTH domain. *Proc Natl Acad Sci* **111:** 13834–13839.

Mauer J, Luo X, Blanjoie A, Jiao X, Grozhik AV, Patil DP, Linder B, Pickering BF, Vasseur JJ, Chen Q, et al. 2017. Reversible methylation of m6Am in the 5′ cap controls mRNA stability. *Nature* **541:** 371–375.

* Merrick WC, Pavitt GD. 2018. Protein synthesis initiation in eukaryotic cells. *Cold Spring Harb Perspect Biol* doi: 10.1101/cshperspect.a033092.

Meyer K, Jaffrey SR. 2017. Rethinking m6A readers, writers, and erasers. *Annu Rev Cell Dev Biol* **33:** 319–342.

Meyer KD, Saletore Y, Zumbo P, Elemento O, Mason CE, Jaffrey SR. 2012. Comprehensive analysis of mRNA methylation reveals enrichment in 3′ UTRs and near stop codons. *Cell* **149:** 1635–1646.

Meyer KD, Patil DP, Zhou J, Zinoviev A, Skabkin MA, Elemento O, Pestova TV, Qian SB, Jaffrey SR. 2015. 5′ UTR m6A promotes cap-independent translation. *Cell* **163:** 999–1010.

Nallagatla SR, Toroney R, Bevilacqua PC. 2011. Regulation of innate immunity through RNA structure and the protein kinase PKR. *Curr Opin Struct Biol* **21:** 119–127.

Noeske J, Wasserman MR, Terry DS, Altman RB, Blanchard SC, Cate JH. 2015. High-resolution structure of the *Escherichia coli* ribosome. *Nat Struct Mol Biol* **22:** 336–341.

Patil DP, Chen CK, Pickering BF, Chow A, Jackson C, Guttman M, Jaffrey SR. 2016. m6A RNA methylation promotes XIST-mediated transcriptional repression. *Nature* **537:** 369–373.

* Robichaud N, Sonenberg N, Ruggero D, Schneider RJ. 2018. Translational control in cancer. *Cold Spring Harb Perspect Biol* doi: 10.1101/cshperspect.a032896.

* Rodnina MV. 2018. Translation in prokaryotes. *Cold Spring Harb Perspect Biol* doi: 10.1101/cshperspect.a032664.

Roost C, Lynch SR, Batista PJ, Qu K, Chang HY, Kool ET. 2015. Structure and thermodynamics of N^6-methyladenosine in RNA: A spring-loaded base modification. *J Am Chem Soc* **137:** 2107–2115.

Roundtree IA, Evans ME, Pan T, He C. 2017. Dynamic RNA modifications in gene expression regulation. *Cell* **169:** 1187–1200.

Schwartz S, Bernstein DA, Mumbach MR, Jovanovic M, Herbst RH, Leon-Ricardo BX, Engreitz JM, Guttman M, Satija R, Lander ES, et al. 2014. Transcriptome-wide mapping reveals widespread dynamic-regulated pseudouridylation of ncRNA and mRNA. *Cell* **159:** 148–162.

Shi H, Wang X, Lu Z, Zhao BS, Ma H, Hsu PJ, Liu C, He C. 2017. YTHDF3 facilitates translation and decay of N^6-methyladenosine-modified RNA. *Cell Res* **27:** 315–328.

Shubina M, Musinova Y, Sheval E. 2016. Nucleolar methyltransferase fibrillarin: Evolution of structure and functions. *Biochemistry (Moscow)* **81:** 941–950.

Slobodin B, Han R, Calderone V, Vrielink JA, Loayza-Puch F, Elkon R, Agami R. 2017. Transcription impacts the efficiency of mRNA translation via co-transcriptional N^6-adenosine methylation. *Cell* **169:** 326–337.

Somme J, Van Laer B, Roovers M, Steyaert J, Versées W, Droogmans L. 2014. Characterization of two homologous 2′-O-methyltransferases showing different specificities for their tRNA substrates. *RNA* **20:** 1257–1271.

Squires JE, Patel HR, Nousch M, Sibbritt T, Humphreys DT, Parker BJ, Suter CM, Preiss T. 2012. Widespread occurrence of 5-methylcytosine in human coding and noncoding RNA. *Nucleic Acids Res* **40:** 5023–5033.

Stoilov P, Rafalska I, Stamm S. 2002. YTH: A new domain in nuclear proteins. *Trends Biochem Sci* **27:** 495–497.

Svitkin YV, Cheng YM, Chakraborty T, Presnyak V, John M, Sonenberg N. 2017. N1-methyl-pseudouridine in mRNA enhances translation through eIF2α-dependent and independent mechanisms by increasing ribosome density. *Nucleic Acids Res* **45:** 6023–6036.

Tang C, Klukovich R, Peng H, Wang Z, Yu T, Zhang Y, Zheng H, Klungland A, Yan W. 2017. ALKBH5-dependent m^{6}A demethylation controls splicing and stability of long 3′-UTR mRNAs in male germ cells. *Proc Natl Acad Sci* **115:** E325–E333.

Theler D, Dominguez C, Blatter M, Boudet J, Allain FH. 2014. Solution structure of the YTH domain in complex with *N^6*-methyladenosine RNA: A reader of methylated RNA. *Nucleic Acids Res* **42:** 13911–13919.

Wang Y, Ludwig J, Schuberth C, Goldeck M, Schlee M, Li H, Juranek S, Sheng G, Micura R, Tuschl T, et al. 2010. Structural and functional insights into 5′-ppp RNA pattern recognition by the innate immune receptor RIG-I. *Nat Struct Mol Biol* **17:** 781–787.

Wang X, Lu Z, Gomez A, Hon GC, Yue Y, Han D, Fu Y, Parisien M, Dai Q, Jia G, et al. 2014. *N^6*-methyladenosine-dependent regulation of messenger RNA stability. *Nature* **505:** 117–120.

Wang X, Zhao BS, Roundtree IA, Lu Z, Han D, Ma H, Weng X, Chen K, Shi H, He C. 2015. *N^6*-methyladenosine modulates messenger RNA translation efficiency. *Cell* **161:** 1388–1399.

Wei CM, Gershowitz A, Moss B. 1976. 5′-Terminal and internal methylated nucleotide sequences in HeLa cell mRNA. *Biochemistry* **15:** 397–401.

* Wek RC. 2018. Role of eIF2α kinases in translational control and adaptation to cellular stress. *Cold Spring Harb Perspect Biol* doi: 10.1101/cshperspect.a032870.

Xiao W, Adhikari S, Dahal U, Chen YS, Hao YJ, Sun HY, Li A, Ping XL, Lai WY, et al. 2016. Nuclear m^{6}A reader YTHDC1 regulates mRNA splicing. *Mol Cell* **61:** 507–519.

Xu C, Wang X, Liu K, Roundtree IA, Tempel W, Li Y, Lu Z, He C, Min J. 2014. Structural basis for selective binding of m^{6}A RNA by the YTHDC1 YTH domain. *Nat Chem Biol* **10:** 927–929.

Yang X, Yang Y, Sun BF, Chen YS, Xu JW, Lai WY, Li A, Wang X, Bhattarai DP, Xiao W, et al. 2017a. 5-methylcytosine promotes mRNA export—NSUN2 as the methyltransferase and ALYREF as an m5C reader. *Cell Res* **27:** 606–625.

Yang Y, Fan X, Mao M, Song X, Wu P, Zhang Y, Jin Y, Yang Y, Chen LL, Wang Y, et al. 2017b. Extensive translation of circular RNAs driven by *N^6*-methyladenosine. *Cell Res* **27:** 626–641.

Yildirim I, Kierzek E, Kierzek R, Schatz GC. 2014. Interplay of LNA and 2′-O-methyl RNA in the structure and thermodynamics of RNA hybrid systems: A molecular dynamics study using the revised AMBER force field and comparison with experimental results. *J Phys Chem B* **118:** 14177–14187.

You C, Dai X, Wang Y. 2017. Position-dependent effects of regioisomeric methylated adenine and guanine ribonucleosides on translation. *Nucleic Acids Res* **45:** 9059–9067.

Yu J, Chen M, Huang H, Zhu J, Song H, Zhu J, Park J, Ji SJ. 2017. Dynamic m^{6}A modification regulates local translation of mRNA in axons. *Nucleic Acids Res* **46:** 1412–1423.

Zhang Z, Theler D, Kaminska KH, Hiller M, de la Grange P, Pudimat R, Rafalska I, Heinrich B, Bujnicki JM, Allain FH, et al. 2010. The YTH domain is a novel RNA binding domain. *J Biol Chem* **285:** 14701–14710.

Zhang C, Samanta D, Lu H, Bullen JW, Zhang H, Chen I, He X, Semenza GL. 2016a. Hypoxia induces the breast cancer stem cell phenotype by HIF-dependent and ALKBH5-mediated m^{6}A-demethylation of NANOG mRNA. *Proc Natl Acad Sci* **113:** E2047–E2056.

Zhang C, Zhi WI, Lu H, Samanta D, Chen I, Gabrielson E, Semenza GL. 2016b. Hypoxia-inducible factors regulate pluripotency factor expression by ZNF217- and ALKBH5-mediated modulation of RNA methylation in breast cancer cells. *Oncotarget* **7:** 64527–64542.

Zhang S, Zhao BS, Zhou A, Lin K, Zheng S, Lu Z, Chen Y, Sulman EP, Xie K, Bogler O, et al. 2017. m^{6}A demethylase ALKBH5 maintains tumorigenicity of glioblastoma stem-like cells by sustaining FOXM1 expression and cell proliferation program. *Cancer Cell* **31**.

Zhao BS, Wang X, Beadell AV, Lu Z, Shi H, Kuuspalu A, Ho RK, He C. 2017. m^{6}A-dependent maternal mRNA clearance facilitates zebrafish maternal-to-zygotic transition. *Nature* **542:** 475–478.

Zheng G, Dahl JA, Niu Y, Fedorcsak P, Huang CM, Li CJ, Vagbo CB, Shi Y, Wang WL, Song SH, et al. 2013. ALKBH5 is a mammalian RNA demethylase that impacts RNA metabolism and mouse fertility. *Mol Cell* **49:** 18–29.

Zhong S, Li H, Bodi Z, Button J, Vespa L, Herzog M, Fray RG. 2008. MTA is an Arabidopsis messenger RNA adenosine methylase and interacts with a homolog of a sex-specific splicing factor. *Plant Cell* **20:** 1278–1288.

Zhou J, Wan J, Gao X, Zhang X, Jaffrey SR, Qian SB. 2015. Dynamic m^{6}A mRNA methylation directs translational control of heat shock response. *Nature* **526:** 591–594.

Zhu T, Roundtree IA, Wang P, Wang X, Wang L, Sun C, Tian Y, Li J, He C, Xu Y. 2014. Crystal structure of the YTH domain of YTHDF2 reveals mechanism for recognition of *N^6*-methyladenosine. *Cell Res* **24:** 1493–1496.

Züst R, Cervantes-Barragan L, Habjan M, Maier R, Neuman BW, Ziebuhr J, Szretter KJ, Baker SC, Barchet W, Diamond MS. 2011. Ribose 2′-O-methylation provides a molecular signature for the distinction of self and nonself mRNA dependent on the RNA sensor Mda5. *Nat Immunol* **12:** 137–143.

Roles of Long Noncoding RNAs and Circular RNAs in Translation

Marina Chekulaeva and Nikolaus Rajewsky

Berlin Institute for Medical Systems Biology, Max Delbrück Center for Molecular Medicine, 13125 Berlin, Germany

Correspondence: marina.chekulaeva@mdc-berlin.de

Most of the eukaryotic genome is pervasively transcribed, yielding hundreds to thousands of long noncoding RNAs (lncRNAs) and circular RNAs (circRNAs), some of which are well conserved during evolution. Functions have been described for a few lncRNAs and circRNAs but remain elusive for most. Both classes of RNAs play regulatory roles in translation by interacting with messenger RNAs (mRNAs), microRNAs (miRNAs), or mRNA-binding proteins (RBPs), thereby modulating translation in *trans*. Moreover, although initially defined as noncoding, a number of lncRNAs and circRNAs have recently been reported to contain functional open reading frames (ORFs). Here, we review current understanding of the roles played by lncRNAs and circRNAs in protein synthesis and discuss challenges and open questions in the field.

circRNAs: BIOGENESIS AND CHALLENGES IN DETECTION AND FUNCTIONAL STUDIES

Hundreds to thousands of different circular RNAs (circRNAs) are expressed in eukaryotic cells (reviewed in Rybak-Wolf et al. 2015; Salzman 2016). circRNAs represent an important class of RNAs, created by head-to-tail splicing, or backsplicing, that link the 3' end of an exon to the 5' end of an upstream exon (Fig. 1) (Salzman et al. 2012, 2013; Jeck et al. 2013; Memczak et al. 2013). Among other forms of covalent RNA circles are certain RNA genomes (viroids, the hepatitis delta virus) and intron-derived RNAs (reviewed in Lasda and Parker 2014). In addition to canonical splicing factors and signals, backsplicing requires specific intronic sequences that mediate circularization, bringing the downstream donor and upstream acceptor sites into close proximity (Ashwal-Fluss et al. 2014; Zhang et al. 2014). Circularization can be promoted via complementary intronic sequences such as Alu repeats, or through the binding of introns by RNA-binding proteins (RBPs) (Dubin et al. 1995; Jeck et al. 2013; Ashwal-Fluss et al. 2014; Liang and Wilusz 2014; Zhang et al. 2014; Conn et al. 2015; Ivanov et al. 2015). For example, the splicing factor Muscleblind (Mbl) binds to introns flanking the second exon of its own transcript, facilitating circMbl formation (Ashwal-Fluss et al. 2014). It also has been proposed that the RBP Quaking may promote the biogenesis of multiple circRNAs via a similar mechanism by binding to specific sites in the flanking introns (Conn et al. 2015). ADAR1 negatively regulates circRNA

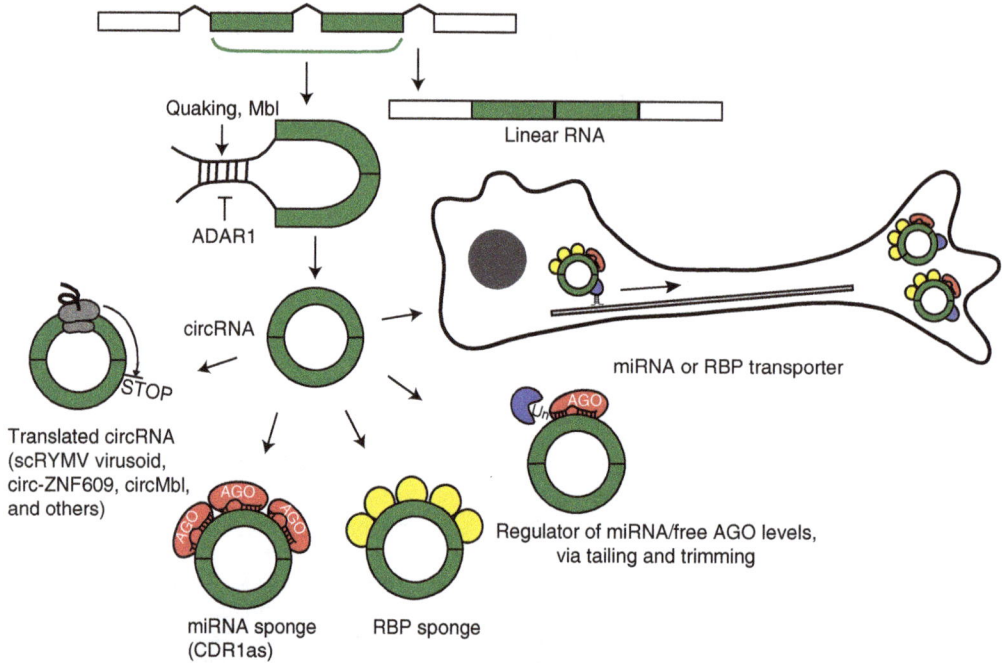

Figure 1. Scheme illustrating biogenesis and regulatory roles of circular RNAs (circRNAs) in translation. circRNAs form as a result of backsplicing, when flanking introns are brought into close proximity because of complementarity or through the binding of introns by specific RBPs. circRNAs play regulatory roles in translation, by acting as sponges and possibly transporting microRNAs (miRNAs) and RNA-binding proteins (RBPs), as well as regulating miRNA levels via tailing and trimming. Some circRNAs contain translatable open reading frames (ORFs). Regulatory proteins are shown, and examples are in parentheses. See text for more details.

biogenesis through A-to-I editing, thereby decreasing RNA base-pairing between the flanking introns (Ivanov et al. 2015).

circRNAs are mostly cytoplasmic (Salzman et al. 2012; Jeck et al. 2013; Memczak et al. 2013; Rybak-Wolf et al. 2015), raising a question as to the mechanism of their nucleocytoplasmic export. Messenger RNA (mRNA) export to the cytoplasm is mediated by the transport/export complex (TREX) that interacts with mRNA primarily via the exon junction complex (EJC), deposited on spliced mRNA (reviewed in Delaleau and Borden 2015). As circRNAs are spliced, their export to the cytoplasm might be mediated via similar mechanisms. As covalent circles, circRNAs are resistant to degradation by RNA exonucleases, and therefore have unusually long half-lives, exceeding 24–48 hours (Jeck et al. 2013; Memczak et al. 2013). That poses another question—how circRNAs are degraded—as it

has been observed that during neuronal development, circRNAs appear to be cleared rapidly (Rybak-Wolf et al. 2015). Curiously, the circRNA CDR1as (cerebellar degeneration-related protein 1 antisense) has an almost perfectly complementary binding site for miR-671, leading to slicing of CDR1as (Hansen et al. 2011; Piwecka et al. 2017). Because of high complementarity of the target, miR-671 is destabilized via a mechanism that involves modification of microRNA (miRNA) ends, adding ("tailing") and removing ("trimming") nucleotides (reviewed in Duchaine and Fabian 2018). These data provide insight into the mechanisms of circRNA decay and release of the bound cargo. Based on the abundance of circRNAs in extracellular vesicles, Lasda and Parker (2014) suggested that formation of such vesicles might serve as an alternative mechanism for circRNA disposal.

Cite this article as *Cold Spring Harb Perspect Biol* doi: 10.1101/cshperspect.a032680

Initially, circRNAs were viewed as splicing artifacts or noise (Cocquerelle et al. 1993; Pasman et al. 1996). This opinion has recently changed in light of a number of findings. First, circRNAs have cell-type-specific expression patterns and are particularly abundant in the brain (Salzman et al. 2013; Rybak-Wolf et al. 2015; Szabo et al. 2015; You et al. 2015). Second, many circRNAs are regulated independently of their linear counterparts, and are well conserved (Jeck et al. 2013; Memczak et al. 2013; Ashwal-Fluss et al. 2014; Rybak-Wolf et al. 2015; You et al. 2015). Also, well-expressed circRNAs have enhanced conservation of nucleotides at third codon positions compared with upstream or downstream codons in bracketing coding sequences (Salzman et al. 2013; Rybak-Wolf et al. 2015; Szabo et al. 2015; You et al. 2015). Finally, biological functions have been reported for some, as discussed in the next section.

The number of circRNAs to which biological functions have been ascribed is limited. One of the reasons is the challenge of depleting circRNAs without affecting their linear counterparts. One method of doing so has been to design small interfering RNAs (siRNAs) that target the circRNA-specific splice junction. However, it is usually not possible to design more than a few siRNAs per junction, which makes it difficult to cope with off-target effects. Uncovering intronic elements that facilitate circRNA biogenesis (Jeck et al. 2013; Ashwal-Fluss et al. 2014; Ivanov et al. 2015) provided a new approach to knocking down circRNAs without affecting linear transcripts. Indeed, disrupting intronic complementary sequences with CRISPR/Cas strongly reduced the levels of circGCN1L1 (Zhang et al. 2016). However, for numerous circRNAs, there are no easily identifiable intronic complementary sequences.

The detection of circRNAs also poses several challenges. They can be identified in RNA sequencing data through reads that map to the backsplicing junction (Glazar et al. 2014; Jeck and Sharpless 2014; Gao et al. 2015). However, some of these reads might be the result of *trans*-splicing, making it important to validate the circularity of individual candidates by treatment with exonucleases to eliminate linear RNAs.

This is also relevant when using vectors that overexpress circRNAs, as they also produce linear transcripts, sometimes as concatemers (Pamudurti et al. 2017), making it challenging to determine which form is responsible for a phenotype.

REGULATORY FUNCTIONS OF circRNAs IN TRANSLATION

The main regulatory function described for circRNAs so far is their action as sponges, which sequester miRNAs (Fig. 1) (Hansen et al. 2013; Memczak et al. 2013). miRNAs function as guides by pairing with partially complementary sites present in their target mRNAs, and subsequently recruiting a complex of proteins that cause mRNA deadenylation, decay, and translational repression (reviewed in Jonas and Izaurralde 2015; Duchaine and Fabian 2018). This means that RNAs with multiple miRNA-binding sites can compete for miRNA binding, leading to the stabilization and translational activation of miRNA-targeted mRNAs. circRNAs are particularly suited for this function because they are abundant and, as covalent circles without a poly(A) tail, they resist miRNA-mediated deadenylation and decay. Two recent studies (Hansen et al. 2013; Memczak et al. 2013) showed that the circRNA CDR1as contains 63 conserved binding sites for miR-7 and depletes miR-7 in neuronal tissues. Indeed, the expression of human CDR1as in zebrafish leads to defects in midbrain development that are reminiscent of an miR-7 knockdown, and a CDR1as knockout mouse showed impaired sensorimotor gating, a deficit associated with neuropsychiatric disorders (Piwecka et al. 2017). Moreover, biochemical miRNA-binding analyses in postmortem brains by "chimera analyses" (Grosswendt et al. 2014) showed that CDR1as is one of the most highly miRNA-bound transcripts in the human and mouse brain. circRNA deriving from the ZNF91 (zinc-finger 91) locus contains 24 miR-23 sites, making it the next best miRNA-sponge candidate (Guo et al. 2014). Another example of a circRNA hypothesized to function as an miRNA sponge is heart-related circRNA (HRCR) (Wang et al. 2016). This has been suggested to play a protective role in cardiac hypertrophy by sequestering miR-223 via its six binding sites. However, bioinformatic analyses suggest that most circRNAs do not function as miRNA sponges, as they do not harbor more AGO2/miRNA-binding sites than linear RNAs (Memczak et al. 2013; Guo et al. 2014; Rybak-Wolf et al. 2015; You et al. 2015). Such a function would also depend on the relative concentrations of the circRNAs and miRNAs in a specific subcellular compartment, but these numbers are difficult to estimate at present, especially in highly polarized cells such as neurons.

It is likely that circRNAs can also function as docking sites for RBPs to sequester them from their target RNAs or mediate their subcellular localization (Fig. 1). Curiously, circRNAs are particularly abundant in synaptosomes (Rybak-Wolf et al. 2015; You et al. 2015; Zappulo et al. 2017), suggesting that they may participate in local RNA regulation. For example, they might transport miRNAs and RBPs to synapses and release them in response to specific stimuli. One possible way of releasing bound factors from a circRNA is if they contain a perfectly complementary site to an miRNA, which can then be cleaved. The roles of circRNAs in the regulation of transcription and splicing are reviewed elsewhere (Ebbesen et al. 2016).

TRANSLATION OF circRNAs

The scanning model of initiation suggested that eukaryotic ribosomes require free 5′ mRNA ends to initiate translation, allowing them to thread onto mRNA like beads on a string (Kozak 1978, 1979; Konarska et al. 1981). This view was challenged by the discovery of internal ribosome entry site (IRES) elements that can directly recruit 40S ribosomal subunits (reviewed in Jackson 2005; Kwan and Thompson 2018). Indeed, an early study by Chen and Sarnow (1995) showed that eukaryotic small ribosomal subunits can initiate translation in vitro from covalent RNA circles containing the encephalomyocarditis virus (EMCV) IRES (Chen and Sarnow 1995). Consistent with this finding, IRES-bearing circRNAs generated from a reporter plasmid via backsplicing produced proteins

(Wang and Wang 2015). Another study suggested that an artificial circRNA with an infinite open reading frame (ORF), that is, lacking a stop codon, can be translated even in the absence of a detectable IRES (Abe et al. 2015).

Several recent studies have reported that endogenously produced circRNAs can be translated in vivo (AbouHaidar et al. 2014; Legnini et al. 2017; Pamudurti et al. 2017; Yang et al. 2017). AbouHaidar et al. (2014) detected the translation product of a circular satellite RNA virusoid associated with rice yellow mottle virus (scRYMV) in a wheat germ in vitro translation system. The protein corresponds to an unusual ORF within scRYMV, combining initiation and termination codons in a UGAUGA sequence.

Another example of circRNA translation is provided by a circular form of the ZNF609 transcript (circ-ZNF609) that regulates myoblast proliferation (Legnini et al. 2017). As circ-ZNF609 contains an ORF that overlaps with the main ORF of linear ZNF609, the investigators tested whether the circular form could be translated. They tagged the endogenous ZNF609 locus with a Flag tag in a way such that the Flag peptide would be produced only from the circular form of the transcript. Mass spectrometry analysis of anti-Flag immunoprecipitates identified a ZNF609-specific peptide, supporting the idea of circ-ZNF609 translation. The mechanism by which this occurs is unclear, but it is ~100-fold less efficient than the translation of its linear counterpart and is dependent on splicing. Importantly, the investigators observed a strong increase in the translation of circ-Flag-ZNF609 on heat-shock stress. Stress leads to eIF4G proteolysis, separating its cap-binding function from its helicase and ribosome-binding activity (reviewed in Holcik and Sonenberg 2005). As a result, cap-dependent translation is inhibited, and the translation of IRES-containing mRNAs is promoted. This result led the investigators to speculate that the translation of circ-ZNF609 might be low by default but activated in response to specific stimuli that inhibit cap-dependent translation.

The study of Pamudurti et al. (2017) searched for ribosome-associated circRNAs (ribo-circRNAs) in the heads of *Drosophila* using ribosome profiling (Riboseq), a technique that can yield a genome-wide snapshot of ribosome footprints on mRNAs (Ingolia et al. 2009, 2018). As circRNAs contain a unique head-to-tail splice junction that is not present in their linear counterparts, they can be distinguished from linear transcripts in RNAseq and Riboseq data. To distinguish reads reflecting bona fide translation from background signals, the authors examined the pattern of Riboseq reads across the circRNA-specific splice junction. A known hallmark of translation is the three-nucleotide (nt) phasing of Riboseq reads, which reflects the codon-by-codon movement of translating ribosomes (Calviello et al. 2016). In addition, accumulation of reads around start and stop codons is often observed for translated ORFs. As start codons and most of the ORF are usually shared between the linear and circular transcripts, Pamudurti et al. (2017) concentrated on the stop codons. Indeed, an enrichment of Riboseq reads around the putative stop codon of the most abundant ribo-circRNA, circMbl, supported the translation of this circRNA. The *mbl* locus produces several circRNAs (Ashwal-Fluss et al. 2014). The authors used targeted mass spectrometry to search for peptides that could be produced from these circRNAs. An analysis of MBL immunoprecipitates identified a peptide that could only be produced by circMbl3, but not by linear Mbl. In findings similar to those of Legnini et al. (2017), Pamudurti et al. (2017) suggested that translation of circMbl is modulated in response to specific stimuli that interfere with cap-dependent translation.

A study by Yang et al. (2017) suggested that the translation of circRNAs might be driven by m^6A RNA modifications (m^6A effects on protein synthesis are reviewed in Peer et al. 2018). The authors generated split GFP reporters that would restore the GFP ORF on circularization. Surprisingly, GFP protein was detected for all of the transfected constructs, independently of whether they contained an IRES element or not. Yang et al. (2017) hypothesized that this is caused by the m^6A motif RRACH (R = G or A; H = A, C, or U) (Csepany et al. 1990; Harper et al. 1990), which had by chance been included in all the plasmids. Mutations of the correspond-

ing motifs reduced levels of GFP. To search for endogenous circRNAs that were potentially translated in human 293 cells, the authors employed three different approaches: (1) identification of m^6A-modified circRNAs; (2) searching for circRNAs associated with polysomes; and (3) looking for peptides matching ORF-containing circRNAs. Each approach resulted in dozens to hundreds of potential candidates, although it is difficult to estimate the degree of overlap between the approaches, due to differences in the sensitivity of the methods used and the usage of different cell lines for different types of analyses. The underlying mechanisms of translation remain mysterious. Based on small hairpin RNA (shRNA) depletion and coimmunoprecipitation experiments, Yang et al. (2017) suggested that the m^6A-binding protein YTH3 recruits eIF4G2/DAP5 to promote cap-independent initiation. This is plausible given a study from Liberman et al. (2015), which showed that eIF4G2, instead of canonical eIF4G, takes part in initiation at a subset of cellular IRESes.

Interestingly, an earlier study of Meyer et al. (2015) used in vitro translation systems to show that m^6A in 5′UTRs (untranslated regions) can indeed initiate translation independently of the cap, via the recruitment of eIF3. However, this mechanism did not involve internal ribosome entry and required free 5′ ends. Meyer et al. (2015) showed this in several types of experiments. First, by incubating m^6A-containing mRNAs bearing two start codons using 40S subunits, eIF1, -1A, -2 and -3, and Met-tRNA$_i^{Met}$, they showed that 48S complexes assemble almost exclusively at the first AUG. This outcome can be easily explained if the preinitiation complex is recruited to the 5′ end and initiates at the first suitable start codon, but it is unlikely to result from internal-entry initiation. Second, when translated in HeLa lysates, an m^6A-modified reporter mRNA with two in-frame start codons produced almost exclusively the longer protein product, which relied on initiation at the first AUG. Finally, the insertion of a stable hairpin at the extreme 5′ end to block 5′ end-dependent initiation inhibited the translation of the m^6A reporter. Interestingly, a similar mode of initiation—independent of the cap, but dependent on the 5′ end—was described for an mRNA reporter carrying an eIF4G-binding domain from the EMCV IRES (Terenin et al. 2013), indicating that simply binding to the 40S subunit is not sufficient for internal initiation—an IRES is also providing for a mechanism for internal entry. These data suggested that m⁶A alone is rather unlikely to drive translation from circRNAs.

Yet another mechanism, involving the recruitment of the YHT-domain containing family protein 1 (YTHDF1), has been proposed to explain the involvement of 3′UTR-located m⁶A in translation (Wang et al. 2015). The authors detected a number of proteins among interactors of YTHDF1, including eIF3 subunits, YBX1, IGF2BP1, G3BP1, and PCBP2, and suggested that they might function collectively to affect the translation of methylated mRNA. So, there is no consensus on the matter whether m⁶A can drive internal initiation and by which mechanisms.

To sum up, several recent studies have provided evidence that some circRNAs can be translated. The evidence includes the presence of ORFs, association with polysomes, enrichment of Riboseq reads around putative stop codons, and identification of peptides corresponding to circRNA sequences based on mass spectrometry. Translated circRNAs possess special properties such as a high stability and independence from the 5′ cap and poly(A) tail that are required for the translation of most eukaryotic mRNAs. Thus, circRNAs could have adapted in ways that lead to translation in special cases where these properties are advantageous. Indeed, reports indicate that the translation of circRNAs is inefficient under normal conditions and is preferentially activated when cap-dependent translation is inhibited (Legnini et al. 2017; Pamudurti et al. 2017). A number of mechanistic questions remain to be resolved through future studies. What drives initiation on circRNAs, and which translation factors are required? If a ribosome terminates translation on a circRNA, does it reinitiate on the same message? Do circRNAs contain ORFs lacking an in-frame termination codon, producing long repetitive proteins as a result of multiple rounds of translation?

How is the translation of circRNAs regulated? The functions of the proteins produced from circRNAs remain unclear as well. One point to note is that many circRNAs share the 5′ part of their potential ORFs with their linear counterparts, which would give them the potential to encode proteins containing only the amino-terminal portion of the protein and to function as dominant negatives.

Rigorous technical procedures will be required to dissect the mechanisms underlying circRNA translation. For example, Pamudurti et al. (2017) observed that circRNA plasmid reporter constructs tend to generate not only circRNAs, but also linear concatemers that encode the same peptides. This makes it important to test the true circularity of the RNA products by using RNase R treatment. It is also essential to include a test for internal initiation, such as a bicistronic assay and/or in vitro circularized RNA, instead of relying exclusively on the cap-independency test (reviewed in Terenin et al. 2017). One must take into account the fact that bicistronic assays are prone to artifacts that result from cryptic promoters or splicing events that produce monocistronic capped mRNAs. Lloyd and colleagues (Van Eden et al. 2004) proposed a stringent test to assess the integrity of bicistronic constructs (Fig. 2): RNAi against the first cistron is performed on cells transfected with either the bicistronic construct or two corresponding monocistronic constructs. In the first case, the production of proteins from both cistrons is expected to be equally reduced. In the second case, the production of the protein produced from the first cistron will be reduced, but the second will remain unaffected. If the result for a bicistronic construct lies somewhere in between (i.e., the product of the second cistron is reduced, but less efficiently than that of the first), this probably means that some amount of monocistronic, possibly capped, mRNA has been produced from the second cistron. Given the fact that tested circRNAs and the corresponding bicistronic reporters have generally produced low amounts of protein, it is particularly important to control for the integrity of the mRNAs that are analyzed. For the same reason, assessing the ratios of two cistrons through

Cite this article as *Cold Spring Harb Perspect Biol* doi: 10.1101/cshperspect.a032680

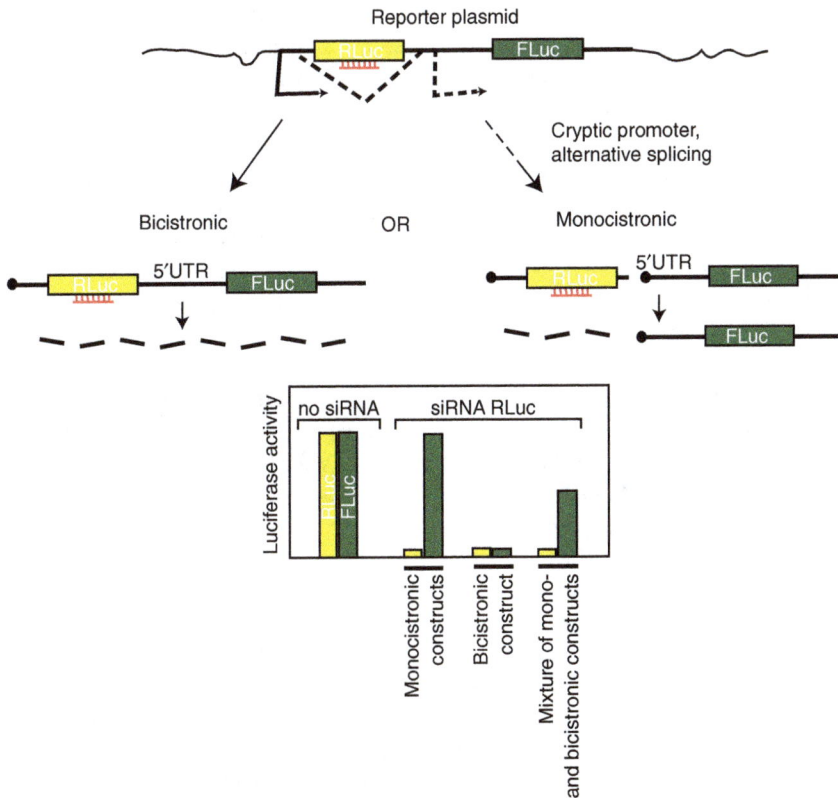

Figure 2. Small interfering RNA (siRNA) test for the integrity of bicistronic reporters. (*Top*) Transcription gives rise to a bicistronic mRNA (*left*) or to two monocistronic mRNAs (*right*), with the RNAi fates depicted. (*Bottom*) Predicted results of luciferase assays. See text for details.

qPCR does not suffice: even if capped monocistronic mRNA represents such a small percentage of bicistronic mRNA that it cannot be detected by qPCR (~1%–2%), it could still be responsible for most of the protein product.

ANNOTATION OF lncRNAs

Long noncoding RNAs (lncRNAs) are broadly defined as noncoding RNAs longer than 200 nt, to distinguish them from transfer RNAs (tRNAs), miRNAs, Piwi-interacting (piRNAs), and other classes of noncoding RNAs (reviewed in Perry and Ulitsky 2016). lncRNAs are classified based on their location relative to protein-coding genes: most are encoded by intergenic regions (termed lincRNAs for long intergenic, or intervening, RNAs), but some overlap with protein-coding genes. Although thousands of

loci in vertebrates produce lncRNAs, biological functions have been ascribed to only a small fraction of them. lncRNAs exhibit lower levels of synthesis, processing, and stability than mRNAs (Mukherjee et al. 2017). They also exhibit a lower level of conservation than protein-coding genes. For example, of ~1000 lncRNAs that are moderately to highly expressed in humans, only ~300 are conserved outside mammals in other vertebrates (Hezroni et al. 2015). Despite the popular belief that lncRNAs are primarily nuclear, recent studies showed that they exhibit a variety of localization patterns, from nuclear to almost exclusively cytoplasmic localization (Derrien et al. 2012; Djebali et al. 2012; Cabili et al. 2015; Mukherjee et al. 2017). Thus, analysis of 31 lncRNAs in three cell lines with single-molecule RNA-FISH showed that ~55% of them were mostly nuclear, ~40% exhibited both nuclear and cytoplasmic localization, and ~5% were mostly cytoplasmic. Consistently, analysis of RNAseq data showed that the nuclear/cytoplasmic ratio is moderately higher for lncRNAs than for mRNAs (Derrien et al. 2012; Djebali et al. 2012; Cabili et al. 2015; Mukherjee et al. 2017). A number of lncRNAs have been implicated in the regulation of cell proliferation, apoptosis, response to stress, and other processes. Based on the mechanisms by which they function, lncRNAs can be loosely divided into three groups (reviewed in Perry and Ulitsky 2016). The first group includes lncRNAs for which only the fact of transcription itself is important, for example, through modification of chromatin in the locus. The second group comprises *cis*-acting lncRNAs that accomplish their functions by attracting *trans*-acting factors to the chromatin locus, thereby modulating transcription. The third group, *trans*-acting lncRNAs, function independently of their transcription sites by recruiting other RNAs or RBPs. Some of these lncRNAs play a role in the regulation of mRNA translation and stability. We discuss this group in more detail below.

REGULATORY ROLES OF lncRNAs IN TRANSLATION

Some lncRNAs sequester and modulate the activity of RBPs and thus affect translation and stability of the mRNAs that these proteins target (Fig. 3). For example, recent studies (Lee et al. 2016; Tichon et al. 2016) showed that the lncRNA NORAD (noncoding RNA activated by DNA damage) harbors 17 binding sites for the mammalian PUF family proteins Pumilio1 and Pumilio2. PUF family proteins are conserved from yeast to animals and plants, in which they deadenylate and repress the translation of their target mRNAs (reviewed in Miller and Olivas 2011). Because Pumilio targets are enriched in genes involved in chromosome segregation, the sequestration of Pumilio by NORAD plays an important role in maintaining genomic stability. The importance of a precise regulation of Pumilio levels can be gleaned from observations that Pumilio1 haploinsufficiency leads to neurodegeneration in mice (Gennarino et al. 2015).

Carrieri et al. (2012) have described a mechanism by which antisense lncRNA can activate translation. Antisense *Uchl1* is a neuron-specific lncRNA transcribed from the opposite strand of the protein-coding gene *Uchl1*. UCHL1 protein is a deubiquitinating enzyme whose loss leads to ataxia and axonal degeneration (Saigoh et al. 1999). Antisense *Uchl1* regulates protein synthesis through the combined activities of two domains: the 5′ antisense region provides specificity through its complementarity to the sense mRNA, and a SINEB2 repeat confers translational activation via an unknown mechanism. The localization of antisense *Uchl1* to the cytoplasm is controlled by signaling pathways such as mammalian target of rapamycin (mTOR) (Carrieri et al. 2012). The authors suggest that antisense *Uchl1* belongs to a new functional class of antisense lncRNAs that act via similar mechanisms involving SINEB2 repeats (Fig. 3).

lncRNAs have also been reported to function as *trans*-acting factors in Staufen-mediated mRNA decay (SMD) (Gong and Maquat 2011). SMD degrades translationally active mRNAs bound by the dsRNA-binding protein Staufen. Staufen-binding sites can be formed through imperfect base-pairing between an Alu element in the 3′UTR of the target mRNA and an Alu element in a lncRNA. For example, the Alu elements within SERPINE1 and FLJ21870 3′UTRs have the potential to base-pair with the lncRNA AF087999 (designated as 1/2-sbsRNA). This finding shows another functional strategy employed by lncRNAs in modulating mRNA metabolism by helping to recruit specific RBPs (Fig. 3).

TRANSLATION OF lncRNAs

By definition, noncoding RNAs lack conserved protein-coding reading frames. Yet, a number of recent studies have reported the translation of RNAs formerly annotated as noncoding (Guttman et al. 2013; Slavoff et al. 2013; Ingolia et al. 2014; Prabakaran et al. 2014; Mackowiak et al. 2015; D'Lima et al. 2017; for reviews, see Andrews and Rothnagel 2014; Saghatelian and Couso 2015). Several approaches have been used to find RNAs with coding potential among

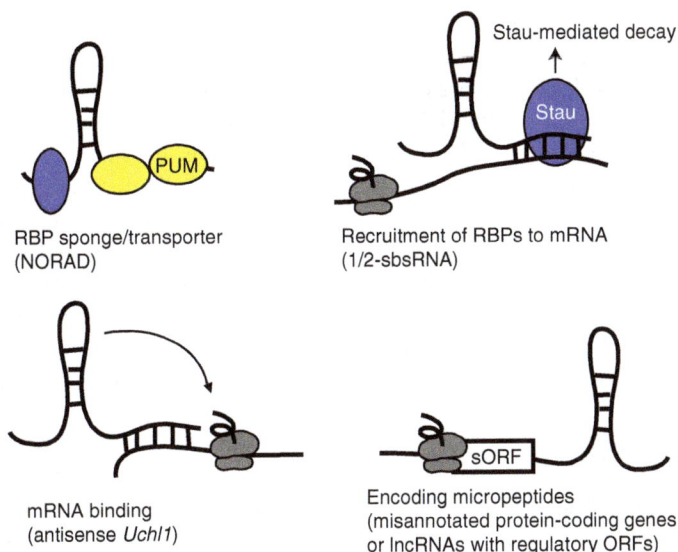

Figure 3. Scheme illustrating regulatory roles of long noncoding RNAs (lncRNAs) in translation. lncRNAs act in sequestering messenger RNA (mRNA)-binding proteins (RBPs), facilitating recruitment of RBPs to mRNAs, or interacting with mRNAs directly. A number of annotated lncRNAs encode micropeptides and thus represent bona fide protein-coding genes. See text for more details. sORF, Short ORF.

sequences previously assumed to be noncoding: (1) a computational search for translatable ORFs and evolutionary conservation analysis; (2) ribosome profiling that uses filters to detect known hallmarks of translation; and (3) mass spectrometry to identify peptides encoded by ORFs that are found exclusively in RNAs that have been annotated as noncoding.

sORFs IN ANNOTATED NONCODING TRANSCRIPTS

An ORF is defined as a continuous stretch of in-frame sense codons that begins with a start codon and ends with a stop codon. However, not all ORFs are translated and defining the parameters that distinguish those that are translated from those that are not has been challenging. Historically, protein-coding genes were discovered using strategies that assumed that true ORFs would be relatively long, usually >100 codons (Claverie 1997). Such strategies were based on the assumptions that (1) shorter ORFs can occur through random processes, and (2) peptides with fewer than 100 amino acids are unlikely to form structures stable enough to perform biological functions. A more recent algorithm, PhyloCSF, is based on conservation and broadens the size range used to predict coding ORFs (Lin et al. 2011; Bazzini et al. 2014). This fits with an increasing body of evidence demonstrating that short ORFs (sORFs) can be translated to produce proteins with important biological functions (reviewed in Andrews and Rothnagel 2014; Saghatelian and Couso 2015).

Such findings have motivated recent studies to search for sORFs among RNAs that have long been annotated as noncoding. An analysis of intergenic regions has identified ~41,000 sORFs in mice (Frith et al. 2006) and between 7159 (Hanada et al. 2007) and 33,809 (Lease and Walker 2006) in *Arabidopsis*. Taking into account additional parameters, such as evolutionary conservation and a sequence's location in transcribed regions, led to predictions that ~5% of these sORFs might be translated (Hanada et al. 2007). Recent work (Mackowiak et al. 2015) offered an integrated computational pipeline for the identification of conserved sORFs in human, mouse, zebrafish, *Drosophila*, and *Caenorhabditis elegans*. By filtering potential amino acid sequences for conservation, the authors identified ∼2000 sORFs in mRNA 3′ and 5′UTRs and in RNAs that had been annotated as noncoding. Cross-checking this prediction with experimental ribosome profiling data has offered evidence in support of translation for 110 sORFs, and peptidomic evidence has been found for 74 sORFs, 36 of which were mapped to sequences annotated as lncRNAs. Interestingly, the peptides they encode are enriched in disordered regions and short linear interaction motifs that match motifs from the ELM (eukaryotic linear motifs) database (Dinkel et al. 2016).

EVIDENCE FROM RIBOSOMAL PROFILING

The computational prediction of ORFs is a powerful technique that is most efficient for conserved ORFs under selective pressure to preserve a specific amino acid sequence that results in a functional protein. However, this approach fails to detect species-specific and very short ORFs. Translation could play a regulatory role by modulating RNA stability, for example, in which case no selective pressure on the protein sequence is expected. Therefore, ribosome profiling, a technique that identifies ribosome-protected RNA fragments (Ingolia et al. 2009), has been used to map ORFs that are translated within RNAs that have been annotated as noncoding. The first studies showed that ribosome occupancy per se is not sufficient to define translated RNAs. For nearly half of the lncRNAs, the density of ribosome footprints was similar to known protein-coding transcripts (Ingolia et al. 2011). To distinguish between ribosome profiling reads that reflect bona fide translation from the background signal, a number of additional criteria were suggested. One approach was based on the fact that ribosomes generally dissociate from mRNA at the end of the coding sequence after translation. Guttman et al. (2013) introduced a ribosome release score (RRS), defined as the ratio between the number of reads within the putative ORF and those within the 3′UTR. For lncRNAs, this produced a median RRS ∼100 times lower than for known protein-coding RNAs, suggesting that lncRNAs are not translated. An important parameter in separating the true ribosome footprints and background reads

resulting from secondary structures and nonribosomal RNPs is the observation that ribosome-protected RNA fragments have a characteristic length of 28–30 nt (Ingolia et al. 2014). To account for this, Ingolia et al. (2014) introduced a fragment length organization similarity score (FLOSS), which reflects how well the lengths of fragments that are detected agree with true 80S ribosome footprints. Several studies further pointed to the 3-nt periodicity of ribosome profiling reads, reflecting the codon-by-codon movement of ribosomes along mRNA during translation (Michel et al. 2012; Bazzini et al. 2014; Duncan and Mata 2014). This property led Calviello et al. (2016) to develop RiboTaper, a computational tool intended to identify translated ORFs with a high degree of confidence. RiboTaper drew on HEK293 cell ribosome profiling data to identify ORFs in 504 noncoding genes, including pseudogenes, antisense, and lncRNAs. Interestingly, these ORFs were generally nonconserved, suggesting that their translation might be playing a regulatory role rather than resulting in the production of stable functional proteins.

PROTEOMIC EVIDENCE AND THE FUNCTIONALITY OF MICROPEPTIDES

Proof that a given ORF is indeed translated is the detection of the corresponding peptides through mass spectrometry. Banfai et al. (2012) analyzed two data sets produced by the Encyclopedia of DNA elements (ENCODE) project involving tandem mass spectrometry (MS/MS) and RNA-seq data for the cell lines K562 and GM12878. The authors detected matching peptides for 69 lncRNAs, amounting to ∼8% of annotated lncRNAs, concluding that most of the cases represented protein-coding genes that had been incorrectly annotated as noncoding. Slavoff et al. (2013) performed peptidomic analysis to detect products of sORFs in human cells. A small fraction of the peptides that were detected matched annotated lincRNAs (8 of 1866). A study by Prabakaran et al. (2014) compared transcriptomic and proteomic data from mouse cortical neurons. This analysis identified 250 unique peptides that mapped to regions that had been annotated as noncoding, including UTRs of known protein-coding genes, pseudogenes, introns, antisense RNAs, and intergenic regions. Of 25 peptides that mapped to intergenic regions, three were also found in the ribosomal profiling data of Ingolia et al. (2011). It should be noted that proteomic approaches are generally less sensitive than transcriptomic because of the PCR-based amplification step in RNAseq protocols. Indeed, even for known protein-coding transcripts, proteomic evidence is normally obtained for only 25%–30% of the transcripts typically detected by RNAseq-based methods. Extrapolating from this observation, the real number of proteins that originate from annotated noncoding RNAs can be expected to be 3–4 times higher than currently confirmed by proteomics studies.

Importantly, some proteins that originated from intergenic regions and RNAs annotated as noncoding were reported to have biological functions. Hanada et al. (2013) overexpressed 473 coding intergenic sORFs in *Arabidopsis*, of which 49 produced visible phenotypes, suggesting that the peptides played some role in plant development. A clear functional example comes from the peptide produced by four tandem sORFs within the *Drosophila tarsal-less* (*tal*) lncRNA (Galindo et al. 2007; Kondo et al. 2007, 2010). Mutations that disrupt the translation of these sORFs lead to an embryonic lethal phenotype similar to mutations in the transcription factor Ovo/Svb (Kondo et al. 2007, 2010). Tal peptides trigger the amino-terminal truncation of the Ovo protein, which converts it from a repressor to an activator. Interestingly, homologs of *tal* have been found in insect species but not in vertebrates (Li et al. 2002). These functional data have led to a recent reclassification of *tal* as a polycistronic mRNA.

The study of Magny et al. (2013) described two peptides, sarcolamban A and B, that are encoded by the putative *Drosophila* noncoding RNA *pncr003:2L*. These peptides are conserved from flies to humans and function as regulators of calcium transport and cardiac muscle contraction. Myoregulin (MLN) and dwarf open reading frame (DWORF) are recently discovered micropeptides encoded by annotated lncRNAs that have turned out to be important for muscle function (Anderson et al. 2015; Nelson et al. 2016). Muscle contraction and relaxation depend on the release of Ca^{2+} from the sarcoplasmic reticulum (SR) and its reuptake by the membrane pump SERCA. MLN functions as a SERCA inhibitor, together with homologous proteins phospholamban (PLN) and sarcolipin (SLN). DWORF, on the other hand, enhances SERCA activity by displacing SERCA inhibitors. Mice lacking MLN show improved performance in exercise, and the deletion of DWORF leads to delays in Ca^{2+} clearance and relaxation.

D'Lima et al. (2017) used proteomics approaches to identify a small protein they called NoBody, for nonannotated P-body dissociating polypeptide. NoBody is encoded by *LINC01420/LOC550643* RNA and is specific to mammals. The protein localizes to P bodies and interacts with components of the decapping machinery. Its overexpression resulted in the dissolution of P bodies, suggesting that NoBody has a function in the regulation of mRNA metabolism.

Curiously, mitochondrial 12S rRNA was also reported to contain a functional sORF encoding a 16-amino-acid peptide MOTS-c (mitochondrial ORF of the 12S rRNA-c) (Lee et al. 2015). MOTS-c inhibits the folate cycle primarily in skeletal muscles, thereby regulating insulin sensitivity and metabolic homeostasis.

To summarize, annotated lncRNAs can be divided into three groups based on their status with regard to translation: (1) actual lncRNAs that do not encode proteins; (2) lncRNAs with conserved sORFs that are protein-coding genes misannotated as lncRNAs because the proteins they encode are shorter than the limit of 100 amino acids used as a benchmark; and (3) lncRNAs with nonconserved sORFs. The question of functions is most interesting and controversial for the last group. A minor but important proportion of such nonconserved sORFs encode functional micropeptides, but the lack of conservation suggests that for most of them their translation is likely to play a regulatory rather than coding role, for example by modulating RNA stability and/or localization. Smith et al. (2014) showed that translation of sORFs from annotated lncRNAs in yeast targets them for nonsense-mediated RNA decay (reviewed in Karousis and Mühlemann 2018). Such sORFs could also represent intermediate steps in de novo protein evolution (Carvunis et al. 2012; Ruiz-Orera et al. 2014), and it is certainly possible that some lncRNAs with nonconserved sORFs play the dual role of encoding functional peptides and regulating the fates of other RNAs.

CONCLUSIONS AND FUTURE PROSPECTS

In conclusion, there is solid evidence that some lncRNAs and circRNAs play regulatory roles in translation. Such functions involve interactions with *trans*-acting translational regulators, such as RBPs and miRNAs. These interactions can have diverse consequences for the sequestration, storage, or intracellular localization of *trans*-acting factors. Moreover, lncRNAs can directly interact with mRNAs and regulate their fates by recruiting specific RBPs. More regulatory mechanisms undoubtedly remain to be discovered.

Besides playing regulatory roles in translation, some circRNAs are translated in vivo, but future work is needed to test whether the products of this translation are functionally important. As circRNAs lack a cap and poly(A) tail, they may have adapted to undergo translation in special situations when cap-dependent translation is inhibited. In the future, it will be important to intensify efforts toward understanding the mechanisms that govern circRNA translation. The translation of sORFs has also been reported for a number of RNAs annotated as lncRNAs. Some represent missannotations of protein-coding genes that undergo corrections when corresponding peptides are found but the generally low conservation of lncRNA sORFs makes it likely that their translation plays more of a regulatory role. Further work is needed to understand the significance of this process. Clearly, the future will see the identification of novel interaction partners for circRNAs and lncRNAs and increasing clarity in defining their functions.

ACKNOWLEDGMENTS

We are grateful to the BIMSB faculty, Sergey Dmitriev (MSU), and Russ Hodge (MDC) for comments on the manuscript. We apologize to authors whose work could not be cited in this review because of size limitations.

REFERENCES

*Reference is also in this collection.

Abe N, Matsumoto K, Nishihara M, Nakano Y, Shibata A, Maruyama H, Shuto S, Matsuda A, Yoshida M, Ito Y, et al. 2015. Rolling circle translation of circular RNA in living human cells. *Sci Rep* **5:** 16435.

AbouHaidar MG, Venkataraman S, Golshani A, Liu B, Ahmad T. 2014. Novel coding, translation, and gene expression of a replicating covalently closed circular RNA of 220 nt. *Proc Natl Acad Sci* **111:** 14542–14547.

Anderson DM, Anderson KM, Chang CL, Makarewich CA, Nelson BR, McAnally JR, Kasaragod P, Shelton JM, Liou J, Bassel-Duby R, et al. 2015. A micropeptide encoded by a putative long noncoding RNA regulates muscle performance. *Cell* **160:** 595–606.

Andrews SJ, Rothnagel JA. 2014. Emerging evidence for functional peptides encoded by short open reading frames. *Nat Rev Genet* **15:** 193–204.

Ashwal-Fluss R, Meyer M, Pamudurti NR, Ivanov A, Bartok O, Hanan M, Evantal N, Memczak S, Rajewsky N, Kadener S. 2014. circRNA biogenesis competes with pre-mRNA splicing. *Mol Cell* **56:** 55–66.

Banfai B, Jia H, Khatun J, Wood E, Risk B, Gundling WE Jr, Kundaje A, Gunawardena HP, Yu Y, Xie L, et al. 2012. Long noncoding RNAs are rarely translated in two human cell lines. *Genome Res* **22:** 1646–1657.

Bazzini AA, Johnstone TG, Christiano R, Mackowiak SD, Obermayer B, Fleming ES, Vejnar CE, Lee MT, Rajewsky N, Walther TC, et al. 2014. Identification of small ORFs in vertebrates using ribosome footprinting and evolutionary conservation. *EMBO J* **33:** 981–993.

Cabili MN, Dunagin MC, McClanahan PD, Biaesch A, Padovan-Merhar O, Regev A, Rinn JL, Raj A. 2015. Localization and abundance analysis of human lncRNAs at single-cell and single-molecule resolution. *Genome Biol* **16:** 20.

Calviello L, Mukherjee N, Wyler E, Zauber H, Hirsekorn A, Selbach M, Landthaler M, Obermayer B, Ohler U. 2016. Detecting actively translated open reading frames in ribosome profiling data. *Nat Methods* **13:** 165–170.

Carrieri C, Cimatti L, Biagioli M, Beugnet A, Zucchelli S, Fedele S, Pesce E, Ferrer I, Collavin L, Santoro C, et al. 2012. Long non-coding antisense RNA controls Uchl1 translation through an embedded SINEB2 repeat. *Nature* **491:** 454–457.

Carvunis AR, Rolland T, Wapinski I, Calderwood MA, Yildirim MA, Simonis N, Charloteaux B, Hidalgo CA, Barbette J, Santhanam B, et al. 2012. Proto-genes and de novo gene birth. *Nature* **487:** 370–374.

Chen CY, Sarnow P. 1995. Initiation of protein synthesis by the eukaryotic translational apparatus on circular RNAs. *Science* **268:** 415–417.

Claverie JM. 1997. Computational methods for the identification of genes in vertebrate genomic sequences. *Hum Mol Genet* **6:** 1735–1744.

 Cite this article as *Cold Spring Harb Perspect Biol* doi: 10.1101/cshperspect.a032680

Cocquerelle C, Mascrez B, Hetuin D, Bailleul B. 1993. Missplicing yields circular RNA molecules. *FASEB J* **7**: 155–160.

Conn SJ, Pillman KA, Toubia J, Conn VM, Salmanidis M, Phillips CA, Roslan S, Schreiber AW, Gregory PA, Goodall GJ. 2015. The RNA binding protein quaking regulates formation of circRNAs. *Cell* **160**: 1125–1134.

Csepany T, Lin A, Baldick CJ Jr, Beemon K. 1990. Sequence specificity of mRNA N^6-adenosine methyltransferase. *J Biol Chem* **265**: 20117–20122.

Delaleau M, Borden KL. 2015. Multiple export mechanisms for mRNAs. *Cells* **4**: 452–473.

Derrien T, Johnson R, Bussotti G, Tanzer A, Djebali S, Tilgner H, Guernec G, Martin D, Merkel A, Knowles DG, et al. 2012. The GENCODE v7 catalog of human long noncoding RNAs: Analysis of their gene structure, evolution, and expression. *Genome Res* **22**: 1775–1789.

Dinkel H, Van Roey K, Michael S, Kumar M, Uyar B, Altenberg B, Milchevskaya V, Schneider M, Kuhn H, Behrendt A, et al. 2016. ELM 2016—Data update and new functionality of the eukaryotic linear motif resource. *Nucleic Acids Res* **44**: D294–D300.

Djebali S, Davis CA, Merkel A, Dobin A, Lassmann T, Mortazavi A, Tanzer A, Lagarde J, Lin W, Schlesinger F, et al. 2012. Landscape of transcription in human cells. *Nature* **489**: 101–108.

D'Lima NG, Ma J, Winkler L, Chu Q, Loh KH, Corpuz EO, Budnik BA, Lykke-Andersen J, Saghatelian A, Slavoff SA. 2017. A human microprotein that interacts with the mRNA decapping complex. *Nat Chem Biol* **13**: 174–180.

Dubin RA, Kazmi MA, Ostrer H. 1995. Inverted repeats are necessary for circularization of the mouse testis Sry transcript. *Gene* **167**: 245–248.

* Duchaine TF, Fabian MR. 2018. Mechanistic insights into microRNA-mediated gene silencing. *Cold Spring Harb Perspect Biol* doi: 10.1101/cshperspect.a032771.

Duncan CD, Mata J. 2014. The translational landscape of fission-yeast meiosis and sporulation. *Nat Struct Mol Biol* **21**: 641–647.

Ebbesen KK, Kjems J, Hansen TB. 2016. Circular RNAs: Identification, biogenesis and function. *Biochim Biophys Acta* **1859**: 163–168.

Frith MC, Forrest AR, Nourbakhsh E, Pang KC, Kai C, Kawai J, Carninci P, Hayashizaki Y, Bailey TL, Grimmond SM. 2006. The abundance of short proteins in the mammalian proteome. *PLoS Genet* **2**: e52.

Galindo MI, Pueyo JI, Fouix S, Bishop SA, Couso JP. 2007. Peptides encoded by short ORFs control development and define a new eukaryotic gene family. *PLoS Biol* **5**: e106.

Gao Y, Wang J, Zhao F. 2015. CIRI: An efficient and unbiased algorithm for de novo circular RNA identification. *Genome Biol* **16**: 4.

Gennarino VA, Singh RK, White JJ, De Maio A, Han K, Kim JY, Jafar-Nejad P, di Ronza A, Kang H, Sayegh LS, et al. 2015. Pumilio1 haploinsufficiency leads to SCA1-like neurodegeneration by increasing wild-type Ataxin1 levels. *Cell* **160**: 1087–1098.

Glazar P, Papavasileiou P, Rajewsky N. 2014. circBase: A database for circular RNAs. *RNA* **20**: 1666–1670.

Gong C, Maquat LE. 2011. lncRNAs transactivate STAU1-mediated mRNA decay by duplexing with $3'$ UTRs via Alu elements. *Nature* **470**: 284–288.

Grosswendt S, Filipchyk A, Manzano M, Klironomos F, Schilling M, Herzog M, Gottwein E, Rajewsky N. 2014. Unambiguous identification of miRNA:target site interactions by different types of ligation reactions. *Mol Cell* **54**: 1042–1054.

Guo JU, Agarwal V, Guo H, Bartel DP. 2014. Expanded identification and characterization of mammalian circular RNAs. *Genome Biol* **15**: 409.

Guttman M, Russell P, Ingolia NT, Weissman JS, Lander ES. 2013. Ribosome profiling provides evidence that large noncoding RNAs do not encode proteins. *Cell* **154**: 240–251.

Hanada K, Zhang X, Borevitz JO, Li WH, Shiu SH. 2007. A large number of novel coding small open reading frames in the intergenic regions of the *Arabidopsis thaliana* genome are transcribed and/or under purifying selection. *Genome Res* **17**: 632–640.

Hanada K, Higuchi-Takeuchi M, Okamoto M, Yoshizumi T, Shimizu M, Nakaminami K, Nishi R, Ohashi C, Iida K, Tanaka M, et al. 2013. Small open reading frames associated with morphogenesis are hidden in plant genomes. *Proc Natl Acad Sci* **110**: 2395–2400.

Hansen TB, Wiklund ED, Bramsen JB, Villadsen SB, Statham AL, Clark SJ, Kjems J. 2011. miRNA-dependent gene silencing involving Ago2-mediated cleavage of a circular antisense RNA. *EMBO J* **30**: 4414–4422.

Hansen TB, Jensen TI, Clausen BH, Bramsen JB, Finsen B, Damgaard CK, Kjems J. 2013. Natural RNA circles function as efficient microRNA sponges. *Nature* **495**: 384–388.

Harper JE, Miceli SM, Roberts RJ, Manley JL. 1990. Sequence specificity of the human mRNA N6-adenosine methylase in vitro. *Nucleic Acids Res* **18**: 5735–5741.

Hezroni H, Koppstein D, Schwartz MG, Avrutin A, Bartel DP, Ulitsky I. 2015. Principles of long noncoding RNA evolution derived from direct comparison of transcriptomes in 17 species. *Cell Rep* **11**: 1110–1122.

Holcik M, Sonenberg N. 2005. Translational control in stress and apoptosis. *Nat Rev Mol Cell Biol* **6**: 318–327.

Ingolia NT, Ghaemmaghami S, Newman JR, Weissman JS. 2009. Genome-wide analysis in vivo of translation with nucleotide resolution using ribosome profiling. *Science* **324**: 218–223.

Ingolia NT, Lareau LF, Weissman JS. 2011. Ribosome profiling of mouse embryonic stem cells reveals the complexity and dynamics of mammalian proteomes. *Cell* **147**: 789–802.

Ingolia NT, Brar GA, Stern-Ginossar N, Harris MS, Talhouarne GJ, Jackson SE, Wills MR, Weissman JS. 2014. Ribosome profiling reveals pervasive translation outside of annotated protein-coding genes. *Cell Rep* **8**: 1365–1379.

* Ingolia NT, Hussmann JA, Weissman JS. 2018. Ribosome profiling: Global views of translation. *Cold Spring Harb Perspect Biol* doi: 10.1101/cshperspect.a032698.

Ivanov A, Memczak S, Wyler E, Torti F, Porath HT, Orejuela MR, Piechotta M, Levanon EY, Landthaler M, Dieterich C, et al. 2015. Analysis of intron sequences reveals hallmarks of circular RNA biogenesis in animals. *Cell Rep* **10**: 170–177.

Jackson RJ. 2005. Alternative mechanisms of initiating translation of mammalian mRNAs. *Biochem Soc Trans* **33**: 1231–1241.

Jeck WR, Sharpless NE. 2014. Detecting and characterizing circular RNAs. *Nat Biotechnol* **32**: 453–461.

Jeck WR, Sorrentino JA, Wang K, Slevin MK, Burd CE, Liu J, Marzluff WF, Sharpless NE. 2013. Circular RNAs are abundant, conserved, and associated with ALU repeats. *RNA* **19**: 141–157.

Jonas S, Izaurralde E. 2015. Towards a molecular understanding of microRNA-mediated gene silencing. *Nat Rev Genet* **16**: 421–433.

* Karousis ED, Mühlemann O. 2018. Nonsense-mediated mRNA decay begins where translation ends. *Cold Spring Harb Perspect Biol* doi: 10.1101/cshperspect.a032862.

Konarska M, Filipowicz W, Domdey H, Gross HJ. 1981. Binding of ribosomes to linear and circular forms of the 5′-terminal leader fragment of tobacco-mosaic-virus RNA. *Eur J Biochem* **114**: 221–227.

Kondo T, Hashimoto Y, Kato K, Inagaki S, Hayashi S, Kageyama Y. 2007. Small peptide regulators of actin-based cell morphogenesis encoded by a polycistronic mRNA. *Nat Cell Biol* **9**: 660–665.

Kondo T, Plaza S, Zanet J, Benrabah E, Valenti P, Hashimoto Y, Kobayashi S, Payre F, Kageyama Y. 2010. Small peptides switch the transcriptional activity of Shavenbaby during *Drosophila* embryogenesis. *Science* **329**: 336–339.

Kozak M. 1978. How do eucaryotic ribosomes select initiation regions in messenger RNA? Cell **15**: 1109–1123.

Kozak M. 1979. Inability of circular mRNA to attach to eukaryotic ribosomes. *Nature* **280**: 82–85.

* Kwan T, Thompson SR. 2018. Noncanonical translation initiation in eukaryotes. *Cold Spring Harb Perspect Biol* doi: 10.1101/cshperspect.a032672.

Lasda E, Parker R. 2014. Circular RNAs: Diversity of form and function. *RNA* **20**: 1829–1842.

Lease KA, Walker JC. 2006. The *Arabidopsis* unannotated secreted peptide database, a resource for plant peptidomics. *Plant Physiol* **142**: 831–838.

Lee C, Zeng J, Drew BG, Sallam T, Martin-Montalvo A, Wan J, Kim SJ, Mehta H, Hevener AL, de Cabo R, et al. 2015. The mitochondrial-derived peptide MOTS-c promotes metabolic homeostasis and reduces obesity and insulin resistance. *Cell Metab* **21**: 443–454.

Lee S, Kopp F, Chang TC, Sataluri A, Chen B, Sivakumar S, Yu H, Xie Y, Mendell JT. 2016. Noncoding RNA NORAD regulates genomic stability by sequestering PUMILIO proteins. *Cell* **164**: 69–80.

Legnini I, Di Timoteo G, Rossi F, Morlando M, Briganti F, Sthandier O, Fatica A, Santini T, Andronache A, Wade M, et al. 2017. circ-ZNF609 is a circular RNA that can be translated and functions in myogenesis. *Mol Cell* **66**: 22–37.e29.

Li B, Dai Q, Li L, Nair M, Mackay DR, Dai X. 2002. Ovol2, a mammalian homolog of *Drosophila* ovo: Gene structure, chromosomal mapping, and aberrant expression in blind-sterile mice. *Genomics* **80**: 319–325.

Liang D, Wilusz JE. 2014. Short intronic repeat sequences facilitate circular RNA production. *Genes Dev* **28**: 2233–2247.

Liberman N, Gandin V, Svitkin YV, David M, Virgili G, Jaramillo M, Holcik M, Nagar B, Kimchi A, Sonenberg N. 2015. DAP5 associates with eIF2β and eIF4AI to promote internal ribosome entry site driven translation. *Nucleic Acids Res* **43**: 3764–3775.

Lin MF, Jungreis I, Kellis M. 2011. PhyloCSF: A comparative genomics method to distinguish protein coding and noncoding regions. *Bioinformatics* **27**: i275–i282.

Mackowiak SD, Zauber H, Bielow C, Thiel D, Kutz K, Calviello L, Mastrobuoni G, Rajewsky N, Kempa S, Selbach M, et al. 2015. Extensive identification and analysis of conserved small ORFs in animals. *Genome Biol* **16**: 179.

Magny EG, Pueyo JI, Pearl FM, Cespedes MA, Niven JE, Bishop SA, Couso JP. 2013. Conserved regulation of cardiac calcium uptake by peptides encoded in small open reading frames. *Science* **341**: 1116–1120.

Memczak S, Jens M, Elefsinioti A, Torti F, Krueger J, Rybak A, Maier L, Mackowiak SD, Gregersen LH, Munschauer M, et al. 2013. Circular RNAs are a large class of animal RNAs with regulatory potency. *Nature* **495**: 333–338.

Meyer KD, Patil DP, Zhou J, Zinoviev A, Skabkin MA, Elemento O, Pestova TV, Qian SB, Jaffrey SR. 2015. 5′ UTR m⁶A promotes cap-independent translation. *Cell* **163**: 999–1010.

Michel AM, Choudhury KR, Firth AE, Ingolia NT, Atkins JF, Baranov PV. 2012. Observation of dually decoded regions of the human genome using ribosome profiling data. *Genome Res* **22**: 2219–2229.

Miller MA, Olivas WM. 2011. Roles of Puf proteins in mRNA degradation and translation. *Wiley Interdiscip Rev RNA* **2**: 471–492.

Mukherjee N, Calviello L, Hirsekorn A, de Pretis S, Pelizzola M, Ohler U. 2017. Integrative classification of human coding and noncoding genes through RNA metabolism profiles. *Nat Struct Mol Biol* **24**: 86–96.

Nelson BR, Makarewich CA, Anderson DM, Winders BR, Troupes CD, Wu F, Reese AL, McAnally JR, Chen X, Kavalali ET, et al. 2016. A peptide encoded by a transcript annotated as long noncoding RNA enhances SERCA activity in muscle. *Science* **351**: 271–275.

Pamudurti NR, Bartok O, Jens M, Ashwal-Fluss R, Stottmeister C, Ruhe L, Hanan M, Wyler E, Perez-Hernandez D, Ramberger E, et al. 2017. Translation of CircRNAs. *Mol Cell* **66**: 9–21.e27.

Pasman Z, Been MD, Garcia-Blanco MA. 1996. Exon circularization in mammalian nuclear extracts. *RNA* **2**: 603–610.

* Peer E, Moshitch-Moshkovitz S, Rechavi G, Dominissini D. 2018. The epitranscriptome in translation regulation. *Cold Spring Harb Perspect Biol* doi: 10.1101/cshperspect.a032623.

Perry RB, Ulitsky I. 2016. The functions of long noncoding RNAs in development and stem cells. *Development* **143**: 3882–3894.

Piwecka M, Glazar P, Hernandez-Miranda LR, Memczak S, Wolf SA, Rybak-Wolf A, Filipchyk A, Klironomos F, Cerda Jara CA, Fenske P, et al. 2017. Loss of a mammalian circular RNA locus causes miRNA deregulation and affects brain function. *Science* **357**: eaam8526.

Prabakaran S, Hemberg M, Chauhan R, Winter D, Tweedie-Cullen RY, Dittrich C, Hong E, Gunawardena J, Steen H, Kreiman G, et al. 2014. Quantitative profiling of peptides from RNAs classified as noncoding. *Nat Commun* **5**: 5429.

Ruiz-Orera J, Messeguer X, Subirana JA, Alba MM. 2014. Long non-coding RNAs as a source of new peptides. *eLife* **3**: e03523.

Rybak-Wolf A, Stottmeister C, Glazar P, Jens M, Pino N, Giusti S, Hanan M, Behm M, Bartok O, Ashwal-Fluss R, et al. 2015. Circular RNAs in the mammalian brain are highly abundant, conserved, and dynamically expressed. *Mol Cell* **58**: 870–885.

Saghatelian A, Couso JP. 2015. Discovery and characterization of smORF-encoded bioactive polypeptides. *Nat Chem Biol* **11**: 909–916.

Saigoh K, Wang YL, Suh JG, Yamanishi T, Sakai Y, Kiyosawa H, Harada T, Ichihara N, Wakana S, Kikuchi T, et al. 1999. Intragenic deletion in the gene encoding ubiquitin carboxy-terminal hydrolase in *gad* mice. *Nat Genet* **23**: 47–51.

Salzman J. 2016. Circular RNA expression: Its potential regulation and function. *Trends Genet* **32**: 309–316.

Salzman J, Gawad C, Wang PL, Lacayo N, Brown PO. 2012. Circular RNAs are the predominant transcript isoform from hundreds of human genes in diverse cell types. *PLoS ONE* **7**: e30733.

Salzman J, Chen RE, Olsen MN, Wang PL, Brown PO. 2013. Cell-type specific features of circular RNA expression. *PLoS Genet* **9**: e1003777.

Slavoff SA, Mitchell AJ, Schwaid AG, Cabili MN, Ma J, Levin JZ, Karger AD, Budnik BA, Rinn JL, Saghatelian A. 2013. Peptidomic discovery of short open reading frame-encoded peptides in human cells. *Nat Chem Biol* **9**: 59–64.

Smith JE, Alvarez-Dominguez JR, Kline N, Huynh NJ, Geisler S, Hu W, Coller J, Baker KE. 2014. Translation of small open reading frames within unannotated RNA transcripts in Saccharomyces cerevisiae. *Cell Rep* **7**: 1858–1866.

Szabo L, Morey R, Palpant NJ, Wang PL, Afari N, Jiang C, Parast MM, Murry CE, Laurent LC, Salzman J. 2015. Statistically based splicing detection reveals neural enrichment and tissue-specific induction of circular RNA during human fetal development. *Genome Biol* **16**: 126.

Terenin IM, Andreev DE, Dmitriev SE, Shatsky IN. 2013. A novel mechanism of eukaryotic translation initiation that is neither m^{7}G-cap-, nor IRES-dependent. *Nucleic Acids Res* **41**: 1807–1816.

Terenin IM, Smirnova VV, Andreev DE, Dmitriev SE, Shatsky IN. 2017. A researcher's guide to the galaxy of IRESs. *Cell Mol Life Sci* **74**: 1431–1455.

Tichon A, Gil N, Lubelsky Y, Havkin Solomon T, Lemze D, Itzkovitz S, Stern-Ginossar N, Ulitsky I. 2016. A conserved abundant cytoplasmic long noncoding RNA modulates repression by Pumilio proteins in human cells. *Nat Commun* **7**: 12209.

Van Eden ME, Byrd MP, Sherrill KW, Lloyd RE. 2004. Demonstrating internal ribosome entry sites in eukaryotic mRNAs using stringent RNA test procedures. *RNA* **10**: 720–730.

Wang Y, Wang Z. 2015. Efficient backsplicing produces translatable circular mRNAs. *RNA* **21**: 172–179.

Wang X, Zhao BS, Roundtree IA, Lu Z, Han D, Ma H, Weng X, Chen K, Shi H, He C. 2015. N^6-methyladenosine modulates messenger RNA translation efficiency. *Cell* **161**: 1388–1399.

Wang K, Long B, Liu F, Wang JX, Liu CY, Zhao B, Zhou LY, Sun T, Wang M, Yu T, et al. 2016. A circular RNA protects the heart from pathological hypertrophy and heart failure by targeting miR-223. *Eur Heart J* **37**: 2602–2611.

Yang Y, Fan X, Mao M, Song X, Wu P, Zhang Y, Jin Y, Yang Y, Chen LL, Wang Y, et al. 2017. Extensive translation of circular RNAs driven by N^6-methyladenosine. *Cell Res* **27**: 626–641.

You X, Vlatkovic I, Babic A, Will T, Epstein I, Tushev G, Akbalik G, Wang M, Glock C, Quedenau C, et al. 2015. Neural circular RNAs are derived from synaptic genes and regulated by development and plasticity. *Nat Neurosci* **18**: 603–610.

Zappulo A, van den Bruck D, Ciolli Mattioli C, Franke V, Imami K, McShane E, Moreno-Estelles M, Calviello L, Filipchyk A, Peguero-Sanchez E, et al. 2017. RNA localization is a key determinant of neurite-enriched proteome. *Nat Commun* **8**: 583.

Zhang XO, Wang HB, Zhang Y, Lu X, Chen LL, Yang L. 2014. Complementary sequence-mediated exon circularization. *Cell* **159**: 134–147.

Zhang Y, Xue W, Li X, Zhang J, Chen S, Zhang JL, Yang L, Chen LL. 2016. The biogenesis of nascent circular RNAs. *Cell Rep* **15**: 611–624.

Noncanonical Translation Initiation in Eukaryotes

Thaddaeus Kwan and Sunnie R. Thompson

Department of Microbiology, University of Alabama at Birmingham, Birmingham, Alabama 35294

Correspondence: sunnie@uab.edu

The vast majority of eukaryotic messenger RNAs (mRNAs) initiate translation through a canonical, cap-dependent mechanism requiring a free 5′ end and 5′ cap and several initiation factors to form a translationally active ribosome. Stresses such as hypoxia, apoptosis, starvation, and viral infection down-regulate cap-dependent translation during which alternative mechanisms of translation initiation prevail to express proteins required to cope with the stress, or to produce viral proteins. The diversity of noncanonical initiation mechanisms encompasses a broad range of strategies and cellular cofactors. Herein, we provide an overview and, whenever possible, a mechanistic understanding of the various noncanonical mechanisms of initiation used by cells and viruses. Despite many unanswered questions, recent advances have propelled our understanding of the scope, diversity, and mechanisms of alternative initiation.

Cap-dependent translation relies on recognition of the 5′ cap by the eukaryotic initiation factor (eIF)4F complex (eIF4E, eIF4G, and eIF4A), which recruits the 43S preinitiation complex ([PIC], 40S, eIF3, eIF1, eIF1A, eIF5, and ternary complex [eIF2•Met-tRNA$_i$•GTP]) to the 5′ end of the messenger RNA (mRNA). Following recruitment, the 43S ribosomal subunit scans the mRNA in a 5′-to-3′ direction until the AUG start codon is recognized and 60S ribosomal subunit joining occurs to make an 80S elongation-competent ribosome (see Merrick and Pavitt 2018). Cap-dependent initiation is down-regulated under conditions of cellular stress (see Wek 2018) or during some viral infections (see Stern-Ginossar et al. 2018). Therefore, viral proteins and cellular proteins that are required to survive or adapt to the cellular stress are synthesized using alternative mechanisms of translation initiation (Fig. 1).

Noncanonical mechanisms of initiation vary with respect to which eIFs are required, whether ribosome scanning and cap recognition are used, and the conditions in which they are favored. For example, ribosomal shunting requires a 5′ cap and all of the canonical initiation factors to load the 40S subunit onto the mRNA before translocating the 40S subunit to a downstream location to initiate translation (Morley and Coldwell 2008). In contrast, the initiation factor requirements for internal ribosome entry sites (IRESs), which require neither a 5′ cap nor a free 5′ end to recruit a ribosome internally to the mRNA, have a range of initiation factor requirements dependent on the type of IRES. For instance, the *Dicistroviridae* intergenic region (IGR) IRES is the simplest studied IRES, requiring no initiation factors (Jan et al. 2003), whereas the hepatitis A viral IRES uses all of the factors needed for canonical translation initiation

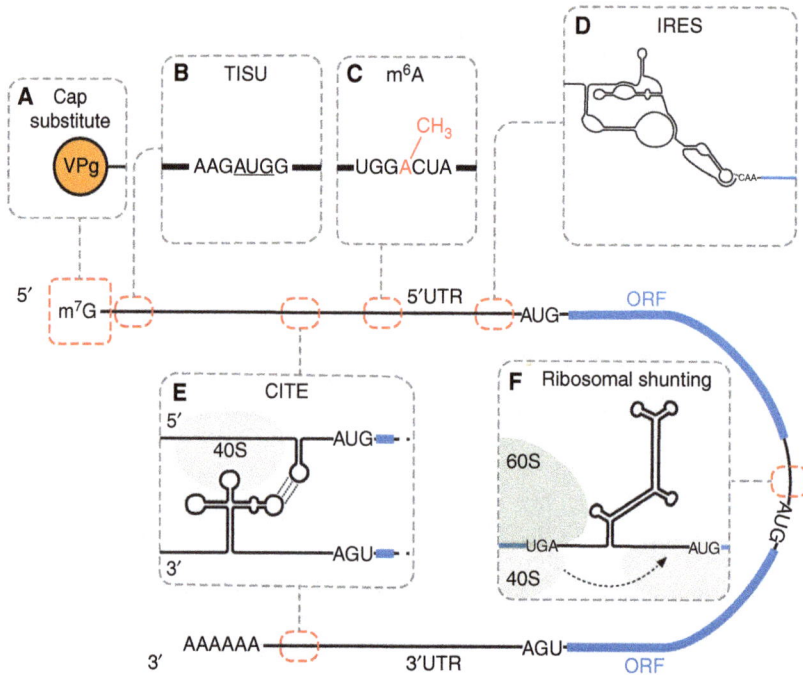

Figure 1. A model conveying the different mechanisms of noncanonical translation initiation. The locations of the *cis*-acting elements (red dashed lines), the open reading frames ([ORFs], blue rectangles), and 5′ and 3′ untranslated regions ([UTRs], black lines) are indicated. (*A*) Viral protein linked to the genome (VPg) rather than a 5′ cap. (*B*) Translation initiator of short 5′UTR (TISU) element with the start codon underlined. (*C*) m^6A modification. (*D*) Internal ribosomal entry site (IRES): the intergenic region (IGR) IRES is shown. (*E*) 3′ Cap-independent translation element (CITE): barley yellow dwarf virus-like translation element (BTE) CITE in the 3′UTR is shown forming a kissing-loop (dotted line) interaction with a stem loop in the 5′UTR. (*F*) Ribosomal shunting: cauliflower mosaic virus (CaMV) ribosomal shunt is shown; the highly structured region promotes shunting of the 40S from the "donor" site to a downstream "acceptor" site for initiation.

including eIF4E (Avanzino et al. 2017). The mechanisms of noncanonical translation initiation that are known are very diverse and differ significantly with respect to how the ribosome is recruited and the mechanism of initiation. Our understanding of these noncanonical mechanisms of initiation has lagged behind our knowledge of cap-dependent translation, largely because of the lack of good genetic and biochemical systems, as most of these noncanonical mechanisms function only in mammalian cells, but not in yeast.

Although many of the mechanisms for noncanonical translation initiation are not well understood, recent advances in the IRES field have shed light on these mechanisms. For example, the first structure of an IRES bound to a ribosome was revealed using cryoelectron microscopy (cryo-EM) (Spahn et al. 2004). Soon afterward, the ribosome-binding domain of an IRES was crystalized, revealing the three-dimensional fold of an IRES (Pfingsten et al. 2006). Previously, the identification of mRNAs that harbor IRESs was cumbersome because of the fact that IRESs must be identified using functional assays to show that translation is cap-independent (Spriggs et al. 2010; Thompson 2012). However, this limitation was largely overcome by using microarrays and next-generation sequencing to identify mRNAs that remained polysome-associated when cap-dependent translation was down-regulated (Johannes et al. 1999; Coldwell et al. 2001; Bushell et al. 2006; Thomas and Johannes 2007). Although these large-scale

Cite this article as *Cold Spring Harb Perspect Biol* doi: 10.1101/cshperspect.a032672

analyses are not sufficient to attribute IRES activity to any of the mRNAs, a recent high-throughput approach validated most of these candidate mRNAs as having bona fide IRES activity in the 5′ untranslated region (5′UTR) (Weingarten-Gabbay et al. 2016). IRESs were first discovered in viruses in 1988 (Jang et al. 1988; Pelletier and Sonenberg 1988), and minimal factors that were required for the activity of the encephalomyocarditis virus (EMCV) IRES were identified in vitro about 8 years later (Pestova et al. 1996a). However, reconstitution of the poliovirus (PV) IRES was more enigmatic because it required additional RNA-binding proteins that were not initiation factors (Sweeney et al. 2014). mRNA modifications have also been implicated in directing noncanonical initiation (see Meyer et al. 2015; Zhou et al. 2015; Peer et al. 2018). Furthermore, the ribosome, which was once considered to be a large protein synthesis machine that merely translated mRNAs into proteins, is becoming better appreciated for its roles in regulation of initiation and selection of mRNAs based on ribosomal RNA modifications and nonessential ribosomal proteins, such as eS25, eL38, and eL40 (Jack et al. 2011; Hertz et al. 2013; Lee et al. 2013; Ban et al. 2014; Meyer et al. 2015; Xue et al. 2015; Yang et al. 2017). This review focuses on how these discoveries have contributed to our understanding of noncanonical mechanisms of translation initiation.

INTERNAL RIBOSOME ENTRY SITES (IRESs)

IRESs are RNA elements that can recruit a ribosome internally to the mRNA rather than at the 5′ end. IRESs are diverse, ranging in size from 9 to >1000 nucleotides and can be unstructured or contain stable secondary and tertiary structures. IRESs were first discovered in PV and EMCV, members of the *Picornaviridae* family (Jang et al. 1988; Pelletier and Sonenberg 1988). Later, IRESs were identified in cellular mRNAs that encode proteins related to cellular stress, survival, angiogenesis, and mitosis (Yang and Sarnow 1997; Walters and Thompson 2016). In 1995, Chen and Sarnow presented compelling evidence that the ribosome is recruited internally to the mRNA by showing that a circular RNA

(no 5′ end or 5′ cap) containing the EMCV IRES but no stop codons could initiate translation and translate around the circular RNA multiple times. Inclusion of a thrombin cleavage site encoded by the circular RNA showed that on addition of thrombin the polypeptide product that was over 106 kDa was reduced to a 40 kDa protein, demonstrating that the circular RNA was intact and that the large protein generated was due to the ribosome transiting around the circular RNA multiple times and was not a protein aggregate (Chen and Sarnow 1995). This also was one of the first studies to challenge the idea that the 5′ end of the mRNA had to be threaded through the 40S subunit. Together, these results showed that ribosomes can be recruited onto mRNAs internally and do not require a free 5′ end or 5′ cap. Since then, the field has continued to amass a large body of evidence to show that both viral and cellular mRNAs can contain IRESs and they are now widely recognized as a cap-independent mechanism of initiation (Carter et al. 2000; Schneider et al. 2001; Ruggero 2013). The significance of IRESs and other noncanonical mechanisms of eukaryotic translation initiation during viral infections, cellular stress, and survival has suggested a major role for translational control during tumorigenesis and viral infection (Walters and Thompson 2016; Lacerda et al. 2017).

Classification of Viral IRESs

Viral IRESs are classified based on their structure, initiation factor requirements, and mechanism of initiation (Table 1). Picornaviral IRESs comprise five of the eight classes of viral IRESs. Type I IRESs were the first to be discovered and are present in enteroviruses (e.g., PV, enterovirus 71) and rhinoviruses (human rhinovirus type 2). Type I IRESs are approximately 490 nucleotides long (Pelletier and Sonenberg 1988) and require eIFs, eIF2, eIF3, eIF4A, and the central domain of eIF4G, as well as the IRES *trans*-acting factor (ITAF) poly(C)-binding protein 2 (PCBP2) (Sweeney et al. 2014). eIF1A and eIF4B are not required, but strongly stimulate type I IRES activity (Sweeney et al. 2014). For type I IRESs, the 40S ribosomal subunit is re-

Table 1. Classification of viral internal ribosome entry sites (IRESs)

		Requirements			
		eIFs	ITAFs	Examples	Notable features
Picornaviral IRES subtypes	Type I (entero-/rhinovirus)	eIF2 eIF3 eIF4A eIF4Gm	PTB PCBP1 PCBP2	Poliovirus EV71 BEV Coxsackievirus B3 HRV2	Ribosomal scanning downstream of ribosome binding to the AUG
	Type II (cardio-/apthovirus)	eIF2 eIF3 eIF4A eIF4B eIF4Gm	PTB ITAF$_{45}$	EMCV FMDV TMEV	Initiates translation at the site of 40S recruitment, 40S does not scan
	Type IV (HAV-like)	eIF2 eIF3 eIF4A eIF4B eIF4E eIF4G	PTB PCBP2 La	HAV	Requires eIF4E but not binding to the 5′ cap
	Type V (Aichivirus)	eIF2 eIF3 eIF4A eIF4B eIF4Gm	PTB DHX29	Aichivirus	Does not require eIF1 and eIF1A
	Type III (HCV-like)	eIF2 eIF3	None	AEV PTV-1 SPV-9	
	HCV-like	eIF2 eIF3	None	HCV CSFV	Direct binding to 40S ribosomal subunit; translates under conditions with low ternary complex
	Dicistrovirus	None	None	CrPV DCV IAV	Initiates translation in the absence of any eIFs in the A-site at a non-AUG codon
				PSIV	Can translate in a +1 reading frame (ORFx)

Continued

Table 1. *Continued*

| | Requirements | | | |
	eIFs	ITAFs	Examples	Notable features
Retrovirus	eIF4A eIF4G eI5A	hnRNPA1 hRIP DDX3	HIV-1 HTLV-1 SIV FIV	
Herpesvirus	eIF4F	PTB PCBP1 KSHV-ORF57	KSHV-vFLIP MDV	vFLIP IRES located in ORF of vCyclin

eIF, Eukaryotic initiation factor; ITAF, IRES *trans*-acting factor; BEV, bovine enterovirus; HRV2, human rhinovirus serotype 2; TMEV, Theiler's murine encephalymyelitis virus; HAV, hepatitis A virus; SPV-9, simian picornavirus 9; CSFV, classical swine fever virus; DCV, *drosophila* C virus; IAV, influenza A virus; PSIV, *Plautia stali* intestine virus; HIV, human immunodeficiency virus; HTLV-1, human T-cell leukemia virus type 1; SIV, simian immunodeficiency virus; FIV, feline immunodeficiency virus; KSHV, Kaposi sarcoma herpesvirus; MDV, Marek's disease virus; CrPV, cricket paralysis virus.

cruited to an upstream AUG before scanning the RNA in a 5′-to-3′ direction to initiate translation at a downstream start codon. eIF1 facilitates correct start codon recognition for type I IRESs, consistent with its role in cap-dependent initiation (Kaminski et al. 2010; Sweeney et al. 2014; see Merrick and Pavitt 2018).

Type II IRESs, found in cardio- and aphthoviruses (e.g., EMCV and foot and mouth disease virus [FMDV]), are structurally different from type I IRESs, yet both classes have similar eIF requirements (Pestova et al. 1996a,b; Lomakin et al. 2000). A defining difference between type I and type II IRESs is that no scanning occurs downstream from the type II IRES despite recruitment of the ribosome to a similar AUG motif (Sweeney et al. 2014). Instead, this AUG serves as the authentic start codon for type II IRESs and neither scanning nor eIF1 is required (Pestova et al. 1996a). However, the FMDV IRES is an exception in that protein synthesis can also occur at a downstream start codon, AUG2, which is dependent on eIF1 and likely involves scanning, meaning that this IRES has characteristics of both type I and type II IRESs (Andreev et al. 2007). This is an example of the complexity of IRES-mediated translation and how the presence or absence of an initiation factor can modulate initiation codon selection for an IRES.

The type III class of IRESs consists of the flavivirus IRESs, the hepatitis C virus (HCV)

IRES, and picornaviral IRESs such as porcine teschovirus 1 (PTV-1) and avian encephalomyelitis virus (AEV) (Pisarev et al. 2004; Bakhshesh et al. 2008). The HCV IRES is the prototype for this class of IRESs and initiates translation by binding to the solvent side of the 40S ribosomal subunit. This binding is stabilized by initiation factor eIF3 and base-pairing between the domain IIId region of the HCV IRES and the loop region of helix 26 of the 18S rRNA (Fraser and Doudna 2007; Matsuda and Mauro 2014). Although methionyl-transfer RNA (Met-tRNA$_i$), is required for HCV start codon recognition (Pestova et al. 1998; Fraser and Doudna 2007) in vitro, the HCV IRES has been shown to function under conditions where the ternary complex is limiting, such as when eIF2 is inactivated via phosphorylation (Gonzalez-Almela et al. 2018). In addition, it has been shown that eIF2 is displaced from the initiating complex on HCV, suggesting that HCV does not use it even if it is available (Jaafar et al. 2016). The initiating Met-tRNA$_i$ can be brought in independently of eIF2, but how this occurs has been very controversial (Terenin et al. 2008; Jaafar et al. 2016; Gonzalez-Almela et al. 2018). This illustrates the importance of confirming in vitro findings in vivo, because in vitro experiments allow a great deal of control over the components present but may not reflect what is occurring under physiological conditions. It now appears that eIF2D or eIF2A are not required for HCV initiation in vivo as previous findings had suggested, but rather that eIF1A and possibly eIF5B are important for selecting (by codon–anticodon base pairing) and stabilizing Met-tRNA$_i$ binding to the HCV 48S complex in Huh-7 cells under conditions where eIF2α is phosphorylated (Jaafar et al. 2016; Gonzalez-Almela et al. 2018). It has been proposed that under stress conditions when eIF2α is phosphorylated, excess levels of Met-tRNA$_i$ would be sufficient for recruitment to the HCV 48S complexes in the absence of a dedicated delivery factor (Jaafar et al. 2016). Consistent with this model, the 40S subunit has been reported to bind to the HCV IRES with a dissociation rate that is insignificant on the time scale of initiation (Fuchs et al. 2015).

The most elegant viral IRESs are the IGR IRESs found in the *Dicistroviridae* family (Sasaki et al. 1998; Sasaki and Nakashima 1999; Wilson et al. 2000; Nakashima and Uchiumi 2009). These IRESs are <200 nucleotides in length and consist of a three-domain, multiple stem-loop structure containing three pseudoknots. The defining mechanistic feature of the IGR-IRES in dicistroviruses is that they can initiate translation in the absence of other initiation factors and also do not make use of a canonical AUG start codon (Sasaki and Nakashima 2000; Wilson et al. 2000; Jan and Sarnow 2002; Deniz et al. 2009). This shows that an RNA structure can independently recruit and manipulate ribosomes to initiate translation, as described next.

Mechanism of Internal Initiation of the *Dicistroviridae* IGR IRESs

The molecular mechanisms for how a *cis*-acting RNA element recruits a 40S subunit internally to the mRNA, loads the mRNA into the mRNA channel, recognizes the start codon, and promotes 60S subunit joining is not understood for most IRESs. After all, 10 to 13 initiation factors are required for cap-dependent initiation, yet the IGR IRES is able to bind directly to 40S subunits in the absence of any initiation factors and form translationally competent 80S complexes on addition of 60S subunits (Jan et al. 2003; Pestova and Hellen 2003).

The ~190 nucleotide cricket paralysis virus (CrPV) IGR IRES forms an RNA structure with three pseudoknots (PKI, PKII, and PKIII) that span all three tRNA-binding sites of the ribosome (Fernandez et al. 2014). PKI mimics an anticodon stem loop bound to a codon and occupies the A site, which is recognized by the universally conserved nucleotides in the decoding center that monitor accurate tRNA selection (G530, A1492, and A1493 in *Escherichia coli* 16S rRNA) (Schuler et al. 2006; Fernandez et al. 2014). PKII and PKIII extend out toward the E site and are required for affinity of the IRES to the 40S ribosomal subunit (Jan and Sarnow 2002; Pestova and Hellen 2003; Murray et al. 2016). The IGR IRES binds to the intersubunit surface of the 40S subunit with high affinity (5 to 20 nM), resembling a pretranslocation conformation of the ribosome (Spahn et al. 2004; Schuler et al. 2006; Jang et al. 2009; Landry et al. 2009; Jang and Jan 2010; Jack et al. 2011; Fernandez et al. 2014).

For the first amino acid to be decoded, PKI must be translocated to the P site by eukaryotic elongation factor 2 (eEF2), resulting in the first codon that is decoded being positioned in the A site (Fernandez et al. 2014; Koh et al. 2014). After aminoacyl-tRNA (aa-tRNA) binding to the A-site codon, a second translocation event must occur (without peptidyl transfer) to move PKI to the E site and the A site aa-tRNA to the P site (Fernandez et al. 2014; Muhs et al. 2015). The resulting complex resembles the normal 80S initiation complex that contains Met-tRNA$_i$ in the P site and an empty A site. Thus, it appears that the CrPV IGR IRES manipulates the ribosome to forego classical initiation and begin with elongation.

Although the CrPV IGR IRES has many properties that make it unique, it provides insights into mechanisms that are applicable to other IRESs. For example, the HCV IRES can also bind to 40S subunits in the absence of any initiation factors. Both HCV and the IGR IRES induce a similar conformational change in the 40S subunit on binding that resembles an "open" conformation for mRNA loading (Spahn et al. 2001, 2004). This conformational change resembles the eIF1- and eIF1A-induced open confor-

Cite this article as *Cold Spring Harb Perspect Biol* doi: 10.1101/cshperspect.a032672

mation of the 43S PIC, whereby the mRNA channel is open to allow for mRNA loading (Passmore et al. 2007). Another common requirement for structurally and functionally diverse IRESs is ribosomal protein S25 (eS25/RPS25). Although the role of eS25 is not known, its requirement by both viral and cellular IRESs suggests that they may share a common eS25-dependent mechanism (Hertz et al. 2013). Given the structural differences and eIF requirements, there will likely be some differences in how diverse IRESs initiate translation; however, what is interesting is how these seemingly diverse IRESs appear to have some common mechanisms for initiating translation.

Cellular IRESs

An estimated 10% of mammalian mRNAs have IRESs (Martinez-Salas et al. 2012; Walters and Thompson 2016; Weingarten-Gabbay et al. 2016). These mRNAs encode proteins important

during early development, mitosis, and cellular stresses such as hypoxia, apoptosis, starvation, heat shock, tumorigenesis, and viral infection (Table 2) (Johannes and Sarnow 1998; Johannes et al. 1999; Coldwell et al. 2001; Holcik 2004; Bushell et al. 2006; Dresios et al. 2006; Xue et al. 2015; Vaklavas et al. 2016; Walters and Thompson 2016). Our understanding of cellular IRESs lags behind that of viral IRESs because of the fact that many of these IRES-containing mRNAs can be translated cap-dependently but switch to IRES-mediated initiation under cellular stress. To accommodate the switch from cap-dependent to cap-independent initiation, cellular IRESs are generally less structured than viral IRESs and can even be modular, meaning that the IRES activity is distributed throughout the 5′UTR. In fact, some cellular IRESs, such as the Gtx IRES, base pair with a region on the 18S ribosomal RNA reminiscent of Shine–Dalgarno sequences (Dresios et al. 2006). The complementary ribosomal sequence is located in the prox-

Table 2. Mammalian cellular internal ribosomal entry site (IRES)

Cellular pathway	Name	References
Amino acid starvation	Cationic amino acid transporter 1 (CAT-1)	Fernandez et al. 2001
	Sodium-coupled amino acid transporter (SNAT2)	Gaccioli et al. 2006
Nutrient signaling	Human insulin receptor (HIR)	Spriggs et al. 2009
hypoxia	Hypoxia inducible factor 1α (HIF-1α)	Schepens et al. 2005
	Vascular endothelial growth factor (VEGF)	Morris et al. 2010
	Fibroblast growth factor 2 (FGF2)	Bonnal et al. 2003
Apoptosis survival	Apoptotic protease activating factor 1 (Apaf-1)	Sella et al. 1999
	B-cell lymphoma 2 (Bcl-2)	Marash et al. 2008
	BAG family molecular chaperone regulator 1 (Bag1)	Pickering et al. 2004
	Cellular inhibitor of apoptosis 1 (cIAP1)	Graber et al. 2010
	X-linked inhibitor of apoptosis (XIAP)	Riley et al. 2010
Oncogene	c-myc	Le Quesne et al. 2001
	c-Jun	Blau et al. 2012
Mitosis	p58 PITSLRE	Cornelis et al. 2000
	Cyclin-dependent kinase 1 (CDK1)	Marash et al. 2008
	p120 catenin (p120)	Silvera et al. 2009
DNA damage response	p53	Yang et al. 2006
	Serine hydroxymethyltransferase 1 (SHMT1)	Fox et al. 2009
	Replication protein A2 (RPA2)	Yin et al. 2013
Differentiation	Homeobox transcription factor (Hox)	Xue et al. 2015
	Platelet-derived growth factor 2 (PDGF2)	Sella et al. 1999
	Runt-related transcription factor 1 (Runx1)	Pozner et al. 1999
	Lymphoid enhancer-binding factor 1 (LEF-1)	Jimenez et al. 2005
Cold shock	Cold inducible RNA-binding protein (CIRP)	Al Fageeh and Smales 2009

imity of the classical Shine–Dalgarno sequence. However, unlike the anti-Shine–Dalgarno sequence, which is single-stranded, this sequence is in a region of helix 26 that is predicted to be base-paired (Deforges et al. 2015). More studies are required to understand the mechanism (Chappell et al. 2000; Dresios et al. 2006; Panopoulos and Mauro 2008). Conversely, some cellular IRESs (such as Bag-1 and c-myc) are structurally complex like viral IRESs (Le Quesne et al. 2001; Pickering et al. 2004; Baird et al. 2007; Spriggs et al. 2008).

An example of an extensively studied cellular protein that is translated by an IRES-dependent mechanism of initiation is the apoptotic protease activating factor 1 (Apaf-1) protein. Upon binding cytochrome c and dATP, Apaf-1 forms an oligomeric apoptosome that cleaves the procaspase 9 protein to stimulate the caspase cascade that commits the cell to apoptosis. Apaf-1 is important for programmed cell death during early development and knockout mice die early in embryogenesis because of a malformation of the brain caused by reduced apoptosis (Cecconi et al. 1998; Yoshida et al. 1998). As expected, the Apaf-1 IRES is most active in neuronal cells (25-fold over reporter activity with no IRES activity). This is a result, in part, of the higher affinity of the nPTB (neuronal polypyrimidine tract-binding protein 1) ITAF compared with the nonneuronal PTB found in other cell types (i.e., HEK293T cells, which show only sixfold more activity than no-IRES controls) (Coldwell et al. 2000; Mitchell et al. 2003; Andreev et al. 2009). Studies that questioned whether Apaf-1 has an IRES were all performed in a cell line known not to support robust Apaf-1 IRES activity (Andreev et al. 2009). The Apaf-1 mRNA is expressed in all tissues except skeletal muscle, but it is poorly translated through a cap-dependent mechanism of initiation, suggesting that most of the Apaf-1 initiation occurs through the IRES (Burgess et al. 1999; Coldwell et al. 2000). In a bicistronic reporter assay (described below) the Apaf-1 IRES was more active than the human rhinovirus IRES but less active than the c-myc IRES (Coldwell et al. 2000). Preventing expression of the first cistron by inserting a hairpin upstream of it did not diminish IRES activity,

demonstrating that expression of the second cistron was not caused by readthrough from the first cistron (Coldwell et al. 2000). In fact, insertion of a stable hairpin upstream of the 5′UTR in a monocistronic reporter reduced Apaf-1 mRNA translation only 1.5-fold compared with cap-dependent translation, which was inhibited 200-fold (Coldwell et al. 2000). We note that studies disputing the existence of the Apaf-1 IRES are not inconsistent with data from other groups demonstrating its IRES activity; rather, it is the interpretation of the data that differs. Thus, it is important for researchers to publish their raw data, not just normalized data, and to assay IRES activity in correct cell types and under conditions in which the endogenous protein is made.

IRES *Trans*-Acting Factors

ITAFs bind to IRESs and function as RNA chaperones to remodel the RNA to enhance or repress IRES activity. Both viral and cellular IRESs require ITAFs, excluding the *Dicistroviridae* IGR IRESs (Jan et al. 2003; Pestova and Hellen 2003; Deniz et al. 2009). Interestingly, although translation occurs in the cytoplasm, most ITAFs are either nuclear proteins or proteins that cycle between the nucleus and cytoplasm. The first identified ITAF was La autoantigen (La), which prevents initiation at aberrant start codons and enhances translation of the PV IRES (Meerovitch et al. 1993).

La enhances the activity of the HCV, coxsackievirus, and several cellular IRESs, such as binding immunoglobulin protein (BiP) and X-linked inhibitor of apoptosis (XIAP) (Holcik and Korneluk 2000; Kim et al. 2001; Ray and Das 2002; Costa-Mattioli et al. 2004). Some ITAF mRNAs, such as those encoding La and upstream of N-ras (Unr), use IRESs for their translation (Carter and Sarnow 2000; Cornelis et al. 2005). Other notable ITAFs reported to interact with multiple IRESs include poly-C-binding protein 2 (PCBP2), and the heterogeneous nuclear ribonucleoprotein (hnRNP) family of RNA-binding proteins (Hunt et al. 1999; Fujimura et al. 2008). Whereas the majority of ITAFs are reported to stimulate IRES-mediated translation, the RNA-binding proteins

AU-rich-binding factor (AUF1), KH-type splicing regulatory protein (KHSRP), and double-stranded RNA-binding nuclear protein 76 (DRBP76) negatively modulate IRES activity (Merrill and Gromeier 2006; Lin et al. 2009, 2014; Cathcart et al. 2013; Kung et al. 2017). Taken together, ITAFs play an important role in tissue and host tropism for viruses as well as a regulatory role for cellular IRESs, which may only be active in certain cell types or under certain conditions, such as hypoxia or apoptosis (Coldwell et al. 2000; Mitchell et al. 2003; Jahan et al. 2013).

Identification of Putative IRESs

As has been described above, viral and cellular IRESs are tremendously diverse in their sequence, size, structure and mechanism of initiation, and the *trans*-acting factors used. This requires validation on an individual basis rather than identification bioinformatically (Baird et al. 2006). Potential IRESs have been identified by determining which mRNAs remain or become associated with polysomes under conditions in which cap-dependent translation is down-regulated (Johannes et al. 1999; Qin and Sarnow 2004; Bushell et al. 2006). A recent study took a high-throughput approach by synthesizing 55,000, 210-nucleotide long oligos from cellular and viral mRNAs that have been previously reported to either have IRES activity or be polysome-associated in one of the stress conditions known to down-regulate cap-dependent initiation. The results confirmed that many of the mRNAs that remain polysome-associated under stress have IRESs (Weingarten-Gabbay et al. 2016).

The most important consideration for identifying and validating putative IRESs is that there is no perfect assay. Rather, establishing a new IRES requires a preponderance of evidence to rule out the possibility that translation could be occurring in any other fashion. The assays and controls have been discussed at length previously (Thompson 2012) and only a short summary will be provided here. The bicistronic reporter assay is commonly used as the first assay toward evaluating a putative IRES. Briefly, a DNA plasmid encoding a bicistronic reporter in which the

putative IRES sequence has been inserted between two reporter genes (most commonly *Renilla* and firefly luciferase) is transfected into cells followed by monitoring of the encoded reporter activity. Expression of the first cistron uses a cap-dependent mechanism of initiating translation, whereas expression of the second cistron is driven by IRES-mediated translation. IRES activity is indicated by expression of the second cistron relative to positive and negative controls and corrected for transfection differences by co-transfection of a separate cap-dependent reporter plasmid. Assays to establish IRES activity should be performed in vivo in cells because translation extracts such as the rabbit reticulocyte lysate can translate promiscuously. Whereas the bicistronic assay itself is simple, the major challenge in the field is that a series of stringent controls are required to show that expression of the second cistron is not dependent on a cryptic promoter, cryptic splicing, or readthrough from the upstream cistron (Bert et al. 2006; Baranick et al. 2008; Thompson 2012). In addition, if studies are performed in vitro there need to be controls for RNA cleavage events that would lead to cap-independent initiation because some in vitro translation extracts can translate indiscriminately. As important as the controls are for identifying an IRES, the same stringency needs to be applied to studies that have claimed that previously identified IRESs have cryptic promoter activity. Specifically, controls need to be in place to show that the promoter activity is coming from the IRES and not the vector backbone (Cobbold et al. 2008). It is also important to note that the bicistronic assay is an artificial setting and that in this context IRESs may have cryptic promoter or splicing activity, but this does not show that they are not IRESs, only that the bicistronic IRES assay cannot be used to establish IRES activity. After all, if an endogenous mRNA has this RNA element in the 5'UTR, then the question still remains as to whether it can function as an IRES even if it cannot be shown in a bicistronic context.

Transfection of a bicistronic reporter as an mRNA to bypass transcription in the nucleus avoids issues such as cryptic promoters and splice sites that might produce false positives. However, transfection of an RNA reporter only works for a subset of cellular IRESs, suggesting that a nuclear experience may be required for assembling ITAFs or adding RNA modifications (Stoneley et al. 2000; Semler and Waterman 2008). The greater dependence of cellular IRESs on ITAFs compared with viral IRESs may be caused by their functional differences, because most viral IRESs are highly structured and are solely translated by an IRES-dependent mechanism, rather than switching back and forth between cap-dependent and IRES-mediated initiation. In contrast, many cellular mRNAs that have an IRES need to be translated both cap-dependently and cap-independently, depending on the cellular conditions (Walters and Thompson 2016).

Better assays are needed to identify and validate IRESs. One possible mechanism that needs to be explored further stems from the observation that viral and cellular IRESs depend on eS25, whereas cap-dependent translation does not (Hertz et al. 2013). Thus, if expression of a downstream open reading frame (ORF) from a bicistronic reporter were caused by a cryptic promoter, splicing, or readthrough, then its expression would not be affected by the loss of eS25 from the cells. Moreover, this also provides us with another way to examine whether endogenous messages are using IRES-mediated translation under a given cellular condition by determining whether polysome association of an mRNA depends on eS25, as shown for p53 (Hertz et al. 2013).

CAP-INDEPENDENT TRANSLATION ELEMENTS (CITEs)

The majority of plant viral RNAs lack a 5′ cap and poly(A) tail but are nevertheless translated efficiently by the host cell translational machinery. A major mechanism behind this phenomenon is the CITE, which was first discovered in satellite tobacco necrosis virus (STNV). CITEs are located in the 3′UTR and recruit the 40S subunit by binding some component of the eIF4F complex while using RNA–RNA interactions between the 5′UTR and 3′UTR that circularize the viral RNA. Ribosomes assemble at or near the 5′-end and scan to the start codon (Simon and Miller 2013; Miras et al. 2017). Although CITEs have only been found in the 3′UTR, they can still promote initiation when they are moved to the 5′UTR or into the coding region (Meulewaeter et al. 1998).

CITEs vary widely in sequence and structure and have been classified into seven categories: translation enhancer domain (TED), barley yellow dwarf virus-like translation element (BTE), Panicum mosaic virus-like translation enhancer (PTE), I-shaped structures (ISS), Y-shaped structures (YSS), T-shaped structures (TSS), and CABYV-Xinjiang-like translation element (CXTE) (Table 3). These have been described in detail previously (Simon and Miller 2013; Miras et al. 2017). Almost all classes of 3′-CITEs bind to some component of the eIF4F complex and circularize with the 5′UTR. For example, BTE binds preferentially to eIF4G and recruits the 40S subunit to the 3′UTR before transferring it to the 5′UTR where translation initiates (Treder et al. 2008; Sharma et al. 2015). However, PTE binds preferentially to eIF4E and TED, ISS and YSS CITEs require intact eIF4F for translation (Gazo et al. 2004; Truniger et al. 2008; Nicholson et al. 2010; Wang et al. 2011). An unusual case is the turnip crinkle virus that has a TSS CITE in the 3′UTR that interacts with the 60S ribosomal subunit instead of forming RNA–RNA interactions with the 5′UTR. Circularization of the RNA occurs through 60S binding with the 40S subunit that is bound to the 5′UTR (Stupina et al. 2008). In the case of pea enation mosaic virus 2 (PEMV2), there are three CITE structures in the 3′UTR: kissing loop (kl)-TSS, PTE, and TSS. The kl-TSS is responsible for the RNA–RNA interaction with the 5′UTR, whereas the PTE binds eIF4E to facilitate translation initiation (Du et al. 2017). More studies are required to understand the mechanism of initiation for CITEs and whether they share similarities or not. Also, it is unknown how CITEs compete with cap-dependent initiation.

N^6-METHYLADENOSINE

The most common posttranscriptional modification, methylation of adenine residues to generate N^6-methyladenosine (m^6A), affects

Table 3. Classification of cap-independent translation elements (CITEs)

Type	Examples	Notable features
TED (STNV)	STNV PLPV PCLPV	Binds eIF4F Apical loop interacts with hairpin in 5′UTR
BTE (BYDV)	*Alphanecrovirus* *Betanecrovirus* *Dianthovirus* *Luteovirus* *Umbravirus*	Binds eIF4G, eIF4E-independent Up to six stem loops can radiate from central hub
PTE (PMV)	PMV PEMV2 SCV	Binds eIF4E Base-pairing interactions (dotted lines) form pseudoknot
ISS (MNSV)	MNSV MNeSV	Binds intact eIF4F Apical loop interacts with hairpin in 5′UTR
YSS (TBSV)	TBSV CIRV	Requires eIF4F/eIFis04F 5′ hairpin interacts with hairpin in 5′UTR
TSS (TCV)	TCV PEMV2	TCV TSS competes with tRNA for P-site binding in 60S
CXTE (MNSV)	MNSV	eIF4E-independent

PLPV, Pelargonium line pattern virus; PCLPV, pelargonium chlorotic ring pattern virus; BYDV, barley yellow dwarf virus; PMV, panicum mosaic virus; SCV, strawberry crinkle virus; MNSV, melon necrotic spot virus; MNeSV, maize necrotic streak virus; TBSV, tomato bushy stunt virus; CIRV, carnation Italian ringspot virus; TCV, turnip crinkle virus.

mRNA turnover and translation efficiency (see Peer et al. 2018). Although the majority of m⁶A modifications are localized to the coding region and 3′UTR, it was recently discovered that modifications located in the 5′UTR can mediate cap-independent translation initiation, although a free 5′ end is still required (Meyer et al. 2015). Importantly, 5′UTR m⁶A modifications are upregulated in response to heat shock by relocalization of the reader protein, YTHDF2, to the nucleus upon heat shock, which prevents the eraser protein, FTO, from removing m⁶A modifications (Meyer et al. 2015; Zhou et al. 2015). m⁶A promotes selective cap-independent initiation on certain mRNAs, such as HSP70, which depends on eIF3 binding to m⁶A residues to recruit the 40S subunit (Meyer et al. 2015). In contrast, another study showed that m⁶A can promote internal initiation of endogenous circular RNAs (circRNA) in cells and that the RNAs are polysome-associated (Yang et al. 2017). m⁶A-mediated translation is still a relative newcomer to the field (see Chekulaeva and Rajewsky 2018), and what is known has only been characterized in single reports that need to be reproduced. Additional studies are required to show that m⁶A modifications are required for noncanonical mechanisms of initiation and to reveal the mechanism for m⁶A-mediated initiation.

RIBOSOMAL SHUNTING

Ribosomal shunting occurs when ribosomes bypass or shunt over parts of the 5′UTR before reaching the start codon. Ribosomes are loaded onto the mRNA in a cap-dependent manner with recruitment of 40S subunits to the 5′ end of a capped mRNA followed by limited scanning to reach a "donor" site at which point the 40S jumps or "shunts" over major parts of a complex leader and lands at a downstream start codon (acceptor site) to initiate translation. Ribosomal shunting has been studied in cauliflower mosaic virus (CaMV) and adenovirus and is also used by a number of other plant and animal viruses (Stern-Ginossar et al. 2018). Additionally, cellular mRNAs such as BACE1, cIAP2, and HSP70 use ribosomal shunting to initiate translation

under stress (Yueh and Schneider 2000; Rogers et al. 2004; Sherrill and Lloyd 2008; Koh et al. 2013).

Ribosome shunting allows viruses to expand their coding capacity by translating downstream ORFs, whereas both viral and cellular mRNAs use shunting to selectively translate mRNAs when cap-dependent translation is down-regulated. It has been proposed that shunting proceeds when the helicase of eIF4F is inactive because little to no unwinding is required (Yueh and Schneider 1996). The criteria for establishing that an mRNA is using ribosomal shunting is that its translation is cap-dependent, yet upstream AUGs and stem-loop structures have minimal or no effect on translation and its 5′UTR does not contain an IRES. Interestingly, many but not all mRNAs that use ribosomal shunting have a short ORF (sORF) that is translated; on termination, instead of the ribosome disassociating from the mRNA, it remains associated and proceeds to shunt or bypass other sORFs and secondary structures to reinitiate at a downstream ORF (Latorre et al. 1998; Yueh and Schneider 2000; Racine and Duncan 2010). Ribosomal bypass or shunting over entire regions of the 5′UTR was shown by insertion of a start codon in a good Kozak consensus in the bypassed sequences. This did not result in translation from this start codon, suggesting that the ribosome was not scanning through this region (Futterer et al. 1993). Additionally, two different RNAs were used to encode the donor site (where the ribosome shunts from) and the pyrimidine-rich acceptor site (where the ribosome lands); when these RNAs were annealed together through complementary sequences, translation still occurred at the downstream ORF, suggesting that the ribosome did not scan through the bypassed region because scanning would have separated the two RNAs (Futterer et al. 1993).

For those mRNAs that have an sORF that is translated, the 40S subunit is retained following termination through base-pairing of sequences in the 5′UTR (adenovirus and hsp70) with the 18S rRNA 3′ hairpin or through binding to eIF3 (CaMV), which remains associated with the 40S after initiation (Ryabova and Hohn 2000; Yueh and Schneider 2000; Morley and Coldwell 2008).

Furthermore, the 40S ribosomal protein, eS25, which is required for IRES-mediated initiation, is also required for ribosomal shunting, although its role in shunting is not understood (Hertz et al. 2013). Many questions still remain about how ribosomes land and reinitiate downstream of the donor sites and which initiation factors or *trans*-acting factors are required for reinitiation.

TRANSLATION INITIATOR OF SHORT 5′UTR (TISU)

A cap-dependent, scanning-free mechanism of translation initiation in mRNAs with extremely short 5′UTRs (<30 nucleotides) requires a TISU element. TISUs are located within 5 to 30 nucleotides of the cap and have a median length of 12 nucleotides (Elfakess and Dikstein 2008). The TISU consensus sequence, SAASAUGGCGGC (S = C or G), differs from the Kozak consensus sequence (RCCAUGG, R = A or G) for start codon recognition (Elfakess and Dikstein 2008). Approximately 4% of mRNAs contain TISUs and encode proteins with housekeeping functions such as mitochondrial activities, energy metabolism, and protein synthesis (Elfakess and Dikstein 2008). TISU-containing mRNAs are specifically resistant to inhibition of protein synthesis caused by energy stress during glucose deprivation (Sinvani et al. 2015).

TISUs require eIF4F, which seems problematic given that recognition of an AUG close to the cap would clash with 43S PIC recruitment proximal to the cap. However, a key step in TISU initiation is release of eIF4F upon AUG recognition (Sinvani et al. 2015). During canonical cap-dependent initiation, eIF1 discriminates against AUGs proximal to the 5′ end, favoring 48S complex assembly on an AUG further downstream (Pestova and Kolupaeva 2002). However, eIF1 is not required for TISU start codon recognition as it did not inhibit translation for a TISU AUG located close to the 5′ cap, nor did it induce leaky scanning to bypass the TISU AUG (Elfakess et al. 2011; Dikstein 2012). Yet, eIF1 and eIF1A are required for TISU initiation. Most likely they are required to induce an open conformation (disruption of h18 and h34 and formation of a

new interaction between h16 and uS3) of the 43S PIC for mRNA loading (Haimov et al. 2015; Sharifulin et al. 2016).

The ability of TISUs to prevent scanning of the 48S subunit following cap-dependent recruitment to the mRNA is caused by sequence-specific interactions with ribosomal proteins (uS3 and eS10) positioned in or near the entry channel (Haimov et al. 2017). The interactions between eIF1A and uS3 and eS10 have been suggested to arrest or prevent scanning by displacing helix 16 (h16), which induces the closed conformation and arrests scanning of the 40S. Taken together, this suggests that a TISU element near the 5′ end of the mRNA can recruit a 43S PIC in close proximity to the cap by displacing the cap-binding complex (eIF4F) and immediately arrest scanning on mRNA binding and AUG recognition.

CONCLUSION

Noncanonical translation plays an important role in gene expression during cellular stress. Transcriptomic analysis of mRNA levels correlates poorly with proteomic analysis of their expression, suggesting that there is a large amount of regulation at the posttranscriptional level (Gygi et al. 1999; Ideker et al. 2001; Griffin et al. 2002). Likely, the role of noncanonical translation has been underappreciated in cancer, early development, and cellular stress conditions. A better understanding of the contributions and mechanisms of noncanonical translation is critical for understanding gene expression. Overall, researchers in the translation field have described the various mechanisms of initiation as different or unique. However, given some of the similarities between ribosomal shunting and IRES-mediated translation, such as ribosome landing occurring at a pyrimidine-rich site, base-pairing with the 18S rRNA, and the requirement for eS25, it is tempting to speculate that there could be some common mechanisms shared between IRESs and ribosomal shunts (Morley and Coldwell 2008; Hertz et al. 2013). In addition, some noncanonical mechanisms use a tRNA-like structure to recruit or assemble ribosomes (Colussi et al. 2014; Butcher and Jan 2016). The mechanisms for how CITES, ribosomal shunting, TISU, IRESs, and m⁶A manipulate ribosomes to load the mRNA and initiate translation are largely unknown. As the mechanisms underlying noncanonical initiation are elucidated, both differences as well as similarities will be revealed.

REFERENCES

*Reference is also in this collection.

Al-Fageeh MB, Smales CM. 2009. Cold-inducible RNA binding protein (CIRP) expression is modulated by alternative mRNAs. *RNA* **15:** 1164–1176.

Andreev DE, Fernandez-Miragall O, Ramajo J, Dmitriev SE, Terenin IM, Martinez-Salas E, Shatsky IN. 2007. Differential factor requirement to assemble translation initiation complexes at the alternative start codons of foot-and-mouth disease virus RNA. *RNA* **13:** 1366–1374.

Andreev DE, Dmitriev SE, Terenin IM, Prassolov VS, Merrick WC, Shatsky IN. 2009. Differential contribution of the m⁷G-cap to the 5′ end-dependent translation initiation of mammalian mRNAs. *Nucleic Acids Res* **37:** 6135–6147.

Avanzino BC, Fuchs G, Fraser CS. 2017. Cellular cap-binding protein, eIF4E, promotes picornavirus genome restructuring and translation. *Proc Natl Acad Sci* **114:** 9611–9616.

Baird SD, Turcotte M, Korneluk RG, Holcik M. 2006. Searching for IRES. *RNA* **12:** 1755–1785.

Baird SD, Lewis SM, Turcotte M, Holcik M. 2007. A search for structurally similar cellular internal ribosome entry sites. *Nucleic Acids Res* **35:** 4664–4677.

Bakhshesh M, Groppelli E, Willcocks MM, Royall E, Belsham GJ, Roberts LO. 2008. The picornavirus avian encephalomyelitis virus possesses a hepatitis C virus-like internal ribosome entry site element. *J Virol* **82:** 1993–2003.

Ban N, Beckmann R, Cate JH, Dinman JD, Dragon F, Ellis SR, Lafontaine DL, Lindahl L, Liljas A, Lipton JM, et al. 2014. A new system for naming ribosomal proteins. *Curr Opin Struct Biol* **24:** 165–169.

Baranick BT, Lemp NA, Nagashima J, Hiraoka K, Kasahara N, Logg CR. 2008. Splicing mediates the activity of four putative cellular internal ribosome entry sites. *Proc Natl Acad Sci* **105:** 4733–4738.

Bert AG, Grepin R, Vadas MA, Goodall GJ. 2006. Assessing IRES activity in the HIF-1α and other cellular 5′ UTRs. *RNA* **12:** 1074–1083.

Blau L, Knirsh R, Ben-Dror I, Oren S, Kuphal S, Hau P, Proescholdt M, Bosserhoff AK, Vardimon L. 2012. Aberrant expression of c-Jun in glioblastoma by internal ribosome entry site (IRES)-mediated translational activation. *Proc Natl Acad Sci* **109:** E2875–E2884.

Bonnal S, Schaeffer C, Creancier L, Clamens S, Moine H, Prats AC, Vagner S. 2003. A single internal ribosome entry site containing a G quartet RNA structure drives fibroblast growth factor 2 gene expression at four alternative translation initiation codons. *J Biol Chem* **278:** 39330–39336.

Burgess DH, Svensson M, Dandrea T, Gronlund K, Hammarquist F, Orrenius S, Cotgreave IA. 1999. Human skeletal muscle cytosols are refractory to cytochrome c-dependent activation of type-II caspases and lack APAF-1. *Cell Death Differ* **6:** 256–261.

Bushell M, Stoneley M, Kong YW, Hamilton TL, Spriggs KA, Dobbyn HC, Qin X, Sarnow P, Willis AE. 2006. Polypyrimidine tract binding protein regulates IRES-mediated gene expression during apoptosis. *Mol Cell* **23:** 401–412.

Butcher SE, Jan E. 2016. tRNA-mimicry in IRES-mediated translation and recoding. *RNA Biol* **13:** 1068–1074.

Carter MS, Sarnow P. 2000. Distinct mRNAs that encode La autoantigen are differentially expressed and contain internal ribosome entry sites. *J Biol Chem* **275:** 28301–28307.

Carter MS, Kuhn KM, Sarnow P. 2000. Cellular internal ribosome entry site elements and the use of cDNA microarrays in their investigation. In *Translational control of gene expression* (ed. Sonenberg N, et al.), pp. 615–635. Cold Spring Harbor Laboratory Press, Cold Spring Harbor, NY.

Cathcart AL, Rozovics JM, Semler BL. 2013. Cellular mRNA decay protein AUF1 negatively regulates enterovirus and human rhinovirus infections. *J Virol* **87:** 10423–10434.

Cecconi F, Alvarez-Bolado G, Meyer BI, Roth KA, Gruss P. 1998. Apaf1 (CED-4 homolog) regulates programmed cell death in mammalian development. *Cell* **94:** 727–737.

Chappell SA, Edelman GM, Mauro VP. 2000. A 9-nt segment of a cellular mRNA can function as an internal ribosome entry site (IRES) and when present in linked multiple copies greatly enhances IRES activity. *Proc Natl Acad Sci* **97:** 1536–1541.

* Chekulaeva M, Rajewsky N. 2018. Roles of lncRNAs and circRNAs in translation. *Cold Spring Harb Perspect Biol* doi: 10.1101/cshperspect.a032680.

Chen CY, Sarnow P. 1995. Initiation of protein synthesis by the eukaryotic translational apparatus on circular RNAs. *Science* **268:** 415–417.

Cobbold LC, Spriggs KA, Haines SJ, Dobbyn HC, Hayes C, de Moor CH, Lilley KS, Bushell M, Willis AE. 2008. Identification of internal ribosome entry segment (IRES)-trans-acting factors for the Myc family of IRESs. *Mol Cell Biol* **28:** 40–49.

Coldwell MJ, Mitchell SA, Stoneley M, MacFarlane M, Willis AE. 2000. Initiation of Apaf-1 translation by internal ribosome entry. *Oncogene* **19:** 899–905.

Coldwell MJ, deSchoolmeester ML, Fraser GA, Pickering BM, Packham G, Willis AE. 2001. The p36 isoform of BAG-1 is translated by internal ribosome entry following heat shock. *Oncogene* **20:** 4095–4100.

Colussi TM, Costantino DA, Hammond JA, Ruehle GM, Nix JC, Kieft JS. 2014. The structural basis of transfer RNA mimicry and conformational plasticity by a viral RNA. *Nature* **511:** 366–369.

Cornelis S, Bruynooghe Y, Denecker G, Van Huffel S, Tinton S, Beyaert R. 2000. Identification and characterization of a novel cell cycle-regulated internal ribosome entry site. *Mol Cell* **5:** 597–605.

Cornelis S, Tinton SA, Schepens B, Bruynooghe Y, Beyaert R. 2005. UNR translation can be driven by an IRES element that is negatively regulated by polypyrimidine tract binding protein. *Nucleic Acids Res* **33**: 3095–3108.

Costa-Mattioli M, Svitkin Y, Sonenberg N. 2004. La autoantigen is necessary for optimal function of the poliovirus and hepatitis C virus internal ribosome entry site in vivo and in vitro. *Mol Cell Biol* **24**: 6861–6870.

Deforges J, Locker N, Sargueil B. 2015. mRNAs that specifically interact with eukaryotic ribosomal subunits. *Biochimie* **114**: 48–57.

Deniz N, Lenarcic EM, Landry DM, Thompson SR. 2009. Translation initiation factors are not required for *Dicistroviridae* IRES function in vivo. *RNA* **15**: 932–946.

Dikstein R. 2012. Transcription and translation in a package deal: The TISU paradigm. *Gene* **491**: 1–4.

Dresios J, Chappell SA, Zhou W, Mauro VP. 2006. An mRNA-rRNA base-pairing mechanism for translation initiation in eukaryotes. *Nat Struct Mol Biol* **13**: 30–34.

Du Z, Alekhina OM, Vassilenko KS, Simon AE. 2017. Concerted action of two 3′ cap-independent translation enhancers increases the competitive strength of translated viral genomes. *Nucleic Acids Res* **45**: 9558–9572.

Elfakess R, Dikstein R. 2008. A translation initiation element specific to mRNAs with very short 5′UTR that also regulates transcription. *PLoS ONE* **3**: e3094.

Elfakess R, Sinvani H, Haimov O, Svitkin Y, Sonenberg N, Dikstein R. 2011. Unique translation initiation of mRNAs-containing TISU element. *Nucleic Acids Res* **39**: 7598–7609.

Fernandez J, Yaman I, Mishra R, Merrick WC, Snider MD, Lamers WH, Hatzoglou M. 2001. Internal ribosome entry site-mediated translation of a mammalian mRNA is regulated by amino acid availability. *J Biol Chem* **276**: 12285–12291.

Fernandez IS, Bai XC, Murshudov G, Scheres SH, Ramakrishnan V. 2014. Initiation of translation by cricket paralysis virus IRES requires its translocation in the ribosome. *Cell* **157**: 823–831.

Fox JT, Shin WK, Caudill MA, Stover PJ. 2009. A UV-responsive internal ribosome entry site enhances serine hydroxymethyltransferase 1 expression for DNA damage repair. *J Biol Chem* **284**: 31097–31108.

Fraser CS, Doudna JA. 2007. Structural and mechanistic insights into hepatitis C viral translation initiation. *Nat Rev Microbiol* **5**: 29–38.

Fuchs G, Petrov AN, Marceau CD, Popov LM, Chen J, O'Leary SE, Wang R, Carette JE, Sarnow P, Puglisi JD. 2015. Kinetic pathway of 40S ribosomal subunit recruitment to hepatitis C virus internal ribosome entry site. *Proc Natl Acad Sci* **112**: 319–325.

Fujimura K, Kano F, Murata M. 2008. Identification of PCBP2, a facilitator of IRES-mediated translation, as a novel constituent of stress granules and processing bodies. *RNA* **14**: 425–431.

Futterer J, Kiss-Laszlo Z, Hohn T. 1993. Nonlinear ribosome migration on cauliflower mosaic virus 35S RNA. *Cell* **73**: 789–802.

Gaccioli F, Huang CC, Wang C, Bevilacqua E, Franchi-Gazzola R, Gazzola GC, Bussolati O, Snider MD, Hatzoglou M. 2006. Amino acid starvation induces the SNAT2 neutral amino acid transporter by a mechanism that involves

eukaryotic initiation factor 2α phosphorylation and cap-independent translation. *J Biol Chem* **281**: 17929–17940.

Gazo BM, Murphy P, Gatchel JR, Browning KS. 2004. A novel interaction of Cap-binding protein complexes eukaryotic initiation factor (eIF) 4F and eIF(iso)4F with a region in the 3′-untranslated region of satellite tobacco necrosis virus. *J Biol Chem* **279**: 13584–13592.

Gonzalez-Almela E, Williams H, Sanz MA, Carrasco L. 2018. The initiation factors eIF2, eIF2A, eIF2D, eIF4A, and eIF4G are not involved in translation driven by hepatitis C virus IRES in human cells. *Front Microbiol* **9**: 207.

Graber TE, Baird SD, Kao PN, Mathews MB, Holcik M. 2010. NF45 functions as an IRES *trans*-acting factor that is required for translation of cIAP1 during the unfolded protein response. *Cell Death Differ* **17**: 719–729.

Griffin TJ, Gygi SP, Ideker T, Rist B, Eng J, Hood L, Aebersold R. 2002. Complementary profiling of gene expression at the transcriptome and proteome levels in *Saccharomyces cerevisiae*. *Mol Cell Proteomics* **1**: 323–333.

Gygi SP, Rochon Y, Franza BR, Aebersold R. 1999. Correlation between protein and mRNA abundance in yeast. *Mol Cell Biol* **19**: 1720–1730.

Haimov O, Sinvani H, Dikstein R. 2015. Cap-dependent, scanning-free translation initiation mechanisms. *Biochim Biophys Acta* **1849**: 1313–1318.

Haimov O, Sinvani H, Martin F, Ulitsky I, Emmanuel R, Tamarkin-Ben-Harush A, Vardy A, Dikstein R. 2017. Efficient and accurate translation initiation directed by TISU involves RPS3 and RPS10e binding and differential eukaryotic initiation factor 1A regulation. *Mol Cell Biol* **37**: e00150.

Hertz MI, Landry DM, Willis AE, Luo G, Thompson SR. 2013. Ribosomal protein S25 dependency reveals a common mechanism for diverse internal ribosome entry sites and ribosome shunting. *Mol Cell Biol* **33**: 1016–1026.

Holcik M. 2004. Targeting translation for treatment of cancer—A novel role for IRES? *Curr Cancer Drug Targets* **4**: 299–311.

Holcik M, Korneluk RG. 2000. Functional characterization of the X-linked inhibitor of apoptosis (XIAP) internal ribosome entry site element: Role of La autoantigen in XIAP translation. *Mol Cell Biol* **20**: 4648–4657.

Hunt SL, Hsuan JJ, Totty N, Jackson RJ. 1999. unr, a cellular cytoplasmic RNA-binding protein with five cold-shock domains, is required for internal initiation of translation of human rhinovirus RNA. *Genes Dev* **13**: 437–448.

Ideker T, Thorsson V, Ranish JA, Christmas R, Buhler J, Eng JK, Bumgarner R, Goodlett DR, Aebersold R, Hood L. 2001. Integrated genomic and proteomic analyses of a systematically perturbed metabolic network. *Science* **292**: 929–934.

Jaafar ZA, Oguro A, Nakamura Y, Kieft JS. 2016. Translation initiation by the hepatitis C virus IRES requires eIF1A and ribosomal complex remodeling. *eLife* **5**: e21198.

Jack K, Bellodi C, Landry DM, Niederer RO, Meskauskas A, Musalgaonkar S, Kopmar N, Krasnykh O, Dean AM, Thompson SR, et al. 2011. rRNA pseudouridylation defects affect ribosomal ligand binding and translational fidelity from yeast to human cells. *Mol Cell* **44**: 660–666.

Jahan N, Wimmer E, Mueller S. 2013. Polypyrimidine tract binding protein-1 (PTB1) is a determinant of the tissue and host tropism of a human rhinovirus/poliovirus chimera PV1(RIPO). *PLoS ONE* **8:** e60791.

Jan E, Sarnow P. 2002. Factorless ribosome assembly on the internal ribosome entry site of cricket paralysis virus. *J Mol Biol* **324:** 889–902.

Jan E, Kinzy TG, Sarnow P. 2003. Divergent tRNA-like element supports initiation, elongation, and termination of protein biosynthesis. *Proc Natl Acad Sci* **100:** 15410–15415.

Jang CJ, Jan E. 2010. Modular domains of the *Dicistroviridae* intergenic internal ribosome entry site. *RNA* **16:** 1182–1195.

Jang SK, Krausslich HG, Nicklin MJ, Duke GM, Palmenberg AC, Wimmer E. 1988. A segment of the 5′ nontranslated region of encephalomyocarditis virus RNA directs internal entry of ribosomes during in vitro translation. *J Virol* **62:** 2636–2643.

Jang CJ, Lo MC, Jan E. 2009. Conserved element of the dicistrovirus IGR IRES that mimics an E-site tRNA/ribosome interaction mediates multiple functions. *J Mol Biol* **387:** 42–58.

Jimenez J, Jang GM, Semler BL, Waterman ML. 2005. An internal ribosome entry site mediates translation of lymphoid enhancer factor-1. *RNA* **11:** 1385–1399.

Johannes G, Sarnow P. 1998. Cap-independent polysomal association of natural mRNAs encoding c-myc, BiP, and eIF4G conferred by internal ribosome entry sites. *RNA* **4:** 1500–1513.

Johannes G, Carter MS, Eisen MB, Brown PO, Sarnow P. 1999. Identification of eukaryotic mRNAs that are translated at reduced cap binding complex eIF4F concentrations using a cDNA microarray. *Proc Natl Acad Sci* **96:** 13118–13123.

Kaminski A, Poyry TA, Skene PJ, Jackson RJ. 2010. Mechanism of initiation site selection promoted by the human rhinovirus 2 internal ribosome entry site. *J Virol* **84:** 6578–6589.

Kim YY, Back SH, Rho J, Lee SH, Jang SK. 2001. La autoantigen enhances translation of BiP mRNA. *Nucleic Acids Res* **29:** 5009–5016.

Koh DC, Edelman GM, Mauro VP. 2013. Physical evidence supporting a ribosomal shunting mechanism of translation initiation for BACE1 mRNA. *Translation (Austin)* **1:** e24400.

Koh CS, Brilot AF, Grigorieff N, Korostelev AA. 2014. Taura syndrome virus IRES initiates translation by binding its tRNA-mRNA-like structural element in the ribosomal decoding center. *Proc Natl Acad Sci* **111:** 9139–9144.

Kung YA, Hung CT, Chien KY, Shih SR. 2017. Control of the negative IRES *trans*-acting factor KHSRP by ubiquitination. *Nucleic Acids Res* **45:** 271–287.

Lacerda R, Menezes J, Romao L. 2017. More than just scanning: The importance of cap-independent mRNA translation initiation for cellular stress response and cancer. *Cell Mol Life Sci* **74:** 1659–1680.

Landry DM, Hertz MI, Thompson SR. 2009. RPS25 is essential for translation initiation by the *Dicistroviridae* and hepatitis C viral IRESs. *Genes Dev* **23:** 2753–2764.

Latorre P, Kolakofsky D, Curran J. 1998. Sendai virus Y proteins are initiated by a ribosomal shunt. *Mol Cell Biol* **18:** 5021–5031.

Lee AS, Burdeinick-Kerr R, Whelan SP. 2013. A ribosome-specialized translation initiation pathway is required for cap-dependent translation of vesicular stomatitis virus mRNAs. *Proc Natl Acad Sci* **110:** 324–329.

Le Quesne JP, Stoneley M, Fraser GA, Willis AE. 2001. Derivation of a structural model for the c-myc IRES. *J Mol Biol* **310:** 111–126.

Lin JY, Li ML, Shih SR. 2009. Far upstream element binding protein 2 interacts with enterovirus 71 internal ribosomal entry site and negatively regulates viral translation. *Nucleic Acids Res* **37:** 47–59.

Lin JY, Li ML, Brewer G. 2014. mRNA decay factor AUF1 binds the internal ribosomal entry site of enterovirus 71 and inhibits virus replication. *PLoS ONE* **9:** e103827.

Lomakin IB, Hellen CU, Pestova TV. 2000. Physical association of eukaryotic initiation factor 4G (eIF4G) with eIF4A strongly enhances binding of eIF4G to the internal ribosomal entry site of encephalomyocarditis virus and is required for internal initiation of translation. *Mol Cell Biol* **20:** 6019–6029.

Marash L, Liberman N, Henis-Korenblit S, Sivan G, Reem E, Elroy-Stein O, Kimchi A. 2008. DAP5 promotes cap-independent translation of Bcl-2 and CDK1 to facilitate cell survival during mitosis. *Mol Cell* **30:** 447–459.

Martinez-Salas E, Pineiro D, Fernandez N. 2012. Alternative mechanisms to initiate translation in eukaryotic mRNAs. *Comp Funct Genomics* **2012:** 391546.

Matsuda D, Mauro VP. 2014. Base pairing between hepatitis C virus RNA and 18S rRNA is required for IRES-dependent translation initiation in vivo. *Proc Natl Acad Sci* **111:** 15385–15389.

Meerovitch K, Svitkin YV, Lee HS, Lejbkowicz F, Kenan DJ, Chan EK, Agol VI, Keene JD, Sonenberg N. 1993. La autoantigen enhances and corrects aberrant translation of poliovirus RNA in reticulocyte lysate. *J Virol* **67:** 3798–3807.

* Merrick WC, Pavitt GD. 2018. Protein synthesis initiation in eukaryotic cells. *Cold Spring Harb Perspect Biol* doi: 10.1101/cshperspect.a033092.

Merrill MK, Gromeier M. 2006. The double-stranded RNA binding protein 76:NF45 heterodimer inhibits translation initiation at the rhinovirus type 2 internal ribosome entry site. *J Virol* **80:** 6936–6942.

Meulewaeter F, Van Montagu M, Cornelissen M. 1998. Features of the autonomous function of the translational enhancer domain of satellite tobacco necrosis virus. *RNA* **4:** 1347–1356.

Meyer KD, Patil DP, Zhou J, Zinoviev A, Skabkin MA, Elemento O, Pestova TV, Qian SB, Jaffrey SR. 2015. 5′ UTR m^{6}A promotes cap-independent translation. *Cell* **163:** 999–1010.

Miras M, Miller WA, Truniger V, Aranda MA. 2017. Noncanonical translation in plant RNA viruses. *Front Plant Sci* **8:** 494.

Mitchell SA, Spriggs KA, Coldwell MJ, Jackson RJ, Willis AE. 2003. The Apaf-1 internal ribosome entry segment attains the correct structural conformation for function via interactions with PTB and unr. *Mol Cell* **11:** 757–771.

Morley SJ, Coldwell MJ. 2008. A cunning stunt: An alternative mechanism of eukaryotic translation initiation. *Sci Signal* **1**: pe32.

Morris MJ, Negishi Y, Pazsint C, Schonhoft JD, Basu S. 2010. An RNA G-quadruplex is essential for cap-independent translation initiation in human VEGF IRES. *J Am Chem Soc* **132**: 17831–17839.

Muhs M, Hilal T, Mielke T, Skabkin MA, Sanbonmatsu KY, Pestova TV, Spahn CM. 2015. Cryo-EM of ribosomal 80S complexes with termination factors reveals the translocated cricket paralysis virus IRES. *Mol Cell* **57**: 422–432.

Murray J, Savva CG, Shin BS, Dever TE, Ramakrishnan V, Fernandez IS. 2016. Structural characterization of ribosome recruitment and translocation by type IV IRES. *eLife* **5**: e13567.

Nakashima N, Uchiumi T. 2009. Functional analysis of structural motifs in dicistroviruses. *Virus Res* **139**: 137–147.

Nicholson BL, Wu B, Chevtchenko I, White KA. 2010. Tombusvirus recruitment of host translational machinery via the 3′ UTR. *RNA* **16**: 1402–1419.

Panopoulos P, Mauro VP. 2008. Antisense masking reveals contributions of mRNA-rRNA base pairing to translation of Gtx and FGF2 mRNAs. *J Biol Chem* **283**: 33087–33093.

Passmore LA, Schmeing TM, Maag D, Applefield DJ, Acker MG, Algire MA, Lorsch JR, Ramakrishnan V. 2007. The eukaryotic translation initiation factors eIF1 and eIF1A induce an open conformation of the 40S ribosome. *Mol Cell* **26**: 41–50.

* Peer E, Moshitch-Moshkovitz S, Rechavi G, Dominissini D. 2018. The epitranscriptome in translation regulation. *Cold Spring Harb Perspect Biol* doi: 10.1101/cshperspect.a032623.

Pelletier J, Sonenberg N. 1988. Internal initiation of translation of eukaryotic mRNA directed by a sequence derived from poliovirus RNA. *Nature* **334**: 320–325.

Pestova TV, Hellen CU. 2003. Translation elongation after assembly of ribosomes on the Cricket paralysis virus internal ribosomal entry site without initiation factors or initiator tRNA. *Genes Dev* **17**: 181–186.

Pestova TV, Kolupaeva VG. 2002. The roles of individual eukaryotic translation initiation factors in ribosomal scanning and initiation codon selection. *Genes Dev* **16**: 2906–2922.

Pestova TV, Hellen CU, Shatsky IN. 1996a. Canonical eukaryotic initiation factors determine initiation of translation by internal ribosomal entry. *Mol Cell Biol* **16**: 6859–6869.

Pestova TV, Shatsky IN, Hellen CU. 1996b. Functional dissection of eukaryotic initiation factor 4F: The 4A subunit and the central domain of the 4G subunit are sufficient to mediate internal entry of 43S preinitiation complexes. *Mol Cell Biol* **16**: 6870–6878.

Pestova TV, Shatsky IN, Fletcher SP, Jackson RJ, Hellen CU. 1998. A prokaryotic-like mode of cytoplasmic eukaryotic ribosome binding to the initiation codon during internal translation initiation of hepatitis C and classical swine fever virus RNAs. *Genes Dev* **12**: 67–83.

Pfingsten JS, Costantino DA, Kieft JS. 2006. Structural basis for ribosome recruitment and manipulation by a viral IRES RNA. *Science* **314**: 1450–1454.

Pickering BM, Mitchell SA, Spriggs KA, Stoneley M, Willis AE. 2004. Bag-1 internal ribosome entry segment activity is promoted by structural changes mediated by poly (rC) binding protein 1 and recruitment of polypyrimidine tract binding protein 1. *Mol Cell Biol* **24**: 5595–5605.

Pisarev AV, Chard LS, Kaku Y, Johns HL, Shatsky IN, Belsham GJ. 2004. Functional and structural similarities between the internal ribosome entry sites of hepatitis C virus and porcine teschovirus, a picornavirus. *J Virol* **78**: 4487–4497.

Pozner A, Goldenberg D, Negreanu V, Le SY, Elroy-Stein O, Levanon D, Groner Y. 2000. Transcription-coupled translation control of AML1/RUNX1 is mediated by cap- and internal ribosome entry site-dependent mechanisms. *Mol Cell Biol* **20**: 2297–2307.

Qin X, Sarnow P. 2004. Preferential translation of internal ribosome entry site-containing mRNAs during the mitotic cycle in mammalian cells. *J Biol Chem* **279**: 13721–13728.

Racine T, Duncan R. 2010. Facilitated leaky scanning and atypical ribosome shunting direct downstream translation initiation on the tricistronic S1 mRNA of avian reovirus. *Nucleic Acids Res* **38**: 7260–7272.

Ray PS, Das S. 2002. La autoantigen is required for the internal ribosome entry site-mediated translation of Coxsackievirus B3 RNA. *Nucleic Acids Res* **30**: 4500–4508.

Riley A, Jordan LE, Holcik M. 2010. Distinct 5′ UTRs regulate XIAP expression under normal growth conditions and during cellular stress. *Nucleic Acids Res* **38**: 4665–4674.

Rogers GW Jr., Edelman GM, Mauro VP. 2004. Differential utilization of upstream AUGs in the β-secretase mRNA suggests that a shunting mechanism regulates translation. *Proc Natl Acad Sci* **101**: 2794–2799.

Ruggero D. 2013. Translational control in cancer etiology. *Cold Spring Harb Perspect Biol* **5**: a012336.

Ryabova LA, Hohn T. 2000. Ribosome shunting in the cauliflower mosaic virus 35S RNA leader is a special case of reinitiation of translation functioning in plant and animal systems. *Genes Dev* **14**: 817–829.

Sasaki J, Nakashima N. 1999. Translation initiation at the CUU codon is mediated by the internal ribosome entry site of an insect picorna-like virus in vitro. *J Virol* **73**: 1219–1226.

Sasaki J, Nakashima N. 2000. Methionine-independent initiation of translation in the capsid protein of an insect RNA virus. *Proc Natl Acad Sci* **97**: 1512–1515.

Sasaki J, Nakashima N, Saito H, Noda H. 1998. An insect picorna-like virus, *Plautia stali* intestine virus, has genes of capsid proteins in the 3′ part of the genome. *Virology* **244**: 50–58.

Schepens B, Tinton SA, Bruynooghe Y, Beyaert R, Cornelis S. 2005. The polypyrimidine tract-binding protein stimulates HIF-1α IRES-mediated translation during hypoxia. *Nucleic Acids Res* **33**: 6884–6894.

Schneider R, Agol VI, Andino R, Bayard F, Cavener DR, Chappell SA, Chen JJ, Darlix JL, Dasgupta A, Donze O, et al. 2001. New ways of initiating translation in eukaryotes. *Mol Cell Biol* **21**: 8238–8246.

Schuler M, Connell SR, Lescoute A, Giesebrecht J, Dabrowski M, Schroeer B, Mielke T, Penczek PA, Westhof E, Spahn CM. 2006. Structure of the ribosome-bound cricket paralysis virus IRES RNA. *Nat Struct Mol Biol* **13:** 1092–1096.

Sella O, Gerlitz G, Le SY, Elroy-Stein O. 1999. Differentiation-induced internal translation of c-*sis* mRNA: Analysis of the *cis* elements and their differentiation-linked binding to the hnRNP C protein. *Mol Cell Biol* **19:** 5429–5440.

Semler BL, Waterman ML. 2008. IRES-mediated pathways to polysomes: Nuclear versus cytoplasmic routes. *Trends Microbiol* **16:** 1–5.

Sharifulin DE, Bartuli YS, Meschaninova MI, Ven'yaminova AG, Graifer DM, Karpova GG. 2016. Exploring accessibility of structural elements of the mammalian 40S ribosomal mRNA entry channel at various steps of translation initiation. *Biochim Biophys Acta* **1864:** 1328–1338.

Sharma SD, Kraft JJ, Miller WA, Goss DJ. 2015. Recruitment of the 40S ribosome subunit to the 3′-untranslated region (UTR) of a viral mRNA, via the eIF4 complex, facilitates cap-independent translation. *J Biol Chem* **290:** 11268–11281.

Sherrill KW, Lloyd RE. 2008. Translation of cIAP2 mRNA is mediated exclusively by a stress-modulated ribosome shunt. *Mol Cell Biol* **28:** 2011–2022.

Silvera D, Arju R, Darvishian F, Levine PH, Zolfaghari L, Goldberg J, Hochman T, Formenti SC, Schneider RJ. 2009. Essential role for eIF4GI overexpression in the pathogenesis of inflammatory breast cancer. *Nat Cell Biol* **11:** 903–908.

Simon AE, Miller WA. 2013. 3′ cap-independent translation enhancers of plant viruses. *Annu Rev Microbiol* **67:** 21–42.

Sinvani H, Haimov O, Svitkin Y, Sonenberg N, Tamarkin-Ben-Harush A, Viollet B, Dikstein R. 2015. Translational tolerance of mitochondrial genes to metabolic energy stress involves TISU and eIF1-eIF4GI cooperation in start codon selection. *Cell Metab* **21:** 479–492.

Spahn CM, Kieft JS, Grassucci RA, Penczek PA, Zhou K, Doudna JA, Frank J. 2001. Hepatitis C virus IRES RNA-induced changes in the conformation of the 40s ribosomal subunit. *Science* **291:** 1959–1962.

Spahn CM, Jan E, Mulder A, Grassucci RA, Sarnow P, Frank J. 2004. Cryo-EM visualization of a viral internal ribosome entry site bound to human ribosomes: The IRES functions as an RNA-based translation factor. *Cell* **118:** 465–475.

Spriggs KA, Stoneley M, Bushell M, Willis AE. 2008. Reprogramming of translation following cell stress allows IRES-mediated translation to predominate. *Biol Cell* **100:** 27–38.

Spriggs KA, Cobbold LC, Ridley SH, Coldwell M, Bottley A, Bushell M, Willis AE, Siddle K. 2009. The human insulin receptor mRNA contains a functional internal ribosome entry segment. *Nucleic Acids Res* **37:** 5881–5893.

Spriggs KA, Bushell M, Willis AE. 2010. Translational regulation of gene expression during conditions of cell stress. *Mol Cell* **40:** 228–237.

* Stern-Ginossar N, Thompson SR, Mathews MB, Mohr I. 2018. Translational control in virus-infected cells. *Cold Spring Harb Perspect Biol* doi: 10.1101/cshperspect. a033001.

Stoneley M, Subkhankulova T, Le Quesne JP, Coldwell MJ, Jopling CL, Belsham GJ, Willis AE. 2000. Analysis of the c-myc IRES; a potential role for cell-type specific trans-acting factors and the nuclear compartment. *Nucleic Acids Res* **28:** 687–694.

Stupina VA, Meskauskas A, McCormack JC, Yingling YG, Shapiro BA, Dinman JD, Simon AE. 2008. The 3′ proximal translational enhancer of Turnip crinkle virus binds to 60S ribosomal subunits. *RNA* **14:** 2379–2393.

Sweeney TR, Abaeva IS, Pestova TV, Hellen CU. 2014. The mechanism of translation initiation on type 1 picornavirus IRESs. *EMBO J* **33:** 76–92.

Terenin IM, Dmitriev SE, Andreev DE, Shatsky IN. 2008. Eukaryotic translation initiation machinery can operate in a bacterial-like mode without eIF2. *Nat Struct Mol Biol* **15:** 836–841.

Thomas JD, Johannes GJ. 2007. Identification of mRNAs that continue to associate with polysomes during hypoxia. *RNA* **13:** 1116–1131.

Thompson SR. 2012. So you want to know if your message has an IRES? *Wiley Interdiscip Rev RNA* **3:** 697–705.

Treder K, Kneller EL, Allen EM, Wang Z, Browning KS, Miller WA. 2008. The 3′ cap-independent translation element of Barley yellow dwarf virus binds eIF4F via the eIF4G subunit to initiate translation. *RNA* **14:** 134–147.

Truniger V, Nieto C, Gonzalez-Ibeas D, Aranda M. 2008. Mechanism of plant eIF4E-mediated resistance against a Carmovirus (*Tombusviridae*): Cap-independent translation of a viral RNA controlled in *cis* by an (a)virulence determinant. *Plant J* **56:** 716–727.

Vaklavas C, Grizzle WE, Choi H, Meng Z, Zinn KR, Shrestha K, Blume SW. 2016. IRES inhibition induces terminal differentiation and synchronized death in triple-negative breast cancer and glioblastoma cells. *Tumour Biol* **37:** 13247–13264.

Walters B, Thompson SR. 2016. Cap-independent translational control of carcinogenesis. *Front Oncol* **6:** 128.

Wang Z, Parisien M, Scheets K, Miller WA. 2011. The cap-binding translation initiation factor, eIF4E, binds a pseudoknot in a viral cap-independent translation element. *Structure* **19:** 868–880.

Weingarten-Gabbay S, Elias-Kirma S, Nir R, Gritsenko AA, Stern-Ginossar N, Yakhini Z, Weinberger A, Segal E. 2016. Comparative genetics. Systematic discovery of cap-independent translation sequences in human and viral genomes. *Science* **351:** aad4939.

* Wek RC. 2018. Role of eIF2α kinases in translational control and adaptation to cellular stress. *Cold Spring Harb Perspect Biol* doi: 10.1101/cshperspect.a032870.

Wilson JE, Pestova TV, Hellen CU, Sarnow P. 2000. Initiation of protein synthesis from the A site of the ribosome. *Cell* **102:** 511–520.

Xue S, Tian S, Fujii K, Kladwang W, Das R, Barna M. 2015. RNA regulons in Hox 5′ UTRs confer ribosome specificity to gene regulation. *Nature* **517:** 33–38.

Yang Q, Sarnow P. 1997. Location of the internal ribosome entry site in the 5′ non-coding region of the immunoglobulin heavy-chain binding protein (BiP) mRNA: Evidence for specific RNA-protein interactions. *Nucleic Acids Res* **25:** 2800–2807.

Yang DQ, Halaby MJ, Zhang Y. 2006. The identification of an internal ribosomal entry site in the 5′-untranslated region of p53 mRNA provides a novel mechanism for the regulation of its translation following DNA damage. *Oncogene* **25:** 4613–4619.

Yang Y, Fan X, Mao M, Song X, Wu P, Zhang Y, Jin Y, Yang Y, Chen LL, Wang Y, et al. 2017. Extensive translation of circular RNAs driven by N^6-methyladenosine. *Cell Res* **27:** 626–641.

Yin JY, Dong ZZ, Liu RY, Chen J, Liu ZQ, Zhang JT. 2013. Translational regulation of RPA2 via internal ribosomal entry site and by eIF3a. *Carcinogenesis* **34:** 1224–1231.

Yoshida H, Kong YY, Yoshida R, Elia AJ, Hakem A, Hakem R, Penninger JM, Mak TW. 1998. Apaf1 is required for mitochondrial pathways of apoptosis and brain development. *Cell* **94:** 739–750.

Yueh A, Schneider RJ. 1996. Selective translation initiation by ribosome jumping in adenovirus-infected and heat-shocked cells. *Genes Dev* **10:** 1557–1567.

Yueh A, Schneider RJ. 2000. Translation by ribosome shunting on adenovirus and hsp70 mRNAs facilitated by complementarity to 18S rRNA. *Genes Dev* **14:** 414–421.

Zhou J, Wan J, Gao X, Zhang X, Jaffrey SR, Qian SB. 2015. Dynamic m^6A mRNA methylation directs translational control of heat shock response. *Nature* **526:** 591–594.

Repeat-Associated Non-ATG Translation in Neurological Diseases

Tao Zu,[1,2] Amrutha Pattamatta,[1,2] and Laura P.W. Ranum[1,2,3,4]

[1]Center for Neuro-Genetics, University of Florida, Gainesville, Florida 32610

[2]Departments of Molecular Genetics and Microbiology, University of Florida, Gainesville, Florida 32610

[3]Departments of Neurology, College of Medicine, University of Florida, Gainesville, Florida 32610

[4]Genetics Institute, University of Florida, Gainesville, Florida 32610

Correspondence: ranum@ufl.edu

More than 40 different neurological diseases are caused by microsatellite repeat expansions that locate within translated or untranslated gene regions, including 5′ and 3′ untranslated regions (UTRs), introns, and protein-coding regions. Expansion mutations are transcribed bidirectionally and have been shown to give rise to proteins, which are synthesized from three reading frames in the absence of an AUG initiation codon through a novel process called repeat-associated non-ATG (RAN) translation. RAN proteins, which were first described in spinocerebellar ataxia type 8 (SCA8) and myotonic dystrophy type 1 (DM1), have now been reported in a growing list of microsatellite expansion diseases. This article reviews what is currently known about RAN proteins in microsatellite expansion diseases and experiments that provide clues on how RAN translation is regulated.

Microsatellite repeats, or short repetitive stretches of DNA containing 2–10 nucleotides are common in the human genome. A subset of these sequences has been shown to be unstable, and when expanded too many times, can cause disease. Microsatellite expansions are known to cause >40 different neurodegenerative diseases, including Huntington disease (HD), fragile X tremor ataxia syndrome (FXTAS), C9ORF72 amyotrophic lateral sclerosis (ALS)/frontotemporal dementia (FTD), and spinocerebellar ataxia types 1, 2, 3, 6, 7, 8, 10, 12, and 17 (Ranum and Cooper 2006; Orr and Zoghbi 2007; Lopez Castel et al. 2010).

A framework for understanding the pathogenic mechanisms of these diseases has been built on inferring the mechanisms of these diseases based on the position of the mutations within their corresponding genes. For example, for diseases such as HD, in which the mutation is translated as a glutamine stretch that is part of a long open reading frame (ORF), most research has been focused on how the mutant protein, which contains a long polyGln repeat tract, disturbs various cellular functions that result in cellular toxicity and disease. In contrast, the deleterious effects of mutant expansion RNAs or RNA gain-of-function effects have been the primary mechanistic focus for diseases characterized by expansion RNAs located outside canonical ORFs. Mutant RNAs produced from these noncoding expansion mutations have been shown to accumulate as nuclear RNA foci that sequester RNA-binding proteins (RBPs) and lead to a loss of their normal function (Ranum and Cooper 2006; Scotti and Swanson 2016). The myotonic dystrophies are the classic example of this type of disorder. For example, the myotonic dystrophy type 1 (DM1) and type 2 (DM2) mutations, which express CUG or CCUG expansion RNAs, respectively, sequester the muscleblind-like (MBNL) family of RNA-binding proteins within the nucleus as RNA foci. This sequestration prevents MBNL proteins from binding to their normal RNA-processing targets, and the resulting MBNL loss-of-function leads to RNA-processing abnormalities including dysregulation of alternative splicing and changes in polyadenylation (Ranum and Cooper 2006; Scotti and Swanson 2016).

The discovery of repeat-associated non-ATG (RAN) translation challenged these models because it showed that repeat expansion mutations can express unexpected repetitive proteins in all three reading frames without an AUG initiation codon (Zu et al. 2011). RAN translation, which was initially discovered in spinocerebellar ataxia type 8 (SCA8) and DM1, has now been reported in seven different repeat expansion diseases, including *C9orf72* ALS/FTD, FXTAS, HD, SCA31, and DM2 (Zu et al. 2011, 2013, 2017; Ash et al. 2013; Mori et al. 2013b; Todd et al. 2013; Banez-Coronel et al. 2015; Ishiguro et al. 2017). In addition to RAN translation, the bidirectional transcription of microsatellite expansion mutations (Cho et al. 2005; Moseley et al. 2006; Ladd et al. 2007; Rudnicki et al. 2007; Batra et al. 2010; Ranum et al. 2010) adds further mechanistic complexity to these disorders, and raises the possibility that one or both expansion transcripts and up to six RAN proteins contribute to many of these diseases (Fig. 1). For the translation field, the discovery that expansion mutations express proteins without a canonical AUG-initiation codon in multiple reading frames raises mechanistic questions about how RAN proteins are initiated and whether these mutant proteins can be therapeutically targeted (Cleary and Ranum 2017).

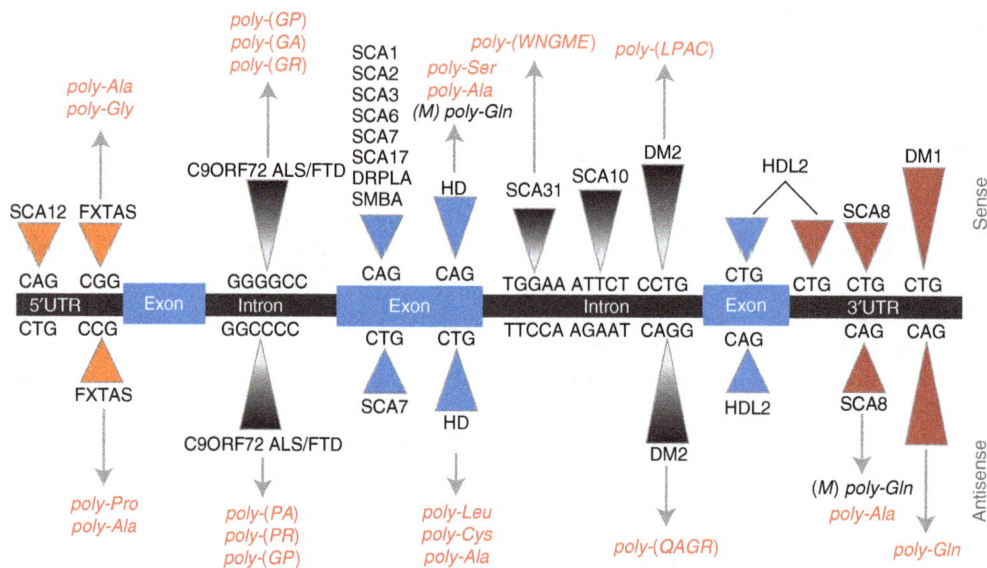

Figure 1. Bidirectional transcription and repeat-associated non-ATG (RAN) translation in repeat expansion disorders. Antisense transcripts and RAN proteins have been observed in a growing number of microsatellite expansions, including spinocerebellar ataxia type 8 (SCA8), myotonic dystrophy type 1 (DM1), myotonic dystrophy type 2 (DM2), fragile X tremor ataxia syndrome (FXTAS), *C9ORF72* amyotrophic lateral sclerosis (ALS)/frontotemporal dementia (FTD), Huntington disease (HD), and spinocerebellar ataxia type 31 (SCA31). RAN proteins that have been identified in various expansion diseases and confirmed in patient tissues are shown in red.

Cite this article as *Cold Spring Harb Perspect Biol* doi: 10.1101/cshperspect.a033019

RAN TRANSLATION

The Discovery of RAN Translation in SCA8 and DM1

SCA8 is an autosomal dominant, neurodegenerative disease caused by a CTG•CAG repeat expansion. In 2006, it was recognized that the SCA8 expansion mutation is bidirectionally transcribed and produces both CUG (*ATXN8OS*) and CAG (*ATXN8*) expansion RNAs, which are expressed in opposite directions across the expansion mutation (Moseley et al. 2006). The CUG expansion transcripts form RNA foci, which have been shown to sequester muscleblind proteins and are thought to cause RNA gain-of-function problems (Daughters et al. 2009). In addition, the expanded CAG *ATXN8* transcript expressed in the opposite direction produces a nearly pure polyGln protein from an unusual short ORF that contains an AUG-initiation codon directly upstream of the CAG repeat (Moseley et al. 2006).

To understand the relative contributions of the effects of the *ATXN8* encoded polyGln protein (Moseley et al. 2006), Ranum and colleagues attempted to block its expression by mutating the upstream ATG initiation codon. Unexpectedly, this mutation did not block expression of the polyGln protein in transfected cells (Zu et al. 2011). This finding raised the question of whether proteins could be made in other reading frames. To answer this question, CAG expansion expression vectors with two stop codons in each reading frame upstream of the CAG tract were generated to ensure AUG initiation could not occur in any of the reading frames. Additionally, epitope tags in each of the three reading frames downstream from the repeat were used to tag proteins expressed from any of the three frames. Surprisingly, homopolymeric polyGln, polyAla, and polySer proteins were expressed in all three reading frames even when the transcripts lacked AUG or near-cognate AUG codons in any reading frame (Zu et al. 2011). Subsequent analyses of CAG expansion transcripts isolated from polyribosomes showed no evidence that editing of the repeat RNAs had introduced an AUG start codon (Zu et al. 2011).

Additional cell-culture experiments showed that hairpin-forming CAG and CUG expansions express RAN proteins without an AUG initiation codon (Zu et al. 2011). Steady-state levels of RAN proteins were differentially affected by repeat length in transfected cells with no RAN protein accumulation detectable with 20 CAG repeats or less, polyGln accumulation evident at ≥ 42 repeats, polySer at ≥ 58, and polyAla at ≥ 73 repeats (Zu et al. 2011). Because repeat length is often correlated with anticipation or worsening of disease, it is possible that the accumulation of RAN proteins in multiple reading frames with longer repeats exacerbates disease. Finally, Zu et al. developed antibodies to detect the carboxy-terminal regions of these proteins in vivo and showed that RAN proteins are expressed in vivo in both mouse and human disease tissue (Zu et al. 2011). In SCA8, a novel polyAla RAN protein was shown to accumulate in Purkinje cells and in DM1 a novel polyGln RAN protein expressed from the antisense CAG expansion transcript was detected in heart, skeletal muscle, and blood.

In summary, CAG and CUG microsatellite expansion RNAs undergo a novel form of translation that leads to the production of homopolymeric expansion proteins from all three reading frames without an AUG-initiation codon. The RAN proteins predicted by cell-culture experiments accumulate in vivo in both SCA8 and DM1 disease tissue.

RAN Proteins in *C9ORF72* ALS/FTD

In 2011, a GGGGCC (G_4C_2) hexanucleotide repeat expansion located in intron 1 of the *chromosome 9 open reading frame 72* (*C9ORF72*) gene was reported by two research groups as the most frequent known genetic cause of both familial ALS and FTD (*C9ORF72* ALS/FTD) (DeJesus-Hernandez et al. 2011; Renton et al. 2011). This discovery was the first to link ALS and FTD to a larger group of microsatellite expansion mutation disorders. Although, initially, haploinsufficiency of the C9ORF72 protein and gain-of-function mechanisms of the GGGGCC expansion RNA were proposed in 2013, several investigators reported that, similar to SCA8 and

DM1, the *C9ORF72* expansion mutation also undergoes RAN translation (Ash et al. 2013; Mori et al. 2013b; Zu et al. 2013). In this case, the hexanucleotide G_4C_2 expansion mutations form RNA foci and also produce dipeptide repeat proteins (DPRs) with repetitive glycine-proline (GP), glycine-alanine (GA), and glycine-arginine (GR) repeat motifs in the sense direction. Also, similar to SCA8, DM1 and a number of other expansion diseases, this mutation was found to be expressed in the antisense direction and these antisense transcripts also produce three additional RAN proteins: proline-alanine (PA), proline-arginine (PR), and GP (Gendron et al. 2013; Mori et al. 2013a; Zu et al. 2013). Immunological evidence using antibodies directed at the predicted carboxy-terminal regions shows that proteins made in five of the six reading frames include unique carboxy-terminal ends that extend beyond the repeat tract (Zu et al. 2013). Although the contribution of the C9-RAN proteins in disease is not yet fully understood, there is substantial evidence that all six C9-RAN proteins are toxic and accumulate in *C9ORF72* ALS/FTD autopsy tissue as cytoplasmic aggregates in multiple regions of the brain that are affected by the disease. These regions include the frontal cortex, hippocampus, and spinal cord (Ash et al. 2013; Gendron et al. 2013; Mackenzie et al. 2013; Mori et al. 2013a,b; Zu et al. 2013; Mackenzie 2016).

RAN Translation across CGG•CCG Expansion in FXTAS

Expansion of a CGG repeat tract in the 5′ untranslated region (UTR) of the *FMR1* gene causes two distinct diseases, depending on the length of the repeat. Larger expansions (>200 repeats) cause transcriptional silencing of the *FMR1* gene resulting in fragile X syndrome. In contrast, shorter alleles within the premutation range (55–200 repeats) cause FXTAS, a late-onset neurodegenerative disorder (Nelson et al. 2013).

Todd et al. (2013) reported RAN proteins are expressed in FXTAS. This work was initiated because these authors noticed puzzling green fluorescence protein (GFP)-positive aggregates

in a fly model expressing 90 CGG repeats but not GFP alone. The CGG repeats are located in what was assumed to be a noncoding 5′UTR upstream of the GFP reporter. The formation of these aggregates was later explained after a series of experiments showed that FXTAS CGG expansion mutations containing 90 repeats undergo RAN translation and that the resulting poly (Gly)$_{90}$-GFP fusion protein forms GFP-positive aggregates. Using mammalian cell culture, Todd et al. went on to show that CGG expansions lead to the production of polyGly-GFP and polyAla-GFP fusion proteins in the absence of an ATG initiation codon. RAN translation in the polyAla reading frame is length-dependent, occurring with constructs containing 88 but not 30 CGG repeats. In contrast, expression of the polyGly protein occurred with 88, 50, and 30 CGGs. Additional experiments showed initiation of translation begins upstream of the CGG repeat in the polyGly reading frame and, in contrast to SCA8, DM1 and *C9ORF72* ALS/FTD repeat expansions, requires near-cognate initiation codons (Kearse et al. 2016). The requirement for near-cognate codons upstream of the repeat in the polyGly frame may explain the expression of polyGly protein, even at relatively short expansion sizes. Finally, the authors showed that polyGly aggregates accumulate in both model systems and human FXTAS brains using several custom carboxy-terminal antibodies (Todd et al. 2013). More recently, Krans et al. (2016) showed that homopolymeric proteins are also expressed in the antisense direction, generating polyPro, polyArg, and polyAla in cell culture, and provided evidence that at least two of the antisense proteins (polyPro and polyAla) accumulate in human brains.

Huntington Disease

HD is a progressive neurodegenerative disorder characterized by severe movement, cognitive, and behavioral changes. The disease is caused by a CAG•CTG expansion located within a long ORF of the *huntingtin* (*HTT*) gene. Although previous work on HD showed both sense CAG and antisense CUG expansion RNAs are expressed in HD (Rudnicki et al. 2007; Ross and

Cite this article as *Cold Spring Harb Perspect Biol* doi: 10.1101/cshperspect.a033019

Tabrizi 2011), most research has focused on understanding the deleterious effects of the mutant huntingtin protein, which contains a long poly-Gln expansion repeat tract.

Recently, Banez-Coronel et al. (2015) showed that both sense and antisense expansion transcripts undergo RAN translation and that additional homopolymeric RAN proteins (poly-Ala, polySer, polyLeu, and polyCys) accumulate in HD patient brains. These HD-RAN proteins accumulate in brain regions most affected in HD, including the striatum, and the severely affected cerebellar tissue from juvenile-onset cases. Brain regions with robust RAN protein accumulation show pathologic features of HD, including apoptosis, neuroinflammation, and white matter abnormalities. Cell-culture studies show that HD-RAN proteins decrease viability and increase death of neural cells. Additionally, these authors showed that repeat length differentially affects RAN protein aggregation and accumulation with increased polyAla levels and aggregation of polySer occurring in cells expressing >52 repeats, which are typically associated with severe juvenile HD. Additionally, it is interesting to note that polySer accumulation is first detectable in cells expressing ≥35 repeats, which is close to the repeat length at which HD can manifest (Ross and Tabrizi 2011).

In summary, four novel RAN proteins accumulate in HD and they can be expressed across repeats in multiple reading frames even when the repeat is located within a canonical ORF.

RAN Translation in Tetra- and Pentanucleotide Expansion Disorders

More recently, RAN proteins have been reported in tetra- and pentanucleotide repeat diseases. Ishiguro et al. (2017) reported the accumulation of a UGGAA expansion-encoded Trp-Asn-Gly-Met-Glu pentapeptide repeat protein (PPR) in a *Drosophila* SCA31 model and SCA31 human autopsy brains. Additionally, overexpression of several RNA-binding proteins (TDP-43, FUS, and hnRNPA2B1) reduce PPR protein levels in *Drosophila*, suggesting that these RBPs may play a role in RNA quality control and regulation of repeat-associated translation.

In DM2, Zu et al. (2017) showed that the intronic CCTG•CAGG expansion mutation located in the *cellular nucleic acid binding protein* (*CNBP*) gene produces tetrapeptide expansion proteins in both the sense (leucine-proline-alanine-cysteine [LPAC]) and antisense (glutamine-alanine-glycine-arginine [QAGR]) directions. Codon-replacement studies show that these proteins are toxic to neural cells independent of RNA gain-of-function effects. Additionally, the authors provide evidence that RNA gain-of-function and RAN mechanisms are linked by showing that nuclear sequestration of CCUG expansion transcripts into RNA foci decreases steady-state levels of LPAC RAN protein by preventing export of CCUG expansion RNAs to the cytoplasm.

RAN Protein Toxicity

RAN proteins with homopolymeric, dipeptide, tetrapeptide, and pentapeptide repeat motifs have now been reported in seven different microsatellite expansion diseases. Although the role of individual RAN proteins in each of these disorders is not yet clear, there is a growing body of evidence that RAN proteins expressed in multiple contexts are toxic to cells independent of RNA gain-of-function effects and, when overexpressed, can induce neurodegeneration in model systems. For example, codon-replacement strategies have shown that RAN proteins are toxic independent of RNA gain-of-function effects in mammalian cell systems for SCA8 (Zu et al. 2011), HD (Banez-Coronel et al. 2015), and DM2 (Zu et al. 2017) repeat expansions. Using mouse models with or without a near-cognate ACG codon in the 5'UTR, Sellier et al. (2017) showed that mice expressing both the polyGly RAN protein and CGG expansion RNAs develop behavioral deficits not found in FXTAS animals expressing CGG expansion RNAs alone. The mice that express both the CGG RNA and the polyGly protein contain a near-cognate ACG codon, which allows polyGly expression; in contrast, mice in which the ACG has been deleted express only the CGG RNA.

The most extensive data on the biophysical properties and toxicity of RAN proteins come from a large number of studies of *C9ORF72* ALS/FTD. In several different in vitro and in vivo overexpression experiments, GA, GR, and PR DPRs have been reported to be toxic by multiple investigators. GA has the biophysical properties to aggregate and form amyloid-like fibrils that stain positive with Congo red and thioflavin-T (May et al. 2014; Chang et al. 2016). Additionally, mass spectrometry has shown that p62, Unc119, ubiquilin-1,2, and some proteosomal subunits are found in GA complexes (May et al. 2014; Schludi et al. 2017). Neuronal cell culture studies show GA overexpression causes impaired dendritic branching, endoplasmic reticulum (ER) stress, and apoptosis (May et al. 2014; Zhang et al. 2016). AAV-mediated overexpression of GA DPR causes cognitive and motor deficits in mice (Chew et al. 2015; Zhang et al. 2016). In contrast, several other studies have reported minimal toxicity of GA in primary neuronal cell culture and *Drosophila* models (Mizielinska et al. 2014; Wen et al. 2014). In fact, several independent studies performed using a non-hairpin-forming alternative codon strategy to distinguish RNA gain-of-function effects from RAN protein toxicity, showed that PR and GR are the most toxic in the *Drosophila* eye (Mizielinska et al. 2014; Wen et al. 2014; Freibaum et al. 2015; Lee et al. 2016). Toxicity of GR and PR has been attributed to (1) nucleolar localization in overexpression systems (Kwon et al. 2014; Tao et al. 2015); (2) association with low-complexity domain (LCD) proteins that cause phase separation into membraneless organelles (Lee et al. 2016; Lin et al. 2016); and (3) plugging the nuclear pore and impairing nucleocytoplasmic transport (Freibaum et al. 2015; Jovicic et al. 2015; Zhang et al. 2015; Shi et al. 2017).

Although a number of independent investigators provide strong support that specific DPRs can be toxic when overexpressed, additional studies that recapitulate the temporal and spatial expression of these proteins in mammalian systems will be needed to understand their impact on the disease. Recently, four research groups developed bacterial artificial chromosome (BAC) mouse models to better understand *C9ORF72* ALS/FTD (O'Rourke et al. 2015; Peters et al. 2015; Jiang et al. 2016; Liu et al. 2016).

Further studies in which RAN translation can be blocked or modulated without affecting the levels of the RNAs in these models will be useful for sorting out the role of RAN proteins in disease.

MECHANISM OF RAN TRANSLATION

RAN translation is now known to occur across a variety of different repeat motifs including CAG, CUG, CGG, CCG GGGGCC, CCCCGG, and, most recently, CCUG and CAGG for DM2 and UGGAA for SCA31 (Fig. 1) (Cleary and Ranum 2017; Ishiguro et al. 2017; Zu et al. 2017). Microsatellite expansion mutations in these diseases are present in ORFs, 3'UTRs, 5'UTRs, and intronic regions. Although the mechanisms whereby RAN proteins are produced are still poorly understood, several common themes are emerging, including the importance of repeat length, RNA structure, RNA-binding proteins, and translation initiation by canonical and noncanonical mechanisms.

Repeat Length and Repeat Sequence

It has been shown by several investigators that RAN translation is repeat length-dependent with the accumulation of RAN proteins more likely to occur for expansions with longer repeat tracts. This length dependence applies across multiple diseases including SCA8, HD, DM1, DM2, and *C9ORF72* ALS/FTD and for FXTAS in the polyAla reading frame (Zu et al. 2011, 2013, 2017; Todd et al. 2013; Banez-Coronel et al. 2015; Cleary and Ranum 2017). In general, longer repeats in these diseases are more likely to result in the accumulation of a cocktail of RAN proteins expressed from multiple reading frames and also more likely to result in increased disease severity.

A curious observation from studies of RAN translation across CAG•CTG expansions in HD and CCTG•CAGG expansion mutations in DM2 is that RAN translation and subsequent RAN protein accumulation are cell-type-dependent in some, but not all, reading frames, suggesting the possibility that RAN translation in these cells is modulated by sequence-specific effects such as secondary structures and near-cognate initiation codons. Additionally, cell-specific expression of specialized initiation factors that favor RAN translation could cause cell-type-specific expression of specific RAN proteins (Banez-Coronel et al. 2015; Zu et al. 2017).

Previous studies using CAG repeats with different 5′ flanking sequences cloned from the SCA3, HD, DM1, or HDL2 loci, showed that flanking sequences can affect RAN protein expression (Zu et al. 2011). Additionally, these investigators showed that insertion of a TAG stop immediately preceding the CAG repeat tract in the glutamine frame prevented translation of polyGln but not of polyAla or polySer. These data suggest translation initiation in the polyGln reading frame occurs upstream of the repeat in this construct. Similarly, Todd et al. (2013) and Scoles et al. (2015) have shown that 5′ and 3′ flanking sequences affect RAN translation in FXTAS and SCA2, respectively. These data suggested that RAN translation can be influenced by both the repeat expansion motif and flanking sequences.

Hairpin and G-Quadruplex RNA Structures

Single-stranded CAG, CUG, CGG, CCUG, and G_4C_2 repeats have all been reported to form stable secondary structures including hairpins and G-quadruplexes (Tian et al. 2000; Sobczak et al. 2003; Mirkin 2007; Fratta et al. 2012; Krzyzosiak et al. 2012; Reddy et al. 2013; Haeusler et al. 2014). When RAN translation was first discovered for hairpin-forming CAG repeats, Zu et al. (2011) hypothesized that hairpin formation is required for RAN translation. Consistent with this prediction, these authors showed hairpin-forming CAG, but not non-hairpin-forming CAA repeat constructs, undergo RAN translation.

In *C9ORF72* ALS/FTD, it has also been shown that codon replacement RNAs, which encode various DPRs but do not form secondary RNA structures, do not undergo RAN translation but rather require an ATG initiation codon for expression (Mizielinska et al. 2014). Because G_4C_2 and CGG repeats can form both G-quadruplex and hairpin structures (Fratta et al. 2012;

Reddy et al. 2013; Haeusler et al. 2014), it is possible that the transition and/or equilibrium between G-quadruplex and hairpin conformations may play a role in RAN translation across these expansions.

Additionally, SCA31 is distinct from other expansion diseases previously reported to undergo RAN translation in two important ways: (1) the repeat expansion mutation is not GC-rich and not predicted to form hairpin structures; (2) repeated iterations of the UGGAA repeat contain AUG initiation codons throughout the repeat tract (e.g., UGGAAUGGAAUG GAAUGGAA...) in all three reading frames (Ishiguro et al. 2017). It will be interesting to determine the frequency of RAN protein aggregates in SCA31. It will also be interesting to determine whether AUG and RAN mechanisms are both at play in this repeat expansion disorder and if one or multiple AUG initiation codons embedded within the repeat tract are used.

Cap-Dependent Initiation across CGG Repeats in *FXTAS*

Kearse et al. (2016) recently showed that translation of the FMR polyGly initiates preferentially at close-cognate start codons upstream of the repeat, in a repeat-length-independent manner. In contrast, translation in the polyAla reading frame for FXTAS is length dependent and does not have a similar close-cognate initiation codon. Additionally, using cell-free in vitro translation protocols, these authors show that the polyGly and polyAla reading frames require an m^7G cap, the eIF4A helicase, and 40S ribosomal scanning. These findings support a model involving canonical preinitiation complex loading and ribosome scanning and suggest ribosome stalling caused by the CGG RNA secondary structure facilitates translation in both the polyAla and polyGly reading frames (Fig. 2). Additionally, mass spectrometry performed by Sellier et al. (2017) on the FMR polyGly protein showed that initiation occurs at an upstream ACG close-cognate start codon and that the amino-terminal amino acid is a methionine in mammalian cells. These data suggest that for FXTAS, a canonical protein initiation mechanism is used in the

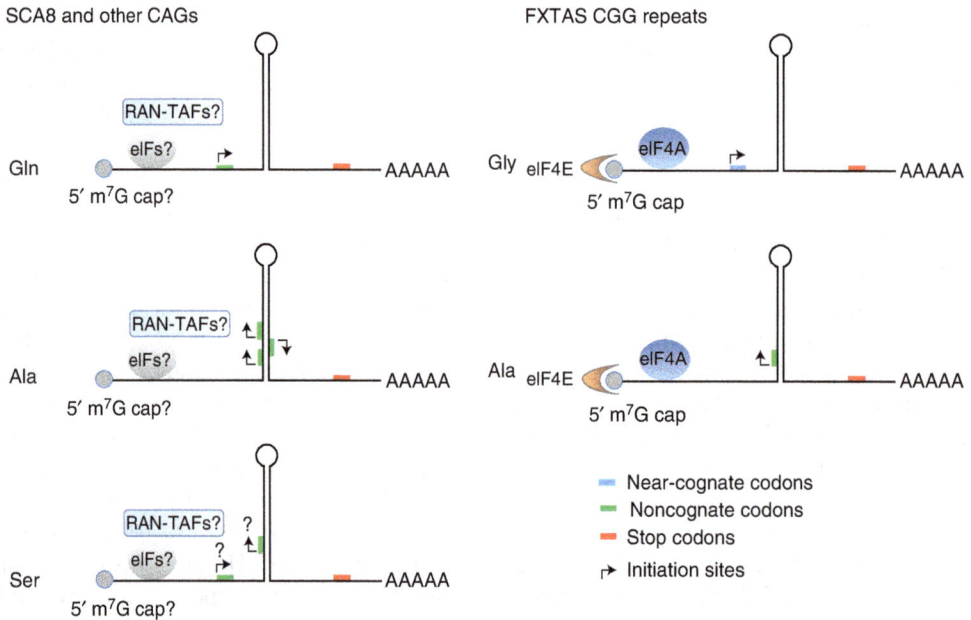

Figure 2. Repeat-associated non-ATG (RAN) translation of CAG and CGG repeats. Model of RAN translation showing repeat-length dependence and preference for hairpin or secondary RNA structures. RAN translation can occur in multiple reading frames and, depending on the disease locus, at start sites upstream of the repeats or at multiple sites within the repeat tract. Experiments on CAG repeats suggest that translation initiation in the polyGln frame occurs upstream of the repeat tract, although initiation in the polyAla reading frame occurs throughout the repeat. Translation of FXTAS polyGly and polyAla have been shown to be cap-dependent and require eIF4E and eIF4A. RAN-TAFs, putative RAN-translation associated factors.

polyGly reading frame but with the use of a close-cognate initiation codon. Additional studies in FXTAS will be required to determine the specific initiation site or sites within the polyAla reading frame.

For CAG•CTG expansions, several experiments showed that, although RAN translation is highly permissive across repeats in mammalian cell culture, RAN translation in a variety of cell-free translation systems, including rabbit reticulocyte lysates (RRLs), occurs less robustly and only in reading frames with close-cognate initiation codons (Zu et al. 2011). These data indicate that RRLs do not necessarily recapitulate the more highly permissive RAN translation, which occurs in mammalian cells. Because the sequence flanking the repeat tracts differs dramatically between disease loci and for individual reading frames, it will be important to determine what mechanistic features are shared between different repeat expansion diseases.

INTERNAL RIBOSOME ENTRY SITES (IRES)

Previous studies showed that translational initiation is affected by secondary structures downstream from start codons (Kozak 1989). In addition, hairpin structures downstream from start codons at suboptimal sites can enhance translational initiation efficiency, possibly by slowing the 40S ribosomal subunit and enhancing the time for interaction between the Met-tRNA$_i$ anticodon and the suboptimal AUGs or close-cognate AUG-like start site (Kochetov et al. 2007).

Some viral mRNAs containing highly structured RNA elements at the 5′UTR have been shown to recruit initiation factors and ribosomal subunits to IRESs (Cullen 2009; Sonenberg and Hinnebusch 2009). Additionally, type IV IRESs in the cricket paralysis virus (CrPV) can facilitate translation initiation without eIFs or Met-tRNA$_i^{Met}$. In these cases, eukaryotic elongation factors 1 and 2 (eEF1 and eEF2) are involved in translation initiation. The PKI hairpin of the CrPV IRES functions as a $tRNA_i^{Met}$ to initiate translation at non-AUG or noncognate AUG codons, including GCU or GCA that encode an alanine (Hellen and Sarnow 2001; Hellen 2009; Fernandez et al. 2014). Because RAN translation most often occurs across repeat expansions that form stable secondary structures, an IRES-like mechanism has also been proposed (Zu et al. 2011; Cleary and Ranum 2017). Mass spectrometry for CAG repeat expansions suggests RAN translation in the GCA/ alanine frame initiates at multiple GCA codons in the repeats without incorporating an amino-terminal methionine (Fig. 2) (Zu et al. 2011). Interestingly, some repeat expansion-binding proteins, such as CUGBP1 and hnRNPs (Timchenko et al. 1996; Conlon et al. 2016), can serve as IRES translation associated factors (ITAFs) (Fox and Stover 2009). Taken together, these data suggest that RAN translation of specific repeats may involve IRES-like mechanisms.

RNA-Binding Proteins and RAN Translation

RNA foci and RAN proteins are found in a growing number of diseases including SCA8, *C9ORF72* ALS/FTD, and now DM2. This raises the question of whether RNA gain-of-function and RAN mechanisms are linked.

DM2 is caused by an intronic CCTG expansion in the *CNBP* gene (Liquori et al. 2001). The clinical similarities between DM1 and DM2, the formation of CUG and CCUG repeat-containing nuclear RNA foci, and similar MBNL/CELF protein-related RNA-processing changes provide strong support that RNA gain-of-function and corresponding MBNL protein loss-of-function contribute to DM2 (Liquori et al. 2001; Cleary and Ranum 2013). In a recent study, Zu et al. (2017) showed that in DM2, antisense CAGG transcripts are expressed and elevated compared with antisense RNA levels found in control DM2 autopsy tissue. Additionally, poly (LPAC) and poly(QAGR) tetrapeptide RAN proteins produced from sense and antisense transcripts accumulate in human DM2 brains.

Similar to RAN proteins for several other repeat expansion mutations, steady-state levels of the LPAC and QAGR proteins increased in a repeat length-dependent manner in transfected cells. RNA foci and LPAC RAN protein expression are inversely correlated, and steady-state levels of RAN proteins expressed from CCUG, CUG, and CAG expansion RNAs are reduced by Mbnl1 overexpression. These data suggest a nuclear sequestration model of disease. In this model, initially, RNA-binding proteins sequester expansion RNAs in the nucleus and play a protective role by preventing cytoplasmic RAN translation. During later stages of the disease, nuclear sequestration fails and cytoplasmic expansion RNAs produce RAN proteins that exacerbate disease (Zu et al. 2017). Because RNA foci and RAN proteins are found in a growing number of diseases (SCA8, DM1, DM2, FXTAS, and *C9ORF72* ALS/FTD), nuclear sequestration of expansion RNAs by specific RNA-binding proteins into nuclear foci may be a general protective mechanism.

An additional puzzle is how transcripts of intronic expansion mutations like the DM2 and *C9ORF72* expansions make it out into the cytoplasm to be translated in the first place. A recent study reported that increased binding of GGGGCC-expanded *C9ORF72* repeats to the nuclear export adaptor SRSF1 may increase nuclear export through the nuclear RNA export factor 1 (NXF1) pathway (Hautbergue et al. 2017). Additionally, intron retention, which is increasingly recognized in a wide range of mammalian transcripts, and also in microsatellite expansion diseases (Gipson et al. 2013; Sathasivam et al. 2013; Braunschweig et al. 2014; Niblock et al. 2016), may provide a plausible alternative mechanism that would allow intronic expansion mutations to be retained and exported into the cytoplasm where they could undergo RAN translation. Finally, disruptions in the nuclear pore complex, which have been reported in both *C9ORF72* ALS/FTD and HD (Freibaum et al. 2015; Jovicic et al. 2015; Zhang et al. 2015; Grima et al. 2017), may provide another route for the accumulation of expansion RNAs in the cytoplasm and RAN protein expression.

CONCLUSIONS AND PERSPECTIVES

RAN proteins have now been reported in seven out of more than 40 diseases known to be caused by microsatellite expansion mutations. Although there has been substantial progress in this field, there remain an even larger number of questions and future experiments. First, additional studies that rigorously define where these proteins begin and end are needed. These studies face two major challenges. Currently, mass spectrometry is challenging because of the repetitive nature of these proteins, and it is not clear that cell-free in vitro systems will accurately recapitulate the highly permissive initiation of RAN proteins in cells. Second, it is important to understand the molecular factors that are required for RAN translation and also to understand whether the mechanisms vary depending on the availability or absence of close-cognate codons and RNA secondary structures. The cell-type differences in the accumulation patterns of RAN proteins may reflect the requirements for various cellular factors that may differ depending on the sequence of both the repeat expansion and flanking sequence. Alternatively, the patchy accumulation of RAN protein aggregates seen in human autopsy tissue (Zu et al. 2013) may result from prion-like seeding and spreading mechanisms or localized differences in RAN protein production or turnover. It is also possible that stress pathways, known to play a critical role in canonical translational regulation, may also play a role in RAN protein diseases. Third, it is critical to understand the role that RAN proteins have in disease. Although many studies show that various RAN proteins are toxic when overexpressed, it is less clear whether and how these proteins contribute to disease when expressed in the spatial and temporal patterns and levels that are found in disease. Finding ways to target RAN proteins will be important for understanding their role in disease.

Understanding the mechanisms and impact of RAN proteins in neurological diseases is important and may lead to therapies that can turn off a pathogenic pathway common to a large number of neurodegenerative disorders. Moreover, because nearly half of the human genome consists of repetitive DNA elements, it is possible that RAN translation of repeats in normal mammalian genomes may lead to the production of an abundant category of novel proteins. Elucidating the mechanisms underlying RAN translation has the potential to shift the paradigm and our current views on the complexity of the proteome and how perturbations at this novel level impact fundamental aspects of biology.

ACKNOWLEDGMENTS

This work is supported by the National Institutes of Health (R01 NS098819, R37 NS040389, and P01 NS058901), Target ALS, CHDI, National Ataxia Foundation, ALS Association, Packard Foundation, and the Muscular Dystrophy Association.

REFERENCES

Ash PE, Bieniek KF, Gendron TF, Caulfield T, Lin WL, Dejesus-Hernandez M, van Blitterswijk MM, Jansen-West K, Paul JW III, Rademakers R, et al. 2013. Unconventional translation of C9ORF72 GGGGCC expansion generates insoluble polypeptides specific to c9FTD/ALS. *Neuron* **77:** 639–646.

Banez-Coronel M, Ayhan F, Tarabochia AD, Zu T, Perez BA, Tusi SK, Pletnikova O, Borchelt DR, Ross CA, Margolis RL, et al. 2015. RAN translation in Huntington disease. *Neuron* **88:** 667–677.

Batra R, Charizanis K, Swanson MS. 2010. Partners in crime: Bidirectional transcription in unstable microsatellite disease. *Hum Mol Genet* **19:** R77–R82.

Braunschweig U, Barbosa-Morais NL, Pan Q, Nachman EN, Alipanahi B, Gonatopoulos-Pournatzis T, Frey B, Irimia M, Blencowe BJ. 2014. Widespread intron retention in mammals functionally tunes transcriptomes. *Genome Res* **24:** 1774–1786.

Chang YJ, Jeng US, Chiang YL, Hwang IS, Chen YR. 2016. The glycine-alanine dipeptide repeat from C9orf72 hexanucleotide expansions forms toxic amyloids possessing cell-to-cell transmission properties. *J Biol Chem* **291:** 4903–4911.

Chew J, Gendron TF, Prudencio M, Sasaguri H, Zhang YJ, Castanedes-Casey M, Lee CW, Jansen-West K, Kurti A, Murray ME, et al. 2015. *C9ORF72* repeat expansions in mice cause TDP-43 pathology, neuronal loss, and behavioral deficits. *Science* **348:** 1151–1154.

Cho DH, Thienes CP, Mahoney SE, Analau E, Filippova GN, Tapscott SJ. 2005. Antisense transcription and heterochromatin at the DM1 CTG repeats are constrained by CTCF. *Mol Cell* **20:** 483–489.

Cleary JD, Ranum LP. 2013. Repeat-associated non-ATG (RAN) translation in neurological disease. *Hum Mol Genet* **22:** R45–R51.

Cleary JD, Ranum LP. 2017. New developments in RAN translation: Insights from multiple diseases. *Curr Opin Genet Dev* **44**: 125–134.

Conlon EG, Lu L, Sharma A, Yamazaki T, Tang T, Shneider NA, Manley JL. 2016. The C9ORF72 GGGGCC expansion forms RNA G-quadruplex inclusions and sequesters hnRNP H to disrupt splicing in ALS brains. *eLife* **5**: e17820.

Cullen BR. 2009. Viral RNAs: Lessons from the enemy. *Cell* **136**: 592–597.

Daughters RS, Tuttle DL, Gao W, Ikeda Y, Moseley ML, Ebner TJ, Swanson MS, Ranum LP. 2009. RNA gain-of-function in spinocerebellar ataxia type 8. *PLoS Genet* **5**: e1000600.

DeJesus-Hernandez M, Mackenzie IR, Boeve BF, Boxer AL, Baker M, Rutherford NJ, Nicholson AM, Finch NA, Flynn H, Adamson J, et al. 2011. Expanded GGGGCC hexanucleotide repeat in noncoding region of *C9ORF72* causes chromosome 9p-linked FTD and ALS. *Neuron* **72**: 245–256.

Fernandez IS, Bai XC, Murshudov G, Scheres SH, Ramakrishnan V. 2014. Initiation of translation by cricket paralysis virus IRES requires its translocation in the ribosome. *Cell* **157**: 823–831.

Fox JT, Stover PJ. 2009. Mechanism of the internal ribosome entry site-mediated translation of serine hydroxymethyltransferase 1. *J Biol Chem* **284**: 31085–31096.

Fratta P, Mizielinska S, Nicoll AJ, Zloh M, Fisher EM, Parkinson G, Isaacs AM. 2012. *C9orf72* hexanucleotide repeat associated with amyotrophic lateral sclerosis and frontotemporal dementia forms RNA G-quadruplexes. *Sci Rep* **2**: 1016.

Freibaum BD, Lu Y, Lopez-Gonzalez R, Kim NC, Almeida S, Lee KH, Badders N, Valentine M, Miller BL, Wong PC, et al. 2015. GGGGCC repeat expansion in *C9orf72* compromises nucleocytoplasmic transport. *Nature* **525**: 129–133.

Gendron TF, Bieniek KF, Zhang YJ, Jansen-West K, Ash PE, Caulfield T, Daughrity L, Dunmore JH, Castanedes-Casey M, Chew J, et al. 2013. Antisense transcripts of the expanded *C9ORF72* hexanucleotide repeat form nuclear RNA foci and undergo repeat-associated non-ATG translation in c9FTD/ALS. *Acta Neuropathol* **126**: 829–844.

Gipson TA, Neueder A, Wexler NS, Bates GP, Housman D. 2013. Aberrantly spliced HTT, a new player in Huntington's disease pathogenesis. *RNA Biol* **10**: 1647–1652.

Grima JC, Daigle JG, Arbez N, Cunningham KC, Zhang K, Ochaba J, Geater C, Morozko E, Stocksdale J, Glatzer JC, et al. 2017. Mutant Huntingtin disrupts the nuclear pore complex. *Neuron* **94**: 93–107.e106.

Haeusler AR, Donnelly CJ, Periz G, Simko EA, Shaw PG, Kim MS, Maragakis NJ, Troncoso JC, Pandey A, Sattler R, et al. 2014. *C9orf72* nucleotide repeat structures initiate molecular cascades of disease. *Nature* **507**: 195–200.

Hautbergue GM, Castelli LM, Ferraiuolo L, Sanchez-Martinez A, Cooper-Knock J, Higginbottom A, Lin YH, Bauer CS, Dodd JE, Myszczynska MA, et al. 2017. SRSF1-dependent nuclear export inhibition of *C9ORF72* repeat transcripts prevents neurodegeneration and associated motor deficits. *Nat Commun* **8**: 16063.

Hellen CU. 2009. IRES-induced conformational changes in the ribosome and the mechanism of translation initiation by internal ribosomal entry. *Biochim Biophys Acta* **1789**: 558–570.

Hellen CU, Sarnow P. 2001. Internal ribosome entry sites in eukaryotic mRNA molecules. *Genes Dev* **15**: 1593–1612.

Ishiguro T, Sato N, Ueyama M, Fujikake N, Sellier C, Kanegami A, Tokuda E, Zamiri B, Gall-Duncan T, Mirceta M, et al. 2017. Regulatory role of RNA chaperone TDP-43 for RNA misfolding and repeat-associated translation in SCA31. *Neuron* **94**: 108–124.e107.

Jiang J, Zhu Q, Gendron TF, Saberi S, McAlonis-Downes M, Seelman A, Stauffer JE, Jafar-Nejad P, Drenner K, Schulte D, et al. 2016. Gain of toxicity from ALS/FTD-linked repeat expansions in C9ORF72 is alleviated by antisense oligonucleotides targeting GGGGCC-containing RNAs. *Neuron* **90**: 535–550.

Jovicic A, Mertens J, Boeynaems S, Bogaert E, Chai N, Yamada SB, Paul JW III, Sun S, Herdy JR, Bieri G, et al. 2015. Modifiers of *C9orf72* dipeptide repeat toxicity connect nucleocytoplasmic transport defects to FTD/ALS. *Nat Neurosci* **18**: 1226–1229.

Kearse MG, Green KM, Krans A, Rodriguez CM, Linsalata AE, Goldstrohm AC, Todd PK. 2016. CGG repeat-associated non-AUG translation utilizes a cap-dependent scanning mechanism of initiation to produce toxic proteins. *Mol Cell* **62**: 314–322.

Kochetov AV, Palyanov A, Titov II, Grigorovich D, Sarai A, Kolchanov NA. 2007. AUG_hairpin: Prediction of a downstream secondary structure influencing the recognition of a translation start site. *BMC Bioinformatics* **8**: 318.

Kozak M. 1989. Context effects and inefficient initiation at non-AUG codons in eucaryotic cell-free translation systems. *Mol Cell Biol* **9**: 5073–5080.

Krans A, Kearse MG, Todd PK. 2016. Repeat-associated non-AUG translation from antisense CCG repeats in fragile X tremor/ataxia syndrome. *Ann Neurol* **80**: 871–881.

Krzyzosiak WJ, Sobczak K, Wojciechowska M, Fiszer A, Mykowska A, Kozlowski P. 2012. Triplet repeat RNA structure and its role as pathogenic agent and therapeutic target. *Nucleic Acids Res* **40**: 11–26.

Kwon I, Xiang S, Kato M, Wu L, Theodoropoulos P, Wang T, Kim J, Yun J, Xie Y, McKnight SL. 2014. Poly-dipeptides encoded by the *C9orf72* repeats bind nucleoli, impede RNA biogenesis, and kill cells. *Science* **345**: 1139–1145.

Ladd PD, Smith LE, Rabaia NA, Moore JM, Georges SA, Hansen RS, Hagerman RJ, Tassone F, Tapscott SJ, Filippova GN. 2007. An antisense transcript spanning the CGG repeat region of *FMR1* is upregulated in premutation carriers but silenced in full mutation individuals. *Hum Mol Genet* **16**: 3174–3187.

Lee KH, Zhang P, Kim HJ, Mitrea DM, Sarkar M, Freibaum BD, Cika J, Coughlin M, Messing J, Molliex A, et al. 2016. C9orf72 dipeptide repeats impair the assembly, dynamics, and function of membrane-less organelles. *Cell* **167**: 774–788.e717.

Lin Y, Mori E, Kato M, Xiang S, Wu L, Kwon I, McKnight SL. 2016. Toxic PR poly-dipeptides encoded by the C9orf72 repeat expansion target LC domain polymers. *Cell* **167**: 789–802.e712.

Liquori CL, Ricker K, Moseley ML, Jacobsen JF, Kress W, Naylor SL, Day JW, Ranum LP. 2001. Myotonic dystrophy type 2 caused by a CCTG expansion in intron 1 of *ZNF9*. *Science* **293**: 864–867.

Liu Y, Pattamatta A, Zu T, Reid T, Bardhi O, Borchelt DR, Yachnis AT, Ranum LP. 2016. *C9orf72* BAC mouse model with motor deficits and neurodegenerative features of ALS/FTD. *Neuron* **90**: 521–534.

Lopez Castel A, Cleary JD, Pearson CE. 2010. Repeat instability as the basis for human diseases and as a potential target for therapy. *Nat Rev Mol Cell Biol* **11**: 165–170.

Mackenzie IR. 2016. The role of dipeptide-repeat protein pathology in *C9orf72* mutation cases. *Neuropathol Appl Neurobiol* **42**: 217–219.

Mackenzie IR, Arzberger T, Kremmer E, Troost D, Lorenzl S, Mori K, Weng SM, Haass C, Kretzschmar HA, Edbauer D, et al. 2013. Dipeptide repeat protein pathology in *C9ORF72* mutation cases: Clinico-pathological correlations. *Acta Neuropathol* **126**: 859–879.

May S, Hornburg D, Schludi MH, Arzberger T, Rentzsch K, Schwenk BM, Grasser FA, Mori K, Kremmer E, Banzhaf-Strathmann J, et al. 2014. *C9orf72* FTLD/ALS-associated Gly-Ala dipeptide repeat proteins cause neuronal toxicity and Unc119 sequestration. *Acta Neuropathol* **128**: 485–503.

Mirkin SM. 2007. Expandable DNA repeats and human disease. *Nature* **447**: 932–940.

Mizielinska S, Gronke S, Niccoli T, Ridler CE, Clayton EL, Devoy A, Moens T, Norona FE, Woollacott IO, Pietrzyk J, et al. 2014. *C9orf72* repeat expansions cause neurodegeneration in *Drosophila* through arginine-rich proteins. *Science* **345**: 1192–1194.

Mori K, Arzberger T, Grasser FA, Gijselinck I, May S, Rentzsch K, Weng SM, Schludi MH, van der Zee J, Cruts M, et al. 2013a. Bidirectional transcripts of the expanded *C9orf72* hexanucleotide repeat are translated into aggregating dipeptide repeat proteins. *Acta Neuropathol* **126**: 881–893.

Mori K, Weng SM, Arzberger T, May S, Rentzsch K, Kremmer E, Schmid B, Kretzschmar HA, Cruts M, Van Broeckhoven C, et al. 2013b. The *C9orf72* GGGGCC repeat is translated into aggregating dipeptide-repeat proteins in FTLD/ALS. *Science* **339**: 1335–1338.

Moseley ML, Zu T, Ikeda Y, Gao W, Mosemiller AK, Daughters RS, Chen G, Weatherspoon MR, Clark HB, Ebner TJ, et al. 2006. Bidirectional expression of CUG and CAG expansion transcripts and intranuclear polyglutamine inclusions in spinocerebellar ataxia type 8. *Nat Genet* **38**: 758–769.

Nelson DL, Orr HT, Warren ST. 2013. The unstable repeats —Three evolving faces of neurological disease. *Neuron* **77**: 825–843.

Niblock M, Smith BN, Lee YB, Sardone V, Topp S, Troakes C, Al-Sarraj S, Leblond CS, Dion PA, Rouleau GA, et al. 2016. Retention of hexanucleotide repeat-containing intron in *C9orf72* mRNA: Implications for the pathogenesis of ALS/FTD. *Acta Neuropathol Commun* **4**: 18.

O'Rourke JG, Bogdanik L, Muhammad AK, Gendron TF, Kim KJ, Austin A, Cady J, Liu EY, Zarrow J, Grant S, et al. 2015. *C9orf72* BAC transgenic mice display typical pathologic features of ALS/FTD. *Neuron* **88**: 892–901.

Orr HT, Zoghbi HY. 2007. Trinucleotide repeat disorders. *Annu Rev Neurosci* **30**: 575–621.

Peters OM, Cabrera GT, Tran H, Gendron TF, McKeon JE, Metterville J, Weiss A, Wightman N, Salameh J, Kim J, et al. 2015. Human *C9ORF72* hexanucleotide expansion reproduces RNA foci and dipeptide repeat proteins but not neurodegeneration in BAC transgenic mice. *Neuron* **88**: 902–909.

Ranum LP, Cooper TA. 2006. RNA-mediated neuromuscular disorders. *Annu Rev Neurosci* **29**: 259–277.

Ranum LP, Daughters RS, Tuttle DL, Gao W, Ikeda Y, Moseley ML, Ebner T, Swanson MS. 2010. Double the trouble: Bidirectional expression of the SCA8 CAG/CTG expansion mutation - evidence for RNA and protein gain of function effects. *Rinsho Shinkeigaku* **50**: 982–983.

Reddy K, Zamiri B, Stanley SY, Macgregor RB Jr, Pearson CE. 2013. The disease-associated r(GGGGCC)n repeat from the *C9orf72* gene forms tract length-dependent uni- and multimolecular RNA G-quadruplex structures. *J Biol Chem* **288**: 9860–9866.

Renton AE, Majounie E, Waite A, Simon-Sanchez J, Rollinson S, Gibbs JR, Schymick JC, Laaksovirta H, van Swieten JC, Myllykangas L, et al. 2011. A hexanucleotide repeat expansion in *C9ORF72* is the cause of chromosome 9p21-linked ALS-FTD. *Neuron* **72**: 257–268.

Ross CA, Tabrizi SJ. 2011. Huntington's disease: From molecular pathogenesis to clinical treatment. *Lancet Neurol* **10**: 83–98.

Rudnicki DD, Holmes SE, Lin MW, Thornton CA, Ross CA, Margolis RL. 2007. Huntington's disease–like 2 is associated with CUG repeat-containing RNA foci. *Ann Neurol* **61**: 272–282.

Sathasivam K, Neueder A, Gipson TA, Landles C, Benjamin AC, Bondulich MK, Smith DL, Faull RL, Roos RA, Howland D, et al. 2013. Aberrant splicing of *HTT* generates the pathogenic exon 1 protein in Huntington disease. *Proc Natl Acad Sci* **110**: 2366–2370.

Schludi MH, Becker L, Garrett L, Gendron TF, Zhou Q, Schreiber F, Popper B, Dimou L, Strom TM, Winkelmann J, et al. 2017. Spinal poly-GA inclusions in a *C9orf72* mouse model trigger motor deficits and inflammation without neuron loss. *Acta Neuropathol* **134**: 241–254.

Scoles DR, Ho MH, Dansithong W, Pflieger LT, Petersen LW, Thai KK, Pulst SM. 2015. Repeat associated non-AUG translation (RAN translation) dependent on sequence downstream of the ATXN2 CAG repeat. *PLoS ONE* **10**: e0128769.

Scotti MM, Swanson MS. 2016. RNA mis-splicing in disease. *Nat Rev Genet* **17**: 19–32.

Sellier C, Buijsen RA, He F, Natla S, Jung L, Tropel P, Gaucherot A, Jacobs H, Meziane H, Vincent A, et al. 2017. Translation of expanded CGG repeats into FMRpolyG is pathogenic and may contribute to fragile X tremor ataxia syndrome. *Neuron* **93**: 331–347.

Shi KY, Mori E, Nizami ZF, Lin Y, Kato M, Xiang S, Wu LC, Ding M, Yu Y, Gall JG, et al. 2017. Toxic PR$_n$ poly-dipeptides encoded by the *C9orf72* repeat expansion block nuclear import and export. *Proc Natl Acad Sci* **114**: E1111–E1117.

Sobczak K, de Mezer M, Michlewski G, Krol J, Krzyzosiak WJ. 2003. RNA structure of trinucleotide repeats associated with human neurological diseases. *Nucleic Acids Res* **31**: 5469–5482.

 Cite this article as *Cold Spring Harb Perspect Biol* doi: 10.1101/cshperspect.a033019

Sonenberg N, Hinnebusch AG. 2009. Regulation of translation initiation in eukaryotes: Mechanisms and biological targets. *Cell* **136**: 731–745.

Tao Z, Wang H, Xia Q, Li K, Li K, Jiang X, Xu G, Wang G, Ying Z. 2015. Nucleolar stress and impaired stress granule formation contribute to *C9orf72* RAN translation-induced cytotoxicity. *Hum Mol Genet* **24**: 2426–2441.

Tian B, White RJ, Xia T, Welle S, Turner DH, Mathews MB, Thornton CA. 2000. Expanded CUG repeat RNAs form hairpins that activate the double-stranded RNA-dependent protein kinase PKR. *RNA* **6**: 79–87.

Timchenko LT, Miller JW, Timchenko NA, DeVore DR, Datar KV, Lin L, Roberts R, Caskey CT, Swanson MS. 1996. Identification of a (CUG)$_n$ triplet repeat RNA-binding protein and its expression in myotonic dystrophy. *Nucleic Acids Res* **24**: 4407–4414.

Todd PK, Oh SY, Krans A, He F, Sellier C, Frazer M, Renoux AJ, Chen KC, Scaglione KM, Basrur V, et al. 2013. CGG repeat-associated translation mediates neurodegeneration in fragile X tremor ataxia syndrome. *Neuron* **78**: 440–455.

Wen X, Tan W, Westergard T, Krishnamurthy K, Markandaiah SS, Shi Y, Lin S, Shneider NA, Monaghan J, Pandey UB, et al. 2014. Antisense proline-arginine RAN dipeptides linked to *C9ORF72*-ALS/FTD form toxic nuclear aggregates that initiate in vitro and in vivo neuronal death. *Neuron* **84**: 1213–1225.

Zhang K, Donnelly CJ, Haeusler AR, Grima JC, Machamer JB, Steinwald P, Daley EL, Miller SJ, Cunningham KM, Vidensky S, et al. 2015. The *C9orf72* repeat expansion disrupts nucleocytoplasmic transport. *Nature* **525**: 56–61.

Zhang YJ, Gendron TF, Grima JC, Sasaguri H, Jansen-West K, Xu YF, Katzman RB, Gass J, Murray ME, Shinohara M, et al. 2016. *C9ORF72* poly(GA) aggregates sequester and impair HR23 and nucleocytoplasmic transport proteins. *Nat Neurosci* **19**: 668–677.

Zu T, Gibbens B, Doty NS, Gomes-Pereira M, Huguet A, Stone MD, Margolis J, Peterson M, Markowski TW, Ingram MA, et al. 2011. Non-ATG-initiated translation directed by microsatellite expansions. *Proc Natl Acad Sci* **108**: 260–265.

Zu T, Liu Y, Banez-Coronel M, Reid T, Pletnikova O, Lewis J, Miller TM, Harms MB, Falchook AE, Subramony SH, et al. 2013. RAN proteins and RNA foci from antisense transcripts in *C9ORF72* ALS and frontotemporal dementia. *Proc Natl Acad Sci* **110**: E4968–E4977.

Zu T, Cleary JD, Liu Y, Banez-Coronel M, Bubenik JL, Ayhan F, Ashizawa T, Xia G, Clark HB, Yachnis AT, et al. 2017. RAN translation regulated by muscleblind proteins in myotonic dystrophy type 2. *Neuron* **95**: 1292–1305.e1295.

The Interplay between the RNA Decay and Translation Machinery in Eukaryotes

Adam M. Heck[1,2] and Jeffrey Wilusz[1,2]

[1]Department of Microbiology, Immunology and Pathology, Colorado State University, Fort Collins, Colorado 80525

[2]Program in Cell & Molecular Biology, Colorado State University, Fort Collins, Colorado 80525

Correspondence: jeffrey.wilusz@colostate.edu

RNA decay plays a major role in regulating gene expression and is tightly networked with other aspects of gene expression to effectively coordinate post-transcriptional regulation. The goal of this work is to provide an overview of the major factors and pathways of general messenger RNA (mRNA) decay in eukaryotic cells, and then discuss the effective interplay of this cytoplasmic process with the protein synthesis machinery. Given the transcript-specific and fluid nature of mRNA stability in response to changing cellular conditions, understanding the fundamental networking between RNA decay and translation will provide a foundation for a complete mechanistic understanding of this important aspect of cell biology.

RNA degradation plays a major role in regulating both the quantity and quality of gene expression in the cell. The abundance of an RNA transcript is a reflection of both its rate of synthesis and degradation. There are numerous examples in the literature regarding the regulation of gene expression associated with major cellular responses occurring at the level of differential RNA stability (Chen and Shyu 2017). In addition, there is a growing body of evidence to suggest that cross-communication or buffering occurs between transcription and RNA decay that contributes to the homeostasis of gene expression (Chavèz et al. 2016). The process of translating messenger RNA (mRNA) is also intimately intertwined with the fate of the transcript (Radhakrishnan and Green 2016). Thus, understanding the interplay or networking between the RNA decay machinery and other aspects of

gene expression is needed to further elucidate the depth of regulation that occurs in cells. The goal of this work is to provide an up-to-date overview of the factors and mechanisms of general mRNA decay and then focus on the interplay of this process with the translational machinery. An overview of key enzymes and factors involved in eukaryotic mRNA decay can be found in Table 1.

THE MAJOR STEPS AND MECHANISMS OF mRNA DECAY IN EUKARYOTES

Step One: Deadenylation/Removal of the Poly(A) Tail

Deadenylation is the initial, and often considered rate-limiting step of the traditional exonucleolytic decay pathways. Deadenylases are

Table 1. An overview of key enzymes and factors involved in eukaryotic mRNA decay

Protein name	Full name	Function
Deadenylation		
PAN2	Poly(A)-nuclease deadenylation subunit 2	Enzymatically active subunit of the PAN2/3 complex that is responsible for the preliminary poly(A) trimming
PAN3	Poly(A)-nuclease deadenylation subunit 3	Cofactor and regulatory subunit of the PAN2/3 complex that is responsible for the preliminary poly(A) trimming
CNOT6 (CCR4)	CCR4-NOT complex subunit 6	3' to 5' exonuclease that is the major catalytically active component of the CCR4-NOT complex that performs most of mRNA deadenylation
CNOT1 (NOT1)	CCR4-NOT complex subunit 1	Scaffolding component of the CCR4-NOT deadenylation complex
CNOT7 (CAF1/ POP2)	CCR4-NOT complex subunit 7	Minor deadenylase in the CCR4-NOT complex; its function is somewhat redundant to CNOT6 (CCR4)
PARN	Poly(A)-specific ribonuclease	3' to 5' exoribonuclease that interacts with the 5' cap and preferentially degrades poly(A) tails
Decapping		
DCP2	Decapping protein 2	Catalytically active component of the decapping complex that removes the 5' mRNA cap through hydrolysis reaction
DCP1	Decapping protein 1	Primary cofactor for DCP2; it enhances hydrolysis activity of DCP2 and interacts with other proteins to recruit DCP1/2 complex to mRNA substrates
PATL1 (Pat1)	PAT1 homolog protein 1	Scaffolding protein in deadenylation-dependent decapping; it connects decapping and deadenylation machinery
Lsm1-7	Sm-like protein complex	Identifies and associates with deadenylated transcripts; it interacts with PALT1 to recruit the DCP1/2 complex
DDX6 (Dhh1)	DEAD-box helicase 6	RNA helicase enzyme that interacts with other decapping factors and is thought to remodel the transcript to promote efficient decapping complex assembly
LSM14 (Scd6)	Sm-like protein 14	Serve as a scaffold for assembly and activation of DCP1/2 complex
EDC1,2,3	Enhancer of mRNA-decapping proteins	Serve as a scaffold for assembly and activation of DCP1/2 complex
5' to 3' decay		
XRN1	5'-3' exoribonuclease	Major 5' to 3' exoribonuclease that preferentially degrades RNAs with a 5' monophosphate
3' to 5' decay		
Exo9	Exosome complex (9-subunit)	The core of the exosome that consists of nine different protein subunits
RRP44	Exosome complex exonuclease RRP44	Catalytically active subunit of the exosome
SKI7	Superkiller protein 7	GTP-binding protein that recruits the exosome to nonpoly(A) mRNAs; it is involved in several, including NSD, NMD, and antiviral activity
DCPS	Decapping scavenger enzyme	Catalyzes the cleavage of the residual 5' cap from short oligonucleotide sequence following 3' to 5' decay
DIS3L2	DIS3-like exonuclease 2	3' to 5' exoribonuclease that specifically recognizes polyuridylated RNAs and mediates their degradation
TUT4/7	Terminal uridyltransferase 4/7	Uridyltransferases that mediate terminal uridylation of mRNAs with short poly(A) tails, thus facilitating their degradation

Continued

Cite this article as *Cold Spring Harb Perspect Biol* doi: 10.1101/cshperspect.a032839

Table 1. *Continued*

Protein name	Full name	Function
Endonucleolytic decay		
MCPIP1	Regnase 1/endoribonuclease ZC3H12A	Endonuclease that facilitates mRNA decay and is linked to various biological functions, including immune and inflammatory responses
HRSP12	Heat-responsive protein 12	Protein with endonuclease activity that is involved in activity and formation of the GMD complex
IRE-1	Inositol-requiring enzyme 1	Endonuclease that targets a series of mRNAs connected to several cellular processes
RNase L	Ribonuclease L or ribonuclease 4	Inducible endoribonuclease associated with antiviral response

mRNA, Messenger RNA; NSD, nonstop decay; NMD, nonsense-mediated decay; GMD, glucocorticoid receptor-mediated decay.

recruited to RNA substrates by a variety of RNA-binding proteins (RBPs) and complexes (e.g., Du et al. 2016; Stowell et al. 2016; Yamaji et al. 2017), including the translation initiation factor transporter 4E-T (Nishimura et al. 2015). There are two enzymatic complexes that carry out the majority of cytoplasmic deadenylation in eukaryotes (Wahle and Winkler 2013). The poly(A) nuclease (PAN) PAN2/PAN3 complex is thought to be responsible for preliminary poly(A) trimming (Wolf and Passmore 2014). PAN2 functions as the catalytically active subunit, whereas PAN3 is critical in recruitment of PAN2 to the poly(A) tail and efficient poly(A) tail shortening (Jonas et al. 2014; Schäfer et al. 2014; Wolf et al. 2014). There are two isoforms of PAN3 (3S and 3L) that can, respectively, activate or repress PAN2 deadenylation activity (Chen et al. 2017). Subsequently, the bulk of deadenylation is performed by the CCR4-NOT complex (Collart and Panasenko 2017). While several proteins comprise this complex, the three key proteins appear to be NOT1, CCR4 (CNOT6), and CAF1/POP2 (CNOT7). NOT1 serves as the structural backbone of the complex (Basquin et al. 2012), whereas CCR4 and CAF1/POP2 are the catalytically active deadenylases (Maryati et al. 2015). Recent evidence suggests that CCR4 and CAF1/POP2 deadenylases can also remove terminal non-adenosine residues if the complex is anchored to the RNA substrate by upstream adenosine tracts (Niinuma et al. 2016). Finally, in addition to PAN2/PAN3 and CCR4-NOT, eukaryotic cells also contain a number of additional deadenylases that can influence poly(A) tail length (Skeparnias et al. 2017) and play roles in small RNA biogenesis. Of these, poly(A) ribonuclease (PARN), the most extensively studied, can interact with the 5′ cap of the mRNA substrate to enhance enzymatic activity/processivity (Virtanen et al. 2013; Niedzwiecka et al. 2016). Interestingly, mutations in PARN are associated with human diseases, including dyskeratosis congenita, bone marrow failure, and hypomyelination (Dhanraj et al. 2015; Mason and Bessler 2015). Additionally, ANGEL1, ANGEL2, and the circadian rhythm–associated Nocturnin enzyme are distant homologs of CCR4 that are thought to target specific mRNAs for poly(A) tail shortening (Kojima et al. 2015).

Step Two: Decapping/Removal of the 5′ ^{7m}Gppp Modification to Generate a 5′ Monophosphate

In addition to a key role in promoting translation initiation, the 7-methylguanosine (m^7G) cap on mRNAs also serves to protect transcripts from highly active 5′ to 3′ exoribonucleases that are present in both the nucleus and cytoplasm of eukaryotic cells (Jones et al. 2012). Thus, the process of decapping is a crucial step in the decay of many mRNAs, and can be either deadenylation-dependent or, in certain cases, deadenylation-independent when decay is initiated by alternative mechanisms that do not involve the poly(A) tail as outlined below. The majority of decapping activity is thought to be provided by DCP2, but recent work has identified seven additional NUDIX proteins (e.g., NUDT3 and NUDT16) as well as DXO proteins that act on distinct RNA subsets (Grudzien-Nogalska and Kiledjian 2017) and alternative cap structures (Jiao et al. 2017). In addition to sequence context, reversible methylation of the adenosine adjacent to the cap can have a large influence on decapping rates and overall mRNA stability (Mauer et al. 2017). The well-characterized DCP2 enzyme requires several additional cofactors in vivo to effectively access mRNA substrates as well as attain an optimally active enzymatic conformation (Wurm et al. 2017). Thus, there is a rather elaborate orchestration involved in DCP2 RNA targeting/activation. Decapping cofactors include DCP1, the Lsm1-7 complex, PATL1 (Pat1), DDX6 (Dhh1), LSM14 (Scd6), and a set of EDC proteins. The DCP2/DCP1 complex (Valkov et al. 2017) catalyzes hydrolysis of the cap structure, but generally interacts with mRNA caps with only weak affinity (Ziemniak et al. 2016). The Lsm1-7 complex (Wilusz and Wilusz 2013) associates with the deadenylated transcript and interacts with PATL1 to promote DCP1/2 recruitment (Chowdhury et al. 2014). The Lsm1-7-PAT1 complex can also be usurped by viruses to promote translation (Jungfleisch et al. 2015). EDC1, 2, and 3 proteins, as well as the related LSM14/Scd6 protein, serve as a scaffold for assembly and activation of DCP1/2 enzyme (Fromm et al. 2012; Charenton et al. 2016). The RNA helicase DDX6 (Dhh1) is an evolutionarily conserved factor that interacts with other decapping activators and likely remodels the transcript for efficient complex assembly (Sharif et al. 2013). Finally, in addition to this multifaceted set of factors involved in direct interaction of the decapping complex with RNAs, numerous RBPs as well as terminal RNA modifications also play key roles in the initial identification of mRNAs targeted for decapping. Terminal uridylation of an mRNA or RNA fragment, for example, can effectively trigger/stimulate decapping (Song and Kiledjian 2007; Rissland and Norbury 2009). Finally, the 7 kDa microprotein NoBody has recently been reported to associate with the decapping complex and correlates well with mRNA decay processes

(D'Lima et al. 2016). Therefore, in addition to conventional protein factors, small polypeptides also appear to play a role in the assembly and function of the decapping complex.

Step Three (Option 1): 5′ to 3′ Exoribonucleolytic Decay in the Cytoplasm

XRN1 is the primary 5′ to 3′ cytoplasmic exonuclease and is well conserved throughout eukaryotes (Jones et al. 2012). XRN1 preferentially degrades RNAs with a 5′ monophosphate end—precisely corresponding to the 5′ termini left following mRNA decapping or most endonucleolytic RNA cleavages. XRN1 is a highly processive enzyme that generally can only be stalled by rather elaborate RNA structures, for example, three helix junction knot-like structures found in flavivirus RNAs (Chapman et al. 2014). Interestingly, there is growing evidence that XRN1 plays a key role in overall RNA decay and homeostasis of gene expression (Chavèz et al. 2016). Repression of XRN1 by stalling on flavivirus structures results in a concomitant repression of decapping and deadenylation in the infected cell (Moon et al. 2012, 2015). In addition, buffering effects between mRNA transcription and degradation rates appear to be mediated by XRN1 in some fashion (Haimovich et al. 2013; Sun et al. 2013).

Step Three (Option 2): 3′ to 5′ Exonucleolytic Decay in the Cytoplasm

3′ to 5′ decay in the cytoplasm is performed primarily by a multiunit complex called the exosome (Zinder and Lima 2017). The core of the exosome (Exo9) consists of nine protein subunits that are highly conserved throughout eukaryotes, but nucleolytically inactive. Exo9 binds to the catalytically active subunit RRP44 to form the active exosome (Exo10) (Kowalinski et al. 2016; Zinder et al. 2016). Deadenylation generates an unprotected 3′ end of an mRNA, which the exosome can interact with to initiate 3′ to 5′ decay. The exosome is capable of performing specific functions based on association with different cofactors. For example, the GTP-binding protein SKI7 (Kowalinski et al. 2015)

Cite this article as *Cold Spring Harb Perspect Biol* doi: 10.1101/cshperspect.a032839

recruits the exosome to mRNAs with abnormal translation termination, thereby enabling non-stop decay (NSD) (Graille and Séraphin 2012). Furthermore, the RNA-induced silencing complex (RISC) recruits the exosome to degrade the 5′ product of Ago2-mediated cleavages (Orban and Izaurralde 2005). Following 3′ to 5′ decay, the m^7G cap is removed from the remaining short oligonucleotide by the scavenger decapping enzyme DCPS (Liu et al. 2008) along with FHIT/Aph1 (Taverniti and Seraphin 2015). From a biological perspective, DCPS mutations have been associated with human neurological diseases (Ng et al. 2015) and DCPS inhibitors are being developed as potential therapeutics for spinal muscular atrophy (Gopalsamy et al. 2017). DCPS has also been implicated in micro-RNA (miRNA) turnover (Bosse et al. 2013), further illustrating the overall importance of this enzyme and impact on multiple pathways in cell biology.

In addition to the exosome, 3′ to 5′ decay of coding and noncoding RNAs can be accomplished by DIS3L2 (Łabno et al. 2016; Pirouz et al. 2016; Ustianenko et al. 2016). DIS3L2 preferentially recognizes mRNAs with short poly(U)-tracts at their 3′ end that are added by poly(U) polymerases (PUPs) like TUT4/7 (Viegas et al. 2015). Intriguingly, DIS3L2 has been linked to apoptosis (Thomas et al. 2015), and mutations in DIS3L2 are associated with Perlman syndrome (Pashler et al. 2016). These observations further underscore the biological and developmental importance of this 3′-to-5′ exoribonuclease.

An Alternative Way to Create Exoribonuclease-Sensitive mRNAs: Endonucleolytic Decay

The generation of an accessible 5′ monophosphate or 3′ hydroxyl for an exoribonuclease is the key step in turnover of the body of the mRNA (Fig. 1). In addition to deadenylation and decapping, an appreciation of the extent that endoribonucleases contribute to general mRNA decay has significantly increased over the years. In addition to the well-characterized cleavage events that occur during miRNA and small-interfering RNA (siRNA)-mediated RNA interference (RNAi) (Park and Shin 2014), there are numerous other endoribonucleases that target and cleave mRNAs in a regulated fashion for subsequent exonucleolytic decay. Because endoribonuclease activity significantly modulates the fate of an mRNA, these enzymes are tightly regulated. In fact, cells contain a cysteine-rich ribonuclease inhibitor (RNH1) to protect them from the plethora of unwanted RNase A type ribonucleases secreted by cells in their environment (Thomas et al. 2016). Four of the major endoribonucleases that play significant roles in the regulation of mRNA stability in mammalian cells are highlighted below.

MCPIP1/Regnase 1 is a CCCH zinc finger–containing protein that plays a major role in regulating cellular mRNA decay, including transcripts encoding components of the immune system (Uehata and Takeuchi 2017), iron homeostasis (Yoshinaga et al. 2017), and body fat accumulation (Habacher et al. 2016). Regnase 1 knockout mice develop anemia, severe systemic inflammation, and produce high amounts of autoantibodies. In terms of immune regulation, a key mRNA target of Regnase 1 appears to be the cytokine interleukin (IL)-6 (Uehata and Takeuchi 2017). Regnase 1 activity is regulated by phosphorylation, particularly through IκB kinases in response to immune stimulation (Iwasaki et al. 2011).

The endoribonuclease HRSP12 plays a major role in the formation and activity of the protein complex that mediates glucocorticoid receptor-mediated decay (GMD), a recently described decay pathway that plays a wide role in cell biology (Park et al. 2016). Although it is well known to be a DNA-binding transcription factor, glucocorticoid receptor also binds in a sequence-selective fashion to RNA and recruits PNCR2 and UPF1 to target the transcript for decay (Cho et al. 2015).

Finally, recent work has expanded the biological impact of two well-studied endoribonucleases, IRE-1 and RNase L. In addition to its contributions to endoplasmic reticulum stress and the unfolded protein response (Moore and Hollien 2015), IRE-1 selectively targets a series of mRNAs and plays a key role in a variety of

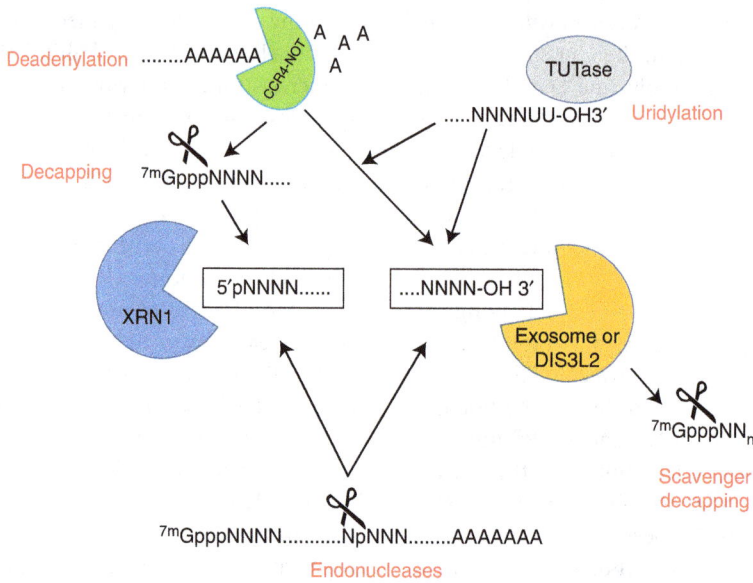

Figure 1. The multiple pathways for exonucleases to gain access to RNAs targeted for degradation. 5′-3′ exoribonucleases (XRN1) require a 5′ monophosphate, whereas 3′-5′ exoribonucleases (exosome and DIS3L2) require an accessible 3′ hydroxyl. A 5′ monophosphate can be generated in a regulated fashion by the process of decapping, which can be deadenylation (poly(A) tail shortening) dependent or deadenylation independent. Deadenylation itself generates an accessible 3′ hydroxyl for exoribonucleases. Poly(U) polymerases (also called TUTases) can uridylate the 3′ end of RNA targets to increase DIS3L2 exonuclease accessibility. In the 3′-5′ exonuclease pathway, the scavenger decapping enzyme DCPS acts on short capped oligonucleotides to promote full degradation. Finally, rather than remodeling the natural 5′ and 3′ ends of the target mRNA, endoribonucleases, including the RNA-induced silencing complex (RISC) complex of the RNA interference (RNAi) pathway, can cleave a transcript internally and generate fragments with 5′ monophosphate and 3′ hydroxyl ends for exonucleolytic decay.

systems, including dendritic cell biology, glucose-responsive insulin secretion, and energy regulation in obesity (Eletto et al. 2016; Shan et al. 2017; Tavernier et al. 2017). RNase L, an inducible endoribonuclease known best for its role in antiviral responses (Cooper et al. 2015), also selectively targets cellular mRNAs to influence overall gene expression (Brennan-Laun et al. 2014; Rath et al. 2015).

INTERPLAY BETWEEN THE TRANSLATIONAL AND RNA DECAY MACHINERIES

Introduction/Evidence for Co-Translational Decay

The definitive function of an mRNA is to produce protein through translation. Building on the advances made over the last two decades in characterizing the pathways of mRNA decay, a full mechanistic understanding of how mRNA turnover is regulated and integrated into overall gene expression is one of the next milestones in the field. Given the extensive networking that occurs between many processes in RNA biology, it is not surprising that the process of translation is intertwined in several ways with RNA decay. Thus, in this section, we focus on co-translational mRNA decay (Fig. 2) and its impact on gene expression.

Numerous early observations suggest that the translation of an mRNA can affect its degradation rate in mammalian cells. Pioneering work in the 1980s and early 1990s on a variety of oncogene-encoding mRNAs, including c-myc (Dani et al. 1984; Linial et al. 1985; Laird-Offringa et al. 1990), c-fos (Rahmsdorf et al. 1987), jun (Ryseck et al. 1988), and myb

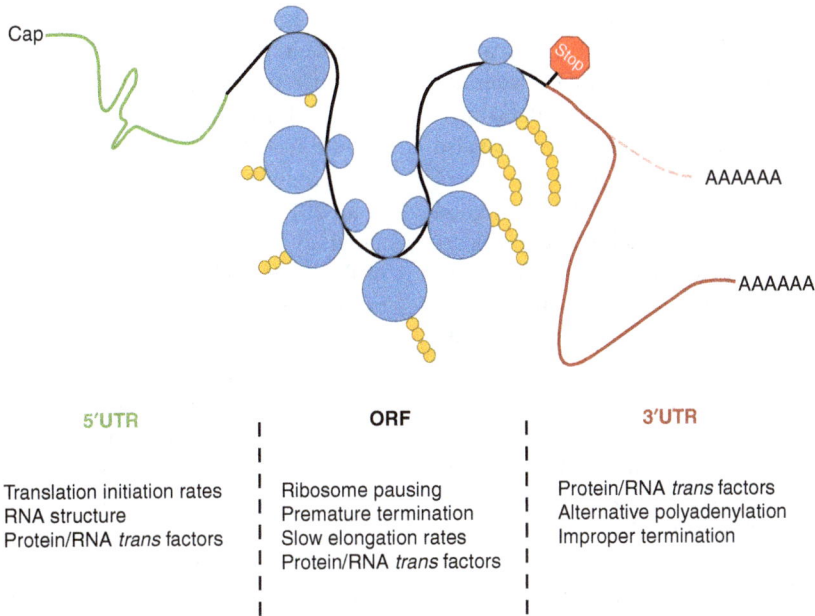

Figure 2. Numerous regulatory points exist throughout the transcript to modulate co-translational messenger RNA (mRNA) decay. A brief overview of the factors and the location on the transcript that can influence the rate of mRNA decay in association with ribosome loading, elongation, and termination. In the 5′ untranslated region (UTR) (green), RNA structure, initiation rates, and RNA-binding protein (RBP)/RNA factors acting in *trans* can co-regulate mRNA decay and translation. As ribosomes (blue) read through the open reading frame (ORF) (black), elongation speed or lack thereof (ribosome pausing) is the primary feature that links active translation and mRNA decay. Events such as premature termination, RNA modification, or protein/RNA *trans* factor interaction can alter ribosomal speed, thus influencing mRNA stability. Finally, the 3′UTR (red) can modulate mRNA decay and translation via interactions with other protein/RNA factors, different 3′UTR isoforms that can arise by alternative polyadenylation (APA), or via improper termination, which can result from a defective ribosome or lack of a stop codon. The cell uses all of these regulatory mechanisms to ensure that accurate and efficient translation is achieved.

(Thompson et al. 1986), indicated that blocking translation, using small molecule drugs or the insertion of stem-loop structures in the 5′ untranslated region (UTR) to stop ribosome scanning, resulted in stabilization of these normally very short-lived mRNAs. The phenomenon was not limited to oncogene-related transcripts, as mRNAs encoding growth factors (Aharon and Schneider 1993), histones (Stimac et al. 1983), enzymes (Altus and Nagamine 1991), and receptors (Koeller et al. 1991) could also be stabilized by translational repression. In fact, an early global analysis using differential phage plaque hybridization extended the connection between mRNA translation and stability by identifying >80 mRNAs whose stability was influenced by translational repression (Almendral et al. 1988).

The changes in mRNA stability observed using global translational inhibitors such as cycloheximide, puromycin, or pactamycin must be interpreted with some caution, because a portion of the effects could be the result of indirect or "*trans*" effects. It is, however, abundantly clear that translation can directly influence the stability of an mRNA. A variety of observations support this contention, including the use of transcript-directed stem-loop structures in mRNAs (Aharon and Schneider 1993), the association of the context of the start codon with stability of the yeast PGK1 mRNA (LaGrandeur and Parker 1999), as well as the extensive literature on premature stop codons influencing transcript stability via nonsense-mediated decay (NMD) (Celik et al. 2017b; Karousis and Mühlemann 2017). In addition, extensive ribosome pausing (e.g., no-go decay [NGD] [Simms et al. 2017]) and incorrect ribosome termination (i.e., NSD [Klauer and van Hoof 2012]) also make a substantial contribution to the stability of an mRNA. There is also an inherent complexity between the interface of active translation and mRNA stability. A good example of this is the observation that the placement of a premature termination codon close to the AUG initiation site fails to elicit NMD, a phenomenon referred to as the AUG proximity effect (Pereira et al. 2015).

There is a growing body of evidence that supports the contention that a large portion of mRNA decay occurs co-translationally. For example, the decapping step of the major 5′–3′ decay pathway in yeast has been shown to take place directly on polysomes (Hu et al. 2009, 2010). In addition, altering growth conditions can cause the decapping activators DHH1 and PAT1 to relocalize to polysomes and target mRNAs for degradation (Drummond et al. 2011). The DHH1 RNA helicase has also been shown to slow ribosome movement and trigger mRNA decapping (Sweet et al. 2012). Finally, viral endonucleases can also be directly recruited to translating mRNAs and target them for degradation (Covarrubias et al. 2011).

Co-translational mRNA decay has also been shown to have practical applications. It is problematic to definitively assess ribosome movement/dynamics in living cells because of concerns about changes that can occur during requisite in vitro processing steps. By sequencing 5′ monophosphorylated decay intermediates, Steinmetz and colleagues have used co-translational mRNA decay as a readout for the dynamics of ribosomal movement on mRNAs in a variety of yeast systems (Pelechano et al. 2015, 2016). The mRNA decay intermediates showed a three-nucleotide pattern of periodicity, consistent with the co-translational decay model in which XRN1 follows the last translating ribosome on an mRNA.

Biases in codon usage are clear in all organisms, but the impact of codon choice on the rate of translation elongation and associated mRNA stability was not fully clear. This was because of conflicting data from ribosome profiling studies

and an associated surprising lack of sensitivity because of a variety of technical biases that can influence data interpretation (Yu et al. 2015). Using cell-free translation assays, Liu and colleagues recently showed a clear association between codon preference and elongation rates (Yu et al. 2015). The elongation rate of ribosomes along an mRNA clearly influences the stability of the associated transcript via the DEAD-box protein Dhh1, which serves as a sensor for ribosome velocity as well as recruiting the mRNA decay machinery (Radhakrishnan et al. 2016). The impact of codon optimality on mRNA stability was revealed by demonstrating that substitution of optimal codons with synonymous but suboptimal ones impacts the rate of ribosome elongation and causes mRNA destabilization (Presnyak et al. 2015). This observation appears to be universal, as codon optimality has been shown to influence RNA decay in *Escherichia coli* (Boel et al. 2016) and ribosome pausing on a transcript also results in increased rates of mRNA decay in plants (Merret et al. 2015). In eukaryotes, the DEAD-box protein DDX6 (which is known as Dhh1 in yeast) appears to be a sensor for ribosome speed and connects translation elongation rates to RNA decay (Radhakrishnan et al. 2016). DDX6/Dhh1 can both interact with the ribosome as well as interact with and regulate mRNA deadenylation and decapping factors (Coller et al. 2001), making it an attractive candidate for a factor that can couple slow ribosome progression with enhanced mRNA decay.

Multiple facets associated with the translation of an mRNA can influence its decay rate (Neymotin et al. 2016). The following sections will provide examples of how different parts of an mRNA, the 5′UTR, the open reading frame (ORF), and the 3′UTR can influence co-translational mRNA decay.

Examples of 5′UTR Impacts on Both Translation and RNA Decay

Hepatitis C Virus

Internal ribosome entry sites (IRESs), commonly found in viral RNAs but also present in some eukaryotic transcripts (Weingarten-Gabbay et

Cite this article as *Cold Spring Harb Perspect Biol* doi: 10.1101/cshperspect.a032839

al. 2016), are highly structured regions in the 5′UTR that promote translation initiation in a cap-independent fashion (Dai et al. 2015). The 5′UTR of hepatitis C virus (HCV), which contains a well-characterized IRES (Khawaja et al. 2015), is thought to also hijack aspects of the cellular RNAi machinery to stabilize its own RNA and increase translation of viral products. microRNA-122 (miR-122) is highly abundant in the liver (Chang et al. 2004), the target tissue of HCV. Binding of the seed sequence and additional nucleotides at the 3′ end of miR-122 to multiple regions in the 5′ terminus of the HCV genome, just upstream of the IRES, stimulates HCV translation (Roberts et al. 2011; Wilson et al. 2011) and enhances RNA stability (Shimakami et al. 2012). Furthermore, it is the association with the miR-122 RISC complex, and not simply double-stranded RNA (dsRNA) formation between the miRNA and the HCV RNA that results in HCV RNA accumulation (Zhang et al. 2012). These studies, coupled with the discovery that the HCV 5′UTR is capable of stalling XRN1 (Moon et al. 2015), led to a model where HCV RNA recruits the miRISC complex, through interactions with miR-122 just upstream of the IRES, to act as a barrier against XRN1 accessing the naturally uncapped HCV RNA. This enhanced RNA stability enables accumulation of HCV RNA and increased viral translation. The usurping of the miRNA machinery by the 5′UTR has recently been generalized to another member of *Flaviviridae*, bovine viral diarrhea virus (BVDV) and its interaction with miR-17 (Scheel et al. 2016).

Protein–RNA Interactions with the 5′UTR that Play Dual Roles in Stability and Translation

RBPs are one of the key factors involved in posttranscriptional gene regulation. RBP interactions with the 5′UTR of target mRNAs can alter RNA degradation kinetics as well as affect translation by interfacing with initiation factors. HuR and AUF1 are well-studied RBPs that are classically known to regulate RNA decay through AU-rich elements. HuR is strongly associated with mRNA stabilization, whereas AUF1 is typically associated with mRNA destabilization (Gram-

matikakis et al. 2017; White et al. 2017). HuR and AUF1, along with the RISC component Ago2, have been implicated in regulating translation of a viral IRES element. HuR and Ago2 binding promotes translational activity of the enterovirus 71 IRES, whereas AUF1 binding represses it (Lin et al. 2015). Similarly, HuR enhances IRES-mediated translation of the X chromosome–linked inhibitor of apoptosis (XIAP) (Durie et al. 2011), whereas Ago2 associated with miR-10a promotes translation of ribosomal protein mRNAs (Ørom et al. 2008). Conversely, HuR can repress IRES-mediated translation of several transcripts (Kullmann et al. 2002; Meng et al. 2005; Yeh et al. 2008). Although the precise mechanism remains uncharacterized, it is postulated that HuR may block assembly of the IRES-associated ribosome–IRES interacting factor (ITAF) complex or impede the complex from initiating translation. Moreover, HuR significantly delays cap-dependent scanning by the 43S preinitiation complex (Meng et al. 2005), resulting in repressed translation and increased degradation of the target mRNA. These studies, along with others in the field, show the complexity associated with RBP-mediated translation/mRNA decay co-regulation, and add support to the notion that RBP–mRNA interactions cannot be defined by a simplistic linear model.

DDX6—A Key Developmental Regulator of Translation and RNA Decay

DDX6 is a member of the DEAD-box family of helicases that is well conserved throughout eukaryotes and has been implicated in mRNA translation and decay (Ostareck et al. 2014), including miRNA-mediated translational silencing (Kuzuoğlu-Öztürk et al. 2016). DDX6 interacts with the CNOT1 component of the CCR4-NOT complex, which interacts with TNRC6/GW182 (Chen et al. 2014; Rouya et al. 2014). An interaction between DDX6 and 4E-transporter (4E-T) (which interacts with eIF4E), in coordination with the CCR4-NOT complex, mediates miRNA-dependent translational repression (Ozgur et al. 2015; Kamenska et al. 2016). 4E-T has also been implicated in physically linking the two termini of mRNAs targeted for decay via interactions with eIF4E at the 5′ cap and the CCR4-NOT deadenylase complex at the 3′ end (Nishimura et al. 2015). The biological relevance of DDX6 coordinating co-translational decay has recently been shown by the ability of progenitor stem cells to modulate stem cell maintenance and differentiation. DDX6 was shown to be up-regulated during neuronal differentiation, and it interacts with the neuronal cell-fate determinant protein TRIM32 to stimulate activity of the miRNA Let-7a (Nicklas et al. 2015), which has previously been shown to help induce neuronal differentiation (Schwamborn et al. 2009). Conversely, in epidermal progenitor cells, DDX6 is believed to control a gene expression profile that inhibits differentiation while maintaining proliferation. Premature differentiation is prevented by destabilizing *KLF4* mRNA through interactions with the 5′UTR, whereas DDX6 forms a complex with YBX1 and eIF4E to promote translation of several self-renewal and proliferation transcripts (Wang et al. 2015). DDX6 is also used/usurped in a variety of ways by viruses (Ostareck et al. 2014), making it an interesting multifunctional protein regulator/organizer in the cytoplasm. This further shows the impressive networking in post-transcriptional gene regulation.

Examples of ORF-Associated Impacts on mRNA Stability

Codon Optimality/Translational Velocity and RNA Decay

Whereas the effect of codon optimality and mRNA stability was described in the introduction to this section, it is useful to emphasize several aspects. First, the impact of codon optimality on translational speed and mRNA decay rates appears to be a conserved phenomenon. In addition to fungi and plants, higher eukaryotes also show a connection between the use of optimal codons and RNA decay, including zebrafish, *Xenopus*, mammals, and *Drosophila* (Bazzini et al. 2016; Chen and Coller 2016; Mishima and Tomari 2016; Zhao et al. 2017). Second, the DHH1 factor, which has previously been shown to slow ribosome movement to enhance co-translational decapping (Sweet et al. 2012), has

emerged as a key factor in monitoring codon optimality of mRNAs in yeast (Radhakrishnan et al. 2016). Next, even NMD rates can be influenced by codon selection (Celik et al. 2017a). This observation clearly shows the connection between the quality control of gene expression and ribosome elongation. Finally, recent global analyses have suggested that both codon optimality and properties of adjacent codons contribute to influence mRNA decay rates, indicating that pairs or perhaps strings of suboptimal codons can differentially regulate mRNA turnover (Harigaya and Parker 2017).

No-Go Decay

NGD is one of several quality-control mechanisms eukaryotic cells use to recognize and eliminate defective mRNAs during the translation process. Specifically, the NGD machinery releases ribosomes in a prolonged elongation stall by targeting the mRNA for endonucleolytic cleavage and subsequent canonical mRNA decay (Simms et al. 2017). Surveillance mechanisms like NGD are critical because the aberrant mRNA transcripts remove ribosomes from the translating pool, which places a significant burden on the cell. Dom34 and Hbs1 are the primary components that make up the NGD machinery (Passos et al. 2009; Chen et al. 2010). The Dom34/Hbs1 complex, in conjunction with ABCE1, promotes ribosome dissociation and nascent peptide transfer RNA (tRNA) drop off (Shoemaker et al. 2010; Pisareva et al. 2011). While it has been reported that NGD results in the endonucleolytic cleavage of the aberrant mRNA (Doma and Parker 2006; Tsuboi et al. 2012), the enzyme(s) that performs the cleavage remains elusive. Reported causes of ribosome stalling that initiate NGD include structured RNA (e.g., stem loops [Doma and Parker 2006]), tracts of positively charged amino acids (Dimitrova et al. 2009), and defective ribosomes (Cole et al. 2009). Although NGD is generally considered a surveillance mechanism for aberrant translation, there are observations of natural NGD having a biological function in *Drosophila* (Xi et al. 2005; Passos et al. 2009) and *Arabidopsis* (Onouchi et al. 2005), suggesting

 Cite this article as *Cold Spring Harb Perspect Biol* doi: 10.1101/cshperspect.a032839

that cells may use NGD to regulate gene expression in certain scenarios.

The Exon Junction Complex (EJC) and Its Impact on General Translation/RNA Stability

The EJC has been intimately associated with mRNA quality control and NMD (Woodward et al. 2017). An excellent source for more information on mRNA quality control and NMD can be found in Karousis and Mühlemann (2017). However, the function of this nuclear-initiated mRNA mark is broad, and it includes an impact on translation and conventional mRNA decay. The core EJC protein MLN51 can bind to eIF3 and promote translation of mRNAs (Chazal et al. 2013). However, additional core EJC proteins also appear to influence translation, including PYM (Diem et al. 2007) and Y14 (Lee et al. 2009). Curiously, the Y14 component of the EJC has been suggested to bind the mRNA 5′ cap as well as the primary decapping enzyme DCP2 through different domains (Chuang et al. 2013). Mutating Y14 so that it cannot bind the cap abrogates its ability to support translation, but does not impede the degradation of target mRNAs (Chuang et al. 2016). This nicely illustrates the interplay of the EJC with cytoplasmic post-transcriptional processes in determining the propensity of the mRNA to be translated as well as degraded.

Examples of 3′UTR Impacts on Both Translation and RNA Decay

Alternative 3′UTRs and Combinatorial Actions of RBPs on 3′UTR Elements that Co-Regulate Decay/Translation

It is widely accepted that the 3′UTRs of eukaryotic mRNAs contain a large proportion of the regulatory elements and binding motifs used by cells to regulate post-transcriptional gene expression. So it is logical to think that an additional mode of regulation that cells can use is to generate transcript isoforms with different 3′UTRs (Bava et al. 2013). This process is termed alternative cleavage and polyadenylation or alternative polyadenylation (APA) and can have profound effects on mRNA translation efficiency and decay (Mayr 2016; Tian and Manley 2017). In a simplified model, shortening of the 3′UTR induced by APA sites can alter RNA structure or eliminate miRNA or RBP-binding sites, which positively or negatively affect mRNA stability and translation. As miRNA-mediated regulation is explained in detail in Fabian (2017), we will touch on a few notable RBPs that co-regulate decay and translation via interactions with the 3′UTR. As mentioned previously, HuR is a widely studied RBP that promotes both mRNA stabilization as well as translation by preventing the interaction of repressive factors with the mRNA via competition for common binding regions (Lal et al. 2004; Kim et al. 2015; Liu et al. 2015; Blackinton and Keene 2016; Bose et al. 2016; Zhang et al. 2017). However, several studies have also implicated HuR in destabilization of transcripts through recruitment of other RBPs (Chang et al. 2010) and the miRISC complex (Kim et al. 2009). Similar to HuR, the Pumilio or PUF family of proteins is present in all eukaryotes (Zamore et al. 1997) and modulates mRNA fate through 3′UTR interactions (Morris et al. 2008; Hafner et al. 2010). PUF proteins are potent repressors that can accelerate deadenylation (Weidmann et al. 2014) through interactions with POP2 (Goldstrohm et al. 2006) and inhibit translation by blocking polypeptide elongation (Friend et al. 2012). PUF proteins play a regulatory role in a wide variety of biological functions, including development (Datla et al. 2014; Mak et al. 2016), cell proliferation (Naudin et al. 2017), stress response (Russo and Olivas 2015), and neurological processes (Sharifnia and Jin 2014; Gennarino et al. 2015; Arvola et al. 2017). Whereas HuR and the PUF family of proteins are two of the most prolific examples of how RBPs can interact with the 3′UTR to influence mRNA translation and stability, there are numerous other 3′UTR-binding RBPs the cell uses to regulate gene expression globally or on a subset of genes.

Nonstop Decay

Finally, NSD is generally considered a quality-control mechanism that targets ribosomes stalled at the 3′ end of mRNAs, which is often caused by lack of a stop codon as a result of premature transcriptional termination and polyadenylation (Ito-Harashima et al. 2007; Simms et al. 2017). Similar to NGD discussed above, the potential loss of functional ribosomes because of build-up on aberrant mRNAs and subsequent impact on cell survival makes NSD a crucial surveillance mechanism. A premature polyadenylation event on a transcript results in ribosome stalling as it tries to decode the poly(A) tail into poly(Lys) residues. Consistent with this notion, poly(A) tracts within the ORF (Ito-Harashima et al. 2007; Kuroha et al. 2010) as well as electrostatic interactions between positively charged lysine residues and the negatively charged tunnel of the ribosome (Lu and Deutsch 2008; Dimitrova et al. 2009) have been shown to elicit ribosome stalling. Several studies have implied the ribosome along with the GTPase SKI7 are responsible for the initial recognition of mRNAs with aberrantly placed poly(A) tracts (Araki et al. 2001; van Hoof et al. 2002; Akimitsu et al. 2007; Arthur et al. 2015; Garzia et al. 2017; Juszkiewicz and Hedge 2017). Whereas SKI7 is primarily associated with NSD, it is a paralog of the NGD factor HBS1, and SKI7 deletion strains in yeast are complemented by introducing HBS1 (van Hoof 2005). Furthermore, Dom34/Hbs1 has been linked to dissociation of ribosomes stalled in the 3′UTR of mRNAs (Tsuboi et al. 2012; Saito et al. 2013), suggesting functional overlap of the SKI7 protein and Dom34/Hbs1 complex. Conversely, a recent study hypothesized a competitive mechanism of action between SKI7 and the Dom34/Hbs1 complex for interactions with ribosomes in NSD (Horikawa et al. 2016). This suggests that the relationship between these two elements is more complicated than originally thought and requires further studies.

CONCLUDING REMARKS/FUTURE DIRECTIONS

In summary, whereas the processes of translation and mRNA decay are naturally competing for the same parts of the mRNA, available evidence indicates that translation and general mRNA decay are not independent processes, but are intricately networked. A variety of interesting questions remain to be answered. Although aspects of the 5′–3′ decay pathway are linked to translational dynamics on the targeted mRNA, what about 3′–5′ decay by the exosome and or DIS3L2? How complex is the interplay between the ribosome and the mRNA decay machinery in terms of protein–protein interactions and the range of cofactor requirements? Although ribosome elongation rates appear to influence mRNA decay, is the relationship mutual? In other words, does RNA decay influence ribosome dynamics on the targeted mRNA? Does the mechanism/route of translation initiation influence mRNA decay rates as well as the rate of elongation? Are changes in mRNA stability as a result of slower elongation rates through suboptimal codons also associated with the NGD pathway and Dom34/HBS1 complex, or do these stability changes simply reflect alterations in conventional mRNA decay pathways? How extensively are associations between translation and decay regulated at the developmental level, and are they perhaps dysregulated/disconnected in cancer cells to promote altered gene expression profiles and remove controls over cellular gene expression? Finally, from a pathogen perspective, how do arboviruses that flip-flop between growth in insect vectors and vertebrate hosts deal with major differences in codon optimality between the two eukaryotic environments and maintain the stability of their RNA transcripts? Elucidating these and other aspects of the interplay between the fundamental translation apparatus and mRNA decay machinery should yield many interesting insights in the future.

ACKNOWLEDGMENTS

We thank members of the Wilusz laboratories for helpful discussions. A.M.H. is supported in part through the National Science Foundation Research Traineeship (NSF-NRT) award #1450032 and an NSF Graduate Research Fellowship. RNA stability work in the Wilusz laboratory is supported by the National Institutes of Health (NIH) awards GM114217, AI123136, and AI130497, as well as a grant from the College of Veterinary Medicine and Biomedical Sciences (CVMBS) College Research Council to J.W.

REFERENCES

*Reference is also in this collection.

Aharon T, Schneider RJ. 1993. Selective destabilization of short-lived mRNAs with the granulocyte-macrophage colony-stimulating factor AU-rich 3' noncoding region is mediated by a cotranslational mechanism. *Mol Cell Biol* 13: 1971–1980.

Akimitsu N, Tanaka J, Pelletier J. 2007. Translation of non-STOP mRNA is repressed post-initiation in mammalian cells. *EMBO J* 26: 2327–2338.

Almendral JM, Sommer D, Macdonald-Bravo H, Burckhardt J, Perera J, Bravo R. 1988. Complexity of the early genetic response to growth factors in mouse fibroblasts. *Mol Cell Biol* 8: 2140–2148.

Altus MS, Nagamine Y. 1991. Protein synthesis inhibition stabilizes urokinase-type plasminogen activator mRNA. Studies in vivo and in cell-free decay reactions. *J Biol Chem* 266: 21190–21196.

Araki Y, Takahashi S, Kobayashi T, Kajiho H, Hoshino S, Katada T. 2001. Ski7p G protein interacts with the exosome and the Ski complex for 3'-to-5' mRNA decay in yeast. *EMBO J* 20: 4684–4693.

Arthur L, Pavlovic-Djuranovic S, Smith-Koutmou K, Green R, Szczesny P, Djuranovic S. 2015. Translational control by lysine-encoding A-rich sequences. *Sci Adv* 1: e1500154.

Arvola RM, Weidmann CA, Tanaka Hall TM, Goldstrohm AC. 2017. Combinatorial control of messenger RNAs by Pumilio, Nanos and brain tumor proteins. *RNA Biol* doi: 10.1080/15476286.2017.1306168.

Atkinson GC, Baldauf SL, Hauryliuk V. 2008. Evolution of nonstop, no-go and nonsense-mediated mRNA decay and their termination factor-derived components. *BMC Evol Biol* 8: 290.

Basquin J, Roudko V, Rode M, Basquin C, Séraphin B, Conti E. 2012. Architecture of the nuclease module of the yeast Ccr4-Not complex: The Not1-Caf1-Ccr4 interaction. *Mol Cell* 48: 207–218.

Bava FA, Eliscovich C, Ferreira PG, Miñana B, Ben-Dov C, Guigó R, Valcárcel J, Méndez R. 2013. CPEB1 coordinates alternative 3'-UTR formation with translational regulation. *Nature* 495: 121–125.

Bazzini AA, del Viso F, Moreno-Mateos MA, Johnstone TG, Vejnar CE, Qin Y, Yao J, Khokha MK, Giraldez AJ. 2016. Codon identity regulates mRNA stability and translation efficiency during the maternal-to-zygotic transition. *EMBO J* 35: 2087–2103.

Blackinton JG, Keene JD. 2016. Functional coordination and HuR-mediated regulation of mRNA stability during T cell activation. *Nucleic Acids Res* 44: 426–436.

Boel G, Letso R, Neely H, Price WN, Wong KH, Su M, Luff JD, Valecha M, Everett JK, Acton TB, et al. 2016. Codon

influence on protein expression in *E. coli* correlates with mRNA levels. *Nature* 529: 358–363.

Bose S, Tholanikunnel TE, Reuben A, Tholanikunnel BG, Spicer EK. 2016. Regulation of nucleolin expression by miR-194, miR-206, and HuR. *Mol Cell Biochem* 417: 141–153.

Bosse GD, Ruegger S, Ow MC, Vasquez-Rifo A, Rondeau EL, Ambros VR, Grobhans H, Simard MJ. 2013. The decapping scavenger enzyme DCS-1 controls microRNA levels in *Caenorhabditis elegans*. *Mol Cell* 50: 281–287.

Brennan-Laun SE, Li XL, Ezelle HJ, Venkataraman T, Blackshear PJ, Wilson GM, Hassel BA. 2014. RNase L attenuates mitogen-stimulated gene expression via transcriptional and post-transcriptional mechanisms to limit the proliferative response. *J Biol Chem* 289: 33629–33643.

Celik A, Baker R, He F, Jacobson A. 2017a. High-resolution profiling of NMD targets in yeast reveals translational fidelity as a basis for substrate selection. *RNA* 23: 735–748.

Celik A, He F, Jacobson A. 2017b. NMD monitors translational fidelity 24/7. *Curr Genet* doi: 10.1007/s00294-017-0709-4.

Chang J, Nicolas E, Marks D, Sander C, Lerro A, Buendia MA, Xu C, Mason WS, Moloshok T, Bort R, et al. 2004. miR-122, a mammalian liver-specific microRNA, is processed from hcr mRNA and may downregulate the high affinity cationic amino acid transporter CAT-1. *RNA Biol* 1: 106–113.

Chang N, Yi J, Guo G, Liu X, Shang Y, Tong T, Cui Q, Zhan M, Gorospe M, Wang W. 2010. HuR uses AUF1 as a cofactor to promote p16INK4 mRNA decay. *Mol Cell Biol* 30: 3875–3886.

Chapman EG, Costantino DA, Rabe JL, Moon SL, Wilusz J, Nix JC, Kieft JS. 2014. the structural basis of pathogenic subgenomic flavivirus RNA (sfRNA) production. *Science* 344: 307–310.

Charenton C, Taverniti V, Gaudon-Plesse C, Back R, Seraphin B, Graille M. 2016. Structure of the active form of Dcp1–Dcp2 decapping enzyme bound to m^{7}GDP and its Edc3 activator. *Nat Struct Mol Biol* 23: 982–986.

Chavèz S, Garcia-Martinez J, Delgado-Ramos L, Pèrez-Ortin JE. 2016. The importance of controlling mRNA turnover during cell proliferation. *Curr Genet* 62: 701–710.

Chazal PE, Daguenet E, Wendling C, Ulryck N, Tomasetto C, Sarguei B, Le Hir H. 2013. EJC core component MLN51 interacts with eIF3 and activates translation. *Proc Natl Acad Sci* 110: 5903–5908.

Chen Y, Coller J. 2016. A universal code for mRNA stability? *Trends Genet* 32: 687–688.

Chen C, Shyu A. 2017. Emerging themes in regulation of global mRNA turnover in *cis*. *Trends Biochem Sci* 42: 16–27.

Chen L, Muhlrad D, Hauryliuk V, Cheng Z, Lim MK, Shyp V, Parker R, Song H. 2010. Structure of the Dom34/Hbs1 complex and implications for no-go decay. *Nat Struct Mol Biol* 17: 1233–1240.

Chen Y, Boland A, Kuzuoğlu-Öztürk D, Bawankar P, Loh B, Chang CT, Weichenrieder O, Izaurralde E. 2014. A DDX6-CNOT1 complex and W-binding pockets in CNOT9 reveal direct links between miRNA target recognition and silencing. *Mol Cell* 54: 737–750.

Chen C, Zhang Y, Xiang Y, Han L, Shyu A. 2017. Antagonistic actions of two human Pan3 isoforms on global mRNA turnover. *RNA* doi: 10.1261/rna.061556.117.

Cho H, Park OH, Park J, Ryu I, Kim J, Ko J, Kim YK. 2015. Glucocorticoid receptor interacts with PNRC2 in a ligand-dependent manner to recruit UPF1 for rapid mRNA degradation. *Proc Natl Acad Sci* 112: 1540–1549.

Chowdhury A, Kalurupalle S, Tharun S. 2014. Pat1 contributes to the RNA binding activity of the Lsm1-7–Pat1 complex. *RNA* 20: 1465–1475.

Chuang TW, Chang WL, Lee KM, Tarn WY. 2013. The RNA-binding protein Y14 inhibits mRNA decapping and modulates processing body formation. *Mol Biol Cell* 24: 1–13.

Chuang TW, Lee K, Lou Y, Lu C, Tarn WY. 2016. A point mutation in the exon junction complex factor Y14 disrupts its function in mRNA cap binding and translation enhancement. *J Biol Chem* 291: 8565–8574.

Cole SE, LaRiviere FJ, Merrikh CN, Moore MJ. 2009. A convergence of rRNA and mRNA quality control pathways revealed by mechanistic analysis of nonfunctional rRNA decay. *Mol Cell* 34: 440–450.

Collart MA, Panasenko OO. 2017. The Ccr4-Not complex: Architecture and structural insights. In *Sub-cellular biochemistry*, Vol. 83, pp. 349–379. Springer, New York.

Coller JM, Tucker M, Sheth U, Valencia-Sanchez MA, Parker R. 2001. The DEAD box helicase, Dhh1p, functions in mRNA decapping and interacts with both the decapping and deadenylase complexes. *RNA* 7: 1717–1727.

Cooper DA, Banerjee S, Chakrabarti A, García-Sastre A, Hesselberth JR, Silverman RH, Barton DJ. 2015. RNase L targets distinct sites in influenza A virus RNAs. *J Virol* 89: 2764–2776.

Covarrubias S, Gaglia MM, Kumar GR, Wong W, Jackson AO, Glaunsinger BA. 2011. Coordinated destruction of cellular messages in translation complexes by the γ herpesvirus host shutoff factor and the mammalian exonuclease Xrn1. *PLoS Pathog* 7: e1002339.

Dai W, Ma W, Li Q, Tao Y, Ding P, Zhu R, Jin J. 2015. The 5′-UTR of DDB2 harbors an IRES element and upregulates translation during stress conditions. *Gene* 573: 57–63.

Dani C, Blanchard JM, Piechaczyk M, El Sabouty S, Marty L, Jeanteur P. 1984. Extreme instability of myc mRNA in normal and transformed human cells. *Proc Natl Acad Sci* 81: 7046–7050.

Datla US, Scovill NC, Brokamp AJ, Kim E, Asch AS, Lee MH. 2014. Role of PUF-8/PUF protein in stem cell control, sperm-oocyte decision and cell fate reprogramming. *J Cell Physiol* 229: 1306–1311.

Dhanraj S, Gunja SMR, Deveau AP, Nissbeck M, Boonyawat B, Coombs AJ, Renieri A, Mucciolo M, Marozza A, Buoni S, et al. 2015. Bone marrow failure and developmental delay caused by mutations in poly(A)-specific ribonuclease (*PARN*). *J Med Genet* 52: 738–748.

Diem MD, Chan CC, Younis I, Dreyfuss G. 2007. PYM binds the cytoplasmic exon-junction complex and ribosomes to enhance translation of spliced mRNAs. *Nat Struct Mol Biol* 14: 1173–1179.

Dimitrova LN, Kuroha K, Tatematsu T, Inada T. 2009. Nascent peptide-dependent translation arrest leads to Not4p-mediated protein degradation by the proteasome. *J Biol Chem* 284: 10343–10352.

D'Lima NG, Ma J, Winkler L, Chu Q, Loh KH, Corpuz EO, Budnik BA, Lykke-Andersen J, Saghatelian A, Slavoff SA. 2016. A human microprotein that interacts with the mRNA decapping complex. *Nat Chem Biol* 13: 174–180.

Doma MK, Parker R. 2006. Endonucleolytic cleavage of eukaryotic mRNAs with stalls in translation elongation. *Nature* 440: 561–564.

Drummond SP, Hildyard J, Firczuk H, Reamtong O, Li N, Kannambath S, Claydon AJ, Beynon RJ, Eyers CE, McCarthy JEG. 2011. Diauxic shift-dependent relocalization of decapping activators Dhh1 and Pat1 to polysomal complexes. *Nucleic Acids Res* 39: 7764–7774.

Du H, Zhao Y, He J, Zhang Y, Xi H, Liu M, Ma J, Wu L. 2016. YTHDF2 destabilizes m6A-containing RNA through direct recruitment of the CCR4/NOT deadenylase complex. *Nat Commun* 7: 1262–1266.

Durie D, Lewis SM, Liwak U, Kisilewicz M, Gorospe M, Holcik M. 2011. RNA-binding protein HuR mediates cytoprotection through stimulation of XIAP translation. *Oncogene* 30: 1460–1469.

Eletto D, Boyle S, Argon Y. 2016. PDIA6 regulates insulin secretion by selectively inhibiting the RIDD activity of IRE1. *FASEB J* 30: 653–665.

Ernoult-Lange M, Baconnais S, Harper M, Minshall N, Souquere S, Boudier T, Bénard M, Andrey P, Pierron G, Kress M, et al. 2012. Multiple binding of repressed mRNAs by the P-body protein Rck/p54. *RNA* 18: 1702–1715.

* Fabian M. 2017. microRNAs. *Cold Spring Harb Perspect Biol* doi: 10.1101/cshperspect.a032771.

Friend K, Campbell ZT, Cooke A, Kroll-Conner P, Wickens MP, Kimble J. 2012. A conserved PUF–Ago–eEF1A complex attenuates translation elongation. *Nat Struct Mol Biol* 19: 176–183.

Fromm SA, Truffault V, Kamenz J, Braun JE, Hoffmann NA, Izaurralde E, Sprangers R. 2012. The structural basis of Edc3- and Scd6-mediated activation of the Dcp1:Dcp2 mRNA decapping complex. *EMBO J* 31: 279–290.

Garzia A, Jafarnejad SM, Meyer C, Chapat C, Gogakos T, Morozov P, Amiri M, Shapiro M, Molina H, Tuschl T, et al. 2017. The E3 ubiquitin ligase and RNA-binding protein ZNF598 orchestrates ribosome quality control of premature polyadenylated mRNAs. *Nat Commun* 8: 16056.

Gennarino V, Singh RK, White JJ, De Maio A, Han K, Kim JY, Jafar-Nejad P, di Ronza A, Kang H, Sayegh LS, et al. 2015. Pumilio1 Haploinsufficiency leads to SCA1-like neurodegeneration by increasing wild-type Ataxin1 levels. *Cell* 160: 1087–1098.

Goldstrohm AC, Hook BA, Seay DJ, Wickens M. 2006. PUF proteins bind Pop2p to regulate messenger RNAs. *Nat Struct Mol Biol* 13: 533–539.

Gopalsamy A, Narayanan A, Liu S, Parikh MD, Kyne RE, Fadeyi O, Tones MA, Cherry JJ, Nabhan JF, LaRosa G, et al. 2017. Design of potent mRNA decapping scavenger enzyme (DcpS) inhibitors with improved physicochemical properties to investigate the mechanism of therapeutic benefit in spinal muscular atrophy (SMA). *J Med Chem* 60: 3094–3108.

Graille M, Séraphin B. 2012. Surveillance pathways rescuing eukaryotic ribosomes lost in translation. *Nat Rev Mol Cell Biol* **13:** 727–735.

Grammatikakis I, Abdelmohsen K, Gorospe M. 2017. Posttranslational control of HuR function. *Wiley Interdiscip Rev RNA* **8:** e1372.

Grudzien-Nogalska E, Kiledjian M. 2017. New insights into decapping enzymes and selective mRNA decay. *Wiley Interdiscip Rev RNA* **8:** e1379.

Guydosh NR, Green R. 2017. Translation of poly(A) tails leads to precise mRNA cleavage. *RNA* **23:** 749–761.

Habacher C, Guo Y, Venz R, Kumari P, Neagu A, Gaidatzis D, Harvald E, Færgeman N, Gut H, Ciosk R. 2016. Ribonuclease-mediated control of body fat. *Dev Cell* **39:** 359–369.

Hafner M, Landthaler M, Burger L, Khorshid M, Hausser J, Berninger P, Rothballer A, Ascano M, Jungkamp A-C, Munschauer M, et al. 2010. Transcriptome-wide Identification of RNA-binding protein and microRNA target sites by PAR-CLIP. *Cell* **141:** 129–141.

Haimovich G, Medina DA, Causse SZ, Garber M, Millan-Zambrano G, Barkai O, Chavez S, Perez-Ortin JE, Darzacq X, Choder M. 2013. Gene expression is circular: Factors for mRNA degradation also foster mRNA synthesis. *Cell* **153:** 1000–1011.

Harigaya Y, Parker R. 2017. The link between adjacent codon pairs and mRNA stability. *BMC Genomics* **18:** 364.

Horikawa W, Endo K, Wada M, Ito K. 2016. Mutations in the G-domain of Ski7 cause specific dysfunction in nonstop decay. *Sci Rep* **6:** 29295.

Hu W, Sweet TJ, Chamnongpol S, Baker KE, Coller J. 2009. Co-translational mRNA decay in *Saccharomyces cerevisiae*. *Nature* **461:** 225–229.

Hu W, Petzold C, Coller J, Baker KE. 2010. Nonsense-mediated mRNA decapping occurs on polyribosomes in *Saccharomyces cerevisiae*. *Nat Struct Mol Biol* **17:** 244–247.

Huch S, Nissan T. 2014. Interrelations between translation and general mRNA degradation in yeast. *Wiley Interdiscip Rev RNA* **5:** 747–763.

Inada T, Aiba H. 2005. Translation of aberrant mRNAs lacking a termination codon or with a shortened 3′-UTR is repressed after initiation in yeast. *EMBO J* **24:** 1584–1595.

Ito-Harashima S, Kuroha K, Tatematsu T, Inada T. 2007. Translation of the poly(A) tail plays crucial roles in nonstop mRNA surveillance via translation repression and protein destabilization by proteasome in yeast. *Genes Dev* **21:** 519–524.

Iwasaki H, Takeuchi O, Teraguchi S, Matsushita K, Uehata T, Kuniyoshi K, Satoh T, Saitoh T, Matsushita M, Standley DM, et al. 2011. The IκB kinase complex regulates the stability of cytokine-encoding mRNA induced by TLR-IL-1R by controlling degradation of regnase-1. *Nat Immunol* **12:** 1167–1175.

Jamar NH, Kritsiligkou P, Grant CM. 2017. The nonstop decay mRNA surveillance pathway is required for oxidative stress tolerance. *Nucleic Acids Res* **45:** 6881–6893.

Jangra RK, Yi M, Lemon SM. 2010. DDX6 (Rck/p54) is required for efficient hepatitis C virus replication but not for internal ribosome entry site-directed translation. *J Virol* **84:** 6810–6824.

Jiao X, Doamekpor SK, Bird JG, Nickels BE, Tong L, Hart RP, Kiledjian M. 2017. 5′ end nicotinamide adenine dinucleotide cap in human cells promotes RNA decay through DXO-mediated deNADding. *Cell* **168:** 1015–1027.e10.

Jonas S, Christie M, Peter D, Bhandari D, Loh B, Huntzinger E, Weichenrieder O, Izaurralde E. 2014. An asymmetric PAN3 dimer recruits a single PAN2 exonuclease to mediate mRNA deadenylation and decay. *Nat Struct Mol Biol* **21:** 599–608.

Jones CI, Zabolotskaya MV, Newbury SF. 2012. The 5′→3′ exoribonuclease XRN1/Pacman and its functions in cellular processes and development. *Wiley Interdiscip Rev RNA* **3:** 455–468.

Jungfleisch J, Chowdhury A, Alves-Rodrigues I, Tharun S, Diez J. 2015. The Lsm1-7-Pat1 complex promotes viral RNA translation and replication by differential mechanisms. *RNA* **21:** 1469–1479.

Juszkiewicz S, Hegde RS. 2017. Initiation of quality control during poly(A) translation requires site-specific ribosome ubiquitination. *Mol Cell* **65:** 743–750.

Kamenska A, Simpson C, Vindry C, Broomhead H, Bénard M, Ernoult-Lange M, Lee BP, Harries LW, Weil D, Standart N. 2016. The DDX6-4E-T interaction mediates translational repression and P-body assembly. *Nucleic Acids Res* **44:** 6318–6334.

* Karousis ED, Mühlemann O. 2017. NMD begins where translation ends. *Cold Spring Harb Perspect Biol* doi: 10.1101/cshperspect.a032862.

Kaufman DR, Papillon J, Larose L, Iwawaki T, Cybulsky AV. 2017. Deletion of inositol-requiring enzyme-1α in podocytes disrupts glomerular capillary integrity and autophagy. *Mol Biol Cell* **28:** 1636–1651.

Khawaja A, Vopalensky V, Pospisek M. 2015. Understanding the potential of hepatitis C virus internal ribosome entry site domains to modulate translation initiation via their structure and function. *Wiley Interdiscip Rev RNA* **6:** 211–224.

Kim HH, Kuwano Y, Srikantan S, Lee EK, Martindale JL, Gorospe M. 2009. HuR recruits let-7/RISC to repress c-Myc expression. *Genes Dev* **23:** 1743–1748.

Kim Y, Noren Hooten N, Dluzen DF, Martindale JL, Gorospe M, Evans MK. 2015. Posttranscriptional regulation of the inflammatory marker C-reactive protein by the RNA-binding protein HuR and microRNA 637. *Mol Cell Biol* **35:** 4212–4221.

Klauer AA, van Hoof A. 2012. Degradation of mRNAs that lack a stop codon: A decade of nonstop progress. *Wiley Interdiscip Rev RNA* **3:** 649–660.

Kobayashi K, Ishitani R, Nureki O. 2013. Recent structural studies on Dom34/aPelota and Hbs1/aEF1α: Important factors for solving general problems of ribosomal stall in translation. *Biophysics (Oxf)* **9:** 131–140.

Koeller DM, Horowitz JA, Casey JL, Klausner RD, Harford JB. 1991. Translation and the stability of mRNAs encoding the transferrin receptor and c-fos. *Proc Natl Acad Sci* **88:** 7778–7782.

Kojima S, Gendreau KL, Sher-Chen EL, Gao P, Green CB. 2015. Changes in poly(A) tail length dynamics from the loss of the circadian deadenylase Nocturnin. *Sci Rep* **5:** 17059.

Kowalinski E, Schuller A, Green R, Conti E. 2015. *Saccharomyces cerevisiae* Ski7 is a GTP-binding protein adopting the characteristic conformation of active translational GTPases. *Structure* **23:** 1336–1343.

Kowalinski E, Kögel A, Ebert J, Reichelt P, Stegmann E, Habermann B, Conti E. 2016. Structure of a cytoplasmic 11-subunit RNA exosome complex. *Mol Cell* **63:** 125–134.

Kullmann M, Göpfert U, Siewe B, Hengst L. 2002. ELAV/Hu proteins inhibit p27 translation via an IRES element in the p27 5′UTR. *Genes Dev* **16:** 3087–3099.

Kuroha K, Akamatsu M, Dimitrova L, Ito T, Kato Y, Shirahige K, Inada T. 2010. Receptor for activated C kinase 1 stimulates nascent polypeptide-dependent translation arrest. *EMBO Rep* **11:** 956–961.

Kuzuoğlu-Öztürk D, Bhandari D, Huntzinger E, Fauser M, Helms S, Izaurralde E. 2016. miRISC and the CCR4/NOT complex silence mRNA targets independently of 43S ribosomal scanning. *EMBO J* **35:** 1186–1203.

Łabno A, Warkocki Z, Kuliński T, Krawczyk PS, Bijata K, Tomecki R, Dziembowski A. 2016. Perlman syndrome nuclease DIS3L2 controls cytoplasmic noncoding RNAs and provides surveillance pathway for maturing snRNAs. *Nucleic Acids Res* **44:** 10437–10453.

LaGrandeur T, Parker R. 1999. The cis acting sequences responsible for the differential decay of the unstable MFA2 and stable PGK1 transcripts in yeast include the context of the translational start codon. *RNA* **5:** 420–433.

Laird-Offringa IA, de Wit CL, Elfferich P, van der Eb AJ. 1990. Poly(A) tail shortening is the translation-dependent step in c-*myc* mRNA degradation. *Mol Cell Biol* **10:** 6132–6140.

Lal A, Mazan-Mamczarz K, Kawai T, Yang X, Martindale JL, Gorospe M. 2004. Concurrent versus individual binding of HuR and AUF1 to common labile target mRNAs. *EMBO J* **23:** 3092–3102.

Lee HC, Choe J, Chi SG, Kim YK. 2009. Exon junction complex enhances translation of spliced mRNAs at multiple steps. *Biochem Biophys Res Commun* **384:** 334–340.

Li X, Ohmori T, Irie K, Kimura Y, Suda Y, Mizuno T, Irie K. 2016. Different regulations of ROM2 and LRG1 expression by Ccr4, Pop2, and Dhh1 in the *Saccharomyces cerevisiae* cell wall integrity pathway. *mSphere* **1:** e00250–16.

Lin JY, Brewer G, Li ML. 2015. HuR and Ago2 bind the internal ribosome entry site of enterovirus 71 and promote virus translation and replication. *PLoS ONE* **10:** e0140291.

Linial M, Gunderson N, Groudine M. 1985. Enhanced transcription of c-myc in bursal lymphoma cells requires continuous protein synthesis. *Science* **230:** 1126–1132.

Liu SW, Rajagopal V, Patel SS, Kiledjian M. 2008. Mechanistic and kinetic analysis of the DcpS scavenger decapping enzyme. *J Biol Chem* **283:** 16427–16436.

Liu L, Ouyang M, Rao JN, Zou T, Xiao L, Chung HK, Wu J, Donahue JM, Gorospe M, Wang JY. 2015. Competition between RNA-binding proteins CELF1 and HuR modulates MYC translation and intestinal epithelium renewal. *Mol Biol Cell* **26:** 1797–1810.

Lu J, Deutsch C. 2008. Electrostatics in the ribosomal tunnel modulate chain elongation rates. *J Mol Biol* **384:** 73–86.

Mak W, Fang C, Holden T, Dratver MB, Lin H. 2016. An important role of Pumilio 1 in regulating the development of the mammalian female germline. *Biol Reprod* **94:** 134.

Maryati M, Airhihen B, Winkler GS. 2015. The enzyme activities of Caf1 and Ccr4 are both required for deadenylation by the human Ccr4-Not nuclease module. *Biochem J* **469:** 169–176.

Mason PJ, Bessler M. 2015. mRNA deadenylation and telomere disease. *J Clin Invest* **125:** 1796–1798.

Mathys H, Basquin J, Ozgur S, Czarnocki-Cieciura M, Bonneau F, Aartse A, Dziembowski A, Nowotny M, Conti E, Filipowicz W. 2014. Structural and biochemical insights to the role of the CCR4-NOT complex and DDX6 ATPase in microRNA repression. *Mol Cell* **54:** 751–765.

Mauer J, Luo X, Blanjoie A, Jiao X, Grozhik AV, Patil DP, Linder B, Pickering BF, Vasseur JJ, Chen Q, et al. 2017. Reversible methylation of m6Am in the 5′ cap controls mRNA stability. *Nature* **541:** 371–375.

Mayr C. 2016. Evolution and biological roles of alternative 3′UTRs. *Trends Cell Biol* **26:** 227–237.

Meng Z, King PH, Nabors LB, Jackson NL, Chen CY, Emanuel PD, Blume SW. 2005. The ELAV RNA-stability factor HuR binds the 5′-untranslated region of the human IGF-IR transcript and differentially represses cap-dependent and IRES-mediated translation. *Nucleic Acids Res* **33:** 2962–2979.

Merret R, Nagarajan VK, Carpentier MC, Park S, Favory JJ, Descombin J, Picart C, Charng YY, Green PJ, Deragon JM, et al. 2015. Heat-induced ribosome pausing triggers mRNA co-translational decay in *Arabidopsis thaliana*. *Nucleic Acids Res* **43:** 4121–4132.

Mishima Y, Tomari Y. 2016. Codon usage and 3′ UTR length determine maternal mRNA stability in Zebrafish. *Mol Cell* **61:** 874–885.

Moon SL, Anderson JR, Kumagai Y, Wilusz CJ, Akira S, Khromykh AA, Wilusz J. 2012. A noncoding RNA produced by arthropod-borne flaviviruses inhibits the cellular exoribonuclease XRN1 and alters host mRNA stability. *RNA* **18:** 2029–2040.

Moon SL, Blackinton JG, Anderson JR, Dozier MK, Dodd BJT, Keene JD, Wilusz CJ, Bradrick SS, Wilusz J. 2015. XRN1 stalling in the 5′ UTR of hepatitis C virus and bovine viral diarrhea virus is associated with dysregulated host mRNA stability. *PLOS Pathog* **11:** e1004708.

Moore K, Hollien J. 2015. Ire1-mediated decay in mammalian cells relies on mRNA sequence, structure, and translational status. *Mol Biol Cell* **26:** 2873–2884.

Morris AR, Mukherjee N, Keene JD. 2008. Ribonomic analysis of human Pum1 reveals cis-trans conservation across species despite evolution of diverse mRNA target sets. *Mol Cell Biol* **28:** 4093–4103.

Mugler CF, Hondele M, Heinrich S, Sachdev R, Vallotton P, Koek AY, Chan LY, Weis K. 2016. ATPase activity of the DEAD-box protein Dhh1 controls processing body formation. *eLife* **5:** e18746.

Naudin C, Hattabi A, Michelet F, Miri-Nezhad A, Benyoucef A, Pflumio F, Guillonneau F, Fichelson S, Vigon I, Dusanter-Fourt I, et al. 2017. PUMILIO/FOXP1 signaling drives expansion of hematopoietic stem/progenitor and leukemia cells. *Blood* **129:** 2493–2506.

Cite this article as *Cold Spring Harb Perspect Biol* doi: 10.1101/cshperspect.a032839

Neymotin B, Ettore V, Gresham D. 2016. Multiple transcript properties related to translation affect mRNA degradation rates in *Saccharomyces cerevisiae*. *G3* doi: 10.1534/g3.116.032276.

Ng CKL, Shboul M, Taverniti V, Bonnard C, Lee H, Eskin A, Nelson SF, Al-Raqad M, Altawalbeh S, Séraphin B, et al. 2015. Loss of the scavenger mRNA decapping enzyme DCPS causes syndromic intellectual disability with neuromuscular defects. *Hum Mol Genet* **24:** 3163–3171.

Nicklas S, Okawa S, Hillje A-L, González-Cano L, Del Sol A, Schwamborn JC. 2015. The RNA helicase DDX6 regulates cell-fate specification in neural stem cells via miRNAs. *Nucleic Acids Res* **43:** 2638–2654.

Niedzwiecka A, Nilsson P, Worch R, Stepinski J, Darzynkiewicz E, Virtanen A. 2016. Molecular recognition of mRNA 5′ cap by 3′ poly(A)-specific ribonuclease (PARN) differs from interactions known for other cap-binding proteins. *Biochim Biophys Acta* **1864:** 331–345.

Niinuma S, Fukaya T, Tomari Y. 2016. CCR4 and CAF1 deadenylases have an intrinsic activity to remove the post-poly(A) sequence. *RNA* **22:** 1550–1559.

Nishimura T, Padamsi Z, Fakim H, Milette S, Dunham WH, Gingras AC, Fabian MR. 2015. The eIF4E-binding protein 4E-T is a component of the mRNA decay machinery that bridges the 5′ and 3′ termini of target mRNAs. *Cell Rep* **11:** 1425–1436.

Onouchi H, Nagami Y, Haraguchi Y, Nakamoto M, Nishimura Y, Sakurai R, Nagao N, Kawasaki D, Kadokura Y, Naito S. 2005. Nascent peptide-mediated translation elongation arrest coupled with mRNA degradation in the CGS1 gene of *Arabidopsis*. *Genes Dev* **19:** 1799–1810.

Orban TI, Izaurralde E. 2005. Decay of mRNAs targeted by RISC requires XRN1, the Ski complex, and the exosome. *RNA* **11:** 459–469.

Ørom UA, Nielsen FC, Lund AH. 2008. MicroRNA-10a binds the 5′UTR of ribosomal protein mRNAs and enhances their translation. *Mol Cell* **30:** 460–471.

Ostareck DH, Naarmann-de Vries IS, Ostareck-Lederer A. 2014. DDX6 and its orthologs as modulators of cellular and viral RNA expression. *Wiley Interdiscip Rev RNA* **5:** 659–678.

Ozgur S, Basquin J, Kamenska A, Filipowicz W, Standart N, Conti E. 2015. Structure of a human 4E-T/DDX6/CNOT1 complex reveals the different interplay of DDX6-binding proteins with the CCR4-NOT complex. *Cell Rep* **13:** 703–711.

Park JH, Shin C. 2014. MicroRNA-directed cleavage of targets: Mechanism and experimental approaches. *BMB Rep* **47:** 417–423.

Park OH, Park J, Yu M, An HT, Ko J, Kim YK. 2016. Identification and molecular characterization of cellular factors required for glucocorticoid receptor-mediated mRNA decay. *Genes Dev* **30:** 2093–2105.

Pashler AL, Towler BP, Jones CI, Newbury SF. 2016. The roles of the exoribonucleases DIS3L2 and XRN1 in human disease. *Biochem Soc Trans* **44:** 1377–1384.

Passos DO, Doma MK, Shoemaker CJ, Muhlrad D, Green R, Weissman J, Hollien J, Parker R. 2009. Analysis of Dom34 and its function in no-go decay. *Mol Biol Cell* **20:** 3025–3032.

Pelechano V, Wei W, Steinmetz LM. 2015. Widespread cotranslational RNA decay reveals ribosome dynamics. *Cell* **161:** 1400–1412.

Pelechano V, Wei W, Steinmetz LM. 2016. Genome-wide quantification of 5′-phosphorylated mRNA degradation intermediates for analysis of ribosome dynamics. *Nat Protoc* **11:** 359–376.

Pereira FJC, Teixeira A, Kong J, Barbosa C, Silva AL, Marques-Ramos A, Liebhaber SA, Romão L. 2015. Resistance of mRNAs with AUG-proximal nonsense mutations to nonsense-mediated decay reflects variables of mRNA structure and translational activity. *Nucleic Acids Res* **43:** 6528–6544.

Pirouz M, Du P, Munafò M, Gregory R. 2016. Dis3l2-mediated decay is a quality control pathway for noncoding RNAs. *Cell Rep* **16:** 1861–1873.

Pisareva VP, Skabkin MA, Hellen CU, Pestova TV, Pisarev AV. 2011. Dissociation by Pelota, Hbs1 and ABCE1 of mammalian vacant 80S ribosomes and stalled elongation complexes. *EMBO J* **30:** 1804–1817.

Presnyak V, Alhusaini N, Chen YH, Martin S, Morris N, Kline N, Olson S, Weinberg D, Baker KE, Graveley BR, et al. 2015. Codon optimality is a major determinant of mRNA stability. *Cell* **160:** 1111–1124.

Radhakrishnan A, Green R. 2016. Connections underlying translation and mRNA stability. *J Mol Biol* **428:** 3558–3564.

Radhakrishnan A, Chen YH, Martin S, Alhusaini N, Green R, Coller J. 2016. The DEAD-Box protein Dhh1p couples mRNA decay and translation by monitoring codon optimality. *Cell* **167:** 122–132.e9.

Rahmsdorf HJ, Schönthal A, Angel P, Litfin M, Rüther U, Herrlich P. 1987. Posttranscriptional regulation of c-fos mRNA expression. *Nucleic Acids Res* **15:** 1643–1659.

Rath S, Donovan J, Whitney G, Chitrakar A, Wang W, Korennykh A. 2015. Human RNase L tunes gene expression by selectively destabilizing the microRNA-regulated transcriptome. *Proc Natl Acad Sci* **112:** 15916–15921.

Rissland OS, Norbury CJ. 2009. Decapping is preceded by 3′ uridylation in a novel pathway of bulk mRNA turnover. *Nat Struct Mol Biol* **16:** 616–623.

Roberts APE, Lewis AP, Jopling CL. 2011. miR-122 activates hepatitis C virus translation by a specialized mechanism requiring particular RNA components. *Nucleic Acids Res* **39:** 7716–7729.

Rouya C, Siddiqui N, Morita M, Duchaine TF, Fabian MR, Sonenberg N. 2014. Human DDX6 effects miRNA-mediated gene silencing via direct binding to CNOT1. *RNA* **20:** 1398–1409.

Russo J, Olivas WM. 2015. Conditional regulation of Puf1p, Puf4p, and Puf5p activity alters YHB1 mRNA stability for a rapid response to toxic nitric oxide stress in yeast. *Mol Biol Cell* **26:** 1015–1029.

Ryseck RP, Hirai SI, Yaniv M, Bravo R. 1988. Transcriptional activation of c-jun during the G_0/G_1 transition in mouse fibroblasts. *Nature* **334:** 535–537.

Saito S, Hosoda N, Hoshino S. 2013. The Hbs1-Dom34 protein complex functions in Non-stop mRNA decay in mammalian cells. *J Biol Chem* **288:** 17832–17843.

Schäfer IB, Rode M, Bonneau F, Schüssler S, Conti E. 2014. The structure of the Pan2–Pan3 core complex reveals cross-talk between deadenylase and pseudokinase. *Nat Struct Mol Biol* 21: 591–598.

Scheel TKH, Luna JM, Liniger M, Nishiuchi E, Rozen-Gagnon K, Shlomai A, Auray G, Gerber M, Fak J, Keller I, et al. 2016. A broad RNA virus survey reveals both miRNA dependence and functional sequestration. *Cell Host Microbe* 19: 409–423.

Schwamborn JC, Berezikov E, Knoblich JA. 2009. The TRIM-NHL protein TRIM32 activates microRNAs and prevents self-renewal in mouse neural progenitors. *Cell* 136: 913–925.

Serman A, Le Roy F, Aigueperse C, Kress M, Dautry F, Weil D. 2007. GW body disassembly triggered by siRNAs independently of their silencing activity. *Nucleic Acids Res* 35: 4715–4727.

Shan B, Wang X, Wu Y, Xu C, Xia Z, Dai J, Shao M, Zhao F, He S, Yang L, et al. 2017. The metabolic ER stress sensor IRE1a suppresses alternative activation of macrophages and impairs energy expenditure in obesity. *Nat Immunol* 18: 519–529.

Sharif H, Ozgur S, Sharma K, Basquin C, Urlaub H, Conti E. 2013. Structural analysis of the yeast Dhh1/Pat1 complex reveals how Dhh1 engages Pat1, Edc3 and RNA in mutually exclusive interactions. *Nucleic Acids Res* 41: 8377–8390.

Sharifnia P, Jin Y. 2014. Regulatory roles of RNA binding proteins in the nervous system of *C. elegans*. *Front Mol Neurosci* 7: 100.

Shimakami T, Yamane D, Jangra RK, Kempf BJ, Spaniel C, Barton DJ, Lemon SM. 2012. Stabilization of hepatitis C virus RNA by an Ago2-miR-122 complex. *Proc Natl Acad Sci* 109: 941–946.

Shoemaker CJ, Eyler DE, Green R. 2010. Dom34:Hbs1 promotes subunit dissociation and peptidyl-tRNA drop-off to initiate no-go decay. *Science* 330: 369–372.

Simms CL, Thomas EN, Zaher HS. 2017. Ribosome-based quality control of mRNA and nascent peptides. *Wiley Interdiscip Rev RNA* 8: e1366.

Skeparnias I, Anastasakis D, Shaukat AN, Grafanaki K, Stathopoulos C. 2017. Expanding the repertoire of deadenylases. *RNA Biol* 1–6.

Song MG, Kiledjian M. 2007. 3′ Terminal oligo U-tract-mediated stimulation of decapping. *RNA* 13: 2356–2365.

Stimac E, Groppi VE, Coffino P. 1983. Increased histone mRNA levels during inhibition of protein synthesis. *Biochem Biophys Res Commun* 114: 131–137.

Stowell JAW, Webster MW, Kögel A, Wolf J, Shelley KL, Passmore LA. 2016. Reconstitution of targeted deadenylation by the Ccr4-Not complex and the YTH domain protein Mmi1. *Cell Rep* 17: 1978–1989.

Sun M, Schwalb B, Pirkl N, Maier KC, Schenk A, Failmezger H, Tresch A, Cramer P. 2013. Global analysis of eukaryotic mrna degradation reveals Xrn1-dependent buffering of transcript levels. *Mol Cell* 52: 52–62.

Sun S, Zhang X, Lyu L, Li X, Yao S, Zhang J. 2016. Autotaxin expression is regulated at the post-transcriptional level by the RNA-binding proteins HuR and AUF1. *J Biol Chem* 291: 25823–25836.

Sweet T, Kovalak C, Coller J, Kiktev D, Inge-Vechtomov S. 2012. The DEAD-Box protein Dhh1 promotes decapping by slowing ribosome movement. *PLoS Biol* 10: e1001342.

Tavernier SJ, Osorio F, Vandersarren L, Vetters J, Vanlangenakker N, Van Isterdael G, Vergote K, De Rycke R, Parthoens E, van de Laar L, et al. 2017. Regulated IRE1-dependent mRNA decay sets the threshold for dendritic cell survival. *Nat Cell Biol* 19: 698–710.

Taverniti V, Seraphin B. 2015. Elimination of cap structures generated by mRNA decay involves the new scavenger mRNA decapping enzyme Aph1/FHIT together with DcpS. *Nucleic Acids Res* 43: 482–492.

Thomas MP, Liu X, Whangbo J, McCrossan G, Sanborn KB, Basar E, Walch M, Lieberman J. 2015. Apoptosis triggers specific, rapid, and global mRNA decay with 3′ uridylated intermediates degraded by DIS3L2. *Cell Rep* 11: 1079–1089.

Thomas SP, Kim E, Kim J-S, Raines RT. 2016. Knockout of the ribonuclease inhibitor gene leaves human cells vulnerable to secretory ribonucleases. *Biochemistry* 55: 6359–6362.

Thompson CB, Challoner PB, Neiman PE, Groudine M. 1986. Expression of the c-*myb* proto-oncogene during cellular proliferation. *Nature* 319: 374–380.

Tian B, Manley JL. 2017. Alternative polyadenylation of mRNA precursors. *Nat Rev Mol Cell Biol* 18: 18–30.

Tsuboi T, Kuroha K, Kudo K, Makino S, Inoue E, Kashima I, Inada T. 2012. Dom34:Hbs1 plays a general role in quality-control systems by dissociation of a stalled ribosome at the 3′ end of aberrant mRNA. *Mol Cell* 46: 518–529.

Turner JD, Vernocchi S, Schmitz S, Muller CP. 2014. Role of the 5′-untranslated regions in post-transcriptional regulation of the human glucocorticoid receptor. *Biochim Biophys Acta* 1839: 1051–1061.

Uehata T, Takeuchi O. 2017. Regnase-1 is an endoribonuclease essential for the maintenance of immune homeostasis. *J Interf Cytokine Res* 37: 220–229.

Ustianenko D, Pasulka J, Feketova Z, Bednarik L, Zigackova D, Fortova A, Zavolan M, Vanacova S. 2016. TUT-DIS3L2 is a mammalian surveillance pathway for aberrant structured non-coding RNAs. *EMBO J* 35: 2179–2191.

Valkov E, Jonas S, Weichenrieder O. 2017. *Mille viae* in eukaryotic mRNA decapping. *Curr Opin Struct Biol* 47: 40–51.

van Hoof A. 2005. Conserved functions of yeast genes support the duplication, degeneration and complementation model for gene duplication. *Genetics* 171: 1455–1461.

van Hoof A, Frischmeyer PA, Dietz HC, Parker R. 2002. Exosome-mediated recognition and degradation of mRNAs lacking a termination codon. *Science* 295: 2262–2264.

Viegas SC, Silva IJ, Apura P, Matos RG, Arraiano CM. 2015. Surprises in the 3′-end: "U" can decide too! *FEBS J* 282: 3489–3499.

Virtanen A, Henriksson N, Nilsson P, Nissbeck M. 2013. Poly(A)-specific ribonuclease (PARN): An allosterically regulated, processive and mRNA cap-interacting deadenylase. *Crit Rev Biochem Mol Biol* 48: 192–209.

Wahle E, Winkler GS. 2013. RNA decay machines: Deadenylation by the Ccr4/Not and Pan2/Pan3 complexes. *Biochim Biophys Acta - Gene Regul Mech* 1829: 561–570.

Wang Y, Arribas-Layton M, Chen Y, Lykke-Andersen J, Sen GLL. 2015. DDX6 orchestrates mammalian progenitor function through the mRNA degradation and translation pathways. *Mol Cell* **60:** 118–130.

Weidmann CA, Raynard NA, Blewett NH, Van Etten J, Goldstrohm AC. 2014. The RNA binding domain of Pumilio antagonizes poly-adenosine binding protein and accelerates deadenylation. *RNA* **20:** 1298–1319.

Weingarten-Gabbay S, Elias-Kirma S, Nir R, Gritsenko AA, Stern-Ginossar N, Yakhini Z, Weinberger A, Segal E. 2016. Systematic discovery of cap-independent translation sequences in human and viral genomes. *Science* **351:** aad4939.

White EJF, Matsangos AE, Wilson GM. 2017. AUF1 regulation of coding and noncoding RNA. *Wiley Interdiscip Rev RNA* **8:** e1393.

Wilson JA, Zhang C, Huys A, Richardson CD. 2011. Human Ago2 is required for efficient microRNA 122 regulation of hepatitis C virus RNA accumulation and translation. *J Virol* **85:** 2342–2350.

Wilusz CJ, Wilusz J. 2013. Lsm proteins and Hfq. *RNA Biol* **10:** 592–601.

Wolf J, Passmore LA. 2014. mRNA deadenylation by Pan2-Pan3. *Biochem Soc Trans* **42:** 184–187.

Wolf J, Valkov E, Allen MD, Meineke B, Gordiyenko Y, McLaughlin SH, Olsen TM, Robinson CV, Bycroft M, Stewart M, et al. 2014. Structural basis for Pan3 binding to Pan2 and its function in mRNA recruitment and deadenylation. *EMBO J* **33:** 1514–1526.

Woodward LA, Mabin JW, Gangras P, Singh G. 2017. The exon junction complex: A lifelong guardian of mRNA fate. *Wiley Interdiscip Rev RNA* **8:** e1411.

Wurm JP, Holdermann I, Overbeck JH, Mayer PHO, Sprangers R. 2017. Changes in conformational equilibria regulate the activity of the Dcp2 decapping enzyme. *Proc Natl Acad Sci* **114:** 6034–6039.

Xi R, Doan C, Liu D, Xie T. 2005. Pelota controls self-renewal of germline stem cells by repressing a Bam-independent differentiation pathway. *Development* **132:** 5365–5374.

Yamaji M, Jishage M, Meyer C, Suryawanshi H, Der E, Yamaji M, Garzia A, Morozov P, Manickavel S, McFarland HL, et al. 2017. DND1 maintains germline stem cells via recruitment of the CCR4–NOT complex to target mRNAs. *Nature* **543:** 568–572.

Yeh CH, Hung LY, Hsu C, Le SY, Lee PT, Liao WL, Lin YT, Chang WC, Tseng JT. 2008. RNA-binding protein HuR interacts with thrombomodulin 5′ untranslated region and represses internal ribosome entry site-mediated translation under IL-1β treatment. *Mol Biol Cell* **19:** 3812–3822.

Yoshinaga M, Nakatsuka Y, Vandenbon A, Ori D, Uehata T, Tsujimura T, Suzuki Y, Mino T, Takeuchi O. 2017. Regnase-1 maintains iron homeostasis via the degradation of transferrin receptor 1 and prolyl-hydroxylase-domain-containing protein 3 mRNAs. *Cell Rep* **19:** 1614–1630.

Yu TX, Rao JN, Zou T, Liu L, Xiao L, Ouyang M, Cao S, Gorospe M, Wang JY. 2013. Competitive binding of CUGBP1 and HuR to occludin mRNA controls its translation and modulates epithelial barrier function. *Mol Biol Cell* **24:** 85–99.

Yu CH, Dang Y, Zhou Z, Wu C, Zhao F, Sachs MS, Liu Y. 2015. Codon usage influences the local rate of translation elongation to regulate cotranslational protein folding. *Mol Cell* **59:** 744–754.

Zamore PD, Williamson JR, Lehmann R. 1997. The Pumilio protein binds RNA through a conserved domain that defines a new class of RNA-binding proteins. *RNA* **3:** 1421–1433.

Zhang C, Huys A, Thibault PA, Wilson JA. 2012. Requirements for human Dicer and TRBP in microRNA-122 regulation of HCV translation and RNA abundance. *Virology* **433:** 479–488.

Zhang Z, Huang A, Zhang A, Zhou C. 2017. HuR promotes breast cancer cell proliferation and survival via binding to CDK3 mRNA. *Biomed Pharmacother* **91:** 788–795.

Zhao F, Yu C, Liu Y. 2017. Codon usage regulates protein structure and function by affecting translation elongation speed in *Drosophila* cells. *Nucleic Acids Res* doi: 10.1093/nar/gkx501.

Zhou M, Bail S, Plasterer HL, Rusche J, Kiledjian M. 2015. DcpS is a transcript-specific modulator of RNA in mammalian cells. *RNA* **21:** 1306–1312.

Zhuang R, Rao JN, Zou T, Liu L, Xiao L, Cao S, Hansraj NZ, Gorospe M, Wang JY. 2013. miR-195 competes with HuR to modulate stim1 mRNA stability and regulate cell migration. *Nucleic Acids Res* **41:** 7905–7919.

Ziemniak M, Mugridge JS, Kowalska J, Rhoads RE, Gross JD, Jemielity J. 2016. Two-headed tetraphosphate cap analogs are inhibitors of the Dcp1/2 RNA decapping complex. *RNA* **22:** 518–529.

Zinder JC, Lima CD. 2017. Targeting RNA for processing or destruction by the eukaryotic RNA exosome and its cofactors. *Genes Dev* **31:** 88–100.

Zinder JC, Wasmuth E V, Lima CD. 2016. Nuclear RNA exosome at 3.1 Å reveals substrate specificities, RNA paths, and allosteric inhibition of Rrp44/Dis3. *Mol Cell* **64:** 734–745.

Nonsense-Mediated mRNA Decay Begins Where Translation Ends

Evangelos D. Karousis and Oliver Mühlemann

Department of Chemistry and Biochemistry, University of Bern, CH-3012 Bern, Switzerland

Correspondence: oliver.muehlemann@dcb.unibe.ch

Nonsense-mediated mRNA decay (NMD) is arguably the best-studied eukaryotic messenger RNA (mRNA) surveillance pathway, yet fundamental questions concerning the molecular mechanism of target RNA selection remain unsolved. Besides degrading defective mRNAs harboring premature termination codons (PTCs), NMD also targets many mRNAs encoding functional full-length proteins. Thus, NMD impacts on a cell's transcriptome and is implicated in a range of biological processes that affect a broad spectrum of cellular homeostasis. Here, we focus on the steps involved in the recognition of NMD targets and the activation of NMD. We summarize the accumulating evidence that tightly links NMD to translation termination and we further discuss the recruitment and activation of the mRNA degradation machinery and the regulation of this complex series of events. Finally, we review emerging ideas concerning the mechanistic details of NMD activation and the potential role of NMD as a general surveyor of translation efficacy.

Originally conceived as a quality control pathway that recognizes and specifically degrades aberrant messenger RNAs (mRNAs) with premature termination codons (PTCs), it has meanwhile become clear that nonsense-mediated mRNA decay (NMD) contributes to posttranscriptional gene regulation in a way that goes far beyond quality control. Exploring in which biological contexts NMD-mediated gene regulation plays an important role is a relatively new but rapidly expanding area of research that has been covered in recent reviews (Nasif et al. 2017; Nickless et al. 2017). Here, we summarize our current understanding regarding the molecular mechanism of NMD, with the focus on data obtained from mammalian systems. Despite more than 25 years of research

and a wealth of biochemical data characterizing interactions between different NMD factors, their enzymatic functions and posttranslational modifications, the mechanism and criteria for selection of an mRNA for the NMD pathway are still not well understood. The slow progress in deciphering the mechanism of NMD can at least partially be attributed to the lack of a suitable in vitro system that faithfully recapitulates the key steps of NMD. Nevertheless, work from many laboratories during the last few years has provided compelling evidence that NMD is tightly coupled to the process of translation termination. During translation termination, it is decided whether the translated mRNA shall remain intact and serve as a template for additional rounds of translation or

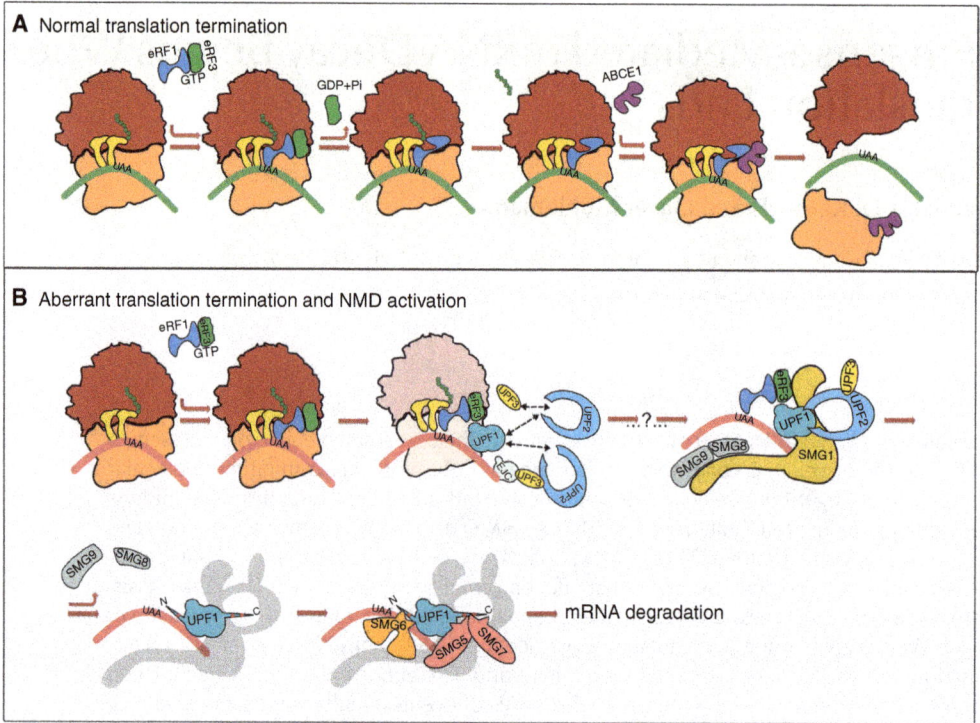

Figure 1. Schematic illustration of the sequential events taking place in normal translation termination and aberrant translation termination resulting in the activation of nonsense-mediated mRNA decay (NMD). (*A*) The presence of a termination codon (TC) in the A site of the ribosome provides the signal for translation termination that is marked by recognition of the TC by eukaryotic release factor (eRF)1 and eRF3, a conformational change of eRF1 that facilitates hydrolysis of the terminal peptidyl-transfer RNA (tRNA) bond and the release of the nascent peptide. Subsequently, release of the ribosome from the messenger RNA (mRNA) is facilitated by ABCE1. (*B*) When translation termination occurs in an unfavorable environment, UPF1 is activated by UPF2 and/or UPF3 that are free in the cytoplasm or positioned nearby on an exon junction complex (EJC). UPF1 activation involves its phosphorylation by SMG1, whose kinase activity is controlled by SMG8 and SMG9. The fate of the stalled ribosomes in the context of NMD activation remains unclear. Phosphorylated UPF1 (indicated by red dots) recruits the NMD-specific SMG6 endonuclease, which cleaves the targeted mRNA nearby the TC and the SMG5/SMG7 heterodimer that stimulates the exonucleolytic degradation of the mRNA by recruiting the CCR4/NOT complex. For clarity, a part of the NMD-activating complex (SMG1, UPF2, and UPF3) is depicted in gray to highlight the factors triggering mRNA degradation in the *lower* part of the panel.

whether it shall be degraded by the NMD pathway (He and Jacobson 2015). In a nutshell, the current view is that NMD ensues when ribosomes at nonsense codons (hereafter called termination codon [TC]) fail to terminate correctly. Because of the tight link between NMD and translation termination, we begin this review with a brief overview of eukaryotic translation termination. For a more detailed review of the mechanism of translation termination, see Hellen (2018).

THE MECHANISM OF EUKARYOTIC TRANSLATION TERMINATION AND RIBOSOME RECYCLING

Translation termination is signaled by the presence of one of the three TCs in the A site of the ribosome. Canonical translation termination is marked by three key events: (1) proper recognition of the termination signal, (2) hydrolysis of the terminal peptidyl-tRNA bond and release of the nascent peptide, and (3) dissociation of the ribosome into its 60S and 40S subunits (Fig. 1A)

Cite this article as *Cold Spring Harb Perspect Biol* doi: 10.1101/cshperspect.a032862

(Dever and Green 2012; Simms et al. 2017). In comparison to the initiation and elongation steps, the step of translation termination is less well studied and accordingly much less is known regarding its molecular mechanism. Instead of cognate aminoacylated (aa) transfer RNAs (tRNAs) that get recruited to the A site of the ribosome during elongation, eukaryotic release factor 1 (eRF)1 binds the A site when it harbors one of the three TCs. On terminating ribosomes, eRF1 is found as a ternary complex with the GTPase eRF3 and GTP. After its recruitment to the A site, GTP hydrolysis by eRF3 stimulates a large conformational change in eRF1 that enhances polypeptide release by engaging the active site of the ribosome. Despite the extended conformational change of the middle and carboxy-terminal parts of eRF1, the amino-terminal part of the protein interacts stably with the TC throughout the process (Alkalaeva et al. 2006; Becker et al. 2012; Eyler et al. 2013; Brown et al. 2015; Shao et al. 2016).

Following GTP hydrolysis, eRF3 dissociates from the termination complex, allowing for the subsequent interaction of eRF1 with the ABC-type ATPase ABCE1 (Rli1 in yeast), a factor that stimulates the recycling of the ribosome by splitting the ribosomal subunits. ABCE1 contains two nucleotide-binding domains (NBDs) and a unique amino-terminal FeS cluster domain aligned by two diamagnetic $[4Fe-4S]^{2+}$ clusters. ATP hydrolysis by ABCE1 causes extended conformational changes that provide the mechanical force leading to the dissociation of the 60S from the 40S ribosomal subunit (Pisarev et al. 2010; Becker et al. 2012). Structural studies showed that an initial closure of the NBDs positions the FeS cluster domain toward eRF1, exerting an immediate force that destabilizes the intersubunit interactions (Heuer et al. 2017). Cross-linking and mass spectrometry approaches showed that after ribosomal splitting, ABCE1 remains bound to the translational GTPase-binding site of the small ribosomal subunit, establishing major contacts with the S24e ribosomal protein mainly through NBD1 and FeS cluster domains. Structural data suggest that ABCE1 performs a tweezer-like movement that positions the FeS cluster domain in a cleft

between S12 and rRNA on the small subunit (Kiosze-Becker et al. 2016). Additionally, ABCE1 associates with the 43S preinitiation complex (consisting of eIF1A, eIF2, eIF3, and initiator tRNA) but its role in translation initiation remains to be elucidated (Heuer et al. 2017). Biochemical data suggest that ABCE1 also accelerates the rate of the peptide release by eRF1 in an ATP-independent manner, providing a functional link between translation termination and ribosome recycling (Shoemaker and Green 2011).

Most of the information concerning translation termination originates from structural studies and in vitro assays. The isolation of translation termination intermediates can be achieved by using purified components in the presence of nucleotide analogs that trap specific steps (Shao et al. 2016) or mutated release factors (Alkalaeva et al. 2006; Brown et al. 2015). The biochemical dissection of translation termination was facilitated by the establishment of a reconstituted system with mammalian components containing purified ribosomal subunits, translation factors, and aminoacylated tRNAs. This in vitro translation system made possible the stepwise investigation of the configuration of termination complexes by using the toeprinting technique, which allows the identification of ribosomal positions on a reporter mRNA at single-nucleotide resolution (Alkalaeva et al. 2006). However, the establishment of similar approaches by using mammalian cell lysates, as well as studies of translation termination in living cells, remains a challenge in the field. The development of techniques that allow studies in living cells such as ribosome profiling and live cell imaging already provides valuable insight into the dynamics of translation termination (Wang et al. 2016; Shirokikh et al. 2017).

TRANSLATION-COUPLED RNA DEGRADATION PATHWAYS

Apart from synthesizing proteins, translation also acts as a quality-control checkpoint for template mRNAs, newly synthesized proteins, and the translation machinery itself. As discussed above, accurate translation termination occurs in the presence of an in-frame TC and requires the engagement of eRF1, which acts as a TC decoding factor, and of the translational GTPase eRF3 in the A site of the ribosome. For mRNAs lacking a TC (nonstop decay) and for those on which ribosomes stall inside an open reading frame (ORF) caused by strong secondary structures or the lack of a cognate tRNA (no-go decay), the rescue of otherwise trapped ribosomes is essential (see Heck and Wilusz 2018). In such cases, the roles of the decoding factor and the GTPase are performed by the HBS1-like protein (HBS1L; Uniprot Q9Y450) and protein Pelota homolog (PELO; Uniprot Q9BRX2; Dom34 in yeast), respectively. When the HBS1L/PELO complex binds a stalled ribosome, degradation of the corresponding mRNA is induced by an endonucleolytic cleavage. These ribosome rescue mechanisms may occur at any point during the translation cycle and require the recruitment of specific factors into the A site of the ribosome. Unlike no-go and nonstop mRNA decay that depend on specialized termination factors, NMD relies on the canonical termination factors eRF1 and eRF3 (Naeger et al. 1992; Dever and Green 2012; Simms et al. 2017).

NMD IS LINKED TO INEFFICIENT TRANSLATION TERMINATION

That NMD depends on translation is well documented (Carter et al. 1995; Thermann et al. 1998). It was initially suggested that only mRNAs associated with the nuclear cap-binding complex (CBC) were sensitive to NMD, leading to the idea that NMD can target mRNAs only during the first round of translation, the so-called pioneer round of translation (Maquat et al. 2010). However, it was later shown that NMD also targets eIF4E-bound mRNAs (Durand and Lykke-Andersen 2013; Rufener and Mühlemann 2013), consistent with data from Saccharomyces cerevisiae (Maderazo et al. 2003; Gao et al. 2005). This led to a revised, unified NMD model, according to which NMD can be triggered at any aberrant translation termination event during any round of translation (He and Jacobson 2015). This model posits that the likelihood for a ribosome to ter-

minate correctly primarily depends on the availability of termination stimulating factors in the immediate vicinity of the TC. For classical NMD targets with a PTC located in a messenger ribonucleoprotein particle (mRNP) environment lacking such factors needed for proper translation termination, the first ribosome attempting to terminate at this PTC will most likely trigger NMD. However, the probability for a terminating ribosome to activate NMD at many other TCs located in more termination-friendly environments would be smaller and the NMD may therefore ensue only after several ribosomes have already terminated properly. This latter scenario might apply for most endogenous NMD targets and explain their mostly moderate NMD-dependent mRNA level reduction (Colombo et al. 2017).

In contrast to no-go and nonstop mRNA decay, which both use the specialized release factors HBS1L and PELO to free the stalled ribosomal subunits, translation termination on NMD-sensitive mRNAs appears to operate with the common release factors eRF1 and eRF3. What, then, is the difference between a termination event that activates NMD and one that does not? Evidence from S. cerevisiae and rabbit reticulocyte lysate indicates that ribosomes reside longer at the TC of NMD-sensitive compared with NMD-insensitive mRNAs (Amrani et al. 2004; Peixeiro et al. 2011). Based on these data, it was proposed that kinetically slow translation termination caused by the lack of termination-promoting factors may be the signal triggering NMD (reviewed in Kervestin and Jacobson 2012; He and Jacobson 2015). The so-far best-characterized termination-promoting factor is poly(A)-binding protein (PABP). Experiments using a fully reconstituted mammalian in vitro translation termination system showed that under limiting concentrations of release factors, PABP directly stimulates translation termination by enhancing the recruitment of eRF3 and eRF1 to the ribosome in a GTP- and peptide release-independent manner (Ivanov et al. 2016). Evidence from both yeast and mammalian cells suggests that NMD ensues when the termination occurs in an mRNP environment that is spatially distant from the PAPB. Accordingly, NMD can be antagonized when cytoplasmic PABP (Pab1p in yeast, PABPC1 in mammals) is artificially tethered into the vicinity of an otherwise NMD-triggering TC (Amrani et al. 2004; Behm-Ansmant et al. 2007; Eberle et al. 2008; Ivanov et al. 2008; Silva et al. 2008; Singh et al. 2008) and the efficiency of suppression is correlated with the physical distance between the TC and the tethered PABP. Somewhat unexpectedly, given the reported interaction of eRF1 with the carboxy-terminal domain of PABP, it was further shown that the carboxyl terminus of PABPC1 is dispensable for suppression of NMD, whereas RRMs 1–3 are sufficient and RRMs 1–2 are necessary. Furthermore, abrogation of the PABPC1–eIF4G interaction reduces NMD suppression by PABPC1 and tethering of eIF4G downstream from an NMD-eliciting TC also suppresses NMD (Fatscher et al. 2014; Joncourt et al. 2014). Although the nature of the termination-promoting activity is still unknown, this finding suggests that it might lie in the formation of the circular structure that the mRNP is thought to adopt by eIF4G interacting with both the 5′ cap-bound eIF4E and the 3′ poly(A) tail-bound PABP (Wells et al. 1998).

UPF1, one of the key NMD factors, has been reported to associate with eRF1 and eRF3 (Czaplinski et al. 1998; Wang et al. 2001; Kashima et al. 2006), fueling the hypothesis that UPF1 antagonizes PABPC1 at termination events that trigger NMD (Kervestin and Jacobson 2012). There is evidence from coimmunoprecipitation experiments that UPF1 and PABPC1 compete for association with eRF3 (Singh et al. 2008). However, a very recent study using recombinant proteins in in vitro experiments showed that this interaction is most likely indirect, bridged by UPF3B (Neu-Yilik et al. 2017). In addition, interactions of eIF3 with hyperphosphorylated UPF1 (Isken et al. 2008) and with UPF2 (Morris et al. 2007) have also been detected, but their roles in NMD are not clear.

KEY EVENTS IN NMD ACTIVATION: THE PIVOTAL ROLE OF UPF1

NMD activation relies on the formation of dynamic protein complexes on the target mRNA

that eventually lead to the degradation of the mRNA. The first NMD factors were identified by genetic screens in *S. cerevisiae* (Leeds et al. 1991, 1992) and *Caenorhabditis elegans* (Pulak and Anderson 1993), and mammalian homologs have been subsequently identified by homology searches (Applequist et al. 1997; Lykke-Andersen et al. 2000; Denning et al. 2001; Serin et al. 2001; Chiu et al. 2003; Reichenbach et al. 2003). More recently, additional screens in *C. elegans* and human cells have identified several additional proteins that appear to be needed for NMD (Longman et al. 2007; Anastasaki et al. 2011; Alexandrov et al. 2017; Hoque et al. 2017); but, apart from the RNA helicase DHX34 (Longman et al. 2013; Hug and Caceres 2014; Melero et al. 2016), their role in NMD is not yet known. Three of the initially identified NMD factors are conserved from yeast to humans: UPF1, UPF2, and UPF3 (of which mammals have two very similar genes, *UPF3A* and *UPF3B*) (Maquat 2004). SMG1, SMG5, SMG6, and SMG7, in contrast, appear to be present only in metazoans (Chen et al. 2008). As outlined below, some of the NMD factors are constituents of more than one of the biochemically characterized protein complexes that form along the pathway of NMD and also have multiple functions in NMD, which complicates the determination of the temporal order of events and ultimately the understanding the molecular mechanism of NMD.

UPF1, an ATP-Dependent RNA Helicase, Is at the Heart of NMD

UPF1 is an ATP-dependent RNA helicase of the helicase superfamily 1 (Fairman-Williams et al. 2010) and a key factor in NMD. UPF1 seems to play a central role in substrate selection and it forms the hub for assembling other NMD factors and recruiting the RNA decay factors. Its helicase activity is tightly regulated and essential for NMD in both yeast and humans (Weng et al. 1998; Franks et al. 2010). ATP binding to the ATP-binding pocket located between the two RecA domains (Rec1A and Rec2A) reduces the affinity of UPF1 for RNA and induces the catalytically active conformation (Bhattacharya et al. 2000; Cheng et al. 2007). Mammalian UPF1 contains three domains that regulate its helicase activity: an amino-terminal domain rich in cysteines and histidines (CH domain); 1B and 1C subdomains embedded within the Rec1A domain; and a low-structured carboxy-terminal SQ-rich domain (Cheng et al. 2007; Chakrabarti et al. 2011; Fiorini et al. 2013). Additionally, UPF1 is regulated by phosphorylation of multiple SQ and TQ motifs located in the amino- and carboxy-terminal parts of the protein and this phosphorylation/dephosphorylation cycle is essential for NMD in metazoans (Yamashita et al. 2001; Ohnishi et al. 2003; Okada-Katsuhata et al. 2012). UPF1 phosphorylation is catalyzed by SMG1, a phosphatidylinositol 3-kinase-related protein kinase (Yamashita 2013).

Is UPF1 Binding Sufficient to Mark mRNAs for Degradation?

The questions of when and where UPF1 binds mRNAs and whether mere binding of UPF1 is sufficient to mark RNAs for degradation have long been investigated and have given rise to contradictory findings. In support of specific UPF1 recruitment to NMD-targeted transcripts, UPF1 has been found significantly enriched on PTC-containing mRNAs compared with PTC-less mRNAs at steady state (Johansson et al. 2007; Johns et al. 2007; Silva et al. 2008; Hwang et al. 2010; Kurosaki et al. 2014; Lee et al. 2015). On the other hand, CLIP-seq and immunoprecipitation experiments in human and mouse cells showed that UPF1 is considerably enriched on the 3′ untranslated region (UTR) of mRNAs independently of whether or not they are targeted by NMD (Hurt et al. 2013; Zund et al. 2013), suggesting that UPF1 binding to RNA is not the discriminatory step in NMD target identification. Furthermore, it appears that UPF1 associates with mRNA before translation starts at seemingly random positions on the RNA and that translating ribosomes subsequently displace UPF1 from the ORF, resulting in the majority of UPF1 being detected on 3′UTRs at steady state (Fig. 2) (Hogg and Goff 2010; Hurt et al. 2013; Zund et al. 2013). Based on the available data, it is possible that transla-

tion affects UPF1-RNA association in two independent ways: (1) translation removes UPF1 from the CDS; and (2) UPF1 gets recruited by release factors (Ivanov et al. 2008) and in yeast cells also by ribosomal proteins (Min et al. 2013) to the 3′UTR of NMD targets. The ATPase activity of UPF1 was shown to be required for releasing UPF1 from nontarget mRNAs (Lee et al. 2015), and there is evidence that UPF1 phosphorylation occurs specifically on NMD targets (Kurosaki et al. 2014). Thus, a plausible scenario is that UPF1 initially binds all mRNAs and that improper/delayed translation termination triggers phosphorylation of the remaining RNA-bound UPF1, eventually activating NMD, whereas ATP hydrolysis results in the release of UPF1 from mRNAs that undergo proper translation termination. In this working model, UPF1 functions as a time bomb that destroys the mRNA if not removed quickly enough, an idea that was postulated some time ago (Hilleren and Parker 1999).

SMG1 Complex Regulates UPF1 Phosphorylation

In mammalian cells, the SMG1-mediated UPF1 phosphorylation is essential for NMD (Yamashita et al. 2001; Okada-Katsuhata et al. 2012) and appears to occur in a translation-dependent manner (Yamashita et al. 2001, 2009; Yamashita 2013). At steady state, most UPF1 of an immortalized cell is hypophosphorylated (Chiu et al. 2003; Okada-Katsuhata et al. 2012). The kinase activity of SMG1 is regulated by SMG8 and SMG9, two proteins forming a complex with SMG1 that keeps SMG1 in a kinase-inactive conformation (Yamashita et al. 2009; Arias-Palomo et al. 2011; Fernandez et al. 2011; Melero et al. 2014). SMG8 inhibits SMG1 activity in vitro but enhances UPF1 phosphorylation in vivo (Yamashita et al. 2009), indicating that in addition to regulating the kinase activity of SMG1, SMG8 also functions in recruiting the SMG1 complex to UPF1. UPF1 can be coimmunoprecipitated from human cells in a complex with SMG1, eRF1, and eRF3 (the so-called SURF complex [Kashima et al. 2006]). These interactions support the view that UPF1 phosphorylation occurs in

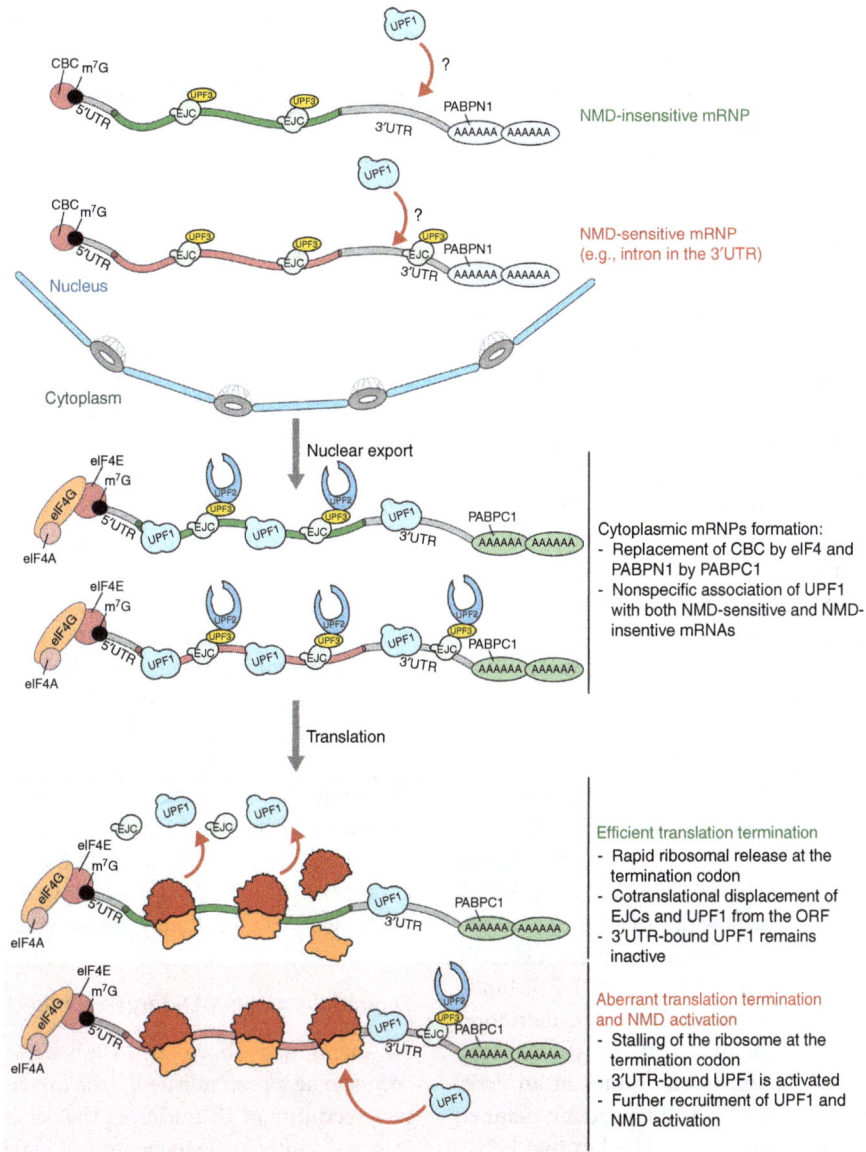

Figure 2. Distribution and molecular architecture of factors that affect nonsense-mediated mRNA decay (NMD) on NMD-sensitive and NMD-insensitive messenger RNAs (mRNAs) at various stages. NMD-insensitive (green) and NMD-sensitive (red) messenger ribonucleoprotein particles (mRNPs) are loaded with UPF3 bound to exon junction complexes (EJCs) before export to the cytoplasm, and it is possible (although not known) that some UPF1 also associates with mRNA in the nucleus. In the cytoplasm, UPF1 is bound to all mRNAs independent of their sensitivity to NMD. UPF2 interacts with UPF3 and the nuclear cap-binding complex (CBC) and the nuclear poly(A)-binding protein (PABPN)1 are replaced by their cytoplasmic counterparts eIF4F (composed of eIF4E, G, and A) and PABPC1, respectively. During translation, UPF1 and EJCs are displaced from the coding regions of the mRNA by elongating ribosomes, while they remain associated on the 3′ untranslated region (UTR). In the context of normal translation termination, ribosome release is very efficient, and the corresponding mRNAs persist. When translation termination is not efficient, however, NMD ensues by UPF1 activation and/or recruitment of additional UPF1 molecules to the site of termination. The presence of an EJC downstream from the termination codon (TC) increases the concentration of UPF2 and UPF3 in the vicinity of UPF1, which is thought to facilitate the activation of NMD.

the context of translation termination (Fig. 1B) (Kashima et al. 2006; Kurosaki et al. 2014). UPF2 (see below) and DHX34 (Hug and Caceres 2014) contribute to the activation of SMG1-catalyzed UPF1 phosphorylation.

UPF2 and UPF3 Are Involved in NMD Activation

UPF2 is conserved in eukaryotes and functions as a ring-like adaptor protein between UPF1 and UPF3 (Serin et al. 2001; Chamieh et al. 2008; Melero et al. 2012). It consists of three MIF4G (middle fragment of eIF4G) domains, with the first two domains providing structural support (Clerici et al. 2009) and the third domain interacting with UPF3B (Kadlec et al. 2004). UPF2's interaction with the CH domain of UPF1 induces a large conformational change of this domain that triggers the helicase activity of UPF1 by switching it from an RNA-clamping to an RNA-unwinding mode (Fig. 1B) (Chakrabarti et al. 2011). In addition to its interaction with UPF1 and UPF3B, UPF2 has been reported to bind to the DEAD-box helicase Dbp6 in yeast (DDX51 in *Homo sapiens*) (Fourati et al. 2014) and to SMG1 and ribosomal proteins of both subunits (Lopez-Perrote et al. 2016). UPF2 also directly interacts with eRF3 through a carboxy-terminal region that is essential for UPF3B binding but does not interfere with UPF1 binding (Lopez-Perrote et al. 2016). Furthermore, UPF2 forms a complex with SURF, allowing its direct association to ribosomes in an exon junction complex (EJC)-independent manner (Lopez-Perrote et al. 2016). The fact that UPF2 appears to associate with some NMD-related components simultaneously and with others in a mutually exclusive way implies an important role for UPF2 in orchestrating the order of events leading to NMD activation (Lopez-Perrote et al. 2016). Somewhat surprisingly, given its central role in NMD, degradation of a subset of NMD targets has been reported to occur independently of UPF2 (Gehring et al. 2005).

UPF3 is the least conserved of the three core NMD factors (Culbertson and Leeds 2003). At steady state, it predominately resides in the nucleus (Serin et al. 2001), but it is thought to

shuttle with the mRNP to the cytoplasm in which NMD takes place (Trcek et al. 2013). Rather than binding RNA, the conserved amino-terminal putative RNA recognition motif (RRM) appears to interact with UPF2 (Kadlec et al. 2004). In addition, the three EJC core factors eIF4A3, MAGOH, and Y14 interact through a composite binding site with a short motif located at the carboxyl terminus of UPF3 (Chamieh et al. 2008). Thus, UPF3 and UPF2 bridge UPF1 to the EJC, providing a molecular link between NMD and the EJC (Chamieh et al. 2008). UPF3 and UPF2 together were also shown to stimulate the helicase activity of UPF1 (Chamieh et al. 2008). Mammals contain two UPF3 paralogs, UPF3A and UPF3B (UPF3B is located on the X chromosome and is also called UPF3X). UPF3A and UPF3B are similar in their amino acid sequence and both associate with other NMD factors but UPF3A is less efficient in destabilizing NMD reporter mRNAs in tethering assays (Lykke-Andersen et al. 2000; Kunz et al. 2006; Chan et al. 2009; Hug and Caceres 2014). In mice, the conditional knockout of UPF3A even enhanced the NMD response of a fraction of endogenous NMD targets, suggesting that UPF3A may function as an antagonist of UPF3B by sequestering UPF2 from the NMD machinery (Shum et al. 2016).

Degradation of NMD-Targeted mRNAs

As mentioned above, UPF1 phosphorylation seems to be a prerequisite for the direct or indirect recruitment of nucleases that degrade the targeted mRNA (Lejeune et al. 2003; Isken et al. 2008). Endonucleolytic cleavage in the vicinity of the NMD-eliciting TC is mediated by SMG6 and the SMG5/SMG7 heterodimer recruits exonucleolytic decay activities (Mühlemann and Lykke-Andersen 2010). SMG6 is a metazoan-specific NMD factor that confers its endonuclease activity through a triad of catalytically active Asp residues in its carboxy-terminal PIN domain (Huntzinger et al. 2008; Eberle et al. 2009). In immortalized human cells (HeLa and HEK-293), the bulk of NMD targets appears to be degraded by SMG6 (Boehm et al. 2014; Lykke-Andersen et al. 2014; Schmidt et al. 2014). The unprotected RNA ends generated by the endo-nucleolytic cleavage are then subjected to exonu-cleolytic degradation by the 5′–3′exonuclease XRN1 and by 3′–5′ endonucleases (Gatfield and Izaurralde 2004; Huntzinger et al. 2008; Eberle et al. 2009; Franks et al. 2010). Notably, SMG6 interacts with phosphorylated T28 of UPF1 (Okada-Katsuhata et al. 2012) and addi-tionally in a phosphorylation-independent man-ner with the unique stalk protruding from the Rec1A domain of UPF1 (Chakrabarti et al. 2014; Nicholson et al. 2014). In tethering exper-iments, SMG6 requires UPF1 and SMG1 to de-grade a reporter transcript, implying that the SMG6 endonuclease might adopt its active con-formation only when bound to phosphorylated UPF1 (Nicholson et al. 2014).

SMG5 and SMG7 form a highly stable het-erodimer (Jonas et al. 2013) that binds to phos-phorylated S1096 of UPF1 in vivo (Ohnishi et al. 2003; Okada-Katsuhata et al. 2012) and triggers RNA degradation when tethered to reporter transcripts (Unterholzner and Izaurralde 2004; Cho et al. 2013). This activity requires the carboxy-terminal proline-rich region of SMG7, which was shown to interact with CNOT8 (POP2), the catalytic subunit of the CCR4-NOT deadenylase complex (Loh et al. 2013). Tethered SMG7 has further been reported to induce reporter transcript decay by DCP2-me-diated decapping and XRN1-mediated 5′-to-3′ exonucleolytic degradation (Unterholzner and Izaurralde 2004). These factors are presumably recruited indirectly via the CCR4-NOT com-plex. There is ample evidence that SMG5/ SMG7 and SMG6 RNA degradation pathways act in, at least partially, a redundant way. Co-depletion of SMG6 with SMG5 or SMG7 inhib-its NMD much more effectively than depletion of each of the factors alone (Luke et al. 2007; Jonas et al. 2013; Metze et al. 2013; Colombo et al. 2017). A transcriptome-wide approach re-vealed that SMG6 and SMG7 target essentially the same sets of transcripts (Colombo et al. 2017), even though individual mRNAs may have preferences for either SMG6- or SMG7-mediated decay (Ottens et al. 2017).

Human UPF1 also interacts with mRNA decay factors independently of SMG5/SMG7,

namely, with the decapping factors DCP1, DCP2, and PNRC2 (Lykke-Andersen 2002; Le-jeune et al. 2003; Fenger-Gron et al. 2005; Isken et al. 2008; Cho et al. 2009; Lai et al. 2012). PNRC2 binds directly to DCP1 and UPF1 (Lai et al. 2012; Cho et al. 2013). The published data regarding whether or not SMG5 interacts with PNRC2 and is involved in the PNRC2-depen-dent decay pathway are contradictory and fur-ther investigation is needed (Cho et al. 2013; Loh et al. 2013). It is noteworthy that many NMD factors colocalize with decapping and 5′–3′ exo-nucleolytic degradation components in process-ing bodies (P-bodies) (Fukuhara et al. 2005; Durand et al. 2007; Franks et al. 2010) but the formation of detectable P-bodies in cells is not required for NMD (Stalder and Mühlemann 2009).

NMD-INDUCING FEATURES

As outlined above, the current model posits that NMD is stimulated when the TC occurs in a microenvironment of the mRNP that is unfa-vorable for translation termination. Assuming that aberrant mRNAs with PTCs are degraded by the same mechanism as endogenous NMD targets with full-length ORFs, one would expect to find common features that render mRNAs susceptible to NMD. Transcriptome-wide ap-proaches to identify NMD targets in cells of different species revealed that the majority of NMD-sensitive transcripts do not contain PTCs but are ordinary mRNAs coding for seem-ingly full-length functional proteins (Rehwinkel et al. 2006; Lykke-Andersen and Jensen 2015; Colombo et al. 2017). Despite the fact that in most of these studies it was not possible to distinguish between direct and indirect targets, there is ample evidence that NMD can target both normal and erroneous transcripts. Among the NMD-inducing features, the presence of the 3′-most exon–exon junction >50 nt down-stream from the TC is the feature with the stron-gest predictive value for NMD susceptibility (Colombo et al. 2017). PTCs resulting from mutations in the ORF or from aberrant or alter-native splicing, as well as genes with an intron in the 3′UTR mostly belong to this class of NMD targets. In addition, long 3′UTRs (>1000 nt in mammalian cells) (Buhler et al. 2006), the presence of actively translated short upstream ORFs (uORFs) (Hurt et al. 2013), or selenocysteine codons (UGA) in cells grown in the absence of selenium (Moriarty et al. 1998) are also features that can—but not always do—trigger NMD (Fig. 3). uORF translation often inhibits translation of the main ORF, either constitutively or in response to stress (Young and Wek 2016). Under such circumstances, ribosomes terminate at the TC of the uORF with usually several EJCs remaining bound further downstream on the mRNA, which creates an NMD-promoting translation termination environment. Despite all these empirically determined features, it has so far remained impossible to computationally predict NMD targets with high confidence.

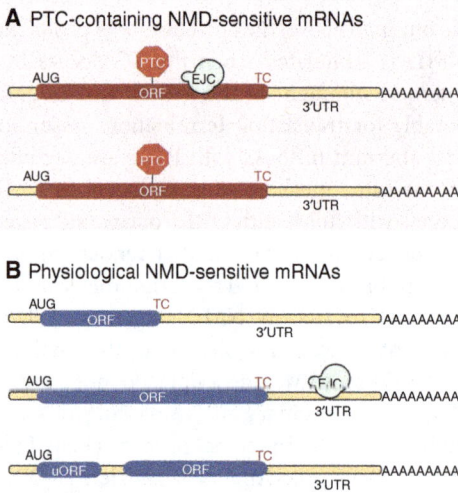

Figure 3. Features that can render aberrant and physiological mRNAs sensitive to nonsense-mediated mRNA decay (NMD). (*A*) Messenger RNAs (mRNAs) in which the open reading frame (ORF) is truncated caused by the presence of a premature termination codon (PTC) are typically sensitive to NMD. The presence of one or several exon junction complexes (EJCs) downstream from the PTC is an important NMD-enhancing feature. (*B*) In mRNAs with an intact ORF coding for a full-length protein, long 3′ untranslated regions (UTRs), EJCs located in the 3′UTR, or short upstream ORFs (uORFs) can also trigger NMD.

Based on the aforementioned ability of PABP to suppress NMD, it has been suggested that the physical distance between the TC and the poly(A) tail might be the crucial feature determining NMD sensitivity (Stalder and Mühlemann 2008). Because such information about the 3D architecture of mRNPs is currently not available, this TC-to-poly(A) tail distance is virtually impossible to measure and predict.

The EJC as an NMD Enhancer

EJCs are multiprotein complexes organized around a central core of four proteins: eIF4A3, MLN51, and MAGOH/Y14, a heterodimer that interacts with UPF3B (Buchwald et al. 2010). The EJC core binds tightly to RNA without apparent sequence specificity, generally ∼24 nt upstream of spliced exon–exon junctions during the course of splicing (Le Hir et al. 2000a,b). However, a transcriptome-wide mapping of eIF4A3 in human cells indicated that not all splicing events result in the deposition of an EJC and a significant fraction of eIF4A3 was detected on mRNA in noncanonical positions (Sauliere et al. 2012; Singh et al. 2012). EJCs are deposited on mRNA in the nucleus during splicing and accompany the mature mRNAs to the cytoplasm, in which those EJCs located in ORFs are thought to be removed by the first ribosome translating the mRNA (Fig. 2). EJCs located in the 3′UTR of an mRNP >30 nt downstream from the TC in contrast will remain bound to the mRNA (Le Hir et al. 2016). The presence of EJCs in mRNPs was shown to activate translation mediated by an interaction of MLN51 with eIF3 subunits a and d (Chazal et al. 2013). Those EJCs that remain bound to the mRNA after translation has begun (i.e., those bound >30 nt downstream of the TC) strongly promote the activation of NMD on the corresponding mRNAs (Buhler et al. 2006; Metze et al. 2013) by recruiting NMD factors UPF3, UPF2, and/or SMG6 to the mRNP (Le Hir et al. 2001; Gehring et al. 2005; Kashima et al. 2010). A similar NMD-promoting effect was observed by tethering EJC components to the 3′UTR of reporter transcripts (Lykke-Andersen et al. 2000; Gehring et al. 2003; Palacios et al. 2004). The presence

Cite this article as *Cold Spring Harb Perspect Biol* doi: 10.1101/cshperspect.a032862

of EJCs in 3′UTRs was reported to enhance the efficiency of SMG6-mediated endonucleolysis without altering the cleavage site (Boehm et al. 2014). This enhanced SMG6-mediated NMD activity can be attributed to the interaction of SMG6 with the EJC via two conserved EJC-binding motifs (EBMs) in the amino-terminal portion of SMG6 (Kashima et al. 2010). Interestingly, the SMG6 EBMs interact with the same surface of the EJC as UPF3B (Kashima et al. 2010), suggesting that these interactions are mutually exclusive and may therefore occur sequentially.

Long 3′UTRs

That mRNAs with a long 3′UTR can be targeted by NMD has been observed in yeast, plants, and animals (Pulak and Anderson 1993; Muhlrad and Parker 1999; Buhler et al. 2006; Kertesz et al. 2006; Eberle et al. 2008; Singh et al. 2008; Hogg and Goff 2010; Hurt et al. 2013; Boehm et al. 2014; Colombo et al. 2017) and, accordingly, insertion of long 3′UTRs into reporter constructs can render them NMD-sensitive (Huang et al. 2011; Yepiskoposyan et al. 2011). Because the physical distance between TC and PABP strongly affects the half-life of NMD reporter transcripts (Eberle et al. 2008), it is very likely that those long 3′UTRs that render mRNAs sensitive to NMD adopt an extended 3′UTR structure in which the poly(A) tail is too far away from the TC to promote proper translation termination. Consistent with this idea, intramolecular secondary structures that bring a distal poly(A) tail into close proximity with an otherwise NMD-triggering TC effectively antagonize NMD, even in the context of EJC-enhanced NMD (Eberle et al. 2008). Although the NMD-suppressing effect of PABP is well documented (Amrani et al. 2004; Behm-Ansmant et al. 2007; Eberle et al. 2008; Ivanov et al. 2008; Silva et al. 2008; Singh et al. 2008), the underlying mechanism remains elusive. Interestingly, several endogenous mRNAs with long 3′UTRs were found to be immune to NMD (Weil and Beemon 2006; Hogg and Goff 2010; Yepiskoposyan et al. 2011; Hurt et al. 2013). A subset of these NMD-insensitive mRNAs with long 3′UTRs

contains AU-rich elements within the first 200 nt downstream from the TC, which were shown to be necessary and sufficient to inhibit NMD (Toma et al. 2015). However, the mechanism by which this NMD protection is conferred is not known. Another well-studied example of an NMD-resistant transcript with a long 3′UTR is the unspliced viral RNA of Rous sarcoma virus (RSV). RSV harbors a 400-nt RNA stability element (RSE) located immediately downstream from the *gag* TC, which is required to confer NMD immunity on the viral RNA (Weil and Beemon 2006). It has been found that the polypyrimidine tract-binding protein 1 (PTBP1) binds to the RSE and thereby blocks UPF1 association with this region (Ge et al. 2016). Insertion of PTBP1-binding sites into NMD reporter constructs downstream of the TC rendered these transcripts NMD-insensitive (Ge et al. 2016).

OPEN QUESTIONS

Despite the fact that NMD has been an area of intensive research for the past two decades, many points still remain unclear. A long-standing open question is whether all eukaryotes have NMD. Trypanosomes have homologs of UPF1 and UPF2 but UPF1 depletion showed no effect on the RNA level of genes that were predicted NMD targets, leading the investigators to conclude that a classical NMD pathway might be absent in *Trypanosoma brucei* (Delhi et al. 2011). In contrast, NMD appears to function in *Giardia lamblia*, another parasitic protist, despite the lack of recognizable homologs of UPF2 and UPF3 (Chen et al. 2008). Although UPF3 seems to be absent from most protists, a UPF3 homolog has been found in *Paramecium tetraurelia* and it was shown to play a role in NMD (Contreras et al. 2014). Interestingly, however, this UPF3 variant lacks the motifs required for interacting with the EJC, indicating that in these early-branched eukaryotes, NMD might function independently of EJCs, similar to *S. cerevisiae*, which lacks EJCs altogether. Consistently, NMD in *Tetrahymena thermophila* has recently been shown to be EJC-independent (Tian et al. 2017). In addition, this study identified a novel protozoa-specific nuclease that is responsible for degrading many NMD targets, a Smg6-like bacterial YacP nuclease (NYN) domain-containing protein (Smg6L). These recent findings corroborate the view that NMD in different species can occur by several different RNA decay pathways and that the evolutionarily conserved part may be confined to the substrate RNA selection, in which UPF1 plays a key function.

Recent ribosome profiling data in *S. cerevisiae* revealed that NMD-targeted mRNAs that lack PTCs according to their annotated sequences tend to show elevated rates of out-of-frame translation (Celik et al. 2017a). Such out-of-frame translation would lead to premature translation termination, which could explain why these mRNAs are NMD targets. It has therefore been proposed that NMD should be considered as a probabilistic quality-control pathway that continually detects errors throughout the translation process (Celik et al. 2017b). This is an interesting novel concept that needs to be further investigated.

Apart from understanding the mechanistic details that govern NMD activation in the context of inefficient translation termination, it would also be interesting to assess whether interconnections between NMD and other translation-dependent mRNA degradation mechanisms exist. Preliminary evidence supports the notion of such cross talk at least with regard to the mRNA degradation machinery involved. When endonucleolytic cleavage occurs upstream of a PTC, the resulting 5′RNA fragment was stabilized on depletion of nonstop decay components Pelota and Hbs1 in *Drosophila* cells (Hashimoto et al. 2017). Along the same lines, it was shown in *C. elegans* that the degradation of PTC-containing mRNAs as well as endogenous NMD targets is coupled to nonstop decay (Arribere and Fire 2018). This implies that the same mRNA could be targeted by more than one translation-dependent RNA surveillance pathway, allowing for more efficient RNA degradation.

It is evident from this review that many fundamental questions regarding the NMD phenomenon are still unsolved. Notably, it is currently not even clear whether what is commonly described as NMD constitutes one defined biochemical pathway or maybe several different pathways with requirements for an overlapping set of proteins. Thus, we are still far away from a comprehensive understanding of the biological functions of NMD, its underpinning molecular mechanism(s), and the evolutionary forces that led to the emergence of NMD.

ACKNOWLEDGMENTS

The research in the laboratory of O.M. is supported by the National Center of Competence in Research (NCCR) RNA & Disease funded by the Swiss National Science Foundation (SNSF), by the SNSF Grant 31003A-162986, and by the canton of Bern.

REFERENCES

*Reference is also in this collection.

Alexandrov A, Shu MD, Steitz JA. 2017. Fluorescence amplification method for forward genetic discovery of factors in human mRNA degradation. *Mol Cell* **65:** 191–201.

Alkalaeva EZ, Pisarev AV, Frolova LY, Kisselev LL, Pestova TV. 2006. In vitro reconstitution of eukaryotic translation reveals cooperativity between release factors eRF1 and eRF3. *Cell* **125:** 1125–1136.

Amrani N, Ganesan R, Kervestin S, Mangus DA, Ghosh S, Jacobson A. 2004. A faux 3′-UTR promotes aberrant termination and triggers nonsense-mediated mRNA decay. *Nature* **432:** 112–118.

Anastasaki C, Longman D, Capper A, Patton EE, Caceres JF. 2011. Dhx34 and Nbas function in the NMD pathway and are required for embryonic development in zebrafish. *Nucleic Acids Res* **39:** 3686–3694.

Applequist SE, Selg M, Raman C, Jack HM. 1997. Cloning and characterization of HUPF1, a human homolog of the *Saccharomyces cerevisiae* nonsense mRNA-reducing UPF1 protein. *Nucleic Acids Res* **25:** 814–821.

Arias-Palomo E, Yamashita A, Fernandez IS, Nunez-Ramirez R, Bamba Y, Izumi N, Ohno S, Llorca O. 2011. The nonsense-mediated mRNA decay SMG-1 kinase is regulated by large-scale conformational changes controlled by SMG-8. *Genes Dev* **25:** 153–164.

Arribere JA, Fire AZ. 2018. Nonsense mRNA suppression via nonstop decay. *eLife* **7:** e33292.

Becker T, Franckenberg S, Wickles S, Shoemaker CJ, Anger AM, Armache JP, Sieber H, Ungewickell C, Berninghausen O, Daberkow I, et al. 2012. Structural basis of highly conserved ribosome recycling in eukaryotes and archaea. *Nature* **482:** 501–506.

Behm-Ansmant I, Gatfield D, Rehwinkel J, Hilgers V, Izaurralde E. 2007. A conserved role for cytoplasmic poly(A)-

Cite this article as *Cold Spring Harb Perspect Biol* doi: 10.1101/cshperspect.a032862

binding protein 1 (PABPC1) in nonsense-mediated mRNA decay. *EMBO J* **26:** 1591–1601.

Bhattacharya A, Czaplinski K, Trifillis P, He F, Jacobson A, Peltz SW. 2000. Characterization of the biochemical properties of the human Upf1 gene product that is involved in nonsense-mediated mRNA decay. *RNA* **6:** 1226–1235.

Boehm V, Haberman N, Ottens F, Ule J, Gehring NH. 2014. 3′UTR length and messenger ribonucleoprotein composition determine endocleavage efficiencies at termination codons. *Cell Rep* **9:** 555–568.

Brown A, Shao S, Murray J, Hegde RS, Ramakrishnan V. 2015. Structural basis for stop codon recognition in eukaryotes. *Nature* **524:** 493–496.

Buchwald G, Ebert J, Basquin C, Sauliere J, Jayachandran U, Bono F, Le Hir H, Conti E. 2010. Insights into the recruitment of the NMD machinery from the crystal structure of a core EJC-UPF3b complex. *Proc Natl Acad Sci* **107:** 10050–10055.

Buhler M, Steiner S, Mohn F, Paillusson A, Mühlemann O. 2006. EJC-independent degradation of nonsense immunoglobulin-mu mRNA depends on 3′UTR length. *Nat Struct Mol Biol* **13:** 462–464.

Carter MS, Doskow J, Morris P, Li S, Nhim RP, Sandstedt S, Wilkinson MF. 1995. A regulatory mechanism that detects premature nonsense codons in T-cell receptor transcripts in vivo is reversed by protein synthesis inhibitors in vitro. *J Biol Chem* **270:** 28995–29003.

Celik A, Baker R, He F, Jacobson A. 2017a. High-resolution profiling of NMD targets in yeast reveals translational fidelity as a basis for substrate selection. *RNA* **23:** 735–748.

Celik A, He F, Jacobson A. 2017b. NMD monitors translational fidelity 24/7. *Curr Genet* **63:** 1007–1010.

Chakrabarti S, Jayachandran U, Bonneau F, Fiorini F, Basquin C, Domcke S, Le Hir H, Conti E. 2011. Molecular mechanisms for the RNA-dependent ATPase activity of Upf1 and its regulation by Upf2. *Mol Cell* **41:** 693–703.

Chakrabarti S, Bonneau F, Schussler S, Eppinger E, Conti E. 2014. Phospho-dependent and phospho-independent interactions of the helicase UPF1 with the NMD factors SMG5-SMG7 and SMG6. *Nucleic Acids Res* **42:** 9447–9460.

Chamieh H, Ballut L, Bonneau F, Le Hir H. 2008. NMD factors UPF2 and UPF3 bridge UPF1 to the exon junction complex and stimulate its RNA helicase activity. *Nat Struct Mol Biol* **15:** 85–93.

Chan WK, Bhalla AD, Le Hir H, Nguyen LS, Huang L, Gecz J, Wilkinson MF. 2009. A UPF3-mediated regulatory switch that maintains RNA surveillance. *Nat Struct Mol Biol* **16:** 747–753.

Chazal PE, Daguenet E, Wendling C, Ulryck N, Tomasetto C, Sargueil B, Le Hir H. 2013. EJC core component MLN51 interacts with eIF3 and activates translation. *Proc Natl Acad Sci* **110:** 5903–5908.

Chen YH, Su LH, Sun CH. 2008. Incomplete nonsense-mediated mRNA decay in *Giardia lamblia*. *Int J Parasitol* **38:** 1305–1317.

Cheng Z, Muhlrad D, Lim MK, Parker R, Song H. 2007. Structural and functional insights into the human Upf1 helicase core. *EMBO J* **26:** 253–264.

Chiu SY, Serin G, Ohara O, Maquat LE. 2003. Characterization of human Smg5/7a: A protein with similarities to *Caenorhabditis elegans* SMG5 and SMG7 that functions in the dephosphorylation of Upf1. *RNA* **9:** 77–87.

Cho H, Kim KM, Kim YK. 2009. Human proline-rich nuclear receptor coregulatory protein 2 mediates an interaction between mRNA surveillance machinery and decapping complex. *Mol Cell* **33:** 75–86.

Cho H, Han S, Choe J, Park SG, Choi SS, Kim YK. 2013. SMG5-PNRC2 is functionally dominant compared with SMG5-SMG7 in mammalian nonsense-mediated mRNA decay. *Nucleic Acids Res* **41:** 1319–1328.

Clerici M, Mourao A, Gutsche I, Gehring NH, Hentze MW, Kulozik A, Kadlec J, Sattler M, Cusack S. 2009. Unusual bipartite mode of interaction between the nonsense-mediated decay factors, UPF1 and UPF2. *EMBO J* **28:** 2293–2306.

Colombo M, Karousis ED, Bourquin J, Bruggmann R, Mühlemann O. 2017. Transcriptome-wide identification of NMD-targeted human mRNAs reveals extensive redundancy between SMG6- and SMG7-mediated degradation pathways. *RNA* **23:** 189–201.

Contreras J, Begley V, Macias S, Villalobo E. 2014. An UPF3-based nonsense-mediated decay in *Paramecium*. *Res Microbiol* **165:** 841–846.

Culbertson MR, Leeds PF. 2003. Looking at mRNA decay pathways through the window of molecular evolution. *Curr Opin Genet Dev* **13:** 207–214.

Czaplinski K, Ruiz-Echevarria MJ, Paushkin SV, Han X, Weng Y, Perlick HA, Dietz HC, Ter-Avanesyan MD, Peltz SW. 1998. The surveillance complex interacts with the translation release factors to enhance termination and degrade aberrant mRNAs. *Genes Dev* **12:** 1665–1677.

Delhi P, Queiroz R, Inchaustegui D, Carrington M, Clayton C. 2011. Is there a classical nonsense-mediated decay pathway in trypanosomes? *PLoS ONE* **6:** e25112.

Denning G, Jamieson L, Maquat LE, Thompson EA, Fields AP. 2001. Cloning of a novel phosphatidylinositol kinase-related kinase: Characterization of the human SMG-1 RNA surveillance protein. *J Biol Chem* **276:** 22709–22714.

Dever TE, Green R. 2012. The elongation, termination, and recycling phases of translation in eukaryotes. *Cold Spring Harb Perspect Biol* **4:** a013706.

Durand S, Lykke-Andersen J. 2013. Nonsense-mediated mRNA decay occurs during eIF4F-dependent translation in human cells. *Nat Struct Mol Biol* **20:** 702–709.

Durand S, Cougot N, Mahuteau-Betzer F, Nguyen CH, Grierson DS, Bertrand E, Tazi J, Lejeune F. 2007. Inhibition of nonsense-mediated mRNA decay (NMD) by a new chemical molecule reveals the dynamic of NMD factors in P-bodies. *J Cell Biol* **178:** 1145–1160.

Eberle AB, Stalder L, Mathys H, Orozco RZ, Mühlemann O. 2008. Posttranscriptional gene regulation by spatial rearrangement of the 3′ untranslated region. *PLoS Biol* **6:** e92.

Eberle AB, Lykke-Andersen S, Mühlemann O, Jensen TH. 2009. SMG6 promotes endonucleolytic cleavage of nonsense mRNA in human cells. *Nat Struct Mol Biol* **16:** 49–55.

Eyler DE, Wehner KA, Green R. 2013. Eukaryotic release factor 3 is required for multiple turnovers of peptide release catalysis by eukaryotic release factor 1. *J Biol Chem* **288**: 29530–29538.

Fairman-Williams ME, Guenther UP, Jankowsky E. 2010. SF1 and SF2 helicases: Family matters. *Curr Opin Struct Biol* **20**: 313–324.

Fatscher T, Boehm V, Weiche B, Gehring NH. 2014. The interaction of cytoplasmic poly(A)-binding protein with eukaryotic initiation factor 4G suppresses nonsense-mediated mRNA decay. *RNA* **20**: 1579–1592.

Fenger-Gron M, Fillman C, Norrild B, Lykke-Andersen J. 2005. Multiple processing body factors and the ARE binding protein TTP activate mRNA decapping. *Mol Cell* **20**: 905–915.

Fernandez IS, Yamashita A, Arias-Palomo E, Bamba Y, Bartolome RA, Canales MA, Teixido J, Ohno S, Llorca O. 2011. Characterization of SMG-9, an essential component of the nonsense-mediated mRNA decay SMG1C complex. *Nucleic Acids Res* **39**: 347–358.

Fiorini F, Boudvillain M, Le Hir H. 2013. Tight intramolecular regulation of the human Upf1 helicase by its N- and C-terminal domains. *Nucleic Acids Res* **41**: 2404–2415.

Fourati Z, Roy B, Millan C, Coureux PD, Kervestin S, van Tilbeurgh H, He F, Uson I, Jacobson A, Graille M. 2014. A highly conserved region essential for NMD in the Upf2 N-terminal domain. *J Mol Biol* **426**: 3689–3702.

Franks TM, Singh G, Lykke-Andersen J. 2010. Upf1 ATPase-dependent mRNP disassembly is required for completion of nonsense- mediated mRNA decay. *Cell* **143**: 938–950.

Fukuhara N, Ebert J, Unterholzner L, Lindner D, Izaurralde E, Conti E. 2005. SMG7 is a 14-3-3-like adaptor in the nonsense-mediated mRNA decay pathway. *Mol Cell* **17**: 537–547.

Gao Q, Das B, Sherman F, Maquat LE. 2005. Cap-binding protein 1-mediated and eukaryotic translation initiation factor 4E-mediated pioneer rounds of translation in yeast. *Proc Natl Acad Sci* **102**: 4258–4263.

Gatfield D, Izaurralde E. 2004. Nonsense-mediated messenger RNA decay is initiated by endonucleolytic cleavage in *Drosophila*. *Nature* **429**: 575–578.

Ge Z, Quek BL, Beemon KL, Hogg JR. 2016. Polypyrimidine tract binding protein 1 protects mRNAs from recognition by the nonsense-mediated mRNA decay pathway. *eLife* **5**: e11155.

Gehring NH, Neu-Yilik G, Schell T, Hentze MW, Kulozik AE. 2003. Y14 and hUpf3b form an NMD-activating complex. *Mol Cell* **11**: 939–949.

Gehring NH, Kunz JB, Neu-Yilik G, Breit S, Viegas MH, Hentze MW, Kulozik AE. 2005. Exon-junction complex components specify distinct routes of nonsense-mediated mRNA decay with differential cofactor requirements. *Mol Cell* **20**: 65–75.

Hashimoto Y, Takahashi M, Sakota E, Nakamura Y. 2017. Nonstop-mRNA decay machinery is involved in the clearance of mRNA 5′-fragments produced by RNAi and NMD in *Drosophila melanogaster* cells. *Biochem Biophys Res Commun* **484**: 1–7.

He F, Jacobson A. 2015. Nonsense-mediated mRNA decay: Degradation of defective transcripts is only part of the story. *Annu Rev Genet* **49**: 339–366.

* Heck AM, Wilusz J. 2018. The interplay between the RNA decay and translation machinery in eukaryotes. *Cold Spring Harb Perspect Biol* doi: 10.1101/cshperspect.a032839.

* Hellen CUT. 2018. Translation termination and ribosome recycling in eukaryotes. *Cold Spring Harb Perspect Biol* doi: 10.1101/cshperspect.a032656.

Heuer A, Gerovac M, Schmidt C, Trowitzsch S, Preis A, Kotter P, Berninghausen O, Becker T, Beckmann R, Tampe R. 2017. Structure of the 40S-ABCE1 post-splitting complex in ribosome recycling and translation initiation. *Nat Struct Mol Biol* **24**: 453–460.

Hilleren P, Parker R. 1999. mRNA surveillance in eukaryotes: Kinetic proofreading of proper translation termination as assessed by mRNP domain organization? *RNA* **5**: 711–719.

Hogg JR, Goff SP. 2010. Upf1 senses 3′UTR length to potentiate mRNA decay. *Cell* **143**: 379–389.

Hoque M, Park JY, Chang YJ, Luchessi AD, Cambiaghi TD, Shamanna R, Hanauske-Abel HM, Holland B, Pe'ery T, Tian B, et al. 2017. Regulation of gene expression by translation factor eIF5A: Hypusine-modified eIF5A enhances nonsense-mediated mRNA decay in human cells. *Translation (Austin)* **5**: e1366294.

Huang L, Lou CH, Chan W, Shum EY, Shao A, Stone E, Karam R, Song HW, Wilkinson MF. 2011. RNA homeostasis governed by cell type-specific and branched feedback loops acting on NMD. *Mol Cell* **43**: 950–961.

Hug N, Caceres JF. 2014. The RNA helicase DHX34 activates NMD by promoting a transition from the surveillance to the decay-inducing complex. *Cell Rep* **8**: 1845–1856.

Huntzinger E, Kashima I, Fauser M, Sauliere J, Izaurralde E. 2008. SMG6 is the catalytic endonuclease that cleaves mRNAs containing nonsense codons in metazoan. *RNA* **14**: 2609–2617.

Hurt JA, Robertson AD, Burge CB. 2013. Global analyses of UPF1 binding and function reveal expanded scope of nonsense-mediated mRNA decay. *Genome Res* **23**: 1636–1650.

Hwang J, Sato H, Tang Y, Matsuda D, Maquat LE. 2010. UPF1 association with the cap-binding protein, CBP80, promotes nonsense-mediated mRNA decay at two distinct steps. *Mol Cell* **39**: 396–409.

Isken O, Kim YK, Hosoda N, Mayeur GL, Hershey JW, Maquat LE. 2008. Upf1 phosphorylation triggers translational repression during nonsense-mediated mRNA decay. *Cell* **133**: 314–327.

Ivanov PV, Gehring NH, Kunz JB, Hentze MW, Kulozik AE. 2008. Interactions between UPF1, eRFs, PABP and the exon junction complex suggest an integrated model for mammalian NMD pathways. *EMBO J* **27**: 736–747.

Ivanov A, Mikhailova T, Eliseev B, Yeramala L, Sokolova E, Susorov D, Shuvalov A, Schaffitzel C, Alkalaeva E. 2016. PABP enhances release factor recruitment and stop codon recognition during translation termination. *Nucleic Acids Res* **44**: 7766–7776.

Johansson MJ, He F, Spatrick P, Li C, Jacobson A. 2007. Association of yeast Upf1p with direct substrates of the NMD pathway. *Proc Natl Acad Sci* **104**: 20872–20877.

Johns L, Grimson A, Kuchma SL, Newman CL, Anderson P. 2007. *Caenorhabditis elegans* SMG-2 selectively marks mRNAs containing premature translation termination codons. *Mol Cell Biol* **27**: 5630–5638.

Jonas S, Weichenrieder O, Izaurralde E. 2013. An unusual arrangement of two 14-3-3-like domains in the SMG5-SMG7 heterodimer is required for efficient nonsense-mediated mRNA decay. *Genes Dev* **27**: 211–225.

Joncourt R, Eberle AB, Rufener SC, Mühlemann O. 2014. Eukaryotic initiation factor 4G suppresses nonsense-mediated mRNA decay by two genetically separable mechanisms. *PLoS ONE* **9**: e104391.

Kadlec J, Izaurralde E, Cusack S. 2004. The structural basis for the interaction between nonsense-mediated mRNA decay factors UPF2 and UPF3. *Nat Struct Mol Biol* **11**: 330–337.

Kashima I, Yamashita A, Izumi N, Kataoka N, Morishita R, Hoshino S, Ohno M, Dreyfuss G, Ohno S. 2006. Binding of a novel SMG-1-Upf1-eRF1-eRF3 complex (SURF) to the exon junction complex triggers Upf1 phosphorylation and nonsense-mediated mRNA decay. *Genes Dev* **20**: 355–367.

Kashima I, Jonas S, Jayachandran U, Buchwald G, Conti E, Lupas AN, Izaurralde E. 2010. SMG6 interacts with the exon junction complex via two conserved EJC-binding motifs (EBMs) required for nonsense-mediated mRNA decay. *Genes Dev* **24**: 2440–2450.

Kertesz S, Kerenyi Z, Merai Z, Bartos I, Palfy T, Barta E, Silhavy D. 2006. Both introns and long 3′-UTRs operate as *cis*-acting elements to trigger nonsense-mediated decay in plants. *Nucleic Acids Res* **34**: 6147–6157.

Kervestin S, Jacobson A. 2012. NMD: A multifaceted response to premature translational termination. *Nat Rev Mol Cell Biol* **13**: 700–712.

Kiosze-Becker K, Ori A, Gerovac M, Heuer A, Nurenberg-Goloub E, Rashid UJ, Becker T, Beckmann R, Beck M, Tampe R. 2016. Structure of the ribosome post-recycling complex probed by chemical cross-linking and mass spectrometry. *Nat Commun* **7**: 13248.

Kunz JB, Neu-Yilik G, Hentze MW, Kulozik AE, Gehring NH. 2006. Functions of hUpf3a and hUpf3b in nonsense-mediated mRNA decay and translation. *RNA* **12**: 1015–1022.

Kurosaki T, Li W, Hoque M, Popp MW, Ermolenko DN, Tian B, Maquat LE. 2014. A post-translational regulatory switch on UPF1 controls targeted mRNA degradation. *Genes Dev* **28**: 1900–1916.

Lai T, Cho H, Liu Z, Bowler MW, Piao S, Parker R, Kim YK, Song H. 2012. Structural basis of the PNRC2-mediated link between mRNA surveillance and decapping. *Structure* **20**: 2025–2037.

Lee SR, Pratt GA, Martinez FJ, Yeo GW, Lykke-Andersen J. 2015. Target discrimination in nonsense-mediated mRNA decay requires Upf1 ATPase activity. *Mol Cell* **59**: 413–425.

Leeds P, Peltz SW, Jacobson A, Culbertson MR. 1991. The product of the yeast *UPF1* gene is required for rapid turnover of mRNAs containing a premature translational termination codon. *Genes Dev* **5**: 2303–2314.

Leeds P, Wood JM, Lee BS, Culbertson MR. 1992. Gene products that promote mRNA turnover in *Saccharomyces cerevisiae*. *Mol Cell Biol* **12**: 2165–2177.

Le Hir H, Izaurralde E, Maquat LE, Moore MJ. 2000a. The spliceosome deposits multiple proteins 20–24 nucleotides upstream of mRNA exon–exon junctions. *EMBO J* **19**: 6860–6869.

Le Hir H, Moore MJ, Maquat LE. 2000b. Pre-mRNA splicing alters mRNP composition: Evidence for stable association of proteins at exon–exon junctions. *Genes Dev* **14**: 1098–1108.

Le Hir H, Gatfield D, Izaurralde E, Moore MJ. 2001. The exon–exon junction complex provides a binding platform for factors involved in mRNA export and nonsense-mediated mRNA decay. *EMBO J* **20**: 4987–4997.

Le Hir H, Sauliere J, Wang Z. 2016. The exon junction complex as a node of post-transcriptional networks. *Nat Rev Mol Cell Biol* **17**: 41–54.

Lejeune F, Li X, Maquat LE. 2003. Nonsense-mediated mRNA decay in mammalian cells involves decapping, deadenylating, and exonucleolytic activities. *Mol Cell* **12**: 675–687.

Loh B, Jonas S, Izaurralde E. 2013. The SMG5–SMG7 heterodimer directly recruits the CCR4–NOT deadenylase complex to mRNAs containing nonsense codons via interaction with POP2. *Genes Dev* **27**: 2125–2138.

Longman D, Plasterk RH, Johnstone IL, Caceres JF. 2007. Mechanistic insights and identification of two novel factors in the *C. elegans* NMD pathway. *Genes Dev* **21**: 1075–1085.

Longman D, Hug N, Keith M, Anastasaki C, Patton EE, Grimes G, Caceres JF. 2013. DHX34 and NBAS form part of an autoregulatory NMD circuit that regulates endogenous RNA targets in human cells, zebrafish and *Caenorhabditis elegans*. *Nucleic Acids Res* **41**: 8319–8331.

Lopez-Perrote A, Castano R, Melero R, Zamarro T, Kurosawa H, Ohnishi T, Uchiyama A, Aoyagi K, Buchwald G, Kataoka N, et al. 2016. Human nonsense-mediated mRNA decay factor UPF2 interacts directly with eRF3 and the SURF complex. *Nucleic Acids Res* **44**: 1909–1923.

Luke B, Azzalin CM, Hug N, Deplazes A, Peter M, Lingner J. 2007. *Saccharomyces cerevisiae* Ebs1p is a putative ortholog of human Smg7 and promotes nonsense-mediated mRNA decay. *Nucleic Acids Res* **35**: 7688–7697.

Lykke-Andersen J. 2002. Identification of a human decapping complex associated with hUpf proteins in nonsense-mediated decay. *Mol Cell Biol* **22**: 8114–8121.

Lykke-Andersen S, Jensen TH. 2015. Nonsense-mediated mRNA decay: An intricate machinery that shapes transcriptomes. *Nat Rev Mol Cell Biol* **16**: 665–677.

Lykke-Andersen J, Shu MD, Steitz JA. 2000. Human Upf proteins target an mRNA for nonsense-mediated decay when bound downstream of a termination codon. *Cell* **103**: 1121–1131.

Lykke-Andersen S, Chen Y, Ardal BR, Lilje B, Waage J, Sandelin A, Jensen TH. 2014. Human nonsense-mediated RNA decay initiates widely by endonucleolysis and targets snoRNA host genes. *Genes Dev* **28**: 2498–2517.

Maderazo AB, Belk JP, He F, Jacobson A. 2003. Nonsense-containing mRNAs that accumulate in the absence of a functional nonsense-mediated mRNA decay pathway are destabilized rapidly upon its restitution. *Mol Cell Biol* **23**: 842–851.

Maquat LE. 2004. Nonsense-mediated mRNA decay: A comparative analysis of different species. *Current Genomics* **5**: 175–190.

Maquat LE, Tarn WY, Isken O. 2010. The pioneer round of translation: Features and functions. *Cell* **142**: 368–374.

Melero R, Buchwald G, Castano R, Raabe M, Gil D, Lazaro M, Urlaub H, Conti E, Llorca O. 2012. The cryo-EM structure of the UPF–EJC complex shows UPF1 poised toward the RNA 3′ end. *Nat Struct Mol Biol* **19**: 498–505.

Melero R, Uchiyama A, Castano R, Kataoka N, Kurosawa H, Ohno S, Yamashita A, Llorca O. 2014. Structures of SMG1-UPFs complexes: SMG1 contributes to regulate UPF2-dependent activation of UPF1 in NMD. *Structure* **22**: 1105–1119.

Melero R, Hug N, Lopez-Perrote A, Yamashita A, Caceres JF, Llorca O. 2016. The RNA helicase DHX34 functions as a scaffold for SMG1-mediated UPF1 phosphorylation. *Nat Commun* **7**: 10585.

Metze S, Herzog VA, Ruepp MD, Mühlemann O. 2013. Comparison of EJC-enhanced and EJC-independent NMD in human cells reveals two partially redundant degradation pathways. *RNA* **19**: 1432–1448.

Min EE, Roy B, Amrani N, He F, Jacobson A. 2013. Yeast Upf1 CH domain interacts with Rps26 of the 40S ribosomal subunit. *RNA* **19**: 1105–1115.

Moriarty PM, Reddy CC, Maquat LE. 1998. Selenium deficiency reduces the abundance of mRNA for Se-dependent glutathione peroxidase 1 by a UGA-dependent mechanism likely to be nonsense codon-mediated decay of cytoplasmic mRNA. *Mol Cell Biol* **18**: 2932–2939.

Morris C, Wittmann J, Jack HM, Jalinot P. 2007. Human INT6/eIF3e is required for nonsense-mediated mRNA decay. *EMBO Rep* **8**: 596–602.

Mühlemann O, Lykke-Andersen J. 2010. How and where are nonsense mRNAs degraded in mammalian cells? *RNA Biol* **7**: 28–32.

Muhlrad D, Parker R. 1999. Aberrant mRNAs with extended 3′UTRs are substrates for rapid degradation by mRNA surveillance. *RNA* **5**: 1299–1307.

Naeger LK, Schoborg RV, Zhao Q, Tullis GE, Pintel DJ. 1992. Nonsense mutations inhibit splicing of MVM RNA in *cis* when they interrupt the reading frame of either exon of the final spliced product. *Genes Dev* **6**: 1107–1119.

Nasif S, Contu L, Mühlemann O. 2017. Beyond quality control: The role of nonsense-mediated mRNA decay (NMD) in regulating gene expression. *Semin Cell Dev Biol* **75**: 78–87.

Neu-Yilik G, Raimondeau E, Eliseev B, Yeramala L, Amthor B, Deniaud A, Huard K, Kerschgens K, Hentze MW, Schaffitzel C, et al. 2017. Dual function of UPF3B in early and late translation termination. *EMBO J* **36**: 2968–2986.

Nicholson P, Josi C, Kurosawa H, Yamashita A, Mühlemann O. 2014. A novel phosphorylation-independent interaction between SMG6 and UPF1 is essential for human NMD. *Nucleic Acids Res* **42**: 9217–9235.

Nickless A, Bailis JM, You Z. 2017. Control of gene expression through the nonsense-mediated RNA decay pathway. *Cell Biosci* **7**: 26.

Ohnishi T, Yamashita A, Kashima I, Schell T, Anders KR, Grimson A, Hachiya T, Hentze MW, Anderson P, Ohno S. 2003. Phosphorylation of hUPF1 induces formation of mRNA surveillance complexes containing hSMG-5 and hSMG-7. *Mol Cell* **12**: 1187–1200.

Okada-Katsuhata Y, Yamashita A, Kutsuzawa K, Izumi N, Hirahara F, Ohno S. 2012. N- and C-terminal Upf1 phosphorylations create binding platforms for SMG-6 and SMG-5:SMG-7 during NMD. *Nucleic Acids Res* **40**: 1251–1266.

Ottens F, Boehm V, Sibley CR, Ule J, Gehring NH. 2017. Transcript-specific characteristics determine the contribution of endo- and exonucleolytic decay pathways during the degradation of nonsense-mediated decay substrates. *RNA* **23**: 1224–1236.

Palacios IM, Gatfield D, St Johnston D, Izaurralde E. 2004. An eIF4AIII-containing complex required for mRNA localization and nonsense-mediated mRNA decay. *Nature* **427**: 753–757.

Peixeiro I, Inacio A, Barbosa C, Silva AL, Liebhaber SA, Romao L. 2011. Interaction of PABPC1 with the translation initiation complex is critical to the NMD resistance of AUG-proximal nonsense mutations. *Nucleic Acids Res* **40**: 1160–1173.

Pisarev AV, Skabkin MA, Pisareva VP, Skabkina OV, Rakotondrafara AM, Hentze MW, Hellen CU, Pestova TV. 2010. The role of ABCE1 in eukaryotic posttermination ribosomal recycling. *Mol Cell* **37**: 196–210.

Pulak R, Anderson P. 1993. mRNA surveillance by the *Caenorhabditis elegans smg* genes. *Genes Dev* **7**: 1885–1897.

Rehwinkel J, Raes J, Izaurralde E. 2006. Nonsense-mediated mRNA decay: Target genes and functional diversification of effectors. *Trends Biochem Sci* **31**: 639–646.

Reichenbach P, Hoss M, Azzalin CM, Nabholz M, Bucher P, Lingner J. 2003. A human homolog of yeast est1 associates with telomerase and uncaps chromosome ends when overexpressed. *Curr Biol* **13**: 568–574.

Rufener SC, Mühlemann O. 2013. eIF4E-bound mRNPs are substrates for nonsense-mediated mRNA decay in mammalian cells. *Nat Struct Mol Biol* **20**: 710–717.

Sauliere J, Murigneux V, Wang Z, Marquenet E, Barbosa I, Le Tonqueze O, Audic Y, Paillard L, Roest Crollius H, Le Hir H. 2012. CLIP-seq of eIF4AIII reveals transcriptome-wide mapping of the human exon junction complex. *Nat Struct Mol Biol* **19**: 1124–1131.

Schmidt SA, Foley PL, Jeong DH, Rymarquis LA, Doyle F, Tenenbaum SA, Belasco JG, Green PJ. 2014. Identification of SMG6 cleavage sites and a preferred RNA cleavage motif by global analysis of endogenous NMD targets in human cells. *Nucleic Acids Res* **43**: 309–323.

Serin G, Gersappe A, Black JD, Aronoff R, Maquat LE. 2001. Identification and characterization of human orthologues to *Saccharomyces cerevisiae* Upf2 protein and Upf3 protein (*Caenorhabditis elegans* SMG-4). *Mol Cell Biol* **21**: 209–223.

Shao S, Murray J, Brown A, Taunton J, Ramakrishnan V, Hegde RS. 2016. Decoding mammalian ribosome-mRNA states by translational GTPase complexes. *Cell* **167**: 1229–1240.e1215.

Shirokikh NE, Archer SK, Beilharz TH, Powell D, Preiss T. 2017. Translation complex profile sequencing to study the in vivo dynamics of mRNA-ribosome interactions during translation initiation, elongation, and termination. *Nat Protoc* **12**: 697–731.

Shoemaker CJ, Green R. 2011. Kinetic analysis reveals the ordered coupling of translation termination and ribosome recycling in yeast. *Proc Natl Acad Sci* **108**: E1392–E1398.

Shum EY, Jones SH, Shao A, Dumdie J, Krause MD, Chan WK, Lou CH, Espinoza JL, Song HW, Phan MH, et al. 2016. The antagonistic gene paralogs Upf3a and Upf3b govern nonsense-mediated RNA decay. *Cell* **165:** 382–395.

Silva AL, Ribeiro P, Inacio A, Liebhaber SA, Romao L. 2008. Proximity of the poly(A)-binding protein to a premature termination codon inhibits mammalian nonsense-mediated mRNA decay. *RNA* **14:** 563–576.

Simms CL, Thomas EN, Zaher HS. 2017. Ribosome-based quality control of mRNA and nascent peptides. *Wiley Interdiscip Rev RNA* doi: 10.1002/wrna.1366.

Singh G, Rebbapragada I, Lykke-Andersen J. 2008. A competition between stimulators and antagonists of Upf complex recruitment governs human nonsense-mediated mRNA decay. *PLoS Biol* **6:** e111.

Singh G, Kucukural A, Cenik C, Leszyk JD, Shaffer SA, Weng Z, Moore MJ. 2012. The cellular EJC interactome reveals higher-order mRNP structure and an EJC-SR protein nexus. *Cell* **151:** 750–764.

Stalder L, Mühlemann O. 2008. The meaning of nonsense. *Trends Cell Biol* **18:** 315–321.

Stalder L, Mühlemann O. 2009. Processing bodies are not required for mammalian nonsense-mediated mRNA decay. *RNA* **15:** 1265–1273.

Thermann R, Neu-Yilik G, Deters A, Frede U, Wehr K, Hagemeier C, Hentze MW, Kulozik AE. 1998. Binary specification of nonsense codons by splicing and cytoplasmic translation. *EMBO J* **17:** 3484–3494.

Tian M, Yang W, Zhang J, Dang H, Lu X, Fu C, Miao W. 2017. Nonsense-mediated mRNA decay in *Tetrahymena* is EJC independent and requires a protozoa-specific nuclease. *Nucleic Acids Res* **45:** 6848–6863.

Toma KG, Rebbapragada I, Durand S, Lykke-Andersen J. 2015. Identification of elements in human long 3′UTRs that inhibit nonsense-mediated decay. *RNA* **21:** 887–897.

Trcek T, Sato H, Singer RH, Maquat LE. 2013. Temporal and spatial characterization of nonsense-mediated mRNA decay. *Genes Dev* **27:** 541–551.

Unterholzner L, Izaurralde E. 2004. SMG7 acts as a molecular link between mRNA surveillance and mRNA decay. *Mol Cell* **16:** 587–596.

Wang W, Czaplinski K, Rao Y, Peltz SW. 2001. The role of Upf proteins in modulating the translation read-through of nonsense-containing transcripts. *EMBO J* **20:** 880–890.

Wang C, Han B, Zhou R, Zhuang X. 2016. Real-time imaging of translation on single mRNA transcripts in live cells. *Cell* **165:** 990–1001.

Weil JE, Beemon KL. 2006. A 3′UTR sequence stabilizes termination codons in the unspliced RNA of Rous sarcoma virus. *RNA* **12:** 102–110.

Wells SE, Hillner PE, Vale RD, Sachs AB. 1998. Circularization of mRNA by eukaryotic translation initiation factors. *Mol Cell* **2:** 135–140.

Weng Y, Czaplinski K, Peltz SW. 1998. ATP is a cofactor of the Upf1 protein that modulates its translation termination and RNA binding activities. *RNA* **4:** 205–214.

Yamashita A. 2013. Role of SMG-1-mediated Upf1 phosphorylation in mammalian nonsense-mediated mRNA decay. *Genes Cells* **18:** 161–175.

Yamashita A, Ohnishi T, Kashima I, Taya Y, Ohno S. 2001. Human SMG-1, a novel phosphatidylinositol 3-kinase-related protein kinase, associates with components of the mRNA surveillance complex and is involved in the regulation of nonsense-mediated mRNA decay. *Genes Dev* **15:** 2215–2228.

Yamashita A, Izumi N, Kashima I, Ohnishi T, Saari B, Katsuhata Y, Muramatsu R, Morita T, Iwamatsu A, Hachiya T, et al. 2009. SMG-8 and SMG-9, two novel subunits of the SMG-1 complex, regulate remodeling of the mRNA surveillance complex during nonsense-mediated mRNA decay. *Genes Dev* **23:** 1091–1105.

Yepiskoposyan H, Aeschimann F, Nilsson D, Okoniewski M, Mühlemann O. 2011. Autoregulation of the nonsense-mediated mRNA decay pathway in human cells. *RNA* **17:** 2108–2118.

Young SK, Wek RC. 2016. Upstream open reading frames differentially regulate gene-specific translation in the integrated stress response. *J Biol Chem* **291:** 16927–16935.

Zund D, Gruber AR, Zavolan M, Mühlemann O. 2013. Translation-dependent displacement of UPF1 from coding sequences causes its enrichment in 3′UTRs. *Nat Struct Mol Biol* **20:** 936–943.

Riboswitches and Translation Control

Ronald R. Breaker

Department of Molecular, Cellular and Developmental Biology, Department of Molecular Biophysics and Biochemistry, Howard Hughes Medical Institute, Yale University, New Haven, Connecticut 06520-8103

Correspondence: ronald.breaker@yale.edu

A growing collection of bacterial riboswitch classes is being discovered that sense central metabolites, coenzymes, and signaling molecules. Included among the various mechanisms of gene regulation exploited by these RNA regulatory elements are several that modulate messenger RNA (mRNA) translation. In this review, the mechanisms of riboswitch-mediated translation control are summarized to highlight both their diversity and potential ancient origins. These mechanisms include ligand-gated presentation or occlusion of ribosome-binding sites, control of alternative splicing of mRNAs, and the regulation of mRNA stability. Moreover, speculation on the potential for novel riboswitch discoveries is presented, including a discussion on the potential for the discovery of a greater diversity of mechanisms for translation control.

Riboswitches are RNA gene-control structures commonly found in the 5′ untranslated regions (UTRs) of messenger RNAs (mRNAs) where they sense and respond to small molecule or ion ligands (Breaker 2011; Serganov and Nudler 2013; Sherwood and Henkin 2016). In most instances, binding of a target ligand to the aptamer domain of the riboswitch triggers changes in the folding pattern of the expression platform (Fig. 1A), which alters the level of gene expression by one or more of many possible mechanisms. Several diverse mechanisms for riboswitch-mediated gene regulation have been established (Fig. 1B), and herein I will focus primarily on the mechanisms that more directly involve the regulation of mRNA translation.

It is important to note that several other long-studied gene-control systems in bacteria exploit some of the same regulatory mechanisms as do riboswitches. For example, the classic at-tenuation mechanisms first discovered in the early 1970s (Yanofsky 2000) require that the translation machinery report on the abundance of an aminoacylated transfer RNA (tRNA) of interest (e.g., for tryptophan) to regulate expression of the main open reading frame (ORF). This sensory process involves a change in translation speed through a region of repetitive tryptophan codons in an upstream ORF (uORF), which leads to alternative folding of the nascent mRNA transcript and premature transcription termination if the aminoacylated tRNA is abundant. Similarly, tryptophan can be directly sensed by TRAP, which is a protein factor that forms multimers and binds the 5′UTR of its target mRNA when tryptophan is abundant (Antson et al. 1999). These types of gene-control systems, which appear to be quite common, require complex-folded biopolymers (protein or RNA factors) other than the 5′UTR to serve as

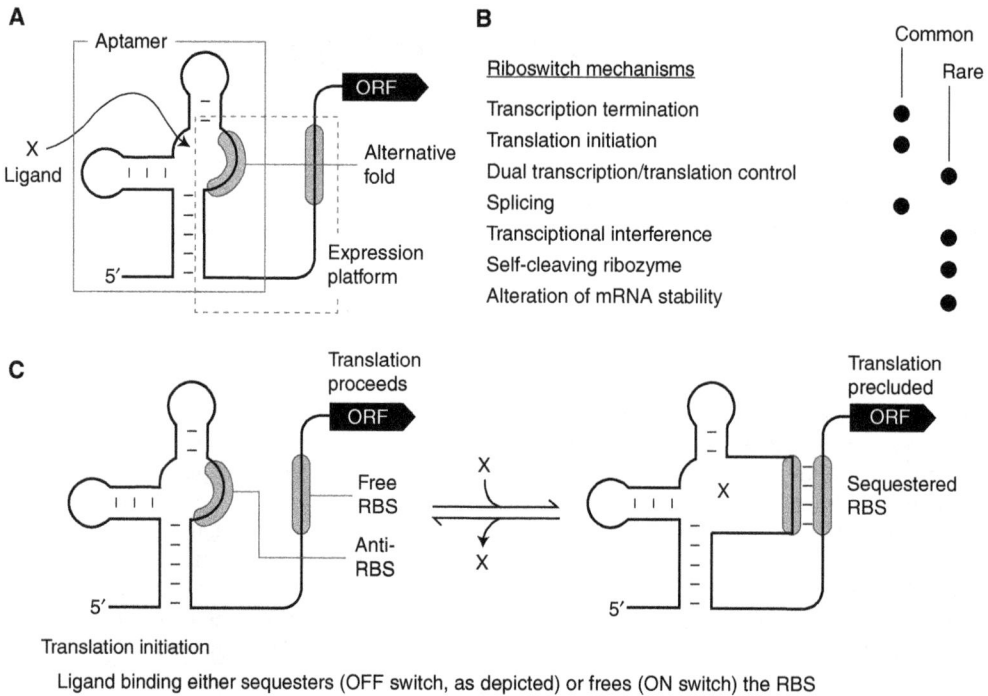

Figure 1. Schematic representations of common riboswitch expression platform arrangements. (*A*) Riboswitches typically carry a single ligand-binding aptamer (gray box) located upstream of (and slightly overlapping) the expression platform (dashed box). Folding changes in the aptamer, brought about by ligand binding, cause folding changes in the expression platform to regulate gene expression by various mechanisms. (*B*) List of experimentally validated or predicted riboswitch gene-control mechanisms. Processes by which the mechanisms highlighted in bold italic font regulate translation and are discussed in the text. (*C*) Schematic representation of a riboswitch that permits translation in the absence of ligand (*left*), but inhibits translation when bound to ligand (*right*). In the model depicted, ligand binding sequesters the ribosome-binding site (RBS) and prevents ribosome binding to the messenger RNA (mRNA). Alternatively, some riboswitch RNAs liberate the RBS on ligand binding to promote ribosome binding and translation.

the sensor for the target ligand. In contrast, riboswitches serve both as the ligand sensor (aptamer) and as the device (expression platform) that communicates the requirement to regulate gene expression. As a result, these systems have been classified separately (Nahvi et al. 2002; Breaker 2011).

Additional examples of regulatory systems that involve 5′UTR structures exist, including T box RNAs, which directly sense non-aminoacylated tRNAs (Green et al. 2010), and RNA thermosensors, which respond to specific temperatures (Kortmann and Narberhaus 2012). These systems are also usually classified separately from riboswitches even though they too use many of the same mechanisms for regulat-

ing gene expression. Specifically, T box RNAs indirectly sense their target amino acids by binding to tRNAs that lack an aminoacyl modification. Although Watson–Crick base-pairing is involved in RNA-RNA (tRNA-T box) complex formation (Grigg et al. 2013), base-pairing appears to be less extensive than that observed for short RNA (sRNA) regulators that bind to 5′UTRs to trigger expression changes (Gottesman and Storz 2011). Interestingly, thermosensors do not need to form a binding site for a ligand at all, given that they respond to changes in temperature. Although all RNA structures can respond to heat as their thermal melting temperatures are approached, biologically relevant thermosensors need to form a metastable

Cite this article as *Cold Spring Harb Perspect Biol* doi: 10.1101/cshperspect.a032797

structure that is precisely tuned to change gene expression at a temperature useful to the cell. This appears to be a relatively easy demand to meet, as RNA thermosensor domains (Grosso-Becera et al. 2015) are typically far simpler than the complex structures formed by riboswitch aptamers (McCown et al. 2017).

The methods used to discover novel classes of riboswitch candidates have had to be modified over the years to continue to yield new findings. This necessity can be understood by recognizing that data hinting that riboswitches indeed existed were derived from bacterial genetics experiments, and that the riboswitches first observed are members of the most common classes. For example, some of the earliest genetic data sets suggesting the existence of riboswitches for lysine (Boy et al. 1979), flavin mononucleotide (FMN) (Gusarov et al. 1997), and adenosylcobalamin (AdoCbl) (Lundrigan et al. 1991) riboswitches were generated and published years before definitive experimental proofs of these riboswitches were made. These three riboswitch classes are among the 10 most abundant classes known. Therefore, they were more likely than rare riboswitch classes to have been encountered by genetics-based mutational screening or other genetic studies.

As each newly found common riboswitch class has been revealed, it becomes increasingly more difficult to uncover evidence pointing to the remaining undiscovered classes. This is because the undiscovered classes tend to be rarer (fewer number of representatives) than those discovered previously (McCown et al. 2017). It became clear that genetic methods would not be effective in searching for additional riboswitch classes. Fortunately, new methods involving comparative sequence analyses that are performed by powerful computer algorithms have been used with much success to search for additional riboswitch classes (e.g., see Barrick et al. 2004; Weinberg et al. 2007, 2010). With improvements in these computation search methods, coupled with the expanding collection of genomic DNA sequence data, it seems certain that many additional riboswitch classes and representatives of many other types of RNA-based gene-control systems will be discovered in the coming years.

More than 38 distinct classes of ligand-binding riboswitches have been experimentally validated to date (Arachchilage et al. 2017; McCown et al. 2017), and it has been proposed (Ames and Breaker 2010; Breaker 2011; McCown et al. 2017) that many hundreds of additional classes remain to be discovered just among the bacterial genomes whose sequences have previously been determined. Thus, it also seems certain that a greater diversity of structures and mechanisms will be uncovered for riboswitches, and some of these might further blur the lines between the various types of RNA-based regulatory systems noted above. Currently, some of the most common mechanisms for riboswitch-mediated gene control involve the direct regulation of translation initiation, or the inhibition of protein production more indirectly by altering mRNA stability or by changing the primary sequence of mRNAs via alternative splicing. The details of some of these riboswitch mechanisms for translation control are discussed below.

DIRECT TRANSLATION CONTROL BY RIBOSWITCHES

After intrinsic transcription terminator control, the next-most-common gene-control mechanism used by many of the known riboswitch classes involves the occlusion or liberation of the ribosome-binding site ([RBS] or Shine–Dalgarno [SD] sequence) (Barrick and Breaker 2007). RBS sequences typically involve a purine-rich segment of approximately six nucleotides located a short distance upstream of the translation start codon, and these small genetic elements are present in ~77% of the protein-coding mRNAs among thousands of bacterial species examined (Omotajo et al. 2015). The RBS sequence can form a Watson–Crick base-pairing interaction with 16S ribosomal RNA in the process of translation initiation (Laursen et al. 2005). Thus, ligand binding to a riboswitch aptamer can direct folding of an expression platform to either display the RBS for access by ribosomes, or block ribosomes from binding to the mRNA (Fig. 1C).

Numerous examples of riboswitch classes that use this direct control of translation are predicted to exist (Barrick and Breaker 2007). However, only a few examples have been experimentally tested by examining individual riboswitch representatives. By using comparative sequence analysis methods, the precise mechanisms for regulating RBS access were divided into three specific types. For the first type, the RBS could serve as an integral part of the aptamer structure (Fig. 1C), and therefore ligand binding precludes ribosome access to yield a riboswitch that functions as a genetic OFF switch in the presence of ligand. Riboswitches for AdoCbl or its close derivatives use this mechanism, as shown at atomic resolution from X-ray crystallography data (Johnson et al. 2012).

A second type has the RBS located at some distance downstream of the aptamer structure. In this case, the RBS does not directly participate in the formation of the aptamer. Therefore, controlling ribosome access must involve alternative folding of structures apart from the aptamer that either hide or openly display the RBS. This mechanism, which was originally proposed for some bacterial thiamin pyrophosphate (TPP) riboswitches (Miranda-Ríos et al. 2001; Rodionov et al. 2002; Winkler et al. 2002), typically involves the use of "anti-RBS" sequences that can base-pair to the RBS sequence in a ligand-regulated manner. As a result, this type of regulation can yield either genetic ON or OFF switches, depending on whether ligand binding to the riboswitch aptamer sequesters the anti-RBS sequence or deploys it. Experimental evidence for TPP riboswitches of this mechanistic type includes (1) ligand-triggered changes in the structural flexibility of the RBS (Winkler et al. 2002; Rentmeister et al. 2007), (2) effects on gene expression with constructs carrying mutations in the anti-RBS sequence (Winkler et al. 2002), (3) monitoring of structural equilibria between translation competent and inactive states (Lang et al. 2007), (4) ligand-dependent modulation of 30S ribosomal subunit binding to mRNA (Ontiveros-Palacios et al. 2008), and (5) bioinformatic discovery of conserved competitor sequences (anti-anti-RBS) for anti-RBS elements (Rodionov et al. 2002; Chauvier et al. 2017).

A third type has the RBS sequence located within an intrinsic terminator stem, such that

formation of the terminator stem will both cease transcription before the ORF is made, and also sequester the RBS if the full-length mRNA has already been produced. This expression platform architecture is very common for TPP riboswitches, as observed via comparative sequence analysis of representatives from diverse bacterial species (Rodionov et al. 2002). Presumably, this dual mechanism is desirable because it offers exceptional regulatory and energy efficiency. For example, if plenty of TPP is available in the cell, ligand binding will terminate transcription before the region corresponding to the adjoining ORF has been fully transcribed. However, if TPP is low in concentration, full-length mRNAs can be made, but then can still permit RBS sequestration by the same TPP-binding interaction if this compound later attains sufficient concentration.

Although the TPP class has been the primary model riboswitch for the study of RBS-sequestration mechanisms, many other riboswitch classes also use similar strategies to control translation. As novel riboswitch classes are discovered, new opportunities become available to assess the types of gene-control mechanisms used by these ligand-binding RNAs. Indeed, bioinformatic analyses of validated riboswitches and riboswitch candidates frequently reveal sequence signatures that indicate their gene-control mechanisms involve RBS availability (e.g., see Vitreschak et al. 2002; Rodionov et al. 2003; Weinberg et al. 2007, 2010, 2017). The basic regulatory framework for each of these riboswitches will likely fall into one of the three general types described above. However, it is clear that regulation of translation initiation can be just the beginning of a more complex gene-control cascade. The two systems described below show the potential for complexity with riboswitches that initially regulate ribosome access to an RBS.

The lysine riboswitch associated with the *lysC* gene of *Escherichia coli* exploits an anti-RBS sequence to preclude translation of the ORF in the presence of its target amino acid (Caron et al. 2012). Elsewhere in this same lysine-bound structure, two adjacent locations for the action of RNase E are presented as non-base-paired RNA. As a result, the mRNA is targeted

Cite this article as *Cold Spring Harb Perspect Biol* doi: 10.1101/cshperspect.a032797

for degradation while simultaneously translation is precluded. In this case, the two mechanisms used to reduce the amount of protein made from the mRNA are not interdependent. It is not essential that translation cease by RBS occlusion for mRNA degradation to proceed, and vice versa. Lysine binding by the riboswitch aptamer is needed only to trigger a difference in the folded state of the RNA that simultaneously triggers the two unrelated events that both independently reduce gene expression.

Another example of possible dual-acting riboswitch mechanisms helps explain a longstanding paradox that was evident when the first riboswitch validation studies began to be published. It was speculated that AdoCbl riboswitches from *E. coli* (Nou and Kadner 2000; Nahvi et al. 2002) and *Salmonella typhimurium* (Nahvi et al. 2004) regulate translation, presumably by an RBS occlusion mechanism like that shown for AdoCbl riboswitches from other species (Johnson et al. 2012). However, why would cells exploit a riboswitch that controls translation of only the first gene in an operon that contains multiple ORFs? This is particularly puzzling for the *S. typhimurium* mRNA that carries the large *cob* operon coding for 25 proteins that function in the long biosynthetic pathway for coenzyme B_{12}. It would be exceedingly wasteful for the riboswitch to regulate the translation of only the first gene by occluding its RBS on AdoCbl binding, but still allow the cell to synthesize the entire mRNA and synthesize all the other 24 proteins encoded by this multi-ORF message.

It was speculated (Link and Breaker 2009; Peters et al. 2009; Roth and Breaker 2009; Breaker 2012) that perhaps the regulation of translation initiation triggers the action of the Rho termination factor (Kriner et al. 2016) that can terminate transcription before the rest of the mRNA can be synthesized. Specifically, inhibition of translation initiation prevents ribosomes from binding to the nascent mRNA, which otherwise would block the access of Rho to permit continued transcription and production of the complete mRNA. Intriguingly, a representative of an Mg^{2+}-II riboswitch and also a representative of an FMN riboswitch have been

reported to regulate Rho-dependent transcription termination (Hollands et al. 2012). However, these riboswitches do not necessarily achieve transcription termination by precluding translation initiation of the downstream ORF. Indeed, the FMN riboswitch example experiences Rho-mediated termination even before the start codon of the ORF is transcribed. Clearly, additional investigations will be needed to more fully explore the mechanisms by which riboswitches directly control translation, and to discover the associated regulatory processes that might ensue when riboswitches sense their cognate ligands.

RIBOZYME REGULATION OF TRANSLATION BY SMALL MOLECULES

Of course, ribosomes, the machines that catalyze the translation process, are directly involved in riboswitch-mediated gene regulation. Therefore, all bacterial riboswitch classes that use a mechanism that sequesters or reveals the ribosome-binding site of an mRNA are regulating ribozyme function by controlling the binding of 16S RNA to its mRNA target. So, in addition to the direct regulation of translation by riboswitches as described in the preceding section, there are several additional examples wherein riboswitches involve the action of ribozymes to indirectly regulate translation. These collaborations between riboswitches and likely ancient ribozyme classes provide intriguing evidence that some modern riboswitches might have their evolutionary origins in the RNA world (Vitreschak et al. 2004; Breaker 2012), a time before proteins emerged to take on many of these same roles.

The first such ribozyme-associated system to be described is the *glmS* riboswitch class (Barrick et al. 2004; McCown et al. 2012), whose members selectively bind the sugar derivative glucosamine-6-phosphate (GlcN6P) to promote RNA-catalyzed RNA strand scission (Winkler et al. 2004; Ferré-D'Amaré 2010). This self-cleaving ribozyme action promotes the rapid ribonuclease-mediated degradation of the adjoining ORF (Collins et al. 2007) coding for the protein enzyme GlmS (glutamine-fructose-6-phosphate amidotransferase). The regulatory logic established by this dual "ribozyme-riboswitch" is quite simple. GlmS proteins catalyze the biosynthesis of GlcN6P, which accumulates to levels that trigger the self-destructing *glmS* ribozyme to cleave the 5′ terminus from the full-length mRNA. However, the mRNA self-cleaves upstream of the coding region (Winkler et al. 2004), which left open the question of precisely how the action of the ribozyme ultimately prevents translation. It was eventually shown that, subsequent to ribozyme-mediated mRNA scission of the 5′UTR, RNase J1 recognizes the trimmed portion of the mRNA carrying the ORF and rapidly destroys this RNA to prevent further translation (Collins et al. 2007).

For the *glmS* ribozyme-riboswitch example, the riboswitch ligand (GlcN6P) is also the cofactor that directly promotes ribozyme-mediated RNA strand scission (Viladoms and Fedor 2012; Bingaman et al. 2017). However, riboswitches can control the action of other ribozymes by more remote mechanisms. In the next section on eukaryotic riboswitches, the regulation of alternative splicing by TPP riboswitches is presented in detail. Specifically, the spliceosomal apparatus that is so important for removing introns in many eukaryotic species is a ribozyme at its core (Fica et al. 2013), and there are several strategies used by fungal and plant riboswitches to regulate alternative splicing.

In addition, there are riboswitches associated with self-splicing ribozymes in bacteria that indirectly regulate translation. Two riboswitch classes are known to exist for the bacterial signaling molecule c-di-GMP (Sudarsan et al. 2008; Lee et al. 2010), the cyclic RNA dinucleotide made from two guanosine nucleotides (Römling et al. 2013). For example, there is a natural tandem arrangement in which a riboswitch aptamer for c-di-GMP was found (Lee et al. 2010) to reside immediately upstream of the 5′ splice site (5′SS) of a group I self-splicing ribozyme (Fig. 2) in the bacterial pathogen *Clostridium difficile* (recently renamed *Clostridioides difficile*) (Lawson et al. 2016). The riboswitch aptamer does not carry a typical expression platform like those that include either a transcription terminator or a structure to obscure an adjacent ribosome-binding site. Rather, ligand binding to the aptamer directly regulates ribozyme-medi-

ated splicing by controlling the structural environment of the 5′SS.

This tandem arrangement between a riboswitch aptamer and a self-splicing ribozyme exploits the typical activities of each RNA device to create a far more sophisticated gene-control apparatus. As expected, the riboswitch aptamer can form two distinct structures that are important for indirectly regulating mRNA translation. When c-di-GMP is bound by the aptamer, the adjacent ribozyme can adopt a structure that presents the proper 5′SS in its active site (Fig. 2A). Thus, the presence of c-di-GMP promotes the fusion of the 5′ and 3′ exons, the complete removal of the ribozyme intron, and the generation of an RBS sequence that permits translation of the processed mRNA (Fig. 2B, bottom) (Chen et al. 2011). In contrast, the absence of c-di-GMP binding allows a reorganization of the aptamer sequence that prevents the use of the 5′SS, and yields a truncated RNA processing product that is missing an RBS (Fig. 2B, top). This truncated mRNA therefore cannot be readily translated.

This tandem aptamer-ribozyme system has two other notable features that increase the genetic complexity shown by such RNA devices. First, the properly spliced mRNA retains the c-di-GMP aptamer, and its ability to form alternative base-pairs based on the presence or absence of c-di-GMP remains intact (Fig. 2C). Therefore, it seems possible that translation of the processed mRNA could yet be regulated by a more conventional riboswitch mechanism of occluding or presenting the RBS, long after the self-splicing ribozyme has been self-extracted. Second, group I ribozymes initiate self-splicing by promoting the nucleophilic attack by a guanosine nucleotide (such as GTP) on the 5′SS, which means that sufficient quantities of both c-di-GMP and GTP are needed to yield translation-ready mRNA products. This means that the tandem system initially functions as a two-input Boolean logic gate, wherein both molecular inputs are needed to promote gene expression (Lee et al. 2010).

Although this particular tandem aptamer-ribozyme architecture is very rare, and is only found in various strains of *C. difficile*, there are

Cite this article as *Cold Spring Harb Perspect Biol* doi: 10.1101/cshperspect.a032797

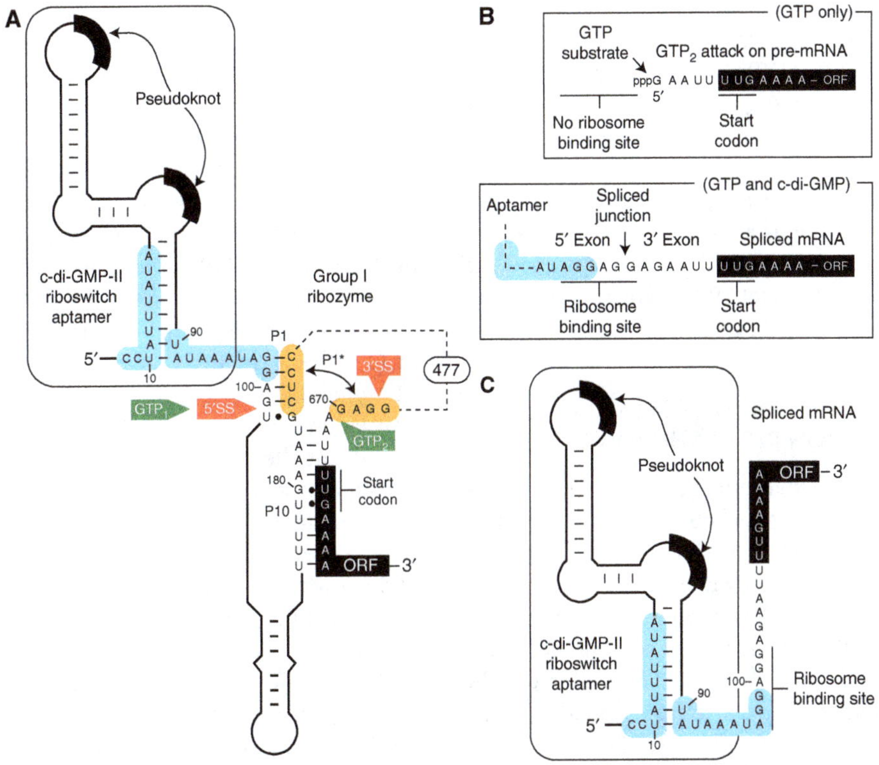

Figure 2. Regulation of self-splicing ribozyme function by a riboswitch aptamer. (*A*) Schematic representation of the 5′ untranslated region (5′UTR) of the CD3246 gene from *Clostridium difficile*, which carries a c-di-GMP-II riboswitch aptamer immediately upstream of a group I self-splicing ribozyme. The binding of c-di-GMP to the aptamer domain requires the structure depicted. This structure promotes ribozyme-catalyzed attack of GTP (called GTP$_1$) at the 5′ splice site (5′SS) and subsequent attack of the 5′ exon's 3′ terminus oxygen on the phosphorus of the 3′ splice site (3′SS) to yield a properly spliced messenger RNA (mRNA) product that can be translated. In the absence of c-di-GMP binding, portions of the aptamer and ribozyme shaded in blue reorganize to form an alternative base-paired structure. This disrupts the P1 stem that is required for proper splicing, and permits the formation of P1* to present an alternative site for GTP attack (called GTP$_2$). GTP attack at this alternative site removes all but four nucleotides located upstream of the unusual UUG start codon for the ORF. (*B*) Two reaction products produced by the tandem aptamer-ribozyme construct depicted in *A*. (*Top*) Processed mRNA product generated by ribozyme action in the absence of c-di-GMP binding. Note that the ribosome-binding site (RBS) has been removed, and thus translation of the downstream open reading frame (ORF) is precluded. (*Bottom*) Processed mRNA product resulting from ribozyme splicing in the presence of c-di-GMP. Splicing creates a strong RBS that permits translation to occur. (*C*) Although the properly spliced mRNA (*B, bottom*) carries an RBS, the riboswitch aptamer presumably can still operate to occlude or reveal this sequence based on the availability of c-di-GMP. Specifically, the nucleotides shaded in blue that alternatively pair to regulate ribozyme splicing activity remain present in the properly spliced mRNA, and presumably can be exploited as an expression platform to continue to regulate gene expression as would a more common riboswitch.

reasons to believe that other tandem arrangements will be discovered. Indeed, we have identified several examples wherein riboswitch aptamers appear in close proximity to group II self-splicing ribozymes. Moreover, there are large numbers of self-cleaving ribozymes that

are present in many bacterial and eukaryotic species (Jimenez et al. 2015). This creates many opportunities for riboswitch aptamers to regulate mRNA processing by controlling the action of ribozymes that cut RNA. Indeed, there exists some evidence that GlcN6P naturally affects the activity of a self-cleaving hepatitis delta virus (HDV)-class ribozyme from bacteria (Passalacqua et al. 2017). This regulation is different than that seen with *glmS* ribozymes that respond to the same compound (Winkler et al. 2004), as the effect is more subtle and appears to be via an allosteric mechanism rather than via the function of GlcN6P as a cofactor for the ribozyme. Moreover, it is not yet clear how this allosteric HDV ribozyme might ultimately regulate translation of the adjacent ORF.

EUKARYOTIC RIBOSWITCHES AND TRANSLATION REGULATION BY SPLICING

Confirmation that certain eukaryotic species exploit riboswitches was obtained by using comparative sequence analysis to uncover homologs of bacterial TPP riboswitches in fungi and plants (Sudarsan et al. 2003). Soon thereafter (Sudarsan et al. 2005), it was recognized that previously observed thiamin-dependent regulation of *thiA* pre-mRNA splicing in fungi (Kubodera et al. 2003) involves a representative of this riboswitch class, and that mutations in the TPP riboswitch were the source of resistance to the antimicrobial compound pyrithiamine in the fungus *Aspergillus oryzae* (Kubodera et al. 2000). These observations provided the first opportunities to examine the mechanisms by which eukaryotic riboswitches regulate gene expression outside of the bacterial and archaeal domains of life.

A detailed mechanism for eukaryotic TPP riboswitch function was first established by examining thiamin-mediated alternative splicing in the fungus *Neurospora crassa* (Cheah et al. 2007). This organism has three TPP riboswitches, each residing in an intron located in the 5′UTR of the *NMT1* mRNA, or disrupting the coding regions of the mRNAs for *THI4* and *NCU01977*. Extensive mutational analysis of riboswitch constructs in vivo revealed that the *NMT1* riboswitch regulates alternative splicing in a TPP-dependent manner by forming alternative RNA structures to sequester or display one of two possible 5′SS.

In the absence of TPP, a portion of the riboswitch aptamer sequence can form Watson–Crick base-pairs with nucleotides adjacent to and including a 5′SS located proximal to the 3′SS (Fig. 3A). Thus, the spliceosomal machinery must choose a distal 5′SS, which yields a processed mRNA that can be efficiently translated to produce the desired NMT1 protein (Fig. 3B). However, in the presence of TPP, the aptamer nucleotides are sequestered, which liberates the proximal 5′SS and yields a processed mRNA with three ORFs, the main coding region for NMT1 plus two uORFs that serve as translational decoys (Fig. 3C). Thus, in the presence of TPP, the alternatively spliced mRNA attracts ribosomes to the uORFs, rather than permitting translation of the main ORF (Cheah et al. 2007).

Similar alternative base-pairing interactions are important for regulation of alternative splicing by another series of fungal TPP riboswitches (Li and Breaker 2013). However, several features of these systems are distinct from that described for the NMT1 gene. First, these riboswitches reside in introns that interrupt the coding region of the pre-mRNA, and so complete removal of the intron is required for translation of the full ORF to occur, as shown for the *N. crassa* gene called *NCU01977*. When TPP is bound by the riboswitch, numerous alternative splicing products result from the apparent random use of numerous possible 3′SS. All these splicing products still carry portions of the intron, which causes premature translation termination. In contrast, the entire intron is removed when TPP concentration in cells is low.

The switching mechanism for riboswitches of the type found in *N. crassa NCU01977* involves alternative pairing between nucleotides (called α) that help form stem P1 when TPP is bound and a distal complementary region (called α′). Although these two complementary segments are separated by ~500 nucleotides, formation of the α-α′ base-paired structure brings the proper 5′SS and 3′SS in proximity to promote complete removal of the intron, and translation of the resulting ORF to produce the full protein product (Li and Breaker 2013).

Regulation of alternative splicing by riboswitches also appears to be a very common mechanism used by eukaryotic riboswitches from diverse species. For example, it has been proposed (Croft et al. 2007) that TPP binding by

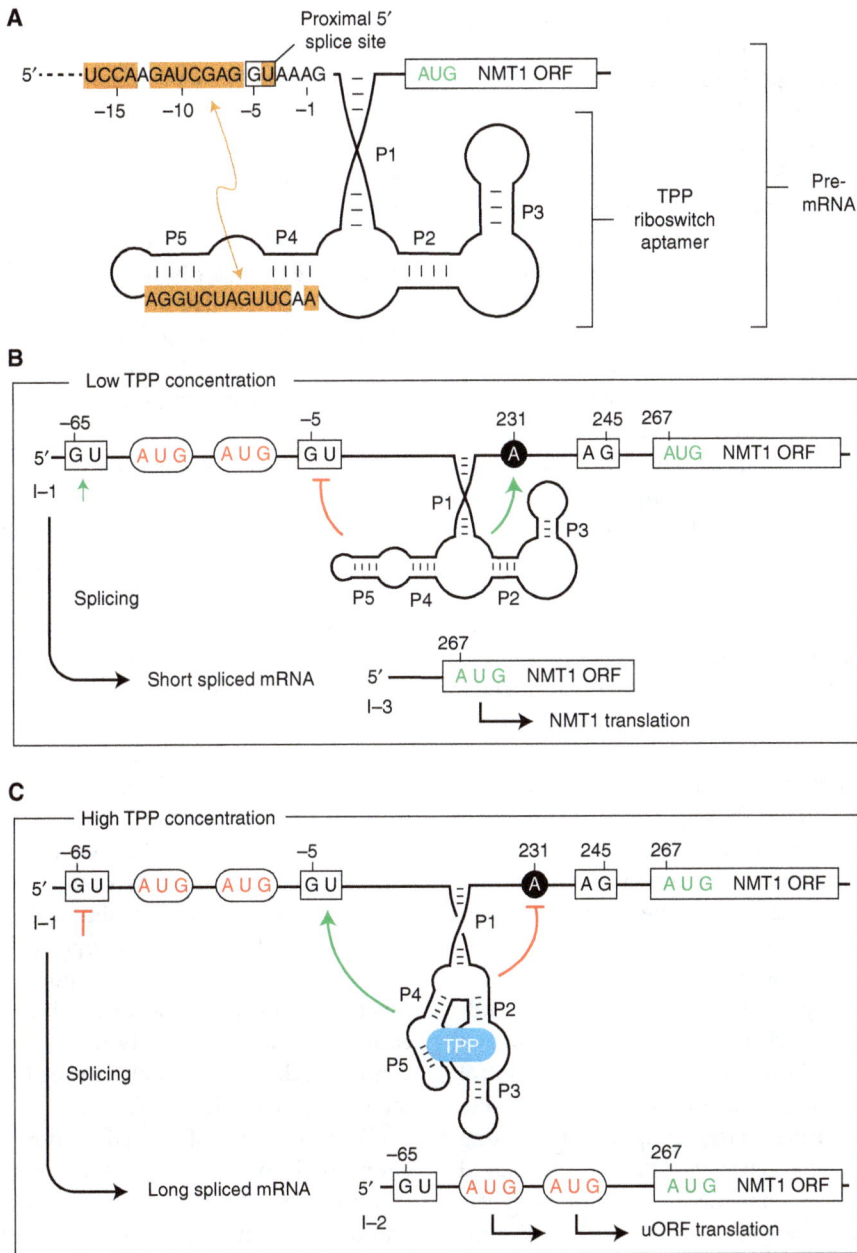

Figure 3. Translation regulation by a fungal thiamin pyrophosphate (TPP) riboswitch that controls alternative splicing. (*A*) Schematic representation of the region of the *Neurospora crassa NMT1* pre–messenger RNA (mRNA) region that includes an intron and TPP riboswitch. The TPP aptamer (including stems P1 through P5) is stabilized on the binding ligand as depicted, or nucleotides from the P4 and P5 region can alternatively base-pair to a region encompassing a 5′SS. (*B*) Results of mRNA splicing when the TPP concentration is low. The alternative base-pairing depicted in A will sequester the proximal 5′SS to promote the use of a distal 5′SS. This produces a short mRNA splicing product wherein the main *NMT1* open reading frame (ORF) can be translated. (*C*) TPP binding to the aptamer prevents occlusion of the proximal 5′SS, which results in a longer mRNA splicing product. This alternatively spliced RNA carries upstream ORFs (uORFs) that are translated instead of the main *NMT1* ORF.

a riboswitch in the *THIC* gene of the green alga *Chlamydomonas reinhardtii* promotes either of two alternative splicing events that lead to incomplete removal of an intron separating coding exons of the pre-mRNA. Translation machinery operating on the resulting processed mRNA yields a truncated protein that presumably is nonfunctional. In the absence of TPP, the intron structure rearranges to promote the use of splice sites resulting in removal of the entire intron, proper fusion of the ORF, and subsequent production of the full-length protein product.

An exceedingly common mechanism for TPP riboswitch-mediated control of gene expression in multicellular plants involves alternative splicing of an intron that resides downstream of the ORF. In this configuration, TPP binding to the riboswitch aptamer of thiamin-related transcripts promotes splicing of an intron in the 3′UTR of the pre-mRNA such that the polyadenylation signal sequence is removed (Bocobza et al. 2007; Wachter et al. 2007). Thus, TPP causes the destabilization of the transcript and reduces mRNA translation. In the absence of TPP, a portion of the riboswitch aptamer can base-pair with the 5′SS to prevent the removal of the intron and the polyadenylation signal. This base-pairing potential is conserved among numerous plant species (Wachter et al. 2007), suggesting that this mechanism of indirect regulation of translation by TPP riboswitch alteration of transcript polyadenylation and mRNA stability is widespread. To date, there are no known eukaryotic riboswitches that directly control translation without first making changes to the mRNA sequence by regulating RNA processing events. However, given the intimate roles played by tRNAs, ribosomal RNAs and various regulatory RNAs, there exist many opportunities for ligand-binding RNAs in eukaryotes to exploit translation control mechanisms other that those described above.

SPECULATION ON REGULATORY MECHANISMS AND CONTRIBUTIONS OF RIBOSWITCHES

It seems likely that there are thousands of distinct bacterial riboswitch classes that have yet to be discovered (Ames and Breaker 2010; Breaker 2011; McCown et al. 2017). Most of these classes will be present among the innumerable bacterial species whose genomes have yet to be sequenced. It has been estimated that ∼85% of the riboswitches encoded by the currently available genomic databases have already been identified (McCown et al. 2017), because members of the more than 38 riboswitch classes that have been discovered to date are so abundant. In other words, the many undiscovered riboswitch classes are so sparsely represented that they form only a small fraction of the total number of individual riboswitches.

If the assumptions that give rise to the numbers above are generally true, we can be confident in estimating the extent to which riboswitches directly regulate translation processes. Updated quantitation of the frequency of various riboswitch gene-control mechanisms is not available, but there are no indications from recent riboswitch discovery efforts to suggest that past tabulations (Barrick and Breaker 2007) are in error. Thus, it is likely that the two most common mechanisms by which riboswitches control gene expression in modern cells will remain (1) transcription termination via the regulated formation of intrinsic transcription terminator stems, followed closely in abundance by (2) translation initiation via the regulated display or occlusion of the ribosome-binding site. All other mechanisms are far rarer, although they each showcase the diversity of possible gene-control mechanisms exploited by regulatory domains made entirely of RNA.

After 15 years of intensive efforts to discover novel riboswitch classes, it is noteworthy that only one riboswitch class (TPP) has been found to be common in some eukaryotic lineages. Furthermore, it is not certain that additional riboswitch classes exist in eukaryotes (Bocobza and Aharoni 2014), particularly given that proteins undoubtedly serve as strong competition for RNAs as organisms evolve new sensors for various targets. However, the massive amount of noncoding RNA in eukaryotes, transcribed either as introns of protein-coding genes or as separately expressed transcripts such as long noncoding RNAs (e.g., Ulitsky

Cite this article as *Cold Spring Harb Perspect Biol* doi: 10.1101/cshperspect.a032797

and Bartel 2013), provides tremendous opportunities for ligand-sensing riboswitches to emerge and operate.

To date, no examples of metabolite- or ion-binding riboswitches have been discovered in humans or indeed any metazoans. If researchers wanted to initiate a concerted campaign to discover more eukaryotic riboswitches, how should they focus their searches? The genomic locations and biochemical functions of eukaryotic TPP riboswitches so far provide a strikingly uniform picture of what other eukaryotic riboswitches might look like if they indeed exist. All known TPP riboswitches in fungi and plants reside in introns, regardless of whether these introns are present in the 5'UTR, in intervening sequences disrupting the ORF, or in the 3'UTR. In each case, they appear to regulate alternative splicing, which ultimately regulates translation of the primary protein product either directly (determine integrity of main ORF) or indirectly (uORF regulation; determine mRNA stability). Eukaryotic TPP riboswitches have yet to be observed in RNA transcripts apart from the mRNAs for thiamin biosynthesis or transport proteins. Therefore, it seems prudent to focus future searches for eukaryotic riboswitches primarily on the alternatively spliced introns of metabolic genes, or genes that are regulated by signaling molecules. The introns of such genes would make excellent locations for metabolite-binding aptamers that selectively respond to their targets, and in the process can be removed by splicing so as to avoid interfering with the protein coding portion of the mRNA.

As noted above, such riboswitches might retain the ancient mechanism of working with RNA-based splicing machinery now present in spliceosomal RNAs to control the information flow from RNA to proteins. Eukaryotes do exploit some RNA-based signaling molecules, such as cAMP, cGMP, and certain cyclic dinucleotides, which have been proposed to reflect ancient signaling compounds from the RNA world (Nelson and Breaker 2017). Perhaps if riboswitches for additional signaling molecules derived from RNA nucleotides are discovered in bacteria, homologs will be uncovered in eukaryotes that respond to these same molecules. How-

ever, the ligands for eukaryotic riboswitches do not necessarily need to be of ancient origin. It is already apparent that eukaryotes largely lack examples of the most common and likely oldest classes of riboswitches being discovered in bacteria. If more complex eukaryotes do exploit riboswitches beyond just those for TPP, they might have more recently acquired the ability to respond to distinctly eukaryotic ligands such as hormones or other signaling molecules important only for multicellular life.

CONCLUDING REMARKS

The abundant information available on riboswitches shows that this mode of gene regulation relies heavily on mechanisms that directly regulate translation, primarily via controlling mRNA binding to ribosomes. However, a diversity of other mechanisms is used by riboswitches to regulate translation even after the full-length mRNA has been synthesized. Most of these processes involve intricate folding of the riboswitch aptamer and its associated expression platform, which must be achieved on a timescale that is relevant to the translation events that come later. These rapid RNA folding and ligand-binding events cannot easily be studied by traditional biochemical or genetic studies that involve examining purified RNAs in bulk or examining mutants in cell culture. Rather, some questions on the details of translation control will remain to be answered until biophysical analyses involving such techniques as fluorescent labeling, single-molecule analyses, and atomic-resolution modeling are fully used (Perez-Gonzalez et al. 2016).

Structural biologists have made truly remarkable progress in revealing the ligand-bound states of nearly all known riboswitch classes (Garst et al. 2011; Peselis and Serganov 2014; McCown et al. 2017). However, riboswitches have evolved to function with highly dynamic structures that can either change their shapes in real time based on ligand-binding events, or perhaps more commonly they make folding pathway decisions based on ligand binding while being actively synthesized by RNA polymerase. To fully understand the structural and kinetic properties of riboswitches, studies that can monitor individual RNAs in real time as they progress through various structural states or forks in their folding pathways will be necessary. Methods that use optical tweezers to evaluate RNA folding during the act of transcription (e.g., Frieda and Block 2012) appear to be particularly promising. However, monitoring similar kinetic processes that involve ribosome docking to a riboswitch-controlled RBS, or spliceosomal particles choosing splice sites whose availabilities are controlled by riboswitch folding, seem daunting.

Regardless of the challenges ahead, it is very important to continue to discover novel riboswitch classes and to examine the mechanistic details of their gene-control actions. This work inevitably expands our understanding of a common mode of gene control in bacteria, and reveals how small ligands can regulate complex processes like alternative splicing in eukaryotes. Furthermore, as the importance of synthetic biology grows in the future, simple and effective engineered genetic switches will be needed and riboswitches might be made to serve roles as designer sensors and switches. In particular, their ability to directly regulate protein synthesis could be exploited to yield long-lived mRNAs that quickly activate or deactivate translation by binding to externally supplied trigger molecules. Finally, each newly discovered natural riboswitch potentially provides a window into our evolutionary past. Even if modern riboswitches do not directly descend from ancient riboswitches, they yield clues that explain how RNAs can fold to sense diverse ligands and how these binding events can interface sometimes directly with modern ribozymes to regulate complex biochemical processes.

ACKNOWLEDGMENTS

R.R.B. thanks all members of the laboratory for many helpful discussions. Riboswitch research in the Breaker laboratory has been supported by the National Institutes of Health (NIH) (GM022778 and DE22340), Yale University, and by investigator support to R.R.B. from the Howard Hughes Medical Institute.

REFERENCES

Ames TD, Breaker RR. 2010. Bacterial riboswitch diversity and analysis. In *The chemical biology of nucleic acids* (ed. Mayer G), Chap. 20. Wiley, Chichester, UK.

Antson AA, Dodson EJ, Dodson G, Greaves RB, Chen X, Gollnick P. 1999. Structure of the *trp* RNA-binding attenuation protein, TRAP, bound to RNA. *Nature* 401: 235–242.

Arachchilage GM, Sherlock ME, Weinberg Z, Breaker RR. 2017. SAM-VI RNAs selectively bind *S*-adenosylmethionine and exhibit similarities to SAM-III riboswitches. *RNA Biol* doi: 10.1080/15476286.2017.1399232.

Barrick JE, Breaker RR. 2007. The distributions, mechanisms, and structures of metabolite-binding riboswitches. *Genome Biol* 8: R239.

Barrick JE, Corbino KA, Winkler WC, Nahvi A, Mandal M, Collins J, Lee M, Roth A, Sudarsan N, Jona I, et al. 2004. New RNA motifs suggest an expanded scope for riboswitches in bacterial genetic control. *Proc Natl Acad Sci* 101: 6421–6426.

Bingaman JL, Zhang S, Stevens DR, Yennawar NH, Hammes-Schiffer S, Bevilacqua PC. 2017. The GlcN6P cofactor plays multiple catalytic roles in the *glmS* ribozyme. *Nat Chem Biol* 13: 439–445.

Bocobza SE, Aharoni A. 2014. Small molecules that interact with RNA: Riboswitch-based gene control and its involvement in metabolic regulation in plants and algae. *Plant J* 79: 693–703.

Bocobza S, Adato A, Mandel T, Shapira M, Nudler E, Aharoni A. 2007. Riboswitch-dependent gene regulation and its evolution in the plant kingdom. *Genes Dev* 21: 2874–2879.

Boy E, Borne F, Patte JC. 1979. Isolation and identification of mutants constitutive for aspartokinase III synthesis in *Escherichia coli* K12. *Biochimie* 61: 1151–1160.

Breaker RR. 2011. Prospects for riboswitch discovery and analysis. *Mol Cell* 43: 867–879.

Breaker RR. 2012. Riboswitches and the RNA World. *Cold Spring Harb Perspect Biol* 4: a003566.

Caron MP, Bastet L, Lussier A, Simoneau-Roy M, Massé E, Lafontaine DA. 2012. Dual-acting riboswitch control of translation initiation and mRNA decay. *Proc Natl Acad Sci* 109: E3444–E3453.

Chauvier A, Picard-Jean F, Berger-Dancause JC, Bastet L, Naghdi MR, Dubé A, Turcotte P, Perreault J, Lafontaine DA. 2017. Transcriptional pausing at the translation start site operates as a critical checkpoint for riboswitch regulation. *Nat Commun* 8: 13892.

Cheah MT, Wachter A, Sudarsan N, Breaker RR. 2007. Control of alternative RNA splicing and gene expression by eukaryotic riboswitches. *Nature* 447: 497–501.

Chen AGY, Sudarsan N, Breaker RR. 2011. Mechanism for gene control by a natural allosteric group I ribozyme. *RNA* 17: 1967–1972.

Collins JA, Irnov I, Baker S, Winkler WC. 2007. Mechanism of mRNA destabilization by the *glmS* ribozyme. *Genes Dev* 21: 3356–3368.

Croft MT, Moulin M, Webb ME, Smith AG. 2007. Thiamine biosynthesis in algae is regulated by riboswitches. *Proc Natl Acad Sci* 104: 20770–20775.

Cite this article as *Cold Spring Harb Perspect Biol* doi: 10.1101/cshperspect.a032797

Ferré-D'Amaré AR. 2010. The *glmS* ribozyme: Use of a small molecule coenzyme by a gene-regulatory RNA. *Q Rev Biophys* **43**: 423–447.

Fica SM, Tuttle N, Novak T, Li N, Lu J, Koodathingal P, Dai Q, Staley JP, Piccirilli JA. 2013. RNA catalyses nuclear pre-pRNA splicing. *Nature* **503**: 229–234.

Frieda KL, Block SM. 2012. Direct observation of cotranscriptional folding in an adenine riboswitch. *Science* **338**: 397–400.

Garst AD, Edwards AL, Batey RT. 2011. Riboswitches: Structures and mechanisms. *Cold Spring Harb Perspect Biol* **3**: a003533.

Gottesman S, Storz G. 2011. Bacterial small RNA regulators: Versatile roles and rapidly evolving variations. *Cold Spring Harb Perspect Biol* **3**: a003798.

Green NJ, Grundy FJ, Henkin TM. 2010. The T box mechanism: tRNA as a regulatory molecule. *FEBS Lett* **584**: 318–324.

Grigg JC, Chen Y, Grundy FJ, Henkin TM, Pollack L, Ke A. 2013. T box RNA decodes both the information content and geometry of tRNA to affect gene expression. *Proc Natl Acad Sci* **110**: 7240–7245.

Grosso-Becera MV, Servín-González L, Soberon-Chávez G. 2015. RNA structures are involved in the thermoregulation of bacterial virulence-associated traits. *Trends Microbiol* **23**: 509–518.

Gusarov II, Kreneva RA, Podcharniaev DA, Iomantas IV, Abalakina EG, Stoinova NV, Perumov DA, Kozlov II. 1997. Riboflavin biosynthetic genes in *Bacillus amyloliquefaciens*: Primary structure, organization and regulation of activity. *Mol Biol (Mosk)* **31**: 446–453.

Hollands K, Proshkin S, Sklyarova S, Epshtein V, Mironov A, Nudler E, Groisman EA. 2012. Riboswitch control of Rho-dependent transcription termination. *Proc Natl Acad Sci* **109**: 5376–5381.

Jimenez RM, Polanco JA, Lupták A. 2015. Chemistry and biology of self-cleaving ribozymes. *Trends Biochem Sci* **40**: 648–661.

Johnson JE Jr, Reyes FE, Polaski JT, Batey RT. 2012. B_{12} cofactors directly stabilize an mRNA regulatory switch. *Nature* **492**: 133–137.

Kortmann J, Narberhaus F. 2012. Bacterial RNA thermometers: Molecular zippers and switches. *Nat Rev Microbiol* **10**: 255–265.

Kriner MA, Sevostyanova A, Groisman EA. 2016. Learning from the leaders: Gene regulation by the transcription termination factor Rho. *Trends Biochem Sci* **41**: 690–699.

Kubodera T, Yamashita N, Nishimura A. 2000. Pyrithiamine resistance gene (*ptrA*) of *Aspergillus oryzae*: Cloning, characterization and application as a dominant selectable marker for transformation. *Biosci Biotechnol Biochem* **64**: 1416–1421.

Kubodera T, Watanabe M, Yoshiuchi K, Yamashita N, Nishimura A, Nakai S, Gomi K, Hanamoto H. 2003. Thiamine-regulated gene expression of *Aspergillus oryzae thiA* requires splicing of the intron containing a riboswitch-like domain in the 5′-UTR. *FEBS Lett* **555**: 516–520.

Lang K, Rieder R, Micura R. 2007. Ligand-induced folding of the *thiM* TPP riboswitch investigated by a structure-based fluorescence spectroscopic approach. *Nucleic Acids Res* **35**: 5370–5378.

Laursen BS, Sørensen HP, Mortensen KK, Sperling-Petersen HU. 2005. Initiation of protein synthesis in bacteria. *Microbiol Mol Biol Rev* **69**: 101–123.

Lawson PA, Citron DM, Tyrrell KL, Finegold SM. 2016. Reclassification of *Clostridium difficile* as *Clostridioides difficile* (Hall and O'Toole 1935) Prévot 1938. *Anaerobe* **40**: 95–99.

Lee ER, Baker JL, Weinberg Z, Sudarsan N, Breaker RR. 2010. An allosteric self-splicing ribozyme triggered by a bacterial second messenger. *Science* **329**: 845–848.

Li S, Breaker RR. 2013. Eukaryotic TPP riboswitch regulation of alternative splicing involving long-distance base pairing. *Nucleic Acids Res* **41**: 3022–3031.

Link KH, Breaker RR. 2009. Engineering ligand-responsive gene-control elements: Lessons learned from natural riboswitches. *Gene Ther* **16**: 1189–1201.

Lundrigan MD, Koster W, Kadner RJ. 1991. Transcribed sequences of the *Escherichia coli btuB* gene control its expression and regulation by vitamin B_{12}. *Proc Natl Acad Sci* **88**: 1479–1483.

McCown PJ, Winkler WC, Breaker RR. 2012. Mechanism and distribution of *glmS* ribozymes. *Methods Mol Biol* **848**: 113–129.

McCown PJ, Corbino KA, Stav S, Sherlock ME, Breaker RR. 2017. Riboswitch diversity and distribution. *RNA* **23**: 995–1011.

Miranda-Ríos J, Navarro M, Soberón M. 2001. A conserved RNA structure (*thi* box) is involved in regulation of thiamin biosynthetic gene expression in bacteria. *Proc Natl Acad Sci* **98**: 9736–9741.

Nahvi A, Sudarsan N, Ebert MS, Zou X, Brown KL, Breaker RR. 2002. Genetic control by a metabolite binding mRNA. *Chem Biol* **9**: 1043–1049.

Nahvi A, Barrick JE, Breaker RR. 2004. Coenzyme B_{12} riboswitches are widespread genetic control elements in prokaryotes. *Nucleic Acids Res* **32**: 143–150.

Nelson JW, Breaker RR. 2017. The lost language of the RNA World. *Sci Signal* **10**: eaam8812.

Nou X, Kadner RJ. 2000. Adenosylcobalamin inhibits ribosome binding to *btuB* RNA. *Proc Natl Acad Sci* **97**: 7190–7195.

Omotajo D, Tate T, Cho H, Choudhary M. 2015. Distribution and diversity of ribosome binding sites in prokaryotic genomes. *BMC Genomics* **16**: 604.

Ontiveros-Palacios N, Smith Am, Grundy FJ, Soberon M, Henkin TM, Miranda-Ríos J. 2008. Molecular basis of gene regulation by the THI-box riboswitch. *Mol Microbiol* **67**: 793–803.

Passalacqua LFM, Jimenez RM, Fong JY, Lupták A. 2017. Allosteric modulation of the *Faecalibacterium prausnitzii* hepatitis delta virus-like ribozyme by glucosamine 6-phosphate: The substrate of the adjacent gene product. *Biochemistry* **56**: 6006–6014.

Perez-Gonzalez C, Grondin JP, Lafontaine DA, Carlos Penedo J. 2016. Biophysical approaches to bacterial gene regulation by riboswitches. *Adv Exp Med Biol* **915**: 157–191.

Peselis A, Serganov A. 2014. Themes and variations in riboswitch structure and function. *Biochim Biophys Acta* **1839:** 908–918.

Peters JM, Mooney RA, Kuan PF, Rowland JL, Keleş S, Landick R. 2009. Rho directs widespread termination of intragenic and stable RNA transcription. *Proc Natl Acad Sci* **106:** 15406–15411.

Rentmeister A, Mayer G, Kuhn N, Famulok M. 2007. Conformational changes in the expression of the *Escherichia coli thiM* riboswitch. *Nucleic Acids Res* **35:** 3713–3722.

Rodionov DA, Vitreschak AG, Mironov AA, Gelfand MS. 2002. Comparative genomics of thiamin biosynthesis in procaryotes: New genes and regulatory mechanisms. *J Biol Chem* **277:** 48949–48959.

Rodionov DA, Vitreschak AG, Mironov AA, Gelfand MS. 2003. Regulation of lysine biosynthesis and transport genes in bacteria: Yet another RNA riboswitch? *Nucleic Acids Res* **31:** 6748–6757.

Römling U, Galperin MY, Gomelsky M. 2013. Cyclic di-GMP: The first 25 years of a universal bacterial second messenger. *Microbiol Mol Biol Rev* **77:** 1–52.

Roth A, Breaker RR. 2009. The structural and functional diversity of metabolite-binding riboswitches. *Annu Rev Biochem* **78:** 305–334.

Serganov A, Nudler E. 2013. A decade of riboswitches. *Cell* **152:** 17–24.

Sherwood AV, Henkin TM. 2016. Riboswitch-mediated gene regulation: Novel RNA architectures dictate gene expression responses. *Annu Rev Microbiol* **70:** 361–374.

Sudarsan N, Barrick JE, Breaker RR. 2003. Metabolite-binding RNA domains are present in the genes of eukaryotes. *RNA* **9:** 644–647.

Sudarsan N, Cohen-Chalamish S, Nakamura S, Emilsson GM, Breaker RR. 2005. Thiamine pyrophosphate riboswitches are targets for the antimicrobial compound pyrithiamine. *Chem Biol* **12:** 1325–1335.

Sudarsan N, Lee ER, Weinberg Z, Moy RH, Kim JN, Link KH, Breaker RR. 2008. Riboswitches in eubacteria sense the second messenger cyclic di-GMP. *Science* **321:** 411–413.

Ulitsky I, Bartel DP. 2013. lincRNAs: Genomics, evolution, and mechanisms. *Cell* **154:** 26–46.

Viladoms J, Fedor MJ. 2012. The *glmS* ribozyme cofactor is a general acid-base catalyst. *J Am Chem Soc* **134:** 19043–19049.

Vitreschak AG, Rodionov DA, Mironov AA, Gelfand MS. 2002. Regulation of riboflavin biosynthesis and transport genes in bacteria by transcriptional and translational attenuation. *Nucleic Acids Res* **30:** 3141–3151.

Vitreschak AG, Rodionov DA, Mironov AA, Gelfand MS. 2004. Riboswitches: The oldest mechanism for the regulation of gene expression? *Trends Genet* **20:** 44–50.

Wachter A, Tunc-Ozdemir M, Grove BC, Green PJ, Shintani DK. 2007. Riboswitch control of gene expression in plants by splicing and alternative $3'$ end processing of mRNAs. **19:** 3437–3450.

Weinberg Z, Barrick JE, Yao Z, Roth A, Kim JN, Gore J, Wang JX, Lee ER, Block KF, Sudarsan N, et al. 2007. Identification of 22 candidate structured RNAs in bacteria using the CMfinder comparative genomics pipeline. *Nucleic Acids Res* **35:** 4809–4819.

Weinberg Z, Wang JX, Bogue J, Yang J, Corbino K, Moy RH, Breaker RR. 2010. Comparative genomics reveals 104 candidate structured RNAs from bacteria, archaea, and their metagenomes. *Genome Biol* **11:** R31.

Weinberg Z, Lünse CE, Corbino KA, Ames TD, Nelson JW, Roth A, Perkins KR, Sherlock ME, Breaker RR. 2017. Detection of 224 candidate structured RNAs by comparative analysis of specific subsets of intergenic regions. *Nucleic Acids Res* **45:** 10811–10823.

Winkler W, Nahvi A, Breaker RR. 2002. Thiamine derivatives bind messenger RNAs directly to regulate bacterial gene expression. *Nature* **419:** 952–956.

Winkler WC, Nahvi A, Roth A, Collins JA, Breaker RR. 2004. Control of gene expression by a natural metabolite-responsive ribozyme. *Nature* **428:** 281–286.

Yanofsky C. 2000. Transcription attenuation: Once viewed as a novel regulatory strategy. *J Bacteriol* **182:** 1–8.

Cite this article as *Cold Spring Harb Perspect Biol* doi: 10.1101/cshperspect.a032797

Mechanistic Insights into MicroRNA-Mediated Gene Silencing

Thomas F. Duchaine[1] and Marc R. Fabian[2,3]

[1]Department of Biochemistry & Goodman Cancer Research Centre, McGill University, Montreal, Quebec H3G 1Y6, Canada

[2]Department of Oncology, McGill University, Montreal, Quebec H3G 1Y6, Canada

[3]Lady Davis Institute, Jewish General Hospital, Montreal, Quebec H3T 1E2, Canada

Correspondence: thomas.duchaine@mcgill.ca; marc.fabian@mcgill.ca

MicroRNAs (miRNAs) posttranscriptionally regulate gene expression by repressing protein synthesis and exert a broad influence over development, physiology, adaptation, and disease. Over the past two decades, great strides have been made toward elucidating how miRNAs go about shutting down messenger RNA (mRNA) translation and promoting mRNA decay.

Here, we review the mechanism of microRNA (miRNA)-mediated gene silencing and describe the most recent transformative discoveries. Focus is placed on miRNA target site recognition, and new insights into the structure of the miRNA-induced silencing complex (miRISC) and miRISC-interacting proteins. Finally, we discuss recent models that have been proposed regarding how the miRISC represses mRNA translation and promotes messenger RNA (mRNA) deadenylation and decay.

THE FUNCTIONS OF miRNAs IN GENE NETWORKS

miRNAs are small, ~21-nucleotide (nt)-long noncoding RNAs that are encoded within the genomes of multicellular organisms and regulate gene expression. At the molecular level, miRNAs base-pair imperfectly with the 3′ untranslated regions (3′UTRs) of targeted mRNAs, from where they engender mRNA translation inhibition and destabilization (Brennecke et al. 2005; Farh et al. 2005; Lim et al. 2005; Stark et al. 2005; Bartel 2009). The functions of miRNAs are thus defined through the identity of their targets. Consequently, the information an miRNA encodes takes on a physiological meaning only when considered as an integral part of gene-regulation networks. The functional impact of miRNAs is both broad and specific; more than 2000 miRNAs have been detected in the human transcriptome, and each miRNA specifically targets one to several dozen mRNAs (Krek et al. 2005; Lall et al. 2006; Baek et al. 2008; Friedman et al. 2009; Fang and Rajewsky 2011). It is estimated that more than 60% of human protein-coding transcripts are targeted by miRNAs (Friedman et al. 2009).

miRNAs play a dual role in gene-regulation networks, in shaping transcriptomes, and in controlling their output in protein synthesis. To perform these global functions, the expression and abundance of miRNAs, those of their target sites in mRNAs, their affinities, and their functional output through regulation of their effector machineries, have all been precisely tuned through evolution (Farh et al. 2005; Stark et al. 2005; Chen and Rajewsky 2006, 2007; Bartel 2009; Friedman et al. 2009). Variations on these parameters between biological conditions and cell types allow miRNAs to serve a diverse array of biological purposes (Ambros et al. 2003a; Ambros 2004; Chen and Rajewsky 2007; Bartel 2009; Sun and Lai 2013; Alberti and Cochella 2017). First, miRNAs can act as rheostats governing the expression of individual genes, by promoting partial but functionally significant repression. This function in fine-tuning gene expression appears pervasive in somatic cells, based on several genome-wide studies (Farh et al. 2005; Selbach et al. 2008; Guo et al. 2010; Fang and Rajewsky 2011). Second, miRNAs play important roles in making developmental decisions, such as steering a daughter cell away from its parental lineage (reviewed in Alberti and Cochella 2017). Examples include control of epidermal seam cell differentiation through regulation of lin-41 mRNA by let-7 miRNA in *Caenorhabditis elegans* (Ecsedi et al. 2015), and the function of miR-155 in controlling macrophage maturation in mammals (Mann et al. 2010). Third, miRNAs can act in a concerted manner over multiple genes in a genetic cascade by toning down a group of genes in a pathway and/or by acting in feedback or feedforward loops. This can have deep impact at the cell, tissue, and organismal levels (Krek et al. 2005; Pelaez and Carthew 2012; Giri and Carthew 2014; Posadas and Carthew 2014; Alberti and Cochella 2017). A well-characterized example is the multilevel functions of lin-4 and let-7 miRNAs in the heterochronic cascade of *C. elegans* (Lee et al. 1993; Pasquinelli et al. 2000; Reinhart et al. 2000; Ambros 2011). In *Drosophila*, miR-7 provides robustness to developmental programs amid stochastic variations in endogenous and exogenous signals (Stark et al.

2005; Li et al. 2009; Alberti and Cochella 2017). Recent evidence highlights the importance of similar functions for miRNAs in controlling stress and aging (Emde and Hornstein 2014; Inukai et al. 2018). Fourth, miRNAs can drive the rapid turnover of large spans of the transcriptome in developmental or other physiological transitions. This is the case during the maternal-to-zygotic transition (MZT) in the zebrafish early embryo, wherein miR-430 instigates the rapid deadenylation and decay of hundreds of maternal mRNA targets (Giraldez et al. 2005, 2006).

Perhaps not surprising in light of their importance in gene-regulation networks, miRNA–mRNA networks are often deregulated in human diseases ranging from heart disease (Xin et al. 2013) to various types of cancer (Croce and Calin 2005; He et al. 2005; Calin and Croce 2006). miRNA expression profiles provide signatures that inform on the pathogenic state of cells and tissues, on the origin of tumor outgrowth, and sometimes even on treatment prognostics (Iorio and Croce 2013; Sun and Lai 2013; Iorio and Croce 2017; Romano et al. 2017).

MATURATION AND LOADING OF miRNAs ONTO ARGONAUTE PROTEINS

The canonical biogenesis pathway of miRNAs (Fig. 1) involves a series of steps governed by the RNase III family enzymes Drosha and Dicer and their associated cofactors (see Ha and Kim 2014 for a review). miRNAs are derived from structured RNA polymerase II–generated primary miRNA transcripts (pri-miRNAs), which encode one or several precursor miRNAs (pre-miRNAs). Drosha and its double-stranded RNA (dsRNA)-binding protein cofactor DGCR8/Pasha process pri-miRNAs and release ∼60- to 75-nt-long hairpin pre-miRNAs, which are exported into the cytoplasm (Lee et al. 2003; Carmell and Hannon 2004). Dicer and a family of dsRNA-binding protein cofactors further process the pre-miRNA yielding a 18- to 21-nt-long RNA duplex consisting of the passenger strand and the guide strand, which is also the mature miRNA (Bernstein et al. 2001; Grishok et al. 2001; Hutvágner et al. 2001; Ketting et al.

Cite this article as *Cold Spring Harb Perspect Biol* doi: 10.1101/cshperspect.a032771

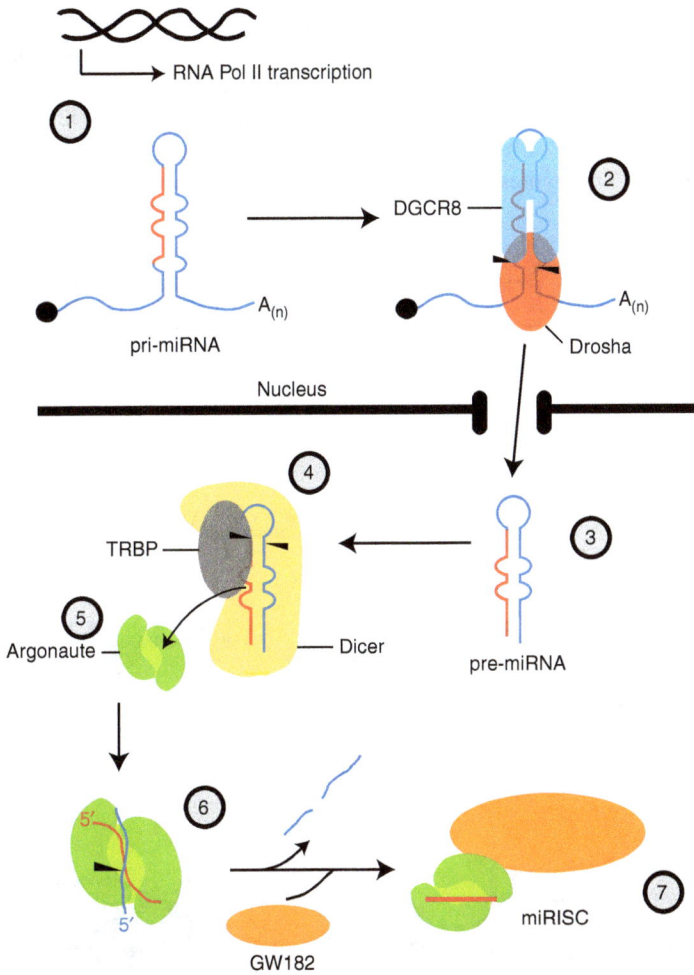

Figure 1. MicroRNA (miRNA) biogenesis and miRNA-induced silencing complex (miRISC) maturation. Primary miRNA (pri-miRNA) transcripts are generated by RNA polymerase II (*1*). Drosha and its double-stranded RNA (dsRNA)-binding protein cofactor DGCR8 bind to a pri-miRNA and release a ~60- to 75-nt-long hairpin precursor (pre)-miRNA (*2*), which is exported from the nucleus to the cytoplasm (*3*). Dicer and a family of dsRNA-binding proteins (TRBPs in mammals) cleave the pre-miRNA (*4*), yielding an 18- to 21-nt-long RNA duplex consisting of the passenger strand (blue), and the guide strand (red), which is also the mature miRNA. The double-stranded miRNA duplex is loaded into an Argonaute protein (*5*), and the passenger strand is either cleaved by Argonaute slicing activity and released (*6*) or released independently of slicing when the two strands are mismatched. The mature single-stranded miRNA (red), Argonaute, and the Argonaute-interacting protein GW182 constitute the core components of miRISC (*7*).

2001; Knight and Bass 2001). Processing by Dicer and loading of the miRNA duplex into an Argonaute protein are closely coordinated processes and occur in a series of transitory (loading) complexes involving both proteins (Gregory et al. 2005; Maniataki and Mourelatos 2005; Kobayashi and Tomari 2016). Within the Argonaute, the miRNA further matures as the passenger strand is either cleaved through the endonuclease activity of the Argonaute (an activity called slicing) and released, or released independently of slicing when the two strands are mismatched (Gregory et al. 2005; Matranga et al. 2005; Miyoshi et al. 2005; Park and Shin 2015). The fully mature,

single-stranded miRNA is stably embedded in
the Argonaute (Fig. 2) wherein it carries out its
function as a guide for silencing activities. Although some species including *Drosophila* and
C. elegans rely on the *sorting* of miRNAs into
functionally specialized Argonautes (DmAGO1

and ALG-1/2, respectively), mammalian species
do not sort miRNAs into a specific Argonaute.
miRNAs are thus found in association with
HsAGO1-4, and HsAGO2 is shared between its
siRNA and miRNA functions (Czech et al. 2009;
Czech and Hannon 2011).

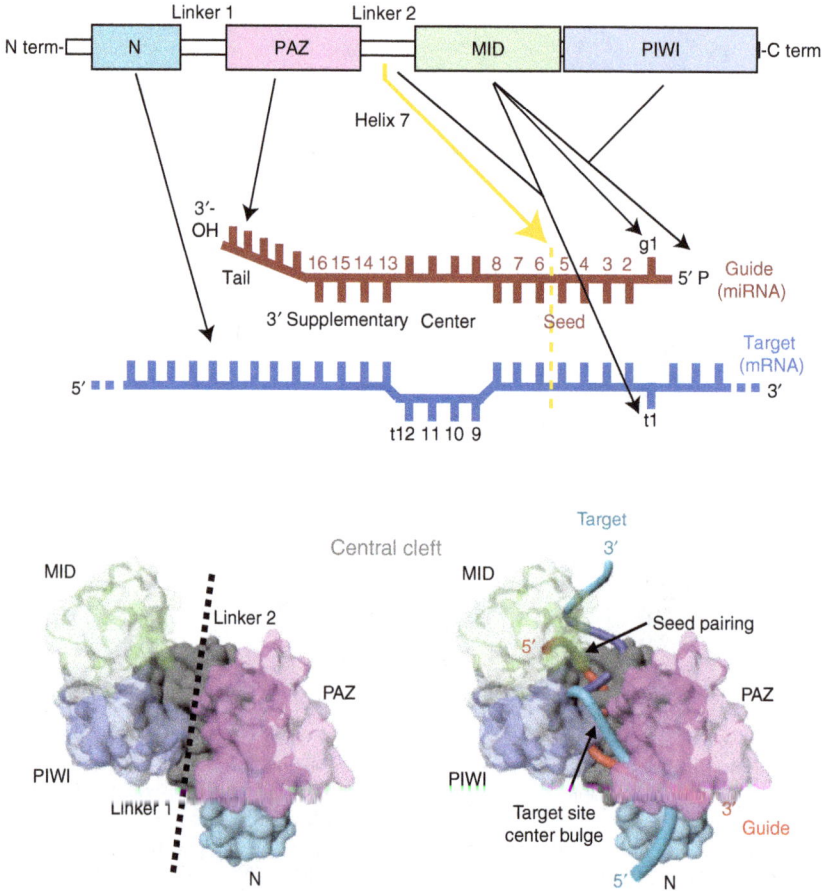

Figure 2. Structure and RNA interactions of Argonautes. (*Top*) The domain distribution and linkers 1 and 2 in
the HsAGO2 primary structure are represented. Key interactions with the guide (the microRNA [miRNA]) and
passenger (the target messenger RNA [mRNA site]) strands are indicated. The extensive interactions of linkers 1
and 2 (L1/L2) with strand backbones in the central cleft are not indicated. Lines are connected when the
interaction occurs at the interface between two domains (PIWI and MID for the guide 5′-phosphate, and
MID and L2 for the target t1 nucleotide). Positioning of helix 7 (encoded in L2) in HsAGO2 and its blocking
of nucleotides g6–8 of the seed are indicated (yellow arrow and dashed line) (Klum et al. 2017). (*Middle*) Linear
representation of the base-pairing of an archetypal miRNA/mRNA pair. Proposed positioning of Seed, Center, 3′
Supplementary, and Tail nucleotides (Wee et al. 2012) are indicated. (*Bottom*) Structure of HsAGO2 bound to
both guide miRNA (red strand), target RNA (blue strand) (*Right* model), and without RNA (*Left* model). Note
that only the dark blue positions of the target RNA were observed in the structure. The path of the target strand in
light blue is extrapolated. Target sequences opposite to Center nucleotides are bulged because of clashes within
the cleft of HsAGO2 in those positions. See text for more details. (The structure [PDB 4W5O] was published in
Schirle et al. 2014 and Sheu-Gruttadauria and MacRae 2017.)

THE miRNA-INDUCED SILENCING COMPLEX

The term RNA-induced silencing complex (RISC) was originally used to describe the assembly of the guide RNA with an Argonaute, the minimal components required to exert the slicing activity associated with RNA interference (RNAi) (Hammond et al. 2000; Hammond et al. 2001; Liu et al. 2004). In miRNA-mediated silencing, the physiological forms of RISC involve more components. The two core proteins of mammalian miRISC are one of the four Argonaute proteins (AGO1–4) and GW182 (glycine-tryptophan [GW] repeat-containing protein of 182 kDa), which directly binds to Argonaute and is essential for recruiting effectors of miRNA-mediated mRNA silencing (Fig. 1). For clarity, herein we distinguish the core components of miRISC (the miRNA, the Argonaute, and GW182 proteins) from proteins and complexes that effect silencing (the translation repression, deadenylation, decapping, and decay cofactors). The amino terminus of mammalian, zebrafish, and *Drosophila* GW182 proteins house numerous conserved GW, WG, or GWG repeats, which form AGO-interacting motifs that bind to two W-binding pockets in AGO proteins (Behm-Ansmant et al. 2006a; El-Shami et al. 2007; Lian et al. 2009; Takimoto et al. 2009; Yao et al. 2011; Pfaff et al. 2013). A recent structural study showed that human GW182 contains three distinct AGO-interacting motifs (motif 1, motif 2, and the Argonaute hook) that bind with high affinity to both HsAGO1 and AGO2 (Elkayam et al. 2017), suggesting that multiple mRNA-bound AGO proteins may recruit a single GW182 molecule to engender mRNA silencing.

miRNA–ARGONAUTE INTERACTIONS IN TARGET RECOGNITION

The functions of miRNAs are inseparable from the structure and interactions of Argonaute proteins (Fig. 2) (Wee et al. 2012; Chandradoss et al. 2015; Salomon et al. 2016; Sheu-Gruttadauria and MacRae 2017). An Argonaute protein folds in a bilobal structure (Song et al. 2004). Two lobes are formed by four distinct domains (amino terminal [N]/PAZ and MID/PIWI) and are separated by a central cleft wherein the guide RNA (the mature miRNA) and the target strand base-pair (Wang et al. 2008b). The guide RNA is held by the Argonaute in an extended conformation. The MID domain binds the 5′ terminal nucleotide of the miRNA (Boland et al. 2011; Frank et al. 2011), and the PAZ domain interacts with the very last (3′ end) nucleotides (Lingel et al. 2003, 2004; Song et al. 2003; Yan et al. 2003; Kidner and Martienssen 2004; Ma et al. 2004). Two linkers (L1 and L2) connect the lobes of the Argonaute and cradle the sugar-phosphate backbone of both guide and target RNAs (Wang et al. 2008a,b; Elkayam et al. 2012; Schirle and MacRae 2012). Argonaute shows sequence preferences for a U or C in the 5′-most nucleotide of the miRNA (position g1, see Fig. 2) and for an A in target sequences facing this position (t1). Binding to nucleotide recognition pockets of the MID domain prevents these nucleotides from base-pairing on target recognition (Schirle et al. 2015). The PIWI domain adopts an RNase H fold within Argonaute (Liu et al. 2004; Song et al. 2004) and carries the catalytic residues for the slicing activity, which is only permitted with extensive or complete base-pairing between the guide and the target RNAs. However, for the vast majority of animal miRNA-binding sites, one or more central residues of the miRNA (g9–g12) are mispaired (Fig. 2). Furthermore, genomic surveys indicate that pairing in these positions is not correlated with miRNA-mediated repression (Grimson et al. 2007). Slicing thus seems primarily to serve a biogenesis purpose in the miRNA pathway, in the cleavage and release of the passenger strand (Bouasker and Simard 2012; Jee et al. 2018). The PIWI domain also carries two tryptophan pockets in its outer structure, which are the binding sites for GW182 (Eulalio et al. 2008, 2009a; Schirle and MacRae 2012; Elkayam et al. 2017).

From within Argonaute, an miRNA is not free to hybridize with an mRNA the way a naked 21-nt RNA molecule would. Argonaute interactions profoundly modify miRNA hybridization kinetics by compartmentalizing the miRNA sequence into subdomains, namely, the 5′ anchor (g1), the seed (g2–g8), the center (g9–g12), the 3′ supplementary (g14–g16), and tail sequences (the 3′-most nucleotides) (Fig. 2) (Wee et al. 2012; Schirle et al. 2014; Salomon et al. 2016). Perhaps the most striking consequence of Argonaute interactions with the miRNA is the unique contribution of the seed nucleotides to target specificity (Lai 2002; Lewis et al. 2003, 2005; Brennecke et al. 2005; Farh et al. 2005). The importance of seed nucleotides became clear early on based on their conservation between paralogous and orthologous miRNA genes (Lagos-Quintana et al. 2001, 2003; Lau et al. 2001; Lee and Ambros 2001; Ambros et al. 2003b; Lim et al. 2003a,b). The functional significance of the seed has since been extensively supported with a conjuncture of conservation, unbiased forward genetics, in vivo and in vitro reporter assays, and structural biology (Lai 2002; Haley and Zamore 2004; Brennecke et al. 2005; Farh et al. 2005; Krek et al. 2005; Lewis et al. 2005; Lim et al. 2005). The positioning of the seed within Argonaute exposes the seed nucleotides (g2–g8) in the central cleft of the protein and enables their base-pairing with mRNAs. Argonaute–miRNA interactions benefit target recognition by preordering the nucleotides of the seed to defray the entropic cost of stacking bases in a readily hybridizable conformation (Nakanishi et al. 2012; Schirle and MacRae 2012; Schirle et al. 2014; Salomon et al. 2016). This accelerates base-pairing of seed nucleotides by >20-fold in comparison to free, disordered nucleic acid oligomers of the same length, to reach rates nearing the limits set by diffusion (Wee et al. 2012; Salomon et al. 2016). Seed pairing with a target site is a two-step process and involves the sequential base-pairing of nucleotides g2–g5 and g6–g8 (Wee et al. 2012; Schirle et al. 2014; Chandradoss et al. 2015; Sheu-Gruttadauria and MacRae 2017). Upon favorable pairing of g2–g5, the helix 7 of Argonaute (formed by L2 residues) is docked away from the guide RNA to give way and enable pairing of nucleotides g6–g8 (Klum et al. 2017). Binding of preordered g2–g5 is rate limiting for target recognition, whereas base-pairing of g6–g8 slows down release from a favorable target site (a pause called *dwelling*).

Another important impact of Argonaute structure on target site binding involves the interactions of the 3′ supplementary nucleotides

with the N and PAZ domains. In target-free Argonaute, nucleotides g14–g18 of the miRNA thread through a narrow channel formed by the N and PAZ domains, wherein they are excluded from base-pairing (Schirle et al. 2014; Sheu-Gruttadauria and MacRae 2017). Recent structures suggest a widening of the N-PAZ channel upon seed pairing to the target, possibly through conformational propagation involving helix 7 (Klum et al. 2017). Widening of the channel repositions the 3′ end of the miRNA and enables base-pairing of g14–g16, which is thought to nucleate an outward helix propagation and contribute to miRISC/target stabilization. Early genomic studies suggested that only about 5% of miRNA targets involve 3′ supplementary pairing of functional significance (Grimson et al. 2007), but this has remained a matter of debate. The most recent structures provide a conceptual framework to interpret studies which suggest that certain miRNAs may rely more on 3′ supplemental nucleotides for target recognition than others (Broughton et al. 2016; Wolter et al. 2017).

FUNCTIONAL INTERACTIONS OF miRNAs AND miRISC IN 3′UTRs

Within 3′UTRs, miRNAs interact with a broad diversity of factors and mechanisms, which determine an mRNA's localization, stability, and translatability (see Mayr 2017 for a recent review). The regulatory functions of a 3′UTR are dictated through the sequences it encodes, its folding structure, and its interactions with miRISC and RNA-binding proteins (RBPs). miRNA target recognition is sensitive to steric hindrance by 3′UTR structures. Sites located near the stop codon (but outside the ribosome footprint) or closer to the poly(A) tail tend to have a more potent effect on silencing, presumably because they are positioned outside of the bulk structures of the 3′UTR. Furthermore, miRNA-binding sites located near A/U-rich regions of 3′UTRs are on average more potent, possibly because of the lower stability of the structures in such regions (Farh et al. 2005; Grimson et al. 2007).

Cooperative interactions between miRNA-binding sites in 3′UTRs are also a major determinant of miRNA-mediated silencing.

Cite this article as *Cold Spring Harb Perspect Biol* doi: 10.1101/cshperspect.a032771

Genome-wide studies in mammals revealed early on that miRNA-binding sites that are located in close proximity to each other in a 3′UTR elicit more potent silencing than sites separated by more than 50 nt (Grimson et al. 2007; Saetrom et al. 2007). In addition, tandem repeats of miRNA-binding sites located in 3′UTRs of reporter mRNAs greatly sensitize artificial miRNA reporter assays (Pillai et al. 2005; Mathonnet et al. 2007; Fabian et al. 2009; Mayya and Duchaine 2015). miRNA-binding site cooperation appears to be a strict requirement in certain biological contexts such as the *C. elegans* embryo (Wu et al. 2010). Two non–mutually exclusive mechanisms have been described to explain miRNA-binding site cooperativity in 3′UTRs. First, cooperative interactions between nearby binding sites can stabilize miRISC binding to the target mRNA (Broderick et al. 2011). Second, cooperation of multiple miRNA-binding sites can promote assembly of the effector machinery, through the recruitment of the CCR4-NOT deadenylase complex to target mRNAs (Flamand et al. 2017; Wu et al. 2017). A deeper understanding of the functional cooperation between miRNA-binding sites warrants further investigation.

Cooperative and competitive interactions between RBPs and miRISC in 3′UTRs also profoundly modulate miRNA-mediated silencing (Bhattacharyya et al. 2006; Huang et al. 2007; Kedde et al. 2007; Elcheva et al. 2009; Takeda et al. 2009; Jafarifar et al. 2011; Kundu et al. 2012; Jens and Rajewsky 2015). For example, the Pumilio RBP can enable miRNA-mediated silencing by improving the accessibility of nearby miRNA-binding sites, presumably by destabilizing local RNA folding structures (Kedde et al. 2010). In an opposite scenario, the RBP Dead end 1 (Dnd1) binds to elements in uridine-rich sequences of 3′UTRs located in close proximity to miRNA-binding sites and inhibits the association of miRISC (Kedde et al. 2007). Finally, several RBPs interact and recruit some of the same effector machineries to mRNAs, in particular, the CCR4-NOT complex. Clarifying how these RBPs functionally interact with miRISC should help to reliably predict the functional output of miRNA-binding sites.

Recently, an elegant single-molecule imaging system has been used to examine the dynamics of miRNA/Argonaute interactions on a 3′UTR mimic (Chandradoss et al. 2015). The results suggested that miRISC could slide along the 3′UTR RNA between miRNA-binding sites. This sliding may represent a *scanning* activity of miRISC that could accelerate target recognition by reducing the collisions to a one-dimensional search along individual 3′UTRs. Even though enticing, the association rates observed were slower than the random-collision, near-diffusion rates observed with other assays (Wee et al. 2012; Salomon et al. 2016). It will be interesting to address the physiological implications of this phenomenon, and how sliding cooperates with or is counteracted by RBPs and 3′UTR folding structures.

GW182-INTERACTING PROTEINS

The carboxy-terminal region of GW182 proteins, termed the silencing domain, is required for GW182 to silence target mRNAs (Figs. 3 and 4). The silencing domain can be subdivided into a number of regions that are predicted to be unstructured, including the middle (Mid) and carboxy-terminal subdomains. These regions flank an RNA-recognition motif (RRM)-like domain, which does not bind RNA because of the presence of an additional α helix at its carboxyl terminus (Eulalio et al. 2009b). The Mid domain can be further subdivided into the M1 and M2 regions, which flank a poly(A)-binding protein (PABP)-interacting motif 2 (PAM2) (Roy et al. 2002). Although miRNA-loaded AGOs require GW182 to silence targeted mRNAs, GW182 alone is not sufficient. It is now well established that GW182 proteins engender miRNA-mediated gene silencing by interacting with a number of proteins, including PABP, the poly(A) nuclease (PAN)2–PAN3 and CCR4–NOT deadenylase complexes, and mRNA decapping factors.

PABP

In addition to binding to the mRNA 3′ poly(A) tail, PABP also interacts with a number of pro-

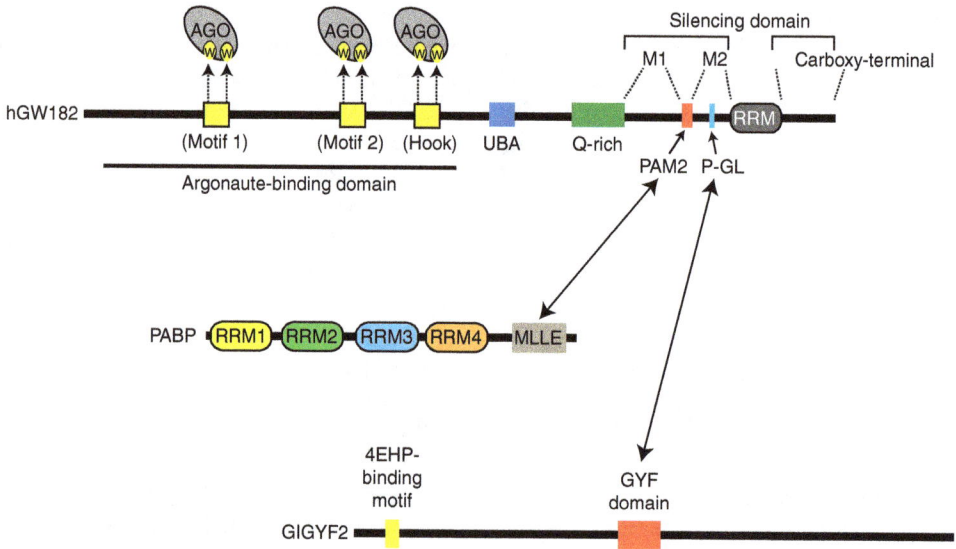

Figure 3. Schematic diagrams of glycine-tryptophan protein of 182 KDa (GW182), poly(A)-binding protein (PABP), and GRB10-interacting GYF protein 2 (GIGYF2). The amino-terminal region of GW182 (the Argonaute-binding domain) contains a number of glycine-tryptophan (GW) repeats, several of which form regions that bind AGO proteins (motif 1, motif 2, and the AGO hook). This region, in addition to the ubiquitin-associated (UBA) and glutamine-rich (Q-rich) domains, is responsible for targeting GW182 to processing (P) bodies. The carboxy-terminal region of mammalian and insect GW182 proteins encompasses the silencing domain, which is the major effector domain by which GW182 engenders miRNA-mediated silencing. It is comprised of the M1 and M2 regions, the PABP association motif 2 (PAM2), the proline-rich PG-L motif, an RNA-recognition motif-like domain (RRM) that cannot bind RNA, and a carboxy-terminal domain. PABP domains include four functional RRMs and a conserved carboxy-terminal MLLE domain that directly binds the mammalian and *Drosophila* GW182 silencing domain via its PAM2 motif. Mammalian GIGYF2 interacts with the PG-L motif in the silencing domain of human GW182 via its GYF domain.

teins that regulate both mRNA translation and mRNA turnover (Derry et al. 2006; Smith et al. 2014). PABP stimulates cap-dependent mRNA translation via its interactions with the 5′ cap–associated translation factor eukaryotic initiation factor (eIF)4G (Kahvejian et al. 2005). In contrast, PABP also associates with PAIP2, which inhibits translation by displacing PABP from the mRNA poly(A) tail (Khaleghpour et al. 2001). PABP can also regulate mRNA decay by directly interacting with the PAN3 subunit of the PAN2–PAN3 deadenylase complex, or the transducer of ERBB2 (Tob), which in turn recruits the CCR4–NOT deadenylase complex to target mRNAs (Siddiqui et al. 2007; Mauxion et al. 2009). PABP also directly binds to GW182 proteins via a stretch of residues in the GW182 carboxy-terminal silencing domain (Fig. 3). Originally referred to as a domain of unknown

function (DUF), this region in the silencing domain is now classified as PABP-interacting motif 2 (PAM2) (Fabian et al. 2009; Jinek et al. 2010; Kozlov et al. 2010). Structurally, mammalian GW182 proteins bind PABP in a manner similar to that of other PABP-interacting proteins (i.e., PAN3, PAIP2, eRF3, and Tob) in that they interact with a PABP carboxy-terminal motif termed the MLLE domain (Jinek et al. 2010; Kozlov et al. 2010). The GW182 PAM2 motif is evolutionarily conserved in mammals, as well as in *Drosophila* where it also binds PABP. In contrast to mammalian and insect GW182 proteins, the *C. elegans* GW182 ortholog AIN-1 lacks a PAM2 motif; nevertheless, it still interacts with PABP, albeit via its amino-terminal RRMs (Kuzuoğlu-Öztürk et al. 2012). Several groups have reported that PABP enhances miRNA-mediated deadenylation in vitro (Fabi-

Cite this article as *Cold Spring Harb Perspect Biol* doi: 10.1101/cshperspect.a032771

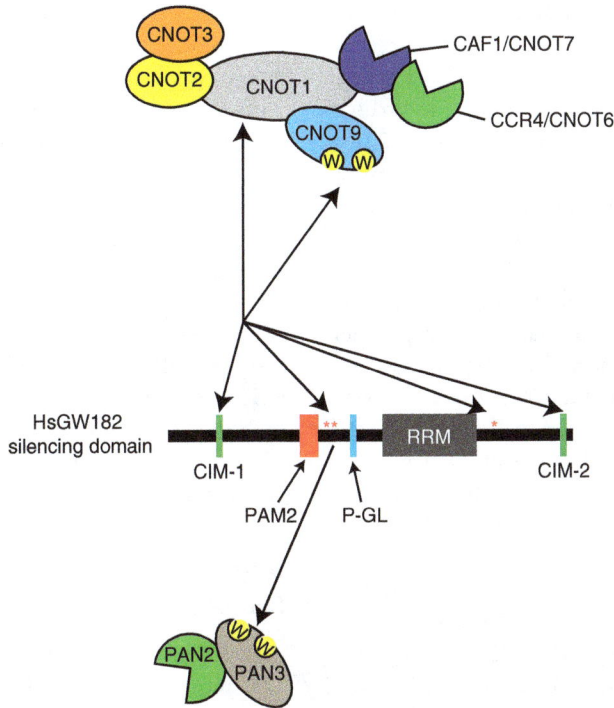

Figure 4. Schematic diagram of the human glycine-tryptophan protein of 182 KDa (GW182) silencing domain and its interactions with the CCR4–NOT and PAN2–PAN3 deadenylase complexes. The GW182 silencing domain reportedly uses two CCR4-interaction motifs (CIM-1 and -2) and a number of conserved tryptophan residues in the M2 region and carboxy-terminal domain to interact with the CCR4–NOT complex. The silencing domain physically contacts both the CNOT1 subunit as well as CNOT9, the latter containing two tryptophan-binding pockets (denoted by W). The PAN3 subunit contains two tryptophan-binding pockets that directly bind to conserved tryptophan residues in the M2 region of the silencing domain. Conserved tryptophans in the silencing domain that have been reported to interact with PAN3 and CNOT subunits are denoted by red asterisks.

an et al. 2009; Flamand et al. 2016) and miRNA-mediated silencing in cells (Huntzinger et al. 2010); however, not all groups have observed the PABP–GW182 interaction being required to promote miRNA-mediated repression (Fukaya and Tomari 2011). PABP and the poly(A) tail promote the association of the miRISC with targeted mRNAs (Moretti et al. 2012), suggesting that the PABP–GW182 interaction promotes GW182 recruitment to AGO-bound, miRNA-targeted mRNAs. Thus, it is possible that the PABP–GW182 interaction plays an important role in promoting miRNA-mediated silencing of mRNAs with long 3'UTRs (Flamand et al. 2016), or with certain miRNA target sites that require PABP to enhance miRISC recruitment.

CCR4–NOT and PAN2–PAN3 Deadenylase Complexes

miRNAs promote the deadenylation of target mRNAs in a number of systems, including human and *Drosophila* cultured cells, zebrafish embryos, as well as mammalian, *Drosophila*, and *C. elegans* cell-free systems (Behm-Ansmant et al. 2006b; Giraldez et al. 2006; Wu et al. 2006; 2010; Wakiyama et al. 2007). miRNA-mediated deadenylation occurs as a biphasic event, which is mediated primarily by the CCR4–NOT deadenylase complex, as well as by the PAN2–PAN3 deadenylase complex (Chen et al. 2009). An array of studies indicates that GW182 proteins act as evolutionarily conserved scaffolds that recruit the CCR4–NOT and PAN2–PAN3 deadenylation complexes to miRNA-targeted mRNAs (Fig. 4) (Braun et al. 2011; Chekulaeva et al. 2011; Fabian et al. 2011; Kuzuoğlu-Öztürk et al. 2012; Christie et al. 2013; Mauri et al. 2017; Wu et al. 2017). Structural and biochemical studies revealed that human GW182 proteins contact several CCR4–NOT complex subunits (Braun et al. 2011; Fabian et al. 2011; Chen et al. 2014; Mathys et al. 2014). The CNOT9 subunit uses two tryptophan-binding pockets to directly bind conserved tryptophans in the GW182 silencing domain (Chen et al. 2014; Mathys et al. 2014). In addition to binding CNOT9, the human GW182 silencing domain directly interacts with the CNOT1 subunit of the CCR4–NOT complex (Fabian et al. 2011). Thus, the GW182 silencing domain most likely makes multiple contacts with the CCR4–NOT complex, by contacting both the CNOT1 and CNOT9 subunits. The GW182 silencing domain also recruits the PAN2–PAN3 heterodimeric complex by interacting with PAN3. This has been reported to occur indirectly via PABP (Chekulaeva et al. 2011; Fabian et al. 2011) or direct binding through a tryptophan-containing pocket in the PAN3 kinase-like domain interacting with the M2 region of the GW182 silencing domain (Braun et al. 2011; Chekulaeva et al. 2011; Christie et al. 2013).

GIGYF2

PABP, the CCR4–NOT and PAN2–PAN3 deadenylase complexes are not the only proteins to directly interact with GW182. The GRB10-interacting GYF protein 2 (GIGYF2) was recently identified as a novel GW182 silencing domain-interacting protein by using a novel proximal biotinylation approach (Split-BioID) to search for miRISC-associating proteins in human cells (Schopp et al. 2017). Coimmunoprecipitation experiments and in vitro pulldowns using recombinant proteins showed that GIGYF2 directly binds to a conserved proline-rich P-GL motif in the Mid region of the GW182 silencing domain (Fig. 3). Knocking down GIGYF2 partially derepressed a tetracycline-inducible miRNA-targeted reporter in HeLa cells. As this derepression oc-

curred at an early time point, before deadenylation (Bethune et al. 2012), GIGYF2 may play a role in translational repression. In keeping with these data, the P-GL motif contributes to tethered GW182-mediated repression in zebrafish embryos (Mishima et al. 2012). Interestingly, GIGYF2 directly binds to the 5' cap–binding protein 4EHP/eIF4E2, which plays a role in translationally repressing select mRNAs (Rom et al. 1998; Cho et al. 2005; Morita et al. 2012; Peter et al. 2017). Moreover, GIGYF2 can silence a luciferase reporter mRNA when artificially tethered to its 3'UTR (Morita et al. 2012). Nevertheless, whereas these data suggest that GIGYF2 interacts with GW182, its exact role in miRNA-mediated repression has yet to be established.

Mechanistic Insights into miRNA-Mediated Translational Repression

Translation initiation begins with eIF4F complex binding to the mRNA 5'cap structure. This complex is comprised of the cap-binding protein eIF4E, the eIF4A RNA helicase, and the scaffolding protein eIF4G, which interact with both eIF4E and eIF4A. Together, these translation initiation factors help recruit the 43S preinitiation complex (PIC) to the mRNA 5' terminus (Gingras et al. 1999; Merrick and Pavitt 2018). eIF4A helicase activity promotes ribosome subunit scanning by unwinding highly structured regions within mRNA 5'UTRs.

A number of early studies suggested that miRNAs translationally repress targeted mRNAs by inhibiting cap-dependent translation at the initiation step, which is currently the most widely supported model (Fig. 5) (Humphreys et al. 2005; Pillai et al. 2005; Mathonnet et al. 2007; Thermann and Hentze 2007). miRNAs were also reported to repress translation postinitiation (i.e., by inhibiting ribosome elongation) (Olsen and Ambros 1999; Nottrott et al. 2006; Petersen et al. 2006). However, transcriptome-wide ribosome profiling experiments, which measure translation efficiencies, do not detect any changes in ribosome elongation rates on miRNA-targeted mRNAs (Guo et al. 2010). Early reports that miRNAs inhibit translation are supported by investigations of the kinetics of

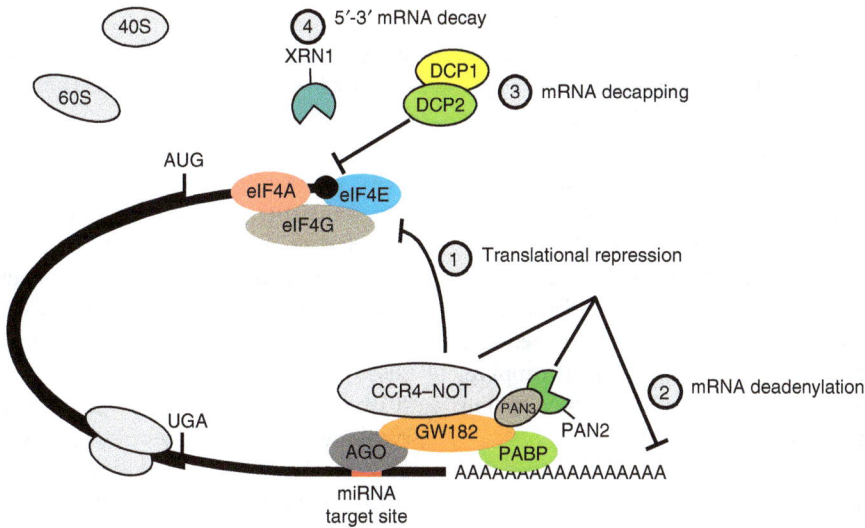

Figure 5. Model of microRNA (miRNA)-mediated gene silencing. Step 1, repression of cap-dependent translation via the CCR4–NOT complex: miRNA-induced silencing complex (miRISC) inhibits translation at the initiation step in a deadenylation-independent manner. Step 2, messenger RNA (mRNA) deadenylation: deadenylation is mediated by the CCR4–NOT complex and, to a lesser extent, the PAN2–PAN3 complex. Step 3, mRNA decapping: following deadenylation, the 5' cap moiety, denoted by a black circle, is removed by the DCP1–DCP2 decapping complex. Step 4, mRNA decay: once the mRNA 5' cap is hydrolyzed, the mRNA is degraded by the XRN1 5'-3' exonuclease.

translational repression and mRNA deadenylation. Studies using fly and mouse cell-free extracts show that translational inhibition can be observed before detectable mRNA deadenylation (Fabian et al. 2009; Zdanowicz et al. 2009). miRNA-mediated translational repression before mRNA deadenylation was also reported in cells (Bethune et al. 2012; Djuranovic et al. 2012) and zebrafish embryos (Bazzini et al. 2012). Although these studies indicate that miRNAs repress mRNA translation, several global studies report that the majority of miRNA-mediated repression can be attributed to mRNA decay (Guo et al. 2010; Eichhorn et al. 2014). How miRNAs facilitate pure translational repression before mRNA destabilization remains a topic of current study, and several groups gained key insights in this field.

miRNA-Mediated Dissociation of Translation Initiation Factors from Target mRNAs

In keeping with repression of translation at the initiation step, several groups have reported that

miRNA-mediated silencing leads to the dissociation of translation initiation factors from targeted mRNAs. RNA pulldown experiments for miRNA- and GW182-targeted reporter mRNAs expressed in *Drosophila* S2 cells showed a decreased association of eIF4E, eIF4G, and PABP with miRNA-targeted mRNAs (Zekri et al. 2013). Similar results were observed in RNA immunoprecipitation experiments performed for miRNA targets in human embryonic kidney (HEK) 293 cells (Rissland et al. 2017) and cell-free extracts that recapitulate miRNA-mediated silencing in vitro (Moretti et al. 2012; Fukao et al. 2014). However, whereas deadenylation should remove PABP from a mRNA poly(A) tail, PABP is still displaced from deadenylation-resistant reporters that are targeted by either miRNAs or tethered GW182 (Zekri et al. 2013; Fukao et al. 2014). Nevertheless, miRNAs efficiently silenced a mRNA that lacked a poly(A) tail but instead had PABP artificially tethered to the mRNA 3' end (Nishihara et al. 2013). Taken together, these data suggest that whereas miRNA-mediated dissociation of PABP may assist in translational repression, it is not the only mode by which miRNAs repress translation. Interestingly, knocking down the decapping factors DCP1 and EDC4 in tandem rescued the association of eIF4E with an miRNA-targeted reporter, even though eIF4G and PABP remained dissociated (Zekri et al. 2013). Thus, the miRNA-mediated silencing machinery may engender the dissociation of PABP and eIF4G before mRNA decapping, whereas eIF4E may be displaced as a result of 5′ cap hydrolysis. The linkages between mRNA translation initiation factors, decapping, and decay are further reviewed in this collection (Heck and Wilusz 2018).

DDX6, 4E-T, and 4EHP

Although the CCR4–NOT complex engenders deadenylation of miRNA targets, several studies showed that it can repress target mRNAs independent of its deadenylase activity (Cooke et al. 2010; Braun et al. 2011; Chekulaeva et al. 2011). In addition, tethering the CCR4–NOT complex can translationally repress deadenylation-resistant reporter mRNAs that contain a stretch of non-A nucleotides following the 3′ poly(A) tail, which suppresses mRNA deadenylation and decay (Kuzuoğlu-Öztürk et al. 2016). How does the CCR4-NOT complex repress mRNA translation? Several early studies showed that knocking down the translational repressor and decapping enhancer protein DDX6 in human cells, or its homolog in flies (Me31B), impairs miRNA-mediated silencing (Chu and Rana 2006; Eulalio et al. 2007). In addition, DDX6 directly binds the MIF4G (middle of eIF4G) domain of CNOT1, the central subunit of the CCR4–NOT complex (Chen et al. 2014; Mathys et al. 2014; Rouya et al. 2014). Moreover, the interaction between DDX6 and CNOT1 enhances miRNA-mediated silencing (Chen et al. 2014; Mathys et al. 2014; Rouya et al. 2014), including silencing of miRNA-targeted reporters where deadenylation is prevented (Mathys et al. 2014; Kuzuoğlu-Öztürk et al. 2016). All in all, these data suggest that DDX6 binding to CNOT1 plays an important role in miRNA-mediated translational repression. Using in vitro ATPase assays, Mathys et al. (2014) showed that DDX6 ATPase

is activated upon binding to CNOT1. Mutational analyses suggest that this plays a role in miRNA-mediated silencing, albeit via an unknown mechanism.

In addition to binding CNOT1, DDX6 directly interacts with several proteins that inhibit translation and/or enhance mRNA decapping and decay. These include EDC3, LSM14, PATL1, and the eIF4E-binding protein, 4E-T, all of which bind to the DDX6 RecA-C domain in a mutually exclusive manner (Tritschler et al. 2009; Jonas and Izaurralde 2013; Sharif et al. 2013; Nishimura et al. 2015; Ozgur et al. 2015; Waghray et al. 2015). The crystal structure of the CNOT1/DDX6/4E-T complex (Ozgur et al. 2015) shows that DDX6, when directly bound to CNOT1, interacts with 4E-T but cannot bind to other binding partners, including EDC3 and PATL1. Notwithstanding that PATL1 cannot directly interact with DDX6 when bound to CNOT1, PATL1 does interact with the CCR4–NOT complex in a DDX6-independent manner (Haas et al. 2010; Ozgur et al. 2010).

How does 4E-T enhance miRNA-mediated gene silencing? 4E-T contains an amino-terminal eIF4E-binding motif, as well as other regions that interact with DDX6, LSM14, PATL1, and UNR (Dostie et al. 2000; Nishimura et al. 2015; Kamenska et al. 2016). Knocking down 4E-T in human cell lines impairs miRNA-mediated silencing of reporter mRNAs (Kamenska et al. 2016; Chapat et al. 2017). Mechanistically, 4E-T translationally represses reporter mRNAs when tethered to their 3′UTRs in human cells and Xenopus oocytes (Kamenska et al. 2014; Waghray et al. 2015). Despite the ability of 4E-T to compete with eIF4G for binding to eIF4E, tethered 4E-T represses mRNA translation in an eIF4E-independent manner (Kamenska et al. 2014; Waghray et al. 2015). Moreover, 4E-T can also interact with the eIF4E homolog 4EHP (also known as eIF4E2), which represses mRNA translation. Recently, 4EHP has been implicated in miRNA-mediated translational repression, as knocking down 4EHP in mammalian cells partially derepressed miRNA-mediated silencing (Chapat et al. 2017; Chen and Gao 2017). 4EHP has also been reported to be essential in miR-145-mediated silencing of DUSP6

Cite this article as *Cold Spring Harb Perspect Biol* doi: 10.1101/cshperspect.a032771

mRNA, a repressor of ERK signaling (Jafarnejad et al. 2018). These data led to a model where a 4E-T/4EHP complex, recruited by CCR4–NOT, represses translation by competing with eIF4E for the 5′ cap (Fig. 6A). However, as 4E-T can bind either eIF4E or 4EHP, it may translationally repress mRNAs whose 5′ cap structures are bound by either eIF4E or 4EHP (Fig. 6B). Although these studies suggest that 4E-T assists in miRNA-mediated translational repression, 4E-T has also been linked to enhancing mRNA decay of CCR4–NOT targets, including the let-7 miRNA target HMGA2 and reporter mRNAs targeted by the AU-rich element-binding protein tristetraprolin (TTP) (Ferraiuolo et al. 2005; Nishimura et al. 2015). Importantly, a 4E-T mutant that cannot interact with eIF4E was defective in promoting the decay of TTP- and miRNA-targeted mRNAs in HeLa cells (Nishimura et al. 2015). Thus, 4E-T may play a role in promoting both early translational repression, as well as the decay of miRNA targets by bringing the CCR4–NOT complex and the decapping machinery into the vicinity of the mRNA 5′ terminus via its interaction with eIF4E or 4EHP.

A Link between the CCR4–NOT Complex and eIF4A

As previously mentioned, the eIF4A translation initiation factor is an ATP-dependent RNA helicase that assists 43S ribosomal scanning by unwinding secondary structures in the mRNA 5′UTR (Sonenberg and Hinnebusch 2009; Merrick and Pavitt 2018). The CCR4–NOT complex was reported to recruit eIF4A2, a vertebrate-specific paralog of eIF4A, and this interaction translationally represses miRNA targets by impairing 43S ribosomal subunit scanning (Meijer et al. 2013). However, these data have been contested in several follow-up studies (Galicia-Vazquez et al. 2015; Kuzuoğlu-Öztürk et al. 2016), one reporting that a human cell line where eIF4A2 has been knocked out displays no noticeable defects in silencing by miRNAs. Although a model wherein miRISC specifically interacts with eIF4A2 to silence target mRNAs has been challenged, two independent studies

reported that miRNA-mediated translational repression targets eIF4A (Fukao et al. 2014; Fukaya et al. 2014). Knocking down eIF4A in *D. melanogaster* S2 cells impairs miRNA-mediated translational repression. Moreover, RNA pulldown and chemical cross-linking experiments in human and fly cell-free extracts that recapitulate miRNA-mediated silencing show that eIF4A is displaced from miRNA-targeted reporter mRNAs (Fukao et al. 2014; Fukaya et al. 2014). Furthermore, the addition of HuD, an eIF4A-binding protein that stimulates translation, to a cell-free extract impairs miRNA-mediated translational repression, whereas a HuD mutant that cannot bind eIF4A does not (Fukao et al. 2014). Taken together, these data suggest that the miRNA machinery interfaces with eIF4A (either directly or indirectly) to repress translation.

In keeping with a link between miRNA-mediated silencing and eIF4A helicase activity, it has been reported that miRNA reporters harboring unstructured 5′UTRs are refractory to miRNA-mediated silencing in a rabbit reticulocyte in vitro translation system (Ricci et al. 2013). However, a recent study (Kuzuoğlu-Öztürk et al. 2016) did not come to the same conclusion. Efficient miRNA-mediated translational repression was observed with luciferase reporters containing short (<10 nt) 5′UTRs, including the translation initiator of short 5′UTR (TISU) that directs cap-dependent translation via a scanning-independent mechanism (Elfakess et al. 2011; Kwan and Thompson 2018). Both the miRISC and the CCR4–NOT complex efficiently translationally repressed these reporters, suggesting that miRNAs inhibit translation initiation at a step other than 43S ribosomal subunit scanning. In addition, varying the 5′ secondary structure had no noticeable impact on translational repression, again demonstrating that miRNA-mediated silencing does not seem to require eIF4A helicase activity (Kuzuoğlu-Öztürk et al. 2016). Although these data indicate that eIF4A helicase activity is not the major target of the miRNA machinery, they do suggest that eIF4A plays an important role in miRNA-mediated translational repression. One possibility is that miRISC-associated factors, including

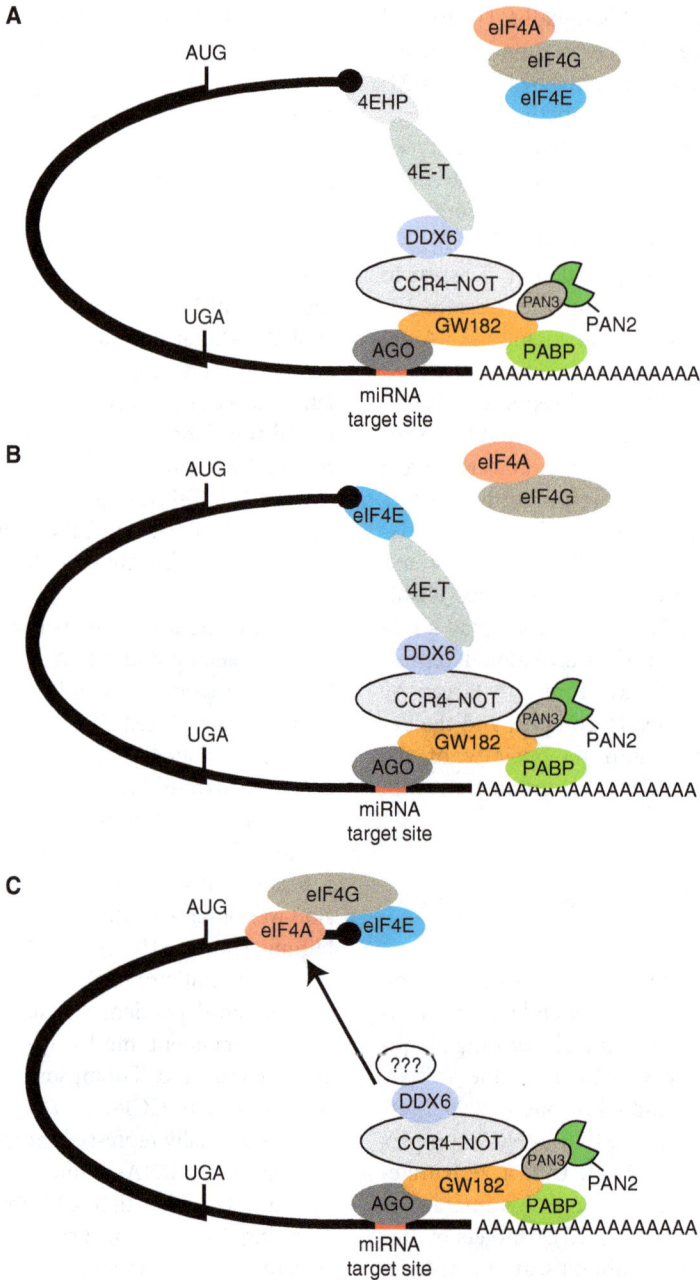

Figure 6. Models for microRNA (miRNA)-mediated translational repression. It is important to note that these models are not mutually exclusive. (*A*) The CCR4–NOT/DDX6 complex recruits 4E-T and 4EHP, the latter functioning to displace eukaryotic initiation factor (eIF)4E from the cap structure and associated translation factors (i.e., eIF4G and eIF4A). (*B*) The CCR4–NOT/DDX6 complex recruits 4E-T, which binds to eIF4E thereby displacing eIF4G. (*C*) The CCR4-NOT/DDX6 complex, potentially with unknown binding partners (denoted by ???), interacts with eIF4A leading to inhibition of translation initiation.

Cite this article as *Cold Spring Harb Perspect Biol* doi: 10.1101/cshperspect.a032771

DDX6, may inhibit translation initiation by physically interacting with eIF4A, rather than by specifically impairing eIF4A helicase activity (Fig. 6C). However, this model remains to be directly tested.

CONCLUSIONS AND FUTURE PERSPECTIVES

These are exciting times in the field of miRNA-mediated silencing. Progress is being made with respect to how the miRISC identifies and interacts with target sites, and on how its interactions direct mRNA translation repression, deadenylation, and decay. Parameters such as cellular miRNA concentrations, their stoichiometric ratio with their mRNA targets, and affinities for target sites are being examined in a quantitative manner. Genome-wide studies have yielded critical information regarding miRNA–mRNA 3′UTR networks. The recent development of libraries capturing the poly(A) tail status in transcriptomes is particularly promising (Subtelny et al. 2014; Rabani et al. 2017; Rissland et al. 2017; Yartseva et al. 2017). Emerging quantitative single-molecule imaging methods consolidate previous models and unveil properties of miRISC that had been missed through batch experiments. Several key aspects of miRNA-mediated silencing remain to be elucidated, however. miRISC recruits the CCR4–NOT deadenylase complex to repress mRNA translation before initiating mRNA deadenylation and decay, but it remains to be established how the CCR4–NOT complex and associated factors bring this about. In addition, whereas several studies indicate that 5′ cap–associated translation initiation factors such as eIF4G and eIF4A are displaced from miRNA-targeted mRNAs before, or in the absence of, mRNA destabilization, the mechanistic underpinnings of this observation remain unknown. Cooperative and/or competitive interplay between RBPs such as TTP and Pumilio, miRISC, and the CCR4–NOT complex also warrants further resolution. Perhaps progress on those topics will explain some of the lingering disparities on the relative contributions of miRNA in driving mRNA translation repression, deadenylation, decapping, and decay.

These topics remain at the front line of current miRNA biology.

ACKNOWLEDGMENTS

We apologize for any directly related work that we have not cited in this review. We thank Dr. Noriko Uetani for her assistance with the graphical design of Figure 2. This work is supported by Canadian Institute of Health Research (CIHR) grants to T.F.D. (MOP-123352) and M.R.F. (MOP-130425), a Natural Sciences and Engineering Research Council of Canada (NSERC) Discovery grant to M.R.F. (RGPIN-2015-03712), Fonds de recherché du Québec-Santé (FRQS) Chercheur-Boursier Senior salary award to T.F.D. and Junior 1 salary award to M.R.F., and CIHR New Investigator award to M.R.F.

REFERENCES

*Reference is also in this collection.

Alberti C, Cochella L. 2017. A framework for understanding the roles of miRNAs in animal development. *Development (Cambridge, England)* **144:** 2548–2559.

Ambros V. 2004. The functions of animal microRNAs. *Nature* **431:** 350–355.

Ambros V. 2011. MicroRNAs and developmental timing. *Curr Opin Genet Dev* **21:** 511–517.

Ambros V, Bartel B, Bartel DP, Burge CB, Carrington JC, Chen X, Dreyfuss G, Eddy SR, Griffiths-Jones S, Marshall M, et al. 2003a. A uniform system for microRNA annotation. *RNA* **9:** 277–279.

Ambros V, Lee RC, Lavanway A, Williams PT, Jewell D. 2003b. MicroRNAs and other tiny endogenous RNAs in *C. elegans. Curr Biol* **13:** 807–818.

Baek D, Villen J, Shin C, Camargo FD, Gygi SP, Bartel DP. 2008. The impact of microRNAs on protein output. *Nature* **455:** 64–71.

Bartel DP. 2009. MicroRNAs: Target recognition and regulatory functions. *Cell* **136:** 215–233.

Bazzini AA, Lee MT, Giraldez AJ. 2012. Ribosome profiling shows that mir-430 reduces translation before causing mRNA decay in zebrafish. *Science* **336:** 233–237.

Behm-Ansmant I, Rehwinkel J, Doerks T, Stark A, Bork P, Izaurralde E. 2006a. mRNA degradation by miRNAs and GW182 requires both CCR4:NOT deadenylase and DCP1:DCP2 decapping complexes. *Genes Dev* **20:** 1885–1898.

Behm-Ansmant I, Rehwinkel J, Izaurralde E. 2006b. MicroRNAs silence gene expression by repressing protein expression and/or by promoting mRNA decay. *Cold Spring Harb Symp Quant Biol* **71:** 523–530.

Bernstein E, Caudy AA, Hammond SM, Hannon GJ. 2001. Role for a bidentate ribonuclease in the initiation step of RNA interference. *Nature* **409:** 363–366.

Bethune J, Artus-Revel CG, Filipowicz W. 2012. Kinetic analysis reveals successive steps leading to miRNA-mediated silencing in mammalian cells. *EMBO Rep* **13:** 716–723.

Bhattacharyya SN, Habermacher R, Martine U, Closs EI, Filipowicz W. 2006. Relief of microRNA-mediated translational repression in human cells subjected to stress. *Cell* **125:** 1111–1124.

Boland A, Huntzinger E, Schmidt S, Izaurralde E, Weichenrieder O. 2011. Crystal structure of the MID-PIWI lobe of a eukaryotic Argonaute protein. *Proc Natl Acad Sci* **108:** 10466–10471.

Bouasker S, Simard MJ. 2012. The slicing activity of miRNA-specific Argonautes is essential for the miRNA pathway in *C. elegans*. *Nucleic Acids Res* **40:** 10452–10462.

Braun JE, Huntzinger E, Fauser M, Izaurralde E. 2011. GW182 proteins directly recruit cytoplasmic deadenylase complexes to miRNA targets. *Mol Cell* **44:** 120–133.

Brennecke J, Stark A, Russell RB, Cohen SM. 2005. Principles of microRNA-target recognition. *PLoS Biol* **3:** e85.

Broderick JA, Salomon WE, Ryder SP, Aronin N, Zamore PD. 2011. Argonaute protein identity and pairing geometry determine cooperativity in mammalian RNA silencing. *Rna* **17:** 1858–1869.

Broughton JP, Lovci MT, Huang JL, Yeo GW, Pasquinelli AE. 2016. Pairing beyond the seed supports microRNA targeting specificity. *Mol Cell* **64:** 320–333.

Calin GA, Croce CM. 2006. MicroRNA-cancer connection: The beginning of a new tale. *Cancer Res* **66:** 7390–7394.

Carmell MA, Hannon GJ. 2004. RNase III enzymes and the initiation of gene silencing. *Nat Struct Mol Biol* **11:** 214–218.

Chandradoss SD, Schirle NT, Szczepaniak M, MacRae IJ, Joo C. 2015. A dynamic search process underlies microRNA targeting. *Cell* **162:** 96–107.

Chapat C, Jafarnejad SM, Matta-Camacho E, Hesketh GG, Gelbart IA, Attig J, Gkogkas CG, Alain T, Stern-Ginossar N, Fabian MR, et al. 2017. Cap-binding protein 4EHP effects translation silencing by microRNAs. *Proc Natl Acad Sci* **114:** 5425–5430.

Chekulaeva M, Mathys H, Zipprich JT, Attig J, Colic M, Parker R, Filipowicz W. 2011. miRNA repression involves GW182-mediated recruitment of CCR4-NOT through conserved W-containing motifs. *Nat Struct Mol Biol* **18:** 1218–1226.

Chen S, Gao G. 2017. MicroRNAs recruit eIF4E2 to repress translation of target mRNAs. *Protein Cell* **8:** 750–761.

Chen K, Rajewsky N. 2006. Deep conservation of micro-RNA-target relationships and 3′UTR motifs in vertebrates, flies, and nematodes. *Cold Spring Harb Symp Quant Biol* **71:** 149–156.

Chen K, Rajewsky N. 2007. The evolution of gene regulation by transcription factors and microRNAs. *Nat Rev Genet* **8:** 93–103.

Chen CY, Zheng D, Xia Z, Shyu AB. 2009. Ago-TNRC6 triggers microRNA-mediated decay by promoting two deadenylation steps. *Nat Struct Mol Biol* **16:** 1160–1166.

Chen Y, Boland A, Kuzuoğlu-Öztürk D, Bawankar P, Loh B, Chang CT, Weichenrieder O, Izaurralde E. 2014. A DDX6–CNOT1 complex and W-binding pockets in CNOT9 reveal direct links between miRNA target recognition and silencing. *Mol Cell* **54:** 737–750.

Cho PF, Poulin F, Cho-Park YA, Cho-Park IB, Chicoine JD, Lasko P, Sonenberg N. 2005. A new paradigm for translational control: Inhibition via 5′-3′ mRNA tethering by Bicoid and the eIF4E cognate 4EHP. *Cell* **121:** 411–423.

Christie M, Boland A, Huntzinger E, Weichenrieder O, Izaurralde E. 2013. Structure of the PAN3 pseudokinase reveals the basis for interactions with the PAN2 deadenylase and the GW182 proteins. *Mol Cell* **51:** 360–373.

Chu CY, Rana TM. 2006. Translation repression in human cells by microRNA-induced gene silencing requires RCK/p54. *PLoS Biol* **4:** e210.

Cooke A, Prigge A, Wickens M. 2010. Translational repression by deadenylases. *J Biol Chem* **285:** 28506–28513.

Croce CM, Calin GA. 2005. miRNAs, cancer, and stem cell division. *Cell* **122:** 6–7.

Czech B, Hannon GJ. 2011. Small RNA sorting: Matchmaking for Argonautes. *Nat Rev Genet* **12:** 19–31.

Czech B, Zhou R, Erlich Y, Brennecke J, Binari R, Villalta C, Gordon A, Perrimon N, Hannon GJ. 2009. Hierarchical rules for Argonaute loading in *Drosophila*. *Mol Cell* **36:** 445–456.

Derry MC, Yanagiya A, Martineau Y, Sonenberg N. 2006. Regulation of poly(A)-binding protein through PABP-interacting proteins. *Cold Spring Harb Symp Quant Biol* **71:** 537–543.

Djuranovic S, Nahvi A, Green R. 2012. miRNA-mediated gene silencing by translational repression followed by mRNA deadenylation and decay. *Science* **336:** 237–240.

Dostie J, Ferraiuolo M, Pause A, Adam SA, Sonenberg N. 2000. A novel shuttling protein, 4E-T, mediates the nuclear import of the mRNA 5′ cap-binding protein, eIF4E. *EMBO J* **19:** 3142–3156.

Ecsedi M, Rausch M, Grosshans H. 2015. The let-7 microRNA directs vulval development through a single target. *Dev Cell* **32:** 335–344.

Eichhorn SW, Guo H, McGeary SE, Rodriguez-Mias RA, Shin C, Baek D, Hsu SH, Ghoshal K, Villen J, Bartel DP. 2014. mRNA destabilization is the dominant effect of mammalian microRNAs by the time substantial repression ensues. *Mol Cell* **56:** 104–115.

Elcheva I, Goswami S, Noubissi FK, Spiegelman VS. 2009. CRD-BP protects the coding region of *βTrCP1* mRNA from miR-183-mediated degradation. *Mol Cell* **35:** 240–246.

Elfakess R, Sinvani H, Haimov O, Svitkin Y, Sonenberg N, Dikstein R. 2011. Unique translation initiation of mRNAs-containing TISU element. *Nucleic Acids Res* **39:** 7598–7609.

Elkayam E, Kuhn CD, Tocilj A, Haase AD, Greene EM, Hannon GJ, Joshua-Tor L. 2012. The structure of human argonaute-2 in complex with miR-20a. *Cell* **150:** 100–110.

Elkayam E, Faehnle CR, Morales M, Sun J, Li H, Joshua-Tor L. 2017. Multivalent recruitment of human Argonaute by GW182. *Mol Cell* **67:** 646–658.e643.

El-Shami M, Pontier D, Lahmy S, Braun L, Picart C, Vega D, Hakimi MA, Jacobsen SE, Cooke R, Lagrange T. 2007.

Reiterated WG/GW motifs form functionally and evolutionarily conserved ARGONAUTE-binding platforms in RNAi-related components. *Genes Dev* **21**: 2539–2544.

Emde A, Hornstein E. 2014. miRNAs at the interface of cellular stress and disease. *EMBO J* **33**: 1428–1437.

Eulalio A, Rehwinkel J, Stricker M, Huntzinger E, Yang SF, Doerks T, Dorner S, Bork P, Boutros M, Izaurralde E. 2007. Target-specific requirements for enhancers of decapping in miRNA-mediated gene silencing. *Genes Dev* **21**: 2558–2570.

Eulalio A, Huntzinger E, Izaurralde E. 2008. GW182 interaction with Argonaute is essential for miRNA-mediated translational repression and mRNA decay. *Nat Struct Mol Biol* **15**: 346–353.

Eulalio A, Helms S, Fritzsch C, Fauser M, Izaurralde E. 2009a. A C-terminal silencing domain in GW182 is essential for miRNA function. *RNA* **15**: 1067–1077.

Eulalio A, Tritschler F, Buttner R, Weichenrieder O, Izaurralde E, Truffault V. 2009b. The RRM domain in GW182 proteins contributes to miRNA-mediated gene silencing. *Nucleic Acids Res* **37**: 2974–2983.

Fabian MR, Mathonnet G, Sundermeier T, Mathys H, Zipprich JT, Svitkin YV, Rivas F, Jinek M, Wohlschlegel J, Doudna JA, et al. 2009. Mammalian miRNA RISC recruits CAF1 and PABP to affect PABP-dependent deadenylation. *Mol Cell* **35**: 868–880.

Fabian MR, Cieplak MK, Frank F, Morita M, Green J, Srikumar T, Nagar B, Yamamoto T, Raught B, Duchaine TF, et al. 2011. miRNA-mediated deadenylation is orchestrated by GW182 through two conserved motifs that interact with CCR4-NOT. *Nat Struct Mol Biol* **18**: 1211–1217.

Fang Z, Rajewsky N. 2011. The impact of miRNA target sites in coding sequences and in 3'UTRs. *PLoS ONE* **6**: e18067.

Farh KK, Grimson A, Jan C, Lewis BP, Johnston WK, Lim LP, Burge CB, Bartel DP. 2005. The widespread impact of mammalian microRNAs on mRNA repression and evolution. *Science* **310**: 1817–1821.

Ferraiuolo MA, Basak S, Dostie J, Murray EL, Schoenberg DR, Sonenberg N. 2005. A role for the eIF4E-binding protein 4E-T in P-body formation and mRNA decay. *J Cell Biol* **170**: 913–924.

Flamand MN, Wu E, Vashisht A, Jannot G, Keiper BD, Simard MJ, Wohlschlegel J, Duchaine TF. 2016. Poly (A)-binding proteins are required for microRNA-mediated silencing and to promote target deadenylation in *C. elegans*. *Nucleic Acids Res* **44**: 5924–5935.

Flamand MN, Gan HH, Mayya VK, Gunsalus KC, Duchaine TF. 2017. A non-canonical site reveals the cooperative mechanisms of microRNA-mediated silencing. *Nucleic Acids Res* **45**: 7212–7225.

Frank F, Fabian MR, Stepinski J, Jemielity J, Darzynkiewicz E, Sonenberg N, Nagar B. 2011. Structural analysis of 5'-mRNA-cap interactions with the human AGO2 MID domain. *EMBO Rep* **12**: 415–420.

Friedman RC, Farh KK, Burge CB, Bartel DP. 2009. Most mammalian mRNAs are conserved targets of microRNAs. *Genome Res* **19**: 92–105.

Fukao A, Mishima Y, Takizawa N, Oka S, Imataka H, Pelletier J, Sonenberg N, Thoma C, Fujiwara T. 2014. MicroRNAs trigger dissociation of eIF4AI and eIF4AII from target mRNAs in humans. *Mol Cell* **56**: 79–89.

Fukaya T, Tomari Y. 2011. PABP is not essential for microRNA-mediated translational repression and deadenylation in vitro. *EMBO J* **30**: 4998–5009.

Fukaya T, Iwakawa HO, Tomari Y. 2014. MicroRNAs block assembly of eIF4F translation initiation complex in *Drosophila*. *Mol Cell* **56**: 67–78.

Galicia-Vazquez G, Chu J, Pelletier J. 2015. eIF4AII is dispensable for miRNA-mediated gene silencing. *RNA* **21**: 1826–1833.

Gingras AC, Raught B, Sonenberg N. 1999. eIF4 initiation factors: Effectors of mRNA recruitment to ribosomes and regulators of translation. *Annu Rev Biochem* **68**: 913–963.

Giraldez AJ, Cinalli RM, Glasner ME, Enright AJ, Thomson JM, Baskerville S, Hammond SM, Bartel DP, Schier AF. 2005. MicroRNAs regulate brain morphogenesis in zebrafish. *Science* **308**: 833–838.

Giraldez AJ, Mishima Y, Rihel J, Grocock RJ, Van Dongen S, Inoue K, Enright AJ, Schier AF. 2006. Zebrafish MiR-430 promotes deadenylation and clearance of maternal mRNAs. *Science* **312**: 75–79.

Giri R, Carthew RW. 2014. MicroRNAs suppress cellular phenotypic heterogeneity. *Cell Cycle* **13**: 1517–1518.

Gregory RI, Chendrimada TP, Cooch N, Shiekhattar R. 2005. Human RISC couples microRNA biogenesis and posttranscriptional gene silencing. *Cell* **123**: 631–640.

Grimson A, Farh KK, Johnston WK, Garrett-Engele P, Lim LP, Bartel DP. 2007. MicroRNA targeting specificity in mammals: Determinants beyond seed pairing. *Mol Cell* **27**: 91–105.

Grishok A, Pasquinelli AE, Conte D, Li N, Parrish S, Ha I, Baillie DL, Fire A, Ruvkun G, Mello CC. 2001. Genes and mechanisms related to RNA interference regulate expression of the small temporal RNAs that control *C. elegans* developmental timing. *Cell* **106**: 23–34.

Guo H, Ingolia NT, Weissman JS, Bartel DP. 2010. Mammalian microRNAs predominantly act to decrease target mRNA levels. *Nature* **466**: 835–840.

Ha M, Kim VN. 2014. Regulation of microRNA biogenesis. *Nat Rev Mol Cell Biol* **15**: 509–524.

Haas G, Braun JE, Igreja C, Tritschler F, Nishihara T, Izaurralde E. 2010. HPat provides a link between deadenylation and decapping in metazoa. *J Cell Biol* **189**: 289–302.

Haley B, Zamore PD. 2004. Kinetic analysis of the RNAi enzyme complex. *Nat Struct Mol Biol* **11**: 599–606.

Hammond SM, Bernstein E, Beach D, Hannon GJ. 2000. An RNA-directed nuclease mediates post-transcriptional gene silencing in *Drosophila* cells. *Nature* **404**: 293–296.

Hammond SM, Boettcher S, Caudy AA, Kobayashi R, Hannon GJ. 2001. Argonaute2, a link between genetic and biochemical analyses of RNAi. *Science* **293**: 1146–1150.

He L, Thomson JM, Hemann MT, Hernando-Monge E, Mu D, Goodson S, Powers S, Cordon-Cardo C, Lowe SW, Hannon GJ, et al. 2005. A microRNA polycistron as a potential human oncogene. *Nature* **435**: 828–833.

* Heck AM, Wilusz J. 2018. The interplay between the RNA decay and translation machinery in eukaryotes. *Cold Spring Harb Perspect Biol* 10.1101/cshperspect.a032839.

Huang J, Liang Z, Yang B, Tian H, Ma J, Zhang H. 2007. Derepression of microRNA-mediated protein translation inhibition by apolipoprotein B mRNA-editing enzyme catalytic polypeptide-like 3G (APOBEC3G) and its family members. *J Biol Chem* **282:** 33632–33640.

Humphreys DT, Westman BJ, Martin DI, Preiss T. 2005. MicroRNAs control translation initiation by inhibiting eukaryotic initiation factor 4E/cap and poly(A) tail function. *Proc Natl Acad Sci* **102:** 16961–16966.

Huntzinger E, Braun JE, Heimstadt S, Zekri L, Izaurralde E. 2010. Two PABPC1-binding sites in GW182 proteins promote miRNA-mediated gene silencing. *EMBO J* **29:** 4146–4160.

Hutvágner G, McLachlan J, Pasquinelli AE, Balint E, Tuschl T, Zamore PD. 2001. A cellular function for the RNA-interference enzyme Dicer in the maturation of the let-7 small temporal RNA. *Science* **293:** 834–838.

Inukai S, Pincus Z, de Lencastre A, Slack FJ. 2018. A micro-RNA feedback loop regulates global microRNA abundance during aging. *RNA* **24:** 159–172.

Iorio MV, Croce CM. 2013. MicroRNA profiling in ovarian cancer. *Methods Mol Biol* **1049:** 187–197.

Iorio MV, Croce CM. 2017. MicroRNA dysregulation in cancer: Diagnostics, monitoring and therapeutics. A comprehensive review. *EMBO Mol Med* **9:** 852.

Jafarifar F, Yao P, Eswarappa SM, Fox PL. 2011. Repression of VEGFA by CA-rich element-binding microRNAs is modulated by hnRNP L. *EMBO J* **30:** 1324–1334.

Jafarnejad SM, Chapat C, Matta-Camacho E, Gelbart IA, Hesketh GG, Arguello M, Garzia A, Kim SH, Attig J, Shapiro M, et al. 2018. Translational control of ERK signaling through miRNA/4EHP-directed silencing. *eLife* **7:** e35034.

Jee D, Yang JS, Park SM, Farmer DT, Wen J, Chou T, Chow A, McManus MT, Kharas MG, Lai EC. 2018. Dual strategies for argonaute2-mediated biogenesis of erythroid miRNAs underlie conserved requirements for slicing in mammals. *Mol Cell* **69:** 265–278.e266.

Jens M, Rajewsky N. 2015. Competition between target sites of regulators shapes post-transcriptional gene regulation. *Nat Rev Genet* **16:** 113–126.

Jinek M, Fabian MR, Coyle SM, Sonenberg N, Doudna JA. 2010. Structural insights into the human GW182–PABC interaction in microRNA-mediated deadenylation. *Nat Struct Mol Biol* **17:** 238–240.

Jonas S, Izaurralde E. 2013. The role of disordered protein regions in the assembly of decapping complexes and RNP granules. *Genes Dev* **27:** 2628–2641.

Kahvejian A, Svitkin YV, Sukarieh R, M'Boutchou MN, Sonenberg N. 2005. Mammalian poly(A)-binding protein is a eukaryotic translation initiation factor, which acts via multiple mechanisms. *Genes Dev* **19:** 104–113.

Kamenska A, Lu WT, Kubacka D, Broomhead H, Minshall N, Bushell M, Standart N. 2014. Human 4E-T represses translation of bound mRNAs and enhances microRNA-mediated silencing. *Nucleic Acids Res* **42:** 3298–3313.

Kamenska A, Simpson C, Vindry C, Broomhead H, Benard M, Ernoult-Lange M, Lee BP, Harries LW, Weil D, Standart N. 2016. The DDX6-4E-T interaction mediates translational repression and P-body assembly. *Nucleic Acids Res* **44:** 6318–6334.

Kedde M, Strasser MJ, Boldajipour B, Oude Vrielink JA, Slanchev K, le Sage C, Nagel R, Voorhoeve PM, van Duijse J, Orom UA, et al. 2007. RNA-binding protein

Dnd1 inhibits microRNA access to target mRNA. *Cell* **131:** 1273–1286.

Kedde M, van Kouwenhove M, Zwart W, Oude Vrielink JA, Elkon R, Agami R. 2010. A Pumilio-induced RNA structure switch in p27-3′ UTR controls miR-221 and miR-222 accessibility. *Nat Cell Biol* **12:** 1014–1020.

Ketting RF, Fischer SE, Bernstein E, Sijen T, Hannon GJ, Plasterk RH. 2001. Dicer functions in RNA interference and in synthesis of small RNA involved in developmental timing in *C. elegans. Genes Dev* **15:** 2654–2659.

Khaleghpour K, Svitkin YV, Craig AW, DeMaria CT, Deo RC, Burley SK, Sonenberg N. 2001. Translational repression by a novel partner of human poly(A) binding protein, Paip2. *Mol Cell* **7:** 205–216.

Kidner CA, Martienssen RA. 2004. Spatially restricted microRNA directs leaf polarity through ARGONAUTE1. *Nature* **428:** 81–84.

Klum SM, Chandradoss SD, Schirle NT, Joo C, MacRae IJ. 2017. Helix-7 in Argonaute2 shapes the microRNA seed region for rapid target recognition. *EMBO J* **37:** 75–88.

Knight SW, Bass BL. 2001. A role for the RNase III enzyme DCR-1 in RNA interference and germ line development in *Caenorhabditis elegans. Science* **293:** 2269–2271.

Kobayashi H, Tomari Y. 2016. RISC assembly: Coordination between small RNAs and Argonaute proteins. *Biochim Biophys Acta* **1859:** 71–81.

Kozlov G, Safaee N, Rosenauer A, Gehring K. 2010. Structural basis of binding of P-body-associated proteins GW182 and ataxin-2 by the Mlle domain of poly(A)-binding protein. *J Biol Chem* **285:** 13599–13606.

Krek A, Grun D, Poy MN, Wolf R, Rosenberg L, Epstein EJ, MacMenamin P, da Piedade I, Gunsalus KC, Stoffel M, et al. 2005. Combinatorial microRNA target predictions. *Nat Genet* **37:** 495–500.

Kundu P, Fabian MR, Sonenberg N, Bhattacharyya SN, Filipowicz W. 2012. HuR protein attenuates miRNA-mediated repression by promoting miRISC dissociation from the target RNA. *Nucleic Acids Res* **40:** 5088–5100.

Kuzuoğlu-Öztürk D, Huntzinger E, Schmidt S, Izaurralde E. 2012. The *Caenorhabditis elegans* GW182 protein AIN-1 interacts with PAB-1 and subunits of the PAN2-PAN3 and CCR4-NOT deadenylase complexes. *Nucleic Acids Res* **40:** 5651–5665.

Kuzuoğlu-Öztürk D, Bhandari D, Huntzinger E, Fauser M, Helms S, Izaurralde E. 2016. miRISC and the CCR4-NOT complex silence mRNA targets independently of 43S ribosomal scanning. *EMBO J* **35:** 1186–1203.

* Kwan T, Thompson SR. Noncanonical translation initiation in eukaryotes. *Cold Spring Harb Perspect Biol* 10.1101/cshperspect.a032672.

Lagos-Quintana M, Rauhut R, Lendeckel W, Tuschl T. 2001. Identification of novel genes coding for small expressed RNAs. *Science* **294:** 853–858.

Lagos-Quintana M, Rauhut R, Meyer J, Borkhardt A, Tuschl T. 2003. New microRNAs from mouse and human. *RNA* **9:** 175–179.

Lai EC. 2002. Micro RNAs are complementary to 3′ UTR sequence motifs that mediate negative post-transcriptional regulation. *Nat Genet* **30:** 363–364.

Lall S, Grun D, Krek A, Chen K, Wang YL, Dewey CN, Sood P, Colombo T, Bray N, Macmenamin P, et al. 2006. A genome-wide map of conserved microRNA targets in *C. elegans. Curr Biol* **16:** 460–471.

Lau NC, Lim LP, Weinstein EG, Bartel DP. 2001. An abundant class of tiny RNAs with probable regulatory roles in *Caenorhabditis elegans. Science* **294:** 858–862.

Lee RC, Ambros V. 2001. An extensive class of small RNAs in *Caenorhabditis elegans. Science* **294:** 862–864.

Lee RC, Feinbaum RL, Ambros V. 1993. The *C. elegans* heterochronic gene *lin-4* encodes small RNAs with antisense complementarity to *lin-14. Cell* **75:** 843–854.

Lee Y, Ahn C, Han J, Choi H, Kim J, Yim J, Lee J, Provost P, Radmark O, Kim S, et al. 2003. The nuclear RNase III Drosha initiates microRNA processing. *Nature* **425:** 415–419.

Lewis BP, Shih IH, Jones-Rhoades MW, Bartel DP, Burge CB. 2003. Prediction of mammalian microRNA targets. *Cell* **115:** 787–798.

Lewis BP, Burge CB, Bartel DP. 2005. Conserved seed pairing, often flanked by adenosines, indicates that thousands of human genes are microRNA targets. *Cell* **120:** 15–20.

Li X, Cassidy JJ, Reinke CA, Fischboeck S, Carthew RW. 2009. A microRNA imparts robustness against environmental fluctuation during development. *Cell* **137:** 273–282.

Lian SL, Li S, Abadal GX, Pauley BA, Fritzler MJ, Chan EK. 2009. The C-terminal half of human Ago2 binds to multiple GW-rich regions of GW182 and requires GW182 to mediate silencing. *RNA* **15:** 804–813.

Lim LP, Glasner ME, Yekta S, Burge CB, Bartel DP. 2003a. Vertebrate microRNA genes. *Science* **299:** 1540.

Lim LP, Lau NC, Weinstein EG, Abdelhakim A, Yekta S, Rhoades MW, Burge CB, Bartel DP. 2003b. The microRNAs of *Caenorhabditis elegans. Genes Dev* **17:** 991–1008.

Lim LP, Lau NC, Garrett-Engele P, Grimson A, Schelter JM, Castle J, Bartel DP, Linsley PS, Johnson JM. 2005. Microarray analysis shows that some microRNAs downregulate large numbers of target mRNAs. *Nature* **433:** 769–773.

Lingel A, Simon B, Izaurralde E, Sattler M. 2003. Structure and nucleic-acid binding of the *Drosophila* Argonaute 2 PAZ domain. *Nature* **426:** 465–469.

Lingel A, Simon B, Izaurralde E, Sattler M. 2004. Nucleic acid 3′-end recognition by the Argonaute2 PAZ domain. *Nat Struct Mol Biol* **11:** 576–577.

Liu J, Carmell MA, Rivas FV, Marsden CG, Thomson JM, Song JJ, Hammond SM, Joshua-Tor L, Hannon GJ. 2004. Argonaute2 is the catalytic engine of mammalian RNAi. *Science* **305:** 1437–1441.

Ma JB, Ye K, Patel DJ. 2004. Structural basis for overhang-specific small interfering RNA recognition by the PAZ domain. *Nature* **429:** 318–322.

Maniataki E, Mourelatos Z. 2005. A human, ATP-independent, RISC assembly machine fueled by pre-miRNA. *Genes Dev* **19:** 2979–2990.

Mann M, Barad O, Agami R, Geiger B, Hornstein E. 2010. miRNA-based mechanism for the commitment of multipotent progenitors to a single cellular fate. *Proc Natl Acad Sci* **107:** 15804–15809.

Mathonnet G, Fabian MR, Svitkin YV, Parsyan A, Huck L, Murata T, Biffo S, Merrick WC, Darzynkiewicz E, Pillai RS, et al. 2007. MicroRNA inhibition of translation initi-ation in vitro by targeting the cap-binding complex eIF4F. *Science* **317:** 1764–1767.

Mathys H, Basquin J, Ozgur S, Czarnocki-Cieciura M, Bonneau F, Aartse A, Dziembowski A, Nowotny M, Conti E, Filipowicz W. 2014. Structural and biochemical insights to the role of the CCR4-NOT complex and DDX6 ATPase in microRNA repression. *Mol Cell* **54:** 751–765.

Matranga C, Tomari Y, Shin C, Bartel DP, Zamore PD. 2005. Passenger-strand cleavage facilitates assembly of siRNA into Ago2-containing RNAi enzyme complexes. *Cell* **123:** 607–620.

Mauri M, Kirchner M, Aharoni R, Ciolli Mattioli C, van den Bruck D, Gutkovitch N, Modepalli V, Selbach M, Moran Y, Chekulaeva M. 2017. Conservation of miRNA-mediated silencing mechanisms across 600 million years of animal evolution. *Nucleic Acids Res* **45:** 938–950.

Mauxion F, Chen CY, Seraphin B, Shyu AB. 2009. BTG/TOB factors impact deadenylases. *Trends Biochem Sci* **34:** 640–647.

Mayr C. 2017. Regulation by 3′-untranslated regions. *Annu Rev Genet* **51:** 171–194.

Mayya VK, Duchaine TF. 2015. On the availability of microRNA-induced silencing complexes, saturation of microRNA-binding sites and stoichiometry. *Nucleic Acids Res* **43:** 7556–7565.

Meijer HA, Kong YW, Lu WT, Wilczynska A, Spriggs RV, Robinson SW, Godfrey JD, Willis AE, Bushell M. 2013. Translational repression and eIF4A2 activity are critical for microRNA-mediated gene regulation. *Science* **340:** 82–85.

* Merrick WC, Pavitt GD. 2018. Protein synthesis initiation in eukaryotic cells. *Cold Spring Harb Perspect Biol* 10.1101/cshperspect.a033092.

Mishima Y, Fukao A, Kishimoto T, Sakamoto H, Fujiwara T, Inoue K. 2012. Translational inhibition by deadenylation-independent mechanisms is central to microRNA-mediated silencing in zebrafish. *Proc Natl Acad Sci* **109:** 1104–1109.

Miyoshi K, Tsukumo H, Nagami T, Siomi H, Siomi MC. 2005. Slicer function of *Drosophila* Argonautes and its involvement in RISC formation. *Genes Dev* **19:** 2837–2848.

Moretti F, Kaiser C, Zdanowicz-Specht A, Hentze MW. 2012. PABP and the poly(A) tail augment microRNA repression by facilitated miRISC binding. *Nat Struct Mol Biol* **19:** 603–608.

Morita M, Ler LW, Fabian MR, Siddiqui N, Mullin M, Henderson VC, Alain T, Fonseca BD, Karashchuk G, Bennett CF, et al. 2012. A novel 4EHP-GIGYF2 translational repressor complex is essential for mammalian development. *Mol Cell Biol* **32:** 3585–3593.

Nakanishi K, Weinberg DE, Bartel DP, Patel DJ. 2012. Structure of yeast Argonaute with guide RNA. *Nature* **486:** 368–374.

Nishihara T, Zekri L, Braun JE, Izaurralde E. 2013. miRISC recruits decapping factors to miRNA targets to enhance their degradation. *Nucleic Acids Res* **41:** 8692–8705.

Nishimura T, Padamsi Z, Fakim H, Milette S, Dunham WH, Gingras AC, Fabian MR. 2015. The eIF4E-binding protein 4E-T is a component of the mRNA decay machinery that bridges the 5′ and 3′ termini of target mRNAs. *Cell Rep* **11:** 1425–1436.

Nottrott S, Simard MJ, Richter JD. 2006. Human let-7a miRNA blocks protein production on actively translating polyribosomes. *Nat Struct Mol Biol* **13:** 1108–1114.

Olsen PH, Ambros V. 1999. The lin-4 regulatory RNA controls developmental timing in *Caenorhabditis elegans* by blocking LIN-14 protein synthesis after the initiation of translation. *Dev Biol* **216:** 671–680.

Ozgur S, Chekulaeva M, Stoecklin G. 2010. Human Pat1b connects deadenylation with mRNA decapping and controls the assembly of processing bodies. *Mol Cell Biol* **30:** 4308–4323.

Ozgur S, Basquin J, Kamenska A, Filipowicz W, Standart N, Conti E. 2015. Structure of a human 4E-T/DDX6/CNOT1 complex reveals the different interplay of DDX6-binding proteins with the CCR4–NOT complex. *Cell Rep* **13:** 703–711.

Park JH, Shin C. 2015. Slicer-independent mechanism drives small-RNA strand separation during human RISC assembly. *Nucleic Acids Res* **43:** 9418–9433.

Pasquinelli AE, Reinhart BJ, Slack F, Martindale MQ, Kuroda MI, Maller B, Hayward DC, Ball EE, Degnan B, Muller P, et al. 2000. Conservation of the sequence and temporal expression of let-7 heterochronic regulatory RNA. *Nature* **408:** 86–89.

Pelaez N, Carthew RW. 2012. Biological robustness and the role of microRNAs: A network perspective. *Curr Top Dev Biol* **99:** 237–255.

Peter D, Weber R, Sandmeir F, Wohlbold L, Helms S, Bawankar P, Valkov E, Igreja C, Izaurralde E. 2017. GIGYF1/2 proteins use auxiliary sequences to selectively bind to 4EHP and repress target mRNA expression. *Genes Dev* **31:** 1147–1161.

Petersen CP, Bordeleau ME, Pelletier J, Sharp PA. 2006. Short RNAs repress translation after initiation in mammalian cells. *Mol Cell* **21:** 533–542.

Pfaff J, Hennig J, Herzog F, Aebersold R, Sattler M, Niessing D, Meister G. 2013. Structural features of Argonaute-GW182 protein interactions. *Proc Natl Acad Sci* **110:** E3770–E3779.

Pillai RS, Bhattacharyya SN, Artus CG, Zoller T, Cougot N, Basyuk E, Bertrand E, Filipowicz W. 2005. Inhibition of translational initiation by Let-7 microRNA in human cells. *Science* **309:** 1573–1576.

Posadas DM, Carthew RW. 2014. MicroRNAs and their roles in developmental canalization. *Curr Opin Genet Dev* **27:** 1–6.

Rabani M, Pieper L, Chew GL, Schier AF. 2017. A massively parallel reporter assay of 3′ UTR sequences identifies in vivo rules for mRNA degradation. *Mol Cell* **68:** 1083–1094.e1085.

Reinhart BJ, Slack FJ, Basson M, Pasquinelli AE, Bettinger JC, Rougvie AE, Horvitz HR, Ruvkun G. 2000. The 21-nucleotide let-7 RNA regulates developmental timing in *Caenorhabditis elegans*. *Nature* **403:** 901–906.

Ricci EP, Limousin T, Soto-Rifo R, Rubilar PS, Decimo D, Ohlmann T. 2013. miRNA repression of translation in vitro takes place during 43S ribosomal scanning. *Nucleic Acids Res* **41:** 586–598.

Rissland OS, Subtelny AO, Wang M, Lugowski A, Nicholson B, Laver JD, Sidhu SS, Smibert CA, Lipshitz HD, Bartel DP. 2017. The influence of microRNAs and poly(A) tail length on endogenous mRNA-protein complexes. *Genome Biol* **18:** 211.

Rom E, Kim HC, Gingras AC, Marcotrigiano J, Favre D, Olsen H, Burley SK, Sonenberg N. 1998. Cloning and characterization of 4EHP, a novel mammalian eIF4E-related cap-binding protein. *J Biol Chem* **273:** 13104–13109.

Romano G, Veneziano D, Acunzo M, Croce CM. 2017. Small non-coding RNA and cancer. *Carcinogenesis* **38:** 485–491.

Rouya C, Siddiqui N, Morita M, Duchaine TF, Fabian MR, Sonenberg N. 2014. Human DDX6 effects miRNA-mediated gene silencing via direct binding to CNOT1. *RNA* **20:** 1398–1409.

Roy G, De Crescenzo G, Khaleghpour K, Kahvejian A, O'Connor-McCourt M, Sonenberg N. 2002. Paip1 interacts with poly(A) binding protein through two independent binding motifs. *Mol Cell Biol* **22:** 3769–3782.

Saetrom P, Heale BS, Snove O Jr, Aagaard L, Alluin J, Rossi JJ. 2007. Distance constraints between microRNA target sites dictate efficacy and cooperativity. *Nucleic Acids Res* **35:** 2333–2342.

Salomon WE, Jolly SM, Moore MJ, Zamore PD, Serebrov V. 2016. Single-molecule imaging reveals that Argonaute reshapes the binding properties of its nucleic acid guides. *Cell* **166:** 517–520.

Schirle NT, MacRae IJ. 2012. The crystal structure of human Argonaute2. *Science* **336:** 1037–1040.

Schirle NT, Sheu-Gruttadauria J, MacRae IJ. 2014. Structural basis for microRNA targeting. *Science* **346:** 608–613.

Schirle NT, Sheu-Gruttadauria J, Chandradoss SD, Joo C, MacRae IJ. 2015. Water-mediated recognition of t1-adenosine anchors Argonaute2 to microRNA targets. *eLife* **4:** e07646.

Schopp IM, Amaya Ramirez CC, Debeljak J, Kreibich E, Skribbe M, Wild K, Bethune J. 2017. Split-BioID a conditional proteomics approach to monitor the composition of spatiotemporally defined protein complexes. *Nat Commun* **8:** 15690.

Selbach M, Schwanhausser B, Thierfelder N, Fang Z, Khanin R, Rajewsky N. 2008. Widespread changes in protein synthesis induced by microRNAs. *Nature* **455:** 58–63.

Sharif H, Ozgur S, Sharma K, Basquin C, Urlaub H, Conti E. 2013. Structural analysis of the yeast Dhh1-Pat1 complex reveals how Dhh1 engages Pat1, Edc3 and RNA in mutually exclusive interactions. *Nucleic Acids Res* **41:** 8377–8390.

Sheu-Gruttadauria J, MacRae IJ. 2017. Structural foundations of RNA silencing by Argonaute. *J Mol Biol* **429:** 2619–2639.

Siddiqui N, Osborne MJ, Gallie DR, Gehring K. 2007. Solution structure of the PABC domain from wheat poly(A)-binding protein: An insight into RNA metabolic and translational control in plants. *Biochemistry* **46:** 4221–4231.

Smith RW, Blee TK, Gray NK. 2014. Poly(A)-binding proteins are required for diverse biological processes in metazoans. *Biochem Soc Trans* **42:** 1229–1237.

Sonenberg N, Hinnebusch AG. 2009. Regulation of translation initiation in eukaryotes: Mechanisms and biological targets. *Cell* **136:** 731–745.

Song JJ, Liu J, Tolia NH, Schneiderman J, Smith SK, Martienssen RA, Hannon GJ, Joshua-Tor L. 2003. The crystal structure of the Argonaute2 PAZ domain reveals an RNA binding motif in RNAi effector complexes. *Nat Struct Biol* **10:** 1026–1032.

Song JJ, Smith SK, Hannon GJ, Joshua-Tor L. 2004. Crystal structure of Argonaute and its implications for RISC slicer activity. *Science* **305:** 1434–1437.

Stark A, Brennecke J, Bushati N, Russell RB, Cohen SM. 2005. Animal microRNAs confer robustness to gene expression and have a significant impact on 3′UTR evolution. *Cell* **123:** 1133–1146.

Subtelny AO, Eichhorn SW, Chen GR, Sive H, Bartel DP. 2014. Poly(A)-tail profiling reveals an embryonic switch in translational control. *Nature* **508:** 66–71.

Sun K, Lai EC. 2013. Adult-specific functions of animal microRNAs. *Nat Rev Genet* **14:** 535–548.

Takeda Y, Mishima Y, Fujiwara T, Sakamoto H, Inoue K. 2009. DAZL relieves miRNA-mediated repression of germline mRNAs by controlling poly(A) tail length in zebrafish. *PLoS ONE* **4:** e7513.

Takimoto K, Wakiyama M, Yokoyama S. 2009. Mammalian GW182 contains multiple Argonaute-binding sites and functions in microRNA-mediated translational repression. *RNA* **15:** 1078–1089.

Thermann R, Hentze MW. 2007. *Drosophila* miR2 induces pseudo-polysomes and inhibits translation initiation. *Nature* **447:** 875–878.

Tritschler F, Braun JE, Eulalio A, Truffault V, Izaurralde E, Weichenrieder O. 2009. Structural basis for the mutually exclusive anchoring of P body components EDC3 and Tral to the DEAD box protein DDX6/Me31B. *Mol Cell* **33:** 661–668.

Waghray S, Williams C, Coon JJ, Wickens M. 2015. *Xenopus* CAF1 requires NOT1-mediated interaction with 4E-T to repress translation in vivo. *RNA* **21:** 1335–1345.

Wakiyama M, Takimoto K, Ohara O, Yokoyama S. 2007. Let-7 microRNA-mediated mRNA deadenylation and translational repression in a mammalian cell-free system. *Genes Dev* **21:** 1857–1862.

Wang Y, Juranek S, Li H, Sheng G, Tuschl T, Patel DJ. 2008a. Structure of an argonaute silencing complex with a seed-containing guide DNA and target RNA duplex. *Nature* **456:** 921–926.

Wang Y, Sheng G, Juranek S, Tuschl T, Patel DJ. 2008b. Structure of the guide-strand-containing argonaute silencing complex. *Nature* **456:** 209–213.

Wee LM, Flores-Jasso CF, Salomon WE, Zamore PD. 2012. Argonaute divides its RNA guide into domains with distinct functions and RNA-binding properties. *Cell* **151:** 1055–1067.

Wolter JM, Le HH, Linse A, Godlove VA, Nguyen TD, Kotagama K, Lynch A, Rawls A, Mangone M. 2017. Evolutionary patterns of metazoan microRNAs reveal targeting principles in the let-7 and miR-10 families. *Genome Res* **27:** 53–63.

Wu L, Fan J, Belasco JG. 2006. MicroRNAs direct rapid deadenylation of mRNA. *Proc Natl Acad Sci* **103:** 4034–4039.

Wu E, Thivierge C, Flamand M, Mathonnet G, Vashisht AA, Wohlschlegel J, Fabian MR, Sonenberg N, Duchaine TF. 2010. Pervasive and cooperative deadenylation of 3′UTRs by embryonic microRNA families. *Mol Cell* **40:** 558–570.

Wu E, Vashisht AA, Chapat C, Flamand MN, Cohen E, Sarov M, Tabach Y, Sonenberg N, Wohlschlegel J, Duchaine TF. 2017. A continuum of mRNP complexes in embryonic microRNA-mediated silencing. *Nucleic Acids Res* **45:** 2081–2098.

Xin M, Olson EN, Bassel-Duby R. 2013. Mending broken hearts: Cardiac development as a basis for adult heart regeneration and repair. *Nat Rev Mol Cell Biol* **14:** 529–541.

Yan KS, Yan S, Farooq A, Han A, Zeng L, Zhou MM. 2003. Structure and conserved RNA binding of the PAZ domain. *Nature* **426:** 468–474.

Yao B, Li S, Lian SL, Fritzler MJ, Chan EK. 2011. Mapping of Ago2-GW182 functional interactions. *Methods Mol Biol* **725:** 45–62.

Yartseva V, Takacs CM, Vejnar CE, Lee MT, Giraldez AJ. 2017. RESA identifies mRNA-regulatory sequences at high resolution. *Nat Methods* **14:** 201–207.

Zdanowicz A, Thermann R, Kowalska J, Jemielity J, Duncan K, Preiss T, Darzynkiewicz E, Hentze MW. 2009. *Drosophila* miR2 primarily targets the m⁷GpppN cap structure for translational repression. *Mol Cell* **35:** 881–888.

Zekri L, Kuzuoğlu-Öztürk D, Izaurralde E. 2013. GW182 proteins cause PABP dissociation from silenced miRNA targets in the absence of deadenylation. *EMBO J* **32:** 1052–1065.

Stress Granules and Processing Bodies in Translational Control

Pavel Ivanov,[1,2,3] Nancy Kedersha,[1,2] and Paul Anderson[1,2]

[1]Division of Rheumatology, Immunology and Allergy, Brigham and Women's Hospital, Boston, Massachusetts 02115

[2]Department of Medicine, Harvard Medical School, Boston, Massachusetts 02115

[3]The Broad Institute of Harvard and M.I.T., Cambridge, Massachusetts 02142

Correspondence: panderson@rics.bwh.harvard.edu

Stress granules (SGs) and processing bodies (PBs) are non-membrane-enclosed RNA granules that dynamically sequester translationally inactive messenger ribonucleoprotein particles (mRNPs) into compartments that are distinct from the surrounding cytoplasm. mRNP remodeling, silencing, and/or storage involves the dynamic partitioning of closed-loop polyadenylated mRNPs into SGs, or the sequestration of deadenylated, linear mRNPs into PBs. SGs form when stress-activated pathways stall translation initiation but allow elongation and termination to occur normally, resulting in a sudden excess of mRNPs that are spatially condensed into discrete foci by protein:protein, protein:RNA, and RNA:RNA interactions. In contrast, PBs can exist in the absence of stress, when specific factors promote mRNA deadenylation, condensation, and sequestration from the translational machinery. The formation and dissolution of SGs and PBs reflect changes in messenger RNA (mRNA) metabolism and allow cells to modulate the proteome and/or mediate life or death decisions during changing environmental conditions.

Tight control of messenger RNA (mRNA) processing, trafficking, degradation, and translation are important in regulating gene expression. These processes are controlled by specific RNA-binding proteins (RBPs) that bind the mRNA within larger complexes called messenger ribonucleoprotein particles (mRNPs) (Mitchell and Parker 2014). In eukaryotes, such mRNPs are often localized to specific cellular compartments, both as a part of mRNA biogenesis under optimal conditions, and as a part of response to changing conditions. Recent data suggest that self-organization of mRNPs into various non-membrane-enclosed subcellular compartments, termed RNA granules, plays critical roles in mRNA metabolism (Shin and Brangwynne 2017). Two of the best-studied RNA granules are stress granules (SGs) and processing bodies (PBs), membraneless cytoplasmic foci formed by the condensation of translationally inactivated mRNPs. Although the composition of sequestered mRNAs and RBPs differs between SGs and PBs (Fig. 1), both RNA granules are linked to translational control events that modulate the proteome and/or influence cell survival. The accumulation and condensation of untranslating

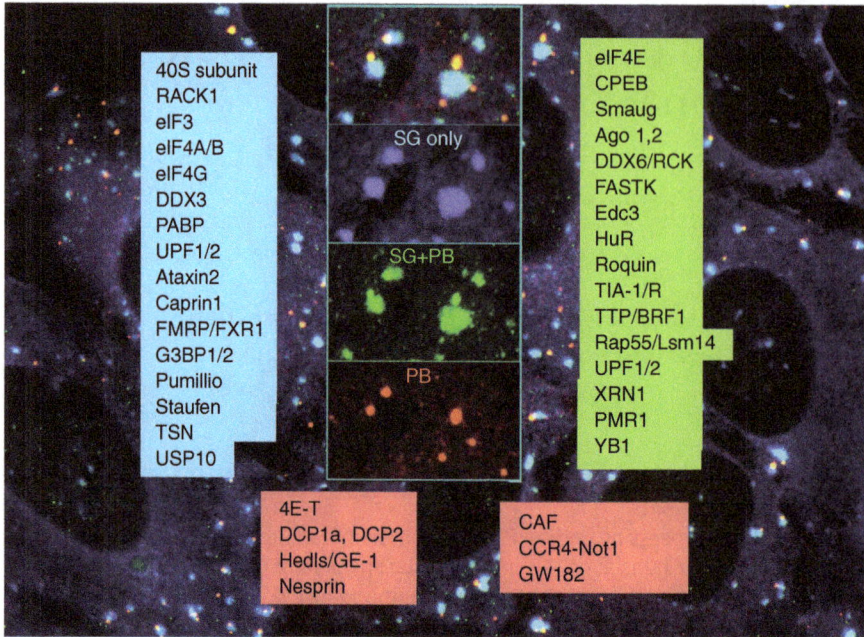

Figure 1. Selected stress granule (SG)- and processing body (PB)-associated proteins. Proteins (partial list) found exclusively in SGs (blue box), in both SGs and PB/GW-bodies (GWBs) (green box), or predominantly in PB/GWBs (red box). Image obtained using arsenite-treated U2OS cells stained for eukaryotic initiation factor 3b (eIF3b) (blue), DCP1a (red), and eIF4E (green).

mRNPs into these discrete cytoplasmic granules are governed by similar events that are intimately connected to various aspects of translational control.

The term "stress granules" was first used to describe phase-dense cytoplasmic particles that appeared in mammalian cells subjected to heat shock. These granules contained various heat-shock proteins (HSPs) (Collier and Schlesinger 1986; Collier et al. 1988), and similar particles were observed in heat-shocked tomato cells (Nover et al. 1983, 1989). Although initial compositional analysis revealed the presence of both HSPs and mRNAs in tomato "heat SGs" (Nover et al. 1983, 1989), later reports clarified that these SGs did not actually contain RNA and thus cannot be classified as RNA granules (Weber et al. 2008). However, before this revised report, the term "stress granules" was also used to describe cytoplasmic foci containing the translational repressor T-cell intracellular antigen 1 (TIA1), the translational enhancer poly(A)-binding protein (PABPC1), and polyadenylated

mRNAs. Colocalization of these factors in discrete cytoplasmic granules was triggered by either heat-shock stress or sodium arsenite-induced oxidative stress (Kedersha et al. 1999). Unlike plant heat-shock granules, these mammalian mRNA-containing stress granules strictly required phosphorylation of eukaryotic translation initiation factor 2α (eIF2α) (Kedersha et al. 1999), thus linking SGs to translational control.

PBs were first described as "XRN1 foci" because of the granular cytoplasmic localization of the exoribonuclease XRN1 (Bashkirov et al. 1997). Subsequent observations revealed that other RNA decay-associated proteins were colocalized in these foci (Ingelfinger et al. 2002; van Dijk et al. 2002; Fenger-Gron et al. 2005; Wilczynska et al. 2005; Yu et al. 2005; Eulalio et al. 2007), leading to their designation as mRNA "processing bodies" (Sheth and Parker 2006). Proteins associated with mRNA silencing, such as the argonautes and glycine-tryptophan protein of 182 KDa (GW182)/trinucleotide repeat containing 6A, were also found in

Cite this article as *Cold Spring Harb Perspect Biol* doi: 10.1101/cshperspect.a032813

organized puncta described as "GW-bodies" (GWBs), which were often coincident with PBs (Eystathioy et al. 2003). For the purposes of this review, we will include GWBs under the umbrella term PBs, but note that they are not identical (reviewed in Stoecklin and Kedersha 2013).

STRESS GRANULES: COMPOSITION AND INITIATION

SGs consist of stalled preinitiation complexes that include small (40S), but not large (60S), ribosomal subunits, translation initiation factors eIF4F, eIF3, and PABP, and polyadenylated mRNAs (reviewed in Anderson and Kedersha 2009). Condensation of stalled preinitiation complexes (PICs) into SGs is mediated by specific RBPs, some of which show sequence-specific binding to mRNAs, and others that interact with the translational machinery. These two components, stalled PICs and SG-nucleating RBPs, together determine a threshold at which SGs form or disperse. Some SG-associated RBPs are shared with PBs, whereas other components are limited to SGs or PBs only. In terms of mRNA, SGs contain poly(A) mRNA, whereas PBs contain largely deadenylated mRNA. Figure 1 shows the SG/PB distribution of some of the best-characterized SG-specific proteins (blue), proteins common to both SGs and PBs (green), and proteins specific to PBs (red).

SGs and PBs are dynamic entities that are in equilibrium with polysomes (Kedersha et al. 2000, 2005). SG-associated mRNPs and RBPs dynamically shuttle between SGs and polysomes. Increasing the pool of translationally inactivated mRNA promotes SG assembly, whereas reducing the pool of untranslated mRNA causes SG disassembly. This balance is reflected by the antagonistic effects of the translation inhibitors emetine or cycloheximide and puromycin (Kedersha et al. 2000). Emetine and cycloheximide inhibit translation elongation, allowing slowly initiating/stalled PICs to complete initiation and be trapped in polysomes, reducing the free mRNP pool available for SGs. In the continued presence of stress, emetine treatment rapidly disassembles SGs, whereas PBs remain (see online Movie 1). In the absence of stress,

PBs are disassembled by 10–40 min of cycloheximide treatment (Andrei et al. 2005), suggesting that their steady-state integrity requires ongoing mRNP input. Conversely, puromycin-induced premature termination of translating mRNAs promotes SG formation (Kedersha et al. 2000). Mitotic cells do not form SGs or PBs, at least in part because of arrested elongation that prevents polysome disassembly (Sivan et al. 2007). The effects of these drugs highlight the link between SGs/PBs and translation, and distinguish them from other RNA granules (see online Movie 1).

SG disassembly occurs when cells adapt to stress, or when stress is removed and normal translational equilibrium is restored. Although SG assembly requires translationally stalled PICs, formation of SGs themselves is not required for translational arrest in the cells; specific stresses or knockdown of specific SG-associated proteins and SG nucleators uncouple SG formation from translational arrest (Ohn et al. 2008; Kedersha et al. 2016). Thus, a separate step beyond translational arrest is required for SG formation. For example, energy starvation induced by cold shock stalls the translational cycle, induces eIF2α phosphorylation, and triggers slow SG formation that takes hours rather than minutes, but these cold-shock SGs rapidly dissolve (5–10 min) when cells are warmed up, although dephosphorylation of eIF2α, polysome formation, and restoration of full translation take several hours (Hofmann et al. 2012). In this case, SG dissolution seems to result from a "decondensation" of mRNPs rather than restored translation that depletes the untranslated pool of mRNPs. Inhibiting the energy-sensing 5′AMP-activated protein kinase (AMPK) prevents cold-shock-induced SG formation, suggesting that as-yet-unidentified AMPK targets regulate the SG condensation event. In support of this idea, ubiquitin-specific peptidase 10 (USP10) is a protein that inhibits SG condensation but not translational arrest (Kedersha et al. 2016), and is both an activator and a substrate of AMPK (Deng et al. 2016).

Although SGs form over the course of minutes to hours, most SG protein components are in much more rapid dynamic equilibrium with the cytosol. Fluorescent recovery after photobleaching (FRAP) analysis indicates that the fluorescence recovery of SG proteins ranges from seconds to minutes. Rapidly shuttling SG RBPs include the translation silencers TIA1, Ras GTP-activating protein-binding protein 1 (G3BP1), tristetraprolin (TTP), and cytoplasmic polyadenylation element-binding protein (CPEB), which recover from bleaching within 10–30 sec. Fluorescence recovery of PABP is somewhat slower (30–60 sec), and recovery of Fas-activated serine-threonine kinase (FASTK) and the fragile-X proteins, fragile-X mental retardation protein (FMRP) and FMRP autosomal homolog 1 (FXR1) is very slow (minutes) (Kedersha et al. 2005; Bley et al. 2015). Thus, proteins that shuttle in and out of SGs show fast, medium, and slow recovery kinetics (Kedersha et al. 2000, 2005; Leung et al. 2006; Mollet et al. 2008). PB components also show a range of kinetics for eIF4E (fast), Lsm1 (medium), and eIF4E-T (slow) (Andrei et al. 2005), and support a model in which SG/PB structure arises from phase-transition-driven events (see below) rather than classic protein-based scaffolding or membrane-mediated partitioning. It must be noted that these FRAP studies used overexpressed, tagged proteins for measurements and used different cell lines; hence, some kinetic differences reported for specific situations may not be universal. As more SG proteins are knocked out and replaced by tagged versions using new technologies, better relative measurements should become available.

STRESS GRANULES AND TRANSLATIONAL MECHANISMS

Stress-induced phosphorylation of eIF2α (phospho-eIF2α) is necessary and sufficient for SG assembly (Fig. 2) (Kedersha et al. 1999, 2002). Early compositional analysis revealed that phospho-eIF2α-induced SGs contain most PIC factors but lack eIF5 and eIF2, proteins necessary for conversion of ribosomal preinitiation complexes into translationally competent ribosomes (Kedersha et al. 2002; Kimball et al. 2003). Mechanistically, stress-induced phospho-eIF2α inhibits translation initiation by depleting the eIF2•GTP-Met-tRNA$_i^{Met}$ ternary complex that

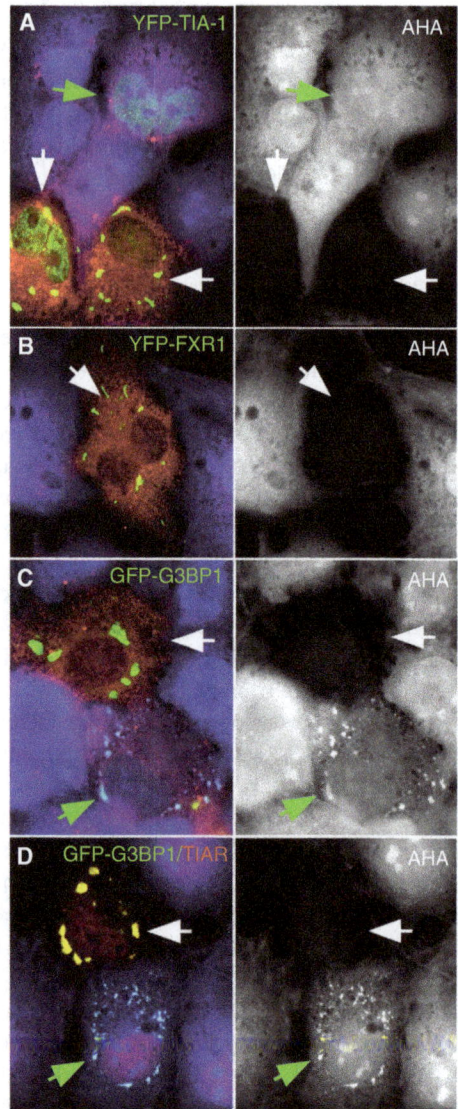

Figure 2. Stress granule (SG) nucleation, eukaryotic initiation factor 2α (eIF2α) phosphorylation, and global translation. Transient expression of (A) SG-nucleating proteins T-cell intracellular antigen 1 (TIA1), or (B) FMRP autosomal homolog 1 (FXR1) in COS7 cells triggers eIF2α phosphorylation (red) and prevents translation (blue in merged figure, shown separately in gray), as shown by labeling with the amino acid analog L-azidohomoalanine (AHA) in some transfectants (white arrows) but not those lacking SGs (green arrow). Overexpression of GTP-activating protein-binding protein 1 (G3BP1) (C,D) can nucleate SGs before/without triggering eIF2α phosphorylation or translational arrest (green arrows). Yellow indicates red/green colocalization.

loads initiator Met-tRNA$_i$^{Met} onto the AUG start codon (reviewed in Jackson et al. 2010; Merrick and Pavitt 2018; Wek 2018). This results in assembly of a translationally stalled 48S initiation complex as shown in Figure 3 (yellow region), and allows elongating ribosomes to "run off" the mRNA, converting polysomes into the closed-loop mRNPs that are the core components of SGs (Kedersha et al. 2002; Kimball et al. 2003). During stress, transcripts bearing elements that evade phospho-eIF2α-mediated translational arrest (e.g., those possessing internal ribosome entry sites or upstream open reading frames) are selectively translated as they produce proteins that protect cells from stress-induced damage. For example, heat shock causes SG formation but HSP70 mRNAs (Kedersha and Anderson 2002) are selectively translated and excluded from SGs. Similarly, HSP90 mRNA is largely excluded from arsenite-induced SGs (Stohr et al. 2006). Thus, SGs are composed of translationally arrested mRNAs released from polysomes and remodeled into mRNPs, but mRNAs that remain actively translated are not remodeled.

Stress and eIF2α phosphorylation are linked by a family of eIF2α kinases that comprise the integrated stress response (ISR), an intracellular surveillance system that monitors different parameters relevant to translation (see Wek 2018). Levels of charged transfer RNAs (tRNAs) are sensed by GCN2 (Wek et al. 1995), redox state and heme availability by HRI (McEwen et al. 2005), endoplasmic reticulum (ER) stress/unfolded protein levels by protein kinase R (PKR)-like ER kinase (PERK) (Harding et al. 2000), and double-stranded RNA (dsRNA) by PKR (Srivastava et al. 1998). Different stresses activate each kinase (reviewed in Donnelly et al. 2013), triggering eIF2α phosphorylation and initiating subsequent SG formation. Deletion or inactivation of any single eIF2α kinase renders cells unresponsive to the corresponding stress, for example, deletion of HRI prevents sodium arsenite-induced eIF2α phosphorylation and subsequent SG formation, but HRI knockout cells still respond to ER stress and PERK activation by both eIF2α phosphorylation and SG formation (Aulas et al. 2017). Knockout cells

from mouse embryos homozygous for non-phosphorylatable mutant eIF2α (eIF2α[S51A], see below) evade stress-induced translational arrest and SG formation (Scheuner et al. 2001; McEwen et al. 2005).

Mechanistically, eIF2α phosphorylation does not directly impact eIF2 function in the preinitiation complex (recognition of an AUG initiation codon on an mRNA). Instead, it inhibits eIF2B, the guanine nucleotide exchange factor that reloads eIF2 with GTP. As eIF2B is limiting in cells, even a small amount of phospho-eIF2α efficiently inhibits eIF2B activity, causing a consequent depletion of the ternary complex that inhibits translation and initiates SG assembly (Jackson et al. 2010; Merrick and Pavitt 2018; Wek 2018). Consequently, pharmacological manipulations targeting eIF2α phosphorylation and eIF2B also modulate SG assembly/disassembly. Integrated stress response inhibitor (ISRIB), a chemical inhibitor of PERK signaling, reverses the inhibitory effects of phospho-eIF2α (Sidrauski et al. 2013, 2015a) by activating eIF2B and antagonizing phospho-eIF2α effects. Treatment with ISRIB reverses the ISR by rapidly restoring translation and causing SG disassembly, despite continuously elevated phospho-eIF2α (Sidrauski et al. 2015a,b). As in the case of emetine-enforced SG disassembly (see online Movie 1), ISRIB treatment disassembles SGs but not PBs (Sidrauski et al. 2015a).

Although stress-induced translational repression triggers synchronous SG assembly in cultured cells, it is less recognized that isolated cells in otherwise healthy cultures can transiently form SGs. Online Movie 2 shows individual cells displaying transient SG formation and resolution while neighboring cells divide and grow normally, suggesting that SG formation occurs during metabolic changes associated with normal growth but is often overlooked. In support of this idea, SG-positive cells have been found in a subset of hypoxia-sensitive neurons in intact brain (DeGracia et al. 2007), in cochlear hair cells exposed to ototoxins (Mangiardi et al. 2004), in individual cells in hypoxic tumors (Moeller et al. 2004), and in virally infected cell cultures (reviewed by Lloyd 2013). In hepatitis C virus (HCV)-infected cultures, individual

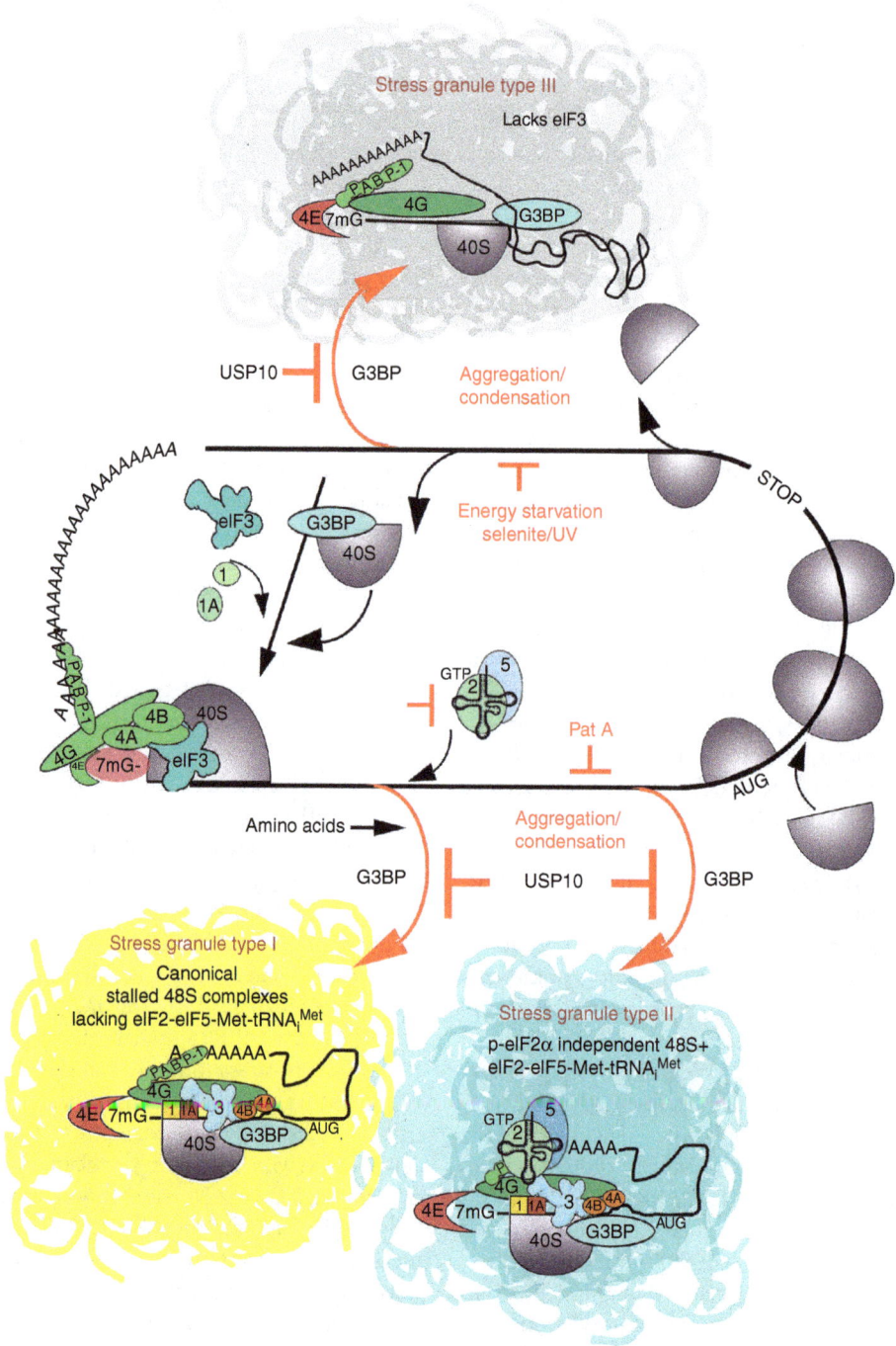

Figure 3. Regulatory stalling points in the translational cycle leading to stress granule (SG) formation. A closed-loop messenger RNA (mRNA) can stall at different points in the translation cycle, resulting in messenger ribonucleoproteins (mRNPs) of different composition eligible for condensation (mediated by GTP-activating protein-binding protein 1 [G3BP]) into SGs. Type I, or canonical SGs, result when eukaryotic initiation factor 2α (eIF2α) phosphorylation inhibits recharging of eIF2•GTP-Met-tRNA$_i^{Met}$, resulting in mRNPs that lack eIF2/5. Type II SGs form when eIF4A activities are inhibited and can be induced in cells lacking phospho-eIF2α, generating SGs that contain eIF2/5. Type III SGs result from xenobiotic stress, and lack eIF3. The mechanism shown here is hypothetical. Not shown are hyperosmotic/G3BP-independent SGs thought to arise from molecular crowding.

Cite this article as *Cold Spring Harb Perspect Biol* doi: 10.1101/cshperspect.a032813

cells show oscillating cycles of SG formation/ disassembly that parallel eIF2α phosphorylation and dephosphorylation, and global protein translation. These cycles result from the opposing effects on translation of the eIF2α kinase PKR and the eIF2α phosphatase protein phosphatase 1 (PP1), respectively (see online Movie 2) (Ruggieri et al. 2012).

In situ metabolic labeling using the methionine analog L-azidohomoserine (AHA) reveals the tight correlation between SG formation, eIF2α phosphorylation, and translational arrest (Fig. 2). However, other noncanonical modes of SG formation are triggered by phospho-eIF2α–independent mechanisms. A model summarizing the major stalling points in the translational cycle linked to SG formation depicts a closed-loop mRNA that can be paused at different checkpoints, initiating polysome disassembly at these points and resulting in SGs of different composition (Fig. 3). Whereas phospho-eIF2α depletes ternary complex to promote the assembly of canonical "type I" SGs lacking eIF2, other agents (pateamine A [Low et al. 2005], hippuristanol [Mazroui et al. 2006], rocaglates [Sadlish et al. 2013]) or lipid mediators (15-deoxy-$\Delta^{12,14}$-prostaglandin J$_2$, or 15-d-PGJ2 [Kim et al. 2007]) inactivate the eIF4A helicase that helps unwind the 5′ untranslated region (UTR) of mRNAs to promote translation initiation. Failure of translation initiation via this mechanism triggers the formation of phospho-eIF2α–independent "type II" SGs that contain eIF2 and eIF5 (Fig. 3, blue), and can be triggered in the absence of phospho-eIF2α. Whereas type I SGs are generally cytoprotective, type II SGs may be cytotoxic (Fujimura et al. 2012; Anderson et al. 2015), although the mechanisms by which these compositionally distinct granules mediate opposing effects on cell survival are not known. Another subclass of SGs lacks eIF3, and is induced by chemotherapeutic drugs such as selenite, UV, or other xenobiotic agents (Fujimura et al. 2012; Anderson et al. 2015). UV-induced SGs do not require phospho-eIF2α but are disassembled by cycloheximide (Moutaoufik et al. 2014). We suggest that these eIF3-negative "type III" SGs may result from stalling before the recycling of terminating 40S subunits into new

43S preinitiation complexes at the 5′ end of the mRNA, a process that requires eIF3 (Pisarev et al. 2007, 2010) and is the commitment step to another round of translation (Fig. 3, top gray-shaded area). Finally, hypertonic conditions trigger assembly of another noncanonical, phospho-eIF2α-independent SG (Bevilacqua et al. 2010; Bounedjah et al. 2012); the mechanism for their formation is related to molecular crowding, but the details are unknown.

Many stresses that trigger the ISR, such as UV irradiation, heat shock or oxidative stress, also activate endogenous RNases such as angiogenin (Lyons et al. 2017), which cleaves tRNAs in their anticodon loops to produce tRNA halves (Fu et al. 2009; Yamasaki et al. 2009) known as tiRNAs (tRNA-derived, stress-induced small RNAs) (Yamasaki et al. 2009). A subset of tiRNAs (5′ tiRNAs derived from tRNA^{Ala} and tRNA^{Cys}) inhibit translation (Emara et al. 2010) by displacing the eIF4F complex from the m^{7}GTP cap, a step that is required for cap-dependent translation (Ivanov et al. 2011a, 2014). Transfected 5′ tiRNA^{Ala} and 5′ tiRNA^{Cys} induce SG assembly that does not require phospho-eIF2α (Emara et al. 2010; Ivanov et al. 2011b), but does require the SG protein YB-1 (Lyons et al. 2016), a cold-shock, domain-containing protein also found in PBs (Yang and Bloch 2007). The mechanisms whereby tiRNAs collaborate with YB1 to induce SGs are unknown, but are under active investigation.

SG/PB CONDENSATION

In addition to stalled PIC mRNPs, SG formation also requires SG-nucleating proteins (Panas et al. 2016). Overexpression of any of these SG nucleators drives "spontaneous" SG assembly (Kedersha and Anderson 2007; Reineke et al. 2015), and knockdown of SG nucleators can impair SG formation. SG nucleating proteins include CAPRIN1, G3BP1, and G3BP2 (collectively referred to as G3BP), TIA1 and TIAR, FMRP, FXR1, and FXR2, TTP, BRF1, FASTK, and CPEB (reviewed in Anderson and Kedersha 2008). Many of these proteins act as translational silencers and display high-affinity, sequence-specific RNA-binding activity, and most require concurrent activation of PKR to trigger eIF2α phosphorylation to induce SGs (Kedersha and Anderson 2007). Figure 2 shows that overexpressed TIA1 (A) or FXR1 (B) nucleates SGs in cells that display both elevated phospho-eIF2α and globally repressed translation (white arrows). In the case of G3BP, overexpressed protein localizes to granule-like regions before activating the eIF2α kinase PKR (Reineke et al. 2012) in a feedforward manner (Fig. 2C,D). Cells lacking G3BP (G3BP-null cells) are unable to assemble mRNPs into SGs, despite overexpression of other SG nucleating proteins (Kedersha et al. 2016); hence, G3BP appears to play a direct role in SG condensation (see below) in addition to activating PKR.

Whereas stalling of translation initiates SG formation by promoting polysome disassembly into PIC mRNPs, the physical condensation of these mRNPs into granules is a separate event that requires G3BP. Two proteins, Caprin1 and USP10, competitively bind G3BP to promote (Solomon et al. 2007) or inhibit SG condensation, respectively (Kedersha et al. 2016). G3BP-null cells are unable to assemble either phospho-eIF2α-dependent type I or eIF4A-dependent type II SGs (Kedersha et al. 2016). SG formation is rescued in these cells by expression of either wild-type G3BP, or G3BP containing a point mutation that renders it unable to bind USP10 or Caprin1. Thus, G3BP is uniquely required for the formation of both type I and II SGs, and its activity is regulated by Caprin1 and USP10, but does not require these proteins to cause SG condensation. In addition, several viruses (Semliki Forest, Chikungunya, and herpes simplex) encode SG-inhibiting proteins that share the same binding motif (FGDF motif) within USP10 that binds to G3BP (Panas et al. 2015). The FGDF motif is sufficient to bind (and in some cases relocalize) G3BP, blocking its ability to mediate SG condensation. The importance of G3BP to SG condensation is further illustrated by the number of other viruses (including poliovirus, encephalomyocarditis virus [EMCV], and Cox-sackievirus) that encode specific proteases that cleave G3BP to inhibit SG formation and disable the host antiviral response (White et al. 2007; Fung et al. 2013; Ng et al. 2013).

PROCESSING BODIES AND TRANSLATIONAL CONTROL

PBs share some components (Fig. 1) and properties with SGs, but differ from SGs in composition, behavior, and proposed functions (reviewed in Decker and Parker 2012; Stoecklin and Kedersha 2013). Shared components include eIF4E, mRNA, and selected RBPs, but PBs lack SG-associated eIF3, PABP, small ribosomal subunits, and many signaling proteins (Fig. 1). Generally, SGs house proteins involved in mRNA translation (discussed above), whereas PBs house proteins associated with mRNA decay (DCP1, DCP2, XRN1, EDC3, hedls/EDC4) or mRNA silencing (GW182, argonautes). This led to the proposal that PBs are sites of active mRNA decay (reviewed in Decker and Parker 2012). However, subsequent studies showed that repressed mRNAs can move from PBs into the cytosol where they reenter translation (Brengues et al. 2005), and disruption of PBs does not prevent global or specialized pathways of mRNA decay (Eulalio et al. 2007). Recent studies confirm that mRNAs segregated into PBs are translationally repressed but not degraded (Hubstenberger et al. 2017) and represent one-fifth of the mRNA transcriptome. The PB-associated subset of mRNAs is enriched for transcripts encoding regulatory proteins, especially regulatory subunits of multiprotein complexes. The sequestration of these mRNAs in PBs allows for their rapid mobilization into the translationally active pool, bypassing transcription, mRNA processing, and nuclear export, and thus enabling local control of PB-localized mRNAs (Hubstenberger et al. 2017). As PBs (and related GWBs) are also distinguished by the inclusion of RNA-induced silencing complex components (argonautes, GW182/TNRC6), at least part of their regulatory control is likely to be miRNA dependent (Eulalio et al. 2008). Importantly, various RBPs cosegregate into PBs with their mRNA targets, providing an alternative mechanism for the transport of specific mRNAs into and out of the PB storage depots to modulate protein production in response to cellular needs (Hubstenberger et al. 2017). A schematic illustrating the relationship

Cite this article as *Cold Spring Harb Perspect Biol* doi: 10.1101/cshperspect.a032813

between polysomes, SGs, and PBs is shown in Figure 4.

SG/PB FORMATION AND THE "LIQUID/ LIQUID PHASE TRANSITION" CONCEPT

As shown in online Movie 2, SGs fuse and show liquid-like behavior, similar to that of other RNA-containing structures (reviewed in Shin and Brangwynne 2017). Although SGs can persist for hours, the residence time of SG proteins (Kedersha et al. 2000, 2005), mRNAs (Mollet et al. 2008), and PB proteins (Andrei et al. 2005) is on the order of seconds to minutes. This dynamic behavior in seemingly stable structures distinguishes SGs and PBs from more static RNA granules and from membrane-bound organelles. SG/PB proteins are highly enriched in intrinsically disordered/low complexity (ID/LC) regions (reviewed in Uversky 2017), which do not assume classic structured domains. These regions are highly flexible, able to assume multiple conformations influenced by posttranslational modifications or templating interactions with other molecules (reviewed in Kedersha et al. 2013; Shin and Brangwynne 2017; Uversky 2017). The high proportion of ID/LC regions in SG-nucleating proteins supports the hypothesis that SG/PB formation is mediated by a liquid/liquid phase separation event (Han et al. 2012; Kato et al. 2012; Weber and Brangwynne 2012) that drives SG and PB components to condense out of the surrounding cytosol. This model suggests that the rapid shutting of ID/LC proteins into and out of SGs may parallel rapid conformational changes in their ID/LC regions.

Most SG-nucleating proteins are ID/LC-rich (reviewed in Kedersha et al. 2013; Panas et al. 2016), consistent with a phase transition model. One such protein is the mRNA decay-promoting protein TTP, which nucleates both SGs and PBs. Phosphorylation of TTP by MAPKAP2 (mitogen-activated protein kinase-activated protein kinase 2) promotes TTP binding to 14–3–3 proteins, and this interaction regulates TTP trafficking by eliminating the ability of TTP to condense into SGs (Stoecklin et al. 2004). As seen in online Movie 3, arsenite-triggered

phosphorylation of TTP results in its rapid export from SGs, but not PBs (Lykke-Andersen and Wagner 2005; Franks and Lykke-Andersen 2007). As 14–3–3 binding to TTP does not inhibit its binding to target mRNAs, regulated binding to 14–3–3 proteins may allow TTP to escort its target mRNAs out of SGs and into PBs (Stoecklin et al. 2004). An attractive model to explain this behavior is that the ID/LC regions of TTP are templated into a fixed conformation on binding to the highly ordered 14–3–3 proteins, "freezing" TTP into a state incompatible with the dynamics required for shuttling into SGs (see online Movie 3).

Most SG-nucleating proteins (TIA-1/TIAR, Caprin1, CPEB1) bind mRNAs in a sequence-specific manner, and likely concentrate and selectively condense their target mRNAs when their translation is stalled. In addition to binding a subset of mRNAs, other SG nucleators such as G3BP1/2 and FMR1/FXR1/2 interact with ribosomal subunits. G3BP binds 40S ribosomal subunits (Simsek et al. 2017) through its RGG region (Kedersha et al. 2016; Simsek et al. 2017). Truncated G3BP lacking the RGG region loses both its ribosome-binding ability and its ability to restore SG assembly to the G3BP-null cells. The G3BP-binding protein Caprin1 is unable to nucleate SG formation in the absence of G3BP, thus it seems likely that G3BP:Caprin1 complexes cooperate to recruit Caprin1-specific transcripts into SGs (Solomon et al. 2007). In contrast to G3BP, the FMR/FXR1/FXR2 proteins bind large ribosomal subunits (Chen et al. 2014; Simsek et al. 2017), and paradoxically their overexpression strongly nucleates SGs despite the fact that SGs lack 60S subunits. Higher-order complexes containing the FMRP/FXR1 proteins and G3BP/Caprin1 proteins have been reported (El Fatimy et al. 2012; Baumgartner et al. 2013) that hint at cooperative effects between these two SG nucleating families, but details are not yet clear. A recent paper notes that these two families show mRNA binding that is regulated by adenosine methylation at position N6 (m^6A). This study finds that m^6A-modified RNA preferentially binds to FMRP/FXR1 and is translationally repressed, whereas G3BP preferentially binds and stabilizes mRNA

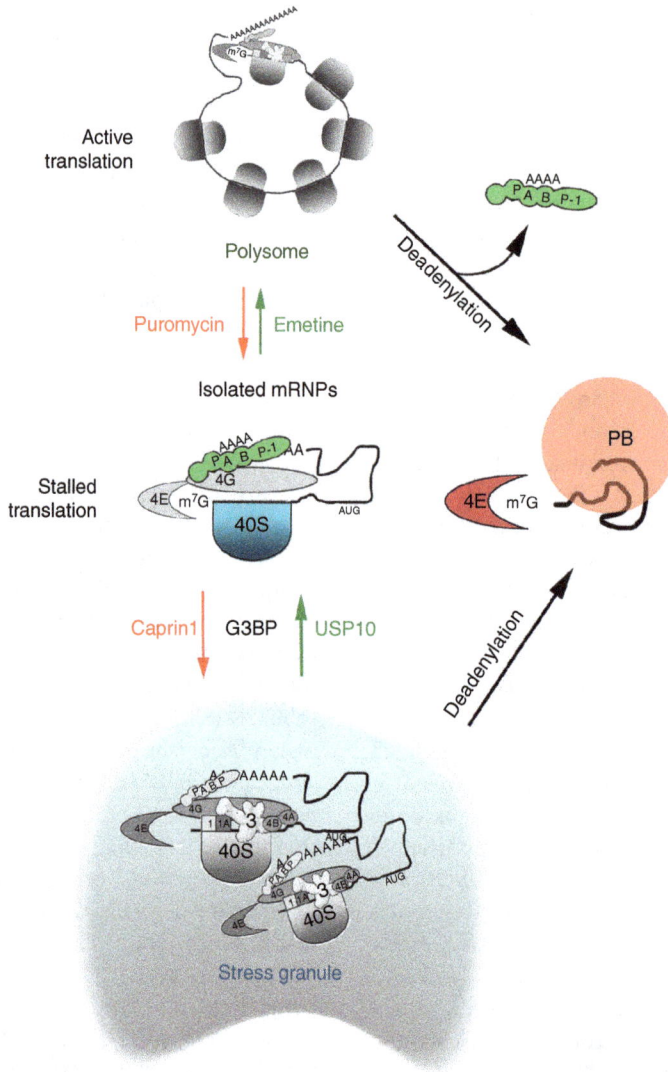

Figure 4. Relationship between polysomes, stress granules (SGs), and processing bodies (PBs). Polysomes are maintained when translation initiation and termination occur at balanced rates on a single messenger RNA (mRNA). When termination occurs more frequently than initiation, polysome disassembly results, resulting in stalled, circularized messenger ribonucleoproteins (mRNPs). Agents that prevent elongation (such as emetine and cycloheximide) inhibit polysome/mRNP conversion, and hence reduce the pool of mRNPs, whereas puromycin promotes premature termination and accelerates the polysome-mRNP conversion. A pool of free mRNPs is necessary, but not sufficient, for SG assembly, which requires the dynamic condensation of mRNPs out of the surrounding cytoplasm. Condensation (and possibly decondensation) requires the mRNA-binding protein G3BP, and is regulated by G3BP-binding proteins Caprin1 (which promotes) and USP10 (which inhibits) the G3BP-mediated condensation of mRNPs into SGs. PBs contain deadenylated mRNPs and/or mRNAs undergoing deadenylation, and PB condensation requires multiple proteins, including EDC4, LSM14, 4-ET, and DDX6 (reviewed in Luo et al. 2018).

that is not m⁶A modified (Edupuganti et al. 2017).

PROPOSED ROLES OF SGs IN CELL SIGNALING

Cellular stress requires rapid reprogramming of RNA metabolism as part of the ISR. SG assembly occurs downstream of stress-induced translational arrest, and the resulting pool of translationally arrested mRNAs constitutes a scaffold for RNA-binding proteins that promote SG condensation. Subsequently, specific mRNA-RBP complexes recruit a variety of signaling molecules that can promote cell survival and adaptation/recovery or induce apoptosis, suggesting that granule condensation may assemble signaling centers that allow cross talk between the ISR and other signaling pathways. The interplay between signaling pathways and SGs occurs at many levels (reviewed in Kedersha et al. 2013). First, signaling events directed by eIF2α kinases regulate cellular translation arrest that initiates SG assembly. Second, stress-modulated posttranslational modifications of SG proteins such as phosphorylation, O-GlcNAc addition, poly-ADP-ribosylation, and arginine methylation directly impact SG assembly and disassembly (reviewed in Ohn and Anderson 2010). Third, SGs modulate signaling cascades during stress by forming transient signaling hubs that coordinate signaling events in both the cytoplasm and the nucleus. Such modulation is achieved by selective sequestration of specific signal transduction-related molecules (such as RACK1, TRAF2, RSK2, etc.) into, or their exclusion from, SGs. As signaling centers, SG formation communicates a "state of emergency," and their transient existence rewires the network of signaling events to control cell fate (Kedersha et al. 2013).

SGs, PBs, AND DISEASE

Given the varied effects that RNA granules have on metabolism and cell signaling, it is not surprising that they are also implicated in many diseases. Defects in SG dynamics are found in cancer, neurodegenerative disease, viral infection, and autoimmune disease. Several recent reviews discuss in detail the roles that RNA granules play in disease pathogenesis (Wolozin 2014; Anderson et al. 2015; Shukla and Parker 2016; Taylor et al. 2016; McCormick and Khaperskyy 2017).

Viruses depend on host translational machinery for their propagation, and several aspects of interactions between cellular translation and viruses directly or indirectly impact dynamics of RNA granules (Stern-Ginossar et al. 2018). In response to different viruses, SG and PB formation can be triggered or repressed (Lloyd 2013). Some viruses physically interact with SGs/PBs or hijack specific SG/PB components to favor viral replication. In some cases, SGs sequester key translation components, thus reducing the availability of ribosomal subunits and associated protein synthesis factors needed for translation of viral transcripts. Some viruses (e.g., EMCV, poliovirus) encode proteases that cleave the key SG nucleator G3BP to enforce SG disassembly and enhance viral replication (White et al. 2007; Fung et al. 2013; Ng et al. 2013). Other viruses (such as influenza A virus) block SG formation via expression of viral proteins (NS1) that directly bind and antagonize the eIF2α kinase PKR, a critical sensor of viral dsRNA (Khaperskyy et al. 2012). Flaviviruses (such as West Nile virus, Dengue virus) hijack multiple SG components (e.g., TIA1, TIAR, G3BP) via their sequestration by viral RNA 3′ stem-loop structures, thus suppressing SG formation (Li et al. 2002; Emara et al. 2008). Semliki Forest virus co-opts the SG nucleator G3BP (Panas et al. 2012), whereas HCV recruits both G3BP and multiple PB components and diverts them into lipid droplet-associated foci containing HCV replication/assembly complexes (Ariumi et al. 2011; Pager et al. 2013).

Interactions between viruses, the translational machinery, and RNA granules directly impact immunity. Virus-induced SGs regulate production of interferons that counteract virus propagation, thus identifying SGs as part of the innate immune system (reviewed in McCormick and Khaperskyy 2017). By preventing SG assembly or dissolving existing SGs, viruses evade the host defense. Several innate immune system-related proteins reside in SGs, including both interferon-inducers (such as PKR, RIG-I, MDA5) and interferon-effectors (RNase L and OASes) (Onomoto et al. 2012). The consequences and importance of colocalization of these factors with SGs for innate immunity are complex and not well understood. SGs may constitute platforms linking immune responses with other cell signaling pathways, or regulate selective production of specific immune factors such as cytokines. However, it is also possible that some SG-associated proteins are directly involved in innate immune responses irrespective of their localization.

SGs are implicated in the pathogenesis of neurodegenerative diseases such as amyotrophic lateral sclerosis, frontotemporal dementia and Alzheimer's disease (reviewed in Wolozin 2012, 2014). The connection between SGs and these diseases comes from histological observations linking SG proteins to intracellular and extracellular protein aggregates that are pathological hallmarks of neurodegeneration (Wolozin 2012, 2014). The finding that mutations in several SG proteins are found in patients with neurodegenerative disease has led to speculation that disease-causing forms of SG proteins (e.g., TDP43, FUS/TLS, hnRNPA1, SMN) modulate SG dynamics in ways that impact neuronal survival (Shukla and Parker 2016; Taylor et al. 2016; Maziuk et al. 2017). Whereas evidence from multiple in vitro and cell culture models using disease-associated mutants of SG-associated proteins supports this hypothesis and shows that SG-related mechanisms can impair neuronal functions leading to neuron loss (Shukla and Parker 2016; Taylor et al. 2016; Maziuk et al. 2017), these studies require further evaluation in humans.

SGs are also associated with cancer (Anderson et al. 2015) and are found in tumors of different histological origins including colorectal cancer, carcinomas, and glioblastomas (Adjibade et al. 2015; El-Naggar et al. 2015; Vilas-Boas et al. 2016). Responding and adapting to stress is important for cancer progression as cancer cells encounter hostile tumor microenvironments characterized by hypoxia, nutrient starvation, high levels of reactive oxygen species

and hyperosmolarity, all of which trigger SG formation. Importantly, SG formation is often found as a part of the response to both radio- and chemotherapies (Moeller et al. 2004; Szaflarski et al. 2016). As SGs are generally prosurvival, their induction by chemotherapy can inadvertently compromise cancer treatment. Therefore, preventing SG assembly may prove to be a valuable therapeutic option in promoting tumor cell death. SGs are also linked to metastasis, where they contribute to cancer cell survival during tumor invasion and dissemination. Mouse xenograft sarcoma models reveal that increased levels of the SG nucleator G3BP1 promote SG formation in sarcoma cells to enhance their survival, resistance to chemotherapy, and metastasis to lungs (Somasekharan et al. 2015). By reducing G3BP1 expression and SG formation, metastasis and tumor invasion are significantly diminished (Somasekharan et al. 2015). Similarly, cancer cells can up-regulate SG assembly and enhance resistance to chemotherapy by producing and secreting the lipid mediator 15-deoxy-Δ12,14-prostaglandin J2 (Grabocka and Bar-Sagi 2016); this stimulates SG formation in an autocrine and paracrine manner to increase adaptation to tumor cell environments (Grabocka and Bar-Sagi 2016).

CONCLUDING REMARKS

Adjusting protein synthesis to adapt to changing conditions is crucial for growth and survival, and requires highly coordinated, rapid, specific, and localized reprogramming of cellular translation. Such regulation requires modulating global translation control as well as selective translation of mRNA subpopulations. Although global translation regulation pathways such as the integrated stress response and the mechanistic target of rapamycin (mTOR)/AMPK nutrient and energy regulatory cascades are understood in some detail, the mechanisms underlying selective translation remain largely unexplored, and RNA granules appear important in these processes. We anticipate that regulatory sequences, modified nucleotides (methylations, editing) (Peer et al. 2018), microRNAs (Duchaine and Fabian 2018), and structures within mRNAs mediate their selective recruitment to or exclusion from the translation machinery and/or RNA granules. Such regulatory mRNA elements may be recognized by specific RBPs that ferry them into RNA granules, and/or dictate their availability for translation. Posttranslational modifications of RBPs and their interactions with other proteins influence their localization, aggregation, and functions; we are only beginning to catalog these so we can ask larger questions.

Very recent studies using new technologies of proximity labeling are finally cataloging the range of proteins and mRNAs that shuttle through SGs and PBs. Using the BirA/BioID (proximity-dependent biotin identification) enzymatic system that biotin-labels interacting proteins over time (several hours), multiple BirA-tagged SG/PB baits were used to identify complex networks of cumulative protein interactors, but the long labeling period was unable to distinguish between stress-specific and normal interactions (Youn et al. 2018). Another study (Markmiller et al. 2018) used a faster labeling system based on the engineered peroxidase APEX2, revealing a 1-min snapshot labeling of proteins in close proximity to APEX2-tagged G3BP. This study detected stress-specific interactions between G3BP and the translational machinery (eIF3, eIF4G), but also revealed a complex network of interactions specific to certain types of stress, interactions that are stress-independent, and many which are cell-type-specific. We eagerly anticipate future studies that marry proximity labeling with techniques such as photoactivatable ribonucleoside-enhanced cross-linking and immunoprecipitation (PAR-CLIP) to obtain a census of SG/PB-associated RNAs under different conditions. Such studies are needed to address global questions such as: How are distinct subsets of mRNAs coordinately regulated in translation? Which signaling pathways regulate cross talk between global and selective translation control mechanisms? How are these integrated with biological outcomes of cell proliferation, motility, and death?

Both SGs and PBs are dynamic, and their contents are in flux. RNA helicases, which are required to facilitate the protein:RNA and RNA:RNA transitions, may play a direct role in re-modeling stalled polysomes into mRNPs eligible for condensation into RNA granules. Specific helicases eIF4A, DDX3, and DDX6 contribute to SG and PB formation and resolution (Bordeleau et al. 2006; Dang et al. 2006; Ohn et al. 2008; Shih et al. 2008; Cencic et al. 2009; Ayache et al. 2015; Iwasaki et al. 2016). These play important functions in translational control as well as signaling (reviewed in Heerma van Voss et al. 2017). Future studies will elucidate their specific roles in SG/PB dynamics.

Physiological and environmental changes often trigger pathophysiological events underlying human disorders, ranging from infectious diseases to cancer and neurodegeneration. Translation control maintains homeostasis in response to environment change to ensure survival. RNA granule formation is both a biomarker of stress-induced translational arrest, and a mechanism for regulating selective translation. SGs influence mRNA sorting and sequestration, potentiate eIF2α phosphorylation, and constitute signaling centers that integrate RNA metabolism with other aspects of cellular physiology. PBs contain a subset of mRNAs, but how these are removed from PBs and translationally activated is unknown. We have only begun to elucidate the mechanisms whereby RNA granule-associated proteins and their client mRNAs fine-tune translation in response to environmental and physiological cues.

ACKNOWLEDGMENTS

We thank members of the Ivanov and Anderson laboratories for helpful discussions. This work is supported by the National Institutes of Health (NS094918 to P.I., GM121410, and GM111700 to P.A.). We apologize to our colleagues whose work could not be mentioned owing to space constraints.

REFERENCES

*Reference is also in this collection.

Adjibade P, St-Sauveur VG, Huberdeau MQ, Fournier MJ, Savard A, Coudert L, Khandjian EW, Mazroui R. 2015. Sorafenib, a multikinase inhibitor, induces formation of stress granules in hepatocarcinoma cells. *Oncotarget* **6:** 43927–43943.

Anderson P, Kedersha N. 2008. Stress granules: The Tao of RNA triage. *Trends Biochem Sci* **33:** 141–150.

Anderson P, Kedersha N. 2009. Stress granules. *Curr Biol* **19:** R397–R398.

Anderson P, Kedersha N, Ivanov P. 2015. Stress granules, P-bodies and cancer. *Biochim Biophys Acta* **1849:** 861–870.

Andrei MA, Ingelfinger D, Heintzmann R, Achsel T, Rivera-Pomar R, Luhrmann R. 2005. A role for eIF4E and eIF4E-transporter in targeting mRNPs to mammalian processing bodies. *RNA* **11:** 717–727.

Ariumi Y, Kuroki M, Kushima Y, Osugi K, Hijikata M, Maki M, Ikeda M, Kato N. 2011. Hepatitis C virus hijacks P-body and stress granule components around lipid droplets. *J Virol* **85:** 6882–6892.

Aulas A, Fay MM, Lyons SM, Achorn CA, Kedersha N, Anderson P, Ivanov P. 2017. Stress-specific differences in assembly and composition of stress granules and related foci. *J Cell Sci* **130:** 927–937.

Ayache J, Benard M, Ernoult-Lange M, Minshall N, Standart N, Kress M, Weil D. 2015. P-body assembly requires DDX6 repression complexes rather than decay or Ataxin2/2L complexes. *Mol Biol Cell* **26:** 2579–2595.

Bashkirov VI, Scherthan H, Solinger JA, Buerstedde JM, Heyer WD. 1997. A mouse cytoplasmic exoribonuclease (mXRN1p) with preference for G4 tetraplex substrates. *J Cell Biol* **136:** 761–773.

Baumgartner R, Stocker H, Hafen E. 2013. The RNA-binding proteins FMR1, rasputin and caprin act together with the UBA protein lingerer to restrict tissue growth in *Drosophila melanogaster*. *PLoS Genet* **9:** e1003598.

Bevilacqua E, Wang X, Majumder M, Gaccioli F, Yuan CL, Wang C, Zhu X, Jordan LE, Scheuner D, Kaufman RJ, et al. 2010. eIF2α phosphorylation tips the balance to apoptosis during osmotic stress. *J Biol Chem* **285:** 17098–17111.

Bley N, Lederer M, Pfalz B, Reinke C, Fuchs T, Glass M, Möller B, Huttelmaier S. 2015. Stress granules are dispensable for mRNA stabilization during cellular stress. *Nucleic Acids Res* **43:** e26.

Bordeleau ME, Cencic R, Lindqvist L, Oberer M, Northcote P, Wagner G, Pelletier J. 2006. RNA-mediated sequestration of the RNA helicase eIF4A by Pateamine A inhibits translation initiation. *Chem Biol* **13:** 1287–1295.

Bounedjah O, Hamon L, Savarin P, Desforges B, Curmi PA, Pastre D. 2012. Macromolecular crowding regulates assembly of mRNA stress granules after osmotic stress: New role for compatible osmolytes. *J Biol Chem* **287:** 2446–2458.

Brengues M, Teixeira D, Parker R. 2005. Movement of eukaryotic mRNAs between polysomes and cytoplasmic processing bodies. *Science* **310:** 486–489.

Cencic R, Carrier M, Galicia-Vazquez G, Bordeleau ME, Sukarieh R, Bourdeau A, Brem B, Teodoro JG, Greger H, Tremblay ML, et al. 2009. Antitumor activity and mechanism of action of the cyclopenta[*b*]benzofuran, silvestrol. *PLoS ONE* **4:** e5223.

Chen E, Sharma MR, Shi X, Agrawal RK, Joseph S. 2014. Fragile X mental retardation protein regulates translation by binding directly to the ribosome. *Mol Cell* **54:** 407–417.

Collier NC, Schlesinger MJ. 1986. The dynamic state of heat shock proteins in chicken embryo fibroblasts. *J Cell Biol* **103:** 1495–1507.

Collier NC, Heuser J, Levy MA, Schlesinger MJ. 1988. Ultrastructural and biochemical analysis of the stress granule in chicken embryo fibroblasts. *J Cell Biol* **106:** 1131–1139.

Dang Y, Kedersha N, Low WK, Romo D, Gorospe M, Kaufman R, Anderson P, Liu JO. 2006. Eukaryotic initiation factor 2α-independent pathway of stress granule induction by the natural product pateamine A. *J Biol Chem* **281:** 32870–32878.

Decker CJ, Parker R. 2012. P-bodies and stress granules: Possible roles in the control of translation and mRNA degradation. *Cold Spring Harb Perspect Biol* **4:** a012286.

DeGracia DJ, Rudolph J, Roberts GG, Rafols JA, Wang J. 2007. Convergence of stress granules and protein aggregates in hippocampal Cornu Ammonis 1 at later reperfusion following global brain ischemia. *Neuroscience* **146:** 562–572.

Deng M, Yang X, Qin B, Liu T, Zhang H, Guo W, Lee SB, Kim JJ, Yuan J, Pei H, et al. 2016. Deubiquitination and activation of AMPK by USP10. *Mol Cell* **61:** 614–624.

Donnelly N, Gorman AM, Gupta S, Samali A. 2013. The eIF2α kinases: Their structures and functions. *Cell Mol Life Sci* **70:** 3493–3511.

* Duchaine TF, Fabian MR. 2018. Mechanistic insights into microRNA-mediated gene silencing. *Cold Spring Harb Perspect Biol* doi: 10.1101/cshperspect.a032771.

Edupuganti RR, Geiger S, Lindeboom RGH, Shi H, Hsu PJ, Lu Z, Wang SY, Baltissen MPA, Jansen P, Rossa M, et al. 2017. N^6-methyladenosine (m^{6}A) recruits and repels proteins to regulate mRNA homeostasis. *Nat Struct Mol Biol* **24:** 870–878.

El Fatimy R, Tremblay S, Dury AY, Solomon S, De Koninck P, Schrader JW, Khandjian EW. 2012. Fragile X mental retardation protein interacts with the RNA-binding protein Caprin1 in neuronal RiboNucleoProtein complexes [corrected]. *PLoS ONE* **7:** e39338.

El-Naggar AM, Veinotte CJ, Cheng H, Grunewald TG, Negri GL, Somasekharan SP, Corkery DP, Tirode F, Mathers J, Khan D, et al. 2015. Translational activation of HIF1α by YB-1 promotes sarcoma metastasis. *Cancer Cell* **27:** 682–697.

Emara MM, Liu H, Davis WG, Brinton MA. 2008. Mutation of mapped TIA-1/TIAR binding sites in the 3′ terminal stem-loop of West Nile virus minus-strand RNA in an infectious clone negatively affects genomic RNA amplification. *J Virol* **82:** 10657–10670.

Emara MM, Ivanov P, Hickman T, Dawra N, Tisdale S, Kedersha N, Hu GF, Anderson P. 2010. Angiogenin-induced tRNA-derived stress-induced RNAs promote stress-induced stress granule assembly. *J Biol Chem* **285:** 10959–10968.

Eulalio A, Behm-Ansmant I, Schweizer D, Izaurralde E. 2007. P-body formation is a consequence, not the cause, of RNA-mediated gene silencing. *Mol Cell Biol* **27:** 3970–3981.

Cite this article as *Cold Spring Harb Perspect Biol* doi: 10.1101/cshperspect.a032813

Eulalio A, Huntzinger E, Izaurralde E. 2008. GW182 inter-action with Argonaute is essential for miRNA-mediated translational repression and mRNA decay. *Nat Struct Mol Biol* **15:** 346–353.

Eystathioy T, Jakymiw A, Chan EK, Seraphin B, Cougot N, Fritzler MJ. 2003. The GW182 protein colocalizes with mRNA degradation associated proteins hDcp1 and hLSm4 in cytoplasmic GW bodies. *RNA* **9:** 1171–1173.

Fenger-Gron M, Fillman C, Norrild B, Lykke-Andersen J. 2005. Multiple processing body factors and the ARE bind-ing protein TTP activate mRNA decapping. *Mol Cell* **20:** 905–915.

Franks TM, Lykke-Andersen J. 2007. TTP and BRF proteins nucleate processing body formation to silence mRNAs with AU-rich elements. *Genes Dev* **21:** 719–735.

Fu H, Feng J, Liu Q, Sun F, Tie Y, Zhu J, Xing R, Sun Z, Zheng X. 2009. Stress induces tRNA cleavage by angiogenin in mammalian cells. *FEBS Lett* **583:** 437–442.

Fujimura K, Sasaki AT, Anderson P. 2012. Selenite targets eIF4E-binding protein-1 to inhibit translation initiation and induce the assembly of non-canonical stress granules. *Nucleic Acids Res* **40:** 8099–8110.

Fung G, Ng CS, Zhang J, Shi J, Wong J, Piesik P, Han L, Chu F, Jagdeo J, Jan E, et al. 2013. Production of a dominant-negative fragment due to G3BP1 cleavage contributes to the disruption of mitochondria-associated protective stress granules during CVB3 infection. *PLoS ONE* **8:** e79546.

Grabocka E, Bar-Sagi D. 2016. Mutant KRAS enhances tu-mor cell fitness by upregulating stress granules. *Cell* **167:** 1803–1813.

Han TW, Kato M, Xie S, Wu LC, Mirzaei H, Pei J, Chen M, Xie Y, Allen J, Xiao G, et al. 2012. Cell-free formation of RNA granules: Bound RNAs identify features and com-ponents of cellular assemblies. *Cell* **149:** 768–779.

Harding HP, Novoa I, Zhang Y, Zeng H, Wek R, Schapira M, Ron D. 2000. Regulated translation initiation controls stress-induced gene expression in mammalian cells. *Mol Cell* **6:** 1099–1108.

Heerma van Voss MR, Vesuna F, Bol GM, Afzal J, Tantravedi S, Bergman Y, Kammers K, Lehar M, Malek R, Ballew M, et al. 2017. Targeting mitochondrial translation by inhib-iting DDX3: A novel radiosensitization strategy for cancer treatment. *Oncogene* **37:** 63–74.

Hofmann S, Cherkasova V, Bankhead P, Bukau B, Stoecklin G. 2012. Translation suppression promotes stress granule formation and cell survival in response to cold shock. *Mol Biol Cell* **23:** 3786–3800.

Hubstenberger A, Courel M, Benard M, Souquere S, Ernoult-Lange M, Chouaib R, Yi Z, Morlot JB, Munier A, Fradet M, et al. 2017. P-body purification reveals the condensation of repressed mRNA regulons. *Mol Cell* **68:** 144–157.

Ingelfinger D, Arndt-Jovin DJ, Luhrmann R, Achsel T. 2002. The human LSm1–7 proteins colocalize with the mRNA-degrading enzymes Dcp1/2 and Xrnl in distinct cytoplas-mic foci. *RNA* **8:** 1489–1501.

Ivanov P, Emara MM, Villen J, Gygi SP, Anderson P. 2011a. Angiogenin-induced tRNA fragments inhibit translation initiation. *Mol Cell* **43:** 613–623.

Ivanov P, Kedersha N, Anderson P. 2011b. Stress puts TIA on TOP. *Genes Dev* **25:** 2119–2124.

Ivanov P, O'Day E, Emara MM, Wagner G, Lieberman J, Anderson P. 2014. G-quadruplex structures contribute to the neuroprotective effects of angiogenin-induced tRNA fragments. *Proc Natl Acad Sci* **111:** 18201–18206.

Iwasaki S, Floor SN, Ingolia NT. 2016. Rocaglates convert DEAD-box protein eIF4A into a sequence-selective trans-lational repressor. *Nature* **534:** 558–561.

Jackson RJ, Hellen CU, Pestova TV. 2010. The mechanism of eukaryotic translation initiation and principles of its regulation. *Nat Rev Mol Cell Biol* **11:** 113–127.

Kato M, Han TW, Xie S, Shi K, Du X, Wu LC, Mirzaei H, Goldsmith EJ, Longgood J, Pei J, et al. 2012. Cell-free formation of RNA granules: Low complexity sequence domains form dynamic fibers within hydrogels. *Cell* **149:** 753–767.

Kedersha N, Anderson P. 2002. Stress granules: Sites of mRNA triage that regulate mRNA stability and translat-ability. *Biochem Soc Trans* **30:** 963–969.

Kedersha N, Anderson P. 2007. Mammalian stress granules and processing bodies. *Methods Enzymol* **431:** 61–81.

Kedersha NL, Gupta M, Li W, Miller I, Anderson P. 1999. RNA-binding proteins TIA-1 and TIAR link the phos-phorylation of eIF-2 α to the assembly of mammalian stress granules. *J Cell Biol* **147:** 1431–1442.

Kedersha N, Cho MR, Li W, Yacono PW, Chen S, Gilks N, Golan DE, Anderson P. 2000. Dynamic shuttling of TIA-1 accompanies the recruitment of mRNA to mammalian stress granules. *J Cell Biol* **151:** 1257–1268.

Kedersha N, Chen S, Gilks N, Li W, Miller IJ, Stahl J, Anderson P. 2002. Evidence that ternary complex (eIF2-GTP-tRNA$_i^{Met}$)-deficient preinitiation complexes are core constituents of mammalian stress granules. *Mol Biol Cell* **13:** 195–210.

Kedersha N, Stoecklin G, Ayodele M, Yacono P, Lykke-An-dersen J, Fritzler MJ, Scheuner D, Kaufman RJ, Golan DE, Anderson P. 2005. Stress granules and processing bodies are dynamically linked sites of mRNP remodeling. *J Cell Biol* **169:** 871–884.

Kedersha N, Ivanov P, Anderson P. 2013. Stress granules and cell signaling: More than just a passing phase? *Trends Biochem Sci* **38:** 494–506.

Kedersha N, Panas MD, Achorn CA, Lyons S, Tisdale S, Hickman T, Thomas M, Lieberman J, McInerney GM, Ivanov P, et al. 2016. G3BP-Caprin1-USP10 complexes mediate stress granule condensation and associate with 40S subunits. *J Cell Biol* **212:** 845–860.

Khaperskyy DA, Hatchette TF, McCormick C. 2012. Influ-enza A virus inhibits cytoplasmic stress granule forma-tion. *FASEB J* **26:** 1629–1639.

Kim WJ, Kim JH, Jang SK. 2007. Anti-inflammatory lipid mediator 15d-PGJ2 inhibits translation through inactiva-tion of eIF4A. *EMBO J* **26:** 5020–5032.

Kimball SR, Horetsky RL, Ron D, Jefferson LS, Harding HP. 2003. Mammalian stress granules represent sites of accu-mulation of stalled translation initiation complexes. *Am J Physiol Cell Physiol* **284:** C273–C284.

Leung AK, Calabrese JM, Sharp PA. 2006. Quantitative anal-ysis of Argonaute protein reveals microRNA-dependent localization to stress granules. *Proc Natl Acad Sci* **103**: 18125–18130.

Li W, Li Y, Kedersha N, Anderson P, Emara M, Swiderek KM, Moreno GT, Brinton MA. 2002. Cell proteins TIA-1 and TIAR interact with the 3′ stem-loop of the West Nile virus complementary minus-strand RNA and facilitate virus replication. *J Virol* **76**: 11989–12000.

Lloyd RE. 2013. Regulation of stress granules and P-bodies during RNA virus infection. *Wiley Interdiscip Rev RNA* **4**: 317–331.

Low WK, Dang Y, Schneider-Poetsch T, Shi Z, Choi NS, Merrick WC, Romo D, Liu JO. 2005. Inhibition of eukaryotic translation initiation by the marine natural product pateamine A. *Mol Cell* **20**: 709–722.

Luo Y, Na Z, Slavoff SA. 2018. P-bodies: Composition, properties, and functions. *Biochemistry* doi: 10.1021/acs.biochem.7b01162.

Lykke-Andersen J, Wagner E. 2005. Recruitment and activation of mRNA decay enzymes by two ARE-mediated decay activation domains in the proteins TTP and BRF-1. *Genes Dev* **19**: 351–361.

Lyons SM, Achorn C, Kedersha NL, Anderson PJ, Ivanov P. 2016. YB-1 regulates tiRNA-induced stress granule formation but not translational repression. *Nucleic Acids Res* **44**: 6949–6960.

Lyons SM, Fay MM, Akiyama Y, Anderson PJ, Ivanov P. 2017. RNA biology of angiogenin: Current state and perspectives. *RNA Biol* **14**: 171–178.

Mangiardi DA, McLaughlin-Williamson K, May KE, Messana EP, Mountain DC, Cotanche DA. 2004. Progression of hair cell ejection and molecular markers of apoptosis in the avian cochlea following gentamicin treatment. *J Comp Neurol* **475**: 1–18.

Markmiller S, Soltanieh S, Server KL, Mak R, Jin W, Fang MY, Luo EC, Krach F, Yang D, Sen A, et al. 2018. Context-dependent and disease-specific diversity in protein interactions within stress granules. *Cell* **172**: 590–604.

Maziuk B, Ballance HI, Wolozin B. 2017. Dysregulation of RNA binding protein aggregation in neurodegenerative disorders. *Front Mol Neurosci* **10**: 89.

Mazroui R, Sukarieh R, Bordeleau ME, Kaufman RJ, Northcote P, Tanaka J, Gallouzi I, Pelletier J. 2006. Inhibition of ribosome recruitment induces stress granule formation independently of eukaryotic initiation factor 2α phosphorylation. *Mol Biol Cell* **17**: 4212–4219.

McCormick C, Khaperskyy DA. 2017. Translation inhibition and stress granules in the antiviral immune response. *Nat Rev Immunol* **17**: 647–660.

McEwen E, Kedersha N, Song B, Scheuner D, Gilks N, Han A, Chen JJ, Anderson P, Kaufman RJ. 2005. Heme-regulated inhibitor kinase-mediated phosphorylation of eukaryotic translation initiation factor 2 inhibits translation, induces stress granule formation, and mediates survival upon arsenite exposure. *J Biol Chem* **280**: 16925–16933.

* Merrick WC, Pavitt GD. 2018. Protein synthesis initiation in eukaryotic cells. *Cold Spring Harb Perspect Biol* doi: 10.1101/cshperspect.a033092.

Mitchell SF, Parker R. 2014. Principles and properties of eukaryotic mRNPs. *Mol Cell* **54**: 547–558.

Moeller BJ, Cao Y, Li CY, Dewhirst MW. 2004. Radiation activates HIF-1 to regulate vascular radiosensitivity in tumors: Role of reoxygenation, free radicals, and stress granules. *Cancer Cell* **5**: 429–441.

Mollet S, Cougot N, Wilczynska A, Dautry F, Kress M, Bertrand E, Weil D. 2008. Translationally repressed mRNA transiently cycles through stress granules during stress. *Mol Biol Cell* **19**: 4469–4479.

Moutaoufik MT, El Fatimy R, Nassour H, Gareau C, Lang J, Tanguay RM, Mazroui R, Khandjian EW. 2014. UVC-induced stress granules in mammalian cells. *PLoS ONE* **9**: e112742.

Ng CS, Jogi M, Yoo JS, Onomoto K, Koike S, Iwasaki T, Yoneyama M, Kato H, Fujita T. 2013. Encephalomyocarditis virus disrupts stress granules, the critical platform for triggering antiviral innate immune responses. *J Virol* **87**: 9511–9522.

Nover L, Scharf KD, Neumann D. 1983. Formation of cytoplasmic heat shock granules in tomato cell cultures and leaves. *Mol Cell Biol* **3**: 1648–1655.

Nover L, Scharf KD, Neumann D. 1989. Cytoplasmic heat shock granules are formed from precursor particles and are associated with a specific set of mRNAs. *Mol Cell Biol* **9**: 1298–1308.

Ohn T, Anderson P. 2010. The role of posttranslational modifications in the assembly of stress granules. *Wiley Interdiscip Rev RNA* **1**: 486–493.

Ohn T, Kedersha N, Hickman T, Tisdale S, Anderson P. 2008. A functional RNAi screen links O-GlcNAc modification of ribosomal proteins to stress granule and processing body assembly. *Nat Cell Biol* **10**: 1224–1231.

Onomoto K, Jogi M, Yoo JS, Narita R, Morimoto S, Takemura A, Sambhara S, Kawaguchi A, Osari S, Nagata K, et al. 2012. Critical role of an antiviral stress granule containing RIG-I and PKR in viral detection and innate immunity. *PLoS ONE* **7**: e43031.

Pager CT, Schutz S, Abraham TM, Luo G, Sarnow P. 2013. Modulation of hepatitis C virus RNA abundance and virus release by dispersion of processing bodies and enrichment of stress granules. *Virology* **435**: 472–484.

Panas MD, Varjak M, Lulla A, Eng KE, Merits A, Karlsson Hedestam GB, McInerney GM. 2012. Sequestration of G3BP coupled with efficient translation inhibits stress granules in Semliki Forest virus infection. *Mol Biol Cell* **23**: 4701–4712.

Panas MD, Schulte T, Thaa B, Sandalova T, Kedersha N, Achour A, McInerney GM. 2015. Viral and cellular proteins containing FGDF motifs bind G3BP to block stress granule formation. *PLoS Pathog* **11**: e1004659.

Panas MD, Ivanov P, Anderson P. 2016. Mechanistic insights into mammalian stress granule dynamics. *J Cell Biol* **215**: 313–323.

* Peer E, Moshitch-Moshkovitz S, Rechavi G, Dominissini D. 2018. The epitranscriptome in translation regulation. *Cold Spring Harb Perspect Biol* doi: 10.1101/cshperspect.a032623.

Pisarev AV, Hellen CU, Pestova TV. 2007. Recycling of eukaryotic posttermination ribosomal complexes. *Cell* **131**: 286–299.

Pisarev AV, Skabkin MA, Pisareva VP, Skabkina OV, Rakotondrafara AM, Hentze MW, Hellen CU, Pestova TV.

2010. The role of ABCE1 in eukaryotic posttermination ribosomal recycling. *Mol Cell* **37:** 196–210.

Reineke LC, Dougherty JD, Pierre P, Lloyd RE. 2012. Large G3BP-induced granules trigger eIF2α phosphorylation. *Mol Biol Cell* **23:** 3499–3510.

Reineke LC, Kedersha N, Langereis MA, van Kuppeveld FJ, Lloyd RE. 2015. Stress granules regulate double-stranded RNA-dependent protein kinase activation through a complex containing G3BP1 and Caprin1. *mBio* **6:** e02486.

Ruggieri A, Dazert E, Metz P, Hofmann S, Bergeest JP, Mazur J, Bankhead P, Hiet MS, Kallis S, Alvisi G, et al. 2012. Dynamic oscillation of translation and stress granule formation mark the cellular response to virus infection. *Cell Host Microbe* **12:** 71–85.

Sadlish H, Galicia-Vazquez G, Paris CG, Aust T, Bhullar B, Chang L, Helliwell SB, Hoepfner D, Knapp B, Riedl R, et al. 2013. Evidence for a functionally relevant rocaglamide binding site on the eIF4A-RNA complex. *ACS Chem Biol* **8:** 1519–1527.

Scheuner D, Song B, McEwen E, Liu C, Laybutt R, Gillespie P, Saunders T, Bonner-Weir S, Kaufman RJ. 2001. Translational control is required for the unfolded protein response and in vivo glucose homeostasis. *Mol Cell* **7:** 1165–1176.

Sheth U, Parker R. 2006. Targeting of aberrant mRNAs to cytoplasmic processing bodies. *Cell* **125:** 1095–1109.

Shih JW, Tsai TY, Chao CH, Wu Lee YH. 2008. Candidate tumor suppressor DDX3 RNA helicase specifically represses cap-dependent translation by acting as an eIF4E inhibitory protein. *Oncogene* **27:** 700–714.

Shin Y, Brangwynne CP. 2017. Liquid phase condensation in cell physiology and disease. *Science* **357:** eaaf4382.

Shukla S, Parker R. 2016. Hypo- and hyper-assembly diseases of RNA-protein complexes. *Trends Mol Med* **22:** 615–628.

Sidrauski C, Acosta-Alvear D, Khoutorsky A, Vedantham P, Hearn BR, Li H, Gamache K, Gallagher CM, Ang KK, Wilson C, et al. 2013. Pharmacological brake-release of mRNA translation enhances cognitive memory. *eLife* **2:** e00498.

Sidrauski C, McGeachy AM, Ingolia NT, Walter P. 2015a. The small molecule ISRIB reverses the effects of eIF2α phosphorylation on translation and stress granule assembly. *eLife* **4:** e05033.

Sidrauski C, Tsai JC, Kampmann M, Hearn BR, Vedantham P, Jaishankar P, Sokabe M, Mendez AS, Newton BW, Tang EL, et al. 2015b. Pharmacological dimerization and activation of the exchange factor eIF2B antagonizes the integrated stress response. *eLife* **4:** e07314.

Simsek D, Tiu GC, Flynn RA, Byeon GW, Leppek K, Xu AF, Chang HY, Barna M. 2017. The mammalian ribo-interactome reveals ribosome functional diversity and heterogeneity. *Cell* **169:** 1051–1065.

Sivan G, Kedersha N, Elroy-Stein O. 2007. Ribosomal slowdown mediates translational arrest during cellular division. *Mol Cell Biol* **27:** 6639–6646.

Solomon S, Xu Y, Wang B, David MD, Schubert P, Kennedy D, Schrader JW. 2007. Distinct structural features of caprin-1 mediate its interaction with G3BP-1 and its induction of phosphorylation of eukaryotic translation initiation factor 2α, entry to cytoplasmic stress granules,

and selective interaction with a subset of mRNAs. *Mol Cell Biol* **27:** 2324–2342.

Somasekharan SP, El-Naggar A, Leprivier G, Cheng H, Hajee S, Grunewald TG, Zhang F, Ng T, Delattre O, Evdokimova V, et al. 2015. YB-1 regulates stress granule formation and tumor progression by translationally activating G3BP1. *J Cell Biol* **208:** 913–929.

Srivastava SP, Kumar KU, Kaufman RJ. 1998. Phosphorylation of eukaryotic translation initiation factor 2 mediates apoptosis in response to activation of the double-stranded RNA-dependent protein kinase. *J Biol Chem* **273:** 2416–2423.

* Stern-Ginossar N, Thompson SR, Mathews MB, Mohr I. 2018. Translational control in virus-infected cells. *Cold Spring Harb Perspect Biol* doi: 10.1101/cshperspect.a033001.

Stoecklin G, Kedersha N. 2013. Relationship of GW/P-bodies with stress granules. *Adv Exp Med Biol* **768:** 197–211.

Stoecklin G, Stubbs T, Kedersha N, Wax S, Rigby WF, Blackwell TK, Anderson P. 2004. MK2-induced tristetraprolin: 14–3–3 complexes prevent stress granule association and ARE-mRNA decay. *EMBO J* **23:** 1313–1324.

Stohr N, Lederer M, Reinke C, Meyer S, Hatzfeld M, Singer RH, Huttelmaier S. 2006. ZBP1 regulates mRNA stability during cellular stress. *J Cell Biol* **175:** 527–534.

Szaflarski W, Fay MM, Kedersha N, Zabel M, Anderson P, Ivanov P. 2016. Vinca alkaloid drugs promote stress-induced translational repression and stress granule formation. *Oncotarget* **7:** 30307–30322.

Taylor JP, Brown RH Jr, Cleveland DW. 2016. Decoding ALS: From genes to mechanism. *Nature* **539:** 197–206.

Uversky VN. 2017. Intrinsically disordered proteins in overcrowded milieu: Membrane-less organelles, phase separation, and intrinsic disorder. *Curr Opin Struct Biol* **44:** 18–30.

van Dijk E, Cougot N, Meyer S, Babajko S, Wahle E, Seraphin B. 2002. Human Dcp2: A catalytically active mRNA decapping enzyme located in specific cytoplasmic structures. *EMBO J* **21:** 6915–6924.

Vilas-Boas FAS, da Silva AM, de Sousa LP, Lima KM, Vago JP, Bittencourt LF, Dantas AE, Gomes DA, Vilela MC, Teixeira MM, et al. 2016. Impairment of stress granule assembly via inhibition of the eIF2α phosphorylation sensitizes glioma cells to chemotherapeutic agents. *J Neurooncol* **127:** 253–260.

Weber SC, Brangwynne CP. 2012. Getting RNA and protein in phase. *Cell* **149:** 1188–1191.

Weber C, Nover L, Fauth M. 2008. Plant stress granules and mRNA processing bodies are distinct from heat stress granules. *Plant J* **56:** 517–530.

* Wek RC. 2018. Role of eIF2α kinases in translational control and adaptation to cellular stress. *Cold Spring Harb Perspect Biol* doi: 10.1101/cshperspect.a032870.

Wek SA, Zhu S, Wek RC. 1995. The histidyl-tRNA synthetase-related sequence in the eIF-2 α protein kinase GCN2 interacts with tRNA and is required for activation in response to starvation for different amino acids. *Mol Cell Biol* **15:** 4497–4506.

White JP, Cardenas AM, Marissen WE, Lloyd RE. 2007. Inhibition of cytoplasmic mRNA stress granule formation by a viral proteinase. *Cell Host Microbe* **2:** 295–305.

Wilczynska A, Aigueperse C, Kress M, Dautry F, Weil D. 2005. The translational regulator CPEB1 provides a link between dcp1 bodies and stress granules. *J Cell Sci* **118:** 981–992.

Wolozin B. 2012. Regulated protein aggregation: Stress granules and neurodegeneration. *Mol Neurodegener* **7:** 56.

Wolozin B. 2014. Physiological protein aggregation run amuck: Stress granules and the genesis of neurodegenerative disease. *Discov Med* **17:** 47–52.

Yamasaki S, Ivanov P, Hu GF, Anderson P. 2009. Angiogenin cleaves tRNA and promotes stress-induced translational repression. *J Cell Biol* **185:** 35–42.

Yang WH, Bloch DB. 2007. Probing the mRNA processing body using protein macroarrays and "autoantigenomics." *RNA* **13:** 704–712.

Youn JY, Dunham WH, Hong SJ, Knight JDR, Bashkurov M, Chen GI, Bagci H, Rathod B, MacLeod G, Eng SWM, et al. 2018. High-density proximity mapping reveals the subcellular organization of mRNA-associated granules and bodies. *Mol Cell* **69:** 517–532.

Yu JH, Yang WH, Gulick T, Bloch KD, Bloch DB. 2005. Ge-1 is a central component of the mammalian cytoplasmic mRNA processing body. *RNA* **11:** 1795–1802.

Cite this article as *Cold Spring Harb Perspect Biol* doi: 10.1101/cshperspect.a032813

Role of eIF2α Kinases in Translational Control and Adaptation to Cellular Stress

Ronald C. Wek

Department of Biochemistry and Molecular Biology, Indiana University School of Medicine, Indianapolis, Indiana 46202–5126

Correspondence: rwek@iu.edu

A central mechanism regulating translation initiation in response to environmental stress involves phosphorylation of the α subunit of eukaryotic initiation factor 2 (eIF2α). Phosphorylation of eIF2α causes inhibition of global translation, which conserves energy and facilitates reprogramming of gene expression and signaling pathways that help to restore protein homeostasis. Coincident with repression of protein synthesis, many gene transcripts involved in the stress response are not affected or are even preferentially translated in response to increased eIF2α phosphorylation by mechanisms involving upstream open reading frames (uORFs). This review highlights the mechanisms regulating eIF2α kinases, the role that uORFs play in translational control, and the impact that alteration of eIF2α phosphorylation by gene mutations or small molecule inhibitors can have on health and disease.

Maintenance of protein homeostasis requires appropriate regulation of translation, as well as protein folding, transport, and degradative processes. Environmental stresses and physiological stimuli can rapidly disrupt protein homeostasis, triggering cell-adaptive responses that are critical to restore the integrity of the proteome. However, the functionality of the adaptive responses can decline or be altered with chronic stress or with aging, leading to diseases that can afflict multiple organs, including the neural system and those contributing to metabolic health. This review addresses the role of translational control in adaptive responses to environmental stresses and the processes by which phosphorylation of the α subunit of eukaryotic initiation factor 2 (P-eIF2α) can modulate translation genome wide to restore protein homeostasis. Key themes in the review will be the mechanisms regulating eIF2α kinases, the role that upstream open reading frames (uORFs) play in translational control, and the impact that altered P-eIF2α levels by gene mutations or small molecule inhibitors can have on health and disease.

PHOSPHORYLATION OF eIF2α DIRECTS TRANSLATION CONTROL

A major mechanism regulating the initiation phase of protein synthesis involves P-eIF2α at serine-51. The eIF2, combined with guanosine triphosphate (GTP), is critical for providing initiator methionyl-transfer RNA (tRNA) (Met-tRNA$_i^{Met}$) to the 43S preinitiation complex that contains the small ribosomal subunit and a myriad of additional translation initiation factors. In the predominant pathway, the preinitiation complex then combines with the 5′-7-methylguanosine "cap" of messenger RNA (mRNA) and scans processively 5′- to 3′- along the leader of the transcript in search of an initiation codon. Complementary binding of the Met-tRNA$_i^{Met}$ to the start codon in the P site of the 40S ribosomal subunit triggers cessation of scanning and hydrolysis of GTP associated with eIF2. Following release of eIF2•GDP (guanosine diphosphate), the large 60S ribosomal subunit then joins to form the 80S ribosome, which carries out the elongation phase of protein synthesis. To facilitate the next round of translation initiation, GDP associated with eIF2 needs to be exchanged for GTP, a process catalyzed by a guanine nucleotide exchange factor, eIF2B. In response to diverse stresses, P-eIF2α alters this translation factor so that it binds tightly to a regulatory portion of eIF2B, thus inhibiting the recycling of eIF2•GDP to the active GTP-bound

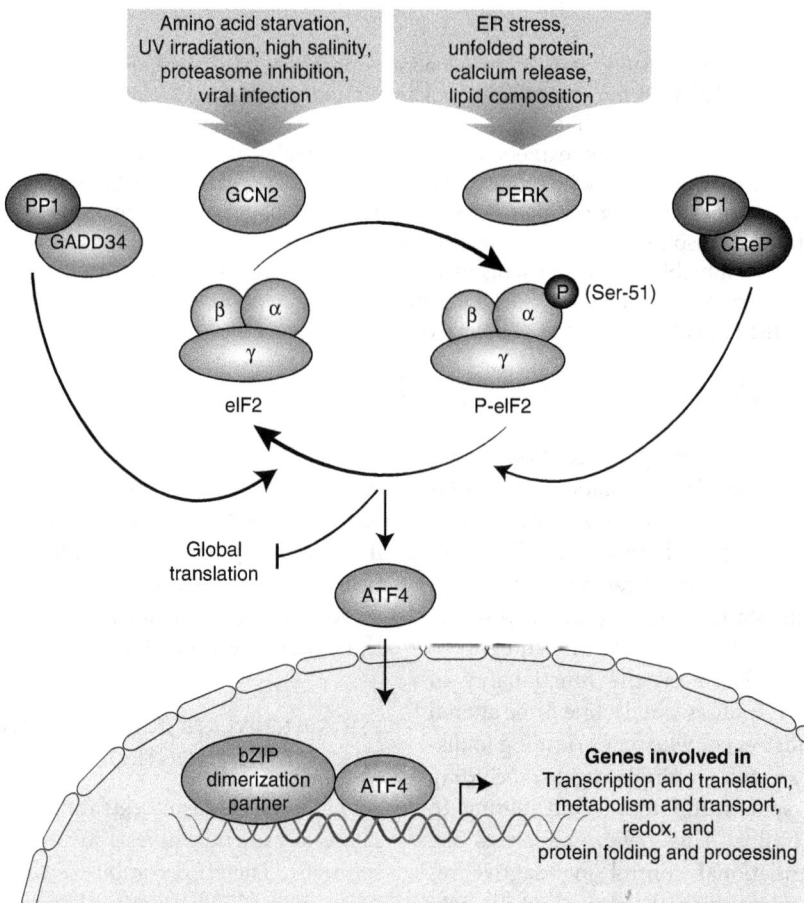

Figure 1. Phosphorylation of the α subunit of eukaryotic initiation factor 2 (eIF2α) regulates global and gene-specific translation. The eIF2α kinases general control nonderepressible 2 (GCN2) and protein kinase R (PKR)-like endoplasmic reticulum (ER) kinase (PERK) are activated by nutritional stress or perturbations in the ER, respectively. Type 1 protein phosphatase complex (PP1c) combines with CReP to dephosphorylate eIF2α during basal conditions and GADD34 in feedback control of the integrated stress response (ISR). Phosphorylation of eIF2α reduces global translation initiation coincident with preferential translation of ATF4, encoding a basic zipper (bZIP) transcriptional activator that dimerizes with other transcript factors to regulate transcription of ISR genes that function in adaptation to stress.

Cite this article as *Cold Spring Harb Perspect Biol* doi: 10.1101/cshperspect.a032870

form (Fig. 1). As a consequence, there is lowered eIF2•GTP and delivery of the Met-tRNA$_i^{Met}$ to ribosomes, culminating in a sharp reduction in global translation initiation.

Repression of translation initiation is an efficient mechanism to conserve energy and nutrients, which are amply consumed by protein synthesis. Furthermore, lowering general translation allow cells to reconfigure gene expression and signaling pathways that optimize stress alleviation. For example, arrest of translational initiation by increased levels of P-eIF2α leads to polysome disassembly that triggers formation of stress granules, which are cytosolic foci of untranslated mRNAs and associated 40S ribosomal subunits and proteins (Kedersha et al. 2013; Ivanov et al. 2017). Stress granules serve as a triage center, sorting incoming messenger ribonucleoproteins for mRNA decay or sequestration for eventual return to the cytoplasm for translation. Therefore, stress granules are critical for reprogramming gene expression. Signaling proteins and enzymes can also be recruited to stress granules, influencing their respective cellular pathways.

Inhibition of global protein synthesis also reshapes the proteome, as proteins that are labile will rapidly be depleted from cells. The biological consequences of these proteomic changes are shown by the activation of nuclear factor κB (NF-κB) in response to accumulation of P-eIF2α and ultraviolet (UV) irradiation (Wu et al. 2004; Jiang and Wek 2005). NF-κB is a transcriptional regulator of genes involved in inflammation, cell proliferation, and apoptosis, and is inhibited by binding to IκBα. Lowered synthesis of IκBα as a consequence of induced P-eIF2α, combined with rapid turnover of IκBα protein, causes a release of IκBα from NF-κB that facilitates NF-κB entry into the nucleus for targeted transcriptional regulation.

FAMILY OF eIF2α KINASES ACTIVATED BY DIFFERENT STRESSES

Coincident with global repression of protein synthesis, select gene transcripts can be resistant or even preferentially translated in response to induced P-eIF2α. An important preferentially translated gene is *ATF4*, which features uORFs embedded in its mRNA that serve as a "bar code" for scanning ribosomes for selective translation (Harding et al. 2000a; Lu et al. 2004; Vattem and Wek 2004). ATF4 is a basic zipper (bZIP) transcription factor of genes involved in nutrient import, metabolism, and alleviation of oxidative stress (Harding et al. 2003). Because P-eIF2α and ATF4 are induced by diverse stresses, this pathway is referred to as the integrated stress response (ISR) (Harding et al. 2003). In mammals, there are four different eIF2α kinases, each containing distinct regulatory domains that serve to sense the cell stress environment through engagement with regulatory ligands and proteins. This review will focus on two of the eIF2α kinase family members, general control nonderepressible 2 (GCN2 or EIFAK4) and protein kinase R (PKR)-like endoplasmic reticulum (ER) kinase (PERK or EIF2AK3), which respond to perturbations in the cytosol and ER, respectively (Fig. 1). The other eIF2α kinases include HRI (EIF2AK1), which primarily functions to balance globin synthesis with heme availability during erythropoiesis, and PKR (EIF2AK2), which participates in the innate immune response to viral infection.

In the example of GCN2, starvation for amino acids enhance P-eIF2α levels and translational control, which quickly limits incorporation of amino acids into nascent polypeptides. In addition to the protein kinase domain, GCN2 has a regulatory region homologous to histidyl-tRNA synthetase (HARS), which binds to uncharged tRNAs that accumulate during deprivation for nutrients (Wek et al. 1989, 1995; Dong et al. 2000). Binding to uncharged tRNA is thought to lead to conformational changes in GCN2 that trigger autophosphorylation and release of inhibitory interactions between the regulatory regions of GCN2 and the kinase domain, resulting in increased P-eIF2α (Lageix et al. 2014, 2015). It should be emphasized that GCN2 can bind to a range of different uncharged tRNAs to monitor the availability of their respective amino acids. Activation of GCN2 also requires GCN1 protein, which binds to the amino-terminal RWD domain of GCN2 and is thought to facilitate GCN2 access to uncharged tRNAs (Marton et al. 1993, 1997). GCN2 can be inhibited by the regulatory protein IMPACT (YIH1), which competes with this eIF2α kinase for its association with GCN1 (Sattlegger et al. 2004; Pereira et al. 2005). Finally, many other stresses have been reported to activate GCN2, including UV irradiation, high salinity, and glucose deprivation (Yang et al. 2000; Goosens et al. 2001; Deng et al. 2002; Zaborske et al. 2009). These stresses also require the function of the HARS-related domain of GCN2, supporting the idea that this mode of regulation is required at least in part for GCN2 activation by stresses not directly linked with amino acid depletion.

PERK is a transmembrane protein situated in the ER, which functions as part of the unfolded protein response (UPR) (Shi et al. 1998; Harding et al. 1999, 2000b). The UPR features both translational and transcriptional gene expression that serves to expand the processing capacity of the ER (Walter and Ron 2011). Regulation of PERK is complex, in part because there are numerous conditions that readily perturb the ER and because the stresses are typically not measured directly but instead are inferred by assessing activation of PERK and the other UPR sensors IRE1 and ATF6. A prevailing model for the regulation of PERK is that the amino-terminal portion of PERK can bind to the ER-resident chaperone BiP (GRP78/HSPA5), maintaining this eIF2α kinase in a repressed conformation (Bertolotti et al. 2000; Ma et al. 2002). Stressful conditions in the ER that disrupt protein folding can trigger the release of the chaperone BiP from PERK, providing for an activated conformation that induces PERK autophosphorylation and P-eIF2α.

The rationale for BiP release from PERK during ER stress is attributed to accumulating unfolded protein in the ER effectively competing for binding with the ER lumenal portion of this eIF2α kinase. BiP dissociation from PERK would be readily reversed when the ER stress is remedied in the cell (Bertolotti et al. 2000). It was generally assumed that BiP bound with PERK through the peptide-binding portion of this ER chaperone. An alternative model has been put forth that the ATPase domain of BiP binds with PERK and this interaction is released

when unfolded protein engages with the canonical peptide-binding domain of BiP (Carrara et al. 2015). Given that BiP is abundant in the ER, it has been argued that the BiP regulatory model of PERK is too coarse for rapid activation of PERK during ER stress (Pincus et al. 2010). For the observed rapid activation of PERK, it has instead been proposed that the lumenal portion of PERK can accommodate direct binding to unfolded protein. This idea is supported by peptide-binding experiments with the UPR sensory protein IRE1 from yeast (Gardner and Walter 2011), but is still unresolved for PERK.

It is noteworthy that PERK has functions independent of its eIF2α kinase activity, as increased cytosolic Ca^{2+} levels can also trigger oligomerization of PERK in the ER, which is suggested to stabilize PERK interactions with the actin-binding protein filamin A (FLNA) (van Vliet et al. 2017). The PERK/FLNA interaction drives F-actin remodeling, facilitating contacts between the ER and plasma membrane that function in the regulation of Ca^{2+} fluxes and lipid signaling. These results indicate that the biological effects attributed to loss of PERK do not always involve dysregulation of eIF2α phosphorylation and translational control.

Enhanced P-eIF2α by GCN2 and PERK are balanced by dephosphorylation by type 1 protein phosphatase complex (PP1c) that is directed to eIF2α via scaffolding proteins GADD34 (PPP1R15A) and CReP (PPP1R15B) (Fig. 1) (Connor et al. 2001; Novoa et al. 2001; Jousse et al. 2003). GADD34 and CReP share sequence similarity in their carboxy-terminal PP1c-anchoring motifs, but have dissimilar regions that serve to engage with eIF2α (Choy et al. 2015). Expression of *GADD34* is enhanced by stress and elevated levels of P-eIF2α and is central for restoration of translation through feedback control of the ISR. CReP functions to maintain lower levels of P-eIF2α during basal conditions. Although expression of *CReP* is suggested to be constitutive, there is potential cross regulation between these two PP1c regulatory proteins (Young et al. 2015). It is also noteworthy that both GADD34 and CReP association with PP1c is stabilized by direct binding with monomeric G-actin, and the abundance and activity of the

Cite this article as *Cold Spring Harb Perspect Biol* doi: 10.1101/cshperspect.a032870

complex and levels of P-eIF2α are responsive to changes in the polymeric status of actin (Chambers et al. 2015; Chen et al. 2015). The involvement of the cytoskeleton as a spatial organizer and regulator of key processes in translation is an emerging theme and may also affect the regulation of GCN2 (Silva et al. 2016). Disruption of the actin cytoskeleton is suggested to facilitate IMPACT release from GCN1, making the activator GCN1 protein more accessible for association with GCN2. GCN2 can then bind uncharged tRNAs more efficiently, leading to enhanced phosphorylation of eIF2α.

Emphasizing the importance of CReP and GADD34 in the implementation and function of the ISR, mice deleted for CReP are growth impaired and deficient for erythropoiesis (Harding et al. 2009). Combined loss of both CReP and GADD34 leads to early embryonic lethality, which can be rescued by expression of a version of eIF2α that that is refractory to phosphorylation. These findings highlight the critical roles that appropriate dephosphorylation of eIF2α plays in regulating translational control in the ISR and in mammalian development.

uORFs IN ISR TRANSLATIONAL CONTROL

In addition to *ATF4*, a number of key ISR regulatory genes affecting diverse cell functions are preferentially translated by mechanisms involving uORFs (Fig. 2). By definition, a uORF encodes at least two amino acid residues followed by a termination codon, which can be fully upstream or overlapping the primary coding sequence (CDS). About half of human genes encode putative uORFs (Iacono et al. 2005; Calvo et al. 2009; Resch et al. 2009). Whereas the mere presence of a predicted uORF does not necessarily indicate that it is translated, a report by Qian and colleagues used ribosome profiling to identify nearly 8000 translation initiation sites upstream of human CDSs (Lee et al. 2012). Their findings suggest that uORF initiation sites also include non-AUG codons, with CUG being the most prominent. There are technical concerns about potential translation artifacts in profiling studies; nonetheless, the prevalence of uORFs is striking among mammalian transcripts. It is im-

portant to note that predicted uORFs are present among mRNAs that are repressed, not affected, or preferentially translated during cellular stress and in the presence of elevated levels of P-eIF2α. Therefore, the presence of a uORF alone is not predictive of whether an mRNA is preferentially translated in response to P-eIF2α induction. Rather, the specific properties of uORFs and their placement and combinations in the 5'-leader of target mRNAs determine translation efficiency in response to P-eIF2α induction (Fig. 3). Additionally, secondary structures in the 5'-leader of mRNAs and RNA-binding proteins can influence the functions of uORFs and translational control.

Typically, uORFs are inhibitory to translation of the downstream CDS. Repression by uORFs can be considerable or more moderate, depending on the degree to which ribosomes initiate translation at the uORF and the ability of the terminating ribosomes to reinitiate translation downstream at a subsequent CDS (Fig. 3). Preferential translation of mRNAs in the ISR involves ribosome bypass or leaky scanning through inhibitory uORFs. How does P-eIF2α allow the ribosomes to proceed through barrier uORFs? The 5'-leader of *ATF4* contains a strong inhibitory uORF2, which overlaps out-of-frame with the *ATF4* CDS, and a short uORF1 that acts as a positive element in *ATF4* translation, promoting downstream reinitiation of translation (Fig. 4A) (Harding et al. 2000a; Lu et al. 2004; Vattem and Wek 2004). Following translation of the 5'-proximal uORF1, 40S ribosomal subunits are thought to be retained on the *ATF4* mRNA and resume scanning. The scanning 40S subunits then rapidly reacquire a new eIF2•GTP•Met-tRNA$_i^{Met}$ ternary complex that is abundant in nonstressed conditions when P-eIF2α is low. As a result, ribosomes initiate translation at the next available CDS, uORF2. Translation of the overlapping out-of-frame uORF2 results in translation termination and ribosome dissociation 3' of the initiation codon of the *ATF4* CDS. Therefore, ATF4 protein levels are reduced and there is lowered transcription of target ISR genes.

During ER stress or nutrient deprivation, induction of P-eIF2α lowers the levels of eIF2•GTP that are required for delivery of Met-tRNA$_i^{Met}$ to

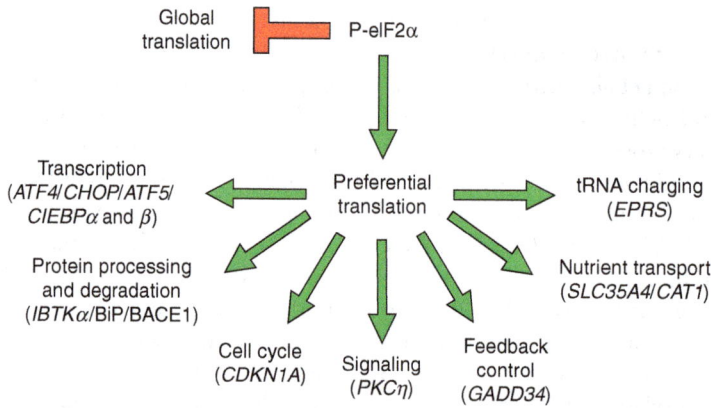

Figure 2. Phosphorylation of the α subunit of eukaryotic initiation factor 2 (P-eIF2α) enhances translation of multiple integrated stress response (ISR) genes by mechanisms involving upstream open reading frames (uORFs). P-eIF2α reduces global protein synthesis concurrent with preferential translation genes involved in diverse cellular functions. Preferential translation of *ATF4, CHOP, GADD34, EPRS,* and *CDKN1A* involves uORFs as described in the text. *IBTKα* (Baird et al. 2014; Willy et al. 2017), *BiP* (Starck et al. 2016), *BACE1* (O'Connor et al. 2008), *PKCη* (Raveh-Amit et al. 2009), *SLC35A4* (Andreev et al. 2015; Sidrauski et al. 2015), and *CAT1* (Yaman et al. 2003) have also been reported to be preferentially translated directly or indirectly by P-eIF2α during stress.

the reinitiating ribosomes. As a consequence, after translation of uORF1, the scanning 40S ribosomal subunit requires more time to acquire a new eIF2 ternary complex needed for recognition of the next initiation codon in the *ATF4* mRNAs. The delay in the acquisition of eIF2 ternary complex allows the 40S ribosomal subunit to scan through the start codon for the inhibitory uORF2 and instead promotes translation initiation at the next available initiation codon, the *ATF4* CDS (Fig. 4A). Increased levels of ATF4 directly enhance adaptive target genes in the ISR (Fig. 1). Translational control by delayed reinitiation was originally described by Hinnebusch and colleagues in budding yeast for the related transcriptional activator GCN4 (Abastado et al. 1991; Hinnebusch 2005).

Given the diverse stress conditions enhancing *ATF4* translation, there may be additional modulators of *ATF4*. For example, another short uORF has been identified upstream of uORF1 in *ATF4* that is occupied by ribosomes in profiling studies. Prior experiments using luciferase reporters fused to 5′-segments of the *ATF4* transcript did not detect any appreciable changes in the induction of translation on ER stress when this upstream uORF was omitted (RC Wek, unpubl.). However, levels of induced *ATF4* trans-

lation measured using reporters transfected into cultured cells are typically lower than those determined for endogenous *ATF4* so there may be additional regulatory features.

Furthermore, mRNA sequences proximal to the 5′ cap can enhance the recruitment of the eIF4E subunit of the cap-binding eIF4F complex and translation efficiency (Keys 2016; Keys and Sabatini 2017). These so-called "juxtaposed sequences" may influence *ATF4* translation and may be critical for loading of the 43S preinitiation complex onto the *ATF4* transcript when eIF2•GTP levels are diminished with increased levels of P-eIF2α and stress. Finally, base modifications in RNA, such as N^6-methyladenosine, may influence the efficiency of ribosome scanning and reinitiation that can affect *ATF4* translation (Meyer et al. 2015; Wang et al. 2015; Zhou et al. 2015).

PREFERENTIAL TRANSLATION BY RIBOSOME BYPASS

A number of transcripts that are preferentially translated in the ISR involve a mechanism featuring a single uORF. One example is *CHOP*, whose translational and transcriptional expression is enhanced by P-eIF2α. Early in the stress

Cite this article as *Cold Spring Harb Perspect Biol* doi: 10.1101/cshperspect.a032870

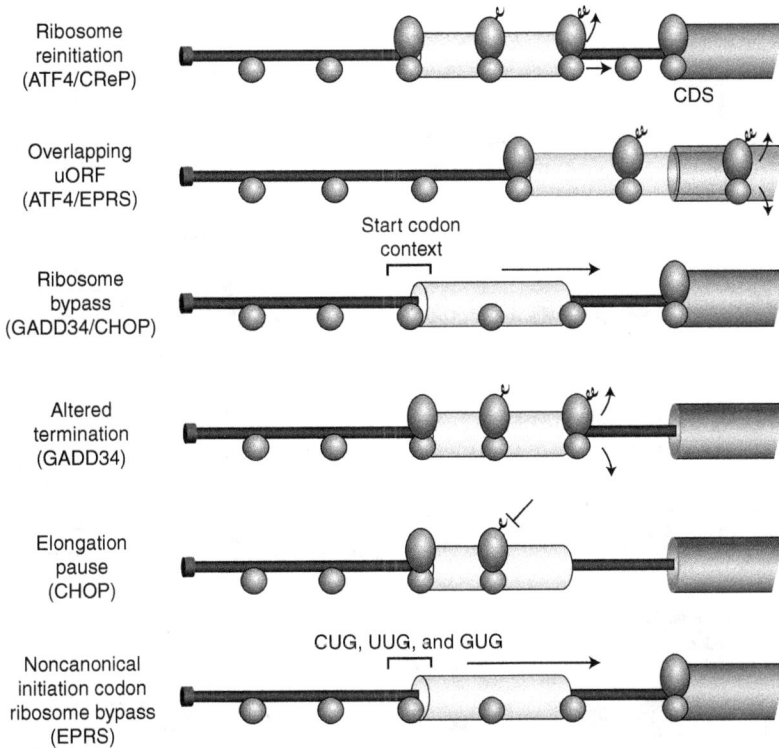

Figure 3. Upstream open reading frames (uORFs) can have different functions in preferential translation in the integrated stress response (ISR). The uORFs and their function are highlighted for the indicated gene transcripts. The 5'-leader of the messenger RNAs (mRNAs) is indicated as a solid line. The coding sequences (CDSs) are indicated by the bar on the far right of each transcript, with uORFs indicated by the light gray bars. Scanning and elongating ribosomes are indicated by the ovals, with small and large ribosomal subunits. Arrows indicate ribosome bypass, reinitiation or termination, and release.

response, CHOP triggers transcription of genes with adaptive functions, including those related to ATF4 (Marciniak et al. 2004; Han et al. 2013). However, with extended stress and sustained P-eIF2α induction, continued *CHOP* expression can trigger expression of genes that elicit apoptosis (Marciniak et al. 2004; Marciniak and Ron 2006; Oslowski and Urano 2011). Thus, CHOP is central to the balance between the adaptive functions of the ISR during acute stress versus induction of cell death during chronic stress conditions.

Preferential translation of *CHOP* features a single uORF that serves to stall elongating ribosomes as judged by experiments with translational reporters and in vitro toeprinting analyses, preventing reinitiation at the downstream *CHOP* CDS (Fig. 3) (Jousse et al. 2001; Palam

et al. 2011; Young et al. 2016b). In response to stress and accumulating P-eIF2α, a subset of scanning ribosomal subunits proceed through the *CHOP* uORF and instead initiate at the CDS. Part of the ability of ribosomes to bypass the uORF in response to increased levels of P-eIF2α involves a less-than-optimal context of the uORF start codon. Emphasizing the importance of the uORF in *CHOP* expression, mutations that prevent translation of the uORF substantially increase the levels of CHOP during both basal and stress conditions and modify the pattern of induction of *CHOP* expression in the ISR (Young et al. 2016b). Elevated CHOP levels sensitize cells to stress, with accelerated apoptosis on cell exposure to ER stress.

Another example of ribosome bypass of a uORF is that of *GADD34*, which contains two

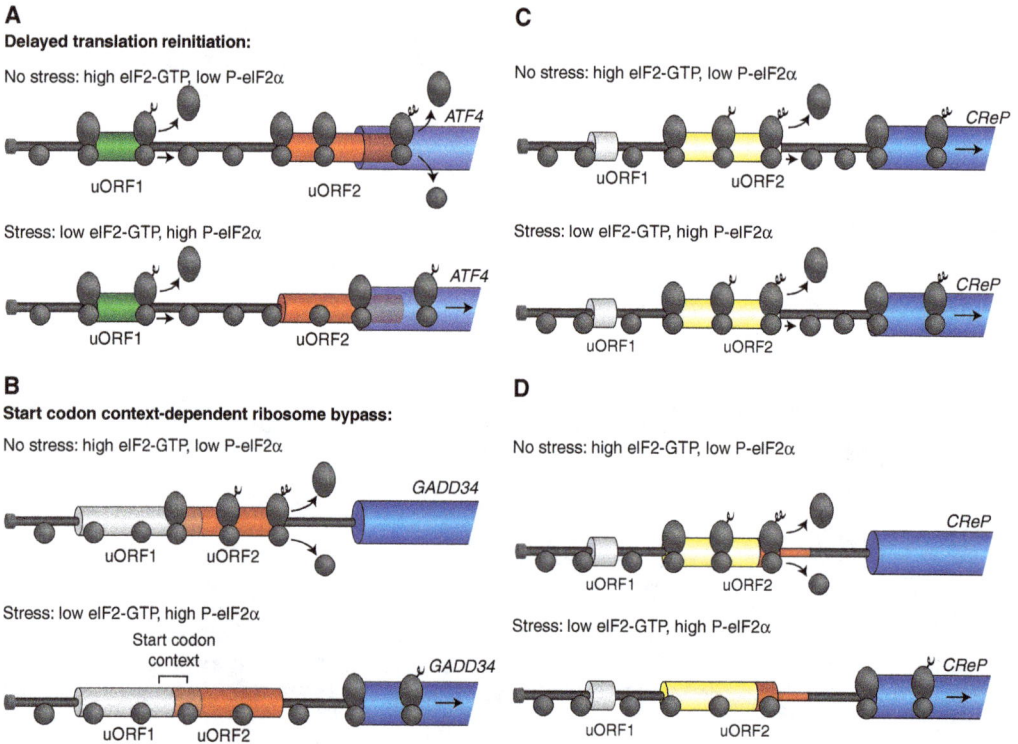

Figure 4. The integrated stress response (ISR) features different translational control mechanisms with upstream open reading frames (uORFs). (*A*) Illustration of the mechanism of *ATF4* delayed translation reinitiation that functions to enhance ATF4 synthesis on phosphorylation of the α subunit of eukaryotic initiation factor 2 (P-eIF2α) and stress. In nonstressed conditions, there are low levels of P-eIF2α and abundant eIF2•GTP (guanosine triphosphate). Following translation of uORF1 (green bar), ribosomes (ovals indicated by large and small subunits) rapidly reacquire new eIF2•GTP•Met-tRNA$_i^{Met}$ and reinitiate at the inhibitory uORF2 (red bar), which overlaps out-of-frame with the *ATF4* coding sequence (CDS) (blue bar). Therefore, there are low levels of ATF4 and its target genes in the absence of stress. In response to stress, enhanced P-eIF2α and low eIF2•GTP delay reinitiation, allowing ribosomes to proceed through uORF2, and instead translate the *ATF4* CDS. (*B*) Translation of *GADD34* involves a fraction of the translating ribosome scanning through an inhibitory uORF2 (red bar) in response to P-eIF2α and stress. The uORF1 (gray bar), which overlaps out-of-frame uORF2, is not well translated and is a modest dampener in the translation of *GADD34*. (*C*) Expression of *CReP* involves a fraction of ribosomes translating uORF2 (red bar) and reinitiating at the *CReP* CDS independent P-eIF2α and stress. Therefore, synthesis of CReP is largely constitutive regardless of stress conditions. The *CReP* ORF1 (gray bar) functions to lower translation of the *CReP* CDS only modestly. (*D*) Substitution of the Pro-Pro-Gly-stop codons and nine nucleotides 3′- of the *GADD34* uORF2 for the corresponding uORF2 regions in the *CReP* transcript (indicated by red portion of the uORF and messenger RNA [mRNA]) leads to lowered translation of the *CReP* hybrid that is preferentially translated in response to stress and P-eIF2α induction (Young et al. 2015).

uORFs (Fig. 4B) (Lee et al. 2009; Young et al. 2015). uORF2 is the primary inhibitor of downstream translation at the *GADD34* CDS and is sufficient to confer preferential translation in response to P-eIF2α. Translation of the Pro-Pro-Gly codons juxtaposed to the termination codon in uORF2 is suggested to block transla-tion reinitiation at the CDS, lowering levels of *GADD34* expression during basal conditions (Young et al. 2015). Preferential translation of *GADD34* in response to stress and elevated P-eIF2α levels occurs by a fraction of the scanning ribosomal subunits bypassing uORF2 by a mechanism involving, at least in part, the poor

Cite this article as *Cold Spring Harb Perspect Biol* doi: 10.1101/cshperspect.a032870

start codon context of this inhibitory uORF. It is interesting to note that translation of *CReP* is resistant to P-eIF2α induction (Andreev et al. 2015; Young et al. 2015). The CReP transcript also has two uORFs that are frequently bypassed even during nonstressed conditions. Furthermore, uORF2 allows for efficient reinitiation of translation at the CDS (Fig. 4C). If the termination codon and 3′-flanking sequences from the GADD34 uORF2 are substituted for those in the *CReP* uORF2, translation of *CReP* becomes induced on stress and with elevated levels of P-eIF2α (Fig. 4D) (Young et al. 2015). This finding emphasizes the importance of precise uORF properties to convey translational control in the ISR.

Preferential translation can also occur via bypass of uORFs with noncanonical initiation codons. Enhanced expression of the bifunctional glutamyl-prolyl tRNA synthetase, EPRS, serves to increase the appropriately charged tRNA pool and prime the cell for resumption of translation once the cellular stress is alleviated (Fig. 3). Two uORFs featuring UUG and CUG initiation codons are considered to be the primary regulators of *EPRS* preferential translation (Young et al. 2016a). An inhibitory uORF with a CUG initiation codon overlaps out-of-frame with the *EPRS* CDS. On the other hand, a uORF featuring UUG terminates upstream of the CDS and allows some of the ribosomes to reinitiate at the downstream *EPRS* CDS (Young et al. 2016a). Both uORFs are bypassed to a moderate extent during basal conditions, with enhanced bypass efficiency during eIF2α-P and stress.

PREFERENTIAL TRANSLATION VARIES BETWEEN STRESSES

There is robust P-eIF2α induction in response to a spectrum of stress conditions, but the pattern of gene-specific translation can be specifically tailored to best adapt to each stress condition. This is noteworthy because the uORFs are embedded in each gene transcript and would not appear to be readily modified for a given stress. Three explanations can be provided for gene-specific translation tailored to a given stress.

First, *ATF4* and other ISR genes subject to preferential translation also have enhanced transcriptional expression in response to ER or nutrient stress, which would increase the amount of mRNA available for translation during the progression of the stress response. However, following exposure to high physiological doses of UV-B or UV-C, there is repressed transcription of *ATF4*, sharply lowering the amount of *ATF4* mRNA that is available for preferential translation (Dey et al. 2010, 2012; Collier et al. 2015). Lowered ATF4 levels also reduce the transcription and, ultimately, the translation of the downstream target gene *CHOP*. Repression of global translation is important for cell survival in response to UV stress, and forced expression of *ATF4* sensitizes cells. Knockdown of *CHOP* suppresses this sensitivity, emphasizing the idea that elevated expression of *CHOP* is detrimental in response to UV irradiation (Collier et al. 2015). The dynamics of changes in both mRNA and translation in response to increased levels of P-eIF2α provide a vehicle to differentially regulate the expression of key ISR genes in response to different stress conditions.

A second explanation features alternative gene promoters and pre-mRNA splicing that can create gene transcripts with different 5′-leaders and uORF configurations, which alter mRNA translation during P-eIF2α induction. For example, the *ATF5α* variant that is controlled by the mechanism of delayed translation initiation is derived from a different promoter than the *ATF5β* variant, which has an expanded collection of uORFs that appear to largely dampen translation (Watatani et al. 2008; Zhou et al. 2008). Both variants express the same CDS, with *ATF5α* participating in the ISR and *ATF5β* mRNA being expressed predominantly during early development (Hansen et al. 2002). Pre-mRNA splicing can alter translation of *CDKN1A* (*p21/WAF1*), which contributes to cell-cycle arrest and increased survival in response to starvation for amino acids. Among the many *CDKN1A* spliced variants in mice that alter the 5′-leader of the gene transcripts, variant 2 features three uORFs that provide for preferential translation in response to induced P-eIF2α (Lehman et al. 2015).

A final explanation for distinct stress-specific programs of preferential translation involves the notion that mRNA translation is spatially organized and regulated in cells and that a given stress can differentially disrupt cell compartments. Nicchitta and colleagues (Reid et al. 2014; Reid and Nicchitta 2015) found that ER-bound ribosomes synthesize a significant fraction of proteins, both those slated to be retained in the cytosol as well as those destined for the secretory pathway. The ER-associated translation system is suggested to be dynamic and be reorganized in response to physiological cues and cellular stresses. In this way, the ER environment for translation may be quite distinct from the cytosol and the influences of P-eIF2α induction may vary, yielding differences in preferential translation.

ISR AND DISEASE

Emphasizing the broad and diverse impact of the ISR, mutations have been identified in ISR genes that afflict distinct tissues and present with different pathologies. For example, nonsense, frameshift, and missense mutations have been reported in *PERK*, leading to Wolcott–Rallison syndrome, which is characterized by neonatal diabetes, osteoporosis, digestive dysfunctions, and hepatic complications, culminating in early death (Delepine et al. 2000; Senée et al. 2004). The inability of PERK to induce translational control in Wolcott–Rallison syndrome leads to disruption of protein homeostasis, especially in specialized secretory tissues that require robust ER secretory processes.

Loss of GCN2 function causes pulmonary disorders, including pulmonary arterial hypertension (PAH), pulmonary veno-occlusive disease (PVOD), and pulmonary capillary hemangiomatosis (PCH) (Best et al. 2014, 2017; Eyries et al. 2014). The rationale for why GCN2 deficiency triggers pulmonary disorders in humans is currently not understood. The lungs are challenged by a variety of inhaled stress agents, including smoke, airborne particles, and microbes. Appropriate induction of the ISR may be critical for cell resistance to these insults and for pulmonary vascular remodeling. Supporting

this idea, ATF4 plays a central role in antioxidation responses and cysteine sufficiency, along with angiogenesis through enhanced expression of vascular endothelial growth factor (VEGF) (Harding et al. 2003; Roybal et al. 2005; Fusakio et al. 2016). GCN2 can also participate in cell proliferation and differentiation, which may be critical for the health of pulmonary tissues and the immune system (Munn et al. 2005; Collier et al. 2017). Finally, it is suggested that loss of GCN2/ATF4 may disrupt signaling through BMPR2 (Eichstaedt et al. 2016). Mutations in *BMPR2* are found in the majority of familial PAH, which segregates as an autosomal dominant with incomplete penetrance.

Missense mutations in *CReP* that destabilize its association with PP1c have been reported to lead to early-onset diabetes, along with growth retardation and microcephaly and learning disabilities, and liver pathologies (Abdulkarim et al. 2015; Kernohan et al. 2015; Mohammad et al. 2016). In islet β cells of the pancreas, loss of *CReP* leads to increased levels of P-eIF2α, which lowers insulin synthesis and secretion and sensitizes these cells to enhanced apoptosis in response to ER stress (Abdulkarim et al. 2015).

Mutations in genes encoding one of the five different subunits of eIF2B lead to vanishing white matter (VWM), or childhood ataxia with central nervous system hypomyelination, which features severe white matter abnormalities, including myelin and cystic degeneration (Leegwater et al. 2001; van der Knaap et al. 2002). The resulting lowered exchange of eIF2•GDP to the GTP-bound form is suggested to lead to some activation of the ISR independent of stress. When combined with stress induction of P-eIF2α, the VWM residue substitutions in eIF2B can enhance the amplitude of the ISR and alter the timing of the response, which is suggested to trigger the destructive features of the ISR (Richardson et al. 2004; Pavitt and Proud 2009).

It is important to note that some VWM mutations do not appear to alter eIF2B interactions with eIF2 or its guanine nucleotide exchange activity. Two related functions have been attributed to eIF2B, which could be adversely effected by VWM mutations (Jennings and Pavitt 2010;

Cite this article as *Cold Spring Harb Perspect Biol* doi: 10.1101/cshperspect.a032870

Jennings et al. 2013, 2017). During translation initiation, GTP associated with eIF2 is hydrolyzed by another translation factor, eIF5, which retains association with eIF2•GDP and thwarts spontaneous release of the nucleotide. eIF2B then promotes release of eIF5 from eIF2 before catalyzing the exchange of eIF2•GDP to the GTP-bound form. Additionally, eIF2B ensures that the phosphorylated version of eIF2 is not included in the eIF2•GTP•Met-tRNA$_i^{Met}$ complex. These additional eIF2B functions may help explain the complex decameric structure of this exchange factor and may be additional targets for disruption by VWM mutations.

In addition to the pathologies resulting from mutations in the ISR genes directly involved in the regulation of P-eIF2α or its effect on eIF2•GTP exchange, there have been reports of genetic disorders that alter stress activation of the eIF2α kinases and the adaptation functions of the ISR. For example, DNAJC3 (P58^{IPK}) is present in the ER lumen and directly aids the chaperone function of BiP by enhancing its ATPase activity and facilitating association of unfolded polypeptides to BiP (Rutkowski et al. 2007; Petrova et al. 2008). Mutations in *DNAJC3* were reported to cause diabetes and widespread neurodegeneration (Synofzik et al. 2014). Loss of DNAJC3 is suggested to disrupt BiP function, increasing the levels of unfolded protein in the ER. This disruption in protein homeostasis can chronically induce PERK phosphorylation of eIF2α, triggering apoptosis. IER3IP1 is another ER protein that is linked with regulation of PERK. Mutations that disrupt *IER3IP1* lead to pathologies related to Wolcott–Rallison syndrome, including neonatal diabetes, microcephalogy, and developmental delays, along with seizures (Abdel-Salam et al. 2012; Shalev et al. 2014). IER3IP1 has a putative G-patch domain found in RNA-associated proteins, and loss of IER3IP1 lowered activation of PERK and the ISR in cultured β islet cells exposed to ER stress, culminating in increased cell death (Sun and Ren 2017). Finally, elevated levels of P-eIF2α have been reported in the diseased brain tissues from Alzheimer's patients and from mouse models of Alzheimer's disease. Genetic deletion of *PERK* lowered the P-eIF2α induction and

translational control, and restored synaptic plasticity and memory in mice that expressed familial Alzheimer's disease-related mutations (Ma et al. 2013; Sossin and Costa-Mattioli 2017). Similar outcomes were observed for deletion of *GCN2* in the Alzheimer's disease model mice, which further supports the idea that aberrant induction of P-eIF2α is an underlying contributor to the pathophysiology of Alzheimer's disease.

THERAPEUTIC TARGETS IN THE ISR

Small molecules have been identified that thwart induction of translational control, or alternatively accentuate the ISR pathway. Those that block P-eIF2α induction and the ISR include PERK and GCN2 inhibitors (Robert et al. 2009; Axten et al. 2012; Harding et al. 2012). For example, the PERK inhibitor GSK2606414 blocks induction of P-eIF2α and interrupts translational control in response to ER stress in cultured cells. This is shown by the application of GSK2606414 to cultured islet β cells subjected to high levels of glucose, which sharply interfered with activation of the PERK portion of the UPR, culminating in rapid accumulation of misfolded insulin protein (Harding et al. 2012). A fluorinated analog of this small molecule, GSK2656157, has been optimized for preclinical development with therapeutic applications for cancer and neurodegenerative disorders (Axten et al. 2013). A cautionary note is that extended exposure to these PERK inhibitors alone can induce P-eIF2α, suggesting compensatory mechanisms that are, at least in part, independent of the ISR (Krishnamoorthy et al. 2014). Supporting this idea, these GSK molecules potently inhibit the protein kinase RIPK1, which functions in the tumor necrosis factor α (TNF-α) pathway, affecting inflammation and cell death (Rojas-Rivera et al. 2017).

ISRIB is another small molecule inhibitor of the ISR that was identified for its ability to block induction of *ATF4* translation in response to ER stress (Sidrauski et al. 2013). ISR does not block P-eIF2α induction per se, but rather stimulates the guanine nucleotide exchange activity of eIF2B, thus compensating for the inhibitory effect of P-eIF2α (Sekine et al. 2015; Sidrauski et al.

2015). As a consequence, ISRIB allows for retention of global translation and thwarts assembly of stress granules in response to stress. Synthesis of specific proteins and synaptic plasticity in the hippocampus are critical for the formation and maintenance of memory. By diminishing the ISR-dependent translation, treatment with ISRIB or genetic alterations that disrupt P-eIF2α induction in mice improve memory in a learning paradigm that requires long-term potentiation (Sidrauski et al. 2013). In contrast, long-term memory is impaired by the small molecule, salubrinal, which prevents PP1c dephosphorylation of P-eIF2α and sustains the ISR and translational control (Costa-Mattioli et al. 2007). Persistent activation of the ISR also occurs in traumatic brain injury, and treatment with ISRIB reverses the cognitive deficits associated with the hippocampus in two different injury models in mice (Chou et al. 2017).

As noted for salubrinal, some small molecules enhance and sustain the ISR. Salubrinal was first discovered in a screen for chemicals that protect cultured cells from pharmacologically induced ER stress (Boyce et al. 2005). Salubrinal affords protection in neurodegenerative model systems that are associated with the induced UPR (Sokka et al. 2007; Reijonen et al. 2008; Saxena et al. 2009; Colla et al. 2012). However, depending on the disease model, sustained P-eIF2α induction by salubrinal can have deleterious consequences (Moreno et al. 2012; Collier et al. 2015). Another strategy for small molecule activation of the ISR involves provoking a defined stress for targeted activation of an eIF2α kinase. Halofuginone is a potent inhibitor of prolyl-tRNA synthetase and rapidly activates GCN2 and the ISR (Keller et al. 2012). Prior treatment with halofuginone induces expression of stress-resistant proteins that protect against subsequent renal and hepatic ischemic injury in a mouse surgery reperfusion model (Peng et al. 2012).

LOOKING TO THE FUTURE: IMPORTANT UNRESOLVED QUESTIONS

As highlighted in this review, P-eIF2α regulates translation of individual gene transcripts that collectively contribute to global changes in protein synthesis and restoration of protein homeostasis. Given the central role of uORFs in the ISR, it is important to identify their mechanistic contributions to repression, resistance, and preferential mRNA translation. What are the varied mechanisms by which uORFs are bypassed in response to P-eIF2α? Furthermore, it is important to establish accurate predictive rules for uORF regulatory functions in translational expression. Many single-nucleotide polymorphisms (SNPs) have been identified in humans that alter potential uORFs (Calvo et al. 2009). Do these genetic variations alter ISR function in health and disease? Finally, how can our knowledge of the ISR be applied to clinical practice? Certainly, small molecules such as ISRIB have great therapeutic potential, but there are also challenges as disruptions in key elements of the ISR have the potential for altering cell adaptation and triggering death processes.

ACKNOWLEDGMENTS

This laboratory is supported by National Institutes of Health (NIH) Grants GM049164, DK109714, AI124723, and the Ralph W. and Grace M. Showalter Research Trust Fund.

REFERENCES

*Reference is also in this collection.

Abastado JP, Miller PF, Jackson BM, Hinnebusch AG. 1991. Suppression of ribosomal reinitiation at upstream open reading frames in amino acid-starved cells forms the basis of GCN4 translational control. *Mol Cell Biol* **11**: 486–496.

Abdel-Salam GM, Schaffer AE, Zaki MS, Dixon-Salazar T, Mostafa IS, Afifi HH, Gleeson JG. 2012. A homozygous IER3IP1 mutation causes microcephaly with simplified gyral pattern, epilepsy, and permanent neonatal diabetes syndrome (MEDS). *Am J Med Genet A* **158A**: 2788–2796.

Abdulkarim B, Nicolino M, Igoillo-Esteve M, Daures M, Romero S, Philippi A, Senee V, Lopes M, Cunha DA, Harding HP, et al. 2015. A missense mutation in PPP1R15B causes a syndrome including diabetes, short stature, and microcephaly. *Diabetes* **64**: 3951–3962.

Andreev DE, O'Connor PB, Fahey C, Kenny EM, Terenin IM, Dmitriev SE, Cormican P, Morris DW, Shatsky IN, Baranov PV. 2015. Translation of 5′ leaders is pervasive in genes resistant to eIF2 repression. *eLife* **4**: e03971.

Axten JM, Medina JR, Feng Y, Shu A, Romeril SP, Grant SW, Li WH, Heerding DA, Minthorn E, Mencken T, et al. 2012. Discovery of 7-methyl-5-(1-{[3-(trifluoromethyl) phenyl]acetyl}-2,3-dihydro-1H-indol-5-yl)-7H-pyrrolo [2,3-d]pyrimidin-4-amine (GSK2606414), a potent and selective first-in-class inhibitor of protein kinase R (PKR)-like endoplasmic reticulum kinase (PERK). *J Med Chem* **55**: 7193–7207.

Axten JM, Romeril SP, Shu A, Ralph J, Medina JR, Feng Y, Li WH, Grant SW, Heerding DA, Minthorn E, et al. 2013. Discovery of GSK2656157: An optimized PERK inhibitor selected for preclinical development. *ACS Med Chem Lett* **4**: 964–968.

Baird TD, Palam LR, Fusakio ME, Willy JA, Davis CM, McClintick JN, Anthony TG, Wek RC. 2014. Selective mRNA translation during eIF2 phosphorylation induces expression of IBTKα. *Mol Biol Cell* **25**: 1686–1697.

Bertolotti A, Zhang Y, Hendershot LM, Harding HP, Ron D. 2000. Dynamic interaction of BiP and ER stress transducers in the unfolded protein response. *Nat Cell Biol* **2**: 326–332.

Best DH, Sumner KL, Austin ED, Chung WK, Brown LM, Borczuk AC, Rosenzweig EB, Bayrak-Toydemir P, Mao R, Cahill BC, et al. 2014. EIF2AK4 mutations in pulmonary capillary hemangiomatosis. *Chest* **145**: 231–236.

Best DH, Sumner KL, Smith BP, Damjanovich-Colmenares K, Nakayama I, Brown LM, Ha Y, Paul E, Morris A, Jama MA, et al. 2017. EIF2AK4 mutations in patients diagnosed with pulmonary arterial hypertension. *Chest* **151**: 821–828.

Boyce M, Bryant KF, Jousse C, Long K, Harding HP, Scheuner D, Kaufman RJ, Ma D, Coen DM, Ron D, et al. 2005. A selective inhibitor of eIF2α dephosphorylation protects cells from ER stress. *Science* **307**: 935–939.

Calvo SE, Pagliarini DJ, Mootha VK. 2009. Upstream open reading frames cause widespread reduction of protein expression and are polymorphic among humans. *Proc Natl Acad Sci* **106**: 7507–7512.

Carrara M, Prischi F, Nowak PR, Kopp MC, Ali MM. 2015. Noncanonical binding of BiP ATPase domain to Ire1 and Perk is dissociated by unfolded protein CH1 to initiate ER stress signaling. *eLife* **4**: e03522.

Chambers JE, Dalton LE, Clarke HJ, Malzer E, Dominicus CS, Patel V, Moorhead G, Ron D, Marciniak SJ. 2015. Actin dynamics tune the integrated stress response by regulating eukaryotic initiation factor 2α dephosphorylation. *eLife* **4**: e04872.

Chen R, Rato C, Yan Y, Crespillo-Casado A, Clarke HJ, Harding HP, Marciniak SJ, Read RJ, Ron D. 2015. G-actin provides substrate-specificity to eukaryotic initiation factor 2α holophosphatases. *eLife* **4**: e04872.

Chou A, Krukowski K, Jopson T, Zhu PJ, Costa-Mattioli M, Walter P, Rosi S. 2017. Inhibition of the integrated stress response reverses cognitive deficits after traumatic brain injury. *Proc Natl Acad Sci* **114**: E6420–E6426.

Choy MS, Yusoff P, Lee IC, Newton JC, Goh CW, Page R, Shenolikar S, Peti W. 2015. Structural and functional analysis of the GADD34:PP1 eIF2α phosphatase. *Cell Rep* **11**: 1885–1891.

Colla E, Coune P, Liu Y, Pletnikova O, Troncoso JC, Iwatsubo T, Schneider BL, Lee MK. 2012. Endoplasmic reticulum stress is important for the manifestations of α-synucleinopathy in vivo. *J Neurosci* **32**: 3306–3320.

Collier AE, Wek RC, Spandau DF. 2015. Translational repression protects human keratinocytes from UVB-induced apoptosis through a discordant eIF2 kinase stress response. *J Invest Dermatol* **135**: 2502–2511.

Collier AE, Wek RC, Spandau DF. 2017. Human keratinocyte differentiation requires translational control by the eIF2α kinase GCN2. *J Invest Dermatol* **137**: 1924–1934.

Connor JH, Weiser DC, Li S, Hallenbeck JM, Shenolikar S. 2001. Growth arrest and DNA damage-inducible protein GADD34 assembles a novel signaling complex containing protein phosphatase 1 and inhibitor 1. *Mol Cell Biol* **21**: 6841–6850.

Costa-Mattioli M, Gobert D, Stern E, Gamache K, Colina R, Cuello C, Sossin W, Kaufman R, Pelletier J, Rosenblum K, et al. 2007. eIF2α phosphorylation bidirectionally regulates the switch from short- to long-term synaptic plasticity and memory. *Cell* **129**: 195–206.

Delepine M, Nicolino M, Barrett T, Golamaully M, Lathrop GM, Julier C. 2000. EIF2AK3, encoding translation initiation factor 2-α kinase 3, is mutated in patients with Wolcott–Rallison syndrome. *Nat Genet* **25**: 406–409.

Deng J, Harding H, Raught B, Gingras A, Berlanga J, Scheuner D, Kaufman R, Ron D, Sonenberg N. 2002. Activation of GCN2 in UV-irradiated cells inhibits translation. *Curr Biol* **12**: 1279–1286.

Dey S, Baird TD, Zhou D, Palam LR, Spandau DF, Wek RC. 2010. Both transcriptional regulation and translational control of ATF4 are central to the integrated stress response. *J Biol Chem* **285**: 33165–33174.

Dey S, Savant S, Teske BF, Hatzoglou M, Calkhoven CF, Wek RC. 2012. Transcriptional repression of ATF4 gene by CCAAT/enhancer-binding protein β (C/EBPβ) differentially regulates integrated stress response. *J Biol Chem* **287**: 21936–21949.

Dong J, Qiu H, Garcia-Barrio M, Anderson J, Hinnebusch AG. 2000. Uncharged tRNA activates GCN2 by displacing the protein kinase moiety from a bipartite tRNA-binding domain. *Mol Cell* **6**: 269–279.

Eichstaedt CA, Song J, Benjamin N, Harutyunova S, Fischer C, Grunig E, Hinderhofer K. 2016. EIF2AK4 mutation as "second hit" in hereditary pulmonary arterial hypertension. *Respir Res* **17**: 141.

Eyries M, Montani D, Girerd B, Perret C, Leroy A, Lonjou C, Chelghoum N, Coulet F, Bonnet D, Dorfmuller P, et al. 2014. EIF2AK4 mutations cause pulmonary veno-occlusive disease, a recessive form of pulmonary hypertension. *Nat Genet* **46**: 65–69.

Fusakio ME, Willy JA, Wang Y, Mirek ET, Al Baghdadi RJ, Adams CM, Anthony TG, Wek RC. 2016. Transcription factor ATF4 directs basal and stress-induced gene expression in the unfolded protein response and cholesterol metabolism in the liver. *Mol Biol Cell* **27**: 1536–1551.

Gardner BM, Walter P. 2011. Unfolded proteins are Ire1-activating ligands that directly induce the unfolded protein response. *Science* **333**: 1891–1894.

Goosens A, Dever TE, Pascual-Ahuir A, Serrano R. 2001. The protein kinase Gcn2p mediates sodium toxicity in yeast. *J Biol Chem* **276**: 30753–30760.

Han J, Back SH, Hur J, Lin YH, Gildersleeve R, Shan J, Yuan CL, Krokowski D, Wang S, Hatzoglou M, et al. 2013. ER-stress-induced transcriptional regulation increases protein synthesis leading to cell death. *Nat Cell Biol* **15**: 481–490.

Hansen MB, Mitchelmore C, Kjaerulff KM, Rasmussen TE, Pedersen KM, Jensen NA. 2002. Mouse *ATF5*: Molecular cloning of two novel mRNAs, genomic organization, and odorant sensory neuron localization. *Genomics* **80**: 344–350.

Harding HP, Zhang Y, Ron D. 1999. Protein translation and folding are coupled by an endoplasmic-reticulum-resident kinase. *Nature* **397**: 271–274.

Harding HP, Novoa I, Zhang Y, Zeng H, Wek R, Schapira M, Ron D. 2000a. Regulated translation initiation controls stress-induced gene expression in mammalian cells. *Mol Cell* **6**: 1099–1108.

Harding HP, Zhang Y, Bertolotti A, Zeng H, Ron D. 2000b. Perk is essential for translation regulation and cell survival during the unfolded protein response. *Mol Cell* **5**: 897–904.

Harding HP, Zhang Y, Zeng H, Novoa I, Lu PD, Calfon M, Sadri N, Yun C, Popko B, Paules R, et al. 2003. An integrated stress response regulates amino acid metabolism and resistance to oxidative stress. *Mol Cell* **11**: 619–633.

Harding HP, Zhang Y, Scheuner D, Chen JJ, Kaufman RJ, Ron D. 2009. Ppp1r15 gene knockout reveals an essential role for translation initiation factor 2α (eIF2α) dephosphorylation in mammalian development. *Proc Natl Acad Sci* **106**: 1832–1837.

Harding HP, Zyryanova AF, Ron D. 2012. Uncoupling proteostasis and development in vitro with a small molecule inhibitor of the pancreatic endoplasmic reticulum kinase, PERK. *J Biol Chem* **287**: 44338–44344.

Hinnebusch AG. 2005. Translational regulation of GCN4 and the general amino acid control of yeast. *Annu Rev Microbiol* **59**: 407–450.

Iacono M, Mignone F, Pesole G. 2005. uAUG and uORFs in human and rodent 5′ untranslated mRNAs. *Gene* **349**: 97–105.

* Ivanov P, Kedersha N, Anderson P. 2017. Stress granules and processing bodies in translational control. *Cold Spring Harb Perspect Biol* doi: 10.1101/cshperspect.a032813.

Jennings MD, Pavitt GD. 2010. eIF5 has GDI activity necessary for translational control by eIF2 phosphorylation. *Nature* **465**: 378–381.

Jennings MD, Zhou Y, Mohammad-Qureshi SS, Bennett D, Pavitt GD. 2013. eIF2B promotes eIF5 dissociation from eIF2•GDP to facilitate guanine nucleotide exchange for translation initiation. *Genes Dev* **27**: 2696–2707.

Jennings MD, Kershaw CJ, Adomavicius T, Pavitt GD. 2017. Fail-safe control of translation initiation by dissociation of eIF2α phosphorylated ternary complexes. *eLife* **6**: e24542.

Jiang HY, Wek RC. 2005. Gcn2 phosphorylation of eIF2α activates NF-κB in response to UV irradiation. *Biochem J* **385**: 371–380.

Jousse C, Bruhat A, Carraro V, Urano F, Ferrara M, Ron D, Fafournoux P. 2001. Inhibition of CHOP translation by a peptide encoded by an open reading frame localized in the chop 5′UTR. *Nucleic Acids Res* **29**: 4341–4351.

Jousse C, Oyadomari S, Novoa I, Lu PD, Zhang H, Harding HP, Ron D. 2003. Inhibition of a constitutive translation initiation factor 2α phosphatase, CReP, promotes survival of stressed cells. *J Cell Biol* **163**: 767–775.

Kedersha N, Ivanov P, Anderson P. 2013. Stress granules and cell signaling: More than just a passing phase? *Trends Biochem Sci* **38**: 494–506.

Keller TL, Zocco D, Sundrud MS, Hendrick M, Edenius M, Yum J, Kim YJ, Lee HK, Cortese JF, Wirth DF, et al. 2012.

Halofuginone and other febrifugine derivatives inhibit prolyl-tRNA synthetase. *Nat Chem Biol* **8**: 311–317.

Kernohan KD, Tetreault M, Liwak-Muir U, Geraghty MT, Qin W, Venkateswaran S, Davila J; Care4Rare Canada Consortium, Holcik M, Majewski J, et al. 2015. Homozygous mutation in the eukaryotic translation initiation factor 2α phosphatase gene, *PPP1R15B*, is associated with severe microcephaly, short stature and intellectual disability. *Hum Mol Genet* **24**: 6293–6300.

Keys HR. 2016. "A multi-level approach to understanding the regulation of translation initiation." PhD thesis, Massachusetts Institute of Technology, Cambridge, MA.

Keys HR, Sabatini DM. 2017. Juxtacap nucleotide sequence modulates eIF4E binding and translation. *bioRxiv* doi: 10.1101/165142.

Krishnamoorthy J, Rajesh K, Mirzajani F, Kesoglidou P, Papadakis AI, Koromilas AE. 2014. Evidence for eIF2α phosphorylation-independent effects of GSK2656157, a novel catalytic inhibitor of PERK with clinical implications. *Cell Cycle* **13**: 801–806.

Lageix S, Rothenburg S, Dever TE, Hinnebusch AG. 2014. Enhanced interaction between pseudokinase and kinase domains in Gcn2 stimulates eIF2α phosphorylation in starved cells. *PLoS Genet* **10**: e1004326.

Lageix S, Zhang J, Rothenburg S, Hinnebusch AG. 2015. Interaction between the tRNA-binding and C-terminal domains of yeast Gcn2 regulates kinase activity in vivo. *PLoS Genet* **11**: e1004991.

Lee YY, Cevallos RC, Jan E. 2009. An upstream open reading frame regulates translation of GADD34 during cellular stresses that induce eIF2α phosphorylation. *J Biol Chem* **284**: 6661–6673.

Lee S, Liu B, Lee S, Huang SX, Shen B, Qian SB. 2012. Global mapping of translation initiation sites in mammalian cells at single-nucleotide resolution. *Proc Natl Acad Sci* **109**: E2424–E2432.

Leegwater PA, Vermeulen G, Konst AA, Naidu S, Mulders J, Visser A, Kersbergen P, Mobach D, Fonds D, van Berkel CG, et al. 2001. Subunits of the translation initiation factor eIF2B are mutant in leukoencephalopathy with vanishing white matter. *Nat Genet* **29**: 383–388.

Lehman SL, Cerniglia GJ, Johannes GJ, Ye J, Ryeom S, Koumenis C. 2015. Translational upregulation of an individual p21Cip1 transcript variant by GCN2 regulates cell proliferation and survival under nutrient stress. *PLoS Genet* **11**: e1005212.

Lu PD, Harding HP, Ron D. 2004. Translation reinitiation at alternative open reading frames regulates gene expression in an integrated stress response. *J Cell Biol* **167**: 27–33.

Ma K, Vattem KM, Wek RC. 2002. Dimerization and release of molecular chaperone inhibition facilitate activation of eukaryotic initiation factor-2 kinase in response to endoplasmic reticulum stress. *J Biol Chem* **277**: 18728–18735.

Ma T, Trinh MA, Wexler AJ, Bourbon C, Gatti E, Pierre P, Cavener DR, Klann E. 2013. Suppression of eIF2α kinases alleviates Alzheimer's disease-related plasticity and memory deficits. *Nat Neurosci* **16**: 1299–1305.

Marciniak SJ, Ron D. 2006. Endoplasmic reticulum stress signaling in disease. *Physiol Rev* **86**: 1133–1149.

Marciniak SJ, Yun CY, Oyadomari S, Novoa I, Zhang Y, Jungreis R, Nagata K, Harding HP, Ron D. 2004. CHOP induces death by promoting protein synthesis and oxidation in the stressed endoplasmic reticulum. *Genes Dev* **18**: 3066–3077.

Marton MJ, Crouch D, Hinnebusch AG. 1993. GCN1, a translational activator of GCN4 in Saccharomyces cerevisiae, is required for phosphorylation of eukaryotic translation initiation factor 2 by protein kinase GCN2. *Mol Cell Biol* **13**: 3541–3556.

Marton MJ, Vazquez de Aldana CR, Qiu H, Chakraburtty K, Hinnebusch AG. 1997. Evidence that GCN1 and GCN20, translational regulators of GCN4, function on elongating ribosomes in activation of eIF2α kinase GCN2. *Mol Cell Biol* **17**: 4474–4489.

Meyer KD, Patil DP, Zhou J, Zinoviev A, Skabkin MA, Elemento O, Pestova TV, Qian SB, Jaffrey SR. 2015. 5′ UTR m⁶A promotes cap-independent translation. *Cell* **163**: 999–1010.

Mohammad S, Wolfe LA, Stobe P, Biskup S, Wainwright MS, Melin-Aldana H, Malladi P, Muenke M, Gahl WA, Whitington PF. 2016. Infantile cirrhosis, growth impairment, and neurodevelopmental anomalies associated with deficiency of PPP1R15B. *J Pediatr* **179**: 144–149.e2.

Moreno JA, Radford H, Peretti D, Steinert JR, Verity N, Martin MG, Halliday M, Morgan J, Dinsdale D, Ortori CA, et al. 2012. Sustained translational repression by eIF2α-P mediates prion neurodegeneration. *Nature* **485**: 507–511.

Munn DH, Sharma MD, Baban B, Harding HP, Zhang Y, Ron D, Mellor AL. 2005. GCN2 kinase in T cells mediates proliferative arrest and anergy induction in response to indoleamine 2,3-dioxygenase. *Immunity* **22**: 633–642.

Novoa I, Zeng H, Harding HP, Ron D. 2001. Feedback inhibition of the unfolded protein response by GADD34-mediated dephosphorylation of eIF2α. *J Cell Biol* **153**: 1011–1022.

O'Connor T, Sadleir KR, Maus E, Velliquette RA, Zhao J, Cole SL, Eimer WA, Hitt B, Bembinster LA, Lammich S, et al. 2008. Phosphorylation of the translation initiation factor eIF2α increases BACE1 levels and promotes amyloidogenesis. *Neuron* **60**: 988–1009.

Oslowski CM, Urano F. 2011. The binary switch that controls the life and death decisions of ER stressed β cells. *Curr Opin Cell Biol* **23**: 207–215.

Palam LR, Baird TD, Wek RC. 2011. Phosphorylation of eIF2 facilitates ribosomal bypass of an inhibitory upstream ORF to enhance CHOP translation. *J Biol Chem* **286**: 10939–10949.

Pavitt GD, Proud CG. 2009. Protein synthesis and its control in neuronal cells with a focus on vanishing white matter disease. *Biochem Soc Trans* **37**: 1298–1310.

Peng W, Robertson L, Gallinetti J, Mejia P, Vose S, Charlip A, Chu T, Mitchell JR. 2012. Surgical stress resistance induced by single amino acid deprivation requires Gcn2 in mice. *Sci Transl Med* **4**: 118ra111.

Pereira CM, Sattlegger E, Jiang HY, Longo BM, Jaqueta CB, Hinnebusch AG, Wek RC, Mello LE, Castilho BA. 2005. IMPACT, a protein preferentially expressed in the mouse brain, binds GCN1 and inhibits GCN2 activation. *J Biol Chem* **280**: 28316–28323.

Petrova K, Oyadomari S, Hendershot LM, Ron D. 2008. Regulated association of misfolded endoplasmic reticu-

lum lumenal proteins with P58/DNAJc3. *EMBO J* **27**: 2862–2872.

Pincus D, Chevalier MW, Aragon T, van Anken E, Vidal SE, El-Samad H, Walter P. 2010. BiP binding to the ER-stress sensor Ire1 tunes the homeostatic behavior of the unfolded protein response. *PLoS Biol* **8**: e1000415.

Raveh-Amit H, Maissel A, Poller J, Marom L, Elroy-Stein O, Shapira M, Livneh E. 2009. Translational control of protein kinase Cη by two upstream open reading frames. *Mol Cell Biol* **29**: 6140–6148.

Reid DW, Nicchitta CV. 2015. Diversity and selectivity in mRNA translation on the endoplasmic reticulum. *Nat Rev Mol Cell Biol* **16**: 221–231.

Reid DW, Chen Q, Tay AS, Shenolikar S, Nicchitta CV. 2014. The unfolded protein response triggers selective mRNA release from the endoplasmic reticulum. *Cell* **158**: 1362–1374.

Reijonen S, Putkonen N, Norremolle A, Lindholm D, Korhonen L. 2008. Inhibition of endoplasmic reticulum stress counteracts neuronal cell death and protein aggregation caused by N-terminal mutant huntingtin proteins. *Exp Cell Res* **314**: 950–960.

Resch AM, Ogurtsov AY, Rogozin IB, Shabalina SA, Koonin EV. 2009. Evolution of alternative and constitutive regions of mammalian 5′UTRs. *BMC Genomics* **10**: 162.

Richardson JP, Mohammad SS, Pavitt GD. 2004. Mutations causing childhood ataxia with central nervous system hypomyelination reduce eukaryotic initiation factor 2B complex formation and activity. *Mol Cell Biol* **24**: 2352–2363.

Robert F, Williams C, Yan Y, Donohue E, Cencic R, Burley SK, Pelletier J. 2009. Blocking UV-induced eIF2α phosphorylation with small molecule inhibitors of GCN2. *Chem Biol Drug Des* **74**: 57–67.

Rojas-Rivera D, Delvaeye T, Roelandt R, Nerinckx W, Augustyns K, Vandenabeele P, Bertrand MJM. 2017. When PERK inhibitors turn out to be new potent RIPK1 inhibitors: Critical issues on the specificity and use of GSK2606414 and GSK2656157. *Cell Death Differ* **24**: 1100–1110.

Roybal CN, Hunsaker LA, Barbash O, Vander Jagt DL, Abcouwer SF. 2005. The oxidative stressor arsenite activates vascular endothelial growth factor mRNA transcription by an ATF4-dependent mechanism. *J Biol Chem* **280**: 20331–20339.

Rutkowski DT, Kang SW, Goodman AG, Garrison JL, Taunton J, Katze MG, Kaufman RJ, Hegde RS. 2007. The role of p58IPK in protecting the stressed endoplasmic reticulum. *Mol Biol Cell* **18**: 3681–3691.

Sattlegger E, Swanson MJ, Ashcraft EA, Jennings JL, Fekete RA, Link AJ, Hinnebusch AG. 2004. YIH1 is an actin-binding protein that inhibits protein kinase GCN2 and impairs general amino acid control when overexpressed. *J Biol Chem* **279**: 29952–29962.

Saxena S, Cabuy E, Caroni P. 2009. A role for motoneuron subtype-selective ER stress in disease manifestations of FALS mice. *Nat Neurosci* **12**: 627–636.

Sekine Y, Zyryanova A, Crespillo-Casado A, Fischer PM, Harding HP, Ron D. 2015. Stress responses. Mutations in a translation initiation factor identify the target of a memory-enhancing compound. *Science* **348**: 1027–1030.

Senée V, Vattem KM, Delépine M, Rainbow L, Haton C, Lecoq A, Shaw N, Robert J-J, Rooman R, Diatloff-Zito C, et al. 2004. Wolcott–Rallison syndrome: Clinical, genetic, and functional study of *EIF2AK3* mutations, and suggestion of genetic heterogeneity. *Diabetes* 53: 1876–1883.

Shalev SA, Tenenbaum-Rakover Y, Horovitz Y, Paz VP, Ye H, Carmody D, Highland HM, Boerwinkle E, Hanis CL, Muzny DM, et al. 2014. Microcephaly, epilepsy, and neonatal diabetes due to compound heterozygous mutations in IER3IP1: Insights into the natural history of a rare disorder. *Pediatr Diabetes* 15: 252–256.

Shi Y, Vattem KM, Sood R, An J, Liang J, Stramm L, Wek RC. 1998. Identification and characterization of pancreatic eukaryotic initiation factor 2 α-subunit kinase, PEK, involved in translation control. *Mol Cell Biol* 18: 7499–7509.

Sidrauski C, Acosta-Alvear D, Khoutorsky A, Vedantham P, Hearn BR, Li H, Gamache K, Gallagher CM, Ang KK, Wilson C, et al. 2013. Pharmacological brake-release of mRNA translation enhances cognitive memory. *eLife* 2: e00498.

Sidrauski C, McGeachy AM, Ingolia NT, Walter P. 2015. The small molecule ISRIB reverses the effects of eIF2α phosphorylation on translation and stress granule assembly. *eLife* 4: e05033.

Silva RC, Sattlegger E, Castilho BA. 2016. Perturbations in actin dynamics reconfigure protein complexes that modulate GCN2 activity and promote an eIF2 response. *J Cell Sci* 129: 4521–4533.

Sokka AL, Putkonen N, Mudo G, Pryazhnikov E, Reijonen S, Khiroug L, Belluardo N, Lindholm D, Korhonen L. 2007. Endoplasmic reticulum stress inhibition protects against excitotoxic neuronal injury in the rat brain. *J Neurosci* 27: 901–908.

* Sossin W, Costa-Mattioli M. 2017. Translational control in the brain in health and disease. *Cold Spring Harb Perspect Biol* doi: 10.1101/cshperspect.a032912.

Starck SR, Tsai JC, Chen K, Shodiya M, Wang L, Yahiro K, Martins-Green M, Shastri N, Walter P. 2016. Translation from the 5′ untranslated region shapes the integrated stress response. *Science* 351: aad3867.

Sun J, Ren D. 2017. IER3IP1 deficiency leads to increased β-cell death and decreased β-cell proliferation. *Oncotarget* 8: 56768–56779.

Synofzik M, Haack TB, Kopajtich R, Gorza M, Rapaport D, Greiner M, Schonfeld C, Freiberg C, Schorr S, Holl RW, et al. 2014. Absence of BiP co-chaperone DNAJC3 causes diabetes mellitus and multisystemic neurodegeneration. *Am J Hum Genet* 95: 689–697.

van der Knaap MS, Leegwater PA, Konst AA, Visser A, Naidu S, Oudejans CB, Schutgens RB, Pronk JC. 2002. Mutations in each of the five subunits of translation initiation factor eIF2B can cause leukoencephalopathy with vanishing white matter. *Ann Neurol* 51: 264–270.

van Vliet AR, Giordano F, Gerlo S, Segura I, Van Eygen S, Molenberghs G, Rocha S, Houcine A, Derua R, Verfaillie T, et al. 2017. The ER stress sensor PERK coordinates ER-plasma membrane contact site formation through interaction with Filamin-A and F-actin remodeling. *Mol Cell* 65: 885–899.

Vattem KM, Wek RC. 2004. Reinitiation involving upstream open reading frames regulates *ATF4* mRNA translation in mammalian cells. *Proc Natl Acad Sci* 101: 11269–11274.

Walter P, Ron D. 2011. The unfolded protein response: From stress pathway to homeostatic regulation. *Science* 334: 1081–1086.

Wang X, Zhao BS, Roundtree IA, Lu Z, Han D, Ma H, Weng X, Chen K, Shi H, He C. 2015. N^6-methyladenosine modulates messenger RNA translation efficiency. *Cell* 161: 1388–1399.

Watatani Y, Ichikawa K, Nakanishi N, Fujimoto M, Takeda H, Kimura N, Hirose H, Takahashi S, Takahashi Y. 2008. Stress-induced translation of ATF5 mRNA is regulated by the 5′-untranslated region. *J Biol Chem* 283: 2543–2553.

Wek RC, Jackson BM, Hinnebusch AG. 1989. Juxtaposition of domains homologous to protein kinases and histidyl-tRNA synthetases in GCN2 protein suggests a mechanism for coupling *GCN4* expression to amino acid availability. *Proc Natl Acad Sci* 86: 4579–4583.

Wek SA, Zhu S, Wek RC. 1995. The histidyl-tRNA synthetase-related sequence in eIF-2 α protein kinase GCN2 interacts with tRNA and is required for activation in response to starvation for different amino acids. *Mol Cell Biol* 15: 4497–4506.

Willy JA, Young SK, Mosley AL, Gawrieh S, Stevens JL, Masuoka HC, Wek RC. 2017. Function of inhibitor of brutons tyrosine kinase isoform α (IBTKα) in nonalcoholic steatohepatitis links autophagy and the unfolded protein response. *J Biol Chem* 292: 14050–14065.

Wu S, Tan M, Hu Y, Wang J-L, Scheuner D, Kaufman RJ. 2004. Ultraviolet light activates NFκB through translation inhibition of IκBa synthesis. *J Biol Chem* 279: 24898–24902.

Yaman I, Fernandez J, Liu H, Caprara M, Komar AA, Koromilas AE, Zhou L, Snider MD, Scheuner D, Kaufman RJ, et al. 2003. The zipper model of translational control: A small upstream ORF is the switch that controls structural remodeling of an mRNA leader. *Cell* 113: 519–531.

Yang R, Wek SA, Wek RC. 2000. Glucose limitation induces *GCN4* translation by activation of Gcn2 protein kinase. *Mol Cell Biol* 20: 2706–2717.

Young SK, Willy JA, Wu C, Sachs MS, Wek RC. 2015. Ribosome reinitiation directs gene-specific translation and regulates the integrated stress response. *J Biol Chem* 290: 28257–28271.

Young SK, Baird TD, Wek RC. 2016a. Translation regulation of the glutamyl-prolyl-tRNA synthetase gene *EPRS* through bypass of upstream open reading frames with noncanonical initiation codons. *J Biol Chem* 291: 10824–10835.

Young SK, Palam LR, Wu C, Sachs MS, Wek RC. 2016b. Ribosome elongation stall directs gene-specific translation in the integrated stress response. *J Biol Chem* 291: 6546–6558.

Zaborske JM, Narasimhan J, Jiang L, Wek SA, Dittmar KA, Freimoser F, Pan T, Wek RC. 2009. Genome-wide analysis of tRNA charging and activation of the eIF2 kinase Gcn2p. *J Biol Chem* 284: 25254–25267.

Zhou D, Palam LR, Jiang L, Narasimhan J, Staschke KA, Wek RC. 2008. Phosphorylation of eIF2 directs ATF5 translational control in response to diverse stress conditions. *J Biol Chem* 283: 7064–7073.

Zhou J, Wan J, Gao X, Zhang X, Jaffrey SR, Qian SB. 2015. Dynamic m^6A mRNA methylation directs translational control of heat shock response. *Nature* 526: 591–594.

Cite this article as *Cold Spring Harb Perspect Biol* doi: 10.1101/cshperspect.a032870

Phosphorylation and Signal Transduction Pathways in Translational Control

Christopher G. Proud

Nutrition & Metabolism, South Australian Health & Medical Research Institute, North Terrace, Adelaide SA5000, Australia; and School of Biological Sciences, University of Adelaide, Adelaide SA5000, Australia

Correspondence: christopher.proud@sahmri.com

Protein synthesis, including the translation of specific messenger RNAs (mRNAs), is regulated by extracellular stimuli such as hormones and by the levels of certain nutrients within cells. This control involves several well-understood signaling pathways and protein kinases, which regulate the phosphorylation of proteins that control the translational machinery. These pathways include the mechanistic target of rapamycin complex 1 (mTORC1), its downstream effectors, and the mitogen-activated protein (MAP) kinase (extracellular ligand-regulated kinase [ERK]) signaling pathway. This review describes the regulatory mechanisms that control translation initiation and elongation factors, in particular the effects of phosphorylation on their interactions or activities. It also discusses current knowledge concerning the impact of these control systems on the translation of specific mRNAs or subsets of mRNAs, both in physiological processes and in diseases such as cancer.

The control of protein synthesis plays key roles in cell growth and proliferation and in many other processes, via shaping the cellular proteome. The importance of translational control is underscored, for example, by the lack of concordance between the transcriptome and the proteome—which, among other factors, reflects the differing efficiencies with which different messenger RNAs (mRNAs) are translated into polypeptides (Schwanhausser et al. 2011; Li et al. 2014)—and the role of dysregulation of translation in human diseases, including cancer (Robichaud et al. 2018a), virus infection (Stern-Ginossar et al. 2018), and many others (Tahmasebi et al. 2018). Furthermore, since mRNA translation consumes a substantial fraction of cellular metabolic energy and an even higher proportion of amino acids, it is crucial that cells match the rate of protein synthesis to the availability of nutrients.

Accordingly, sophisticated control mechanisms exist to allow extracellular stimuli (e.g., hormones, growth factors), intracellular metabolites (essential amino acids, nucleotides) and cues, such as energy status, to regulate protein synthesis. These mechanisms include control of the translation of specific mRNAs, via signaling pathways that impinge on components of the initiation or elongation machinery, which is mainly achieved through protein kinases that phosphorylate and thereby regulate translation initiation or elongation factors, and certain other proteins.

CAVEATS

Studies on the role of specific signaling pathways or their target phosphoproteins rely heavily on the use of small molecule inhibitors, especially of protein kinases, and on testing proteins in which a given phosphorylation site has been mutated to one that cannot be phosphorylated or a negatively charged one, which may mimic a phosphorylated residue. There are important caveats to both approaches. Compounds that inhibit protein kinases are often not specific for a single type of enzyme because the ATP-binding sites of many kinases are similar. For example, the compounds often used as inhibitors of the mitogen-activated protein (MAP) kinase-interacting kinases (MNKs, which phosphorylate eukaryotic initiation factor [eIF]4E), CGP57380 (Tschopp et al. 2000), and cercosporamide (Konicek et al. 2011), affect other protein kinases (Bain et al. 2007; Beggs et al. 2015), several of them more potently than their effects on MNK1 and MNK2. It is, therefore, important to test chemically distinct entities that affect the kinase of interest and, wherever possible, to complement such work with genetic studies using cells in which the enzyme of interest has been knocked out. Genetic studies have been made much more tractable by the advent of clustered regularly interspaced short palindromic repeats (CRISPR)/Cas9 genome editing, which can be applied to cell lines as well as to studies in animals.

Rapamycin is a macrolide compound that is a highly specific inhibitor of the protein kinase complex termed mechanistic (or mammalian) target of rapamycin complex 1 (mTORC1). Rapamycin is specific largely because it does not target the active site of mTOR, but instead binds (together with its partner, FKBP12, an immunophilin) to another domain, thereby partially occluding access to the active site for some protein substrates (Yang et al. 2013). Although in the short term its effects are limited to inhibiting mTORC1, longer-term exposure of cells to rapamycin also impairs mTORC2 signaling by interfering with assembly of this complex (Sarbassov et al. 2006).

The effects of mutating a phosphorylation site of interest must also be treated with caution. Nonphosphorylatable mutants are generally created by mutating the relevant serine or threonine to alanine, which, because it lacks a hydroxyl group, cannot accept a phosphate. However, alanine also differs from serine and threonine in being unable to form H-bonds.

Conversely, phosphomimetic mutants of Ser or Thr to Glu or Asp may sometimes mimic the addition of a phosphate, but since the geometry (tetrahedral vs. planar) and charge (double vs. single) of phospho-Ser/Thr and Glu/Asp differ, phosphomimicry is by no means guaranteed.

mTORC1 SIGNALING

One major pathway that regulates the translational machinery in several ways is the mTOR pathway. mTOR forms two types of multisubunit complexes, mTORC1 and mTORC2. mTORC1 phosphorylates proteins that directly regulate the translational apparatus as well as protein kinases that phosphorylate and regulate translation factors (Saxton and Sabatini 2017).

In addition to mTOR, mTORC1 contains Raptor, mLst8, and Rheb, a GTPase (Ben-Sahra and Manning 2017). Raptor and Rheb are found only in mTORC1, not in mTORC2. Raptor binds substrate proteins, conferring specificity, whereas Rheb activates mTORC1 when it is bound to GTP. Rheb provides a key link to extracellular stimuli through the phosphatidylinositide 3-kinase (PI3K) or classical MAP kinase (extracellular ligand-regulated kinase [ERK]) signaling pathways (Fig. 1) (Saxton and Sabatini 2017). Rheb is regulated by the tuberous sclerosis complex (TSC), which includes the proteins TSC1, TSC2, and TBC1D7 (Dibble et al. 2012). TSC2 acts as a GTPase activator protein for Rheb, promoting its conversion to its inactive GDP-bound state. Protein kinases within or downstream of these pathways, Akt (also called protein kinase B [PKB]), ERK, or the ERK-substrates termed RSKs, phosphorylate the TSC complex and impair its ability to promote GTP hydrolysis on Rheb. This allows stimuli that switch on PI3K or ERK signaling to enhance the levels of Rheb-GTP and thereby activate mTORC1.

Several components of the PI3K and ERK pathways undergo gain-of-function oncogenic mutations in cancers, whereas others are tumor suppressors whose function is lost in some cancers. These pathways, and hence mTORC1 signaling, are constitutively activated in a high

Cite this article as *Cold Spring Harb Perspect Biol* doi: 10.1101/cshperspect.a033050

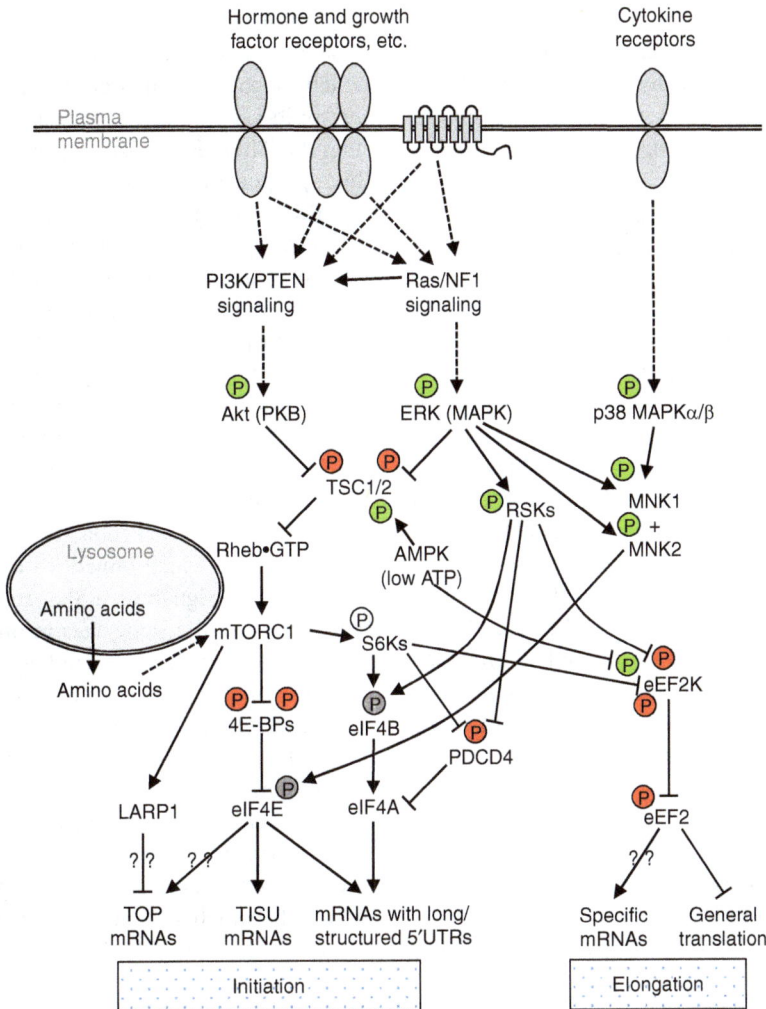

Figure 1. Major signaling connections to the translational machinery. This simplified schematic shows the main signaling links to the translational machinery. Solid lines depict direct links and dashed lines indirect ones (involving multiple components, for example). (P) indicates phosphorylation events, green indicates activation, red indicates an inhibitory phosphorylation, and gray denotes that the functional consequence is unclear. Not all the components or phosphorylation sites mentioned in the text are shown. Question marks denote some of the open questions described in the text.

percentage of human cancers, estimated at 70% or higher (Saxton and Sabatini 2017).

mTORC1 signaling is also activated by amino acids, at the surface of the lysosome (Fig. 1), by sophisticated sensing and signaling mechanisms involving a distinct set of GTPases, the Rags (Wolfson and Sabatini 2017). Interestingly, another nutrient-regulated signaling pathway, the AMP-activated protein kinase (AMPK), also impinges on mTORC1 at the lysosomal

surface, where AMPK can be activated (Lin and Hardie 2017). This happens, for example, when cellular ATP levels fall and levels of ADP, or especially AMP, rise. Recent studies have identified an additional, novel link between intermediary metabolism and the control of AMPK, whereby levels of the glycolytic intermediate fructose 1,6-bisphosphate also control AMPK activity (Lin and Hardie 2017). When AMPK is switched on, it inactivates mTORC1 by phosphorylating TSC2 and/or Raptor, thereby restraining energy-using biosynthetic processes that are normally driven by mTORC1, including protein and ribosome biosynthesis (Ben-Sahra and Manning 2017).

mTORC1 PHOSPHORYLATES PROTEINS THAT REGULATE THE TRANSLATIONAL MACHINERY

Among the many substrates for mTORC1, two classes are particularly relevant for this review. The first class comprises the small phosphoproteins termed eIF4E-binding proteins (4E-BPs) that bind to and sequester eIF4E (Figs. 1 and 2) (Merrick and Pavitt 2018). ("eIF4E" is used here to mean eIF4E1; its paralogs bind to 4E-BPs less strongly [in the case of eIF4E2, also termed eIF4E-homologous protein, 4EHP] or not at all [eIF4E3]) (Joshi et al. 2004). The second class comprises the S6 kinases (S6Ks; Fig. 2), which are named for their ability to phosphorylate ribosomal protein (rp) S6, a component of the small ribosomal subunit.

MAP KINASE SIGNALING

The MAP kinase (ERK) pathway regulates mTORC1, for example, through the phosphorylation of TSC1/2 (Huang and Manning 2008)

or Raptor (Carriere et al. 2011). This pathway is activated by a range of extracellular stimuli and by oncogenic mutations (e.g., in the GTPase Ras or its GTPase-activator protein, neurofibromin 1). Thus, ERK signaling is activated in a high proportion of cancers.

ERK also phosphorylates and activates several other protein kinases, including the RSKs, whose original name (ribosomal S6 kinases) may cause confusion with the S6Ks (Fig. 1). They phosphorylate S6 at a subset of the sites targeted by S6Ks reflecting the fact that S6Ks and RSKs (as well as Akt) act at similar, but not identical, consensus recognition sequences. RSKs and S6Ks phosphorylate rpS6 on Ser235/236, whereas S6Ks additionally phosphorylate Ser240 and Ser244 (Pende et al. 2004), phosphorylation of which provides a useful readout of mTORC1 signaling. RSKs and S6Ks also phosphorylate the same sites in other proteins that regulate translation, including eIF4B (Shahbazian et al. 2006) and the eEF2 kinase (eEF2K) that regulates elongation (Wang et al. 2001).

ERKs are not the only members of the MAP kinase family relevant to control of the translational machinery; others include the p38 MAP kinases, which are turned on by distinct stimuli (such as cytokines) and include p38 MAPKs α and β, which can activate MNK1 (Fig. 1) (Waskiewicz et al. 1997; Wang et al. 1998). p38

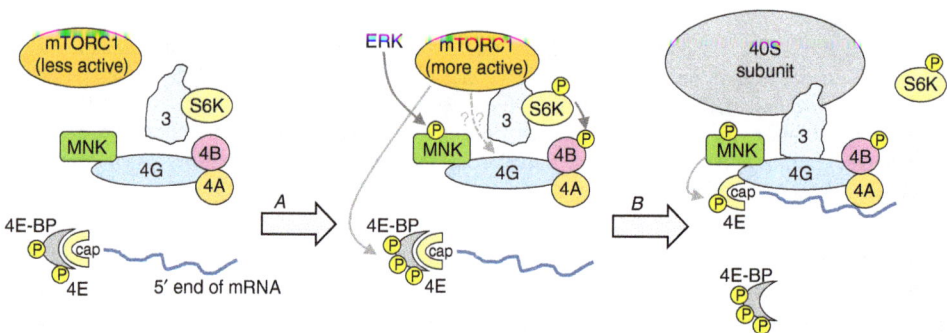

Figure 2. Signaling links to the cap-binding initiation machinery. S6 kinase (S6K) indicates S6 kinases 1 and 2, and MAP kinase-interacting kinase (MNK) denotes MNK1/2. Not all phosphorylation events are depicted. Thin solid gray lines denote direct phosphorylation; the dashed gray line indicates that mechanistic target of rapamycin complex 1 (mTORC1) signaling can promote eukaryotic initiation factor (eIF)4G/eIF3 binding. Step *A*, stimulation of cells (e.g., with insulin or a growth factor), leads to the activation of kinases including mTORC1, MNKs, and S6Ks, leading to Step *B* where different protein:protein interactions occur and assembly of eIF4F, together with eIF3 and 40S ribosomal subunits. Please see the text for further information.

Cite this article as *Cold Spring Harb Perspect Biol* doi: 10.1101/cshperspect.a033050

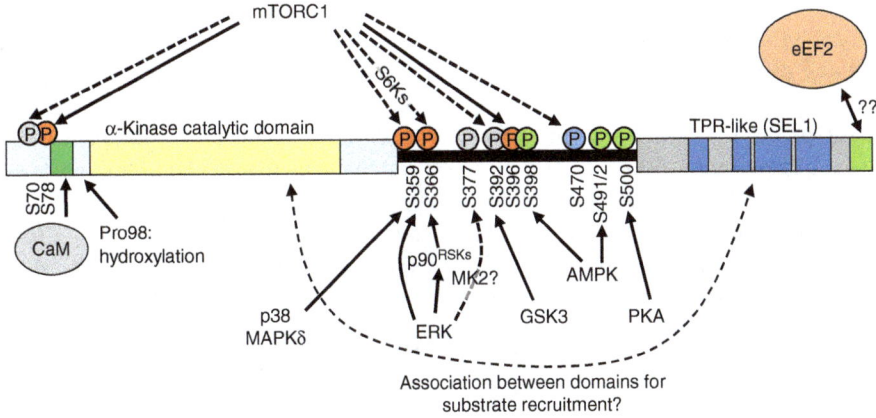

Figure 3. Regulation of eukaryotic elongation factor 2 (eEF2) kinase. The principal domains of eEF2K are depicted schematically, not to scale, including the CaM-binding site and a region at the extreme carboxyl terminus, which is required for eEF2K to phosphorylate eEF2 and may interact directly with that protein. Known phosphorylation sites are shown; orange for inhibitory sites, green for one that activates eEF2K, and gray for ones whose function is not clear or do not affect activity. Solid lines indicate direct phosphorylation; dashed ones denote indirect or unclear connections.

MAPK δ (whose control is poorly understood) also phosphorylates eEF2K (see below and Dever et al. 2018; Fig. 3).

PHOSPHORYLATION OF COMPONENTS OF THE INITIATION MACHINERY

Several eIFs are subject to regulatory phosphorylation. The first example to be discovered was the α subunit of the trimer eIF2. This phosphorylation event inhibits general translation and the inputs to this system are primarily intracellular stresses rather than extracellular stimuli (Wek 2018).

eIF2 is active only when bound (via its γ subunit) to GTP. The GTP is hydrolyzed during initiation. Because regeneration of active eIF2•GTP is slow, an additional factor is required, the GDP-dissociation stimulator protein (Williams et al. 2001), eIF2B, also referred to as a guanine nucleotide exchange factor (GEF). eIF2B is a decamer of five subunits ($\alpha_2\beta_2\gamma_2\delta_2\varepsilon_2$) (Gordiyenko et al. 2014; Wortham et al. 2014). Its largest subunit, ε, which houses the catalytic domain (Gomez et al. 2002), is subject to phosphorylation at several sites. These include Ser540 (in human eIF2Bε), which is a substrate

for GSK3 (Welsh et al. 1998), a kinase that is inactivated by stimuli such as insulin, which stimulate protein synthesis. Because phosphorylation of this site impairs eIF2B's GEF function (Welsh et al. 1998), this link provides a potential mechanism by which agents such as insulin can activate translation initiation (alongside a number of other events discussed below).

mTORC1 can interact with eIF3 (Holz et al. 2005; Harris et al. 2006), as can its substrate S6K1 (Holz et al. 2005), which in turn phosphorylates components of initiation complexes, such as eIF4B (Figs. 1 and 2). mTORC1 signaling also leads to increased binding of eIF3 to eIF4G, which is induced by insulin (Harris et al. 2006). As eIF3 interacts with the 40S subunit, this interaction likely helps to promote recruitment of ribosomes to eIF4F and thus to mRNAs, although it remains unclear how mTOR modulates eIF3/eIF4G binding.

mTORC1 phosphorylates the 4E-BPs, impairing their ability to interact with eIF4E (Figs. 1 and 2). Phosphorylation occurs at multiple sites, Thr37, Thr46, Ser65, and Thr70, in human 4E-BP1. Phosphorylation of the first two sites promotes subsequent phosphorylation at the other sites, which in turn seem to be most important for the release of 4E-BP1 from eIF4E. mTORC1 can phosphorylate all four sites. The properties and regulation of 4E-BP2 and 3 are much less well studied than 4E-BP1. Binding to a 4E-BP prevents eIF4E from interacting with the scaffold protein eIF4G, which in turn interacts with several other proteins involved in translation initiation (Fig. 2) (Gingras et al. 1999). The binding sites on eIF4E for 4E-BPs and eIF4G overlap, and indeed these proteins possess similar linear sequence motifs involved in their binding to eIF4E (Mader et al. 1995). This mechanism therefore allows mTORC1 signaling to promote the formation of eIF4F complexes (Merrick and Pavitt 2018), and thus (through additional protein:protein interactions; Fig. 2) the recruitment of 40S subunits to the 5′ end of mRNAs.

MNK1 and MNK2 phosphorylate eIF4E on Ser209 (Fig. 1), which remains their only validated in vivo substrate (Ueda et al. 2004; Buxade et al. 2008); moreover, MNKs are the only kinases that act on this site. MNKs interact with eIF4G, bringing them close to their substrate eIF4E (Pyronnet et al. 1999).

It should be noted that eIF4E is not, as sometimes stated, always the least abundant initiation factor, although this varies between organisms and cell types (see, for example, Rau et al. 1996; von der Haar and McCarthy 2002; Schwanhausser et al. 2011); some others may occur at lower levels (Merrick and Pavitt 2018). When one copy of the eIF4E gene is deleted in mice, they show no adverse effects (Truitt et al. 2015). However, the 4E-BPs regulate eIF4E availability and it is this parameter, rather than the absolute levels of eIF4E, that can limit the factor's activity.

Rapamycin generally has little effect on the phosphorylation of 4E-BP1, and therefore does not impair eIF4E:eIF4G binding (e.g., Choo et al. 2008; Huo et al. 2012). As one would expect, the phosphorylation of 4E-BP1 is efficiently blocked by compounds that interfere with the catalytic activity of mTOR (mTOR kinase inhibitors; mTOR-KIs). Thus, rapamycin is expected to have less impact on the control of translation initiation by 4E-BPs than mTOR-KIs do. The reason why rapamycin has little effect on the phosphorylation of 4E-BPs reflects

its mode of action. Structural studies indicate that the FKBP12:rapamycin complex partially occludes but does not completely obstruct the active site cleft (Yang et al. 2013), thereby restricting substrate access. Detailed studies suggest that rapamycin inhibits the ability of mTOR to phosphorylate "weak" substrates such as the S6Ks mentioned above, but not "strong" substrates such as 4E-BPs (Kang et al. 2013).

IMPACT OF mTORC1 SIGNALING ON TRANSLATION

The fact that rapamycin only affects phosphorylation of some of the substrates of mTORC1 is important when evaluating the effects of rapamycin versus mTOR-KIs on overall protein synthesis and on the translation of specific mRNAs. Rapamycin has a lesser inhibitory effect compared to the much stronger effect of mTOR-KIs, and they have differential effects on the synthesis of specific proteins (Huo et al. 2012). One example is provided by the 5′-terminal oligopyrimidines (TOP) mRNAs, which possess a run of pyrimidine nucleotides at the extreme 5′ end and include those encoding all cytoplasmic ribosomal proteins and several translation elongation factors (Meyuhas and Kahan 2014). Synthesis of these proteins is impaired by rapamycin and more strongly inhibited by mTOR-KIs. This is presumably because mTOR-KIs inhibit mTORC1 signaling more strongly than rapamycin but one cannot formally rule out a contribution from mTORC2 signaling, which is also inhibited by mTOR-KIs. The regulation of TOP mRNA translation by mTORC1 provides a mechanism by which nutrient-activated signaling can also promote ribosome biogenesis, along with promoting rRNA transcription (Iadevaia et al. 2014) and nucleolar function (Chauvin et al. 2013) to increase the cellular capacity for protein synthesis. The control of this interesting and important set of mRNAs is discussed further below.

Several studies have examined the effect that interfering with mTOR signaling has on mRNA translation and protein synthesis. Such effects could be exerted through 4E-BP-mediated control of the availability of eIF4E; the associated effects on the recruitment of eIF4A (via eIF4G) to the 5′ end of mRNAs; the regulation of eIF4A by eIF4B, which is an indirect target of mTORC1 signaling (see later section); the effect of mTORC1 on the elongation machinery; or other mechanisms.

Larsson et al. (2012) used polysome profiling (i.e., sucrose gradient centrifugation) to assess which mRNAs were associated with larger polysomes. They studied the effects of treatment with rapamycin, PP242 (an mTOR-KI) or metformin, an agent that causes activation of AMPK and therefore inhibits mTORC1 signaling. Because AMPK has diverse substrates (Lin and Hardie 2017), it may impact translation in additional ways that do not occur in response to inhibition of mTOR. The data from their polysome profiling analysis revealed that the three agents they used decreased the association of almost 600 mRNAs with "large" polysomes; some were affected by all stimuli and others by only one or two of them. The great majority of mRNAs affected by PP242 were also influenced by rapamycin (with some exceptions), and two-thirds of them were also affected by metformin, although, interestingly, almost half the mRNAs whose polysomal association was decreased by metformin were not affected by mTOR inhibition, consistent with additional actions of this drug, likely through other AMPK targets. Gene ontology (GO) analysis revealed distinct functional categories of mRNAs that were affected by the different treatments. For example, those selectively affected by PP242 and by metformin, but not rapamycin, included many involved in cell-cycle control, whereas the set affected only by PP242 was enriched in mRNAs encoding proteins involved in mitochondrial functions and mRNA translation; those affected only by metformin included many involved in RNA processing.

Thoreen et al. (2012) and Hsieh et al. (2012) used ribosome profiling technology (Ingolia et al. 2018) to assess which mRNAs were associated with ribosomes under different conditions, in combination with pharmacological inhibition of mTOR/mTORC1 using mTOR-KIs or rapamycin. These studies concluded that mTOR inhibition predominantly reduces ribosome footprints on TOP mRNAs. Hsieh et al.

(2012) also identified several mRNAs for proteins involved in metastasis or invasion as being regulated by mTOR, including Y-box protein 1 (YB1), a posttranscriptional regulator of genes involved in these processes. A distinct feature, a pyrimidine-rich translational element distinct in make-up and position from TOP elements, was present in many mTOR-regulated mRNAs. The importance of this feature for translation initiation, and its mode of action, remain unknown.

Many of the differences in the conclusions drawn by the ribosome profiling studies as compared to the polysome profiling work (Larsson et al. 2012) likely arise for methodological reasons (Gandin et al. 2016). Ribosome profiling provides data on the positions of ribosomes on mRNAs and, as performed by Thoreen et al. (2012) and Hsieh et al. (2012), detects big shifts in ribosome occupancy on highly expressed mRNAs, whereas polysome profiling can detect more modest shifts of mRNAs from larger to smaller polysomes. TOP mRNAs are abundant, and as discussed in Masvidal et al. (2017), next-generation sequencing (NGS) reads corresponding to them therefore tend to make up a high proportion of the data from such ribosome-profiling experiments. High-read "depth" is necessary to obtain quantifiable data for less abundant mRNAs.

Gandin et al. (2016) extended polysome-profiling studies to examine the impacts both of inhibiting mTOR and depleting the helicase eIF4A on the set of mRNAs found in larger polysomes and coupled this with analysis of transcription start sites. This study identified two distinct subsets of "mTOR-sensitive" mRNAs, with quite distinct features in their 5′ untranslated regions (UTRs) (Fig. 1). One subset has long 5′UTRs and includes mRNAs for proteins involved in cell survival (such as that for Mcl1 or survivin [Birc5]) or proliferation (e.g., cyclin D3, ODC). This subset was anticipated given that translation of such mRNAs is expected to require eIF4A's helicase function and thus assembly of the eIF4F complex on them. Consistent with this, the polysomal association of members of this set is also impaired by depleting cells of eIF4A1. The effect of mTOR signaling on the translation of these mRNAs very likely reflects the control of eIF4F assembly by mTOR (through 4E-BPs) and perhaps also the input from mTORC1 via S6Ks to eIF4B (Figs. 1 and 2).

In contrast, mRNAs in the second subset of mTOR-sensitive transcripts possess very short, unstructured 5′UTRs and, presumably as a consequence, their translation is not impaired by depletion or chemical inhibition of eIF4A (Gandin et al. 2016). These mRNAs have been described as containing translation initiator of short 5′UTR (TISU) elements (Elfakess et al. 2011; Kwan and Thompson 2018), although the brevity of the 5′UTR rather than a specific sequence element may be the key feature for their translational control. Translation of these mRNAs requires eIF4E-eIF4G binding and is regulated by 4E-BPs (Fig. 1) (Morita et al. 2013; Sinvani et al. 2015). Many such mRNAs encode proteins involved in mitochondrial function, the vast majority of mitochondrial proteins being encoded by nuclear genes.

The ability of mTORC1 signaling to promote the translation of mRNAs for mitochondrial proteins likely represents part of a concerted program through which this pathway upregulates mitochondrial biogenesis. This program also involves the transcription factor PGC-1α, which drives the transcription of many genes involved in mitochondrial function or biogenesis (Cunningham et al. 2007; Duvel et al. 2010). This mechanism may serve to increase the mitochondrial capacity for ATP production to support the many anabolic processes that are turned on by mTORC1 signaling, such as protein synthesis and ribosome biogenesis.

A LIMITATION OF RIBOSOME AND POLYSOME PROFILING TECHNIQUES

A limitation of both ribosome profiling and polysome profiling approaches concerns the control of elongation. mRNAs whose rate of elongation is slower under a given condition will shift into larger polysomes as ribosomes build up on them, so that in both analyses these mRNAs will appear to be "more translated," whereas in fact the opposite is true. Methods

such as bioorthogonal noncanonical amino acid tagging (BONCAT), involving the labeling of newly synthesized polypeptides with chemical and mass tags (Dieterich et al. 2006), report the rate of accumulation of new polypeptides and thus the rate of synthesis of proteins rather than the association of their mRNAs with ribosomes/polysomes. These approaches have been used to study global proteome dynamics (Schwanhausser et al. 2011), the role of eEF2K in the synthesis of new proteins in primary neurons (Kenney et al. 2016), and the impact of mTOR signaling on the synthesis of specific proteins (Huo et al. 2012). The latter study concluded that mTOR-KIs and rapamycin have differing effects on the synthesis of many proteins. In particular, in agreement with other studies, rapamycin inhibits the synthesis of proteins encoded by 5′TOP mRNAs rather less strongly than mTOR-KIs. Interestingly, the synthesis of certain other proteins shows a similar pattern of inhibition; the mRNAs for some of them may belong to the TOP class although others evidently do not. The fact that the annotation of the 5′ end of mRNAs is often incorrect makes interpretation more difficult, an issue that is addressed by Gandin et al. (2016).

REGULATION OF THE TRANSLATION OF TOP mRNAs

Although the translation (polysomal association) of TOP mRNAs has been known to be promoted by mTORC1 signaling for almost three decades, and despite extensive work in several laboratories, the mechanisms that link mTORC1 signaling to control of TOP mRNA translation still remain to be definitively clarified (Meyuhas and Kahan 2014). At one time, S6K-catalyzed phosphorylation of S6 was thought to be involved but there is now unequivocal evidence against this idea, for example, from data obtained using cells in which S6Ks have been knocked out or the phosphorylation sites in S6 have been mutated (reviewed in Meyuhas and Kahan 2014). The phosphorylation of S6 is discussed in more detail below.

As mTORC1-regulated repressors of translation initiation, the 4E-BPs appear well-placed to regulate TOP mRNA translation, and data show that they do indeed play such a role in response to pharmacological inhibition of mTOR. Thoreen et al. (2012) concluded that, by interfering with eIF4E-eIF4G binding, mTOR inhibitors cause a greater dissociation of eIF4E from TOP mRNAs than from other mRNAs, thereby reducing the translation of the TOP mRNAs. However, several lines of evidence indicate that 4E-BPs are not the key regulators of TOP mRNA translation under conditions such as amino acid deprivation or hypoxia, both of which impair mTORC1 signaling (Meyuhas and Kahan 2014; Miloslavski et al. 2014).

Recent studies have identified a new potential regulator of TOP mRNA translation, La-related protein 1 (LARP1) (Fig. 1), although the data for the role of LARP1 in controlling TOP mRNA translation are conflicting. Tcherkezian et al. (2014) used a proteomic screen to identify proteins that associate with the 5′ cap of mRNAs in an mTORC1-dependent manner; one of them was LARP1, which also binds poly(A)-binding protein (PABP) and mTORC1 itself. Their data indicate that LARP1 is a positive regulator of mRNA translation based on the observation that knockdown of LARP1 reduced polysome levels, suggesting impaired translation initiation. The degree of polysome size reduction of TOP mRNAs was greater than that of mRNAs in general, although the effect was clearly not restricted to TOP messages.

Fonseca et al. (2015) agree that LARP1 interacts with mTORC1 but provide evidence that it acts as a repressor of TOP mRNA translation and interacts with TOP elements. Their data show a "preferential" association of LARP1 with TOP mRNAs and that such association is increased in response to rapamycin or Torin1, an mTOR-KI. In particular, the repression of TOP mRNA translation caused by amino acid starvation is attenuated when LARP1 is knocked down, consistent with LARP1 being a repressor rather than an activator of TOP mRNA translation. It should be noted that the effect of amino acid starvation on TOP mRNA translation also involves the RNA-binding proteins T-cell intracellular antigen 1 (TIA1) and TIA-R (Damgaard and Lykke-Andersen 2011), although, again, this appears not to be a general mechanism for their control.

Further studies have provided additional evidence that LARP1 acts as a repressor of TOP mRNA translation, although the mechanisms proposed differ. Two studies reported that LARP1 directly binds the cap of TOP mRNAs, thus impeding access for eIF4E (Lahr et al. 2017; Philippe et al. 2017). Lahr et al. solved structures of the carboxy-terminal DM15 domain of LARP1, including structures of co-complexes with nucleotides 4-11 of the 5′ end of the TOP mRNA encoding rpS6 (i.e., without the initial cytosine and the next two nucleotides) and m^{7}GTP-C (analog of the extreme 5′ end of a TOP mRNA). This revealed binding pockets in the DM15 domain of LARP1 for the 7-methylated GTP, the first C, and a TOP. Philippe et al. (2017) also reported that LARP1 binds the cap and TOP sequence, that LARP1 represses TOP mRNA translation in vitro, and that this ability is enhanced using LARP1 from cells pretreated with Torin1. A recent report shows that LARP1 is phosphorylated by mTORC1; this causes LARP1 to dissociate from the 5′UTRs of TOP mRNAs and relieves its inhibitory effect on their translation (Hong et al. 2017). Puzzlingly, however, the binding sites for LARP1 on these mRNAs generally did not match the TOP elements at their 5′-termini.

Overall, these data suggest a model whereby (1) LARP1 is regulated by mTORC1 such that its ability to bind TOP mRNAs is somehow enhanced when mTORC1 signaling is inhibited, and (2) association of LARP1 with the cap and 5′ end of a TOP mRNA blocks access for eIF4E and thus ribosomes, leading to impairment of initiation on such mRNAs. It remains unclear precisely how mTORC1 regulates LARP1—presumably by phosphorylation—and given that LARP1 binds Raptor, perhaps direct phosphorylation by mTORC1. LARP1 contains numerous phosphorylation sites, but it is not yet known which are regulated by mTORC1 signaling and which control LARP1's ability to bind TOP mRNAs. It should be noted that overexpression of eIF4E failed to outcompete a repressor activity that represses the translation of TOP mRNAs (Shama et al. 1995), which appears to argue against roles for 4E-BPs or LARP1.

eIF4E PHOSPHORYLATION BY THE MNKs

Phosphorylation of eIF4E was discovered early in studies on translation factor modifications (Duncan et al. 1987) but the function of this regulated modification remains obscure. Phosphorylation occurs at a single site, Ser209 (Flynn and Proud 1995; Joshi et al. 1995), and is not required for animal viability, as mice carrying a homozygous Ser209Ala mutation are viable (Furic et al. 2010). Furthermore, mice lacking both copies of both genes for the relevant kinases, the MNKs, are also viable and fertile (Ueda et al. 2004). However, as described below, there are now a number of examples of the control of the translation of mRNAs requiring phosphorylation of eIF4E and/or MNK function. eIF4E2/4EHP and eIF4E3 do not contain an equivalent of Ser209.

In humans (but apparently not mice), each MNK gene gives rise to two distinct mRNA species caused by alternative splicing (Buxade et al. 2008); this results in two protein isoforms each for MNK1 and MNK2 termed "a" and "b," which differ at their carboxyl termini. All isoforms contain the complete kinase domain and an amino-terminal polybasic region that interacts with eIF4G, which also serves as a nuclear localization signal (Parra-Palau et al. 2003; Scheper et al. 2003). MNK1a also contains a nuclear export signal and is therefore mainly cytoplasmic; MNK2a does not have this feature and is largely nuclear. MNK1a and MNK2a each contain a carboxy-terminal sequence that interacts with MAP kinases; this motif is absent from the shorter MNK1b and MNK2b isoforms. The MAP kinase-binding sequences differ however, such that although MNK1a interacts with ERK and p38 MAP kinase $\alpha\beta$, MNK2a only binds ERK (Waskiewicz et al. 1997). Indeed, unlike MNK1a, MNK2a can bind active phosphorylated ERK, which likely explains its high basal activity. In contrast, in unstimulated cells, MNK1a activity is low but can be markedly increased by stimuli that switch on ERK signaling (e.g., serum, some growth factors, various recep-

tor agonists) or p38 MAP kinase α/β (some cytokines, certain "stress" conditions). The control of MNK1b and MNK2b, which lack the MAP kinase interaction site, is not well understood (see O'Loghlen et al. 2007; Buxade et al. 2008).

MNK signaling intersects with mTORC1 signaling in several ways. First, as noted, MNKs interact with eIF4G, facilitating the phosphorylation of eIF4E (Pyronnet et al. 1999). A consequence of this is that agents that promote the phosphorylation of 4E-BPs and their release from eIF4E, allowing increased eIF4E-eIF4G binding, should facilitate eIF4E phosphorylation. Second, rapamycin treatment can paradoxically enhance eIF4E phosphorylation and the activity of MNK2 (rather than MNK1 [Stead and Proud 2013]). Third, MNKs were recently reported to stimulate signaling through mTORC1 by phosphorylating TELO2, a partner for mTORC1 that promotes interactions between mTORC1 and its substrates (Brown and Gromeier 2017).

WHAT ROLE DOES eIF4E PHOSPHORYLATION PLAY IN CONTROLLING mRNA TRANSLATION?

To date, no comprehensive datasets have been published for the effects of inhibiting phosphorylation of eIF4E on the translation of specific mRNAs of the kind published for mTOR or eIF4A (e.g., Hsieh et al. 2012; Thoreen et al. 2012; Gandin et al. 2016). However, a set of data was reported by Furic et al. (2010) for embryonic fibroblasts from wild-type and eIF4E^{Ser209Ala} knock-in mice; this polysome-profiling study identified 35 mRNAs whose association with polysomes was reduced in the latter cells, implying that eIF4E phosphorylation promotes the initiation of their translation.

It remains unclear how phosphorylation of eIF4E influences the translation of specific transcripts, for example, whether a feature of these mRNAs (and, if so, which feature) confers sensitivity to eIF4E phosphorylation. Several studies, using compounds that inhibit MNKs, MNK-knockout (KO) cells, or cells where the eIF4E phosphorylation site has been changed (Ser$_{209}$Ala), show that MNKs and/or the phosphorylation of eIF4E positively regulate the translation of mRNAs for proteins involved in a range of processes, including migration/invasion, the epithelial–mesenchymal transition (EMT), cell proliferation and survival, immune responses, and circadian rhythms. A handful of other MNK substrates has been reported, two of which, heterogeneous nuclear ribonucleoprotein (hnRNP) A1 and polypyrimidine tract-binding protein-associated splicing factor (PSF) (Buxade et al. 2008), are nucleic acid-binding proteins. Their phosphorylation may affect gene expression, although it remains to be shown that they are indeed substrates for the MNKs in vivo.

Structural data for eIF4E reveal that Ser209 is located near the "channel" through which the mRNA exits the cap-binding site on eIF4E (Marcotrigiano et al. 1997; Matsuo et al. 1997); it is therefore not close to this binding site, or indeed to the region of the surface of eIF4E for 4E-BPs/eIF4G. Biophysical studies (Scheper et al. 2002; Zuberek et al. 2004; Slepenkov et al. 2006) indicate that phosphorylation of eIF4E decreases its affinity for a cap analog or short capped RNAs. The consequences for translational control of the reduced affinity and lower affinity of the eIF4E-cap interaction when eIF4E is phosphorylated remain unclear.

Reflecting both the ability of eIF4E overexpression to transform cells (Lazaris-Karatzas et al. 1990) and the fact that MNKs are activated by the oncogenic Ras/Raf/ERK pathway, initial studies focused on the role of the MNKs and phosphorylation of eIF4E in cancers. This work and other studies revealed requirements for the MNKs and/or Ser209 in eIF4E for cell transformation and tumor progression (Wendel et al. 2004, 2007; Furic et al. 2010; Ueda et al. 2010; Proud 2015). Key to understanding the mechanistic basis of these observations will be to identify which mRNAs' translation is regulated by MNK-mediated phosphorylation of eIF4E. A modest number of candidates has been identified as being positively controlled in this way, including those for Mcl1 and other prosurvival proteins, certain cyclins, chemokines, matrix metalloproteinases, and vascular endothelial growth factor C (VEGF C), a proangiogenic factor (Wendel et al. 2007; Furic et al.

2010, and references therein; Proud 2015; Robichaud et al. 2018b).

PHOSPHORYLATION OF eIF4E CAN INDIRECTLY INFLUENCE THE TRANSCRIPTIONAL MACHINERY

Lim et al. (2013) discovered that eIF4E phosphorylation favors signaling through the Wnt/β-catenin pathway in blast crisis chronic myeloid leukemia. Of particular relevance here is that eIF4E and its phosphorylation promote translation of the mRNA for β-catenin and also, by an unknown mechanism, the translocation of β-catenin into the nucleus where it mediates Wnt regulation of gene transcription.

eIF4E phosphorylation regulates the translation of mRNAs for regulators of transcription in cells of the immune system. For example, the translation of the mRNA for IκBα (inhibitor of κBα), a negative regulator of the key transcription factor NF-κB, is reduced in cells that homozygously express the nonphosphorylatable eIF4E^{Ser209Ala} mutant (Herdy et al. 2012), allowing greater activation of NF-κB and enhanced production of interferon β. In other words, reduced eIF4E phosphorylation can enhance antiviral responses, a device that may be used in virus-infected cells to limit viral spread. However, the regulation of eIF4E phosphorylation during infection differs between viruses (Cuesta et al. 2000; Walsh and Mohr 2004; Stern-Ginossar et al. 2018). The control of the IκBα/NF-κB system by the MNKs may also be important in inflammatory responses (Joshi and Platanias 2014; Bao et al. 2017).

MNKs and/or eIF4E phosphorylation also promote the translation of positive regulators of transcription in immune cells (i.e., interferon regulatory factor [IRF]-8; Xu et al. 2012), which supports the induction of genes involved in host defenses against pathogen infection and the development of lymphoid and myeloid cell lineages. MNKs may also regulate cytokine expression through translational control of the expression of RANTES factor of late-activated T lymphocytes-1 (RFLAT-1), which in turn drives production of the cytokine RANTES/CCL5 (Nikolcheva et al. 2002).

Taken together, these and other data (Buxade et al. 2005; Su et al. 2015; Brace et al. 2017) suggest that the MNKs play multiple, and to some extent opposing, effects in immune and inflammatory responses by modulating the translation of mRNAs for transcriptional regulators.

ROLES FOR eIF4E PHOSPHORYLATION IN NEUROLOGICAL PROCESSES

In neurons of the suprachiasmatic nucleus (SCN), translational control of proteins such as period 1 and 2 (Per1/2) and cryptochrome (Cry) plays a key role in establishing circadian rhythms as part of sophisticated and highly conserved feedback mechanisms. Cao et al. (2015) showed that brief exposure to light activates MNK signaling and promotes the translation of the mRNAs for Per1 and Per2. Interestingly, 4E-BP1 regulates the translation of another mRNA, *Vip1*, in the SCN, where its product, vasoactive intestinal peptide, plays a key role in the SCN in maintaining circadian synchrony (Cao et al. 2013).

The protein Arc plays a crucial role in long-term synaptic plasticity and memory formation. Its levels are subject to rapid regulation via translational control mechanisms that require eIF4E phosphorylation (and therefore MNK1, the main MNK in neurons) as well as other events (Panja et al. 2009, 2014). Although brain-derived neurotrophic factor (BDNF) normally promotes eIF4E-eIF4G binding in synaptoneurosome preparations from cells of the dentate gyrus (part of the hippocampal formation that plays a key role in memory), it fails to do so when MNK1 has been knocked out (Panja and Bramham 2014). BDNF normally causes the release from eIF4E of another partner, cytoplasmic fragile X mental retardation protein (FMRP)-interacting protein 1 (CYFIP1), but does not do so when eIF4E cannot be phosphorylated (e.g., in MNK1-KO cells; see Panja and Bramham 2014). CYFIP1 also binds FMRP mutations, which are the most frequent cause of inherited autism (Hagerman et al. 2017). FMRP binds subsets of mRNAs to repress their translation; thus, the lack of eIF4G-eIF4E binding and the sustained eIF4E-CYFIP1-FMRP association likely work together to prevent engagement of the proper translational program in MNK1-KO cells. In primary neurons from MNK1-KO mice, BDNF fails to activate protein synthesis and, in particular, the synthesis of proteins involved in neurotransmission and synaptic plasticity is impaired. Many of the corresponding mRNAs interact with FMRP (Genheden et al. 2015).

ROLES FOR eIF4E PHOSPHORYLATION IN EMT AND TUMOR METASTASIS

FMRP also binds mRNAs for proteins that are involved in the EMT, a process that is critical for tumor cells to migrate, invade, and metastasize (Luca et al. 2013). Consistent with this, the phosphorylation of eIF4E augments the translation of the mRNAs for SNAIL and matrix metalloproteinase-3 (MMP3), proteins important for EMT and metastasis (Furic et al. 2010; Robichaud et al. 2015). MNKs also positively regulate cell migration and the translation of the mRNA for vimentin, a protein that is characteristic of mesenchymal cells (Beggs et al. 2015).

Two major questions yet to be addressed are (1) how does the phosphorylation of eIF4E, which occurs at a location in eIF4E adjacent only to the extreme $5'$ nucleotides of eIF4E-associated mRNAs, regulate the translation of specific transcripts; and (2) which other mRNAs are translationally regulated through phosphorylation of eIF4E? Published studies have generally focused on only single mRNAs or small numbers of messages.

PHOSPHORYLATION OF OTHER TRANSLATION INITIATION FACTORS

In mammalian cells, several other initiation factors are phosphorylated, including eIF4G, eIF4B and several subunits of eIF3, some of which are modified at high stoichiometry, suggesting that at least some may play roles in the control of translation (Andaya et al. 2014). However, information on the consequences of the phosphorylation of these translation factors on protein synthesis remains sparse, and discussion here focuses on examples where there is some information on the functional consequences of phosphorylation of specific sites.

Mammalian eIF3 consists of 13 different subunits, some of which form a "core" complex (Merrick and Pavitt 2018). The phosphorylation of eIF3f at Ser46 by cyclin-dependent kinase 11 (CDK11^{P46}) (Shi et al. 2009) is reported to enhance its association with the core complex. However, the consequences of this for translation are unclear. Phosphorylation of Ser183 in eIF3h is implicated in oncogenesis; overexpressing eIF4h is oncogenic, whereas expressing a Ser-to-Ala mutant of this phosphorylation site is not. In contrast, when Ser183 is mutated to Asp, eIF3H retains its oncogenic properties (Zhang et al. 2008).

The genomes of yeast (*Saccharomyces cerevisiae*) and mammals encode two different forms of eIF4G; here, the mammalian proteins are termed eIF4GI and eIF4GII (rather confusingly, and they are curated as eIF4G1 and eIF4G3). Human eIF4GI contains more than a dozen phosphorylated residues (Andaya et al. 2014), although the functional significance of most of them is not known. Protein kinase Cα phosphorylates eIF4GI at Thr1186 in the linker region between two of eIF4GI's domains and this phosphorylation promotes binding of eIF4GI to MNK1 even though the MNK1-binding site is at the far carboxyl terminus of eIF4GI (i.e., not adjacent to Thr1186; Dobrikov et al. 2011). This effect would be expected to promote the phosphorylation of eIF4E by MNK1. The linker region also contains phosphorylation sites that are stimulated by serum in a rapamycin-sensitive manner (Raught et al. 2000); their function is unclear.

eIF4GI is also phosphorylated on Ser896 by a stress-activated protein kinase termed p21-activated kinase 2 (Pak2), and this modification impairs cap-dependent translation (Ling et al. 2005).

eIF4GII undergoes phosphorylation during mitosis and eIF4GII hyperphosphorylation is associated with its reduced association with eIF4E (Pyronnet et al. 2001). This effect may contribute to the impairment of cap-dependent mRNA translation, which occurs during mitosis. In budding yeast, both isoforms of eIF4G are phosphorylated during glucose starvation by the protein kinase Ksp1 (Chang and Huh 2018). Interestingly, data obtained using a mutant of Tif4631 (encoding eIF4GI in yeast) indicate a role for phosphorylation of eIF4GI in modulating the stability of mRNAs encoding enzymes of glycolysis when cells are deprived of glucose (Chang and Huh 2018).

Further studies are clearly needed to determine the regulation and functional consequences of other known phosphorylation events in eIF4G isoforms, eIF3 subunits, and other translation factors.

PHOSPHORYLATION OF ELONGATION FACTORS

In mammalian cells, there are two main translation elongation factors, eEF1A and eEF2. A third multimeric factor, eEF1B, acts as a GEF for eEF1A, to regenerate active eEF1A•GTP, the form of this factor that is competent to bind aminoacyl-tRNAs and recruit them to the ribosomal A site (Dever et al. 2018). eEF2 is required for the translocation step of elongation where the ribosome moves by one codon relative to the mRNA and the tRNA carrying the nascent chain shifts from the A site to the P site.

eEF1A and all four subunits of eEF1B are subject to phosphorylation, with CDKs phosphorylating eIF1Bβ and δ (Sasikumar et al. 2012). However, the role of phosphorylation in regulating the activities of eEF1A and eEF1B and in modulating translation elongation remains to be established.

eEF2 is subject to phosphorylation at Thr56 (residue 57 of its encoded sequence, before the initial methionine is removed [Price et al. 1991; Redpath et al. 1993]). This impairs its binding to ribosomes and thus its activity, but in a nondominant way; phosphorylated eEF2 is inactive and removed from the active pool of this factor but it does not interfere with the function of nonphosphorylated eEF2. Thus, phosphorylation of eEF2 slows elongation, as shown by "transit time" measurements (Redpath et al. 1996a); this should conserve energy (the equivalent of at least four ATPs being needed to add each amino acid to a polypeptide) and, of course, amino acids.

eEF2K IS AN ATYPICAL PROTEIN KINASE AND IS SUBJECT TO DIVERSE REGULATORY INPUTS

Phosphorylation of Thr56 in eEF2 is catalyzed by the specific and unusual kinase, eEF2K (Ryazanov et al. 1988; Mitsui et al. 1993; Redpath et al. 1996b). Such phosphorylation is entirely lost in cells lacking eEF2K, showing it is the only enzyme involved (although AMPK had previously been reported to phosphorylate Thr56 [Hong-Brown et al. 2007]). AMPK can phosphorylate and activate eEF2K (Browne et al. 2004; Johanns et al. 2017), providing a mechanism by which low energy levels can slow down rates of elongation and thus reduce the consumption of ATP/GTP.

eEF2K is not a member of the main protein kinase superfamily comprising almost 500 genes but is one of only six α-kinases present in mammals (Ryazanov et al. 1997). The catalytic domains of α-kinases show no sequence homology with the main superfamily (Drennan and Ryazanov 2004), although there are similarities at the 3D structural level (Yamaguchi et al. 2001; Drennan and Ryazanov 2004; Ye et al. 2010; Galan et al. 2014). eEF2K activity is normally dependent on Ca^{2+}-ions and calmodulin (CaM), and it is activated in contracting muscle where Ca^{2+}-ions trigger contraction. Because contraction uses a great deal of ATP, coordinating eEF2K activity with contraction could help ensure that contraction is not affected by reduced ATP levels caused by protein synthesis. Consistent with this model, eEF2 phosphorylation rises very quickly in muscle after the onset of exercise (Rose et al. 2005).

eEF2K is also subjected to other regulatory inputs linked to nutrients, extracellular stimuli, or other cues (Fig. 3) (Liu and Proud 2016). In brief, eEF2K is switched off by signaling through mTORC1, which promotes the phosphorylation of eEF2K at several sites including one (Ser78) near its CaM-binding site, which inhibits activation by CaM (Browne and Proud 2004). Another inhibitory site, Ser366 (Wang et al. 2001), is a target for S6Ks (which are activated by mTORC1). Ser78 and also Ser396 are direct substrates for mTORC1, whereas other sites appear to be regulated indirectly by this pathway. Signaling through the MAP kinase pathway also inhibits eEF2K, via direct phosphorylation by ERK at Ser359 and by the RSKs at Ser366 (Fig. 3) (Wang et al. 2001, 2014).

The CaM-binding site also acts as a locus for regulation of eEF2K by hypoxia, because a neighboring proline is subject to oxygen-dependent hydroxylation that impairs eEF2K activity (Fig. 3) (Moore et al. 2015). This binding site contains several histidines, whose protonation at pH values associated with acidosis enhances CaM binding and thus eEF2K activity (Xie et al. 2015).

Taken together, these inputs allow eEF2K to be activated under conditions where energy requirements are high or energy is depleted, where amino acids are in short supply, or where levels of oxygen (needed to efficiently generate ATP in many cell types) are inadequate (Fig. 1). Inhibiting elongation rather than initiation when nutrients are scarce could have two advantages: (1) by preserving polysomes, rather than causing their disaggregation, it will allow translation to resume quickly once conditions improve; and (2) the continued association of mRNAs with polysomes may protect them against degradation. Conversely, under favorable conditions such as the presence of anabolic hormones or growth factors, eEF2K is switched off to remove a potential limitation on elongation rates.

WHAT IS THE ROLE OF CONTROL OF ELONGATION?

Early studies indicated that chemical inhibition of elongation could, paradoxically, enhance the synthesis of certain proteins (Fig. 1) (Walden and Thach 1986; Scheetz et al. 2000). eEF2K regulates the synthesis of certain proteins in neurons (Kenney et al. 2016), where it undergoes transient activation in response to stimulation of cells with a neurotransmitter receptor antagonist and controls the synthesis of microtubule-associated proteins. It remains unclear how slowing elongation promotes translation of some mRNAs; one suggestion is that for "weak" mRNAs (ones for which initiation is inefficient and thus limiting), slowing elongation results in initiation no longer being the limiting event, enabling these transcripts to compete better in recruiting ribosomes. Two aspects of this model are puzzling: (1) slow moving ribosomes near the beginning of the coding region would be expected to impede access for initiating ribosomes, and (2) it also requires an adequate supply of free ("spare") ribosomes that could be recruited onto these mRNAs when elongation is slowed. Conversely, the levels of certain short-lived proteins, such as some antiapoptotic regulators, may drop when elongation is inhibited; this mechanism is suggested to explain the role of eEF2K in ensuring oocyte quality by promoting the apoptotic elimination of defective cells (Chu et al. 2014).

Several lines of evidence indicate that eEF2K helps to protect cancer cells against nutrient deprivation or other insults including some chemotherapeutic agents (Kenney et al. 2014). Mice lacking eEF2K or bearing a "kinase-dead" mutation in its gene are viable and fertile under vivarium conditions, indicating that eEF2K is not an essential gene (Chu et al. 2014; Moore et al. 2015). Thus, eEF2K appears to be an attractive target for novel anticancer agents. However, the picture is clouded by the observation that eEF2K actually restrains the development of colon tumors in a well-characterized mouse model (Faller et al. 2014), raising the prospect that inhibiting it might promote development of some tumors. This "double-edged sword" situation also applies to other pathways that are considered or used as targets in anticancer therapy, such as autophagy (White and DiPaola 2009) and indeed mTORC1 (Palm et al. 2015), inhibition of which can exert both pro- and antitumor effects depending on the stage in tumor development or the setting. Inhibition of eEF2K may be most effective for treating established solid tumors that are poorly vascularized. Reported connections between eEF2K and other diseases have recently been reviewed (Liu and Proud 2016).

PHOSPHORYLATION OF RIBOSOMAL PROTEIN S6

Several ribosomal proteins are subject to phosphorylation but by far the most attention has focused on S6; that said, the function of its phos-

phorylation is still not understood. As mentioned above, S6Ks phosphorylate several sites in the carboxyl terminus of S6, whereas RSKs act on two of them. Increased S6 phosphorylation is seen under conditions where cell growth and/or proliferation are enhanced, giving rise to the early idea that S6 phosphorylation promoted these outcomes. Indeed, mice lacking S6K1 (Shima et al. 1998) or fruit-flies lacking the *Drosophila* ortholog (Montagne et al. 1999) are smaller than control animals, revealing a role for S6K signaling in cell and organismal growth.

One possible explanation for these observations would be that phosphorylation of S6 promotes protein synthesis and/or the translation of specific mRNAs whose products are important for cell growth. One such class of mRNAs, once thought to be controlled by S6Ks, is the TOP mRNAs, which are well placed to control cell growth as they encode components of the translational machinery (Meyuhas and Kahan 2014). However, their association with active polysomes is still reduced by rapamycin in S6K-KO cells (Pende et al. 2004) and in cells where the phosphorylation sites in S6 have been mutated to nonphosphorylatable alanine residues (Ruvinsky et al. 2005).

Further studies on cells in which the phosphorylation sites in S6 are mutated revealed faster rates of cell division and protein synthesis, and smaller cell size, which was not further decreased by rapamycin (Ruvinsky et al. 2005). Thus, S6 phosphorylation appears to be a positive regulator of cell size but not of general protein synthesis or TOP mRNA translation. Intriguingly, pancreatic β cells from mice expressing nonphosphorylatable S6 are resistant to transformation by active Akt (which, among the effects, turns on mTORC1 signaling) and also to genotoxic or "proteotoxic" stress (Wittenberg et al. 2016). Probing the molecular mechanisms that underlie these effects should reveal much-needed insights into the function(s) of S6 phosphorylation.

OTHER S6K SUBSTRATES RELEVANT TO TRANSLATION

S6Ks phosphorylate a number of other substrates relevant to mRNA translation. These include eEF2K (Figs. 1 and 3) and eIF4B, which interacts with eIF4F and eIF4G and promotes the helicase activity of its eIF4A subunit (Fig. 2) (Raught et al. 2004; Shahbazian et al. 2006). The S6K site in eIF4B, Ser422, is also phosphorylated by the RSKs, downstream of MAP kinase (ERK) signaling (Shahbazian et al. 2006); it, and a second site Ser406, are also targets for the related kinase Akt (van Gorp et al. 2009). The effects of phosphorylation of eIF4B at this site, Ser422, are unclear, with one report concluding that phosphorylation impairs translation initiation (Raught et al. 2004) and a second that it enhances the association of eIF4B with eIF3 (Shahbazian et al. 2006). Mutation of either Ser406 or Ser422 to alanine abolished the ability of eIF4B to promote translation initiation when overexpressed (van Gorp et al. 2009), suggesting that both sites positively regulate eIF4B function (Shahbazian et al. 2006; van Gorp et al. 2009).

S6Ks also phosphorylate the programmed cell death 4 protein (PDCD4), promoting its release from eIF4A (and its degradation), thereby allowing eIF4A to bind eIF4G (Yang et al. 2003; Dorrello et al. 2006). The availability of small-molecule inhibitors of eIF4A has permitted studies on the impact of eIF4A function on the translation of specific mRNAs in mammalian cells (Rubio et al. 2014; Wolfe et al. 2014), revealing roles in tumor biology.

Both eIF4B (Figs. 1 and 2) and PDCD4 are also phosphorylated by RSKs (Shahbazian et al. 2006; Galan et al. 2014). S6Ks also regulate other proteins involved in RNA biology, including SKAR (S6K1 Aly/REF-like substrate), which plays a role in splicing and whose phosphorylation enhances the translation of newly made, spliced mRNAs, perhaps by recruiting S6K (Ma et al. 2008).

S6Ks also help promote the transcriptional program of ribosome biogenesis, an effect that seems to require phosphorylation of S6; again, the mechanism is not known (Chauvin et al. 2013). This role of the S6Ks in the function of the nucleolus, where ribosomes are assembled, may well be related to the observations linking S6Ks with cell size, as ribosome biogenesis is a key factor controlling cell growth (Rudra and Warner 2004).

WHAT DON'T WE KNOW?

Recent years have seen substantial progress is identifying the links between signaling pathways and the translational machinery, such as phosphorylation events. Further examples remain to be discovered. Our understanding of the impact of signaling pathways on the translation of specific mRNAs is greatly helped by the advent of new technologies such as ribosome profiling and stable-isotope labeling methods to monitor new protein synthesis. However, further work is clearly needed to elucidate the molecular mechanisms by which events such as the phosphorylation of eIF4E, or the control of elongation, modulate the synthesis of specific proteins. Conversely, it is unclear how the features of subsets of mRNAs, such as TISU elements, confer translational control. mTOR signaling modulates translation of many mRNAs that are not TOP messages; how does it do this?

One study has indicated that the MNKs negatively modulate the expression of eIF4E3 and that loss of MNK function is associated with enhanced eIF4E3-dependent translation of a subset of mRNAs (Landon et al. 2014). Although this is just a single report, it does raise the possibility that signaling pathways modulate the use of isoforms of translation factors. For example, the roles of eIF4E2, which has been reported to be involved both in repressing and promoting translation of different subsets of mRNAs (Cho et al. 2005; Uniacke et al. 2012; Chapat et al. 2017; Jafarnejad et al. 2018), and of eIF4E3 remain to be fully elucidated. What role do signaling events play in modulating their effects on gene expression?

Control of translation under different conditions almost certainly involves the combined effects of several signaling pathways and control of multiple translation factors. How do these inputs work together to modulate protein synthesis?

LARP1 appears to provide an example of an RNA-binding protein that is regulated by phosphorylation and controls the translation of a specific subset of mRNAs (TOP mRNAs). Do other RNA-binding proteins serve similar roles, linking signaling pathways to regulation of the translation of certain mRNAs?

Cite this article as *Cold Spring Harb Perspect Biol* doi: 10.1101/cshperspect.a033050

Lastly, the signaling pathways that control the translational machinery are dysregulated in diseases such as cancer. How can modulating these pathways or the translational components they control be exploited to prevent, manage, or treat human disease?

ACKNOWLEDGMENTS

Research in the author's laboratory is funded by the National Health & Medical Research Council and the South Australian Health & Medical Research Institute (SAHMRI). There are no competing interests to disclose. I thank Oded Meyuhas (Jerusalem) and Kirk Jensen, Xuemin Wang, and Jianling Xie (SAHMRI) for their helpful comments.

REFERENCES

*Reference is also in this collection.

Andaya A, Villa N, Jia W, Fraser CS, Leary JA. 2014. Phosphorylation stoichiometries of human eukaryotic initiation factors. *Int J Mol Sci* **15**: 11523–11538.

Bain J, Plater L, Elliott M, Shpiro N, Hastie CJ, McLauchlan H, Klevernic I, Arthur JS, Alessi DR, Cohen P. 2007. The selectivity of protein kinase inhibitors: A further update. *Biochem J* **408**: 297–315.

Bao Y, Wu X, Chen J, Hu X, Zeng F, Cheng J, Jin H, Lin X, Chen LF. 2017. Brd4 modulates the innate immune response through Mnk2-eIF4E pathway-dependent translational control of IκBα. *Proc Natl Acad Sci* **114**: E3993–E4001.

Beggs JE, Tian S, Jones GG, Xie J, Iadevaia V, Jenei V, Thomas GJ, Proud CG. 2015. The MAP kinase-interacting kinases regulate cell migration, vimentin expression and eIF4E/CYFIP1 binding. *Biochem J* **467**: 63–76.

Ben-Sahra I, Manning BD. 2017. mTORC1 signaling and the metabolic control of cell growth. *Curr Opin Cell Biol* **45**: 72–82.

Brace PT, Tezera LB, Bielecka MK, Mellows T, Garay D, Tian S, Rand L, Green J, Jogai S, Steele AJ, et al. 2017. Mycobacterium tuberculosis subverts negative regulatory pathways in human macrophages to drive immunopathology. *PLoS Pathog* **13**: e1006367.

Brown MC, Gromeier M. 2017. MNK controls mTORC1: Substrate association through regulation of TELO2 binding with mTORC1. *Cell Rep* **18**: 1444–1457.

Browne GJ, Proud CG. 2004. A novel mTOR-regulated phosphorylation site in elongation factor 2 kinase modulates the activity of the kinase and its binding to calmodulin. *Mol Cell Biol* **24**: 2986–2997.

Browne GJ, Finn SG, Proud CG. 2004. Stimulation of the AMP-activated protein kinase leads to activation of eukaryotic elongation factor 2 kinase and to its phosphorylation at a novel site, serine 398. *J Biol Chem* **279**: 12220–12231.

Buxade M, Parra JL, Rousseau S, Shpiro N, Marquez R, Morrice N, Bain J, Espel E, Proud CG. 2005. The Mnks are novel components in the control of TNFα biosynthesis and phosphorylate and regulate hnRNP A1. *Immunity* **23**: 177–189.

Buxade M, Parra-Palau JL, Proud CG. 2008. The Mnks: MAP kinase-interacting kinases (MAP kinase signal-integrating kinases). *Front Biosci* **13**: 5359–5373.

Cao R, Robinson B, Xu H, Gkogkas C, Khoutorsky A, Alain T, Yanagiya A, Nevarko T, Liu AC, Amir S, et al. 2013. Translational control of entrainment and synchrony of the suprachiasmatic circadian clock by mTOR/4E-BP1 signaling. *Neuron* **79**: 712–724.

Cao R, Gkogkas CG, de ZN, Blum ID, Yanagiya A, Tsukumo Y, Xu H, Lee C, Storch KF, Liu AC, et al. 2015. Light-regulated translational control of circadian behavior by eIF4E phosphorylation. *Nat Neurosci* **18**: 855–862.

Carriere A, Romeo Y, Acosta-Jaquez HA, Moreau J, Bonneil E, Thibault P, Fingar DC, Roux PP. 2011. ERK1/2 phosphorylate Raptor to promote Ras-dependent activation of mTOR complex 1 (mTORC1). *J Biol Chem* **286**: 567–577.

Chang Y, Huh WK. 2018. Ksp1-dependent phosphorylation of eIF4G modulates post-transcriptional regulation of specific mRNAs under glucose deprivation conditions. *Nucleic Acids Res* doi: 10.1093/nar/gky1097.

Chapat C, Jafarnejad SM, Matta-Camacho E, Hesketh GG, Gelbart IA, Attig J, Gkogkas CG, Alain T, Stern-Ginossar N, Fabian MR, et al. 2017. Cap-binding protein 4EHP effects translation silencing by microRNAs. *Proc Natl Acad Sci* **114**: 5425–5430.

Chauvin C, Koka V, Nouschi A, Mieulet V, Hoareau-Aveilla C, Dreazen A, Cagnard N, Carpentier W, Kiss T, Meyuhas O, et al. 2013. Ribosomal protein S6 kinase activity controls the ribosome biogenesis transcriptional program. *Oncogene* **33**: 474–483.

Cho PF, Poulin F, Cho-Park YA, Cho-Park IB, Chicoine JD, Lasko P, Sonenberg N. 2005. A new paradigm for translational control: Inhibition via 5′-3′ mRNA tethering by Bicoid and the eIF4E cognate 4EHP. *Cell* **121**: 411–423.

Choo AY, Yoon SO, Kim SG, Roux PP, Blenis J. 2008. Rapamycin differentially inhibits S6Ks and 4E-BP1 to mediate cell-type-specific repression of mRNA translation. *Proc Natl Acad Sci* **105**: 17414–17419.

Chu HP, Liao Y, Novak JS, Hu Z, Merkin JJ, Shymkiv Y, Braeckman BP, Dorovkov MV, Nguyen A, Clifford PM, et al. 2014. Germline quality control: eEF2K stands guard to eliminate defective oocytes. *Dev Cell* **28**: 561–572.

Cuesta R, Xi Q, Schneider RJ. 2000. Adenovirus-specific translation by displacement of kinase Mnk1 from cap-initiation complex eIF4F. *EMBO J* **19**: 3465–3474.

Cunningham JT, Rodgers JT, Arlow DH, Vazquez F, Mootha VK, Puigserver P. 2007. mTOR controls mitochondrial oxidative function through a YY1-PGC-1α transcriptional complex. *Nature* **450**: 736–740.

Damgaard CK, Lykke-Andersen J. 2011. Translational co-regulation of 5′TOP mRNAs by TIA-1 and TIAR. *Genes Dev* **25**: 2057–2068.

* Dever TE, Dinman JD, Green R. 2018. Translation elongation and recoding in eukaryotes. *Cold Spring Harb Perspect Biol* doi: 10.1101/cshperspect.a032649.

Dibble CC, Elis W, Menon S, Qin W, Klekota J, Asara JM, Finan PM, Kwiatkowski DJ, Murphy LO, Manning BD. 2012. TBC1D7 is a third subunit of the TSC1-TSC2 complex upstream of mTORC1. *Mol Cell* 47: 535–546.

Dieterich DC, Link AJ, Graumann J, Tirrell DA, Schuman EM. 2006. Selective identification of newly synthesized proteins in mammalian cells using bioorthogonal noncanonical amino acid tagging (BONCAT). *Proc Natl Acad Sci* 103: 9482–9487.

Dobrikov M, Dobrikova E, Shveygert M, Gromeier M. 2011. Phosphorylation of eukaryotic translation initiation factor 4G1 (eIF4G1) by protein kinase Cα regulates eIF4G1 binding to Mnk1. *Mol Cell Biol* 31: 2947–2959.

Dorrello NV, Peschiaroli A, Guardavaccaro D, Colburn NH, Sherman NE, Pagano M. 2006. S6K1- and βTRCP-mediated degradation of PDCD4 promotes protein translation and cell growth. *Science* 314: 467–471.

Drennan D, Ryazanov AG. 2004. α-Kinases: Analysis of the family and comparison with conventional protein kinases. *Prog Biophys Mol Biol* 85: 1–32.

Duncan R, Milburn SC, Hershey JW. 1987. Regulated phosphorylation and low abundance of HeLa cell initiation factor eIF-4F suggest a role in translational control. Heat shock effects on eIF-4F. *J Biol Chem* 262: 380–388.

Duvel K, Yecies JL, Menon S, Raman P, Lipovsky AI, Souza AL, Triantafellow E, Ma Q, Gorski R, Cleaver S, et al. 2010. Activation of a metabolic gene regulatory network downstream of mTOR complex 1. *Mol Cell* 39: 171–183.

Elfakess R, Sinvani H, Haimov O, Svitkin Y, Sonenberg N, Dikstein R. 2011. Unique translation initiation of mRNAs-containing TISU element. *Nucleic Acids Res* 39: 7598–7609.

Faller WJ, Jackson TJ, Knight JR, Ridgway RA, Jamieson T, Karim SA, Jones C, Radulescu S, Huels DJ, Myant KB, et al. 2014. mTORC1-mediated translational elongation limits intestinal tumour initiation and growth. *Nature* 517: 497–500.

Flynn A, Proud CG. 1995. Serine 209, not serine 53, is the major site of phosphorylation in initiation factor eIF-4E in serum-treated Chinese hamster ovary cells. *J Biol Chem* 270: 21684–21688.

Fonseca BD, Zakaria C, Jia JJ, Graber TE, Svitkin Y, Tahmasebi S, Healy D, Hoang HD, Jensen JM, Diao IT, et al. 2015. La-related protein 1 (LARP1) represses terminal oligopyrimidine (TOP) mRNA translation downstream of mTOR complex 1 (mTORC1). *J Biol Chem* 290: 15996–16020.

Furic L, Rong L, Larsson O, Koumakpayi IH, Yoshida K, Brueschke A, Petroulakis E, Robichaud N, Pollak M, Gaboury LA, et al. 2010. eIF4E phosphorylation promotes tumorigenesis and is associated with prostate cancer progression. *Proc Natl Acad Sci* 107: 14134–14139.

Galan JA, Geraghty KM, Lavoie G, Kanshin E, Tcherkezian J, Calabrese V, Jeschke GR, Turk BE, Ballif BA, Blenis J, et al. 2014. Phosphoproteomic analysis identifies the tumor suppressor PDCD4 as a RSK substrate negatively regulated by 14-3-3. *Proc Natl Acad Sci* 111: E2918–E2927.

Gandin V, Masvidal L, Hulea L, Gravel SP, Cargnello M, McLaughlan S, Cai Y, Balanathan P, Morita M, Rajakumar A, et al. 2016. nanoCAGE reveals 5′ UTR features that define specific modes of translation of functionally related MTOR-sensitive mRNAs. *Genome Res* 26: 636–648.

Genheden M, Kenney JW, Johnston HE, Manousopoulou A, Garbis SD, Proud CG. 2015. BDNF stimulation of protein synthesis in cortical neurons requires the MAP kinase-interacting kinase MNK1. *J Neurosci* 35: 972–984.

Gingras A-C, Raught B, Sonenberg N. 1999. eIF4 translation factors: Effectors of mRNA recruitment to ribosomes and regulators of translation. *Annu Rev Biochem* 68: 913–963.

Gomez E, Mohammad SS, Pavitt GD. 2002. Characterization of the minimal catalytic domain within eIF2B: The guanine-nucleotide exchange factor for translation initiation. *EMBO J* 21: 5292–5301.

Gordiyenko Y, Schmidt C, Jennings MD, Matak-Vinkovic D, Pavitt GD, Robinson CV. 2014. eIF2B is a decameric guanine nucleotide exchange factor with a γ2ε2 tetrameric core. *Nat Commun* 5: 3902.

Hagerman RJ, Berry-Kravis E, Hazlett HC, Bailey DB Jr, Moine H, Kooy RF, Tassone F, Gantois I, Sonenberg N, Mandel JL, et al. 2017. Fragile X syndrome. *Nat Rev Dis Primers* 3: 17065.

Harris TE, Chi A, Shabanowitz J, Hunt DF, Rhoads RE, Lawrence JC Jr. 2006. mTOR-dependent stimulation of the association of eIF4G and eIF3 by insulin. *EMBO J* 25: 1659–1668.

Herdy B, Jaramillo M, Svitkin YV, Rosenfeld AB, Kobayashi M, Walsh D, Alain T, Sean P, Robichaud N, Topisirovic I, et al. 2012. Translational control of the activation of transcription factor NF-κB and production of type I interferon by phosphorylation of the translation factor eIF4E. *Nat Immunol* 13: 543–550.

Holz MK, Ballif BA, Gygi SP, Blenis J. 2005. mTOR and S6K1 mediate assembly of the translation preinitiation complex through dynamic protein interchange and ordered phosphorylation events. *Cell* 123: 569–580.

Hong-Brown LQ, Brown CR, Huber DS, Lang CH. 2007. Alcohol regulates eukaryotic elongation factor 2 phosphorylation via an AMP-activated protein kinase-dependent mechanism in C2C12 skeletal myocytes. *J Biol Chem* 282: 3702–3712.

Hong S, Freeberg MA, Han T, Kamath A, Yao Y, Fukuda T, Suzuki T, Kim JK, Inoki K. 2017. LARP1 functions as a molecular switch for mTORC1-mediated translation of an essential class of mRNAs. *eLife* 6: e25237.

Hsieh AC, Liu Y, Edlind MP, Ingolia NT, Janes MR, Sher A, Shi EY, Stumpf CR, Christensen C, Bonham MJ, et al. 2012. The translational landscape of mTOR signalling steers cancer initiation and metastasis. *Nature* 485: 55–61.

Huang J, Manning BD. 2008. The TSC1-TSC2 complex: A molecular switchboard controlling cell growth. *Biochem J* 412: 179–190.

Huo Y, Iadevaia V, Yao Z, Kelly I, Cosulich S, Guichard S, Foster LJ, Proud CG. 2012. Stable isotope-labelling analysis of the impact of inhibition of the mammalian target of rapamycin on protein synthesis. *Biochem J* 444: 141–151.

Iadevaia V, Liu R, Proud CG. 2014. mTORC1 signaling controls multiple steps in ribosome biogenesis. *Semin Cell Dev Biol* **36**: 113–120.

* Ingolia NT, Hussmann JA, Weissman JS. 2018. Ribosome profiling: Global views of translation. *Cold Spring Harb Perspect Biol* doi: 10.1101/cshperspect.a032698.

Jafarnejad SM, Chapat C, Matta-Camacho E, Gelbart IA, Hesketh GG, Arguello M, Garzia A, Kim SH, Attig J, Shapiro M, et al. 2018. Translational control of ERK signaling through miRNA/4EHP-directed silencing. *eLife* **7**: e35034.

Johanns M, Pyr Dit Ruys S, Houddane A, Vertommen D, Herinckx G, Hue L, Proud CG, Rider MH. 2017. Direct and indirect activation of eukaryotic elongation factor 2 kinase by AMP-activated protein kinase. *Cell Signal* **36**: 212–221.

Joshi S, Platanias LC. 2014. Mnk kinase pathway: Cellular functions and biological outcomes. *World J Biol Chem* **5**: 321–333.

Joshi B, Cai AL, Keiper BD, Minich WB, Mendez R, Beach CM, Stolarski R, Darzynkiewicz E, Rhoads RE. 1995. Phosphorylation of eukaryotic protein synthesis initiation factor 4E at serine 209. *J Biol Chem* **270**: 14597–14603.

Joshi B, Cameron A, Jagus R. 2004. Characterization of mammalian eIF4E-family members. *Eur J Biochem* **271**: 2189–2203.

Kang SA, Pacold ME, Cervantes CL, Lim D, Lou HJ, Ottina K, Gray NS, Turk BE, Yaffe MB, Sabatini DM. 2013. mTORC1 phosphorylation sites encode their sensitivity to starvation and rapamycin. *Science* **341**: 1236566.

Kenney JW, Moore CE, Wang X, Proud CG. 2014. Eukaryotic elongation factor 2 kinase, an unusual enzyme with multiple roles. *Adv Biol Regul* **55**: 15–27.

Kenney JW, Genheden M, Moon KM, Wang X, Foster LJ, Proud CG. 2016. Eukaryotic elongation factor 2 kinase regulates the synthesis of microtubule-related proteins in neurons. *J Neurochem* **136**: 276–284.

Konicek BW, Stephens JR, McNulty AM, Robichaud N, Peery RB, Dumstorf CA, Dowless MS, Iversen PW, Parsons S, Ellis KE, et al. 2011. Therapeutic inhibition of MAP kinase interacting kinase blocks eukaryotic initiation factor 4E phosphorylation and suppresses outgrowth of experimental lung metastases. *Cancer Res* **71**: 1849–1857.

* Kwan T, Thompson SR. 2018. Noncanonical translation initiation in eukaryotes. *Cold Spring Harb Perspect Biol* doi: 10.1101/cshperspect.a032672.

Lahr RM, Fonseca BD, Ciotti GE, Al-Ashtal HA, Jia JJ, Niklaus MR, Blagden SP, Alain T, Berman AJ. 2017. La-related protein 1 (LARP1) binds the mRNA cap, blocking eIF4F assembly on TOP mRNAs. *eLife* **6**: e24146.

Landon AL, Muniandy PA, Shetty AC, Lehrmann E, Volpon L, Houng S, Zhang Y, Dai B, Peroutka R, Mazan-Mamczarz K, et al. 2014. MNKs act as a regulatory switch for eIF4E1 and eIF4E3 driven mRNA translation in DLBCL. *Nat Commun* **5**: 5413.

Larsson O, Morita M, Topisirovic I, Alain T, Blouin MJ, Pollak M, Sonenberg N. 2012. Distinct perturbation of the translatome by the antidiabetic drug metformin. *Proc Natl Acad Sci* **109**: 8977–8982.

Lazaris-Karatzas A, Montine KS, Sonenberg N. 1990. Malignant transformation by a eukaryotic initiation factor subunit that binds to mRNA 5′ cap. *Nature* **345**: 544–547.

Li JJ, Bickel PJ, Biggin MD. 2014. System wide analyses have underestimated protein abundances and the importance of transcription in mammals. *Peer J* **2**: e270.

Lim S, Saw TY, Zhang M, Janes MR, Nacro K, Hill J, Lim AQ, Chang CT, Fruman DA, Rizzieri DA, et al. 2013. Targeting of the MNK-eIF4E axis in blast crisis chronic myeloid leukemia inhibits leukemia stem cell function. *Proc Natl Acad Sci* **110**: E2298–E2307.

Lin SC, Hardie DG. 2017. AMPK: Sensing glucose as well as cellular energy status. *Cell Metab* **27**: 299–313.

Ling J, Morley SJ, Traugh JA. 2005. Inhibition of cap-dependent translation via phosphorylation of eIF4G by protein kinase Pak2. *EMBO J* **24**: 4094–4105.

Liu R, Proud CG. 2016. Eukaryotic elongation factor 2 kinase as a drug target in cancer, and in cardiovascular and neurodegenerative diseases. *Acta Pharmacol Sin* **37**: 285–294.

Luca R, Averna M, Zalfa F, Vecchi M, Bianchi F, La FG, Del NF, Nardacci R, Bianchi M, Nuciforo P, et al. 2013. The fragile X protein binds mRNAs involved in cancer progression and modulates metastasis formation. *EMBO Mol Med* **5**: 1523–1536.

Ma XM, Yoon SO, Richardson CJ, Julich K, Blenis J. 2008. SKAR links pre-mRNA splicing to mTOR/S6K1-mediated enhanced translation efficiency of spliced mRNAs. *Cell* **133**: 303–313.

Mader S, Lee H, Pause A, Sonenberg N. 1995. The translation initiation factor eIF-4E binds to a common motif shared by the translation factor eIF-4γ and the translational repressors 4E-binding proteins. *Mol Cell Biol* **15**: 4990–4997.

Marcotrigiano J, Gingras A-C, Sonenberg N, Burley SK. 1997. Co-crystal structure of the messenger RNA 5′ cap-binding protein (eIF4E) bound to 7-methyl-GDP. *Cell* **89**: 951–961.

Masvidal L, Hulea L, Furic L, Topisirovic I, Larsson O. 2017. mTOR-sensitive translation: Cleared fog reveals more trees. *RNA Biol* **14**: 1–7.

Matsuo H, Li HJ, McGuire AM, Fletcher CM, Gingras A-C, Sonenberg N, Wagner G. 1997. Structure of translation factor eIF4E bound to m^7GDP and interaction with 4E-binding protein. *Nat Struct Biol* **4**: 717–724.

* Merrick WC, Pavitt GD. 2018. Protein synthesis initiation in eukaryotic cells. *Cold Spring Harb Perspect Biol* doi: 10.1101/cshperspect.a033092.

Meyuhas O, Kahan T. 2014. The race to decipher the top secrets of TOP mRNAs. *Biochim Biophys Acta* **1849**: 801–811.

Miloslavski R, Cohen E, Avraham A, Iluz Y, Hayouka Z, Kasir J, Mudhasani R, Jones SN, Cybulski N, Ruegg MA, et al. 2014. Oxygen sufficiency controls TOP mRNA translation via the TSC-Rheb-mTOR pathway in a 4E-BP-independent manner. *J Mol Cell Biol* **6**: 255–266.

Mitsui K, Brady M, Palfrey HC, Nairn AC. 1993. Purification and characterization of calmodulin-dependent protein kinase III from rabbit reticulocytes and rat pancreas. *J Biol Chem* **268**: 13422–13433.

Montagne J, Stewart MJ, Stocker H, Hafen E, Kozma SC, Thomas G. 1999. *Drosophila* S6 kinase: A regulator of cell size. *Science* **285:** 2126–2129.

Moore CE, Mikolajek H, Regufe da MS, Wang X, Kenney JW, Werner JM, Proud CG. 2015. Elongation factor 2 kinase is regulated by proline hydroxylation and protects cells during hypoxia. *Mol Cell Biol* **35:** 1788–1804.

Morita M, Gravel SP, Chenard V, Sikstrom K, Zheng L, Alain T, Gandin V, Avizonis D, Arguello M, Zakaria C, et al. 2013. mTORC1 controls mitochondrial activity and biogenesis through 4E-BP-dependent translational regulation. *Cell Metab* **18:** 698–711.

Nikolcheva T, Pyronnet S, Chou SY, Sonenberg N, Song A, Clayberger C, Krensky AM. 2002. A translational rheostat for RFLAT-1 regulates RANTES expression in T lymphocytes. *J Clin Invest* **110:** 119–126.

O'Loghlen A, Gonzalez VM, Jurado T, Salinas M, Martin ME. 2007. Characterization of the activity of human MAP kinase-interacting kinase Mnk1b. *Biochim Biophys Acta* **1773:** 1416–1427.

Palm W, Park Y, Wright K, Pavlova NN, Tuveson DA, Thompson CB. 2015. The utilization of extracellular proteins as nutrients is suppressed by mTORC1. *Cell* **162:** 259–270.

Panja D, Bramham CR. 2014. BDNF mechanisms in late LTP formation: A synthesis and breakdown. *Neuropharmacology* **76:** 664–676.

Panja D, Dagyte G, Bidinosti M, Wibrand K, Kristiansen AM, Sonenberg N, Bramham CR. 2009. Novel translational control in Arc-dependent long term potentiation consolidation in vivo. *J Biol Chem* **284:** 31498–31511.

Panja D, Kenney JW, D'Andrea L, Zalfa F, Vedeler A, Wibrand K, Fukunaga R, Bagni C, Proud CG, Bramham CR. 2014. Two-stage translational control of dentate gyrus LTP consolidation is mediated by sustained BDNF-TrkB signaling to MNK. *Cell Rep* **9:** 1430–1445.

Parra-Palau JL, Scheper GC, Wilson ML, Proud CG. 2003. Features in the N and C termini of the MAPK-interacting kinase Mnk1 mediate its nucleocytoplasmic shuttling. *J Biol Chem* **278:** 44197–44204.

Pende M, Um SH, Mieulet V, Sticker M, Goss VL, Mestan J, Mueller M, Fumagalli S, Kozma SC, Thomas G. 2004. S6K1$^{-/-}$/S6K2$^{-/-}$ mice exhibit perinatal lethality and rapamycin-sensitive 5′-terminal oligopyrimidine mRNA translation and reveal a mitogen-activated protein kinase-dependent S6 kinase pathway. *Mol Cell Biol* **24:** 3112–3124.

Philippe L, Vasseur JJ, Debart F, Thoreen CC. 2017. La-related protein 1 (LARP1) repression of TOP mRNA translation is mediated through its cap-binding domain and controlled by an adjacent regulatory region. *Nucleic Acids Res* **16:** 1457–1469.

Price NT, Redpath NT, Severinov KV, Campbell DG, Russell JM, Proud CG. 1991. Identification of the phosphorylation sites in elongation factor-2 from rabbit reticulocytes. *FEBS Lett* **282:** 253–258.

Proud CG. 2015. Mnks, eIF4E phosphorylation and cancer. *Biochim Biophys Acta* **1849:** 766–773.

Pyronnet S, Imataka H, Gingras AC, Fukunaga R, Hunter T, Sonenberg N. 1999. Human eukaryotic translation initiation factor 4G (eIF4G) recruits Mnk1 to phosphorylate eIF4E. *EMBO J* **18:** 270–279.

Pyronnet S, Dostie J, Sonenberg N. 2001. Suppression of cap-dependent translation in mitosis. *Genes Dev* **15:** 2083–2093.

Rau M, Ohlmann T, Morley SJ, Pain VM. 1996. A re-evaluation of the cap-binding translation factor, eIF4E, as a rate-limiting factor for initiation of translation in reticulocyte lysate. *J Biol Chem* **271:** 8983–8990.

Raught B, Gingras AC, Gygi SP, Imataka H, Morino S, Gradi A, Aebersold R, Sonenberg N. 2000. Serum-stimulated, rapamycin-sensitive phosphorylation sites in the eukaryotic translation initiation factor 4GI. *EMBO J* **19:** 434–444.

Raught B, Peiretti F, Gingras AC, Livingstone M, Shahbazian D, Mayeur GL, Polakiewicz RD, Sonenberg N, Hershey JW. 2004. Phosphorylation of eucaryotic translation initiation factor 4B Ser422 is modulated by S6 kinases. *EMBO J* **23:** 1761–1769.

Redpath NT, Price NT, Severinov KV, Proud CG. 1993. Regulation of elongation factor-2 by multisite phosphorylation. *Eur J Biochem* **213:** 689–699.

Redpath NT, Foulstone EJ, Proud CG. 1996a. Regulation of translation elongation factor-2 by insulin via a rapamycin-sensitive signalling pathway. *EMBO J* **15:** 2291–2297.

Redpath NT, Price NT, Proud CG. 1996b. Cloning and expression of cDNA encoding protein synthesis elongation factor-2 kinase. *J Biol Chem* **271:** 17547–17554.

Robichaud N, del Rincon SV, Huor B, Alain T, Petruccelli LA, Hearnden J, Goncalves C, Grotegut S, Spruck CH, Furic L, et al. 2015. Phosphorylation of eIF4E promotes EMT and metastasis via translational control of SNAIL and MMP-3. *Oncogene* **34:** 2032–2042.

* Robichaud N, Sonenberg N, Ruggero D, Schneider RJ. 2018a. Translational control in cancer. *Cold Spring Harb Perspect Biol* doi: 10.1101/cshperspect.a032896.

Robichaud N, Hsu BE, Istomine R, Alvarez F, Blagih J, Ma EH, Morales SV, Dai DL, Li G, Souleimanova M, et al. 2018b. Translational control in the tumor microenvironment promotes lung metastasis: Phosphorylation of eIF4E in neutrophils. *Proc Natl Acad Sci* **115:** E2202–E2209.

Rose AJ, Broholm C, Kiilerich K, Finn SG, Proud CG, Rider MH, Richter EA, Kiens B. 2005. Exercise rapidly increases eukaryotic elongation factor 2 phosphorylation in skeletal muscle of men. *J Physiol* **569:** 223–228.

Rubio CA, Weisburd B, Holderfield M, Arias C, Fang E, DeRisi JL, Fanidi A. 2014. Transcriptome-wide characterization of the eIF4A signature highlights plasticity in translation regulation. *Genome Biol* **15:** 476.

Rudra D, Warner JR. 2004. What better measure than ribosome synthesis? *Genes Dev* **18:** 2431–2436.

Ruvinsky I, Sharon N, Lerer T, Cohen H, Stolovich-Rain M, Nir T, Dor Y, Zisman P, Meyuhas O. 2005. Ribosomal protein S6 phosphorylation is a determinant of cell size and glucose homeostasis. *Genes Dev* **19:** 2199–2211.

Ryazanov AG, Natapov PG, Shestakova EA, Severin FF, Spirin AS. 1988. Phosphorylation of the elongation factor 2: The fifth Ca^{2+}/calmodulin-dependent system of protein phosphorylation. *Biochimie* **70:** 619–626.

Ryazanov AG, Ward MD, Mendola CE, Pavur KS, Dorovkov MV, Wiedmann M, Erdjument-Bromage H, Tempst P, Parmer TG, Prostko CR, et al. 1997. Identification of a

Cite this article as *Cold Spring Harb Perspect Biol* doi: 10.1101/cshperspect.a033050

new class of protein kinases represented by eukaryotic elongation factor-2 kinase. *Proc Natl Acad Sci* **94:** 4884–4889.

Sarbassov dD, Ali SM, Sengupta S, Sheen JH, Hsu PP, Bagley AF, Markhard AL, Sabatini DM. 2006. Prolonged rapamycin treatment inhibits mTORC2 assembly and Akt/PKB. *Mol Cell* **22:** 159–168.

Sasikumar AN, Perez WB, Kinzy TG. 2012. The many roles of the eukaryotic elongation factor 1 complex. *Wiley Interdiscip Rev RNA* **3:** 543–555.

Saxton RA, Sabatini DM. 2017. mTOR signaling in growth, metabolism, and disease. *Cell* **169:** 361–371.

Scheetz AJ, Nairn AC, Constantine-Paton M. 2000. NMDA-mediated control of protein synthesis at developing synapses. *Nat Neurosci* **3:** 1–7.

Scheper GC, van Kollenburg B, Hu J, Luo Y, Goss DJ, Proud CG. 2002. Phosphorylation of eukaryotic initiation factor 4E markedly reduces its affinity for capped mRNA. *J Biol Chem* **277:** 3303–3309.

Scheper GC, Parra JL, Wilson M, van KB, Vertegaal AC, Han ZG, Proud CG. 2003. The N and C termini of the splice variants of the human mitogen-activated protein kinase-interacting kinase Mnk2 determine activity and localization. *Mol Cell Biol* **23:** 5692–5705.

Schwanhausser B, Busse D, Li N, Dittmar G, Schuchhardt J, Wolf J, Chen W, Selbach M. 2011. Global quantification of mammalian gene expression control. *Nature* **473:** 337–342.

Shahbazian D, Roux PP, Mieulet V, Cohen MS, Raught B, Taunton J, Hershey JW, Blenis J, Pende M, Sonenberg N. 2006. The mTOR/PI3K and MAPK pathways converge on eIF4B to control its phosphorylation and activity. *EMBO J* **25:** 2781–2791.

Shama S, Avni D, Frederickson RM, Sonenberg N, Meyuhas O. 1995. Overexpression of initiation factor eIF-4E does not relieve the translational repression of ribosomal protein mRNAs in quiescent cells. *Gene Expr* **4:** 241–252.

Shi J, Hershey JW, Nelson MA. 2009. Phosphorylation of the eukaryotic initiation factor 3f by cyclin-dependent kinase 11 during apoptosis. *FEBS Lett* **583:** 971–977.

Shima H, Pende M, Chen Y, Fumagalli S, Thomas G, Kozma SC. 1998. Disruption of the $p70^{S6k}/p85^{S6k}$ gene reveals a small mouse phenotype and a new functional S6 kinase. *EMBO J* **17:** 6649–6659.

Sinvani H, Haimov O, Svitkin Y, Sonenberg N, Tamarkin-Ben-Harush A, Viollet B, Dikstein R. 2015. Translational tolerance of mitochondrial genes to metabolic energy stress involves TISU and eIF1-eIF4GI cooperation in start codon selection. *Cell Metab* **21:** 479–492.

Slepenkov SV, Darzynkiewicz E, Rhoads RE. 2006. Stopped-flow kinetic analysis of eIF4E and phosphorylated eIF4E binding to cap analogs and capped oligoribonucleotides: Evidence for a one-step binding mechanism. *J Biol Chem* **281:** 14927–14938.

Stead RL, Proud CG. 2013. Rapamycin enhances eIF4E phosphorylation by activating MAP kinase-interacting kinase 2a (Mnk2a). *FEBS Lett* **587:** 2623–2628.

* Stern-Ginossar N, Thompson SR, Mathews MB, Mohr I. 2018. Translational control in virus-infected cells. *Cold Spring Harb Perspect Biol* doi: 10.1101/cshperspect. a033001.

Su X, Yu Y, Zhong Y, Giannopoulou EG, Hu X, Liu H, Cross JR, Ratsch G, Rice CM, Ivashkiv LB. 2015. Interferon-γ regulates cellular metabolism and mRNA translation to potentiate macrophage activation. *Nat Immunol* **16:** 838–849.

Tahmasebi S, Khoutorsky A, Mathews MB, Sonenberg N. 2018. Translation deregulation in human disease. *Nature Rev Mol Cell Biol* (in press).

Tcherkezian J, Cargnello M, Romeo Y, Huttlin EL, Lavoie G, Gygi SP, Roux PP. 2014. Proteomic analysis of cap-dependent translation identifies LARP1 as a key regulator of 5′TOP mRNA translation. *Genes Dev* **28:** 357–371.

Thoreen CC, Chantranupong L, Keys HR, Wang T, Gray NS, Sabatini DM. 2012. A unifying model for mTORC1-mediated regulation of mRNA translation. *Nature* **485:** 109–113.

Truitt ML, Conn CS, Shi Z, Pang X, Tokuyasu T, Coady AM, Seo Y, Barna M, Ruggero D. 2015. Differential requirements for eIF4E dose in normal development and cancer. *Cell* **162:** 59–71.

Tschopp C, Knauf U, Brauchle M, Zurini M, Ramage P, Glueck D, New L, Han J, Gram H. 2000. Phosphorylation of eIF-4E on Ser 209 in response to mitogenic and inflammatory stimuli is faithfully detected by specific antibodies. *Mol Cell Biol Res Commun* **3:** 205–211.

Ueda T, Watanabe-Fukunaga R, Fukuyama H, Nagata S, Fukunaga R. 2004. Mnk2 and Mnk1 are essential for constitutive and inducible phosphorylation of eukaryotic initiation factor 4E but not for cell growth or development. *Mol Cell Biol* **24:** 6539–6549.

Ueda T, Sasaki M, Elia AJ, Chio II, Hamada K, Fukunaga R, Mak TW. 2010. Combined deficiency for MAP kinase-interacting kinase 1 and 2 (Mnk1 and Mnk2) delays tumor development. *Proc Natl Acad Sci* **107:** 13984–13990.

Uniacke J, Holterman CE, Lachance G, Franovic A, Jacob MD, Fabian MR, Payette J, Holcik M, Pause A, Lee S. 2012. An oxygen-regulated switch in the protein synthesis machinery. *Nature* **486:** 126–129.

van Gorp AG, van der Vos KE, Brenkman AB, Bremer A, van den Broek N, Zwartkruis F, Hershey JW, Burgering BM, Calkhoven CF, Coffer PJ. 2009. AGC kinases regulate phosphorylation and activation of eukaryotic translation initiation factor 4B. *Oncogene* **28:** 95–106.

von der Haar T, McCarthy JE. 2002. Intracellular translation initiation factor levels in *Saccharomyces cerevisiae* and their role in cap-complex function. *Mol Microbiol* **46:** 531–544.

Walden WE, Thach RE. 1986. Translational control of gene expression in a normal fibroblast. Characterization of a subclass of mRNAs with unusual kinetic properties. *Biochemistry* **25:** 2033–2041.

Walsh D, Mohr I. 2004. Phosphorylation of eIF4E by Mnk-1 enhances HSV-1 translation and replication in quiescent cells. *Genes Dev* **18:** 660–672.

Wang X, Flynn A, Waskiewicz AJ, Webb BLJ, Vries RG, Baines IA, Cooper J, Proud CG. 1998. The phosphorylation of eukaryotic initiation factor eIF4E in response to phorbol esters, cell stresses and cytokines is mediated by distinct MAP kinase pathways. *J Biol Chem* **273:** 9373–9377.

Wang X, Li W, Williams M, Terada N, Alessi DR, Proud CG. 2001. Regulation of elongation factor 2 kinase by p90[RSK1] and p70 S6 kinase. *EMBO J* **20:** 4370–4379.

Wang X, Regufe da MS, Liu R, Moore CE, Xie J, Lanucara F, Agarwala U, Pyr Dit RS, Vertommen D, Rider MH, et al. 2014. Eukaryotic elongation factor 2 kinase activity is controlled by multiple inputs from oncogenic signaling. *Mol Cell Biol* **34:** 4088–4103.

Waskiewicz AJ, Flynn A, Proud CG, Cooper JA. 1997. Mitogen-activated kinases activate the serine/threonine kinases Mnk1 and Mnk2. *EMBO J* **16:** 1909–1920.

* Wek RC. 2018. Role of eIF2α kinases in translational control and adaptation to cellular stress. *Cold Spring Harb Perspect Biol* doi: 10.1101/cshperspect.a032870.

Welsh GI, Miller CM, Loughlin AJ, Price NT, Proud CG. 1998. Regulation of eukaryotic initiation factor eIF2B: Glycogen synthase kinase-3 phosphorylates a conserved serine which undergoes dephosphorylation in response to insulin. *FEBS Lett* **421:** 125–130.

Wendel HG, De Stanchina E, Fridman JS, Malina A, Ray S, Kogan S, Cordon-Cardo C, Pelletier J, Lowe SW. 2004. Survival signalling by Akt and eIF4E in oncogenesis and cancer therapy. *Nature* **428:** 332–337.

Wendel HG, Silva RL, Malina A, Mills JR, Zhu H, Ueda T, Watanabe-Fukunaga R, Fukunaga R, Teruya-Feldstein J, Pelletier J, et al. 2007. Dissecting eIF4E action in tumorigenesis. *Genes Dev* **21:** 3232–3237.

White E, DiPaola RS. 2009. The double-edged sword of autophagy modulation in cancer. *Clin Cancer Res* **15:** 5308–5316.

Williams DD, Price NT, Loughlin AJ, Proud CG. 2001. Characterisation of the initiation factor eIF2B complex from mammalian cells as a GDP-dissociation stimulator protein. *J Biol Chem* **276:** 24697–24703.

Wittenberg AD, Azar S, Klochendler A, Stolovich-Rain M, Avraham S, Birnbaum L, Binder Gallimidi A, Katz M, Dor Y, Meyuhas O. 2016. Phosphorylated ribosomal protein S6 is required for Akt-driven hyperplasia and malignant transformation, but not for hypertrophy, aneuploidy and hyperfunction of pancreatic β-cells. *PLoS ONE* **11:** e0149995.

Wolfe AL, Singh K, Zhong Y, Drewe P, Rajasekhar VK, Sanghvi VR, Mavrakis KJ, Jiang M, Roderick JE, Van der Meulen J, et al. 2014. RNA G-quadruplexes cause eIF4A-dependent oncogene translation in cancer. *Nature* **513:** 65–70.

Wolfson RL, Sabatini DM. 2017. The dawn of the age of amino acid sensors for the mTORC1 pathway. *Cell Metab* **26:** 301–309.

Wortham NC, Martinez M, Gordiyenko Y, Robinson CV, Proud CG. 2014. Analysis of the subunit organization of the eIF2B complex reveals new insights into its structure and regulation. *FASEB J* **28:** 2225–2237.

Xie J, Mikolajek H, Pigott CR, Hooper KJ, Mellows T, Moore CE, Mohammed H, Werner JM, Thomas GJ, Proud CG. 2015. Molecular mechanism for the control of eukaryotic elongation factor 2 kinase by pH: Role in cancer cell survival. *Mol Cell Biol* **35:** 1805–1824.

Xu H, Zhu J, Smith S, Foldi J, Zhao B, Chung AY, Outtz H, Kitajewski J, Shi C, Weber S, et al. 2012. Notch-RBP-J signaling regulates the transcription factor IRF8 to promote inflammatory macrophage polarization. *Nat Immunol* **13:** 642–650.

Yamaguchi H, Matsushita M, Nairn AC, Kuriyan J. 2001. Crystal structure of the atypical protein kinase domain of a TRP channel with phosphotransferase activity. *Mol Cell* **7:** 1047–1057.

Yang HS, Jansen AP, Komar AA, Zheng X, Merrick WC, Costes S, Lockett SJ, Sonenberg N, Colburn NH. 2003. The transformation suppressor Pdcd4 is a novel eukaryotic translation initiation factor 4A binding protein that inhibits translation. *Mol Cell Biol* **23:** 26–37.

Yang H, Rudge DG, Koos JD, Vaidialingam B, Yang HJ, Pavletich NP. 2013. mTOR kinase structure, mechanism and regulation. *Nature* **497:** 217–223.

Ye Q, Crawley SW, Yang Y, Cote GP, Jia Z. 2010. Crystal structure of the α-kinase domain of *Dictyostelium* myosin heavy chain kinase A. *Sci Signal* **3:** ra17.

Zhang L, Smit-McBride Z, Pan X, Rheinhardt J, Hershey JW. 2008. An oncogenic role for the phosphorylated h-subunit of human translation initiation factor eIF3. *J Biol Chem* **283:** 24047–24060.

Zuberek J, Jemielity J, Jablonowska A, Stepinski J, Dadlez M, Stolarski R, Darzynkiewicz E. 2004. Influence of electric charge variation at residues 209 and 159 on the interaction of eIF4E with the mRNA 5′ terminus. *Biochemistry* **43:** 5370–5379.

Translational Control in Cancer

Nathaniel Robichaud,[1] Nahum Sonenberg,[1] Davide Ruggero,[2] and Robert J. Schneider[3]

[1]Goodman Cancer Research Centre and Department of Biochemistry, McGill University, Montreal, Quebec H3A 1A3, Canada

[2]Helen Diller Family Comprehensive Cancer Center, and Departments of Urology and of Cellular and Molecular Pharmacology, University of California, San Francisco, California 94158

[3]NYU School of Medicine, Alexandria Center for Life Science, New York, New York 10016

Correspondence: nahum.sonenberg@mcgill.ca; robert.schneider@nyumc.org

The translation of messenger RNAs (mRNAs) into proteins is a key event in the regulation of gene expression. This is especially true in the cancer setting, as many oncogenes and transforming events are regulated at this level. Cancer-promoting factors that are translationally regulated include cyclins, antiapoptotic factors, proangiogenic factors, regulators of cell metabolism, prometastatic factors, immune modulators, and proteins involved in DNA repair. This review discusses the diverse means by which cancer cells deregulate and reprogram translation, and the resulting oncogenic impacts, providing insights into the complexity of translational control in cancer and its targeting for cancer therapy.

ON THE IMPORTANCE OF TRANSLATIONAL CONTROL IN CANCER

Considerable resources are dedicated to messenger RNA (mRNA) translation in normal proliferating cells. Up to 20% of cellular energy is used for protein synthesis, as compared with 15% for transcription and DNA replication, and 20% for various cation pumps (Buttgereit and Brand 1995). Moreover, the majority of transcription is directed to the synthesis of ribosomal RNA (rRNA) and mRNAs encoding ribosomal proteins, further increasing the dedication of cellular energy to mRNA translation, making it the most energy-demanding cellular process (Rolfe and Brown 1997). The rapid and continuous proliferation of highly malignant cancers requires continuous protein synthesis and increased ribosome content, further increas-ing the energy consumption directed to protein synthesis (Silvera et al. 2010). Most tumor cells are under physiological stresses such as hypoxia and nutritional deprivation that down-regulate mRNA translation in normal cells but become uncoupled from regulation as a part of the transformation process, further stressing the cell.

In this article, we review how cancer cells hijack the translational machinery for their sustained proliferation, survival, and metastasis (spread) to distant tissue sites, and how the changes in activity and expression of distinct translation factors confer cancer-specific translation of mRNAs. A brief overview of the scanning mechanism of translation initiation and the factors involved in this process is presented in Figure 1, as discussed at length elsewhere (Kwan and Thompson 2018; Merrick and Pavitt 2018).

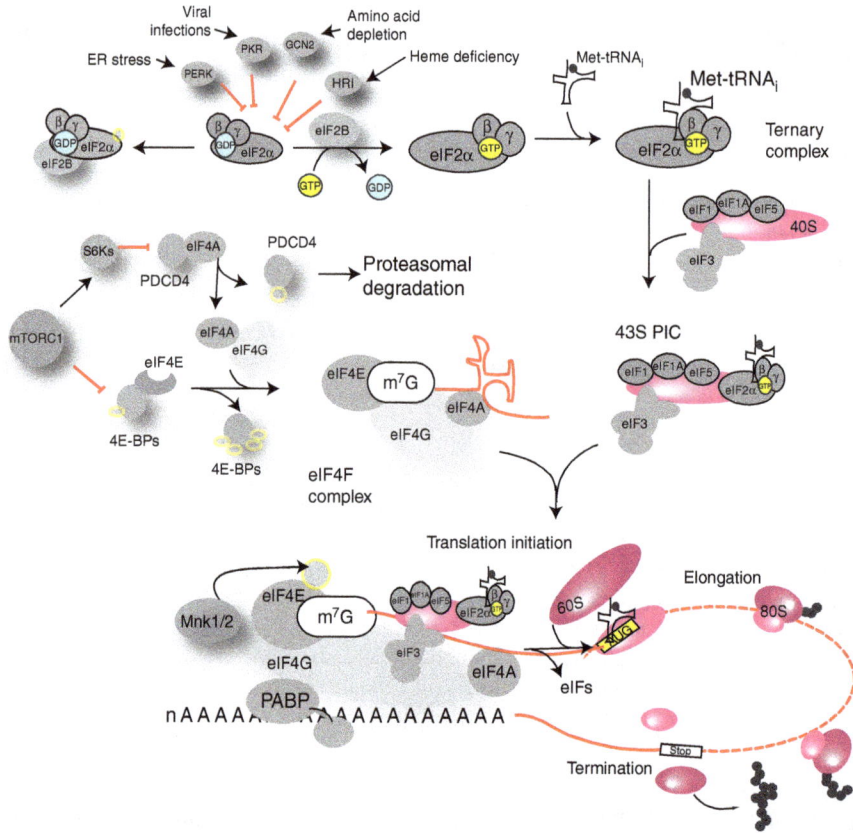

Figure 1. Overview of translation initiation. Initiation proceeds via a scanning mechanism, whereby the 40S ribosomal subunit is recruited to the 5′ extremity of the messenger RNA (mRNA) and scans the 5′ untranslated region (UTR) of the mRNA toward its 3′ end. When the anticodon of the initiator methionyl-transfer RNA (tRNA) (Met-tRNA$_i$) base-pairs with the start codon, the 60S subunit is recruited and elongation begins with the sequential addition of amino acids until a stop codon is reached and termination occurs. (*Top*) Formation of the ternary complex (TC): eukaryotic initiation factor (eIF)2—composed of α, β, and γ subunits—GTP and Met-tRNA$_i$. eIF2α can be phosphorylated by protein kinase R (PKR), PERK, GCN2, or HRI, responding to different stresses such as double-stranded RNA, misfolded proteins, amino acid deficiency, and heme deficiency, respectively. Phosphorylation of eIF2α leads to stabilization of the GDP-loaded complex with the guanine nucleotide exchange factor eIF2B and reduced cycling to the active, GTP-bound TC, resulting in inhibition of global protein synthesis and active translation of upstream open reading frame (uORF)-containing mRNAs such as ATF4. TCs associate with the 40S ribosome and other initiation factors, forming the preinitiation complex (PIC). (*Middle*) The eIF4F complex consists of the cap-binding protein eIF4E, the scaffolding protein eIF4G, and the eIF4A helicase. The eIF4E-binding proteins (4E-BP1/2/3) sequester eIF4E and prevent its binding to eIF4G. Similarly, PDCD4 sequesters eIF4A. Phosphorylation downstream from mammalian target of rapamycin (mTOR) alleviates the inhibitory activity of PDCD4 and 4E-BPs, allowing for eIF4F complex formation, recruitment of the 43S PIC and the initiation of translation. In addition, eIF4E can be phosphorylated by the MNKs.

MECHANISMS OF DEREGULATED AND SELECTIVE mRNA TRANSLATION IN CANCER

Mathematical modeling has predicted that, given constant rates of ribosome elongation on mRNAs and limited translation initiation factors, the translation of specific mRNAs for which ribosome recruitment is inefficient will be disproportionately affected by changes in the level or activity of initiation factors (Lodish 1974). In principle, this basic model helps ex-

Cite this article as *Cold Spring Harb Perspect Biol* doi: 10.1101/cshperspect.a032896

plain a number of aspects of translational deregulation in cancer cells discussed here, including how changes in factors involved in the translation of most mRNAs can yield specific advantages. This model predicts that cancer cells are particularly well equipped to promote angiogenesis, survival, and proliferation, even in times of high physiological stress. This can be achieved through a variety of complex molecular alterations that increase selective translation of poorly translated mRNAs, including increased expression or availability of certain translation initiation factors, and increased activity of signaling pathways regulating them (reviewed in Silvera et al. 2010; Ruggero 2013; Bhat et al. 2015; Truitt and Ruggero 2016; de la Parra et al. 2017). Some of the many ways by which this can be achieved are described below and summarized in Figure 2.

Gain/Loss of Initiation Factors

Aberrant expression of translation initiation factors was the first mechanism to be identified by which cancer cells deregulate translation,

shown originally by the ability of overexpressed eukaryotic initiation factor (eIF)4E to transform NIH 3T3 cells in vitro (Lazaris-Karatzas et al. 1990). Several initiation factors were subsequently found to be overexpressed in human cancers, as discussed below and shown in Figure 2.

eIF4F Complex Formation

Ribosomes are recruited to the 5′ end of the mRNA via the eIF4F complex, which consists of eIF4E, eIF4G, and eIF4A (Fig. 1). The enzymatic component, eIF4A, provides the helicase activity required to unwind secondary structures present in mRNA 5′ untranslated regions (UTRs), a process that is vastly enhanced by binding to eIF4G and eIF4E (Feoktistova et al. 2013). Considering that oncogenic mRNAs tend to have long and stable 5′UTRs, they are particularly sensitive to the activity of eIF4A and the formation of eIF4F (Kozak 1987; Chu and Pelletier 2015; Gandin et al. 2016). All three eIF4F subunits can be deregulated in cancer cells, their genomic loci have all been shown to be amplified in human tumors, and they are all targets of

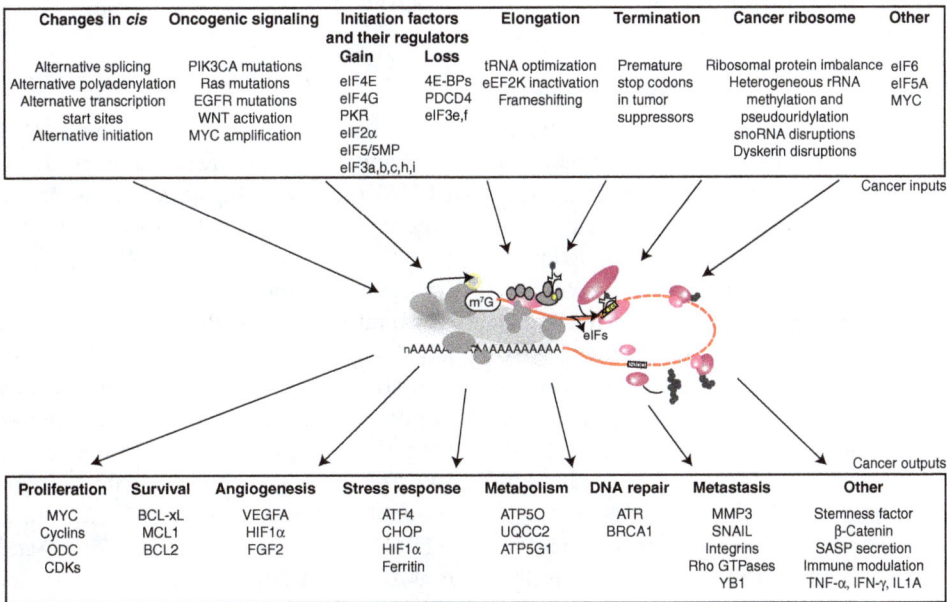

Changes in *cis*	Oncogenic signaling	Initiation factors and their regulators		Elongation	Termination	Cancer ribosome	Other
		Gain	**Loss**				
Alternative splicing	PIK3CA mutations	eIF4E	4E-BPs	tRNA optimization	Premature	Ribosomal protein imbalance	eIF6
Alternative polyadenylation	Ras mutations	eIF4G	PDCD4	eEF2K inactivation	stop codons	Heterogeneous rRNA	eIF5A
Alternative transcription	EGFR mutations	PKR	eIF3e,f	Frameshifting	in tumor	methylation and	MYC
start sites	WNT activation	eIF2α			suppressors	pseudouridylation	
Alternative initiation	MYC amplification	eIF5/5MP				snoRNA disruptions	
		eIF3a,b,c,h,i				Dyskerin disruptions	

Cancer inputs

Cancer outputs

Proliferation	Survival	Angiogenesis	Stress response	Metabolism	DNA repair	Metastasis	Other
MYC	BCL-xL	VEGFA	ATF4	ATP5O	ATR	MMP3	Stemness factor
Cyclins	MCL1	HIF1α	CHOP	UQCC2	BRCA1	SNAIL	β-Catenin
ODC	BCL2	FGF2	HIF1α	ATP5G1		Integrins	SASP secretion
CDKs			Ferritin			Rho GTPases	Immune modulation
						YB1	TNF-α, IFN-γ, IL1A

Figure 2. Cancer inputs and outputs. Summarized view of the oncogenic lesions feeding into the translational machinery (cancer inputs, *top*) and of the resulting advantages conferred by aberrant translation, with examples of regulated mRNAs (cancer outputs, *bottom*). (*Center*) Schematic of translation, as shown in Figure 1.

the MYC oncoprotein (Silvera et al. 2010; Ruggero 2013; Bhat et al. 2015). eIF4E and eIF4G act as classical oncogenes, their overexpression resulting in transformation in vitro (in cell culture) and in vivo (in animals) (Lazaris-Karatzas et al. 1990; Fukuchi-Shimogori et al. 1997; Ruggero et al. 2004; Silvera et al. 2009). Importantly, it was recently shown by using a mouse model of haploinsufficient levels of eIF4E (loss of one allele) that a 50% reduction of eIF4E levels does not limit general protein synthesis and embryonic development in the eIF4E$^{+/-}$ mouse, which is quite remarkably viable. However, eIF4E$^{+/-}$ cells and mice are highly resistant to cellular transformation and tumorigenicity, even when driven by a powerful oncogene such as *Hras-V12* (Truitt et al. 2015). eIF4E overexpression in cancer cells is therefore highly specific for oncogenic transformation.

Translation initiation in cancer cells can also be regulated by phosphorylation of eIF4F components. For example, eIF4E phosphorylation by MNK1 and MNK2 kinases has been shown to promote tumor development and dissemination (Topisirovic et al. 2004; Wendel et al. 2007; Furic et al. 2010; Robichaud et al. 2015; Proud 2018), and is elevated in human lung, breast, and prostate cancers, among others (Fan et al. 2009; Graff et al. 2009; Ramon y Cajal et al. 2014). Multiple phosphorylation sites exist on eIF4G, not all of which have known functions, although Ser1186 phosphorylation is thought to regulate MNK recruitment, and thus phosphorylation of eIF4E (Raught et al. 2000; Dobrikov et al. 2011). MNK-mediated phosphorylation of eIF4E has also been shown to be involved in translational reprogramming that drives resistance to tamoxifen in estrogen-receptor-positive breast cancer (Geter et al. 2017). The mechanism underlying translational regulation by eIF4E phosphorylation is poorly understood, but is thought to involve initiation factor recycling (Scheper and Proud 2002; Proud 2018). This hypothesis is based on the requirement of eIF4G and eIF3 for MNK recruitment and the decreased affinity of eIF4E for the cap resulting from its phosphorylation (Pyronnet et al. 1999; Scheper et al. 2002; Slepenkov et al. 2006; Walsh and Mohr 2014).

Another regulatory mechanism involves sequestration of initiation factors to prevent eIF4F complex formation. Thus, eIF4A can be sequestered by the tumor suppressor programmed cell death 4 (PDCD4), the loss of which is associated with cancer cell invasion and poor patient survival in some cancers (Proud 2018). The exact role of PDCD4 in cancer remains to be established and better characterized (Yang et al. 2003, 2004; Wang et al. 2008; Meric-Bernstam et al. 2012; Modelska et al. 2015). Using a similar mechanism, the 4E-BPs, which compete with eIF4G for binding to eIF4E, are thought to act as tumor suppressors by inhibiting cap-dependent translation (Alain et al. 2012). 4E-BP expression can be lost, as in pancreatic cancer, or its function impaired by inhibitory phosphorylation (Martineau et al. 2014). In contrast, 4E-BP expression is increased in the setting of stage III nonmetastatic esophageal, breast, and prostate cancers, in which it is proposed to oppose metastasis, but lead to the development of large locally advanced tumors (Salehi and Mashayekhi 2006; Braunstein et al. 2007; Coleman et al. 2009; Graff et al. 2009). Thus, the role of the 4E-BPs in cancer may be more complex than originally thought.

Ternary Complex Formation

The ternary complex (TC) is composed of eIF2, GTP, and the initiator methionine transfer RNA (tRNA) (Fig. 1). Deregulated TC formation in cancer cells is a complex issue that has led to different, occasionally conflicting findings regarding the role of eIF2α phosphorylation (Koromilas 2015). On the one hand, it is generally thought that increased eIF2α phosphorylation grants cancer cells a heightened ability to respond to stress conditions encountered along the path to malignancy, by promoting the translation of upstream open reading frame (uORF)-containing stress-response mRNAs such as ATF4 (Robichaud and Sonenberg 2017; Sendoel et al. 2017; Wek 2018). Accordingly, overexpression of eIF2α or one of its kinases, protein kinase R (PKR), has been shown to promote transformation in some contexts, although the mechanism remains unclear (Wang et al. 1999;

Cite this article as *Cold Spring Harb Perspect Biol* doi: 10.1101/cshperspect.a032896

Rosenwald et al. 2001; Kim et al. 2002; Rosenwald et al. 2003; Ye et al. 2010). On the other hand, long-term eIF2α phosphorylation promotes apoptosis, and has prompted research into the development of cancer therapies that promote the activity of eIF2α kinases or the inhibition of eIF2α phosphatases (Schewe and Aguirre-Ghiso 2009; Denoyelle et al. 2012; Hamamura et al. 2014). These results indicate that the outcome of eIF2α phosphorylation in cancer cells is highly context specific, perhaps related to disease site or underlying driver mutations, and may change over time (Silvera et al. 2010).

Additional ways to modulate TC activity in cancer cells have been less explored but include overexpression of eIF5 or its mimic proteins (MPs), 5MP1 and 5MP2. When present in excess, these proteins can bind to eIF2 and sequester it from the 40S ribosome (Singh et al. 2006, 2011). Similar to eIF2α phosphorylation, eIF2 binding by eIF5 or the 5MPs reduces global protein synthesis but enhances translation of uORF-containing mRNAs, including ATF4 (Kozel et al. 2016). This mechanism appears to be important for the malignant properties of some cancer types such as fibrosarcoma and salivary mucoepidermoid carcinoma (Li et al. 2009).

eIF3, Connecting eIF4F and Preinitiation Complexes

eIF3 is a multisubunit complex that binds directly to eIF4G, bridging it to the preinitiation complex (PIC) (Fig. 1), thus connecting mRNAs with the 40S ribosomal subunit and allowing scanning to occur (Hinnebusch et al. 2016; Merrick and Pavitt 2018). Increased eIF3 levels should promote bridging mRNA to the PIC, and therefore increase the rate of translation initiation. This appears to occur when the a, b, and c subunits of eIF3 are overexpressed, resulting in increased levels of whole eIF3 complex, increased global protein synthesis, and increased translation of oncogenic transcripts (Zhang et al. 2007). However, studies on eIF3 present a more complex picture. Indeed, when overexpressed or silenced in immortalized cells, certain individual subunits of eIF3 display oncogenic

properties, while other subunits behave oppositely, as tumor suppressors (Hershey 2015). Such confusing results may arise from the nontranslational roles of certain eIF3 subunits, such as eIF3a, which has been reported to bind to components of the cytoskeleton (MacDonald et al. 1999; Lin et al. 2001), or eIF3f and eIF3i, which have been proposed to regulate signal transduction pathways (Wang et al. 2013; Lee et al. 2016). New roles for eIF3 in translation have also been reported. These include binding to mRNA structures in the 5′UTR of such cancer-relevant mRNAs as c-Jun and Btg1 (Lee et al. 2015), or even directly to the cap of the c-Jun mRNA (Lee et al. 2016), opening new avenues of research on eIF3-dependent translation in cancer.

Translation Elongation and Termination

Although much of the scientific literature has been focused on translation initiation, an understanding of oncogenic changes in elongation and termination is emerging as well. For example, a dominant role for the loss of inhibitory regulation of elongation via eukaryotic elongation factor 2 phosphorylation by its kinase (eEF2K) has been shown for intestinal tumor formation (Faller et al. 2015; Proud 2018). Furthermore, the increased availability of specific tRNA isoaccepting species in cancer cells appears to play a role in tumorigenesis (Gingold et al. 2014). Indeed, the speed of amino acid incorporation during the elongation phase is dependent on the availability of the corresponding charged tRNA (Novoa and Ribas de Pouplana 2012). Several studies have reported distinct translation programs in which proliferating undifferentiated cells and cancer cells express tRNAs optimized to correspond to the codon usage of pro-proliferative mRNAs (Pavon-Eternod et al. 2009; Gingold et al. 2014; Topisirovic and Sonenberg 2014). Hence, in cancer cells, the repertoire of available tRNAs is thought to be reprogrammed such that the species required for the translation of oncogenic mRNAs are present at sufficient levels. In addition, elongation can be deregulated in cancer via programmed -1 ribosomal frameshifting (-1 RPF), a process by which sequence elements force elongating ribosomes back by one base, leading to frameshifts, premature stop codons, and nonsense-mediated mRNA decay (NMD) (Dever et al. 2018; Karousis and Mühlemann 2018). This mechanism may explain the oncogenic role of, for example, silent mutations inducing frameshifting in tumor suppressors (Sulima et al. 2017).

An area that is underexplored is aberrant or altered regulation of termination in the cancer setting. However, termination at premature stop codons can be a cancer driver if it occurs as a result of somatic mutations in tumor-suppressor genes (Bordeira-Carrico et al. 2012), resulting in NMD of the corresponding transcript (Karousis and Mühlemann 2018). NMD can be prevented by using aminoglycosides or small molecule drugs that promote readthrough of premature stop codons. Clinical introduction of such a small molecule inhibitor for the treatment of Duchenne muscular dystrophy, known as Translarna (ataluren) (Welch et al. 2007; Finkel et al. 2013) could be useful in the oncology setting, although its level of efficacy remains to be established.

Two initiation factors with confusing, multiple roles in mRNA translation are also associated with altered translational regulation in cancer cells. One is eIF6, a ribosomal subunit anti-association factor that prevents aberrant interactions between the 40S and 60S ribosomal subunits. eIF6 must be displaced from the ribosome for the final step of 60S ribosome biosynthesis in the nucleolus, and it can promote 80S ribosome disassembly in the cytosol by preventing the reassociation of post-termination 60S ribosomes, thereby impairing further rounds of initiation with prolonged sequestration (Ceci et al. 2003; Brina et al. 2015). Thus, eIF6 may play multiple roles in deregulating translation. Aberrant eIF6 expression has been observed in colorectal, and head and neck cancers, in which it accumulates in the nucleolus (Sanvito et al. 2000; Rosso et al. 2004). In contrast, reduced eIF6 levels have been shown to prevent oncogene-induced transformation and delay lymphomagenesis (Gandin et al. 2008; Miluzio et al. 2011).

The second translation factor that has multiple oncogenic activities is eIF5A. eIF5A was originally described as an initiation factor important for the formation of the first peptide bond during elongation, but has since been found to play a role in the elongation of poorly translated tripeptide regions in mRNAs: prolines, glycines, and/or basic residues (Benne et al. 1978; Gutierrez et al. 2013; Mathews and Hershey 2015; Pelechano and Alepuz 2017; Dever et al. 2018). More general roles in elongation and translation termination have also been proposed based on the accumulation of stalled ribosomes in cells lacking eIF5A in yeast (Pelechano and Alepuz 2017; Schuller et al. 2017). There is considerable interest in eIF5A, as both of its isoforms are overexpressed in a variety of cancers, including pancreatic, hepatic, colon, lung, and ovarian, and have been linked to the metastatic capacity of cancer cells (reviewed in Mathews and Hershey 2015). As the only known mammalian protein containing a hypusine modification, eIF5A is an attractive target, since its activity can be abrogated by inhibiting the enzymes catalyzing the hypusination (Nakanishi and Cleveland 2016).

Changes in 5′ and 3′ UTR Length and Composition in Cancer Cells

Sequence and structural motifs present in mRNAs determine their intrinsic translational efficiency and their ability to be regulated by *trans*-acting factors such as microRNAs (miRNAs), RNA-binding proteins, and initiation factors (Truitt and Ruggero 2016; Duchaine and Fabian 2018). These motifs tend to be overrepresented in oncogenic mRNAs, conferring tight translational regulation (Kozak 1987). Furthermore, mutations in these noncoding motifs have been found to significantly modulate the expression of proto-oncogenes (Diederichs et al. 2016). Increased secondary structure in the 5′UTR was one of the first *cis*-acting elements to be identified that affect the rate or efficiency of cap-dependent mRNA translation initiation (Sonenberg et al. 1981). Subsequently, studies have shown that oncogenic mRNAs typically possess stable 5′UTR structures and, accordingly, show greater dependence on eIF4F. These include mRNAs encoding MYC, ornithine de-

Cite this article as *Cold Spring Harb Perspect Biol* doi: 10.1101/cshperspect.a032896

carboxylase (ODC), a number of cyclins and cyclin-dependent kinases, among others (Darveau et al. 1985; Grens and Scheffler 1990; Rosenwald et al. 1993).

mRNA sequence elements found in the 5′ and 3′UTRs other than length and stability can also regulate the efficiency of translation. For example, enhanced dependence on eIF4E, but not eIF4A, was shown for the translation initiator of the short 5′UTR (TISU) element found in certain mRNAs (Elfakess et al. 2011; Kwan and Thompson 2018). This is in contrast to mRNAs containing internal ribosome entry sites (IRESs) that are cap-independent, but more highly dependent on eIF4G and eIF4A (Komar and Hatzoglou 2011). mRNAs can contain alternative initiation codons and inhibitory ORFs upstream of the canonical initiating AUG, which can severely hamper normal start site identification by the 43S PIC. These sequence elements are enriched in oncogenic transcripts (Kozak 1987). However, under periods of stress, including oncogenic stress (hypoxia and nutritional deprivation), some mRNAs with uORFs show selectively increased translation resulting from increased eIF2α phosphorylation (Hinnebusch et al. 2016; Wek 2018). Additionally, structural or sequence motifs in some 5′UTRs mediate the recruitment of RNA-binding proteins that modulate mRNA translation. One well-investigated example is that of the transforming growth factor β (TGF-β)-activated translation (BAT) element, which regulates the translation of certain mRNAs involved in the epithelial-to-mesenchymal transition (EMT) that promotes cell migration (Chaudhury et al. 2010). Finally, binding sites for miRNAs are particularly common motifs that affect translation, in addition to mRNA stability, as are AU-rich motifs. All of these elements have been thoroughly reviewed elsewhere (Komar and Hatzoglou 2011; Jonas and Izaurralde 2015; Wurth and Gebauer 2015; Hinnebusch et al. 2016; Truitt and Ruggero 2016) and are described by Duchaine and Fabian (2018).

The majority of identified *cis*-acting elements reduce mRNA translation efficiency, and are found in various combinations in individual oncogenic mRNAs, probably to govern and attenuate translation of mRNAs that have the capacity to transform cells when overexpressed (Mayr and Bartel 2009; Dieudonne et al. 2015). Nevertheless, cancer cells often develop mechanisms that bypass this stringent control. For example, the genome-wide shortening of 5′ and/or 3′UTRs is quite common in cellular transformation, which eliminates the suppressive RNA motifs (Mayr and Bartel 2009; Dieudonne et al. 2015). Alternative transcription start sites downstream from the 5′UTR translation regulatory element are one such reported mechanism. For example, initiating transcription downstream from two inhibitory uORFs in *MDM2* results in increased translation of the *MDM2* mRNA and inhibition of p53 (Landers et al. 1997). Similarly, translation inhibitory elements in the 3′UTR are sometimes eliminated by alternative polyadenylation in cancer cells, producing shorter 3′UTRs lacking, for instance, miRNA-binding sites and AU-rich elements that promote rapid mRNA decay (Mayr 2016).

Oncogenic Signaling

Most physiological signals, including stresses, nutritional and growth factor stimulation, metabolic functions, and endocrine factors, among others, are integrated through the translational machinery (Robichaud and Sonenberg 2017). The mammalian target of rapamycin (mTOR) in particular plays a crucial role in translation regulatory signaling (Fig. 1) by phosphorylating the 4E-BPs to allow eIF4F complex formation, and ribosomal protein S6 kinase (S6K), which, in turn, regulates eIF4A via PDCD4 and eIF4B, as well as eEF2K, to alleviate inhibition of elongation (Raught et al. 2004; Dorrello et al. 2006; Faller et al. 2015; Proud 2018). Many of the most commonly mutated genes across many cancer types encode key proteins that regulate signaling pathways impinging on translation, including *PIK3CA*, *KRAS*, *PTEN*, *APC*, and *EGFR*, among others (Kandoth et al. 2013). Particularly interesting is the case of the *MYC* oncogene, which promotes almost all aspects of translation (reviewed in van Riggelen et al. 2010). Although mRNA translation is far from the only effector of these signaling pathways, its importance is shown by the central role for "oncogenic translation" activity required to initiate and maintain the transformed phenotype (Truitt and Ruggero 2016). For example, rapid inhibition of translation following inhibition of upstream kinases such as AKT and KRAS occurs before any transcriptional changes in models of glioblastoma (Rajasekhar et al. 2003). Consequently, deregulation of the translation machinery in response to cancer drug treatment may well be an early driver of drug resistance as a means of maintaining and/or reprogramming cancer cell protein synthesis (Ilic et al. 2011; Zindy et al. 2011; Alain et al. 2012; Boussemart et al. 2014; Cope et al. 2014; Musa et al. 2016).

Is There a Cancer Ribosome?

It has long been discussed whether there are cancer-specific modifications in ribosomes themselves that could promote cancer-specific mRNA translational reprogramming. Discovery of "ribosomopathies," a family of syndromes caused by inherited mutations in genes encoding ribosomal proteins and their regulators that are characterized by initial defects in hematopoiesis, followed by increased cancer susceptibility, supports this hypothesis (Sulima et al. 2017). However, the mechanism underlying the oncogenic effects of these mutations is unclear.

It is noteworthy that the stoichiometry of ribosomal proteins and rRNA modifications, such as methylation and pseudouridylation, varies in cancer cells (Truitt and Ruggero 2016). This suggests that individual ribosomes may possess unique modifications that alter their ability to translate certain mRNAs with respect to others. Whether such "cancer ribosomes" could promote the selective translation of oncogenic mRNAs and/or restrict the synthesis of tumor suppressors remains to be established. However, in support of this model is the finding that disruption of dyskerin, the enzyme-catalyzing pseudouridylation, or of small nucleolar RNAs (snoRNAs) that guide dyskerin to rRNA sites, are common in many cancers and can impair the translation of mRNAs encoding

critical tumor suppressors, such as p53 and p27 (Truitt and Ruggero 2016). Most recently, the existence of ribosome heterogeneity has recently been shown in embryonic stem cells, in which a distinct subset of mRNAs is preferentially translated by different ribosome pools (Shi et al. 2017).

Alternatively, the link between ribosomal proteins, ribosomopathies, and cancer has been attributed to nontranslational roles of components of the translation machinery, mainly p53 stabilization (Gentilella et al. 2015; Zhou et al. 2015b). Thus, a subribosomal complex composed of the 5S rRNA and ribosomal proteins RPL5 and RPL11, binds to and sequesters MDM2, resulting in p53 stabilization and cell-cycle arrest (Gentilella et al. 2015). The complex forms when deregulated ribosome biogenesis leads to the imbalance of ribosomal components, as is the case in ribosomopathies (Gentilella et al. 2015). It is thought that somatic loss of p53 signaling allows hematopoietic cells to escape from the cell cycle arrest caused by defective ribosome biosynthesis, resulting in the cancer predisposition associated with ribosomopathies (Sulima et al. 2017).

SELECTIVE ONCOGENIC ADVANTAGES OF DEREGULATED TRANSLATION

Proliferation and Apoptosis

A large body of research has shown translational regulation of antiapoptotic factors, cyclins, and cyclin-dependent kinases (Sonenberg 1994; Silvera et al. 2010; Ruggero 2013; Teng et al. 2013; de la Parra et al. 2017). Although the role of translation in cell proliferation and survival has historically been the focus of much of the research in this field, other hallmarks of cancer are also regulated at the level of translation, as delineated below (Fig. 2).

Angiogenesis

Tumor angiogenesis is an ongoing process of continuous remodeling to accommodate tumor growth and is promoted by a variety of translational mechanisms. The mRNAs encoding two

Cite this article as *Cold Spring Harb Perspect Biol* doi: 10.1101/cshperspect.a032896

major regulators of angiogenesis, VEGFA and HIF1α, are translated via a variety of mechanisms that ensure cancer cells' ability to adapt to hypoxia. Thus, the translation of *VEGFA* and *HIF1α* mRNAs can be promoted by both cap-dependent and cap-independent mechanisms, via the use of IRESs, uORFs, and possibly other noncanonical regulatory elements (Lang et al. 2002; Braunstein et al. 2007; Bastide et al. 2008; Arcondeguy et al. 2013). Although their translation is associated with increased eIF4E expression in human tumors (Nathan et al. 1997; Scott et al. 1998; Dodd et al. 2015), the complex translational regulation of *VEGFA* and *HIF1α* mRNAs allows for their translation to be maintained even in profound hypoxia and nutrient deprivation. Interestingly, HIF1α binds to the *EIF4E* promoter to promote its transcription, suggesting the possibility that the response to hypoxia could switch from an initial cap-independent mechanism to a cap-dependent one (Yi et al. 2013).

Stress Responses

In addition to hypoxia, cancer cells must modulate translation in the face of a variety of other stresses (Young and Wek 2016; Robichaud and Sonenberg 2017). Interestingly, the responses to diverse stressors share common regulatory mechanisms. Thus, up to 49% of the transcriptome, and essentially all mRNAs translated under stress conditions, are regulated by eIF2α phosphorylation, as they have been reported to include uORFs. These mRNAs disproportionately encode proteins involved in pathways that allow cancer cells to adapt to their environment (Calvo et al. 2009; Andreev et al. 2015; Young and Wek 2016; Wek 2018). Furthermore, IRESs, mRNA methylation, and a variety of noncanonical mechanisms of translation initiation maintain protein synthesis in the face of various stresses that inhibit cap-dependent translation (Meyer et al. 2015; Zhou et al. 2015a; Lacerda et al. 2017; Robichaud and Sonenberg 2017). How specific subsets of mRNAs are selectively translated in response to each stress is not well understood. Another issue requiring further investigation relates to

the fact that inhibition of general translation associated with eIF2α phosphorylation, if persistent, eventually causes cell death (Young and Wek 2016; Robichaud and Sonenberg 2017). Cancer cells may partially escape apoptosis because of the fact that eIF2α phosphorylation promotes the translation of factors promoting its dephosphorylation, resulting in a feedback inhibitory loop (Andreev et al. 2015).

Emerging Oncogenic Advantages of Deregulated Translation

Considering that most cancer deaths are caused by metastatic dissemination, a key emerging concept is the ability of cancer cells to deregulate the translation of prometastatic factors such as matrix metalloproteases, integrins, transcription factors involved in the EMT, and GTPases involved in migration (Silvera et al. 2009; Nasr et al. 2013; Fujimura et al. 2015; Pinzaglia et al. 2015; Robichaud et al. 2015). The importance of cancer-cell-specific translation in the maintenance of cellular energy balance is also becoming clearer, as energy status and protein synthesis are regulated reciprocally to achieve an equilibrium (Proud 2006; Morita et al. 2013; Gandin et al. 2016; Miluzio et al. 2016; Robichaud and Sonenberg 2017). In addition, the interplay between translation and reactive oxygen species (ROS) in cancer cells has recently been revealed. Thus, components of the translation machinery are particularly sensitive to cysteine oxidation by ROS (Chio et al. 2016), whereas the mRNAs encoding key antioxidant proteins possess a motif termed cytosine-enriched regulator of translation (CERT) that confers translation regulation in response to increased eIF4E expression levels (Truitt et al. 2015). Finally, deregulated translation can also promote the expression of proteins involved in DNA repair such as BRCA1, thus enabling the escape from oncogene-induced senescence and resistance to DNA-damaging agents (Badura et al. 2012; Avdulov et al. 2015; Musa et al. 2016). Protein synthesis thus provides a crucial means for cancer cells to disrupt a variety of processes important for all steps of tumor biology.

CONSIDERING HETEROGENEOUS CELL POPULATIONS IN TUMORS

Cancer Stem Cells

Translational regulation in stem cells has emerged as an important new area of study, especially given the low transcriptional activity of these cells in embryonic and hematopoietic systems (see Teixeira and Lehmann 2018). Thus, the 4E-BPs are required to limit translation and ensure the maintenance of hematopoietic and embryonic stem cells (Signer et al. 2016; Tahmasebi et al. 2016), whereas eIF2α phosphorylation promotes the maintenance of muscle stem cells (Zismanov et al. 2016). In the cancer setting, tumor-initiating cells in a mouse skin cancer model display reduced protein synthesis and aberrant uORF translation, linking their stem-like state to eIF2α phosphorylation (Blanco et al. 2016). Phosphorylation of eIF4E has also been implicated in the stem-like phenotype by promoting the synthesis of stem cell maintenance factors such as β-catenin (Altman et al. 2013; Lim et al. 2013; Bell et al. 2016). These studies suggest that cancer stem-like cells may be characterized by low protein synthesis rates compared with most cancer cells caused by hypophosphorylation of 4E-BPs and/or hyperphosphorylation of eIF2α. The spectrum of inhibitors of translation to which cancer stem cells respond may therefore be different than that of bulk tumor cells, which must be taken into consideration for potential cancer therapies.

Immune Cells

The importance of translational regulation in immune cell populations is evident from the immune-suppressive effects of inhibitors such as rapamycin (Martel et al. 1977; So et al. 2016). Indeed, mTOR activity in T cells regulates translation, metabolic reprogramming, differentiation, lineage determination, and activation (Bjur et al. 2013; Araki et al. 2017; Linke et al. 2017; Yoo et al. 2017). The above-cited studies notwithstanding, little is known regarding translation regulatory events specific to the many different immune cell populations, particularly in the cancer-stroma context. eIF4E phos-phorylation appears to be important for the synthesis of cancer-relevant cytokines and chemokines, such as tumor necrosis factor α (TNF-α) and interferon γ (IFN-γ) (Andersson and Sundler 2006; Rowlett et al. 2008; Herdy et al. 2012; Salvador-Bernáldez et al. 2017), and for the development of experimental auto-immune encephalomyelitis (Gorentla et al. 2013). In neutrophils, eIF4E phosphorylation promotes survival and metastatic dissemination in a mouse model of breast cancer (Robichaud et al. 2018). Thus, translation in a variety of immune cell populations can regulate various aspects of tumor biology.

CONCLUDING REMARKS

As we ponder the remarkable and numerous ways in which translational control can be usurped in cancer biology, we are left to discover exciting and promising paths to therapeutic interventions. The possibility that benefits can be derived from targeting translation in both the cancer and immune compartments, as appears to be the case for MNK inhibitors, should improve patient outcome. These findings are particularly meaningful considering the clinical development of MNK inhibitors in immune-oncology (ClinicalTrials.gov identifiers: NCT03258398, NCT02439346). Whether other means of inhibiting translation may be similarly useful or even more potent is an intriguing direction that remains to be investigated (Chu and Pelletier 2018). Indeed, studying the effect of newly developed translation-targeting drugs on the tumor microenvironment should help better predict their clinical benefit, and identify useful therapeutic combinations.

ACKNOWLEDGMENTS

This work is by supported by grants to N.S. from The Susan G. Komen Breast Cancer Foundation (IIR12224057) and CIHR (MOP-7214 and FND-148423). N.S. is a Howard Hughes Medical Institute Senior International Scholar. N.R. is supported by the Vanier Canada Graduate Scholarship (267839) and Rosalind Goodman

Cite this article as *Cold Spring Harb Perspect Biol* doi: 10.1101/cshperspect.a032896

Commemorative Scholarship (GCRC). R.J.S. is supported by the National Institutes of Health (1R01CA178509, 1R01CA207893), the New York State Department of Health (DOH01-ROWLEY-2015-00035), and the Breast Cancer Research Foundation (BCRF-16-143). D.R. is supported by the National Institutes of Health (R01 CA140456, R01 CA184624, R01 CA154916, and R01NS089868).

REFERENCES

*Reference is also in this collection.

Alain T, Morita M, Fonseca BD, Yanagiya A, Siddiqui N, Bhat M, Zammit D, Marcus V, Metrakos P, Voyer LA, et al. 2012. eIF4E/4E-BP ratio predicts the efficacy of mTOR targeted therapies. *Cancer Res* 72: 6468–6476.

Altman JK, Szilard A, Konicek BW, Iversen PW, Kroczynska B, Glaser H, Sassano A, Vakana E, Graff JR, Platanias LC. 2013. Inhibition of Mnk kinase activity by cercosporamide and suppressive effects on acute myeloid leukemia precursors. *Blood* 121: 3675–3681.

Andersson K, Sundler R. 2006. Posttranscriptional regulation of TNFα expression via eukaryotic initiation factor 4E (eIF4E) phosphorylation in mouse macrophages. *Cytokine* 33: 52–57.

Andreev DE, O'Connor PB, Fahey C, Kenny EM, Terenin IM, Dmitriev SE, Cormican P, Morris DW, Shatsky IN, Baranov PV. 2015. Translation of 5′ leaders is pervasive in genes resistant to eIF2 repression. *eLife* 4: e03971.

Araki K, Morita M, Bederman AG, Konieczny BT, Kissick HT, Sonenberg N, Ahmed R. 2017. Translation is actively regulated during the differentiation of CD8+ effector T cells. *Nat Immunol* 18: 1046–1057.

Arcondeguy T, Lacazette E, Millevoi S, Prats H, Touriol C. 2013. VEGF-A mRNA processing, stability and translation: A paradigm for intricate regulation of gene expression at the post-transcriptional level. *Nucleic Acids Res* 41: 7997–8010.

Avdulov S, Herrera J, Smith K, Peterson M, Gomez-Garcia JR, Beadnell TC, Schwertfeger KL, Benyumov AO, Manivel JC, Li S, et al. 2015. eIF4E threshold levels differ in governing normal and neoplastic expansion of mammary stem and luminal progenitor cells. *Cancer Res* 75: 687–697.

Badura M, Braunstein S, Zavadil J, Schneider RJ. 2012. DNA damage and eIF4G1 in breast cancer cells reprogram translation for survival and DNA repair mRNAs. *Proc Natl Acad Sci* 109: 18767–18772.

Bastide A, Karaa Z, Bornes S, Hieblot C, Lacazette E, Prats H, Touriol C. 2008. An upstream open reading frame within an IRES controls expression of a specific VEGF-A isoform. *Nucleic Acids Res* 36: 2434–2445.

Bell JB, Eckerdt FD, Alley K, Magnusson LP, Hussain H, Bi Y, Arslan AD, Clymer J, Alvarez AA, Goldman S, et al. 2016. MNK inhibition disrupts mesenchymal glioma stem cells and prolongs survival in a mouse model of glioblastoma. *Mol Cancer Res* 14: 984–993.

Benne R, Brown-Luedi ML, Hershey JW. 1978. Purification and characterization of protein synthesis initiation factors eIF-1, eIF-4C, eIF-4D, and eIF-5 from rabbit reticulocytes. *J Biol Chem* 253: 3070–3077.

Bhat M, Robichaud N, Hulea L, Sonenberg N, Pelletier J, Topisirovic I. 2015. Targeting the translation machinery in cancer. *Nat Rev Drug Discov* 14: 261–278.

Bjur E, Larsson O, Yurchenko E, Zheng L, Gandin V, Topisirovic I, Li S, Wagner CR, Sonenberg N, Piccirillo CA. 2013. Distinct translational control in CD4+ T cell subsets. *PLoS Genet* 9: e1003494.

Blanco S, Bandiera R, Popis M, Hussain S, Lombard P, Aleksic J, Sajini A, Tanna H, Cortes-Garrido R, Gkatza N, et al. 2016. Stem cell function and stress response are controlled by protein synthesis. *Nature* 534: 335–340.

Bordeira-Carrico R, Pego AP, Santos M, Oliveira C. 2012. Cancer syndromes and therapy by stop-codon read-through. *Trends Mol Med* 18: 667–678.

Boussemart L, Malka-Mahieu H, Girault I, Allard D, Hemmingsson O, Tomasic G, Thomas M, Basmadjian C, Ribeiro N, Thuaud F, et al. 2014. eIF4F is a nexus of resistance to anti-BRAF and anti-MEK cancer therapies. *Nature* 513: 105–109.

Braunstein S, Karpisheva K, Pola C, Goldberg J, Hochman T, Yee H, Cangiarella J, Arju R, Formenti SC, Schneider RJ. 2007. A hypoxia-controlled cap-dependent to cap-independent translation switch in breast cancer. *Mol Cell* 28: 501–512.

Brina D, Miluzio A, Ricciardi S, Biffo S. 2015. eIF6 anti-association activity is required for ribosome biogenesis, translational control and tumor progression. *Biochim Biophys Acta* 1849: 830–835.

Buttgereit F, Brand MD. 1995. A hierarchy of ATP-consuming processes in mammalian cells. *Biochem J* 312: 163–167.

Calvo SE, Pagliarini DJ, Mootha VK. 2009. Upstream open reading frames cause widespread reduction of protein expression and are polymorphic among humans. *Proc Natl Acad Sci* 106: 7507–7512.

Ceci M, Gaviraghi C, Gorrini C, Sala LA, Offenhauser N, Marchisio PC, Biffo S. 2003. Release of eIF6 (p27BBP) from the 60S subunit allows 80S ribosome assembly. *Nature* 426: 579–584.

Chaudhury A, Hussey GS, Ray PS, Jin G, Fox PL, Howe PH. 2010. TGF-β-mediated phosphorylation of hnRNP E1 induces EMT via transcript-selective translational induction of Dab2 and ILEI. *Nat Cell Biol* 12: 286–293.

Chio II, Jafarnejad SM, Ponz-Sarvise M, Park Y, Rivera K, Palm W, Wilson J, Sangar V, Hao Y, Ohlund D, et al. 2016. NRF2 Promotes tumor maintenance by modulating mRNA translation in pancreatic cancer. *Cell* 166: 963–976.

Chu J, Pelletier J. 2015. Targeting the eIF4A RNA helicase as an anti-neoplastic approach. *Biochim Biophys Acta* 1849: 781–791.

* Chu J, Pelletier J. 2018. Translating therapeutics. *Cold Spring Harb Perspect Biol* doi: 10.1101/cshperspect.a032995.

Coleman LJ, Peter MB, Teall TJ, Brannan RA, Hanby AM, Honarpisheh H, Shaaban AM, Smith L, Speirs V, Verghese ET, et al. 2009. Combined analysis of eIF4E and 4E-binding protein expression predicts breast cancer survival and estimates eIF4E activity. *Br J Cancer* 100: 1393–1399.

Cope CL, Gilley R, Balmanno K, Sale MJ, Howarth KD, Hampson M, Smith PD, Guichard SM, Cook SJ. 2014. Adaptation to mTOR kinase inhibitors by amplification of eIF4E to maintain cap-dependent translation. *J Cell Sci* **127:** 788–800.

Darveau A, Pelletier J, Sonenberg N. 1985. Differential efficiencies of in vitro translation of mouse c-myc transcripts differing in the 5′ untranslated region. *Proc Natl Acad Sci* **82:** 2315–2319.

de la Parra C, Walters BA, Geter P, Schneider RJ. 2017. Translation initiation factors and their relevance in cancer. *Curr Opin Genet Dev* **48:** 82–88.

Denoyelle S, Chen T, Chen L, Wang Y, Klosi E, Halperin JA, Aktas BH, Chorev M. 2012. In vitro inhibition of translation initiation by *N,N*′-diarylureas—Potential anticancer agents. *Bioorg Med Chem Lett* **22:** 402–409.

* Dever TE, Dinman JD, Green R. 2018. Translation elongation and recoding in eukaryotes. *Cold Spring Harb Perspect Biol* doi: 10.1101/cshperspect.a032649.

Diederichs S, Bartsch L, Berkmann JC, Frose K, Heitmann J, Hoppe C, Iggena D, Jazmati D, Karschnia P, Linsenmeier M, et al. 2016. The dark matter of the cancer genome: Aberrations in regulatory elements, untranslated regions, splice sites, non-coding RNA and synonymous mutations. *EMBO Mol Med* **8:** 442–457.

Dieudonne FX, O'Connor PB, Gubler-Jaquier P, Yasrebi H, Conne B, Nikolaev S, Antonarakis S, Baranov PV, Curran J. 2015. The effect of heterogeneous transcription start sites (TSS) on the translatome: Implications for the mammalian cellular phenotype. *BMC Genomics* **16:** 986.

Dobrikov M, Dobrikova E, Shveygert M, Gromeier M. 2011. Phosphorylation of eukaryotic translation initiation factor 4G1 (eIF4G1) by protein kinase Cα regulates eIF4G1 binding to Mnk1. *Mol Cell Biol* **31:** 2947–2959.

Dodd KM, Yang J, Shen MH, Sampson JR, Tee AR. 2015. mTORC1 drives HIF-1α and VEGF-A signalling via multiple mechanisms involving 4E-BP1, S6K1 and STAT3. *Oncogene* **34:** 2239–2250.

Dorrello NV, Peschiaroli A, Guardavaccaro D, Colburn NH, Sherman NE, Pagano M. 2006. S6K1- and βTRCP-mediated degradation of PDCD4 promotes protein translation and cell growth. *Science* **314:** 467–471.

* Duchaine TF, Fabian MR. 2018. Mechanistic insights into microRNA-mediated gene silencing. *Cold Spring Harb Perspect Biol* doi: 10.1101/cshperspect.a032771.

Elfakess R, Sinvani H, Haimov O, Svitkin Y, Sonenberg N, Dikstein R. 2011. Unique translation initiation of mRNAs-containing TISU element. *Nucleic Acids Res* **39:** 7598–7609.

Faller WJ, Jackson TJ, Knight JR, Ridgway RA, Jamieson T, Karim SA, Jones C, Radulescu S, Huels DJ, Myant KB, et al. 2015. mTORC1-mediated translational elongation limits intestinal tumour initiation and growth. *Nature* **517:** 497–500.

Fan S, Ramalingam SS, Kauh J, Xu Z, Khuri FR, Sun SY. 2009. Phosphorylated eukaryotic translation initiation factor 4 (eIF4E) is elevated in human cancer tissues. *Cancer Biol Ther* **8:** 1463–1469.

Feoktistova K, Tuvshintogs E, Do A, Fraser CS. 2013. Human eIF4E promotes mRNA restructuring by stimulating eIF4A helicase activity. *Proc Natl Acad Sci* **110:** 13339–13344.

Finkel RS, Flanigan KM, Wong B, Bonnemann C, Sampson J, Sweeney HL, Reha A, Northcutt VJ, Elfring G, Barth J, et al. 2013. Phase 2a study of ataluren-mediated dystrophin production in patients with nonsense mutation Duchenne muscular dystrophy. *PLoS ONE* **8:** e81302.

Fujimura K, Choi S, Wyse M, Strnadel J, Wright T, Klemke R. 2015. Eukaryotic translation initiation factor 5A (EIF5A) regulates pancreatic cancer metastasis by modulating RhoA and Rho-associated kinase (ROCK) protein expression levels. *J Biol Chem* **290:** 29907–29919.

Fukuchi-Shimogori T, Ishii I, Kashiwagi K, Mashiba H, Ekimoto H, Igarashi K. 1997. Malignant transformation by overproduction of translation initiation factor eIF4G. *Cancer Res* **57:** 5041–5044.

Furic L, Rong L, Larsson O, Koumakpayi IH, Yoshida K, Brueschke A, Petroulakis E, Robichaud N, Pollak M, Gaboury LA, et al. 2010. eIF4E phosphorylation promotes tumorigenesis and is associated with prostate cancer progression. *Proc Natl Acad Sci* **107:** 14134–14139.

Gandin V, Miluzio A, Barbieri AM, Beugnet A, Kiyokawa H, Marchisio PC, Biffo S. 2008. Eukaryotic initiation factor 6 is rate-limiting in translation, growth and transformation. *Nature* **455:** 684–688.

Gandin V, Masvidal L, Hulea L, Gravel SP, Cargnello M, McLaughlan S, Cai Y, Balanathan P, Morita M, Rajakumar A, et al. 2016. nanoCAGE reveals 5′ UTR features that define specific modes of translation of functionally related MTOR-sensitive mRNAs. *Genome Res* **26:** 636–648.

Gentilella A, Kozma SC, Thomas G. 2015. A liaison between mTOR signaling, ribosome biogenesis and cancer. *Biochim Biophys Acta* **1849:** 812–820.

Geter PA, Ernlund AW, Bakogianni S, Alard A, Arju R, Giashuddin S, Gadi A, Bromberg J, Schneider RJ. 2017. Hyperactive mTOR and MNK1 phosphorylation of eIF4E confer tamoxifen resistance and estrogen independence through selective mRNA translation reprogramming. *Genes Dev* **31:** 2235–2249.

Gingold H, Tehler D, Christoffersen NR, Nielsen MM, Asmar F, Kooistra SM, Christophersen NS, Christensen LL, Borre M, Sorensen KD, et al. 2014. A dual program for translation regulation in cellular proliferation and differentiation. *Cell* **158:** 1281–1292.

Gorentla BK, Krishna S, Shin J, Inoue M, Shinohara ML, Grayson JM, Fukunaga R, Zhong XP. 2013. Mnk1 and 2 are dispensable for T cell development and activation but important for the pathogenesis of experimental autoimmune encephalomyelitis. *J Immunol* **190:** 1026–1037.

Graff JR, Konicek BW, Lynch RL, Dumstorf CA, Dowless MS, McNulty AM, Parsons SH, Brail LH, Colligan BM, Koop JW, et al. 2009. eIF4E activation is commonly elevated in advanced human prostate cancers and significantly related to reduced patient survival. *Cancer Res* **69:** 3866–3873.

Grens A, Scheffler IE. 1990. The 5′- and 3′-untranslated regions of ornithine decarboxylase mRNA affect the translational efficiency. *J Biol Chem* **265:** 11810–11816.

Gutierrez E, Shin BS, Woolstenhulme CJ, Kim JR, Saini P, Buskirk AR, Dever TE. 2013. eIF5A promotes translation of polyproline motifs. *Mol Cell* **51:** 35–45.

Hamamura K, Minami K, Tanjung N, Wan Q, Koizumi M, Matsuura N, Na S, Yokota H. 2014. Attenuation of malignant phenotypes of breast cancer cells through eIF2α-mediated downregulation of Rac1 signaling. *Intl J Oncol* **44:** 1980–1988.

Herdy B, Jaramillo M, Svitkin YV, Rosenfeld AB, Kobayashi M, Walsh D, Alain T, Sean P, Robichaud N, Topisirovic I, et al. 2012. Translational control of the activation of transcription factor NF-κB and production of type I interferon by phosphorylation of the translation factor eIF4E. *Nat Immunol* **13:** 543–550.

Hershey JW. 2015. The role of eIF3 and its individual subunits in cancer. *Biochim Biophys Acta* **1849:** 792–800.

Hinnebusch AG, Ivanov IP, Sonenberg N. 2016. Translational control by 5′-untranslated regions of eukaryotic mRNAs. *Science* **352:** 1413–1416.

Ilic N, Utermark T, Widlund HR, Roberts TM. 2011. PI3K-targeted therapy can be evaded by gene amplification along the MYC-eukaryotic translation initiation factor 4E (eIF4E) axis. *Proc Natl Acad Sci* **108:** E699–E708.

Jonas S, Izaurralde E. 2015. Towards a molecular understanding of microRNA-mediated gene silencing. *Nat Rev Genet* **16:** 421–433.

Kandoth C, McLellan MD, Vandin F, Ye K, Niu B, Lu C, Xie M, Zhang Q, McMichael JF, Wyczalkowski MA, et al. 2013. Mutational landscape and significance across 12 major cancer types. *Nature* **502:** 333–339.

* Karousis ED, Mühlemann O. 2018. Nonsense-mediated mRNA decay begins where translation ends. *Cold Spring Harb Perspect Biol* doi: 10.1101/cshperspect.a032862.

Kim SH, Gunnery S, Choe JK, Mathews MB. 2002. Neoplastic progression in melanoma and colon cancer is associated with increased expression and activity of the interferon-inducible protein kinase, PKR. *Oncogene* **21:** 8741–8748.

Komar AA, Hatzoglou M. 2011. Cellular IRES-mediated translation: The war of ITAFs in pathophysiological states. *Cell Cycle* **10:** 229–240.

Koromilas AE. 2015. Roles of the translation initiation factor eIF2α serine 51 phosphorylation in cancer formation and treatment. *Biochim Biophys Acta* **1849:** 871–880.

Kozak M. 1987. An analysis of 5′-noncoding sequences from 699 vertebrate messenger RNAs. *Nucleic Acids Res* **15:** 8125–8148.

Kozel C, Thompson B, Hustak S, Moore C, Nakashima A, Singh CR, Reid M, Cox C, Papadopoulos E, Luna RE, et al. 2016. Overexpression of eIF5 or its protein mimic 5MP perturbs eIF2 function and induces ATF4 translation through delayed re-initiation. *Nucleic Acids Res* **44:** 8704–8713.

* Kwan T, Thompson SR. 2018. Noncanonical translation initiation in eukaryotes. *Cold Spring Harb Perspect Biol* doi: 10.1101/cshperspect.a032672.

Lacerda R, Menezes J, Romao L. 2017. More than just scanning: The importance of cap-independent mRNA translation initiation for cellular stress response and cancer. *Cell Mol Life Sci* **74:** 1659–1680.

Landers JE, Cassel SL, George DL. 1997. Translational enhancement of *mdm2* oncogene expression in human tumor cells containing a stabilized wild-type p53 protein. *Cancer Res* **57:** 3562–3568.

Lang KJ, Kappel A, Goodall GJ. 2002. Hypoxia-inducible factor-1α mRNA contains an internal ribosome entry site that allows efficient translation during normoxia and hypoxia. *Mol Biol Cell* **13:** 1792–1801.

Lazaris-Karatzas A, Montine KS, Sonenberg N. 1990. Malignant transformation by a eukaryotic initiation factor subunit that binds to mRNA 5′ cap. *Nature* **345:** 544–547.

Lee AS, Kranzusch PJ, Cate JH. 2015. eIF3 targets cell-proliferation messenger RNAs for translational activation or repression. *Nature* **522:** 111–114.

Lee AS, Kranzusch PJ, Doudna JA, Cate JH. 2016. eIF3d is an mRNA cap-binding protein that is required for specialized translation initiation. *Nature* **536:** 96–99.

Li S, Chai Z, Li Y, Liu D, Bai Z, Situ Z. 2009. BZW1, a novel proliferation regulator that promotes growth of salivary muocepodermoid carcinoma. *Cancer Lett* **284:** 86–94.

Lim S, Saw TY, Zhang M, Janes MR, Nacro K, Hill J, Lim AQ, Chang CT, Fruman DA, Rizzieri DA, et al. 2013. Targeting of the MNK-eIF4E axis in blast crisis chronic myeloid leukemia inhibits leukemia stem cell function. *Proc Natl Acad Sci* **110:** E2298–E2307.

Lin L, Holbro T, Alonso G, Gerosa D, Burger MM. 2001. Molecular interaction between human tumor marker protein p150, the largest subunit of eIF3, and intermediate filament protein K7. *J Cell Biochem* **80:** 483–490.

Linke M, Fritsch SD, Sukhbaatar N, Hengstschlager M, Weichhart T. 2017. mTORC1 and mTORC2 as regulators of cell metabolism in immunity. *FEBS Lett* **591:** 3089–3103.

Lodish HF. 1974. Model for the regulation of mRNA translation applied to haemoglobin synthesis. *Nature* **251:** 385–388.

MacDonald JI, Verdi JM, Meakin SO. 1999. Activity-dependent interaction of the intracellular domain of rat trkA with intermediate filament proteins, the β-6 proteasomal subunit, Ras-GRF1, and the p162 subunit of eIF3. *J Mol Neurosci* **13:** 141–158.

Martel RR, Klicius J, Galet S. 1977. Inhibition of the immune response by rapamycin, a new antifungal antibiotic. *Can J Physiol Pharmacol* **55:** 48–51.

Martineau Y, Azar R, Muller D, Lasfargues C, El Khawand S, Anesia R, Pelletier J, Bousquet C, Pyronnet S. 2014. Pancreatic tumours escape from translational control through 4E-BP1 loss. *Oncogene* **33:** 1367–1374.

Mathews MB, Hershey JW. 2015. The translation factor eIF5A and human cancer. *Biochim Biophys Acta* **1849:** 836–844.

Mayr C. 2016. Evolution and biological roles of alternative 3′UTRs. *Trends Cell Biol* **26:** 227–237.

Mayr C, Bartel DP. 2009. Widespread shortening of 3′UTRs by alternative cleavage and polyadenylation activates oncogenes in cancer cells. *Cell* **138:** 673–684.

Meric-Bernstam F, Chen H, Akcakanat A, Do KA, Lluch A, Hennessy BT, Hortobagyi GN, Mills GB, Gonzalez-Angulo AM. 2012. Aberrations in translational regulation are associated with poor prognosis in hormone receptor-positive breast cancer. *Breast Cancer Res* **14:** R138.

* Merrick WC, Pavitt GD. 2018. Protein synthesis initiation in eukaryotic cells. *Cold Spring Harb Perspect Biol* doi: 10.1101/cshperspect.a033092.

Meyer KD, Patil DP, Zhou J, Zinoviev A, Skabkin MA, Elemento O, Pestova TV, Qian SB, Jaffrey SR. 2015. 5′ UTR m[6]A promotes cap-independent translation. *Cell* **163**: 999–1010.

Miluzio A, Beugnet A, Grosso S, Brina D, Mancino M, Campaner S, Amati B, de Marco A, Biffo S. 2011. Impairment of cytoplasmic eIF6 activity restricts lymphomagenesis and tumor progression without affecting normal growth. *Cancer Cell* **19**: 765–775.

Miluzio A, Ricciardi S, Manfrini N, Alfieri R, Oliveto S, Brina D, Biffo S. 2016. Translational control by mTOR-independent routes: How eIF6 organizes metabolism. *Biochem Soc Trans* **44**: 1667–1673.

Modelska A, Turro E, Russell R, Beaton J, Sbarrato T, Spriggs K, Miller J, Graf S, Provenzano E, Blows F, et al. 2015. The malignant phenotype in breast cancer is driven by eIF4A1-mediated changes in the translational landscape. *Cell Death Dis* **6**: e1603.

Morita M, Gravel SP, Chenard V, Sikstrom K, Zheng L, Alain T, Gandin V, Avizonis D, Arguello M, Zakaria C, et al. 2013. mTORC1 controls mitochondrial activity and biogenesis through 4E-BP-dependent translational regulation. *Cell Metab* **18**: 698–711.

Musa F, Alard A, David-West G, Curtin JP, Blank SV, Schneider RJ. 2016. Dual mTORC1/2 Inhibition as a novel strategy for the resensitization and treatment of platinum-resistant ovarian cancer. *Mol Cancer Ther* **15**: 1557–1567.

Nakanishi S, Cleveland JL. 2016. Targeting the polyamine-hypusine circuit for the prevention and treatment of cancer. *Amino Acids* **48**: 2353–2362.

Nasr Z, Robert F, Porco JA Jr, Muller WJ, Pelletier J. 2013. eIF4F suppression in breast cancer affects maintenance and progression. *Oncogene* **32**: 861–871.

Nathan CO, Carter P, Liu L, Li BD, Abreo F, Tudor A, Zimmer SG, De Benedetti A. 1997. Elevated expression of eIF4E and FGF-2 isoforms during vascularization of breast carcinomas. *Oncogene* **15**: 1087–1094.

Novoa EM, Ribas de Pouplana L. 2012. Speeding with control: Codon usage, tRNAs, and ribosomes. *Trends Genet* **28**: 574–581.

Pavon-Eternod M, Gomes S, Geslain R, Dai Q, Rosner MR, Pan T. 2009. tRNA over-expression in breast cancer and functional consequences. *Nucleic Acids Res* **37**: 7268–7280.

Pelechano V, Alepuz P. 2017. eIF5A facilitates translation termination globally and promotes the elongation of many non polyproline-specific tripeptide sequences. *Nucleic Acids Res* **45**: 7326–7338.

Pinzaglia M, Montaldo C, Polinari D, Simone M, La Teana A, Tripodi M, Mancone C, Londei P, Benelli D. 2015. eIF6 over-expression increases the motility and invasiveness of cancer cells by modulating the expression of a critical subset of membrane-bound proteins. *BMC Cancer* **15**: 131.

Proud CG. 2006. Regulation of protein synthesis by insulin. *Biochem Soc Trans* **34**: 213–216.

* Proud CG. 2018. Phosphorylation and signal transduction pathways in translational control. *Cold Spring Harb Perspect Biol* doi: 10.1101/cshperspect.a033050.

Pyronnet S, Imataka H, Gingras AC, Fukunaga R, Hunter T, Sonenberg N. 1999. Human eukaryotic translation initi-

ation factor 4G (eIF4G) recruits mnk1 to phosphorylate eIF4E. *EMBO J* **18**: 270–279.

Rajasekhar VK, Viale A, Socci ND, Wiedmann M, Hu X, Holland EC. 2003. Oncogenic Ras and Akt signaling contribute to glioblastoma formation by differential recruitment of existing mRNAs to polysomes. *Mol Cell* **12**: 889–901.

Ramon y Cajal S, De Mattos-Arruda L, Sonenberg N, Cortes J, Peg V. 2014. The intra-tumor heterogeneity of cell signaling factors in breast cancer: p4E–BP1 and peIF4E are diffusely expressed and are real potential targets. *Clin Transl Oncol* **16**: 937–941.

Raught B, Gingras AC, Gygi SP, Imataka H, Morino S, Gradi A, Aebersold R, Sonenberg N. 2000. Serum-stimulated, rapamycin-sensitive phosphorylation sites in the eukaryotic translation initiation factor 4GI. *EMBO J* **19**: 434–444.

Raught B, Peiretti F, Gingras AC, Livingstone M, Shahbazian D, Mayeur GL, Polakiewicz RD, Sonenberg N, Hershey JW. 2004. Phosphorylation of eucaryotic translation initiation factor 4B Ser422 is modulated by S6 kinases. *EMBO J* **23**: 1761–1769.

Robichaud N, Sonenberg N. 2017. Translational control and the cancer cell response to stress. *Curr Opin Cell Biol* **45**: 102–109.

Robichaud N, del Rincon SV, Huor B, Alain T, Petruccelli LA, Hearnden J, Goncalves C, Grotegut S, Spruck CH, Furic L, et al. 2015. Phosphorylation of eIF4E promotes EMT and metastasis via translational control of SNAIL and MMP-3. *Oncogene* **34**: 2032–2042.

Robichaud N, Hsu BE, Istomine R, Alvarez F, Blagih J, Ma EH, Morales SV, Dai DL, Li G, Souleimanova M, et al. 2018. Translational control in the tumor microenvironment promotes lung metastasis: Phosphorylation of eIF4E in neutrophils. *Proc Natl Acad Sci* **16**: 937–941.

Rolfe DF, Brown GC. 1997. Cellular energy utilization and molecular origin of standard metabolic rate in mammals. *Physiol Rev* **77**: 731–758.

Rosenwald IB, Lazaris-Karatzas A, Sonenberg N, Schmidt EV. 1993. Elevated levels of cyclin D1 protein in response to increased expression of eukaryotic initiation factor 4E. *Mol Cell Biol* **13**: 7358–7363.

Rosenwald IB, Hutzler MJ, Wang S, Savas L, Fraire AE. 2001. Expression of eukaryotic translation initiation factors 4E and 2α is increased frequently in bronchioloalveolar but not in squamous cell carcinomas of the lung. *Cancer* **92**: 2164–2171.

Rosenwald IB, Wang S, Savas L, Woda B, Pullman J. 2003. Expression of translation initiation factor eIF-2α is increased in benign and malignant melanocytic and colonic epithelial neoplasms. *Cancer* **98**: 1080–1088.

Rosso P, Cortesina G, Sanvito F, Donadini A, Di Benedetto B, Biffo S, Marchisio PC. 2004. Overexpression of p27BBP in head and neck carcinomas and their lymph node metastases. *Head Neck* **26**: 408–417.

Rowlett RM, Chrestensen CA, Nyce M, Harp MG, Pelo JW, Cominelli F, Ernst PB, Pizarro TT, Sturgill TW, Worthington MT. 2008. MNK kinases regulate multiple TLR pathways and innate proinflammatory cytokines in macrophages. *Am J Physiol Gastrointest Liver Physiol* **294**: G452–G459.

Ruggero D. 2013. Translational control in cancer etiology. *Cold Spring Harb Perspect Biol* 5: a012336.

Ruggero D, Montanaro L, Ma L, Xu W, Londei P, Cordon-Cardo C, Pandolfi PP. 2004. The translation factor eIF-4E promotes tumor formation and cooperates with c-Myc in lymphomagenesis. *Nat Med* 10: 484–486.

Salehi Z, Mashayekhi F. 2006. Expression of the eukaryotic translation initiation factor 4E (eIF4E) and 4E-BP1 in esophageal cancer. *Clin Biochem* 39: 404–409.

Salvador-Bernáldez M, Mateus SB, Del Barco Barrantes I, Arthur SC, Martínez-A C, Nebreda AR, Salvador JM. 2017. p38α regulates cytokine-induced IFNγ secretion via the Mnk1/eIF4E pathway in Th1 cells. *Immunol Cell Biol* 95: 814–823.

Sanvito F, Vivoli F, Gambini S, Santambrogio G, Catena M, Viale E, Veglia F, Donadini A, Biffo S, Marchisio PC. 2000. Expression of a highly conserved protein, p27BBP, during the progression of human colorectal cancer. *Cancer Res* 60: 510–516.

Scheper GC, Proud CG. 2002. Does phosphorylation of the cap-binding protein eIF4E play a role in translation initiation? *Eur J Biochem* 269: 5350–5359.

Scheper GC, van Kollenburg B, Hu J, Luo Y, Goss DJ, Proud CG. 2002. Phosphorylation of eukaryotic initiation factor 4E markedly reduces its affinity for capped mRNA. *J Biol Chem* 277: 3303–3309.

Schewe DM, Aguirre-Ghiso JA. 2009. Inhibition of eIF2α dephosphorylation maximizes bortezomib efficiency and eliminates quiescent multiple myeloma cells surviving proteasome inhibitor therapy. *Cancer Res* 69: 1545–1552.

Schuller AP, Wu CC, Dever TE, Buskirk AR, Green R. 2017. eIF5A functions globally in translation elongation and termination. *Mol Cell* 66: 194–205.e195.

Scott PA, Smith K, Poulsom R, De Benedetti A, Bicknell R, Harris AL. 1998. Differential expression of vascular endothelial growth factor mRNA vs protein isoform expression in human breast cancer and relationship to eIF-4E. *Br J Cancer* 77: 2120–2128.

Sendoel A, Dunn JG, Rodriguez EH, Naik S, Gomez NC, Hurwitz B, Levorse J, Dill BD, Schramek D, Molina H, et al. 2017. Translation from unconventional 5′ start sites drives tumour initiation. *Nature* 541: 494–499.

Shi Z, Fujii K, Kovary KM, Genuth NR, Rost HL, Teruel MN, Barna M. 2017. Heterogeneous ribosomes preferentially translate distinct subpools of mRNAs genome-wide. *Mol Cell* 67: 71–83.e77.

Signer RA, Qi L, Zhao Z, Thompson D, Sigova AA, Fan ZP, DeMartino GN, Young RA, Sonenberg N, Morrison SJ. 2016. The rate of protein synthesis in hematopoietic stem cells is limited partly by 4E-BPs. *Genes Dev* 30: 1698–1703.

Silvera D, Arju R, Darvishian F, Levine PH, Zolfaghari L, Goldberg J, Hochman T, Formenti SC, Schneider RJ. 2009. Essential role for eIF4GI overexpression in the pathogenesis of inflammatory breast cancer. *Nat Cell Biol* 11: 903–908.

Silvera D, Formenti SC, Schneider RJ. 2010. Translational control in cancer. *Nat Rev Cancer* 10: 254–266.

Singh CR, Lee B, Udagawa T, Mohammad-Qureshi SS, Yamamoto Y, Pavitt GD, Asano K. 2006. An eIF5/eIF2 complex antagonizes guanine nucleotide exchange by eIF2B during translation initiation. *EMBO J* 25: 4537–4546.

Singh CR, Watanabe R, Zhou D, Jennings MD, Fukao A, Lee B, Ikeda Y, Chiorini JA, Campbell SG, Ashe MP, et al. 2011. Mechanisms of translational regulation by a human eIF5-mimic protein. *Nucleic Acids Res* 39: 8314–8328.

Slepenkov SV, Darzynkiewicz E, Rhoads RE. 2006. Stopped-flow kinetic analysis of eIF4E and phosphorylated eIF4E binding to cap analogs and capped oligoribonucleotides: Evidence for a one-step binding mechanism. *J Biol Chem* 281: 14927–14938.

So L, Lee J, Palafox M, Mallya S, Woxland CG, Arguello M, Truitt ML, Sonenberg N, Ruggero D, Fruman DA. 2016. The 4E-BP-eIF4E axis promotes rapamycin-sensitive growth and proliferation in lymphocytes. *Sci Signal* 9: ra57.

Sonenberg N. 1994. Regulation of translation and cell growth by eIF-4E. *Biochimie* 76: 839–846.

Sonenberg N, Guertin D, Cleveland D, Trachsel H. 1981. Probing the function of the eucaryotic 5′ cap structure by using a monoclonal antibody directed against cap-binding proteins. *Cell* 27: 563–572.

Sulima SO, Hofman IJF, De Keersmaecker K, Dinman JD. 2017. How ribosomes translate cancer. *Cancer Discov* 7: 1069–1087.

Tahmasebi S, Jafarnejad SM, Tam IS, Gonatopoulos-Pournatzis T, Matta-Camacho E, Tsukumo Y, Yanagiya A, Li W, Atlasi Y, Caron M, et al. 2016. Control of embryonic stem cell self-renewal and differentiation via coordinated alternative splicing and translation of YY2. *Proc Natl Acad Sci* 113: 12360–12367.

* Teixeira FK, Lehmann R. 2018. Translational control during developmental transitions. *Cold Spring Harb Perspect Biol* doi: 10.1101/cshperspect.a032987.

Teng T, Thomas G, Mercer CA. 2013. Growth control and ribosomopathies. *Curr Opin Genet Dev* 23: 63–71.

Topisirovic I, Sonenberg N. 2014. Distinctive tRNA repertoires in proliferating versus differentiating cells. *Cell* 158: 1238–1239.

Topisirovic I, Ruiz-Gutierrez M, Borden KL. 2004. Phosphorylation of the eukaryotic translation initiation factor eIF4E contributes to its transformation and mRNA transport activities. *Cancer Res* 64: 8639–8642.

Truitt ML, Ruggero D. 2016. New frontiers in translational control of the cancer genome. *Nat Rev Cancer* 16: 288–304.

Truitt ML, Conn CS, Shi Z, Pang X, Tokuyasu T, Coady AM, Seo Y, Barna M, Ruggero D. 2015. Differential requirements for eIF4E dose in normal development and cancer. *Cell* 162: 59–71.

van Riggelen J, Yetil A, Felsher DW. 2010. MYC as a regulator of ribosome biogenesis and protein synthesis. *Nat Rev Cancer* 10: 301–309.

Walsh D, Mohr I. 2014. Coupling 40S ribosome recruitment to modification of a cap-binding initiation factor by eIF3 subunit e. *Genes Dev* 28: 835–840.

Wang S, Rosenwald IB, Hutzler MJ, Pihan GA, Savas L, Chen JJ, Woda BA. 1999. Expression of the eukaryotic translation initiation factors 4E and 2α in non-Hodgkin's lymphomas. *Am J Pathol* 155: 247–255.

Wang X, Wei Z, Gao F, Zhang X, Zhou C, Zhu F, Wang Q, Gao Q, Ma C, Sun W, et al. 2008. Expression and prognostic significance of PDCD4 in human epithelial ovarian carcinoma. *Anticancer Res* **28:** 2991–2996.

Wang YW, Lin KT, Chen SC, Gu DL, Chen CF, Tu PH, Jou YS. 2013. Overexpressed-eIF3I interacted and activated oncogenic Akt1 is a theranostic target in human hepatocellular carcinoma. *Hepatology* **58:** 239–250.

* Wek RC. 2018. Role of eIF2α kinases in translational control and adaptation to cellular stress. *Cold Spring Harb Perspect Biol* doi: 10.1101/cshperspect.a032870.

Welch EM, Barton ER, Zhuo J, Tomizawa Y, Friesen WJ, Trifillis P, Paushkin S, Patel M, Trotta CR, Hwang S, et al. 2007. PTC124 targets genetic disorders caused by nonsense mutations. *Nature* **447:** 87–91.

Wendel HG, Silva RL, Malina A, Mills JR, Zhu H, Ueda T, Watanabe-Fukunaga R, Fukunaga R, Teruya-Feldstein J, Pelletier J, et al. 2007. Dissecting eIF4E action in tumorigenesis. *Genes Dev* **21:** 3232–3237.

Wurth L, Gebauer F. 2015. RNA-binding proteins, multifaceted translational regulators in cancer. *Biochim Biophys Acta* **1849:** 881–886.

Yang HS, Jansen AP, Komar AA, Zheng X, Merrick WC, Costes S, Lockett SJ, Sonenberg N, Colburn NH. 2003. The transformation suppressor Pdcd4 is a novel eukaryotic translation initiation factor 4A binding protein that inhibits translation. *Mol Cell Biol* **23:** 26–37.

Yang HS, Cho MH, Zakowicz H, Hegamyer G, Sonenberg N, Colburn NH. 2004. A novel function of the MA-3 domains in transformation and translation suppressor Pdcd4 is essential for its binding to eukaryotic translation initiation factor 4A. *Mol Cell Biol* **24:** 3894–3906.

Ye J, Kumanova M, Hart LS, Sloane K, Zhang H, De Panis DN, Bobrovnikova-Marjon E, Diehl JA, Ron D, Koumenis C. 2010. The GCN2-ATF4 pathway is critical for tumour cell survival and proliferation in response to nutrient deprivation. *EMBO J* **29:** 2082–2096.

Yi T, Papadopoulos E, Hagner PR, Wagner G. 2013. Hypoxia-inducible factor-1α (HIF-1α) promotes cap-dependent translation of selective mRNAs through up-regulating initiation factor eIF4E1 in breast cancer cells under hypoxia conditions. *J Biol Chem* **288:** 18732–18742.

Yoo HS, Lee K, Na K, Zhang YX, Lim HJ, Yi T, Song SU, Jeon MS. 2017. Mesenchymal stromal cells inhibit CD25 expression via the mTOR pathway to potentiate T-cell suppression. *Cell Death Dis* **8:** e2632.

Young SK, Wek RC. 2016. Upstream open reading frames differentially regulate gene-specific translation in the integrated stress response. *J Biol Chem* **291:** 16927–16935.

Zhang L, Pan X, Hershey JW. 2007. Individual overexpression of five subunits of human translation initiation factor eIF3 promotes malignant transformation of immortal fibroblast cells. *J Biol Chem* **282:** 5790–5800.

Zhou J, Wan J, Gao X, Zhang X, Jaffrey SR, Qian SB. 2015a. Dynamic m⁶A mRNA methylation directs translational control of heat shock response. *Nature* **526:** 591–594.

Zhou X, Liao WJ, Liao JM, Liao P, Lu H. 2015b. Ribosomal proteins: Functions beyond the ribosome. *J Mol Cell Biol* **7:** 92–104.

Zindy P, Berge Y, Allal B, Filleron T, Pierredon S, Cammas A, Beck S, Mhamdi L, Fan L, Favre G, et al. 2011. Formation of the eIF4F translation-initiation complex determines sensitivity to anticancer drugs targeting the EGFR and HER2 receptors. *Cancer Res* **71:** 4068–4073.

Zismanov V, Chichkov V, Colangelo V, Jamet S, Wang S, Syme A, Koromilas AE, Crist C. 2016. Phosphorylation of eIF2α is a translational control mechanism regulating muscle stem cell quiescence and self-renewal. *Cell Stem Cell* **18:** 79–90.

Translational Control during Developmental Transitions

Felipe Karam Teixeira[1] and Ruth Lehmann[2]

[1]Department of Genetics, University of Cambridge, Cambridge CB2 3EH, United Kingdom

[2]Howard Hughes Medical Institute (HHMI) and Kimmel Center for Biology and Medicine of the Skirball Institute, Department of Cell Biology, New York University School of Medicine, New York, New York 10016

Correspondence: fk319@cam.ac.uk; ruth.lehmann@med.nyu.edu

The many steps of gene expression, from the transcription of a gene to the production of its protein product, are well understood. Yet, transcriptional regulation has been the focal point for the study of gene expression during development. However, quantitative studies reveal that messenger RNA (mRNA) levels are not necessarily good predictors of the respective proteins' levels in a cell. This discrepancy is, at least in part, the result of developmentally regulated, translational mechanisms that control the spatiotemporal regulation of gene expression. In this review, we focus on translational regulatory mechanisms mediating global transitions in gene expression: the shift from the maternal to the embryonic developmental program in the early embryo and the switch from the self-renewal of stem cells to differentiation in the adult.

The control of fate transitions is essential during development, the process by which an entire organism is built from a single totipotent cell, and for homeostasis in adults, in which stem cells sustain tissue integrity by contributing newly differentiated progeny cells. In both cases, changes in cellular fate rely on changes in gene expression, which instruct transitioning cells with new directives and provide the means to remodel cellular structure and to achieve a new identity. Not surprisingly, gene expression is regulated at practically every level during development and stem cell differentiation, from transcription and splicing of pre-messenger RNAs (mRNAs), through mRNA localization and decay, to translation and protein maturation, modification, and degradation.

From plants to mammals, accumulating evidence indicates that translational control plays critical roles during organismal development, tissue homeostasis, and tumorigenesis (Buszczak et al. 2014; Simsek and Barna 2017). Although global quantification of gene expression in different cell systems shows a good correlation between mRNA and protein levels, mRNA levels seem poor predictors for protein abundance during cell state transitions, particularly during early embryonic development (Peshkin et al. 2015; Liu et al. 2016). Yet, the study of spatiotemporal regulation of gene expression and its function in controlling fate transitions has been mostly limited to chromatin-based and transcriptional mechanisms. Among other reasons, this biased attention was facilitated by the advent

---

Cite this article as *Cold Spring Harb Perspect Biol* doi: 10.1101/cshperspect.a032987

of new technologies to systematically study changes in mRNA accumulation and chromatin modifications at the genome level. Indeed, until recently, the study of translational changes occurring in a given developmental transition was limited to small sets of individually tested candidate genes. With the advance of techniques able to tease apart translational changes from fluctuations in mRNA accumulation at the genome level and at nucleotide-resolution—such as ribosome profiling (Ingolia et al. 2009, 2018; Brar and Weissman 2015)—it is now possible to comprehensively visualize changes in translation during development. This has led to the realization of a much broader scale at which translation affects fate decisions and has considerably expanded our mechanistic understanding.

Historically, the most studied examples of translational control during development have been described at developmental periods during which transcriptional input is minimal, such as during oogenesis and early embryogenesis. In the absence of transcriptional activity, gene expression control is restricted to later regulatory steps, most notably translation, mRNA localization, and mRNA and protein decay. Indeed, a large amount of genetic, molecular, and biochemical evidence, mostly gathered from flies and frogs, has provided an impressive detailed picture of the multitude of posttranscriptional mechanisms driving development. More recently, genome-wide approaches revealed the widespread importance of certain regulatory mechanisms and allowed researchers to uncover new layers of regulation and to interrogate other important developmental and cell fate transitions, including stem cell differentiation and cancer.

Here, we review the knowledge gathered from studies conducted in animal models on how translational control plays essential roles during two critical developmental processes relying on cell fate transitions: embryonic development and adult stem cell differentiation. We argue that, as the processes of posttranscriptional regulation are closer to the final protein product than the process of mRNA transcription, this type of regulation may be preferred and advantageous during rapid cell fate transitions. Finally, we discuss how recent studies revealed new mechanisms controlling translation during development.

TRANSLATIONAL CONTROL DURING EMBRYO DEVELOPMENT

With little or no transcriptional input, the late stages of egg maturation and the first steps of embryo development rely exclusively on maternally provided gene product pools (mRNAs and proteins) to lay down the basic structures on which a new organism will be built. Far from being static, these pools rapidly evolve during the course of early development to both drive and adapt to transitioning states, creating a complex and synchronized ballet of dynamically changing mRNA and protein levels. At these developmental stages, sophisticated gene expression modules are in place to spatially and temporally regulate the expression of proteins required to control growth, establish the embryonic axes, and define new cellular fates (Fig. 1). The dependency on maternally provided pools ceases once the zygotic genome is activated and transcription is reestablished in a process known as the maternal-to-zygotic transition (MZT) (Fig. 1A) (Lee et al. 2014). Although the length of the transcriptional quiescence phase varies depending on the species, reflecting evolutionary and developmental strategies, its pervasiveness in animals and plants suggests a widespread requirement for posttranscriptional regulatory mechanisms during early embryo development (Baroux et al. 2008; Tadros and Lipshitz 2009).

mRNA Localization and the Spatial Control of Translation

The most compelling and well-characterized examples of spatially regulated expression of proteins at the subcellular level were delineated by work focusing on embryonic axis specification in *Drosophila*. In this system, the anteroposterior and dorsoventral embryonic axes are predetermined during oogenesis by the targeted intracellular localization of mRNAs encoding key factors that are at the top of cascades establishing embryonic protein gradients. At first, an active and polarized microtubule-dependent mecha-

 Cite this article as *Cold Spring Harb Perspect Biol* doi: 10.1101/cshperspect.a032987

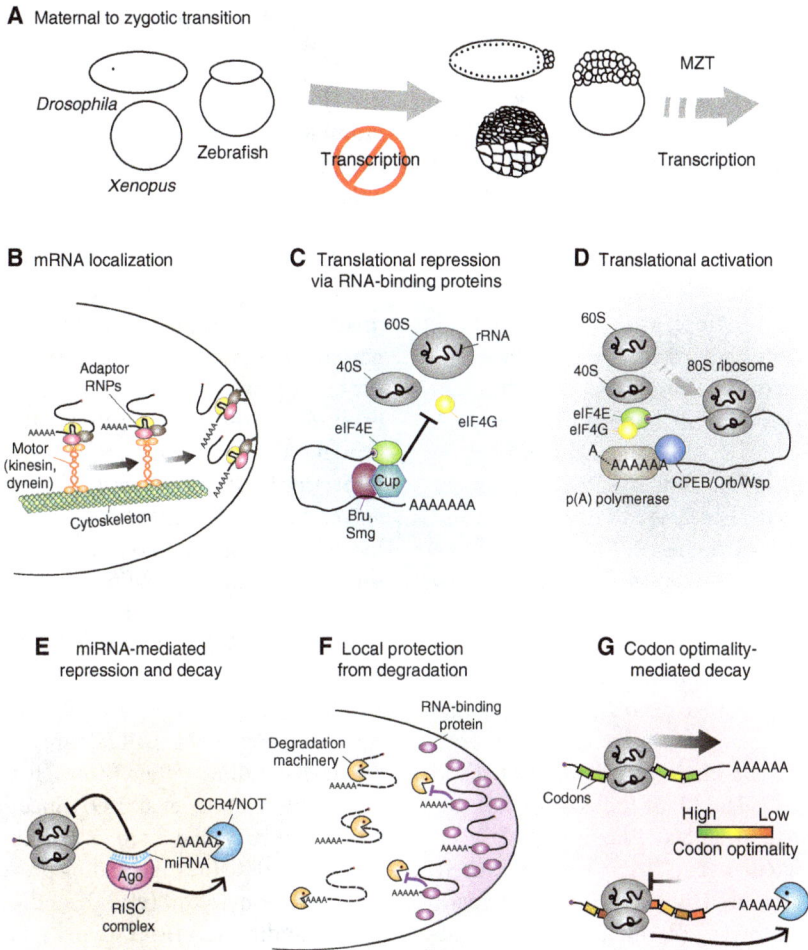

Figure 1. Mechanisms of translational control during embryo development. Examples of modules regulating protein synthesis during the first steps of embryo development, a period during which the zygotic genome is mostly transcriptionally quiescent. (*A*) Maternal-to-zygotic transition (MZT). (*Left*) Fertilized *Drosophila* (*top*), *Xenopus* (*bottom*), and zebrafish (*right*) embryos are transcriptionally silent and contain maternally synthesized gene products that sustain development and lay down the basic structures and embryonic axes. (*Right*) During the MZT, maternal products are largely degraded, and zygotic transcription ensues. This allows asynchrony in cell divisions and morphogenetic cell movements of gastrulation needed for body plan building. (*B*) Messenger RNA (mRNA) localization is mediated by RNA-binding proteins that recognize secondary structures present in the 3′ untranslated region (UTR) of target mRNAs. Assembled adaptor ribonucleoprotein complexes (RNPs) are recruited to molecular motors and delivered to specific intracellular domains. (*C*) Translational repression can be actively established on targets by RNA-binding proteins—such as Bruno (Bru) and Smaug (Smg)—that recognize sequences within the transcript 3′UTRs and recruit other factors to block translation initiation. (*D*) Translational activation is associated with the recruitment and activity of poly(A) [p(A)] polymerases and interactions between the poly(A)-binding protein and translation eukaryotic initiation factor (eIF)4G and eIF4E on target mRNA transcripts. (*E*) Base complementarity-targeted microRNA regulation is mediated by the RNA-induced silencing complex (RISC) and induces large-scale translational repression and CCR4/NOT-dependent target mRNA decay during the MZT. (*F*) Local protection of evenly distributed mRNAs against bulk degradation, which is mediated by localized RNA-binding proteins recognizing specific motifs in the 3′UTR of target mRNAs leads to spatially regulated expression at the subcellular level. (*G*) Slower ribosome translocation rates over transcripts containing suboptimal codons trigger mRNA deadenylation and decay during embryo development. rRNA, ribosomal RNA; miRNA, microRNA.

nism (minus-end-directed), driven by dynein (Bullock and Ish-Horowicz 2001; Dienstbier et al. 2009), and dependent on Bicaudal-D and on the RNA-binding protein Egalitarian (Egl), delivers master-regulator mRNAs—namely, *oskar* (*osk*), *bicoid* (*bcd*), and *gurken* (*grk*)—from transcribing nurse cells to the transcriptionally quiescent oocyte. Once inside the oocyte, *osk* mRNA-containing particles are brought to the posterior side of the *Drosophila* oocyte through a kinesin-dependent mechanism (Brendza et al. 2000; Zimyanin et al. 2008; Lehmann 2016; Gaspar et al. 2017). Concurrently, *bcd* and *grk* mRNAs are actively delivered to the anterior and anterodorsal cortexes, respectively, through the action of dynein motors (plus-end directed) (Davidson et al. 2016; Trovisco et al. 2016). Localization specificity is mostly provided by sequences and secondary structures present in the 3′ untranslated region (UTR) of target mRNAs, which are in turn recognized by double-stranded RNA-binding proteins like Egl and Staufen (Stau) (Fig. 1B) (St Johnston et al. 1991, 1992; Micklem et al. 2000; Jambor et al. 2011, 2014; Le Bouteiller et al. 2013). mRNA–protein interactions required for transport are normally established in the cytoplasm, and form ribonucleoprotein complexes (RNPs) that may contain multiple mRNA copies per particle and are charged into the microtubule network (Jambor et al. 2011; Little et al. 2015; Trcek et al. 2015; Gaspar et al. 2017). Nevertheless, meaningful protein–mRNA interactions can equally well be established inside the transcribing nuclei (Chin and Lecuyer 2017). In the best-studied example, *osk* mRNA splicing and the interaction of *osk* mRNA with the exon junction complex (EJC) are critical for its final posterior oocyte localization (Hachet and Ephrussi 2004; Ghosh et al. 2012; Marchand et al. 2012). This indicates that at least in some cases, target mRNA recognition is initiated before nuclear export (Chin and Lecuyer 2017).

Regardless of the mechanism required for RNP assembly and recruitment to molecular motors, the general role for localized mRNA in determining embryo axial polarity seems to be widespread in animals. In fact, the importance of localized mRNA in defining cellular polarity goes far beyond the above-described example, as it has been implicated in the polarization of migrating cells, in the establishment of apicobasal structure, and in asymmetric cell divisions (reviewed in Medioni et al. 2012; Chin and Lecuyer 2017; Moor et al. 2017).

Very important for the successful establishment of a protein gradient, localized mRNAs have to be translationally silenced during transport. Frequently, translational silencing is actively established on targets by RNA-binding proteins that recognize sequences within the transcript UTRs. In contrast to mechanisms involved in recruitment of mRNAs to molecular motors, which seem to rely on secondary-structure RNA motifs, translational repression is regulated by sequence-specific RNA-binding proteins that directly interact with motifs in the 3′UTR of the target RNA (Fig. 1C). For instance, specific motifs in the *osk* 3′UTR (known as Bruno response elements [BREs]) are recognized by the RNA-binding protein Bruno (Bru) (Kim-Ha et al. 1995; Webster et al. 1997; Snee et al. 2008), which in turn is thought to recruit Cup. The Cup protein binds to the cap-binding protein eukaryotic initiation factor (eIF)4E and antagonizes eIF4E binding to eIF4G, thereby blocking *osk* translation (Wilhelm et al. 2003; Nakamura et al. 2004; Nelson et al. 2004; Zappavigna et al. 2004; Chekulaeva et al. 2006). The fact that Cup can interact with eIF4E in an RNA-independent manner further shows the key role of sequence-specific RNA-binding proteins such as Bru in specifying the set of translationally regulated mRNAs (Nakamura et al. 2004). Similarly, Bru binds to the *grk* 3′UTR and, together with another RNA-binding protein (Squid), is thought to repress *grk* translation during transport (Norvell et al. 1999; Saunders and Cohen 1999; Filardo and Ephrussi 2003; Caceres and Nilson 2009; Weil et al. 2012). In addition to repressive mechanisms relying on protein–protein interactions, it is nonetheless likely that translational repression during transport may also be accomplished by physical exclusion of the translation initiation apparatus, either as a result of the fact that target mRNAs are compacted into large RNPs (Chekulaeva et al. 2006) or because of the preferential enrichment of translational activators only at the final destination (Davidson et al. 2016).

Much less understood are the mechanisms by which localized mRNAs become translationally active. Generally, poly(A) polymerases are recruited to target mRNA transcripts during translational activation or derepression (Fig. 1D) (Castagnetti and Ephrussi 2003; Norvell et al. 2015). For instance, *osk* and *grk* translation is activated during oogenesis and requires the Oo18 RNA-binding protein (Orb, the homolog of the *Xenopus* cytoplasmic polyadenylation element-binding protein [CPEB]), a protein that is enriched inside the oocyte and that can directly associate with the poly(A) polymerases (Chang et al. 1999, 2001; Castagnetti and Ephrussi 2003; Weil et al. 2012; Norvell et al. 2015). Although *grk* and *osk* activation differ in many other mechanistic aspects, such as the fact that Orb directly recognizes sequences in the *osk* 3′UTR while the physical link between Orb and *grk* has not been reported, the poly(A) tails of both *grk* and *osk* mRNAs increase in length during activation. Indeed, translational activation and poly(A) polymerase activity are often correlated, although translation of maternally deposited mRNA per se does not necessarily require an increase in poly(A) tail length during early embryogenesis (Eichhorn et al. 2016). Such is the case of *bcd* mRNA, which is silenced during oogenesis and activated during embryogenesis. Global analysis of mRNA poly(A) traits in *Drosophila* oocytes and early embryos revealed that *bcd* mRNA translational activation is not associated with an increase in poly(A) tail length, suggesting that its temporal regulation is defined by the release of a repressive state rather than the promotion of translational activation and poly(A) polymerase activity (Eichhorn et al. 2016).

During oogenesis, regulated translation of localized *osk* mRNA at the posterior cortex of the forming egg nucleates a specialized cytoplasm—known as germ plasm—and is required for the entrapment of a set of mRNAs passively moving in the cytoplasmic stream created during oogenesis (Lehmann 2016). Among these is *nos*, a master determinant of posterior fate, as well as many other mRNAs required for the formation of germ cells at the embryonic posterior cortex,

including *pgc* and *gcl* (Wang and Lehmann 1991; Gavis and Lehmann 1992, 1994; Little et al. 2015; Trcek et al. 2015). Posterior localization is determined by sequences and secondary structures present in their 3′UTRs, even though mechanistically the movement of mRNAs does not include their recruitment to molecular motors or the assembly of large RNPs (Crucs et al. 2000; Forrest et al. 2004; Rangan et al. 2009; Little et al. 2015; Trcek et al. 2015). Regardless of such differences, translation of germ plasm-enriched mRNAs is also tightly regulated, with most of them being repressed during oogenesis and individually activated during embryogenesis (Rangan et al. 2009). For instance, *nos* mRNA translation is repressed during oogenesis by Glorund and Smaug (Smg), two RNA-binding proteins that interact with the *nos* 3′UTR (Smibert et al. 1999; Kalifa et al. 2006; Tamayo et al. 2017). Similarly to Bru, Smg interacts with Cup, which then binds to eIF4E and inhibits *nos* translation initiation by precluding the eIF4E–eIF4G interaction and formation of the preinitiation complex at the 5′ end of *nos* mRNA (Nelson et al. 2004; Andrews et al. 2011).

Transcript-specific translational regulation may be achieved via association of mRNAs within large, membraneless RNP granules. For example, at the *Drosophila* oocyte anterior cortex, translationally active *grk* is preferentially distributed at the edges of RNP granules, although translationally repressed *bcd* is found both in the center and at the edge of the granule (Weil et al. 2012). This correlation is not observed for mRNAs localizing to the germ granules at the posterior pole. Here, multiple copies of a single RNA organize into homotypic RNA clusters, which occupy fixed locations with respect to the protein germ granule. This spatial organization, however, does not reflect translational activity (Trcek et al. 2015). In line with the importance of such posttranscriptional mechanisms for embryonic development, 71% of the 2314 maternally deposited mRNAs tested in *Drosophila* embryos were shown to be subcellularly localized (Lecuyer et al. 2007). Beyond embryos, accumulating evidence indicates that mRNA localization is more prevalent than expected and may regulate other aspects of cellular biology, including cell signaling, neuronal axon growth, and dendritic plasticity (reviewed in Chin and Lecuyer 2017).

Translational Efficiency and mRNA Decay

Thousands of maternally transcribed mRNAs are deposited into the transcriptionally quiescent oocyte during oogenesis and the vast majority of these mRNAs are found evenly distributed in the egg cytoplasm. To drive the first steps of embryonic development, their translation is temporally—and in some cases even spatially—regulated. For instance, prominent developmental transitions such as oocyte maturation and egg activation, as well as the activation of the zygotic genome during the MZT, are characterized by large-scale changes in gene expression (Fig. 1A) (reviewed in Laver et al. 2015; Yartseva and Giraldez 2015). Mechanistically, these changes are mostly achieved by the bulk regulation of translation efficiency and mRNA decay, although global changes coexist with the more targeted regulatory mechanisms described in the previous section.

During the stages involving oocyte maturation and egg activation in which meiotic divisions proceed toward completion, regulation of gene expression is mostly achieved at the translational level, with minimal fluctuation at the mRNA accumulation level (Su et al. 2007; Tadros et al. 2007; Thomsen et al. 2010; Chen et al. 2011; Kronja et al. 2014; Eichhorn et al. 2016). In Drosophila, ~1350 out of ~5800 mRNAs assessed by ribosome profiling were shown to be under translational control during egg activation, with 70% of those being translationally activated and 30% repressed (Kronja et al. 2014). Interestingly, most of the observed changes in translational efficiency depend on the activity of the serine/threonine kinase PAN GU (Tadros et al. 2007; Kronja et al. 2014), although the details of how this is achieved and which PAN GU substrates are involved in translational regulation are not yet fully understood. Given that translational changes during egg activation are mainly determined by changes in mRNA poly(A) tail length, which are dictated by the opposing activities of deadenylases and the

poly(A) polymerase Wispy (Benoit et al. 2008; Cui et al. 2013; Subtelny et al. 2014), it is possible that PAN GU acts directly upstream of such modifying enzymes (Eichhorn et al. 2016).

In contrast to the prevalence of translational regulation during oocyte maturation and egg activation, MZT is largely accompanied by widespread changes in mRNA levels (affecting 30%–40% of maternal mRNAs) that are mainly driven by mRNA decay mechanisms (reviewed in Laver et al. 2015; Yartseva and Giraldez 2015). Maternal mRNA clearance starts before zygotic transcriptional activation but is intensified on its onset. Accumulating evidence indicates that mRNA destabilization is often preceded by translational inhibition, and that both types of regulation are mediated by the same machineries involving the deadenylation of target mRNAs (Fig. 1E). For instance, the RNA-binding protein Smaug mediates most of the maternal mRNA translational repression and decay before zygotic activation in Drosophila (Nelson et al. 2004; Tadros et al. 2007; Benoit et al. 2009; Chen et al. 2014). Indeed, PAN GU activates Smaug mRNA translation during egg activation (Tadros et al. 2007), with Smaug protein accumulating during early embryogenesis but decreasing soon after the MZT (Benoit et al. 2009). Smaug binds to secondary structure RNA motifs (Smaug recognition elements [SREs]) (Dahanukar et al. 1999; Smibert et al. 1999) within target mRNAs, and mediates either decay by recruiting the CCR4/NOT deadenylase complex or translational repression by interacting with Cup (Nelson et al. 2004; Laver et al. 2015; Eichhorn et al. 2016). However, it is still unclear what determines the recruitment of either downstream effector.

On genome activation, zygotically transcribed microRNAs play a crucial role in translational control and maternal mRNA clearance in many animal species (Fig. 1F). This is notably the case for miR-430 in zebrafish, miR-309 in Drosophila, and miR-427 in Xenopus, which are responsible for destabilization of hundreds of transcripts during early embryogenesis (Giraldez et al. 2006; Bushati et al. 2008; Lund et al. 2009). Mechanistically, microRNA specificity is determined by base complementarity with target mRNAs (Duchaine and Fabian

Cite this article as Cold Spring Harb Perspect Biol doi: 10.1101/cshperspect.a032987

2018), but the mode of action can vary between translational repression and target mRNA cleavage and decay. Interestingly, careful analysis of *miR-403*-mediated regulation in zebrafish using genome-wide approaches revealed that targeting initially leads to translation initiation, which is then followed by mRNA deadenylation and decay (Bazzini et al. 2012).

In parallel to the regulation mediated by RNA-binding proteins and microRNAs, recent work shed light on a new mechanism by which the process of translation elongation influences mRNA stability (see Dever et al. 2018) and dictates mRNA decay during the MZT in *Drosophila*, zebrafish, *Xenopus*, and mouse (Presnyak et al. 2015; Bazzini et al. 2016; Mishima and Tomari 2016). Genome-wide approaches indicate that mRNAs encoding open reading frames (ORFs) enriched with uncommon, suboptimal codons decay faster than those enriched with optimal codons. "Codon optimality" is mostly defined by transfer RNA (tRNA) availability, and slower ribosome translocation rates over transcripts containing suboptimal codons trigger mRNA deadenylation by the CCR4/NOT complex, promoting decay (Fig. 1G). Interestingly, deadenylation activity is increased when suboptimal codons are placed closer to the stop codon and in mRNAs containing shorter 3'UTRs (Mishima and Tomari 2016), suggesting that physical proximity between the ribosome and the poly(A) tail may be required for the decay. The molecular mechanism by which ribosome elongation speed is interpreted and how this connects to the decay machinery remain to be determined. Nevertheless, analysis of mRNA decay kinetics during zebrafish embryo development indicates that the magnitude of codon composition-mediated mRNA decay regulation is comparable to that observed for microRNA regulation (Bazzini et al. 2016). In addition, a third mechanism, which can promote the degradation of transcripts enriched for the RNA modification N^6-methyladenosine (m^6A) (Wang et al. 2014; Peer et al. 2018), complements microRNA- and codon composition-mediated decay during the MZT in zebrafish embryos (Zhao et al. 2017).

Finally, translational regulation of evenly distributed mRNAs is not solely regulated in time but can also be differentially regulated in space. A notable example is provided by the uniformly distributed *hunchback* mRNA (*hb*) in *Drosophila* (Lehmann and Nusslein-Volhard 1987; Tautz 1988; Irish et al. 1989). Like thousands of other mRNAs, *hb* is maternally provided and only translated during early embryogenesis. *hb* translation is inhibited at the posterior and forms a protein gradient that is required during the MZT for regulating the expression of genes defining the anterior–posterior embryonic axis. *hb* translation repression is mediated by the sequence-specific binding of Pumilio (Pum) and its partners Nanos and Brain tumor (Brat) to the 3'UTR of *hb* mRNA (Irish et al. 1989; Wharton and Struhl 1991; Murata and Wharton 1995; Chagnovich and Lehmann 2001; Sonoda and Wharton 2001). Mechanistically, Nanos modulates Pum target specificity and binding to the Nanos response element (NRE) found at the 3'UTRs (Murata and Wharton 1995; Weidmann et al. 2016). In the case of *hb* mRNA, Nanos also recruits Brat, forming the NRE complex (Sonoda and Wharton 2001). Brat then interacts with the eIF4E-homologous protein (4EHP), which binds to the mRNA 5' cap structure and negatively regulates translation by blocking eIF4E binding and formation of the preinitiation complex at the cap (Cho et al. 2006; Arvola et al. 2017). Spatially, this regulation is defined by Nanos protein distribution because of the posterior localization and local translation of *nanos* mRNA (Gavis and Lehmann 1992). Therefore, the spatial regulation of evenly distributed mRNAs may ultimately depend on an initial subcellular enrichment of translational repressors.

TRANSLATIONAL CONTROL DURING STEM CELL MAINTENANCE AND DIFFERENTIATION

Throughout development and adult life, precursor cells transit between quiescent, self-renewing, and differentiating states in a spatiotemporally coordinated fashion to promote embryonic development, tissue formation, and later to sustain organismal homeostasis. Although particularities exist in different stem cell lineages and developmental stages, transitions are fundamentally driven and accompanied by changes in gene expression, including translational control (Buszczak et al. 2014). For instance, many stem cell lineages present lower global protein synthesis rates in comparison to their immediate differentiating daughters. This was observed both in vitro (Sampath et al. 2008; Ingolia et al. 2011) and in stem cell systems in vivo, the latter of which includes *Drosophila* adult germline stem cells (Sanchez et al. 2016), mouse adult hematopoietic stem cells (Signer et al. 2014, 2016), mouse adult neural stem cells (NSCs) (Llorens-Bobadilla et al. 2015), mouse adult skeletal muscle stem cells (or satellite cells) (Zismanov et al. 2016), and mouse adult hair follicle stem cells (Fig. 2) (Blanco et al. 2016). In the studied cases, changes in protein synthesis rate during self-renewal and differentiation were mediated by regulatory mechanisms rather than being a consequence of changes in cell size, proliferation rate, or cell-cycle status (Sampath et al. 2008; Buszczak et al., 2014; Signer et al. 2014; Blanco et al. 2016; Sanchez et al. 2016).

Genetic experiments using mutations affecting translational control indicate that changes in protein synthesis rates are required and sufficient to modulate self-renewal and differentiation capacities in many stem cell systems. For instance, mutations inducing a global reduction in protein synthesis favor the accumulation of undifferentiated cells at the expense of differentiation. This is the case for *Nsun2*-deficient mouse hair follicle stem cells and male germline stem cells (Blanco et al. 2011; Hussain et al. 2013), for knockdowns of mechanistic target of rapamycin (mTOR) or transformation/transcription domain-associated protein (TRRAP)—both belonging to the phosphatidylinositol 3-kinase-related kinase (PIKK) family—in *Drosophila* female germline stem cells (Sanchez et al. 2016), and for mTOR-inhibited cultured mouse embryonic stem cells (Bulut-Karslioglu et al. 2016). Similarly, phosphorylation of eIF2α decreases protein synthesis rates and promotes mouse skeletal muscle stem cell self-renewal rather than differentiation (Zismanov et al. 2016). Muscle stem cells unable to phosphorylate eIF2α exit the quiescent state and differentiate. Furthermore, pharmacological treatment inhibiting eIF2α dephosphorylation promotes the regenerative capacity of muscle stem cells (Zismanov et al. 2016). Consistent with this, a genetically or chemically induced increase in protein synthesis, for instance by ectopically activating the mTOR pathway in stem cells, induces premature differentiation and stem cell depletion in many of the above-mentioned systems (Yilmaz et al. 2006; Zhang et al. 2006; Castilho et al. 2009; Sun et al. 2010; Bonaguidi et al. 2011; Deng et al. 2015). The central role for pro-

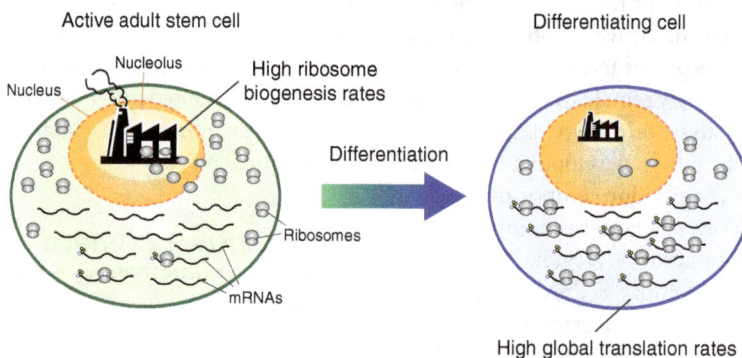

Figure 2. Translational control during stem cell maintenance and differentiation. Actively dividing stem cells (*left*) present lower global protein synthesis rates (fewer ribosomes attached to messenger RNAs [mRNAs]) in comparison to their immediate differentiating daughters (*right*). Yet, ribosomal RNA synthesis and ribosome biogenesis (shown by a black factory) are up-regulated compared with their differentiated progeny. This has been observed in several stem cells, such as *Drosophila* adult female germline and mouse embryonic stem cells, indicating that ribosome biogenesis and protein synthesis rates may be uncoupled in such stem cell systems.

Cite this article as *Cold Spring Harb Perspect Biol* doi: 10.1101/cshperspect.a032987

tein synthesis in this process is exemplified by the fact that a hypomorphic mutation affecting the ribosomal protein Rpl24 is sufficient to counterbalance the activation of the mTOR pathway in mouse hematopoietic stem cells, restoring normal self-renewal and differentiation functions (Signer et al. 2014). Yet, strong translation abrogation by means of affecting ribosome biogenesis can lead to stem cell loss as well as anomalous differentiation (Fichelson et al. 2009; LaFever et al. 2010; Le Bouteiller et al. 2013; Zhang et al. 2014; Sanchez et al. 2016). This shows how fine-tuning mechanisms, rather than drastic changes in protein synthesis, balance self-renewal and differentiation functions, possibly in a lineage-specific manner (Buszczak et al. 2014).

The mechanisms by which the protein synthesis rate is limited in wild-type stem cells are still poorly understood, but current evidence suggests that it may be achieved by different means depending on the specific stem cell lineage. For instance, translation inhibition in mouse adult hematopoietic stem cells and in mouse embryonic stem cells is at least partially mediated by the regulation of the eukaryotic initiation factor 4E-binding proteins 4E-BP1 and 4E-BP2 (Signer et al. 2016; Tahmasebi et al. 2016). In these systems, 4E-BP proteins are preferentially hypophosphorylated; this favors 4E-BP binding to the eIF4E, leading to inhibition of translation initiation. As a consequence, the translation of certain subsets of target mRNAs becomes more strongly inhibited (Roux and Topisirovic 2012; Morita et al. 2013; Proud 2018). In mouse skeletal muscle stem cells, on the other hand, high levels of eIF2α phosphorylation, another fulcrum of translation initiation regulation (Hinnebusch and Lorsch 2012; Merrick and Pavitt 2018; Wek 2018), are responsible for limiting protein synthesis (Zismanov et al. 2016). In contrast to these systems, mouse NSCs seem to use ribosome biogenesis regulation rather than translation initiation to modulate protein synthesis; here, stem cells express lower levels of ribosomal subunits in comparison to their immediate daughters (Llorens-Bobadilla et al. 2015). This type of regulation may not be generally used in other stem cell systems,

however. For instance, ribosomal RNA synthesis and ribosome biogenesis are up-regulated in *Drosophila* adult germline stem cells compared with their differentiated progeny; yet, stem cells show lower protein synthesis rates (Fig. 2) (Neumuller et al. 2008; Zhang et al. 2014; Sanchez et al. 2016; reviewed in Brombin et al. 2015). It remains to be determined how ribosome biogenesis and protein synthesis rates are uncoupled in some stem cell systems. Recurrent observations in several independent systems suggest that active, yet unknown processes limit protein synthesis specifically in stem cells (such as the above-mentioned regulation of translation initiation) and/or conversely boost it in differentiating progeny. Given that differentiation relies on higher protein synthesis rates, it has been proposed that a surfeit of ribosome biogenesis in stem cells may prime daughter cells with enough ribosomes required for efficient differentiation (Ingolia et al. 2011; Sanchez et al. 2016).

In parallel to the global regulation of the translation machinery, specific translational regulators drive fate transitions during adult stem cell self-renewal and differentiation, with many examples being described in the *Drosophila* female germline stem cell system. In this system, a series of RNA-binding proteins initially identified through genetic screens delineates a regulatory cascade that is both required and sufficient for self-renewal and differentiation (reviewed in Slaidina and Lehmann 2014). These include conserved translational repressors, such as Nanos and Pum (Forbes and Lehmann 1998; Tsuda et al. 2003; Miller and Olivas 2011), as well as the DExH-box RNA helicase Benign gonial cell neoplasm (Bgcn) (Bailey et al. 2017), the TRIM-NHL tumor suppressor proteins Mei-P26 and Brat (Neumuller et al. 2008; Harris et al. 2011; Insco et al. 2012), and the cytoplasmic RNA-binding Fox 1 (Rbfox1) (Carreira-Rosario et al. 2016). It is important to note that, in addition to its function as translational repressor, Mei-P26 also regulates translation by directly interacting with Ago1 and inhibiting the microRNA pathway (Neumuller et al. 2008).

By regulating each other's expression through mutual translational repression, the translational repressor factors establish a self-limiting regulatory network that promotes precise fate transitions during female germline stem cell differentiation. Similar to the mechanisms described in embryos, translational silencing during germline stem cell self-renewal and differentiation is generally mediated by repressor complexes that bind to sequences within the target mRNA 3′UTRs. In germline stem cells, for instance, translational repressors Nanos and Pum cooperate to promote self-renewal by inhibiting the translation of differentiation factors including Brat and Mei-P26, likely through the recruitment of the CCR4/NOT decay machinery to target mRNAs (Harris et al. 2011; Joly et al. 2013). In differentiating daughter cells, the expression of the differentiation factor Bag-of-marbles (Bam) leads to the assembly of a complex containing the sex determination factor Sex-lethal (Sxl) and translational repressors Bgcn and Mei-P26 (Li et al. 2009, 2013; Chau et al. 2012). Sxl and Mei-P26 are able to interact with the *Nanos* mRNA 3′UTR and repress its translation in differentiating daughter cells (Chau et al. 2012; Li et al. 2012, 2013), although the mechanism by which this repression is mediated remains to be determined. In the absence of Nanos, Pum is free to interact with Brat, and the Pum–Brat complex inhibits the translation of other self-renewal factors in differentiating cells, including transcription factors Myc and Mad (Harris et al. 2011). Interestingly, in more advanced differentiation stages, cytoplasmic Rbfox1 is expressed and binds to a sequence-specific motif in *Pum* mRNA 3′UTR, thereby repressing Pum translation and ensuring the progression to terminally differentiated states (Tastan et al. 2010; Carreira-Rosario et al. 2016).

Finally, it has recently been proposed that changes in tRNA concentrations may account for some of the gene expression differences observed in proliferating and differentiating cells (Gingold et al. 2014). Given that codon optimality is likely to be determined by tRNA availability (Presnyak et al. 2015; Bazzini et al. 2016), it is to be expected that in addition to a direct effect on translation (Gingold et al. 2014), the observed variation in tRNA abundance in these two states may also induce changes in mRNA stability like those observed during embryogen-

esis (Bazzini et al. 2016). Although it remains to be determined whether tRNA repertoires vary during stem cell fate transitions, it has recently been shown that changes in the expression of specific tRNAs modulate protein expression and metastatic activity during cancer progression through both translational control and mRNA decay (Goodarzi et al. 2016).

CONCLUDING REMARKS

Regulation at the posttranscriptional level is increasingly recognized as an important enabler of cell fate transitions in a variety of developmental and stem cell systems. At a global level, regulation of translational efficiency and RNA or protein turnover may be used to facilitate transitions between cell states by relinquishing entire gene programs to make way for synthesis of a new program. It may not be surprising that these posttranscriptional mechanisms are found so prominently at transitions that require gross turnover, like the transition from the maternally provided gene products of the egg to new gene synthesis by the embryo or during the transition from long-term, self-renewing stem cells to differentiating progeny. While broad, these transitions do not affect all mRNAs and our understanding of the mechanisms of regulation and target selection is still limited. Although this review has focused on 3′UTR-mediated regulation during the early stages of embryogenesis and global regulation at the stem cell–differentiation transition, additional mechanisms mediate translational efficiency via 5′ upstream sequences. These include control via RNA secondary structures, internal ribosome entry sites, upstream ORFs, and specialized ribosomes (Hinnebusch 2005; Chew et al. 2016; Johnstone et al. 2016; Shi et al. 2017; Leppek et al. 2018).

Although the intricacies of transcriptional regulation have long been recognized to modulate gene expression, posttranscriptional mechanisms were thought of as "housekeeping" and permissive. Recent discoveries and technologies challenge this perception and promise opportunities for continued discoveries of fundamental significance. As developmental biology moves toward single-cell level analysis, one of the challenges is the development of technologies to globally assess changes in translational control in individual cells, thereby complementing the recently established techniques (such as single-cell RNA-sequencing) that measure mRNA accumulation levels at single-cell resolution.

ACKNOWLEDGMENTS

We apologize to our colleagues whose work was not cited due to space limitations. F.K.T. is a Wellcome Trust and Royal Society Sir Henry Dale Fellow (206257/Z/17/Z). R.L. is supported by the National Institutes of Health (NIH) R37HD41900 and is a Howard Hughes Medical Institute (HHMI) Investigator.

REFERENCES

*Reference in also in this collection.

Andrews S, Snowflack DR, Clark IE, Gavis ER. 2011. Multiple mechanisms collaborate to repress nanos translation in the *Drosophila* ovary and embryo. *RNA* 17: 967–977.

Arvola RM, Weidmann CA, Tanaka Hall TM, Goldstrohm AC. 2017. Combinatorial control of messenger RNAs by Pumilio, Nanos and brain tumor proteins. *RNA Biol* 14: 1445–1456.

Bailey AS, Batista PJ, Gold RS, Chen YG, de Rooij DG, Chang HY, Fuller MT. 2017. The conserved RNA helicase YTHDC2 regulates the transition from proliferation to differentiation in the germline. *eLife* 6: e26116.

Baroux C, Autran D, Gillmor CS, Grimanelli D, Grossniklaus U. 2008. The maternal to zygotic transition in animals and plants. *Cold Spring Harb Symp Quant Biol* 73: 89–100.

Bazzini AA, Lee MT, Giraldez AJ. 2012. Ribosome profiling shows that miR-430 reduces translation before causing mRNA decay in zebrafish. *Science* 336: 233–237.

Bazzini AA, Del Viso F, Moreno-Mateos MA, Johnstone TG, Vejnar CE, Qin Y, Yao J, Khokha MK, Giraldez AJ. 2016. Codon identity regulates mRNA stability and translation efficiency during the maternal-to-zygotic transition. *EMBO J* 35: 2087–2103.

Benoit P, Papin C, Kwak JE, Wickens M, Simonelig M. 2008. PAP- and GLD-2-type poly(A) polymerases are required sequentially in cytoplasmic polyadenylation and oogenesis in *Drosophila*. *Development* 135: 1969–1979.

Benoit B, He CH, Zhang F, Votruba SM, Tadros W, Westwood JT, Smibert CA, Lipshitz HD, Theurkauf WE. 2009. An essential role for the RNA-binding protein Smaug during the *Drosophila* maternal-to-zygotic transition. *Development* 136: 923–932.

Blanco S, Kurowski A, Nichols J, Watt FM, Benitah SA, Frye M. 2011. The RNA-methyltransferase Misu (NSun2) poises epidermal stem cells to differentiate. *PLoS Genet* 7: e1002403.

Blanco S, Bandiera R, Popis M, Hussain S, Lombard P, Aleksic J, Sajini A, Tanna H, Cortes-Garrido R, Gkatza N, et al. 2016. Stem cell function and stress response are controlled by protein synthesis. *Nature* 534: 335–340.

Bonaguidi MA, Wheeler MA, Shapiro JS, Stadel RP, Sun GJ, Ming GL, Song H. 2011. In vivo clonal analysis reveals self-renewing and multipotent adult neural stem cell characteristics. *Cell* 145: 1142–1155.

Brar GA, Weissman JS. 2015. Ribosome profiling reveals the what, when, where and how of protein synthesis. *Nat Rev Mol Cell Biol* 16: 651–664.

Brendza RP, Serbus LR, Duffy JB, Saxton WM. 2000. A function for kinesin I in the posterior transport of *oskar* mRNA and Staufen protein. *Science* 289: 2120–2122.

Brombin A, Joly JS, Jamen F. 2015. New tricks for an old dog: Ribosome biogenesis contributes to stem cell homeostasis. *Curr Opin Genet Dev* 34: 61–70.

Bullock SL, Ish-Horowicz D. 2001. Conserved signals and machinery for RNA transport in *Drosophila* oogenesis and embryogenesis. *Nature* 414: 611–616.

Bulut-Karslioglu A, Biechele S, Jin H, Macrae TA, Hejna M, Gertsenstein M, Song JS, Ramalho-Santos M. 2016. Inhibition of mTOR induces a paused pluripotent state. *Nature* 540:119–123.

Bushati N, Stark A, Brennecke J, Cohen SM. 2008. Temporal reciprocity of miRNAs and their targets during the maternal-to-zygotic transition in *Drosophila*. *Curr Biol* 18: 501–506.

Buszczak M, Signer RA, Morrison SJ. 2014. Cellular differences in protein synthesis regulate tissue homeostasis. *Cell* 159: 242–251.

Caceres L, Nilson LA. 2009. Translational repression of *gurken* mRNA in the *Drosophila* oocyte requires the hnRNP Squid in the nurse cells. *Dev Biol* 326: 327–334.

Carreira-Rosario A, Bhargava V, Hillebrand J, Kollipara RK, Ramaswami M, Buszczak M. 2016. Repression of Pumilio protein expression by Rbfox1 promotes germ cell differentiation. *Dev Cell* 36: 562–571.

Castagnetti S, Ephrussi A. 2003. Orb and a long poly(A) tail are required for efficient *oskar* translation at the posterior pole of the *Drosophila* oocyte. *Development* 130: 835–843.

Castilho RM, Squarize CH, Chodosh LA, Williams BO, Gutkind JS. 2009. mTOR mediates Wnt-induced epidermal stem cell exhaustion and aging. *Cell Stem Cell* 5: 279–289.

Chagnovich D, Lehmann R. 2001. Poly(A)-independent regulation of maternal hunchback translation in the *Drosophila* embryo. *Proc Natl Acad Sci* 98: 11359–11364.

Chang JS, Tan L, Schedl P. 1999. The *Drosophila* CPEB homolog, Orb, is required for *oskar* protein expression in oocytes. *Dev Biol* 215: 91–106.

Chang JS, Tan L, Wolf MR, Schedl P. 2001. Functioning of the *Drosophila orb* gene in *gurken* mRNA localization and translation. *Development* 128: 3169–3177.

Chau J, Kulnane LS, Salz HK. 2012. Sex-lethal enables germline stem cell differentiation by down-regulating Nanos protein levels during *Drosophila* oogenesis. *Proc Natl Acad Sci* 109: 9465–9470.

Chekulaeva M, Hentze MW, Ephrussi A. 2006. Bruno acts as a dual repressor of *oskar* translation, promoting mRNA oligomerization and formation of silencing particles. *Cell* 124: 521–533.

Chen J, Melton C, Suh N, Oh JS, Horner K, Xie F, Sette C, Blelloch R, Conti M. 2011. Genome-wide analysis of translation reveals a critical role for deleted in azoospermia-like (*Dazl*) at the oocyte-to-zygote transition. *Genes Dev* **25:** 755–766.

Chen L, Dumelie JG, Li X, Cheng MH, Yang Z, Laver JD, Siddiqui NU, Westwood JT, Morris Q, Lipshitz HD, et al. 2014. Global regulation of mRNA translation and stability in the early *Drosophila* embryo by the Smaug RNA-binding protein. *Genome Biol* **15:** R4.

Chew GL, Pauli A, Schier AF. 2016. Conservation of uORF repressiveness and sequence features in mouse, human and zebrafish. *Nat Commun* **7:** 11663.

Chin A, Lecuyer E. 2017. RNA localization: Making its way to the center stage. *Biochim Biophys Acta* **1861:** 2956–2970.

Cho PF, Gamberi C, Cho-Park YA, Cho-Park IB, Lasko P, Sonenberg N. 2006. Cap-dependent translational inhibition establishes two opposing morphogen gradients in *Drosophila* embryos. *Curr Biol* **16:** 2035–2041.

Crucs S, Chatterjee S, Gavis ER. 2000. Overlapping but distinct RNA elements control repression and activation of *nanos* translation. *Mol Cell* **5:** 457–467.

Cui J, Sartain CV, Pleiss JA, Wolfner MF. 2013. Cytoplasmic polyadenylation is a major mRNA regulator during oogenesis and egg activation in *Drosophila*. *Dev Biol* **383:** 121–131.

Dahanukar A, Walker JA, Wharton RP. 1999. Smaug, a novel RNA-binding protein that operates a translational switch in *Drosophila*. *Mol Cell* **4:** 209–218.

Davidson A, Parton RM, Rabouille C, Weil TT, Davis I. 2016. Localized translation of *gurken*/TGF-α mRNA during axis specification is controlled by access to Orb/CPEB on processing bodies. *Cell Rep* **14:** 2451–2462.

Deng Z, Lei X, Zhang X, Zhang H, Liu S, Chen Q, Hu H, Wang X, Ning L, Cao Y, et al. 2015. mTOR signaling promotes stem cell activation via counterbalancing BMP-mediated suppression during hair regeneration. *J Mol Cell Biol* **7:** 62–72.

* Dever TE, Dinman JD, Green R. 2018. Translation elongation and recoding in eukaryotes. *Cold Spring Harb Perspect Biol* doi: 10.1101/cshperspect.a032649.

Dienstbier M, Boehl F, Li X, Bullock SL. 2009. Egalitarian is a selective RNA-binding protein linking mRNA localization signals to the dynein motor. *Genes Dev* **23:** 1546–1558.

* Duchaine TF, Fabian MR. 2018. Mechanistic insights into microRNA-mediated gene silencing. *Cold Spring Harb Perspect Biol* doi: 10.1101/cshperspect.a032771.

Eichhorn SW, Subtelny AO, Kronja I, Kwasnieski JC, Orr-Weaver TL, Bartel DP. 2016. mRNA poly(A)-tail changes specified by deadenylation broadly reshape translation in *Drosophila* oocytes and early embryos. *eLife* **5:** e16955.

Fichelson P, Moch C, Ivanovitch K, Martin C, Sidor CM, Lepesant JA, Bellaiche Y, Huynh JR. 2009. Live-imaging of single stem cells within their niche reveals that a U3snoRNP component segregates asymmetrically and is required for self-renewal in *Drosophila*. *Nat Cell Biol* **11:** 685–693.

Filardo P, Ephrussi A. 2003. Bruno regulates *gurken* during *Drosophila* oogenesis. *Mech Dev* **120:** 289–297.

Forbes A, Lehmann R. 1998. Nanos and Pumilio have critical roles in the development and function of *Drosophila* germline stem cells. *Development* **125:** 679–690.

Forrest KM, Clark IE, Jain RA, Gavis ER. 2004. Temporal complexity within a translational control element in the *nanos* mRNA. *Development* **131:** 5849–5857.

Gaspar I, Sysoev V, Komissarov A, Ephrussi A. 2017. An RNA-binding atypical tropomyosin recruits kinesin-1 dynamically to *oskar* mRNPs. *EMBO J* **36:** 319–333.

Gavis ER, Lehmann R. 1992. Localization of *nanos* RNA controls embryonic polarity. *Cell* **71:** 301–313.

Gavis ER, Lehmann R. 1994. Translational regulation of *nanos* by RNA localization. *Nature* **369:** 315–318.

Ghosh S, Marchand V, Gaspar I, Ephrussi A. 2012. Control of RNP motility and localization by a splicing-dependent structure in *oskar* mRNA. *Nat Struct Mol Biol* **19:** 441–449.

Gingold H, Tehler D, Christoffersen NR, Nielsen MM, Asmar F, Kooistra SM, Christophersen NS, Christensen LL, Borre M, Sorensen KD, et al. 2014. A dual program for translation regulation in cellular proliferation and differentiation. *Cell* **158:** 1281–1292.

Giraldez AJ, Mishima Y, Rihel J, Grocock RJ, Van Dongen S, Inoue K, Enright AJ, Schier AF. 2006. Zebrafish MiR-430 promotes deadenylation and clearance of maternal mRNAs. *Science* **312:** 75–79.

Goodarzi H, Nguyen HCB, Zhang S, Dill BD, Molina H, Tavazoie SF. 2016. Modulated expression of specific tRNAs drives gene expression and cancer progression. *Cell* **165:** 1416–1427.

Hachet O, Ephrussi A. 2004. Splicing of *oskar* RNA in the nucleus is coupled to its cytoplasmic localization. *Nature* **428:** 959–963.

Harris RE, Pargett M, Sutcliffe C, Umulis D, Ashe HL. 2011. Brat promotes stem cell differentiation via control of a bistable switch that restricts BMP signaling. *Dev Cell* **20:** 72–83.

Hinnebusch AG. 2005. Translational regulation of GCN4 and the general amino acid control of yeast. *Annu Rev Microbiol* **59:** 407–450.

Hinnebusch AG, Lorsch JR. 2012. The mechanism of eukaryotic translation initiation: New insights and challenges. *Cold Spring Harb Perspect Biol* **4:** a011544.

Hussain S, Tuorto F, Menon S, Blanco S, Cox C, Flores JV, Watt S, Kudo NR, Lyko F, Frye M. 2013. The mouse cytosine-5 RNA methyltransferase NSun2 is a component of the chromatoid body and required for testis differentiation. *Mol Cell Biol* **33:** 1561–1570.

Ingolia NT, Ghaemmaghami S, Newman JR, Weissman JS. 2009. Genome-wide analysis in vivo of translation with nucleotide resolution using ribosome profiling. *Science* **324:** 218–223.

Ingolia NT, Lareau LF, Weissman JS. 2011. Ribosome profiling of mouse embryonic stem cells reveals the complexity and dynamics of mammalian proteomes. *Cell* **147:** 789–802.

* Ingolia NT, Hussmann JA, Weissman JS. 2018. Ribosome profiling: Global views of translation. *Cold Spring Harb Perspect Biol* doi: 10.1101/cshperspect.a032698.

Insco ML, Bailey AS, Kim J, Olivares GH, Wapinski OL, Tam CH, Fuller MT. 2012. A self-limiting switch based on translational control regulates the transition from proliferation to differentiation in an adult stem cell lineage. *Cell Stem Cell* **11**: 689–700.

Irish V, Lehmann R, Akam M. 1989. The *Drosophila* posterior-group gene *nanos* functions by repressing hunchback activity. *Nature* **338**: 646–648.

Jambor H, Brunel C, Ephrussi A. 2011. Dimerization of *oskar* 3′ UTRs promotes hitchhiking for RNA localization in the *Drosophila* oocyte. *RNA* **17**: 2049–2057.

Jambor H, Mueller S, Bullock SL, Ephrussi A. 2014. A stem-loop structure directs *oskar* mRNA to microtubule minus ends. *RNA* **20**: 429–439.

Johnstone TG, Bazzini AA, Giraldez AJ. 2016. Upstream ORFs are prevalent translational repressors in vertebrates. *EMBO J* **35**: 706–723.

Joly W, Chartier A, Rojas-Rios P, Busseau I, Simonelig M. 2013. The CCR4 deadenylase acts with Nanos and Pumilio in the fine-tuning of Mei-P26 expression to promote germline stem cell self-renewal. *Stem Cell Rep* **1**: 411–424.

Kalifa Y, Huang T, Rosen LN, Chatterjee S, Gavis ER. 2006. Glorund, a *Drosophila* hnRNP F/H homolog, is an ovarian repressor of *nanos* translation. *Dev Cell* **10**: 291–301.

Kim-Ha J, Kerr K, Macdonald PM. 1995. Translational regulation of *oskar* mRNA by bruno, an ovarian RNA-binding protein, is essential. *Cell* **81**: 403–412.

Kronja I, Yuan B, Eichhorn SW, Dzeyk K, Krijgsveld J, Bartel DP, Orr-Weaver TL. 2014. Widespread changes in the posttranscriptional landscape at the *Drosophila* oocyte-to-embryo transition. *Cell Rep* **7**: 1495–1508.

LaFever L, Feoktistov A, Hsu HJ, Drummond-Barbosa D. 2010. Specific roles of target of rapamycin in the control of stem cells and their progeny in the *Drosophila* ovary. *Development* **137**: 2117–2126.

Laver JD, Marsolais AJ, Smibert CA, Lipshitz HD. 2015. Regulation and function of maternal gene products during the maternal-to-zygotic transition in *Drosophila. Curr Top Dev Biol* **113**: 43–84.

Le Bouteiller M, Souilhol C, Beck-Cormier S, Stedman A, Burlen-Defranoux O, Vandormael-Pournin S, Bernex F, Cumano A, Cohen-Tannoudji M. 2013. Notchless-dependent ribosome synthesis is required for the maintenance of adult hematopoietic stem cells. *J Exp Med* **210**: 2351–2369.

Lecuyer E, Yoshida H, Parthasarathy N, Alm C, Babak T, Cerovina T, Hughes TR, Tomancak P, Krause HM. 2007. Global analysis of mRNA localization reveals a prominent role in organizing cellular architecture and function. *Cell* **131**: 174–187.

Lee MT, Bonneau AR, Giraldez AJ. 2014. Zygotic genome activation during the maternal-to-zygotic transition. *Annu Rev Cell Dev Biol* **30**: 581–613.

Lehmann R. 2016. Germ plasm biogenesis—An Oskar-centric perspective. *Curr Top Dev Biol* **116**: 679–707.

Lehmann R, Nusslein-Volhard C. 1987. *hunchback*, a gene required for segmentation of an anterior and posterior region of the *Drosophila* embryo. *Dev Biol* **119**: 402–417.

Leppek K, Das R, Barna M. 2018. Functional 5′ UTR mRNA structures in eukaryotic translation regulation and how to find them. *Nat Rev Mol Cell Biol* **19**: 158–174.

Li Y, Minor NT, Park JK, McKearin DM, Maines JZ. 2009. Bam and Bgcn antagonize *Nanos*-dependent germ-line

stem cell maintenance. *Proc Natl Acad Sci* **106**: 9304–9309.

Li Y, Maines JZ, Tastan OY, McKearin DM, Buszczak M. 2012. Mei-P26 regulates the maintenance of ovarian germline stem cells by promoting BMP signaling. *Development* **139**: 1547–1556.

Li Y, Zhang Q, Carreira-Rosario A, Maines JZ, McKearin DM, Buszczak M. 2013. Mei-p26 cooperates with Bam, Bgcn and Sxl to promote early germline development in the *Drosophila* ovary. *PLoS ONE* **8**: e58301.

Little SC, Sinsimer KS, Lee JJ, Wieschaus EF, Gavis ER. 2015. Independent and coordinate trafficking of single *Drosophila* germ plasm mRNAs. *Nat Cell Biol* **17**: 558–568.

Liu Y, Beyer A, Aebersold R. 2016. On the dependency of cellular protein levels on mRNA abundance. *Cell* **165**: 535–550.

Llorens-Bobadilla E, Zhao S, Baser A, Saiz-Castro G, Zwadlo K, Martin-Villalba A. 2015. Single-cell transcriptomics reveals a population of dormant neural stem cells that become activated upon brain injury. *Cell Stem Cell* **17**: 329–340.

Lund E, Liu M, Hartley RS, Sheets MD, Dahlberg JE. 2009. Deadenylation of maternal mRNAs mediated by miR-427 in *Xenopus laevis* embryos. *RNA* **15**: 2351–2363.

Marchand V, Gaspar I, Ephrussi A. 2012. An intracellular transmission control protocol: Assembly and transport of ribonucleoprotein complexes. *Curr Opin Cell Biol* **24**: 202–210.

Medioni C, Mowry K, Besse F. 2012. Principles and roles of mRNA localization in animal development. *Development* **139**: 3263–3276.

* Merrick WC, Pavitt GD. 2018. Protein synthesis initiation in eukaryotic cells. *Cold Spring Harb Perspect Biol* doi: 10.1101/cshperspect.a033092.

Micklem DR, Adams J, Grunert S, St Johnston D. 2000. Distinct roles of two conserved Staufen domains in *oskar* mRNA localization and translation. *EMBO J* **19**: 1366–1377.

Miller MA, Olivas WM. 2011. Roles of Puf proteins in mRNA degradation and translation. *Wiley Interdiscip Rev RNA* **2**: 471–492.

Mishima Y, Tomari Y. 2016. Codon usage and 3′ UTR length determine maternal mRNA stability in zebrafish. *Mol Cell* **61**: 874–885.

Moor AE, Golan M, Massasa EE, Lemze D, Weizman T, Shenhav R, Baydatch S, Mizrahi O, Winkler R, Golani O, et al. 2017. Global mRNA polarization regulates translation efficiency in the intestinal epithelium. *Science* **357**: 1299–1303.

Morita M, Gravel SP, Chenard V, Sikstrom K, Zheng L, Alain T, Gandin V, Avizonis D, Arguello M, Zakaria C, et al. 2013. mTORC1 controls mitochondrial activity and biogenesis through 4E-BP-dependent translational regulation. *Cell Metab* **18**: 698–711.

Murata Y, Wharton RP. 1995. Binding of pumilio to maternal hunchback mRNA is required for posterior patterning in *Drosophila* embryos. *Cell* **80**: 747–756.

Nakamura A, Sato K, Hanyu-Nakamura K. 2004. *Drosophila* cup is an eIF4E binding protein that associates with Bruno and regulates *oskar* mRNA translation in oogenesis. *Dev Cell* **6**: 69–78.

Nelson MR, Leidal AM, Smibert CA. 2004. *Drosophila* cup is an eIF4E-binding protein that functions in Smaug-mediated translational repression. *EMBO J* **23:** 150–159.

Neumuller RA, Betschinger J, Fischer A, Bushati N, Poernbacher I, Mechtler K, Cohen SM, Knoblich JA. 2008. Mei-P26 regulates microRNAs and cell growth in the *Drosophila* ovarian stem cell lineage. *Nature* **454:** 241–245.

Norvell A, Kelley RL, Wehr K, Schupbach T. 1999. Specific isoforms of squid, a *Drosophila* hnRNP, perform distinct roles in Gurken localization during oogenesis. *Genes Dev* **13:** 864–876.

Norvell A, Wong J, Randolph K, Thompson L. 2015. Wispy and Orb cooperate in the cytoplasmic polyadenylation of localized *gurken* mRNA. *Dev Dyn* **244:** 1276–1285.

* Peer E, Moshitch-Moshkovitz S, Rechavi G, Dominissini D. 2018. The epitranscriptome in translation regulation. *Cold Spring Harb Perspect Biol* doi: 10.1101/cshperspect. a032623.

Peshkin L, Wühr M, Pearl E, Haas W, Freeman RM Jr, Gerhart JC, Klein AM, Horb M, Gygi SP, Kirschner MW. 2015. On the relationship of protein and mRNA dynamics in vertebrate embryonic development. *Dev Cell* **35:** 383–94.

Presnyak V, Alhusaini N, Chen YH, Martin S, Morris N, Kline N, Olson S, Weinberg D, Baker KE, Graveley BR, et al. 2015. Codon optimality is a major determinant of mRNA stability. *Cell* **160:** 1111–1124.

* Proud CG. 2018. Phosphorylation and signal transduction pathways in translational control. *Cold Spring Harb Perspect Biol* doi: 10.1101/cshperspect.a033050.

Rangan P, DeGennaro M, Jaime-Bustamante K, Coux RX, Martinho RG, Lehmann R. 2009. Temporal and spatial control of germ-plasm RNAs. *Curr Biol* **19:** 72–77.

Roux PP, Topisirovic I. 2012. Regulation of mRNA translation by signaling pathways. *Cold Spring Harb Perspect Biol* **4:** a012252.

Sampath P, Pritchard DK, Pabon L, Reinecke H, Schwartz SM, Morris DR, Murry CE. 2008. A hierarchical network controls protein translation during murine embryonic stem cell self-renewal and differentiation. *Cell Stem Cell* **2:** 448–460.

Sanchez CG, Teixeira FK, Czech B, Preall JB, Zamparini AL, Seifert JR, Malone CD, Hannon GJ, Lehmann R. 2016. Regulation of ribosome biogenesis and protein synthesis controls germline stem cell differentiation. *Cell Stem Cell* **18:** 276–290.

Saunders C, Cohen RS. 1999. The role of oocyte transcription, the 5′UTR, and translation repression and derepression in *Drosophila gurken* mRNA and protein localization. *Mol Cell* **3:** 43–54.

Shi Z, Fujii K, Kovary KM, Genuth NR, Rost HL, Teruel MN, Barna M. 2017. Heterogeneous ribosomes preferentially translate distinct subpools of mRNAs genome-wide. *Mol Cell* **67:** 71–83.e77.

Signer RA, Magee JA, Salic A, Morrison SJ. 2014. Haematopoietic stem cells require a highly regulated protein synthesis rate. *Nature* **509:** 49–54.

Signer RA, Qi L, Zhao Z, Thompson D, Sigova AA, Fan ZP, DeMartino GN, Young RA, Sonenberg N, Morrison SJ. 2016. The rate of protein synthesis in hematopoietic stem cells is limited partly by 4E-BPs. *Genes Dev* **30:** 1698–1703.

Simsek D, Barna M. 2017. An emerging role for the ribosome as a nexus for post-translational modifications. *Curr Opin Cell Biol* **45:** 92–101.

Slaidina M, Lehmann R. 2014. Translational control in germline stem cell development. *J Cell Biol* **207:** 13–21.

Smibert CA, Lie YS, Shillinglaw W, Henzel WJ, Macdonald PM. 1999. Smaug, a novel and conserved protein, contributes to repression of *nanos* mRNA translation in vitro. *RNA* **5:** 1535–1547.

Snee M, Benz D, Jen J, Macdonald PM. 2008. Two distinct domains of Bruno bind specifically to the *oskar* mRNA. *RNA Biol* **5:** 1–9.

Sonoda J, Wharton RP. 2001. *Drosophila* brain tumor is a translational repressor. *Genes Dev* **15:** 762–773.

St Johnston D, Beuchle D, Nusslein-Volhard C. 1991. *staufen*, a gene required to localize maternal RNAs in the *Drosophila* egg. *Cell* **66:** 51–63.

St Johnston D, Brown NH, Gall JG, Jantsch M. 1992. A conserved double-stranded RNA-binding domain. *Proc Natl Acad Sci* **89:** 10979–10983.

Su YQ, Sugiura K, Woo Y, Wigglesworth K, Kamdar S, Affourtit J, Eppig JJ. 2007. Selective degradation of transcripts during meiotic maturation of mouse oocytes. *Dev Biol* **302:** 104–117.

Subtelny AO, Eichhorn SW, Chen GR, Sive H, Bartel DP. 2014. Poly(A)-tail profiling reveals an embryonic switch in translational control. *Nature* **508:** 66–71.

Sun P, Quan Z, Zhang B, Wu T, Xi R. 2010. TSC1/2 tumour suppressor complex maintains *Drosophila* germline stem cells by preventing differentiation. *Development* **137:** 2461–2469.

Tadros W, Lipshitz HD. 2009. The maternal-to-zygotic transition: A play in two acts. *Development* **136:** 3033–3042.

Tadros W, Goldman AL, Babak T, Menzies F, Vardy L, Orr-Weaver T, Hughes TR, Westwood JT, Smibert CA, Lipshitz HD. 2007. SMAUG is a major regulator of maternal mRNA destabilization in *Drosophila* and its translation is activated by the PAN GU kinase. *Dev Cell* **12:** 143–155.

Tahmasebi S, Jafarnejad SM, Tam IS, Gonatopoulos-Pournatzis T, Matta-Camacho E, Tsukumo Y, Yanagiya A, Li W, Atlasi Y, Caron M, et al. 2016. Control of embryonic stem cell self-renewal and differentiation via coordinated alternative splicing and translation of YY2. *Proc Natl Acad Sci* **113:** 12360–12367.

Tamayo JV, Teramoto T, Chatterjee S, Hall TMT, Gavis ER. 2017. The *Drosophila* hnRNP F/H homolog Glorund uses two distinct RNA-binding modes to diversify target recognition. *Cell Rep* **19:** 150–161.

Tastan OY, Maines JZ, Li Y, McKearin DM, Buszczak M. 2010. *Drosophila* ataxin 2-binding protein 1 marks an intermediate step in the molecular differentiation of female germline cysts. *Development* **137:** 3167–3176.

Tautz D. 1988. Regulation of the *Drosophila* segmentation gene hunchback by two maternal morphogenetic centres. *Nature* **332:** 281–284.

Thomsen S, Anders S, Janga SC, Huber W, Alonso CR. 2010. Genome-wide analysis of mRNA decay patterns during early *Drosophila* development. *Genome Biol* **11:** R93.

Cite this article as *Cold Spring Harb Perspect Biol* doi: 10.1101/cshperspect.a032987

Trcek T, Grosch M, York A, Shroff H, Lionnet T, Lehmann R. 2015. *Drosophila* germ granules are structured and contain homotypic mRNA clusters. *Nat Commun* **6:** 7962.

Trovisco V, Belaya K, Nashchekin D, Irion U, Sirinakis G, Butler R, Lee JJ, Gavis ER, St Johnston D. 2016. *bicoid* mRNA localises to the *Drosophila* oocyte anterior by random Dynein-mediated transport and anchoring. *eLife* **5:** e17537.

Tsuda M, Sasaoka Y, Kiso M, Abe K, Haraguchi S, Kobayashi S, Saga Y. 2003. Conserved role of nanos proteins in germ cell development. *Science* **301:** 1239–1241.

Wang C, Lehmann R. 1991. Nanos is the localized posterior determinant in *Drosophila*. *Cell* **66:** 637–647.

Wang X, Lu Z, Gomez A, Hon GC, Yue Y, Han D, Fu Y, Parisien M, Dai Q, Jia G, et al. 2014. N^6-methyladenosine-dependent regulation of messenger RNA stability. *Nature* **505:** 117–120.

Webster PJ, Liang L, Berg CA, Lasko P, Macdonald PM. 1997. Translational repressor *bruno* plays multiple roles in development and is widely conserved. *Genes Dev* **11:** 2510–2521.

Weidmann CA, Qiu C, Arvola RM, Lou TF, Killingsworth J, Campbell ZT, Tanaka Hall TM, Goldstrohm AC. 2016. *Drosophila* Nanos acts as a molecular clamp that modulates the RNA-binding and repression activities of Pumilio. *eLife* **5:** e17096.

Weil TT, Parton RM, Herpers B, Soetaert J, Veenendaal T, Xanthakis D, Dobbie IM, Halstead JM, Hayashi R, Rabouille C, et al. 2012. *Drosophila* patterning is established by differential association of mRNAs with P bodies. *Nat Cell Biol* **14:** 1305–1313.

* Wek RC. 2018. Role of eIF2α kinases in translational control and adaptation to cellular stress. *Cold Spring Harb Perspect Biol* doi: 10.1101/cshperspect.a032870.

Wharton RP, Struhl G. 1991. RNA regulatory elements mediate control of *Drosophila* body pattern by the posterior morphogen *nanos*. *Cell* **67:** 955–967.

Wilhelm JE, Hilton M, Amos Q, Henzel WJ. 2003. Cup is an eIF4E binding protein required for both the translational repression of *oskar* and the recruitment of Barentsz. *J Cell Biol* **163:** 1197–1204.

Yartseva V, Giraldez AJ. 2015. The maternal-to-zygotic transition during vertebrate development: A model for reprogramming. *Curr Top Dev Biol* **113:** 191–232.

Yilmaz OH, Valdez R, Theisen BK, Guo W, Ferguson DO, Wu H, Morrison SJ. 2006. Pten dependence distinguishes haematopoietic stem cells from leukaemia-initiating cells. *Nature* **441:** 475–482.

Zappavigna V, Piccioni F, Villaescusa JC, Verrotti AC. 2004. Cup is a nucleocytoplasmic shuttling protein that interacts with the eukaryotic translation initiation factor 4E to modulate *Drosophila* ovary development. *Proc Natl Acad Sci* **101:** 14800–14805.

Zhang J, Grindley JC, Yin T, Jayasinghe S, He XC, Ross JT, Haug JS, Rupp D, Porter-Westpfahl KS, Wiedemann LM, et al. 2006. PTEN maintains haematopoietic stem cells and acts in lineage choice and leukaemia prevention. *Nature* **441:** 518–522.

Zhang Q, Shalaby NA, Buszczak M. 2014. Changes in rRNA transcription influence proliferation and cell fate within a stem cell lineage. *Science* **343:** 298–301.

Zhao BS, Wang X, Beadell AV, Lu Z, Shi H, Kuuspalu A, Ho RK, He C. 2017. m⁶A-dependent maternal mRNA clearance facilitates zebrafish maternal-to-zygotic transition. *Nature* **542:** 475–478.

Zimyanin VL, Belaya K, Pecreaux J, Gilchrist MJ, Clark A, Davis I, St Johnston D. 2008. In vivo imaging of *oskar* mRNA transport reveals the mechanism of posterior localization. *Cell* **134:** 843–853.

Zismanov V, Chichkov V, Colangelo V, Jamet S, Wang S, Syme A, Koromilas AE, Crist C. 2016. Phosphorylation of eIF2α is a translational control mechanism regulating muscle stem cell quiescence and self-renewal. *Cell Stem Cell* **18:** 79–90.

Translational Control in the Brain in Health and Disease

Wayne S. Sossin[1] and Mauro Costa-Mattioli[2]

[1]Montreal Neurological Institute, McGill University, Montreal, Quebec H3A-2B4, Canada

[2]Department of Neuroscience, Memory and Brain Research Center, Baylor College of Medicine, Houston, Texas 77030

Correspondence: wayne.sossin@mcgill.ca; costamat@bcm.edu

Translational control in neurons is crucially required for long-lasting changes in synaptic function and memory storage. The importance of protein synthesis control to brain processes is underscored by the large number of neurological disorders in which translation rates are perturbed, such as autism and neurodegenerative disorders. Here we review the general principles of neuronal translation, focusing on the particular relevance of several key regulators of nervous system translation, including eukaryotic initiation factor 2α (eIF2α), the mechanistic (or mammalian) target of rapamycin complex 1 (mTORC1), and the eukaryotic elongation factor 2 (eEF2). These pathways regulate the overall rate of protein synthesis in neurons and have selective effects on the translation of specific messenger RNAs (mRNAs). The importance of these general and specific translational control mechanisms is considered in the normal functioning of the nervous system, particularly during synaptic plasticity underlying memory, and in the context of neurological disorders.

TRANSLATIONAL CONTROL IN MEMORY

Memory storage is widely thought to have a physical basis in long-lasting changes in synaptic function (Malenka and Bear 2004; Mayford et al. 2012; Neves et al. 2008). For instance, high activity in a given neural pathway can persistently increase the efficacy of synaptic connections in a process called long-term potentiation (LTP). The reverse is also true: reduced activity persistently lowers synaptic efficacy, resulting in long-term depression (LTD). These long-lasting changes in synaptic strength are forms of synaptic plasticity and require de novo protein synthesis (Sutton and Schuman 2006; Costa-Mattioli et al. 2009; Trinh and Klann 2013; Buffington et al. 2014). Pioneering work by Josefa and Louis Flexner showed that short-lasting and longer-lasting memories are differentiated by the requirement for protein synthesis only for the latter (Flexner et al. 1967; Squire and Davis 1981). Thus, the requirement for new protein synthesis underlies long-lasting changes in both synaptic plasticity and memory.

However, despite more than 50 years of research in the field, several important questions remain unanswered: What are the major translational control programs regulating synaptic plasticity and memory storage? Where in the neuron is protein synthesis important for memory formation (cell soma vs. synapse; presynaptic axons vs. postsynaptic dendrites)? What is the nature of the messenger RNAs (mRNAs) whose translation is increased in response to a learning experience? Finally, is the translational control machinery dysregulated in cognitive disorders? If so, by specifically correcting the aberrant translational program, can we restore cognition in these brain disorders?

GENERAL VERSUS SPECIFIC PROTEIN SYNTHESIS

Two aspects of translational control are relevant for synaptic plasticity and memory: general translational control, which determines the overall rate of protein synthesis; and specific translational control, whereby selective regulatory mechanisms allow for translation of specific mRNAs.

When a memory is made, gene expression takes place selectively in a small number of neurons in a given brain area (Guzowski 2002; Reijmers and Mayford 2009). Growing evidence supports the idea that these are the neurons that encode the memory (Josselyn et al. 2015; Tonegawa et al. 2015). Currently, it is not known whether these neurons require (1) a general increase in translation, or (2) the translation of a specific subset of mRNAs to generate the long-lasting synaptic changes that encode memory.

Overall increases in the rate of protein synthesis after learning have been observed in a number of different learning paradigms (Davis and Squire 1984; Kelleher et al. 2004; Batista et al. 2016; Liu and Cline 2016). A general increase in translation rates could be important for the growth and stabilization of new synaptic connections formed in neurons that encode memory (Bailey and Kandel 2008; Xu et al. 2009; Yang et al. 2009; Ryan et al. 2015). The general activation of translation is associated with a number of processes that are discussed below, including decreased phosphorylation of the translation factors eukaryotic initiation factor (eIF)2α and eukaryotic elongation factor 2 (eEF2) as well as activation of mechanistic target of rapamycin complex 1 (mTORC1).

By contrast, changes in the rate of translation of a subset of mRNAs may be sufficient for inducing or stabilizing changes at specific synapses. As we discuss below, the same pathways that regulate general translation can also regulate the synthesis of specific proteins. Thus, it is difficult to differentiate the relative importance of general and specific protein synthesis in the formation of a memory simply by identifying the regulatory pathway involved. Indeed, the two processes may be linked; for example, the specific translation of mRNAs with a 5′ terminal oligopyrimidine tract (TOP mRNAs) (Tsokas et al. 2007; Gobert et al. 2008) promotes general protein synthesis.

DENDRITIC VERSUS SOMATIC PROTEIN SYNTHESIS

Memory is thought to require the strengthening of a specific subset of synapses. The requirement for both protein synthesis and synapse specificity does not necessarily imply a requirement for the activation of translation locally at the synapses. Proteins can also be synthesized in the soma of a neuron and then selectively transported to synapses. Nevertheless, several lines of evidence support the notion that local translation is required for LTP and LTD (Sutton and Schuman 2006). More than 2000 mRNAs are found in synaptic processes (Cajigas et al. 2012), presumably through selective transport of these mRNAs, and increases in protein synthesis can be seen locally at synapses (Aakalu et al. 2001; Wang et al. 2009). Moreover, in some cases, protein synthesis-dependent LTP and LTD can be observed when the soma is physically disconnected from the synapse (Kang and Schuman 1996; Huber et al. 2000; Liu et al. 2003; Cracco et al. 2005; Gelinas and Nguyen 2005), supporting the notion that local translation can mediate long-lasting forms of synaptic plasticity. However, it is not entirely clear whether local translation is used for memory storage. Local application of protein synthesis inhibitors to synaptic regions decreases, but does not completely block, long-lasting forms of synaptic plasticity (Bradshaw et al. 2003). Local protein synthesis may be required to generate high concentrations

Cite this article as *Cold Spring Harb Perspect Biol* doi: 10.1101/cshperspect.a032912

of proteins locally at synapses to support basic synaptic functions (Cajigas et al. 2012), as opposed to the specific production of proteins to induce changes in synaptic plasticity. Indeed, many of the proteins that are known to be locally synthesized near synapses (e.g., calcium-cal-modulin kinase IIα [CAMKIIα], β-actin, and protein kinase Mζ [PKMζ]) have high basal levels at synapses, and for at least one of these, CAMKIIα, the high basal levels require local protein synthesis (Miller et al. 2002). It is an open question how the small resultant change in concentration of these proteins can explain a functional change in synaptic strength. Furthermore, it is currently unknown how many out of the 2000 mRNAs found in synaptic processes are required for memory formation. In addition, so far, we have not been able to identify a translational control mechanism that takes place only at synapses (but not in the soma). Consequently, direct loss- or gain-of function evidence that local translation is required for memory consolidation is still missing. Even if local translation is required for the synaptic plasticity underlying memory, it is still unclear whether this requires selective translational control of a few mRNAs or a general increase in the translation of all transported mRNAs. These are important issues in the field that require further research.

LATE-LTP AND SYNAPTIC TAGGING

Even if local translation produces proteins only near activated synapses, protein diffusion would allow these proteins to be accessible to many synapses situated nearby in the same neuron (Rangaraju et al. 2017). Therefore, it is difficult to envision mechanisms for these proteins to act only at activated synapses in the absence of additional changes that restrict the actions of the new proteins to these synapses. Activation of protein synthesis is thus rarely sufficient to increase synaptic strength in the absence of other changes, a concept formalized in the "synaptic tagging" model (Frey and Morris 1998). In this model, the learning stimulus that generates memory leads to both (1) the translation of plasticity-related proteins (PRPs) in the neuron, and

(2) the formation of a "synaptic tag," which marks the activated synapse. The synaptic tag determines that the PRPs act only at the activated synapse. The production of PRPs is a conceptual framework that supports the necessity of specific protein synthesis for synaptic plasticity underlying memory but does not predict whether the PRPs are made locally or in the soma (Frey and Morris 1998).

There have been many suggestions to explain the molecular basis of the synaptic tag (Martin and Kosik 2002), including alterations in the actin cytoskeleton (Ramachandran and Frey 2009), slots for neurotransmitter-gated ion channels (Granger et al. 2013), and scaffold proteins that link persistently active kinases to their substrates (Hu et al. 2017). Candidates for the PRPs include persistently active kinases, such as truncated persistently active forms of protein kinase C, termed PKM (Sajikumar et al. 2005) and phosphorylated, constitutively active CAMKIIα (Sajikumar et al. 2007; Sanhueza and Lisman 2013). Other candidates include neurotransmitter-gated ion channels and immediate early genes such as Homer and activity-regulated cytoskeleton-associated protein (Arc) (Martin and Kosik 2002). PKMζ (Muslimov et al. 2004; Eom et al. 2014), CAMKIIα (Ouyang et al. 1999; Aakalu et al. 2001), neurotransmitter-gated channels (Ju et al. 2004), and Arc (Steward et al. 2014) have all been shown to be synthesized locally at or near the synapse after stimuli that produce long-lasting synaptic changes and memory, whereas Homer is synthesized only in the soma (Okada et al. 2009).

PRESYNAPTIC VERSUS POSTSYNAPTIC PROTEIN SYNTHESIS

The vast majority of the work on neuronal protein synthesis has focused on local translation in the dendrite, the postsynaptic region of the neuron. These studies are driven by the assumption that ribosomes travel down dendrites, but not down axons to the presynaptic region of the cell. Indeed, the main evidence against presynaptic protein synthesis is the failure to visualize polysomes in presynaptic endings in electron microscopic (EM) studies (Twiss and Fainzilber

2009). However, emerging evidence supports the notion that presynaptic protein synthesis is also important for synaptic plasticity. First, many studies have shown that during development, local translation in axonal growth cones is important for axon guidance, and in mature neurons local translation in the axon is important for response to injury (Jung et al. 2012). In GABAergic neurons, presynaptic protein synthesis is required for long-term plasticity (Younts et al. 2016). In invertebrates, many forms of plasticity depend on presynaptic protein synthesis (Martin et al. 2000; Costa-Mattioli et al. 2009). More recently, surveys of mRNAs found in the synaptic region have shown abundant presence of mRNAs for presynaptic proteins (Cajigas et al. 2012). Finally, whereas it has been difficult to identify polyribosomes at presynaptic sites by EM (Buxbaum et al. 2014), ribosomes and polysomes have been found in presynaptic endings from squid by electron spectroscopy and [^{3}H]-leucine labeling (Crispino et al. 1997). In future experiments, it would be interesting to determine whether the same translational control mechanisms operate in presynaptic and postsynaptic compartments.

eIF2α-MEDIATED TRANSLATION: A MEMORY SWITCH

As discussed above, making new proteins is a critical step required for the generation of long-lasting memories. In particular, the eIF2α signaling pathway plays a dominant role in regulating memory formation. Phosphorylation of the α subunit of eIF2 at Ser51 is tightly regulated by four kinases: (1) the heme-regulated kinase HRI; (2) the double-strand RNA-dependent protein kinase R (PKR); (3) the PKR-like endoplasmic reticulum kinase (PERK); and (4) the highly conserved kinase general control nonderepressible 2 (GCN2). Dephosphorylation of eIF2α is regulated by two phosphatase complexes: (1) the catalytic subunit protein phosphatase 1 (PP1) and the regulatory subunit PPP1R15A/GADD34, and (2) PP1 and the regulatory protein PPP1R15B/CReP (Fig. 1A) (Merrick and Pavitt 2018; Wek 2018). Behavioral training leads to a decrease in the phosphorylation of eIF2α in the hippocampus (Costa-Mattioli et al. 2007), a brain structure crucially required for memory formation (Squire et al. 2015). Accordingly, heterozygous eIF2α$^{S/A}$ knock-in mice (where the single phosphorylation site at serine 51 is replaced by alanine in one allele), with reduced phosphorylation of eIF2α in the hippocampus, show enhanced memory in a variety of behavioral tasks (Costa-Mattioli et al. 2007), indicating that eIF2α phosphorylation normally serves as a memory repressor. Consistent with the results of the eIF2α$^{S/A}$ knock-in mice, genetic deletion of any of the eIF2α kinases expressed in the brain, GCN2, PKR, or PERK, enhances memory formation (Costa-Mattioli et al. 2005, 2007; Zhu et al. 2011; Stern et al. 2013; Ounallah-Saad et al. 2014). However, using conditional knockout mice in which PERK is removed only in the forebrain, deficits in behavior including increased repetitive behaviors and behavioral inflexibility were observed (Trinh et al. 2012). We speculate that this may be because PERK phosphorylates targets other than eIF2α and/or mod-

Figure 1. Eukaryotic initiation factor (eIF)2α bidirectionally regulates the two major forms of plasticity, long-term potentiation (LTP) and long-term depression (LTD) in the brain. (A) Regulation of translation by eIF2α phosphorylation. (B) Increased eIF2α phosphorylation prevents long-lasting LTP induced by four trains of high-frequency stimulation (arrows), but facilitates metabotropic glutamate receptors (mGLuRs)-LTD. The strength of the synapse is measured by postsynaptic potentials (PSPs), or excitatory postsynaptic currents (ePSCs) normalized to the strength before induction of LTP or mGLuR-LTD, respectively. (C) Decreased eIF2α phosphorylation facilitates LTP induced by one high frequency stimulation but blocks mGLuR-induced LTD. The strength of the synapse is measured by PSPs or ePSCs normalized to the strength before induction of LTP or mGLuR-LTD, respectively. PKR, Protein kinase R; GCN2, general control nonderepressible 2; PERK, PKR-like endoplasmic reticulum kinase; PP1, protein phosphatase 1; mRNA, messenger RNA; uORF, upstream open reading frames; UTR, untranslated region.

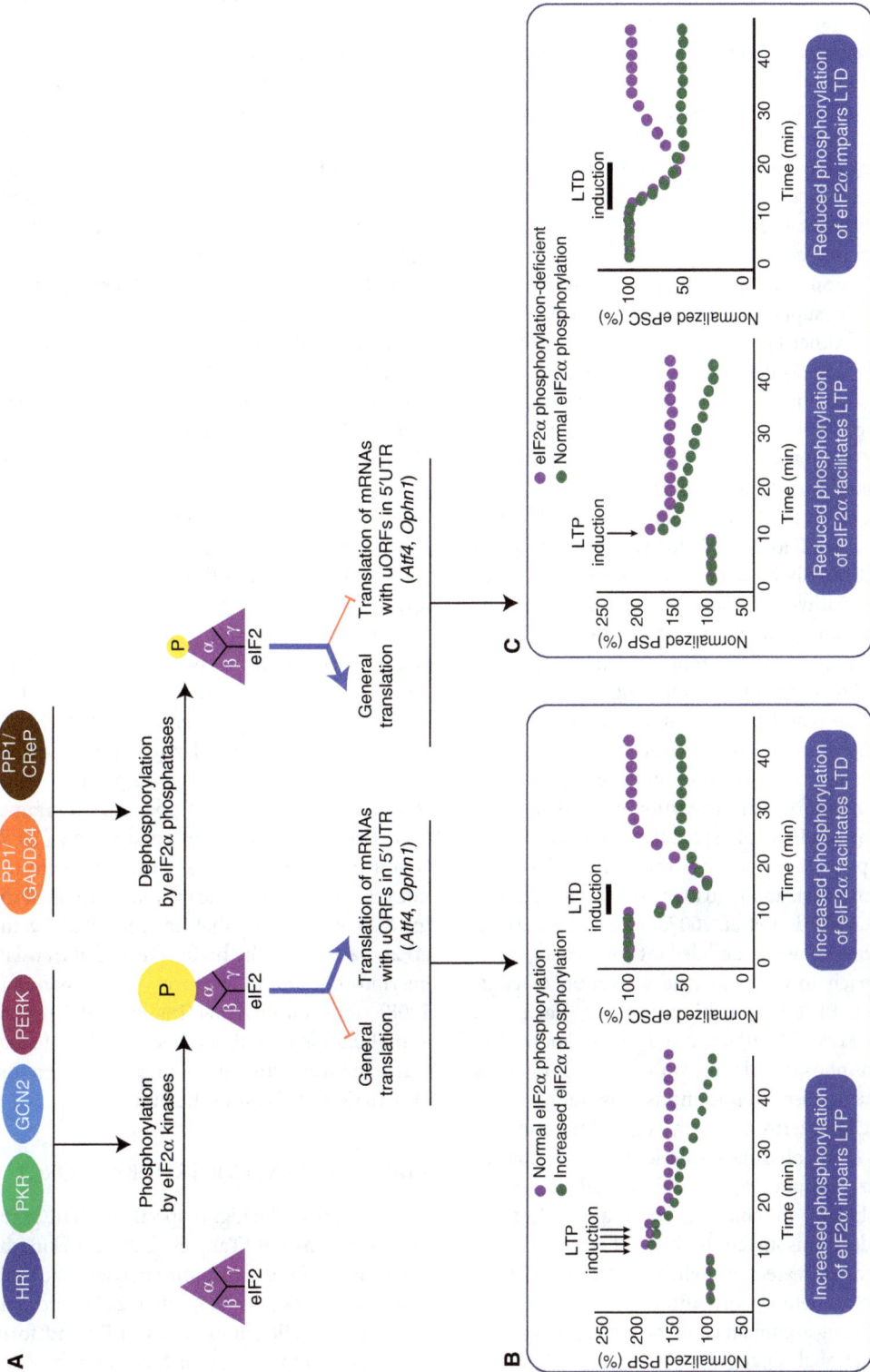

Figure 1. (*See legend on preceding page.*)

ulates calcium dynamics (Zhu et al. 2016), because these perturbations have not been seen in the heterozygous eIF2α$^{S/A}$ knock-in mice. Indeed, several targets of PERK other than eIF2α that may regulate synaptic plasticity have recently been identified including CREB (Sen et al. 2017), calcineurin (Wang et al. 2013), and endoplasmic reticulum (ER)-mitochondrial tethering (van Vliet et al. 2016).

Pharmacological experiments also support the notion that eIF2α phosphorylation is a memory suppressor. Pharmacological inhibition of either (1) PKR with a specific inhibitor (PKRi), or (2) derepression of the translational program controlled by eIF2α with a recently discovered small-molecule inhibitor (integrated stress response inhibitor [ISRIB]), enhances memory formation in both mice and chicks (Zhu et al. 2011; Sidrauski et al. 2013; Stern et al. 2013; Batista et al. 2016), highlighting the evolutionarily conserved role of the eIF2α signaling pathway in learning and memory. Thus, convergent genetic and pharmacological manipulations provide strong evidence that derepression of the translational program controlled by eIF2α is sufficient to enhance memory.

The opposite is also true: increased eIF2α phosphorylation in the brain impairs long-term memory. Administration of Sal003, an inhibitor of eIF2α phosphatases, increases eIF2α phosphorylation in both mice and chicks and prevents their ability to form long-term memory (Costa-Mattioli et al. 2007; Batista et al. 2016). Moreover, using an elegant phamacogenetic approach to allow specific activation of engineered PKR in the hippocampus, Jiang and coworkers (2010) showed that increasing eIF2α phosphorylation only in CA1 neurons of the mouse hippocampus was sufficient to block long-term memory. Importantly, in this study, the pharmacogenetic manipulation of eIF2α phosphorylation did not affect general translation. In contrast, local application of low-dose anisomycin to the CA1 region significantly decreased overall translation without blocking memory formation. Thus, in this model, a strong argument could be made that specific translational changes downstream of eIF2α phosphorylation are required for memory, as opposed to a requirement for a general increase in translation.

During LTP, eIF2α phosphorylation is reduced (Costa-Mattioli et al. 2005, 2007). However, an increase in eIF2α phosphorylation occurs during a protein synthesis–dependent form of LTD associated with activation of metabotropic glutamate receptors, termed mGluR-LTD (Di Prisco et al. 2014; Trinh et al. 2014). Indeed, genetic reduction of eIF2α phosphorylation (or treatment with ISRIB) enhances LTP (Costa-Mattioli et al. 2007; Zhu et al. 2011; Huang et al. 2016; Placzek et al. 2016) but blocks mGluR-LTD (Fig. 1B) (Di Prisco et al. 2014). By contrast, increased eIF2α phosphorylation impairs LTP (Costa-Mattioli et al. 2007; Jiang et al. 2010; Huang et al. 2016) but induces mGluR-LTD (Fig. 1C) (Di Prisco et al. 2014). Taken together, eIF2α phosphorylation bidirectionally controls two major forms of synaptic plasticity (LTP and LTD) and is thus crucial for memory formation.

Some of the proteins regulated by phosphorylation of eIF2α during memory formation have been identified, including ATF4 (Costa-Mattioli et al. 2007) and oligophrenin-1 (OPHN1) (Di Prisco et al. 2014). Similar to other situations where eIF2α regulation is important, these mRNAs contain upstream open reading frames (uORFs) in their 5′untranslated region (UTR) (Hinnebusch et al. 2016). However, there has not yet been a systematic examination of all the mRNAs in neurons that are regulated by the eIF2α pathway in the brain. The use of translating ribosome affinity purification (Heiman et al. 2008), ribosome profiling (Ingolia et al. 2018), or a combination of both methods, will help to elucidate the translational state regulated by phosphorylation of eIF2α in the brain.

TRANSLATIONAL CONTROL BY mTORC1

Treatment with the highly specific mTORC1 inhibitor, rapamycin (Tang et al. 2002; Bekinschtein et al. 2007) or direct pharmacogenetic inhibition of mTORC1 (Stoica et al. 2011) prevents the ability to elicit long-lasting LTP and form long-term memory (Costa-Mattioli and Monteggia 2013). mTORC1 integrates information from various synaptic inputs and is believed to control the formation of memory by regulating translation via S6 kinase (S6K) and eIF4E-binding proteins (4E-BPs) (Sonenberg and Hinnebusch 2009; Proud 2018). S6K regulates translation initiation and elongation through phosphorylation of eukaryotic initiation factor 4B (eIF4B) and eEF2 kinase (eEF2K), respectively. By contrast, 4E-BPs exclusively repress translation initiation through binding to eIF4E. Although inhibition of mTORC1 blocks memory formation, activation of mTORC1 does not enhance memory formation (Graber et al. 2013). This is in contrast to eIF2α phosphorylation, where as described above, memory formation is facilitated when levels of eIF2α are reduced. Gain-of-function activation of mTORC1 by inhibiting upstream suppressors of mTORC1, tuberous sclerosis complex (Tsc) or phosphatase and tensin homolog (Pten), impairs, not enhances, synaptic plasticity and memory formation (von der Brelie et al. 2006; Ehninger et al. 2008; Sperow et al. 2012; Lugo et al. 2013). Similarly, whereas mice lacking 4E-BP2, the major 4E-BP isoform in the brain, show a lowered threshold for the induction of LTP (Banko et al. 2005), under normal learning conditions, 4E-BP2-deficient mice show impaired LTP and long-term memory (Banko et al. 2005). Thus, the correct balance of mTORC1 activation seems critical for normal memory formation.

The cap-binding factor eIF4E is normally inhibited by nonphosphorylated 4E-BPs (Sonenberg and Hinnebusch 2009). In nonneuronal cells, activation of mTORC1 promotes translation rates via phosphorylation of 4E-BPs (Hsieh et al. 2012; Thoreen et al. 2012; Proud 2018). However, in the adult brain, there are only low levels of phosphorylated 4E-BP2 (Bidinosti et al. 2009), the major 4E-BP form. The low levels of phosphorylation are associated with increased deamidation of 4E-BP2, which increases its association with the mTORC1 adaptor raptor, although the significance of this enhanced binding is still unclear (Bidinosti et al. 2009). Thus, it is unclear whether mTORC1 regulates translation in the adult brain through phosphorylation of 4E-BPs. It should also be noted that

mTORC1 regulates a variety of biological processes other than protein synthesis, including autophagy, metabolism, and lipid metabolism biosynthesis (Wullschleger et al. 2006; Laplante and Sabatini 2012). Thus, it is not yet clear whether mTORC1 regulates memory formation by controlling protein synthesis or other biological processes.

REGULATION OF TRANSLATION ELONGATION

Interest in the role of regulating translational elongation in the nervous system has increased in recent years (Taha et al. 2013; Richter and Coller 2015). Although we have a good understanding of how translation initiation of specific mRNAs can be regulated by uORFs, structured 5′UTRs, and interactions between RNA-binding proteins and the initiation machinery (Sossin and Lacaille 2010; Darnell and Richter 2012), the role of elongation in controlling translation of specific transcripts is less clear.

There are two major elongation factors, the GTPases eEF1 and eEF2. eEF1 is comprised of eEF1A and eEF1B; eEF1A recruits aminoacylated transfer RNAs (tRNAs) to the A site in the ribosome and is complexed with eEF1B, its guanine nucleotide exchange factor (GEF) (Sasikumar et al. 2012). The translation of eEF1A mRNA is regulated by a TOP sequence in its 5′UTR. In neurons, eEF1A mRNA is translated locally at dendrites during memory formation and synaptic plasticity (Giustetto et al. 2003; Huang et al. 2005; Tsokas et al. 2005).

In contrast to eEF1, eEF2 is one of the few GTPases that does not require a GEF for activation. eEF2 ratchets the peptidyl-tRNA from the hybrid position on the ribosome after peptide bond formation to the P-site position (Kaul et al. 2011; Dever et al. 2018). There are reports of increased synthesis of eEF2 after induction of synaptic plasticity (Carroll et al. 2004; Takei et al. 2009) and eEF2, like eEF1A, is regulated through a TOP motif. Thus, because TOP mRNAs are regulated downstream of the mTORC1 pathway, one mechanism for mTORC1 regulation of elongation could be through increasing levels of elongation factors.

REGULATION OF ELONGATION THROUGH eEF2K

Much attention has been devoted to studying eEF2 phosphorylation at Thr57 by eEF2K (Nairn and Palfrey 1987; Proud 2015; Dever et al. 2018). No other kinase appears to phosphorylate eEF2 at this site, as there is a complete lack of detectable eEF2 phosphorylation when eEF2K is genetically ablated (Heise et al. 2017). Phosphorylation of eEF2 reduces its binding to the ribosome and thus decreases elongation (Proud 2015, 2018). The impact of eEF2 phosphorylation on translation depends on whether (1) the rate of elongation is rate-limiting for translation overall, (2) the levels of eEF2 are rate-limiting for elongation, and (3) the extent of eEF2 phosphorylation. Increasing levels of eEF2 in cultured neurons decreased elongation times and increased protein synthesis, suggesting that eEF2 levels can be rate-limiting in neurons under some circumstances (Takei et al. 2009). However, increasing levels of active eEF2 by eliminating eEF2 phosphorylation through removal of eEF2K did not change the overall rate of protein synthesis in the brain (Heise et al. 2017).

A decrease in eEF2 phosphorylation in the hippocampus occurs after fear conditioning and after induction of long-term increases in synaptic strength in *Aplysia* neurons (Im et al. 2009; McCamphill et al. 2015). Preventing the decrease in eEF2 phosphorylation blocks long-term increases in synaptic strength in *Aplysia* (McCamphill et al. 2017), demonstrating that eEF2 dephosphorylation is required for long-term facilitation. Accordingly, in mice lacking eEF2K and with decreased eEF2 phosphorylation, long-term fear memory is impaired (Heise et al. 2017).

eEF2 phosphorylation is regulated by mTORC1 (Carroll et al. 2004; Inamura et al. 2005; McCamphill et al. 2015) through a conserved S6K phosphorylation site in eEF2K that inactivates the kinase (Wang et al. 2001; Weatherill et al. 2011). This site can also be phosphorylated by ERK through RSK2 (Wang et al. 2001).

Increased eEF2 phosphorylation has been shown in a number of synaptic plasticity and learning procedures, including novel taste learning (Belelovsky et al. 2005), LTP in the dentate gyrus (Panja et al. 2009), mGluR-LTD (Park et al. 2008), and a form of intermediate facilitation in *Aplysia* (McCamphill et al. 2015). Preventing the increase in eEF2 phosphorylation blocks both mGluR-LTD (Park et al. 2008) and intermediate facilitation in *Aplysia* (McCamphill et al. 2015), but eEF2 phosphorylation does not appear to be required for LTP in the dentate gyrus (Panja et al. 2009). It is not clear whether the increased eEF2 phosphorylation inhibits overall translation elongation rates in these situations, but it has been shown to be required for the increased translation of specific mRNAs including CAMKIIα, brain-derived neurotrophic factor (BDNF), microtubule-associated protein (Map)1B and other cytoskeletal proteins (Fig. 2) (Scheetz et al. 2000; Davidkova and Carroll 2007; Park et al. 2008; Verpelli et al. 2010; Kenney et al. 2016; Heise et al. 2017).

There are two proposed mechanisms by which increased eEF2 phosphorylation could lead to the up-regulation of translation of specific mRNAs. When elongation is slow, the ribosome is slow to clear the initiation site; this decreases the rate of initiation on well-translated mRNAs, freeing rate-limiting initiation factors (reviewed in Sossin and Lacaille 2010). The second proposed mechanism involves eEF2 phosphorylation, which plays a role in the release of stalled polysomes. This proposal arose because of a correlation between synaptic plasticity events that specifically require eEF2 phosphorylation and those that are mediated by initiation-independent translation (i.e., forms of plasticity that are blocked by elongation inhibitors and resistant to translation inhibitors; see McCamphill et al. 2015). These forms of plasticity include mGluR-LTD in hippocampal neurons (Graber et al. 2013) and intermediate facilitation in *Aplysia* neurons (McCamphill et al. 2015).

Stalled polysomes have been suggested as an important mechanism for regulated transport of mRNAs to the synapse (Sossin and DesGroseillers 2006). mRNA initiation takes place in the cell soma; at some point elongation is stalled, and the stalled polysome is packaged into an RNA granule that is then transported to the syn-

Cite this article as *Cold Spring Harb Perspect Biol* doi: 10.1101/cshperspect.a032912

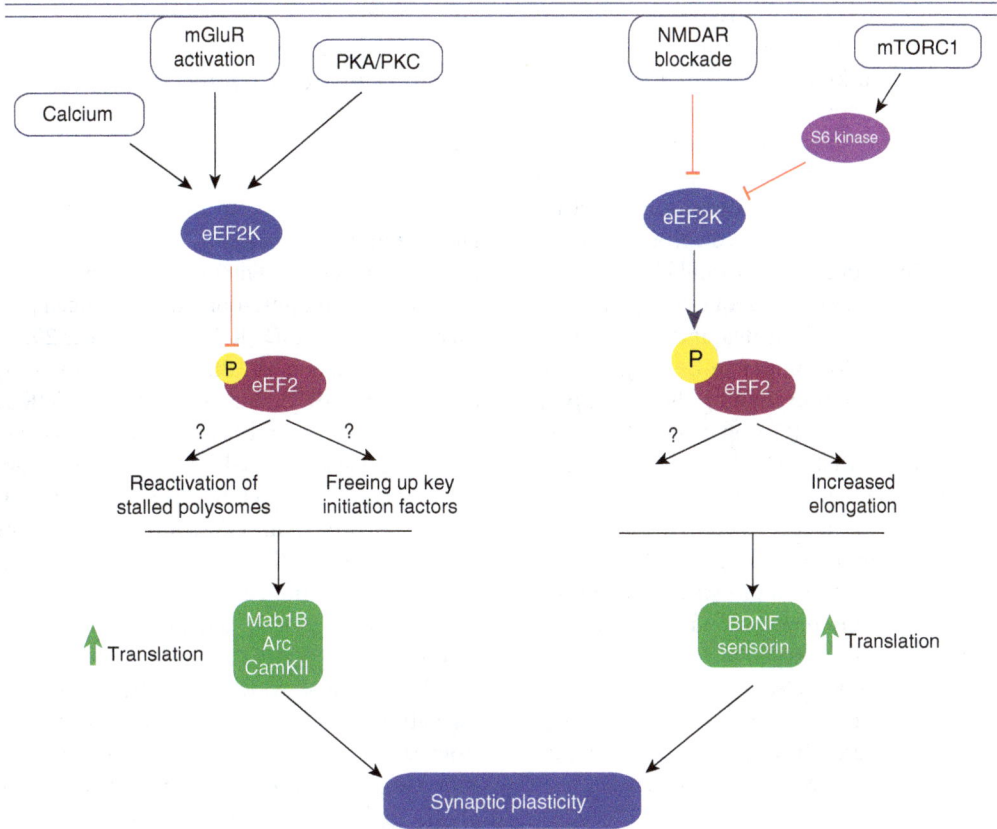

Figure 2. Upstream and downstream of eukaryotic elongation factor 2 (eEF2) phosphorylation. In neurons, eukaryotic elongation factor 2 kinase (eEF2K) can be activated by calcium influx, metabotropic glutamate receptor (mGLuR) activation, or protein kinases. The resultant increase in eEF2 phosphorylation increases translation of selective transcripts either by freeing up key initiation factors or by reactivating stalled polysomes. The increased synthesis of these proteins is linked to changes in synaptic strength (synaptic plasticity). In contrast, blockade of the N-methyl-D-aspartate receptor (NMDAR) or activation of the S6 kinase downstream of target of rapamycin complex 1 (TORC1) leads to a decrease in the activity of eEF2K. The resultant decrease in eEF2 phosphorylation causes an overall increased elongation rate with specific increases in the synthesis of several secreted factors, such as brain-derived neurotrophic factor (BDNF) in vertebrates and sensorin in *Aplysia*. The increased synthesis of these factors is linked to protein synthesis-dependent increases in synaptic strength (synaptic plasticity). PKA, Protein kinase A; PKC, protein kinase C.

apse. After the stall is relieved, translation from previously stalled polysomes is dependent only on elongation, but not initiation, similar to the forms of plasticity that require eEF2 phoshorylation noted above. The fragile X mental retardation protein (FMRP), the protein lost in fragile X syndrome in humans, is associated with stalled polysomes (Mazroui et al. 2002; Darnell et al. 2011; El Fatimy et al. 2016). FMRP can stall polysomes through direct binding to ribosomal RNA (Chen et al. 2014). The role of FMRP in stalling

polysomes is consistent with the role of FMRP in regulating mGLuR-LTD (Hou et al. 2006; Nosyreva and Huber 2006; Graber et al. 2017). However, how phosphorylation of eEF2 regulates stalled polysomes is not immediately clear. One possibility is that the inactive phosphorylated form of eEF2 plays a role in stalling, and removal of this inactive form through eEF2 dephosphorylation is required for unstalling.

If mRNAs are transported to synapses in stalled polysomes, these polysomes would need to be protected from the ribosome-associated quality-control pathway that rescues stalled polysomes (Buskirk and Green 2017). Indeed, this rescue pathway is particularly important when levels of rare charged tRNAs are low, and the absence of this rescue pathway leads to neuronal degeneration (Ishimura et al. 2014). This may indicate that ribosomes are a limiting resource in neurons, which would be consistent with stalled polysomes, sequestering many of the ribosomes in dendrites and thus making the number of ribosomes rate-limiting for basal translation. A study examining changes in polysomes after learning (Ostroff et al. 2017) identified polysomes that are both sensitive and resistant to the initiation inhibitor 4EGI-1, which blocks the association of eIF4E and eIF4G, which is required for cap-dependent translation (Moerke et al. 2007). In this study, the sensitive and resistant polysomes were differentially distributed at distinct types of synapses (Ostroff et al. 2017). It was unclear in this study, however, whether the 4EGI-1-resistant polysomes were stalled or initiated in a cap-independent (and thus 4EGI-1-independent) manner.

TRANSLATIONAL CONTROL AND NEURODEVELOPMENTAL DISORDERS

Genetic perturbations impinging on translational control mechanisms have been associated with developmental disorders of the nervous system. In the last few years, it has been shown that specific translational control mechanisms are perturbed in these cases. The questions to be addressed are (1) whether restoration of translation reverses the cognitive decline in these disorders, and (2) whether drugs can be developed to target translation in the treatment of brain disorders.

PERTURBED TRANSLATIONAL CONTROL IN NEUROLOGICAL DISORDERS

How dysregulation of the signaling pathways impinging on translation contributes to the pathophysiology of several neurodegenerative diseases and neurodevelopmental disorders has been recently reviewed (Buffington et al.

2014; Huber et al. 2015; Kapur et al. 2017). One important context of translational regulation in neurological disorders is autism spectrum disorders (ASDs). The importance of translational control in ASD is underscored by the many neurological disorders in which mTORC1 activity is perturbed as a result of single gene mutations in mTORC1 upstream regulators (Pten, Tsc, or Fmr1). Indeed, in all three cases, enhanced translation rates have been postulated to cause ASD (Kelleher and Bear 2008). However, deletion of Pten or Tsc not only activates mTORC1, but also alters mTORC2 activity (Laplante and Sabatini 2012). Because mTORC2 is also required for LTP and mnemonic processes (Huang et al. 2013), it is currently unclear which mTOR complex drives the ASD pathology in these TORopathies. In addition, as we mentioned above, mTORC1 regulates other translation-independent processes (Proud 2018).

The strongest evidence that translation upregulation leads to ASD-like behaviors emerged from studies of mice lacking 4E-BPs or overexpressing eIF4E. In both models, increased eIF4E-mediated translation leads to ASD-like behaviors (Gkogkas et al. 2013; Santini et al. 2013). Moreover, agents that modestly decrease translation rates, like rapamycin, metformin, or an inhibitor of the eIF4E kinase, MNK1, reverse ASD-like behaviors in FMRP-deficient mice (Busquets-Garcia et al. 2013; Gkogkas et al. 2014; Gantois et al. 2017), supporting the notion that enhanced overall translation contributes to the pathology associated with fragile X syndrome. Although most of these studies rescued the phenotypes in Fmr1-deficient mice by repressing initiation, in one study, reduction of the RNA-binding protein CPEB rescued pathology in Fmr1-deficient mice and this was associated with normalization of translation elongation rates (Udagawa et al. 2013).

These data suggest that increased translation may cause neuronal dysfunction in the fragile X syndrome. Accordingly, deletion of 4E-BP2 leads to an increase in the synthesis of the synaptic adhesion protein neuroligin-1, and preventing this increase ameliorates autistic-like behaviors in this rodent model (Gkogkas et al.

Cite this article as *Cold Spring Harb Perspect Biol* doi: 10.1101/cshperspect.a032912

2013). Matrix metalloprotease 9 (Mmp9) is overexpressed in fragile X mice and decreasing its translation through reducing eIF4E phosphorylation rescues several autistic-like behaviors in this model (Gkogkas et al. 2014).

The second context in which translation dysregulation has been observed is neurodegenerative disease. In this case, a decrease in translation rates was reported in Alzheimer's disease (AD) (Langstrom et al. 1989), likely because of the activation of the integrated stress response (ISR), wherein eIF2α phosphorylation is a central component (Wek 2018). Increased phosphorylation of eIF2α has been observed in AD (Chang et al. 2002a,b; Hoozemans et al. 2005; Page et al. 2006), traumatic brain injury (Dash et al. 2015; Chou et al. 2017), and prion disease (Moreno et al. 2012). It is currently unclear whether translation dysregulation underlies, or is a downstream effector of, neurodegenerative disease. Nevertheless, increasing translation could ameliorate the disease. Indeed, both genetic reduction of eIF2α phosphorylation (PKR or PERK knockout mice and GADD34-overexpressing vectors) or pharmacological correction of the abnormal translational program controlled by eIF2α (e.g., using ISRIB and PKR inhibitors) reverses the cognitive decline associated with AD (Lourenco et al. 2013; Ma et al. 2013; Segev et al. 2015), traumatic brain injury (Chou et al. 2017), as well as the pathology associated with prion disease (Moreno et al. 2012).

Recently, eEF2 phosphorylation has also been reported to be elevated both in mouse models of AD and AD patients (Ma et al. 2014). Inhibiting eEF2K restored LTP that had been inhibited by β amyloid (Ma et al. 2014).

mTOR complexes have also been proposed to be targets for neurodegenerative diseases (Santos et al. 2011), although the proposed role for mTOR in neurodegenerative disease has mainly been through mTOR regulation of autophagy as opposed to direct regulation of protein synthesis.

Thus, perturbed eIF2α phosphorylation (and possibly eEF2 phosphorylation), is responsible, at least in part, for some of the cognitive and cellular aspects underlying neurodegenerative disorders. Neurodegenerative diseases are also caused by expansions of nucleotide repeats in proteins and these may also have an important contribution from dysregulated translation centered on those nucleotide repeats (Zu et al. 2018), further emphasizing the importance of translational control in neuronal disease.

CONCLUDING REMARKS

Regulation of mRNA translation initiation and elongation is essential for synaptic plasticity and memory formation. Moreover, alterations in translation rates resulting from mutations in signaling pathways regulating translation or activation of the ISR contribute to pathogenesis in a variety of neurological disorders. One of the most intriguing questions in the memory field, the identification of the nature of the proteins that are synthesized in response to a memory paradigm, remains largely unanswered. Given that memory is encoded in a small subset of neurons in a particular brain area, answering this question will require state-of-the-art intersectional molecular genetic approaches combined with translating ribosome affinity purification (TRAP) in the specific neurons that encode memories.

A better understanding of translational control in brain processes will require a full dissection at the cell-type level. The formation of memory is not only mediated by principal (pyramidal) neurons in a given brain area. In fact, other neuronal cell types, such as inhibitory or dopaminergic neurons, are crucially involved in memory formation. However, understanding the regulation of translation in these cell types is just beginning (Ran et al. 2009; Placzek et al. 2016).

From a therapeutic standpoint, it would be interesting to determine whether some of the very promising genetic manipulations blocking or activating translation can be replicated using pharmacology. Pharmacological agents that modulate translation rates by regulating the integrated stress response, such as PKRi and ISRIB, have been identified and initial results suggest the usefulness of these drugs in improving cognition (Zhu et al. 2011; Sidrauski et al. 2013; Chou et al. 2017). Their efficacy in multiple mouse models of disease remains to be examined before considering whether these drugs can be used to treat human disease. Drugs that act on other targets, including downstream effectors of TORC1, regulators of stalled polysomes, and inhibitors and activators of eEF2K are also interesting candidates for ameliorating disease.

REFERENCES

*Reference is also in this collection.

Aakalu G, Smith WB, Nguyen N, Jiang C, Schuman EM. 2001. Dynamic visualization of local protein synthesis in hippocampal neurons. *Neuron* 30: 489–502.

Bailey CH, Kandel ER. 2008. Synaptic remodeling, synaptic growth and the storage of long-term memory in *Aplysia*. *Prog Brain Res* 169: 179–198.

Banko JL, Poulin F, Hou L, DeMaria CT, Sonenberg N, Klann E. 2005. The translation repressor 4E-BP2 is critical for eIF4F complex formation, synaptic plasticity, and memory in the hippocampus. *J Neurosci* 25: 9581–9590.

Batista G, Johnson JL, Dominguez E, Costa-Mattioli M, Pena JL. 2016. Translational control of auditory imprinting and structural plasticity by eIF2α. *eLife* 5: e17197.

Bekinschtein P, Katche C, Slipczuk LN, Igaz LM, Cammarota M, Izquierdo I, Medina JH. 2007. mTOR signaling in the hippocampus is necessary for memory formation. *Neurobiol Learn Mem* 87: 303–307.

Belelovsky K, Elkobi A, Kaphzan H, Nairn AC, Rosenblum K. 2005. A molecular switch for translational control in taste memory consolidation. *Eur J Neurosci* 22: 2560–2568.

Bidinosti M, Ran I, Sanchez-Carbente MR, Martineau Y, Gingras AC, Gkogkas C, Raught B, Bramham CR, Sossin WS, Costa-Mattioli M, et al. 2009. Postnatal deamidation of 4E-BP2 in brain enhances its association with raptor and alters kinetics of excitatory synaptic transmission. *Mol Cell* 37: 797–808.

Bradshaw KD, Emptage NJ, Bliss TV. 2003. A role for dendritic protein synthesis in hippocampal late LTP. *Eur J Neurosci* 18: 3150–3152.

Buffington SA, Huang W, Costa-Mattioli M. 2014. Translational control in synaptic plasticity and cognitive dysfunction. *Annu Rev Neurosci* 37: 17–38.

Buskirk AR, Green R. 2017. Ribosome pausing, arrest and rescue in bacteria and eukaryotes. *Philos Trans R Soc Lond B Biol Sci* 372: 20160183.

Busquets-Garcia A, Gomis-Gonzalez M, Guegan T, Agustin-Pavon C, Pastor A, Mato S, Perez-Samartin A, Matute C, de la Torre R, Dierssen M, et al. 2013. Targeting the endocannabinoid system in the treatment of fragile X syndrome. *Nat Med* 19: 603–607.

Buxbaum AR, Wu B, Singer RH. 2014. Single β-actin mRNA detection in neurons reveals a mechanism for regulating its translatability. *Science* 343: 419–422.

Cajigas IJ, Tushev G, Will TJ, tom Dieck S, Fuerst N, Schuman EM. 2012. The local transcriptome in the synaptic neuropil revealed by deep sequencing and high-resolution imaging. *Neuron* 74: 453–466.

Carroll M, Warren O, Fan X, Sossin WS. 2004. 5-HT stimulates eEF2 dephosphorylation in a rapamycin-sensitive manner in *Aplysia* neurites. *J Neurochem* 90: 1464–1476.

Chang RC, Suen KC, Ma CH, Elyaman W, Ng HK, Hugon J. 2002a. Involvement of double-stranded RNA-dependent protein kinase and phosphorylation of eukaryotic initiation factor-2α in neuronal degeneration. *J Neurochem* 83: 1215–1225.

Chang RC, Wong AK, Ng HK, Hugon J. 2002b. Phosphorylation of eukaryotic initiation factor-2α (eIF2α) is associated with neuronal degeneration in Alzheimer's disease. *Neuroreport* 13: 2429–2432.

Chen E, Sharma MR, Shi X, Agrawal RK, Joseph S. 2014. Fragile X mental retardation protein regulates translation by binding directly to the ribosome. *Mol Cell* 54: 407–417.

Chou A, Krukowski K, Jopson T, Zhu PJ, Costa-Mattioli M, Walter P, Rosi S. 2017. Inhibition of the integrated stress response reverses cognitive deficits after traumatic brain injury. *Proc Natl Acad Sci* 114: E6420–E6426.

Costa-Mattioli M, Monteggia LM. 2013. mTOR complexes in neurodevelopmental and neuropsychiatric disorders. *Nat Neurosci* 16: 1537–1543.

Costa-Mattioli M, Gobert D, Harding H, Herdy B, Azzi M, Bruno M, Bidinosti M, Ben Mamou C, Marcinkiewicz E, Yoshida M, et al. 2005. Translational control of hippocampal synaptic plasticity and memory by the eIF2α kinase GCN2. *Nature* 436: 1166–1173.

Costa-Mattioli M, Gobert D, Stern E, Gamache K, Colina R, Cuello C, Sossin W, Kaufman R, Pelletier J, Rosenblum K, et al. 2007. eIF2α phosphorylation bidirectionally regulates the switch from short- to long-term synaptic plasticity and memory. *Cell* 129: 195–206.

Costa-Mattioli M, Sossin WS, Klann E, Sonenberg N. 2009. Translational control of long-lasting synaptic plasticity and memory. *Neuron* 61: 10–26.

Cracco JB, Serrano P, Moskowitz SI, Bergold PJ, Sacktor TC. 2005. Protein synthesis-dependent LTP in isolated dendrites of CA1 pyramidal cells. *Hippocampus* 15: 551–556.

Crispino M, Kaplan BB, Martin R, Alvarez J, Chun JT, Benech JC, Giuditta A. 1997. Active polysomes are present in the large presynaptic endings of the synaptosomal fraction from squid brain. *J Neurosci* 17: 7694–7702.

Darnell JC, Richter JD. 2012. Cytoplasmic RNA-binding proteins and the control of complex brain function. *Cold Spring Harb Perspect Biol* 4: a012344.

Darnell JC, Van Driesche SJ, Zhang C, Hung KY, Mele A, Fraser CE, Stone EF, Chen C, Fak JJ, Chi SW, et al. 2011. FMRP stalls ribosomal translocation on mRNAs linked to synaptic function and autism. *Cell* 146: 247–261.

Dash PK, Hylin MJ, Hood KN, Orsi SA, Zhao J, Redell JB, Tsvetkov AS, Moore AN. 2015. Inhibition of eukaryotic initiation factor 2α phosphatase reduces tissue damage and improves learning and memory after experimental traumatic brain injury. *J Neurotrauma* 32: 1608–1620.

Cite this article as *Cold Spring Harb Perspect Biol* doi: 10.1101/cshperspect.a032912

Davidkova G, Carroll RC. 2007. Characterization of the role of microtubule-associated protein 1B in metabotropic glutamate receptor-mediated endocytosis of AMPA receptors in hippocampus. *J Neurosci* **27:** 13273–13278.

Davis HP, Squire LR. 1984. Protein synthesis and memory: A review. *Psychol Bull* **96:** 518–559.

* Dever TE, Dinman JD, Green R. 2018. Translation elongation and recoding in eukaryotes. *Cold Spring Harb Perspect Biol* doi: 10.1101/cshperspect.a032649.

Di Prisco GV, Huang W, Buffington SA, Hsu CC, Bonnen PE, Placzek AN, Sidrauski C, Krnjevic K, Kaufman RJ, Walter P, et al. 2014. Translational control of mGluR-dependent long-term depression and object-place learning by eIF2α. *Nat Neurosci* **17:** 1073–1082.

Ehninger D, Han S, Shilyansky C, Zhou Y, Li W, Kwiatkowski DJ, Ramesh V, Silva AJ. 2008. Reversal of learning deficits in a Tsc2$^{+/-}$ mouse model of tuberous sclerosis. *Nat Med* **14:** 843–848.

El Fatimy R, Davidovic L, Tremblay S, Jaglin X, Dury A, Robert C, De Koninck P, Khandjian EW. 2016. Tracking the Fragile X mental retardation protein in a highly ordered neuronal ribonucleoparticles population: A link between stalled polyribosomes and RNA granules. *PLoS Genet* **12:** e1006192.

Eom T, Muslimov IA, Tsokas P, Berardi V, Zhong J, Sacktor TC, Tiedge H. 2014. Neuronal BC RNAs cooperate with eIF4B to mediate activity-dependent translational control. *J Cell Biol* **207:** 237–252.

Flexner LB, Flexner JB, Roberts RB. 1967. Memory in mice analyzed with antibiotics. Antibiotics are useful to study stages of memory and to indicate molecular events which sustain memory. *Science* **155:** 1377–1383.

Frey U, Morris RG. 1998. Synaptic tagging: Implications for late maintenance of hippocampal long-term potentiation. *Trends Neurosci* **21:** 181–188.

Gantois I, Khoutorsky A, Popic J, Aguilar-Valles A, Freemantle E, Cao R, Sharma V, Pooters T, Nagpal A, Skalecka A, et al. 2017. Metformin ameliorates core deficits in a mouse model of fragile X syndrome. *Nat Med* **23:** 674–677.

Gelinas JN, Nguyen PV. 2005. β-Adrenergic receptor activation facilitates induction of a protein synthesis-dependent late phase of long-term potentiation. *J Neurosci* **25:** 3294–3303.

Giustetto M, Hegde AN, Si K, Casadio A, Inokuchi K, Pei W, Kandel ER, Schwartz JH. 2003. Axonal transport of eukaryotic translation elongation factor 1α mRNA couples transcription in the nucleus to long-term facilitation at the synapse. *Proc Natl Acad Sci* **100:** 13680–13685.

Gkogkas CG, Khoutorsky A, Ran I, Rampakakis E, Nevarko T, Weatherill DB, Vasuta C, Yee S, Truitt M, Dallaire P, et al. 2013. Autism-related deficits via dysregulated eIF4E-dependent translational control. *Nature* **493:** 371–377.

Gkogkas CG, Khoutorsky A, Cao R, Jafarnejad SM, Prager-Khoutorsky M, Giannakas N, Kaminari A, Fragkouli A, Nader K, Price TJ, et al. 2014. Pharmacogenetic inhibition of eIF4E-dependent Mmp9 mRNA translation reverses fragile X syndrome-like phenotypes. *Cell Rep* **9:** 1742–1755.

Gobert D, Topolnik L, Azzi M, Huang L, Badeaux F, Desgroseillers L, Sossin WS, Lacaille JC. 2008. Forskolin induction of late-LTP and up-regulation of 5′ TOP mRNAs translation via mTOR, ERK, and PI3K in hippocampal pyramidal cells. *J Neurochem* **106:** 1160–1174.

Graber TE, Hebert-Seropian S, Khoutorsky A, David A, Yewdell JW, Lacaille JC, Sossin WS. 2013. Reactivation of stalled polyribosomes in synaptic plasticity. *Proc Natl Acad Sci* **110:** 16205–16210.

Graber TE, Freemantle E, Anadolu M, Hebert-Seropian S, MacAdam R, Shin U, Hoang HD, Alain T, Lacaille JC, Sossin WS. 2017. UPF1 governs synaptic plasticity through association with a STAU2 RNA granule. *J Neurosci* **37:** 9116–9131.

Granger AJ, Shi Y, Lu W, Cerpas M, Nicoll RA. 2013. LTP requires a reserve pool of glutamate receptors independent of subunit type. *Nature* **493:** 495–500.

Guzowski JF. 2002. Insights into immediate-early gene function in hippocampal memory consolidation using antisense oligonucleotide and fluorescent imaging approaches. *Hippocampus* **12:** 86–104.

Heiman M, Schaefer A, Gong S, Peterson JD, Day M, Ramsey KE, Suarez-Farinas M, Schwarz C, Stephan DA, Surmeier DJ, et al. 2008. A translational profiling approach for the molecular characterization of CNS cell types. *Cell* **135:** 738–748.

Heise C, Taha E, Murru L, Ponzoni L, Cattaneo A, Guarnieri FC, Montani C, Mossa A, Vezzoli E, Ippolito G, et al. 2017. eEF2K/eEF2 pathway controls the excitation/inhibition balance and susceptibility to epileptic seizures. *Cereb Cortex* **27:** 2226–2248.

Hinnebusch AG, Ivanov IP, Sonenberg N. 2016. Translational control by 5′-untranslated regions of eukaryotic mRNAs. *Science* **352:** 1413–1416.

Hoozemans JJ, Veerhuis R, Van Haastert ES, Rozemuller JM, Baas F, Eikelenboom P, Scheper W. 2005. The unfolded protein response is activated in Alzheimer's disease. *Acta Neuropathol* **110:** 165–172.

Hou L, Antion MD, Hu D, Spencer CM, Paylor R, Klann E. 2006. Dynamic translational and proteasomal regulation of fragile X mental retardation protein controls mGluR-dependent long-term depression. *Neuron* **51:** 441–454.

Hsieh AC, Liu Y, Edlind MP, Ingolia NT, Janes MR, Sher A, Shi EY, Stumpf CR, Christensen C, Bonham MJ, et al. 2012. The translational landscape of mTOR signalling steers cancer initiation and metastasis. *Nature* **485:** 55–61.

Hu J, Ferguson L, Adler K, Farah CA, Hastings MH, Sossin WS, Schacher S. 2017. Selective erasure of distinct forms of long-term synaptic plasticity underlying different forms of memory in the same postsynaptic neuron. *Curr Biol* **27:** 1888–1899.e1884.

Huang F, Chotiner JK, Steward O. 2005. The mRNA for elongation factor 1α is localized in dendrites and translated in response to treatments that induce long-term depression. *J Neurosci* **25:** 7199–7209.

Huang W, Zhu PJ, Zhang S, Zhou H, Stoica L, Galiano M, Krnjevic K, Roman G, Costa-Mattioli M. 2013. mTORC2 controls actin polymerization required for consolidation of long-term memory. *Nat Neurosci* **16:** 441–448.

Huang W, Placzek AN, Viana Di Prisco G, Khatiwada S, Sidrauski C, Krnjevic K, Walter P, Dani JA, Costa-Mattioli M. 2016. Translational control by eIF2α phosphorylation regulates vulnerability to the synaptic and behavioral effects of cocaine. *eLife* **5:** e12052.

Huber KM, Kayser MS, Bear MF. 2000. Role for rapid dendritic protein synthesis in hippocampal mGluR-dependent long-term depression. *Science* **288:** 1254–1257.

Huber KM, Klann E, Costa-Mattioli M, Zukin RS. 2015. Dysregulation of mammalian target of rapamycin signaling in mouse models of autism. *J Neurosci* **35:** 13836–13842.

Im HI, Nakajima A, Gong B, Xiong X, Mamiya T, Gershon ES, Zhuo M, Tang YP. 2009. Post-training dephosphorylation of eEF-2 promotes protein synthesis for memory consolidation. *PLoS ONE* **4:** e7424.

Inamura N, Nawa H, Takei N. 2005. Enhancement of translation elongation in neurons by brain-derived neurotrophic factor: Implications for mammalian target of rapamycin signaling. *J Neurochem* **95:** 1438–1445.

* Ingolia NT, Hussmann JA, Weissman JS. 2018. Ribosome profiling: Global views of translation. *Cold Spring Harb Perspect Biol* doi: 10.1101/cshperspect.a032698.

Ishimura R, Nagy G, Dotu I, Zhou H, Yang XL, Schimmel P, Senju S, Nishimura Y, Chuang JH, Ackerman SL. 2014. RNA function. Ribosome stalling induced by mutation of a CNS-specific tRNA causes neurodegeneration. *Science* **345:** 455–459.

Jiang Z, Belforte JE, Lu Y, Yabe Y, Pickel J, Smith CB, Je HS, Lu B, Nakazawa K. 2010. eIF2α phosphorylation-dependent translation in CA1 pyramidal cells impairs hippocampal memory consolidation without affecting general translation. *J Neurosci* **30:** 2582–2594.

Josselyn SA, Kohler S, Frankland PW. 2015. Finding the engram. *Nat Rev Neurosci* **16:** 521–534.

Ju W, Morishita W, Tsui J, Gaietta G, Deerinck TJ, Adams SR, Garner CC, Tsien RY, Ellisman MH, Malenka RC. 2004. Activity-dependent regulation of dendritic synthesis and trafficking of AMPA receptors. *Nat Neurosci* **7:** 244–253.

Jung H, Yoon BC, Holt CE. 2012. Axonal mRNA localization and local protein synthesis in nervous system assembly, maintenance and repair. *Nat Rev Neurosci* **13:** 308–324.

Kang H, Schuman EM. 1996. A requirement for local protein synthesis in neurotrophin-induced hippocampal synaptic plasticity. *Science* **273:** 1402–1406.

Kapur M, Monaghan CE, Ackerman SL. 2017. Regulation of mRNA translation in neurons—A matter of life and death. *Neuron* **96:** 616–637.

Kaul G, Pattan G, Rafeequi T. 2011. Eukaryotic elongation factor-2 (eEF2): Its regulation and peptide chain elongation. *Cell Biochem Funct* **29:** 227–234.

Kelleher RJ III, Bear MF. 2008. The autistic neuron: Troubled translation? *Cell* **135:** 401–406.

Kelleher RJ III, Govindarajan A, Jung HY, Kang H, Tonegawa S. 2004. Translational control by MAPK signaling in long-term synaptic plasticity and memory. *Cell* **116:** 467–479.

Kenney JW, Genheden M, Moon KM, Wang X, Foster LJ, Proud CG. 2016. Eukaryotic elongation factor 2 kinase regulates the synthesis of microtubule-related proteins in neurons. *J Neurochem* **136:** 276–284.

Langstrom NS, Anderson JP, Lindroos HG, Winblad B, Wallace WC. 1989. Alzheimer's disease-associated reduction of polysomal mRNA translation. *Brain Res Mol Brain Res* **5:** 259–269.

Laplante M, Sabatini DM. 2012. mTOR signaling in growth control and disease. *Cell* **149:** 274–293.

Liu HH, Cline HT. 2016. Fragile X mental retardation protein is required to maintain visual conditioning-induced behavioral plasticity by limiting local protein synthesis. *J Neurosci* **36:** 7325–7339.

Liu K, Hu JY, Wang D, Schacher S. 2003. Protein synthesis at synapse versus cell body: Enhanced but transient expression of long-term facilitation at isolated synapses. *J Neurobiol* **56:** 275–286.

Lourenco MV, Clarke JR, Frozza RL, Bomfim TR, Forny-Germano L, Batista AF, Sathler LB, Brito-Moreira J, Amaral OB, Silva CA, et al. 2013. TNF-α mediates PKR-dependent memory impairment and brain IRS-1 inhibition induced by Alzheimer's β-amyloid oligomers in mice and monkeys. *Cell Metab* **18:** 831–843.

Lugo JN, Smith GD, Morrison JB, White J. 2013. Deletion of PTEN produces deficits in conditioned fear and increases fragile X mental retardation protein. *Learn Mem* **20:** 670–673.

Ma T, Trinh MA, Wexler AJ, Bourbon C, Gatti E, Pierre P, Cavener DR, Klann E. 2013. Suppression of eIF2α kinases alleviates Alzheimer's disease-related plasticity and memory deficits. *Nat Neurosci* **16:** 1299–1305.

Ma T, Chen Y, Vingtdeux V, Zhao H, Viollet B, Marambaud P, Klann E. 2014. Inhibition of AMP-activated protein kinase signaling alleviates impairments in hippocampal synaptic plasticity induced by amyloid β. *J Neurosci* **34:** 12230–12238.

Malenka RC, Bear MF. 2004. LTP and LTD: An embarrassment of riches. *Neuron* **44:** 5–21.

Martin KC, Kosik KS. 2002. Synaptic tagging—Who's it? *Nat Rev Neurosci* **3:** 813–820.

Martin KC, Barad M, Kandel ER. 2000. Local protein synthesis and its role in synapse-specific plasticity. *Curr Opin Neurobiol* **10:** 587–592.

Mayford M, Siegelbaum SA, Kandel ER. 2012. Synapses and memory storage. *Cold Spring Harb Perspect Biol* **4:** a005751.

Mazroui R, Huot ME, Tremblay S, Filion C, Labelle Y, Khandjian EW. 2002. Trapping of messenger RNA by Fragile X mental retardation protein into cytoplasmic granules induces translation repression. *Hum Mol Genet* **11:** 3007–3017.

McCamphill PK, Farah CA, Anadolu MN, Hoque S, Sossin WS. 2015. Bidirectional regulation of eEF2 phosphorylation controls synaptic plasticity by decoding neuronal activity patterns. *J Neurosci* **35:** 4403–4417.

McCamphill PK, Ferguson L, Sossin WS. 2017. A decrease in eukaryotic elongation factor 2 phosphorylation is required for local translation of sensorin and long-term facilitation in *Aplysia*. *J Neurochem* **142:** 246–259.

* Merrick WC, Pavitt GD. 2018. Protein synthesis initiation in eukaryotic cells. *Cold Spring Harb Perspect Biol* doi: 10.1101/chperspect.a033092.

Miller S, Yasuda M, Coats JK, Jones Y, Martone ME, Mayford M. 2002. Disruption of dendritic translation of CaMKIIα impairs stabilization of synaptic plasticity and memory consolidation. *Neuron* **36:** 507–519.

Moerke NJ, Aktas H, Chen H, Cantel S, Reibarkh MY, Fahmy A, Gross JD, Degterev A, Yuan J, Chorev M, et al. 2007. Small-molecule inhibition of the interaction between the translation initiation factors eIF4E and eIF4G. *Cell* **128:** 257–267.

Moreno JA, Radford H, Peretti D, Steinert JR, Verity N, Martin MG, Halliday M, Morgan J, Dinsdale D, Ortori CA, et al. 2012. Sustained translational repression by eIF2α-P mediates prion neurodegeneration. *Nature* **485:** 507–511.

Muslimov IA, Nimmrich V, Hernandez AI, Tcherepanov A, Sacktor TC, Tiedge H. 2004. Dendritic transport and localization of protein kinase Mζ mRNA: Implications for molecular memory consolidation. *J Biol Chem* **279:** 52613–52622.

Nairn AC, Palfrey HC. 1987. Identification of the major M_r 100,000 substrate for calmodulin-dependent protein kinase III in mammalian cells as elongation factor-2. *J Biol Chem* **262:** 17299–17303.

Neves G, Cooke SF, Bliss TV. 2008. Synaptic plasticity, memory and the hippocampus: A neural network approach to causality. *Nat Rev Neurosci* **9:** 65–75.

Nosyreva ED, Huber KM. 2006. Metabotropic receptor-dependent long-term depression persists in the absence of protein synthesis in the mouse model of fragile X syndrome. *J Neurophysiol* **95:** 3291–3295.

Okada D, Ozawa F, Inokuchi K. 2009. Input-specific spine entry of soma-derived Vesl-1S protein conforms to synaptic tagging. *Science* **324:** 904–909.

Ostroff LE, Botsford B, Gindina S, Cowansage KK, LeDoux JE, Klann E, Hoeffer C. 2017. Accumulation of polyribosomes in dendritic spine heads, but not bases and necks, during memory consolidation depends on cap-dependent translation initiation. *J Neurosci* **37:** 1862–1872.

Ounallah-Saad H, Sharma V, Edry E, Rosenblum K. 2014. Genetic or pharmacological reduction of PERK enhances cortical-dependent taste learning. *J Neurosci* **34:** 14624–14632.

Ouyang Y, Rosenstein A, Kreiman G, Schuman EM, Kennedy MB. 1999. Tetanic stimulation leads to increased accumulation of Ca^{2+}/calmodulin-dependent protein kinase II via dendritic protein synthesis in hippocampal neurons. *J Neurosci* **19:** 7823–7833.

Page G, Rioux Bilan A, Ingrand S, Lafay-Chebassier C, Pain S, Perault Pochat MC, Bouras C, Bayer T, Hugon J. 2006. Activated double-stranded RNA-dependent protein kinase and neuronal death in models of Alzheimer's disease. *Neuroscience* **139:** 1343–1354.

Panja D, Dagyte G, Bidinosti M, Wibrand K, Kristiansen AM, Sonenberg N, Bramham CR. 2009. Novel translational control in Arc-dependent long term potentiation consolidation in vivo. *J Biol Chem* **284:** 31498–31511.

Park S, Park JM, Kim S, Kim JA, Shepherd JD, Smith-Hicks CL, Chowdhury S, Kaufmann W, Kuhl D, Ryazanov AG, et al. 2008. Elongation factor 2 and fragile X mental retardation protein control the dynamic translation of Arc/Arg3.1 essential for mGluR-LTD. *Neuron* **59:** 70–83.

Placzek AN, Prisco GV, Khatiwada S, Sgritta M, Huang W, Krnjevic K, Kaufman RJ, Dani JA, Walter P, Costa-Mattioli M. 2016. eIF2α-mediated translational control regulates the persistence of cocaine-induced LTP in midbrain dopamine neurons. *eLife* **5:** e17517.

Proud CG. 2015. Regulation and roles of elongation factor 2 kinase. *Biochem Soc Trans* **43:** 328–332.

* Proud CG. 2018. Phosphorylation and signal transduction pathways in translational control. *Cold Spring Harb Perspect Biol* doi: 10.1101/cshperspect.a033050.

Ramachandran B, Frey JU. 2009. Interfering with the actin network and its effect on long-term potentiation and synaptic tagging in hippocampal CA1 neurons in slices in vitro. *J Neurosci* **29:** 12167–12173.

Ran I, Laplante I, Bourgeois C, Pepin J, Lacaille P, Costa-Mattioli M, Pelletier J, Sonenberg N, Lacaille JC. 2009. Persistent transcription- and translation-dependent long-term potentiation induced by mGluR1 in hippocampal interneurons. *J Neurosci* **29:** 5605–5615.

Rangaraju V, Tom Dieck S, Schuman EM. 2017. Local translation in neuronal compartments: How local is local? *EMBO Rep* **18:** 693–711.

Reijmers L, Mayford M. 2009. Genetic control of active neural circuits. *Front Mol Neurosci* **2:** 27.

Richter JD, Coller J. 2015. Pausing on polyribosomes: Make way for elongation in translational control. *Cell* **163:** 292–300.

Ryan TJ, Roy DS, Pignatelli M, Arons A, Tonegawa S. 2015. Memory. Engram cells retain memory under retrograde amnesia. *Science* **348:** 1007–1013.

Sajikumar S, Navakkode S, Sacktor TC, Frey JU. 2005. Synaptic tagging and cross-tagging: The role of protein kinase Mζ in maintaining long-term potentiation but not long-term depression. *J Neurosci* **25:** 5750–5756.

Sajikumar S, Navakkode S, Frey JU. 2007. Identification of compartment- and process-specific molecules required for "synaptic tagging" during long-term potentiation and long-term depression in hippocampal CA1. *J Neurosci* **27:** 5068–5080.

Sanhueza M, Lisman J. 2013. The CaMKII/NMDAR complex as a molecular memory. *Mol Brain* **6:** 10.

Santini E, Huynh TN, MacAskill AF, Carter AG, Pierre P, Ruggero D, Kaphzan H, Klann E. 2013. Exaggerated translation causes synaptic and behavioural aberrations associated with autism. *Nature* **493:** 411–415.

Santos RX, Correia SC, Cardoso S, Carvalho C, Santos MS, Moreira PI. 2011. Effects of rapamycin and TOR on aging and memory: Implications for Alzheimer's disease. *J Neurochem* **117:** 927–936.

Sasikumar AN, Perez WB, Kinzy TG. 2012. The many roles of the eukaryotic elongation factor 1 complex. *Wiley Interdiscip Rev RNA* **3:** 543–555.

Scheetz AJ, Nairn AC, Constantine-Paton M. 2000. NMDA receptor-mediated control of protein synthesis at developing synapses. *Nat Neurosci* **3:** 211–216.

Segev Y, Barrera I, Ounallah-Saad H, Wibrand K, Sporild I, Livne A, Rosenberg T, David O, Mints M, Bramham CR, et al. 2015. PKR inhibition rescues memory deficit and ATF4 overexpression in ApoE ε4 human replacement mice. *J Neurosci* **35**: 12986–12993.

Sen T, Gupta R, Kaiser H, Sen N. 2017. Activation of PERK elicits memory impairment through inactivation of CREB and downregulation of PSD95 after traumatic brain injury. *J Neurosci* **37**: 5900–5911.

Sidrauski C, Acosta-Alvear D, Khoutorsky A, Vedantham P, Hearn BR, Li H, Gamache K, Gallagher CM, Ang KK, Wilson C, et al. 2013. Pharmacological brake-release of mRNA translation enhances cognitive memory. *eLife* **2**: e00498.

Sonenberg N, Hinnebusch AG. 2009. Regulation of translation initiation in eukaryotes: Mechanisms and biological targets. *Cell* **136**: 731–745.

Sossin WS, DesGroseillers L. 2006. Intracellular trafficking of RNA in neurons. *Traffic* **7**: 1581–1589.

Sossin WS, Lacaille JC. 2010. Mechanisms of translational regulation in synaptic plasticity. *Curr Opin Neurobiol* **20**: 450–456.

Sperow M, Berry RB, Bayazitov IT, Zhu G, Baker SJ, Zakharenko SS. 2012. Phosphatase and tensin homologue (PTEN) regulates synaptic plasticity independently of its effect on neuronal morphology and migration. *J Physiol* **590**: 777–792.

Squire LR, Davis HP. 1981. The pharmacology of memory: A neurobiological perspective. *Annu Rev Pharmacol Toxicol* **21**: 323–356.

Squire LR, Genzel L, Wixted JT, Morris RG. 2015. Memory consolidation. *Cold Spring Harb Perspect Biol* **7**: a021766.

Stern E, Chinnakkaruppan A, David O, Sonenberg N, Rosenblum K. 2013. Blocking the eIF2α kinase (PKR) enhances positive and negative forms of cortex-dependent taste memory. *J Neurosci* **33**: 2517–2525.

Steward O, Farris S, Pirbhoy PS, Darnell J, Driesche SJ. 2014. Localization and local translation of Arc/Arg3.1 mRNA at synapses: Some observations and paradoxes. *Front Mol Neurosci* **7**: 101.

Stoica L, Zhu PJ, Huang W, Zhou H, Kozma SC, Costa-Mattioli M. 2011. Selective pharmacogenetic inhibition of mammalian target of rapamycin complex I (mTORC1) blocks long-term synaptic plasticity and memory storage. *Proc Natl Acad Sci* **108**: 3791–3796.

Sutton MA, Schuman EM. 2006. Dendritic protein synthesis, synaptic plasticity, and memory. *Cell* **127**: 49–58.

Taha E, Gildish I, Gal-Ben-Ari S, Rosenblum K. 2013. The role of eEF2 pathway in learning and synaptic plasticity. *Neurobiol Learn Mem* **105**: 100–106.

Takei N, Kawamura M, Ishizuka Y, Kakiya N, Inamura N, Namba H, Nawa H. 2009. Brain-derived neurotrophic factor enhances the basal rate of protein synthesis by increasing active eukaryotic elongation factor 2 levels and promoting translation elongation in cortical neurons. *J Biol Chem* **284**: 26340–26348.

Tang SJ, Reis G, Kang H, Gingras AC, Sonenberg N, Schuman EM. 2002. A rapamycin-sensitive signaling pathway contributes to long-term synaptic plasticity in the hippocampus. *Proc Natl Acad Sci* **99**: 467–472.

Thoreen CC, Chantranupong L, Keys HR, Wang T, Gray NS, Sabatini DM. 2012. A unifying model for mTORC1-mediated regulation of mRNA translation. *Nature* **485**: 109–113.

Tonegawa S, Liu X, Ramirez S, Redondo R. 2015. Memory engram cells have come of age. *Neuron* **87**: 918–931.

Trinh MA, Klann E. 2013. Translational control by eIF2α kinases in long-lasting synaptic plasticity and long-term memory. *Neurobiol Learn Mem* **105**: 93–99.

Trinh MA, Kaphzan H, Wek RC, Pierre P, Cavener DR, Klann E. 2012. Brain-specific disruption of the eIF2α kinase PERK decreases ATF4 expression and impairs behavioral flexibility. *Cell Rep* **1**: 676–688.

Trinh MA, Ma T, Kaphzan H, Bhattacharya A, Antion MD, Cavener DR, Hoeffer CA, Klann E. 2014. The eIF2α kinase PERK limits the expression of hippocampal metabotropic glutamate receptor-dependent long-term depression. *Learn Mem* **21**: 298–304.

Tsokas P, Grace EA, Chan P, Ma T, Sealfon SC, Iyengar R, Landau EM, Blitzer RD. 2005. Local protein synthesis mediates a rapid increase in dendritic elongation factor 1A after induction of late long-term potentiation. *J Neurosci* **25**: 5833–5843.

Tsokas P, Ma T, Iyengar R, Landau EM, Blitzer RD. 2007. Mitogen-activated protein kinase upregulates the dendritic translation machinery in long-term potentiation by controlling the mammalian target of rapamycin pathway. *J Neurosci* **27**: 5885–5894.

Twiss JL, Fainzilber M. 2009. Ribosomes in axons—Scrounging from the neighbors? *Trends Cell Biol* **19**: 236–243.

Udagawa T, Farny NG, Jakovcevski M, Kaphzan H, Alarcon JM, Anilkumar S, Ivshina M, Hurt JA, Nagaoka K, Nalavadi VC, et al. 2013. Genetic and acute CPEB1 depletion ameliorate fragile X pathophysiology. *Nat Med* **19**: 1473–1477.

van Vliet AR, Garg AD, Agostinis P. 2016. Coordination of stress, Ca²⁺, and immunogenic signaling pathways by PERK at the endoplasmic reticulum. *Biol Chem* **397**: 649–656.

Verpelli C, Piccoli G, Zibetti C, Zanchi A, Gardoni F, Huang K, Brambilla D, Di Luca M, Battaglioli E, Sala C. 2010. Synaptic activity controls dendritic spine morphology by modulating eEF2-dependent BDNF synthesis. *J Neurosci* **30**: 5830–5842.

von der Brelie C, Waltereit R, Zhang L, Beck H, Kirschstein T. 2006. Impaired synaptic plasticity in a rat model of tuberous sclerosis. *Eur J Neurosci* **23**: 686–692.

Wang X, Li W, Williams M, Terada N, Alessi DR, Proud, CG. 2001. Regulation of elongation factor 2 kinase by p90(RSK1) and p70 S6 kinase. *EMBO J* **20**: 4370–4379.

Wang DO, Kim SM, Zhao Y, Hwang H, Miura SK, Sossin WS, Martin KC. 2009. Synapse- and stimulus-specific local translation during long-term neuronal plasticity. *Science* **324**: 1536–1540.

Wang R, McGrath BC, Kopp RF, Roe MW, Tang X, Chen G, Cavener DR. 2013. Insulin secretion and Ca²⁺ dynamics in β-cells are regulated by PERK (EIF2AK3) in concert with calcineurin. *J Biol Chem* **288**: 33824–33836.

Cite this article as *Cold Spring Harb Perspect Biol* doi: 10.1101/cshperspect.a032912

Weatherill DB, McCamphill PK, Pethoukov E, Dunn TW, Fan X, Sossin WS. 2011. Compartment-specific, differential regulation of eukaryotic elongation factor 2 and its kinase within *Aplysia* sensory neurons. *J Neurochem* **117:** 841–855.

* Wek RC. 2018. Role of eIF2α kinases in translational control and adaptation to cellular stress. *Cold Spring Harb Perspect Biol* doi: 10.1101/cshperspect.a032870.

Wullschleger S, Loewith R, Hall MN. 2006. TOR signaling in growth and metabolism. *Cell* **124:** 471–484.

Xu T, Yu X, Perlik AJ, Tobin WF, Zweig JA, Tennant K, Jones T, Zuo Y. 2009. Rapid formation and selective stabilization of synapses for enduring motor memories. *Nature* **462:** 915–919.

Yang G, Pan F, Gan WB. 2009. Stably maintained dendritic spines are associated with lifelong memories. *Nature* **462:** 920–924.

Younts TJ, Monday HR, Dudok B, Klein ME, Jordan BA, Katona I, Castillo PE. 2016. Presynaptic protein synthesis is required for long-term plasticity of GABA release. *Neuron* **92:** 479–492.

Zhu PJ, Huang W, Kalikulov D, Yoo JW, Placzek AN, Stoica L, Zhou H, Bell JC, Friedlander MJ, Krnjevic K, et al. 2011. Suppression of PKR promotes network excitability and enhanced cognition by interferon-γ-mediated disinhibition. *Cell* **147:** 1384–1396.

Zhu S, McGrath BC, Bai Y, Tang X, Cavener DR. 2016. PERK regulates Gq protein-coupled intracellular Ca^{2+} dynamics in primary cortical neurons. *Mol Brain* **9:** 87.

* Zu T, Pattamatta A, Ranum LPW. 2018. RAN translation in neurological diseases. *Cold Spring Harb Perspect Biol* doi: 10.1101/cshperspect.a033019.

Translational Control in Virus-Infected Cells

Noam Stern-Ginossar,[1] Sunnie R. Thompson,[2] Michael B. Mathews,[3] and Ian Mohr[4]

[1]Department of Molecular Genetics, Weizmann Institute of Science, Rehovot 76100, Israel

[2]Department of Microbiology, University of Alabama at Birmingham, Birmingham, Alabama 35294

[3]Department of Medicine, Rutgers New Jersey Medical School, Newark, New Jersey 07103

[4]Department of Microbiology, New York University School of Medicine, New York, New York 10016

Correspondence: noam.stern-ginossar@weizmann.ac.il; sunnie@uab.edu; mathews@njms.rutgers.edu; ian.mohr@nyumc.org

As obligate intracellular parasites, virus reproduction requires host cell functions. Despite variations in genome size and configuration, nucleic acid composition, and their repertoire of encoded functions, all viruses remain unconditionally dependent on the protein synthesis machinery resident within their cellular hosts to translate viral messenger RNAs (mRNAs). A complex signaling network responsive to physiological stress, including infection, regulates host translation factors and ribosome availability. Furthermore, access to the translation apparatus is patrolled by powerful host immune defenses programmed to restrict viral invaders. Here, we review the tactics and mechanisms used by viruses to appropriate control over host ribosomes, subvert host defenses, and dominate the infected cell translational landscape. These not only define aspects of infection biology paramount for virus reproduction, but continue to drive fundamental discoveries into how cellular protein synthesis is controlled in health and disease.

A defining attribute of virus infection biology is its absolute reliance on the host translation machinery to produce the polypeptides needed for virus reproduction. This feature cannot be overstated, as virus replication and spread depend on conscripting host ribosomes to translate viral messenger RNAs (mRNAs). Failure to engage ribosomes would have dire consequences for virus reproduction and evolution. To ensure translation of their mRNAs, virus-encoded functions dominate cell signaling pathways that control the host protein synthesis machinery. Commandeering these regulatory circuits preserves the functionality of components that recruit ribosomes to viral mRNAs. Although many host cell intrinsic immune defenses target translation factors to incapacitate the protein synthesis apparatus of the infected cell, viral factors have evolved to limit host antiviral responses. Exploiting virus model systems continues to reveal fundamental parameters governing how mRNA translation is controlled in infected and uninfected cells. Here, we review molecular interactions between select plant and animal viruses and their hosts that regulate protein synthesis in infected cells, and recent developments in the field are highlighted.

A PRIMER ON VIROLOGY AND VIRUS REPRODUCTION STRATEGIES

Viruses are vastly diverse, and although a comprehensive discussion of virus biology is not our purpose, an overview of select principles is warranted. Viruses are obligate intracellular parasites with varied lifestyles. Whereas acute infections are often cytolytic and destroy the cell, persistent infections can be tolerated. Some viruses establish permanent infections, in which viruses continually replicate or, alternatively, remain latent within cells and reproduce only periodically. These distinct reproduction strategies differentially impact the host translation machinery and translated mRNA landscape. Viruses replicate within and may remodel the cytoplasm or nucleus. Not all infections cause clinical disease, and infection outcome varies among organisms or cell types and is dependent on immune status. Finally, viruses have diverse genome structures comprised of single- or double-stranded (ds) DNA or RNA, and RNA virus genomes can be composed of single or multiple nucleic acid segments. Viral genome size varies from less than 10 Kb for small RNA viruses to greater than 200 Kb for large human DNA viruses and megabase genomes of DNA viruses that infect *Acanthamoeba* (Schultz et al. 2017). Notwithstanding genome size, viruses are under selective pressure to optimize coding capacity, and tactics like polyprotein processing, where proteases generate multiple proteins from one open reading frame (ORF; e.g., in picornaviruses and flaviviruses) and recoding (e.g., frameshifting as in retrovirus Gag-Pol) are frequently observed (Jan et al. 2016; Atkins et al. 2016).

Genome structure largely informs mRNA biogenesis and mRNA features impact translation. Single-strand RNA virus genomes having an identical polarity to mRNA, termed plus (+)-strand RNA viruses, function as mRNA and are directly translated on infection. RNA virus genomes of opposite polarity, designated minus (−)-strand RNA viruses, and dsRNA virus genomes require an RNA-dependent RNA polymerase to produce mRNA. Discrete structural elements within 5′ and 3′ untranslated regions (UTRs) support RNA virus genome replication and mRNA translation. Viral mRNA 5′ ends may be uncapped, protein-linked, modified by viral capping enzymes to contain a 5′ methyl-7-GTP (m^7GTP) cap, or derived from host mRNA 5′ terminal fragments that are naturally capped (Decroly et al. 2012). Uncapped viral mRNAs

deploy specialized genome elements to recruit 40S ribosomal subunits that support translation even when canonical cap-dependent translation is impaired (Kwan and Thompson 2018). Polyadenylated 3′ ends are generally template-encoded for RNA viruses or replaced by an element that recruits a viral 3′ terminal binding protein (Poon et al. 1999; Deo et al. 2002; Kempf and Barton 2015). After converting (+)-strand RNA genomes into dsDNA using reverse transcriptase, retroviruses produce mRNA using the host RNA polymerase II transcriptional machinery. Whereas most DNA viruses replicate within the nucleus and use host enzymes for mRNA biogenesis, others, including poxviruses and asfarviruses, replicate in the cytoplasm, encoding their own transcription, capping, and 3′ end processing machinery (Van Etten et al. 2010).

STRESS RESPONSES AND INFECTION BIOLOGY

Controlling gene expression by mRNA translation enables swift responses to environmental and physiological insults, including infection. Indeed, many host proteins and mechanisms regulating protein synthesis in response to stress in uninfected cells (reviewed by Wek 2018) play significant roles during infection. Likewise, features of translational control in virus-infected cells resemble and perhaps were co-opted from host cell stress responses.

Host Defenses and Antiviral Immunity

While hijacking ribosomes enables virus protein production, it also is a vulnerability exploited by the host. Significantly, many innate host defenses limit mRNA access to ribosomes by selectively or globally impairing the translation machinery. The struggle between viruses and their hosts to control ribosomes occupies the front lines in the innate immune response to virus infection.

By detecting conserved viral nucleic acid features, including uncapped single-stranded RNA, dsRNA, or cytoplasmic DNA, host sentinel pattern recognition receptors trigger type I interferon (IFN) production (Chiang and Gack 2017). The resulting IFN-induced host proteins

 Cite this article as *Cold Spring Harb Perspect Biol* doi: 10.1101/cshperspect.a033001

establish an "antiviral state" that is not permissive for virus reproduction. Many IFN-induced proteins impact translation and ribosome recruitment to mRNA (Diamond and Farzan 2013; Schoggins 2014). Central among these are PKR, a eukaryotic initiation factor (eIF)2α kinase, and oligoadenylate synthetase (OAS), which are both activated by dsRNA. Importantly, dsRNA accumulates in cells infected with RNA or DNA viruses. Although some viruses have dsRNA genomes, dsRNA is an obligatory intermediate in RNA virus genome replication and can be present as secondary structures in single-strand RNA. dsRNA also accumulates in cells infected with DNA viruses, which produce mRNAs from op-

posing strands of the viral genome. Upon activation by dsRNA, PKR phosphorylates the α subunit of eIF2 (Fig. 1). This inactivates eIF2, a GTP-regulated, three-subunit complex that loads 40S ribosomal subunits with initiator methionyl-transfer RNA (Met-tRNA$_i^{Met}$) (Merrick and Pavitt 2018; Wek 2018). Normally, GTP hydrolysis is stimulated by eIF5 following start codon recognition enabling 60S subunit joining. Subsequent eIF2•GDP recycling to the GTP-bound form requires guanine nucleoside exchange factor (GEF) eIF2B (Merrick and Pavitt 2018; Wek 2018). However, phosphorylated eIF2 binds tightly to eIF2B, inhibiting its GEF activity (Fig. 1). Because eIF2B is limiting, small changes in phosphorylat-

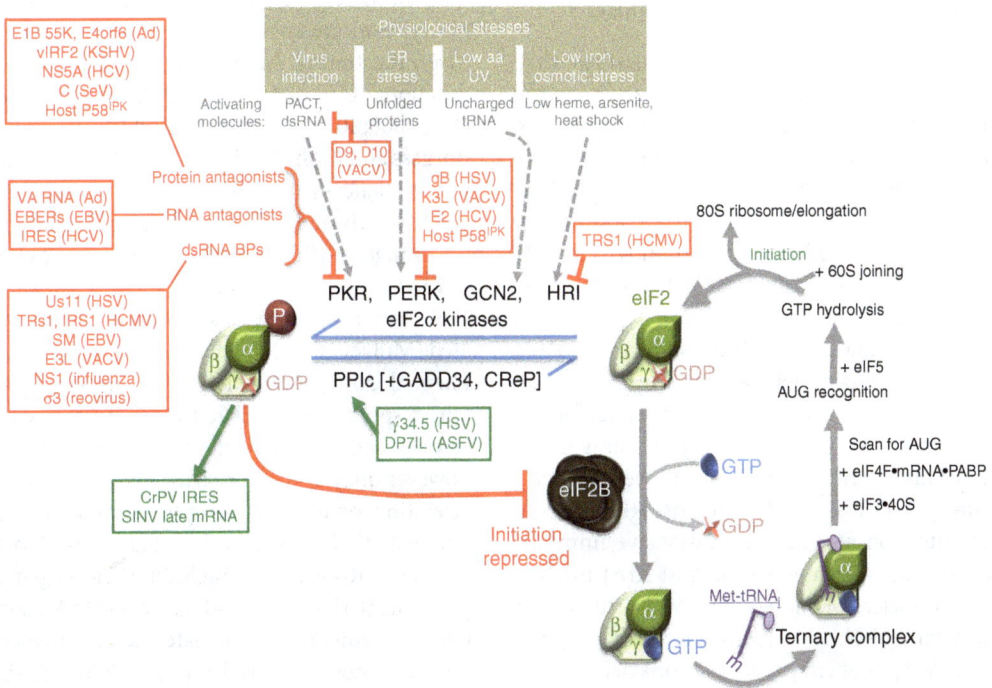

Figure 1. Targeting stress-responsive host defenses by viral functions controls eukaryotic initiation factor (eIF)2α phosphorylation and translation initiation. Composed of three subunits (α,β,γ), eIF2 is a GTP-binding translation initiation factor that loads eIF3-bound 40S ribosomal subunits with methionyl-transfer RNA (Met-tRNA$_i$) (*right* panel). Subsequent mRNA recruitment, AUG recognition, and eIF5-stimulated GTP hydrolysis is followed by 60S subunit joining and translation initiation by the 80S ribosome. Recycling inactive eIF2•GDP to the active GTP-bound form requires the GEF eIF2B. Phosphorylation of eIF2α on S51 blocks initiation by binding to and inhibiting eIF2B, preventing GDP-GTP exchange. Four eIF2 kinases, each of which is activated by specific molecules that accumulate in response to a discrete physiological stress, and the protein phosphatase 1 catalytic subunit (PP1c), partnered with an inducible (GADD34) or constitutively active (CReP) regulatory subunit, control eIF2α phosphorylation. Viral functions that activate (green), respond to (green), or repress (red) the indicated host effectors are shown. aa, amino acid.

ed eIF2α have a large impact on protein synthesis and globally inhibit initiation.

Whereas OAS is similarly activated by dsRNA, it interferes with protein synthesis via a different mechanism. OAS synthesizes short oligoadenylate (OA) polymers with a $2'-5'$ linkage that stimulate RNase L, an endogenous ribonuclease whose activation is dependent on OA (Sadler and Williams 2008; Han et al. 2014). Active RNase L degrades ribosomal RNA (rRNA) and mRNA, inactivating both the translation machinery and its templates. RNase L cleavage products, which contain $2'-3'$ cyclic phosphorylated termini, also stimulate inflammasome activation (Chakrabarti et al. 2015).

Viral Tactics to Counter Host Defenses

Viruses have acquired countermeasures that neutralize host dsRNA-activated defenses or allow their replication despite them. Coronavirus and rotavirus-encoded phosphodiesterases attack OA, limiting its accumulation and RNase L activation (Zhang et al. 2013). In addition, a picornavirus-encoded RNA and protein inhibit RNase L (Townsend et al. 2008; Sorgeloos et al. 2013), and Rift Valley fever virus (RVFV) triggers PKR degradation (Mudhasani et al. 2016). dsRNA-binding proteins encoded by a variety of viruses shield dsRNA from host sensors, foiling PKR and OAS activation (Fig. 1), and some physically interact with PKR to prevent its activation (Jan et al. 2016). Surprisingly, dsRNA accumulation and dsRNA-responsive immune effectors are regulated by the host Xrn1 mRNA exoribonuclease (Burgess and Mohr 2015). By accelerating RNA decay, viral enzymes like poxvirus decapping enzymes or coronavirus nsp15 endonuclease, which presumably produce Xrn1 substrates, limit dsRNA accumulation (Burgess and Mohr 2015; Liu et al. 2015; Deng et al. 2017; Kindler et al. 2017). Alternatively, an adenovirus (Ad) small RNA, VA RNA$_I$, binds PKR but is not large enough to support PKR dimerization, which is needed for activation (Mathews and Shenk 1991; Launer-Felty et al. 2015).

Broader strategies protect eIF2α from phosphorylation not only by PKR, but by the other stress-activated eIF2α kinases PERK, HRI, and GCN2 (Fig. 1) (Pavio et al. 2003; Berlanga et al. 2006; Won et al. 2012; Vincent et al. 2017). Examples include the poxvirus-encoded eIF2α pseudo-substrate K3L (Sood et al. 2000; Seo et al. 2008), the HCMV TRS1 protein that binds and inhibits PKR and HRI (Hakki et al. 2006; Vincent et al. 2017), and a phosphatase regulatory subunit encoded by herpes simplex virus (HSV)-1 (γ34.5) and African swine fever virus (ASFV) DP71L, which recruits the host protein phosphatase 1α catalytic subunit to eIF2α where it counteracts the activity of eIF2α kinases (Rojas et al. 2015; Barber et al. 2017). Induction of host p58^{IPK} by infection can also limit PKR, PERK, and GCN2 activation (Goodman et al. 2011; Roobol et al. 2015). Multiple functions that act synergistically to restrict eIF2α phosphorylation are encoded by HSV-1 and poxviruses, illustrating the importance of preserving functional eIF2 (Mulvey et al. 2003, 2007; Seo et al. 2008; Rice et al. 2011; Sciortino et al. 2013; Burgess and Mohr 2015; Liu et al. 2015). Mutant viruses deficient in functions that subvert host dsRNA-activated defenses often have an altered host range (Haller et al. 2014; Carpentier et al. 2016; Peng et al. 2016), are hypersensitive to IFNs and in some cases are profoundly attenuated (Mulvey et al. 2004; White and Jacobs 2012; Liu et al. 2015). Activation of the unfolded protein response and PERK by hepatitis C virus (HCV) and vesicular stomatitis virus (VSV), however, may also favor virus replication by accelerating type I IFN receptor degradation to attenuate IFN responses (Liu et al. 2009). Finally, several RNA viruses including cricket paralysis virus (CrPV) and Sindbis virus (SINV) contain cis-elements that obviate the requirement for eIF2 altogether (Wilson et al. 2000; Spahn et al. 2004; Kerr et al. 2016; Sanz et al. 2017). Although in vitro studies suggested a mechanism in which eIF2A or eIF2D supply Met-tRNA$_i^{Met}$, recent work using knockout cell lines shows that these factors are dispensable (Sanz et al. 2017; Gonzales-Almela et al. 2018). The role of RNA structures and contribution of other proteins to eIF2-independent initiation on CrPV and SINV remain to be established.

Stress responses, including those resulting from infection, impact mRNA and translation factor subcellular distribution. Stalled initiation complexes containing messenger ribonucleoproteins (mRNPs), mRNA, 40S subunits, initiation factors, and RNA-binding proteins (RBPs) accumulate in phase-dense, non-membrane-bound aggregates called stress granules (SGs) (Protter and Parker 2016; Ivanov et al. 2018). Although eIF2α phosphorylation often accompanies SG formation, implying a reversible way of regulating initiation, SG accumulation can be inhibited or viral mRNAs released from SGs regardless of eIF2α phosphorylation (Montero et al. 2008; Qin et al. 2011). SG core components can directly interface with viral RNA genomes or mRNAs, repressing their translation (Albornoz et al. 2014). Host cytoplasmic RNA sensors including MDA5, RIG-I, and PKR can be recruited to SGs in response to eIF2α phosphorylation, suggesting SGs are platforms for viral RNA detection (Reineke and Lloyd 2015; Oh et al. 2016; Sánchez-Aparicio et al. 2017). Indeed, many viruses interfere with SG formation (Emara and Brinton 2007; White et al. 2007; Finnen et al. 2014; Khaperskyy et al. 2014; Dauber et al. 2016; Humoud et al. 2016; Nelson et al. 2016; Rabouw et al. 2016; Amorim et al. 2017; Basu et al. 2017; Choudhury et al. 2017; Khong et al. 2017; Le Sage et al. 2017). In cells infected with Middle East respiratory syndrome (MERS) coronavirus, Kaposi sarcoma–associated herpesvirus (KSHV), or HSV-1, this is achieved in part by preventing PKR activation (Rabouw et al. 2016; Sharma et al. 2017; Burgess and Mohr 2018). Even though SG formation is inhibited by dengue virus (DENV) and Zika virus, host protein synthesis remains suppressed, uncoupling these processes (Roth et al. 2017). Depleting key SG components like G3BP reduces Zika virus replication, however, indicating that host SG components may be repurposed to benefit virus reproduction without SG formation (Hou et al. 2017). Whereas SGs can be dynamic platforms for staging host antiviral responses (Tsai and Lloyd 2014; McCormick and Khaperskyy 2017), SG formation is stimulated by rabies virus (Nikolic et al. 2016) and enhances respiratory syncytial virus replication (Lindquist et al. 2010). The rationale underlying this seemingly opposite strategy is unknown.

Remodeling Host mRNA Translation in Infected Cells

By interfering with host translation, which overwhelmingly is cap-dependent, viruses antagonize host defenses. In its extreme form, termed "host shut-off," virus infection impairs ongoing host protein synthesis, limiting production of host defense molecules and allowing viral mRNAs to better compete for limiting factors (Mohr and Sonenberg 2012; Mohr 2016). Different mechanisms account for host shut-off (Fig. 2). Picornaviruses suppress host protein synthesis by inhibiting cap-dependent translation and often target eIF4F, a multisubunit initiation factor comprised of the cap-binding protein eIF4E, the eIF4G scaffold, and the DEAD-box-containing RNA helicase eIF4A, needed to load 40S subunits onto m^{7}GTP-capped mRNAs. Poliovirus 2A proteinase cleaves eIF4G, separating the eIF4E-binding domain from the eIF3-binding region (Gradi et al. 1998), whereas 3C proteinases cleave the poly(A)-binding protein (PABP) (Rivera and Lloyd 2008; Kobayashi et al. 2012a). Cleavage of eIF4G by poliovirus or group A rhinovirus 2A protease is stimulated by eIF4E (Aumayr et al. 2017; Avanzino et al. 2017). During encephalomyocarditis virus (EMCV) infection, hypophosphorylated 4E-BP1 repressor accumulates, inhibiting eIF4E binding to eIF4G and limiting eIF4F assembly (Gingras et al. 1996). Enterovirus 71 (EV 71) instead induces host microRNA miR-141 expression, which reduces eIF4E abundance to suppress host protein synthesis (Ho et al. 2010).

Remodeling the host mRNA pool available for translation by cytoplasmic ribosomes is another shut-off mechanism. By encoding virus factors that stimulate mRNA turnover, viral mRNAs, which are actively transcribed in acutely infected cells, dominate the mRNA pool. Accelerating mRNA decay also sculpts viral mRNA populations, facilitating temporal viral gene expression transitions (Jan et al. 2016). Viruses that produce mRNAs with m^{7}GTP-capped 5′ termini, like herpesviruses and poxviruses, rely on mRNA decay to liberate host ribosomes without impairing cap-dependent translation. Viral functions subsequently can stimulate

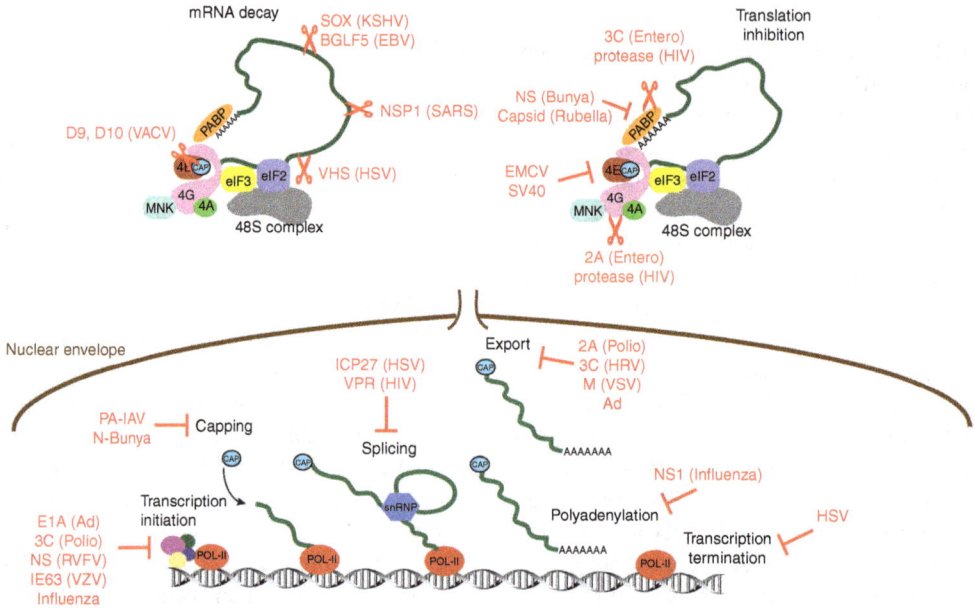

Figure 2. Viruses use multiple mechanisms to block host messenger RNA (mRNA) translation. Cellular mRNA biogenesis starts in the nucleus where RNA polymerase II generates primary transcripts that undergo several processing steps including 5′ capping, splicing, and polyadenylation before functional mRNAs are produced. Mature mRNAs are exported to the cytoplasm where they are translated. Initiation of cap-dependent mRNA translation typically requires recruitment of the multisubunit eukaryotic translation initiation factor (eIF)4F, which consists of the cap-binding protein eIF4E, the RNA helicase eIF4A, and the adaptor protein eIF4G. Binding of eIF4G to poly(A)-binding protein (PABP) mediates communication between 5′ and 3′ ends. Virus-encoded factors that degrade (*scissors*) cellular proteins and mRNAs or repress cellular functions are shown.

cap-dependent translation by subverting cell signaling (Walsh et al. 2005, 2008). To generate substrates for mRNA decay, some viruses encode endoribonucleases (influenza, HSV-1, coronavirus, KSHV) (Kamitani et al. 2009; Read 2013; Abernathy and Glaunsinger 2015; Khaperskyy et al. 2016), whereas others like poxviruses encode decapping enzymes (Parrish and Moss 2007). The HSV-1 endoribonuclease vhs interacts with eIF4B and eIF4H to target mRNA cleavage (Read 2013), the KSHV SOX protein directs host RNA decay components, including the exoribonuclease XRN1, to translating ribosomes (Abernathy and Glaunsinger 2015), and the coronavirus nuclease nsp1 associates with 40S ribosomal subunits (Kamitani et al. 2009). Viruses can also disrupt host mRNA biogenesis and nuclear export to remodel the host mRNA pool (Faria et al. 2005; Rutkowski et al. 2015; Gong et al. 2016).

Not all viruses impair ongoing host translation during their acute reproductive cycle, however. Notably, host protein synthesis proceeds in human cytomegalovirus (HCMV)-infected cells. Activation of the mechanistic target of rapamycin complex 1 (mTORC1) (see Proud 2018) by the viral UL38 protein promotes cap-dependent translation and stimulates synthesis of factors encoded by mRNAs containing a 5′ terminal oligopyrimidine (5′-TOP; see Proud 2018) sequence element including PABP (McKinney et al. 2012, 2014). Higher PABP levels allow HCMV to overcome a host response that increases PABP-interacting protein 2 (Paip2) abundance, which inhibits PABP binding to eIF4G and poly(A) RNA (Fig. 3). Interfering with the HCMV-induced PABP increase results in restriction of virus replication by Paip2, highlighting an unexpected antiviral role for Paip2 in host-cell-intrinsic defenses (McKinney et al.

Cite this article as *Cold Spring Harb Perspect Biol* doi: 10.1101/cshperspect.a033001

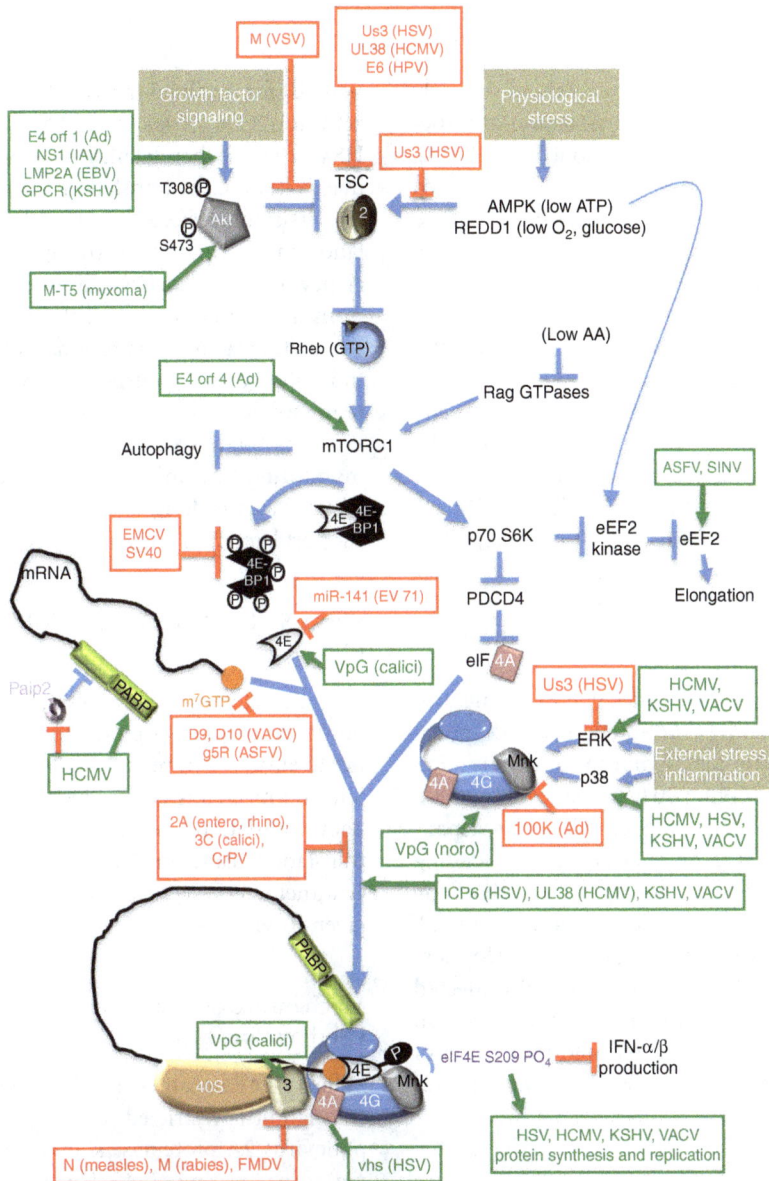

Figure 3. Subverting mechanistic target of rapamycin complex 1 (mTORC1) signaling to control translation in virus-infected cells. Occupying a nexus of intersecting signaling networks, the cellular ser/thr kinase mTORC1 plays a critical role stimulating anabolic programs, like protein synthesis, and repressing catabolic outcomes like autophagy. Briefly, growth factor–induced Akt phosphorylation (T308, S473) and activation represses TSC1/2, which, in turn, allows Rheb-GTP to activate mTORC1. Nutrient, energy, amino acid (aa), or oxygen insufficiency (physiological stress) all repress mTORC1 through discrete effectors. Signaling through mTORC1 allows swift changes in translational output in response to differing environmental and physiological inputs by controlling initiation and elongation. Initiation is stimulated by phosphorylating and inactivating 4E-BP translational repressor family members (e.g., 4E-BP1), which bind the cap-binding protein eIF4E to prevent its interaction with eIF4G. Through its substrate p70S6K, mTORC1 controls the DEAD-box-containing RNA helicase eIF4A, which together with eukaryotic initiation factor (eIF)4E and eIF4G comprises the multisubunit initiation factor eIF4F. Regulated eIF4F assembly controls cap-dependent mRNA translation as eIF4F recruits 40S subunits to the mRNA capped 5′ end. By repressing eEF2 kinase, mTORC1 stimulates eEF2 and elongation. The impact of mTORC1 activation on translation initiation and elongation is shown, as are viral factors that stimulate (green) and repress (red) the indicated cellular effectors. SV40, simian virus 40; calici, calicivirus; noro, norovirus; entero, enterovirus; rhino, rhinovirus.

2013). Avian herpesvirus microRNAs similarly repress Paip2 to stimulate translation of a specific viral mRNA (Tahiri-Alaoui et al. 2014). Additional host genes, like the IFN-induced C19orf66 (RyDEN) may also target PABP to restrict both DNA and RNA viruses, including DENV (Suzuki et al. 2016).

Regulating eIF4E phosphorylation impacts selective host mRNA translation and modulates IFN production, influencing host defenses. Coronaviruses, flaviviruses, noroviruses, and many large DNA viruses promote eIF4F assembly, positioning the eIF4G-associated kinase Mnk1 proximal to its substrate eIF4E (Fig. 3) (Mizutani et al. 2004; Walsh and Mohr 2004, 2006; Walsh et al. 2005, 2008; Royall et al. 2015; Roth et al. 2017; Proud 2018). This stimulates eIF4E S209 phosphorylation and translation of IκB mRNA, which encodes an NF-κB inhibitor (Herdy et al. 2012). Besides counteracting NF-κB-dependent IFN production, eIF4E phosphorylation stimulates viral mRNA translation (Walsh and Mohr 2004; Walsh et al. 2008). Significantly, eIF4E phosphorylation is dependent on eIF3 subunit e, showing how loading eIF3-bound 40S marks cap-bound eIF4F by phosphorylation to regulate selective mRNA translation (Walsh and Mohr 2014). Unphosphorylated eIF4E accumulates in cells infected with adenovirus and many RNA viruses (Jan et al. 2016), resulting in reduced translation of IκB-encoding mRNA, NF-κB activation, and IFN production.

Harnessing Stress Responses to Control Viral Persistence

Unlike acute infections, different viral lifestyles necessitate specialized interactions with host cells. Some, like herpesviruses, establish lifelong latency where virus reproduction is suppressed in specific cell types. Periodically, latent infections reactivate and reenter the productive growth cycle, which allows virus reproduction and spread to new hosts. Reactivation is influenced by host defenses and physiological stress responses induced by disrupting homeostasis. KSHV stimulates eIF4F assembly on inducible reactivation, and eIF4E phosphorylation is needed for optimal production of the lytic activator RTA (Arias et al. 2009). In contrast, HSV-1 establishes latency in neurons, where inactivation of the translation repressor 4E-BP1 by persistent mTORC1-dependent phosphorylation in response to neurotrophic factor sufficiency promotes latency (Camarena et al. 2010). Conversely, preventing 4E-BP1 phosphorylation stimulates reactivation presumably by restricting translation of eIF4E-responsive mRNAs that are repressed by 4E-BP1 (Kobayashi et al. 2012b). Whereas the mRNA targets remain unknown, monitoring host mRNA translation can gauge homeostasis in long-term, latent infections to support latency or trigger virus reproduction.

VIRAL STRATEGIES TO CAPTURE RIBOSOMES

The absolute dependence of viruses on host ribosomes for protein production demands tactics to ensure their recruitment to viral mRNAs. Shut-off mechanisms suppress host cap-dependent translation and effectively remodel the translation-ready mRNA pool. Ribosome recruitment, however, cannot be left to chance given its vital role in virus reproduction.

Mechanisms of Cap-Independent Translation Used by Viruses

When eIF4F becomes limiting, cap-dependent translation is restricted. Cleavage of eIF4G by poliovirus 2A proteinase shows a severe example of inactivating eIF4F. Uncapped poliovirus mRNA contains a specialized, cis-acting RNA internal ribosome entry site (IRES) in the 5′UTR. In contrast to canonical cap-dependent initiation where 40S ribosomes load onto the mRNA 5′ end, highly structured IRES elements enable ribosome recruitment to specific internal sites within viral RNA (Kwan and Thompson 2018). IRESs were first discovered in poliovirus and EMCV where they allow translation initiation of uncapped, single-stranded (+)-sense RNA viral genomes immediately upon release into the cytoplasm (Jang et al. 1988; Pelletier and Sonenberg 1988). Subsequently, IRESs have been identified in many virus genomes includ-

Cite this article as *Cold Spring Harb Perspect Biol* doi: 10.1101/cshperspect.a033001

ing retroviruses like the human immunodeficiency virus (HIV)-1, which uses an IRES to express the Gag protein late in infection when cap-dependent initiation is impaired (Brasey et al. 2003; Amorim et al. 2014; Carvajal et al. 2016). Some DNA virus genomes like KSHV even contain IRES elements to facilitate translation of polycistronic mRNA (Othman et al. 2014).

Although they share the general capacity to direct initiation when canonical cap-dependent translation mechanisms are suppressed, different IRESs are structurally distinct and show varied functional requirements for eukaryotic initiation factors to load 40S subunits (Table 1) (Jan et al. 2016; Kwan and Thompson 2018). For example, the poliovirus IRES requires the eIF4G carboxy-terminal fragment, and selectively recruits 40S subunits to viral mRNA in a bona fide cap-independent manner (Sweeney et al. 2014). Even though the hepatitis A virus (HAV) RNA genome lacks an m^7GTP cap, it contains the only IRES found to require the cap-binding protein eIF4E (Table 1). Stimulation of the HAV IRES by eIF4E is independent of m^7GTP cap binding. Instead, eIF4E increases the affinity of eIF4G for several picornavirus IRESs, including HAV, and promotes IRES restructuring by the eIF4A helicase (Avanzino et al. 2017). In contrast, some IRESs, such as those in the CrPV intergenic region (IGR) and HCV, recruit 40S subunits without eIFs (Jan and Sarnow 2002; Pestova and Hellen 2003). Whereas the HCV IRES requires eIF3 and eIF2 to initiate translation (Fraser and Doudna 2007), the CrPV IGR can directly assemble 80S ribosomes and initiate translation in the absence of canonical initiation factors and even the initiator tRNA (Thompson et al. 2001; Jan et al. 2003; Pestova and Hellen 2003; Deniz et al. 2009).

In an alternative strategy, some RNA viruses like norovirus and calicivirus rely on a $5'$ terminal protein covalently linked to the $5'$ end of the

Table 1. Major types of IRESs

IRES	Examples	Required initiation factors[a]	References
Picornaviruses			
I. Enterovirus	Poliovirus, EV71, bovine enterovirus	eIF1A, eIF2, eIF3, eIF4A, eIF4B, central domain eIF4G, ITAFs (PCBP1/2, PTB)	Martinez-Salas et al. 2015
II. Cardio-/apthovirus	EMCV, FMDV	eIF2, eIF3, eIF4A, eIF4B, central domain eIF4G, ITAF PTB)	Martinez-Salas et al. 2015
III. HAV-like	Hepatitis A virus	eIF2, eIF3, eIF4A, eIF4B, eIF4E, eIF4G, ITAF (PTB, PCBP2)	Martinez-Salas et al. 2015
IV. HCV-like picornaviruses	Simian picornavirus type 9 (SPV9), porcine teschovirus I (PTV-1), Avian encephalomyelitis virus	eIF2, eIF3	Pisarev et al. 2004; Chard et al. 2006; Martinez-Salas et al. 2015
V. Aichivirus	Aichivirus	eIF2, eIF3, eIF4A, central domain eIF4G, ITAF (DHX29, PTB)	Yu et al. 2011
HCV-like	HCV, CSFV	eIF2, eIF3, binds to 40S directly	Fraser and Doudna 2007
Dicistrovirus intragenic (IGR)	CrPV, DCV, IAPV, PSIV	No initiation factors, binds 40S and assembles 80s directly	Hertz and Thompson 2011b
Retrovirus	HIV-1, HTLV-1, SIV, FIV	Central domain eIF4G, hypusine-eIF5A, eIF4A, likely other eIFs, ITAF (hnRNPA1, hRIP, DDX3), viral proteins (Rev)	Ohlmann et al. 2014; Plank et al. 2014; Caceres et al. 2016
KSHV vFLIP	KSHV vFLIP	eIF4A, eIF4G, eIF4E, eIF3, eIF2	Othman et al. 2014

[a]ITAFs: IRES transacting factors.

genome to recruit eIF4E, eIF4G, and/or eIF3 (Daughenbaugh et al. 2003; Leen et al. 2016). Different *cis* elements mediating long-range RNA interactions have been identified in plant (+)-strand RNA viruses (barley yellow dwarf virus, maize necrotic streak virus, carnation Italian ringspot virus, tomato bushy stunt virus, saguaro cactus virus). These cap-independent translation enhancers (CITEs) reside within 3′UTRs where they recruit 40S ribosomes directly or via eIF4F subunits eIF4E or eIF4G (Simon and Miller 2013). Functional base-pairing subsequently repositions 3′-CITE-loaded ribosomes proximal to 5′UTR structural elements (Nicholson and White 2014). Multiple 3′-CITEs in pea enation mosaic virus recruit both 40S and 60S subunits (Gao et al. 2014, 2017). By adopting alternate conformations induced by viral RNA polymerase binding, CITEs also repress translation to facilitate RNA replication (Le et al. 2017).

Strategies and Mechanisms Used by Viruses that Produce m⁷GTP-Capped mRNAs

Viral mRNAs with a 5′-m⁷GTP cap and 3′-poly(A) tail are indistinguishable from host mRNAs and are typically translated by cap-dependent processes. A number of different strategies that are used to commandeer control over the translational machinery by dominating host cell signaling pathways (reviewed by Proud 2018) have been defined (Fig. 3). Herpesviruses and poxviruses constitutively stimulate mTORC1 to ensure that the 4E-BP1 translational repressor is inactivated (Moorman et al. 2008; Walsh et al. 2008; Arias et al. 2009; Chuluunbaatar et al. 2010). By targeting the host tuberous sclerosis complex (TSC), the HSV1 Us3 ser/thr kinase not only enforces mTORC1 activation by mimicking the cellular kinase Akt (Fig. 3), but also subverts host AMPK-dependent responses to energy insufficiency and supports virus replication during stress (Vink et al. 2017). In addition to inactivating 4E-BP1, these viruses rely on diverse mechanisms to promote assembly of eIF4F, the multisubunit initiation factor that recruits 40S ribosomes to m⁷GTP-capped mRNA. These include an eIF4G-binding protein encoded by HSV-1 (ICP6) that stimulates eIF4E binding to

eIF4G, promotes eIF4E phosphorylation, and stimulates viral mRNA translation (Walsh and Mohr 2006). To capitalize on limiting cytoplasmic PABP availability in HSV-1-infected cells, the viral ICP27 RNA-binding protein may stimulate initiation downstream of cap binding by harnessing PABP and eIF4G in a mechanism similar to that used by the cellular regulator Dazl (Deleted in azoospermia-like) (Smith et al. 2017). HCMV increases the overall abundance of eIFs and PABP. Moreover, preventing the virus-induced rise in PABP levels reduces eIF4F complex assembly and virus growth (McKinney et al. 2012). Instead of manipulating translation factor abundance, vaccinia virus (VACV) and ASFV, which replicate in the cytoplasm, increase the effective local concentration of eIFs by sequestering them within discrete replication compartments (Katsafanas and Moss 2007; Walsh et al. 2008; Castelló et al. 2009; Zaborowska et al. 2012; Desmet et al. 2014).

Following ribosome loading on capped viral mRNAs, AUG start codons are usually identified by 5′UTR scanning (Merrick and Pavitt 2018). However, some viruses use *cis* elements that allow ribosomes to bypass 5′UTR segments and resume AUG scanning further downstream. This nonlinear ribosome translocation, called ribosome shunting, is used to produce heat shock proteins in uninfected cells subjected to stress and involves base-pairing with 18S rRNA and mRNA *cis* elements (Yueh and Schneider 2000). In adenovirus-infected cells, ribosome shunting requires a *cis*-acting tripartite leader RNA sequence in late mRNAs, the virus-encoded 100K protein and eIF4G (Xi et al. 2004). Shunting also occurs in mRNAs encoded by cauliflower mosaic virus, human papillomavirus (HPV), DNA pararetrovirus, and a picorna-like virus (Remm et al. 1999; Pooggin et al. 2012).

Controlling Elongation

In comparison with initiation, our understanding of whether viruses control translation elongation is much less developed. Elongation factors are repurposed to function in RNA replication by many bacterial and plant viruses (Takeshita and Tomita 2012; Li et al. 2013). ASFV and SINV concentrate eEF2 within virus replication compartments (Castelló et al. 2009; Sanz et al. 2009). Whereas mTORC1 activation by many viruses likely inhibits eEF2K (Fig. 3), stimulating elongation, the impact on infection biology remains largely unknown. Finally, expression of prokaryote-like elongation factors encoded by the giant mimivirus changes in response to nutrient availability (Silva et al. 2015), hinting that viral translation factors may supplant or supplement those resident in the host.

TACKLING THE RIBOSOME DIRECTLY

Viruses depend on host ribosomes regardless of whether they use cap-dependent or noncanonical translation initiation mechanisms. Whereas ribosome loading onto mRNAs in eukaryotes is typically reliant on initiation factors, a more direct role for the ribosome itself in regulating translation is emerging.

Dicistrovirus IRESs

Dicistroviruses are (+)-sense, single-stranded RNA viruses with two ORFs that are translated by two different IRESs. The 5′-IRES drives expression of the nonstructural proteins at a low level early in infection, whereas the IGR IRES drives robust structural protein expression late in infection (Garrey et al. 2010). IGR IRESs assemble 80S ribosomes without assistance from eIFs and initiate from the ribosomal A site at a non-AUG codon by priming the ribosome for elongation (Sasaki and Nakashima 2000; Wilson et al. 2000; Jan and Sarnow 2002; Jan et al. 2003; Pestova et al. 2004; Cevallos and Sarnow 2005; Yamamoto et al. 2007; Deniz et al. 2009; Zhang et al. 2016). By dissociating eIF4G from eIF4E to impair host protein synthesis, dicistroviruses subvert host ribosomes to translate viral IRES-containing RNAs (Garrey et al. 2010).

Ribosomal Proteins Control Translation in Virus-Infected Cells

Through discrete functions and modifications of specific protein subunits, ribosomes themselves selectively control viral and host mRNA translation (Fig. 4). Indeed, significant insight into how

ribosome composition impacts translation has been gleaned from viral models. Two 40S subunit-associated proteins (eS25, RACK1) dispensable for cap-dependent initiation, but required for IRES activity, have been identified (Landry et al. 2009; Hertz and Thompson 2011a; Majzoub et al. 2014; Olivares et al. 2014; Carvajal et al. 2016). Whereas RACK1 is required by specific IRESs (dicistroviral 5′-IRES, HCV), it is not compulsory for the IGR IRES (Majzoub et al. 2014). In contrast, eS25 is essential for 40S subunit recruitment by numerous IRESs, including the dicistrovirus IGR and picornavirus, flavivirus, retrovirus, and cellular IRESs (Landry et al. 2009; Hertz et al. 2013; Olivares et al. 2014; Carvajal et al. 2016). Unexpectedly, eS25 is also important for ribosomal shunting, suggesting a potential shared mechanism between these two noncanonical initiation strategies (Hertz et al. 2013). Although RACK1 is not essential for cap-dependent translation of cellular mRNAs, poxvirus late mRNAs, which have 5′UTR adenosine tracts, require phosphorylated RACK1 for translation. By expressing a viral kinase that phosphorylates a loop region on RACK1, poxviruses remodel host ribosomes to confer a translational advantage on viral mRNAs that contain adenosine repeats in their 5′UTR (Jha et al. 2017). Consistent with this, adenosine-rich leaders enhance translation in plants where the RACK1 loop region is naturally negatively charged, suggesting that poxviruses have exploited a conserved mechanism to preferentially translate certain capped mRNAs (Jha et al. 2017).

A 60S subunit protein (eL40/RPL40) is required for VSV, rabies, and measles virus mRNA translation (Lee et al. 2013). These viruses produce capped, polyadenylated transcripts like cellular mRNAs, yet viral mRNAs are efficiently translated by an unknown mechanism even though host protein synthesis is suppressed. A screen for eL40-dependent mRNAs in yeast revealed that 7% of the yeast mRNAs required eL40, suggesting that VSV has usurped a highly conserved mechanism for translating a subset of cap-dependent mRNAs (Lee et al. 2013).

Dengue virus NS1 protein associates with ribosomal proteins from both the small and large subunits (Cervantes-Salazar et al. 2015).

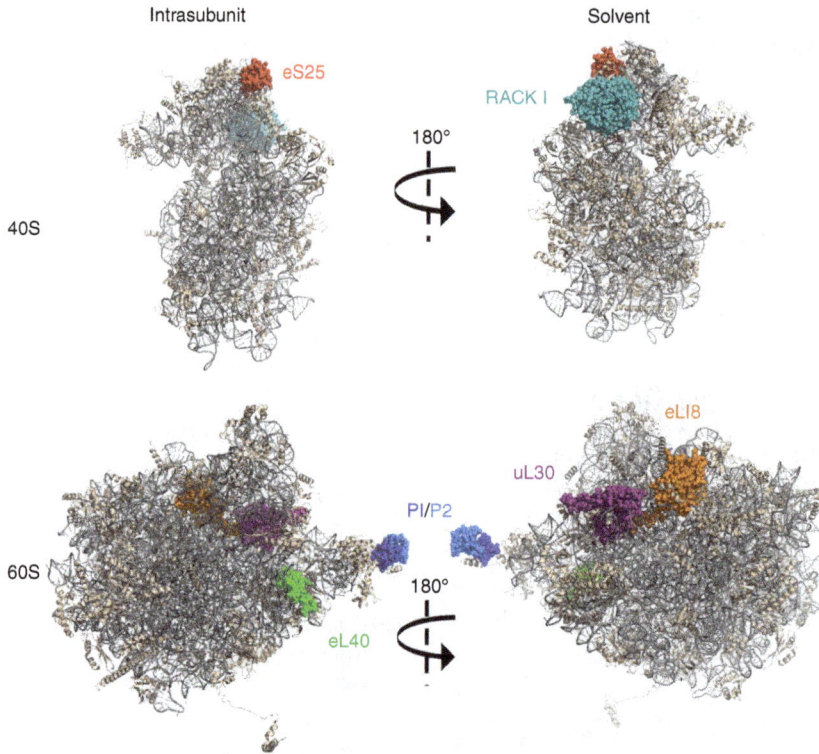

Figure 4. Ribosomal proteins are required for translation of viral RNAs. Locations of ribosomal proteins on 40S and 60S subunits that are important for viral messenger RNA (mRNA) translation are shown on the ribosome structure derived from PDB entry 4v88. (Figure based on data in Ben-Shem et al. 2011.) Both intrasubunit (*left*) and solvent surfaces (*right*) are shown. Ribosomal RNA (rRNA) (gray) and ribosomal proteins (tan) are depicted with the proteins implicated in viral mRNA translation in color: eS25 (red), RACK1 (turquoise), eL40 (green), uL30 (purple), eL18 (orange), and P1/P2 (dark blue/light blue).

Whereas the DENV RNA genome has a capped 5′ end, it lacks a poly(A) tail (Edgil et al. 2006). Two 60S ribosomal subunit proteins, eL18 and uL30 (previously called RPL18 and RPL7; Ban et al. 2014), were shown to relocalize with NS1 following DENV infection. Whether they play a direct role in translation is not known; however, knockdown of eL18 reduced viral titers by one log (Cervantes-Salazar et al. 2015). P1/P2 (RPLP1/2) are 60S subunit proteins that are part of the ribosome L1 stalk. They are also required for DENV protein expression early in infection (before replication) (Campos et al. 2017). Because the P1/P2 complex only associates with 80S ribosomes (not free subunits) and is important for eEF2 recruitment and GTPase activity, whether it effects elongation rather than

initiation needs to be evaluated (Uchiumi et al. 2002; Bautista-Santos and Zinker 2014).

NEW INSIGHTS FROM GENOME-WIDE STRATEGIES

Recent progress using genome-wide, high-throughput technologies has reshaped our understanding of translational control. Besides uncovering new principles and mechanisms, these methodologies provide a global, unbiased view of the infected cell translational landscape.

Genome Annotation

Virus genomes contain densely packed coding information that is often accessed via specialized translation strategies. Ribosome profiling (RP) (sequencing ribosome-protected fragments) allows global translation analysis and direct, experimental annotation of translation events (Ingolia et al. 2009, 2018). It couples classic nuclease footprinting with deep sequencing and indicates ribosome position with single-nucleotide resolution. Besides identifying precise boundaries of translated regions, the three-nucleotide footprint periodicity (reflecting translocation steps) indicates which reading frame is being decoded. Translational start sites can be mapped using conditions that preferentially capture initiating ribosomes (Ingolia et al. 2011; Lee et al. 2012; Stern-Ginossar et al. 2012). These approaches allow transcriptome annotation in areas translated in two overlapping reading frames, a frequent occurrence in virus genomes. Two herpesvirus genomes (HCMV, KSHV) (Stern-Ginossar et al. 2012; Arias et al. 2014), VACV (Yang et al. 2015), and a murine coronavirus (Irigoyen et al. 2016) have been annotated using RP, identifying numerous novel, mostly short ORFs translated upstream of, or within, known virus coding regions. Ribosome pause sites were identified in coronavirus A59, but their significance remains unknown (Irigoyen et al. 2016). RP also accurately measures virus gene expression kinetics throughout infection. During EMCV infection, RP exposed a temporally regulated frameshifting event. Although negligible early in infection, frameshifting efficiency increased to 70% at late time points, suggesting a new mechanism to modulate relative levels of EMCV structural and nonstructural proteins (Napthine et al. 2017).

Host Shut-Off

Application of genome-wide technologies to investigate host shut-off has clarified our understanding of underlying mechanisms and their relative contribution to impairing host translation. Concurrent measurement of translation and mRNA levels throughout influenza A virus (IAV), VACV, and coronavirus infections revealed that genome-wide changes to the host translation landscape are primarily driven by remodeling the mRNA pool (Bercovich-Kinori

et al. 2016; Irigoyen et al. 2016; Dai et al. 2017). Although many viruses encode endoribonucleases, their specificity varies. In α (HSV-1) and γ (Epstein–Barr virus [EBV], KSHV) herpesvirus subfamily members, viral and cellular mRNAs are degraded, whereas viral mRNAs are spared by coronavirus and IAV (Rivas et al. 2016). A degenerate sequence motif was identified that enables cleavage of numerous RNA targets by a γ-herpesvirus endoribonuclease (Gaglia et al. 2015). Finally, transcription termination of cellular, but not viral, genes was unexpectedly disrupted by HSV-1, broadening our perception of the range of processes that potentially impact host shut-off (Rutkowski et al. 2015). Furthermore, γ-herpesvirus-induced mRNA degradation reduced host but not viral mRNA transcription, suggesting this cellular feedback mechanism empowers viral gene expression (Abernathy et al. 2015).

Infected Cell Translational Landscape

The notion of host shut-off as a blunt, indiscriminate instrument to halt host gene expression has been revised by genome-wide studies. While host shut-off curtails host antiviral responses, impairing overall cellular protein production could adversely impact virus reproduction. Precisely how the infected cell translational landscape impacts viral propagation is just beginning to be shown. Significantly, cellular mRNAs encoding critical maintenance functions like oxidative phosphorylation are spared from host shut-off and translated in IAV- and VACV-infected cells (Bercovich-Kinori et al. 2016; Dai et al. 2017). Although ongoing production of these proteins requires stable mRNA expression in IAV-infected cells, translation of mRNAs encoding oxidative phosphorylation effector proteins is stimulated by VACV. Continuous oxidative phosphorylation is important for viral propagation in both cases. Discrimination among targets during host shut-off is likely greater than previously anticipated. Similarly, numerous host mRNAs escape KSHV-mediated shut-off and affect viral pathogenesis (Glaunsinger and Ganem 2004). A different tactic was observed in HCMV-infected cells, where host protein synthesis is not globally suppressed, so it was assumed that host protein synthesis proceeded uninterrupted. Using polysome profiling (McKinney et al. 2014) and RP (Tirosh et al. 2015), HCMV was shown to dramatically reshape the infected cell translation landscape. This translational reprogramming is dependent on mTOR activation, and expression of the virus-encoded UL38 mTORC1 activator in uninfected cells in part recapitulates these translational alterations. Importantly, interfering with the virus-induced activation of cellular mRNA translation can limit or enhance HCMV growth (McKinney et al. 2014). These examples show how genome-wide methodologies have revised our understanding of how viruses facilitate selective translation of host mRNAs needed for infection.

Synthetic Genome Recoding

Coding genes are biased in the relative frequencies of codons specifying the same amino acid. In some organisms, codon bias reflects optimization for specific tRNAs and elongation rates (Gingold and Pilpel 2011). Some synonymous codon pairs are used more or less frequently than expected, a phenomenon termed codon pair bias (Yarus and Folley 1985) that may influence translation efficiency (Tats et al. 2008). Indeed, altering codon pair frequencies reduced virus replication without affecting protein sequence (Coleman et al. 2008; Mueller et al. 2010; Martrus et al. 2013; Yang et al. 2013; Le Nouen et al. 2014). As the resulting virus attenuation depends on numerous mutations, reversion is markedly reduced, providing an exciting opportunity for live-attenuated vaccine development (Wimmer et al. 2009). However, in addition to codon pair bias and translation efficiency, other viral genome features, such as suppression of CpG and UpA dinucleotide frequencies, and variations in the propensity to mutate, which generate differential access to protein sequence space can contribute to the reduced replication phenotypes (Lauring et al. 2013; Tulloch et al. 2014; Kunec and Osterrieder 2016). Therefore, precisely how codon usage impacts virus attenuation and the relative contribution of translation repression requires further investigation.

RNA MODIFICATION AND INFECTION BIOLOGY

Although noncoding RNAs contain numerous modifications, only a few have been found in mRNAs. By far the most prevalent internal modified base on mRNAs is a methyl group on the N^6 position of adenosine (m^6A) (Yue et al. 2015; Peer et al. 2018).

Modulating Viral mRNA Metabolism by m^6A

The m^6A modification occurs predominantly in the nucleus and is mediated by the enzyme methyltransferase-like 3 (METTL3) together with METTL14, WTAP, KIAA1429, and RBM15/RBM15B. Following nuclear export, m^6A is recognized by cytoplasmic "reader" proteins, YTHDF1, YTHDF2, and YTHDF3 (Yue et al. 2015; Peer et al. 2018). m^6A modification reportedly influences mRNA splicing, export, translation and stability, all of which may impact virus biology.

Advances in genome-wide m^6A mapping on mRNAs and identification of the m^6A machinery fueled investigations into how this modification influences virus gene expression. m^6A modification was reported to enhance HIV-1 replication by regulating viral mRNA nuclear export or by stimulating viral gene expression (Kennedy et al. 2016; Lichinchi et al. 2016a; Tirumuru et al. 2016). YTHDF m^6A-reader proteins bind HIV-1 RNA at m^6A sites but their suggested function varies from promoting viral transcript abundance and translation to suppressing reverse transcription. Several m^6A-modified regions were identified in flaviviruses, such as HCV, Zika virus, DENV, yellow fever virus, and West Nile virus (Gokhale et al. 2016; Lichinchi et al. 2016b). Because these viruses replicate exclusively within the cytoplasm, the m^6A methyltransferase machinery is apparently active in this compartment. Knockdown of the host m^6A machinery enhances HCV and Zika virus production. HCV mRNA translation and replication are unaffected, but m^6A inhibits HCV RNA packaging into viral particles (Gokhale et al. 2016). Although the mechanism(s) through which m^{6}A act remain unclear, virus model systems could help clarify how m^{6}A controls gene expression.

Discriminating Self versus Nonself mRNAs

Most eukaryotic mRNAs contain a Cap-0 (m^{7}GpppN) structure with a methyl group at the guanine N-7 position. In higher eukaryotes, the mRNA cap is further modified by ribose 2'-O-methylation on the first and sometimes second cap-proximal nucleotides, resulting in Cap-1 (m^{7}GpppNmN) or Cap-2 (m^{7}GpppNmNm). Although 2'-O-methylation does not affect translation directly, it provides a molecular signature of "self." Hence mRNAs that lack 2'-O-methylation are marked as "nonself," triggering type I IFN production (Zust et al. 2011), which induces transcription of IFN-stimulated genes (ISGs). Among the most highly upregulated ISGs are those encoding IFIT (IFN-induced proteins with tetratricopeptide repeats) proteins (Fensterl and Sen 2015). Viral 2'-O-methyl-transferases encoded by coronavirus, flaviviruses, and VACV prevent recognition by IFIT1 (Daffis et al. 2010; Szretter et al. 2012; Menachery et al. 2014) and cap-proximal structural elements in α virus Cap-0 mRNA restrict IFIT1 action (Hyde et al. 2014). Mechanistically, IFIT1 can sequester eIF4F, inhibiting viral Cap-0 mRNA translation (Habjan et al. 2013). However, N1 methylation (Cap-1) may be insufficient to protect all mRNAs from IFIT1, as the second N^2 methylation (Cap-2) and secondary structure may also impact IFIT1 activity (Daugherty et al. 2016; Young et al. 2016; Abbas et al. 2017). Most mammals encode several IFIT proteins with varying affinities for distinct RNA ligands (Hyde and Diamond 2015), suggesting that additional RNA determinants are recognized. Besides competing with eIF4E for cap binding, IFIT1 binds eIF3 subunits, preventing ribosome recruitment, and may inhibit 48S complex formation (Hyde and Diamond 2015). Because IFITs alter translation in uninfected cells (Guo et al. 2000) and increase following infection, IFITs might remodel the infected cell-translation landscape.

CODING CAPACITY

To maximize their genome coding capacity, viruses use multiple strategies, including leaky scanning, polyproteins, reinitiation, translational bypass (hopping), readthrough of stop codons, and programmed ribosomal frameshifting (PRF) (Fig. 5).

Coding and Recoding Strategies

PRF is a recoding event that shifts the reading frame of a translating ribosome one or two nucleotides in the + or − direction (Caliskan et al. 2015; Dever et al. 2018). PRF frequencies range from 1% to 80%, but are typically low. A 7-nucleotide slippery site sequence, the spacer region, and structural barriers that stall the ribosome, impact PRF efficiency. Retroviruses generate high levels of Gag and low levels of a Gag-Pol fusion from the same transcript via PRF or by readthrough of a stop codon (Hung et al. 1998; Csibra et al. 2014). Perturbations in this ratio affect viral assembly, RNA packaging and maturation. Mimiviruses use both PRF and readthrough to encode a polypeptide chain release factor homolog (Jeudy et al. 2012). West Nile virus, which encodes a single polyprotein, uses an efficient -1 PRF (∼30% to 70%) to generate additional structural proteins, which increases virus levels in birds and mosquitoes for efficient transmission (Melian et al. 2010, 2014). *Dicistroviridae* IGR IRESs induce a +1 frameshift during initiation by extending pseudoknot I by one base pair to translate a short protein, ORFx (Ren et al. 2012). In addition, +1 frameshifting in IAV gene segment 3, which encodes the viral RNA polymerase subunit PA, produces the PA-X endonuclease that coordinates host shut-off (Jagger et al. 2012). Rarer -2 PRF events, first reported in porcine reproductive and respiratory syndrome virus (PRRSV, an RNA virus), also occur (Fang et al. 2012). PRRSV -2 frameshifting is responsive to a viral protein transactivator that binds to a C-rich region downstream of the PRF (Li et al. 2014). Efficient +1/−2 PRF within repetitive sequence elements encoded by the DNA viruses EBV and KSHV also generates alternative reading frame versions of their re-

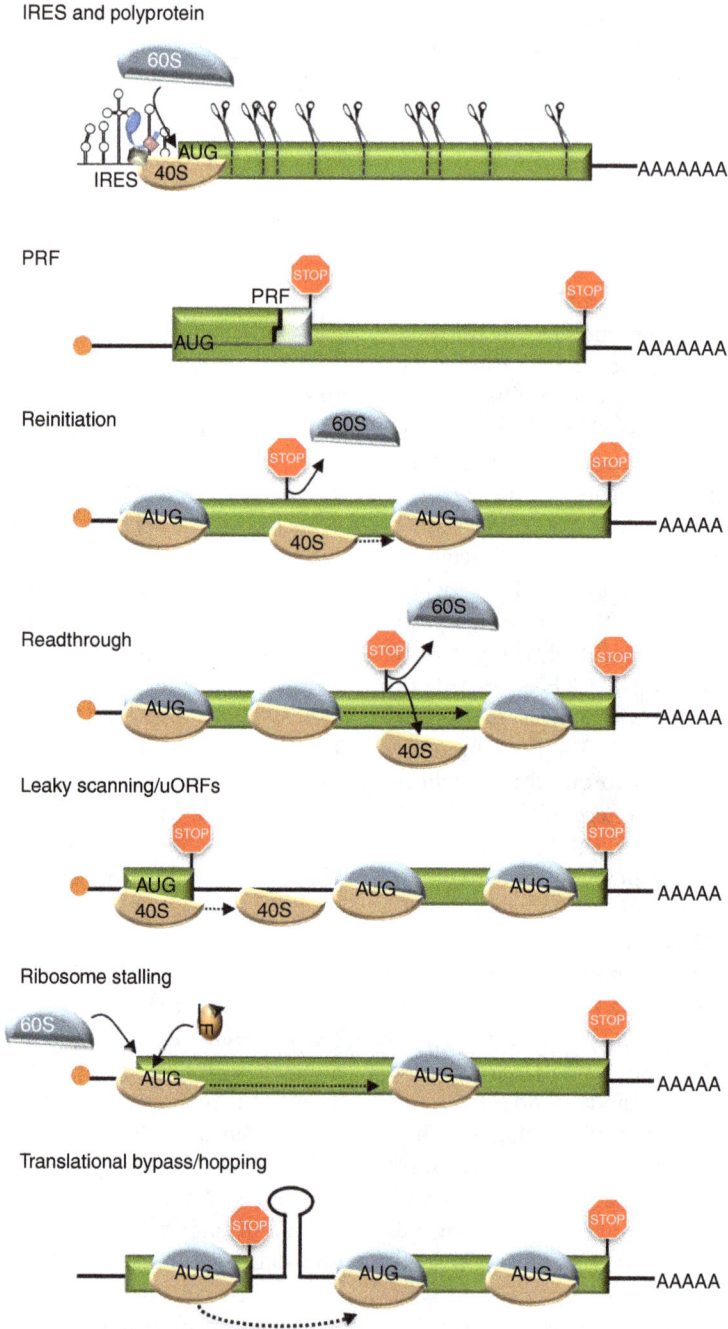

Figure 5. Strategies to maximize viral genome coding capacity. Internal ribosome entry site (IRES)-driven polyprotein production, programmed ribosome frameshifting (PRF), reinitiation, stop-codon readthrough, leaky scanning, ribosome stalling, and translational bypass/ribosome hopping are shown in the cartoon. (See the text for a detailed description.) Stop codons (stop signs), proteolytic cleavage sites (scissors), 5′ cap structure (orange circle), open reading frames (green), ribosome subunits (brown and blue), and initiation factor loaded with transfer RNA (tRNA) are shown.

spective latency proteins EBNA1 and LANA1, although their functions remain unknown (Kwun et al. 2014).

Polyproteins, which are proteolytically processed by viral and cellular proteases, are used by many (+)-stranded RNA viruses and allow for a single viral RNA to encode several different proteins. Alternatively, some DNA and RNA viruses express multiple proteins from a polycistronic transcript using a reinitiation mechanism (ribosomal termination-reinitiation or stop–start) (Wise et al. 2011; Kronstad et al. 2013, 2014; Royall and Locker 2016). For example, caliciviruses recruit ribosomes to the viral RNA 5′ end to translate an upstream ORF (uORF). Calicivirus RNAs contain a termination upstream ribosomal binding site (TURBS) that retains the 40S subunit after termination. By base pairing with 18S rRNA, TURBS facilitates efficient translation reinitiation at a nearby AUG or non-AUG codon (Luttermann and Meyers 2014; Royall and Locker 2016). In vitro, reinitiation by a terminated 40S ribosomal subunit requires eIF2, 1, and 1A but not eIF3 (Zinoviev et al. 2015).

To produce a single polypeptide from two discrete ORFs, bacteriophage T4 relies on a translational bypass or ribosome hopping. After translating a glycine codon that precedes a stop codon, the 80S ribosome "hops" over a stretch of sequences and resumes translation at a downstream glycine codon. This hopping mechanism likely requires a compact structure in the gap region to bring together the two glycine codons in the translating ribosome and to prevent the release factor from entering the A site (Todd and Walter 2013).

uORFs

Some DNA and RNA viral polycistronic transcripts contain upstream start codons that can translate short uORFs. The uORFs are a barrier to efficient initiation at a downstream ORF. Viruses modulate downstream ORF expression by leaky scanning, which transpires when the 40S subunit scans past an upstream AUG in a weak or moderate Kozak consensus sequence context before initiating at a downstream AUG (Wise

et al. 2011; Kronstad et al. 2013, 2014). For example, members of the triple gene block (TGB) superfamily of movement proteins encoded by many plant viruses, which allow cell-to-cell movement and vascular spread needed to cause disease, are often produced by leaky scanning along a single mRNA transcript that contains overlapping ORFs (Lezzhov et al. 2015; Miras et al. 2017).

Ribosome Stalling

During cap-dependent initiation, the start codon is recognized when the Met-tRNA$_i^{Met}$ base-pairs with the start codon. When eIF2 is phosphorylated, levels of ternary complex (eIF2•GTP•Met-tRNA$_i^{Met}$) are reduced, which shuts down protein synthesis globally. Alphaviruses use a stable RNA stem-loop structure located within their coding sequence to stall the ribosome on the initiation codon of the 26S mRNA when ternary complex levels are low and rely on an as-yet-unknown initiation mechanism to avoid a major antiviral defense (Toribio et al. 2016, 2018). Once the 60S subunit joins the stalled ribosome, it is released to translate the viral RNA.

CONCLUDING REMARKS

The categorical requirement for host ribosomes to translate viral mRNAs continues to provide powerful opportunities to investigate how protein synthesis is regulated. Indeed, exploiting virus model systems has defined fundamental features of the cellular protein synthesis machinery, how it is regulated in uninfected cells, and how it responds to physiological stress. Insights gleaned from investigating viral mechanisms have implications for understanding how different forms of acute and chronic stress impact translational control of gene expression in health and disease. As virus reproduction is dependent on protein synthesis, the identity and roles of leading molecular actors during infection have been revealed. RNA structural elements and modifications detected by host sentinel molecules coordinate powerful antiviral responses intended to restrict virus access to the translational apparatus. Further insights can be expected from functional analysis of the genes encoding certain translation system components in giant *Acanthamoeba* viruses (Abrahão et al. 2017; Schultz et al. 2017).

Virus–host interactions that control protein synthesis could potentially be valuable targets for new antiviral therapies and/or lead to new vaccines and biological treatments. Included among these are virus-encoded functions that counter host dsRNA-dependent, antiviral responses. By capitalizing on this biology, a replicating HSV-1 missing the ICP34.5 eIF2α phosphatase subunit, but producing a viral dsRNA-binding protein that inhibits PKR and OAS, was developed into the first oncolytic virus immunotherapy approved by U.S. and European regulatory agencies (Taneja et al. 2001; U.S. Food and Drug Administration 2015; European Medicines Agency 2016; Ribas et al. 2017). Potent innate defenses in normal cells, including IFN-induced functions like PKR and OAS, effectively limit replication of the oncolytic HSV-1 lacking the virus-encoded phosphatase subunit. However, impaired cell-intrinsic immune responses in cancer cells support preferential virus reproduction and spread through tumor tissue. This new class of biological therapeutics for cancer was an unexpected outgrowth of understanding and manipulating a translational control mechanism in virus-infected cells. New fundamental mechanisms of initiation and regulation have been revealed. In particular, roles for RNA modifications and the ribosome itself in regulating translation have emerged with discrete ribosomal proteins selectively controlling translation. Furthermore, application of genome-wide methodologies has revealed surprising insights into how host shut-off is achieved and how viruses that do not impair host protein synthesis globally impact the host translational landscape. While the underlying mechanisms remain unknown, viral systems continue to be instrumental in delineating how ribosomes differentially recruit or exclude messages and how discrete modifications of ribosomes and mRNAs comprehensively shape translation in response to physiological and environmental stress.

REFERENCES

*Reference is also in this collection.

Abbas YM, Laudenbach BT, Martinez-Montero S, Cencic R, Habjan M, Pichlmair A, Damha MJ, Pelletier J, Nagar B. 2017. Structure of human IFIT1 with capped RNA reveals adaptable mRNA binding and mechanisms for sensing N1 and N2 ribose 2′-O methylations. *Proc Natl Acad Sci* **114:** E2106–E2115.

Abernathy E, Glaunsinger B. 2015. Emerging roles for RNA degradation in viral replication and antiviral defense. *Virology* **479–480:** 600–608.

Abernathy E, Gilbertson S, Alla R, Glaunsinger B. 2015. Viral nucleases induce an mRNA degradation-transcription feedback loop in mammalian cells. *Cell Host Microbe* **18:** 243–253.

Abrahão JS, Araújo R, Colson P, La Scola B. 2017. The analysis of translation-related gene set boosts debates around origin and evolution of mimiviruses. *PLoS Genet* **13:** e1006532.

Albornoz A, Carletti T, Corazza G, Marcello A. 2014. The stress granule component TIA-1 binds tick-borne encephalitis virus RNA and is recruited to peri-nuclear sites of viral replication to inhibit viral translation. *J Virol* **88:** 6611–6622.

Amorim R, Costa SM, Cavaleiro NP, da Silva EE, da Costa LJ. 2014. HIV-1 transcripts use IRES-initiation under conditions where cap-dependent translation is restricted by poliovirus 2A protease. *PLoS ONE* **9:** e88619.

Amorim R, Temzi A, Griffin BD, Mouland AJ. 2017. Zika virus inhibits eIF2α-dependent stress granule assembly. *PLoS Negl Trop Dis* **11:** e0005775.

Arias C, Walsh D, Harbell J, Wilson AC, Mohr I. 2009. Activation of host translational control pathways by a viral developmental switch. *PLOS Pathog* **5:** e1000334.

Arias C, Weisburd B, Stern-Ginossar N, Mercier A, Madrid AS, Bellare P, Holdorf M, Weissman JS, Ganem D. 2014. KSHV 2.0: A comprehensive annotation of the Kaposi's sarcoma-associated herpesvirus genome using next-generation sequencing reveals novel genomic and functional features. *PLoS Pathog* **10:** e1003847.

Atkins JF, Loughran G, Bhatt PR, Firth AE, Baranov PV. 2016. Ribosomal frameshifting and transcriptional slippage: From genetic steganography and cryptography to adventitious use. *Nucleic Acids Res* **44:** 7007–7008.

Aumayr M, Schrempf A, Uzulmez O, Olek KM, Skern T. 2017. Interaction of 2A protease of human rhinovirus genetic group A with eIF4E is required for eIF4G cleavage during infection. *Virology* **511:** 123–134.

Avanzino BC, Fuchs G, Fraser CS. 2017. Cellular cap-binding protein, eIF4E, promotes picornavirus genome restructuring and translation. *Proc Natl Acad Sci* **114:** 9611–9616.

Ban N, Beckmann R, Cate JH, Dinman JD, Dragon F, Ellis SR, Lafontaine DL, Lindahl L, Liljas A, Lipton JM, et al. 2014. A new system for naming ribosomal proteins. *Curr Opin Struct Biol* **24:** 165–169.

Barber C, Netherton C, Goatley L, Moon A, Goodbourn S, Dixon L. 2017. Identification of residues within the African swine fever virus DP71L protein required for dephosphorylation of translation initiation factor eIF2α and inhibiting activation of pro-apoptotic CHOP. *Virology* **504:** 107–113.

Basu M, Courtney SC, Brinton MA. 2017. Arsenite-induced stress granule formation is inhibited by elevated levels of reduced glutathione in West Nile virus-infected cells. *PLoS Pathog* **13:** e1006240.

Bautista-Santos A, Zinker S. 2014. The P1/P2 protein heterodimers assemble to the ribosomal stalk at the moment when the ribosome is committed to translation but not to the native 60S ribosomal subunit in *Saccharomyces cerevisiae*. *Biochemistry* **53:** 4105–4112.

Ben-Shem A, Garreau de Loubresse N, Melnikov S, Jenner L, Yusupova G, Yusupov M. 2011. The structure of the eukaryotic ribosome at 3.0 Å resolution. *Science* **334:** 1524–1529.

Bercovich-Kinori A, Tai J, Gelbart IA, Shitrit A, Ben-Moshe S, Drori Y, Itzkovitz S, Mandelboim M, Stern-Ginossar N. 2016. A systematic view on influenza induced host shutoff. *eLife* **5:** e18311.

Berlanga JJ, Ventoso I, Harding HP, Deng J, Ron D, Sonenberg N, Carrasco L, de Haro C. 2006. Antiviral effect of the mammalian translation initiation factor 2α kinase GCN2 against RNA viruses. *EMBO J* **25:** 1730–1740.

Brasey A, Lopez-Lastra M, Ohlmann T, Beerens N, Berkhout B, Darlix JL, Sonenberg N. 2003. The leader of human immunodeficiency virus type 1 genomic RNA harbors an internal ribosome entry segment that is active during the G2/M phase of the cell cycle. *J Virol* **77:** 3939–3949.

Burgess HM, Mohr I. 2015. Cellular 5′-3′ mRNA exonuclease Xrn1 controls double-stranded RNA accumulation and anti-viral responses. *Cell Host Microbe* **17:** 332–344.

Burgess HM, Mohr I. 2018. Defining the role of stress granules in innate immune suppression by the HSV-1 endoribonuclease VHS. *J Virol* doi:10.1128/JVI.00829-18.

Caceres CJ, Angulo J, Contreras N, Pino K, Vera-Otarola J, Lopez-Lastra M. 2016. Targeting deoxyhypusine hydroxylase activity impairs cap-independent translation initiation driven by the 5′untranslated region of the HIV-1, HTLV-1, and MMTV mRNAs. *Antiviral Res* **134:** 192–206.

Caliskan N, Peske F, Rodnina MV. 2015. Changed in translation: mRNA recoding by -1 programmed ribosomal frameshifting. *Trends Biochem Sci* **40:** 265–274.

Camarena V, Kobayashi M, Kim JY, Roehm P, Perez R, Gardiner J, Wilson AC, Mohr I, Chao MV. 2010. Nature and duration of growth factor signaling through receptor tyrosine kinases regulates HSV-1 latency in neurons. *Cell Host Microbe* **8:** 320–330.

Campos RK, Wong B, Xie X, Lu YF, Shi PY, Pompon J, Garcia-Blanco MA, Bradrick SS. 2017. RPLP1 and RPLP2 are essential flavivirus host factors that promote early viral protein accumulation. *J Virol* **91:** e01706-16.

Carpentier KS, Esparo NM, Child SJ, Geballe AP. 2016. A single amino acid dictates protein kinase R susceptibility to unrelated viral antagonists. *PLoS Pathog* **12:** e1005966.

Carvajal F, Vallejos M, Walters B, Contreras N, Hertz MI, Olivares E, Caceres CJ, Pino K, Letelier A, Thompson SR, et al. 2016. Structural domains within the HIV-1 mRNA and the ribosomal protein S25 influence cap-independent translation initiation. *FEBS J* **283:** 2508–2527.

Castelló A, Quintas A, Sánchez EG, Sabina P, Nogal M, Carrasco L, Revilla Y. 2009. Regulation of host translational machinery by African swine fever virus. *PLOS Pathog* **5:** e1000562.

Cervantes-Salazar M, Angel-Ambrocio AH, Soto-Acosta R, Bautista-Carbajal P, Hurtado-Monzon AM, Alcaraz-Estrada SL, Ludert JE, Del Angel RM. 2015. Dengue virus NS1 protein interacts with the ribosomal protein RPL18: This interaction is required for viral translation and replication in Huh-7 cells. *Virology* **484:** 113–126.

Cevallos RC, Sarnow P. 2005. Factor-independent assembly of elongation-competent ribosomes by an internal ribosome entry site located in an RNA virus that infects penaeid shrimp. *J Virol* **79:** 677–683.

Chakrabarti A, Banerjee S, Franchi L, Loo YM, Gale M Jr, Nunez G, Silverman RH. 2015. RNase L activates the NLRP3 inflammasome during viral infections. *Cell Host Microbe* **17:** 466–477.

Chard LS, Bordeleau ME, Pelletier J, Tanaka J, Belsham GJ. 2006. Hepatitis C virus-related internal ribosome entry sites are found in multiple genera of the family *Picornaviridae*. *J Gen Virol* **87:** 927–936.

Chiang C, Gack MU. 2017. Post-translational control of intracellular pathogen sensing pathways. *Trends Immunol* **38:** 39–52.

Choudhury P, Bussiere L, Miller CL. 2017. Mammalian orthoreovirus factories modulate stress granule protein localization by interaction with G3BP1. *J Virol* doi: 10.1128/ JVI.01298-17.

Chuluunbaatar U, Roller R, Feldman ME, Brown S, Shokat KM, Mohr I. 2010. Constitutive mTORC1 activation by a herpesvirus Akt surrogate stimulates mRNA translation and viral replication. *Genes Dev* **24:** 2627–2639.

Coleman JR, Papamichail D, Skiena S, Futcher B, Wimmer E, Mueller S. 2008. Virus attenuation by genome-scale changes in codon pair bias. *Science* **320:** 1784–1787.

Csibra E, Brierley I, Irigoyen N. 2014. Modulation of stop codon read-through efficiency and its effect on the replication of murine leukemia virus. *J Virol* **88:** 10364–10376.

Daffis S, Szretter KJ, Schriewer J, Li J, Youn S, Errett J, Lin TY, Schneller S, Zust R, Dong H, et al. 2010. 2′-O-methylation of the viral mRNA cap evades host restriction by IFIT family members. *Nature* **468:** 452–456.

Dai A, Cao S, Dhungel P, Luan Y, Liu Y, Xie Z, Yang Z. 2017. Ribosome profiling reveals translational upregulation of cellular oxidative phosphorylation mRNAs during vaccinia virus-induced host shutoff. *J Virol* **91:** e01858.

Dauber B, Poon D, Dos Santos T, Duguay BA, Mehta N, Saffran HA, Smiley JR. 2016. The herpes simplex virus virion host shutoff protein enhances translation of viral true late mRNAs independently of suppressing protein kinase R and stress granule formation. *J Virol* **90:** 6049–6057.

Daughenbaugh KF, Fraser CS, Hershey JW, Hardy ME. 2003. The genome-linked protein VPg of the Norwalk virus binds eIF3, suggesting its role in translation initiation complex recruitment. *EMBO J* **22:** 2852–2859.

Daugherty MD, Schaller AM, Geballe AP, Malik HS. 2016. Evolution-guided functional analyses reveal diverse antiviral specificities encoded by IFIT1 genes in mammals. *eLife* **5:** e14228.

Decroly E, Ferron F, Lescar J, Canard B. 2012. Conventional and unconventional mechanisms for capping viral mRNA. *Nat Rev Microbiol* **10:** 51–65.

Deng X, Hackbart M, Mettelman RC, O'Brien A, Mielech AM, Yi G, Kao CC, Baker SC. 2017. Coronavirus nonstructural protein 15 mediates evasion of dsRNA sensors and limits apoptosis in macrophages. *Proc Natl Acad Sci* **114:** E4251–E4260.

Deniz N, Lenarcic EM, Landry DM, Thompson SR. 2009. Translation initiation factors are not required for *Dicistroviridae* IRES function in vivo. *RNA* **15:** 932–946.

Deo RC, Groft CM, Rajashankar KR, Burley SK. 2002. Recognition of the rotavirus mRNA 3′ consensus by an asymmetric NSP3 homodimer. *Cell* **108:** 71–81.

Desmet EA, Anguish LJ, Parker JS. 2014. Virus-mediated compartmentalization of the host translational machinery. *MBio* **5:** e01463.

* Dever TE, Dinman JD, Green R. 2018. Translation elongation and recoding in eukaryotes. *Cold Spring Harb Perspect Biol* doi: 10.1101/cshperspect.a032649.

Diamond MS, Farzan M. 2013. The broad-spectrum antiviral functions of IFIT and IFITM proteins. *Nat Rev Immunol* **13:** 46–57.

Edgil D, Polacek C, Harris E. 2006. Dengue virus utilizes a novel strategy for translation initiation when cap-dependent translation is inhibited. *J Virol* **80:** 2976–2986.

Emara MM, Brinton MA. 2007. Interaction of TIA-1/TIAR with West Nile and dengue virus products in infected cells interferes with stress granule formation and processing body assembly. *Proc Natl Acad Sci* **104:** 9041–9046.

European Medicines Agency. 2016. Imlygic, www.ema.europa.eu.

Fang Y, Treffers EE, Li Y, Tas A, Sun Z, van der Meer Y, de Ru AH, van Veelen PA, Atkins JF, Snijder EJ, et al. 2012. Efficient -2 frameshifting by mammalian ribosomes to synthesize an additional arterivirus protein. *Proc Natl Acad Sci* **109:** E2920–E2928.

Faria PA, Chakraborty P, Levay A, Barber GN, Ezelle HJ, Enninga J, Arana C, van Deursen J, Fontoura BM. 2005. VSV disrupts the Rae1/mrnp41 mRNA nuclear export pathway. *Mol Cell* **17:** 93–102.

Fensterl V, Sen GC. 2015. Interferon-induced Ifit proteins: Their role in viral pathogenesis. *J Virol* **89:** 2462–2468.

Finnen RL, Hay TJ, Dauber B, Smiley JR, Banfield BW. 2014. The herpes simplex virus 2 virion-associated ribonuclease vhs interferes with stress granule formation. *J Virol* **88:** 12727–12739.

Fraser CS, Doudna JA. 2007. Structural and mechanistic insights into hepatitis C viral translation initiation. *Nat Rev Microbiol* **5:** 29–38.

Gaglia MM, Rycroft CH, Glaunsinger BA. 2015. Transcriptome-wide cleavage site mapping on cellular mRNAs reveals features underlying sequence-specific cleavage by the viral ribonuclease SOX. *PLoS Pathog* **11:** e1005305.

Gao F, Simon AE. 2017. Differential use of 3′CITEs by the subgenomic RNA of Pea enation mosaic virus 2. *Virology* **510:** 194–204.

Gao F, Kasprzak WK, Szarko C, Shapiro BA, Simon AE. 2014. The 3′ untranslated region of pea enationmosaic virus contains two T-shaped, ribosome-binding, cap-independent translation enhancers. *J Virol* **88:** 11696–1712.

Garrey JL, Lee YY, Au HH, Bushell M, Jan E. 2010. Host and viral translational mechanisms during cricket paralysis virus infection. *J Virol* **84:** 1124–1138.

Gingold H, Pilpel Y. 2011. Determinants of translation efficiency and accuracy. *Mol Syst Biol* **7:** 481.

Gingras AC, Svitkin Y, Belsham GJ, Pause A, Sonenberg N. 1996. Activation of the translational suppressor 4E-BP1 following infection with encephalomyocarditis virus and poliovirus. *Proc Natl Acad Sci* **93:** 5578–5583.

Glaunsinger B, Ganem D. 2004. Highly selective escape from KSHV-mediated host mRNA shutoff and its implications for viral pathogenesis. *J Exp Med* **200:** 391–398.

Gokhale NS, McIntyre AB, McFadden MJ, Roder AE, Kennedy EM, Gandara JA, Hopcraft SE, Quicke KM, Vazquez C, Willer J, et al. 2016. N^6-methyladenosine in flaviviridae viral RNA genomes regulates infection. *Cell Host Microbe* **20:** 654–665.

Gong D, Kim YH, Xiao Y, Du Y, Xie Y, Lee KK, Feng J, Farhat N, Zhao D, Shu S, et al. 2016. A herpesvirus protein selectively inhibits cellular mRNA nuclear export. *Cell Host Microbe* **20:** 642–653.

Gonzales-Almela E, Williams H, Sanz MA, Carrasco L. 2018. The initiation factors eIF2, eIF2A, eIF2D, eIF4A and eIF4G are not involved in translation driven by hepatitis C virus IRES in human cells. *Front Microbiol* **9:** 207.

Goodman AG, Tanner BC, Chang ST, Esteban M, Katze MG. 2011. Virus infection rapidly activates the P58(IPK) pathway, delaying peak kinase activation to enhance viral replication. *Virology* **417:** 27–36.

Guo J, Hui DJ, Merrick WC, Sen GC. 2000. A new pathway of translational regulation mediated by eukaryotic initiation factor 3. *EMBO J* **19:** 6891–6899.

Gradi A, Svitkin YV, Imataka H, Sonenberg N. 1998. Proteolysis of human eukaryotic translation initiation factor eIF4GII, but not eIF4GI, coincides with the shutoff of host protein synthesis after poliovirus infection. *Proc Natl Acad Sci* **95:** 11089–11094.

Habjan M, Hubel P, Lacerda L, Benda C, Holze C, Eberl CH, Mann A, Kindler E, Gil-Cruz C, Ziebuhr J, et al. 2013. Sequestration by IFIT1 impairs translation of 2′O-unmethylated capped RNA. *PLoS Pathog* **9:** e1003663.

Hakki M, Marshall EE, De Niro KL, Geballe AP. 2006. Binding and nuclear relocalization of protein kinase R by human cytomegalovirus TRS1. *J Virol* **80:** 11817–11826.

Haller SL, Peng C, McFadden G, Rothenburg S. 2014. Poxviruses and the evolution of host range and virulence. *Infect Genet Evol* **21:** 15–40.

Han Y, Donovan J, Rath S, Whitney G, Chitrakar A, Korennykh A. 2014. Structure of human RNase L reveals the basis for regulated RNA decay in the IFN response. *Science* **343:** 1244–1248.

Herdy B, Jaramillo M, Svitkin YV, Rosenfeld AB, Kobayashi M, Walsh D, Alain T, Sean P, Robichaud N, Topisirovic I, et al. 2012. Translational control of the activation of transcription factor NF-κB and production of type I interferon by phosphorylation of the translation factor eIF4E. *Nat Immunol* **13:** 543–550.

Hertz MI, Thompson SR. 2011a. In vivo functional analysis of the *Dicistroviridae* intergenic region internal ribosome entry sites. *Nucleic Acids Res* **39:** 7276–7288.

Hertz MI, Thompson SR. 2011b. Mechanism of translation initiation by *Dicistroviridae* IGR IRESs. *Virology* **411:** 355–361.

Hertz MI, Landry DM, Willis AE, Luo G, Thompson SR. 2013. Ribosomal protein S25 dependency reveals a common mechanism for diverse internal ribosome entry sites and ribosome shunting. *Mol Cell Biol* **33:** 1016–1026.

Ho BC, Yu SL, Chen JJ, Chang SY, Yan BS, Hong QS, Singh S, Kao CL, Chen HY, Su KY, et al. 2010. Enterovirus-induced miR-141 contributes to shutoff of host protein translation by targeting the translation initiation factor eIF4E. *Cell Host Microbe* **9:** 58–69.

Hou S, Kumar A, Xu Z, Airo AM, Stryapunina I, Wong CP, Branton W, Tchesnokov E, Götte M, Power C, et al. 2017. Zika virus hijacks stress granule proteins and modulates the host stress response. *J Virol* doi: 10.1128/JVI.00474-17.

Humoud MN, Doyle N, Royall E, Willcocks MM, Sorgeloos F, van Kuppeveld F, Roberts LO, Goodfellow IG, Langereis MA, Locker N. 2016. Feline calicivirus infection disrupts assembly of cytoplasmic stress granules and induces g3bp1 cleavage. *J Virol* **90:** 6489–6501.

Hung M, Patel P, Davis S, Green SR. 1998. Importance of ribosomal frameshifting for human immunodeficiency virus type 1 particle assembly and replication. *J Virol* **72:** 4819–4824.

Hyde JL, Diamond MS. 2015. Innate immune restriction and antagonism of viral RNA lacking 2-*O* methylation. *Virology* **479–480:** 66–74.

Hyde JL, Gardner CL, Kimura T, White JP, Liu G, Trobaugh DW, Huang C, Tonelli M, Paessler S, Takeda K, et al. 2014. A viral RNA structural element alters host recognition of nonself RNA. *Science* **343:** 783–787.

Ingolia NT, Ghaemmaghami S, Newman JR, Weissman JS. 2009. Genome-wide analysis in vivo of translation with nucleotide resolution using ribosome profiling. *Science* **324:** 218–223.

Ingolia NT, Lareau LF, Weissman JS. 2011. Ribosome profiling of mouse embryonic stem cells reveals the complexity and dynamics of mammalian proteomes. *Cell* **147:** 789–802.

* Ingolia NT, Hussmann JA, Weissman JS. 2018. Ribosome profiling: Global views of translation. *Cold Spring Harb Perspect Biol* doi: 10.1101/cshperspect.a032698.

Irigoyen N, Firth AE, Jones JD, Chung BY, Siddell SG, Brierley I. 2016. High- resolution analysis of coronavirus gene expression by RNA sequencing and ribosome profiling. *PLoS Pathog* **12:** e1005473.

* Ivanov P, Kedersha N, Anderson P. 2018. Stress granules and processing bodies in translational control. *Cold Spring Harb Perspect Biol* doi: 10.1101/cshperspect.a032813.

Jagger BW, Wise HM, Kash JC, Walters KA, Wills NM, Xiao YL, Dunfee RL, Schwartzman LM, Ozinsky A, Bell GL, et al. 2012. An overlapping protein-coding region in influenza A virus segment 3 modulates the host response. *Science* **337:** 199–204.

Jan E, Sarnow P. 2002. Factorless ribosome assembly on the internal ribosome entry site of cricket paralysis virus. *J Mol Biol* **324:** 889–902.

Jan E, Kinzy TG, Sarnow P. 2003. Divergent tRNA-like element supports initiation, elongation, and termination of

protein biosynthesis. *Proc Natl Acad Sci* **100:** 15410–15415.

Jan E, Mohr I, Walsh D. 2016. A cap-to-tail guide to mRNA translation strategies in virus-infected cells. *Ann Rev Virol* **3:** 283–307.

Jang SK, Krausslich HG, Nicklin MJ, Duke GM, Palmenberg AC, Wimmer E. 1988. A segment of the 5′ nontranslated region of encephalomyocarditis virus RNA directs internal entry of ribosomes during in vitro translation. *J Virol* **62:** 2636–2643.

Jeudy S, Abergel C, Claverie JM, Legendre M. 2012. Translation in giant viruses: A unique mixture of bacterial and eukaryotic termination schemes. *PLoS Genet* **8:** e1003122.

Jha S, Rollins MG, Fuchs G, Procter DJ, Hall EA, Cozzolino K, Sarnow P, Savas JN, Walsh D. 2017. Trans-kingdom mimicry underlies ribosome customization by a poxvirus kinase. *Nature* **546:** 651–655.

Kamitani W, Huang C, Narayanan K, Lokugamage KG, Makino S. 2009. A two-pronged strategy to suppress host protein synthesis by SARS coronavirus Nsp1 protein. *Nat Struct Mol Biol* **16:** 1134–1140.

Katsafanas GC, Moss B. 2007. Colocalization of transcription and translation within cytoplasmic poxvirus factories coordinates viral expression and subjugates host functions. *Cell Host Microbe* **2:** 221–228.

Kempf BJ, Barton DJ. 2015. Picornavirus RNA polyadenylation by 3D(pol), the viral RNA-dependent RNA polymerase. *Virus Res* **206:** 3–11.

Kennedy EM, Bogerd HP, Kornepati AV, Kang D, Ghoshal D, Marshall JB, Poling BC, Tsai K, Gokhale NS, Horner SM, et al. 2016. Posttranscriptional m^{6}A editing of HIV-1 mRNAs enhances viral gene expression. *Cell Host Microbe* **19:** 675–685.

Kerr CH, Ma ZW, Jang CJ, Thompson SR, Jan E. 2016. Molecular analysis of the factorless internal ribosome entry site in Cricket Paralysis virus infection. *J Sci Rep* **6:** 37319.

Khaperskyy DA, Emara MM, Johnston BP, Anderson P, Hatchette TF, McCormick C. 2014. Influenza a virus host shutoff disables antiviral stress-induced translation arrest. *PLoS Pathog* **10:** e1004217.

Khaperskyy DA, Schmaling S, Larkins-Ford J, McCormick C, Gaglia MM. 2016. Selective degradation of host RNA polymerase II transcripts by influenza A virus PA-X host shutoff protein. *PLoS Pathog* **12:** e1005427.

Khong A, Kerr CH, Yeung CH, Keatings K, Nayak A, Allan DW, Jan E. 2017. Disruption of stress granule formation by the multifunctional cricket paralysis virus 1a protein. *J Virol* **91.**

Kindler E, Gil-Cruz C, Spanier J, Li Y, Wilhelm J, Rabouw HH, Züst R, Hwang M, V'kovski P, Stalder H, et al. 2017. Early endonuclease-mediated evasion of RNA sensing ensures efficient coronavirus replication. *PLoS Pathog* **13:** e1006195.

Kobayashi M, Arias C, Garabedian A, Palmenberg AC, Mohr I. 2012a. Site-specific cleavage of the host poly(A) binding protein by the encephalomyocarditis virus 3C proteinase stimulates viral replication. *J Virol* **86:** 10686–10694.

Kobayashi M, Wilson AC, Chao MV, Mohr I. 2012b. Control of viral latency in neurons by axonal mTOR signaling and the 4E-BP translation repressor. *Genes Dev* **26:** 1527–1532.

Kronstad LM, Brulois KF, Jung JU, Glaunsinger BA. 2013. Dual short upstream open reading frames control translation of a herpesviral polycistronic mRNA. *PLoS Pathog* **9:** e1003156.

Kronstad LM, Brulois KF, Jung JU, Glaunsinger BA. 2014. Reinitiation after translation of two upstream open reading frames (ORF) governs expression of the ORF35-37 Kaposi's sarcoma-associated herpesvirus polycistronic mRNA. *J Virol* **88:** 6512–6518.

Kunec D, Osterrieder N. 2016. Codon pair bias is a direct consequence of dinucleotide bias. *Cell Rep* **14:** 55–67.

* Kwan T, Thompson SR. 2018. Noncanonical translation initiation in eukaryotes. *Cold Spring Harb Perspect Biol* doi: 10.1101/cshperspect.a032672.

Kwun HJ, Toptan T, Ramos da Silva S, Atkins JF, Moore PS, Chang Y. 2014. Human DNA tumor viruses generate alternative reading frame proteins through repeat sequence recoding. *Proc Natl Acad Sci* **111:** E4342–E4349.

Landry DM, Hertz MI, Thompson SR. 2009. RPS25 is essential for translation initiation by the *Dicistroviridae* and hepatitis C viral IRESs. *Genes Dev* **23:** 2753–2764.

Launer-Felty K, Wong CJ, Cole JL. 2015. Structural analysis of adenovirus VAI RNA defines the mechanism of inhibition of PKR. *Biophys J* **108:** 748–757.

Lauring AS, Frydman J, Andino R. 2013. The role of mutational robustness in RNA virus evolution. *Nat Rev Microbiol* **11:** 327–336.

Le MT, Kasprzak WK, Kim T, Gao F, Young MY, Yuan X, Shapiro BA, Seog J, Simon AE. 2017. Folding behavior of a T-shaped, ribosome-binding translation enhancer implicated in a wide-spread conformational switch. *eLife* **13:** e22883.

Lee S, Liu B, Lee S, Huang SX, Shen B, Qian SB. 2012. Global mapping of translation initiation sites in mammalian cells at single-nucleotide resolution. *Proc Natl Acad Sci* **109:** E2424–E2432.

Lee AS, Burdeinick-Kerr R, Whelan SP. 2013. A ribosome-specialized translation initiation pathway is required for cap-dependent translation of vesicular stomatitis virus mRNAs. *Proc Natl Acad Sci* **110:** 324–329.

Leen EN, Sorgeloos F, Correia S, Chaudhry Y, Cannac F, et al. 2016. A conserved interaction between a C-terminal motif in norovirus VPg and the HEAT-1 domain of eIF4G is essential for translation initiation. *PLOS Pathog* **12:** e1005379.

Le Nouen C, Brock LG, Luongo C, McCarty T, Yang L, Mehedi M, Wimmer E, Mueller S, Collins PL, Buchholz UJ, et al. 2014. Attenuation of human respiratory syncytial virus by genome-scale codon-pair deoptimization. *Proc Natl Acad Sci* **111:** 13169–13174.

Le Sage V, Cinti A, McCarthy S, Amorim R, Rao S, Daino GL, Tramontano E, Branch DR, Mouland AJ. 2017. Ebola virus VP35 blocks stress granule assembly. *Virology* **502:** 73–83.

Lezzhov AA, Gushchin VA, Lazareva EA, Vishnichenko VK, Morozov SY, Solovyev AG. 2015. Translation of the shallot virus X TGB3 gene depends on non-AUG initiation and leaky scanning. *J Gen Virol* **96:** 3159–3164.

Li D, Wei T, Abbott CM, Harrich D. 2013. The unexpected roles of eukaryotic translation elongation factors in RNA virus replication and pathogenesis. *Microbiol Mol Biol Rev* **77:** 253–266.

Li Y, Treffers EE, Napthine S, Tas A, Zhu L, Sun Z, Bell S, Mark BL, van Veelen PA, van Hemert MJ, et al. 2014. Transactivation of programmed ribosomal frameshifting by a viral protein. *Proc Natl Acad Sci* **111:** E2172–E2181.

Lindquist ME, Lifland AW, Utley TJ, Santangelo PJ, Crowe JE Jr. 2010. Respiratory syncytial virus induces host RNA stress granules to facilitate viral replication. *J Virol* **84:** 12274–12284.

Lichinchi G, Gao S, Saletore Y, Gonzalez GM, Bansal V, Wang Y, Mason CE, Rana TM. 2016a. Dynamics of the human and viral m⁶A RNA methylomes during HIV-1 infection of T cells. *Nat Microbiol* **1:** 16011.

Lichinchi G, Zhao BS, Wu Y, Lu Z, Qin Y, He C, Rana TM. 2016b. Dynamics of human and viral RNA methylation during Zika virus infection. *Cell Host Microbe* **20:** 666–673.

Liu J, HuangFu WC, Kumar KG, Qian J, Casey JP, Hamanaka RB, Grigoriadou C, Aldabe R, Diehl JA, Fuchs SY. 2009. Virus-induced unfolded protein response attenuates antiviral defenses via phosphorylation-dependent degradation of the type I interferon receptor. *Cell Host Microbe* **5:** 72–83.

Liu SW, Katsafanas GC, Liu R, Wyatt LS, Moss B. 2015. Poxvirus decapping enzymes enhance virulence by preventing the accumulation of dsRNA and the induction of innate antiviral responses. *Cell Host Microbe* **17:** 320–331.

Luttermann C, Meyers G. 2014. Two alternative ways of start site selection in human norovirus reinitiation of translation. *J Biol Chem* **289:** 11739–11754.

Majzoub K, Hafirassou ML, Meignin C, Goto A, Marzi S, Fedorova A, Verdier Y, Vinh J, Hoffmann JA, Martin F, et al. 2014. RACK1 controls IRES-mediated translation of viruses. *Cell* **159:** 1086–1095.

Martinez-Salas E, Francisco-Velilla R, Fernandez-Chamorro J, Lozano G, Diaz-Toledano R. 2015. Picornavirus IRES elements: RNA structure and host protein interactions. *Virus Res* **206:** 62–73.

Martrus G, Nevot M, Andres C, Clotet B, Martinez MA. 2013. Changes in codon-pair bias of human immunodeficiency virus type 1 have profound effects on virus replication in cell culture. *Retrovirology* **10:** 78.

Mathews MB, Shenk T. 1991. Adenovirus virus-associated RNA and translational control. *J Virol* **65:** 5657–5682.

McCormick C, Khaperskyy DA. 2017. Translation inhibition and stress granules in the antiviral immune response. *Nat Rev Immunol* doi: 10.1038/nri.2017.63.

McKinney C, Perez C, Mohr I. 2012. Poly(A) binding protein abundance regulates eukaryotic translation initiation factor 4F assembly in human cytomegalovirus-infected cells. *Proc Natl Acad Sci* **109:** 5627–5632.

McKinney C, Yu D, Mohr I. 2013. A new role for the cellular PABP repressor Paip2 as an innate restriction factor capable of limiting productive cytomegalovirus replication. *Genes Dev* **27:** 1809–1820.

McKinney C, Zavadil J, Bianco C, Shiflett L, Brown S, Mohr I. 2014. Global reprogramming of the cellular translational landscape facilitates cytomegalovirus replication. *Cell Rep* **6:** 9–17.

Melian EB, Hinzman E, Nagasaki T, Firth AE, Wills NM, Nouwens AS, Blitvich BJ, Leung J, Funk A, Atkins JF, et al. 2010. NS1′ of flaviviruses in the Japanese encephalitis virus serogroup is a product of ribosomal frameshifting and plays a role in viral neuroinvasiveness. *J Virol* **84:** 1641–1647.

Melian EB, Hall-Mendelin S, Du F, Owens N, Bosco-Lauth AM, Nagasaki T, Rudd S, Brault AC, Bowen RA, Hall RA, et al. 2014. Programmed ribosomal frameshift alters expression of West Nile virus genes and facilitates virus replication in birds and mosquitoes. *PLoS Pathog* **10:** e1004447.

Menachery VD, Yount BL Jr, Josset L, Gralinski LE, Scobey T, Agnihothram S, Katze MG, Baric RS. 2014. Attenuation and restoration of severe acute respiratory syndrome coronavirus mutant lacking 2′-O-methyltransferase activity. *J Virol* **88:** 4251–4264.

* Merrick WC, Pavitt GD. 2018. Protein synthesis initiation in eukaryotic cells. *Cold Spring Harb Perspect Biol* doi: 10.1101/cshperspect.a033092.

Miras M, Miller WA, Truniger V, Aranda MA. 2017. Noncanonical translation in plant RNA viruses. *Front Plant Sci* **8:** 494.

Mizutani T, Fukushi S, Saijo M, Kurane I, Morikawa S. 2004. Phosphorylation of p38 MAPK and its downstream targets in SARS coronavirus-infected cells. *Biochem Biophys Res Commun* **319:** 1228–1234.

Mohr I. 2016. Closing in on the causes of host shutoff. *eLife* **5:** e20755.

Mohr I, Sonenberg N. 2012. Host translation at the nexus of infection and immunity. *Cell Host Microbe* **12:** 470–483.

Montero H, Rojas M, Arias CF, Lopez S. 2008. Rotavirus infection induces the phosphorylation of eIF2α but prevents the formation of stress granules. *J Virol* **82:** 1496–1504.

Moorman NJ, Cristea IM, Terhune SS, Rout MP, Chait BT, Shenk T. 2008. Human cytomegalovirus protein UL38 inhibits host cell stress responses by antagonizing the tuberous sclerosis protein complex. *Cell Host Microbe* **3:** 253–262.

Mudhasani R, Tran JP, Retterer C, Kota KP, Whitehouse CA, Bavari S. 2016. Protein kinase R degradation is essential for Rift Valley fever virus infection and is regulated by SKP1-CUL1-F-box(SCF)$^{FBXW11-NSs}$ E3 ligase. *PLOS Pathog* **12:** e1005437.

Mueller S, Coleman JR, Papamichail D, Ward CB, Nimnual A, Futcher B, Skiena S, Wimmer E. 2010. Live attenuated influenza vaccines by computer-aided rational design. *Nat Biotechnol* **28:** 723–726.

Mulvey M, Poppers J, Sternberg D, Mohr I. 2003. Regulation of eIF2α phosphorylation by different functions that act during discrete phases in the herpes simplex virus type 1 life cycle. *J Virol* **77:** 10917–10928.

Mulvey M, Camarena V, Mohr I. 2004. Full resistance of HSV-1 infected primary human cells to interferon α require both the Us11 and γ134.5 gene products. *J Virol* **78:** 10193–10196.

Mulvey M, Arias C, Mohr I. 2007. Maintenance of endoplasmic reticulum (ER) homeostasis in herpes simplex virus

type 1-infected cells through the association of a viral glycoprotein with PERK, a cellular ER stress sensor. *J Virol* **81:** 3377–3390.

Napthine S, Ling R, Finch LK, Jones JD, Bell S, Brierley I, Firth AE. 2017. Protein-directed ribosomal frameshifting temporally regulates gene expression. *Nat Commun* **8:** 15582.

Nelson EV, Schmidt KM, Deflubé LR, Doğanay S, Banadyga L, Olejnik J, Hume AJ, Ryabchikova E, Ebihara H, Kedersha N, et al. 2016. Ebola virus does not induce stress granule formation during infection and sequesters stress granule proteins within viral inclusions. *J Virol* **90:** 7268–7268.

Nicholson BL, White KA. 2014. Functional long-range RNA–RNA interactions in positive-strand RNA viruses. *Nat Rev Microbiol* **12:** 493–504.

Nikolic J, Civas A, Lama Z, Lagaudrière-Gesbert C, Blondel D. 2016. Rabies virus infection induces the formation of stress granules closely connected to the viral factories. *PLoS Pathog* **12:** e1005942.

Oh SW, Onomoto K, Wakimoto M, Onoguchi K, Ishidate F, Fujiwara T, Yoneyama M, Kato H, Fujita T. 2016. Leader-containing uncapped viral transcript activates RIG-I in antiviral stress granules. *PLoS Pathog* **12:** e1005444.

Ohlmann T, Mengardi C, Lopez-Lastra M. 2014. Translation initiation of the HIV-1 mRNA. *Translation (Austin)* **2:** e960242.

Olivares E, Landry DM, Caceres CJ, Pino K, Rossi F, Navarrete C, Huidobro-Toro JP, Thompson SR, Lopez-Lastra M. 2014. The 5′ untranslated region of the human T-cell lymphotropic virus type 1 mRNA enables cap-independent translation initiation. *J Virol* **88:** 5936–5955.

Othman Z, Sulaiman MK, Willcocks MM, Ulryck N, Blackbourn DJ, Sargueil B, Roberts LO, Locker N. 2014. Functional analysis of Kaposi's sarcoma-associated herpesvirus vFLIP expression reveals a new mode of IRES-mediated translation. *RNA* **20:** 1803–1814.

Parrish S, Moss B. 2007. Characterization of a second vaccinia virus mRNA-decapping enzyme conserved in poxviruses. *J Virol* **81:** 12973–12978.

Pavio N, Romano PR, Graczyk TM, Feinstone SM, Taylor DR. 2003. Protein synthesis and endoplasmic reticulum stress can be modulated by the hepatitis C virus envelope protein E2 through the eukaryotic initiation factor 2α kinase PERK. *J Virol* **77:** 3578–3585.

* Peer E, Moshitch-Moshkovitz S, Rechavi G, Dominissini D. 2018. The epitranscriptome in translation regulation. *Cold Spring Harb Perspect Biol* doi: 10.1101/cshperspect.a032623.

Pelletier J, Sonenberg N. 1988. Internal initiation of translation of eukaryotic mRNA directed by a sequence derived from poliovirus RNA. *Nature* **334:** 320–325.

Peng C, Haller SL, Rahman MM, McFadden G, Rothenburg S. 2016. Myxoma virus M156 is a specific inhibitor of rabbit PKR but contains a loss-of-function mutation in Australian virus isolates. *Proc Natl Acad Sci* **113:** 3855–3860.

Pestova TV, Hellen CU. 2003. Translation elongation after assembly of ribosomes on the Cricket paralysis virus internal ribosomal entry site without initiation factors or initiator tRNA. *Genes Dev* **17:** 181–186.

Pestova TV, Lomakin IB, Hellen CU. 2004. Position of the CrPV IRES on the 40S subunit and factor dependence of IRES/80S ribosome assembly. *EMBO Rep* **5**: 906–913.

Pisarev AV, Chard LS, Kaku Y, Johns HL, Shatsky IN, Belsham GJ. 2004. Functional and structural similarities between the internal ribosome entry sites of hepatitis C virus and porcine teschovirus, a picornavirus. *J Virol* **78**: 4487–4497.

Plank TD, Whitehurst JT, Cencic R, Pelletier J, Kieft JS. 2014. Internal translation initiation from HIV-1 transcripts is conferred by a common RNA structure. *Translation (Austin)* **2**: e27694.

Pooggin MM, Rajeswaran R, Schepetilnikov MV, Ryabova LA. 2012. Short ORF-dependent ribosome shunting operates in an RNA picorna-like virus and a DNA pararetrovirus that cause rice tungro disease. *PLOS Pathog* **8**: e1002568.

Poon LL, Pritlove DC, Fodor E, Brownlee GG. 1999. Direct evidence that the poly(A) tail of influenza A virus mRNA is synthesized by reiterative copying of a U track in the virion RNA template. *J Virol* **73**: 3473–3476.

Protter DS, Parker R. 2016. Principles and properties of stress granules. *Trends Cell Biol* **26**: 668–679.

* Proud CG. 2018. Phosphorylation and signal transduction pathways in translational control. *Cold Spring Harb Perspect Biol* doi: 10.1101/cshperspect.a033050.

Qin Q, Carroll K, Hastings C, Miller CL. 2011. Mammalian orthoreovirus escape from host translational shutoff correlates with stress granule disruption and is independent of eIF2α phosphorylation and PKR. *J Virol* **85**: 8798–8810.

Rabouw HH, Langereis MA, Knaap RC, Dalebout TJ, Canton J, Sola I, Enjuanes L, Bredenbeek PJ, Kikkert M, de Groot RJ, et al. 2016. Middle East respiratory coronavirus accessory protein 4a inhibits PKR-mediated antiviral stress responses. *PLoS Pathog* **12**: e1005982.

Read GS. 2013. Virus-encoded endonucleases: Expected and novel functions. *RNA* **4**: 693–708.

Reineke LC, Lloyd RE. 2015. The stress granule protein G3BP1 recruits protein kinase R to promote multiple innate immune antiviral responses. *J Virol* **89**: 2575–2589.

Remm M, Remm A, Ustav M. 1999. Human papillomavirus type 18 E1 protein is translated from polycistronic mRNA by a discontinuous scanning mechanism. *J Virol* **73**: 3062–3070.

Ren Q, Wang QS, Firth AE, Chan MM, Gouw JW, Guarna MM, Foster LJ, Atkins JF, Jan E. 2012. Alternative reading frame selection mediated by a tRNA-like domain of an internal ribosome entry site. *Proc Natl Acad Sci* **109**: E630–E639.

Ribas A, Dummer R, Puzanov I, VanderWalde A, Andtbacka RHI, Michielin O, Olszanski AJ, Malvehy J, Cebon J, Fernandez E, et al. 2017. Oncolytic virotherapy promotes intratumoral T cell infiltration and improves anti-PD-1 immunotherapy. *Cell* **170**: 1109–1119.e10.

Rice AD, Turner PC, Embury JE, Moldawer LL, Baker HV, Moyer RW. 2011. Roles of vaccinia virus genes E3L and K3L and host genes PKR and RNase L during intratracheal infection of C57BL/6 mice. *J Virol* **85**: 550–567.

Rivas HG, Schmaling SK, Gaglia MM. 2016. Shutoff of host gene expression in influenza A virus and herpesviruses: Similar mechanisms and common themes. *Viruses* **8**: 102.

Rivera CI, Lloyd RE. 2008. Modulation of enteroviral proteinase cleavage of poly(A)-binding protein (PABP) by conformation and PABP-associated factors. *Virology* **375**: 59–72.

Rojas M, Vasconcelos G, Dever TE. 2015. An eIF2α-binding motif in protein phosphatase 1 subunitGADD34 and its viral orthologs is required to promote dephosphorylation of eIF2α. *Proc Natl Acad Sci* **112**: E3466–E3475.

Roobol A, Roobol J, Bastide A, Knight JR, Willis AE, Smales CM. 2015. p58IPK is an inhibitor of the eIF2α kinase GCN2 and its localization and expression underpin protein synthesis and ER processing capacity. *Biochem J* **465**: 213–225.

Roth H, Magg V, Uch F, Mutz P, Klein P, Haneke K, Lohmann V, Bartenschlager R, Fackler OT, Locker N, et al. 2017. Flavivirus infection uncouples translation suppression from cellular stress responses. *MBio* **8**: e00488.

Royall E, Locker N. 2016. Translational control during calicivirus infection. *Viruses* **8**: 104.

Royall E, Doyle N, Abdul-Wahab A, Emmott E, Morley SJ, Goodfellow I, Roberts LO, Locker N. 2015. Murine norovirus 1 (MNV1) replication induces translational control of the host by regulating eIF4E activity during infection. *J Biol Chem* **290**: 4748–4758.

Rutkowski AJ, Erhard F, L'Hernault A, Bonfert T, Schilhabel M, Crump C, Rosenstiel P, Efstathiou S, Zimmer R, Friedel CC, et al. 2015. Widespread disruption of host transcription termination in HSV-1 infection. *Nat Commun* **6**: 7126.

Sadler AJ, Williams BR. 2008. Interferon-inducible antiviral effectors. *Nat Rev Immunol* **8**: 559–568.

Sánchez-Aparicio MT, Ayllón J, Leo-Macias A, Wolff T, García-Sastre A. 2017. Subcellular localizations of RIG-I, TRIM25, and MAVS complexes. *J Virol* **91**.

Sanz MA, Castelló A, Ventoso I, Berlanga JJ, Carrasco L. 2009. Dual mechanism for the translation of subgenomic mRNA from Sindbis virus in infected and uninfected cells. *PLoS ONE* **4**: e4772.

Sanz MA, Almela EG, Carrasco L. 2017. Translation of sindbis subgenomic mRNA is independent of eIF2, eIF2A and eIF2D. *Sci Rep* **7**: 43876.

Sasaki J, Nakashima N. 2000. Methionine-independent initiation of translation in the capsid protein of an insect RNA virus. *Proc Natl Acad Sci* **97**: 1512–1515.

Schoggins JW. 2014. Interferon-stimulated genes: Roles in viral pathogenesis. *Curr Opin Virol* **6**: 40–46.

Schultz F, Yutin N, Ivanova NN, Ortega DR, Lee TK, Vierheilig J, Daims H, Horn M, Wagner M, Jensen GJ, et al. 2017. Giant viruses with an expanded complement of translation system components. *Science* **356**: 82–85.

Sciortino MT, Parisi T, Siracusano G, Mastino A, Taddeo B, Roizman B. 2013. The virion host shutoff RNase plays a key role in blocking the activation of protein kinase R in cells infected with herpes simplex virus 1. *J Virol* **87**: 3271–3276.

Seo EJ, Liu F, Kawagishi-Kobayashi M, Ung TL, Cao C, Dar AC, Sicheri F, Dever TE. 2008. Protein kinase PKR mutants resistant to the poxvirus pseudosubstrate K3L protein. *Proc Natl Acad Sci* **105**: 16894–16899.

Sharma NR, Majerciak V, Kruhlak MJ, Zheng ZM. 2017. KSHV inhibits stress granule formation by viral ORF57 blocking PKR activation. *PLoS Pathog* **13:** e1006677.

Silva LC, Almeida GM, Assis FL, Albarnaz JD, Boratto PV, Dornas FP, Andrade KR, La Scola B, Kroon EG, da Fonseca FG, et al. 2015. Modulation of the expression of mimivirus-encoded translation-related genes in response to nutrient availability during *Acanthamoeba castellanii* infection. *Front Microbiol* **6:** 539.

Simon AE, Miller WA. 2013. 3′ cap-independent translation enhancers of plant viruses. *Annu Rev Microbiol* **67:** 21–42.

Smith RWP, Anderson RC, Larralde O, Smith JWS, Gorgoni B, Richardson WA, Malik P, Graham SV, Gray NK. 2017. Viral and cellular mRNA-specific activators harness PABP and eIF4G to promote translation initiation downstream of cap binding. *Proc Natl Acad Sci* **114:** 6310–6315.

Sood R, Porter AC, Ma K, Quilliam LA, Wek RC. 2000. Pancreatic eukaryotic initiation factor-2α kinase (PEK) homologues in humans, *Drosophila melanogaster* and *Caenorhabditis elegans* that mediate translational control in response to endoplasmic reticulum stress. *Biochem J* **346:** 281–293.

Sorgeloos F, Jha BK, Silverman RH, Michiels T. 2013. Evasion of antiviral innate immunity by Theiler's virus L* protein through direct inhibition of RNase L. *PLOS Pathog* **9:** e1003474.

Spahn CM, Jan E, Mulder A, Grassucci RA, Sarnow P, Frank J. 2004. Cryo-EM visualization of a viral internal ribosome entry site bound to human ribosomes: The IRES functions as an RNA-based translation factor. *Cell* **118:** 465–475.

Stern-Ginossar N, Weisburd B, Michalski A, Le VT, Hein MY, Huang SX, Ma M, Shen B, Qian SB, Hengel H, et al. 2012. Decoding human cytomegalovirus. *Science* **338:** 1088–1093.

Suzuki Y, Chin WX, Han Q, Ichiyama K, Lee CH, Eyo ZW, Takahashi H, Takahashi C, Tan BH, Hishiki T, et al. 2016. Characterization of RyDEN (C19orf66) as an interferon stimulated cellular inhibitor against dengue virus replication. *PLoS Pathog* **12:** e1005357.

Sweeney TR, Abaeva IS, Pestova TV, Hellen CU. 2014. The mechanism of translation initiation on Type 1 picornavirus IRESs. *EMBO J* **33:** 76–92.

Szretter KJ, Daniels BP, Cho H, Gainey MD, Yokoyama WM, Gale M Jr, Virgin HW, Klein RS, Sen GC, Diamond MS. 2012. 2′-O methylation of the viral mRNA cap by West Nile virus evades ifit1-dependent and -independent mechanisms of host restriction in vivo. *PLoS Pathog* **8:** e1002698.

Tahiri-Alaoui A, Zhao Y, Sadigh Y, Popplestone J, Kgosana L, Smith LP, Nair V. 2014. Poly(A) binding protein 1 enhances cap-dependent translation initiation of neurovirulence factor from avian herpesvirus. *PLoS ONE* **9:** e114466.

Takeshita D, Tomita K. 2012. Molecular basis for RNA polymerization by Qβ replicase. *Nat Struct Mol Biol* **9:** 229–237.

Taneja S, MacGregor J, Markus S, Ha S, Mohr I. 2001. Enhanced antitumor efficacy of a herpes simplex virus mutant isolated by genetic selection in cancer cells. *Proc Natl Acad Sci* **98:** 8804–8808.

Tats A, Tenson T, Remm M. 2008. Preferred and avoided codon pairs in three domains of life. *BMC Genomics* **9:** 463.

Thompson SR, Gulyas KD, Sarnow P. 2001. Internal initiation in *Saccharomyces cerevisiae* mediated by an initiator tRNA/eIF2-independent internal ribosome entry site element. *Proc Natl Acad Sci* **98:** 12972–12977.

Tirosh O, Cohen Y, Shitrit A, Shani O, Le-Trilling VT, Trilling M, Friedlander G, Tanenbaum M, Stern-Ginossar N. 2015. The transcription and translation landscapes during human cytomegalovirus infection reveal novel host-pathogen interactions. *PLoS Pathog* **11:** e1005288.

Tirumuru N, Zhao BS, Lu W, Lu Z, He C, Wu L. 2016. N^6-methyladenosine of HIV-1 RNA regulates viral infection and HIV-1 Gag protein expression. *eLife* **5:** e15528.

Todd GC, Walter NG. 2013. Secondary structure of bacteriophage T4 gene 60 mRNA: Implications for translational bypassing. *RNA* **19:** 685–700.

Toribio R, Diaz-Lopez I, Boskovic J, Ventoso I. 2016. An RNA trapping mechanism in Alphavirus mRNA promotes ribosome stalling and translation initiation. *Nucleic Acids Res* **44:** 4368–4380.

Toribio R, Diaz-Lopez I, Boskovic J, Ventoso I. 2018. Translation initiation of αvirus mRNA reveals new insights into the topology of the 48S initiation complex. *Nucleic Acids Res* doi: 10.1093/nar/gky071.

Townsend HL, Jha BK, Han JQ, Maluf NK, Silverman RH, Barton DJ. 2008. A viral RNA competitively inhibits the antiviral endoribonuclease domain of RNase L. *RNA* **14:** 1026–1036.

Tsai WC, Lloyd RE. 2014. Cytoplasmic RNA granules and viral infection. *Ann Rev Virol* **1:** 147–170.

Tulloch F, Atkinson NJ, Evans DJ, Ryan MD, Simmonds P. 2014. RNA virus attenuation by codon pair deoptimisation is an artefact of increases in CpG/UpA dinucleotide frequencies. *eLife* **3:** e04531.

Uchiumi T, Honma S, Nomura T, Dabbs ER, Hachimori A. 2002. Translation elongation by a hybrid ribosome in which proteins at the GTPase center of the *Escherichia coli* ribosome are replaced with rat counterparts. *J Biol Chem* **277:** 3857–3862.

U.S. Food and Drug Administration. 2015. Imlygic, www.fda.gov.

Van Etten JL, Lane LC, Dunigan DD. 2010. DNA viruses: The really big ones (giruses). *Annu Rev Microbiol* **64:** 83–99.

Vincent HA, Ziehr B, Moorman NJ. 2017. Mechanism of protein kinase R inhibition by human cytomegalovirus pTRS1. *J Virol* **91.**

Vink EI, Smiley JR, Mohr I. 2017. Subversion of host responses to energy insufficiency by Us3 supports herpes simplex virus replication during stress. *J Virol* **91:** e00295.

Walsh D, Mohr I. 2004. Phosphorylation of eIF4E by Mnk-1 enhances HSV-1 translation and replication in quiescent cells. *Genes Dev* **18:** 660–672.

Walsh D, Mohr I. 2006. Assembly of an active translation initiation factor complex by a viral protein. *Genes Dev* **20:** 461–472.

Walsh D, Mohr I. 2014. Coupling 40S ribosome recruitment to modification of a cap-binding initiation factor by eIF3 subunit e. *Genes Dev* **28:** 835–840.

Walsh D, Perez C, Notary J, Mohr I. 2005. Regulation of the translation initiation factor eIF4F by multiple mechanisms in human cytomegalovirus-infected cells. *J Virol* **79:** 8057–8064.

Walsh D, Arias C, Perez C, Halladin D, Escandon M, Ueda T, Watanabe-Fukunaga R, Fukunaga R, Mohr I. 2008. EukaryoticEukk translation initiation factor 4F architectural alterations accompany translation initiation factor redistributions in poxvirus-infected cells. *Mol Cell Biol* **28:** 2648–2658.

* Wek RC. 2018. Role of eIF2α kinases and translational control and adaptation to cellular stress. *Cold Spring Harb Perspect Biol* doi: 10.1101/cshperspect.a032870.

White SD, Jacobs BL. 2012. The amino terminus of the vaccinia virus E3 protein is necessary to inhibit the interferon response. *J Virol* **86:** 5895–5904.

White JP, Cardenas AM, Marissen WE, Lloyd RE. 2007. Inhibition of cytoplasmic mRNA stress granule formation by a viral proteinase. *Cell Host Microbe* **2:** 295–305.

Wilson JE, Pestova TV, Hellen CU, Sarnow P. 2000. Initiation of protein synthesis from the A site of the ribosome. *Cell* **102:** 511–520.

Wimmer E, Mueller S, Tumpey TM, Taubenberger JK. 2009. Synthetic viruses: A new opportunity to understand and prevent viral disease. *Nat Biotechnol* **27:** 1163–1172.

Wise HM, Barbezange C, Jagger BW, Dalton RM, Gog JR, Curran MD, Taubenberger JK, Anderson EC, Digard P. 2011. Overlapping signals for translational regulation and packaging of influenza A virus segment 2. *Nucleic Acids Res* **39:** 7775–7790.

Won S, Eidenschenk C, Arnold CN, Siggs OM, Sun L, Brandl K, Mullen TM, Nemerow GR, Moresco EM, Beutler B. 2012. Increased susceptibility to DNA virus infection in mice with a GCN2 mutation. *J Virol* **86:** 1802–1808.

Xi Q, Cuesta R, Schneider RJ. 2004. Tethering of eIF4G to adenoviral mRNAs by viral 100k protein drives ribosome shunting. *Genes Dev* **18:** 1997–2009.

Yamamoto H, Nakashima N, Ikeda Y, Uchiumi T. 2007. Binding mode of the first aminoacyl-tRNA in translation initiation mediated by *Plautia stali* intestine virus internal ribosome entry site. *J Biol Chem* **282:** 7770–7776.

Yang C, Skiena S, Futcher B, Mueller S, Wimmer E. 2013. Deliberate reduction of hemagglutinin and neuraminidase expression of influenza virus leads to an ultraprotective live vaccine in mice. *Proc Natl Acad Sci* **110:** 9481–9486.

Yang Z, Cao S, Martens CA, Porcella SF, Xie Z, Ma M, Shen B, Moss B. 2015. Deciphering poxvirus gene expression by RNA sequencing and ribosome profiling. *J Virol* **89:** 6874–6886.

Yarus M, Folley LS. 1985. Sense codons are found in specific contexts. *J Mol Biol* **182:** 529–540.

Young DF, Andrejeva J, Li X, Inesta-Vaquera F, Dong C, Cowling VH, Goodbourn S, Randall RE. 2016. Human IFIT1 inhibits mRNA translation of rubulaviruses but not other members of the paramyxoviridae family. *J Virol* **90:** 9446–9456.

Yu Y, Sweeney TR, Kafasla P, Jackson RJ, Pestova TV, Hellen CU. 2011. The mechanism of translation initiation on Aichivirus RNA mediated by a novel type of picornavirus IRES. *EMBO J* **30:** 4423–4436.

Yue Y, Liu J, He C. 2015. RNA N^6-methyladenosine methylation in post-transcriptional gene expression regulation. *Genes Dev* **29:** 1343–1355.

Yueh A, Schneider RJ. 2000. Translation by ribosome shunting on adenovirus and hsp70 mRNAs facilitated by complementarity to 18S rRNA. *Genes Dev* **14:** 414–421.

Zaborowska I, Kellner K, Henry M, Meleady P, Walsh D. 2012. Recruitment of host translation initiation factor eIF4G by the vaccinia virus ssDNA binding protein I3. *Virology* **425:** 11–22.

Zhang R, Jha BK, Ogden KM, Dong B, Zhao L, Elliott R, Patton JT, Silverman RH, Weiss SR. 2013. Homologous $2',5'$-phosphodiesterases from disparate RNA viruses antagonize antiviral innate immunity. *Proc Natl Acad Sci* **110:** 13114–13119.

Zhang H, Ng MY, Chen Y, Cooperman BS. 2016. Kinetics of initiating polypeptide elongation in an IRES-dependent system. *eLife* **5:** e13429.

Zinoviev A, Hellen CU, Pestova TV. 2015. Multiple mechanisms of reinitiation on bicistronic calicivirus mRNAs. *Mol Cell* **57:** 1059–1073.

Zust R, Cervantes-Barragan L, Habjan M, Maier R, Neuman BW, Ziebuhr J, Szretter KJ, Baker SC, Barchet W, Diamond MS, et al. 2011. Ribose $2'$-O-methylation provides a molecular signature for the distinction of self and non-self mRNA dependent on the RNA sensor Mda5. *Nat Immunol* **12:** 137–143.

Therapeutic Opportunities in Eukaryotic Translation

Jennifer Chu[1] and Jerry Pelletier[1,2,3]

[1]Department of Biochemistry, McGill University, Montreal, Quebec H3G 1Y6, Canada

[2]Department of Oncology, McGill University, Montreal, Quebec H3G 1Y6, Canada

[3]Rosalind and Morris Goodman Cancer Research Center, McGill University, Montreal, Quebec H3G 1Y6, Canada

Correspondence: jerry.pelletier@mcgill.ca

The ability to block biological processes with selective small molecules provides advantages distinct from most other experimental approaches. These include rapid time to onset, swift reversibility, ability to probe activities in manners that cannot be accessed by genetic means, and the potential to be further developed as therapeutic agents. Small molecule inhibitors can also be used to alter expression and activity without affecting the stoichiometry of interacting partners. These tenets have been especially evident in the field of translation. Small molecule inhibitors were instrumental in enabling investigators to capture short-lived complexes and characterize specific steps of protein synthesis. In addition, several drugs that are the mainstay of modern antimicrobial drug therapy are potent inhibitors of prokaryotic translation. Currently, there is much interest in targeting eukaryotic translation as decades of research have revealed that deregulated protein synthesis in cancer cells represents a targetable vulnerability. In addition to being potential therapeutics, small molecules that manipulate translation have also been shown to influence cognitive processes such as memory. In this review, we focus on small molecule modulators that target the eukaryotic translation initiation apparatus and provide an update on their potential application to the treatment of disease.

A large collection of chemical probes targeting various steps of prokaryotic and eukaryotic translation has been identified and characterized over the last ~50 years (Pestka 1977; Vazquez 1979; Pelletier and Peltz 2007; Malina et al. 2012). Early efforts in this endeavor were primed by the realization that prokaryotic and eukaryotic translations differ sufficiently to enable selective targeting, from which emerged some of the most effective antimicrobial therapies (aminoglycosides, tetracyclines, macrolides, chloramphenicol, clindamycin, spectinomycin, streptogramins, lincosamides, oxazolindinones, etc.). In the last 15 years, there has been a resurgence of interest in targeting translation driven by a number of medical needs. One is the multidrug resistance crisis currently facing the field of medical microbiology and prompting an urgent need to identify new antibiotics. Another is the realization that targeting the translation apparatus in parasites, such as malaria, offers a broad therapeutic window (Baragaña et al. 2015). A third is the recognition that translational control is frequently usurped in human can-

Cite this article as *Cold Spring Harb Perspect Biol* doi: 10.1101/cshperspect.a032995

cers and represents a druggable vulnerability (Silvera et al. 2010; Stumpf and Ruggero 2011; Bhat et al. 2015; Pelletier et al. 2015; Chu et al. 2016a). Our goal in this review is to visit ongoing developments aimed at targeting the two main regulatory nodes of eukaryotic translation initiation: the eIF4F-mediated ribosome recruitment step and the modulation of eIF2 activity. eIF4F and eIF2 link translational output to changes in extra- and intracellular cues and are critical actors in maintaining cellular homeostasis. For a detailed review of the roles of eIF4F and eIF2 in translation, the reader is referred to Merrick and Pavitt (2018).

THERAPEUTIC TARGETS IN TRANSLATIONAL CONTROL

Messenger RNA (mRNA) Activation by eIF4F

Cap-dependent translation is mediated by eIF4F, a heterotrimeric complex comprising eIF4E, eIF4A, and eIF4G, with the essential eIF4E cap-binding protein being rate-limiting for translation (Fig. 1). Under conditions of growth restriction, the bulk of eIF4E is sequestered away from eIF4F into translationally inactive complexes by eIF4E-binding proteins (there

are three isoforms), leading to reduced cap-dependent translation (Pause et al. 1994). The phosphoinositide 3-kinase (PI3K)/mechanistic target of rapamycin (mTOR) pathway is critical to cell growth and survival and responds to a multitude of extra- and intracellular cues. Increased signaling flux leads to elevated 4E-BP phosphorylation and dissociation of eIF4E:4E-BP complexes. Because 4E-BPs and eIF4G share a common binding site on eIF4E, 4E-BP phosphorylation leads to increased eIF4F complex formation. This results in elevated translation of some mRNAs, several of which have been implicated in tumor onset and maintenance (e.g., c-MYC, cyclins, ornithine decarboxylase [ODC], BCL-2, MCL-1, BCL-X_L, osteopontin, survivin, vascular endothelial growth factor [VEGF], fibroblast growth factors [FGF]) (Colina et al. 2008; Mills et al. 2008; Petroulakis et al. 2009; Dowling et al. 2010; Bhat et al. 2015; Chu et al. 2016a). There are a number of strategies that have been explored to abrogate eIF4F function, including disruption of the eIF4E:eIF4G association, perturbation of the eIF4E:cap interaction, and inhibition of eIF4A and eIF4G activity (Fig. 1).

In mammals, eIF4E activity is also regulated through phosphorylation of Ser 209 by MNK1

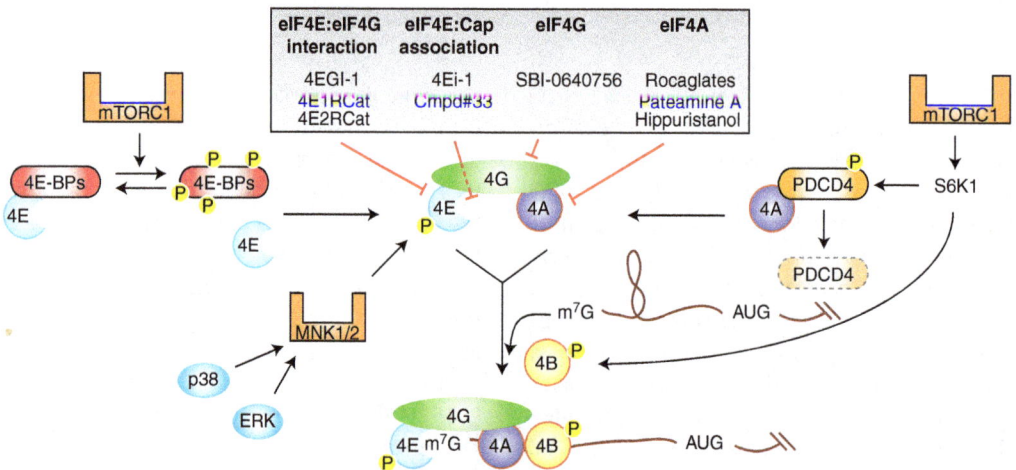

Figure 1. Schematic diagram outlining the regulation of eIF4F assembly by mechanistic target of rapamycin (mTOR), eIF4E phosphorylation by MNK1/2, and messenger (mRNA) recruitment by eIF4F. The targets of known inhibitors are denoted. mTOR complex 1 (mTORC1) is a complex composed of mTOR, Raptor, MLST8, PRAS40, and DEPTOR and links nutrient/redox/energy levels to translation initiation.

Cite this article as *Cold Spring Harb Perspect Biol* doi: 10.1101/cshperspect.a032995

(mitogen activated protein kinase [MAPK]-interacting kinases) and MNK2, two kinases that function downstream from the p38 MAPK and MEK-ERK (MAPK/ERK kinase-extracellular signal-regulated kinase) signaling cascades (Waskiewicz et al. 1999). In response to stress (p38 MAPK) or mitogens (MEK/ERK) (Cargnello and Roux 2011), the MNKs bind to eIF4G and phosphorylate eIF4E (Pyronnet et al. 1999). This causes preferential translation stimulation of a subset of mRNAs—several of which are known to be required for tumor maintenance (e.g., MCL1, MMP3, SNAIL) (Wendel et al. 2004; Furic et al. 2010; Robichaud et al. 2015).

The association and activity of eIF4A, the helicase subunit of eIF4F, is also affected by mTOR signaling flux through the S6 kinases (S6K1 and S6K2). Activation of S6K1 by mTOR leads to phosphorylation and degradation of programmed cell death 4 (PDCD4), an eIF4A-binding partner that represses translation (Fig. 1) (Yang et al. 2003a). Loss of PDCD4 results in increased eIF4A assembly into the eIF4F complex. S6K also phosphorylates the eIF4A auxiliary factor, eIF4B, which is an RNA-binding protein that promotes initiation complex formation and stimulates eIF4A helicase activity (Raught et al. 2004; Holz et al. 2005).

Ternary Complex Formation

The second major regulatory checkpoint in translation initiation is ternary complex (TC) formation (Fig. 2). Here, the initiator Met-tRNA$_i^{Met}$, eIF2, and guanosine triphosphate (GTP) form a TC that is recruited to 40S ribosomal subunits. Upon initiation codon recognition, GTP is hydrolyzed and eIF2-guanosine diphosphate (GDP) is released from the ribosome in complex with eIF5. The eIF2B guanine nucleotide exchange factor is required to exchange GDP for GTP as well as to displace eIF5 from eIF2 to enable its participation in subsequent rounds of initiation (Jennings and Pavitt 2014). In response to stress, the eIF2α subunit is phosphorylated at Ser51 and this increases its affinity for the limiting amounts of eIF2B, leading to sequestration of the latter and reduced eIF2-GDP/eIF2-GTP recycling (Jennings and Pavitt 2014). Decreasing TC availability hinders global translation, while permitting stimulation of translation of a set of mRNAs encoding regulators of the stress response (e.g., GCN4 in yeast, ATF4 in humans), which is an event dependent on appropriately positioned short, upstream open reading frames (ORFs) (see Wek 2018). There are four known kinases that phosphorylate eIF2α: (1) GCN2

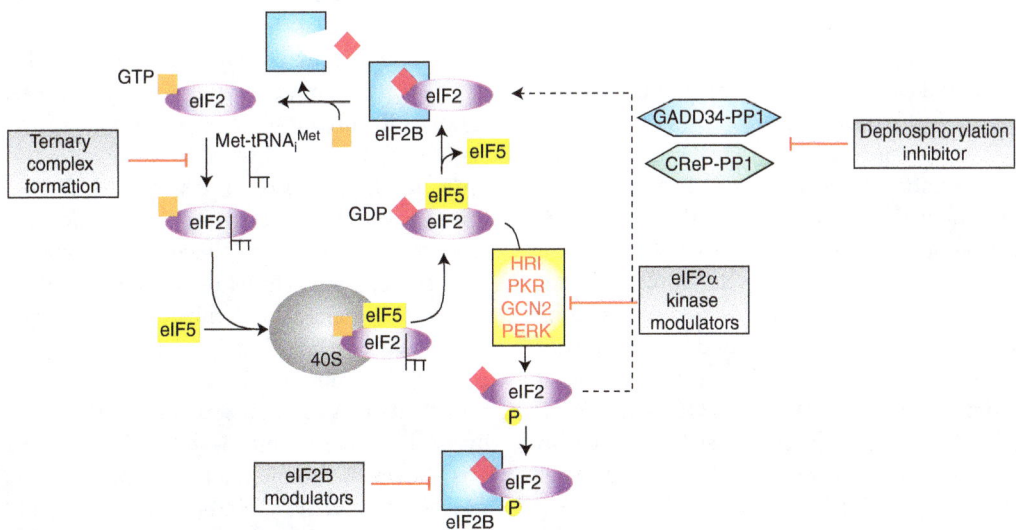

Figure 2. Schematic diagram outlining the regulation of eIF2 recycling and the different steps known to be modulated by small molecules.

(general control nonderepressible), (2) HRI (heme-regulated inhibitor), (3) PKR (double-stranded RNA-dependent protein kinase), and (4) PERK (PKR-like endoplasmic reticulum kinase). Dephosphorylation of eIF2α is mediated by protein phosphatase 1 (PP1) containing either GADD34 or CReP (Fig. 2). Small molecules targeting the eIF2 regulatory node have been identified that interfere with TC formation, modulate eIF2α kinase activity, block dephosphorylation of phospho-eIF2α, and perturb eIF2B activity (Fig. 2).

TRANSLATIONAL CONTROL AND DISEASE

Cancer

The discovery that translation is dysregulated in cancer cells emerged from early studies documenting a small (~6%) change (increase and decrease) in unique mRNA species associated with polyribosomes from resting compared with growing cells (Williams and Penman 1975; Getz et al. 1976). The subsequent demonstration that eIF4E overexpression could drive tumor initiation (Lazaris-Karatzas et al. 1990) put translational control on the cancer research map and opened a new therapeutic avenue. Subsequent studies highlighted the pervasive extent to which deregulated translation contributes to tumor initiation and maintenance (Robert and Pelletier 2009). First, changes in expression levels and activity of a number of initiation factors have been documented in human cancer cells and tumors, many of which are associated with tumor progression and poor prognosis (Silvera et al. 2010; Chu et al. 2016a). The nonoverlapping nature of the associated changes in mRNA translation suggests a complex layer of gene-expression regulation that remains to be systematically explored. Second, eIF4F activity is often altered by signaling pathways found activated in cancer (PI3K/mTOR, MEK-ERK, p38 MAPK) and this event has been linked to increased cell proliferation and survival (Bhat et al. 2015). Third, the quintessential c-MYC proto-oncogene regulates transcription of the genes encoding eIF2α, eIF4E, eIF4A1, and eIF4G1 (Rosenwald et al. 1993; Lin et al. 2008).

Because c-MYC is a well-characterized eIF4F-responsive mRNA (Lin et al. 2012; Chu et al. 2016b), this establishes a positive feedback loop. Given that MYC occupies a pinnacle position in the cellular gene expression hierarchy and is among the top amplified genes across all human cancer types (Beroukhim et al. 2010), one would expect the MYC-eIF4F loop to be tightly regulated in normal cells. Fourth, mRNA sequence variations that alter translational efficiency have been associated with tumor initiation or maintenance. These include point mutations in the 5′ leader region of c-MYC, BRCA1, and CDKN2A that affect protein output, as well as usage of alternative splice sites or transcription start sites that lead to significant nucleotide changes in the 5′ leader region (Meric and Hunt 2003). Last, increased eIF4F activity, altered eIF4E/4E-BP1 ratios, and persistent eIF4F complex levels can impart resistance to chemotherapy (e.g., doxorubicin [Wendel et al. 2004], dexamethasone [Robert et al. 2014], cisplatin, and antimitotic microtubule stabilizers [Zhou et al. 2011]) and to therapies targeting the PI3K/mTOR (Mallya et al. 2014) and MAPK pathways (Ilic et al. 2011; Zindy et al. 2011; Alain et al. 2012; Boussemart et al. 2014; Cope et al. 2014; Mallya et al. 2014). Collectively, these studies indicate that changes in translational control can have profound consequences for normal cellular physiology.

A number of in vitro studies have highlighted the value of targeting the eIF4E and eIF4A subunits to curtail tumor cell survival. (1) Early experiments using antisense approaches to suppress eIF4E expression in cell culture showed that this would block the growth of transformed cells, reduce soft agar colonization, increase tumor latency periods in xenograft models, reduce metastatic potential, and enhance chemosensitivity in vitro (Rinker-Schaeffer et al. 1993; Graff et al. 1995; Thumma et al. 2015). (2) Sequestration of eIF4E by ectopic expression of 4E-BP1 or by peptides harboring the eIF4E-binding site was shown to inhibit tumor cell proliferation (Rousseau et al. 1996; Herbert et al. 2000) and block eIF4E- and MYC-driven transformation ex vivo and in vivo in xenograft models (Jiang et al. 2003; Lynch et al. 2004). A fusion peptide containing the eIF4E-binding site linked to a gonadotropin receptor agonist displayed antineoplastic activity when delivered by intraperitoneal injection to an ovarian xenograft model, with no detectable toxicity (Ko et al. 2009). (3) Antisense suppression of eIF4A or sequestration of eIF4A by ectopic expression of PDCD4 in tumor cells reduced cellular proliferation (Eberle et al. 2002), blocked transformation (Yang et al. 2003b), and delayed tumor onset in a chemically induced murine skin tumor model (Jansen et al. 2005). Delivery of an expression vector driving expression of PDCD4 to H-Ras12V mice with liver cancer significantly reduced tumor burden (Kim et al. 2013).

Manipulating TC availability as an anticancer approach has not been as extensively assessed. One problem is that eIF2α phosphorylation can have either prosurvival consequences or trigger cell death, depending on context (Bi et al. 2005; Koromilas and Mounir 2013). For example, expression of a nonphosphorylatable eIF2α mutant (S51A) is capable of transforming NIH 3T3 cells (Donze et al. 1995). Phosphorylation of eIF2α is observed in response to treatment with different proteasome inhibitors and is associated with an apoptotic response (Jiang and Wek 2005a). However, phosphorylation of eIF2α in response to ultraviolet (UV) irradiation (mediated by GCN2) facilitates nuclear factor (NF)-κB activation and leads to a prosurvival response (Jiang and Wek 2005b). Thus, the issue of context has to be better understood to rationalize how best to manipulate this pathway to achieve a proapoptotic response when applied to the treatment of cancer cells.

Cognitive Functions

eIF2B exists as a dimer of two pentamers of five subunits (α–ε) (Kashiwagi et al. 2016). The α, β, and δ subunits make up the regulatory subcomplex, whereas the γ and ε subunits are responsible for the catalytic activity. Mutations in any of the five eIF2B subunits can lead to a progressive loss of brain white matter known as leukoencephalopathy with vanishing white matter (VWM) or childhood ataxia with central nervous system hypomyelination (CACH) (Ka-

shiwagi et al. 2016). These mutations impact on eIF2B assembly and/or affect eIF2B: eIF2 interactions, both of which diminish TC formation and reduce translational output. This disorder highlights the critical role that eIF2B plays in nervous system development and maintenance.

Perturbing TC formation through modifying eIF2α phosphorylation status can also affect cognitive function. In eIF2$^{+/S51A}$ mice in which eIF2α phosphorylation is reduced, long-term potentiation and memory are enhanced (Costa-Mattioli et al. 2007). Genetic deletion of PERK or GCN2 in mouse models of Alzheimer's disease (AD) alleviates deficits in protein synthesis, synaptic plasticity, and memory (Ma et al. 2013). Compounds that alter eIF2α phosphorylation in the brain also impact cognitive functions. Minocycline (a semisynthetic tetracycline) can attenuate eIF2α phosphorylation in an AD mouse model and this is associated with improved learning and memory (Choi et al. 2007). Trazodone hydrochloride and dibenzoylmethane, two compounds identified for their ability to reverse eIF2α phosphorylation, can block neurodegeneration, restore memory deficits, and improve survival in prion-disease mice (Halliday et al. 2017). These and additional studies have implicated the eIF2α kinases in cognitive functions and offer new and exciting avenues to potentially treat neurodegenerative disorders as well as alter cognitive functions (Trinh and Klann 2013).

Dysregulated translation has also been implicated as a core deficit in fragile X syndrome, with hyperactivation of mTORC1 and MAPK signaling associated with behavioral issues, learning deficits, and developmental delays (Hou et al. 2006; Gkogkas et al. 2014). Remarkably, metformin treatment of Fmr1$^{-/-}$ mice ameliorated the core deficits in this model and this effect was associated with normalization of ERK signaling and reduced eIF4E phosphorylation (Gantois et al. 2017).

IS TARGETING TRANSLATION INITIATION SAFE?

Targeting translation as an antineoplastic approach is already in the clinic, as exemplified by the use of omacetaxine mepesuccinate, a semisynthetic formulation of homoharringtonine (HHT), against tyrosine kinase inhibitor-resistant chronic myeloid leukemia (CML). HHT inhibits formation of the first peptide bond (Garreau de Loubresse et al. 2014) and, therefore, is quite different from a targeted therapy in which only one tumor cell dependency is blocked. A significant percent of patients with CML show a major cytogenetic response (i.e., reduction in Philadelphia chromosome-positive cells) when administered omacetaxine mepesuccinate (Alvandi et al. 2014). Several adverse reactions were reported (Alvandi et al. 2014), including some that were considered serious but manageable. HHT will block synthesis of all proteins in transformed and normal cells and the selectivity of HHT for CML versus normal cells likely stems from inhibition of synthesis of short-lived oncoproteins to which tumor cells have become addicted, such as MCL-1 and c-MYC (Chen et al. 2006; Robert et al. 2009a).

Unlike the global inhibitory effects on translation exerted in cells treated with HHT, suppression of eIF4E is expected to lead to a more selective reduction in translation of a subset of mRNAs. The preclinical data to date suggest that partial suppression of eIF4E at the organismal level is well tolerated. In mice engineered to express inducible eIF4E short hairpin RNAs (shRNAs) (*sh4E/rtTA*) in all tissues, transient eIF4E suppression affected only a subset of normal regenerating cells (e.g., immature/undifferentiated intestinal crypts), but was otherwise well tolerated and the aforementioned effects were rapidly and completely reversible (Lin et al. 2012). Systemic suppression of *sh4E/rtTA* mice also carrying an Eμ-Myc translocation significantly delayed B-cell tumor onset (Lin et al. 2012). Similarly, eIF4E$^{+/-}$ mice showed no developmental defects and had normal body weights and survival (Truitt et al. 2015). When eIF4E$^{+/-}$ mice were crossed to a KRas^{G12D}-driven lung cancer model, eIF4E$^{+/-}$/KRas^{G12D} mice showed reduced tumor burden compared with eIF4E$^{+/+}$/KRas^{G12D} mice (Truitt et al. 2015). Telling are the consequences of suppressing eIF4E expression systemically using antisense oligonucleotides (ASOs). Here, an eIF4E ASO that targets the 3′ untranslated region (LY2275796; GenBank: M15353, nucleotides 1285–1304) was developed and tested in vivo (Graff et al. 2007). When administered at 50 mg/kg thrice weekly in an MDA-MB-231 breast cancer xenograft model, a 64% reduction in eIF4E protein levels was documented in tumors and was associated with a pronounced suppression of tumor outgrowth. At 25 mg/kg thrice weekly (after an initial loading dose of 50 mg/kg) in a PC3 prostate cancer xenograft model, a 56% reduction in eIF4E protein levels was documented in tumors and was associated with a significant reduction in tumor outgrowth. No change in mean body weight was observed in the MDA-MB-231 or PC3 models. Lower concentrations of LY2275796 (5 mg/kg) had no effect on PC3 tumor outgrowth and failed to reduce eIF4E proteins in the tumors. Normal mice dosed with LY2275796 (40 mg/kg twice weekly; 80 mg/kg/wk) showed an 80% decrease in eIF4E protein levels in the liver (the site where ASOs preferentially accumulate) with no appreciable change in liver and spleen histology, body weight, or liver transaminase levels detected (Graff et al. 2007).

Two clinical trials with LY2275796 have been undertaken but they failed to satisfactorily address the issue of effectiveness of eIF4E suppression. A phase I trial primarily sought to determine an appropriate delivery dose for LY2275796 in patients (Hong et al. 2011). In this study, LY2275796 was delivered for 3 consecutive days followed by weekly maintenance doses (Hong et al. 2011). The maximum tolerated dose was designated as 1000 mg based on clinical, pharmacokinetic, and pharmacodynamics observations with patients having manageable toxicities (grade 1/2). This dose was reported to cause an 80% reduction in eIF4E mRNA levels (as determined using a branched DNA assay), but this could not be attributed to a specific effect of LY2275796 because expression of housekeeping genes was also reduced (by 64%). In pre- and posttreatment biopsy samples, cytoplasmic eIF4E protein levels were assessed by immunohistochemistry and reported to be reduced by ~25%. Cytoplasmic VEGF and nuclear cyclin D1 protein levels (two eIF4F-responsive mRNAs) were unaffected or reduced

~25%, respectively. The best response observed was seven patients out of 22 that achieved stable disease for a minimum of 6 weeks. We note that in this study, in which the median patient age was 58.5 years (and based on an average weight of 76 kg (see shortsupport.org/Research/Papers/NHANES.pdf), patients received ~52 mg/kg/wk (first week) and 13 mg/kg/wk (maintenance dose) of LY2275796, which is significantly lower than the doses used to achieve an antitumor response in the aforementioned mouse xenograft studies.

A second clinical trial looked at combining the 4E-ASO with irinotecan in solid tumors and irinotecan-refractory colorectal cancer (Duffy et al. 2016). The combination of LY2275796 (1000 mg/wk; 13 mg/kg/wk) and irinotecan was not well tolerated because of chronic low-grade toxicities (Duffy et al. 2016) and appearance of more severe toxicities (grade 3/4). This likely stemmed from an LY2275796: irinotecan drug interaction that increased the half-life of irinotecan and its metabolites leading to an increased drug exposure. Analysis of post- versus pretreated tumor biopsies found a reduction (~10%–75%) in eIF4E mRNA levels in 5/9 of patients, but no change in eIF4E protein levels (as assessed by immunohistochemistry). Overall, although the preclinical studies are encouraging, we cannot make general conclusions regarding the effectiveness of suppressing eIF4E and more robust and reliable approaches are needed to inhibit eIF4E/eIF4F activity in the clinical setting.

INHIBITING eIF4E

Targeting the eIF4E–Cap Interaction

Cap analogs have been used for the last 35 years to inhibit eIF4E cap-binding activity (primarily in in vitro biochemical assays), but assessment of their potential as antineoplastic agents is hindered by poor cell permeability and in vivo stability (Wagner et al. 2000). Cocrystal structures of eIF4E with cap analogs have defined the key features critical for cap recognition (Marcotrigiano et al. 1997, 1999; Niedzwiecka et al. 2002; Tomoo et al. 2002). These include the

delocalized positive charge of the 7-methyl guanosine ring, which forms cation-π stacking interactions with Trp56 and Trp102, a hydrogen-bonding interaction between the guanine carbonyl group and the backbone amide of Trp102, hydrogen bond engagement of guanine N1 and N2 with Glu103, and a hydrogen bonding network between phosphates, and mediated through water, with Arg112, Arg157, and Lys162 (Fig. 3A). This latter feature explains the higher binding affinity of m^{7}GTP for eIF4E relative to m^{7}GDP (39-fold) and m^{7}GMP (407-fold). The eIF4E cap-binding pocket has three cavities that can accommodate extensions from the N7 position, and this explains why compounds that harbor bulky N7 substituents, like N7-benzyl GMP and 7-(p-fluorobenzyl)-GMP, still function as cap analogs (Brown et al. 2007). This insight was used to design a novel guanine analog (Cmpd #33) that showed a 147-fold increase in binding affinity for eIF4E compared with m^{7}GMP (Fig. 3B) (Chen et al. 2012b). Unfortunately, Cmpd #33 suffers from poor cellular permeability.

To overcome this liability, Wagner and colleagues (Wagner et al. 2000) developed 4Ei-1, an N7-benzyl GMP tryptamine phosphoramidate pronucelotide (Fig. 3B) that is cell permeable and converts to the biologically active compound, 7-benzyl guanosine monophosphate in cells (Ghosh et al. 2009). 4Ei-1 has been shown to block the epithelial-to-mesenchymal transition in zebrafish (Ghosh et al. 2009) and in lung epithelial cells by inhibiting translation of Snail1 (Smith et al. 2015). It also chemosensitizes lung cancer cells to gemcitabine (Li et al. 2013) and mesothelioma cells to pemetrexed (Chen et al. 2014).

Targeting the eIF4E:eIF4G Interaction

Two independently conducted high throughput screens led to the discovery of three eIF4E:eIF4G interaction inhibitors: 4EGI-1, 4E1RCat, and 4E2RCat (Fig. 3C) (Moerke et al. 2007; Cencic et al. 2011a). Compound 4EGI-1 (which exists as two interconverting isomeric forms, E and Z) binds to eIF4E at a location removed from the eIF4G-binding site and causes local conforma-

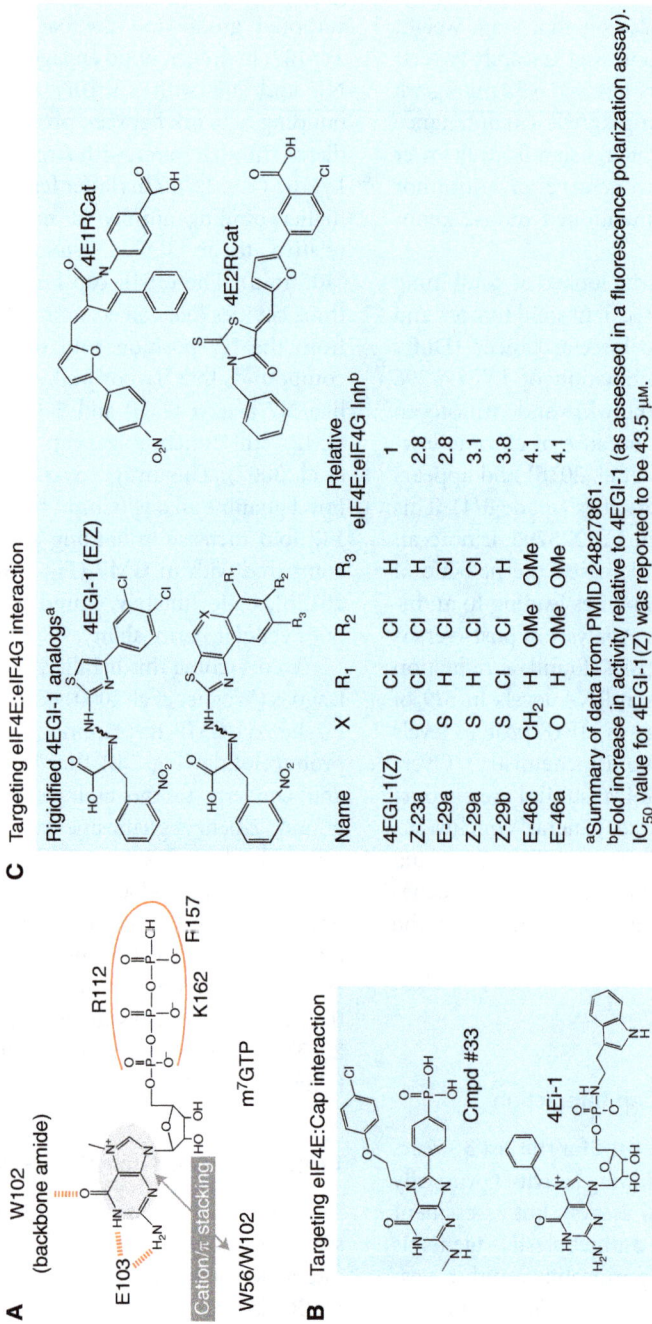

Figure 3. Small molecule inhibitors of eIF4E. (*A*) Interactions between m^{7}GTP and eIF4E are denoted. (*B*) Chemical inhibitors of eIF4E:cap interaction. (*C*) Chemical inhibitors of eIF4E:eIF4G interaction.

tional changes that lead to disengagement of eIF4G from eIF4E (Papadopoulos et al. 2014). An unexpected property of 4EGI-1 is that it increases 4E-BP1 binding to eIF4E in cells (Moerke et al. 2007). 4EGI-1 inhibits expression of known eIF4F-responsive mRNAs, such as cyclins D1 and E, c-MYC, and BCL-2 (Moerke et al. 2007), and blocks the hypoxia-induced gene-activation program (Yi et al. 2013). It selectively kills transformed cells more than nontransformed cells (Moerke et al. 2007; Tamburini et al. 2009) and shows efficacy in several cancer models, including U87 glioma, breast, and melanoma xenograft models (Chen et al. 2012a; Descamps et al. 2012; Wu et al. 2016) and against multiple myeloma (Descamps et al. 2012) and melanoma cell lines (Mahalingam et al. 2014). 4EGI-1 also impairs mitochondrial function (Yang et al. 2015). This compound has been used in studies to probe the role of eIF4F in memory, autism, and viral replication (Hoeffer et al. 2011; McMahon et al. 2011; Gkogkas et al. 2013; Santini et al. 2013). Rigidified variants of 4EGI-1 have been synthesized and characterized, with some capable of better inhibiting the eIF4E:eIF4G interaction than the parental compound (Fig. 3C) (Mahalingam et al. 2014).

4E1RCat and 4E2RCat prevent the association of both eIF4G and 4EBP1 with eIF4E (Fig. 3C) (Cencic et al. 2011a). 4E1RCat dampened protein synthesis by ~30% in Jurkat cells (Cencic et al. 2011a). 4E1RCat is pharmacologically active in mice and was shown to synergize with doxorubicin in a mouse lymphoma model, prolonging tumor-free survival compared with single-agent-treated cohorts (Cencic et al. 2011a). 4E1RCat blocks mitosis-restricted protein synthesis (Shuda et al. 2015) and decreases MCL-1 levels in HL-1 cardiomyocytes (Arnold et al. 2014). 4E1RCat and 4E2RCat have been used to block coronavirus replication (Cencic et al. 2011b).

Targeting eIF4A

There are two related eIF4A homologs in mammals, eIF4A1 and eIF4A2, which share 90% identity at the amino acid level (Nielsen and Trachsel 1988; Galicia-Vazquez et al. 2012). Accurate quantitation of the two proteins has not been extensively performed across a large number of cell or tissue types. At the mRNA level, eIF4A1 is generally more abundantly expressed with notable exceptions being in fetal brain, heart, and kidney and adult brain, ovary, and skeletal muscle (Galicia-Vazquez et al. 2012). eIF4A is an abundant factor (about three to six molecules/ribosome) with the majority (~90%) existing outside of the eIF4F complex. Both eIF4A1 and eIF4A2 can assemble into the eIF4F complex. eIF4A1 suppression is not well tolerated, whereas eIF4A2 can be completely eliminated in NIH 3T3 fibroblast cells using CRISPR/Cas9 with no detectable effects on protein synthesis rates, proliferation, and viability. An analysis of the viability of 398 cancer cells lines in which eIF4A1 or eIF4A2 expression was suppressed using shRNAs (21 shRNAs targeting eIF4A1 and 19 shRNAs targeting eIF4A2 were used in this study) is consistent with eIF4A1, but not eIF4A2, being essential (McDonald et al. 2017). Three natural products, pateamine A (PatA), hippuristanol, and rocaglate family members have been identified as selective and potent eIF4A inhibitors.

Pateamine A

PatA is a naturally occurring metabolite isolated from the marine sponge *Mycale hentscheli* and was originally found to have potent cytotoxic activity against P388 leukemia cells (half-maximal inhibitory concentration, $IC_{50} = 0.27$ nM) (Northcote et al. 1991). Initial efforts to identify the molecular target of PatA involved affinity chromatography of total cell extracts prepared from HL60 cells using immobilized PatA, in which cytokeratin, tubulin, eIF4A1, eIF4A2, and eIF4A3 (DDX48) were retained (Bordeleau et al. 2005). In a separate set of experiments, also using immobilized PatA, eIF4A1 and serine/threonine kinase receptor associated protein (STRAP) were captured from RKO colorectal carcinoma cell extracts (Low et al. 2005). Overexpression of eIF4A1, but not STRAP, in HeLa cells increased the PatA IC_{50} fivefold suggesting that inhibition of eIF4A is primarily responsible for PatA's ability to block cell proliferation (Low et al. 2005). In addition to inhibiting translation through targeting eIF4A1 and/or eIF4A2, PatA also impedes nonsense-mediated mRNA decay by targeting eIF4A3, a core component of the exon junction complex (Dang et al. 2006). PatA is not a pan-helicase inhibitor as it does not block Ded1p helicase activity, in vitro splicing reactions, or cellular DNA and RNA synthesis (Bordeleau et al. 2005). Moreover, PatA does not inhibit prokaryotic translation (Bordeleau et al. 2005).

With respect to its mechanism of action, PatA increases the ATP hydrolysis rate and RNA-binding ability of free eIF4A (Bordeleau et al. 2005; Low et al. 2005), but not if eIF4A is present in the eIF4F complex (Bordeleau et al. 2005), suggesting that binding of eIF4A to eIF4G might occlude PatA binding. Exposure of cells to PatA reduces the levels of eIF4A1 found within the eIF4F complex (Low et al. 2005). On the other hand, high concentrations of PatA (10 μM) have been reported to stimulate the interaction between eIF4A and eIF4B (Low et al. 2005), but this may be an indirect effect mediated by a bridging RNA fragment (Bordeleau et al. 2006b). Taken together, the current body of evidence depicts PatA as a compound that forces nonspecific eIF4A-RNA engagements, depleting eIF4A from the eIF4F complex with concomitant inhibition of cap-dependent translation (Bordeleau et al. 2006b). PatA blocks 48S complex formation in vitro and PatA-exposed cells have reduced polysomes (Bordeleau et al. 2005). The hepatitis C virus internal ribosome entry site (HCV IRES), which does not require any of the eIF4F subunits for initiation, is significantly more resistant to inhibition by PatA (Bordeleau et al. 2005; Low et al. 2005). As an inhibitor of translation, PatA also induces the assembly of stress granules (Dang et al. 2006; Mazroui et al. 2006). Although the PatA-binding site on eIF4A has yet to be determined, it may be located within the eIF4A amino-terminal domain (NTD) because a carboxy-terminal deletion of eIF4A (Δ246–406) is still capable of binding PatA (Low et al. 2007).

The total chemical synthesis of PatA has been reported and second-generation analogs

have been prepared (Romo et al. 2004; Kuznetsov et al. 2009; Low et al. 2014). One analog, DMDA-PatA, has comparable cytotoxicity and activity toward eIF4A as PatA. PatA is an irreversible inhibitor of translation, possibly caused by the presence of a Michael addition site on this compound (Bordeleau et al. 2005; Low et al. 2005) and analogs lacking this feature are no longer cytotoxic (Low et al. 2014). PatA and DMDA-PatA inhibit proliferation and induce apoptosis in many tumor cell lines at nanomolar concentrations ex vivo as well as in melanoma xenograft models (Northcote et al. 1991; Hood et al. 2001; Low et al. 2005; Kuznetsov et al. 2009). DMDA-PatA is not a substrate for multidrug resistance protein 1 (MDR1) (Kuznetsov et al. 2009), a protein in which overexpression is associated with intrinsic and acquired drug resistance (Hodges et al. 2011).

Hippuristanol

Hippuristanol is a polyoxygenated steroid first isolated from the gorgonian *Isis hippuris* (Higa and Tanaka 1981; Higa et al. 1981). It prevents eIF4A from interacting with RNA without affecting ATP binding (Bordeleau et al. 2006a). The binding site for hippuristanol resides in the eIF4A carboxy-terminal domain (CTD) and involves amino acids spanning, and adjacent to, motifs V and VI, which are two regions that are conserved among DEAD-box RNA helicases and that participate in RNA and ATP binding (Bordeleau et al. 2006a; Lindqvist et al. 2008). The amino acids flanking motifs V and VI are not conserved among other DEAD-box helicases, providing an explanation for the specificity of hippuristanol toward eIF4A.

Hippuristanol has been used to link eIF4A-dependent initiation events to long-term synaptic potentiation (Ran et al. 2009; Hoeffer et al. 2013), herpes simplex virus 1 host shutoff (Dauber et al. 2011; Radtke et al. 2013), and influenza virus mRNA translation (Yánguez et al. 2011). Hippuristanol shows single-agent activity against DBA/MC fibrosarcoma cells (Higa et al. 1981), multiple myeloma cells (Robert et al. 2014), Burkitt lymphoma cells (Cencic et al. 2013), adult T-cell leukemia cells in vitro

Cite this article as *Cold Spring Harb Perspect Biol* doi: 10.1101/cshperspect.a032995

and in vivo (Tsumuraya et al. 2011), and against lymphocytic leukemia P-388 tumors in mice (Higa et al. 1981). Combining hippuristanol with vemurafenib (anti-B-Raf therapy) (Boussemart et al. 2014), ABT-737 (Cencic et al. 2013), doxorubicin (Cencic et al. 2013), or dexamethasone (Robert et al. 2014) overcomes resistance to the initial therapy, an effect that is also obtained with rocaglates (see below), implying a role for eIF4A-dependent translation in drug resistance.

Several synthetic routes to hippuristanol have been reported (Li et al. 2009; Ravindar et al. 2010, 2011; Somaiah et al. 2014). These provided some insight into structure–activity relationships (SAR) with respect to translation inhibition (Li et al. 2009; Somaiah et al. 2014) and cellular cytotoxicity (Li et al. 2009). However, our understanding of hippuristanol SAR is still limited because of an inability to synthetically access functionality at a large number of locations on this molecule. To date, no analog has been made that has equivalent or greater potency toward inhibiting translation than hippuristanol.

Rocaglates

Rocaglates originated from plants of the *Aglaia* genus. They interact directly with eIF4A as determined by affinity chromatography (Chambers et al. 2013), differential scanning fluorimetry (Chu et al. 2016b), and cellular thermal shift assay (CETSA) (Chu et al. 2016b). A significant body of work has shown that rocaglates possess single-agent activity in a variety of cancer models, reverse the chemoresistance phenotype in a number of settings (including toward PI3K/mTOR and RAS/MAPK pathway inhibitors), and also affect tumor-supportive processes such as angiogenesis (reviewed in Bhat et al. 2015; Chu and Pelletier 2015). Importantly, through the generation of an eIF4A rocaglate-resistant allele using CRISPR/Cas9, the in vivo antiproliferative property of this class of compounds toward tumor cells has been shown to be a direct consequence of eIF4A1 inhibition (Chu et al. 2016b).

There has been significant interest in identifying mRNA features that modulate rocaglate

sensitivity. Ribosome profiling analysis of MDA-MB-231 breast cancer cells treated for 2 h with 25 nM silvestrol yielded modest repression of global translation, with only 3.3% of translating mRNAs being markedly inhibited (Rubio et al. 2014). These mRNAs had 5′ leaders with longer, structured sequences and relatively low overall guanine-cytosine (GC) content compared with nonresponsive mRNAs (Rubio et al. 2014). Ribosome profiling analysis of KOPT-K1 T acute lymphoblastic leukemia cells exposed to 25 nM silvestrol for 45 min also identified 5′ leader length as a silvestrol responsive feature (Wolfe et al. 2014). In addition, ~1/3 of silvestrol-responsive mRNAs harbored a putative G-quadruplex, a feature previously shown to confer rocaglate responsiveness (Cencic et al. 2009). A third study examined the consequences on translation in HEK 293 cells treated with the rocaglate, RocA (Iwasaki et al. 2016), and found that RocA-responsiveness did not correlate with the thermodynamic stability of the 5′ leader or the presence of G-quadruplexes (Iwasaki et al. 2016). Using a Bind-n-Seq approach, a high-throughput method by which to identify in vitro protein–RNA interactions, revealed that RocA leads to clamping of eIF4A onto RNA-containing polypurine sequences (4–6 nts in length). In vitro experiments revealed that this clamping, when it occurs on mRNAs, can lead to reduced initiation events at start codons residing downstream from the polypurine stretch, suggesting impaired ribosome scanning. Rocaglates can also block eIF4F activity because cells exposed to these have reduced levels of eIF4A in the eIF4F complex, an effect that will block 48S pre-initiation complex formation (Bordeleau et al. 2008; Cencic et al. 2009).

Targeting eIF4G

eIF4G remains a largely unexplored drug target. Virtual docking experiments aimed at identifying novel inhibitors of AKT led to the discovery of BI-69A11, a compound that unexpectedly was also found to bind to eIF4G1 following mass spectrometry analysis of proteins from UACC903 cells retained using biotinylated BI-69A11 (Gaitonde et al. 2009; Feng et al. 2015).

The pursuit of analogs with better pharmacokinetic properties led to the development of SBI-0640756 (SBI-756) (Feng et al. 2015). Isolation of eIF4F by cap-analog affinity chromatography from mouse embryonic fibroblast (MEF) cell extracts revealed partial loss of eIF4G in SBI-756-treated cells. SBI-756 inhibits AKT and NF-κB activity and blocks growth of NF1-mutant, NRas-mutant, and BRAF-resistant melanoma cells (Feng et al. 2015). Nontransformed fibroblasts are threefold less sensitive than A375 or UACC903 melanoma cells to the cytotoxic effects of SBI-756. It will be important to determine to what extent these biological effects are a consequence of eIF4G perturbation versus other activities that this compound may have.

INHIBITING eIF2

Perturbing eIF2 Activity

Blocking Dephosphorylation of Phospho-eIF2α

Ser/Thr protein phosphatase 1 can be recruited to dephosphorylate eIF2α by either GADD34 or CReP, the two proteins shown to blunt the endoplasmic reticulum (ER) stress response (Novoa et al. 2001; Jousse et al. 2003). Because GADD34 enhances the toxicity of ER stress and GADD34$^{-/-}$ mice develop normally (Marciniak et al. 2004), it was postulated that GADD34 inhibition might be of therapeutic benefit. To this end, a screen was undertaken in search of compounds (19,000 molecules tested) that could protect rat PC12 cells against ER stress-induced apoptosis (Boyce et al. 2005). Salubrinal, and a more potent and soluble derivative (Sal003) (Robert et al. 2006), were found to prevent the dephosphorylation of phospho-eIF2α, likely by blocking the conserved PP1-binding domain of GADD34 and CReP (Boyce et al. 2005; Fullwood et al. 2012). These compounds have been used to show protective effects in a number of neurodegenerative disease models driven by ER stress that include (1) amyloid β-protein aggregation (Hoozemans et al. 2009; Lee et al. 2010); (2) mutant Huntingtin aggregation (Reijonen et al. 2008); (3) misexpression of α-synuclein seen in Parkinson's disease (Smith et al. 2005); and (4) polyubiquitinated protein build up in a transgenic model of amyotrophic lateral sclerosis (Saxena et al. 2009). These results are difficult to reconcile with those of genetic experiments in which loss of PERK in mice expressing a familial AD-related mutation in APP and PSEN1 prevented deficits in synaptic plasticity and spatial memory associated with this model (Ma et al. 2013). It is possible that the differences are caused by salubrinal having additional activity (which has not been systematically assessed), to the timing of changes in eIF2α phosphorylation status (e.g., loss of PERK before onset of AD pathology versus maintenance of phospho-eIF2α levels during ER stress), or to differences in the models used. Salubrinal has been shown capable of resensitizing multiple myeloma cells to bortezomib (Schewe and Aguirre-Ghiso 2009). Current compound liabilities include low solubility and poor bioavailability (Long et al. 2005).

Guanabenz is a drug used to treat hypertension and acts as an α2-adrenergic agonist. It binds to GADD34, but not CReP, and disrupts its interaction with PP1α. Guanabenz is capable of rescuing cells from protein misfolding stress (Tsaytler et al. 2011) and blocks the proliferation, survival, and invasion of 4T1 and MDA-MB-231 mammary tumor cells (Hamamura et al. 2014). These results are exciting given the clinical experience with Guanabez and its known safety profile and are encouraging for future strategies aimed at pharmacologically modulating ER stress.

Blocking eIF2α Function

Another manner by which eIF2 function can be blocked is highlighted by experiments with NSC119889 and NSC119893, two fluorescein derivatives that prevent the binding of Met-tRNA$_i^{Met}$ to eIF2 (Robert et al. 2006). These compounds were used to show a reduced requirement of the HCV IRES for TCs (Robert et al. 2006); the HCV IRES can use eIF5B, eIF3, and eIF2A for translation initiation under stress (Terenin et al. 2008; Kim et al. 2011).

Targeting eIF2B

Using a luciferase reporter-based screen, Sidrauski et al. (2012) identified ISRIB (integrated stress response inhibitor), a compound that inhibits PERK-mediated activation of ATF4 translation. The *trans*-isomer ISRIB-A1 is 100-fold more potent than its *cis*-isomer (Sidrauski et al. 2015a). ISRIB-A1 does not diminish PERK phosphorylation (if anything, an increase is observed) nor does it affect eIF2α phosphorylation in ER-stressed cells (Sidrauski et al. 2015a). Yet, global translation rates are sustained in ISRIB-treated cells in the face of ER stress or when the eIF2α S51D phosphomimetic allele is introduced into cells. In fact, ISRIB blocks the integrated stress response (ISR) triggered by all eIF2α kinases (Sidrauski et al. 2012). Memory consolidation and enhancement are limited by the ISR and ISRIB-treated mice display enhanced spatial and fear-associated learning (Sidrauski et al. 2012). ISRIB completely reverses the translational effects elicited by eIF2α phosphorylation and induces no major change in mRNA translation or levels in treated cells (Sidrauski et al. 2015b). At the molecular level, ISRIB likely overcomes the effects of eIF2α phosphorylation by acting on eIF2B. ISRIB causes dimerization of eIF2B, enhances the thermostability of the eIF2B δ subunit in the CETSA assay, and enhances the GEF activity of eIF2B (Sidrauski et al. 2015a). Amino acid changes within the amino terminus of the eIF2B δ subunit confer ISRIB resistance (Sekine et al. 2015). Analogs with higher potencies are currently sought after and, one such compound, ISRIB-A17, has been found to be 10-fold more potent than the parental ISRIB-A1 compound (Sidrauski et al. 2015a; Hearn et al. 2016).

Modulating eIF2α Kinase Activity

Many compounds have been identified that cause an increase in phospho-eIF2α levels, but data are lacking regarding the specificity of these compounds and there is a paucity of genetic information linking their biological activity to changes in phospho-eIF2α levels. For example, the flavonoids, genisten, and quercetin, are cytotoxic to tumor cells in vitro and induce eIF2α phosphorylation at high concentrations (100 μM) (Ito et al. 1999). However, supporting genetic data are needed to link changes in phospho-eIF2α levels to the biological activity of these (and other) compounds, especially because genisten (Yan et al. 2010) and quercetin (Boly et al. 2011) are known tyrosine and serine/threonine kinase inhibitors and exert profound effects on the phosphoproteome.

1. Heme-regulated inhibitor (HRI). There are two known modulators of HRI activity: aminopyrazolindane and *N,N′*-diarylureas. Aminopyrazolindane was identified as an inhibitor from a screen reporting on HRI kinase activity (Garcia-Villegas et al. 2009). This compound suffers from low bioavailability and very rapid clearance in rats (Garcia-Villegas et al. 2009). A screen of 102,000 small molecules identified *N,N′*-diarylureas as a class of compounds capable of activating HRI and inducing eIF2α phosphorylation (Chen et al. 2011; Denoyelle et al. 2012). *N,N′*-diarylureas interact directly with HRI, as determined by nuclear magnetic resonance (NMR) and by drug affinity–responsive target stability (DARTS), an assay that assesses small molecule binding to target by monitoring protection against proteolytic degradation (Chen et al. 2011). *N,N′*-diarylureas reduce proliferation of tumor cells, which is an effect linked to enhanced eIF2α phosphorylation because ectopic expression of eIF2α^{S51A} rescued the proliferative block. A subsequent screen of a ~1900 member *N,N′*-disubstituted urea library identified a compound (*N*-aryl, *N′*-cyclohexylphenoxyurea) that significantly increased eIF2α phosphorylation and CHOP synthesis (Chen et al. 2013). We note that because the diarylurea scaffold is a widely used framework for the design of receptor tyrosine kinase (RTK) inhibitors (Shan et al. 2016), it will be important to link the biological activity of new compounds to changes in eIF2α phosphorylation.

2. PKR. There have been a number of PKR inhibitors described in the literature. The compound 2-amino purine was the first reported

inhibitor of PKR, but is neither very potent nor selective (Huang and Schneider 1990). Another inhibitor (Cmpd #16) was isolated by the Beal Laboratory (Jammi et al. 2003) from a collection of ATP site inhibitors and shown to be capable of rescuing a translational block imposed in vitro using a recombinant glutathione S-transferase (GST) fusion containing the PKR kinase domain. This compound is able to rescue translation inhibition caused by PKR in cells (Eley et al. 2007). In a separate study, a collection of kinase inhibitors was screened for their ability to block PKR activity (Bryk et al. 2011). Pharmacophore modeling of these hits (and of those from the Beal Laboratory screen [Jammi et al. 2003]), followed by an in silico screen of 15 million compounds identified several novel compounds that showed activity toward PKR, although their selectivity for PKR over other kinases remains to be established.

3. GCN2. An autophosphorylation assay was used to identify GCN2 kinase domain inhibitors. From a collection of kinase inhibitors, three compounds (indirubine-3′-monoxime, SP600124, and staurosporine) were identified (Robert et al. 2009b). Evaluation of 14 different indirubine-3′-monoxime analogs identified Syk1 (spleen tyrosine kinase inhibitor) as having equipotent activity as the parental compound. Inhibition of GCN2 in cells with these compounds showed a protective effect against UV-induced eIF2α phosphorylation (Deng et al. 2002).

4. PERK. A phenotype-based screen was implemented based on ectopic expression of a CHOP-luciferase fusion mRNA in CHO cells designed to report on ISR induction (Harding et al. 2005). Two hits were identified from this screen, TGD31 and TGD45, both of which were subsequently benzoylated to improve cell permeability, leading to the generation of TGD31 and TG45BZ. Treatment of cells with TG45BZ causes ER stress and activation of PERK, but not of GCN2 or PKR (HRI activation could not be assessed because of its low expression level in fibro-

blasts). The ISR induced by these compounds is thought to be promoted by activation of PERK as well as additional contributing mechanisms, because ISR markers were also induced in TGD45BZ-treated PERK$^{-/-}$ MEFs (Harding et al. 2005).

Structure-guided in silico docking was used to identify a different set of compounds with activity toward PERK, but kinome-wide assessment of their activities is lacking (Wang et al. 2010). In an assay assessing eIF2α phosphorylation by GST-PERK, a theinopyrimidine-containing compound was identified from a collection of kinase inhibitors and inhibited PERK with an IC$_{50}$ of 11 nM (Axten et al. 2012). A series of analogs was generated, and informed by PERK-inhibitor cocrystal structures, GSK2606414 was developed. Kinome profiling showed extremely high selectivity for PERK and xenograft studies showed that this compound could reduce pancreatic tumor growth with no notable side effects (Axten et al. 2012). Medicinal chemistry optimization led to the development of GSK2656157, displaying an IC$_{50}$ of 0.9 nM against PERK and impressive selectivity; among 300 kinases tested against GSK2656157, the most potent off-target activity was against HRI (IC$_{50}$ of 460 nM) (Atkins et al. 2013; Axten et al. 2013). This compound reduced tumor outgrowth in xenograft settings established from pancreatic and multiple myeloma cells, as well as reduced angiogenesis and vascular perfusion. GSK2606414 and GSK2656157 have recently been shown to inhibit RIPK1 and induce RIPK1-kinase dependent cell death in a PERK-independent manner, indicating an off-target biological activity (Rojas-Rivera et al. 2017).

FUTURE PERSPECTIVE

A number of exciting opportunities remain for the discovery of small molecules targeting translation initiation. The interaction between eIF4A and eIF4G and the association between eIF4G and MNK1/2 have not been exploited for compound discovery. In addition, compounds that discriminate between eIF4A1, eIF4A2, eIF4G1, and eIF4G3 activity would be useful to tease out potential differences among these factors in the translation process.

Recently, eIF2A, a factor that binds Met-tRNA$_i$ (but not GTP) and interacts with 40S ribosomes in a codon-dependent manner, has been identified as an essential factor required for tumor initiation in squamous cell carcinoma (Sendoel et al. 2017). It is thought that eIF2A mediates initiation at upstream ORFs (uORFs), imparting a profound effect on the cellular proteome. eIF2A$^{-/-}$ mice are viable (Golovko et al. 2016), indicating that eIF2A inhibitors might affect normal cells minimally and produce few or minor side effects, while significantly impacting on tumor cell survival.

New compounds targeting other translation factors would be enormously helpful as tools to define whether these play critical roles in normal and abnormal physiology. For example, the multifactorial complex eIF3 has been implicated as a translational activator or repressor by binding to specific RNA elements, has cap-binding activity, promotes reinitiation, and overexpression of some of its subunits is oncogenic (Zhang et al. 2007; Roy et al. 2010; Hershey 2015; Lee et al. 2015, 2016; Kumar et al. 2016). A better understanding of the role of eIF3 in the initiation process may provide insight into the molecular basis of different translation initiation pathways. Likewise, the role that eIF4B plays in preinitiation complex recruitment, in stimulating eIF4A helicase activity, and unraveling its putative eIF4A-independent function in translation still remain to be elucidated. Hence, future progress in this field holds promise to shed light on the fundamental process of translation and offers the exciting possibility of uncovering novel therapeutic strategies and compounds.

ACKNOWLEDGMENTS

Research on this topic in J.P.'s laboratory has been made possible through support from the Canadian Institutes of Health Research and the Richard and Edith Strauss Canada Foundation.

REFERENCES

*Reference is also in this collection.

Alain T, Morita M, Fonseca BD, Yanagiya A, Siddiqui N, Bhat M, Zammit D, Marcus V, Metrakos P, Voyer LA, et al. 2012. eIF4E/4E-BP ratio predicts the efficacy of mTOR targeted therapies. *Cancer Res* 72: 6468–6476.

Alvandi F, Kwitkowski VE, Ko CW, Rothmann MD, Ricci S, Saber H, Ghosh D, Brown J, Pfeiler E, Chikhale E, et al. 2014. U.S. Food and Drug Administration approval summary: Omacetaxine mepesuccinate as treatment for chronic myeloid leukemia. *Oncologist* 19: 94–99.

Arnold N, Koppula PR, Gul R, Luck C, Pulakat L. 2014. Regulation of cardiac expression of the diabetic marker microRNA miR-29. *PLoS ONE* 9: e103284.

Atkins C, Liu Q, Minthorn E, Zhang SY, Figueroa DJ, Moss K, Stanley TB, Sanders B, Goetz A, Gaul N, et al. 2013. Characterization of a novel PERK kinase inhibitor with antitumor and antiangiogenic activity. *Cancer Res* 73: 1993–2002.

Axten JM, Medina JR, Feng Y, Shu A, Romeril SP, Grant SW, Li WH, Heerding DA, Minthorn E, Mencken T, et al. 2012. Discovery of 7-methyl-5-(1-{[3-(trifluoromethyl)phenyl]acetyl}-2,3-dihydro-1*H*-indol-5-yl)-7*H*-pyrrolo[2,3-*d*]pyrimidin-4-amine (GSK2606414), a potent and selective first-in-class inhibitor of protein kinase R (PKR)-like endoplasmic reticulum kinase (PERK). *J Med Chem* 55: 7193–7207.

Axten JM, Romeril SP, Shu A, Ralph J, Medina JR, Feng Y, Li WH, Grant SW, Heerding DA, Minthorn E, et al. 2013. Discovery of GSK2656157: An optimized PERK inhibitor selected for preclinical development. *ACS Med Chem Lett* 4: 964–968.

Baragaña B, Hallyburton I, Lee MC, Norcross NR, Grimaldi R, Otto TD, Proto WR, Blagborough AM, Meister S, Wirjanata G, et al. 2015. A novel multiple-stage antimalarial agent that inhibits protein synthesis. *Nature* 522: 315–320.

Beroukhim R, Mermel CH, Porter D, Wei G, Raychaudhuri S, Donovan J, Barretina J, Boehm JS, Dobson J, Urashima M, et al. 2010. The landscape of somatic copy-number alteration across human cancers. *Nature* 463: 899–905.

Bhat M, Robichaud N, Hulea L, Sonenberg N, Pelletier J, Topisirovic I. 2015. Targeting the translation machinery in cancer. *Nat Rev Drug Discov* 14: 261–278.

Bi M, Naczki C, Koritzinsky M, Fels D, Blais J, Hu N, Harding H, Novoa I, Varia M, Raleigh J, et al. 2005. ER stress-regulated translation increases tolerance to extreme hypoxia and promotes tumor growth. *EMBO J* 24: 3470–3481.

Boly R, Gras T, Lamkami T, Guissou P, Serteyn D, Kiss R, Dubois J. 2011. Quercetin inhibits a large panel of kinases implicated in cancer cell biology. *Int J Oncol* 38: 833–842.

Bordeleau ME, Matthews J, Wojnar JM, Lindqvist L, Novac O, Jankowsky E, Sonenberg N, Northcote P, Teesdale-Spittle P, Pelletier J. 2005. Stimulation of mammalian translation initiation factor eIF4A activity by a small molecule inhibitor of eukaryotic translation. *Proc Natl Acad Sci* 102: 10460–10465.

Bordeleau ME, Mori A, Oberer M, Lindqvist L, Chard LS, Higa T, Belsham GJ, Wagner G, Tanaka J, Pelletier J.

2006a. Functional characterization of IRESes by an inhibitor of the RNA helicase eIF4A. *Nat Chem Biol* **2**: 213–220.

Bordeleau ME, Cencic R, Lindqvist L, Oberer M, Northcote P, Wagner G, Pelletier J. 2006b. RNA-mediated sequestration of the RNA helicase eIF4A by pateamine A inhibits translation initiation. *Chem Biol* **13**: 1287–1295.

Bordeleau ME, Robert F, Gerard B, Lindqvist L, Chen SM, Wendel HG, Brem B, Greger H, Lowe SW, Porco JA Jr, et al. 2008. Therapeutic suppression of translation initiation modulates chemosensitivity in a mouse lymphoma model. *J Clin Invest* **118**: 2651–2660.

Boussemart L, Malka-Mahieu H, Girault I, Allard D, Hemmingsson O, Tomasic G, Thomas M, Basmadjian C, Ribeiro N, Thuaud F, et al. 2014. eIF4F is a nexus of resistance to anti-BRAF and anti-MEK cancer therapies. *Nature* **513**: 105–109.

Boyce M, Bryant KF, Jousse C, Long K, Harding HP, Scheuner D, Kaufman RJ, Ma D, Coen DM, Ron D, et al. 2005. A selective inhibitor of eIF2α dephosphorylation protects cells from ER stress. *Science* **307**: 935–939.

Brown CJ, McNae I, Fischer PM, Walkinshaw MD. 2007. Crystallographic and mass spectrometric characterisation of eIF4E with N7-alkylated cap derivatives. *J Mol Biol* **372**: 7–15.

Bryk R, Wu K, Raimundo BC, Boardman PE, Chao P, Conn GL, Anderson E, Cole JL, Duffy NP, Nathan C, et al. 2011. Identification of new inhibitors of protein kinase R guided by statistical modeling. *Bioorg Med Chem Lett* **21**: 4108–4114.

Cargnello M, Roux PP. 2011. Activation and function of the MAPKs and their substrates, the MAPK-activated protein kinases. *Microbiol Mol Biol Rev* **75**: 50–83.

Cencic R, Carrier M, Galicia-Vázquez G, Bordeleau ME, Sukarieh R, Bourdeau A, Brem B, Teodoro JG, Greger H, Tremblay ML, et al. 2009. Antitumor activity and mechanism of action of the cyclopenta[*b*]benzofuran, silvestrol. *PLoS ONE* **4**: e5223.

Cencic R, Hall DR, Robert F, Du Y, Min J, Li L, Qui M, Lewis I, Kurtkaya S, Dingledine R, et al. 2011a. Reversing chemoresistance by small molecule inhibition of the translation initiation complex eIF4F. *Proc Natl Acad Sci* **108**: 1046–1051.

Cencic R, Desforges M, Hall DR, Kozakov D, Du Y, Min J, Dingledine R, Fu H, Vajda S, Talbot PJ, et al. 2011b. Blocking eIF4E–eIF4G interaction as a strategy to impair coronavirus replication. *J Virol* **85**: 6381–6389.

Cencic R, Robert F, Galicia-Vázquez G, Malina A, Ravindar K, Somaiah R, Pierre P, Tanaka J, Deslongchamps P, Pelletier J. 2013. Modifying chemotherapy response by targeted inhibition of eukaryotic initiation factor 4A. *Blood Cancer J* **3**: e128.

Chambers JM, Lindqvist LM, Webb A, Huang DC, Savage GP, Rizzacasa MA. 2013. Synthesis of biotinylated episilvestrol: Highly selective targeting of the translation factors eIF4AI/II. *Org Lett* **15**: 1406–1409.

Chen R, Gandhi V, Plunkett W. 2006. A sequential blockade strategy for the design of combination therapies to overcome oncogene addiction in chronic myelogenous leukemia. *Cancer Res* **66**: 10959–10966.

Chen T, Ozel D, Qiao Y, Harbinski F, Chen L, Denoyelle S, He X, Zvereva N, Supko JG, Chorev M, et al. 2011. Chem-

ical genetics identify eIF2α kinase heme-regulated inhibitor as an anticancer target. *Nat Chem Biol* **7**: 610–616.

Chen L, Aktas BH, Wang Y, He X, Sahoo R, Zhang N, Denoyelle S, Kabha E, Yang H, Freedman RY, et al. 2012a. Tumor suppression by small molecule inhibitors of translation initiation. *Oncotarget* **3**: 869–881.

Chen X, Kopecky DJ, Mihalic J, Jeffries S, Min X, Heath J, Deignan J, Lai S, Fu Z, Guimaraes C, et al. 2012b. Structure-guided design, synthesis, and evaluation of guanine-derived inhibitors of the eIF4E mRNA–cap interaction. *J Med Chem* **55**: 3837–3851.

Chen T, Takrouri K, Hee-Hwang S, Rana S, Yefidoff-Freedman R, Halperin J, Natarajan A, Morisseau C, Hammock B, Chorev M, et al. 2013. Explorations of substituted urea functionality for the discovery of new activators of the heme-regulated inhibitor kinase. *J Med Chem* **56**: 9457–9470.

Chen EZ, Jacobson BA, Patel MR, Okon AM, Li S, Xiong K, Vaidya AJ, Bitterman PB, Wagner CR, Kratzke RA. 2014. Small-molecule inhibition of oncogenic eukaryotic protein translation in mesothelioma cells. *Invest New Drugs* **32**: 598–603.

Choi Y, Kim HS, Shin KY, Kim EM, Kim M, Kim HS, Park CH, Jeong YH, Yoo J, Lee JP, et al. 2007. Minocycline attenuates neuronal cell death and improves cognitive impairment in Alzheimer's disease models. *Neuropsychopharmacology* **32**: 2393–2404.

Chu J, Pelletier J. 2015. Targeting the eIF4A RNA helicase as an anti-neoplastic approach. *Biochim Biophys Acta* **1849**: 781–791.

Chu J, Cargnello M, Topisirovic I, Pelletier J. 2016a. Translation initiation factors: Reprogramming protein synthesis in cancer. *Trends Cell Biol* **26**: 918–933.

Chu J, Galicia-Vázquez G, Cencic R, Mills JR, Katigbak A, Porco JA Jr, Pelletier J. 2016b. CRISPR-mediated drug-target validation reveals selective pharmacological inhibition of the RNA helicase, eIF4A. *Cell Rep* **15**: 2340–2347.

Colina R, Costa-Mattioli M, Dowling RJ, Jaramillo M, Tai LH, Breitbach CJ, Martineau Y, Larsson O, Rong L, Svitkin YV, et al. 2008. Translational control of the innate immune response through IRF-7. *Nature* **452**: 323–328.

Cope CL, Gilley R, Balmanno K, Sale MJ, Howarth KD, Hampson M, Smith PD, Guichard SM, Cook SJ. 2014. Adaptation to mTOR kinase inhibitors by amplification of eIF4E to maintain cap-dependent translation. *J Cell Sci* **127**: 788–800.

Costa-Mattioli M, Gobert D, Stern E, Gamache K, Colina R, Cuello C, Sossin W, Kaufman R, Pelletier J, Rosenblum K, et al. 2007. eIF2α phosphorylation bidirectionally regulates the switch from short- to long-term synaptic plasticity and memory. *Cell* **129**: 195–206.

Dang Y, Kedersha N, Low WK, Romo D, Gorospe M, Kaufman R, Anderson P, Liu JO. 2006. Eukaryotic initiation factor 2α-independent pathway of stress granule induction by the natural product pateamine A. *J Biol Chem* **281**: 32870–32878.

Dauber B, Pelletier J, Smiley JR. 2011. The herpes simplex virus 1 vhs protein enhances translation of viral true late mRNAs and virus production in a cell type-dependent manner. *J Virol* **85**: 5363–5373.

Deng J, Harding HP, Raught B, Gingras AC, Berlanga JJ, Scheuner D, Kaurman RJ, Ron D, Sonenberg N. 2002.

Activation of GCN2 in UV-irradiated cells inhibits translation. *Curr Biol* **12:** 1279–1286.

Denoyelle S, Chen T, Chen L, Wang Y, Klosi E, Halperin JA, Aktas BH, Chorev M. 2012. In vitro inhibition of translation initiation by *N,N'*-diarylureas—Potential anti-cancer agents. *Bioorg Med Chem Lett* **22:** 402–409.

Descamps G, Gomez-Bougie P, Tamburini J, Green A, Bouscary D, Maïga S, Moreau P, Le Gouill S, Pellat-Deceunynck C, Amiot M. 2012. The cap-translation inhibitor 4EGI-1 induces apoptosis in multiple myeloma through Noxa induction. *Br J Cancer* **106:** 1660–1667.

Donze O, Jagus R, Koromilas AE, Hershey JW, Sonenberg N. 1995. Abrogation of translation initiation factor eIF-2 phosphorylation causes malignant transformation of NIH 3T3 cells. *EMBO J* **14:** 3828–3834.

Dowling RJ, Topisirovic I, Alain T, Bidinosti M, Fonseca BD, Petroulakis E, Wang X, Larsson O, Selvaraj A, Liu Y, et al. 2010. mTORC1-mediated cell proliferation, but not cell growth, controlled by the 4E-BPs. *Science* **328:** 1172–1176.

Duffy AG, Makarova-Rusher OV, Ulahannan SV, Rahma OE, Fioravanti S, Walker M, Abdullah S, Raffeld M, Anderson V, Abi-Jaoudeh N, et al. 2016. Modulation of tumor eIF4E by antisense inhibition: A phase I/II translational clinical trial of ISIS 183750-an antisense oligonucleotide against eIF4E-in combination with irinotecan in solid tumors and irinotecan-refractory colorectal cancer. *Int J Cancer* **139:** 1648–1657.

Eberle J, Fecker LF, Bittner JU, Orfanos CE, Geilen CC. 2002. Decreased proliferation of human melanoma cell lines caused by antisense RNA against translation factor eIF-4A1. *Br J Cancer* **86:** 1957–1962.

Eley HL, Russell ST, Tisdale MJ. 2007. Attenuation of muscle atrophy in a murine model of cachexia by inhibition of the dsRNA-dependent protein kinase. *Br J Cancer* **96:** 1216–1222.

Feng Y, Pinkerton AB, Hulea L, Zhang T, Davies MA, Grotegut S, Cheli Y, Yin H, Lau E, Kim H, et al. 2015. SBI-0640756 attenuates the growth of clinically unresponsive melanomas by disrupting the eIF4F translation initiation complex. *Cancer Res* **75:** 5211–5218.

Fullwood MJ, Zhou W, Shenolikar S. 2012. Targeting phosphorylation of eukaryotic initiation factor-2α to treat human disease. *Prog Mol Biol Transl Sci* **106:** 75–106.

Furic L, Rong L, Larsson O, Koumakpayi IH, Yoshida K, Brueschke A, Petroulakis E, Robichaud N, Pollak M, Gaboury LA, et al. 2010. eIF4E phosphorylation promotes tumorigenesis and is associated with prostate cancer progression. *Proc Natl Acad Sci* **107:** 14134–14139.

Gaitonde S, De SK, Tcherpakov M, Dewing A, Yuan H, Riel-Mehan M, Krajewski S, Robertson G, Pellecchia M, Ronai Z. 2009. BI-69A11-mediated inhibition of AKT leads to effective regression of xenograft melanoma. *Pigment Cell Melanoma Res* **22:** 187–195.

Galicia-Vazquez G, Cencic R, Robert F, Agenor AQ, Pelletier J. 2012. A cellular response linking eIF4AI activity to eIF4AII transcription. *RNA* **18:** 1373–1384.

Gantois I, Khoutorsky A, Popic J, Aguilar-Valles A, Freemantle E, Cao R, Sharma V, Pooters T, Nagpal A, Skalecka A, et al. 2017. Metformin ameliorates core deficits in a mouse model of fragile X syndrome. *Nat Med* **23:** 674–677.

Garcia-Villegas R, Lopez-Alvarez LE, Arni S, Rosenbaum T, Morales MA. 2009. Identification and functional characterization of the promoter of the mouse sodium-activated sodium channel Na_x gene (Scn7a). *J Neurosci Res* **87:** 2509–2519.

Garreau de Loubresse N, Prokhorova I, Holtkamp W, Rodnina MV, Yusupova G, Yusupov M. 2014. Structural basis for the inhibition of the eukaryotic ribosome. *Nature* **513:** 517–522.

Getz MJ, Elder PK, Benz EW Jr, Stephens RE, Moses HL. 1976. Effect of cell proliferation on levels and diversity of poly(A)-containing mRNA. *Cell* **7:** 255–265.

Ghosh B, Benyumov AO, Ghosh P, Jia Y, Avdulov S, Dahlberg PS, Peterson M, Smith K, Polunovsky VA, Bitterman PB, et al. 2009. Nontoxic chemical interdiction of the epithelial-to-mesenchymal transition by targeting cap-dependent translation. *ACS Chem Biol* **4:** 367–377.

Gkogkas CG, Khoutorsky A, Ran I, Rampakakis E, Nevarko T, Weatherill DB, Vasuta C, Yee S, Truitt M, Dallaire P, et al. 2013. Autism-related deficits via dysregulated eIF4E-dependent translational control. *Nature* **493:** 371–377.

Gkogkas CG, Khoutorsky A, Cao R, Jafarnejad SM, Prager-Khoutorsky M, Giannakas N, Kaminari A, Fragkouli A, Nader K, Price TJ, et al. 2014. Pharmacogenetic inhibition of eIF4E-dependent Mmp9 mRNA translation reverses fragile X syndrome-like phenotypes. *Cell Rep* **9:** 1742–1755.

Golovko A, Kojukhov A, Guan BJ, Morpurgo B, Merrick WC, Mazumder B, Hatzoglou M, Komar AA. 2016. The eIF2A knockout mouse. *Cell Cycle* **15:** 3115–3120.

Graff JR, Boghaert ER, De Benedetti A, Tudor DL, Zimmer CC, Chan SK, Zimmer SG. 1995. Reduction of translation initiation factor 4E decreases the malignancy of *ras*-transformed cloned rat embryo fibroblasts. *Int J Cancer* **60:** 255–263.

Graff JR, Konicek BW, Vincent TM, Lynch RL, Monteith D, Weir SN, Schwier P, Capen A, Goode RL, Dowless MS, et al. 2007. Therapeutic suppression of translation initiation factor eIF4E expression reduces tumor growth without toxicity. *J Clin Invest* **117:** 2638–2648.

Halliday M, et al. 2017. Repurposed drugs targeting eIF2α-P-mediated translational repression prevent neurodegeneration in mice. *Brain* **140:** 1768–1783.

Hamamura K, Minami K, Tanjung N, Wan Q, Koizumi M, Matsuura N, Na S, Yokota H. 2014. Attenuation of malignant phenotypes of breast cancer cells through eIF2α-mediated downregulation of Rac1 signaling. *Int J Oncol* **44:** 1980–1988.

Harding HP, Zhang Y, Khersonsky S, Marciniak S, Scheuner D, Kaufman RJ, Javitt N, Chang YT, Ron D. 2005. Bioactive small molecules reveal antagonism between the integrated stress response and sterol-regulated gene expression. *Cell Metab* **2:** 361–371.

Hearn BR, Jaishankar P, Sidrauski C, Tsai JC, Vedantham P, Fontaine SD, Walter P, Renslo AR. 2016. Structure–activity studies of bis-O-arylglycolamides: Inhibitors of the integrated stress response. *ChemMedChem* **11:** 870–880.

Herbert TP, Fahraeus R, Prescott A, Lane DP, Proud CG. 2000. Rapid induction of apoptosis mediated by peptides that bind initiation factor eIF4E. *Curr Biol* **10:** 793–796.

Hershey JW. 2015. The role of eIF3 and its individual subunits in cancer. *Biochim Biophys Acta* **1849:** 792–800.

Higa T, Tanaka J. 1981. 18-Oxygenated polyfunctional ste-roids from the gorgonian *Isis hippuris*. *Tetrahed Lett* **22**: 2777–2780.

Higa T, Tanaka J, Tsukitani Y, Kikuchi H. 1981. Hippuris-tanols, cytotoxic polyoxygenated steroids from the gorgo-nian *Isis hippuris*. *Chem Lett* **10**: 1647–1650.

Hodges LM, Markova SM, Chinn LW, Gow JM, Kroetz DL, Klein TE, Altman RB. 2011. Very important pharmaco-gene summary: ABCB1 (MDR1, P-glycoprotein). *Pharmacogenet Genomics* **21**: 152–161.

Hoeffer CA, Cowansage KK, Arnold EC, Banko JL, Moerke NJ, Rodriguez R, Schmidt EK, Klosi E, Chorev M, Lloyd RE, et al. 2011. Inhibition of the interactions between eukaryotic initiation factors 4E and 4G impairs long-term associative memory consolidation but not recon-solidation. *Proc Natl Acad Sci* **108**: 3383–3388.

Hoeffer CA, Santini E, Ma T, Arnold EC, Whelan AM, Wong H, Pierre P, Pelletier J, Klann E. 2013. Multiple compo-nents of eIF4F are required for protein synthesis-depen-dent hippocampal long-term potentiation. *J Neurophysiol* **109**: 68–76.

Holz MK, Ballif BA, Gygi SP, Blenis J. 2005. mTOR and S6K1 mediate assembly of the translation preinitiation complex through dynamic protein interchange and ordered phos-phorylation events. *Cell* **123**: 569–580.

Hong DS, Kurzrock R, Oh Y, Wheler J, Naing A, Brail L, Callies S, André V, Kadam SK, Nasir A, et al. 2011. A phase 1 dose escalation, pharmacokinetic, and pharma-codynamic evaluation of eIF-4E antisense oligonucleo-tide LY2275796 in patients with advanced cancer. *Clin Cancer Res* **17**: 6582–6591.

Hood KA, West LM, Northcote PT, Berridge MV, Miller JH. 2001. Induction of apoptosis by the marine sponge (*My-cale*) metabolites, mycalamide A and pateamine. *Apopto-sis* **6**: 207–219.

Hoozemans JJ, van Haastert ES, Nijholt DA, Rozemuller AJ, Eikelenboom P, Scheper W. 2009. The unfolded protein response is activated in pretangle neurons in Alzheimer's disease hippocampus. *Am J Pathol* **174**: 1241–1251.

Hou L, Antion MD, Hu D, Spencer CM, Paylor R, Klann E. 2006. Dynamic translational and proteasomal regulation of fragile X mental retardation protein controls mGluR-dependent long-term depression. *Neuron* **51**: 441–454.

Huang JT, Schneider RJ. 1990. Adenovirus inhibition of cel-lular protein synthesis is prevented by the drug 2-amino-purine. *Proc Natl Acad Sci* **87**: 7115–7119.

Ilic N, Utermark T, Widlund HR, Roberts TM. 2011. PI3K-targeted therapy can be evaded by gene amplification along the MYC-eukaryotic translation initiation factor 4E (eIF4E) axis. *Proc Natl Acad Sci* **108**: E699–E708.

Ito T, Warnken SP, May WS. 1999. Protein synthesis inhibi-tion by flavonoids: Roles of eukaryotic initiation factor 2α kinases. *Biochem Biophys Res Commun* **265**: 589–594.

Iwasaki S, Floor SN, Ingolia NT. 2016. Rocaglates convert DEAD-box protein eIF4A into a sequence-selective trans-lational repressor. *Nature* **534**: 558–561.

Jammi NV, Whitby LR, Beal PA. 2003. Small molecule in-hibitors of the RNA-dependent protein kinase. *Biochem Biophys Res Commun* **308**: 50–57.

Jansen AP, Camalier CE, Colburn NH. 2005. Epidermal expression of the translation inhibitor programmed cell death 4 suppresses tumorigenesis. *Cancer Res* **65**: 6034–6041.

Jennings MD, Pavitt GD. 2014. A new function and com-plexity for protein translation initiation factor eIF2B. *Cell Cycle* **13**: 2660–2665.

Jiang H, Coleman J, Miskimins R, Miskimins WK. 2003. Expression of constitutively active 4EBP-1 enhances p27^Kip1 expression and inhibits proliferation of MCF7 breast cancer cells. *Cancer Cell Int* **3**: 2.

Jiang HY, Wek RC. 2005a. Phosphorylation of eIF2α reduces protein synthesis and enhances apoptosis in response to proteasome inhibition. *J Biol Chem* **280**: 14189–14202.

Jiang HY, Wek RC. 2005b. GCN2 phosphorylation of eIF2α activates NF-κB in response to UV irradiation. *Biochem J* **385**: 371–380.

Jousse C, Oyadomari S, Novoa I, Lu P, Zhang Y, Harding HP, Ron D. 2003. Inhibition of a constitutive translation ini-tiation factor 2α phosphatase, CReP, promotes survival of stressed cells. *J Cell Biol* **163**: 767–775.

Kashiwagi K, Takahashi M, Nishimoto M, Hiyama TB, Higo T, Umehara T, Sakamoto K, Ito T, Yokoyama S. 2016. Crystal structure of eukaryotic translation initiation factor 2B. *Nature* **531**: 122–125.

Kim JH, Park SM, Park JH, Keum SJ, Jang SK. 2011. eIF2A mediates translation of hepatitis C viral mRNA under stress conditions. *EMBO J* **30**: 2454–2464.

Kim YK, Minai-Tehrani A, Lee JH, Cho CS, Cho MH, Jiang HL. 2013. Therapeutic efficiency of folated poly(ethylene glycol)-chitosan-graft-polyethylenimine-*Pdcd4* complexes in H-*ras*12V mice with liver cancer. *Int J Manomed* **8**: 1489–1498.

Ko SY, Guo H, Barengo N, Naora H. 2009. Inhibition of ovarian cancer growth by a tumor-targeting peptide that binds eukaryotic translation initiation factor 4E. *Clin Cancer Res* **15**: 4336–4347.

Koromilas AE, Mounir Z. 2013. Control of oncogenesis by eIF2α phosphorylation: Implications in PTEN and PI3K-Akt signaling and tumor treatment. *Future Oncol* **9**: 1005–1015.

Kumar P, Hellen CU, Pestova TV. 2016. Toward the mech-anism of eIF4F-mediated ribosomal attachment to mam-malian capped mRNAs. *Genes Dev* **30**: 1573–1588.

Kuznetsov G, Xu Q, Rudolph-Owen L, Tendyke K, Liu J, Towle M, Zhao N, Marsh J, Agoulnik S, Twine N, et al. 2009. Potent in vitro and in vivo anticancer activities of des-methyl, des-amino pateamine A, a synthetic analogue of marine natural product pateamine A. *Mol Cancer Ther* **8**: 1250–1260.

Lazaris-Karatzas A, Montine KS, Sonenberg N. 1990. Malig-nant transformation by a eukaryotic initiation factor sub-unit that binds to mRNA 5' cap. *Nature* **345**: 544–547.

Lee DY, Lee KS, Lee HJ, Kim DH, Noh YH, Yu K, Jung HY, Lee SH, Lee JY, Youn YC, et al. 2010. Activation of PERK signaling attenuates Aβ-mediated ER stress. *PLoS ONE* **5**: e10489.

Lee AS, Kranzusch PJ, Cate JH. 2015. eIF3 targets cell-pro-liferation messenger RNAs for translational activation or repression. *Nature* **522**: 111–114.

Lee AS, Kranzusch PJ, Doudna JA, Cate JH. 2016. eIF3d is an mRNA cap-binding protein that is required for special-ized translation initiation. *Nature* **536**: 96–99.

Cite this article as *Cold Spring Harb Perspect Biol* doi: 10.1101/cshperspect.a032995

Li W, Dang Y, Liu JO, Yu B. 2009. Expeditious synthesis of hippuristanol and congeners with potent antiproliferative activities. *Chemistry* **15**: 10356–10359.

Li S, Jia Y, Jacobson B, McCauley J, Kratzke R, Bitterman PB, Wagner CR. 2013. Treatment of breast and lung cancer cells with a N-7 benzyl guanosine monophosphate tryptamine phosphoramidate pronucleotide (4Ei-1) results in chemosensitization to gemcitabine and induced eIF4E proteasomal degradation. *Mol Pharm* **10**: 523–531.

Lin CJ, Cencic R, Mills JR, Robert F, Pelletier J. 2008. c-Myc and eIF4F are components of a feedforward loop that links transcription and translation. *Cancer Res* **68**: 5326–5334.

Lin CJ, Nasr Z, Premsrirut PK, Proco JA Jr, Hippo Y, Lowe SW, Pelletier J. 2012. Targeting synthetic lethal interactions between Myc and the eIF4F complex impedes tumorigenesis. *Cell Rep* **1**: 325–333.

Lindqvist L, Oberer M, Reibarkh M, Cencic R, Bordeleau ME, Vogt E, Marintchev A, Tanaka J, Fagotto F, Altmann M, et al. 2008. Selective pharmacological targeting of a DEAD box RNA helicase. *PLoS ONE* **3**: e1583.

Long K, Boyce M, Lin H, Yuan J, Ma D. 2005. Structure–activity relationship studies of salubrinal lead to its active biotinylated derivative. *Bioorg Med Chem Lett* **15**: 3849–3852.

Low WK, Dang Y, Schneider-Poetsch T, Shi Z, Choi NS, Merrick WC, Romo D, Liu JO. 2005. Inhibition of eukaryotic translation initiation by the marine natural product pateamine A. *Mol Cell* **20**: 709–722.

Low WK, Dang Y, Bhat S, Romo D, Liu JO. 2007. Substrate-dependent targeting of eukaryotic translation initiation factor 4A by pateamine A: Negation of domain–linker regulation of activity. *Chem Biol* **14**: 715–727.

Low WK, Li J, Zhu M, Kommaraju SS, Shah-Mittal J, Hull K, Liu JO, Romo D. 2014. Second-generation derivatives of the eukaryotic translation initiation inhibitor pateamine A targeting eIF4A as potential anticancer agents. *Bioorg Med Chem* **22**: 116–125.

Lynch M, Fitzgerald C, Johnston KA, Wang S, Schmidt EV. 2004. Activated eIF4E-binding protein slows G1 progression and blocks transformation by c-*myc* without inhibiting cell growth. *J Biol Chem* **279**: 3327–3339.

Ma T, Trinh MA, Wexler AJ, Bourbon C, Gatti E, Pierre P, Cavener DR, Klann E. 2013. Suppression of eIF2α kinases alleviates Alzheimer's disease-related plasticity and memory deficits. *Nat Neurosci* **16**: 1299–1305.

Mahalingam P, Takrouri K, Chen T, Sahoo R, Papadopoulos E, Chen L, Wagner G, Aktas BH, Halperin JA, Chorev M. 2014. Synthesis of rigidified eIF4E/eIF4G inhibitor-1 (4EGI-1) mimetic and their in vitro characterization as inhibitors of protein–protein interaction. *J Med Chem* **57**: 5094–5111.

Malina A, Mills JR, Pelletier J. 2012. Emerging therapeutics targeting mRNA translation. *Cold Spring Harb Perspect Biol* **4**: a012377.

Mallya S, Fitch BA, Lee JS, So L, Janes MR, Fruman DA. 2014. Resistance to mTOR kinase inhibitors in lymphoma cells lacking 4EBP1. *PLoS ONE* **9**: e88865.

Marciniak SJ, Yun CY, Oyadomari S, Novoa I, Zhang Y, Jungreis R, Nagata K, Harding HP, Ron D. 2004. CHOP induces death by promoting protein synthesis and oxidation in the stressed endoplasmic reticulum. *Genes Dev* **18**: 3066–3077.

Marcotrigiano J, Gingras AC, Sonenberg N, Burley SK. 1997. Cocrystal structure of the messenger RNA 5′ cap-binding protein (eIF4E) bound to 7-methyl-GDP. *Cell* **89**: 951–961.

Marcotrigiano J, Gingras AC, Sonenberg N, Burley SK. 1999. Cap-dependent translation initiation in eukaryotes is regulated by a molecular mimic of eIF4G. *Mol Cell* **3**: 707–716.

Mazroui R, Sukarieh R, Bordeleau ME, Kaufman RJ, Northcote P, Tanaka J, Gallouzi I, Pelletier J. 2006. Inhibition of ribosome recruitment induces stress granule formation independently of eukaryotic initiation factor 2α phosphorylation. *Mol Biol Cell* **17**: 4212–4219.

McDonald ER III, de Weck A, Schlabach MR, Billy E, Mavrakis KJ, Hoffman GR, Belur D, Castelletti D, Frias E, Gampa K, et al. 2017. Project DRIVE: A compendium of cancer dependencies and synthetic lethal relationships uncovered by large-scale, deep RNAi screening. *Cell* **170**: 577–592 e510.

McMahon R, Zaborowska I, Walsh D. 2011. Noncytotoxic inhibition of viral infection through eIF4F-independent suppression of translation by 4EGi-1. *J Virol* **85**: 853–864.

Meric F, Hunt KK. 2003. Translation initiation in cancer: A novel target for therapy. *Mol Canc Ther* **1**: 971–979.

* Merrick WC, Pavitt GD. 2018. Protein synthesis initiation in eukaryotic cells. *Cold Spring Harb Perspect Biol* doi: 10.1101/cshperspect.a033092.

Mills JR, Hippo Y, Robert F, Chen SM, Malina A, Lin CJ, Trojahn U, Wendel HG, Charest A, Bronson RT, et al. 2008. mTORC1 promotes survival through translational control of Mcl-1. *Proc Natl Acad Sci* **105**: 10853–10858.

Moerke NJ, Aktas H, Chen H, Cantel S, Reibarkh MY, Fahmy A, Gross JD, Degterev A, Yuan J, Chorev M, et al. 2007. Small-molecule inhibition of the interaction between the translation initiation factors eIF4E and eIF4G. *Cell* **128**: 257–267.

Niedzwiecka A, Marcotrigiano J, Stepinski J, Jankowska-Anyszka M, Wyslouch-Cieszynska A, Dadlez M, Gingras AC, Mak P, Darzynkiewicz E, Sonenberg N, et al. 2002. Biophysical studies of eIF4E cap-binding protein: Recognition of mRNA 5′ cap structure and synthetic fragments of eIF4G and 4E-BP1 proteins. *J Mol Biol* **319**: 615–635.

Nielsen PJ, Trachsel H. 1988. The mouse protein synthesis initiation factor 4A gene family includes two related functional genes which are differentially expressed. *EMBO J* **7**: 2097–2105.

Northcote PT, Blunt JW, Munro MHG. 1991. Pateamine—A potent cytotoxin from the New Zealand marine sponge, *Mycale* sp. *Tetrahedron Lett* **32**: 6411–6414.

Novoa I, Zeng H, Harding HP, Ron D. 2001. Feedback inhibition of the unfolded protein response by GADD34-mediated dephosphorylation of eIF2α. *J Cell Biol* **153**: 1011–1022.

Papadopoulos E, Jenni S, Kabha E, Takrouri KJ, Yi T, Salvi N, Luna RE, Gavathiotis E, Mahalingam P, Arthanari H, et al. 2014. Structure of the eukaryotic translation initiation factor eIF4E in complex with 4EGI-1 reveals an allosteric mechanism for dissociating eIF4G. *Proc Natl Acad Sci* **111**: E3187–E3195.

Pause A, Belsham GJ, Gingras AC, Donzé O, Lin TA, Lawrence JC Jr, Sonenberg N. 1994. Insulin-dependent stimulation of protein synthesis by phosphorylation of a regulator of 5′-cap function. *Nature* **371**: 762–767.

Pelletier J, Peltz SW. 2007. *Therapeutic opportunities in translation*, pp. 855–895. Cold Spring Harbor Laboratory Press, New York.

Pelletier J, Graff J, Ruggero D, Sonenberg N. 2015. Targeting the eIF4F translation initiation complex: A critical nexus for cancer development. *Cancer Res* **75**: 250–263.

Pestka S. 1977. *Inhibitors of protein synthesis*. Academic, New York.

Petroulakis E, Parsyan A, Dowling RJ, LeBacquer O, Martineau Y, Bidinosti M, Larsson O, Alain T, Rong L, Mamane Y, et al. 2009. p53-dependent translational control of senescence and transformation via 4E-BPs. *Cancer Cell* **16**: 439–446.

Pyronnet S, Imataka H, Gingras AC, Fukunaga R, Hunter T, Sonenberg N. 1999. Human eukaryotic translation initiation factor 4G (eIF4G) recruits mnk1 to phosphorylate eIF4E. *EMBO J* **18**: 270–279.

Radtke K, English L, Rondeau C, Leib D, Lippé R, Desjardins M. 2013. Inhibition of the host translation shutoff response by herpes simplex virus 1 triggers nuclear envelope-derived autophagy. *J Virol* **87**: 3990–3997.

Ran I, Laplante I, Bourgeois C, Pépin J, Lacaille P, Costa-Mattioli M, Pelletier J, Sonenberg N, Lacaille JC. 2009. Persistent transcription- and translation-dependent long-term potentiation induced by mGluR1 in hippocampal interneurons. *J Neurosci* **29**: 5605–5615.

Raught B, Peiretti F, Gingras AC, Livingstone M, Shahbazian D, Mayeur GL, Polakiewicz RD, Sonenberg N, Hershey JW. 2004. Phosphorylation of eucaryotic translation initiation factor 4B Ser422 is modulated by S6 kinases. *EMBO J* **23**: 1761–1769.

Ravindar K, Reddy MS, Lindqvist L, Pelletier J, Deslongchamps P. 2010. Efficient synthetic approach to potent antiproliferative agent hippuristanol via Hg(II)-catalyzed spiroketalization. *Org Lett* **12**: 4420–4423.

Ravindar K, Reddy MS, Lindqvist L, Pelletier J, Deslongchamps P. 2011. Synthesis of the antiproliferative agent hippuristanol and its analogues via Suárez cyclizations and Hg(II)-catalyzed spiroketalizations. *J Org Chem* **76**: 1269–1284.

Reijonen S, Putkonen N, Norremolle A, Lindholm D, Korhonen L. 2008. Inhibition of endoplasmic reticulum stress counteracts neuronal cell death and protein aggregation caused by N-terminal mutant huntingtin proteins. *Exp Cell Res* **314**: 950–960.

Rinker-Schaeffer CW, Graff JR, De Benedetti A, Zimmer SG, Rhoads RE. 1993. Decreasing the level of translation initiation factor 4E with antisense RNA causes reversal of *ras*-mediated transformation and tumorigenesis of cloned rat embryo fibroblasts. *Int J Cancer* **55**: 841–847.

Robert F, Pelletier J. 2009. Translation initiation: A critical signalling node in cancer. *Expert Opin Ther Targets* **13**: 1279–1293.

Robert F, Kapp LD, Khan SN, Acker MG, Kolitz S, Kazemi S, Kaufman RJ, Merrick WC, Koromilas AE, Lorsch JR, et al. 2006. Initiation of protein synthesis by hepatitis C virus is refractory to reduced eIF2·GTP·Met-tRNA$_i^{Met}$ ternary complex availability. *Mol Biol Cell* **17**: 4632–4644.

Robert F, Carrier M, Rawe S, Chen S, Lowe S, Pelletier J. 2009a. Altering chemosensitivity by modulating translation elongation. *PLoS ONE* **4**: e5428.

Robert F, Williams C, Yan Y, Donohue E, Cencic R, Burley SK, Pelletier J. 2009b. Blocking UV-induced eIF2α phosphorylation with small molecule inhibitors of GCN2. *Chem Biol Drug Des* **74**: 57–67.

Robert F, Roman W, Bramoullé A, Fellmann C, Roulston A, Shustik C, Porco JA Jr, Shore GC, Sebag M, Pelletier J. 2014. Translation initiation factor eIF4F modifies the dexamethasone response in multiple myeloma. *Proc Natl Acad Sci* **111**: 13421–13426.

Robichaud N, del Rincon SV, Huor B, Alain T, Petruccelli LA, Hearnden J, Goncalves C, Grotegut S, Spruck CH, Furic L, et al. 2015. Phosphorylation of eIF4E promotes EMT and metastasis via translational control of SNAIL and MMP-3. *Oncogene* **34**: 2032–2042.

Rojas-Rivera D, Delvaeye T, Roelandt R, Nerinckx W, Augustyns K, Vandenabeele P, Bertrand MJM. 2017. When PERK inhibitors turn out to be new potent RIPK1 inhibitors: Critical issues on the specificity and use of GSK2606414 and GSK2656157. *Cell Death Differ* **24**: 1100–1110.

Romo D, Choi NS, Li S, Buchler I, Shi Z, Liu JO. 2004. Evidence for separate binding and scaffolding domains in the immunosuppressive and antitumor marine natural product, pateamine A: Design, synthesis, and activity studies leading to a potent simplified derivative. *J Am Chem Soc* **126**: 10582–10588.

Rosenwald IB, Rhoads DB, Callanan LD, Isselbacher KJ, Schmidt EV. 1993. Increased expression of eukaryotic translation initiation factors eIF-4E and eIF-2 α in response to growth induction by c-*myc*. *Proc Natl Acad Sci* **90**: 6175–6178.

Rousseau D, Gingras AC, Pause A, Sonenberg N. 1996. The eIF4E-binding proteins 1 and 2 are negative regulators of cell growth. *Oncogene* **13**: 2415–2420.

Roy B, Vaughn JN, Kim BH, Zhou F, Gilchrist MA, Von Arnim AG. 2010. The h subunit of eIF3 promotes reinitiation competence during translation of mRNAs harboring upstream open reading frames. *RNA* **16**: 748–761.

Rubio CA, Weisburd B, Holderfield M, Arias C, Fang E, DeRisi JL, Fanidi A. 2014. Transcriptome-wide characterization of the eIF4A signature highlights plasticity in translation regulation. *Genome Biol* **15**: 476.

Santini E, Huynh TN, MacAskill AF, Carter AG, Pierre P, Ruggero D, Kaphzan H, Klann E. 2013. Exaggerated translation causes synaptic and behavioural aberrations associated with autism. *Nature* **493**: 411–415.

Saxena S, Cabuy E, Caroni P. 2009. A role for motoneuron subtype-selective ER stress in disease manifestations of FALS mice. *Nat Neurosci* **12**: 627–636.

Schewe DM, Aguirre-Ghiso JA. 2009. Inhibition of eIF2α dephosphorylation maximizes bortezomib efficiency and eliminates quiescent multiple myeloma cells surviving proteasome inhibitor therapy. *Cancer Res* **69**: 1545–1552.

Sekine Y, Zyryanova A, Crespillo-Casado A, Fischer PM, Harding HP, Ron D. 2015. Stress responses. Mutations in a translation initiation factor identify the target of a memory-enhancing compound. *Science* **348**: 1027–1030.

Sendoel A, Dunn JG, Rodriguez EH, Naik S, Gomez NC, Hurwitz B, Levorse J, Dill BD, Schramek D, Molina H, et al. 2017. Translation from unconventional 5′ start sites drives tumour initiation. *Nature* **541**: 494–499.

Shan Y, Wang C, Zhang L, Wang J, Wang M, Dong Y. 2016. Expanding the structural diversity of diarylureas as multitarget tyrosine kinase inhibitors. *Bioorg Med Chem* **24**: 750–758.

Shuda M, Velásquez C, Cheng E, Cordek DG, Kwun HJ, Chang Y, Moore PS. 2015. CDK1 substitutes for mTOR kinase to activate mitotic cap-dependent protein translation. *Proc Natl Acad Sci* **112**: 5875–5882.

Sidrauski C, Acosta-Alvear D, Khoutorsky A, Vedantham P, Hearn BR, Li H, Gamache K, Gallagher CM, Ang KK, Wilson C, et al. 2012. Pharmacological brake-release of mRNA translation enhances cognitive memory. *eLife* **2**: e00498.

Sidrauski C, Tsai JC, Kampmann M, Hearn BR, Vedantham P, Jaishankar P, Sokabe M, Mendez AS, Newton BW, Tang EL, et al. 2015a. Pharmacological dimerization and activation of the exchange factor eIF2B antagonizes the integrated stress response. *eLife* **4**: e07314.

Sidrauski C, McGeachy AM, Ingolia NT, Walter P. 2015b. The small molecule ISRIB reverses the effects of eIF2α phosphorylation on translation and stress granule assembly. *eLife* **4**: e05033.

Silvera D, Formenti SC, Schneider RJ. 2010. Translational control in cancer. *Nat Rev Cancer* **10**: 254–266.

Smith WW, Jiang H, Pei Z, Tanaka Y, Morita H, Sawa A, Dawson VL, Dawson TM, Ross CA. 2005. Endoplasmic reticulum stress and mitochondrial cell death pathways mediate A53T mutant α-synuclein-induced toxicity. *Hum Mol Genet* **14**: 3801–3811.

Smith KA, Zhou B, Avdulov S, Benyumov A, Peterson M, Liu Y, Okon A, Hergert P, Braziunas J, Wagner CR, et al. 2015. Transforming growth factor-β1 induced epithelial mesenchymal transition is blocked by a chemical antagonist of translation factor eIF4E. *Sci Rep* **5**: 18233.

Somaiah R, Ravindar K, Cencic R, Pelletier J, Deslongchamp P. 2014. Synthesis of the antiproliferative agent hippuristanol and its analogues from hydrocortisone via Hg(II)-catalyzed spiroketalization: Structure–activity relationship. *J Med Chem* **57**: 2511–2523.

Stumpf CR, Ruggero D. 2011. The cancerous translation apparatus. *Curr Opin Genet Dev* **21**: 474–483.

Tamburini J, Green AS, Bardet V, Chapuis N, Park S, Willems L, Uzunov M, Ifrah N, Dreyfus F, Lacombe C, et al. 2009. Protein synthesis is resistant to rapamycin and constitutes a promising therapeutic target in acute myeloid leukemia. *Blood* **114**: 1618–1627.

Terenin IM, Dmitriev SE, Andreev DE, Shatsky IN. 2008. Eukaryotic translation initiation machinery can operate in a bacterial-like mode without eIF2. *Nat Struct Mol Biol* **15**: 836–841.

Thumma SC, Jacobson BA, Patel MR, Konicek BW, Franklin MJ, Jay-Dixon J, Sadiq A, De A, Graff JR, Kratzke RA. 2015. Antisense oligonucleotide targeting eukaryotic translation initiation factor 4E reduces growth and enhances chemosensitivity of non-small-cell lung cancer cells. *Cancer Gene Ther* **22**: 396–401.

Tomoo K, Shen X, Okabe K, Nozoe Y, Fukuhara S, Morino S, Ishida T, Taniguchi T, Hasegawa H, Terashima A, et al. 2002. Crystal structures of 7-methylguanosine 5′-triphosphate (m⁷GTP)- and P¹-7-methylguanosine-P³-adeno-sine-5′,5′-triphosphate (m⁷GpppA)-bound human full-length eukaryotic initiation factor 4E: Biological importance of the C-terminal flexible region. *Biochem J* **362**: 539–544.

Trinh MA, Klann E. 2013. Translational control by eIF2α kinases in long-lasting synaptic plasticity and long-term memory. *Neurobiol Learn Mem* **105**: 93–99.

Truitt ML, Conn CS, Shi Z, Pang X, Tokuyasu T, Coady AM, Seo Y, Barna M, Ruggero D. 2015. Differential requirements for eIF4E dose in normal development and cancer. *Cell* **162**: 59–71.

Tsaytler P, Harding HP, Ron D, Bertolotti A. 2011. Selective inhibition of a regulatory subunit of protein phosphatase 1 restores proteostasis. *Science* **332**: 91–94.

Tsumuraya T, Ishikawa C, Machijima Y, Nakachi S, Senba M, Tanaka J, Mori N. 2011. Effects of hippuristanol, an inhibitor of eIF4A, on adult T-cell leukemia. *Biochem Pharmacol* **81**: 713–722.

Vazquez D. 1979. Inhibitors of protein biosynthesis. *Mol Biol Biochem Biophys* **30**: 1–312.

Wagner CR, Iyer VV, McIntee EJ. 2000. Toward the in vivo delivery of antiviral and anticancer nucleotides. *Med Res Rev* **20**: 417–451.

Wang H, Blais J, Ron D, Cardozo T. 2010. Structural determinants of PERK inhibitor potency and selectivity. *Chem Biol Drug Des* **76**: 480–495.

Waskiewicz AJ, Johnson JC, Penn B, Mahalingam M, Kimball SR, Cooper JA. 1999. Phosphorylation of the cap-binding protein eukaryotic translation initiation factor 4E by protein kinase Mnk1 in vivo. *Mol Cell Biol* **19**: 1871–1880.

* Wek RC. 2018. Role of eIF2α kinases in translational control and adaptation to cellular stress. *Cold Spring Harb Perspect Biol* doi: 10.1101/cshperspect.a032870.

Wendel HG, De Stanchina E, Fridman JS, Malina A, Ray S, Kogan S, Coron-Cardo C, Pelletier J, Lowe SW. 2004. Survival signalling by Akt and eIF4E in oncogenesis and cancer therapy. *Nature* **428**: 332–337.

Williams JG, Penman S. 1975. The messenger RNA sequences in growing and resting mouse fibroblasts. *Cell* **6**: 197–206.

Wolfe AL, Singh K, Zhong Y, Drewe P, Rajasekhar VK, Sanghvi VR, Mavrakis KJ, Jiang M, Roderick JE, Van der Meulen J, et al. 2014. RNA G-quadruplexes cause eIF4A-dependent oncogene translation in cancer. *Nature* **513**: 65–70.

Wu M, Zhang C, Li XJ, Liu Q, Wanggou S. 2016. Anti-cancer effect of cap-translation inhibitor 4EGI-1 in human glioma U87 cells: Involvement of mitochondrial dysfunction and ER stress. *Cell Physiol Biochem* **40**: 1013–1028.

Yan GR, Xiao CL, He GW, Yin XF, Chen NP, Cao Y, He QY. 2010. Global phosphoproteomic effects of natural tyrosine kinase inhibitor, genistein, on signaling pathways. *Proteomics* **10**: 976–986.

Yang HS, Jansen AP, Komar AA, Zheng X, Merrick WC, Costes S, Lockett SJ, Sonenberg N, Colburn NH. 2003a. The transformation suppressor Pdcd4 is a novel eukaryotic translation initiation factor 4A binding protein that inhibits translation. *Mol Cell Biol* **23**: 26–37.

Yang HS, Knies JL, Stark C, Colburn NH. 2003b. Pdcd4 suppresses tumor phenotype in JB6 cells by inhibiting AP-1 transactivation. *Oncogene* **22:** 3712–3720.

Yang X, Dong QF, Li LW, Huo JL, Li PQ, Fei Z, Zhen HN. 2015. The cap-translation inhibitor 4EGI-1 induces mitochondrial dysfunction via regulation of mitochondrial dynamic proteins in human glioma U251 cells. *Neurochem Int* **90:** 98–106.

Yángüez E, Castello A, Welnowska E, Carrasco L, Goodfellow I, Nieto A. 2011. Functional impairment of eIF4A and eIF4G factors correlates with inhibition of influenza virus mRNA translation. *Virology* **413:** 93–102.

Yi T, Papadopoulos E, Hagner PR, Wagner G. 2013. Hypoxia-inducible factor-1α (HIF-1α) promotes cap-dependent translation of selective mRNAs through upregulating initiation factor eIF4E1 in breast cancer cells

under hypoxia conditions. *J Biol Chem* **288:** 18732–18742.

Zhang L, Pan X, Hershey JW. 2007. Individual overexpression of five subunits of human translation initiation factor eIF3 promotes malignant transformation of immortal fibroblast cells. *J Biol Chem* **282:** 5790–5800.

Zhou FF, Yan M, Guo GF, Wang F, Qiu HJ, Zheng FM, Zhang Y, Liu Q, Zhu XF, Xia LP. 2011. Knockdown of eIF4E suppresses cell growth and migration, enhances chemosensitivity and correlates with increase in Bax/Bcl-2 ratio in triple-negative breast cancer cells. *Med Oncol* **28:** 1302–1307.

Zindy P, Bergé Y, Allal B, Filleron T, Pierredon S, Cammas A, Beck S, Mhamdi L, Fan L, Favre G, et al. 2011. Formation of the eIF4F translation-initiation complex determines sensitivity to anticancer drugs targeting the EGFR and HER2 receptors. *Cancer Res* **71:** 4068–4073.

Cite this article as *Cold Spring Harb Perspect Biol* doi: 10.1101/cshperspect.a032995

Index